AF758390

Noah's Ravens

LIFE OF THE PAST

James O. Farlow, editor

NOAH'S RAVENS

Interpreting the Makers of Tridactyl Dinosaur Footprints

JAMES O. FARLOW

With Contributions by DAN COROIAN
and PHILIP J. CURRIE

INDIANA UNIVERSITY PRESS

This book is a publication of

Indiana University Press
Office of Scholarly Publishing
Herman B Wells Library 350
1320 East 10th Street
Bloomington, Indiana 47405 USA

iupress.indiana.edu

© 2018 by James O. Farlow
All rights reserved

No part of this book may be reproduced or utilized in any form or by any means, electronic or mechanical, including photocopying and recording, or by any information storage and retrieval system, without permission in writing from the publisher.

The paper used in this publication meets the minimum requirements of the American National Standard for Information Sciences—Permanence of Paper for Printed Library Materials, ANSI Z39.48-1992.

Manufactured in the United States of America

Cataloging information is available from the Library of Congress.

ISBN 978-0-253-02725-2 (hdbk.)
ISBN 978-0-253-03716-9 (web PDF)

1 2 3 4 5 23 22 21 20 19 18

Contents

C

Acknowledgments

I HAVE BENEFITED OVER THE YEARS FROM THE ASSISTANCE, access to specimens (or data or art), hospitality and/or logistical support, and/or wise counsel of numerous individuals, some of whom are now deceased, but not forgotten. I thank Laura Abraczinskas, Thomas Adams, David and Margaret Akers, Herculano Alvarenga, Don Baird, Billy Paul and Pam Baker, Matteo Belvedere, Mike Brett-Surman, Dan Brinkman, Lisa Buckley, Pierre Bultynck, Ralph Chapman, Sandra Chapman, Luis Chiappe, Walter Coombs, Raymond Coory, Tim Corey, Rodolfo Coria, Nick Czaplewski, Ann Darrow, Ben Dattilo, Kyle Davies, Lisa Davis, Federico Degrange, Carl Denham, Everett and Carolyn Deschner, Sarah Doyle, Jack Driscoll, Ruth Elsey, Annelise Folie, Jed Freels, Georges Gand, Jose García-Ramos, Rob Gaston, Steve Gatesy, Patrick Getty, Whit Gibbons, Gerard Gierliński, Brian Gill, David Gillette, Tammy Gordon, Tim Hamley, Øvind Hammer, Amy Henrici, Tom Holtz, Jack Horner, Jim Kirkland, Richard Krueger, Glen Kuban, Cory Kumagai, Wann Langston, Peter Larson, Young-Nam Lee, Giuseppe Leonardi, Martin Lockley, Andy Main, Anthony Maltese, Cliff and Row Manuel, Rich McCrea, Jack McIntosh, Carl Mehling, Christian Meyer, Cliff Miles, Colin Miskelly, Ralph Molnar, George Mustoe, Andrea Oettl, Paul Olsen, John Ostrom, Frank Paladino, Greg Paul, Felix Pérez-Lorente, Laura Piñuela, Raymond Pippert, Philip Powell, Emma Rainforth, Alan Resetar, Nathan Robinson, John Ruben, Paul Sereno, Kirby Siber, Boyd Simpson, Jana Sizemore, Matt Smith, Mark Staton, Cub Stephens, Glenn Storrs, Cecilia Succar, Clarence Tennis, Alan Tennyson, Cotter Tharin, Tony Thulborn, Doug Townsend, Kate Wellspring, Jim Whitcraft, Trevor Worthy, Joanna Wright, Yvonne Zubovic, and Kristof Zyskowski. It was a particular pleasure to work with Dan Coroian and Phil Currie on the two chapters of the book that they coauthored. There may be others whose names have slipped from my memory, and if so, I apologize.

I was assisted in my zoo and laboratory work by student research assistants, most notably Jana (McClain) Benson and Kate Shearer. They put up with countless hours of mud, bad-tempered big birds, and tedious tracing and measurement of footprint shapes.

As will be apparent in the chapters that follow, I owe a great debt to many "libraries" of biodiversity, the zoos and natural history museums with whose denizens and specimens I have worked. I especially want to thank the staff of the Fort Wayne Children's Zoo, in Indiana, where for several years I routinely collected footprints of emus. I could not have asked for greater cooperation and assistance with my research than I received from Mark Weldon and other zoo personnel.

I am also grateful to a more literal library, specifically the interlibrary loan section at Indiana University–Purdue University Fort Wayne (now Purdue University Fort Wayne), whose staff helped me track down many an old and/or obscure publication.

This research has been supported by grants from the American Chemical Society Petroleum Research Fund, American Philosophical Society, Dinosaur Society, Indiana University, Purdue University Fort Wayne (and the late Patricia Farrell in particular), the National Geographic Society, and the National Science Foundation.

Annette Richter and Peter Falkingham read the entire manuscript, rescuing me from an embarrassing number of bloopers and infelicities of expression, and offering many constructive criticisms. I greatly appreciate this service, and hope that their efforts were reasonably painless! Bob Sloan, Alan Bower, Gary Dunham, and Peggy Solic of Indiana University Press assisted at various stages in the development of this book, sometimes forcing changes on me about which I initially grumbled, but later appreciated.

Most of all, I thank my wife, Karen. She has put up with my absences during field work and research trips to countless museums, and occasionally with interesting things "temporarily" stored in our home freezer. I can express my appreciation for her no better than by quoting an ancient source of wisdom: "A woman's beauty lights up a man's face, and there is nothing he desires more. If kindness and humility mark her speech, her husband is more fortunate than other men" (Sirach 36: 27–28, New Revised Standard Version).

Noah's Ravens

Introduction: Noah's Ravens

1

And it happened, at the end of forty days, that Noah opened the window of the ark he had made. And he sent out the raven and it went forth to and fro until the waters should dry up from the earth.

Genesis 8:6–7 (Alter 2004: 46)

About the year 1802, (possibly a year earlier or later,) Mr. Pliny Moody of South Hadley, in Massachusetts, then a boy, turned up with a plough upon his father's farm in that place, a stone, containing in relief five tracks . . . and it was put down as a door-step . . . and the neighbors used facetiously to remark to Mr. Moody, that he must have heavy poultry that could make such tracks on stone. After Mr. Moody (junior) had left home for school or college, Dr. Elihu Dwight of South Hadley purchased this stone, because it contained these tracks. . . . Dr. Dwight used pleasantly to remark to his visitors, that *these were probably the tracks of Noah's raven.*

Edward Hitchcock (1844: 297; italics in original text)

IN ONE OF THE OLDEST STORIES IN WORLD LITERATURE (Finkel 2014)—possibly based on some real natural disaster in the ancient Near East (cf. Ryan and Pittman 1998; Montgomery 2012)—God or the gods become(s) fed up with humanity, and resolve(s) to wash nearly the lot of them away in a global flood. Fortunately, the hero of the story—variously named Ziusudra, Atrahasis, Utnapishti(m), or, in the most familiar version of the legend, Noah—has found favor with his deity, and is told to build a huge boat to save a remnant of people and animals from the coming watery catastrophe. Once the flood finally begins to ebb, the hero releases a raven, which recognizes the improving weather and takes off for parts unknown, apparently heading west—a bird which turns out to be a highly motivated aeronautical distance traveler.

Because several millennia later, and thousands of miles away, a young farmer in western Massachusetts discovered a sequence of fossil footprints of some three-toed, bipedal animal (fig. I.1). The slab was eventually bought by a neighbor, Dr. Dwight, who wryly joked (see epigraph above) that these were the tracks of none other than Noah's raven.

These and other fossil footmarks of the Connecticut Valley eventually came to the attention of the Reverend Edward Hitchcock, a pioneering American geologist whose landmark studies of what were eventually recognized as non-avian dinosaur tracks launched research on dinosaurs in America more generally (Colbert 1968; Dean 1969; Steinbock 1989; Bakker 2004; Pemberton et al. 2007; Farlow et al. 2012), although Hitchcock thought that his trackmakers had, for the most part, been flightless birds. To honor Dr. Dwight's quip, and to credit that legendary keeper of the world's biggest ocean-going zoo (and thereby acknowledge the importance of zoos for this study), I extend the historic nickname from the maker of a single trackway to the makers of trackways of tridactyl bipedal dinosaurs more generally, and so collectively dub them: Noah's Ravens. (I thank Ben Dattilo for helping me with the idea for doing this.)

DEEP IN THE HEART OF TEXAS

About 60 miles southwest of Fort Worth, Texas, in Somervell County near the little town of Glen Rose, the Paluxy River, a beautiful, free-flowing stream (something of an oddity in a state obsessed with damming such things), makes a hairpin turn to the north, and then back again to the south, in Dinosaur Valley State Park (DVSP). At the beginning of the twentieth century, a truant schoolboy discovered fossilized dinosaur footprints in the rocky bed of a tributary of the Paluxy (Jasinski 2008). Soon thereafter other tracksites were found in the bed of the Paluxy itself, including huge prints made by sauropods. All of these tracks eventually came to the attention of the remarkable fossil hunter R. T. Bird of the American Museum of Natural History in New York (Bird 1985), who in 1940 collected portions of trackways of both a sauropod and a carnivorous dinosaur that may have been following it. DVSP was created by the state of Texas to protect the most important of the local dinosaur tracksites (fig. I.2).

I became involved in studying Early Cretaceous dinosaur footprints of central Texas by accident. As a graduate student at Yale University, I had obviously known about Bird's collection of the Paluxy River tracks, and I was mildly interested in the much older Early Jurassic prints of the Connecticut

I.1. Amherst College 16/2, *Anomoepus scambus*, the sequence of footprints found by Pliny Moody and waggishly said to have been made by Noah's raven. *Photograph used by permission of the Beneski Museum of Natural History at Amherst College, courtesy of the Trustees of Amherst College.*

Valley. However, my advisor, the late, great John Ostrom, specifically warned me against having anything to do with dinosaur footprints. Apart from what tracks might show about dinosaur locomotion and behavior, they were, in John's opinion, a dead-end line of inquiry.

In the early 1980s I received an unexpected letter from an old friend. I had spent the summer of 1972 at the Savannah River Ecology Laboratory near Aiken, South Carolina, working as part of a team studying the phenomenally fast growth of local slider turtles in waters heated by discharge from nuclear reactors of the Savannah River Site. My roommate that summer, Mark Staton, was doing his own study of snakes on the site. In addition to our research projects, we had great times herping together at SREL and other places in the Carolinas.

Mark later married into a family that owned the F^6 Ranch in Kimble County, Texas. Around 1980 a flood had uncovered dinosaur tracks in the bed of a creek on the ranch, in rocks a bit younger than those of the Paluxy River. Mark remembered my interest in dinosaurs, and so wrote to me asking if I wanted to come study the newly exposed footprints. I figured it would be a nice, short-term project. Consequently another faculty member at the college where I then taught and I led a student group to the ranch, where we enjoyed the hospitality of Dave and Margaret Akers (the ranch owners) and had a splendid time measuring, photographing, and casting the tracks.

The footprints at the ranch were all of the three-toed variety, and spanned a modest range of sizes. Interestingly enough, some of the dinosaurs seemed to have been running, probably at a fairly fast clip (Farlow 1981). Some of the prints were comparable in size to the larger footmarks studied by Edward Hitchcock in New England, but others, like many of the tridactyl prints at Dinosaur Valley State Park, were much bigger. I spent a lot of time by myself at the F^6 Ranch tracksite, thinking about the footprints. What kinds of dinosaurs made them? They were all obviously made by bipedal dinosaurs, and I assumed that most of the trackmakers had been theropods (mostly, but not entirely, meat-eaters), but I wondered if some of the prints might have been made by bipedal ornithopods (plant-eaters)—but I was by no means certain that I could tell the difference. The prints that could reasonably be ascribed to theropods included both large and

Facing, **I.2.** Trackways of large bipedal dinosaurs, probably theropods, in the dolostone bed (Glen Rose Formation, Blue Hole Ballroom) of the Paluxy River in Dinosaur Valley State Park, Glen Rose, Texas (Farlow et al. 2015). The individual footprints are about 50 cm long. This is one of several spectacular occurrences of dinosaur tracks in the park (Farlow et al. 2015).

small footmarks. Were the smaller prints made by younger individuals of the same species responsible for the big tracks? And what name should be used to describe all of these footprints? I remembered that R. T. Bird had collected dinosaur tracks from the Paluxy River nearly half a century earlier. I figured that after my group finished working at the ranch, I could go to the literature, see what Bird and subsequent workers had decided about such footprints, use their work to interpret what I saw at the ranch, write the project up, and be done with footprints.

Except that when I went to the literature on Cretaceous dinosaur footprints from Texas, there just wasn't much. Ellis W. Shuler (1917, 1935, 1937) of Southern Methodist University had written short articles about the original finds in the Glen Rose area, and Bird (1939, 1941, 1944, 1954) wrote some short popular articles about his own collecting efforts. There were other short papers describing track occurrences in other areas of central Texas (see Bird [1985: 214] for citations of early work, and Farlow et al. [2006, 2015] for citations of more recent publications). The most technical paper about the footprints had been published by Wann Langston, Jr. (1974) of the University of Texas. None of these publications, however, really addressed all of the questions that had come to mind as I worked on the trackways at the ranch. If my questions were going to be answered, I'd have to do it myself. And so my short-term project became a study of the dinosaur tracks of Texas that has lasted more than 30 years.

In his early studies of the tridactyl dinosaur footprints from the Paluxy River area, Shuler (1917, 1935) cautiously suggested that one of the ichnogeneric names originally applied by Hitchcock to some of the Early Jurassic prints from eastern North America, *Eubrontes* (cf. Rainforth 2005), might also be suitable for the Lone Star tracks. He therefore named the Glen Rose prints *Eubrontes (?) titanopelopatidus* and *Eubrontes (?) glenrosensis*. Langston (1974), on the other hand, thought that the Texas prints were more like tridactyl footprints from the Early Cretaceous of western Canada, which had been given the ichnogeneric name *Irenesauripus* (Sternberg 1932). So for this if no other reason I would eventually have to spend some time looking at both the Canadian and the eastern North American prints.

At that point I still thought that dinosaur footprints were going to be a minor component of my research effort (I was actually then more interested in the teeth of carnivorous dinosaurs). John Ostrom was right; nobody cared much about dinosaur tracks. While it is true that after Hitchcock's time there had been a few workers studying dinosaur footprints, such as Richard Swann Lull and Donald Baird in the eastern United States, Paul Ellenberger in southern Africa, Giuseppe Leonardi in South America, and Harmut Haubold, Georges Demathieu and Georges Gand in Europe, ichnology was for the most part a nice, quiet, backwater area of research on dinosaurs.

And then the metaphorical dam broke.

It seems that just as I was becoming interested in dinosaur tracks, so were many other folks around the world. From about the early 1980s on, and showing no signs of letting up even as I write this, there has been an explosion of research interest in dinosaur footprints. Among the more active workers have been Martin Lockley in the western United States, M. L. Casanovas Cladellas and Felix Pérez-Lorente in Spain, Georges Gand in France, Tony Thulborn in Australia, and Lida Xing in China, to name just a few. There has been a latter-day deluge, a flood this time of papers and books about dinosaur tracks.

As I tried to keep up with all this work, the same questions that had perplexed me at the F^6 Ranch were never far from my mind. A few papers had a particular impact on my own research.

One of these was by Paul Olsen (1980) of Columbia University's Lamont Doherty Earth Observatory, who had occupied a basement office of Yale's Peabody Museum across the hall from my own when I was a graduate student at Yale. Paul worked on the stratigraphy and paleontology of the Newark Supergroup, a thick sequence of sedimentary rocks in eastern North America formed during the initial stages of continental breakup of Europe and North America. Among the Newark Supergroup fossils he studied were the very kinds of dinosaur footprints Edward Hitchcock had researched. In his 1980 paper Paul presented an interesting graph of footprints attributed to theropods; he plotted the amount by which the middle toe (digit III) protruded forward beyond the limits of the two peripheral toes (what I will term the digit III projection in this book) against the length of the footprint behind this point (which I will characterize as backfoot length). The graph showed a continuous trend (fig. I.3), which Paul suggested might indicate that the prints constituted an ontogenetic sequence. In later work (Olsen et al. 1998) Paul backed off a bit from this interpretation, but that 1980 graph planted the seed of a question in my mind: What would an ontogenetic sequence of tridactyl footprints, from those of very small, young individuals to those of large adults, look like? How much change in footprint shape would there be?

In 1992 Robert Weems of the U.S. Geological Survey extended Olsen's approach by considering the width across the tips of the two peripheral toes as well as digit III projection and backfoot length. In so doing Weems was able to separate the Newark Supergroup tridactyl footprints attributed to saurischian dinosaurs (theropods and, possibly, prosauropods) into distinct groups. This suggested that a multivariate

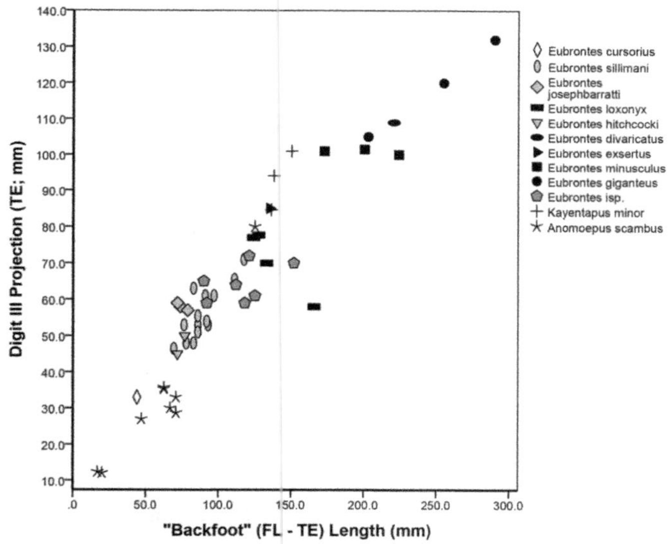

I.3. Graph of digit III projection (also known as the toetip extension [TE]) against backfoot length (overall footprint length minus the toetip extension [FL–TE]) in tridactyl dinosaur footprints from the Newark Supergroup of eastern North America. Olsen (1980) pioneered the use of such graphs. The data plotted here are from my own measurements and represent either mean values for footprints within trackways or individual measurements of "singleton" footprint specimens. The names applied to specimens largely follow the work of Weems (1987, 1992) and Rainforth (2005); see chapter 8 for details.

approach could be fruitfully applied to the recognition of different kinds of tracks and trackmakers.

This impression was bolstered by the third of the papers that had the greatest influence on me. J. J. Moratalla and his colleagues (1988*a*) at the Universidad Autonoma in Madrid took a series of measurements that had traditionally been used to distinguish footprints attributed to theropods from prints attributed to ornithopods, and showed that a multivariate analysis using these parameters did indeed recognize the same distinct groups of footprints. This was no small result, because it showed that a quantitative analysis picked out the same shape categories that had previously been identified more qualitatively. The differences in footprint shape were real, and more than mere pseudo-patterns like faces or animal shapes seen in clouds.

Moratalla et al. (1988*a*) were not the only workers to apply statistical analyses to the discrimination of different morphotypes of tridactyl dinosaur footprints. French workers, most notably Georges Demathieu and Georges Gand, had actively pursued this approach; see, for example, Demathieu (1990), Demathieu et al. (2002), and Gand et al. (2007), and earlier studies cited therein. The research of the American, French, and Spanish workers convinced me of the usefulness of quantitative approaches to discriminating among footprint morphotypes. As will become apparent much later in this book, Demathieu's 1990 paper had particular relevance to my own research.

But it seemed to me that the approach of Moratalla et al. (1988*a*) had one limitation. How did one know, in the first place, that the traditional morphological characters that had been thought to distinguish theropod from ornithopod footprints really were features specific to theropods and ornithopods?

And this brings me to the single publication that probably had the greatest impact of all on my thinking: Donald Baird's classic 1957 study of the Newark Supergroup footprints from Milford, New Jersey. One statement therein stood out: "I submit that the characters most diagnostic for the classification of footprints as such, as well as most useful for comparison with skeletal remains, are those which reflect the bony structure of the foot" (Baird 1957: 469). Baird's contention that one should pay especially close attention to footprint characters that could be correlated with features of foot skeletons struck me like the proverbial thunderbolt with its . . . well . . . "obviousness." If you can't tell the foot skeletons of bipedal dinosaurs apart, are you likely to be able to tell their footprints apart? As zoological categories, theropods and ornithopods are defined on the basis of skeletal criteria. It seemed to me that there needed to be explicit criteria for correlating between footprint shapes and footprint skeletal proportions (Buckley 2015 seems to have been thinking along similar lines). I have been looking at foot skeletons ever since.

BONES, BIG DUMB BIRDS, AND ALLIGATORS

By the mid-1980s I had concluded that, as prescient as John Ostrom had been about other aspects of dinosaur paleontology, he had been wrong about studying dinosaur footprints. I had begun working on the Early Cretaceous tracks from Texas in earnest. At this time I was contacted by the producers of an educational film series about the importance of mathematics for solving scientific and other problems. They wanted to do a segment about my work on Texas dinosaur footprints, but they wanted to film something livelier than static footprints in rock. Could I think of something?

I did: Our local Fort Wayne Children's Zoo had ostriches. Birds are bipedal archosaurs and, as Ostrom's research was showing, most likely a group within dinosaurs, and the only one to survive to the present. Ostriches are the biggest extant birds, and flightless to boot, and they can be far from static when properly motivated. So we filmed ostriches trotting around a holding pen at the zoo, and I measured their trackways and made casts of their footprints (Farlow 1989).

Ostriches, however, have an odd two-toed foot, and so they are not the best analogs for comparing with tridactyl non-avian dinosaurs. Kevin Padian and Paul Olsen (1989) had done some experimental work with footprint formation by rheas, which have nice, three-toed feet, and their project and my ostrich film shoot got me thinking about doing some more of this kind of research. As it happened, the Fort Wayne Children's Zoo had opened an Australian wildlife exhibit, among the denizens of which were individuals of another species of big, flightless, tridactyl bird: emus. So I embarked on a massive study of footprint and trackway formation by these ratites, chasing birds from hatchlings to adults across a variety of substrates, and making a very large collection of plaster casts of their footprints. I also collected footprints from many other species of ground birds at zoos around the country for comparison with my emu tracks. I wanted to quantify variability in footprint shape within emus, assess the extent to which footprint shape changed as emus grew and as a function of substrate conditions, and determine how well footprint shape could be used to tell different species of bird trackmakers apart.

At the same time I began visiting natural history museums and universities in the United States and around the world. I was collecting measurements of the bones in foot skeletons of dinosaurs, but also of living and extinct species of ground birds and crocodylians. I wanted to see how well pedal proportions could distinguish among species or more inclusive groups of animals. If one couldn't distinguish between the skeletons of an allosaur and a tyrannosaur, for instance, there was a fair chance that one wouldn't be able to tell their footprints apart on the basis of print shape alone. I also wanted to determine in what features of foot shape different kinds of non-avian dinosaurs differed from one another, and to what extent.

I reasoned, though, that the number of reasonably complete foot skeletons of Mesozoic dinosaurs might be limited, and so I spent as much or more time measuring foot skeletons of ground birds. This would be a parallel approach to my study of bird tracks, asking the same question asked of the bird prints and of the dinosaur foot skeletons: To what extent can foot skeletons (whether of birds or of non-avian dinosaurs) and footprints (of ground birds) be told apart?

One thing just led to another. I was disappointed that many of the footprints I collected of small ground birds lacked detail. So I tried something else: feet of study skins in ornithology collections, which showed surprisingly good preservation of the soft tissues of the feet. Another source of data!

Following the rationale of the extant phylogenetic bracket (Bryant and Russell 1992; Witmer 1995), I thought it important to look at intraspecific and interspecific variability in pedal proportions of that other major clade of extant archosaurs, the crocodylians. These reptiles might not be as phylogenetically close to non-avian dinosaurs as birds are, but, as with other aspects of their biology (Brazaitis and Watanabe 2011), what is observed about foot shape variability in crocs might nonetheless shed light on what was going on in Mesozoic dinosaurs. I looked at crocodylian foot skeletons, but also at the configuration of the surface features of the underside of the hindfoot in intact alligators and pickled specimens of other species. This led me on minor adventures, to the Rockefeller Wildlife Refuge in Louisiana, and back to the Savannah River Ecology Laboratory, where I took part in capturing live alligators of various sizes to measure their feet and other body parts.

BACK TO THE ORIGINAL NOAH'S RAVENS

Even though throughout this research I have been primarily interested in tridactyl footprints from the Early Cretaceous of Texas, I nonetheless thought it important to spend a great deal of time examining the classic Early Jurassic Newark Supergroup tracks made famous by the likes of Edward Hitchcock, Richard Swann Lull, Donald Baird, Robert Weems, Paul Olsen, and Emma Rainforth. Shuler had thought the Glen Rose tracks might be aptly described as a form of *Eubrontes*, and so it seemed necessary for me to gain some familiarity with that ichnogenus. This resulted in my working with Peter Galton on the spectacular exposures of large tridactyl footprints at Dinosaur State Park in Rocky Hill, Connecticut (Farlow and Galton 2003)—ironically, the same footprints that John Ostrom had warned me against getting involved with when I was a graduate student.

There was, however, a more important reason for my interest in the dinosaur tracks from eastern North America. As time went on, I found that I was becoming less interested in the question of what ichnotaxonomic label to apply to these or my tracks from Texas than in the matter of how well they served as proxies for the biodiversity of the dinosaurs that made them. Could I make a reasonable estimate of how many different kinds of trackmakers there were, and what the affinities of those trackmakers were? On the basis of my research on feet and footprints of dinosaurs, birds, and crocodylians I thought I was getting a handle on how to approach such questions, but I needed a test case on which to try my methods. The Newark Supergroup tridactyl dinosaur footprints include some of the best-preserved dinosaur tracks in the world (as also recognized by Demathieu 1990: 98), so they seemed to me to be ideal for my purposes. If I can't say anything intelligent about the biodiversity encapsulated in these beautiful footprints, I probably won't be able to say much about tracks from any other footprint faunas.

And that is what this book is all about. In it I will present the results of my comparative studies of dinosaur, bird, and crocodylian foot skeletons, crocodylian and bird intact feet, and ground bird footprints. I will then use the results of this work, as applied to a sample of the Newark Supergroup footprints themselves, to make an interpretation of the minimum number of kinds of dinosaurs responsible for the footprints, and try to identify the groups of dinosaurs to which the trackmakers belonged.

I will not, however, attempt any kind of systematic revision of the nomenclature applied to the Newark Supergroup prints. That is a separate, albeit closely related, matter from the topic of this book. It would be an undertaking comparable in magnitude to what I present here, and I have only one lifetime to devote to such things! Besides, there are other persons already working on the ichnotaxonomy of eastern North American dinosaur tracks, and I think them better qualified than I to tackle issues of footprint naming and classification. Having said that, I do hope, of course, that what I have done in this study will be of some use to other workers, not just to students of the Newark Supergroup footprints, but to dinosaur ichnologists more generally.

My original intent was to publish this work in a series of papers, the more customary way of doing things in science. But the more I worked on the project, the more the individual topics seemed to intermingle, every topic having ramifications for every other topic. It seemed harder to know how to separate them out, and I finally gave up. It all felt like, to borrow a phrase from an impeccable source, "one long argument." So here it all is.

I have tried, throughout this book, to make my arguments and interpretations data driven, rather than reflecting any *a priori* theoretical stance. I have, in fact, on more than one occasion been mightily surprised by what the data were telling me, and had to change my mind about what I thought was going on.

WHY SO MANY TABLES AND GRAPHS?

This book is not, I regret to say, light bedtime reading (although some sections might constitute a sure-fire cure for insomnia). To understand its arguments fully, the reader will repeatedly have to flip back and forth from table to table, or figure to figure, both within and across chapters and in appendices, and may be overwhelmed by the multitude of tables and graphs that are presented. The need for so many different presentations of data arises from two unavoidable problems that we will repeatedly encounter: the incompleteness of many specimens, both of footprints and foot skeletons, due to the vagaries of preservation, and the great size range of specimens—from small fry like *Bambiraptor* to monsters like *Tyrannosaurus*—being compared.

In the three main toes of the pedal skeleton of a typical bipedal dinosaur, there are three phalanges on digit II, four on digit III, and five on digit IV. Ideally we would like to be able to obtain measurements of the lengths and widths of each of these phalanges. Unfortunately, many dinosaur foot skeletons, as preserved, lack some of the toe bones, or the preservation of those bones is such that some of the measurements we want cannot be made.

Similar problems arise with footprints. We would like to have measurements of the lengths and widths of the individual digital pads, and lengths of the claw marks, as well as other parameters, but in many cases the preservation of the print is such that some of these data cannot be obtained.

Then there is the problem of the huge range of absolute sizes in the foot skeletons and footprints of bipedal dinosaurs. If we want to compare shapes of feet and prints, we need to be able to take these size differences into account by some kind of scaling—otherwise, big specimens will tend to be more like other big specimens, and small specimens like other small specimens, even if they differ considerably in proportions.

The most informative approach would be to scale measurements in such a way that all of the size and shape parameters of a foot or a footprint are taken into account in creating the scaling factor. However, the number of foot skeletons for which all of the measurements of individual toe bones can be made, or the number of footprints in which all of the measurements of interest can be made, is much smaller than the number of foot skeletons or footprints in which only some of the desired measurements can be taken. Consequently using a scaling factor that is based on all the possible measurements would limit us to a rather small sample size of specimens in the analysis, which might give misleading results.

We will therefore constantly be forced to choose between many measurements that can be made on only a few specimens, or a small number of measurements that can be made on many specimens. I will attempt to tack between Scylla and Charybdis by using many different scales, ranging from simple bivariate comparisons at one extreme, to more complicated multivariate scales at the other, to see if different ways of scaling parameters yield consistent results. Because I will often be analyzing the same features of foot or footprint shape in many different ways, I will have to employ many tables and graphs of data in our analyses.

I will try to make all this no more painful than necessary, and so this is a good place to describe a procedure I will use

Table I.1. Key of abbreviations to museum collections and other institutions

Abbreviations	Institutions	Abbreviations	Institutions
AC, ACM, ACM.ICH	Beneski Museum of Natural History, Amherst College	MSU	Michigan State University Museum
AM	Auckland War Memorial Museum	MUCPv	Museo de Ciencias Naturales, Universidad Nacional del Comahue
AMNH	American Museum of Natural History	NCSM	North Carolina Museum of Natural Sciences
ANSP	Philadelphia Academy of Natural Sciences, Drexel University	NHMUK	Natural History Museum, London
AU	University of Arkansas Museum	NMNZ, DM	Museum of New Zealand Te Papa Tongarewa
BHI	Black Hills Institute of Geological Research	OMNH, OU	Sam Noble Oklahoma Museum of Natural History
BNMH, BMRP	Burpee Museum of Natural History	PU	Princeton University
BYU	Brigham Young University Earth Science Museum	PVSJ	Museo de Ciencias Naturales, Universidad Nacional de San Juan, Argentina
CEUM	Utah State University Eastern Prehistoric Museum	QM	Queensland Museum (Brisbane, Australia)
CITES	Convention on International Trade in Endangered Species	QVM	Queen Victoria Museum and Art Gallery (Launceston, Tasmania)
CM	Carnegie Museum of Natural History	RMDRC	Rocky Mountain Dinosaur Resource Center, Woodland Park, Colorado
CM AV	Canterbury Museum	ROM	Royal Ontario Museum
CMN	Canadian Museum of Nature	RTMP, TMP	Royal Tyrrell Museum of Paleontology
CMNH	Cleveland Museum of Natural History	RWR	Rockefeller Wildlife Refuge, Grand Chenier, Louisiana
DGM	Divisão de Geologia e Mineralogia do Departamento Nacional da Produção Mineral, Rio de Janeiro	SCMG	Sheridan College Geology Museum, Sheridan, Wyoming
DMNH	Denver Museum of Nature and Science	SGM	Ministere de l'Energie des Mines, Rabat, Morocco
DNM	Dinosaur National Monument	SMA	Sauriermuseum Aathal, Switzerland
FIP	Florida Institute of Paleontology	SMM	Science Museum of Minnesota
FMNH	Field Museum of Natural History	SMP	State Museum of Pennsylvania
FWMSH	Fort Worth Museum of Science and History	SMU	Southern Methodist University
GI, IGM	Geological Institute, Mongolian Academy of Sciences	SREL	Savannah River Ecology Laboratory, Aiken, South Carolina
GM	Geiseltal Museum, Halle, Germany	TMM	Texas Vertebrate Paleontology Collections at the University of Texas
HMN	Museum für Naturkunde, Berlin	TPII	Thanksgiving Point Institute (North American Museum of Ancient Life, Lehi, Utah)
IRSNB	Institut Royal des Sciences Naturelles de Belgique	UCMP	University of California Museum of Paleontology
LACM	Natural History Museum of Los Angeles County	UF	Florida Museum of Natural History
LHPV	Lang Hao Institute for Paleontology, Hohhot, Nei Mongol Autonomous Region, China	UMA	University of Massachusetts Amherst Natural History Collections
MACN	Museo Argentino de Ciencias Naturales "Bernardino Rivadavia"	UMMZ	University of Michigan Museum of Zoology
MCF-PVPH	Museo Municiap Carmen Funes Paleontología de Vertebrados, Plaza Huincul, Argentina	UMNH	Utah Museum of Natural History
MCZ	Museum of Comparative Zoology, Harvard University	UNSM	University of Nebraska State Museum
MMP	Museo Municipal de Ciencias Naturales, Lorenzo Scaglia, Mar del Plata, Argentina	USNM	National Museum of Natural History, Smithsonian Institution
MNA	Museum of Northern Arizona	WO	Waitomo Caves Museum, New Zealand
MOR	Museum of the Rockies	WWU	Western Washington University
MPCA	Museo Carlos Ameghino Cipolletti, Río Negro, Argentina	YORYM	Yorkshire Museum (York Museums Trust)
MPC-D	Paleontological Center, Mongolian Academy of Sciences, Ulaanbaatar	YPM VP	Peabody Museum of Natural History, Yale University
MSF	Sauriermuseum-Frick, Switzerland	YPM VPPU	Peabody Museum of Natural History, Yale University—specimen formerly in collection of Princeton University

throughout this book to make it more user-friendly. There will be many tables of data, or tables providing results of analyses, in most of the chapters. Some of these tables will have results that are more immediately critical for understanding the text than others; such tables will appear in the main text of the chapter. Many other tables, however, are either less immediately critical, or are so long and complicated that including them in the main text of the chapter would be disruptive. Such tables will, however, present useful supplementary material that supports the conclusions of the chapter, and so need to appear somewhere. I will relegate such "second-order" tables to the Appendix at the end of the book.

At many places through this book I will refer to particular specimens in museum or other collections. A key to the institutional acronyms/abbreviations appears in table I.1.

OK, that's it for the introduction. It's show time.

1 Intraspecific and Interspecific Variability in Pedal Phalangeal and Digital Dimensions and Proportions in Non-avian Dinosaurs, Birds, and Crocodylians

I often say that when you can measure what you are speaking about, and express it in numbers, you know something about it; but when you cannot measure it, when you cannot express it in numbers, your knowledge is of a meagre and unsatisfactory kind.

William Thomson (Lord Kelvin), *Popular Lectures* 1: 73, 1883

REGRETTABLY, NON-AVIAN DINOSAURS ARE EXTINCT. WE can no longer watch such animals in the act of making footprints, and so it is no longer possible to be certain that footprints we think might have been made by dinosaurs of the same species really were. We must take an indirect approach. Knowing what the intact sole of the foot of dinosaurs was like would help here; dinosaur "mummies" and other remarkably preserved specimens sometimes provide such information (e.g., Brown 1916; Cuesta et al. 2015; Wang X. et al. 2017), but there aren't enough of them.

Non-avian dinosaur species are usually named on the basis of skeletal material. We can look at foot skeletons of non-avian dinosaurs and other archosaurs. If pedal intraspecific size and shape variability turned out to be comparable among non-avian dinosaurs, birds, and crocodylians, this would bolster our confidence that the same would have been true for their intact feet and footprints—although we must always keep in mind that ascertaining that skeletons of extinct animals like dinosaurs were members of the same species is itself a challenging problem.

The present chapter examines size and shape variability of foot skeletons of extant crocodylian and bird species that are unambiguously members of the same species, as well as variability of foot skeletons of what are thought to be members of the same species of several kinds of extinct birds and non-avian dinosaurs. The question is, essentially, how different in size and shape of foot can skeletons be and still belong to the same species? And is the amount of variability among species within a group greater than that within a species? We will consider several different measures of variability, themselves of highly variable sophistication, for comparing intraspecific and interspecific variability in pedal proportions.

MATERIALS AND METHODS

I collected measurements of the phalangeal skeletons of numerous specimens of crocodylians, birds, and non-avian dinosaurs from collections in North America, Europe, Australia, and New Zealand (table A1.1; figs. 1.1–1.12); these measurements are used in this and later chapters. As in earlier studies (Farlow and Lockley 1993; Farlow and Chapman 1997; Farlow 2001; Smith and Farlow 2003; Farlow et al. 2012, 2013, 2014), I measured non-ungual phalangeal lengths (fig. 1.1) along the medial and lateral sides of the toe bone, parallel to the long axis of the bone, from about the dorsoventral midpoint (or the most concave point, if different) of the concave proximal cotyla (terminology of Baumel et al. 1979) to the dorsoventral midpoint (or most distal point, if different) of the convex distal articular trochlea of the bone. If it was not possible to measure phalanx lengths on both the medial and lateral sides of the bone, whichever length could be measured was used instead of the mean length. Non-ungual proximal widths were taken as the greatest transverse distance across the proximal face of the phalanx, perpendicular to the long axis of the bone. Non-ungual phalanx distal widths were taken as the greatest transverse distance across the distal trochlea, perpendicular to the long axis of the bone. Ungual lengths were measured in a straight line, again on both medial and lateral sides (where possible) of the bone, from the dorsoventral midpoint (or the most distal point) of the concave proximal cotyla to the distal tip of the ungual, located near the midline of the bone.

There were certain unavoidable problems in the database. Because there are numerous phalanges in an archosaurian foot, it is common (particularly for non-avian dinosaurs) for fossil specimens to have incomplete or poorly preserved feet. This is less of a problem for foot skeletons of extant or subfossil birds. For specimens of extant crocodylians, however, I only measured lengths of phalanges that remained bound together in articulation by dried soft tissues; I was less confident of my ability to correctly identify loose phalanges of crocodylians in specimen boxes than of my ability to make such identifications for bird and dinosaur specimens. Another problem for most crocodylian feet in osteological collections is that the horny claw does not easily separate from the ungual, making accurate measurements of unguals impossible, and so crocodylian unguals were not used in the present study.

For many dinosaur specimens it was not possible to measure phalangeal lengths on both the medial and lateral sides of the bone, either because the foot was still embedded in matrix or because of poor preservation. This could potentially introduce "bogus" variability within a sample for a particular taxon, if phalanx lengths of some specimens used in an analysis were means of both medial and lateral lengths, while lengths of other specimens were only medial or lateral lengths. This problem would likely be greater for phalanges of digits II and IV (the peripheral toes of functionally tridactyl dinosaurs), and less for the more symmetrical digit III phalanges.

If a given specimen had complete sets of phalanges for both feet, I commonly used measurements for only one foot. If, however, both feet were incomplete, I often combined measurements from both feet to create data for a composite, "synthetic" foot of that individual. Where possible, I took the average of medial and lateral lengths of each phalanx as its phalangeal length, but used just the medial or the lateral length if this was the only measurement of phalangeal length that could be made; I call these averages "blended" phalangeal measurements. As already noted, employing data for feet in which some, but not all, phalangeal lengths were means of medial and lateral lengths could introduce artifacts in calculations of intraspecific or other kinds of within-group variability in phalangeal size and shape, a possibility that will be considered in this chapter for the dinosaur species with the largest sample of feet, *Iguanodon bernissartensis*. In some fossil specimens the tips of unguals had broken off; if I thought I could accurately estimate ungual length within a few millimeters, or if I thought that the measured ungual length was a reasonable approximation of its true length, I went ahead and used the ungual length in my analyses. Because of this postmortem damage to unguals, however, variability in ungual lengths may be artificially higher in comparison with variability in lengths of non-ungual phalanges.

Most measurements of phalanx or digit lengths and widths were made to the nearest millimeter. For small phalanges of young crocodylians, however, lengths were measured to the nearest 0.5 or even 0.1 mm.

In some analyses, data just for the larger pedal phalanges (II1–3, III1–4, IV1, IV2, IV5) were analyzed, with phalanges II1, II2, III1–III3, IV1, and IV2 (here designated the "big seven" phalanges) given particular attention; phalanges IV3 and IV4 were not so treated, but their measurements were obviously used in analyses of overall digit lengths. Overall, cumulative digit length was calculated as the summed lengths of the phalanges of each digit. Digit lengths were examined both with and without the unguals.

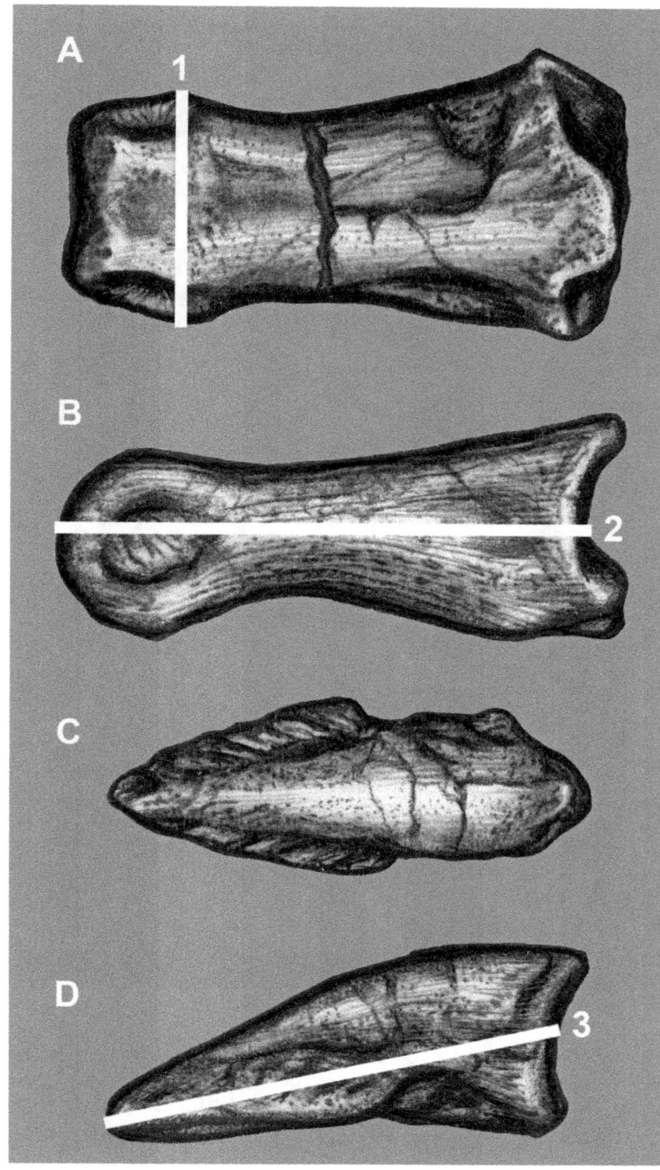

1.1. Measurements of pedal phalanges. *A, B,* Non-ungual phalanx in dorsal (*A*) and side (*B*) views. Measurement 1 = distal width; measurement 2 = phalanx length. *C, D,* Ungual phalanx in dorsal (*C*) and side (*D*) views. Measurement 3 = ungual length. Proximal widths of non-ungual phalanges and unguals are not drawn on the figure but were measured as the maximum transverse width across the proximal articular end of the bone. *Drawing by Emma Schachner, taken from Farlow et al. (2014).*

Apart from issues of completeness and quality of preservation of foot skeletons, the biggest issue in the present study was the sample size of feet for each species, which was limited by what was available in museum collections or (in the case of alligators) what I was able to collect myself. Consequently, my most detailed analyses were for alligators (*Alligator mississippiensis*), emus (*Dromaius novaehollandiae*), ostriches (*Struthio camelus*), rheas (*Rhea americana* and

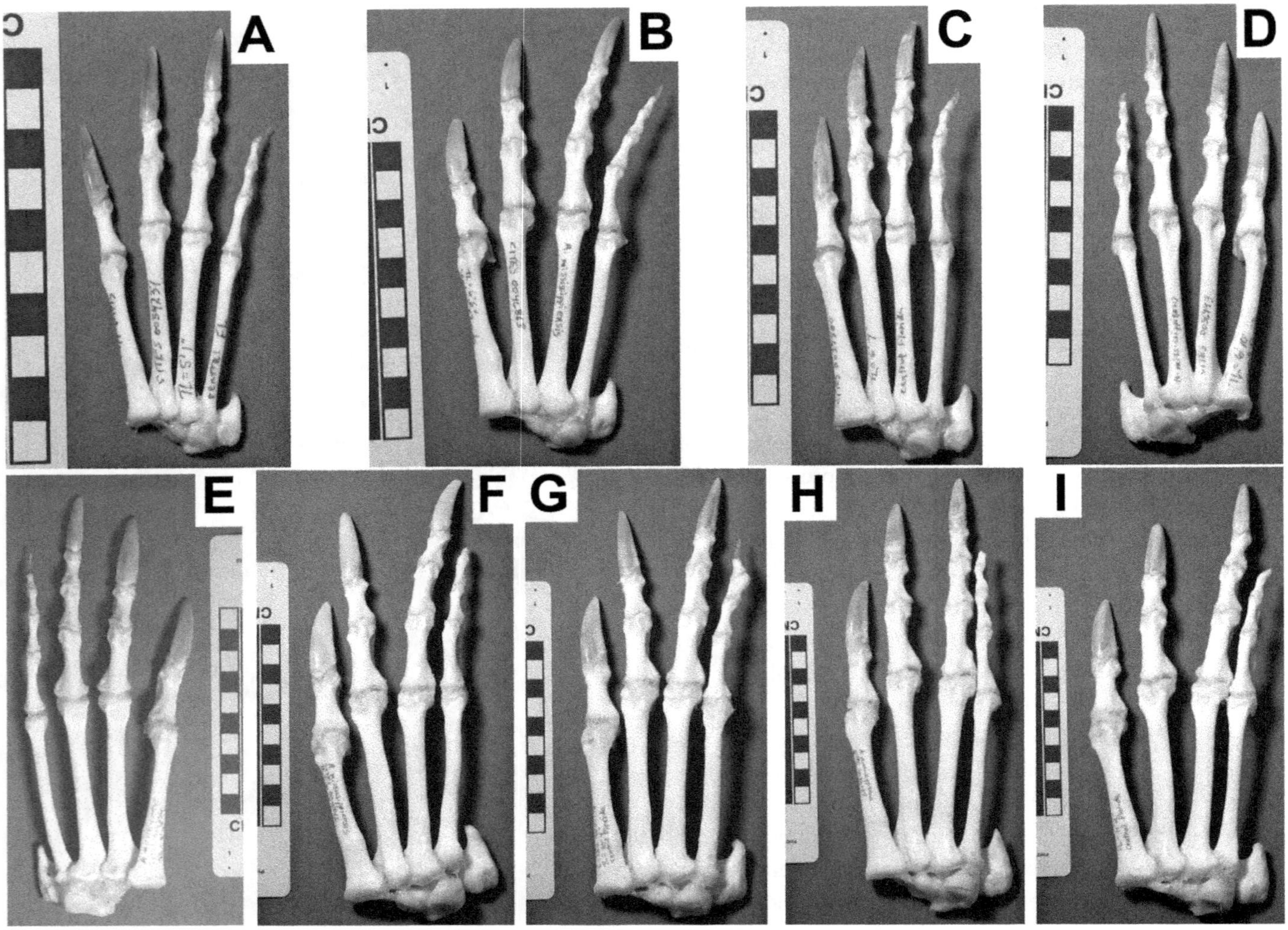

1.2. Series of foot skeletons (*A–C, F–I* = rights; *D, E* = lefts) of legally hunted specimens of *Alligator mississippiensis* from central Florida; the same feet also were measured while intact, before being defleshed (chap. 5) (some of the same specimens are illustrated as intact feet in fig. 5.2). Specimens are arranged in order of increasing total length (TL) of the intact alligator. Scale bar in centimeters. *A*, CITES 0054237; TL = 155 cm. *B*, CITES 0046863; TL = 198 cm. *C*, CITES 0027580; TL = 201 cm. *D*, CITES 0056493; TL = 208 cm. *E*, CITES 0043361; TL = 272 cm. *F*, CITES 0041015; TL = 282 cm. *G*, CITES 0042178; TL = 315 cm. *H*, CITES 0040187; TL = 325 cm. *I*, CITES 0041602; TL = 340 cm.

Pterocnemia [assigned to *Rhea* by Sibley and Monroe (1990), but not Dickinson (2003) or Clements (2007)] *pennata*), kiwi (*Apteryx australis*), moa (particularly *Anomalopteryx didiformis, Pachyornis elephantopus*, and *Dinornis robustus*), and the non-avian dinosaurs *Coelophysis bauri*, *Tyrannosaurus rex* (provisionally including *Nanotyrannus lancensis*), and *Iguanodon bernissartensis*, forms represented by a reasonable (but not huge) number of specimens.

Some of these taxa deserve further mention. Early in my work I recognized that moa (Dinornithiformes) could be of particular interest for this project. The nine generally recognized species constitute an adaptive radiation of big flightless birds that lived in prehistoric New Zealand (Cooper et al. 1992, 2001; Worthy and Holdaway 2002; Bunce et al. 2003, 2009; Huynen et al. 2003; Baker et al. 2005; Worthy 2005; Allentoft and Rawlence 2012; Worthy and Scofield 2012; Brassey et al. 2013; Olson and Turvey 2013; Bishop 2015; Angst and Buffetaut 2017; Mayr 2017), an area roughly comparable in size, interestingly enough, to the collected depositional basins of the Newark Supergroup of eastern North America during the Early Jurassic (home to the historically important footprint fauna of Edward Hitchcock). As subfossil birds, moa have only recently become extinct, and so were sure to be represented by larger sample sizes of foot skeletons than non-avian dinosaurs. In addition, the moa would have been as close in size to non-avian dinosaurs as any other avian clade. For all these reasons they have been given particular emphasis in this study.

In collecting measurements of moa foot skeletons in museum collections in New Zealand, Britain, and the United

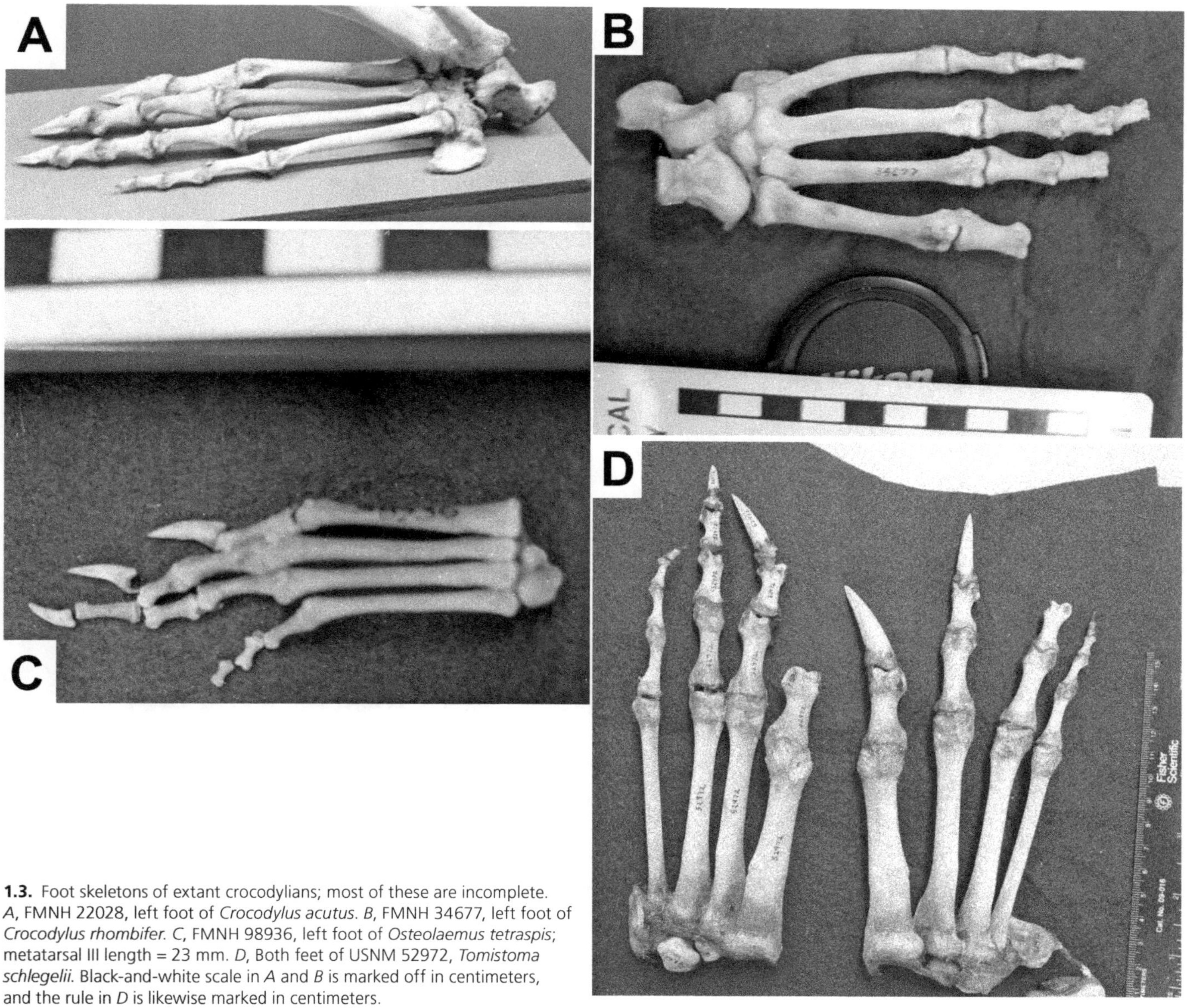

1.3. Foot skeletons of extant crocodylians; most of these are incomplete. *A*, FMNH 22028, left foot of *Crocodylus acutus*. *B*, FMNH 34677, left foot of *Crocodylus rhombifer*. *C*, FMNH 98936, left foot of *Osteolaemus tetraspis*; metatarsal III length = 23 mm. *D*, Both feet of USNM 52972, *Tomistoma schlegelii*. Black-and-white scale in *A* and *B* is marked off in centimeters, and the rule in *D* is likewise marked in centimeters.

States, I generally used the museum specimen identifications associated with those specimens, supplemented by discussions with moa specialists (particularly Trevor H. Worthy). Allentoft et al. (2010) compared species and specimen identifications based on morphology with identifications based on ancient mitochondrial DNA, and found fairly good agreement between molecular and morphological species identifications of moa in their sample; discrepancies involved *Euryapteryx curtus* vs. *Pachyornis elephantopus*, *Emeus crassus* vs. *P. elephantopus*, *E. curtus* vs. *E. crassus*, or juvenile birds identified as emeids vs. *Dinornis robustus*. One specimen in my sample (Canterbury Museum CM AV 8622) that was identified on the basis of morphology as being *E. curtus* was assigned by Allentoft et al. (2010) on the basis of DNA to *P. elephantopus*. In analyses of intraspecific variability in pedal proportions of *P. elephantopus* reported in this chapter, results will be reported both excluding and including CM AV 8622.

As the potential maker of tridactyl dinosaur footprints represented by the largest sample size, *Iguanodon bernissartensis* figures prominently in this chapter. Consequently some comment about the nature and taxonomy of these specimens is in order.

My sample of *I. bernissartensis* and related forms comprises skeletons (table 1.1) in the collections of the Institut Royal des Sciences Naturelles de Belgique (IRSNB) and London's Natural History Museum (NHMUK; formerly BMNH) (Norman 1980, 1986, 2010, 2012, 2014; Paul 2007, 2008*b*, 2010; Verdú et al. 2017). Norman (1987*a*: 247) speculated that the *Iguanodon* assemblage at Bernissart accumulated over "an

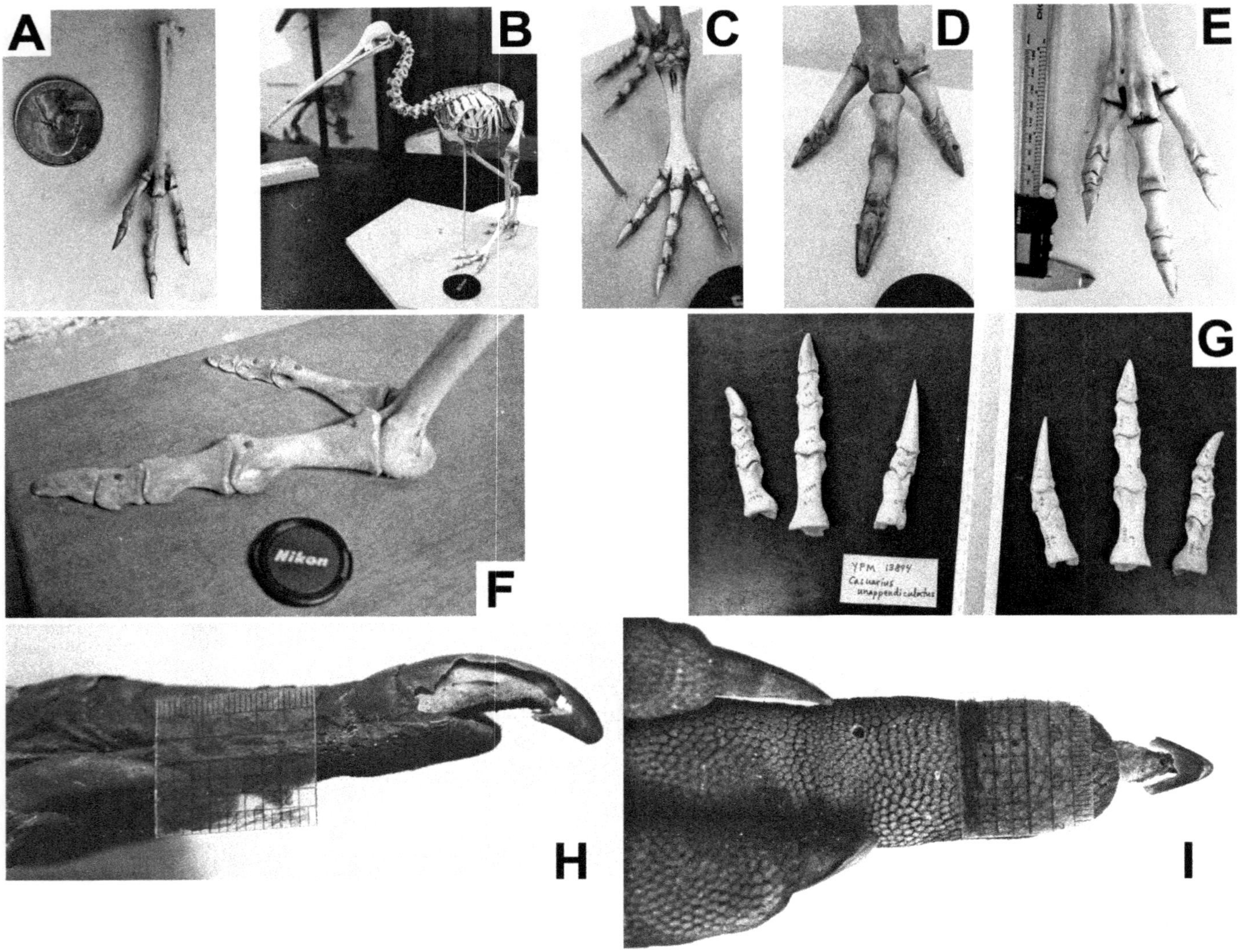

1.4. Foot skeletons of modern palaeognathous birds. *A*, Left foot of USNM 345018, elegant crested tinamou (*Eudromia elegans*); American quarter provides scale. *B*, *C*, Mounted skeleton and left foot of kiwi (*Apteryx* sp.), University of Illinois Museum; 35-mm camera lens cap provides scale in these photographs and in *D* and *F*. *D*, Left foot of greater rhea (*Rhea americana*), University of Illinois. *E*, Right foot of USNM 288184, lesser rhea (*Pterocnemia* [or *Rhea*] *pennata*); calipers are marked off in centimeters and inches. *F*, Right foot of MSU OR.8537, ostrich (*Struthio camelus*); courtesy Michigan State University Museum. *G*, Left and right pedal phalanges of YPM ORN 13894, northern cassowary (*Casuarius unappendiculatus*); total digit III length = 155 mm. *H*, *I*, Distal end of digit III of MSU OR.8778, emu (*Dromaius novaehollandiae*) in medial (*H*) and ventral (*I*) views; transparent scale marked off in centimeters is placed against the specimen. Note that the bony ungual terminates well proximal to the tip of the horny claw. *Courtesy Michigan State University Museum.*

appreciable period of time (?10-100 years)." This is a short enough time that it is unlikely that there would have been any evolutionary change in pedal shape in the species present, and so I will treat the Bernissart *Iguanodon* specimens as an essentially contemporaneous assemblage.

The Bernissart *Iguanodon* specimens fall into three categories, as recognized by Norman (1980, 1986): three "subadult" individuals of *I. bernissartensis*, several adult individuals of the same species (but obviously we cannot be certain about the sexual maturity of these animals in the way that is possible for, say, extant American alligators—another species that figures prominently in this book), and at least two specimens of a second species. Although the two putative species have been interpreted as sexual dimorphs of a single species (Carpenter 1999), Norman (1986) provisionally concluded that they were probably distinct species.

At the time I examined the specimen, NHMUK R1829 was labeled in the Natural History Museum (London)'s collection as *I. mantelli*. Norman (1986) regarded this taxon as an objective junior synonym of *I. anglicus*, which in turn is a dubious taxon based on teeth (Norman 1986; Charig and Chapman 1998, 2000). *I. mantelli* was therefore not an appropriate name for the small, gracile form (IRSNB R 57 [old 1551]) of *Iguanodon* at Bernissart, even though it had previously been identified under this name. Norman (1986) concluded that IRSNB R 57 was appropriately assigned to I.

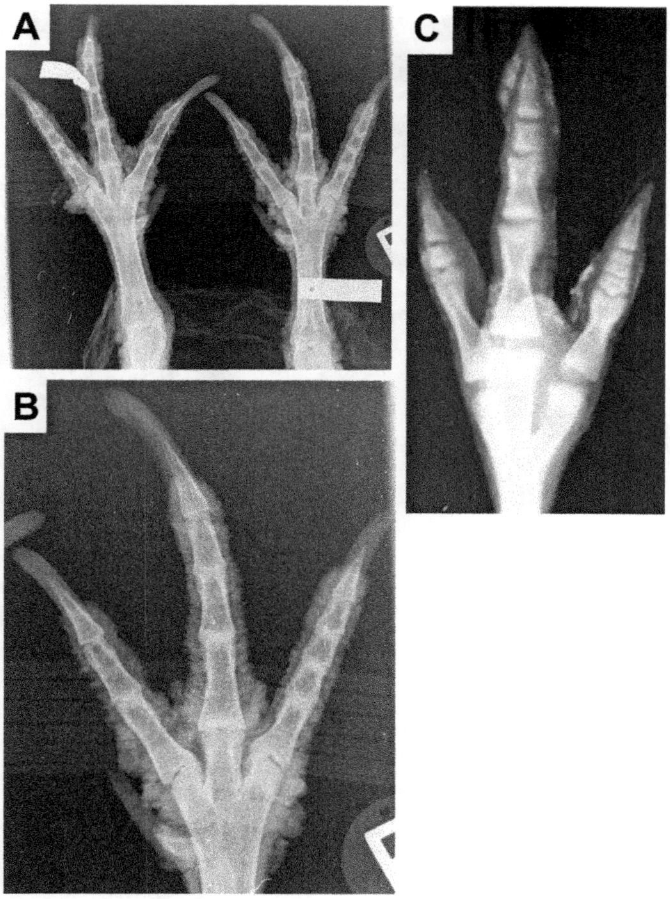

1.5. Relationship between the foot skeleton and soft tissues of the foot in two extant paleognathous birds. *A*, *B*, Both feet (*A*) and detail of right foot (*B*) of a kiwi (*Apteryx* sp.). *C*, Right foot of *Rhea americana*. In the rhea, the non-ungual phalanges distal to the first phalanx of digits II and IV are very short. In digit II the joint between phalanges II1 and II2, and also that between II2 and II3, are incorporated within a single digital pad, as are the non-ungual phalanges of digit IV and the base of ungual IV5. The joint between phalanges III1 and III2 corresponds to a digital pad just under midway along the middle toe, and the joint between III2 and III3, and sometimes also that between III3 and III4, are incorporated in a single large distal toe pad. The tips of the claws of the toes extend somewhat beyond the tips of the underlying unguals. (See fig. 6.35*F*–6.35*J* for photographs of the soles of the feet of rhea study skins.) In the kiwi, at least some of the digital pads seem to coincide with joints between phalanges, but interpretation of their relationships is complicated by the presence of interpad spaces between digital pads (fig. 6.36). The horny claws in this kiwi specimen extend considerably beyond the limits of the unguals, but this was a captive bird, so the claws may have been unusually long.

Table 1.1. Identification of *Iguanodon* specimens employed in this study. IRSNB = Institut Royal des Sciences Naturelles de Belgique; NHMUK = Natural History Museum (London). The IRSNB specimens have often been described and figured in other publications under slightly different catalog numbers; see table A1.1 for a correlation between the older and current numbers.

Taxon	Specimens
I. bernissartensis (subadult)	IRSNB Vert-5144-1726, Vert-5144-1729, Vert-5144-1730
I. bernissartensis (adult)	IRSNB R 51, R 52, R 53, R 54, R 56, Vert-5144-1562, Vert-5144-1639, Vert-5144-1657, Vert-5144-1710, Vert-5144-1712, Vert-5144-1714, Vert-5144-1715, Vert-5144-1716, Vert-5144-1723, Vert-5144-1724, Vert-5144-1725; NHMUK R2506
Mantellisaurus atherfieldensis	IRSNB R 57; NHMUK R1829
Iguanodon sp.	NHMUK R1863

atherfieldensis. Although he did not assign NHMUK R1829 to *I. atherfieldensis* in his 1986 monograph, Norman (2012, 2014) indicated that *atherfieldensis* would be the appropriate species name for the specimen.

Paul (2007) went even further than that, first assigning IRSNB R 57, the Bernissart specimen of *I. atherfieldensis*, to a new genus, such that the full name of the species would be *Mantellisaurus atherfieldensis*, and subsequently (Paul 2008b, 2010) assigning this specimen to a new genus and species, *Dollodon bampingi*. Norman (2010, 2012, 2014) accepted the validity of *Mantellisaurus* as the generic name for specimens previously assigned to *Iguanodon atherfieldensis*, an interpretation that will be followed here (see Lomax and Tamura 2014 for a summary of current thinking about the status of these and other Early Cretaceous iguanodonts from Europe and Britain).

The interpretation that the Bernissart iguanodontians constitute two species is perhaps provided further support by the taphonomic occurrences of the two forms at different localities. At Bernissart, *Iguanodon bernissartensis* dominates the assemblage, while at another mass-accumulation locality (Nehden in Germany; Norman 1987*a*) *Mantellisaurus atherfieldensis* numerically dominates. Although one could imagine scenarios in which the sexes of a single species were segregated by habitat or behavior, it seems simpler to interpret these differences in relative abundance as reflecting separation of different species.

A further complication for the present study that must be acknowledged—and has already been discussed for *Iguanodon*—stems from the fact that dinosaur specimens assigned to the same species often come from different stratigraphic levels within the formation in which they occur, such that the individual animals may have lived at times separated by thousands of years, hundreds of thousands of years, or even longer intervals of time. This raises two obvious questions. First: is the variability seen among specimens attributed to a particular species over a long time interval going to be the same as what one would have observed from a sample of individuals that lived at the same time? The latter, of course, would be more relevant to assessing the amount of intraspecific variability seen in measurements of footprints from a particular tracksite.

The second troubling question is whether specimens collected over the entire interval of a formation that accumulated over a long time scale should even be considered

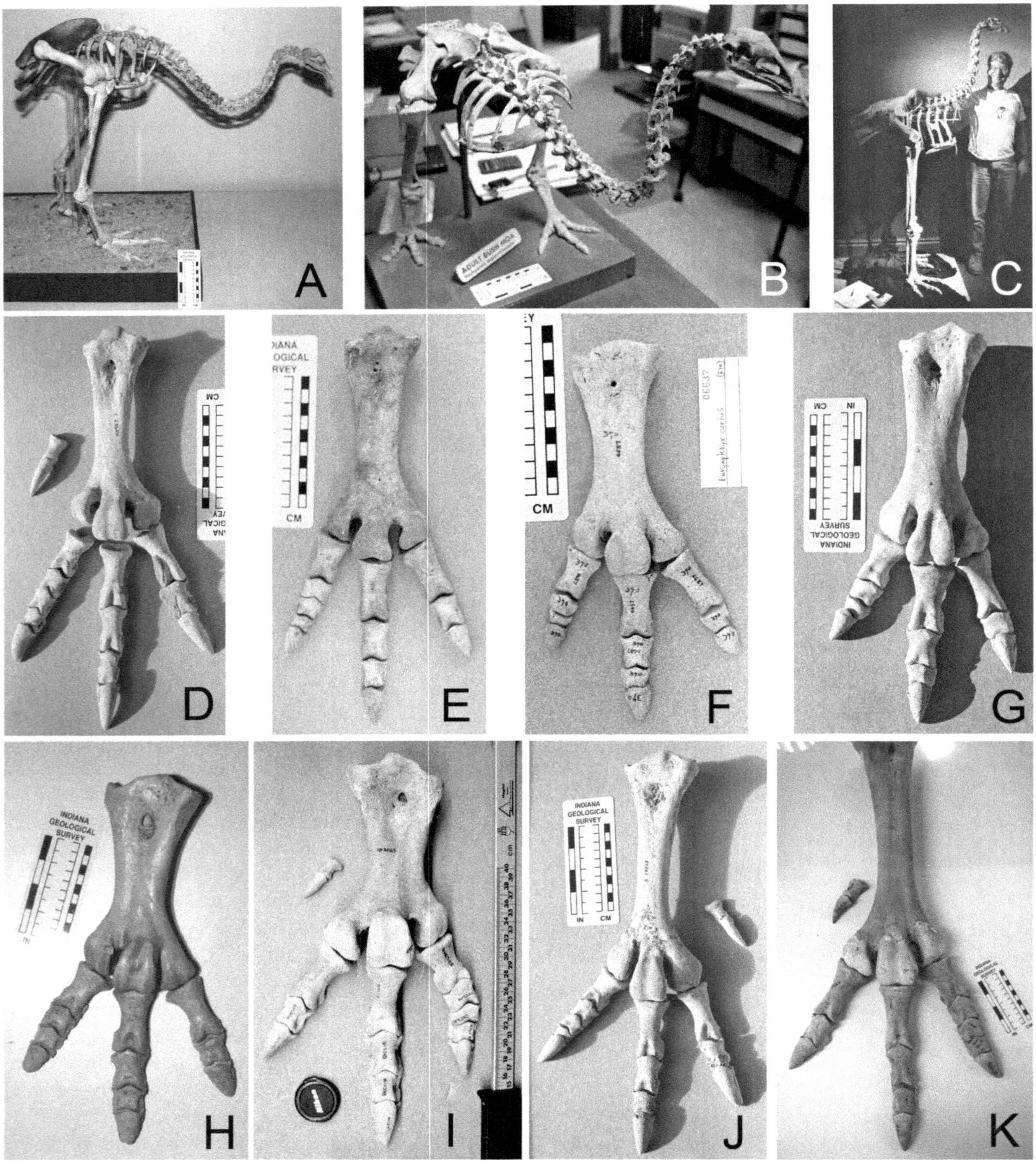

1.6. Moa and their feet; the black-and-white scale in the photographs is marked off in centimeters and inches. Digit I is not shown in all specimens. *A*, NMNZ S.023700, mounted skeleton of *Megalapteryx didinus* from collection of Museum of New Zealand Te Papa Tongarewa. *B*, Mounted composite skeleton of *Pachyornis geranoides* (formerly *P. mappini*), Waitomo Caves Museum. *C*, Mounted skeleton of YORYM: 2004.20.a, *Dinornis robustus* (the individual on the *left*), York Museums Trust (Yorkshire Museum). *D*, NMNZ S. 025657, right foot of *Megalapteryx didinus* from collection of Museum of New Zealand Te Papa Tongarewa; digit I is mistakenly placed beside digit IV in this photograph. *E*, AM LB5551, right foot of *Anomalopteryx didiformis*. *F*, AM LB6637, right foot of *Euryapteryx curtus*. *G*, NMNZ S.000469, right foot of *Emeus crassus* from collection of Museum of New Zealand Te Papa Tongarewa. *H*, NMNZ S.031552, right foot of *Euryapteryx curtus* (formerly *E. gravis* and *E. geranoides*) from collection of Museum of New Zealand Te Papa Tongarewa. *I*, CM AV 8383, left foot of *Pachyornis elephantopus*. Note conspicuous medial curvature of digit IV when the phalanges are articulated. *J*, *K*, *Dinornis robustus* from collection of Museum of New Zealand Te Papa Tongarewa. *J*, Right foot of NMNZ S.028225. *K*, Left foot of NMNZ S.034088.

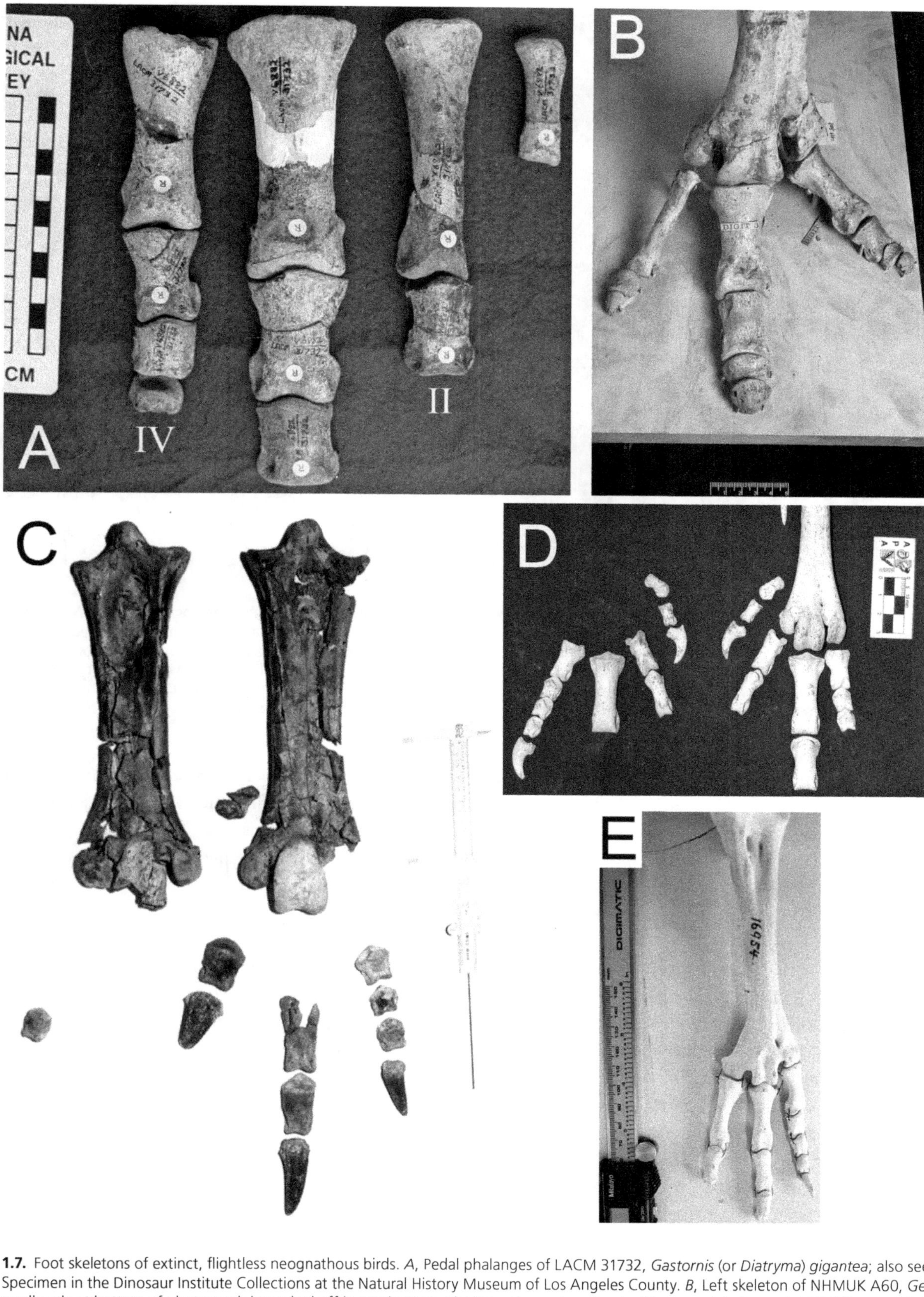

1.7. Foot skeletons of extinct, flightless neognathous birds. *A*, Pedal phalanges of LACM 31732, *Gastornis* (or *Diatryma*) *gigantea*; also see figure 6.33*E*. Specimen in the Dinosaur Institute Collections at the Natural History Museum of Los Angeles County. *B*, Left skeleton of NHMUK A60, *Genyornis newtoni*; small scale at bottom of photograph is marked off in centimeters. Photograph courtesy Sandra Chapman (photograph NHMUK copyright). *C*, Foot skeletons of DGM 1418, *Paraphysornis brasiliensis* (Alvarenga et al. 2011). Photograph courtesy of Herculano Alvarenga; handle of calipers is marked off in centimeter increments. *D*, Right and left pedal phalanges and distal left tarsometatarsus of MMP 5050, *Llallawavis scagliai* (Degrange et al. 2015); photograph courtesy Federico Degrange. *E*, USNM 16954, cast of left foot of *Raphus cucullatus* (cf. Parish 2013; Claessens et al. 2015); handle of calipers is marked off in centimeter and inch increments.

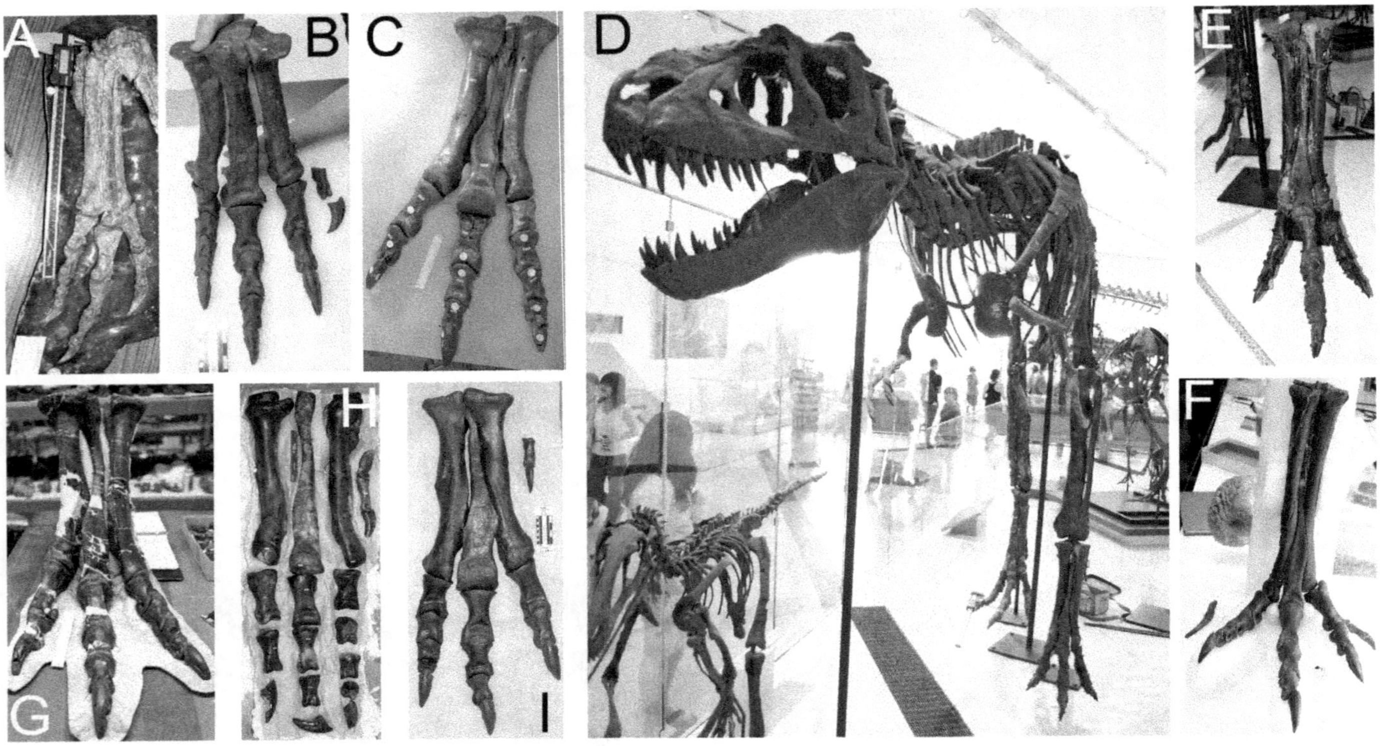

1.8. Large theropod dinosaurs and their feet; digit I is not shown in all specimens. A variety of scales are used in these photographs: large calipers (40.5 cm long) in *A*; black-and-white scale marked off in centimeters and inches in *B* and *I*; 6-inch ruler in *C*, *G*, and *H*; large calipers with handles marked off in centimeters in *E*; and tape measure marked off in centimeters in *F*. *A*, Right foot of TMM 43646-1, *Dilophosaurus wetherilli*. *B*, Cast of right foot of SMA 0005 ("Big Al II"), *Allosaurus fragilis*. *C*, Left foot of MOR 657, *Albertosaurus* sp. The composition of digit IV (particularly the first phalanx) is questionable. *D–F*, *Gorgosaurus libratus*. *D*, Mounted skeleton, and *E*, left foot of ROM 1247. *F*, Right foot of FMNH PR 2211. *G*, Left foot of MOR 590, *Daspletosaurus* sp. *H*, Left foot of CMN 350, *Daspletosaurus torosus* (digit I is beside digit IV). *I*, Cast of right foot of BHI 6230 ("Wyrex"), *Tyrannosaurus rex*.

to be members of the same species. To cite a rather charismatic example, should all specimens of *Triceratops* from the Hell Creek Formation be assigned to a single species, or is there enough difference between earlier and later specimens within the formation that they should be assigned to different species (Scannella et al. 2014)?

These are questions that I cannot address, except for a practical consideration. The sample size of specimens attributed to individual dinosaur species is already small enough as to make realistic assessments of intraspecific variability at best tentative. Distributing these specimens into multiple species would greatly aggravate this problem. Consequently I will provisionally accept the current assignments of specimens to their various species, with the understanding that the resulting inferences about intraspecific variability may not be accurate; presumably my measures of intraspecific variability in that case would overestimate the true measures of intraspecific variability.

Before we leave the subject of Hell Creek Formation dinosaur, another matter that must be considered is what to do with *Nanotyrannus lancensis*. Should this be regarded as a species in its own right (Bakker et al. 1988; Currie 2003*a*; Larson 2013; Persons and Currie 2016; Schmerge and Rothschild 2016) or an immature form of *Tyrannosaurus rex* (Carr 1999; Carr and Williamson 2004; Brusatte et al. 2016)? I will consider the effects of both options.

Another taxon that requires special comment is the big allosaur *Saurophaganax maximus*. The pedal phalanges that I measured in this study came from the Stovall/WPA quarry Kenton Pit 1 (Nicholas Czaplewski, personal communication). When I measured the phalanges, and made my own identification of which bone was which phalanx, I was under the impression that the bones had been found in association, so that I could be reasonably confident that they represent a single individual animal, and I treated them as such in some studies (e.g. Farlow 2001; Farlow et al. 2013). Since then I have learned that my impression about the association of these bones was incorrect. It is very likely, instead, that the *Saurophaganax* foot skeleton for which I give measurements in table A1.1 is a composite: too bad, but that's the way the tootsie tumbles. However, the news may not be entirely bad. The overall distribution of *Saurophaganax* bones in the quarry suggests that they came from two individuals of about the same size (Kyle Davies, personal communication), and so the pedal phalanges together may approximate the proportions of the foot skeleton of a single animal. Because of

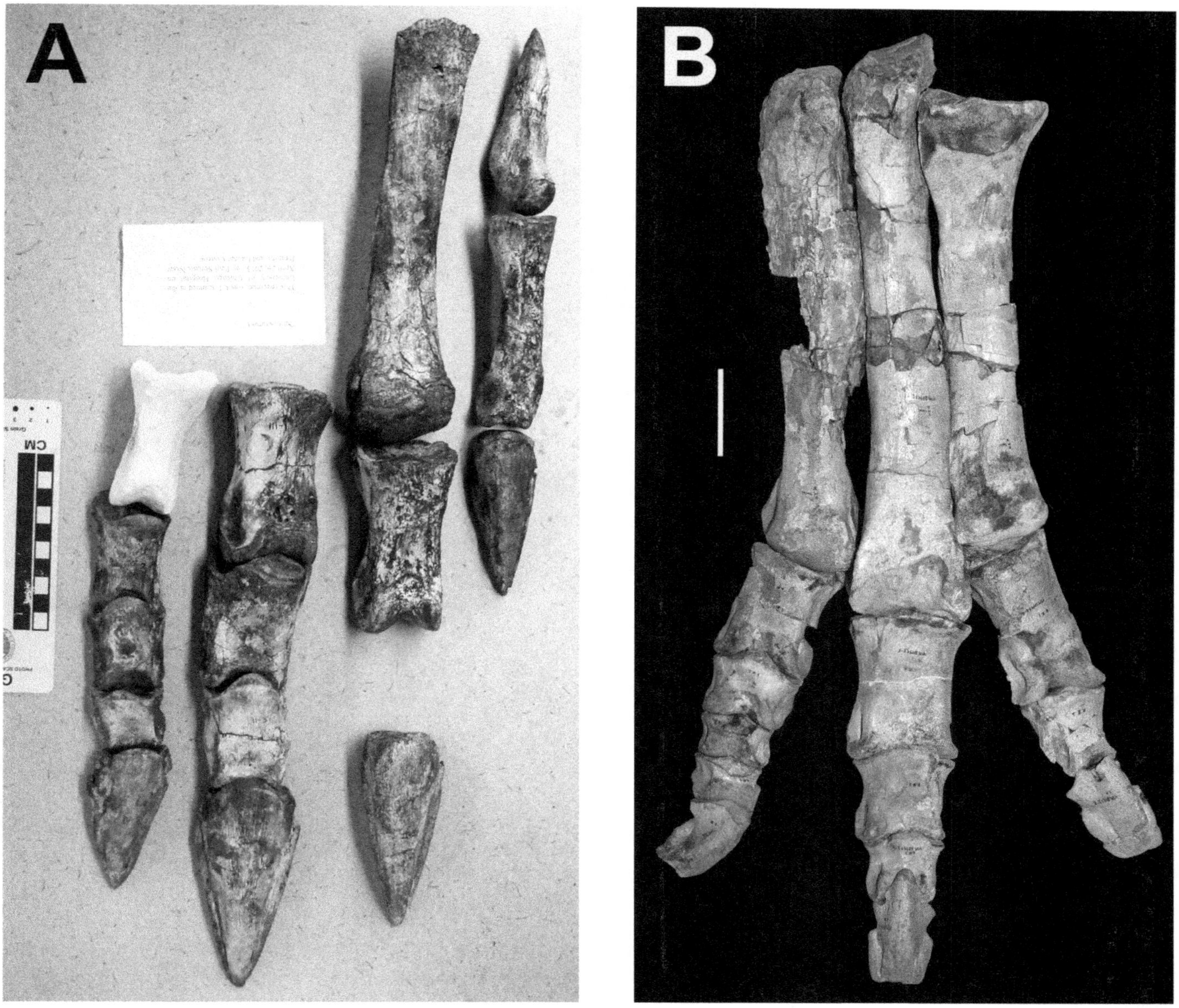

1.9. Foot skeletons of atypical large theropods. *A*, *Spinosaurus aegyptiacus* Specimen C, right foot (Ibrahim et al. 2014). Phalanx II2 is missing from this specimen. Phalanx IV1 is now lost but was cast before being lost. Note the long, broad unguals, with conspicuous "wings"; compare these with the unguals of ornithomimosaurs (fig. 1.10*A*, 1.10*B*) and small to medium-sized ornithischians (fig. 1.11). *B*, Right foot of the huge ornithomimosaur *Deinocheirus mirificus* MPC-D 100/127; scale bar is 10 cm (Lee et al. 2014). Note the broad unguals with squared-off distal tips. *Photograph courtesy Young-Nam Lee.*

the interest in foot skeletons of large theropods for this book, I will therefore continue to present data for *Saurophaganax* under this assumption, but the reader should keep in mind the possible error of doing so.

INTRASPECIFIC SIZE VARIABILITY

Size variability of phalangeal and digital lengths was analyzed for seven extant and extinct archosaurian species (tables 1.2–1.8). I restricted this analysis, to the extent that this was possible, to adult individuals of each species. For alligators (table 1.2), data for animals of total length of 183 cm or more, a common criterion for sexual maturity in *A. mississippiensis* (chapter 5), were selected. I did not have a measurement of overall body size, or a minimum size associated with sexual maturity, for any of the other species, and most obviously lacked such information for the extinct forms. I eliminated from consideration individuals of such species that were considerably smaller than others of the same species and those whose phalangeal articular surfaces had a conspicuously porous appearance of the kind associated with immaturity. Data are reported for specimens for which a complete set of measurements could be made (listwise treatment), and also where the number of specimens for which measurements of

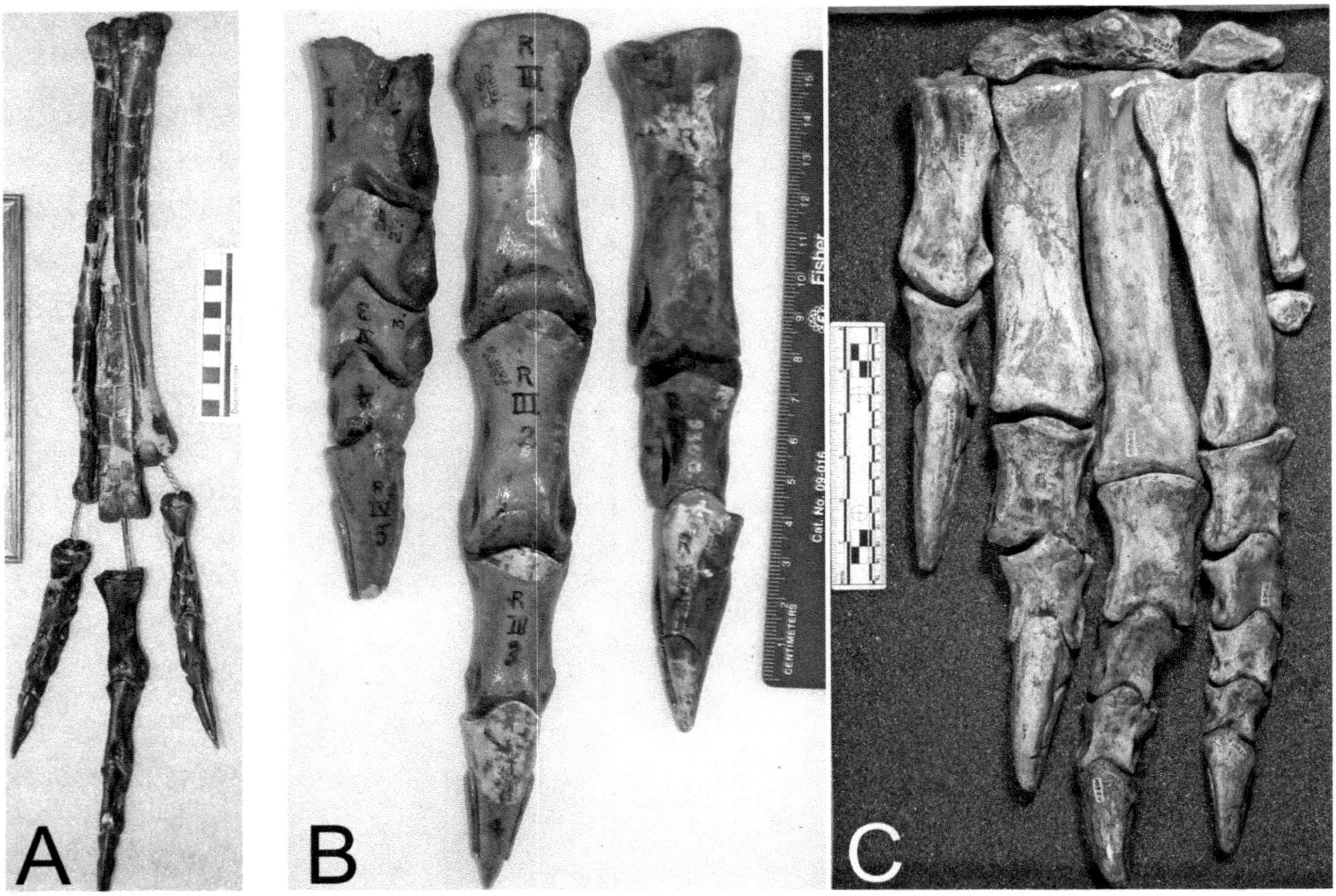

1.10. Miscellaneous saurischian feet; centimeter scale in all photographs. *A*, Right foot of ROM 797 (formerly 5163), *Ornithomimus edmontonicus* (formerly *Dromiceiomimus brevetertius*). *B*, Pedal phalanges of right foot of CMN 930, *Struthiomimus altus*. *C*, USNM V 10924, cast of left foot of *Plateosaurus longiceps*. *Image copyright Smithsonian Institution, all rights reserved.*

a particular bone was greater than the number of specimens for which a complete set of measurements could be made (in which case a larger number of cases is reported).

Iguanodon bernissartensis, the only non-avian dinosaur examined, was given special treatment (table 1.8). Although specimens in the NHMUK have been assigned to this species (table 1.1), I considered intraspecific size variability in pedal element lengths only for the Bernissart sample, regarding as immature individuals the three "subadult" animals recognized by Norman (1980, 1986)—although I obviously had no information about the reproductive status of any of these dinosaurs. (It is worth noting, however, that dinosaurs, like alligators, seem to have reached sexual maturity at body sizes much smaller than those at which they became skeletally mature; cf. Sander 2000; Schweitzer et al. 2005; Erickson et al. 2007, 2009; Klein and Sander 2008; Lee and Werning 2008; Giebeler and Werner 2011; Sander et al. 2011; Lee et al. 2013; Hone et al. 2016). For individuals regarded as mature, intraspecific size variability of pedal elements was examined in three ways: (1) for "synthetic" and "blended" feet (see above for the distinction), with the number of data cases varying with the number of dinosaurs for which each element measurement could be made; (2) for synthetic and blended feet, with the analysis restricted to those individuals for which all pedal measurements could be made, resulting in the same data cases being used in all determinations of element size variability; and (3) for synthetic but not blended feet, with measurements restricted to lengths taken along the medial or lateral sides of elements, and the number of data cases varying with the number of dinosaurs for which each element measurement could be made.

INTRASPECIFIC AND INTERSPECIFIC SIZE-RELATED CHANGES IN PEDAL PROPORTIONS

To see if there were intraspecific ontogenetic or interspecific allometric changes in phalangeal proportions, I determined reduced major axis (RMA) equations (Imbrie 1956; Rayner

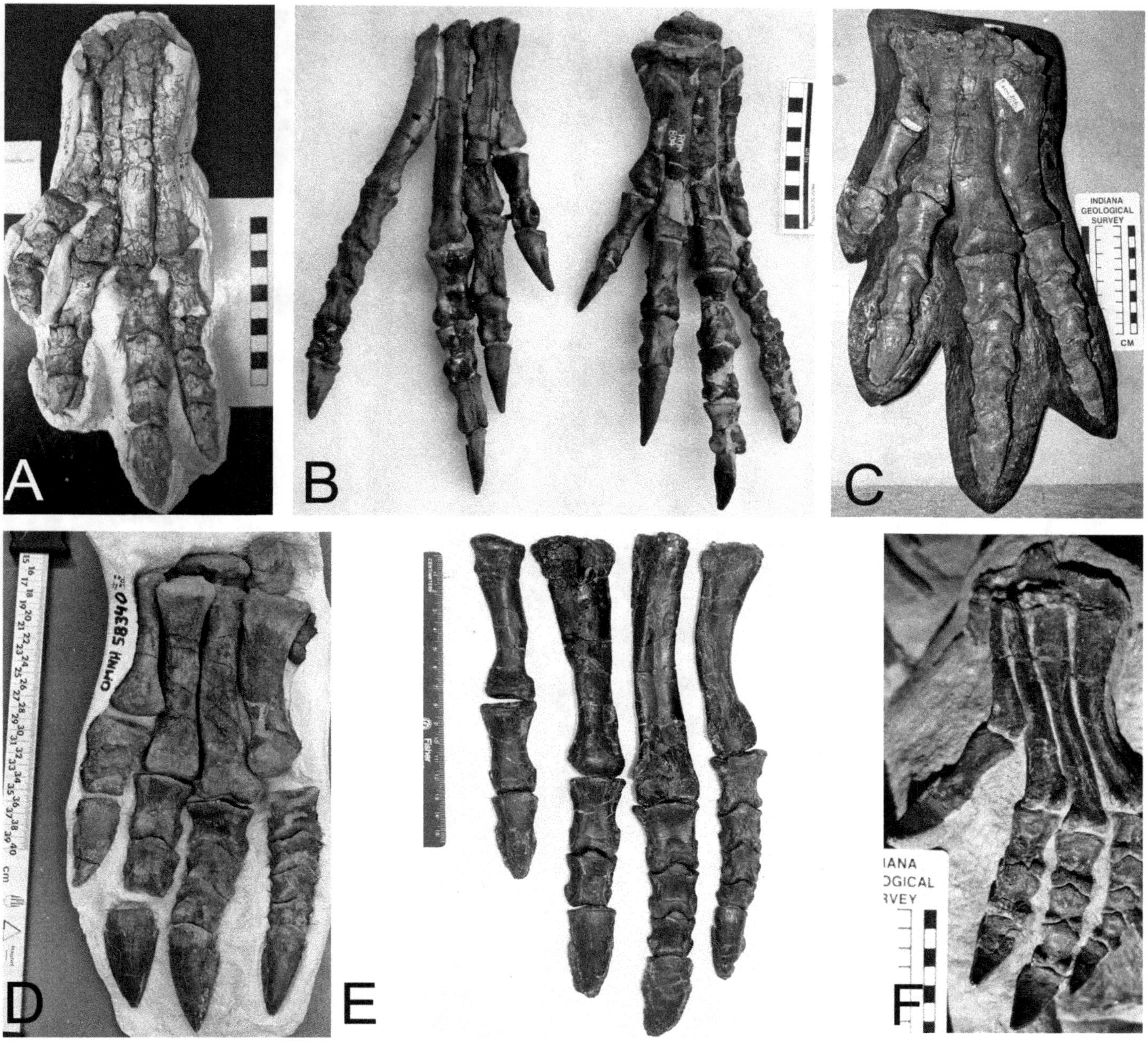

1.11. Foot skeletons of small to medium-sized ornithischians; centimeter scales throughout. *A*, Left foot of SMU 73181, the Proctor Lake ornithopod. *B*, Both feet of ROM 804, *Parksosaurus warreni*. *C*, Left foot of LACM 33542, *Thescelosaurus neglectus*. Specimen in the Dinosaur Institute Collections at the Natural History Museum of Los Angeles County. *D*, OMNH 58340, left foot of *Tenontosaurus* cf. *tilletti*. *E*, CMN 8889, left foot of *Leptoceratops gracilis*. *F*, TMP 82.11.1, right foot (ventral view) of *Montanoceratops* sp. Note the general similarity between these feet and those of a prosauropod (fig. 1.10*C*) and even the huge theropod *Spinosaurus* (fig. 1.9*A*).

1985; Leduc 1987) for bivariate and multivariate relationships between selected log-transformed values, and then tested whether the 95% confidence intervals for the equation slopes excluded 1, the slope for isometric relationships. I hypothesized that interspecific comparisons were more likely to show size-related changes in pedal phalangeal or digital proportions than series of within-species feet of different sizes. Consequently, I expected to see either more interspecific than intraspecific allometric relationships, or at least more extreme interspecific than intraspecific allometric relationships, in pedal and digital proportions.

Several versions of such analyses were done. The simplest, of course, involved comparisons between only two variables. For alligators and other crocodylians, where I had good measurements or proxies of overall animal size (total length, femur length [Farlow and Britton 2000; Farlow et al. 2005])

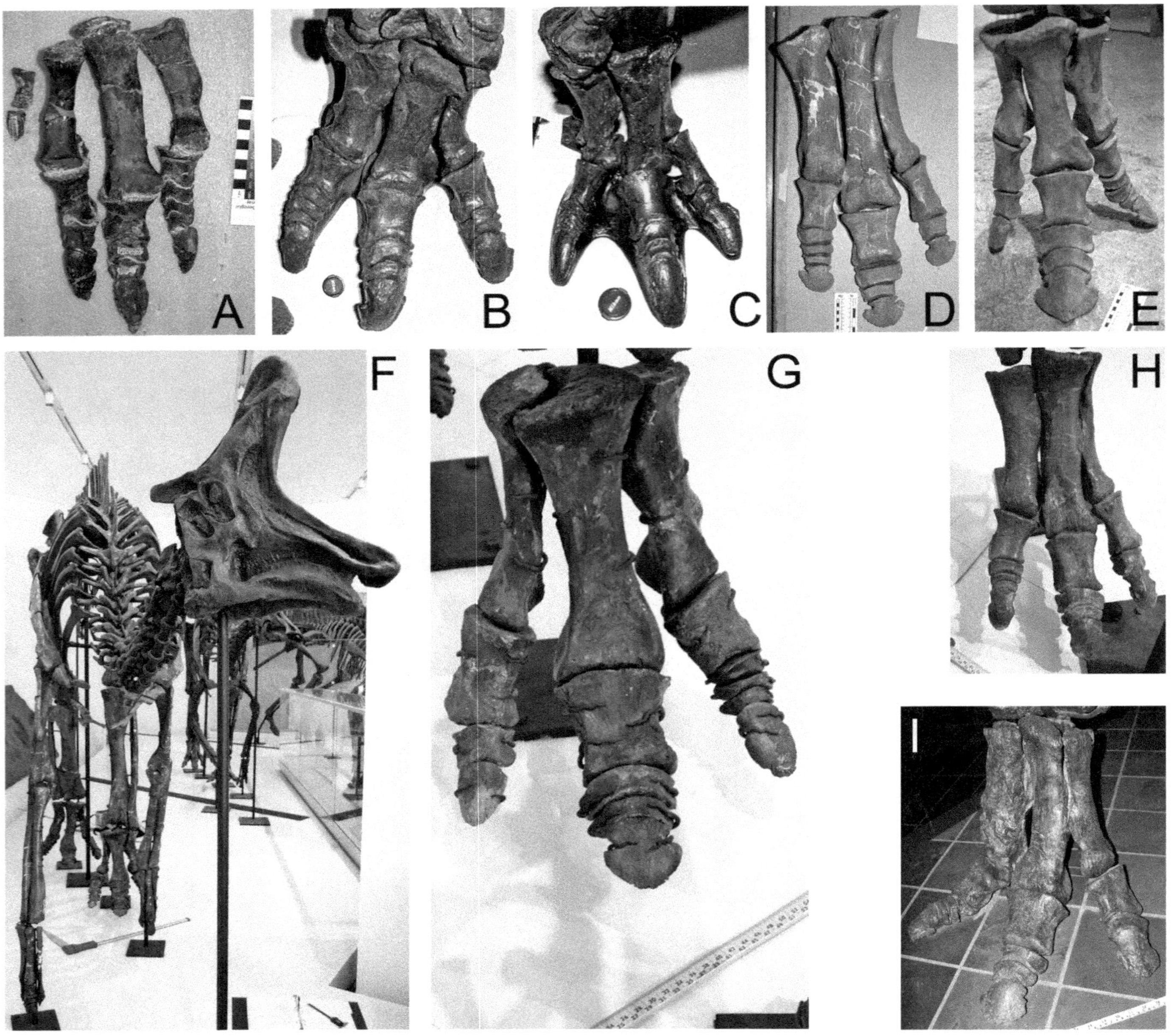

1.12. Large ornithopods and their feet. *A*, Left foot of CEUM 52457, *Camptosaurus* sp. *B*, Right foot of IRSNB R 51, *Iguanodon bernissartensis*. *C*, Right foot of IRSNB R 57, *Mantellisaurus* (formerly *Iguanodon*) *atherfieldensis*. *D*, Right foot of MOR 794, *Brachylophosaurus canadensis*. *E*, Left foot of DMNH EPV.1493, *Edmontosaurus annectens*. *F*, *G*, ROM 1218, *Lambeosaurus lambei*. *F*, Mounted skeleton. *G*, Left foot. *H*, Right foot of ROM 845, *Corythosaurus casuarius*. *I*) Right foot of CMN 8501, *Hypacrosaurus altispinus*. Scale bar in *A*, *D*, and *E* marked off in centimeters; caliper handles in *G* and *H* marked off in centimeters; steel tape in *J* marked off in inches and centimeters; 35-mm lens cap provides scale in *B* and *C*.

for several specimens, I considered the relationship between animal size and several measures of hindfoot size (metatarsal length, the mean of the log-transformed "big seven" phalanx lengths, digit III length excluding the ungual). For crocodylians and ground birds I examined intraspecific and interspecific relationships between metatarsal III or tarsometatarsal length and proxies for the size of the digital portion of the foot (digit III length excluding the ungual, log-transformed "big seven" mean), between III1L and the lengths of other "big seven" phalanges, and between the log-transformed length of each of the "big seven" phalanges and the mean of the remaining log-transformed six "big seven" phalanges.

I searched for allometric relationships between phalangeal and digital lengths in selected species for which I had data covering a substantial part of the size range of the species (alligators and two moa species), although only in the case of *Alligator mississippiensis* did I have data for very small individuals. In addition, I did interspecific analyses for extant crocodylians, two moa species together, and all moa species. For the crocodylian and "all moa" samples, I did alternative analyses in which all specimens in my sample were used (disregarding the fact that I had very different sample sizes for different species), and "one-per-species" treatments in which I used only one specimen (generally the largest for which data were available) for each species.

For those bivariate or multivariate relationships that were significant (95% confidence interval (CI) for the slope excluded 1), I made a further distinction between relationships that I regarded as "only barely" allometric (upper confidence limit [CI] of the RMA slope was between 0.95 and 1.00 [in the case of negative allometry], or lower CI of the RMA slope was between 1.00 and 1.05 [positive allometry]) and those that showed stronger allometry (upper CI < 0.95, or lower CI > 1.05), and placed more emphasis on the latter. This was done in part to mitigate the problem that, with many individual analyses being run, the chances increase that some relationships would incorrectly be regarded as allometric that in fact were not. I also tried to allow for spurious allometry by putting greater emphasis on those shape comparisons where more than one of a class of comparisons all appeared to be allometric (e.g. different measures of foot size as a function of total length or total length in crocodylians; table A1.2).

INTRASPECIFIC AND INTERSPECIFIC MEASURES OF SHAPE VARIABILITY

Shape variability in foot skeletons obviously is not necessarily related to size alone. We hope, after all, to be able to tell feet (and footprints) of different species apart on the basis of pedal proportions. Consequently I used several measures of foot shape to compare variability within and across species, after factoring out the effects of overall size.

Maximum/Minimum Ratios of Scaled Parameters. A simple, quick-and-dirty measurement of variability of phalangeal and digital proportions was calculated by standardizing individual phalanx lengths against the length of phalanx III1, and the lengths of digits II and IV (with or without unguals) against the length of digit III (with or without the ungual). The procedure involved selecting a common length of phalanx III1 or digit III length across all the specimens in my sample, and multiplying the lengths of all other phalanges or digits by a scaling factor that would keep the ratio of the length of the scaled parameter (lengths of all phalanges other than III1, or the lengths of digits II and IV) to that of the scaling parameter (that is, phalanx III1 length or digit III length) the same as the ratio of the raw (unscaled) parameter length to that of the actual III1 or digit III length of the specimen. The scaled phalanx or digit length is therefore a kind of bivariate proportional comparison. Variability of the scaled parameter within and across species serves as a measure of how variable the proportions of the two pedal parameters are.

The simplest measure of variability of the scaled parameter is the ratio of the maximum value of the scaled parameter to the minimum value of the scaled parameter, both within a species and across species. The larger the value of the maximum/minimum ratio, the more variable the proportions of the scaled parameter as compared with the scaling parameter. One might expect (or hope) that the maximum/minimum ratio of the scaled parameter will be greater for comparisons across species than for comparisons within species.

Given the small sample sizes of complete dinosaur foot skeletons, this simple comparison of proportional variability has some appeal; for example, only two phalanges (phalanx III1 and the phalanx being considered) must be preserved in order for the foot to be included in the sample. Thus the sample size of specimens that can be compared is maximized.

On the other hand, with the small sample sizes with which we must deal, it is possible that a greater value of the maximum/minimum ratio in an interspecific sample than in an intraspecific sample might reflect nothing more than a larger number of specimens, without real differences between/among species in the phalangeal or digital proportions being investigated. This possibility was evaluated in two ways. First, using data for the moa *Dinornis robustus* (a species of very large ground bird represented by the largest number of foot skeletons in my sample), I randomly selected 5, 10, or 15 cases (repeating this procedure 15 times for each treatment) of scaled parameters from the sample of 24 foot skeletons for which all of the "big seven" phalanges were preserved (a "listwise" treatment of the data). The same procedure was used for the overall lengths of digits II and IV (excluding the ungual) scaled against the overall length of digit III (excluding the ungual).

Maximum/minimum values of each scaled parameter were calculated from the cases randomly selected in each trial. The median value of the maximum/minimum ratio for each scaled parameter in the 15 trials in each treatment was then taken as a proxy for variability in the maximum/minimum ratio; higher values of the median indicate that the sampling procedure results in greater disparity between the smallest and largest values of each scaled parameter.

The second way of evaluating the maximum/minimum values of scaled parameters was similar to the first, but involved single-species vs. two-species or multispecies comparisons. In this test, a specified number of data cases of scaled parameters were randomly selected (repeated in either

15 or 30 replicates for each such treatment) from the pooled sample of two or more species. The number of cases randomly selected was the same as the number of cases for the single species with which the pooled sample was compared. The median value of the maximum/minimum ratio of each scaled parameter for the pooled two-species or multispecies sample across the 15 or 30 replicates was compared with the single value of the maximum/minimum ratio for the sample of scaled values of that parameter for the single species. If the median value of the ratio across the 15 or 30 replicates is greater than the value of the ratio for the same number of data cases as the individual species, this means that in more than half of the replicates the ratio of the pooled sample was greater than that of the individual species, indicating greater shape variability in the pooled sample than in the individual species. As an additional way of comparing within-species and across-species variability, I also tabulated the percentage of the 15 or 30 interspecific trials in which the maximum/minimum ratio of the scaled parameter for the randomly selected cases was greater than the single-species value of this ratio. This kind of analysis was used to compare pedal proportions of *Alligator mississippiensis* with those of crocodylians more generally, and also to compare intraspecific with interspecific samples of moa feet.

The moa listwise sample, representing the largest number of specimens in my database, was particularly interesting for this analysis. In these trials, 11, 17, or 24 cases (corresponding to the number of specimens in my sample) for the three species *Pachyornis elephantopus* (excluding CM AV 8622, the specimen that Allentoft et al. [2010] assigned to this species on molecular as opposed to morphological grounds), *Anomalopteryx didiformis*, and *Dinornis robustus*, respectively, were randomly selected (with 15 repetitions) from the entire moa sample, and the median values of the maximum/minimum ratio of the scaled parameters were compared with those of each of the three species. In addition, the percentage of 15 trials in which the maximum/minimum of the scaled parameter exceeded that for each of the three species was tabulated. More will be said about this procedure later in this chapter when explaining the results of these trials.

For moa and other groups, I compared maximum/minimum values of scaled pedal phalangeal and digital parameters for a variety of single-species, two-species, and multispecies sets of data. In contrast to the preceding listwise tests, in these other comparisons data were not limited to foot skeletons in which all of the pedal phalanges of interest were preserved (and so included data for incomplete foot skeletons). Consequently the number of specimens of each species or interspecific grouping could vary from one scaled parameter to the next.

As a final test of whether maximum/minimum ratios of scaled parameters were capable of providing a quick-and-dirty, seat-of-the-pants metric of shape variability capable of distinguishing single-species from interspecific samples, I performed analyses of covariance (ANCOVAs) of lengths of the non-ungual phalanges of digits II–IV across moa species, with the length of phalanx III1 serving as the covariate. These results were then compared alongside evaluations of the extent to which interspecific samples of scaled phalangeal proportions in moa were greater than those for single species.

Coefficient of Variation of Scaled Phalangeal Parameters. Maximum/minimum ratios of scaled parameters are equivalent to the range of values of scaled parameters, and thus reflect the extreme values of each scaled parameter. A measure of the dispersion of values of scaled parameters that will be less affected by extreme values is the coefficient of variation (CV; calculated as [standard deviation/mean] × 100) of the scaled values.

Range and Standard Deviation of GM-Scaled Parameters. A more sophisticated way of scaling parameters is to log transform all of the parameters of interest, take the mean of the log-transformed values, and subtract that mean from each of the log-transformed parameters. Thus for the "big seven" phalanges, the mean of the logarithm of all the phalanx lengths was subtracted from the logarithm of each phalanx length to scale that phalanx length. Similarly, the three digit lengths were log transformed, and the mean of the three log-transformed values was subtracted from the logarithm of each digit length. This procedure is related to the calculation of the geometric mean (the arithmetic mean of log-transformed data is the log of the geometric mean of the untransformed data, D; Townsend, personal communication), and so it will be designated "GM scaling." Unfortunately, because measurements of all of the relevant parameters in a foot skeleton must be available for this kind of scaling to be done, GM scaling comes at the cost of reducing sample sizes for those taxa (particularly for many extinct forms).

GM scaling unavoidably results in some negative values for scaled parameters, because many of the log-transformed measurements that go into calculating the mean of the log-transformed parameters during scaling obviously will take smaller values than the mean. To avoid negative numbers, instead of reporting maximum/minimum values of GM-scaled parameters, I will use the ranges of GM-scaled values; and instead of calculating the coefficient of variation of GM-scaled parameters, I will stick with the standard deviation.

Coefficient of Relative Dispersion About the Reduced Major Axis (D_d). Imbrie (1956) described this parameter as expressing "the amount of shape variation as a proportion of the average shape attained by the sample" (Imbrie 1956: 241). As its name implies D_d measures how much points are scattered around a line (the reduced major axis) that is fitted to a bivariate set of data. It is calculated as:

$$D_d = 100 \times \sqrt{\{[2 \times (1 - r) \times (\text{SD } X^2 + \text{SD } Y^2)] / (\text{mean } X^2 + \text{mean } Y^2)\}}$$

where r is the correlation coefficient, SD is the standard deviation, and X and Y are the two variables of interest. X will be described as the "independent variable" throughout this study, although strictly speaking, there is neither an independent nor a dependent variable in an RMA analysis. Another way of describing D_d is to think of it as a two-dimensional version of the coefficient of variation.

Because D_d measures variability a bit differently than the above-described measures of variability, I calculate it for many more intraspecific and interspecific samples than some of the other parameters. For the most part, I calculate D_d using log-transformed values of data (in chapter 6, dealing with emu footprints, D_d is calculated using both raw and log-transformed data). I will do simple bivariate comparisons, but for the "big seven" phalanges I will also do a bivariate comparison after first doing multivariate GM scaling. For each of the "big seven" phalanges, I compute the mean of the other "big seven" phalanges, excluding the phalanx under consideration. D_d is then calculated for the relationship between the log-transformed length of the phalanx of interest and the mean of the log-transformed values of the remaining six phalanx lengths.

Interspecific Variability in Pedal Proportions. To investigate how the number of species, and the kinds of different species, in an interspecific sample affect all of the above-described measures of proportional variability, I created two-species and multispecies samples that were both unrealistic and realistic. Unrealistic samples pooled species from different times and geographic regions that never would coexist in the same fauna. The idea here was that such extremely unrealistic samples should give some notion about the upper limits of foot shape variability—of whether it is even possible to distinguish diverse interspecific samples from less diverse interspecific assemblages, or the latter from intraspecific samples. So I would, for example, combine all foot skeletons of *Allosaurus* with all foot skeletons of *Tyrannosaurus*, or foot skeletons of all tyrannosaurids in my sample with those of *Tyrannosaurus rex* alone, to see if the shape variability parameters described above showed higher values for pooled two or more species samples as opposed to those for the single species.

The more realistic samples were taken from my moa data, representing species that lived at the same time in the same geographic region. Thus I compared pooled samples of all moa species, a variety of combinations of two or more species, and those of single species.

In the interests of space, I did not examine all possible species combinations, and for those species combinations that I did consider, I did not always look at all of the different ways of measuring pedal shape variability. Instead, in this chapter I present what I hope is a judicious sampling of the ways I could have analyzed my data, generally using what seemed the most interesting species combinations and the species combinations that had enough specimens in the sample to make analysis worthwhile.

RESULTS

Intraspecific Size Variability

Alligator mississippiensis. My American alligator sample (table 1.2) included individuals in museum collections as well as some of the same unfortunate reptiles killed during the annual autumn hunt in Florida, whose intact feet I measured (chapter 5) and then skeletonized for osteological measurement. As already mentioned, I was unable to measure ungual lengths in most crocodylian skeletons.

The coefficient of variation for individual phalanges of sexually mature animals ranges from 19% to 22%, and the maximum/minimum ratio from 1.6 to 1.9. The first phalanx of digits II–IV seems to be more variable than the second phalanx. Aggregate toe lengths for digits II and III (excluding the ungual; digit IV was not considered because of the small number of specimens for which all digit IV phalanges were available) had CVs of about 22% and maximum/minimum ratios of 1.8.

Birds. Emus (table 1.3) and greater rheas (table 1.4) generally show less intraspecific variability in pedal element lengths than alligators, with CVs generally less than 10%, but with some CV values for more distal pedal phalanges (II2, III3, III4, or IV2) sometimes taking larger values. Maximum/minimum ratios are likewise usually less than for alligators, with values typically 1.1–1.5, with the larger values again being associated with distal phalanx lengths.

Variability of putatively adult specimens of the three moa species, expressed in terms of CV or the maximum/minimum ratio (tables 1.5–1.7), tends to be greater than for

Table 1.2. Variability of phalangeal and digital measurements (in millimeters) in mature *Alligator mississippiensis*. Total length of individuals measured = 198–340 cm. Data are reported for specimens for which a complete set of measurements could be made (listwise treatment; *N* = 8), and also where the number of specimens for which measurements of a particular bone was greater than the number of specimens for which a complete set of measurements could be made (in which case a larger number of cases is reported). Lengths of unguals and of the last two phalanges of digit IV are not reported because of small sample size. Phalanx abbreviations are expressed in terms of the particular digit (roman numeral) and the particular phalanx within that digit (arabic numeral); thus, I1 is the first phalanx of digit I, III2 is the second phalanx of digit III, etc. L = length; CV = coefficient of variation.

Parameter	Number of specimens	Minimum	Maximum	Mean (CV)	Maximum/minimum ratio
I1L	8	19	34	25.7 (21.8)	1.79
	9	19	34	26.0 (20.4)	1.79
II1L	8	22	40	30.4 (22.3)	1.82
II2L	8	14	24	18.7 (20.7)	1.71
III1L	8	23	41	32.3 (22.0)	1.78
	9	23	41	32.3 (20.5)	1.78
III2L	8	15	26	20.9 (20.7)	1.73
	9	15	26	20.9 (19.4)	1.73
III3L	8	10	19	15.3 (21.8)	1.90
IV1L	8	20	36	28.3 (21.3)	1.80
IV2L	8	13	21	16.5 (18.9)	1.62
IIL excluding ungual	8	36	64	49.1 (21.6)	1.78
IIIL excluding ungual	8	48	86	68.4 (21.5)	1.79

Table 1.3. Variability of phalangeal and digital measurements (in millimeters) in emus (*Dromaius novaehollandiae*). Data are reported for individuals interpreted as mature (phalanx III1 length at least 50 mm). Data are reported for specimens for which a complete set of measurements could be made (listwise treatment; *N* = 11), and also where the number of specimens for which measurements of a particular bone was greater than the number of specimens for which a complete set of measurements could be made (in which a larger number of cases is reported). CV = coefficient of variation; parameter abbreviations as in table 1.2.

Parameter	Number of specimens	Minimum	Maximum	Mean (CV)	Maximum/minimum ratio
II1L	11	44	55	49.0 (7.58)	1.25
	12	42	55	48.4 (8.42)	1.31
II2L	11	12	15	13.5 (8.96)	1.25
	12	10	15	13.3 (11.7)	1.50
II3L	11	23	29	26.7 (7.50)	1.26
III1L	11	50	64	57.4 (7.20)	1.28
	12	50	64	57.0 (7.25)	1.28
III2L	11	33	39	36.2 (5.22)	1.18
	12	33	39	36.0 (5.30)	1.18
III3L	11	16	21	18.8 (8.17)	1.31
	12	16	21	18.6 (9.02)	1.31
III4L	11	26	36	30.7 (10.9)	1.38
IV1L	11	37	45	40.8 (6.65)	1.22
	12	37	45	40.6 (6.68)	1.22
IV2L	11	13	17	15.0 (8.43)	1.31
	12	13	17	14.9 (8.31)	1.31
IV5L	11	21	27	23.6 (8.73)	1.29
Digit II length excluding ungual	11	57	69	62.5 (6.76)	1.21
	12	52	69	61.7 (8.19)	1.33
Digit II length	11	81	98	89.3 (6.28)	1.21
Digit III length excluding ungual	11	101	122	112.4 (5.91)	1.21
	12	101	122	111.6 (6.17)	1.21
Digit III length	11	131	158	143.1 (6.34)	1.21
Digit IV length excluding ungual	11	64	78	70.6 (6.28)	1.22
	12	64	78	70.3 (6.19)	1.22
Digit IV length	11	87	105	94.3 (6.51)	1.21

Table 1.4. Variability of phalangeal and digital measurements (in mm) in greater rheas (*Rhea americana*). Data are reported for individuals interpreted as mature (phalanx III1 length at least 35 mm). Data are reported for specimens for which a complete set of measurements could be made (listwise treatment; *N* = 8), and also where the number of specimens for which measurements of a particular bone was greater than the number of specimens for which a complete set of measurements could be made (in which case a larger number of cases is reported). CV = coefficient of variation; parameter abbreviations as in table 1.2.

Parameter	Number of specimens	Minimum	Maximum	Mean (CV)	Maximum/minimum ratio
II1L	8	40	45	42.7 (4.29)	1.13
	9	38	45	42.2 (5.53)	1.18
II2L	8	7	13	8.3 (24.0)	1.86
	9	7	13	8.1 23.4)	1.86
II3L	8	20	23	21.9 (5.15)	1.15
III1L	8	38	43	40.6 (3.93)	1.13
	9	37	43	40.2 (4.78)	1.16
III2L	8	22	26	24.9 (5.86)	1.18
	9	22	26	24.6 (6.79)	1.18
III3L	8	10	14	11.3 (11.4)	1.40
	9	10	14	11.1 (11.4)	1.40
III4L	8	28	31	29.9 (5.20)	1.11
	9	26	31	29.4 (6.60)	1.19
IV1L	8	29	35	32.3 (6.37)	1.21
	9	29	35	32.0 (6.44)	1.21
IV2L	8	6	8	6.5 (11.6)	1.33
	9	5	8	6.3 (13.7)	1.60
IV5L	8	18	26	21.9 (10.5)	1.44
Digit II length excluding ungual	8	47	54	51.0 (4.45)	1.15
	9	45	54	50.3 (5.79)	1.20
Digit II length	8	70	75	72.9 (2.37)	1.07
Digit III length excluding ungual	8	70	81	76.7 (4.55)	1.16
	9	69	81	75.9 (5.49)	1.17
Digit III length	8	98	111	106.6 (3.72)	1.13
	9	95	111	105.3 (5.09)	1.17
Digit IV length excluding ungual	8	45	51	47.9 (4.79)	1.13
	9	41	51	47.1 (6.67)	1.24
Digit IV length	8	66	74	69.7 (4.18)	1.12

the two extant ratites, especially in the giant moa *Dinornis robustus* (table 1.7), but even in the giant moa CV values are less than for alligators (although maximum/minimum ratios of *D. robustus* are comparable to those for alligators). The greater variability of the moa than of the extant birds may reflect uncertainties about the size at which specimens became mature, the huge sexual size dimorphism in moa (Bunce et al. 2003; Allentoft et al. 2010), the fact that the moa sample comes from subfossil assemblages of different ages (with adult bird size varying over time; cf. Worthy and Holdaway 2002), or some combination of these factors.

In contrast to alligators, there is a clear tendency in the birds for distal phalanges (sometimes, but not always, including the ungual) in a toe to be more variable in size than more proximal phalanges in the same toe, whether one looks at CV or at maximum/minimum ratios. Somewhat surprisingly, there is also a tendency for aggregate toe lengths that include the ungual to be a bit less variable than aggregate toe lengths without the ungual.

Whether the greater variability of alligator digital dimensions than those of most of the bird species is biologically meaningful is uncertain. It could simply reflect the exigencies of specimen availability and the fact that I had a defined size criterion for maturity in alligators that I did not have for birds. On the other hand, non-avian reptiles reach sexual maturity at a relatively smaller body mass (in comparison with the average asymptotic size attained) than do birds, in which age at sexual maturity is close to the final adult size (Ruben 1995; Erickson et al. 2007; Lee and Werning 2008). Although my data are inadequate for this test, intraspecific measurements of pedal dimensions of ground birds across the size range of animals known to be sexually mature might well show a smaller size variability than in crocodylians, but how one might ever hope to apply this to ascertaining the maturity of the makers of fossilized footprints is far from clear.

Table 1.5. Variability of phalangeal and digital measurements (in mm) in the moa *Anomalopteryx didiformis*. Data are reported for individuals interpreted as mature (phalanx III1 length at least 40 mm). Data are reported for specimens for which a complete set of measurements could be made (listwise treatment; $N = 14$), and also where the number of specimens for which measurements of a particular bone was greater than the number of specimens for which a complete set of measurements could be made (in which case a larger number of cases is reported). CV = coefficient of variation; parameter abbreviations as in table 1.2.

Parameter	Number of specimens	Minimum	Maximum	Mean (CV)	Maximum/minimum ratio
II1L	14	38	51	46.2 (9.30)	1.34
	21	38	51	46.2 (9.43)	1.34
II2L	14	19	28	24.6 (11.4)	1.47
	21	19	35	24.5 (15.1)	1.84
II3L	14	28	46	37.4 (15.7)	1.64
	17	28	46	38.3 (15.3)	1.64
III1L	14	43	57	51.9 (9.10)	1.33
	22	43	57	51.9 (8.80)	1.33
III2L	14	25	34	31.1 (8.98)	1.36
	19	25	37	31.2 (10.6)	1.48
III3L	14	19	28	22.6 (11.6)	1.47
	18	17	28	22.0 (13.4)	1.65
III4L	14	33	48	38.9 (12.4)	1.45
	17	31	48	38.8 (12.9)	1.55
IV1L	14	32	43	38.6 (8.65)	1.34
	21	31	43	38.1 (9.42)	1.39
IV2L	14	14	20	17.4 (9.48)	1.43
	19	14	24	17.4 (13.3)	1.71
IV5L	14	25	39	31.6 (11.8)	1.56
	18	23	40	32.2 (15.0)	1.74
Digit II length excluding ungual	14	58	79	70.8 (9.63)	1.36
	20	58	83	70.6 (10.7)	1.43
Digit II length	14	93	125	108.2 (10.4)	1.34
	16	93	125	109.9 (10.4)	1.34
Digit III length excluding ungual	14	87	119	105.6 (8.90)	1.37
	18	87	119	104.7 (9.27)	1.37
Digit III length	14	123	167	144.5 (8.88)	1.36
	16	120	167	142.6 (9.43)	1.39
Digit IV length excluding ungual	14	69	93	83.1 (8.45)	1.35
	17	66	93	82.2 (9.24)	1.41
Digit IV length	14	98	132	114.7 (8.39)	1.35
	16	89	132	113.7 (10.0)	1.48

Iguanodon bernissartensis. My initial concern notwithstanding, "blended" phalangeal lengths do not seem to be consistently more variable than lengths based on only one side of the toe. CV for individual phalanges of putatively mature individuals of *I. bernissartensis* ranges from 4% to 17%, with most values being less than 13% (table 1.8); maximum/minimum ratios are mostly 1.5 or less. Phalangeal size variability is more like what is seen in the birds than in alligators, but much should not be made of that, given our ignorance about the size at which *I. bernissartensis* became sexually mature (although it was likely significantly less than the asymptotic adult body mass), and also the small number of individual dinosaurs in my sample. In contrast with the bird species examined, there is no consistent tendency for distal phalanges to be more variable in size than more proximal phalanges of the same toe.

CVs of aggregate digit lengths of mature *I. bernissartensis* range from 3% to 9% (again comparable to what I found for the ground bird species examined in this study), with the length of digit IV (excluding the ungual) the most variable measurement. Maximum/minimum ratios ranged from 1.1 to 1.5.

Intraspecific Size Variability in Phalanges and Digits: Conclusions. Yablokov (1974: appendix table 6) reported that CVs of foot bones of mammalian species (presumably adult individuals) commonly range from 2% to 11%, and CVs of phalanges 3% to 9%. Intraspecific size variability data for the phalanges and digits of emus and rheas and *I. bernissartensis* mostly are in the same ballpark, and moa species somewhat more variable, but the alligator data are considerably more variable. Again, it is possible that the difference between

Table 1.6. Variability of phalangeal and digital measurements (in mm) in the moa *Pachyornis elephantopus*. Data are reported for individuals interpreted as mature (phalanx III1 length at least 67 mm). The first and second rows of data are for specimens assigned to this species on morphological grounds; the first row of data comprises specimens for which a complete set of measurements could be made (listwise treatment), and the second row is for all specimens for which measurements of that particular bone could be made. The third row of data is non-listwise like the second, but includes measurements of a specimen (CM AV 8622; previously identified as *Euryapteryx curtus*) that Allentoft et al. (2010) identified on molecular grounds as belonging to *P. elephantopus*. CV = coefficient of variation; parameter abbreviations as in table 1.2.

Parameter	Number of specimens	Minimum	Maximum	Mean (CV)	Maximum/minimum ratio
	9	63	73	68.1 (4.84)	1.16
II1L	11	63	80	69.6 (6.84)	1.27
	12	63	80	69.5 (6.57)	1.27
	9	22	32	28.0 (12.8)	1.45
II2L	11	22	50	30.0 (24.5)	2.27
	12	22	50	29.7 (24.0)	2.27
	9	47	55	50.6 (4.96)	1.17
II3L	10	47	55	50.6 (4.68)	1.17
	11	45	55	50.1 (5.61)	1.22
	9	67	80	73.7 (5.92)	1.19
III1L	12	67	88	75.3 (7.46)	1.31
	13	67	88	75.1 (7.20)	1.31
	9	36	45	41.0 (7.21)	1.25
III2L	11	36	45	40.7 (6.69)	1.25
	12	36	45	40.5 (6.70)	1.25
	9	21	32	27.0 (14.6)	1.52
III3L	11	21	32	27.3 (13.1)	1.51
	12	21	32	27.1 (12.8)	1.51
	9	47	60	55.4 (7.55)	1.28
III4L	11	47	60	55.3 (7.11)	1.28
	12	44	60	54.3 (9.13)	1.36
	9	51	59	54.8 (4.63)	1.16
IV1L	12	51	64	56.2 (6.48)	1.25
	13	51	64	56.1 (6.20)	1.25
	9	16	25	19.9 (13.6)	1.56
IV2L	11	16	25	19.6 (13.1)	1.56
	12	16	25	19.5 (12.8)	1.56
	9	38	45	42.3 (4.72)	1.18
IV5L	11	38	45	42.6 (4.48)	1.18
	12	37	45	42.2 (5.79)	1.22
	9	88	105	96.1 (6.34)	1.19
Digit II length excluding ungual	11	88	130	99.6 (11.6)	1.48
	12	88	130	99.2 (11.2)	1.48
	9	135	158	146.7 (5.38)	1.17
Digit II length	10	135	158	147.2 (5.18)	1.17
	11	135	158	146.5 (5.22)	1.17
	9	126	156	141.7 (6.72)	1.24
Digit III length excluding ungual	11	126	156	142.1 (6.11)	1.24
	12	126	156	141.6 (5.97)	1.24
	9	177	216	197.1 (5.97)	1.22
Digit III length	11	177	216	197.4 (5.49)	1.22
	12	177	216	195.9 (5.86)	1.22
	9	94	120	105.3 (7.09)	1.28
Digit IV length excluding ungual	10	94	120	105.2 (6.70)	1.28
	11	87	120	103.5 (8.36)	1.38
	9	132	165	147.7 (6.31)	1.25
Digit IV length	10	132	165	147.7 (5.95)	1.25
	11	124	165	145.5 (7.54)	1.33

Table 1.7. Variability of phalangeal and digital measurements (in mm) in the moa *Dinornis robustus*. Data are reported for individuals interpreted as mature (phalanx III1 length at least 70 mm). Data are reported for specimens for which a complete set of measurements could be made (listwise treatment; $N = 22$), and also where the number of specimens for which measurements of a particular bone was greater than the number of specimens for which a complete set of measurements could be made (in which case a larger number of cases is reported). CV = coefficient of variation; parameter abbreviations as in table 1.2.

Parameter	Number of specimens	Minimum	Maximum	Mean (CV)	Maximum/minimum ratio
II1L	22	68	109	89.8 (10.8)	1.60
	25	68	109	88.4 (11.3)	1.60
II2L	22	33	52	41.4 (11.5)	1.58
	24	31	52	40.8 (12.3)	1.68
II3L	22	46	80	63.6 (11.8)	1.74
	23	46	80	63.1 (12.1)	1.74
III1L	22	73	115	98.5 (11.1)	1.58
	25	73	115	97.1 (11.5)	1.58
III2L	22	39	64	51.1 (12.0)	1.64
	24	39	64	50.8 (11.8)	1.64
III3L	22	24	41	33.3 (14.0)	1.71
	23	24	41	33.3 (13.7)	1.71
III4L	22	48	78	63.7 (12.1)	1.63
	23	48	78	63.1 (12.6)	1.63
IV1L	22	53	82	70.2 (10.6)	1.55
	25	53	82	69.1 (11.2)	1.55
IV2L	22	24	44	32.2 (14.5)	1.83
	24	24	44	32.0 (14.2)	1.83
IV5L	22	43	67	54.0 (10.8)	1.56
	24	43	67	53.7 (10.9)	1.56
Digit II length excluding ungual	22	101	161	131.1 (10.8)	1.59
	24	101	161	130.0 (10.9)	1.59
Digit II length	22	147	237	194.7 (10.4)	1.61
	23	147	237	193.3 (10.8)	1.61
Digit III length excluding ungual	22	141	220	183.0 (11.3)	1.56
	23	141	220	182.6 (11.2)	1.56
Digit III length	22	190	295	246.7 (11.0)	1.55
Digit IV length excluding ungual	22	110	177	148.1 (11.5)	1.61
	24	110	177	147.3 (11.3)	1.61
Digit IV length	22	153	239	202.1 (10.8)	1.56
	24	153	239	201.0 (10.7)	1.56

alligators as opposed to birds and mammals reflects differences in the size ratio of fully grown/just sexually mature individuals in these groups.

Intraspecific vs. Interspecific Size-Related Changes in Pedal Proportions

Crocodylians. Alligators showed clear negative allometry (table A1.2) between measures of pedal size (metatarsal III length, "big seven" mean, digit III length excluding ungual) and measures of animal size (total length, femur length), although many of these were "only barely" allometric, probably due to the small sample sizes (see chapter 5 for analyses of intact foot size vs. alligator size involving a much larger database). Measures of the size of the digital/phalangeal region of the foot ("big seven" mean, digit III length excluding ungual) were negatively allometric with respect to metatarsal III length. Young alligators thus have relatively longer feet and toes than do their elders.

In alligators, the lengths of phalanges I1, II1, II2, III2, III3, and IV2 show at least slight negative allometry with respect to the length of phalanx III1, but the length of IV1 is positively allometric with respect to the length of III1. In multivariate comparisons, in which the length of each of the "big seven" phalanges is compared with the mean of the log-transformed "big seven" phalanges other than itself (a proxy for overall size of the digital portion of the foot), phalanges II2 and III3 show at least slight negative allometry, phalanges III1 and IV1 show at least slight positive allometry, and phalanges II1, III2, and IV2 seem to be isometric. This suggests that some of the more distal non-ungual phalanges become shorter with respect to some of the basal phalanges with increasing alligator size. This would extend the negative allometry seen in comparisons of hindlimb length with respect to body size (chapters 4, 5), of autopodial size with respect to more proximal body parts (total length, femur length), and of the digital portion of the foot relative to the metatarsal region, to comparisons of distal vs. proximal portions within the digital part of the foot.

In comparisons of the overall lengths of digits (excluding unguals) in alligators, digit II may show negative allometry with respect to digit III, while digit IV may show positive allometry with respect to digit II, but isometry with respect to digit III. (Questions of allometry of lengths of the intact toes of alligators, using a much larger database, will be considered in chapter 5.)

As with alligators, negative allometry of pedal dimensions with respect to femur length (measurements of total lengths of the intact animal were unavailable for most crocodylian skeletal specimens), and of the digital portion of the foot with respect to metatarsal III length, are seen in the interspecific crocodylian sample (table A1.2). Of course, alligators make up a substantial portion of the interspecific crocodylian sample, and so the interspecific results may be unduly influenced by that fact. However, negative allometry is also suggested by negative slopes in the "one per species" treatments, even though these do not differ significantly from a value of 1, again probably due to the small number of species for which data were available.

RMA slopes of interspecific bivariate comparisons of phalangeal lengths compared with the length of phalanx III1 are mostly similar to those for alligators when all the available data (ignoring differences in the number of specimens per species) are used (again possibly reflecting the fact that I had more data for American alligators than for any other species). However, some of the RMA slopes in the "one per species" treatment (lengths of I1, II1, and III2) are closer to 1.0 (and most of them are not significantly different from 1.0) than their intraspecific alligator counterparts, contrary to my expectations. Multivariate interspecific comparisons of phalanx lengths with "big seven" means (excluding the phalanges of interest in those means), as with alligators, suggest isometry for the lengths of phalanges II1 and III2, negative allometry for phalanges II2 and III3, and positive allometry for phalanges III1, IV1, and (this last unlike the intraspecific alligator relationship) IV2. "One per species" treatment slopes are comparable in value to those for treatments in which all available data are used, even though they are less often significantly different from 1.

Interspecific comparisons of digit lengths are close in value to their intraspecific alligator counterparts. The length of digit II appears to be isometric with respect to that of digit III (alligators may show very slight negative allometry, although the value of the slope differs little from its interspecific counterpart), while the length of digit IV may be positively allometric with respect to the length of digits II and III. However, in none of the "one per species" treatments is the slope significantly different from 1.

Summarizing, the limited data available for crocodylian species other than the American alligator seem not to support my hypothesis that interspecific relationships are more likely to show allometry than are intraspecific relationships. My data suggest that size-related changes in pedal proportions are generally consistent across crocodylians, regardless of species.

Moa. My sample of moa foot skeletons probably did not include as close to the full range of sizes within a species (cf. Huynen et al. 2014) as did my alligator sample (table A1.3), and so the chances of detecting intraspecific allometry in the two moa species for which I had a reasonable sample size (*Dinornis robustus* and *Anomalopteryx didiformis*)—even if it did exist—were probably less than for alligators. *D. robustus* showed no significant allometry between measures of the size of the digital part of the foot ("big seven" mean, digit III length including or excluding the ungual) and tarsometatarsus length, although the sample size was quite small. *A. didiformis* showed positive allometry between both measures of the length of digit III relative to tarsometatarsus length, but the relationship was significant only for digit III length excluding the ungual.

For *D. robustus*, the only phalangeal or digital relationship suggesting significant allometry was a seemingly positive relationship of the length of phalanx III2 relative to that of phalanx III1, and whether much stock should be placed in this is doubtful, because the length of phalanx III2 showed no allometry relative to the "big seven" mean. However, there was only a twofold greater ratio between the size of the largest and smallest individuals in my *D. robustus* sample, which is not much more than the extreme reversed size dimorphism (150%; females > males) known for this species (Bunce et al. 2003; Allentoft et al. 2010), and so the inability to detect allometry could reflect the lack of very young individuals in my sample.

The size difference among specimens of *A. didiformis* in my sample (ca. 2.5–3×) was somewhat larger than for *D. robustus*, and two phalangeal/digital comparisons suggested possible allometry; the length of phalanx II1 may show negative allometry with respect to the length of phalanx III1, and the length of digit II excluding the ungual may show negative allometry with respect to the length of digit III excluding the ungual. These two results suggest a possible

Table 1.8. Variability in pedal phalangeal and digital measurements in *Iguanodon bernissartensis* from Bernissart, Belgium. Data are reported for individuals interpreted as mature (phalanx III1 length at least 100 mm). Data are summarized for four treatments: (1) Length of the medial side of the phalanx or digit; (2) length of the lateral side of the digit; (3) "synthetic/blended" feet, in which incomplete left and right feet of the same individual were combined to provide a more complete composite foot, and (where possible) medial and lateral lengths of phalanges were averaged; and (4) listwise synthetic/blended feet, in which it was possible to obtain a complete set of measurements for all phalanges (that is, at least one length measurement [medial, lateral, or both] could be made on all phalanges [of either the left or right foot, or combining phalanges of both feet]). All measurements are in millimeters. CV = coefficient of variation; parameter abbreviations as in table 1.2.

Parameter	Treatment	Minimum	Maximum	Mean (CV)	Maximum/minimum ratio	*N*
II1L	Medial	135	188	151.2 (9.23)	1.39	11
	Lateral	109	159	123.5 (11.8)	1.46	11
	Synthetic/blended	116	173	137.3 (10.2)	1.49	12
	Synthetic/blended listwise	125	149	135.9 (5.53)	1.19	8
II2L	Medial	36	59	51.5 (12.7)	1.64	10
	Lateral	38	51	45.1 (8.56)	1.34	11
	Synthetic/blended	37	55	48.9 (9.89)	1.49	13
	Synthetic/blended listwise	41	55	48.7 (8.31)	1.34	8
II3L	Medial	106	136	124.4 (7.65)	1.28	9
	Lateral	107	136	123.9 (7.86)	1.27	8
	Synthetic/blended	107	134	122.8 (7.36)	1.25	11
	Synthetic/blended listwise	111	134	125.5 (6.57)	1.21	8
III1L	Medial	113	136	126.7 (5.65)	1.20	11
	Lateral	108	140	119.9 (8.21)	1.30	9
	Synthetic/blended	111	143	124.4 (6.96)	1.29	15
	Synthetic/blended listwise	115	138	125.6 (5.17)	1.20	8
III2L	Medial	38	44	41.2 (4.95)	1.16	11
	Lateral	37	47	40.6 (6.54)	1.27	11
	Synthetic/blended	39	45	41.6 (5.33)	1.15	13
	Synthetic/blended listwise	39	45	41.5 (5.47)	1.15	8
III3L	Medial	30	40	35.1 (9.54)	1.33	10
	Lateral	30	40	35.3 (8.56)	1.33	10
	Synthetic/blended	31	40	35.9 (7.63)	1.29	13
	Synthetic/blended listwise	31	37	34.4 (5.59)	1.19	8
III4L	Medial	121	151	134.8 (8.29)	1.25	9
	Lateral	101	147	127.3 (12.6)	1.45	9
	Synthetic/blended	101	149	127.6 (10.6)	1.48	11
	Synthetic/blended listwise	101	149	128.5 (11.8)	1.48	8
IV1L	Medial	77	116	94.3 (14.5)	1.51	10
	Lateral	103	136	113.9 (8.21)	1.32	10
	Synthetic/blended	78	136	106.6 (13.6)	1.74	14
	Synthetic/blended listwise	78	117	103.3 (11.8)	1.50	8

tendency for digit II to diminish in size relative to digit III with increasing bird size. However, the length of digit II including the ungual shows no indication of allometry with respect to the length of digit III including the ungual, and the length of phalanx II1 shows no allometry with respect to the "big seven" mean that excludes it. Consequently I see no unambiguous allometry of phalangeal or digital proportions in this species, either, although once again, if I had data for very small chicks, that conclusion might change.

Unlike for crocodylians, I was able to measure widths of phalanges of moa toes. Neither *D. robustus* nor *A. didiformis* showed allometry of the distal width of phalanx III2 (a spot which would be about halfway out the length of the toe) compared with the length of digit III excluding the unugual.

If the data for *D. robustus* and *A. didiformis* are pooled to create an interspecific sample, there are more candidates for allometric relationships than for either species alone (but of course the size range of specimens, and the sample size, both become larger). The three measures of digital foot size are all negatively allometric with respect to tarsometatarsus length. The lengths of phalanges II2, III2, III3, and to a lesser extent IV1 appear to be negatively allometric relative to the length of phalanx III1 (although II2 and IV1 only barely so). The length of digit II is positively allometric (but only barely so) with respect to the length of digit III (both including and excluding the unguals). The lengths of II1, III1, IV1, and IV2 are positively allometric (IV1 and IV2 only barely so) with respect to the "big seven" means that exclude them, and the

Parameter	Treatment	Minimum	Maximum	Mean (CV)	Maximum/minimum ratio	*N*
IV2L	Medial	26	51	33.3 (20.2)	1.96	10
	Lateral	30	53	39.9 (15.1)	1.77	11
	Synthetic/blended	30	52	36.1 (13.9)	1.73	15
	Synthetic/blended listwise	30	39	35.7 (7.59)	1.30	8
IV5L	Medial	104	121	111.2 (5.72)	1.16	9
	Lateral	99	118	106.6 (6.90)	1.19	9
	Synthetic/blended	102	124	109.5 (5.93)	1.22	13
	Synthetic/blended listwise	102	119	108.9 (4.85)	1.17	8
Digit II length excluding ungual	Medial	187	224	204.1 (4.92)	1.20	10
	Lateral	155	197	168.6 (7.59)	1.27	11
	Synthetic/blended	167	210	185.9 (6.65)	1.26	12
	Synthetic/blended listwise	170	202	185.6 (5.54)	1.19	8
Digit II length	Medial	293	342	325.4 (4.86)	1.17	8
	Lateral	274	310	288.9 (4.00)	1.13	8
	Synthetic/blended	274	329	307.0 (4.91)	1.20	10
	Synthetic/blended listwise	301	329	311.1 (3.48)	1.09	8
Digit III length excluding ungual	Medial	195	214	204.4 (3.49)	1.10	10
	Lateral	180	214	197.4 (4.99)	1.19	8
	Synthetic/blended	185	212	201.1 (3.40)	1.15	13
	Synthetic/blended listwise	197	212	201.5 (2.61)	1.08	8
Digit III length	Medial	320	350	338.8 (3.06)	1.09	8
	Lateral	304	347	330.4 (4.50)	1.14	7
	Synthetic/blended	298	348	327.6 (4.81)	1.17	11
	Synthetic/blended listwise	298	348	330.0 (5.07)	1.17	8
Digit IV length excluding ungual	Medial	159	204	178.3 (8.95)	1.28	9
	Lateral	198	238	212.2 (5.13)	1.20	9
	Synthetic/blended	157	230	197.4 (8.82)	1.46	14
	Synthetic/blended listwise	157	208	193.4 (8.08)	1.32	8
Digit IV length	Medial	274	324	294.0 (5.72)	1.18	8
	Lateral	302	341	316.6 (4.00)	1.13	7
	Synthetic/blended	276	343	308.2 (6.22)	1.24	12
	Synthetic/blended listwise	276	317	302.3 (4.28)	1.15	8

lengths of III2 and III3 are negatively allometric with respect to their corresponding "big seven" means.

An even bigger interspecific sample, with an even greater size range, is created by pooling all moa specimens of all species, disregarding differences in the number of specimens representing each species. The three measures of digital foot size are again negatively allometric with respect to tarsometatarsus length (but the two measures of digit III length only barely so). The length of phalanx IV1 is negatively allometric with respect to the length of phalanx III1 (but only barely so), the length of digit IV (both with and without the ungual) and the length of digit II (but only if the ungual is included) are positively allometric with respect to the length of digit III, and the lengths of phalanges II2 and IV2 are positively allometric (but only barely so) with respect to the "big seven" means that exclude them. Interestingly, there are fewer apparently allometric phalangeal/digital relationships in this "all moa" treatment than in the pooled *D. robustus–A. didiformis* sample.

The moa "one per species" sample spans a respectable size range, but with only nine specimens has little chance of showing significant allometric relationships. It is nonetheless interesting to compare the slopes of relationships in the "one per species" treatment with their "all moa" counterparts for consistency. In all comparisons in which the "all moa" treatment suggests significant allometry except one (digit II length including ungual with digit III length including ungual), the "one per species" treatment agrees with the "all moa" treatment in whether the possible allometry is negative or positive.

None of the interspecific comparisons of the distal width of phalanx III2 with the length of digit III excluding the ungual suggests allometry. Just as toe widths do not become proportionally stouter or slimmer with increasing bird size within the two moa species examined, neither does there seem to be any size change in toe stoutness across species.

Summarizing, the limited sample size and size range of moa specimens make statements about intraspecific and interspecific allometry tentative at best. However, unlike alligators vs. crocodylians more generally, there seem to be more possible examples of interspecific allometry among moa than within either of the two moa species for which I had moderately large sample sizes. As with the analogous comparison in alligators (as well as crocodylians more generally), with increasing size across moa species the digital portion of the foot seems to become shorter with respect to the length of the metatarsus. Also like alligators, and across crocodylians, across moa species there may be positive allometry of the length of digit IV relative to other parts of the digital portion of the foot.

Intraspecific vs. Interspecific Variability in Phalangeal and Digital Proportions

Effects of Sample Size on Maximum/Minimum Ratios of Scaled Parameters (table A1.4). I have already explained the procedures used in this section in the methods section of this chapter, but the analyses that follow are complicated enough that briefly restating here what was done, and why it was done, before presenting the results of each analysis may facilitate understanding of those results.

For *Alligator mississippiensis*, there were 19 specimens in which it was possible to measure the lengths of all of the following non-ungual pedal phalanges: I1, II1, II2, III2, III3, IV1, and IV2, as well as the length of phalanx III1, to a common value of which the lengths of all the other phalanx lengths were scaled. Pooling measurements of the 19 alligators with data for all other crocodylian specimens, of all species, in which lengths of all of the phalanges could be measured gives a sample size of 42 individual animals. I compared the ratio of the maximum to the minimum value of each phalanx length, as scaled to the length of phalanx III1, of the intraspecific American alligator sample with the interspecific sample of all crocodylians. This was done two ways. I first looked at the interspecific value of the maximum/minimum ratio for the entire crocodylian sample. Then I did 30 trials in which I randomly selected 19 of the 42 crocodylian cases (the alligator cases were included in the 42 cases from which 19 samples were selected), and calculated the median of the maximum/minimum ratio of the scaled phalanx length across the trials, and determined the percentage of the trials in which the maximum/minimum ratio of the 19 randomly selected cases was greater than the maximum/minimum ratio of the 19 American alligator cases.

For the intraspecific *A. mississippiensis* sample, the maximum/minimum ratio of phalanx lengths scaled to a common value of phalanx III1 length ranged from 1.1 to 1.4, with the larger values associated with the more distal scaled phalanges. For every scaled phalanx length, the maximum/minimum ratio for the interspecific crocodylian sample was greater than the ratio for the intraspecific alligator sample; this was true both for the entire crocodylian sample of 42 cases, as well as for median value of the 30 replicates of 19 randomly selected cases. For phalanges I1, II1, II2, and III2, in fact, every one (100%) of the 30 replicates of 19 cases randomly selected from among the 42 crocodylian data cases had a maximum/minimum ratio of scaled values greater than that of the intraspecific alligator sample, and for phalanx III3 97% of the 30 trials resulted in a ratio greater than the intraspecific alligator sample. For phalanx IV1 the percentage of trials with a maximum/minimum ratio greater than the ratio for American alligators alone dropped to 87%, and for phalanx IV2 to 53%. So it seems that for crocodylians, for most of the non-ungual phalanges, an interspecific sample does indeed show more proportional length variability than a sample drawn just from *A. mississippiensis*.

For the moa *Dinornis robustus* there were 24 specimens in which all the parameters tested could be measured, and so the listwise sample size for this species was 24. Five, 10, or 15 cases were randomly selected from this sample of 24 cases, and each of these three permutations of the trial was carried out 15 times for each pedal parameter of interest. As expected, the median value of the maximum/minimum ratio of each scaled parameter did increase with an increasing number of cases randomly selected out of the possible 24 cases, most notably for the scaled lengths of phalanges III3 and IV2 (the smallest phalanges considered in these tests). For other scaled phalanx lengths, and for digit lengths, the increase in variability with an increase in the number of cases randomly selected was less striking.

Anomalopteryx didiformis was represented by a listwise sample size of 17 foot skeletons. These 17 data cases were pooled with the 24 *D. robustus* cases for a two-species interspecific set of trials. The maximum/minimum ratio of the scaled phalanx lengths in the 41 pooled, two-species sample was at least a bit greater than the maximum/minimum ratio for either of the two species for all of the "big seven" phalanges except IV1; for IV1 the maximum/minimum ratio for the pooled sample was greater than that for *D. robustus*, but the same as for *A. didiformis*.

Thirty trials in which 17 cases were randomly selected from the pooled sample of *D. robustus* and *A. didiformis* cases were carried out for each scaled phalanx length. The median value for each scaled parameter across the 30 replicates was at least somewhat greater than the maximum/minimum ratio for the entire listwise sample of each species for all scaled phalanx lengths except phalanges IV1 and IV2. For phalanx IV1, the median value of the 30 tests was greater than the maximum/minimum ratio for *D. robustus*, but not for *A. didiformis*. For phalanx IV2 the median value was the same as the maximum/minimum ratio for *A. didiformis*, and less than that for *D. robustus*. The value of the maximum/minimum ratio calculated from the 17 cases randomly selected from the two-species sample was greater than the maximum/minimum ratio for the listwise sample of cases for either species alone in more than half of the 30 trials for all phalanx lengths except IV1 (for *A. didiformis*) and IV2 (both species).

These results can be compared with an ANCOVA of the interspecific moa sample (table A1.5). Despite some reservations due to problems of meeting the assumptions of the ANCOVA procedure, in ANCOVAs of the lengths of phalanges II1, II2, III2, IV1, and IV2, with the length of phalanx III1 as the covariate, *A. didiformis* seems to be significantly different from *D. robustus* in the relative proportions of phalanges II2, III2, and III3, but not II1, IV1, and IV2. For phalanges II2, III2, and III3, but not phalanges II1, IV1, and IV2, pooling data for the two species resulted in maximum/minimum ratios of the scaled parameters that were substantially higher than for either of the two species by itself (table A1.3).

The moa *Pachyornis elephantopus* is represented by 11 listwise cases in these tests. These cases were combined with the listwise cases for *A. didiformis*, *D. robustus*, and all other moa species in my sample to create a pooled moa foot skeleton sample of 82 cases. The maximum/minimum ratio of phalanx lengths scaled against that of III1 across the 82 "all moa" cases was greater than that for *D. robustus*, *A. didiformis*, and *P. elephantopus* alone for all of the "big seven" phalanges.

I randomly selected 11, 17, or 24 cases from the pooled "all moa" sample of 82 cases 15 times for each of the "big seven" phalanx lengths. The median value of the maximum/minimum ratio of the scaled parameter calculated from the randomly selected cases was greater than the single-species maximum/minimum ratio for *D. robustus* and *A. didiformis* for all scaled phalanx lengths except IV1 (*A. didiformis*, random 11 and random 17, but not random 24, cases) and IV2 (*A. didiformis*, random 11 cases; *D. robustus*, random 11 and 17 cases).

P. elephantopus usually had more variable phalangeal proportions than the other two moa species. The maximum/minimum ratio of the scaled parameter of this species was greater than the median value of the maximum/minimum ratio of the scaled parameter calculated from the randomly selected cases for all moa species, for at least some versions of the test, for phalanges II1 (random 11, 17, and 24 cases), II2 (random 11 cases), and IV2 (random 11 cases). For phalanges III2, III3, and IV1 the median value of maximum/minimum ratio calculated from the randomly selected cases was equal to or greater than the *P. elephantopus* single-species ratio.

We can look at these results another way. Fifty percent or more of the median values of the maximum/minimum ratio of the scaled parameter, calculated from the randomly selected cases of the scaled parameter, drawing from the entire sample of moa specimens, were greater than the single-species maximum/minimum ratios of the same scaled parameters for the individual species *D. robustus* and *A. didiformis* for all scaled phalangeal lengths (and all three versions of the random selection tests), except in some versions of the tests, for one or the other species, for IV1 and IV2.

In contrast, the version of the test in which 11 cases were randomly selected from the entire set of moa specimens resulted in median values of the maximum/minimum ratio of the scaled parameter that were greater than the single-species value of the maximum/minimum ratio of the scaled parameter for *P. elephantopus* in fewer than 50% of the 15 trials for the scaled lengths of phalanges II1, II2, IV1, and IV2. For the 17-case and 24-case versions of the test, however, the all-moa sample again resulted in median values that were greater than those for *P. elephantopus* alone in half or more of the 15 replicates, just as with the other two moa species. Furthermore, the same was true for all versions of the test for phalanges III2 and III3.

Once again we can compare these results with an ANCOVA across several moa species (table A1.5). *P. elephantopus*, *A. didiformis*, and *D. robustus* have significantly different proportions (with phalanx III1 length the covariate) for all of the test phalanges (II1–2, III2–3, IV1–2) than those for at least one other moa species in all or nearly all comparisons.

The recitation of the preceding results is admittedly rather mind-numbing, but the payoff is this: For moa (as with crocodylians), observing greater values of the maximum/minimum ratio of scaled parameters for interspecific than for intraspecific samples does not seem merely to reflect larger numbers of specimens in the interspecific samples. There are at least some differences across moa species in pedal proportions, although one can be most confident about this conclusion if the sample size of specimens is reasonably large (say, at least 15 specimens).

The good news, then, is that the minimum/maximum ratio of phalanx lengths scaled to a common length of phalanx III1 often really is greater for interspecific than for intraspecific samples, indicating that this very simple, quick-and-dirty bivariate shape comparison might potentially have some use in deciding whether a sample of feet—or with luck footprints—came from a single species or more than one species. Although the ratio clearly is affected by the number of specimens examined, this does not necessarily mean that greater interspecific than intraspecific values of the ratio cannot be trusted. With that in mind, we now consider several intraspecific and interspecific samples in more detail.

Within-Species vs. Across-Species Values of Maximum/Minimum Ratios of Scaled Phalanx and Digit Lengths. Values of the maximum/minimum ratios of lengths of phalanges II1, II2, III2, III3, IV1, and IV2, scaled to a common length of phalanx III1, of crocodylians, ground birds, and non-avian dinosaurs are summarized in table A1.6. Some listwise data from table A1.4 are repeated in table A1.6 for convenience of comparison, but most of the ratios reported in table A1.6 are based on data cases in which only the phalanx of interest, and the phalanx (III1) against whose length the phalanx of interest was scaled, had to be present and measurable in order to be included. Consequently the sample size can vary among the different phalanges, and the sample sizes are often larger than in the listwise treatment.

In most taxa, whether single species or interspecific samples, maximum/minimum ratios of scaled lengths are less for the larger, more proximal (II1, III2, IV1) than for the smaller, more distal (II2, III3, IV2) phalanges. The largest values of the maximum/minimum ratio of scaled phalanx lengths are 1.37 for II1 (*Iguanodon bernissartensis* "synthetic"/"blended" data treatment); 2.04 for II2 (*Pachyornis elephantopus*); 1.50 for III2 (*I. bernissartensis* "synthetic"/"blended" data treatment); 1.55 for III3 (*P. elephantopus*); 1.86 for IV1 (*I. bernissartensis* "synthetic"/"blended" data treatment); and 1.80 for IV2 (*I. bernissartensis* "synthetic"/"blended" data treatment). Thus two forms, one an ornithopod dinosaur and the other a moa, have the most variable phalangeal proportions of the taxa in my sample. Apart from those two species, maximum/minimum ratios of scaled parameters are in the range of 1.1–1.4 for phalanges II1 and IV1, 1.1–1.4 for III2, and 1.1–1.9 for II2, III3, and IV2.

In most interspecific vs. intraspecific comparisons, at least some of the interspecific maximum/minimum ratios are larger than for a single species, as expected. For two-species or probable two-species pooled cases (*Dromaius novaehollandiae* + *Ardeotis kori*, *Rhea americana* + *Pterocnemia pennata*, *Dinornis* spp., *D. robustus* + *A. didiformis*, *Allosaurus* spp., *Edmontosaurus* spp.), maximum/minimum ratios of scaled parameters fall in the range of 1.1–1.2 (II1), 1.2–2.0 (II2), 1.3–1.7 (III2), 1.4–1.9 (III3), 1.1–1.6 (IV1), and 1.3–1.9 (IV2). There is thus no single value that allows one to conclude with certainty that two rather than one species are represented in a sample.

For samples consisting of three or more species, maximum/minimum ratios of scaled phalanx lengths range as follows: 1.1–1.7 (II1), 1.1–3.7 (II2), 1.1–1.9 (III2), 1.3–2.4 (III3), 1.1–1.9 (IV1), and 1.2–4.1 (IV2). There are some interspecific ranges of the ratio whose minimum values remain not much greater than those for single species, but the upper values have increased greatly.

The largest values of the ratio for all of the scaled phalanx lengths except that of IV1 are for the pooled sample of all large tridactyl ground birds measured in this study, which obviously consists of many species (kori bustard, extant ratites, moa). On the other hand, pooling data for allosauroids, tyrannosaurids, and two more basal large theropods (*Dilophosaurus* and *Aucasaurus*) does not increase the maximum/minimum ratios of scaled parameters beyond those for allosauroids alone, and adding data for *Mantellisaurus atherfieldensis* to data for all specimens of *Iguanodon* (the latter possibly including species in addition to *I. bernissartensis*) increases values of the maximum/minimum ratio over those for *I. bernissartensis* alone only for phalanx IV2.

Comparing all of the intraspecific and interspecific values of the maximum/minimum ratio of scaled phalanx lengths of ground birds and non-avian dinosaurs graphically (fig. 1.13*A*, 1.13*B*) shows at least some tendency for the ratio to creep upward as the number of species in the sample increases for all the phalanges examined; the tendency is particularly striking for phalanges II1, III2, III3, and IV2 and less convincing for II2 and IV1. The very high values of the ratio for phalanx IV1 associated with a small number of species are for *I. bernissartensis* (one species, maximum/minimum ratio = 1.86), *Iguanodon* + related forms (two species, ratio = 1.82), and *Allosaurus* spp. (two species, ratio = 1.59). Thus there is a great amount of scatter in these plots. Consequently higher values of the maximum/minimum ratio are not invariably generated by increasing the total number of species included in the sample.

Similar conclusions can be drawn from intraspecific vs. interspecific comparisons of maximum/minimum ratios of the overall lengths of digits II and IV scaled to that of digit III (table A1.7; fig. 1.13*C*, 1.13*D*). The sample sizes in many of these comparisons are substantially smaller than those for phalangeal proportions (especially for extinct forms), simply because all of the phalanges in both of two digits must be preserved and measurable for the specimen to be included as a data case. In some comparisons (all crocodylians vs. *Alligator mississippiensis*; emu + cassowaries [*Casuarius* spp.], emu + kori bustard [*Ardeotis kori*], or emu + rheas vs. emu alone; *D. robustus* + *A. didiformis* vs. *D. robustus* or *A. didiformis* alone; all moa or all ground birds vs. individual ground bird species) the interspecific values of the ratio are noticeably larger than those for individual species for at least

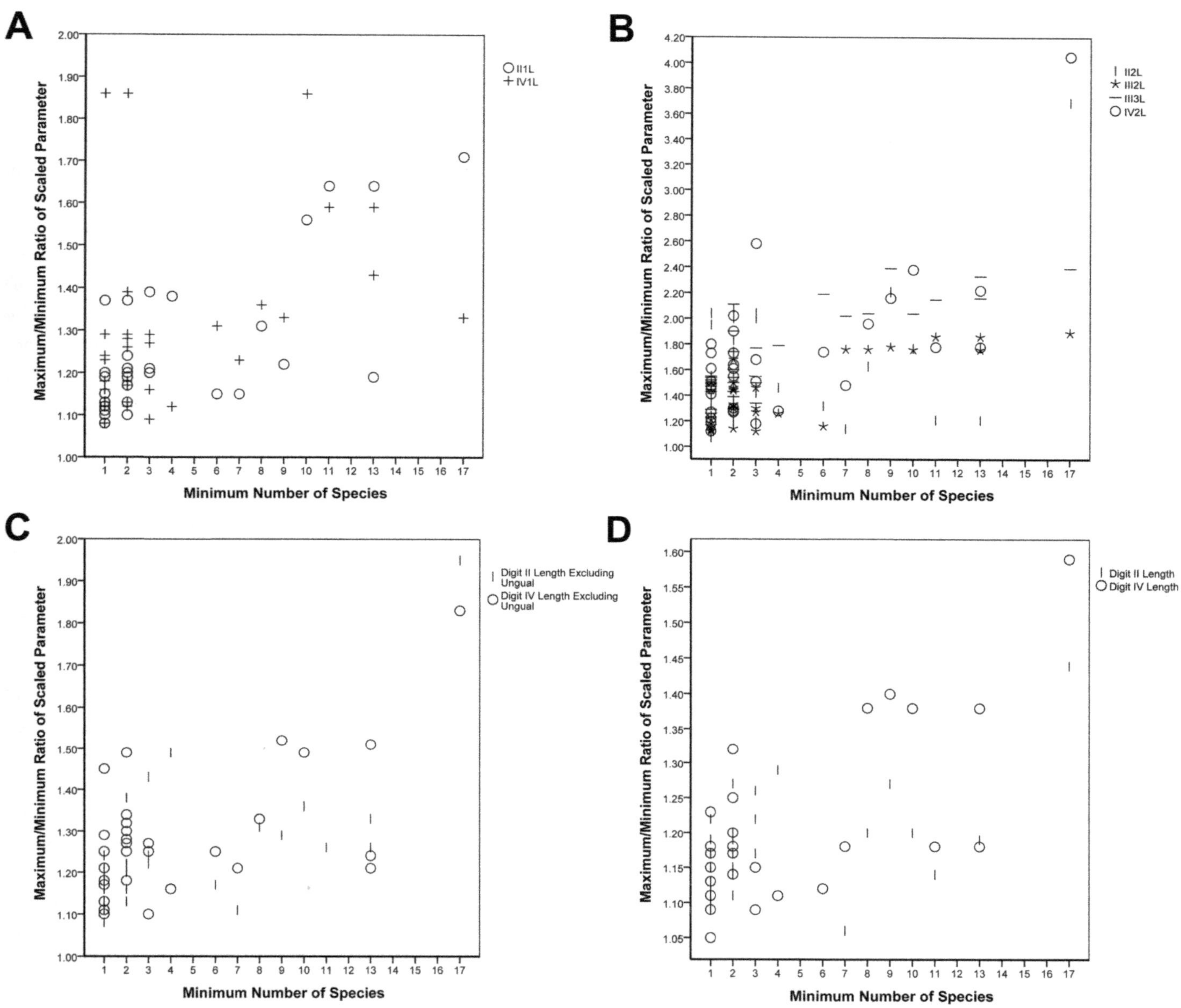

1.13. Relationship between the minimum number of species ("minimum" because of uncertainty about the number of species represented by some of the dinosaur samples) represented in a sample of foot skeletons and the maximum/minimum ratio of values of phalanx or digit lengths scaled to a common length. Data used in these graphs are for most of the intraspecific and interspecific samples of ground birds and non-avian dinosaurs in tables A1.6 and A1.7. *A*, Proximal phalanx lengths scaled to a common value of the length of phalanx III1. *B*, Distal phalanx lengths scaled to a common value of the length of phalanx III1. *C*, Lengths of digits II and IV (excluding the ungual), scaled to a common value of the length of digit III (excluding the ungual). *D*, Lengths of digits II and IV scaled to a common length of digit III.

some scaled digit lengths. However, in other comparisons (*Casuarius* spp. vs. *C. casuarius*; both rhea species vs. *R. americana*; *Pachyornis* spp. vs. *P. elephantopus*; *Dinornis* spp. vs. *D. robustus*; *Iguanodon bernissartensis* + closely related forms vs. *I. bernissartensis* alone) the interspecific values of the maximum/minimum ratio of the scaled digit lengths are slightly or no greater than those for the individual species. Even so, graphic comparisons of all the data for ground birds and non-avian dinosaurs (fig. 1.13*C*, 1.13*D*) show a more striking increase in the maximum/minimum ratio with increasing number of species for scaled digit lengths than for scaled phalangeal lengths.

Of course, one could reasonably argue that, because many of the interspecific comparisons in tables A1.6 and A1.7 and in figure 1.13 pool species from different geographic regions and/or times, they are therefore unrealistic for evaluating the amount of difference one would expect to see among foot skeletons from species of a single fauna. This concern can be addressed by considering species known to have been members of the same fauna, the moa species from

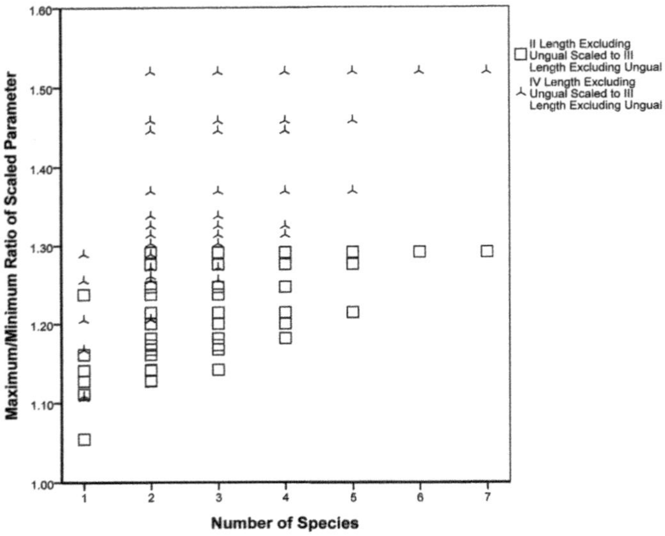

1.14. Relationship between variability in digital proportions and the number of species being compared. The graph shows the ratio of the maximum to the minimum value of the scaled digit length plotted as a function of the number of species in the sample. The species involved are moa from the South Island of New Zealand; to increase the sample size, data for North Island specimens of species that occurred on both islands are included. The minimum number of specimens in the sample for *Dinornis robustus* is 24; for *Anomalopteryx didiformis*, 17; for *Emeus crassus*, 8; for *Euryapteryx curtus*, 9; for *Megalapteryx didinus*, 3; for *Pachyornis elephantopus*, 11 (these analyses include the specimen assigned to this species on molecular grounds by Allentoft et al. 2010); and for *P. australis*, 1. Data for the aggregate lengths of the non-ungual phalanges of digits II and IV, scaled to a common value of the aggregate length of the non-ungual phalanges of digit III, are presented. For all species except *P. australis*, data are presented for the individual species, and for all combinations of pooled data for two, three, four, and five species; the single specimen of *P. australis* is included in the pooled data for all seven species of South Island moa. The number of cases plotted in the graph is sometimes less than the actual number of cases because identical values are superimposed. The various combinations of species in each numerical category of number of pooled species cases (all two-species combinations, all three-species combinations, and so on) show a wide range of values for the minimum/maximum ratio. The maximum value of the ratio does not change beyond that of two-species combinations for higher numbers of species in the comparison, but the minimum value of the ratio creeps steadily upward. The median values of the ratio therefore increase from those for all of the one-species samples (digit II: 1.13; digit IV: 1.19) to those for progressively more species in the comparison, at least through those for five-species combinations (two species: digit II = 1.18, digit IV = 1.30; three species: digit II = 1.23, digit IV = 1.37; four species: digit II = 1.28, digit IV = 1.46; five species: digit II = 1.29, digit IV = 1.52; six species: digit II = 1.29; digit IV = 1.52).

the South Island of New Zealand (Bunce et al. 2009). Figure 1.14 shows such an analysis for the lengths of digits II and IV (excluding the unguals) scaled to a common length of digit III (excluding the ungual). Data are plotted for each of the six moa species represented by more than one specimen, for all species combinations of pooled data for two, three, four, and five species, for all six moa species, and for the six moa species plus a seventh species represented by a single specimen.

Values of the maximum/minimum ratio of scaled digit lengths jump upward from those for the six within-species (one species) samples to those for the various two-species combinations. As the number of species being pooled increases from two through five, the largest value of the maximum/minimum ratio does not increase. This may in part be an artifact generated by the fact that in pooling data for ever more species the same specimens are repeatedly generating the extreme values of the scaled digit lengths. Of greater interest, the smallest value of the maximum/minimum ratio creeps steadily upward. As a result, the median value of the maximum/minimum ratio also moves upward, leveling off at the value for five-species combinations. Consequently the maximum/minimum ratio of the scaled digit lengths does show a tendency to increase as more species are added to the sample, but the amount of the increase becomes less as the number of species pooled becomes progressively larger. The scaled length of digit IV is noticeably more variable than that of digit II.

Inspection of the distribution of values of the maximum/minimum ratio in tables A1.6 and A1.7, and figures 1.13 and 1.14, suggests some rough rules of thumb for deciding whether a collection of foot skeletons is likely to represent a single species, or two or more species. For the "big seven" individual phalanges considered in this analysis (II1–2, III2–3, IV1–2, all scaled to a common length of III1), one should suspect (albeit without complete certainty) that more than one species may be involved if the maximum/minimum ratio of scaled proximal phalanx lengths takes values of 1.5 or more, and for scaled distal phalanx lengths values of 2 or more, although interpreting how many more than one species are present is difficult, maybe even impossible. For the overall lengths of digits II and IV scaled to a common length of digit III, values of the ratio of 1.3 or more (a conservative cut-off, judging from fig. 1.14) likely indicate that more than one species is included in the sample. For scaled digit lengths and scaled lengths of at least phalanx II1, and possibly phalanges III2, III3, and IV2, the higher the value of the maximum/minimum ratio of the scaled parameters, the greater the odds that an increasing number of species is included in the sample. However, the moa data (fig. 1.14) suggest that it may be easier to distinguish between samples drawn from one, as opposed to more than one, species than to ascertain how many more than one species there are in a sample of two or more species.

Within-Species vs. Across-Species Values of the Coefficient of Variation of Scaled Phalangeal Parameters (tables A1.8, A1.9; fig. 1.15). The results of these analyses look so much like those for maximum/minimum ratios of scaled parameters that I will forgo a detailed description here. Suffice it to say that there is a tendency for values of CV to

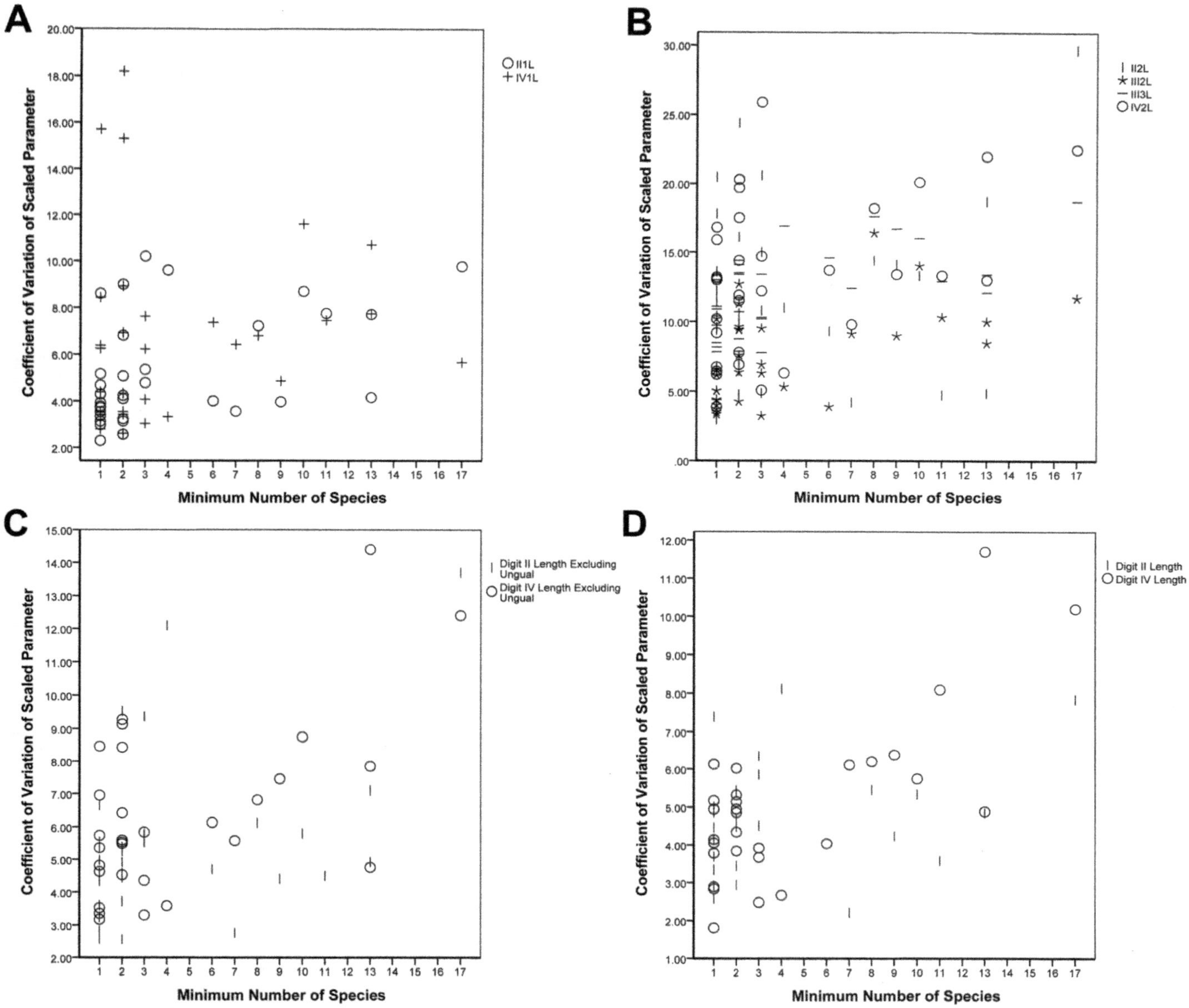

1.15. Relationship between the minimum number of species represented in a sample of foot skeletons and the coefficient of variation of the ratio of values of phalanx or digit lengths scaled to a common length. Data used in these graphs are for most of the intraspecific and interspecific samples of ground birds and non-avian dinosaurs in tables A1.8 and A1.9. *A*, Proximal phalanx lengths scaled to a common value of the length of phalanx III1. *B*, Distal phalanx lengths scaled to a common value of the length of phalanx III1. *C*, Lengths of digits II and IV (excluding the ungual) scaled to a common value of the length of digit III (excluding the ungual). *D*, Lengths of digits II and IV scaled to a common length of digit III.

increase from intraspecific to interspecific samples, and for at least some phalanges, and for overall digit lengths, for CV to become larger as the number of species represented in the sample increases.

Comparing the CV of scaled lengths with the maximum/minimum ratio of scaled lengths of the same within-species and across-species comparisons (figs. 1.13, 1.15, 1.16) shows that the two measures of shape variability are strongly correlated—they tell the same story. However, the relationship is curved rather than linear, probably reflecting the fact that the maximum/minimum ratio is more strongly affected by extreme cases than is CV.

Range and Standard Deviation of GM-Scaled Parameters (tables A1.10, A1.11; fig. 1.17). Because the GM scaling used here requires that all of the "big seven" phalanges, or all of the phalanges in all three digits, be preserved and measurable for a specimen to be included in the analysis, the sample size is necessarily smaller than in simple bivariate scaling in which only two phalanges or two digits must be complete.

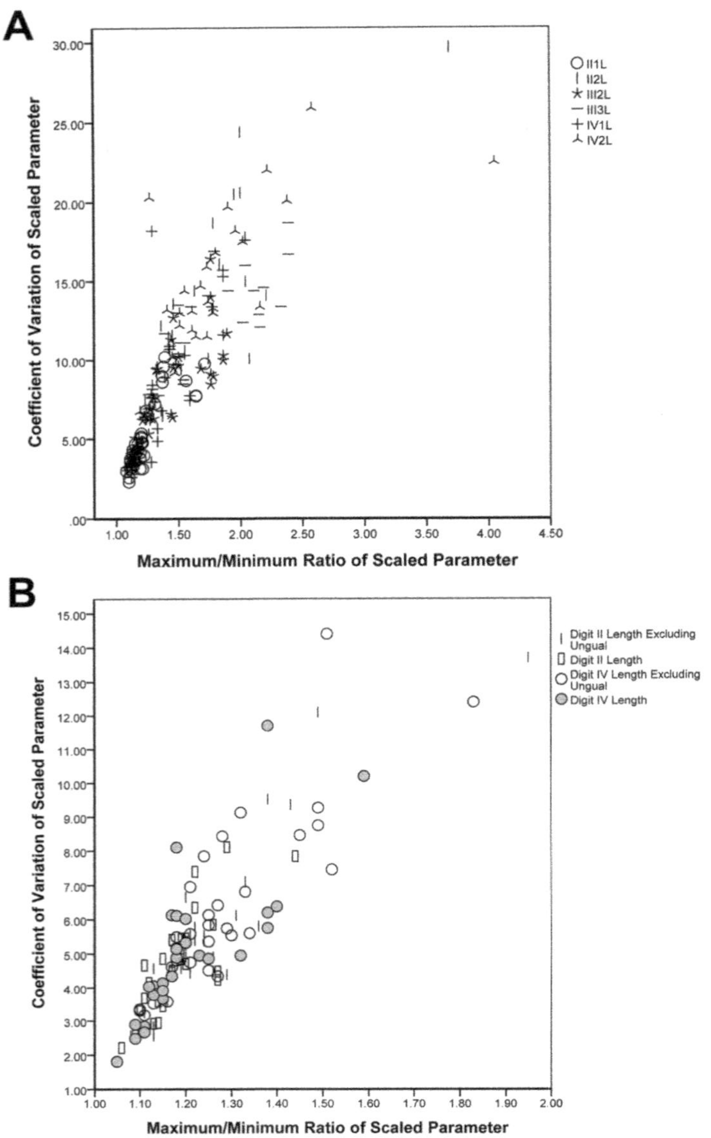
A
Coefficient of Variation of Scaled Parameter
Maximum/Minimum Ratio of Scaled Parameter
II1L
II2L
III2L
III3L
IV1L
IV2L
B
Coefficient of Variation of Scaled Parameter
Maximum/Minimum Ratio of Scaled Parameter
Digit II Length Excluding Ungual
Digit II Length
Digit IV Length Excluding Ungual
Digit IV Length

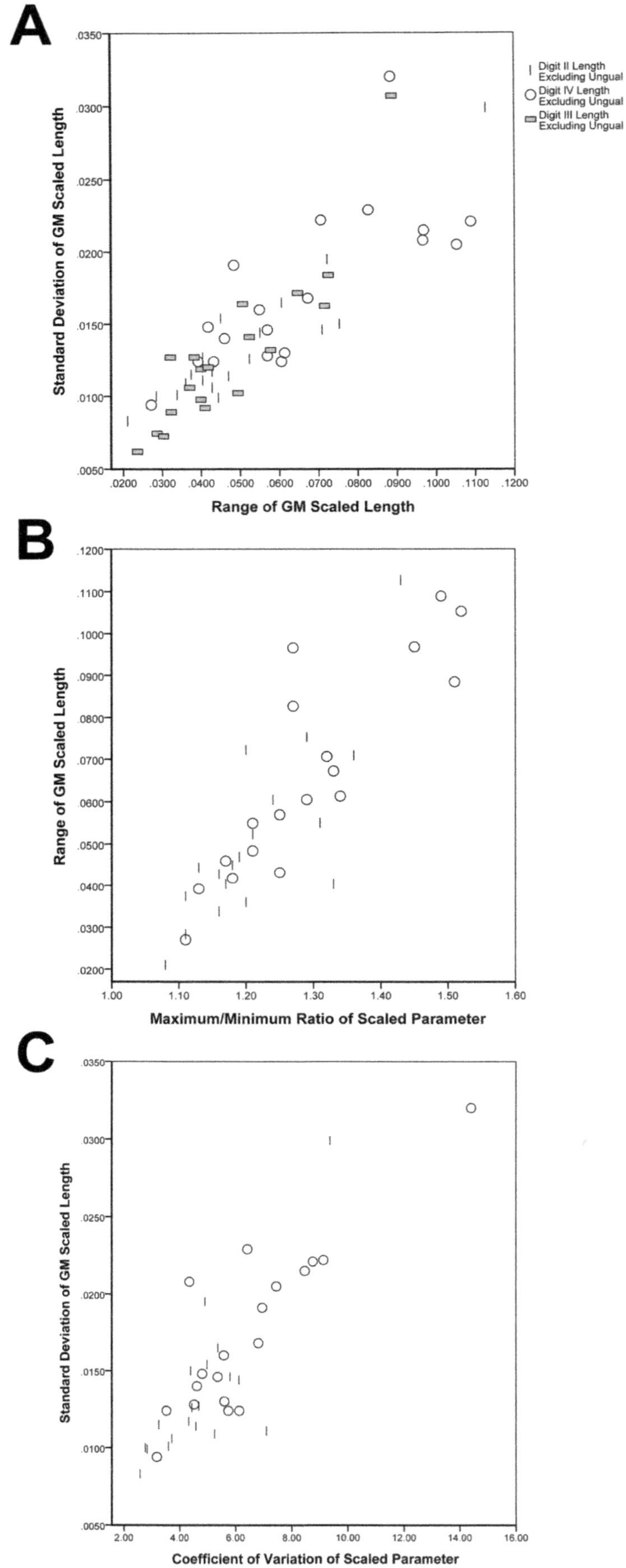
A
Standard Deviation of GM Scaled Length
Range of GM Scaled Length
Digit II Length Excluding Ungual
Digit IV Length Excluding Ungual
Digit III Length Excluding Ungual
B
Range of GM Scaled Length
Maximum/Minimum Ratio of Scaled Parameter
C
Standard Deviation of GM Scaled Length
Coefficient of Variation of Scaled Parameter

Consequently results are reported (tables A1.16, 1.17) for fewer kinds of intraspecific and interspecific samples of GM-scaled variability than for maximum/minimum ratios and coefficients of variation of scaled bivariate parameters.

As with the measures of proportional variability of phalanx lengths already considered, GM-scaled values of the larger, more proximal phalanges (II1, III1, IV1) are generally less variable than GM-scaled values of the smaller, more distal phalanges (II2, III2, and especially III3 and IV2). The GM-scaled length of digit IV (excluding the ungual) is usually more variable than that of digits II or III.

For some comparisons, the range and/or standard deviation of GM-scaled parameters for interspecific samples is/are at least somewhat greater than for intraspecific samples (crocodylians vs. *Alligator mississippiensis* alone, emu + rheas vs. *Dromaius novaehollandiae* or *Rhea americana* alone, *Dinornis robustus* + *Anomalopteryx didiformis* vs. *D. robustus* or *A. didiformis* alone, all moa vs. *D. robustus*, *A. didiformis*, and [for many comparisons] *Pachyornis elephantopus* alone, *Iguanodon bernissartensis* + closely related ornithopods + hadrosaurids vs. *I. bernissartensis* alone). For the two species of rhea combined, GM-scaled measures of variability were generally greater than those for *R. americana* alone for the "big seven" phalanges, but not overall digit lengths (excluding the unguals).

The two parameters of GM-scaled variability (range, standard deviation) are positively correlated (fig. 1.17*A*), and each seems (given the limitations of the data; fig. 1.17*B*, 1.17*C*) to be correlated with its simpler, two-parameter scaled counterpart).

Coefficient of Relative Dispersion About the Reduced Major Axis of Intraspecific and Interspecific Relationships. D_d values of bivariate comparisons of log-transformed phalangeal lengths were calculated for several species represented by a reasonable number (generally 10 or more) of specimens (tables A1.12–A1.16). For comparisons of relative phalanx length, either phalanx III1 was the toe bone against which others were compared, or the length of each of the "big seven" phalanges was compared against the mean of the other six of the "big seven" phalanges excluding itself.

Intraspecific D_d values of both kinds of bivariate comparisons of log-transformed phalangeal lengths are 6 or less (tables A1.12, A1.13; fig. 1.18*A*, 1.18*B*). Distal phalanges in a digit compared with phalanx III1 tend to have higher D_d values than more proximal phalanges. Intraspecific D_d values involving overall digit lengths (tables A1.12, A1.15; fig. 1.18*C*) are 1.6 or less.

At first glance (fig. 1.18*A*–1.18*C*) D_d might seem to change little as the number of species involved in the comparison increases, but this impression is misleading, because some of the interspecific samples did not have enough cases for individual species in those samples for points of those individual species to be plotted in the graphs. In comparisons of interspecific samples for which at least one of the individual species is represented by a reasonable number of specimens (tables A1.13, A1.14–A1.16), the D_d value for the interspecific sample is nearly always greater than that of at least one of the individual species with which it is compared, and in several comparisons is greater than that of all the individual species within the interspecific group. D_d values show a nice correlation with the maximum/minimum ratio of parameters scaled to a common length (fig. 1.18*D*, 1.18*E*).

Summing Up: A Final Set of Comparisons. This chapter began with what could be taken as a rather pompous epigraph about quantitative reasoning. In using it I don't mean to be arrogant, but I nonetheless think that trying to understand in quantitative terms the extent to which foot skeletons differ across potential tridactyl trackmakers is a good first step in trying to understand how much variability we could hope to see among footprints made by different kinds of bipedal dinosaurs.

We have now considered intraspecific vs. interspecific variability of phalangeal and digital proportions using several different measures of variability, both simple and complex: maximum/minimum ratios of phalanx and digital lengths (as scaled either against a single reference phalanx or digit

Facing left, **1.16.** Comparison of two measures of phalangeal and digital shape variability, the maximum/minimum ratio of scaled lengths and the coefficient of variation of scaled lengths. Data plotted here are from the same comparisons graphed in figures 1.13 and 1.15. *A*, Phalanx lengths scaled to a common length of phalanx III1. *B*, Lengths of digits II and IV scaled to a common length of digit III.

Facing right, **1.17.** Relationships involving the lengths of digits II, III, and IV (excluding the ungual) and various intraspecific and interspecific samples of non-avian dinosaurs and ground birds (table A1.11). Digit lengths were log-transformed, and the mean of the three log-transformed digit lengths was subtracted from each of the log-transformed digit lengths to scale the latter (geometric mean [GM] scaling). *A*, Range of GM-scaled digit lengths vs. standard deviation of GM-scaled digit lengths. Most points fall neatly along a linear trend, but there are three outliers, for which the standard deviation is greater than might be expected for the value of the range. For digit II this involves the pooled sample of rhea and emu points, and for digits III and IV the pooled sample of tyrannosaurid and ornithomimosaur points. *B*, *C*, Comparison of variability of GM-scaled lengths of digits II and IV (excluding the ungual) with their simpler counterparts, in which the digit length without the ungual is scaled against the length of digit III without its ungual. These graphs are a bit misleading in that the points for the GM-scaled parameters are generally based on one or a small number (fewer) specimens than their simpler counterparts. *B*, Minimum/maximum ratio of scaled digit length vs. range of GM-scaled digit length. *C*, Coefficient of variation of scaled parameter vs. standard deviation of GM-scaled digit length.

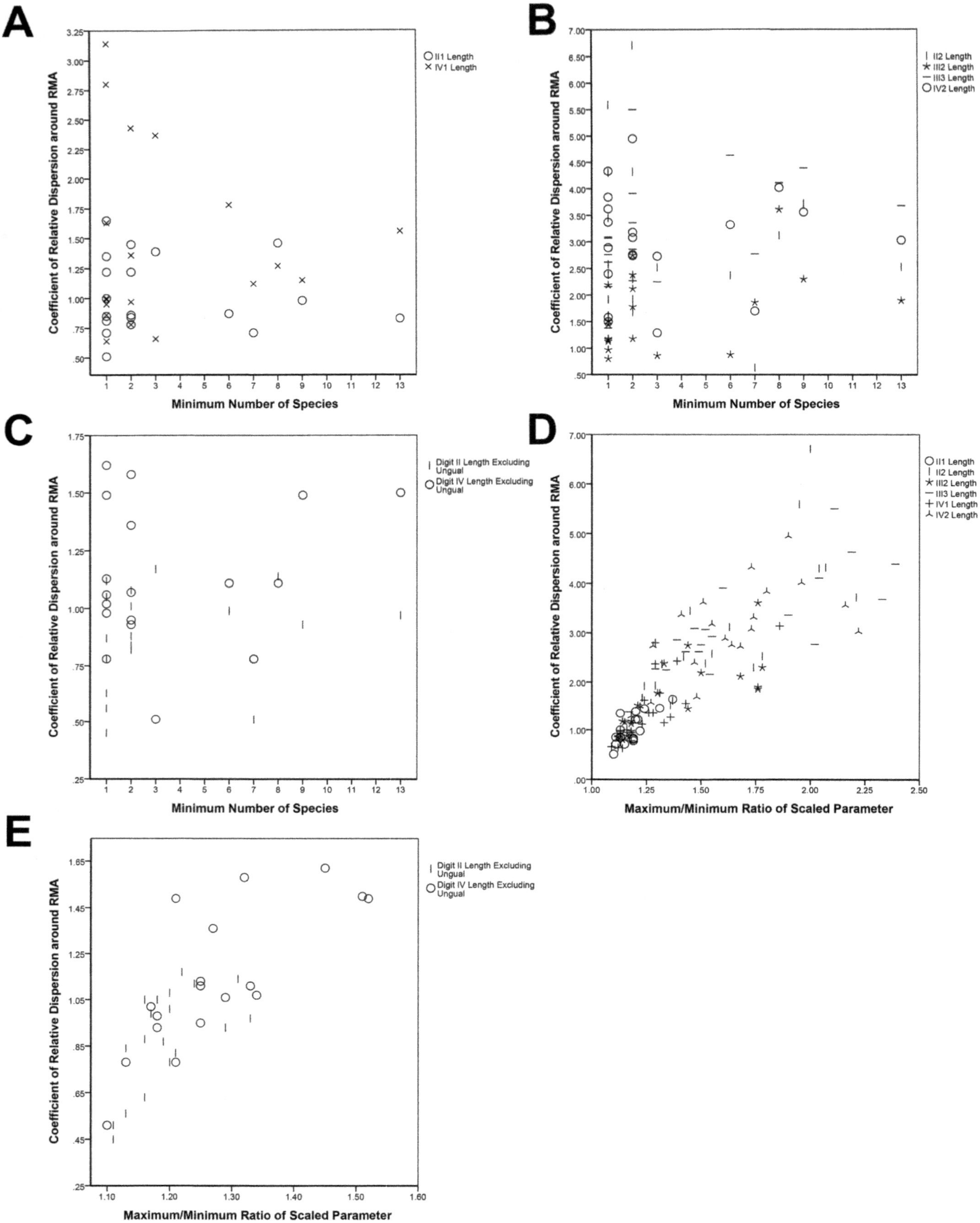
A
B
C
D
E
Coefficient of Relative Dispersion around RMA
Minimum Number of Species
Maximum/Minimum Ratio of Scaled Parameter
II1 Length
IV1 Length
II2 Length
III2 Length
III3 Length
IV2 Length
Digit II Length Excluding Ungual
Digit IV Length Excluding Ungual

length, or GM-scaled lengths); the coefficient of variation of scaled lengths; the range and standard deviation of GM-scaled lengths; and the coefficient of relative dispersion around the reduced major axis. Table A1.17 examines, for multiple samples in which the measure of variability is calculated for a single species, two species, or three or more species, the way in which the median of all of the comparisons of each of these measures of variability changes as the number of species in the sample increases. Whether one considers the proximal or distal phalanges of digits II–IV, or the lengths of digits II–IV, in nearly all cases the median value of the measure of variability increases. That is, if we calculate the measure of variability for several individual species, the median of the value across all those species is less than if the measure is calculated for many two-species comparisons, and less still than if the measure is calculated for multiple samples representing three or more species.

If you have stuck with me through all of this difficult chapter, your eyes may have glazed over with tedium of the many different kinds of intraspecific and interspecific comparisons I have made—and if you thought it was tedious to read, imagine what it was like for me to write! But the good news in all this is that the available data do not give us any reason to think that intraspecific variability in phalangeal or digital proportions in non-avian dinosaurs systematically differs from what is seen in ground birds or crocodylians. That being the case, we are probably also justified in thinking that intraspecific variability in the shape of intact feet, or in footprints made by those feet, would probably have been similar for non-avian dinosaurs, birds, and crocodylians. Consequently the values of D_d and maximum/minimum scaled values of footprint parameters described in later chapters could give us some idea of what to expect among footprints made by the same species of trackmaker.

Furthermore, the results of this chapter indicate that interspecific variability in pedal proportions commonly should exceed intraspecific variability. This raises hope that shape variability in assemblages of footprints made by more than one species of bipedal, tridactyl dinosaur might, in principle, be distinguished from merely intraspecific footprint samples. Unfortunately, determining how many more than one species are present in the assemblage will likely be more difficult. Very likely the best we would be able to do, by comparing digital and other pedal proportions among footprints in a tracksite, is to offer an estimate of the minimum number of trackmaker species responsible.

There are other ways of looking at the skeletal data than those explored in this chapter, and there remain some important questions to consider: If we had an assemblage of tridactyl footprints made by individuals of more than one species, how, or to what extent, could we determine the nature of species responsible for those footmarks? Could we in fact hope to tell them apart? And could we use similarity or difference in footprint shape to say whether the makers of two different footprints were close as opposed to distant relatives? To what extent does overall footprint shape serve as a proxy for phylogenetic propinquity? To such matters we must now turn.

Facing, **1.18.** The coefficient of relative dispersion (D_d) around the reduced major axis (RMA) plotted as a measure of intraspecific vs. interspecific shape diversity in feet of ground birds and non-avian dinosaurs. For comparisons involving individual phalanges, log-transformed phalanx lengths are compared against the log-transformed length of phalanx III1 in calculating D_d. For comparisons involving the lengths of digits II and IV, log-transformed lengths of the digit (excluding the ungual) are compared against the log-transformed length of digit III (excluding the ungual). *A,* D_d of the first phalanges of digits II and IV as a function of the minimum number of species in the sample. *B,* D_d of the distal non-ungual phalanges of digits II and IV as a function of the minimum number of species in the sample. *C,* D_d of the lengths of digits II and IV as a function of the minimum number of species in the sample. *D,* D_d of individual non-ungual phalanges compared with the maximum/minimum ratio of the length of each phalanx scaled against a common length of phalanx III1. *E,* D_d of the lengths of digits II and IV (excluding the ungual) compared with the maximum/minimum ratio of the length of each digit (excluding the ungual) scaled against a common length of digit III (excluding the ungual). In general, D_d shows little or no change as the number of the species in the sample increases, but this measure is clearly correlated with the maximum/minimum ratio of scaled parameters.

2 Pedal Shape and Phylogenetic Relationships

IN THE PREVIOUS CHAPTER WE CONSIDERED SIZE AND shape variability in phalangeal and digital proportions, both within and across taxa, of crocodylians, non-avian dinosaurs, and ground birds. The main question of interest in that chapter was whether the shapes of digital skeletons were more variable when the sample was known to represent more than one species.

In the present chapter I consider a related set of questions. As noted by Tony Thulborn in his classic book about dinosaur footprints, "The shape of the footprints reflects the anatomy of the track-maker's feet and is a particularly important clue because each major group of dinosaurs had its own distinctive pattern of foot structure" (Thulborn 1990: 105).

Do members of the same species have foot skeletons that are more like each other than like those of related species? Are the pedal skeletons of species and genera that are members of more inclusive clades more like those of other members of the same clade than like species and genera in other clades?

Of course, it may be too much to expect that overall phenetic comparisons of foot (or footprint) shape, which probably include a mixture of both primitive and derived pedal characters, would result in topologies of similarity that would match the topological patterns of phylogenetic trees created from synapomorphic characters drawn from the entire cranial and postcranial skeleton. This would seem to be a matter of comparing the proverbial apples and oranges (Buckley [2015] independently came to the same conclusion). On the other hand, the number of morphological features that can be observed, measured, or coded in fossil footprints will often not be large enough to allow much choosiness over the kinds of characters one compares across taxa. Consequently it seems worthwhile at least to take a look at how closely patterns of phylogenetic relationship are tracked (if one pardons the expression) by patterns of overall similarity and dissimilarity in foot and footprint shape among actual or potential trackmakers.

In this chapter I consider these matters for the digital and phalangeal skeletons of non-avian theropods, ornithischian dinosaurs that might have been at least facultatively bipedal, several living and extinct groups of large ground birds, and to a limited extent extant crocodylian species. My data analyses are based mainly on the toes because, in birds and dinosaurs, these are the portions of the foot most likely to be involved in making footprints. However, I also give some consideration to the role of the metatarsal region of the foot in non-avian dinosaurs.

The nucleus of the data set used in these analyses is the same as that used in chapter 1: measurements of the lengths and distal widths of pedal phalanges (mostly of digits II–IV, but some attention will be given to digit I). This data set has been used in a series of publications over the years (Farlow and Lockley 1993; Farlow and Chapman 1997; Farlow 2001; Smith and Farlow 2003; Farlow et al. 2006, 2012, 2013, 2014). Each of these studies has used the data in a different way than earlier papers, and/or has redone previous analyses with additional (and in some cases improved or corrected) data. The same will be done here.

Two versions of the data set will be used. Most of the analyses reported here are based on my own measurements of phalangeal lengths and distal widths, or measurements made by others following my explicit protocols (fig. 1.1). With non-avian dinosaur (and sometimes avian) skeletons, there is always a problem of incomplete measurements of foot material. Most of the time, it is not possible to make all of the desired measurements on all the pedal phalanges; elements have often not been preserved, or have been poorly preserved, such that measurement is impossible. In previous studies, my coauthors and I employed a variety of compromises between having complete measurements of all relevant phalanges (and thus reducing the sample size to those specimens for which this is possible) or reducing the number of phalanges used in the analysis (and thus being able to include more, and incomplete, foot skeletons). In the present chapter I will base most of our analyses on the "best of the best" specimens, those in which every measurement of interest could be made.

For some comparisons, however, I will go to the opposite extreme. For comparisons of relative digit lengths, I also use not only data collected by myself, but also data collected by other workers (particularly Philip J. Currie and contributors to the Online Dinosaur Program), including data published as parts of descriptions of new taxa. Published tables of measurements of pedal phalanges seldom give explicit descriptions of how the measurements were made, and so there is a problem of whether they are compatible with those of other workers. In my experience, this is not a huge problem, except for the unguals. To mitigate this aspect of operator error, and to allow for as large a sample size as possible, this second data set consists of just the lengths of phalanges I1, II1–2, III1–3, and IV1–3.

Analyses of phalangeal lengths and widths will be done using principal component analysis (PCA, with a covariance matrix) of log-transformed measurements. I will also do cluster analyses of scaled phalanx lengths and widths. Scaling will be done by first calculating the mean of all log-transformed phalangeal lengths and widths for each animal's foot, and then subtracting that mean from each log-transformed phalangeal measurement. Clusters will be created using the between-groups method, based on the squared Euclidean distance. Finally, some analyses will involve simple bivariate comparisons.

A complication in both PCAs and cluster analyses arises from the fact that some moa species (*Emeus crassus* and *Euryapteryx curtus*; fig. 1.6*F*–1.6*H*) have a digit IV with only four, as opposed to the usual five, phalanges. As in a previous study (Farlow et al. 2013) I treat the missing phalanx as IV4, and treat the ungual as phalanx IV5. For cluster analyses and PCAs involving birds, instead of using the lengths and distal widths of phalanges IV3 and IV4, I employ the combined lengths of phalanges IV3 plus IV4, and the distal width of phalanx IV3. For PCAs and cluster analyses of non-avian dinosaur feet, however, I use the lengths and distal widths of both phalanges IV3 and IV4.

Because I was less confident of my ability to identify loose phalanges of crocodylians in specimen boxes than toe bones of dinosaurs and birds, my measurements of crocodylian foot skeletons were limited to osteological specimens in which the phalanges remained articulated. A further complication for measuring phalanx lengths is that in most of the crocodylian osteological specimens the horny sheaths of claws remained firmly attached to their phalanges. I therefore restricted my measurements of phalanx lengths to non-ungual phalanges. I measured all such phalanges of digits I (I1), II (II1–2), and III (III1–3), and the first three phalanges of digit IV. Because dried tissues held articulated phalanges in place, it was seldom possible to measure phalangeal widths, and so my comparisons of crocodylian feet are limited to data for phalangeal lengths. Toe widths of alligators will, however, be briefly examined in chapter 4.

RESULTS

Crocodylians. Molecular (Oaks 2011; Zhang et al. 2011) and morphology-based (Brochu et al. 2012) phylogenies of extant crocodylians (Grigg and Kirshner 2015) agree in recognizing two major groups, alligatoroids (alligators and caimans) and crocodyloids (crocodiles). The chief discrepancies between the two kinds of phylogeny involve the placement of true gharials (whether closely related to crocodiles, or basal to the node linking alligatoroids and crocodyloids), and whether false gharials are more closely related to true gharials or to crocodiles.

The number of specimens in my sample, even though drawn from several collections, is limited, and unsurprisingly dominated by *Alligator mississippiensis*. Consequently few conclusions will be drawn from these data, and these only tentative. In a cluster analysis of scaled non-ungual phalanx lengths, 14 of the 17 *A. mississippiensis* specimens are more like each other than like any of the other crocodylians (fig. 2.1). There may also be a tendency for specimens of *Tomistoma schlegelii*, *Paleosuchus trigonatus*, and *Caiman crocodilus* to cluster with conspecifics apart from other species. Beyond that, there seems little correspondence between the dendrogram of pedal phalangeal proportions and the phylogenetic relationships of the extant crocodylian species.

Much the same holds true for the relative lengths of phalanx I1 and the aggregate lengths of II1–2 and IV1–3, all expressed as percentages of the aggregate length of phalanges III1–3 (fig. 2.2). Points for *A. mississippiensis* plot close together, associated with relatively short lengths of phalanx I1 and the non-ungual length of digit II, and a relatively long aggregate length of the first three phalanges of digit IV, compared with the other species in the sample. These other species are a phylogenetically heterogeneous lot, including points for the Chinese alligator, common caiman, crocodiles, and true and false gharials. The data, although scant, suggest that there is very little phylogenetic signal in relative digit lengths, just as we saw for phalangeal proportions. Speculating a bit, one wonders if this could be one manifestation of the relatively low rate of evolutionary change in the crocodylian genome (Green et al. 2014).

Moa. Molecular evidence (Bunce et al. 2009; Allentoft and Rawlence 2012) suggests that *Emeus crassus* and *Euryapteryx curtus* (it is possible that there is more than one species of *Euryapteryx*: Huynen and Lambert 2014) are close relatives, that these two moa species link up with *Anomalopteryx didiformis*, and that those three species join with the species of *Pachyornis* in what can be regarded as the family Emeidae. The two species of *Dinornis* then constitute the

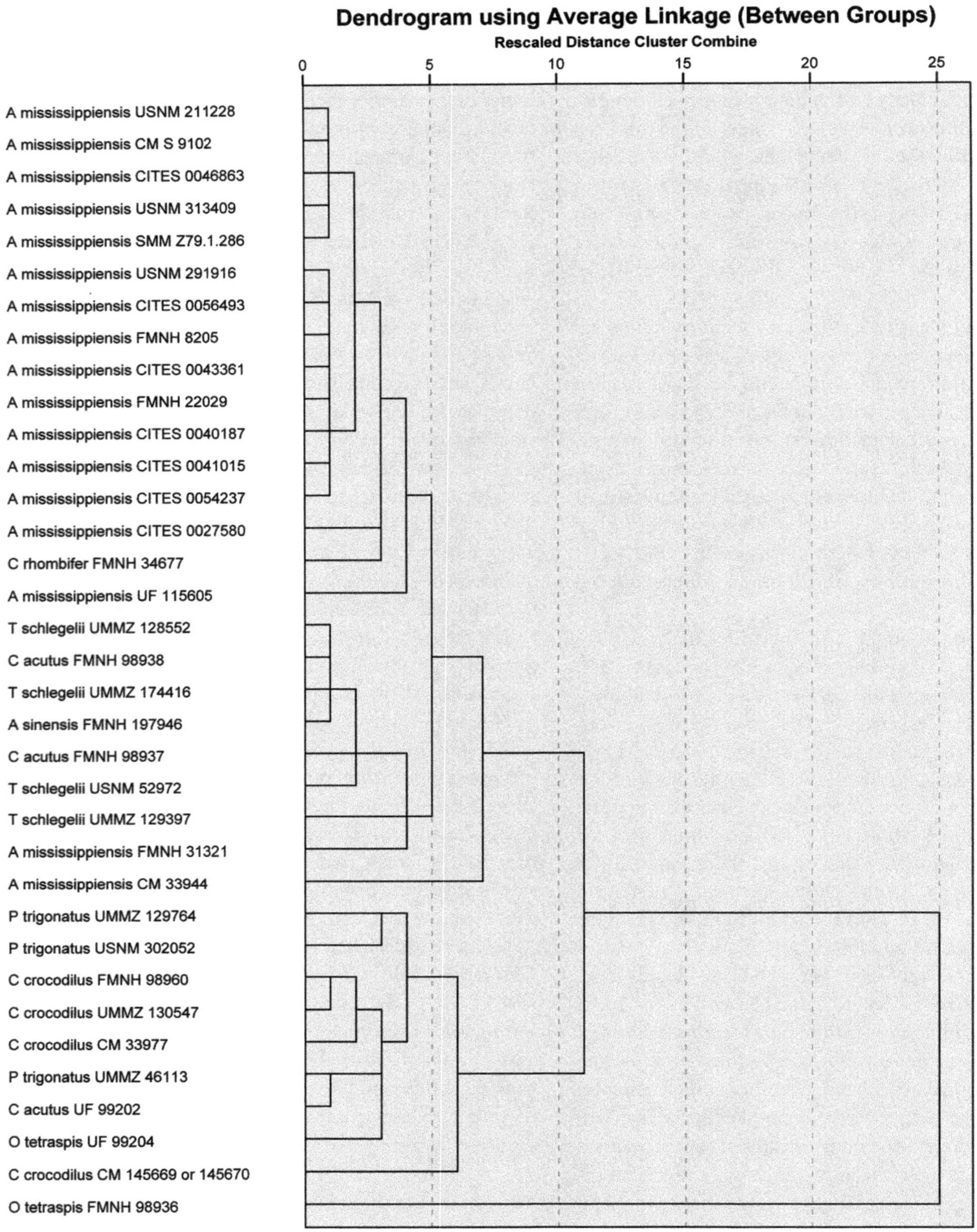

2.1. Cluster analysis of scaled lengths of phalanges I1, II1 and II2, III1–3, and IV1–3 of crocodylians (the USNM museum abbreviation for Smithsonian Institution vertebrate zoology [as opposed to vertebrate paleontology] used here and in fig. 2.10 appears in table A1.1 as NMNH). Lengths were scaled by subtracting the mean of all log-transformed values from each log-transformed value. Three main clusters are generated: a cluster composed almost entirely of American alligators (*Alligator mississippiensis*), but also one specimen of the Cuban crocodile (*Crocodylus rhombifer*); a heterogeneous cluster with cases for false gharial (*Tomistoma schlegelii*), American crocodile (*Crocodylus acutus*), Chinese alligator (*Alligator sinensis*), and one American alligator specimen; and an equally heterogeneous cluster with cases for smooth-fronted caiman (*Paleosuchus trigonatus*), common caiman (*Caiman crocodilus*), American crocodile, and African dwarf crocodile (*Osteolaemus tetraspis*). There seems to be no correspondence between similarity in pedal phalangeal proportions and phylogenetic relationships in these reptiles.

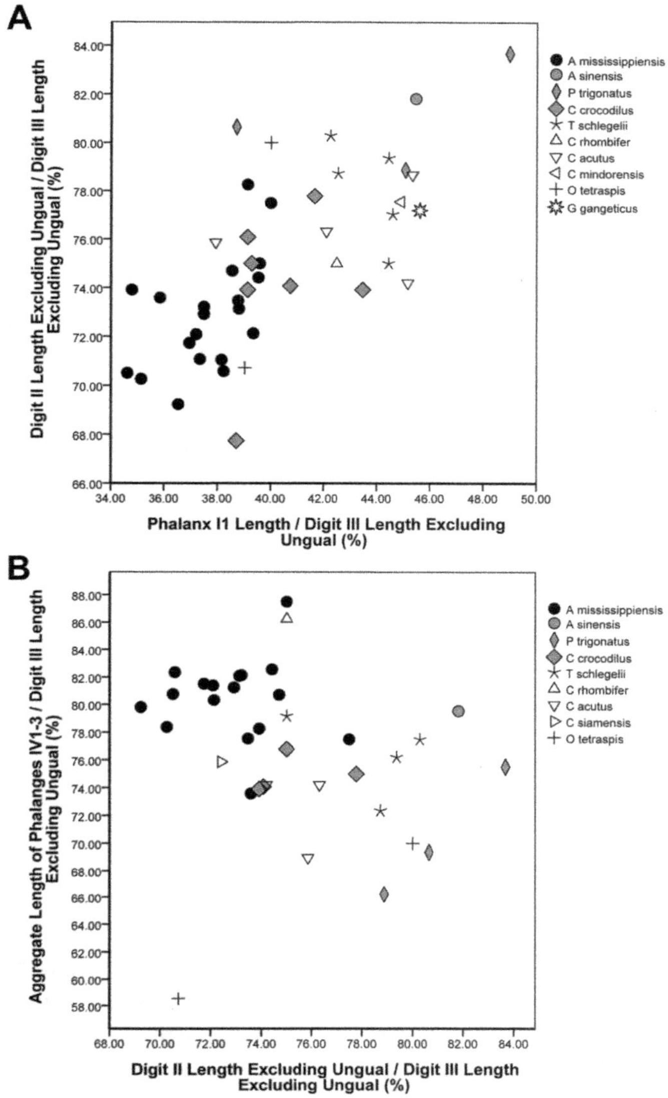

2.2. Relative digit lengths in crocodylians. The lengths of the non-ungual portions of digits I (a single phalanx) and II, and the aggregate lengths of phalanges 1–3 of digit IV, are expressed as percentages of the aggregate lengths of the non-ungual phalanges of digit III. *A*, Digit I vs. digit II. *B*, Digit II vs. digit IV. The American alligator has a relatively short phalanx I and digit II and a relatively long digit IV. The few data points for different species of *Crocodylus* occur close together with intermediate values of the relative lengths of phalanx I1 and digit II, but otherwise there is no clear phylogenetic picture in the distribution of values for the various species.

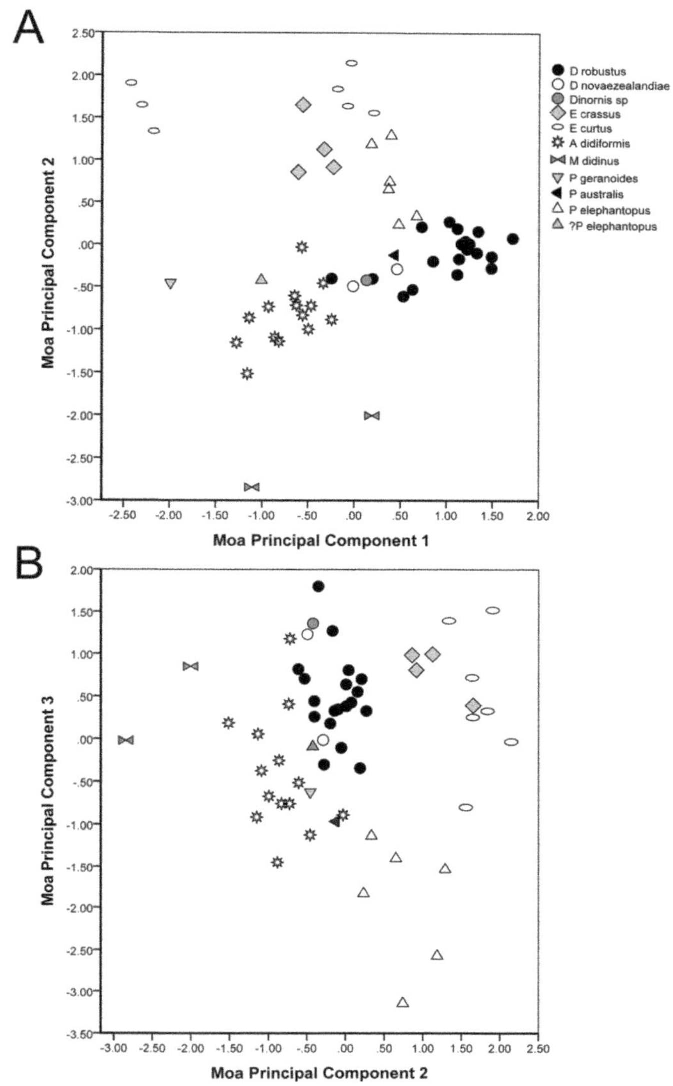

2.3. Principal component analysis (PCA) of log-transformed pedal phalanx lengths and widths of digits II–IV of moa (table 2.1). Principal component 1 (PC1) is mainly associated with overall size. Positive values of PC 2 (*A*, *B*) are associated with relatively broad phalanges and long first phalanges of each digit, whereas negative values are associated with long phalanges other than the first. Positive values of PC3 (*B*) are associated with relatively long first and second phalanges of the digits; negative values are associated with relatively long phalanges distal to the first two, and also broad phalanges. The PCs provide some separation of moa taxa. *Megalapteryx didinus* has very negative values of PC2, whereas *Emeus crassus*, *Euryapteryx curtus*, and *Pachyornis elephantopus* have strongly positive values. Plotting PC3 against PC2 puts *P. elephantopus*, *Megalapteryx*, *Emeus* + *Euryapteryx*, and *Anomalopteryx didiformis* + *Dinornis* spp., into different regions of morphospace.

family Dinornithidae. The emeids and the dinornithids together are the sister clade of *Megalapteryx didinus*. Worthy and Scofield (2012), using morphological data, recovered a single most parsimonious tree in which, once again, *Emeus* and *Euryapteryx* are close relatives. These two forms then link up with *Pachyornis*, the combined *Pachyornis-Emeus-Euryapteryx* then joins with *Megalapteryx*, and *Dinornis* is the outgroup to all the other moa. So the molecular and morphological trees are not completely congruent, but the agreement between them isn't bad.

In the PCA of log-transformed phalanx lengths and widths, almost 90% of the data variance is accounted for by size (table 2.1: principal component [PC] 1). PC2 accounts for about 7% of the variance. Positive values of PC2 are associated with broad phalanges and, to a lesser extent, long first phalanges of digits II–IV. Negative values of PC2 are

Table 2.1. Principal component analysis (using a covariance matrix) of log-transformed measurements of phalangeal lengths and distal widths of digits II–IV in moa (*N* = 58).

Parameter	PC1 loading (raw [rescaled])	PC2 loading (raw [rescaled])	PC3 loading (raw [rescaled])
Phalanx II1 length	0.136 (0.978)	0.017 (0.119)	0.014 (0.103)
Phalanx II1 distal width	0.134 (0.963)	0.033 (0.237)	−0.008 (−0.056)
Phalanx II2 length	0.142 (0.946)	−0.038 (−0.251)	0.009 (0.063)
Phalanx II2 distal width	0.131 (0.965)	0.030 (0.222)	−0.005 (−0.041)
Phalanx II3 length	0.138 (0.953)	−0.023 (−0.162)	−0.012 (−0.086)
Phalanx III1 length	0.137 (0.985)	0.012 (0.086)	0.014 (0.104)
Phalanx III1 distal width	0.136 (0.951)	0.040 (0.282)	−0.004 (−0.029)
Phalanx III2 length	0.129 (0.982)	−0.011 (−0.083)	0.005 (0.036)
Phalanx III2 distal width	0.131 (0.944)	0.040 (0.292)	−0.012 (−0.090)
Phalanx III3 length	0.126 (0.929)	−0.033 (−0.246)	−0.016 (−0.116)
Phalanx III3 distal width	0.131 (0.957)	0.035 (0.252)	−0.009 (−0.065)
Phalanx III4 length	0.124 (0.956)	−0.019 (−0.143)	−0.011 (−0.086)
Phalanx IV1 length	0.129 (0.975)	0.020 (0.150)	0.016 (0.121)
Phalanx IV1 distal width	0.140 (0.974)	0.030 (0.207)	−0.001 (−0.005)
Phalanx IV2 length	0.140 (0.939)	−0.021 (−0.138)	0.042 (0.283)
Phalanx IV2 distal width	0.134 (0.959)	0.034 (0.240)	−0.004 (−0.031)
Combined lengths of phalanges IV3 and IV4	0.176 (0.833)	−0.112 (−0.530)	−0.012 (−0.055)
Phalanx IV3 distal width	0.138 (0.966)	0.023 (0.160)	−0.008 (−0.055)
Phalanx IV5 length	0.131 (0.955)	−0.022 (−0.162)	0.001 (0.011)
Eigenvalues (% of variance)	0.354 (89.510)	0.027 (6.773)	0.004 (0.922)
Cumulative variance explained (%)	89.510	96.284	97.206

Kaiser-Meyer-Olkin measure of sampling adequacy = 0.946; Bartlett's test of sphericity: $x2$ = 3,113.520, $P < .001$.

associated with relatively long phalanges distal to the first phalanx of digits II–IV.

Plotting PC2 against PC1 (fig. 2.3A) unsurprisingly spreads the various moa species along PC1 on the basis of size, separating the two species of giant moa (*Dinornis* spp.) from smaller forms like *Anomalopteryx didiformis*. More interestingly, PC2 separates moa species with proportionately broad toes, and short distal phalanges (most notably *Euryapteryx curtus*, *Emeus crassus* [no doubt related to the fact that these two species have only four, rather than the usual five, phalanges in digit IV], and *Pachyornis elephantopus*), from species with relatively narrow toes and longer distal phalanges (*A. didiformis* and *Megalapteryx didinus*) (fig. 1.6).

Although PC3 accounts for less than 1% of the variance in the data, it seems to separate *P. elephantopus* from the other moa species (fig. 2.3*B*). Positive values of PC3 are associated with relatively long first and second phalanges of digits II–IV, while negative values are associated with long distal phalanges, and also broad phalanges.

A cluster analysis of the scaled phalanx lengths and widths (fig. 2.4) recognizes at least four distinct groups: *M. didinus*, a combined cluster of the two stout-toed moa (*E. crassus* and *E. curtus*), most of the specimens of *P. elephantopus* along with a single nonconformist *A. didiformis*, and a final supercluster containing most specimens of *A. didiformis*, the two *Dinornis* species, and the smaller species of *Pachyornis*. Within this big cluster, there is a tendency for *A. didiformis* to cluster together, but specimens of *Dinornis robustus* link up seemingly at random with the other species in the cluster. This result is quite consistent with the groupings revealed by the PCA. The fact that conspecific specimens tend to be more like each other than like other species is also consistent, of course, with the results of those ghastly comparisons of within-species vs. across-species variability of moa pedal shape reported in chapter 1.

It seems, then, that phalangeal proportions do a modest job of separating out moa species. If absolute size is ignored (even big moa grew up from small chicks: Huynen et al. 2014), phalangeal proportions discriminate four shape groups among moa: *Megalapteryx didinus*, *Emeus crassus* + *Euryapteryx curtus*, *Pachyornis elephantopus*, and then everybody else. If size is also considered (chicks less likely to leave footprints than adults?) *Dinornis* separates nicely from *Anomalopteryx*. The two species of *Dinornis* are probably indistinguishable.

These phenetic groupings are only somewhat consistent with the moa phylogeny (Bunce et al. 2009; Worthy and Scofield 2012). On the basis of pedal phalangeal proportions, *M. didinus* is distinctly different from the other species, the two *Dinornis* species are so similar that their feet cannot be told apart, and *E. curtus* and *E. crassus* are likewise quite similar. On the other hand, phalangeal proportions do not group the three species of *Pachyornis* together apart from other moa. *Anomalopteryx* and *Pachyornis* have phalangeal

2.4. Cluster analysis of scaled phalanx lengths and widths of digits II–IV of moa. Clusters were generated for this and all remaining analyses in this chapter the same way. Lengths and widths were scaled by subtracting the mean of all log-transformed values from each log-transformed value. Clusters were created using the between-groups linkage method, based on the squared Euclidean distance. Five distinct clusters can be seen: a cluster mainly but not entirely composed of specimens of the two species of *Dinornis*; an *Anomalopteryx* cluster; a cluster mainly but not entirely composed of specimens of *Pachyornis elephantopus*; a cluster of *Emeus* + *Euryapteryx*; and a *Megalapteryx* cluster. (Some of the institutional specimen catalog numbers are now somewhat different from those in this figure and in figs. 2.7, 2.10, and 2.17; see table A1.1 for the current catalog numbers.)

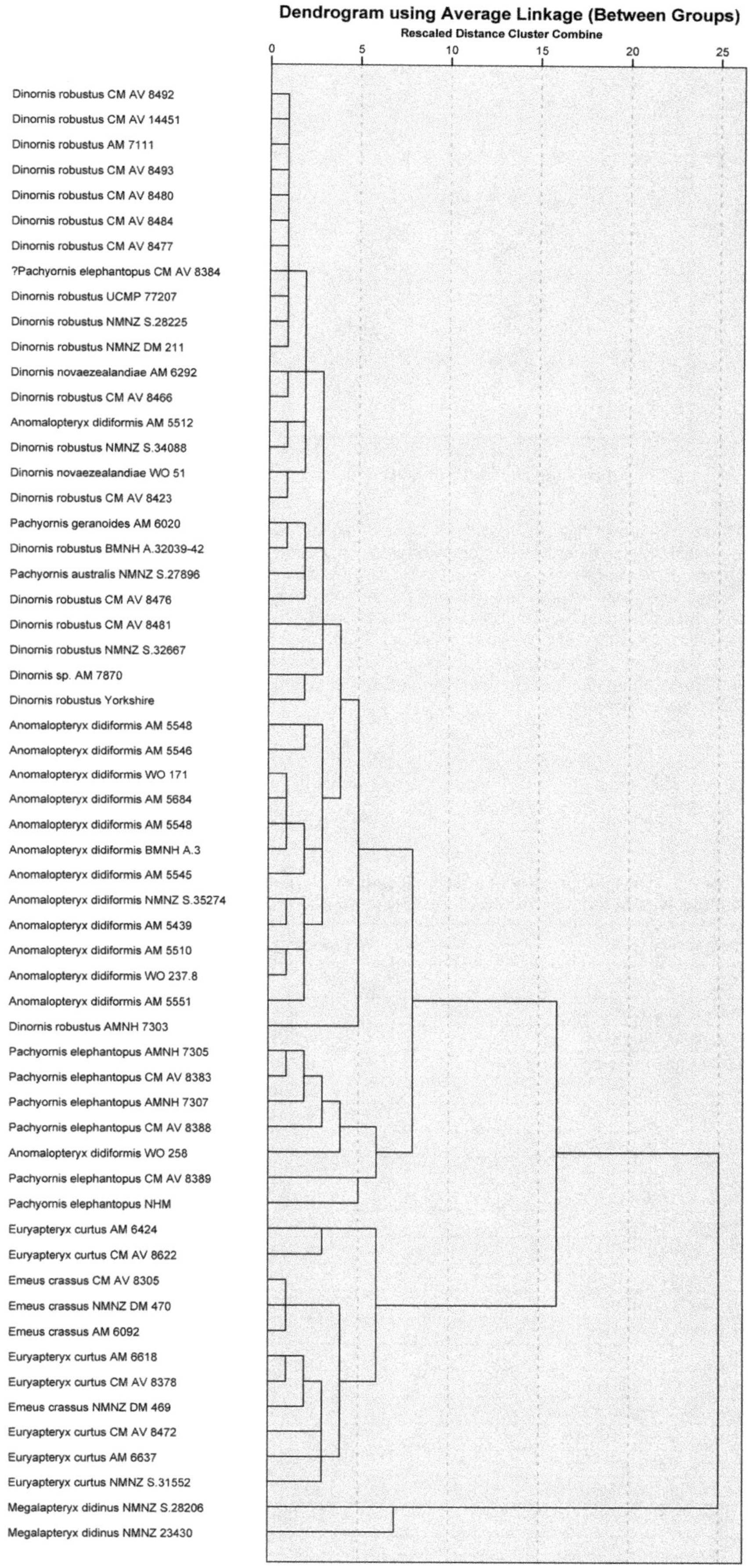

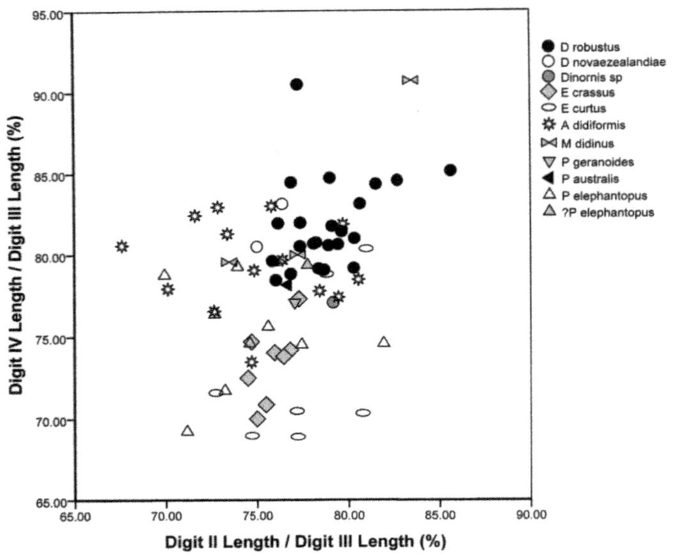

2.5. Relative overall digit lengths of moa. *Dinornis* and *Megalapteryx* have relatively long digits II and IV compared with digit III. *Emeus* and *Euryapteryx*, in contrast, have a relatively short digit II and an even more markedly short digit IV (both taxa have only four as opposed to the usual five phalanges in digit IV). *Pachyornis elephantopus* likewise tends to have relatively short digits II and IV. Many specimens of *Anomalopteryx* have a relatively short digit II, but not so much digit IV. Relative digital lengths thus provide some separation of moa taxa, but less than that afforded by phalangeal proportions.

proportions more like those of *Dinornis* than those of *Emeus* and *Euryapteryx*.

If all we have to compare are relative proportions of the overall lengths of digits II–III (fig. 2.5), we lose some, but not all, of the ability to distinguish among species. *E. crassus* and *E. curtus* have a relatively short digit IV (reflecting the absence of one phalanx in this toe) and to a lesser extent digit II, but are probably indistinguishable from each other. Most specimens of *P. elephantopus* also have relatively short digits II and IV. The remaining species, however, show a great deal of overlap in digital proportions, although *Dinornis* and *Anomalopteryx* show a tendency toward separation on the basis of shorter and longer lengths of digit III, respectively, compared with the lengths of digits II and IV.

Struthioniforms (other than *Struthio*). In this section extant ratites will be treated as a distinct group of ground birds, but there is considerable question as to whether they are in fact a natural group, or if so, how extant species relate to extinct forms, like moa and elephant birds, and to tinamous. Ratite phylogeny has been investigated using both morphological and molecular data (Chojnowski et al. 2008; Hackett et al. 2008; Harshman et al. 2008; Bourdon et al. 2009; Phillips et al. 2010; Johnston 2011; Worthy and Scofield 2012; Smith et al. 2013; Agnolin 2017; Angst and Buffetaut 2017; Grealy et al. 2017; Mayr 2017; Yonezawa et al. 2017); and older references

Table 2.2. Principal component analysis (using a covariance matrix) of log-transformed measurements of phalangeal lengths and distal widths of digits II–IV in extant struthioniforms other than *Struthio* (*N* = 37).

Parameter	PC1 loading (raw [rescaled])	PC2 loading (raw [rescaled])	PC3 loading (raw [rescaled])
Phalanx II1 length	0.164 (0.920)	−0.048 (−0.269)	0.036 (0.204)
Phalanx II1 distal width	0.168 (0.972)	0.000 (0.000)	−0.022 (−0.129)
Phalanx II2 length	0.038 (0.296)	0.083 (0.654)	0.071 (0.554)
Phalanx II2 distal width	0.174 (0.973)	0.003 (0.017)	−0.026 (−0.146)
Phalanx II3 length	0.182 (0.834)	0.086 (0.394)	−0.082 (−0.375)
Phalanx III1 length	0.208 (0.987)	−0.018 (−0.083)	0.016 (0.074)
Phalanx III1 distal width	0.223 (0.979)	−0.032 (−0.140)	0.016 (0.069)
Phalanx III2 length	0.191 (0.992)	0.003 (0.016)	0.009 (0.047)
Phalanx III2 distal width	0.227 (0.979)	−0.030 (−0.130)	0.022 (0.095)
Phalanx III3 length	0.121 (0.811)	0.082 (0.554)	−0.002 (−0.013)
Phalanx III3 distal width	0.222 (0.972)	−0.042 (−0.184)	0.018 (0.079)
Phalanx III4 length	0.159 (0.931)	0.010 (0.060)	−0.023 (−0.134)
Phalanx IV1 length	0.205 (0.977)	−0.035 (−0.169)	0.014 (0.067)
Phalanx IV1 distal width	0.190 (0.980)	−0.031 (−0.160)	−0.007 (−0.038)
Phalanx IV2 length	0.159 (0.797)	0.103 (0.516)	0.045 (0.227)
Phalanx IV2 distal width	0.186 (0.975)	−0.033 (−0.174)	−0.007 (−0.035)
Combined length of phalanges IV3 and IV4	0.063 (0.494)	0.105 (0.824)	0.015 (0.118)
Phalanx IV3 distal width	0.184 (0.975)	−0.023 (−0.120)	−0.011 (−0.056)
Phalanx IV5 length	0.158 (0.934)	0.024 (0.144)	−0.035 (−0.208)
Eigenvalues (% of variance)	0.592 (86.100)	0.054 (7.802)	0.020 (2.919)
Cumulative variance explained (%)	86.100	93.901	96.820

Kaiser-Meyer-Olkin measure of sampling adequacy = 0.878; Bartlett's test of sphericity: $x2$ = 1,889.054, $P < .001$.

cited in these papers). Among the contentious issues are whether kiwi (*Apteryx*) are more closely related to moa, elephant birds, or a clade composed of emus (*Dromaius*) and cassowaries (*Casuarius*) and whether tinamous are a group within the ratite clade. More positively, all analyses group the two extant rheas together, and recognize that emus and cassowaries are most closely related to each other. Some analyses find ostriches (*Struthio*) to be close relatives of rheas, some that ostriches are the outgroup to other extant ratites, and some find ostriches to be the outgroup to a larger ratite grouping that includes moa. Yikes. So calling all of the extant ratites struthioniforms may not be completely appropriate (cf. Mayr 2017), but for our purposes it seems a convenient label, as long as one is aware of its possible shortcomings.

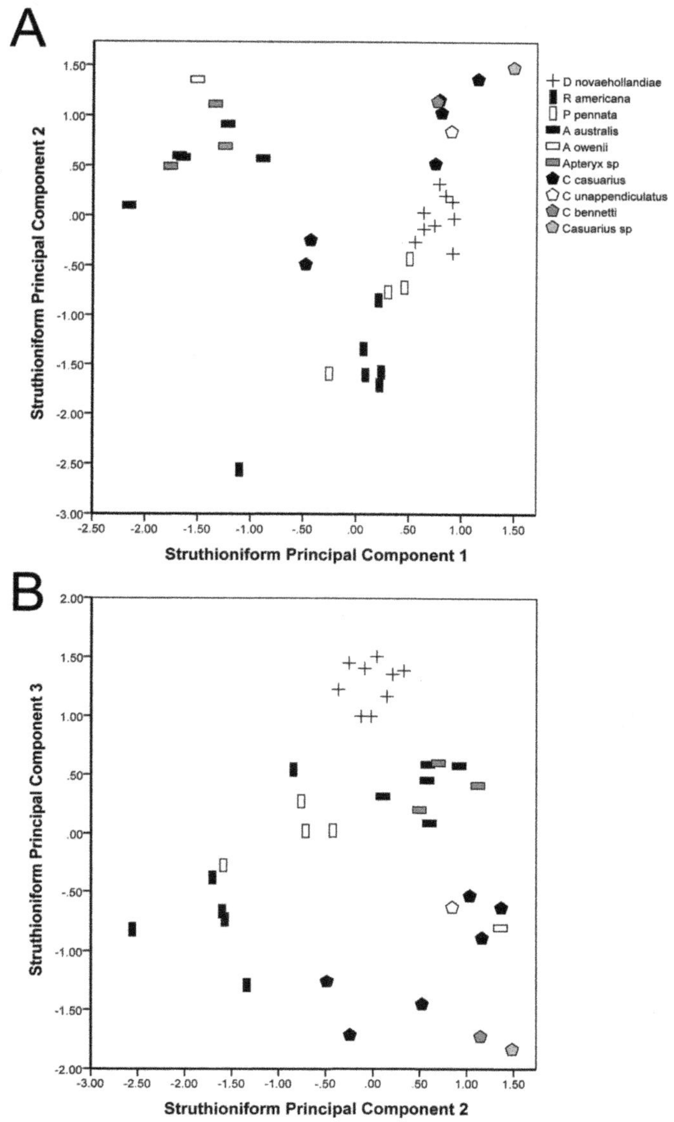

2.6. Principal component analysis (PCA) of log-transformed pedal phalanx lengths and widths of digits II–IV of extant struthioniform birds other than ostriches (table 1.2). Principal component one (PC1) is mainly associated with overall size. Positive values of PC2 (*A*, *B*) are associated with relatively long distal phalanges, whereas negative values are associated with relatively long first phalanges, and also broad phalanges on digits III and IV. Positive values of PC3 (*B*) are associated with relatively long first and second phalanges (especially phalanx II2) and a broad digit III, whereas negative values are associated with relatively long ungual phalanges and broad digits II and IV. The PCs provide substantial separation of taxa. Kiwi (*Apteryx* spp.) are characterized by very positive values of PC2 and (mostly) fairly positive values of PC3. Cassowaries (*Casuarius* spp.) are characterized by slightly negative to very positive values of PC2 but negative values of PC3. Rheas (*Rhea* and *Pterocnemia*) have negative values of PC2 and negative to slightly positive values of PC3. Emus (*Dromaius*) have roughly neutral values of PC2 but very positive values of PC3.

In a PCA of log-transformed phalanx lengths and widths, 86% of the data variance is accounted for by size (table 2.2: PC1). PC2 accounts for about 8% of the variance. Positive values of PC2 are associated with long distal phalanges. Negative values of PC2 are associated with relatively long first phalanges and broad phalanges.

As with moa, plotting PC2 against PC1 (fig. 2.6*A*) spreads the various extant ratite species along PC1 on the basis of size. There are three bands of points, one for kiwi (*Apteryx*), a second for cassowaries (*Casuarius*), and a third for rheas (*Rhea*, *Pterocnemia*) and emus (*Dromaius*). In all three bands, the data suggest increasingly positive values of PC2 with increasing bird size, both across and within species. Kiwi have much higher values of PC2 for the size of their feet than do the other species, cassowaries have higher values of PC2 at any given size than do rheas and emus, and emus have higher values than rheas.

PC3 contrasts species with long first and second phalanges (particularly II2) and a broad digit III (positive values) with species that have long unguals and broad digits II and IV (negative values). Plotting PC3 against PC2 (fig. 2.6*B*) provides additional separation of species. Cassowaries are associated with positive values of PC2 and negative values of PC3. Kiwi likewise have high values of PC2, but most kiwi specimens have slightly positive values of PC3. Emus have particularly positive values of PC3, and plot well away from their close relatives, the cassowaries. The two rhea species range from fairly negative to slightly positive in values of PC3, with all but one specimen of *Rhea americana* showing lower values of PC3 than *Pterocnemia pennata*.

A cluster analysis using scaled phalanx lengths and widths (fig. 2.7) reveals four major clusters, one each for emus, rheas, cassowaries, and kiwi, the last of which form a cluster distinct from the other three. Emus do not link up particularly closely with cassowaries. The four specimens of *P. pennata* cluster together, as do four of the five specimens of *R. americana*. The different species of *Casuarius* and *Apteryx* seem to hook up more willy-nilly, but that may reflect the fact that most species of both genera are represented by a single specimen.

Overall digital length proportions (fig. 2.8) also seem to do a respectable job of separating extant ratites into groups. Kiwi have relatively long digits II and IV, compared with the

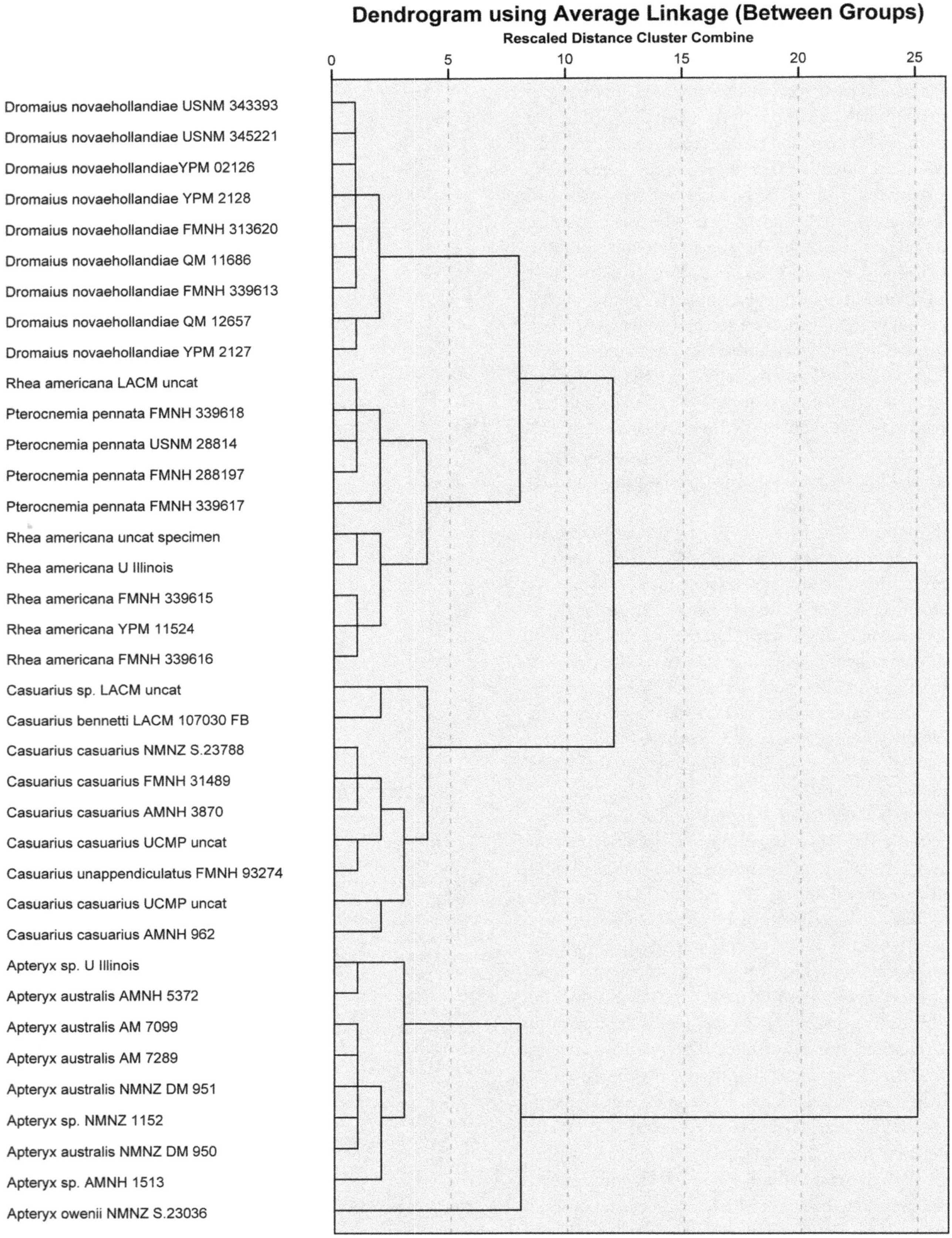

2.7. Cluster analysis of scaled phalanx lengths and widths of digits II–IV of struthioniforms other than ostriches. Four first-order clusters were created, one each for emus, rheas, cassowaries, and kiwi. At successively higher levels, emus link up with rheas; this cluster then joins cassowaries; and emus + rheas + cassowaries finally join with kiwi.

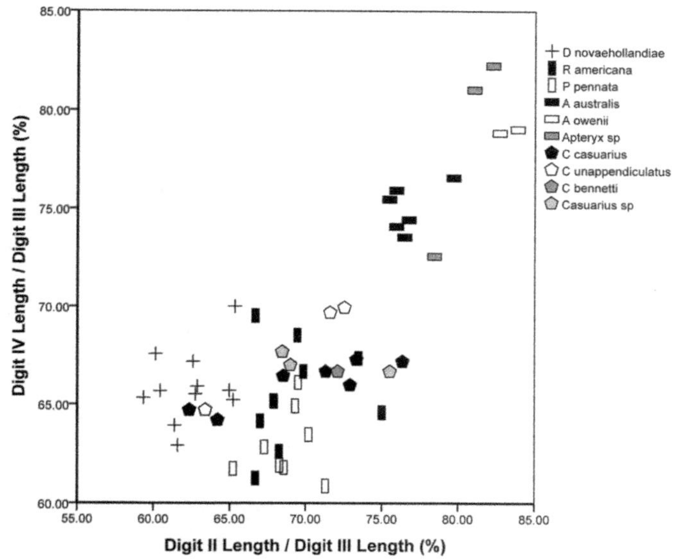

2.8. Relative overall digit lengths of struthioniforms other than ostriches. Kiwi differ markedly from the others in having relatively long digits II and IV compared with digit III. Emus tend to have a relatively shorter digit II than in rheas and cassowaries, whereas rheas show a slight tendency to have a relatively shorter digit IV than in emus and cassowaries.

length of digit III, while emus have relatively shorter digits II and IV. Cassowaries and rheas have a digit IV/digit III length ratio comparable to that of emus, but at the same time have a higher digit II/digit III length ratio. Relative digit lengths seem to do a better job of separating the different genera, at least, of extant ratites than they do for moa.

All Large Birds. Phylogenetic relationships among neognaths are just as controversial as those of paleognaths (Hackett et al. 2008; Harshman et al. 2008; Naish 2012; McCormack et al. 2013; Jarvis et al. 2014; Prum et al., 2015; Mayr 2017; Angst and Buffetaut 2017; Worthy et al. 2017). There is agreement, however, about at least some of the major patterns of relationship. Waterfowl and game birds together form one major clade of neognaths (Galloanserae), which also includes such illustrious extinct big flightless birds as gastornithids (Andors 1992) and dromornithids (Murray and Vickers-Rich 2004). The other big neognath clade (Neoaves) includes all other birds, which include such ground birds of interest to us as bustards, seriemas, phorusrhacids, and the dodo (*Raphus*).

A PCA of big ground birds other than ostriches (III1 length at least 20 mm; table 2.3) shows PC1 (associated with size) to account for more than 91% of the data variance, and PC2 about 4%. Beyond that, this time it seems appropriate

Table 2.3. Principal component analysis (using a covariance matrix) of log-transformed measurements of phalangeal lengths and distal widths of digits II–IV in large (phalanx III1 length at least 20 mm) ground birds of multiple clades (N = 104).

Parameter	PC1 loading (raw [rescaled])	PC2 loading (raw [rescaled])	PC3 loading (raw [rescaled])	PC4 loading (raw [rescaled])
Phalanx II1 length	0.208 (0.961)	−0.034 (−0.156)	−0.002 (−0.008)	0.028 (0.128)
Phalanx II1 distal width	0.278 (0.985)	−0.008 (−0.029)	−0.028 (−0.098)	−0.031 (−0.111)
Phalanx II2 length	0.212 (0.867)	0.096 (0.391)	−0.064 (−0.260)	0.022 (0.088)
Phalanx II2 distal width	0.278 (0.985)	−0.007 (−0.025)	−0.024 (−0.086)	−0.031 (−0.110)
Phalanx II3 length	0.212 (0.916)	0.042 (0.180)	0.069 (0.297)	−0.035 (−0.153)
Phalanx III1 length	0.210 (0.967)	−0.029 (−0.132)	0.024 (0.111)	0.036 (0.165)
Phalanx III1 distal width	0.256 (0.978)	−0.048 (−0.184)	0.001 (0.003)	0.012 (0.048)
Phalanx III2 length	0.174 (0.939)	−0.013 (−0.070)	0.050 (0.269)	0.029 (0.157)
Phalanx III2 distal width	0.257 (0.977)	−0.049 (−0.186)	−0.002 (−0.006)	0.011 (0.040)
Phalanx III3 length	0.173 (0.914)	0.056 (0.294)	0.033 (0.177)	−0.001 (−0.005)
Phalanx III3 distal width	0.266 (0.980)	−0.046 (−0.168)	−0.012 (−0.043)	0.009 (0.034)
Phalanx III4 length	0.200 (0.963)	0.014 (0.070)	0.031 (0.149)	−0.019 (−0.090)
Phalanx IV1 length	0.204 (0.958)	−0.042 (−0.196)	0.018 (0.082)	0.034 (0.157)
Phalanx IV1 distal width	0.281 (0.989)	−0.027 (−0.097)	−0.021 (−0.075)	−0.014 (−0.048)
Phalanx IV2 length	0.232 (0.939)	0.046 (0.187)	−0.003 (−0.011)	0.049 (0.199)
Phalanx IV2 distal width	0.276 (0.985)	−0.029 (−0.103)	−0.023 (−0.084)	−0.021 (−0.077)
Combined length of phalanges IV3 and IV4	0.216 (0.846)	0.129 (0.504)	−0.004 (−0.017)	0.012 (0.049)
Phalanx IV3 distal width	0.281 (0.986)	−0.019 (−0.068)	−0.021 (−0.075)	−0.025 (−0.088)
Phalanx IV5 length	0.218 (0.960)	0.027 (0.120)	0.036 (0.156)	−0.030 (−0.133)
Eigenvalues (% of variance)	1.059 (91.591)	0.046 (4.014)	0.018 (1.599)	0.013 (1.137)
Cumulative variance explained (%)	91.591	95.605	97.204	98.341

Kaiser-Meyer-Olkin measure of sampling adequacy = 0.957; Bartlett's test of sphericity: $x2$ = 6,308.835, $P < .001$.

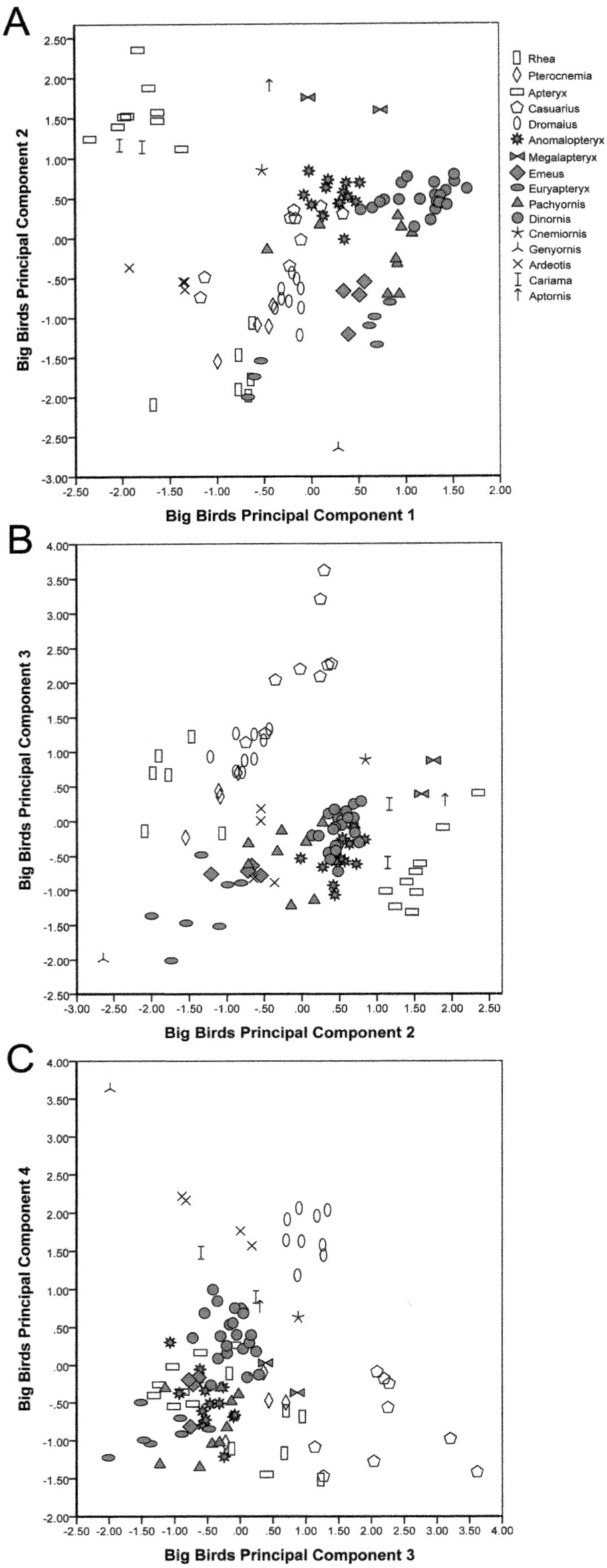
A
B
C
Big Birds Principal Component 1
Big Birds Principal Component 2
Big Birds Principal Component 3
Big Birds Principal Component 4
Rhea
Pterocnemia
Apteryx
Casuarius
Dromaius
Anomalopteryx
Megalapteryx
Emeus
Euryapteryx
Pachyornis
Dinornis
Cnemiornis
Genyornis
Ardeotis
Cariama
Aptornis

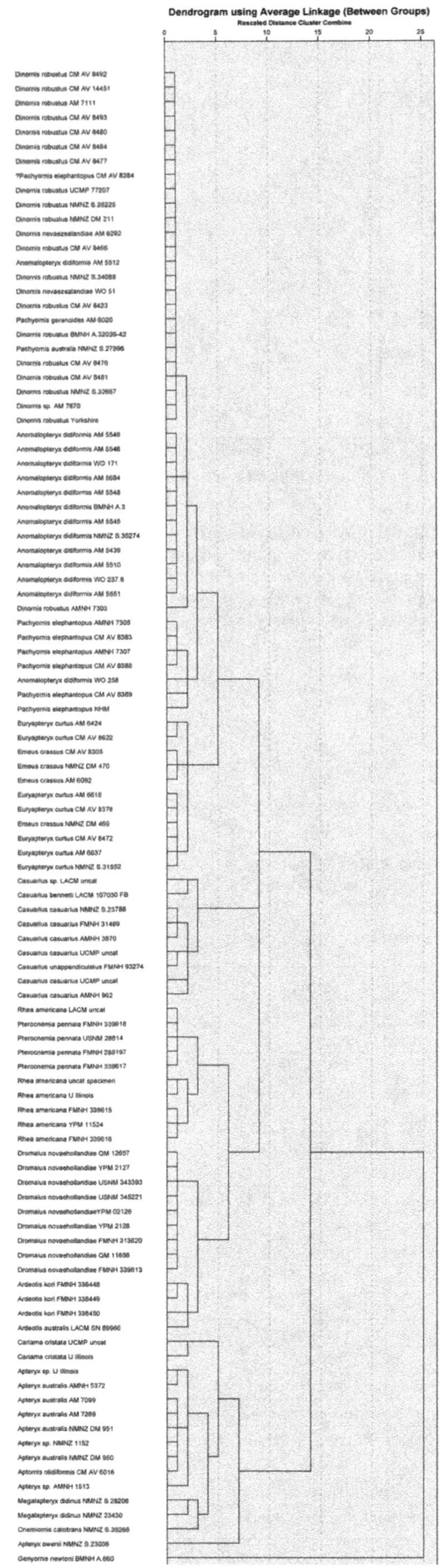
Dendrogram using Average Linkage (Between Groups)
Rescaled Distance Cluster Combine
0
5
10
15
20
25
Dinornis robustus CM AV 8492
Dinornis robustus CM AV 14451
Dinornis robustus AM 7111
Dinornis robustus CM AV 8493
Dinornis robustus CM AV 8480
Dinornis robustus CM AV 8484
Dinornis robustus CM AV 8477
?Pachyornis elephantopus CM AV 8384
Dinornis robustus UCMP 77207
Dinornis robustus NMNZ S.28225
Dinornis robustus NMNZ DM 211
Dinornis novaezealandiae AM 6292
Dinornis robustus CM AV 8466
Anomalopteryx didiformis AM 5512
Dinornis robustus NMNZ S.34088
Dinornis novaezealandiae WO 51
Dinornis robustus CM AV 8423
Pachyornis geranoides AM 6020
Dinornis robustus BMNH A.32035-42
Pachyornis australis NMNZ S.27896
Dinornis robustus CM AV 8476
Dinornis robustus CM AV 8481
Dinornis robustus NMNZ S.32687
Dinornis sp. AM 7870
Dinornis robustus Yorkshire
Anomalopteryx didiformis AM 5546
Anomalopteryx didiformis AM 5546
Anomalopteryx didiformis WO 171
Anomalopteryx didiformis AM 5684
Anomalopteryx didiformis AM 5548
Anomalopteryx didiformis BMNH A.3
Anomalopteryx didiformis AM 5545
Anomalopteryx didiformis NMNZ S.35274
Anomalopteryx didiformis AM 5439
Anomalopteryx didiformis AM 5510
Anomalopteryx didiformis WO 237.8
Anomalopteryx didiformis AM 5551
Dinornis robustus AMNH 7303
Pachyornis elephantopus AMNH 7305
Pachyornis elephantopus CM AV 8383
Pachyornis elephantopus AMNH 7307
Pachyornis elephantopus CM AV 8388
Anomalopteryx didiformis WO 258
Pachyornis elephantopus CM AV 8389
Pachyornis elephantopus NHM
Euryapteryx curtus AM 6424
Euryapteryx curtus CM AV 8622
Emeus crassus CM AV 8305
Emeus crassus NMNZ DM 470
Emeus crassus AM 6092
Euryapteryx curtus AM 6618
Euryapteryx curtus CM AV 8378
Emeus crassus NMNZ DM 469
Euryapteryx curtus CM AV 8472
Euryapteryx curtus AM 6637
Euryapteryx curtus NMNZ S.31552
Casuarius sp. LACM uncat
Casuarius bennetti LACM 107030 FB
Casuarius casuarius NMNZ S.23788
Casuarius casuarius FMNH 31489
Casuarius casuarius AMNH 3870
Casuarius casuarius UCMP uncat
Casuarius unappendiculatus FMNH 93274
Casuarius casuarius UCMP uncat
Casuarius casuarius AMNH 962
Rhea americana LACM uncat
Pterocnemia pennata FMNH 339818
Pterocnemia pennata USNM 28814
Pterocnemia pennata FMNH 288197
Pterocnemia pennata FMNH 339617
Rhea americana uncat specimen
Rhea americana U Illinois
Rhea americana FMNH 339615
Rhea americana YPM 11524
Rhea americana FMNH 339616
Dromaius novaehollandiae QM 12657
Dromaius novaehollandiae YPM 2127
Dromaius novaehollandiae USNM 343393
Dromaius novaehollandiae USNM 345221
Dromaius novaehollandiaeYPM 02126
Dromaius novaehollandiae YPM 2128
Dromaius novaehollandiae FMNH 313620
Dromaius novaehollandiae QM 11656
Dromaius novaehollandiae FMNH 339613
Ardeotis kori FMNH 338448
Ardeotis kori FMNH 338449
Ardeotis kori FMNH 338450
Ardeotis australis LACM SN 89966
Cariama cristata UCMP uncat
Cariama cristata U Illinois
Apteryx sp. U Illinois
Apteryx australis AMNH 5372
Apteryx australis AM 7099
Apteryx australis AM 7289
Apteryx australis NMNZ DM 951
Apteryx sp. NMNZ 1152
Apteryx australis NMNZ DM 950
Aptornis otidiformis CM AV 6016
Apteryx sp. AMNH 1513
Megalapteryx didinus NMNZ S.28206
Megalapteryx didinus NMNZ 23430
Cnemiornis calcitrans NMNZ S.35266
Apteryx oweni NMNZ S.23036
Genyornis newtoni BMNH A.660

to recognize two more components, because PC3 accounts for only a bit more of the data variance than PC4 (1.6% and 1.1%, respectively).

Positive values of PC2 are associated with relatively long distal non-ungual and ungual phalanges, while negative values are associated with long proximal phalanges and broad phalanges. Kiwi, seriemas (*Cariama*), the adzebill (*Aptornis*), and the moa *Megalapteryx* plot along the positive end of PC2, while the dromornithid *Genyornis*, rheas, emus, and the moa *Euryapteryx* and *Emeus* plot near the negative end (fig. 2.9A, 2.9*B*). Other moa, cassowaries, and other ground birds fall around the middle of PC2. Interestingly, bustards (*Ardeotis*) have PC2 values similar to those of emus; throughout this book we will repeatedly see similarities between emus and bustards in foot and footprint shape.

Positive values of PC3 are associated with relatively long unguals and a relatively long digit III, while negative values are most strongly associated with a long phalanx II2, with minor contributions from having broad phalanges. Cassowaries have very positive positions along PC3 (fig. 2.9*B*), due to having a very short II2 (fig. 1.4*G*). Rheas and emus plot fairly positively along PC3, but kiwi are more negative. Most moa, in contrast, have neutral (values close to zero) or negative values of PC3, and separate from struthioniforms other than kiwi along this axis. Indeed, plotting PC3 against PC2 (fig. 2.9*B*) does a remarkably good job of separating moa as a group from extant struthioniforms other than kiwi. *Genyornis* and the moa *Euryapterx* have the most negative values of PC3; both forms have relatively short unguals (figs. 1.6*H*, 1.7*B*); *Euryapteryx* has stout digits, and *Genyornis* a rather broad digit III, but a very slender IV.

Positive values of PC4 are associated with relatively long first and second phalanges of the digits, a relatively long combined length of phalanges IV3 and IV4, and a broad digit III. Negative values are associated with relatively long unguals and broad digits II and IV. *Genyornis* has the most extreme positive value of PC4 (fig. 2.9C). Emus, *Ardeotis*, and *Cariama* are much less positive than *Genyornis* along PC4, but are more positive than the remaining ground birds in the sample.

The striking success with which ground birds segregate into different regions of morphospace along the principal components is also seen in a cluster analysis of scaled phalanx lengths and widths (fig. 2.10). Several distinct low-level or first-order clusters are recognized: (1) most specimens of the two species of giant moa *Dinornis*, along with specimens of *Pachyornis geranoides* and *P. australis*; (2) most specimens of the moa *Anomalopteryx*; (3) most specimens of *Pachyornis elephantopus*; (4) the stout-toed moa *Euryapteryx* and *Emeus*; (5) cassowaries; (6) rheas; (7) emus; (8) bustards (*Ardeotis*); (9) the seriema *Cariama*; (10) most kiwi plus *Aptornis*; and (11) the moa *Megalapteryx*.

Within the *Dinornis* cluster, *D. robustus* and *D. novaezealandiae* seem once again to be indistinguishable. Most, but not all, specimens of *Rhea americana* cluster within the rhea cluster apart from *Pterocnemia pennata*. In the fat-toed moa cluster, the three specimens of *Emeus crassus* group together, but *Euryapteryx curtus* forms two clusters that are no closer to each other than to the *Emeus* group.

At higher-level clustering, the most strikingly positive result is that all moa except *Megalapteryx* group together apart from other big ground birds. After that, the news isn't so good. Emus and bustards form a cluster, which in turn links up with the rhea cluster. The two *Megalapteryx* specimens join with the extinct New Zealand flightless goose *Cnemiornis*, and then with kiwi, and finally seriemas. The moa cluster forms a higher-level cluster with the cassowary, rhea, and emu + bustard cluster, and this eventually joins the seriema + kiwi + *Megalapteryx* + *Cnemiornis* cluster. Finally, *Genyornis* forms a cluster with all the other birds.

It is very encouraging, but once again not that surprising (chapter 1), that conspecific birds tend to be more like each other than like other species. It is also nice to see that most moa tend to be more like other moa than not. Beyond that, the foot shape clusters bear very little resemblance to the known phylogenetic relationships of birds. Paleognaths

Facing left, **2.9.** Principal component analysis (PCA) of log-transformed pedal phalanx lengths and widths of digits II–IV of all large birds (with a phalanx III1 length of at least 20 mm) in our sample (table 1.3). Principal component one (PC1) is, as usual, mainly associated with overall size. Positive values of PC2 (*A, B*) are associated with relatively long distal phalanges (second and third phalanges of digit II; third and fourth phalanges of digit III; second to fifth phalanges of digit IV), whereas negative values are associated with relatively long first phalanges and also with broad phalanges. Positive values of PC3 (*B, C*) are associated with long unguals and a long digit III; negative values are associated with a long phalanx II2 and, to a lesser extent, broad phalanges. Positive values of PC4 (*C*) are associated with long first and second phalanges, a long combined length of phalanges IV3 and IV4, and a broad digit III; negative values are associated with long unguals and relatively broad digits II and IV. Kiwi, *Megalapteryx*, adzebill (*Aptornis*), and seriemas (*Cariama*) have very positive values of PC2, whereas rheas, *Euryapteryx*, and the dromornithid *Genyornis* have very negative values. Struthioniforms other than kiwi (and of course ostriches, which were not included in this analysis) largely segregate from moa in a plot of PC2 against PC3 (*B*), with most struthioniforms showing more positive values of PC3 at any given value of PC2. *Genyornis*, bustards (*Ardeotis*), emus, and seriemas segregate from other large ground birds on the basis of positive to very positive values of PC4 (*C*).

Facing right, **2.10.** Cluster analysis of scaled phalanx lengths and widths of digits II–IV of all large birds (other than ostriches) in my sample. Several first-order clusters emerge, characterized mostly (but not always entirely) by one kind of bird: *Dinornis, Anomalopteryx, Pachyornis elephantopus, Euryapteryx + Emeus, Casuarius*, rheas, *Dromaius, Ardeotis, Cariama*, and *Apteryx*. Some phylogenetic signal is detected, in that moa other than *Megalapteryx* cluster together. However, there is a lot of phylogenetic disorder in the higher-level clusters; see text for details.

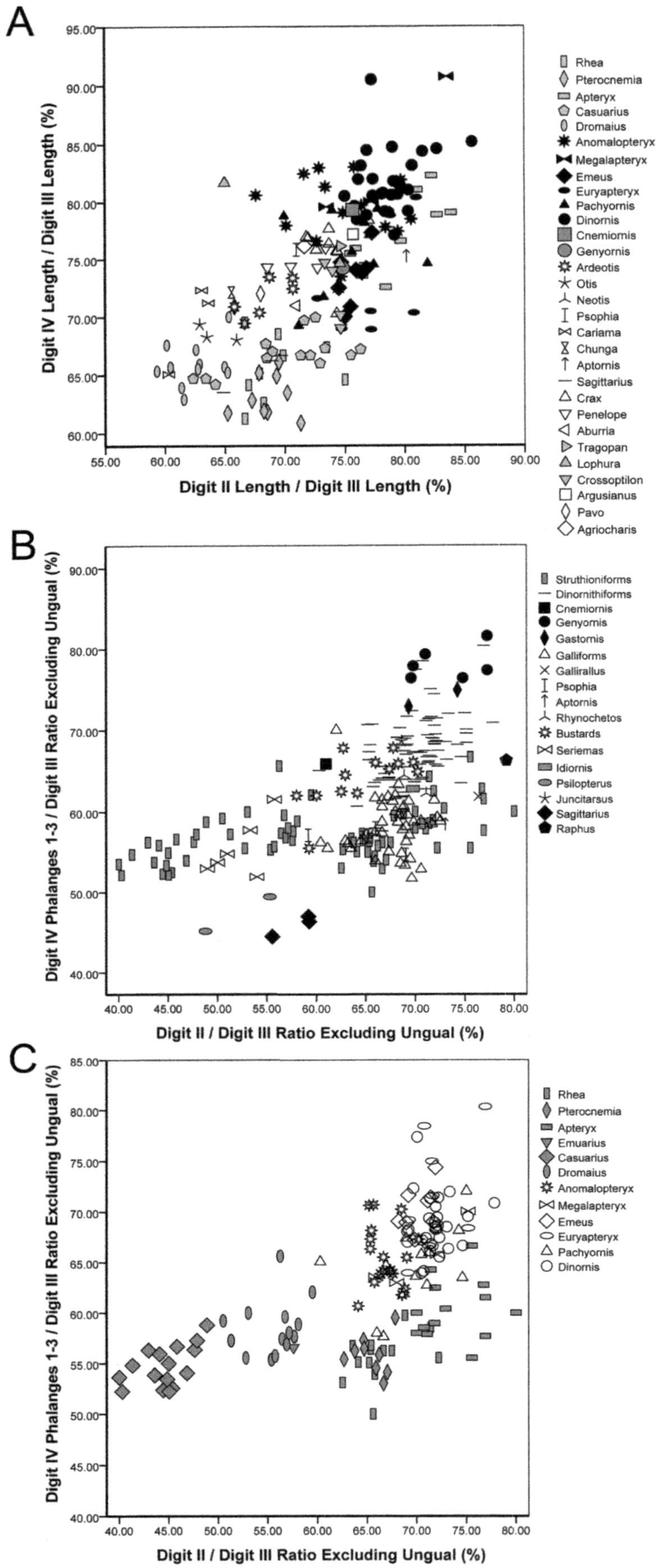

2.11. Relative overall digit lengths of all large ground birds (with a phalanx III1 length of at least 20 mm). *A*, Data from my own measurements only. Some moa (*Dinornis*, *Megalapteryx*, *Anomalopteryx*), kiwi, and the extinct flightless goose *Cnemiornis* tend to have relatively long digits II and IV, whereas most struthioniforms and other ground birds generally have a relatively long digit III. The stout-toed moa *Emeus* and *Euryapteryx* have moderately long relative lengths of digit II, but not digit IV. *B*, *C*, Data from my own measurements, but also from the literature; *B* and *C* differ in which genera or groups are highlighted. To increase sample sizes, and to reduce bogus variability caused by differences in measurements (particularly of ungual lengths) among authors, digit lengths exclude unguals and also the fourth phalanx of digit IV. As with total digit length ratios (*A*), *Genyornis* and moa are seen to have relatively long digits II and IV, as does *Gastornis*. Although there is some overlap with moa, struthioniforms (other than kiwi) and most other large ground birds have relatively shorter digits II and especially digit IV (particularly true of rheas, secretarybird [*Sagittarius*], and the phorusrhacid *Psilopterus*).

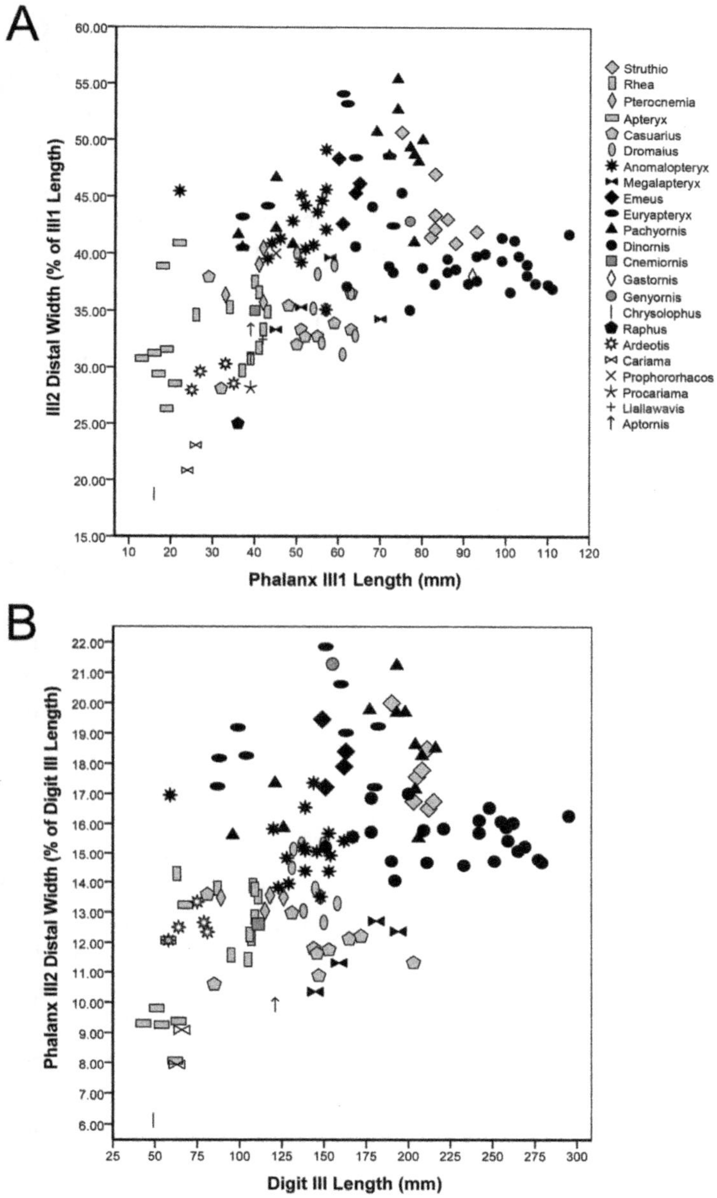

2.12. Relative distal width of phalanx III2 (a bone located about halfway out the length of the digit) in large ground birds. *A*, Phalanx III2 distal width (as a percentage of the length of phalanx III1) plotted as a function of phalanx III1 length. *B*, Phalanx III2 distal width (as a percentage of the aggregate length of digit II) plotted as a function of digit III length. Both graphs show a tendency toward broader digits with increasing size across bird groups.

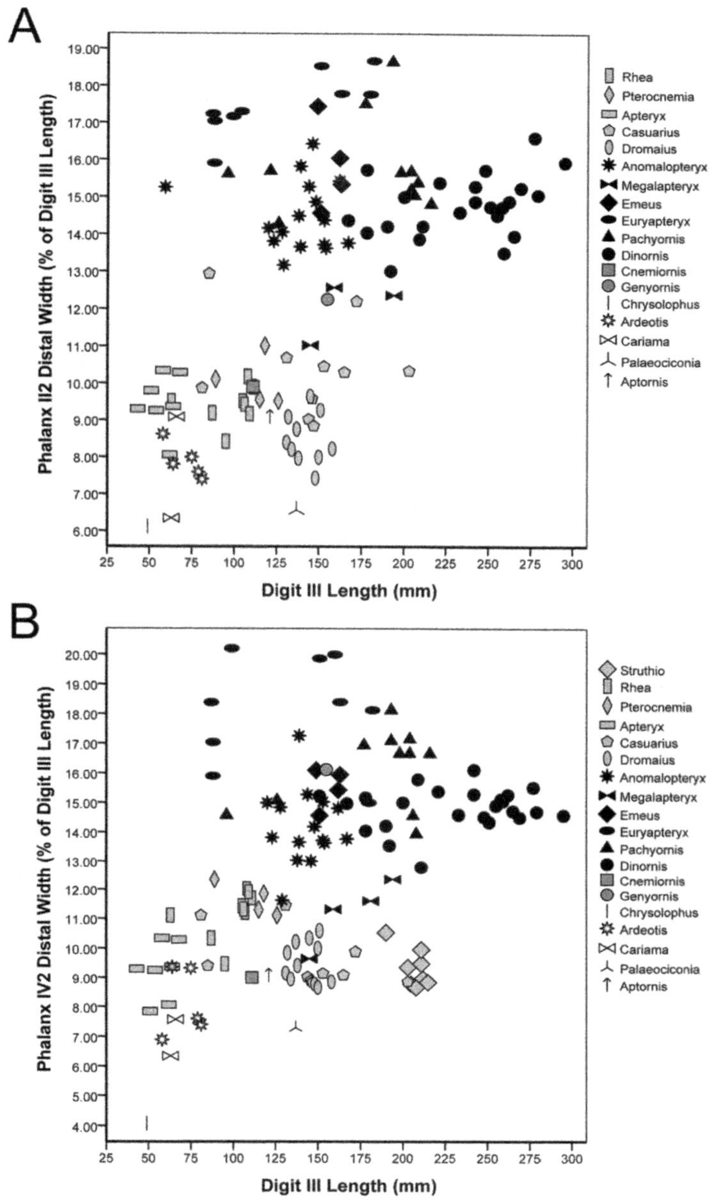

2.13. Relative distal widths of phalanges II2 and IV2 (as a percentage of digit III length) plotted as a function of phalanx III1 length in large ground birds. *A*, Phalanx II2 (data for *Struthio* excluded). *B*, Phalanx IV2. Digits II and IV both show a tendency to become relatively broader in bigger birds, although the trend is spoiled a bit by the very narrow digit IV of the ostrich. (*Struthio*).

join with neognaths (e.g., emus and bustards, *Cariama* and *Cnemiornis* with *Apteryx* and *Megalapteryx*) apart from other paleognaths, *Megalapteryx* wants to be a kiwi, and close relatives like emus and cassowaries don't show it.

Overall digit length proportions (fig. 2.11) provide some separation of groups of large ground birds. Moa, *Genyornis*, and *Gastornis* (specimens formerly known as *Diatryma*) are characterized by relatively long digits II and especially IV, compared with the length of digit III. In contrast, struthioniforms (other than *Apteryx*) and other large ground birds have a relatively longer digit III compared with the lengths of digits II and IV. Secretarybirds (*Sagittarius*) and the phorusrhacid *Psilopterus* have a particularly short digit IV. There is considerable overlap among the fields for struthioniforms, galliforms, and bustards.

Within moa, *Dinornis*, *Anomalopteryx*, and *Megalapteryx* tend to have relatively long digits II and IV compared with digit III, while *Emeus*, *Euryapteryx*, and *Pachyornis* tend to

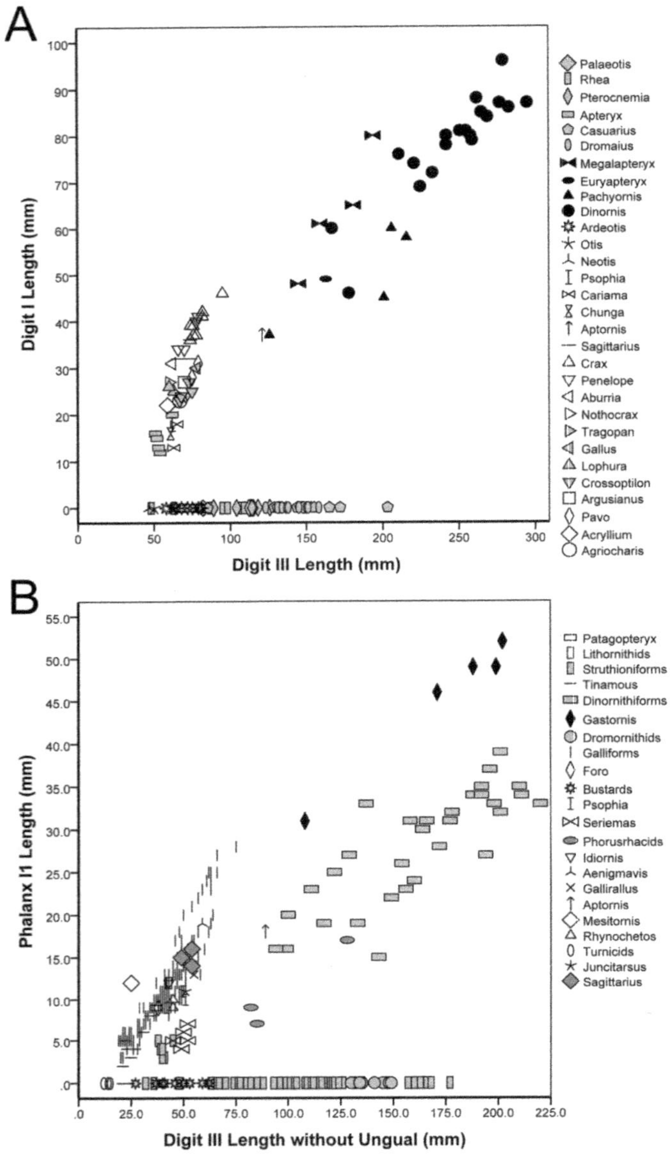

2.14. Digit I length plotted as a function of digit III length in large ground birds. *A*, Total digit lengths (data from my own measurements). *B*, Digit lengths excluding the unguals (my data plus data from the literature and other sources). Some species (extant ratites other than kiwi, bustards) in our sample have completely lost digit I. In those forms in which digit I is present, it appears to become relatively shorter in bigger birds, particularly in moa (note the shallower slope of the relationship in moa than in smaller birds).

have relatively shorter digits II and IV (fig. 2.11*A*). Interestingly, if digit relative lengths are compared without including the ungual (fig. 2.11*B*, 2.11*C*), struthioniforms segregate cleanly along a continuum from forms with very low to high relative lengths of digit II (a twofold range of values), in the order (from low to high values) of cassowaries, emus, rheas (which also have a relatively short digit IV if the unguals are excluded), and finally kiwi.

The relative overall lengths of digits II–IV thus seem in some cases to provide some useful signal for discriminating among groups, and among taxa within groups, of large ground birds. However, there is so much overlap in digit length ratios among groups that are not close relatives that relative digit lengths in feet—and presumably footprints—will frequently be unreliable guides to whether ground birds are close relatives.

Across all large bird groups in our sample, viewed together, there is a suggestion that toes tend to become relatively broader with increasing animal size (figs. 2.12, 2.13), although there is a great deal of scatter in these relationships. In fact, neither moa (table A1.3) nor struthioniforms show any tendency for larger birds within their own group to have relatively broader toes, except for digit III of struthioniforms. The overall trend toward fatter toes in bigger birds is more a matter of moa generally being bigger birds than struthioniforms, and moa also tending to have stouter toes than struthioniforms. *Dinornis* is especially slim-toed for its size, as is *Struthio* for digit IV. In contrast, *Emeus*, *Euryapteryx*, *Anomalopteryx*, *Pachyornis*, and to a lesser extent *Apteryx* have relatively broad digits. *Ardeotis* and *Genyornis* have a relatively broad digit III, but not so much digits II and IV. Relative digit stoutness will be considered again, using slightly different comparisons, in chapters 3 and 4.

A feature of bird foot skeletons not considered thus far is the length of digit I (fig. 2.14), due to the fact that in many species this toe is either absent or so short as to be unlikely to leave an impression in footprints. Digit I is absent (length = zero) in extant ratites other than *Apteryx*, and also in bustards (Johnsgard 1991), among other groups (see chapter 6). Among the birds in our sample in which digit I is present, smaller birds from several clades follow a trend in which the length of digit I increases steeply as digit III becomes longer. Larger birds (*Aptornis*, *Gastornis*, and especially moa) fall away from this trend, and so have a relatively shorter digit I.

Non-avian Dinosaurs. And now we come to the animals of greatest interest for this study. We begin by examining relative lengths of the three main metatarsals, II, III, and IV. Bipedal dinosaurs usually walked with their metatarsals off the ground, the metatarsus acting as a segment of leg length. Consequently the metatarsals would not be expected routinely to impress the substrate over their entire length; only the distal ends of the metatarsals would usually be expected to register in footprints (cf. Farlow et al. 2000; Rainforth and Manzella 2007). However, because the three metatarsals are of unequal length, this could affect the shape of the proximal end of the footprint, and also the relative forward projection of digits II–IV, thereby affecting how far digit III extends beyond the limits of digits II and IV in the footprint (cf. Olsen 1980; Olsen et al. 1998; Weems 1992) and the shape of the footprint "anterior triangle" (cf. Lockley 2009).

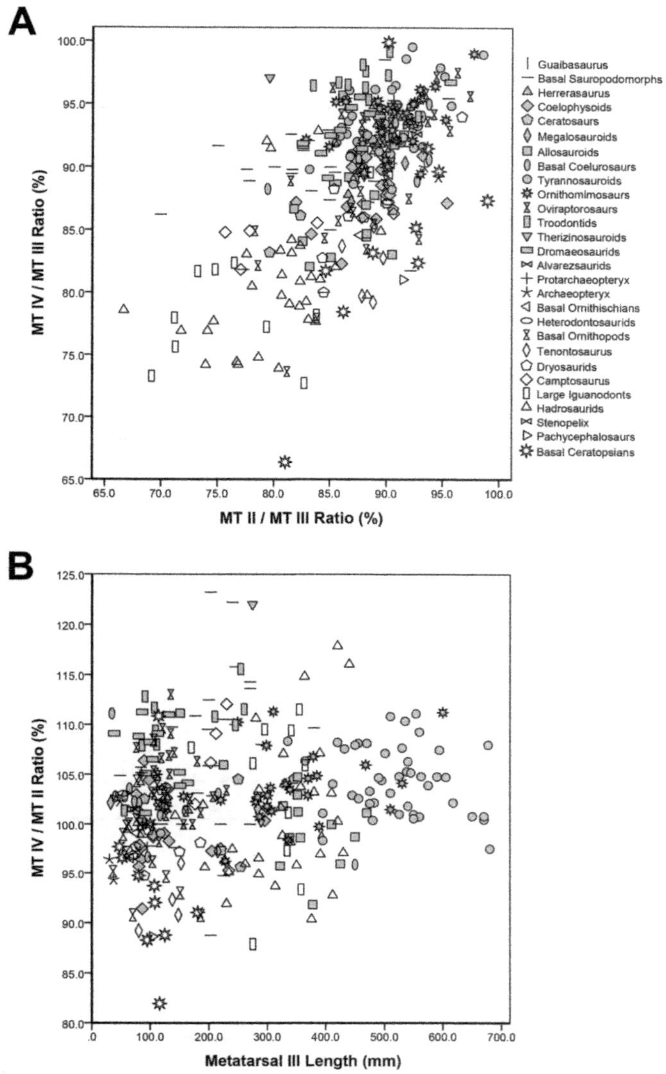

2.15. Metatarsal (MT) ratios in dinosaurs. *A*, Lengths of metatarsals II and IV, both expressed as a percentage of the length of metatarsal III. Note the tendency for theropods to have relatively long outer metatarsals (II and IV) compared with metatarsal III, for basal sauropodomorphs to have a relatively large metatarsal IV/metatarsal III ratio, and for most ornithischians to have relatively short outer metatarsals compared with metatarsal III. *B*, Ratio of metatarsal IV length to metatarsal II length in dinosaurs of differing size (with metatarsal III serving as the proxy for animal size). This comparison shows less separation of saurischians from ornithischians. Among large theropods, tyrannosauroids tend to have a relatively longer metatarsal IV than that in large allosauroids in this comparison. Data from this study, as well as the Online Dinosaur Project, P. J. Currie (personal communication), and also Hulke (1882), Osborn (1899), Broom (1911), Gilmore (1915), van Hoepen (1920*a*, 1920*b*), Haughton (1924), Young (1941*a*, 1941*b*, 1958), Casamiquela (1967), Ostrom (1970), Thulborn (1972), Galton (1973, 1974, 1976, 1981, 1983), Morris (1976), Santa Luca (1984), Galton and van Heerden (1985), Bonaparte (1991), Xu and Zhang (2005), Dalla Vecchia (2009), Hu et al. (2009), Rinehart et al. (2009), Sereno et al. (2009), Lee et al. (2011, 2014), Xu et al. (2011), Zanno et al. (2011), Gao et al (2012), Han et al. (2012), Rauhut et al. (2012 [and Rauhut personal communication]), Cullen et al. (2013), Llandres Serrano et al. (2013), White et al. (2013), Xu et al. (2013), and Choiniere et al. (2014).

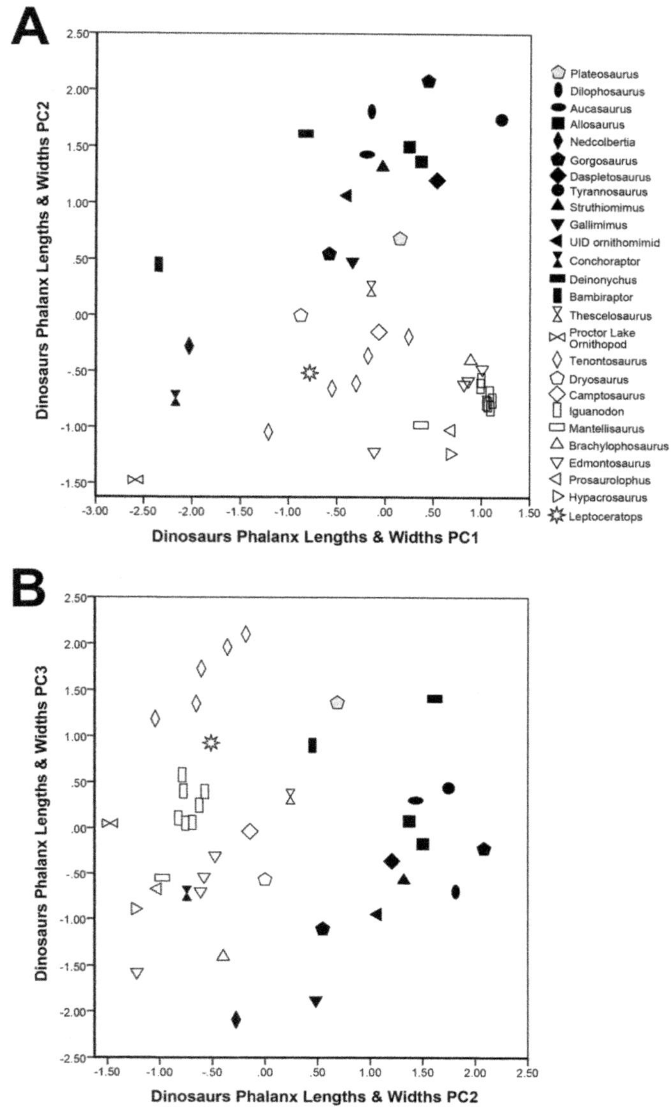

2.16. Principal component analysis (PCA) of log-transformed pedal phalanx lengths and widths of digits II–IV of non-avian dinosaurs (table 2.4). *A*, PC1 vs. PC2. PC1 is again associated with animal size. Positive values of PC2 are associated with relatively long phalanges (especially phalanges positioned along a toe between the first and ungual phalanges), whereas negative values are associated with broad non-ungual phalanges. *B*, PC2 vs. PC3. Positive values of PC3 are associated with long unguals and long distal non-ungual phalanges, whereas negative values of PC3 are associated with relatively long more proximal phalanges. Most theropods separate from ornithischians along PC2; various groups of both theropods and ornithischians spread out along PC3.

Table 2.4. Principal component analysis (using a covariance matrix) of log-transformed measurements of phalangeal lengths and distal widths of digits II–IV in non-avian dinosaurs (*N* = 41).

Parameter	PC1 loading (raw [rescaled])	PC2 loading (raw [rescaled])	PC3 loading (raw [rescaled])
Phalanx II1 length	0.263 (0.944)	0.053 (0.191)	−0.060 (−0.214)
Phalanx II1 distal width	0.341 (0.985)	−0.053 (−0.152)	−0.001 (−0.002)
Phalanx II2 length	0.198 (0.843)	0.119 (0.505)	−0.017 (−0.074)
Phalanx II2 distal width	0.334 (0.975)	−0.065 (−0.190)	0.015 (0.042)
Phalanx II3 length	0.216 (0.930)	0.008 (0.035)	0.073 (0.314)
Phalanx III1 length	0.242 (0.917)	0.073 (0.277)	−0.062 (−0.234)
Phalanx III1 distal width	0.343 (0.986)	−0.042 (−0.122)	−0.025 (−0.073)
Phalanx III2 length	0.149 (0.657)	0.160 (0.707)	−0.032 (−0.144)
Phalanx III2 distal width	0.359 (0.980)	−0.065 (−0.177)	−0.013 (−0.036)
Phalanx III3 length	0.147 (0.688)	0.151 (0.708)	−0.005 (−0.021)
Phalanx III3 distal width	0.351 (0.965)	−0.090 (−0.248)	0.002 (0.005)
Phalanx III4 length	0.238 (0.959)	−0.004 (−0.016)	0.058 (0.234)
Phalanx IV1 length	0.259 (0.960)	0.037 (0.137)	−0.033 (−0.124)
Phalanx IV1 distal width	0.345 (0.990)	−0.035 (−0.100)	−0.017 (−0.047)
Phalanx IV2 length	0.163 (0.742)	0.139 (0.635)	0.029 (0.131)
Phalanx IV2 distal width	0.347 (0.991)	−0.041 (−0.118)	−0.006 (−0.016)
Phalanx IV3 length	0.159 (0.754)	0.126 (0.601)	0.045 (0.214)
Phalanx IV3 distal width	0.347 (0.987)	−0.048 (−0.137)	−0.003 (−0.008)
Phalanx IV4 length	0.169 (0.830)	0.081 (0.398)	0.061 (0.302)
Phalanx IV4 distal width	0.328 (0.983)	−0.052 (−0.156)	0.003 (0.010)
Phalanx IV5 length	0.245 (0.970)	0.006 (0.023)	0.050 (0.197)
Eigenvalues (% of variance)	1.587 (88.691)	0.143 (8.000)	0.029 (1.624)
Cumulative variance explained (%)	88.691	96.691	98.315

Kaiser-Meyer-Olkin measure of sampling adequacy = 0.915; Bartlett's test of sphericity: $\chi 2$ = 2,566.011, < .001.

Comparing the relative lengths of the three metatarsals (fig. 2.15), theropods tend to have relatively longer metatarsals II and IV (compared with the length of metatarsal III) than do ornithischians. Basal sauropodomorphs show considerable overlap with ornithischians (but not so much with theropods) in the metatarsal II/metatarsal III length ratio. Basal sauropodomorphs tend to have a longer metatarsal IV/metatarsal II length ratio than do theropods and ornithischians, although the separation is far from clean. Theropods and ornithischians show nearly complete overlap with each other in the metatarsal IV/metatarsal II length ratio. Within large theropods, there is a slight tendency for tyrannosaurs to have slightly longer metatarsal IV/metatarsal II length ratios than do allosaurs.

On to phalangeal lengths and widths. As with ground birds, PC1 accounts for almost 90% of the variance in log-transformed phalanx lengths and widths, again indicating that PC1 is largely associated with size (table 2.4; fig. 2.16A). Positive values of PC2 are associated with lengths of nonungual phalanges (particularly those other than the first phalanx of each digit), while negative values are associated with broad phalanges. The prosauropod *Plateosaurus* and most theropods have positive values of PC2, while most ornithischians have negative values (fig. 2.16A, 2.16B). Positive values of PC3 are most strongly associated with relatively long unguals, as well as long phalanges 2–4 of digit IV, while negative values are mainly associated with long first phalanges of digits II–IV, and also long second phalanges of digits II and III. There is little separation of theropods and ornithischians as groups along PC3.

Large theropods, whether *Dilophosaurus*, *Allosaurus*, or tyrannosaurids, tend to have strongly positive values of PC2, but more neutral values of PC3. It is interesting, however, that the smaller of two specimens of *Gorgosaurus libratus* has more negative, ornithomimid-like values of PC3 than its larger conspecific.

Dromaeosaurids (*Bambiraptor*, *Deinonychus*), *Plateosaurus*, and the small to medium-sized ornithischians *Tenontosaurus* and *Leptoceratops* have very positive values of PC3, reflecting their whopping big unguals (compared to the rest of their phalanges; figs. 1.10, 1.11). Basal ornithopods (*Thescelosaurus*, the Proctor Lake ornithopod), *Camptosaurus*, and *Iguanodon* show neutral to slightly positive values of PC3, while hadrosaurids (*Brachylophosaurus*, *Edmontosaurus*, *Prosaurolophus*, *Hypacrosaurus*) and the small to mid-sized theropods *Conchoraptor*, *Nedcolbertia* (interpreted as a basal ornithomimosaur by Brownstein [2017]), and *Gallimimus* have fairly negative values of PC3.

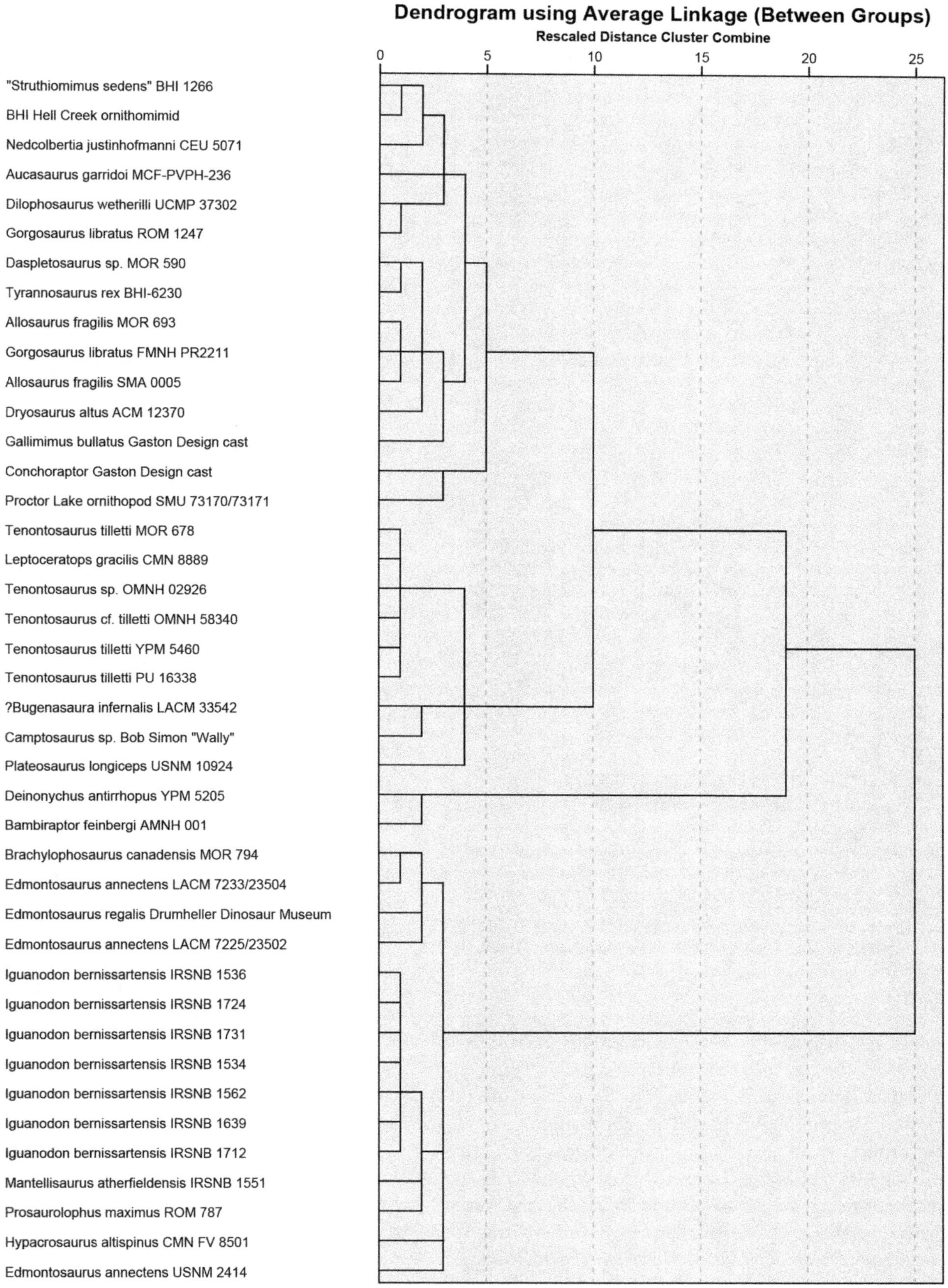

2.17. Cluster analysis of scaled phalanx lengths and widths (digits II–IV) of non-avian dinosaurs. Four major groups are defined: most theropods other than dromaeosaurids, plus *Dryosaurus* and the Proctor Lake ornithopod; some small to medium-sized ornithischians, plus *Plateosaurus*; dromaeosaurids; and large ornithopods.

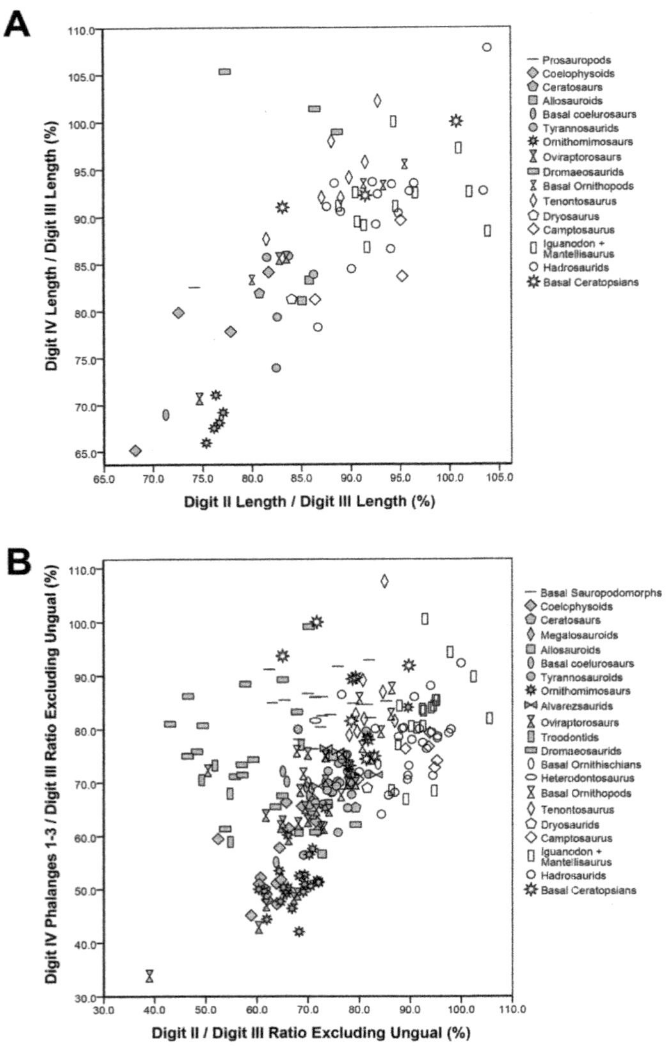

2.18. Relative lengths of digits II–IV in non-avian dinosaurs. *A*, Including unguals in digit lengths; data from this study. *B*, Excluding unguals (and phalanx IV4) to increase sample size; data from this study, and also the literature. Theropods other than dromaeosaurids tend to have relatively short digits II and IV compared with digit III, whereas ornithischians tend to have relatively long digits II and IV, and basal sauropodomorphs a relatively long digit IV.

A cluster analysis of scaled phalanx lengths and widths of digits II–IV (fig. 2.17) picks out four major groups: theropods other than dromaeosaurids along with *Dryosaurus* and the Proctor Lake ornithopod; miscellaneous small to medium-sized ornithischians plus *Plateosaurus*; dromaeosaurids; large ornithopods. Within the cluster of big ornithopods, all the specimens of *Iguanodon bernissartensis* cluster together, a pleasing result (cf. chapter 1). The same is mostly true for specimens of *Edmontosaurus*. However, the hadrosaurid *Prosaurolophus* joins up with *Iguanodon* and *Mantellisaurus* apart from other hadrosaurids (*Edmontosaurus, Brachylophosaurus, Hypacrosaurus*). It may be that the real pattern among big ornithopods is that most of them have feet so similar that one cannot reliably distinguish among them—and, perhaps in consequence, among footprints made by those feet.

Within the cluster of medium-sized ornithischians, the five specimens of *Tenontosaurus* cluster together along with *Leptoceratops*. One specimen of *Thescelosaurus* (labeled ?*Bugenasaura* in the cluster diagram) links with a specimen of *Camptosaurus*. Finding *Plateosaurus* among the small to medium-sized ornithischians may seem jarring, but the similarity of its foot to that of prosauropods has been noted previously (Farlow and Chapman 1997; Farlow et al. 2012; also see figs. 1.10C and 1.11 here).

Within the theropod cluster, there is no tendency for tyrannosaurids to cluster together apart from other large theropods, corroborating results found in earlier studies (Farlow 2001; Farlow et al. 2013). In contrast to previous work, however, ornithomimids show less tendency here to group together, but that could reflect the small number of specimens. *Conchoraptor* is the theropod most different (other than the dromaeosaurids) from other theropods, and clusters first with—of all things—the Proctor Lake ornithopod. This isn't terribly surprising, given the behavior of *Conchoraptor* in the PCA (fig. 2.16*B*), where it fell among points for ornithischians.

As they do with birds, relative digit lengths (fig. 2.18) provide some discrimination among groups of non-avian dinosaurs. Theropods other than dromaeosaurids tend to have relatively short digits II and IV compared with the length of digit III—just the opposite of what was seen in metatarsal length ratios (fig. 2.15A). Among theropods, ornithomimosaurs fairly consistently show particularly short digits II and IV. Dromaeosaurids have a somewhat longer digit II relative length than many other theropods on the basis of total digit length (fig. 2.18A; due to the huge—but strongly curved—digit II ungual), but a particularly short digit II relative length if unguals are excluded (fig. 2.18*B*). Dromaeosaurids and troodontids have a particularly long digit IV relative length compared with other theropods, with values for dromaeosaurids comparable to those for prosauropods (basal sauropodomorphs) and ornithischians. Tyrannosaurs generally show little difference from allosaurs in relative digit lengths.

Ornithischians generally show relatively long digits II and IV, compared with the length of digit III. The same may be true for prosauropods (at least when digit lengths are compared without unguals), especially for digit IV.

Among ornithischians, *Tenontosaurus* fairly consistently has a relatively long digit IV, while *Dryosaurus*, *Camptosaurus*, and basal ornithopods ("hypsilophodontids") tend to have a relatively short digit IV. Even so, there is a great deal

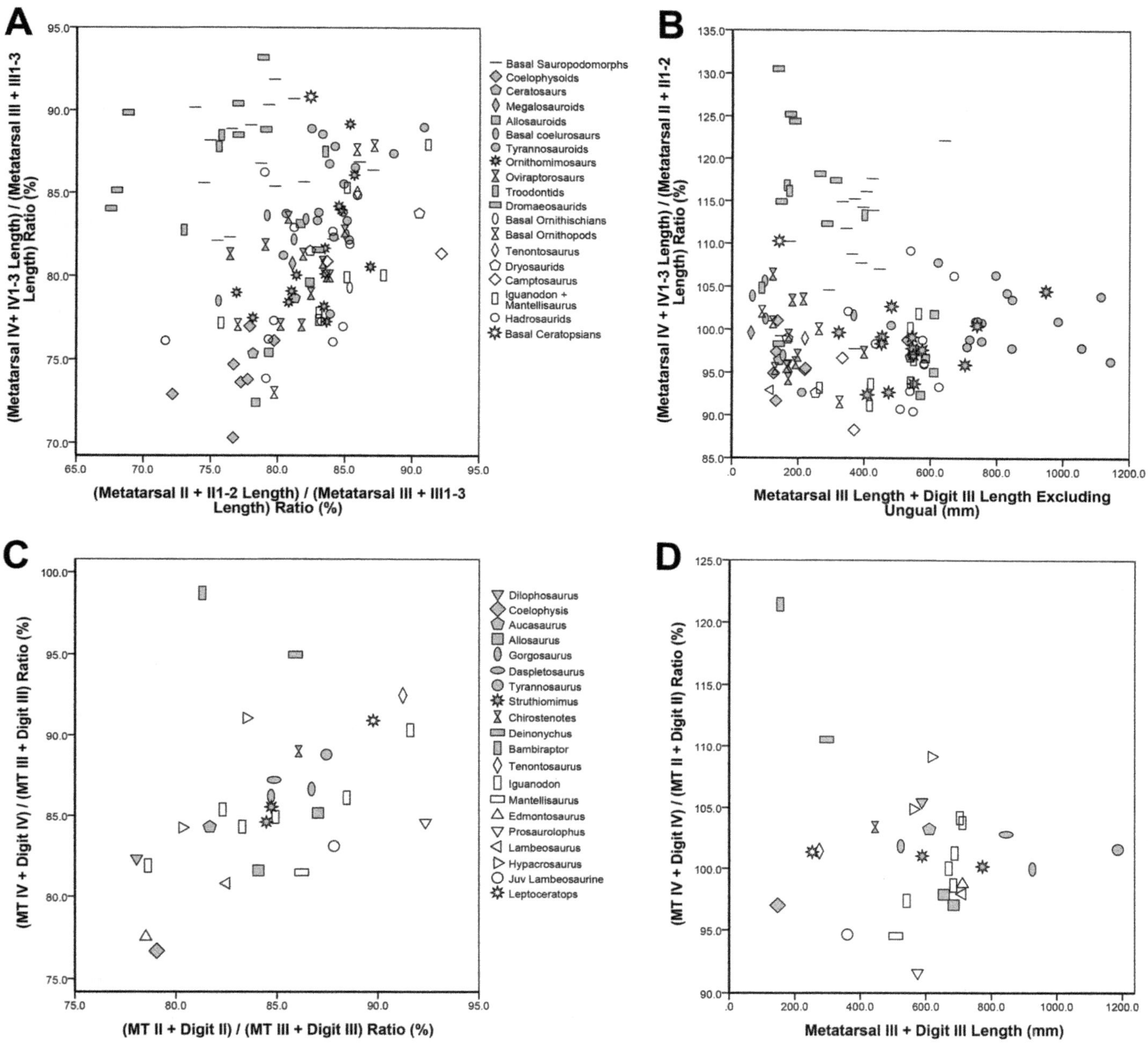

2.19. Ratios of the combined lengths of metatarsals and their digits (both including and excluding unguals). Data from this study and also many of the sources cited for figure 2.15. *A* and *B* have the same symbol key, as do *C* and *D*. *A*, Ratio of the combined length of metatarsal II and digit II (excluding the ungual) to the combined length of metatarsal III and digit III (excluding the ungual) vs. the ratio of the combined length of metatarsal IV and the first three phalanges of digit IV to the combined length of metatarsal III and digit III (excluding the ungual). Most theropods show little separation from ornithischians, but dromaeosaurids and troodontids differ from other theropods, and from ornithischians, in having a relatively long combined metatarsal IV plus digit IV length, designated here as "(metatarsal IV + digit IV)," and a relatively short (metatarsal II + digit II). Basal sauropodomorphs generally separate from theropods—other than dromaeosaurids and troodontids—and ornithischians in having a relatively low (metatarsal II + digit II) / (metatarsal III + digit III) ratio. *B*, Ratio of the combined length of metatarsal IV and the first three phalanges of digit IV to the combined length of metatarsal II and digit II (excluding the ungual) plotted as a function of the combined length of metatarsal III and digit III (excluding the ungual). Troodontids, dromaeosaurids, and most basal sauropodomorphs tend to have a relatively long (metatarsal IV + digit IV) compared with the length of (metatarsal II + digit II), and tyrannosaurs may show a higher ratio than allosaurs. *C*, *D*, Same comparisons as in *A* and *B*, except that all phalanges, including unguals, are included in the digit lengths, which considerably reduces the sample size. However, dromaeosaurids and troodontids continue to show a relatively long (metatarsal IV + digit IV), whether compared with the combined length of (metatarsal III + digit III) or of (metatarsal II + digit II). Tyrannosaurs still seem to show a relatively longer (metatarsal IV + digit IV), compared with (metatarsal II + digit II), than do allosaurs. Basal sauropodomorphs do not appear in *C* and *D* because of lack of data for specimens with measurements for both metatarsals and complete digits.

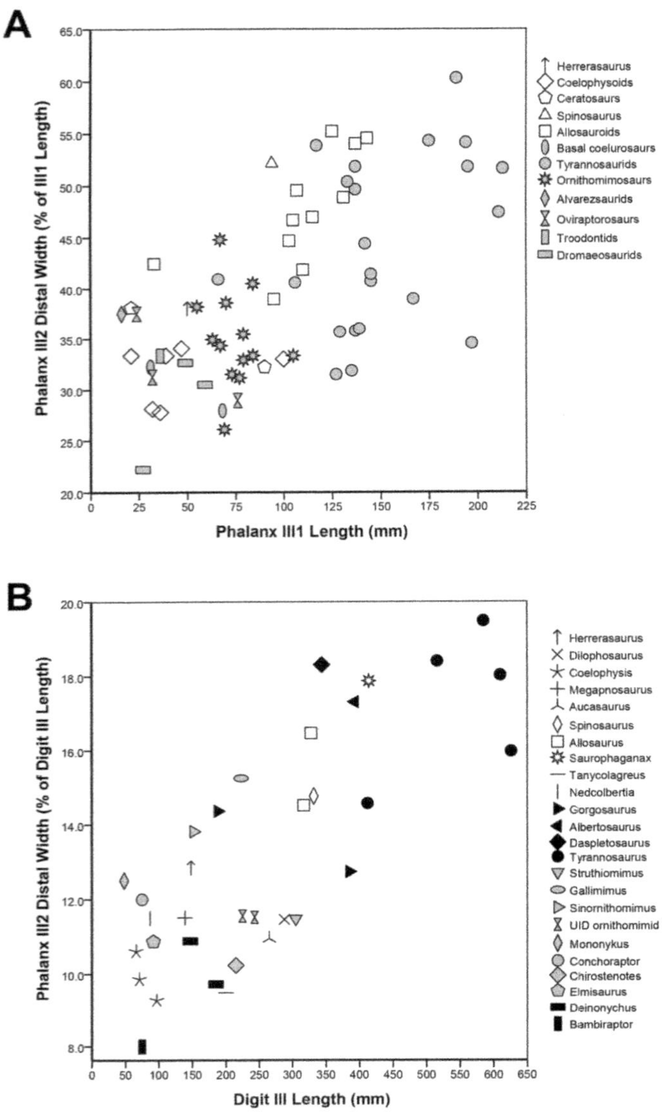

2.20. Relative distal width of phalanx III2 in non-avian theropods. *A*, Phalanx III2 distal width (as a percentage of the length of phalanx III1) plotted as a function of phalanx III1 length. *B*, Phalanx III2 distal width (as a percentage of the aggregate length of digit III) plotted as a function of digit III length. As with ground birds, both graphs show a tendency toward broader digits with increasing size across groups. The point for *Saurophaganax* in *B* and in figure 2.21*B* should be viewed with caution, because the pedal phalanges of this form probably come from more than one animal, albeit from individuals of about the same size.

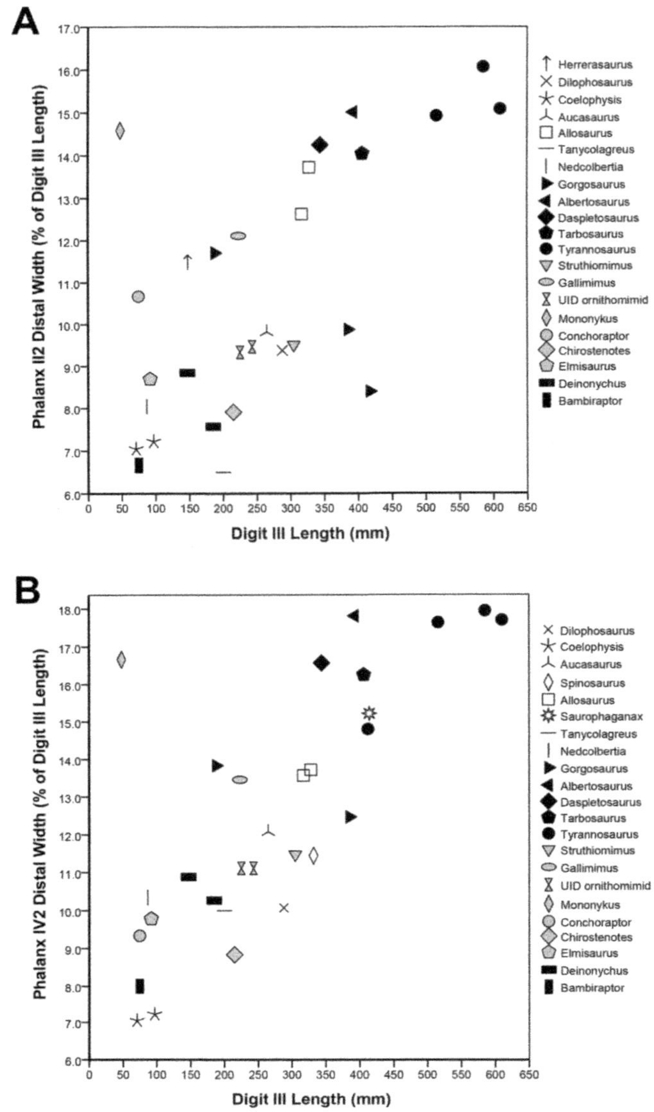

2.21. Relative distal widths of phalanges II2 and IV2 (as a percentage of digit III length) plotted as a function of digit III length in non-avian theropods. *A*, Phalanx II2. *B*, Phalanx IV2. Digits II and IV both show a tendency to become relatively broader in bigger dinosaurs. *Mononykus*, however, has unusually broad digits II and IV for its size.

of overlap among different groups of ornithischians in relative digit lengths. Points for hadrosaurids and *Iguanodon* in particular overlap those of other ornithischian groups. It is therefore unlikely that one would be able confidently to distinguish feet or footprints of one group of ornithischians from another on the basis of the relative overall lengths of digits II–IV alone.

Unfortunately (and very regrettably), the relatively longer metatarsal II/metatarsal III and metatarsal IV/metatarsal III ratios seen in theropods as opposed to ornithopods (fig. 2.15A), and the relatively short digit II/digit III and digit IV/digit III ratios seen in theropods as opposed to ornithopods (fig. 2.18), cancel each other out if we consider the ratios in terms of combined metatarsal and digit lengths (fig. 2.19), thus demolishing a potential criterion for distinguishing theropod from ornithischian feet and footprints. However, dromaeosaurids and troodontids continue to show a relatively long metatarsal IV plus digit IV, and tyrannosaurs may still show a relatively longer metatarsal IV plus digit IV, compared to metatarsal II plus digit II, than do allosaurs.

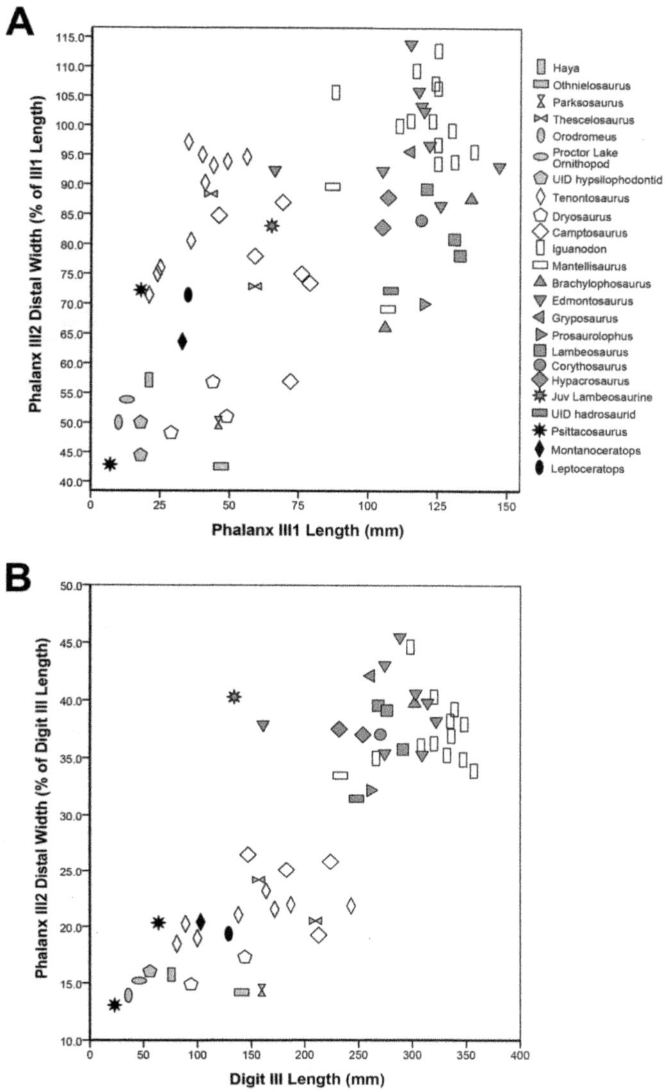

2.22. Relative distal width of phalanx III2 in ornithischians. *A*, Phalanx III2 distal width (as a percentage of the length of phalanx III1) plotted as a function of phalanx III1 length. *B*, Phalanx III2 distal width (as a percentage of the aggregate length of digit III) plotted as a function of digit III length. As with ground birds and non-avian theropods, both graphs show a tendency toward broader digits with increasing size across groups.

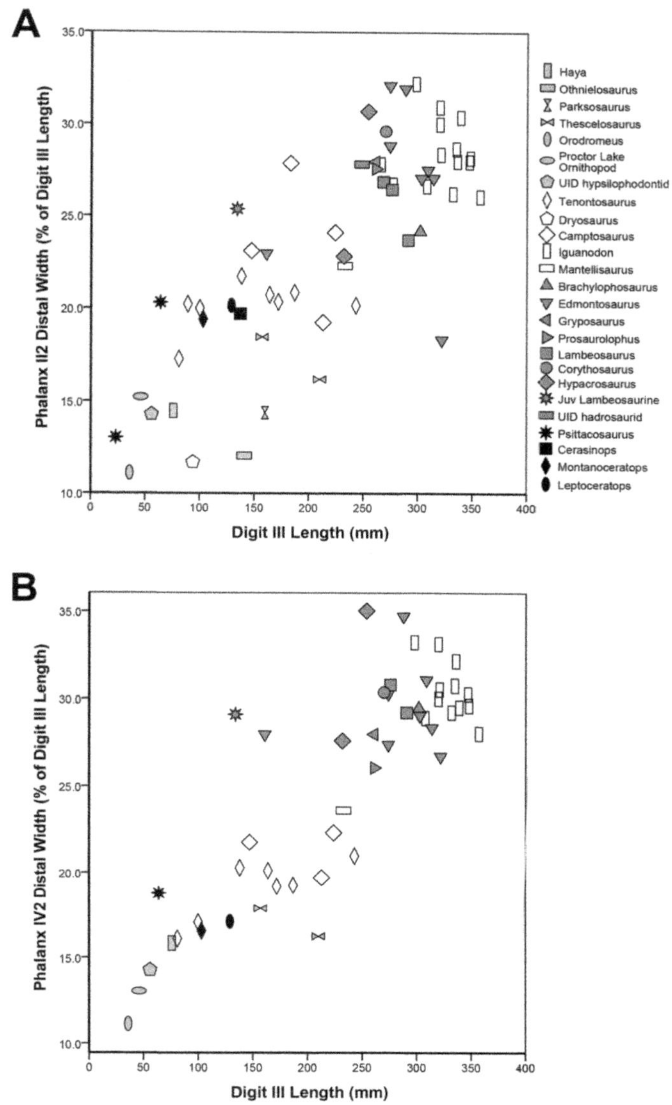

2.23. Relative distal widths of phalanges II2 and IV2 (as a percentage of digit III length), each plotted as a function of digit III length in ornithischians. *A*, Phalanx II2. *B*, Phalanx IV2.

Among non-avian theropods, even more clearly than among ground birds, there is a tendency for larger forms to have relatively broader digits (figs. 2.12, 2.13, 2.20, 2.21). Adding to other peculiarities of its skeleton, *Mononykus* has unusually stout digits II and IV, but not III, for a dinosaur its size.

Ornithischians similarly show a clear tendency for larger forms to have relatively broader digits (figs. 2.22, 2.23). As with ground birds, however, the tendency for toes to become stouter across groups isn't seen so clearly within groups. Thus hadrosaurs show little tendency to increase toe stoutness with increasing animal size across their own group (at least over the size range for which I have data), but hadrosaurs are big dinosaurs, and have relatively fat toes (cf. Moreno et al. 2007), and so contribute to the overall tendency to increase toe stoutness with increasing dinosaur size.

Generalizing across major groups, big non-avian theropods tend to have narrower digits for their size than do ground birds and ornithischians (fig. 2.24). Birds do show considerable overlap with non-avian theropods in this relationship, but the fatter-toed moa are more like ornithischians

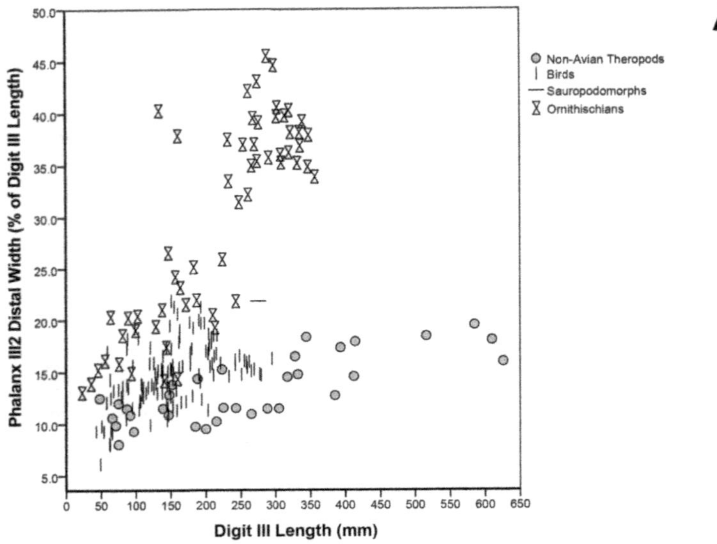

2.24. Relative width of digit III in ground birds, non-avian dinosaurs, and ornithischians. All groups show a tendency for toes to become stouter with increasing animal size, with large non-avian theropods showing the narrowest toes among large forms. Ground birds show extensive overlap with non-avian theropods in this relationship, but some moa have stouter toes than those of theropods with comparable digit III length. Ornithischians have the broadest toes and show very little overlap in the stoutness vs. digit length relationship with non-avian theropods but exhibit more overlap with ground birds. Digits II and IV show a similar pattern to that for digit III.

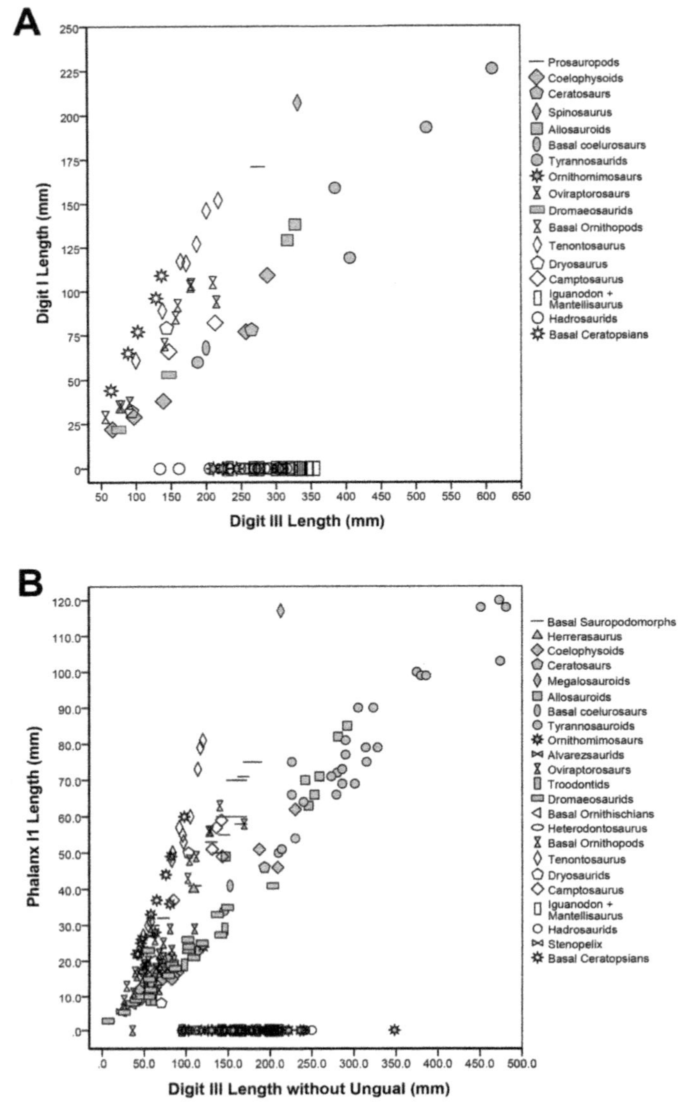

2.25. Relative length of digit I in non-avian dinosaurs. *A*, Total digit lengths (data from my own measurements in this study). *B*, Digit lengths excluding the unguals (my data plus data from the literature and other sources). Some forms—most ornithomimosaurs, large ornithopods—have completely lost digit I (length = 0). In forms that still have a digit I, it is relatively short (compared with the length of digit III) in most theropods (*Spinosaurus* being a glaring exception), but longer in basal sauropodomorphs (prosauropods) and ornithischians (particularly *Tenontosaurus* and basal ceratopsians).

than non-avian theropods in digit stoutness. Large ornithopods have particularly broad toes.

As with ground birds, the relative length of digit I in non-avian dinosaurs is quite variable (fig. 2.25). The pattern is remarkably similar to that for ground birds (fig. 2.14). Hadrosaurs and other large iguanondontoids lack digit I, as do most ornithomimosaurs. Among forms that retain a digit I, most theropods other than *Spinosaurus* (Ibrahim et al. 2014) have a relatively short digit I (making them analogous to moa), while ornithischians and prosauropods have a relatively large digit I. Among ornithischians, *Tenontosaurus* and basal ceratopsians have a particularly long digit I.

Digit I would not be expected routinely to leave a mark in footprints of most theropods, unless they moved through very soft and deep sediments (cf. Gatesy et al. 1999), or placed the entire length of the foot (including the metatarsal region) against the substrate (cf. Hitchcock 1858; Lull 1953; Kuban 1989; Pérez-Lorente 1993, 2015; Boutakiout et al. 2009*b*; Gierliński et al. 2009; Milner et al. 2009; Farlow et al. 2015). I would expect digit I to register a mark in footprints of prosauropods and many small to medium-sized ornithischians (cf. Stanford et al. 2004), unless the dinosaur was doing some kind of interesting dorsiflexion during locomotion—a matter that will be considered in detail for prosauropods in chapter 9.

In the first two chapters we have looked at phalangeal and digital proportions of many actual or potential makers of tridactyl bird and non-avian dinosaur footprints, both within and across species and higher taxa. However, there is another way of examining these data before we have finished with foot skeletons, and to this matter we now turn.

Toe-Tapering Profiles in Non-avian Dinosaurs and Ground Birds

3

With Contributions by Dan Coroian

Theropod footprints comprise tapering digits, often with a V-shaped outline, while ornithopod footprints typically have parallel-sided digits with a more U-shaped outline. . . . These differences in shape tend to be most pronounced in digit III, which was usually planted very firmly into the substrate.

T. Thulborn, *Dinosaur Tracks* (1990: 220)

THULBORN'S OBSERVATION PROMPTED US TO EXAMINE toe-tapering profiles of the foot skeletons of non-avian theropods, bipedal or potentially bipedal ornithischians, and large ground birds to consider the extent to which such profiles can discriminate among different kinds of potential tridactyl footprint-maker. If toe-tapering profiles can tell the foot skeletons apart, they might also work for discriminating among different kinds of tridactyl footprints.

METHODS AND MATERIALS

Toe-tapering profiles (TTPs) were created by plotting the distal width of each pedal phalanx against the cumulative length of the phalanges up to and including that phalanx. Thus the distal width of phalanx II1 was plotted against the length of phalanx II1, and the distal width of phalanx II2 was plotted against the aggregate lengths of phalanges II1 and II2; the distal width of ungual II3 was assigned a value of zero, and was plotted against the cumulative lengths of phalanges II1–II3. A comparable procedure was followed for digits III and IV.

Because many foot skeletons of non-avian dinosaurs have incomplete phalanges owing to the vagaries of preservation, we increased the sample size of specimens by some judicious (we hope) approximations. If the proximal width of the first phalanx (used in calculating "straight-line phalanx lengths"; see below) could not be measured, we used the value of the distal width in its place; this approximation seemed warranted because the first phalanx of a digit shows relatively little decrease in transverse width from its proximal to its distal width. Similarly, if either the medial or lateral length of a phalanx or ungual could not be measured, we assumed that the missing value was the same as the length of the other side of the phalanx. This approximation is less justified than our assumption that the proximal width of a first phalanx is the same as the distal width of the same phalanx, at least for digits II and IV, but it seemed better to do this than to exclude the specimen from the analysis.

If the tip of an ungual was missing a small amount of its material, we either used the actual length measurements of the ungual as preserved, or estimated how much material was missing (as long as this was a very small amount). Deciding whether to approximate the medial and lateral lengths of a slightly incomplete ungual (and thus to include that specimen in the analysis of scaled TTPs), and if used how best to approximate the medial and lateral lengths of that ungual, were ad hoc decisions for each specimen.

For large, non-avian theropods (the dinosaurs of greatest interest to Farlow, given the likely makers of tridactyl dinosaur footprints of the Early Cretaceous Glen Rose Formation of Texas), TTPs were created for all specimens for which the first phalanx, and as many of the progressively more distal phalanges of each of the three main pedal digits (II–IV), could be measured. For many specimens, the profiles for one or more of the digits were incomplete owing to missing phalanges or unguals. Non-ungual phalanges and unguals were measured as in previous chapters (fig. 1.1), and the TTPs using such values are here characterized as "raw" (as opposed to the "straight-line" analyses described below). To remove the effects of size in TTPs, we also scaled raw cumulative digit lengths and associated phalanx widths to a common value, scaling them as percentages of the length of phalanx III1, thus allowing TTPs of forms as big as *Tyrannosaurus* to be compared directly with those of much smaller animals.

For a more restricted set of data cases comprising non-avian theropods, other non-avian dinosaurs, ground birds, and also the dinosaur-like basal suchian *Poposaurus* (cf. Weinbaum and Hungerbühler 2007; Gauthier et al. 2011; Schachner et al. 2011; Bates and Schachner 2012; Farlow et al. 2014) for which all of the length and distal width measurements of all of the phalanges of digits II–IV could be made,

TTPs were also scaled to remove the effects of absolute size in a different way. Each non-ungual phalanx was modeled as a trapezoid, the height of which was calculated from the proximal and distal transverse widths of the bone, along with the mean value of the medial and lateral lengths of the bone. The height of this trapezoid is the "straight-line" length of the non-ungual phalanx. Unguals were modeled as isosceles triangles, the base width of which was taken to be the distal width of the final non-ungual phalanx, and the sides of which were the mean of the medial and lateral lengths of the ungual, as routinely measured. Note that calculating the straight-line lengths of non-ungual phalanges, and even more so unguals, in this way removes a slight artifact that arises in comparing dinosaurs with very broad unguals (such as large ornithopods) with dinosaurs with relatively narrow toes. If a phalanx shows substantial taper from its proximal to its distal end (which unguals, of course, will necessarily do), the standard "raw" length measurement (mean of medial and lateral lengths) will be somewhat larger for a broad-clawed animal than a narrow-clawed animal, even if the straight-line lengths of the unguals of the two animals are the same. Presenting TTPs using straight-line phalanx and ungual lengths removes this artifact.

All straight-line lengths and distal widths of the phalanges and unguals of digits II–IV were then scaled as a percentage of the straight-line length of phalanx III1 (which itself was obviously assigned a value of 100%). Scaled TTPs were then created for digits II–IV. Because all of the measurements had to have been made, or at least approximated, in this more restrictive analysis, there are considerably fewer specimens represented than for TTPs based on "raw" measurements (such as those for large theropods, as described above).

Finally, toe-tapering curves (TTCs) were fitted to the scaled TTPs of digits II–IV for each taxon included in the analysis. If a species was represented by more than one specimen, the mean values of the scaled measurements were used in creating the TTCs. Separate curves were fitted for the non-ungual portions of the TTPs, and also for TTPs including the unguals.

TTCs created from means of values of scaled parameters for each species, or single cases in which only one is available for that species, obviously reduce information about variability within species, but at the same time reduce clutter in the illustrated TTPs, thereby making them easier to interpret. We will discuss TTCs for non-avian theropods, ground birds, and bipedal or potentially bipedal ornithischians, and for comparative purposes also consider TTCs for the dinosaur-like bipedal suchian *Poposaurus gracilis* and the sauropodomorph *Plateosaurus* (in some of the graphs, the specimen is identified as *P. longiceps*, but only out of convenience, because that was the label on the specimen at the time we measured it; the appropriate species name may actually be different (cf. Galton 1984*b*; Moser 2003; Prieto-Márquez and Norell 2011).

RESULTS

Raw Toe-Tapering Profiles for Large Theropods

Raw TTPs for most of the large theropods follow a common pattern (fig. 3.1): a curving, concave-down profile from the first phalanx of a digit through that digit's ungual (curves are not drawn between points for individual feet to avoid clutter in the graphs, which are already quite busy enough). Values of the distal width of a phalanx plotted against the cumulative widths of phalanges up to and including that phalanx for the various taxa generally fall along or close to rays that radiate outward from the origin of the graph, from the first phalanx to the terminal ungual (whose ray parallels the *x*-axis because the distal width of the ungual tip is defined as zero). Most of the curves, if drawn between points for the same foot, would be roughly concentric, suggesting fairly similar TTPs across most taxa, but there is one conspicuous oddball: digits III and IV of *Spinosaurus* (digit II of this specimen is incomplete) show no decline in distal width from one non-ungual phalanx to the next, and may actually increase along the TTP for digit IV. *Spinosaurus* also has a remarkably small foot for so large a theropod dinosaur (cf. Ibrahim et al. 2014).

Scaling digit cumulative lengths and phalanx distal widths to percentages of the length of phalanx III1 (fig. 3.2) continues to show most species with fairly similar concave-down TTPs.

There is a clear tendency for larger-bodied individuals to have stouter toes, regardless of the species or more inclusive clade to which they belong (figs. 2.20, 2.21, 3.2, 3.3), as

Facing left, **3.1.** "Raw" (nonscaled) toe-tapering profiles (TTPs) for large theropods. Many specimens for which data are plotted in these graphs are incomplete, lacking one or more of the more distal phalanges. Points for *Nanotyrannus* are labeled separately from those of (other?) specimens of *Tyrannosaurus* in these and other graphs in this chapter. As elsewhere in this book, *"Camptosaurus amplus"* is regarded as a large theropod, following a suggestion by R. T. Bakker in Galton and Powell (1980) and Galton (2015). Although treated here as from a single individual, the pedal phalanges of *Saurophaganax* likely represent two individuals of about the same size (chapter 1). *A*, Digit II. *B*, Digit III. *C*, Digit IV.

Facing right, **3.2.** Scaled raw toe-tapering profiles (TTPs) for large theropods. All lengths and distal widths are expressed as a percentage of the length of phalanx III1. *A*, Digit II. *B*, Digit III. *C*, Digit IV. *Dilophosaurus*, *Aucasaurus*, and *Nanotyrannus* (the last of these possibly an immature individual of *Tyrannosaurus*) tend to be more slender-toed than other large theropods, whereas *Tyrannosaurus* tends to be broad-toed. *Spinosaurus* is very unusual in showing little or no digit tapering in phalanges proximal to the unguals.

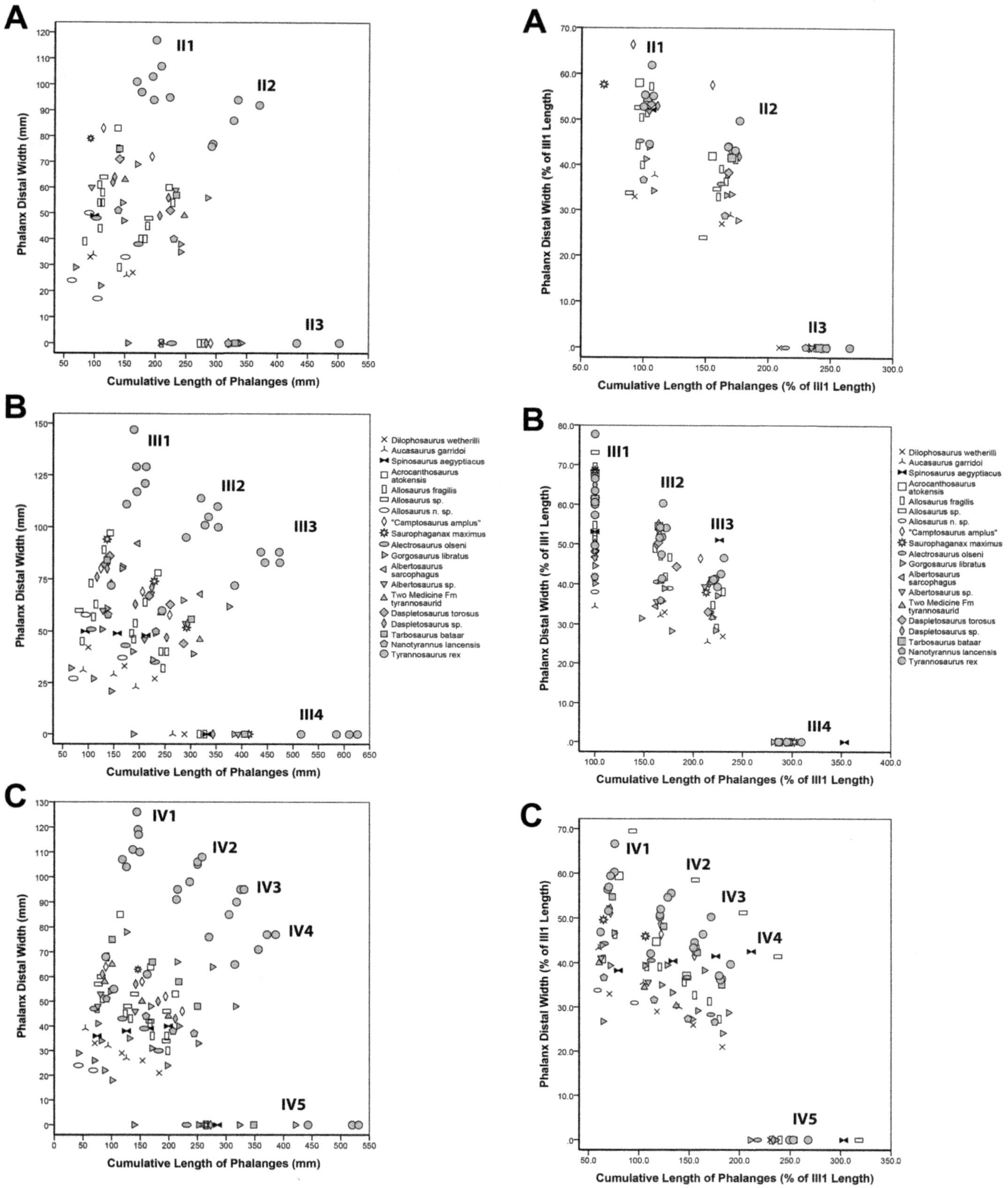
A
Phalanx Distal Width (mm)
Cumulative Length of Phalanges (mm)
II1
II2
II3
A
Phalanx Distal Width (% of III1 Length)
Cumulative Length of Phalanges (% of III1 Length)
B
III1
III2
III3
III4
C
IV1
IV2
IV3
IV4
IV5
Dilophosaurus wetherilli
Aucasaurus garridoi
Spinosaurus aegyptiacus
Acrocanthosaurus atokensis
Allosaurus fragilis
Allosaurus sp.
Allosaurus n. sp.
"Camptosaurus amplus"
Saurophaganax maximus
Alectrosaurus olseni
Gorgosaurus libratus
Albertosaurus sarcophagus
Albertosaurus sp.
Two Medicine Fm tyrannosaurid
Daspletosaurus torosus
Daspletosaurus sp.
Tarbosaurus bataar
Nanotyrannus lancensis
Tyrannosaurus rex

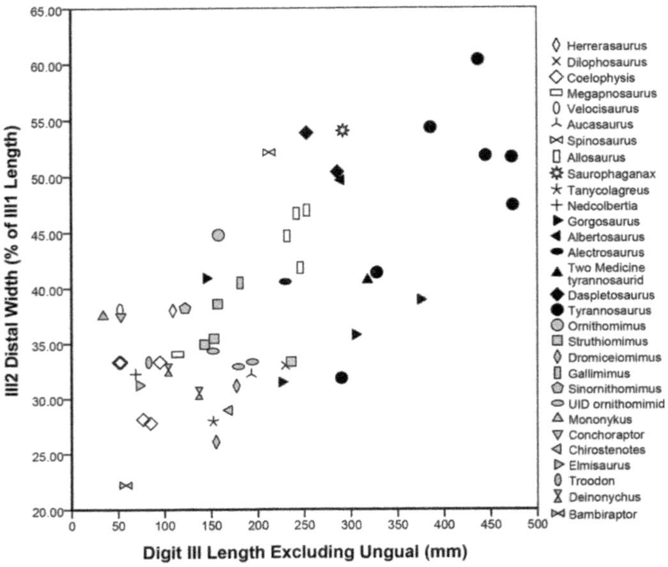

3.3. Phalanx III2 distal width (as a percentage of phalanx III1 length) plotted as a function of digit III length (excluding the ungual) for non-avian theropods of all sizes (cf. fig. 2.20). Note the clear tendency for toes to become stouter with increasing dinosaur size, both across taxa and (at least in *Tyrannosaurus*, particularly if *Nanotyrannus* [the smallest point labeled *Tyrannosaurus* in this graph] is an immature form of the species) possibly within large-bodied species.

previously noted in chapter 2. This is most clearly seen in *Tyrannosaurus*, especially if *Nanotyrannus* is in fact an immature of that taxon. The tendency for large-bodied forms to have stouter toes also means that larger-bodied theropods have steeper TTPs than do smaller forms.

Within taxa there is a considerable spread of values of relative width at each of the joints between phalanges of digits II–IV (again best seen in *Tyrannosaurus*), and also considerable overlap between clades (e.g., allosaurs and tyrannosaurs). TTPs thus seem to be of limited use in discriminating among most taxa of large theropods. *Spinosaurus*, however, remains distinctive. What is particularly evident with the scaled raw plots (fig. 3.2*B*, 3.2*C*) is how much longer the unguals of this dinosaur are, compared with the lengths of the non-ungual portions of the digits (fig. 1.9*A*), than in nearly all other large theropods (note the horizontal distance separating points for the cumulative lengths of the non-ungual phalanges from points for the ungual). In this feature it resembles *Plateosaurus* (fig. 1.10*C*) and ornithischians like *Tenontosaurus* (fig. 1.11*D*).

Scaled Straight-Line Toe-Tapering Profiles and Toe-Tapering Curves

Non-avian Theropods (fig. 3.4). Extending comparisons from large theropods to all non-avian theropods for which all phalanx lengths and distal widths of digits II–IV could

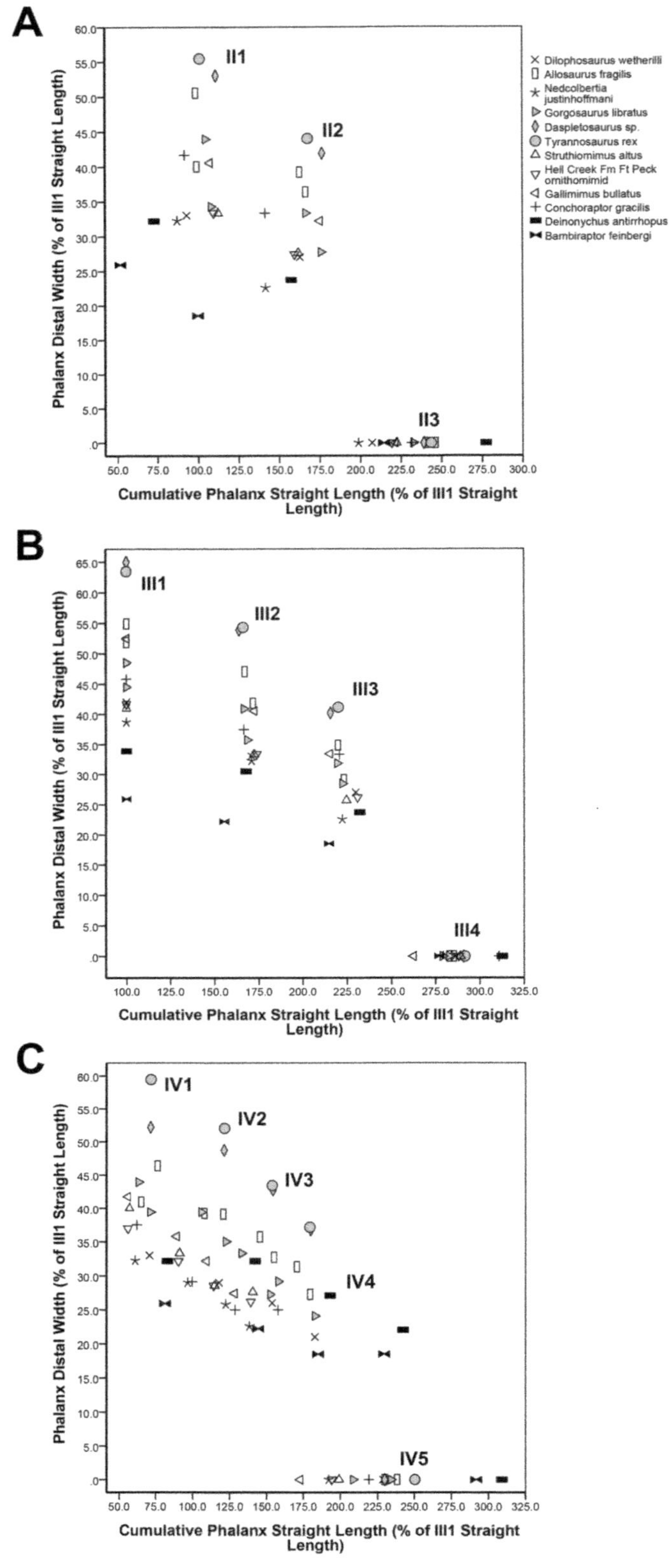

3.4. Scaled straight-line toe-tapering profiles (TTPs) for non-avian theropods of all sizes. *A*, Digit II. *B*, Digit III. *C*, Digit IV. As with scaled raw TTPs, large-bodied theropods (allosaurs, tyrannosaurs) generally have broader digits than those of small-bodied forms. Dromaeosaurids have particularly short scaled non-ungual phalanges on digit II.

A

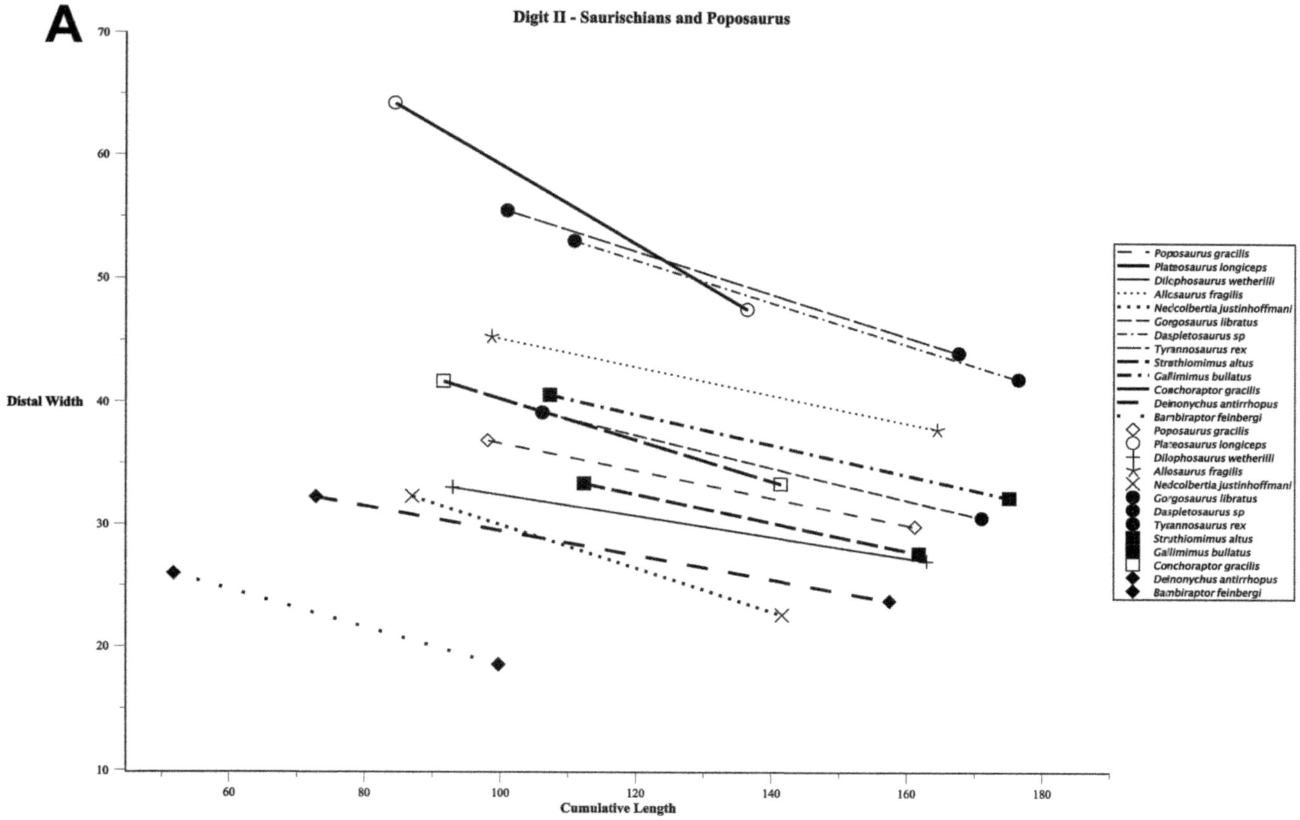

B

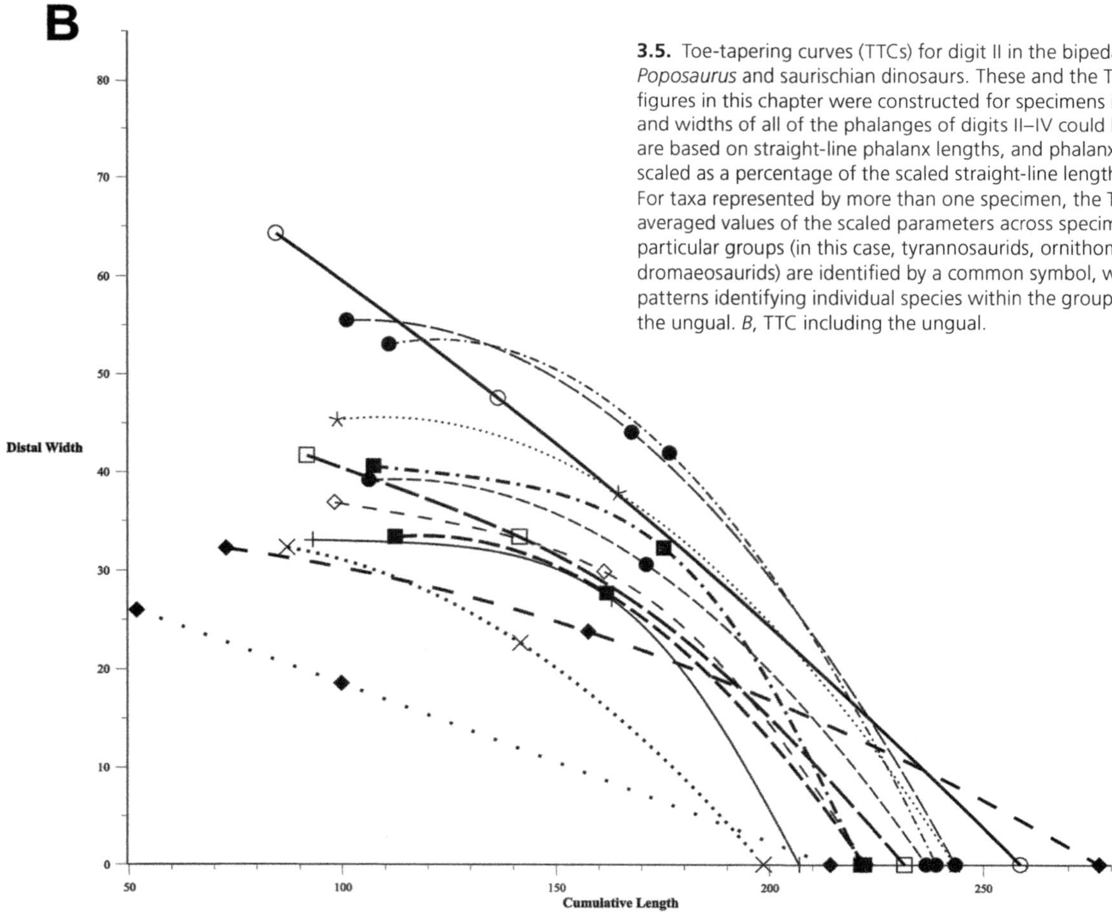

3.5. Toe-tapering curves (TTCs) for digit II in the bipedal suchian *Poposaurus* and saurischian dinosaurs. These and the TTCs in subsequent figures in this chapter were constructed for specimens in which lengths and widths of all of the phalanges of digits II–IV could be measured. TTCs are based on straight-line phalanx lengths, and phalanx distal widths, scaled as a percentage of the scaled straight-line length of phalanx III1. For taxa represented by more than one specimen, the TTC is based on averaged values of the scaled parameters across specimens. Points for particular groups (in this case, tyrannosaurids, ornithomimosaurs, and dromaeosaurids) are identified by a common symbol, with different line patterns identifying individual species within the group. *A*, TTC excluding the ungual. *B*, TTC including the ungual.

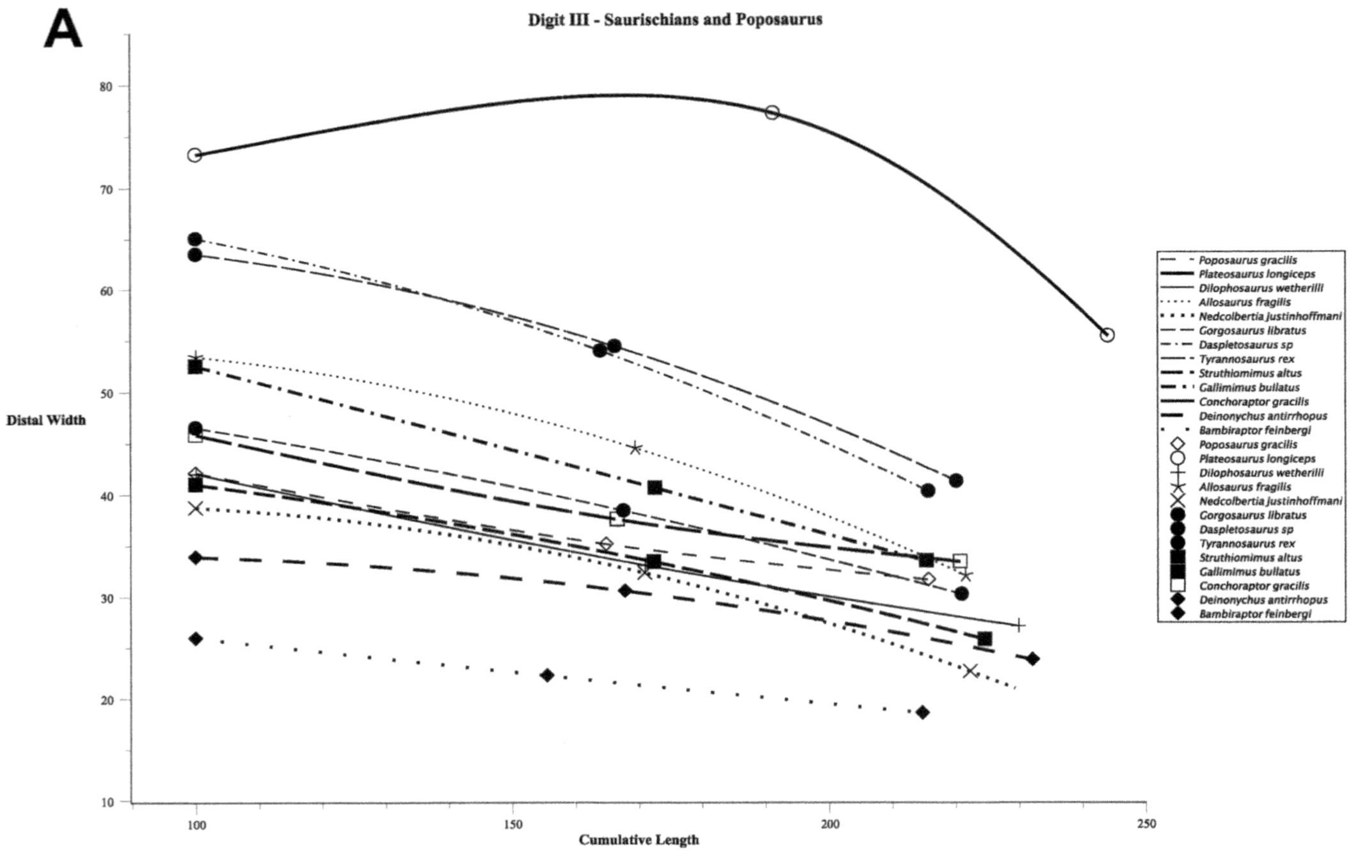

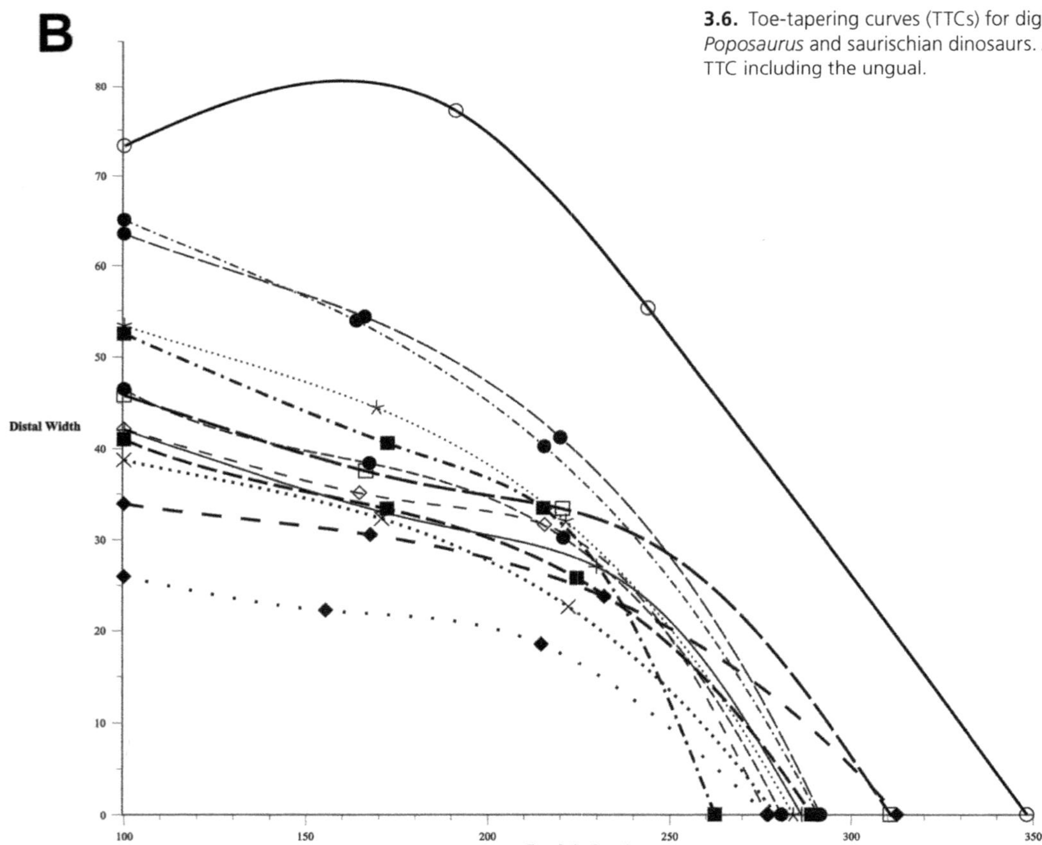

3.6. Toe-tapering curves (TTCs) for digit III in the bipedal suchian *Poposaurus* and saurischian dinosaurs. *A*, TTC excluding the ungual. *B*, TTC including the ungual.

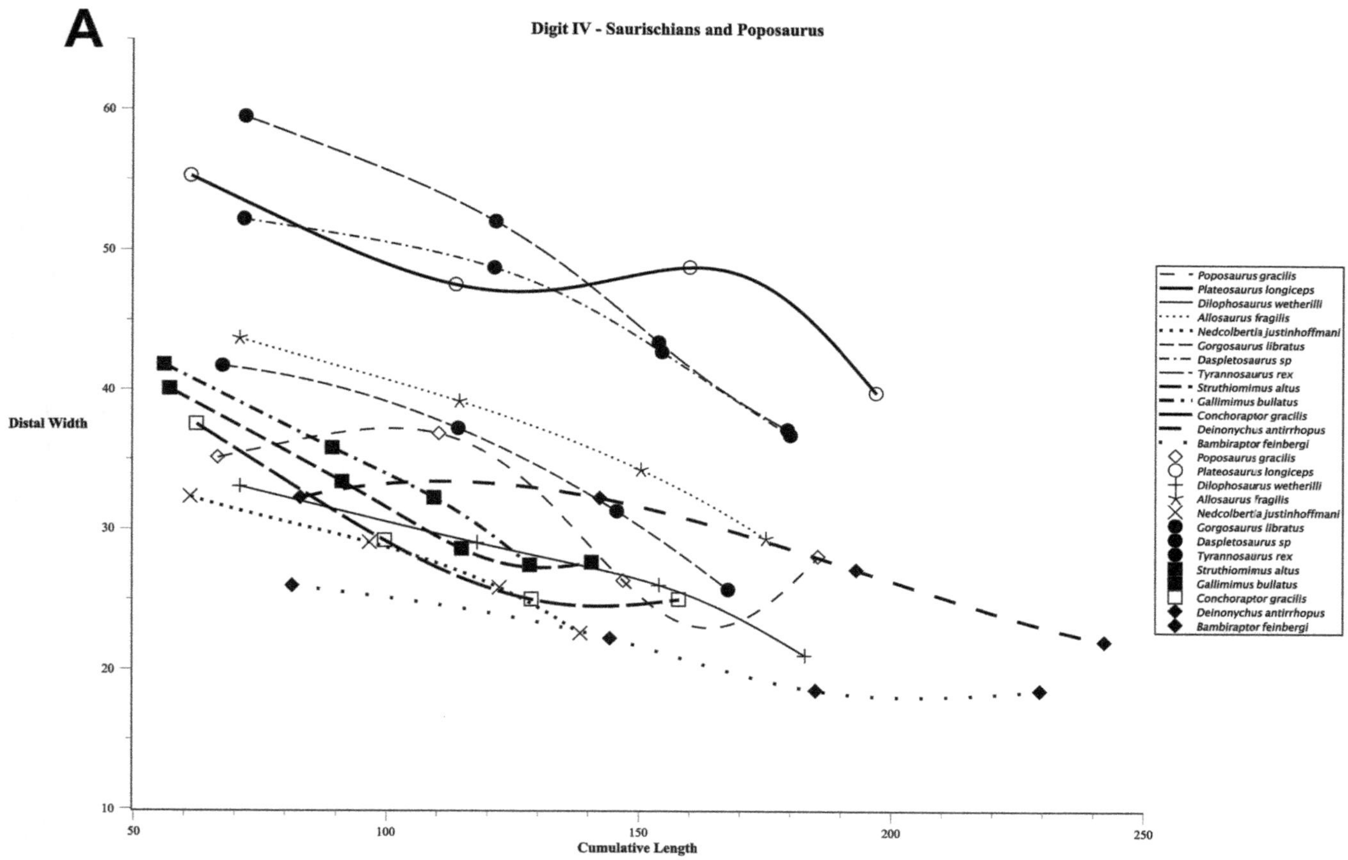

3.7. Toe-tapering curves (TTCs) for digit IV in the bipedal suchian *Poposaurus* and saurischian dinosaurs. *A*, TTC excluding the ungual. *B*, TTC including the ungual.

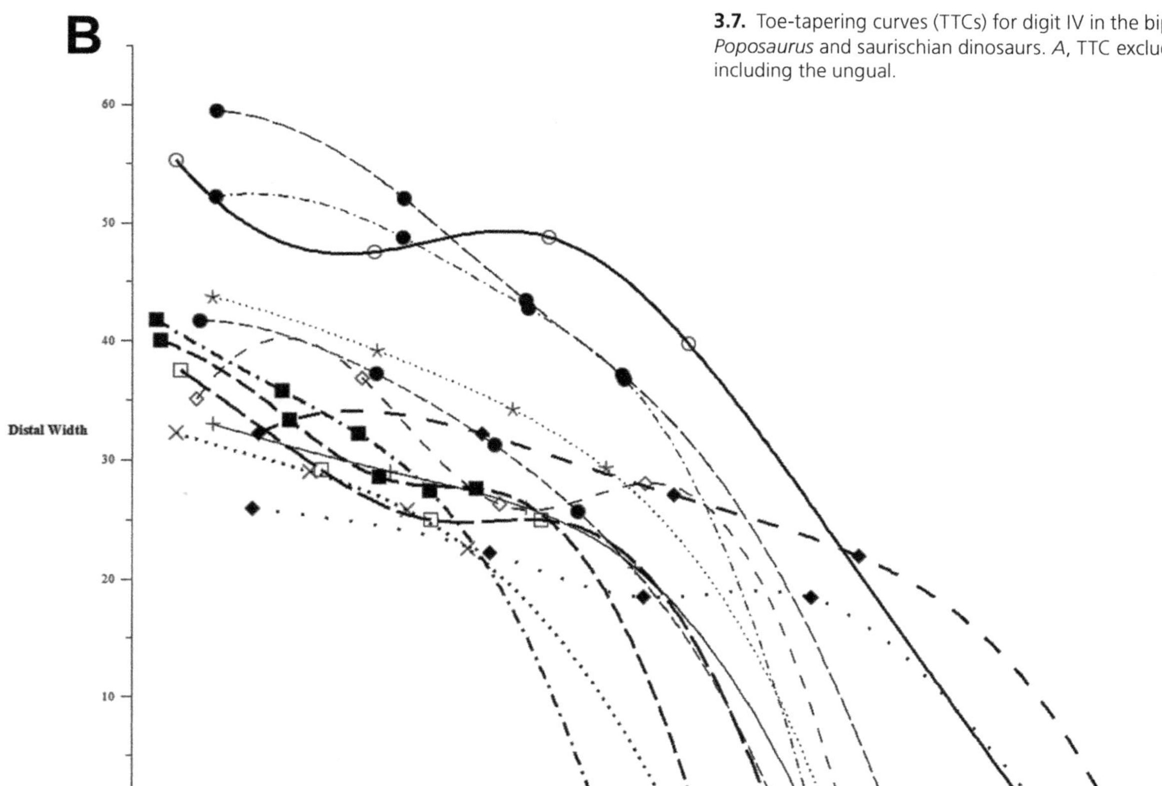

be made reveals few surprises. As with raw TTPs, straight-line TTPs of large-bodied theropods like tyrannosaurs and allosaurs are generally steeper than TTPs of small-bodied theropods like dromaeosaurids and oviraptorosaurs; the maximum distal width of the first phalanx of each toe, as seen in large forms, is about 65% of its straight-line length, and values of scaled distal width decline from there. Rays marking the location of distal ends of non-ungual phalanges along the lengths of the digits are reasonably distinct for digits II and III, but become smeared out for digit IV. Dromaeosaurids (*Deinonychus*, *Bambiraptor*), like *Spinosaurus* (which is not included in these analyses because one of its pedal phalanges is missing), have relatively long unguals on digit II. Dromaeosaurids are also distinctive in having relatively short non-ungual phalanges on digit II, and long non-ungual phalanges on digit IV. In most non-avian theropods the total straight-line relative cumulative lengths of digits II and IV are about 200%–250% of the length of phalanx III1, and digit III is roughly 275%–300% of the length of phalanx III1. In *Deinonychus*, however, all three digits are at least 275% of the length of phalanx III1, and digit IV in *Deinonychus*, and also *Bambiraptor*, has a substantially longer relative cumulative length than that of other non-avian theropods.

Toe-tapering curves (TTCs) for digit II of non-avian theropods that do not include the ungual (fig. 3.5A) obviously have to be straight lines, because only two phalanges are involved in their creation. Lines for most of the non-avian theropods, and also *Poposaurus*, are nearly parallel, with larger forms having relatively broader toes than smaller forms. *Plateosaurus* has a steeper curve than the theropods. With TTCs that include the ungual (fig. 3.5*B*), many species of medium-sized to large theropods (*Dilophosaurus wetherilli*, *Allosaurus fragilis*, ornithomimids, *Gorgosaurus libratus*, and also the bipedal suchian *Poposaurus gracilis*) continue to have very similar shapes. In contrast, TTCs of the two dromaeosaurids (*Bambiraptor feinbergi*, *Deinonychus antirrhopus*) are much flatter, nearly linear, and indicate relatively slender toes, very short first phalanges and—unsurprisingly for digit II of these dinosaurs—very long unguals. *Daspletosaurus* sp. and *Tyrannosaurus rex* have concave-down profiles like most of the other theropods, but because these dinosaurs have relatively stouter non-ungual phalanges their TTCs are steeper. The prosauropod *Plateosaurus longiceps* has a linear and steeply declining profile, and a very long ungual (fig. 1.10C).

Digit III TTCs tell much the same story. Once again, TTCs created without the ungual are linear (fig. 3.6A) or nearly so for the theropods and *Poposaurus*, with larger forms having stouter toes (cf. fig. 3.3). Proportional lengths of the non-ungual phalanges of the dromaeosaurids do not differ so much from those of other theropods as do those of digit II, and cumulative scaled lengths of digit III of the theropods and *Poposaurus* are much the same. *Plateosaurus*, in contrast, has a concave-down profile, stouter toes than the theropods and *Poposaurus*, and relatively longer second and third phalanges.

With TTCs that include the ungual (fig. 3.6*B*), the dromaeosaurids again have flatter TTCs and narrower toes than the other theropods and *Poposaurus*, but (unlike with digit II) without a huge ungual at the end of the toe. *Daspletosaurus* and *Tyrannosaurus* again show steeply declining, concave-down TTCs. *Plateosaurus* again shows a very stout digit that actually increases slightly in distal width from the first to the second phalanx, and once again a relatively long ungual.

Digit IV TTCs, unfortunately, are rather chaotic, whether the ungual is excluded or included in their creation (fig. 3.7). The two dromaeosaurids have a relatively much longer digit IV than do the other theropods, comparable to that of *Plateosaurus*. The ornithomimosaurs and *Nedcolbertia justinhoffmani* (the latter possibly an ornithomimosaur; Brownstein 2017), in contrast, have a relatively short digit IV. *Daspletosaurus* and *Tyrannosaurus* again have steeper TTCs than the other theropods. The more proximal phalanges of digit IV of *Plateosaurus* are comparable in relative distal width to those of *Daspletosaurus* and *Tyrannosaurus*, but the very long ungual of digit IV, as with digits II and III, gives the prosauropod a longer cumulative relative length of this toe compared with those of the tyrannosaurs.

Struthioniforms. Straight-line TTPs for extant ratites (fig. 3.8) are messier than those for non-avian theropods; distinct rays marking the positions of the distal ends of non-ungual phalanges are harder to recognize, especially for digit IV. The ostrich (*Struthio*) stands out from other ratites in having lost digit II, of course, but also in having a relatively broad and short digit III (in comparison with the length of phalanx III1), as well as a relatively narrow and short digit IV (in fact, in none of the ostrich foot skeletons we examined was a fifth phalanx present, which is why figure 3.8C shows digit IV terminating well above the zero value for the distal width of phalanx IV5 that marks the end point of this toe in other species). Digit IV of the ostrich is also unusual in increasing in distal width from the distal end of the first phalanx to the distal end of the second phalanx.

Maximum relative distal widths of the first phalanx of digits II–IV of extant ratites other than ostriches are typically 25%–45% of the length of phalanx III1, putting them in the same general range as the more slender-toed non-avian theropods. Digits II–IV are generally 150%–275% of the length of phalanx III1, while digit III is 225%–350% of the length of phalanx III1.

Cassowaries (*Casuarius* spp.) have relatively short digits. This might seem surprising for digit II, given the long ungual

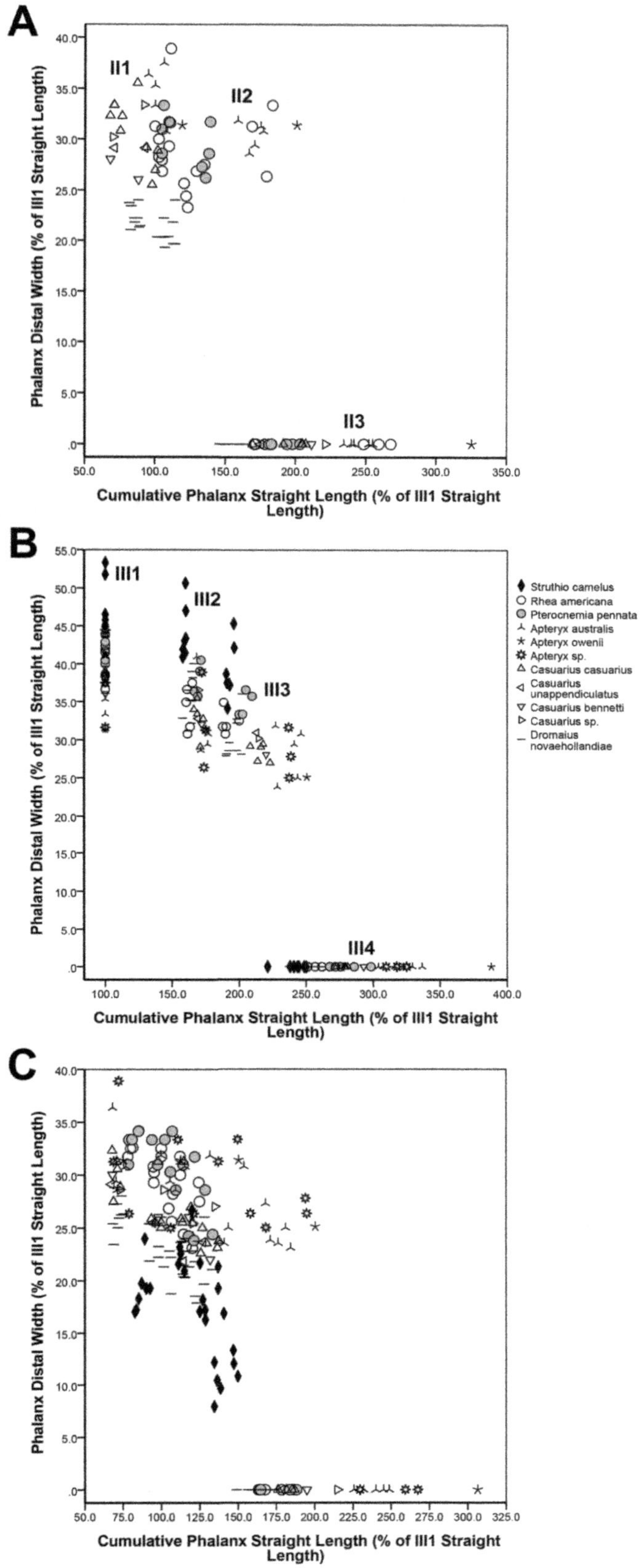

3.8. Scaled straight-line toe-tapering profiles (TTPs) for extant ratites (struthioniforms). *A*, Digit II. *B*, Digit III. *C*, Digit IV.

II3, but cassowaries have particularly short phalanges II1 and II2, compared with the length of phalanx III1. Emus (*Dromaius novaehollandiae*) likewise have relatively short toes that are also fairly narrow. Kiwi (*Apteryx* spp.), in contrast, have relatively long toes compared with the length of phalanx III1. Rheas (*Rhea americana*, *Pterocnemia pennata*), like kiwi, have a relatively long digit II, but unlike kiwi have a relatively short digit IV, and also a shorter digit III than in kiwi. Thus the various groups of extant ratites do differ among themselves in their scaled TTPs, which may therefore have some utility in distinguishing among these bird species.

TTCs of extant ratites are less messy than their corresponding TTPs (probably because they are based on species means, and so plot fewer individual cases). For digit II, the TTCs excluding the ungual again perforce are straight lines (fig. 3.9A). The relatively short first and second phalanges of the three cassowary species, and the very long second phalanges of the kiwi, set them apart at opposite ends from the other ratites. Unlike with non-avian theropods, there is no clear tendency for bigger-bodied species to have stouter toes. One kiwi and one cassowary species show flat TTCs, unlike the other birds in the sample.

Digit II TTCs including the unguals (fig. 3.9*B*) produce nice concave-down profiles for ratites. The three species of *Casuarius* all show relatively short first and second phalanges and (like dromaeosaurids) relatively long unguals. Emus and rheas have TTCs of similar shape, but emus have narrower toes than do rheas, and a shorter cumulative length, due to having a relatively shorter first phalanx than that of rheas. The two kiwi have relatively long non-ungual phalanges and unguals, giving them a relatively long digit II cumulative length.

TTCs for digit III of struthioniforms that exclude the ungual have linear to slightly concave shapes (fig. 3.10A). Kiwi and cassowaries have a relatively slim digit III, while emus, rheas, and especially ostriches have a stouter digit III. Kiwi and cassowaries have a relatively long third phalanx. The main difference in TTCs for digit III that include the ungual (fig. 3.10*B*) involves the relative length of the ungual. The ostrich has a very short ungual, kiwi have very long unguals, and the other ratites are somewhere in between these extremes.

As with non-avian theropods, TTCs for digit IV of struthioniforms do not consistently show nice, well-behaved, concave-down shapes (fig. 3.11). *Struthio* has a relatively long first phalanx, digit IV increases in relative distal width between the first and second phalanx, and digit IV of this species terminates without an ungual. The three *Casuarius* species show complicated patterns of distal width from one phalanx to the next. Kiwi, again, have relatively long phalanges and a very long digit.

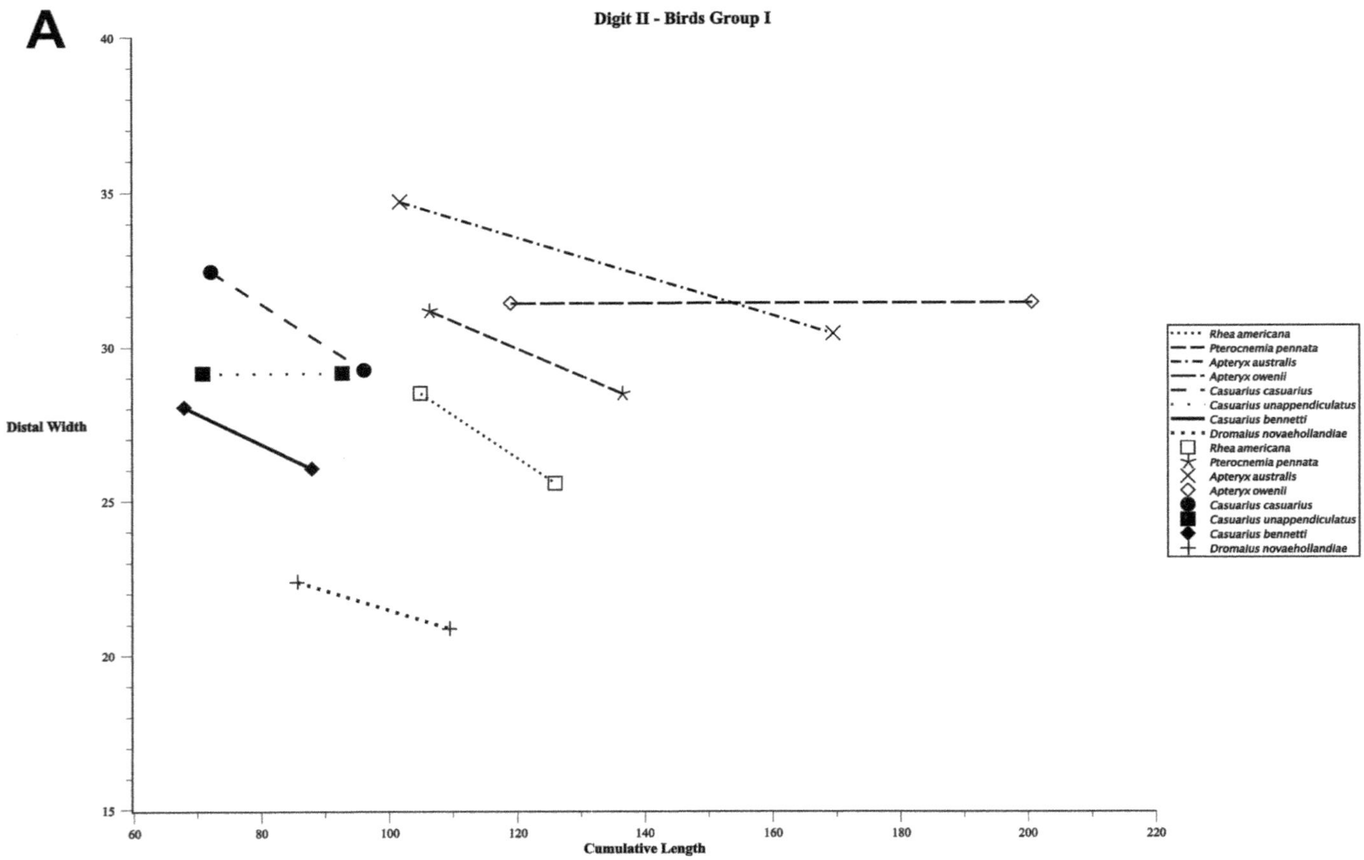

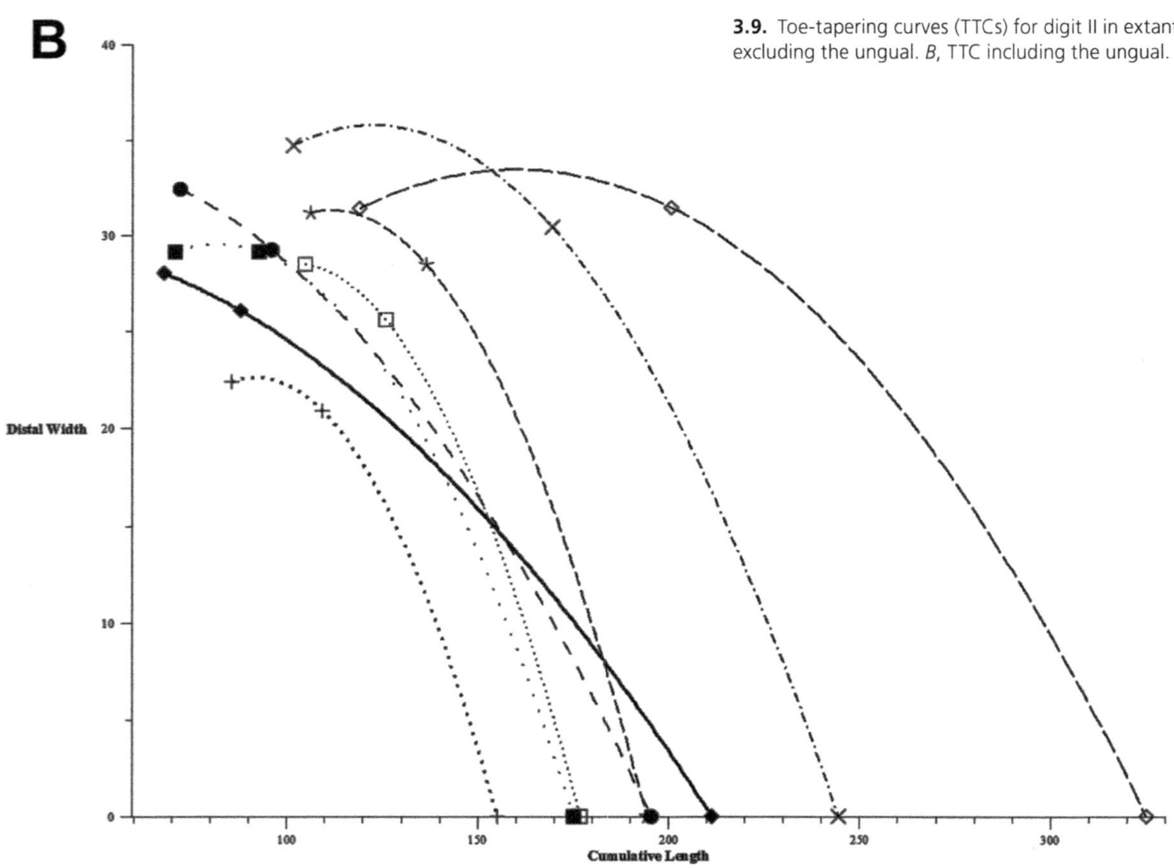

3.9. Toe-tapering curves (TTCs) for digit II in extant ratites. *A*, TTC excluding the ungual. *B*, TTC including the ungual.

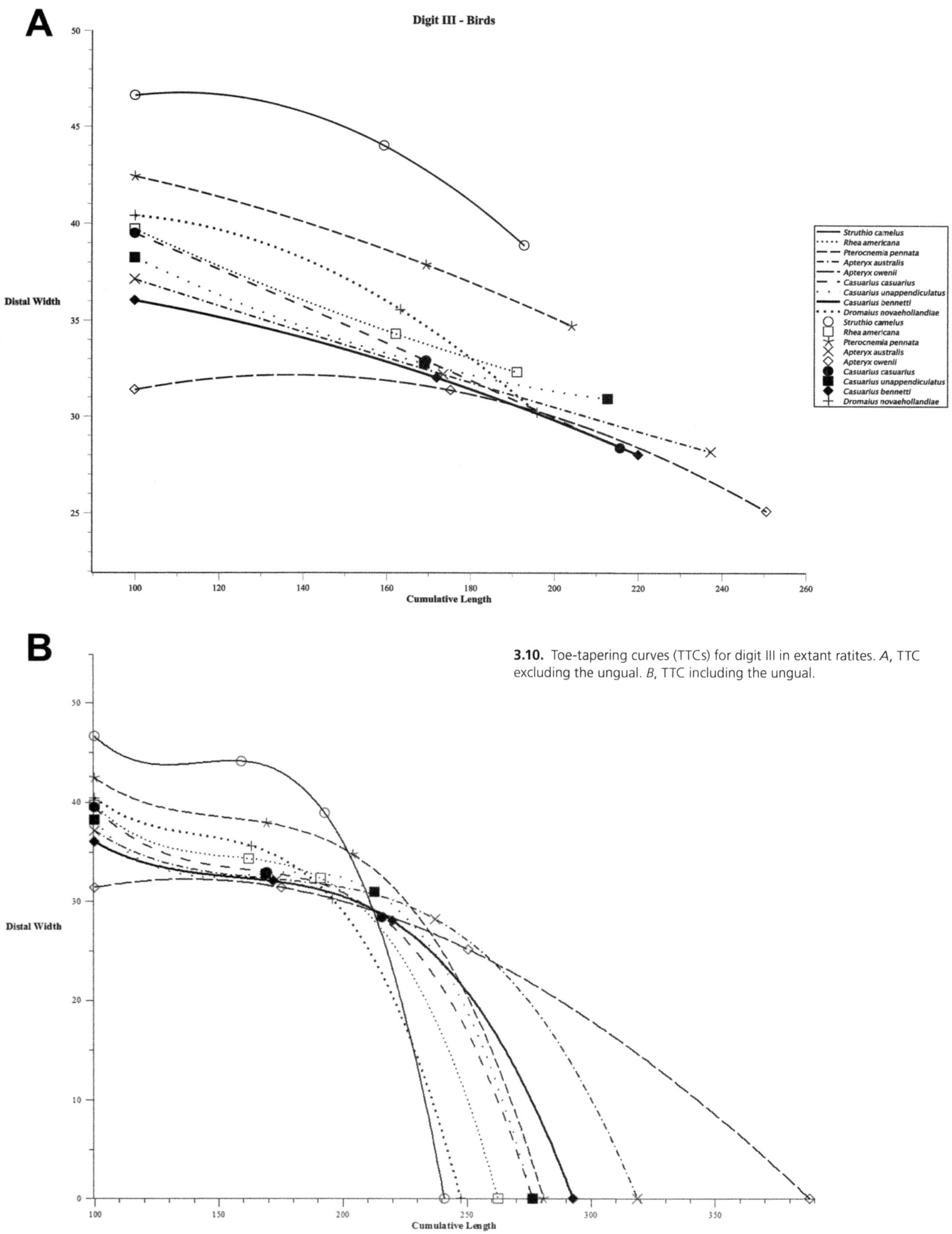

3.10. Toe-tapering curves (TTCs) for digit III in extant ratites. *A*, TTC excluding the ungual. *B*, TTC including the ungual.

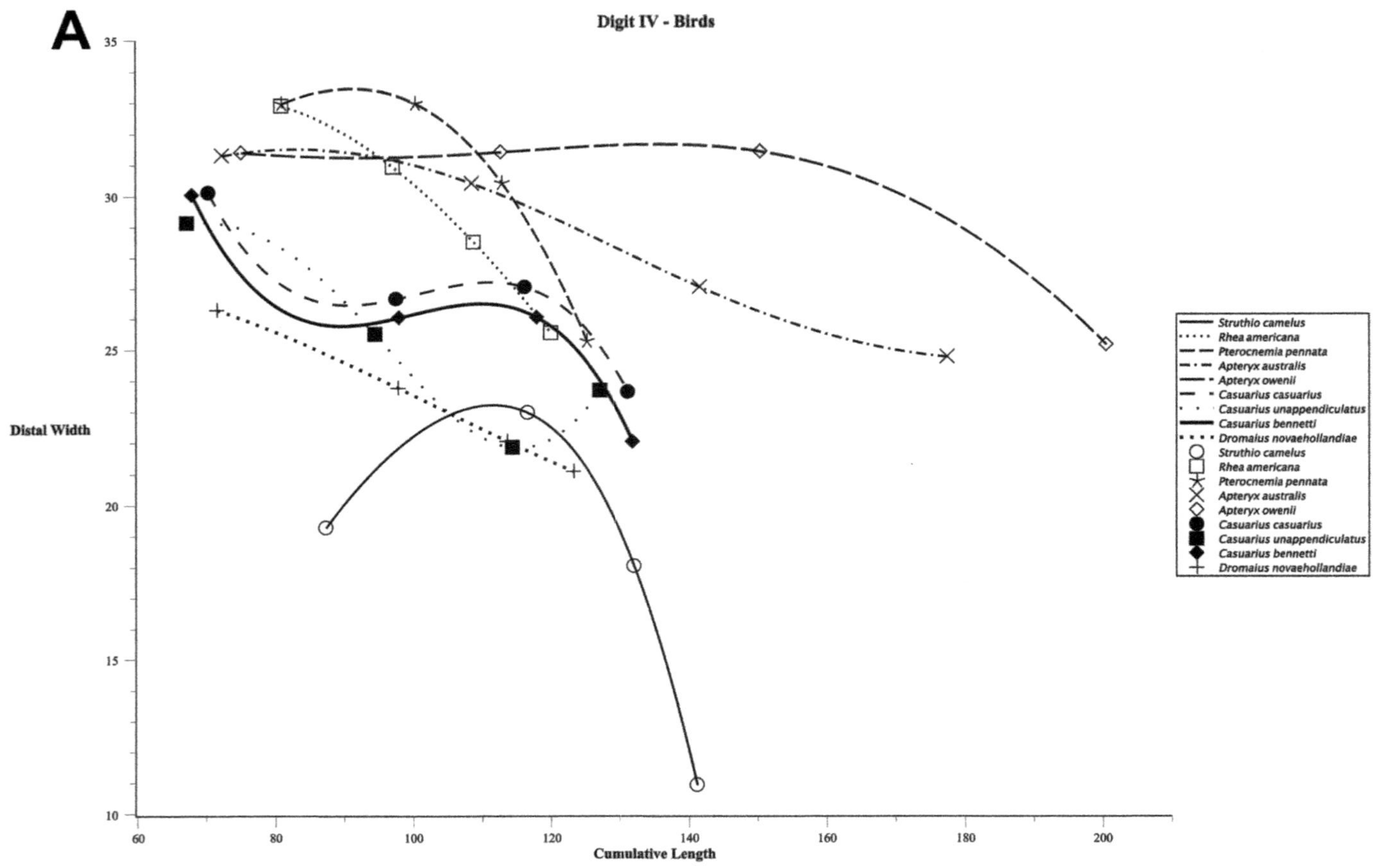

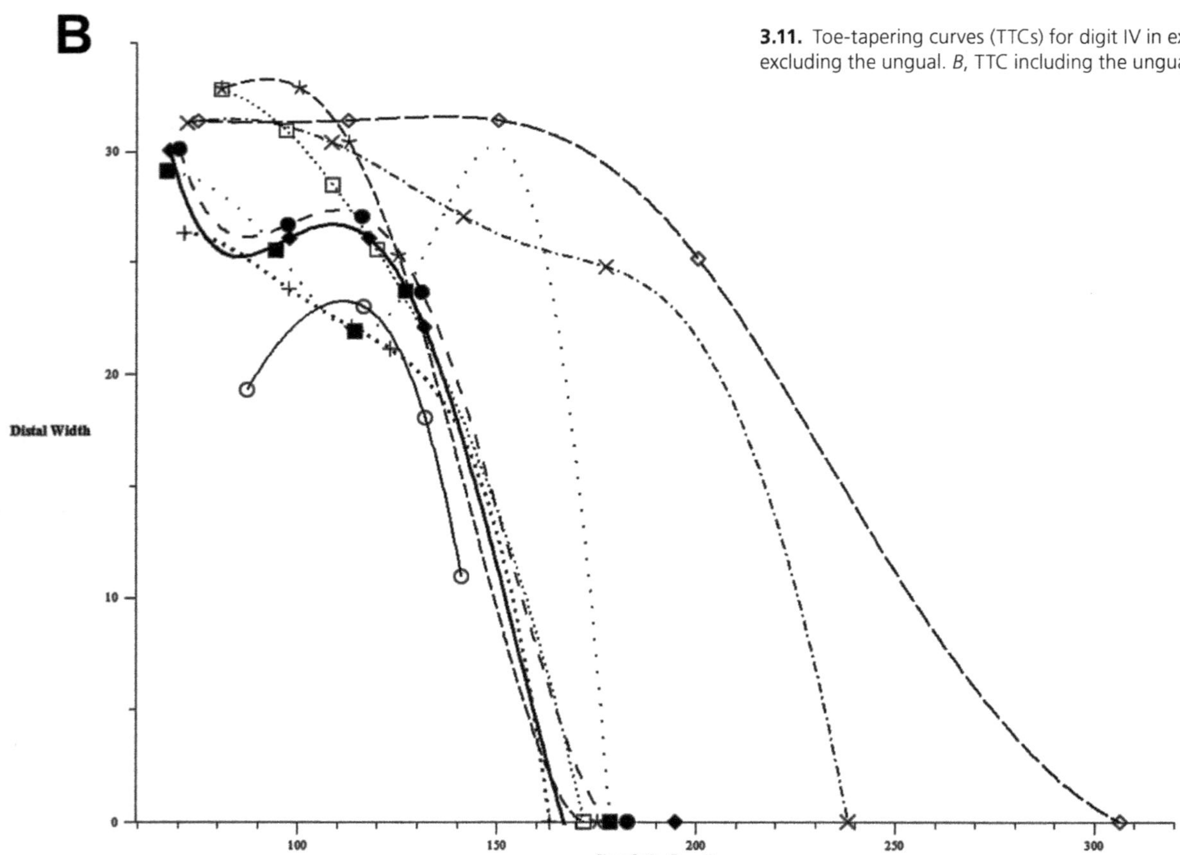

3.11. Toe-tapering curves (TTCs) for digit IV in extant ratites. *A*, TTC excluding the ungual. *B*, TTC including the ungual.

Dinornithiforms. Scaled straight-line TTPs of moa show discrete aggregations of points for the distal ends of the phalanges of digits II and III, but as with struthioniforms, digit IV is much messier (fig. 3.12). Digits II and IV of all species have cumulative scaled lengths ranging about 150%–250% of the length of phalanx III1, and cumulative scaled lengths of digit III of 225%–325% of the length of phalanx III1, much as in struthioniforms and non-avian theropods. Maximum scaled widths of the distal ends of the first phalanx of the three digits range 35%–60% of the length of phalanx III1, and so some moa are broader-toed than most struthioniforms (figs. 2.12, 2.13, 3.8–3.11). Most of the moa species have remarkably similar TTCs, differing mainly (and only moderately) in relative toe widths and digit cumulative lengths (figs. 3.13–3.15). Although there is considerable overlap of points in TTPs across moa species, there are some species that stand apart from others. *Megalapteryx didinus* and the two species of *Dinornis* are rather narrow-toed species, while *Pachyornis elephantopus*, *Euryapteryx curtus*, and *Emeus crassus* are stouter-toed forms. *Megalapteryx didinus* and *Anomalopteryx didiformis* have relatively longer phalanges distal to the first phalanx of all three digits than do other moa; *Megalapteryx* has the relatively longest toes of the moa. Thus TTPs and TTCs have some, albeit limited, ability to discriminate among moa species.

As with non-avian dinosaurs, there is a tendency for digits to become relatively stouter with increasing digit length across the combined sample of struthioniforms, dinornithiforms, and other big ground birds in our sample (figs. 2.12, 2.13, 3.16). *Dinornis* spp., however, depart from this trend, with a nearly flat-lined relationship between relative width of digit III as a function of digit III length excluding the ungual.

Ornithischians. Scaled straight-line TTPs for bipedal or potentially bipedal ornithischians, like those for extant ratites, are rather messy (fig. 3.17), indicating variability among groups in relative phalangeal proportions. *Thescelosaurus*, *Tenontosaurus*, and *Leptoceratops* have relatively long non-ungual phalanges, especially long unguals, and very long cumulative digit relative lengths (digit II ca. 275%–375%, digit III 350%–475%, and digit IV 275%–450% of the length phalanx III1). Large iguanodonts and hadrosaurs, in contrast, have relatively shorter phalanges, with cumulative digit relative lengths 275% or less (digits II and IV) and 300% or less (digit III) of the length of phalanx III1, giving these big herbivores very steep TPPs compared with their smaller kin.

The TTCs are somewhat easier to interpret (figs. 3.18–3.20). *Mantellisaurus atherfieldensis*, *Iguanodon bernissartensis*, and the four hadrosaur species all show fairly steep TTCs and relatively short total digit relative lengths, while *Tenontosaurus tilletti*, *Leptoceratops gracilis*, and to a lesser extent

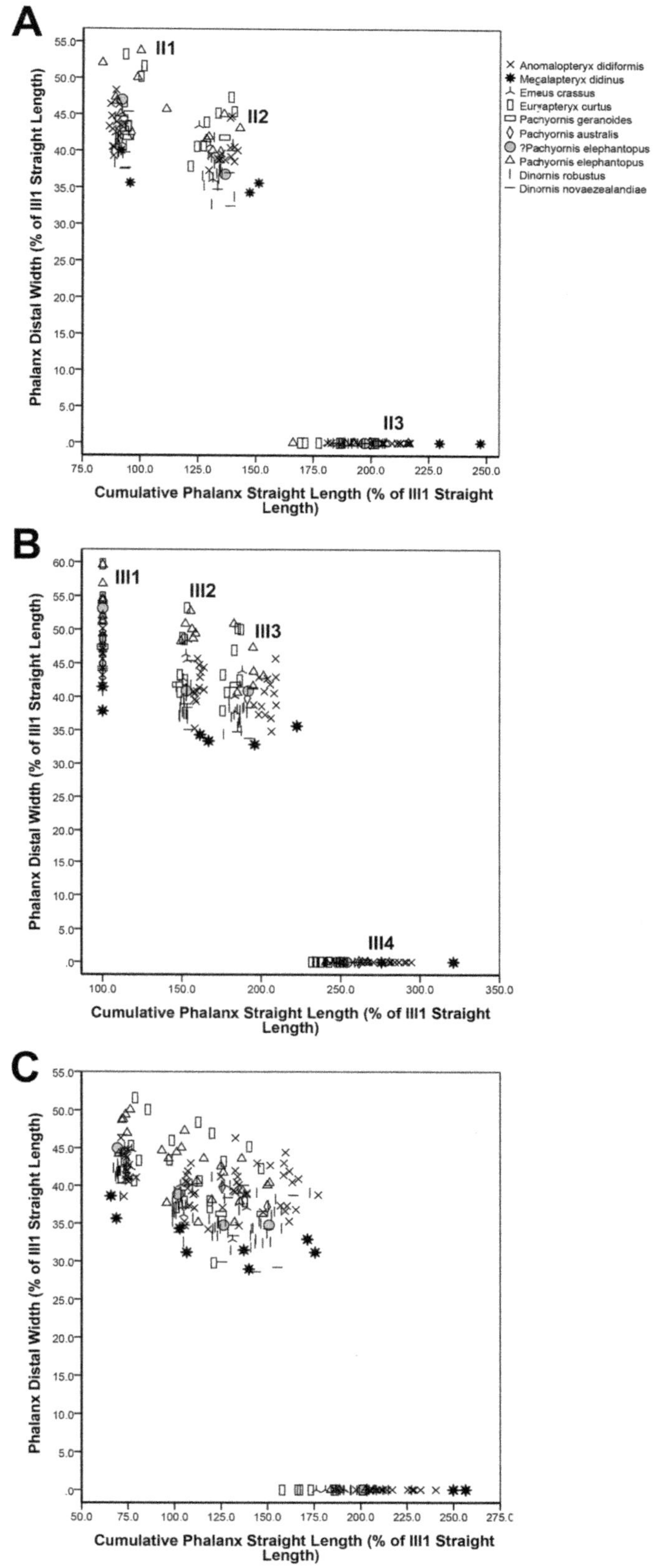

3.12. Scaled straight-line toe-tapering profiles (TTPs) for dinornithiforms (moa). *A*, Digit II. *B*, Digit III. *C*, Digit IV.

A

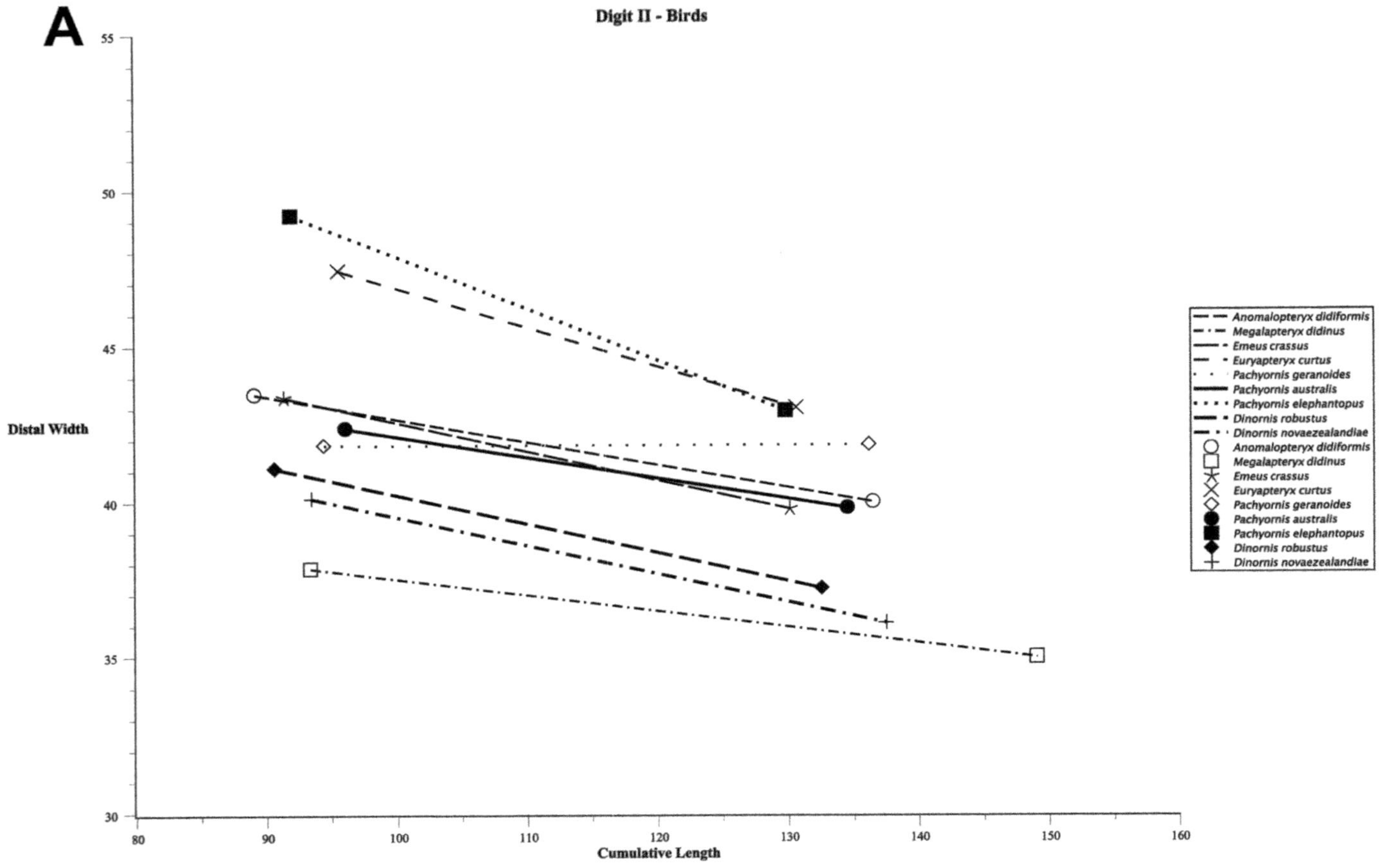

B

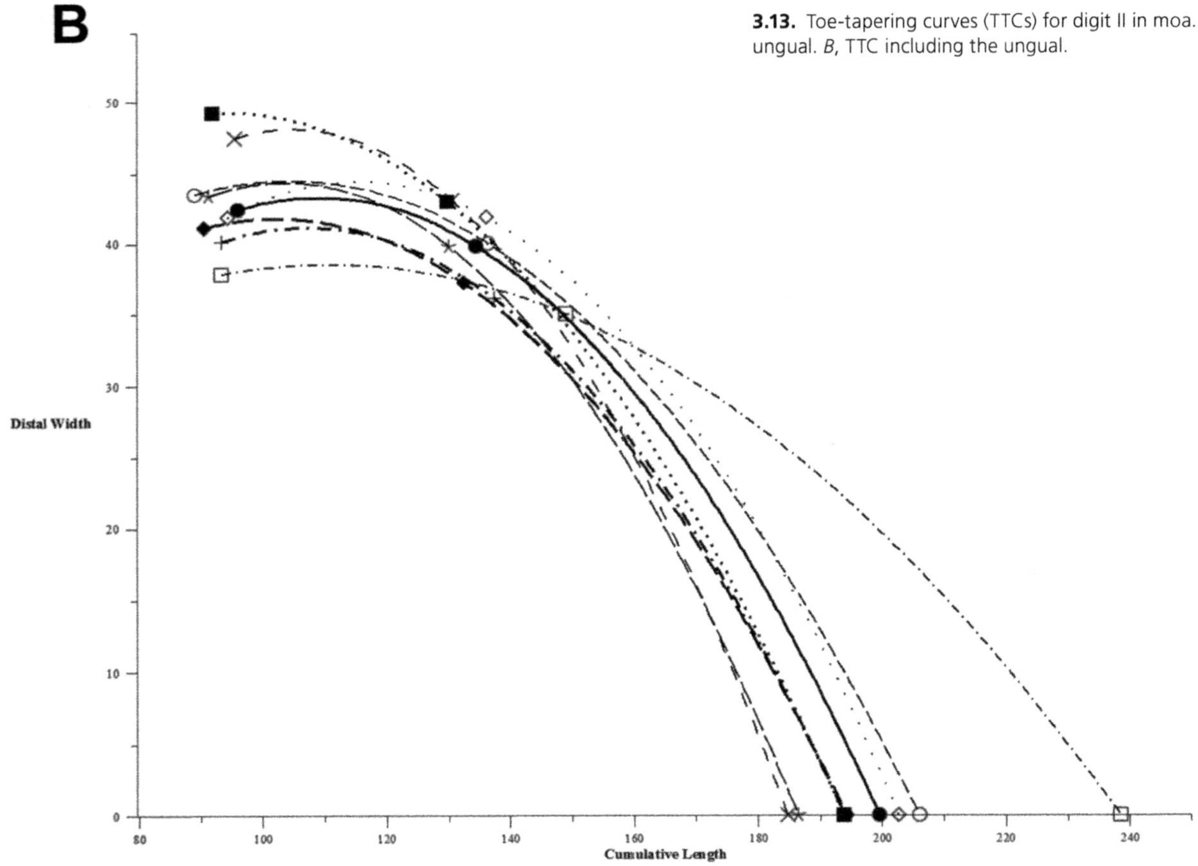

3.13. Toe-tapering curves (TTCs) for digit II in moa. *A*, TTC excluding the ungual. *B*, TTC including the ungual.

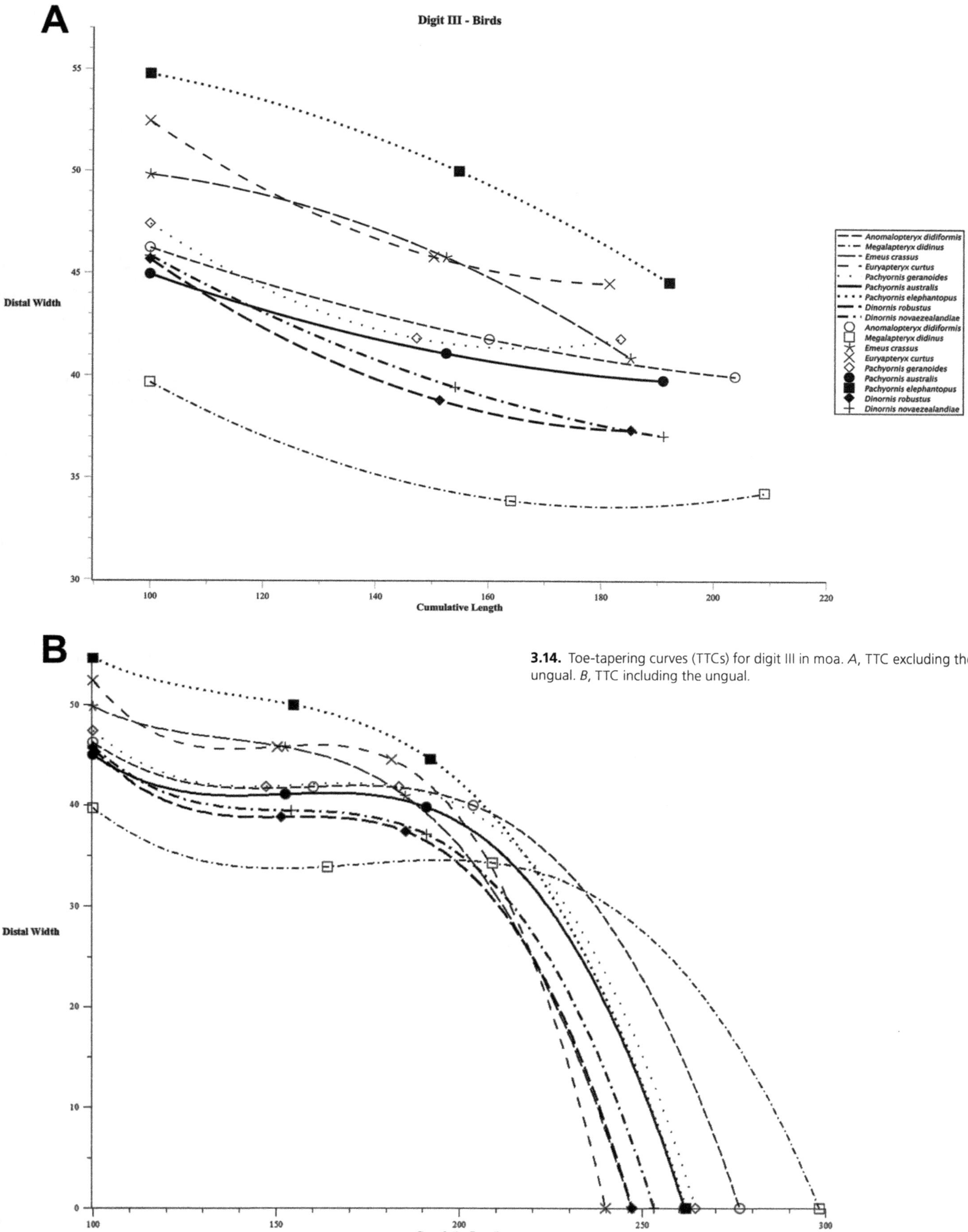

3.14. Toe-tapering curves (TTCs) for digit III in moa. *A*, TTC excluding the ungual. *B*, TTC including the ungual.

A

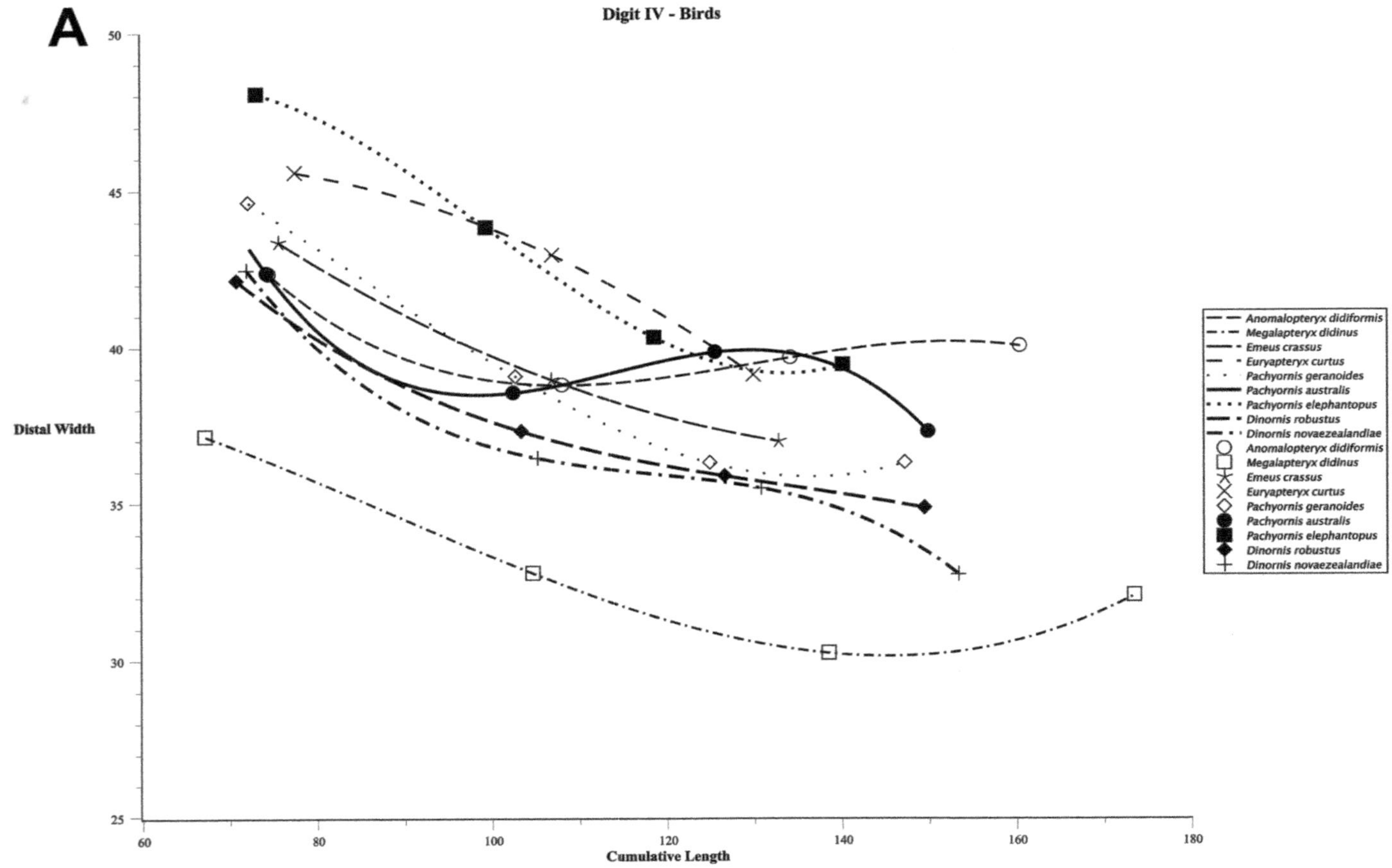

B

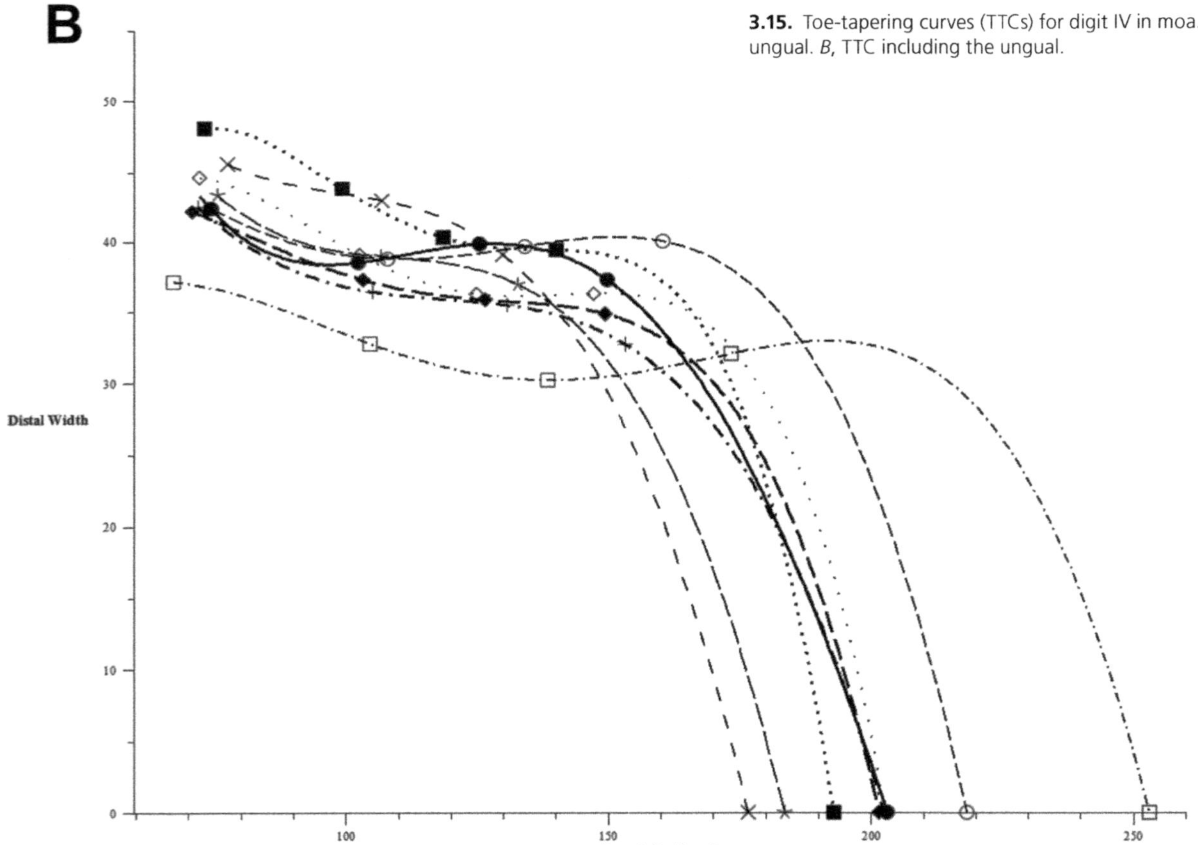

3.15. Toe-tapering curves (TTCs) for digit IV in moa. *A*, TTC excluding the ungual. *B*, TTC including the ungual.

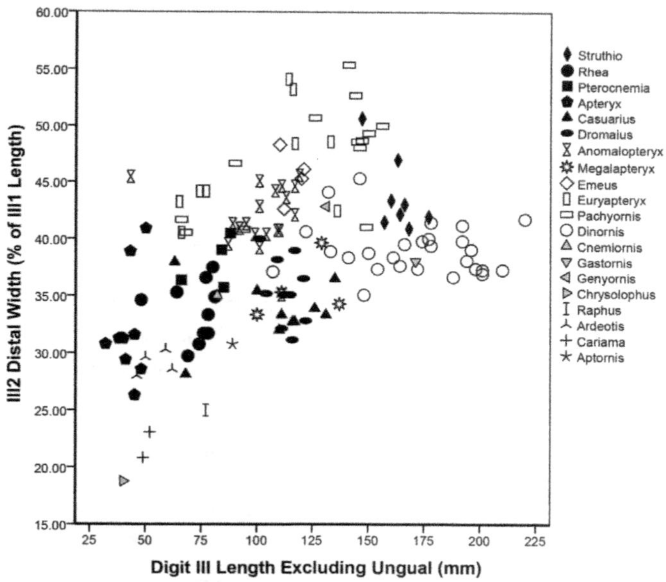

3.16. Phalanx III2 distal width (as a percentage of phalanx III1 length) as a function of digit III length (excluding the ungual) in ground birds. As with non-avian theropods, there is a tendency for toes to become stouter with increasing size across clades, although moa of the genus *Dinornis* have relatively narrow toes for birds so large (cf. figs. 2.12, 2.13).

Thescelosaurus neglectus have flatter TTCs, indicating relatively long non-ungual phalanges that do not decrease much in length distally, and very long unguals. *Camptosaurus* sp. has TTCs for digits II and IV that are more like those for the big ornithopods, while the digit III TTC for *Camptosaurus* is more like that of *Tentontosaurus* and *Leptoceratops*, although digit III of *Camptosaurus* is relatively narrower.

Like ground birds and non-avian theropods, ornithischians show an increase in relative digit width with increasing size (figs. 2.22, 2.23, 3.21). *Tenontosaurus* is consistently one of the thicker-toed forms, and like ornithischians as a group and *Tyrannosaurus* (fig. 3.3), shows a marked (at least in part ontogenetic?) increase in digit stoutness with increasing animal size. Some of the other genera (e.g., *Camptosaurus* and *Edmontosaurus*) show a fair amount of scatter in relative digit width in animals of comparable toe length.

DISCUSSION

Our examination of TTPs and TTCs amplifies the results of analyses of phalangeal proportions of birds and dinosaurs in previous chapters. Several conclusions stand out.

First, it is interesting that in all groups, whether non-avian theropods, ground birds, or bipedal or potentially bipedal ornithischians, there is a strong tendency for larger animals to have relatively stouter toes (as already noted in chap. 2, and which we will see again in chap. 4). This is seen across

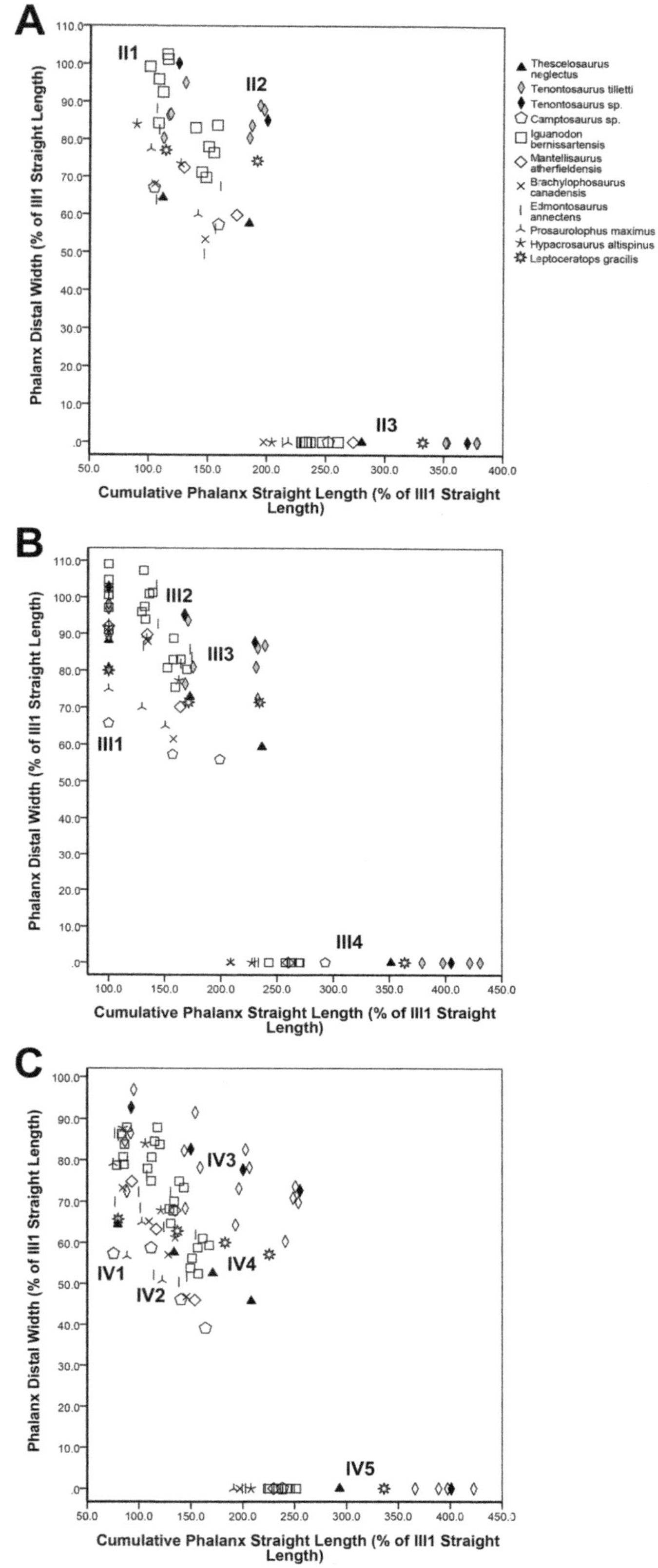

3.17. Scaled straight-line toe-tapering profiles (TTPs) for ornithischians. *A*, Digit II. *B*, Digit III. *C*, Digit IV.

A

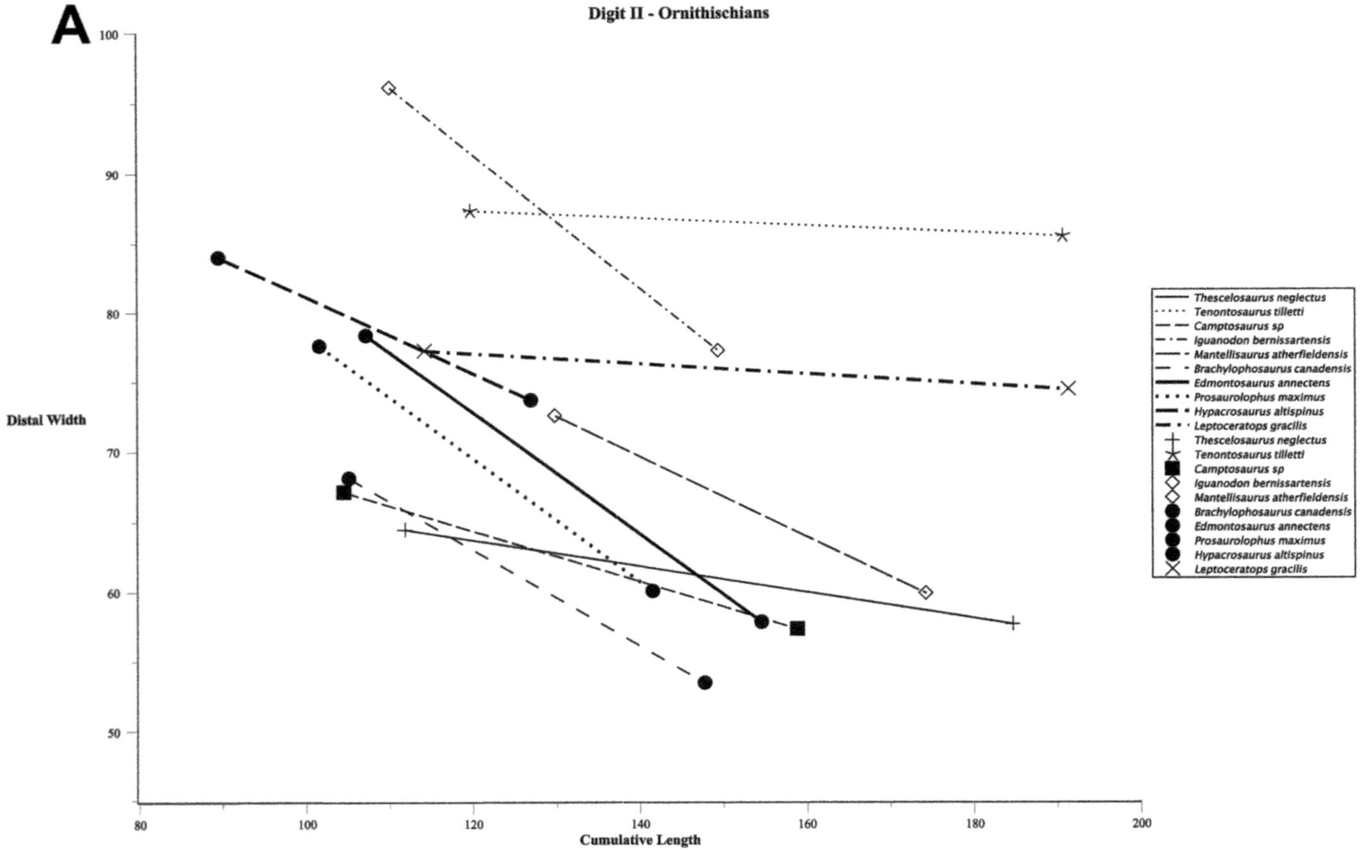

B

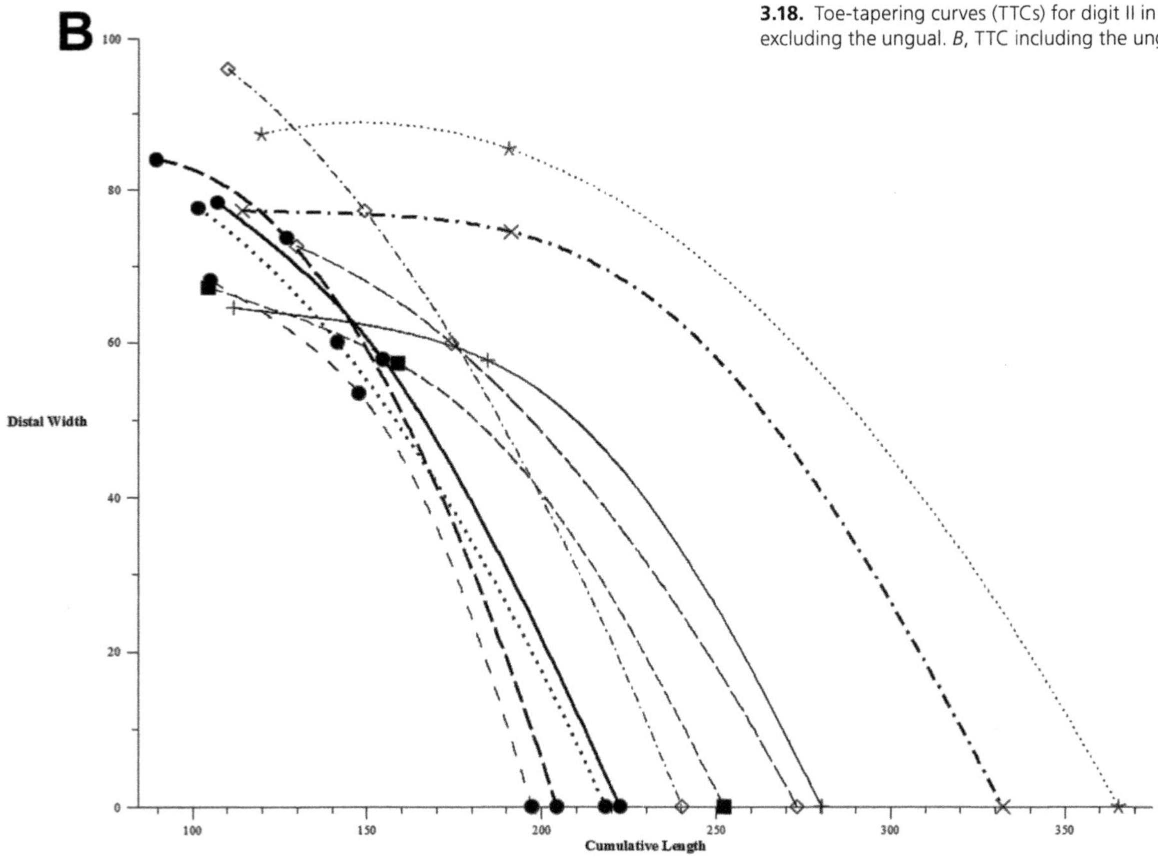

3.18. Toe-tapering curves (TTCs) for digit II in ornithischians. *A*, TTC excluding the ungual. *B*, TTC including the ungual.

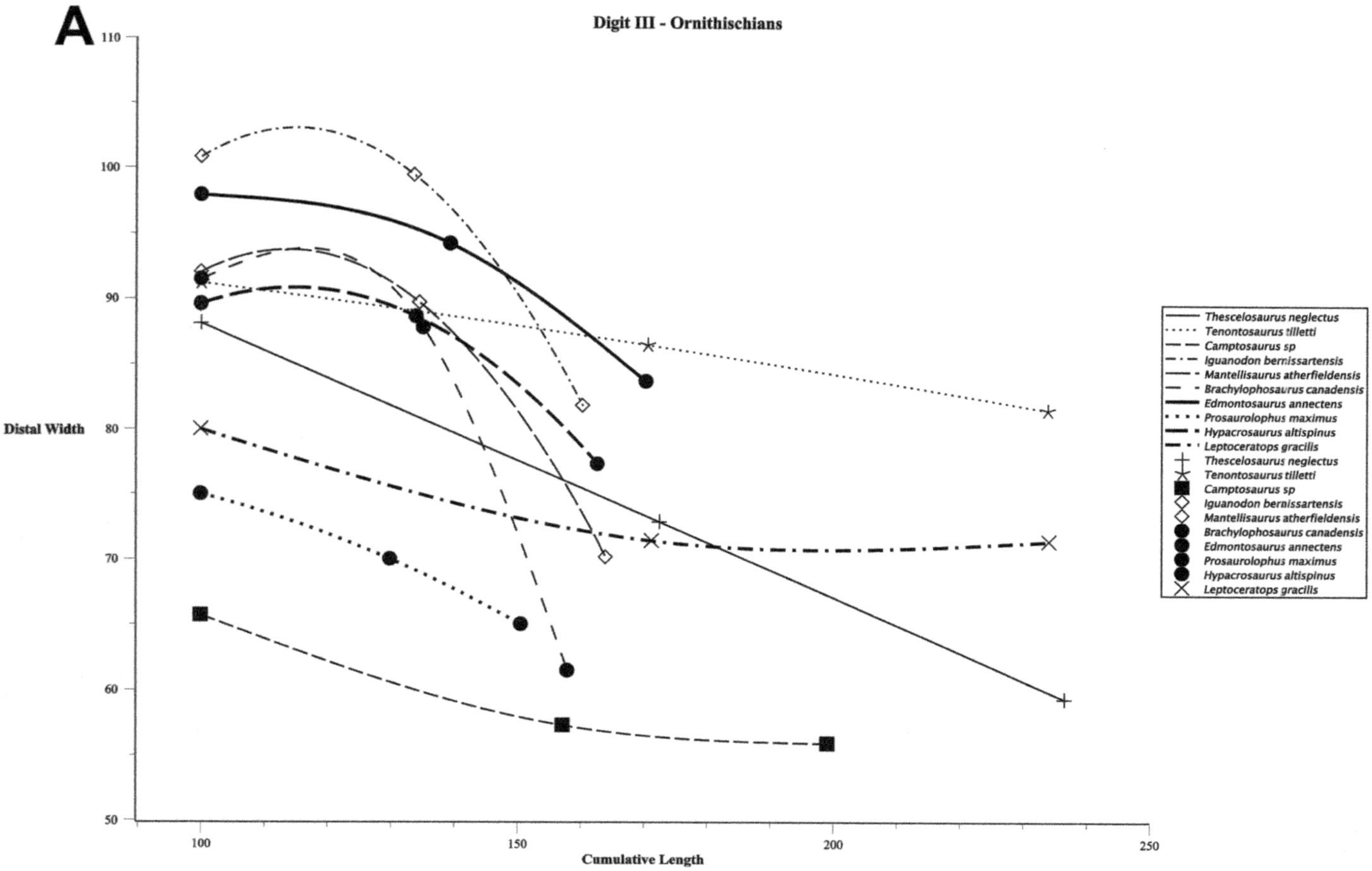

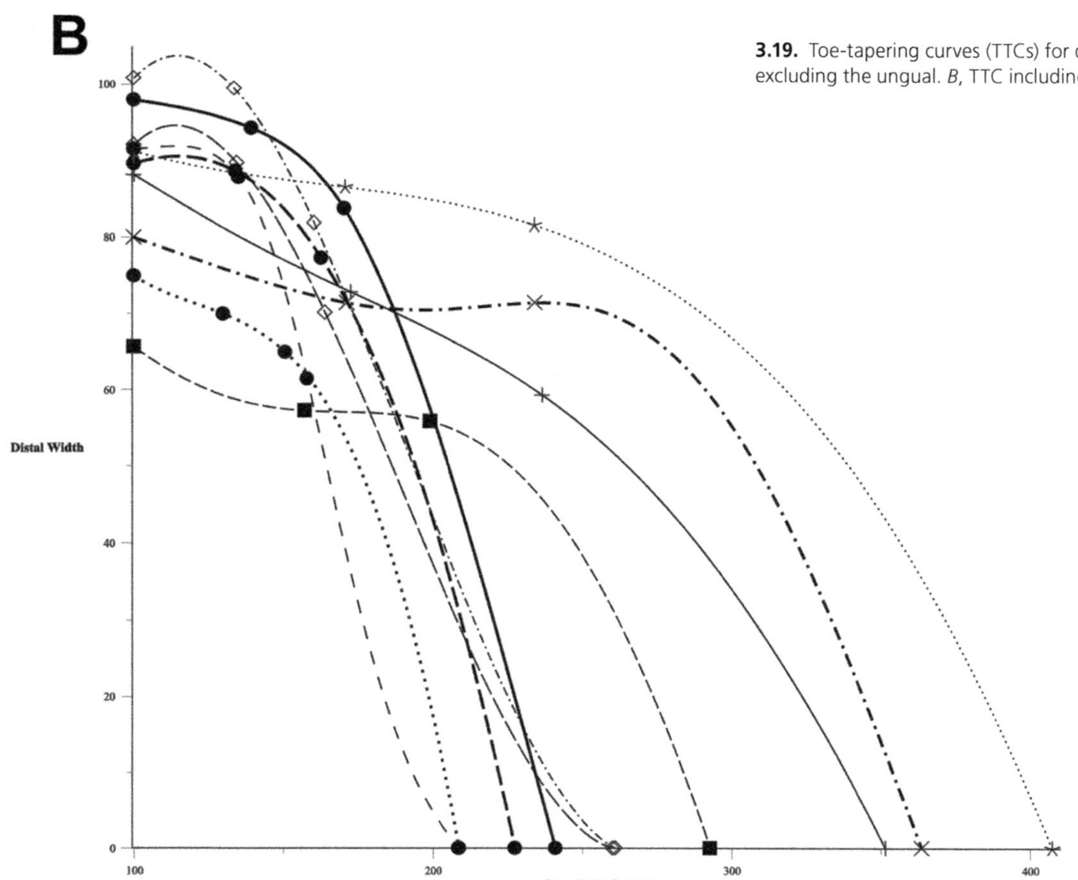

3.19. Toe-tapering curves (TTCs) for digit III in ornithischians. *A*, TTC excluding the ungual. *B*, TTC including the ungual.

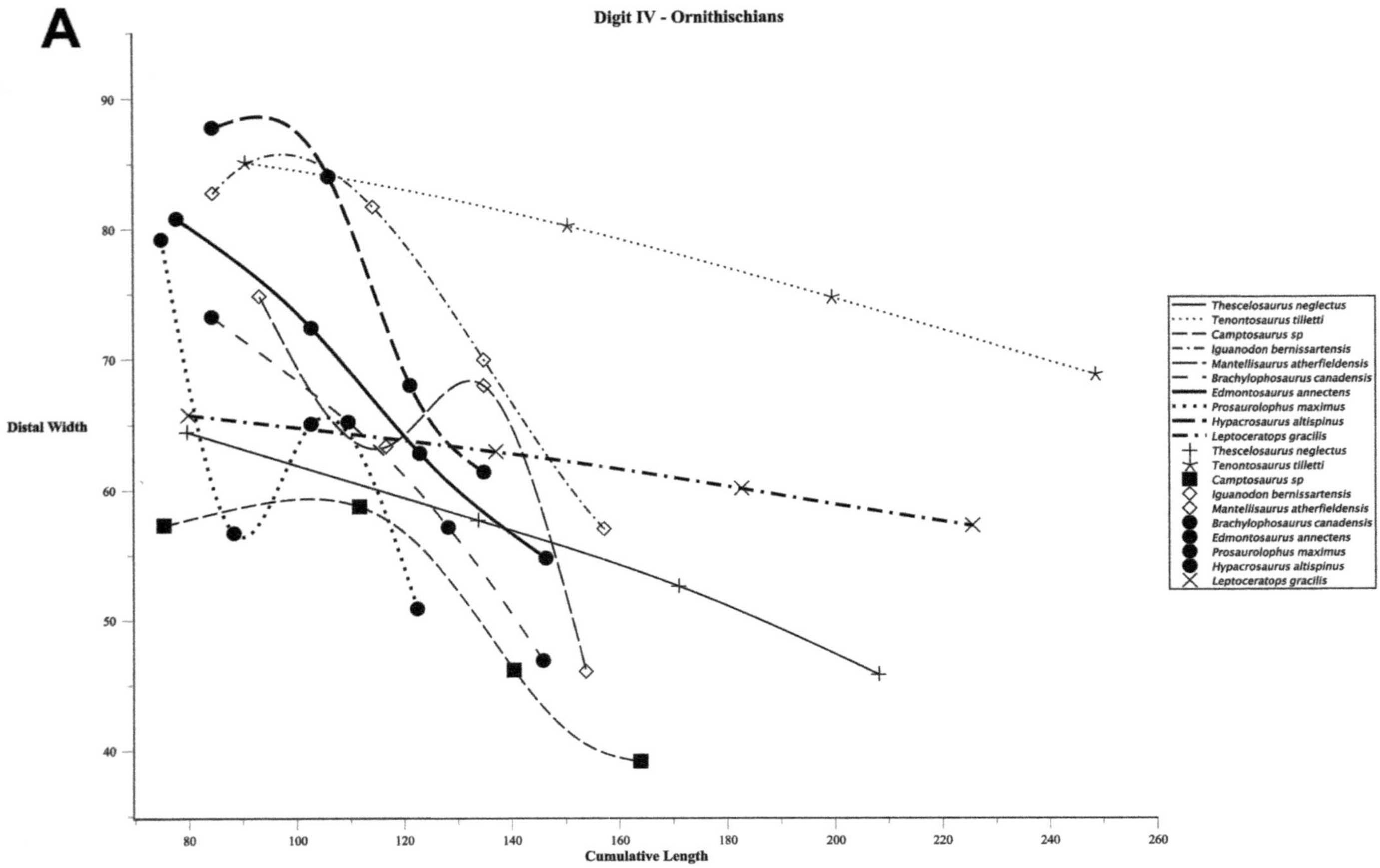

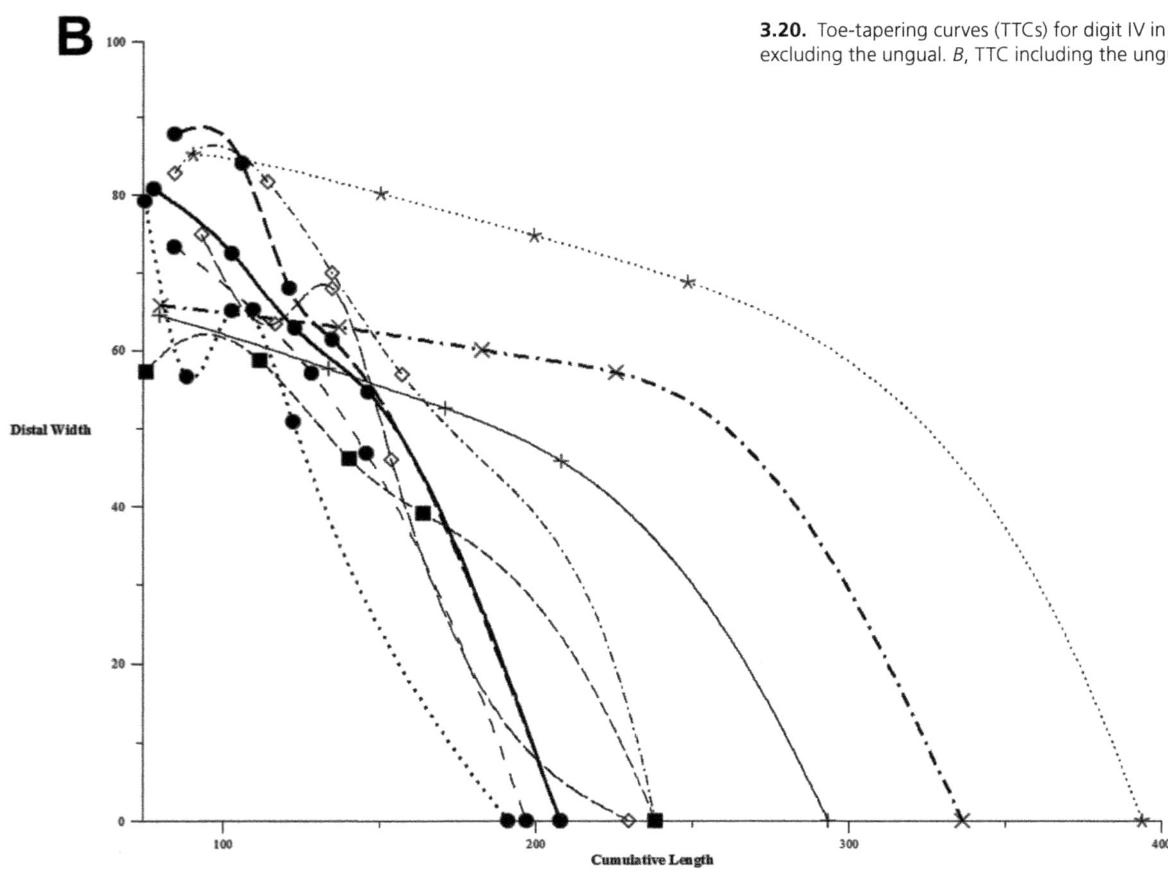

3.20. Toe-tapering curves (TTCs) for digit IV in ornithischians. *A*, TTC excluding the ungual. *B*, TTC including the ungual.

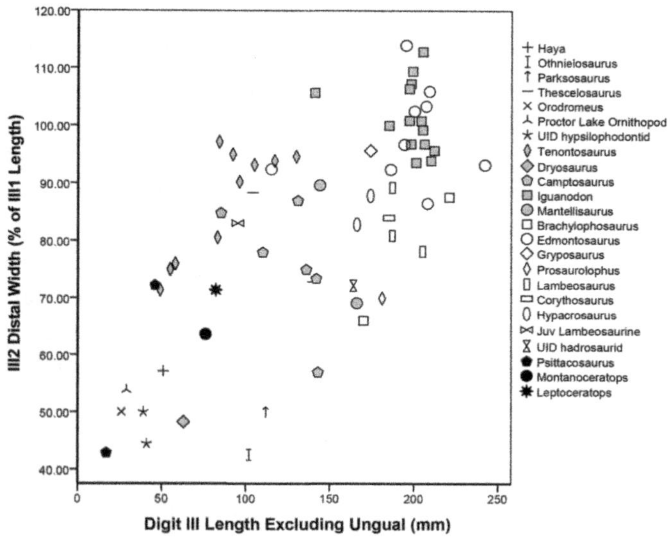

3.21. Phalanx III2 distal width (as a percentage of phalanx III1 length) plotted as a function of digit III length (excluding the ungual) in ornithischians (cf. figs. 2.22, 2.23). As with non-avian theropods and ground birds, there is a tendency for toes to become stouter with increasing size.

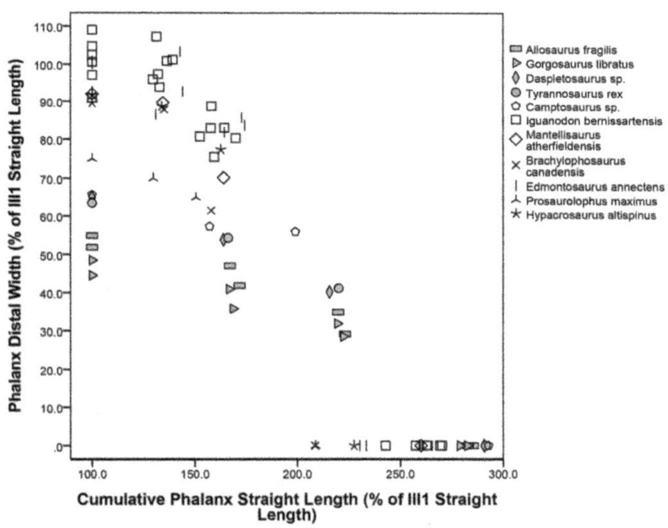

Above, **3.22.** Comparison of digit III toe-tapering profiles (TTPs) of large theropods and large ornithopods; cases represent individual species, so some species are represented by more than one individual. Note the tendency for large ornithopods to show steeper TTPs than those for large theropods.

Right, **3.23.** Overall comparison of scaled toe-tapering profiles (TTPs) for non-avian theropods, ornithischians, and large ground birds (the last including several forms not illustrated in preceding graphs). *A,* Digit II. *B,* Digit III. *C,* Digit IV. Ornithischians fairly consistently have stouter toes than those of birds and non-avian theropods; the latter two groups show considerable overlap in graphed points.

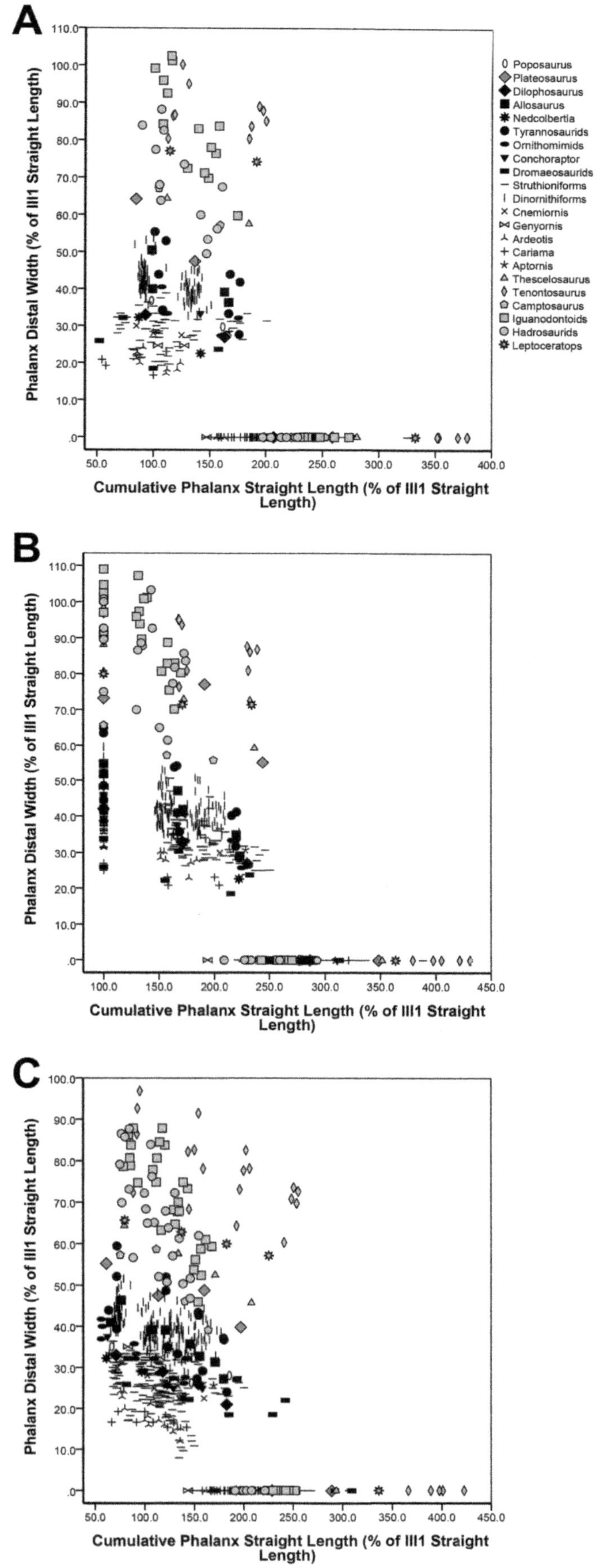

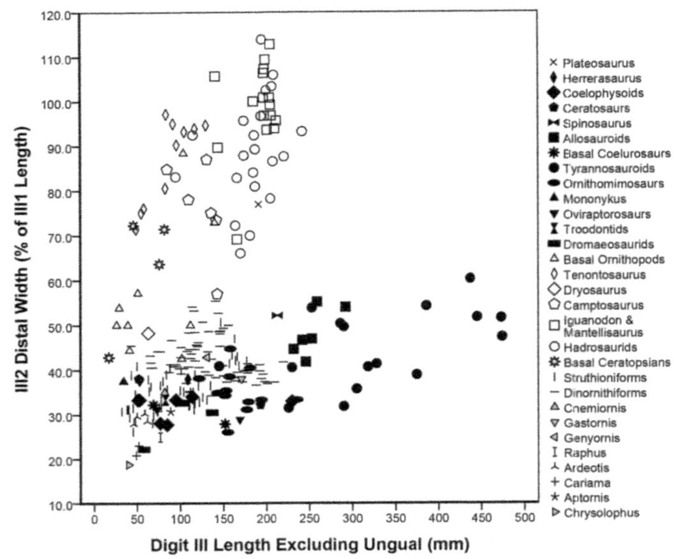

3.24. Comparison of relative widths of digit III (expressed as a percentage of the length of phalanx III1) among non-avian theropods, ground birds, and bipedal or potentially bipedal ornithischians (cf. fig. 2.24). Nearly all ornithischians, and all large ornithopods, have relatively stouter toes than those of birds or non-avian theropods. Non-avian theropods and ground birds show substantial overlap in this relationship.

species, and possibly within species as well. The stouter the toe at its base, the more it will have to decrease distally in order to terminate at the tip of the ungual with a distal width of zero. Consequently, for groups of dinosaurs with basically similar toe shapes (e.g., large theropods), the bigger the animal, the steeper its TTP usually will be (fig. 3.4). However, some forms (e.g., *Spinosaurus*) may show little taper along the lengths of their non-ungual phalanges, instead waiting until the ungual itself before showing much decrease in width.

The increase in relative toe stoutness with increasing dinosaur size will contribute (along with increases in interdigital angle with increasing size) to a feature that has previously been observed in fossil footprints of dinosaurs: an increase in footprint width relative to footprint length with increasing footprint size (cf. Lockley 2009). Our skeletal results corroborate this ichnological prediction.

On the other hand, the difference in toe tapering that Thulborn (1990) suggested should discriminate between theropod and ornithopod footprints regrettably does not receive much support from toe-tapering profiles and curves. TTPs of large ornithopods show as much or more taper than do those of large theropods (fig. 3.22). However, Thulborn's observation still holds true for the very tips of toe impressions, at least for most large theropods (the gigantic ornithomimosaur *Deinocheirus* [fig. 1.9*B*] being a very unusual exception) compared with large ornithopods, due to the blunter ends of the latter (figs.1.8, 1.12).

Some groups are readily discriminated from others on the basis of TTPs. As a group, ornithischians tend to be much stouter-toed than ground birds and non-avian theropods. This is readily seen if TTPs and scaled digit widths for all dinosaurs, non-avian and avian, are compared together (figs. 2.24, 3.23, 3.24). Ground birds and non-avian theropods show considerable overlap in TTPs and scaled digit widths. Large ornithopods have relatively shorter scaled non-ungual phalanx lengths than do most ground birds and non-avian theropods, but have relatively longer scaled ungual lengths (cf. Llandres Serrano et al. 2013) than ground birds and theropods, which causes large ornithopods to have total scaled cumulative digit lengths as long as those of ground birds and non-avian theropods. *Thescelosaurus*, *Tenontosaurus*, and *Leptoceratops* have scaled non-ungual phalanges as long as those of non-avian theropods and ground birds, which, together with the extremely long scaled ungual lengths of these medium-sized ornithischians, results in their having cumulative scaled digit lengths longer than nearly all birds (other than kiwi) and non-avian theropods in our sample.

At the other extreme is the dromornithid *Genyornis*. Like other dromornithids (Murray and Vickers-Rich 2004), this bird has extraordinarily short unguals (fig.1.7*B*), with little or no tapering proximal to the unguals, and total scaled digit lengths of less than 150% (digits II and IV), and less than 200% (digit III), of the length of phalanx III1.

Some forms show distinctly different TTPs and TTCs than other members of the same group. Dromaeosaurids and *Spinosaurus* are very different, but in different ways, from other non-avian theropods. Kiwi, cassowaries, rheas, and emu differ from one another in TTPs and TTCs of one or more digits—and the ostrich is, of course, very different from all other ratites. *Tenontosaurus* and *Leptoceratops* have TTPs and TTCs that readily distinguish them from other ornithischians.

However, there are also similarities across groups. At the largest level of comparison again, TTPs of theropods show considerable overlap with those of ground birds, especially moa (cf. Farlow et al. 2013). Within non-avian theropods, allosaurs and tyrannosaurs have very similar TTPs and TTCs. Moa other than *Megalapteryx* differ from each other only in moderate differences in relative digit width. The various hadrosaurs have feet that are nearly indistinguishable.

Even more interesting, there are some similarities in TTPs between species that are not at all closely related (fig. 3.25). The emu (*Dromaius*) has TTPs almost identical to those of the kori bustard (*Ardeotis kori*), kiwi (*Apteryx* spp.) are very similar to the moa *Megalapteryx*, and the prosauropod *Plateosaurus* is rather similar (at least in digits III and IV) to the ornithischians *Tenontosaurus* and *Leptoceratops*—similar

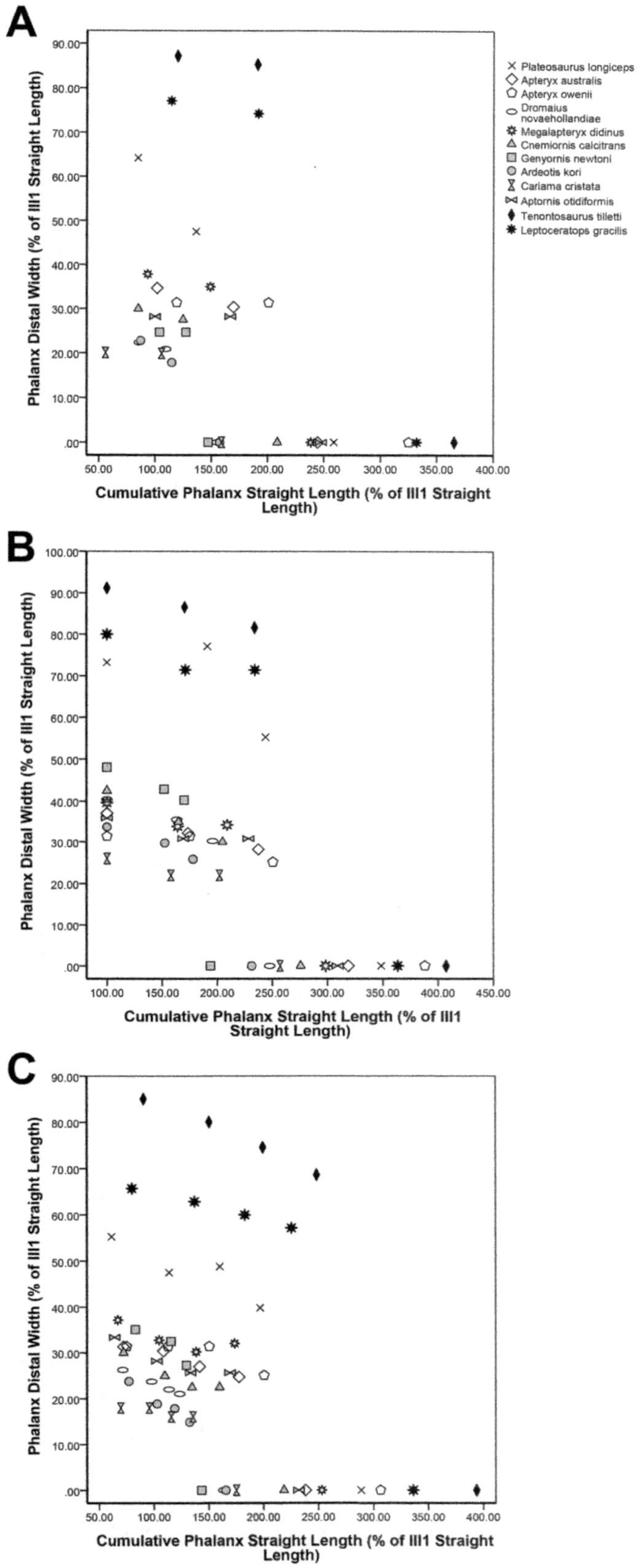

3.25. Convergent toe-tapering profiles (TTPs) between or among species that are not closely related. To reduce clutter, mean values were used for data points graphed for species represented by multiple specimens, rather than data for individual specimens. *A*, Digit II. *B*, Digit III. *C*, Digit IV. Note that the ratite *Dromaius novaehollandiae* has TTPs almost identical to those of the bustard *Ardeotis kori*, that kiwi (*Apteryx* spp.) are very similar in this respect to the moa *Megalapteryx didinus*, and that the sauropodomorph *Plateosaurus* is fairly similar (especially for digits III and IV) to the ornithischians *Tenontosaurus* and *Leptoceratops*.

enough, in fact, that it would not surprise us if a dinosaur trace fossil reminiscent of the early Mesozoic *Otozoum* (generally thought to have been made by prosauropods; Rainforth 2003) were to turn up in the Early Cretaceous.

Our consideration of TTPs, then, supports the conclusions reached in earlier chapters about the efficacy of pedal skeletal proportions in distinguishing among different clades of bipedal or potentially bipedal, tridactyl or functionally tridactyl dinosaurs. The foot skeletons of some groups can indeed be told apart. However, it is difficult to discriminate among species or even higher categories within some groups (e.g., among most large theropods, or among most hadrosaurs), and there are some startling cases of convergence in foot skeletal proportions across groups that are only distantly related.

In our examination of foot skeletons of crocodylians, ground birds, and non-avian dinosaurs in the last few chapters, we have noted some instances of size-related changes in pedal proportions that might be expressed in footprints—most notably, the tendency toward relatively stouter toes seen with increasing size. There are other possible size-related changes in trackmaker body proportions that might impact trackway patterns, though, which we must now consider.

4 Ontogenetic and Across-Species Trends in Hindfoot and Hindlimb Proportions

With Contributions by Philip J. Currie

BECAUSE TRACKWAYS ARE MADE BY THE BUSINESS END of an animal's limb (that is, its foot), the proportions of the segments of the limb and the foot might be expected to have an influence (along with kinematics of the limb, and the interaction of the foot with the recording substrate) on the spacing of footprints in trackways made by that animal. If the portion of the hindlimb of a bipedal animal that is held off the ground is short in comparison with the length of that part of the foot that is routinely placed on the ground, then we might expect the animal to make a trackway in which the ratios of measures of step length to footprint length are less than those made by another animal with a longer leg length compared with its foot length.

In this chapter we explore limb and foot proportions of crocodylians, ground birds, and bipedal or potentially bipedal (cf. Maidment and Barrett 2014) non-avian dinosaurs. Emphasis will be placed on intraspecific, ontogenetic sequences of specimens within species, but we also make comparisons across groups of species. We will mainly be concerned with skeletal proportions, but some soft-tissue data taken from extant alligator feet and bird study skins will also be examined. We also consider the extent to which dinosaur trackway measurements of relative stride length match predictions from skeletal data.

METHODS AND MATERIALS

Because we are considering size-related changes in limb proportions, a variety of size proxies will be used, depending on what data are available. Most of these are self-explanatory, but some require brief explanation. The most frequently used size proxies will be limb bone lengths, either individual bones (most often the femur and third metatarsal [or tarsometatarsus for birds]) or combinations of bones. "Lower" leg length is defined as the sum of the lengths of the tibia and metatarsal III. "Overall" leg length is defined as the sum of the lengths of the femur, tibia, and metatarsal III. While the aggregate length of the femur, tibia, and metatarsal III will likely underestimate the actual length of the limb, due to the presence of cartilage caps at the ends of the bones (Bonnan et al. 2010; Holliday et al. 2010; Tsal et al. 2018), this sum can serve at least as a proxy for overall hindleg length. "Total" leg length is defined as the sum of the lengths of the femur, tibia, metatarsal III, and digit III. Digit III lengths here usually exclude the ungual to increase sample sizes of specimens; spot checks showed that including the ungual doesn't change the relationships described here.

For alligators, total length is measured from the snout to the tail tip of the intact (i.e., not defleshed) animal; head + body length excluding tail will be used for some comparisons.

Rough proxies for the length of the tarsometatarsus and the digital portion of toe III were measured on bird study skins in museum collections (table A4.1). Rough tarsometatarsus length was measured along the dorsal surface of the foot, from the crook of the ankle to the midpoint of the joint between the tarsometatarsus and the first phalanx of digit III (as determined by feel on the study skin foot). "Overall foot digital length" was measured in a straight line along the dorsal surface of digit III, from the midpoint of the joint with the tarsometatarsus to the tip of the claw.

For an ontogenetic sequence of white leghorn chickens described by Manion (1984), one of the size proxies used will be the cube root of body mass.

Skeletal data were obtained from many sources: the published literature (Gilmore 1920; Parks 1920, 1933, 1935; Wellnhofer 1974, 1993*a*; Ostrom 1976, 1978; Santa Luca 1980; Peng 1992; Zhou and Zhang 2001; Kobayashi and Lü 2003; Clarke et al. 2006; You et al. 2006; Zhou et al. 2008, 2014; Hu et al. 2009; Rinehart et al. 2009; Chiappe and Göhlich 2010; O'Connor et al. 2010, 2015; Ji et al. 2011; Han et al. 2012; Li et al. 2012; Rauhut et al. 2012; Sereno 2012; Godefroit et al. 2013; White et al. 2013, 2016; Zhang et al. 2013; Escaso et al. 2014; Foth et al. 2014; Pei et al. 2014; Wang 2014; Wang et al. 2014*a*, 2014*b*, 2016*a*, 2016*b*; Apesteguía et al. 2016; Huang et al. 2016), the Online Dinosaur Project, Farlow's and Currie's measurements, and measurements solicited from colleagues. Our thanks to all who supplied us with data.

We scoured the literature, as well as our own field data and unpublished data supplied by colleagues, for measurements of footprint length and stride length in trackways attributed to non-avian theropod and bipedal (or at least potentially bipedal) basal ornithischians and ornithopods.

We tried to be thorough, eventually collecting measurements of more than 2,000 trackways (table A4.2), but given the huge size of the literature, and the range of international, national, regional, and local publications in which many of the studies have appeared, we would not be surprised to learn that we have missed some. Sometimes different authors apply different names to the same site, and different labels to the same trackways; we tried to avoid duplication of data, but the authors' ineptitude in some of the languages in which studies have been published may have resulted in mistakes.

We usually (but not always) followed the identifications of trackways made by the authors of the studies. It would not surprise us, however, to learn that many trackways attributed to small non-avian theropods had in fact been made by small ornithischians, or vice versa (cf. Farlow et al. 2012, 2014; Castanera et al. 2013*a*; Buckley 2015; Hübner 2016), or even by birds (or, again, vice versa: cf. Gierliński 1996*a*; Gierliński et al. 2017).

When authors reported mean footprint lengths and stride lengths in text or tables, we usually used those values. However, some of our measurements were extracted from trackway photographs or diagrams. We used stride length rather than pace length, even though far more measurements of the latter are published, on the grounds that it is less likely that footprints will mistakenly be assigned to the same trackway when there are at least three of them in sequence. Thus a given trackway had to consist of at least one measurable stride for us to include it.

For ornithischians, we included data for animals thought to have been walking bipedally as well as animals known to have been moving quadrupedally, partly on the grounds that trackways initially attributed to bipeds sometimes turn out to have been made by quadrupeds (cf. Castanera et al. 2013*b*).

Analyses will be based on looking for patterns in bivariate graphs, and by reduced major axis (RMA) analyses of bivariate relationships in known or presumed ontogenetic sequences of specimens. Relationships are judged to be isometric if the 95% confidence interval (CI) of the RMA slope (calculated following Rayner [1985] and Leduc [1987]) includes 1.0, and otherwise are regarded as allometric. Allometric relationships will informally be regarded as "barely so" if the upper CI of the RMA slope is between 0.95 and 1.00, or the lower limit is between 1.00 and 1.05.

There will be some variability in the symbols used to depict particular groups across graphs. In addition, some groups may be lumped together in some graphs, but labeled separately in others. These plotting conventions were imposed by limitations in the number of discrete symbols we were able to accommodate in a single graph, while at the same time keeping some degree of consistency in symbols from one graph to the next. This frequently required some uneasy compromises in symbol selection.

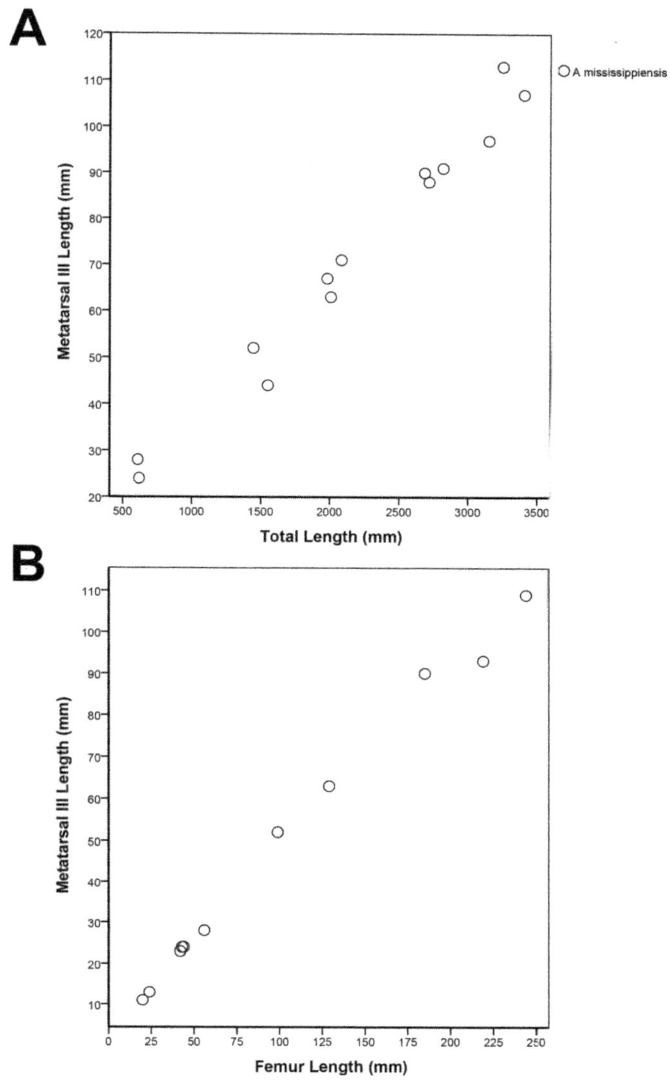

4.1. Length of metatarsal III plotted as a function of total length (*A*) and femur length (*B*) in American alligators.

LIMB PROPORTIONS

Alligators and other crocodylians. Despite the rather low sample size, American alligators show unambiguous negative allometry of skeletal measures of foot length with overall body size and leg length (tables A1.2, 4.1; also see Farlow and Britton 2000). Metatarsal III length is negatively allometric with respect to both total length and femur length (fig. 4.1), albeit only "barely" so. The length of digit III excluding the ungual shows negative allometry with respect to

Table 4.1. Reduced major axis (RMA) relationships between foot and limb bones in crocodylians, non-avian theropods, and ground birds, and between different measurements of study skins in ground birds. All measurements (in millimeters) were log transformed prior to analysis. RMA slopes significantly different than 1 are indicated in **bold**. CI = confidence interval.

Species	Independent variable	Dependent variable	r^2	Number of cases	Slope	95% CI of slope	Source of data
Alligator mississippiensis	Total length	Metatarsal III length + digit III length excluding ungual	0.944	11	**0.828**[a]	0.688–0.997	This study
	Length from snout to base of tail	Pes length	0.965	13	**0.860**[a]	0.758–0.977	Allen et al. (2010)
	Femur length	Metatarsal III length	0.997	11	**0.911**[a]	0.874–0.950	This study
		Digit III length excluding ungual	0.996	6	**0.868**	0.795–0.948	
		Metatarsal III + digit III length excluding ungual	0.997	6	**0.891**[a]	0.825–0.961	
		Pes length	0.972	13	**0.846**[a]	0.763–0.957	Allen et al. (2010)
	Femur + tibia length	Pes length	0.879	13	0.915	0.712–1.176	
	Metatarsal III length	Digit III length excluding ungual	0.989	20	**0.902**[a]	0.856–0.950	This study
	Digit III length excluding ungual	Phalanx III2 distal width	0.966	9	1.167	0.985–1.382	
Coelophysis bauri	Femur length	Metatarsal II length	0.828	17	0.891	0.689–1.151	Rinehart et al. (2009); this study
		Metatarsal III length	0.857	24	0.898	0.748–1.078	
		Metatarsal IV length	0.952	14	0.980	0.850–1.130	
		Digit II length excluding ungual	0.864	10	1.234	0.882–1.725	
		Digit III length excluding ungual	0.975	9	0.995	0.862–1.150	
	Metatarsal II length	Digit II length excluding ungual	0.917	14	**1.472**	1.216–1.783	
	Metatarsal III length	Digit III length excluding ungual	0.915	12	**1.245**[a]	1.001–1.549	
Allosaurus fragilis	Femur length	Metatarsal III length	0.897	11	0.914	0.704–1.187	Holtz (1994); R. Bykowski (pers. comm.); this study
	Femur + tibia length	Metatarsal III length	0.935	10	0.937	0.753–1.165	
Allosaurus spp.	Digit II length excluding ungual	Phalanx II2 distal width	0.989	9	**1.559**	1.419–1.714	J. R. Foster (pers. comm.); this study
Gorgosaurus libratus	Femur length	Metatarsal II length	0.966	11	**0.660**	0.572–0.761	This study
		Metatarsal III length	0.916	18	**0.655**	0.557–0.770	

femur length, and "barely" negative allometry with respect to metatarsal III length (fig. 4.2A–4.2C). The slope of the relationship between digit III length and alligator total length suggests negative allometry, but is not significantly different from isometry. The ratio of metatarsal III length to digit III length (excluding the ungual) changes with increasing overall foot size (fig. 4.2*D*); it appears to increase sharply from very small to somewhat larger animals before flattening out or even decreasing to the biggest individuals.

The combined length of metatarsal III and digit III length excluding the ungual is "barely" negatively allometric with respect to alligator total length and femur length (fig. 4.3A, 4.3*B*). Pes length including the ungual is "barely" negatively allometric with respect to alligator length from the tip of the snout to the base of the tail (fig. 4.3C). The slope of the RMA relationship between pes length and femur + tibia length suggests negative allometry (table 4.1; fig. 4.3*D*), but does not differ significantly from isometry. All told, alligators do seem to grow into their feet and their toes (a topic that will be considered with a larger data set for intact alligator body proportions in chapter 5).

Although data for other crocodylian species are limited, the data suggest that the trends defined for American alligators may also apply across other extant species (figs. 4.2*B*, 4.2*C*, 4.3*B*).

In contrast to foot length relative to measures of alligator size, the limited data indicate possible (but not statistically significant with the data at hand) positive allometry of digit

Species	Independent variable	Dependent variable	r^2	Number of cases	Slope	95% CI of slope	Source of data
Gorgosaurus libratus continued		Metatarsal IV length	0.926	12	**0.640**	0.523–0.783	
		Digit III length excluding ungual	0.927	10	1.083	0.858–1.368	
	Femur + tibia length	Metatarsal III length	0.958	17	**0.730**	0.650–0.820	
		Metatarsal III length + digit III length excluding ungual	0.987	9	0.914	0.825–1.013	
	Metatarsal III length	Digit III length excluding ungual	0.946	9	**1.609**	1.295–1.999	
	(Femur + tibia + metatarsal III) length	Digit III length excluding ungual	0.952	9	**1.307**	1.066–1.602	
Tarbosaurus bataar	Femur length	Metatarsal III length	0.903	9	**0.508**	0.376–0.687	This study
Dinornis robustus	Tarsometatarsus length	Digit III length excluding ungual	0.909	9	0.950	0.710–1.270	This study
Anomalopteryx didiformis	Tarsometatarsus length	Digit III length excluding ungual	0.986	15	**1.130**	1.052–1.214	
Rhea americana study skins	Rough tarsometatarsus length	Rough digit III length	0.943	16	**0.858**[a]	0.745–0.989	
Dromaius noveahollandiae study skins	Rough tarsometatarsus length	Rough digit III length	0.976	31	**0.853**	0.804–0.906	
Domestic chickens	Femur length	Digit III length	0.994	33	**0.872**	0.848–0.897	Manion (1984)
	(Femur + tibiotarsus) length	Tarsometatarsus length	0.997	44	**1.028**[a]	1.010–1.045	
		Tarsometatarsus length + digit III length	0.997	33	**0.931**[a]	0.913–0.950	
	(Femur + tibiotarsus + tarsometatarsus) length: all birds	Digit III length	0.994	33	**0.834**	0.811–0.858	
	(Femur + tibiotarsus + tarsometatarsus) length: limb length ≥ 200 mm	Digit III length	0.938	20	**0.732**	0.644–0.832	
	Tarsometatarsus length	Digit III length	0.994	33	**0.827**	0.804–0.851	

[a] Only barely allometric (upper CI of RMA slope between 0.95 and 1.00, or lower limit between 1.00 and 1.05).

width with respect to digit length (table 4.1; fig. 4.4). Large alligators may become fatter-toed, as also suggested by examination of their intact feet (cf. fig. 5.2).

Birds. Manion's (1984) ontogenetic series of domestic chickens, unlike alligators, shows "barely" positive allometry of the parameter corresponding to metatarsal III length, tarsometatarsus length, with respect to leg length (femur + tibiotarsus length) (table 4.1; fig. 4.5A). In contrast, like alligators, chickens show at least "barely" negative allometry of digit III length with respect to femur length, tarsometatarsus length, and the combined lengths of the femur, tibiotarsus, and tarsometatarsus, and also negative allometry of tarsometatarsus length + digit III length with respect to femur + tibiotarsus length (table 4.1; fig. 4.5*B*, 4.5C). The ratio of overall limb length (femur + tibiotarsus + tarsometatarsus) to digit III length is positively correlated with a linear measure of overall bird size (cube root of body mass; fig.4.6), showing an ontogenetic increase of limb length relative to the digital portion of the foot. So chickens, like alligators, mostly grow into their feet—particularly the digital portion thereof.

Skeletal measures of digit III length and tarsometatarsus length of medium-sized to large species of ground birds in our data set (tables A1.1, 4.1; fig. 4.7A, 4.7*B*) generally are based on too few specimens, and too small a size range (few or no chicks), to say much about ontogenetic changes in shape. Over the size range of measured specimens, the big moa *Dinornis robustus* shows a digit III length/tarsometatarsus

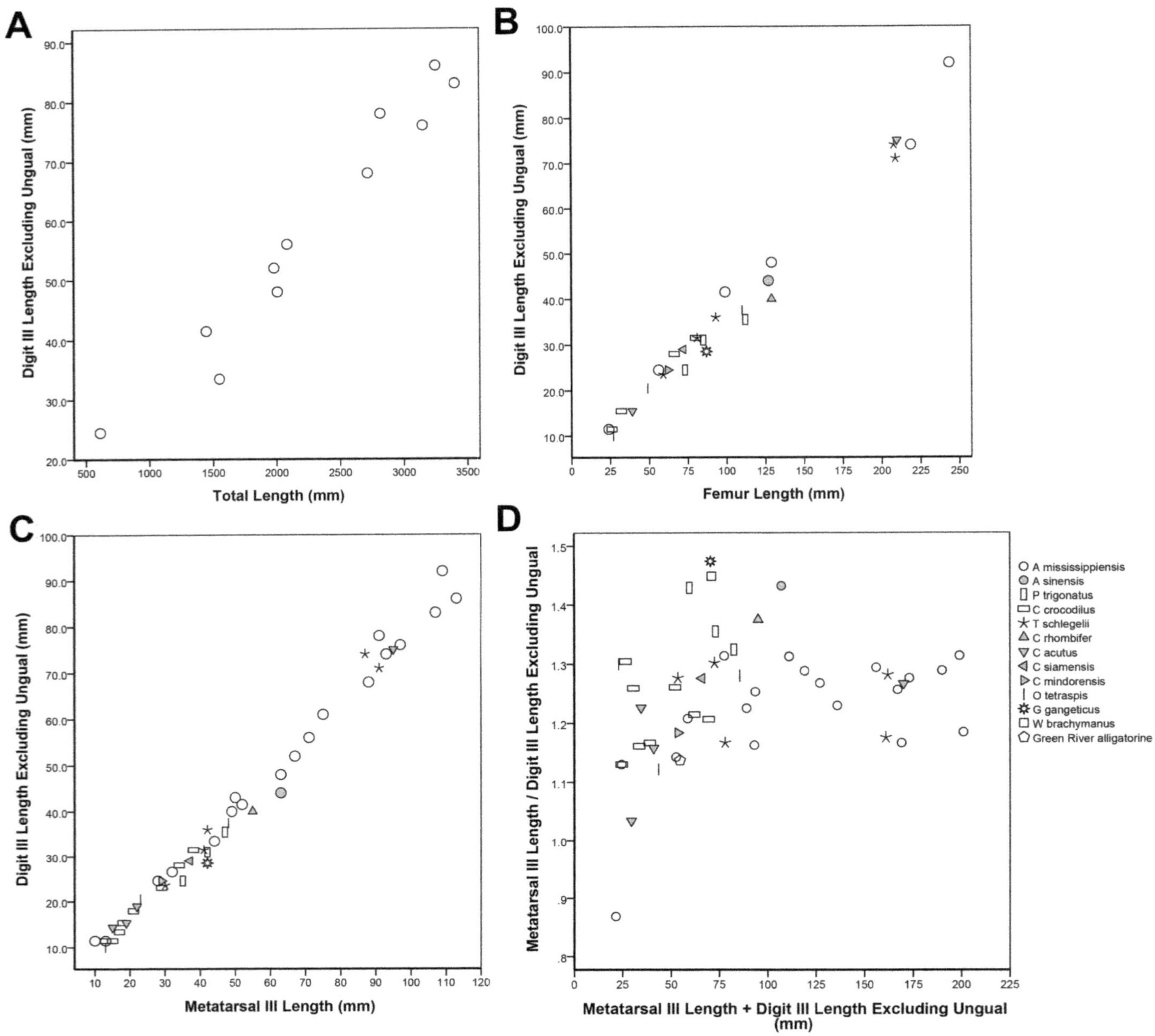

4.2. *A–C*, Length of digit III (excluding the ungual) plotted as a function of three measures of size in American alligators and other crocodylians. *A*, Total length. *B*, Femur length. *C*, Metatarsal III length. *D*, Ratio of metatarsal III length to the length of digit III (excluding the ungual) plotted as a function of total foot length (excluding the ungual).

length relationship that does not differ significantly from isometry, while the smaller moa *Anomalopteryx didiformis* seems to show positive allometry in the same relationship (table 4.1; fig. 4.7*B*). It would be nice to have data for very young individuals of these extinct birds. Across moa species, however, the digital portion of the foot becomes shorter relative to the length of the tarsometatarsus with increasing bird size, as previously discussed (chapter 1; tableA1.3).

Comparing across species, most large extant ratites and their extinct relatives (*Struthio*, *Palaeotis*, *Rhea*, *Pterocnemia*, and *Dromaius*) have a relatively shorter digit III compared with the length of the tarsometatarsus than do moa (*Dinornis*, *Anomalopteryx*, *Megalapteryx*, *Emeus*, *Euryapteryx*, and *Pachyornis*). Cassowaries (*Casuarius*) and kiwi (*Apteryx*) seem not to have received the phylogenetic memo on this matter, however, because they plot among the moa instead of among other extant ratites. In contrast, bustards (*Ardeotis*, *Otis*) fall along the sequence of points defined by most extant ratites. As we shall see in chapter 6, bustards also have very emu-like feet and footprints.

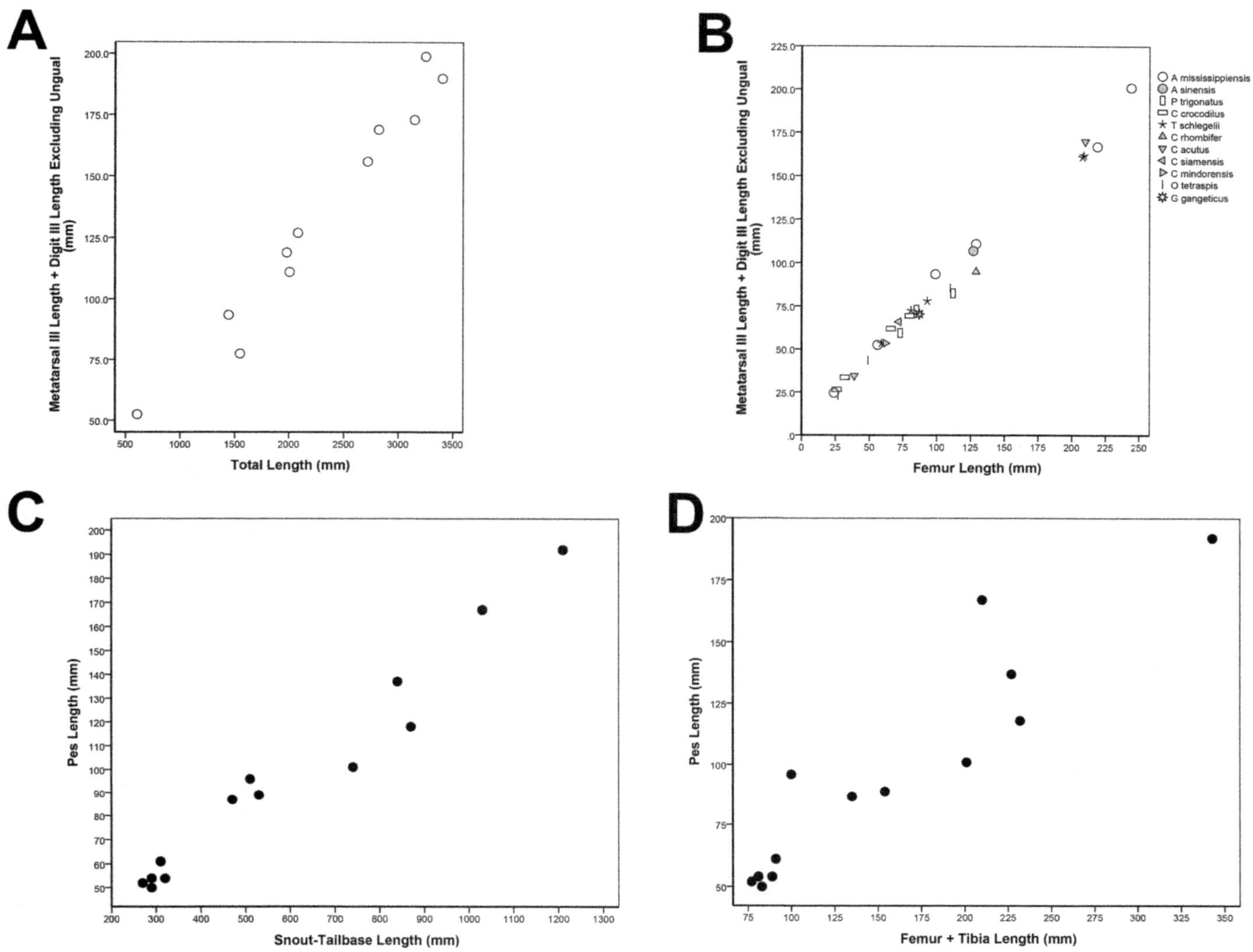

Above, **4.3.** Measures of overall foot length in American alligators and other crocodylians. *A*, Metatarsal III length + digit III length excluding the ungual plotted as a function of total length in American alligators. *B*, Metatarsal III length + digit III length excluding the ungual plotted as a function of femur length among crocodylians. *C*, *D*, Total pes length of American alligators (measured to the tip of the digit III ungual) plotted as a function of *C*, animal length from the tip of the snout to the base of the tail, and *D*, combined length of femur and tibia. *Data in* C *and* D *from Allen et al. (2010).*

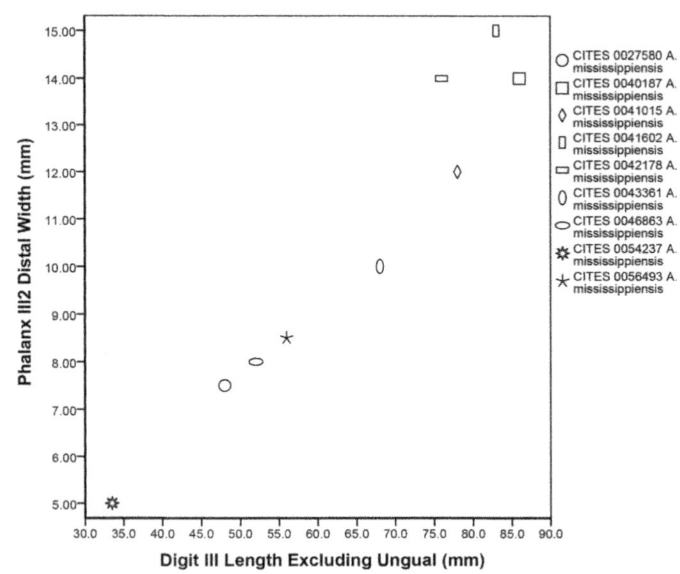

Right, **4.4.** Distal width of the second phalanx of digit III plotted as a function of the length of digit III without the ungual in wild-caught alligators. Individual alligators are identified on the basis of their CITES number.

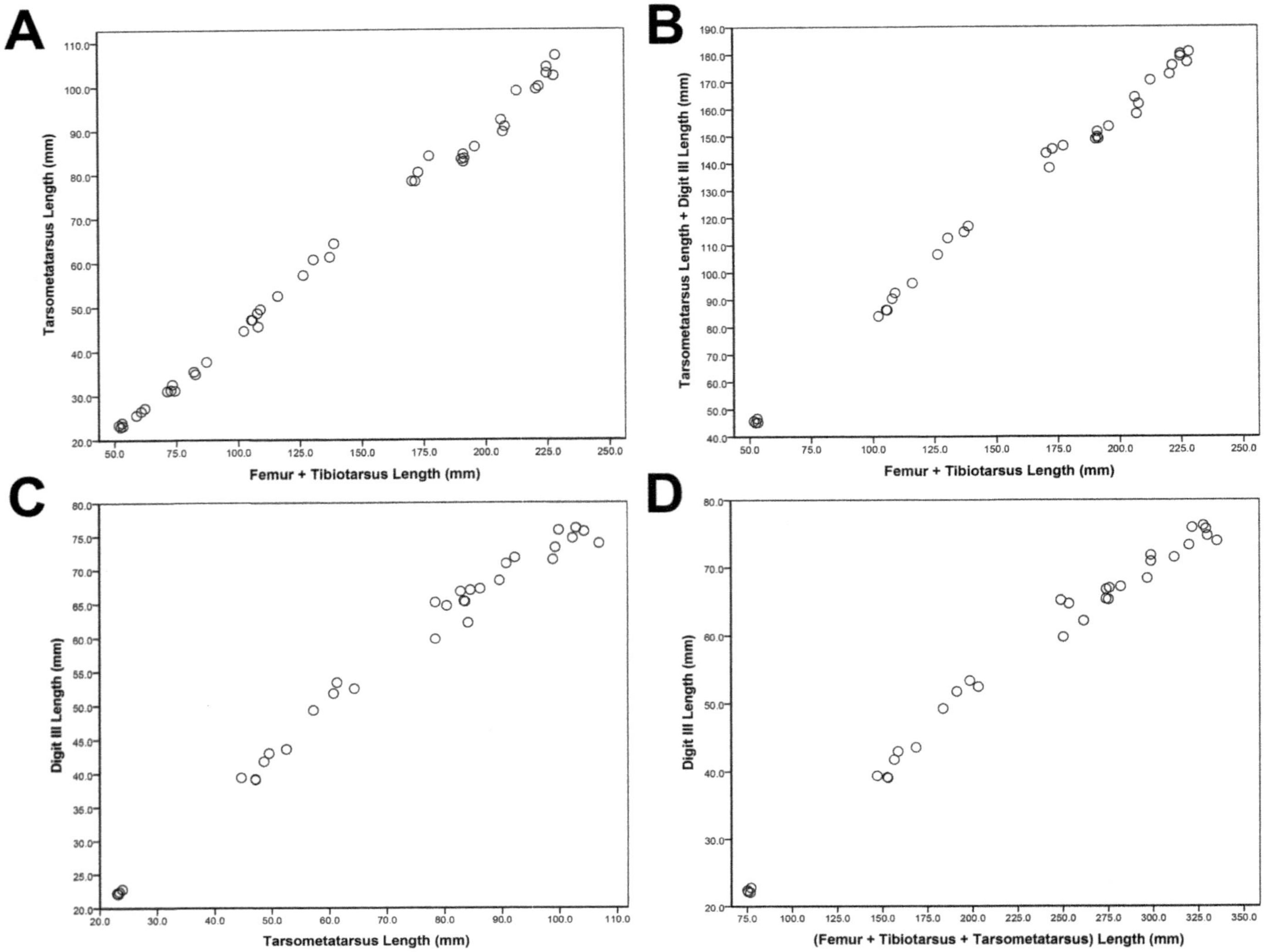

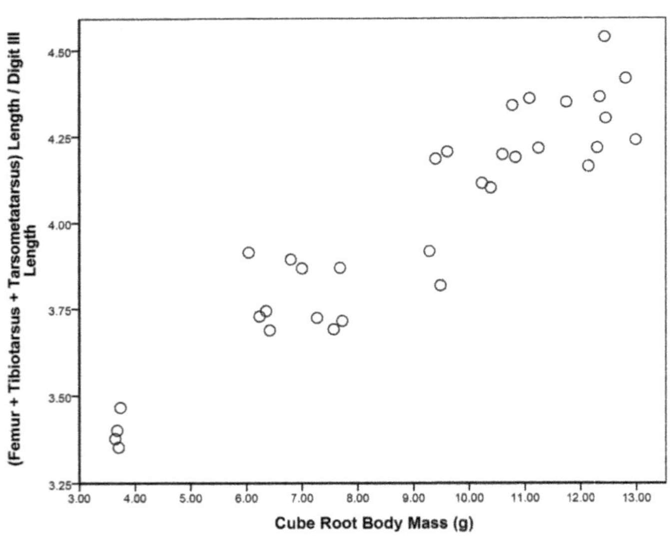

Above, **4.5.** Hindlimb proportions in domestic chickens (white leghorn); data from Manion (1984). *A,* Tarsometatarsus length plotted as a function of the combined length of the femur and tibiotarsus; note tendency for larger birds to have relatively longer tarsometatarsi. *B,* Combined tarsometatarsus and digit III lengths plotted as a function of the combined length of the femur and tibiotarsus. *C,* Digit III length plotted as a function of tarsometatarsus length. *D,* Digit III length plotted as a function of the combined lengths of the femur, tibiotarsus, and tarsometatarsus. In contrast with the tarsometatarsus alone (*A*), all measures of foot length that include digit III show a tendency for decreasing the relative length of the foot with increasing bird size.

Left, **4.6.** Ratio of overall limb length (femur + tibiotarsus + tarsometatarsus) to digit III length as a function of bird size (cube root of body mass) in domestic chickens; data from Manion (1984). As the chickens grow, the length of the leg becomes progressively longer, relative to the length of digit III; the birds "grow into" their toes.

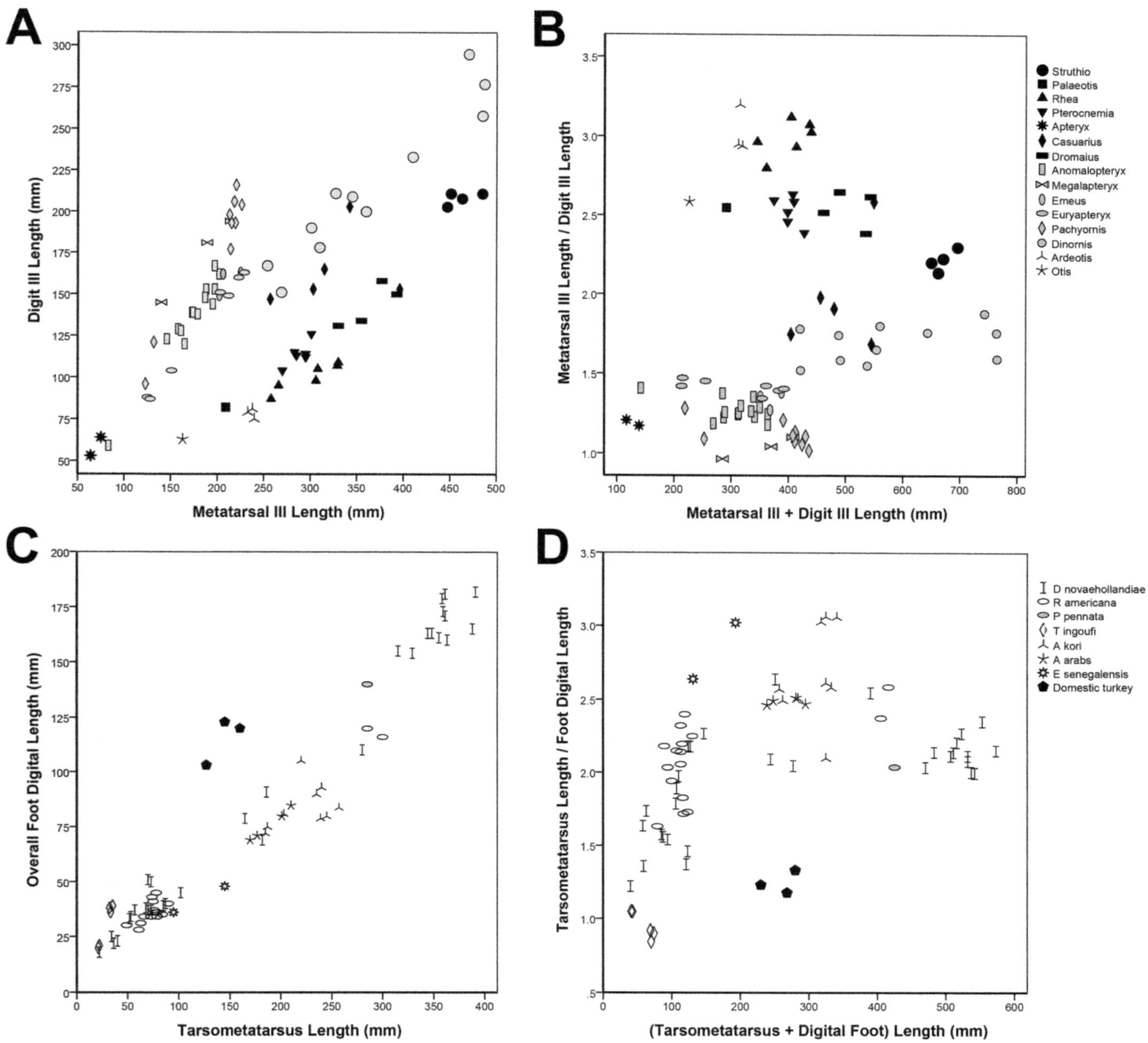

4.7. Digit III length (here including the ungual) as related to tarsometatarsus length in various living and extinct ground birds. *A* and *B* have the same keys, as do *C* and *D*. Full names of bird species in *C* and *D* are as follows: emu, *Dromaius novaehollandiae*; rheas, *Rhea americana* and *Pterocnemia pennata*; tinamou, *Tinamotis ingoufi*; and bustards, *Ardeotis kori, Ardeotis arabs*, and *Eupodotis senegalensis*. *A*, *B*, Measurements based on skeletons (tarsometatarsus length is labeled as "Metatarsal III Length" in both *A* and *B* because these data were extracted from a larger database that included measurements for non-avian dinosaurs); birds identified to genus. *A*, Digit III length vs. tarsometatarsus length. *B*, Ratio of tarsometatarsus length to digit III length plotted as a function of the combined length of the tarsometatarsus and digit III. *C*, *D*, Rough measurements of the length of the tarsometatarsus compared with that of the digital portion of the foot, made on study skins of extant bird species (see text for details). *C*, Length of the digital portion of the foot vs. tarsometatarsus length. *D*, Ratio of tarsometatarsus length to length of the digital portion of the foot plotted as a function of the combined length of the tarsometatarsus and digit III. Note that both emus and greater rheas show a pattern of rapid increase in the ratio from very small to somewhat larger birds, after which the ratio stays constant.

A larger ontogenetic size range of specimens is available from study skins (table A6.26) of greater rheas (*Rhea americana*) and emus (*Dromaius novaehollandiae*). Both species show negative allometry of digit III length with respect to the rough measure of tarsometatarsus length (table 4.1; fig. 4.7C). Bustards (*Ardeotis*, *Eupodotis*) again fall near emu and rheas in this relationship, while domestic turkeys have a relatively short tarsometatarsus. Plotting the ratio of tarsometatarsus length to digit III length against total foot length (fig. 4.7*D*) shows a pattern rather like that for crocodylians (fig. 4.2*D*),

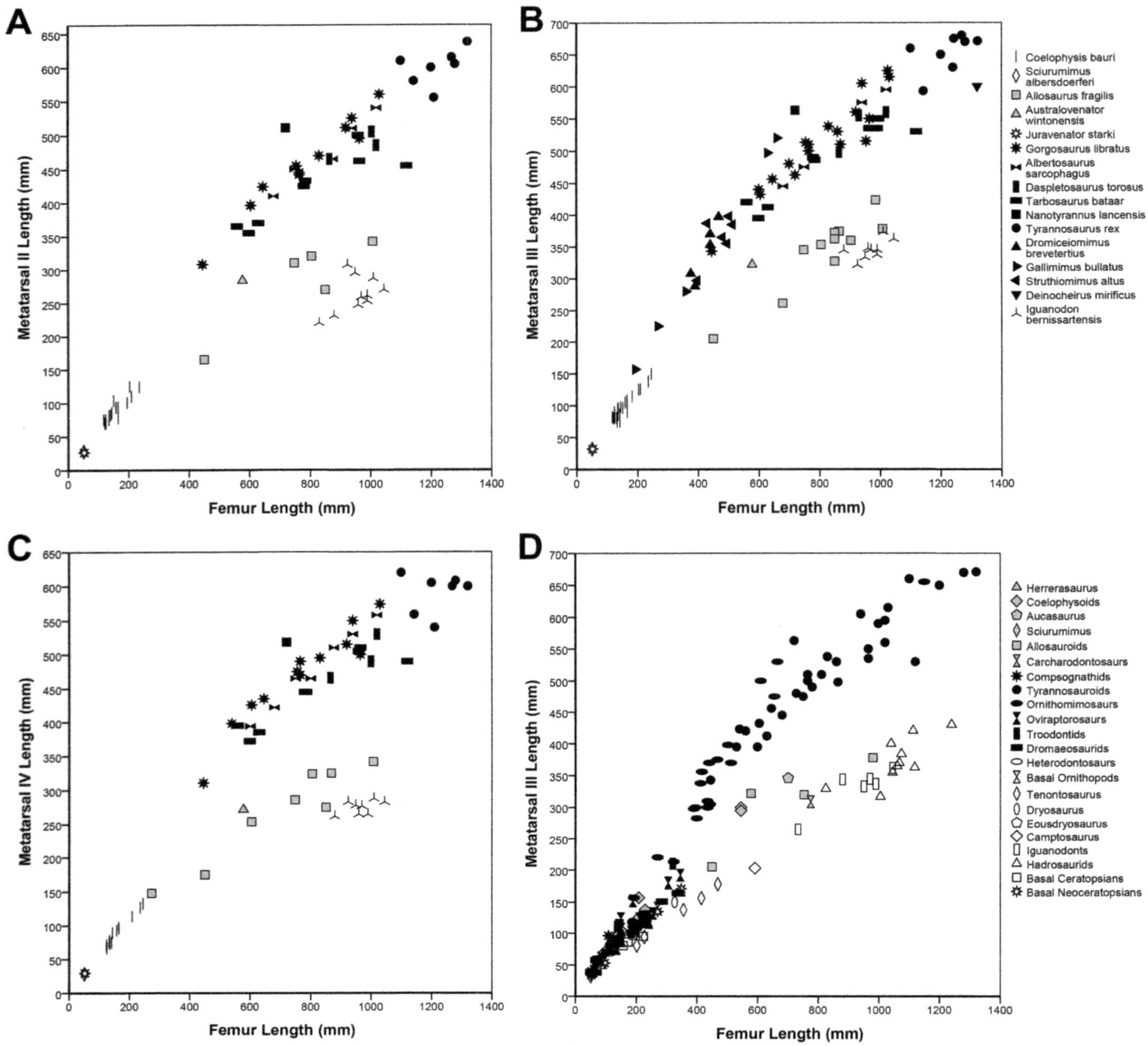

4.8. Metatarsal length plotted as a function of femur length in bipedal or potentially bipedal, non-avian dinosaurs. *A–C*, Metatarsals II, III, and IV of selected dinosaur species; same symbol key applies to all three of these graphs. *Nanotyrannus* is labeled differently than *Tyrannosaurus*, although it is possible that *Nanotyrannus* is an immature form of *Tyrannosaurus*. *D*, Metatarsal III across a wide range of dinosaur groups. Note tendency for metatarsals to become relatively shorter with increasing femur length, both across theropods as a group and within some species (*Gorgosaurus libratus* and *Tarbosaurus bataar*, and possibly *Coelophysis bauri* and *Allosaurus fragilis*). *Dilophosaurus*, *Aucasaurus*, *Allosaurus*, *Australovenator*, and medium-sized to large ornithopods (e.g., *Iguanodon bernissartensis*) have conspicuously shorter metatarsals than those in tyrannosaurs and other coelurosaurs of comparable femur length. In these and all other relationships graphed in this chapter, points for the small theropods *Juravenator* and *Sciurumimus* (a juvenile megalosaur) nearly always overlap.

with the ratio increasing quickly from small to mid-sized emus and rheas, and then flattening out among the biggest birds—possibly even declining a bit in emus.

Thus the available data for ground birds suggest that, as in alligators, the length of the digital portion of the foot (and thus potentially the length of footprints) becomes proportionally shorter as the animal grows from younger to older animals—at least if the sample includes a sufficiently large size range, including very young individuals.

A tendency for pedal digits to become relatively stouter with increasing bird size, at least across species, was noted in chapter 2 (figs. 2.12, 2.13).

Non-avian Dinosaurs. The data set for non-avian dinosaurs is even more bedeviled by problems of small sample sizes and lack of measurements for young individuals of the various species. However, in all the theropod species examined (table 4.1; fig. 4.8) the calculated RMA slope of the relationship between metatarsal length, against femur or femur + tibia length, either indicates (*Gorgosaurus libratus, Tarbosaurus bataar*) or suggests (without being statistically significant) negative allometry (*Allosaurus fragilis, Coelophysis bauri* [except perhaps for metatarsal IV of *Coelophysis*]). The same would likely be true for *Tyrannosaurus rex*, particularly if *Nanotyrannus lancensis* is regarded as an immature form of that species. Wosik et al. (2018) found a weak, nonsignificant negative relationship between metatarsal III length and femur length in the hadrosaurid genus *Edmontosaurus*, which might well have become statistically significant if the number of specimens in their sample (up to 13) had been larger.

The intraspecific RMA slopes of metatarsal length against femur length for *Gorgosaurus libratus* and *Tarbosaurus bataar* are close to the values of interspecific regression slopes of the same relationships in tyrannosaurids (Currie 2003*a*; cf. Persons and Currie 2016). Similarly, the nonsignificant slopes of the relationship between metatarsal III length and femur length in *Coelophysis bauri* and *Allosaurus fragilis* are nearly identical to the interspecific slope of the same relationship in non-arctometatarsalian theropods (Holtz 1994). Most tyrannosaurid specimens plot close together in metatarsal length/femur length relationships (fig. 4.8), although *Nanotyrannus* (?young *Tyrannosaurus*) has relatively long metatarsals, while larger individuals of *Tarbosaurus* have relatively short metatarsals. The tyrannosaurids as a group have much longer metatarsals for a given femur length than do *Coelophysis* and *Allosaurus* (cf. Holtz 1994).

Gorgosaurus libratus shows a significant negative allometric relationship between metatarsal III length and the combined lengths of the femur and the tibia (table 4.1), indicating that relative decline in metatarsus length with increasing animal size applies not just to comparisons with the femur, but also to the leg as a whole. The value of the slope suggests that the same could be true for *Coelophysis bauri*, but the slope does not differ significantly from isometry.

Non-avian theropods thus seem likely to have grown into their metatarsals, as do alligators and chickens, but whether this was true of the digital portions of their toes is uncertain. Digit length data spanning a fair size range across several specimens are available for *Gorgosaurus libratus*, but even the smallest individual in the sample would have been substantially bigger than a hatchling. Although visually the relationship between digit length and femur length or metatarsal length in this species looks like it might flatten out (i.e., relatively shorter digits) at larger animal sizes (figs. 4.9, 4.10), the RMA analyses don't bear this out (table 4.1): Digit III length in this species, if anything, seems to be positively allometric with respect to femur length, leg length, metatarsus length, and leg + metatarsus length; and digit III length clearly becomes proportionally longer relative to leg length across tyrannosaurs (cf. McCrea et al. 2014*a*), and theropod dinosaurs as a whole (fig. 4.11). For *Coelophysis bauri* digit length appears to be isometric with respect to femur length, but positively allometric with respect to metatarsus length (table 4.1).

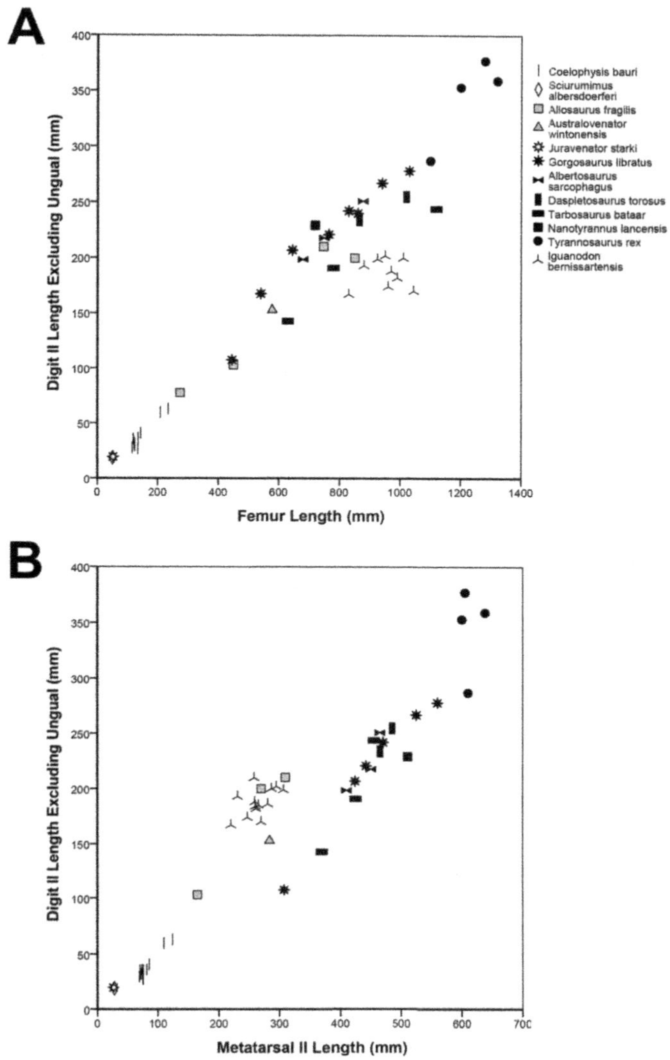

4.9. Length of digit II (excluding the ungual to increase sample sizes) plotted as a function of femur (*A*) or metatarsal (*B*) length in selected non-avian dinosaur species. *Iguanodon* has a distinctly shorter digit II with respect to femur length than that of large theropods. *Allosaurus* and *Iguanodon* have a distinctly longer digit II with respect to metatarsal length than that of tyrannosaurs.

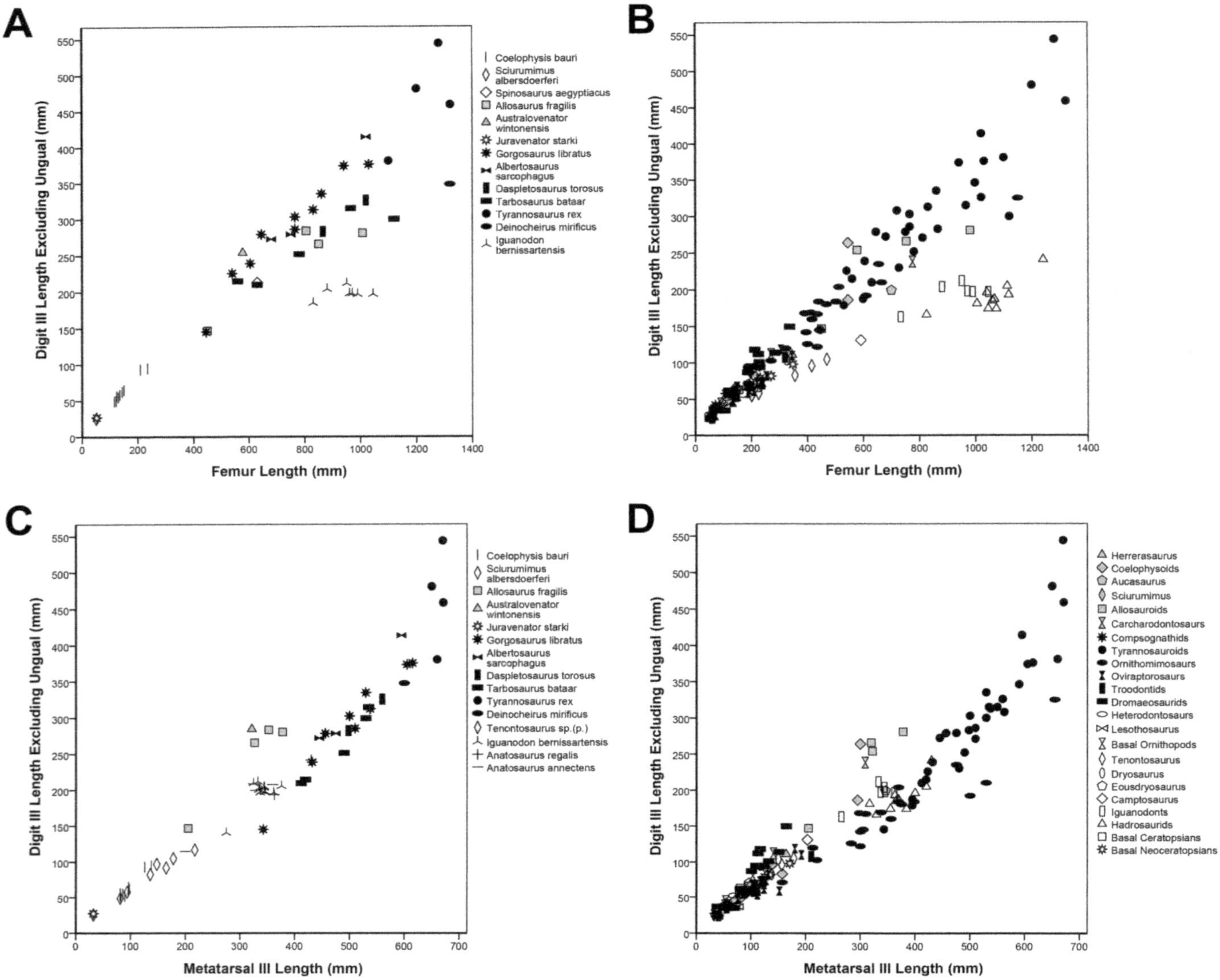

4.10. Relative digit III length (excluding the ungual to increase sample sizes) in bipedal or potentially bipedal dinosaurs. *B* and *D* have the same symbol key. *A*, *B*, Digit III length plotted as a function of femur length in selected non-avian dinosaur species (*A*) and across a wide range of dinosaur groups (*B*). In contrast with the relationship between metatarsal length and femur length (fig. 4.8), digit lengths seem to be isometric with respect to femur length, both within and across theropod species, except in the biggest tyrannosaurs, which show a tendency to increase relative digit III length. For a given femur length, *Iguanodon* and other large ornithopods have shorter digits than those of *Allosaurus* and tyrannosaurs. *C*, *D*, Digit III length excluding the ungual plotted as a function of metatarsal III length in selected non-avian dinosaur species (*C*) and across a wide range of dinosaur groups (*D*). Digit III lengths show positive allometry with respect to metatarsal length in *Gorgosaurus libratus* (table 4.1) and across tyrannosaurs as a group. For a given metatarsal III length, *Aucasaurus*, *Australovenator*, and *Allosaurus* have longer digit IIIs than those of tyrannosaurs and a variety of medium-sized to large ornithopods.

Allosaurus may have a relatively shorter digit III length for a given femur length than do many tyrannosaurs (fig. 4.10A, 4.10*B*), but a longer digit III length for a given metatarsal III length (fig. 4.10*C*, 4.10*D*). These differences appear to cancel each other out if digit III length is compared with overall leg length (fig. 4.11A, 4.11*B*), because *Allosaurus* (fig. 4.11A) falls squarely among the tyrannosaurs in this relationship.

Across theropods as a group, there is a tendency for digits (especially II and III) to become relatively stouter with increasing size (fig. 4.12; also see figs. 2.20–2.21), as also seen in ground birds (figs. 2.12, 2.13). The same appears to be true in feet of *Allosaurus* (mostly *A. fragilis*; table 4.1). So it is possible that, as in alligators, fully grown individuals of large theropods had broader toes than younger/smaller conspecifics.

Small ornithischians overlap non-avian theropods in the femur/metatarsal length relationship (fig. 4.8*D*). Medium-sized to large ornithopods (*Tenontosaurus*, *Iguanodon*, *Mantellisaurus*, hadrosaurs) have considerably shorter metatarsals for a given femur length than do tyrannosaurs and ornithomimosaurs (fig. 4.8A–4.8*D*). In contrast, for a given femur

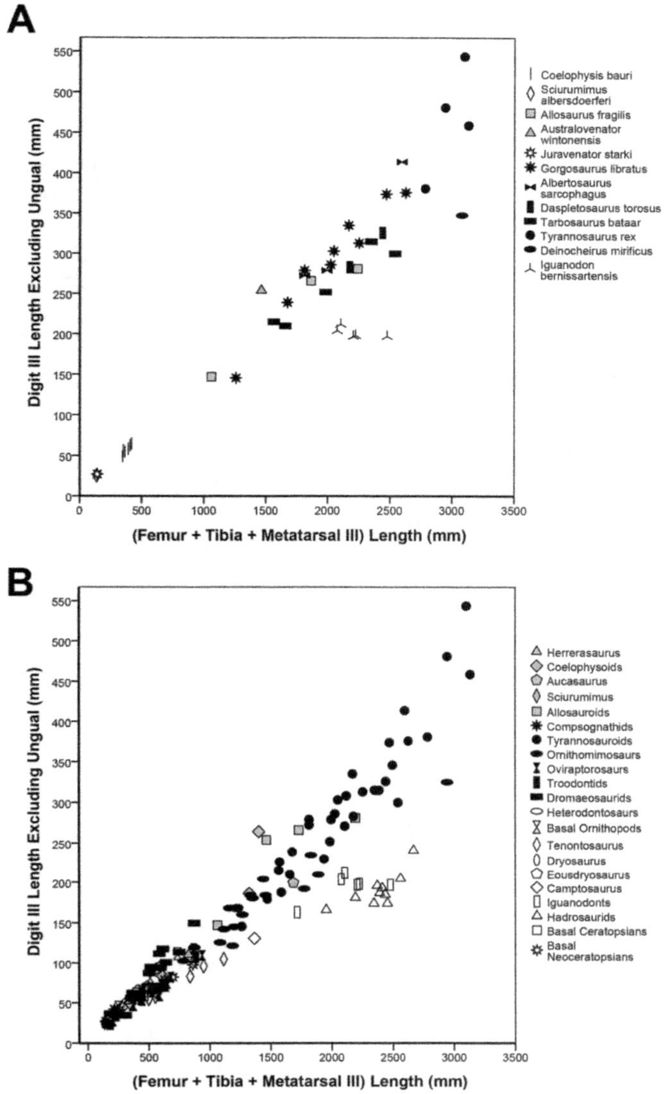

Above, **4.11.** Digit III length (excluding the ungual) plotted as a function of overall leg length (femur + tibia + metatarsal III) in *A*, selected non-avian dinosaur species, and *B*, across a wide range of dinosaur groups. Data points for nearly all theropods in the sample fall along the same trend, which shows a tendency to increase the relative length of digit III with increasing leg length, especially in tyrannosaurs. *Iguanodon* and other large ornithopods have a conspicuously shorter digit III at a given leg length than that of theropods.

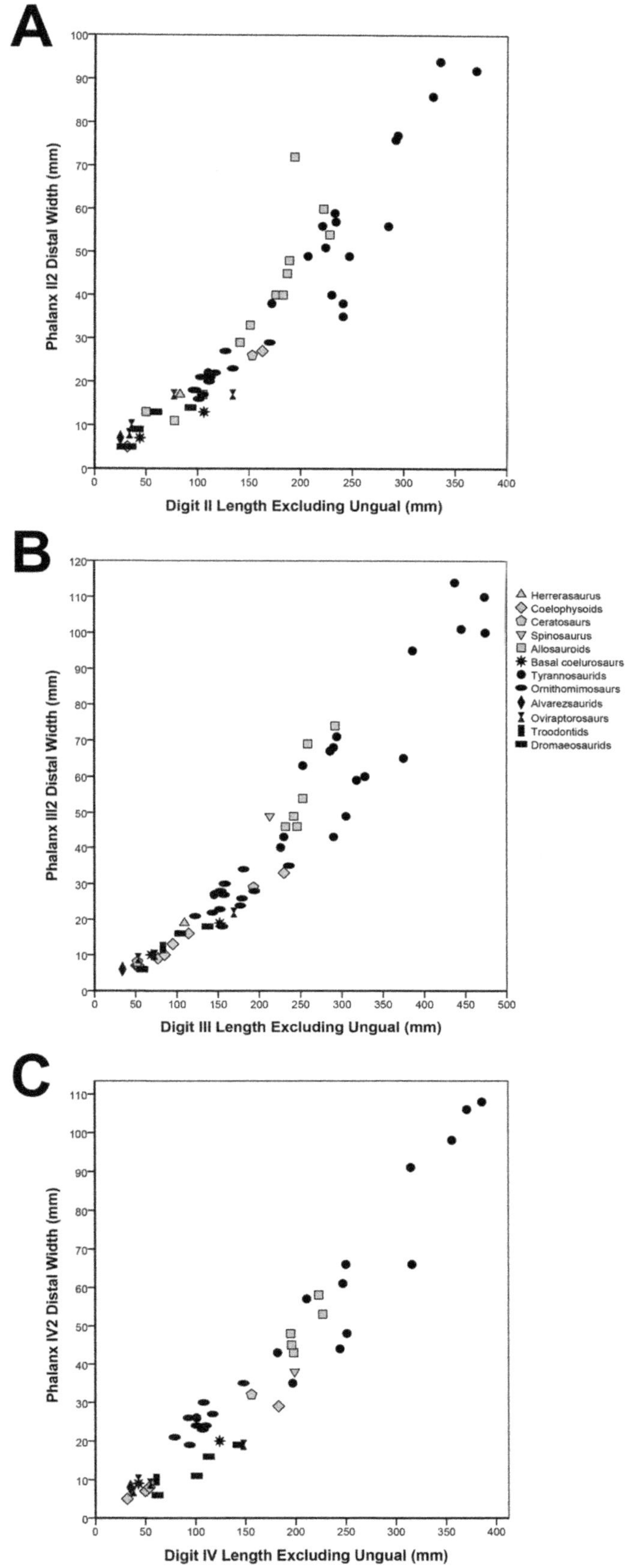

Right, **4.12.** Digit width plotted as a function of digit length in non-avian theropods. Data are for distal widths of the second phalanx (a position roughly halfway out along the total length of the digit when the ungual is included) of digits II–IV; the ungual is excluded in these plots, however, to increase the sample size of specimens. *A*, Digit II. The particularly stout allosauroid point is "*Camptosaurus amplus,*" interpreted here as a theropod, following Galton (2015). *B*, Digit III. *C*, Digit IV. Note the tendency for relatively stouter toes among the larger forms, especially for digits II and III. See figures 2.20 and 2.21 for the same trends illustrated in a different way.

length, big ornithopods have a metatarsal length that is either comparable to, or only a little less than, that of *Allosaurus*. The large ornithopod *Iguanodon bernissartensis*—at least over the size range represented in our data—nearly flat-lines the relationship between metatarsal length and femur length (fig. 4.8*A*–4.8*C*). Examined across the entire range of species, however, ornithopods, like theropods, show a tendency to shorten the metatarsus relative to femur length among larger-bodied forms (cf. Maidment and Barrett 2014).

Smaller ornithischians show a digit III length/femur length ratio comparable to that of non-avian theropods, but medium-sized to large ornithopods have shorter toes for a given femur length than do large theropods (fig. 4.10*A*, 4.10*B*). In contrast, ornithopods of all sizes fall among non-avian theropods along the metatarsal III/digit III relationship (fig. 4.10*C*, 4.10*D*). With increasing size, ornithopods have a proportionally shorter digit III relative to overall leg length than do non-avian theropods (fig. 4.11). Even more than in theropods, ornithischian toes become progressively stouter at large body sizes (figs. 2.22–2.24; cf. Moreno et al. 2007).

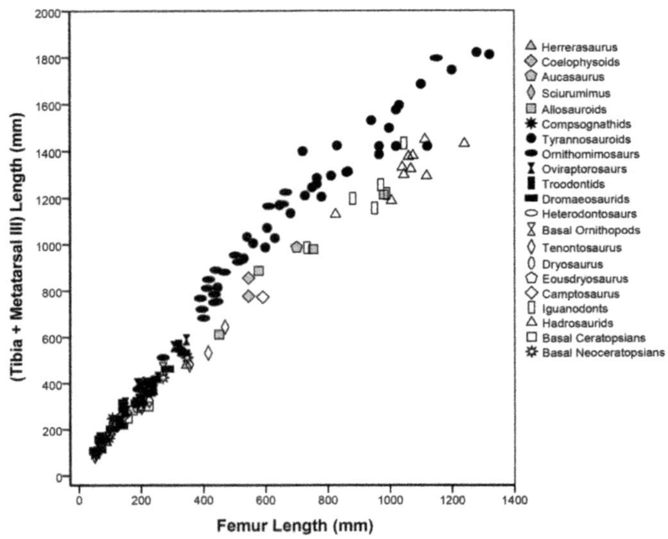

4.13. Lower leg (tibia + metatarsal III) length plotted as a function of femur length across a wide range of dinosaur groups. The relationship curve flattens out in large theropods (especially tyrannosaurs). Ornithopods tend to have a relatively shorter lower leg than that of most theropods, a trend particularly clear in larger forms.

Limb Proportions: Discussion. Because alligators, emus, rheas, and chickens are represented by a larger ontogenetic size range of specimens (more very young/small individuals) than any of the non-avian dinosaurs in our sample, it is possible that the negative allometry between digit III length vs. other parameters of foot and leg size seen in these species might also have occurred in non-avian theropods and ornithischians. The isometry or even positive allometry between digit length vs. metatarsal or leg length observed in the non-avian theropods represented in our sample would then be presumed to apply only to individuals well beyond the hatchling size. On the other hand, negative allometry between digit III length and femur + tibiotarsus + tarsometatarsus length seen across the entire size range of Manion's chickens also applies to just the larger birds (leg + tarsometatarsus length ≥ 200 mm) by themselves (table 4.1), and so the difference between large theropods in our sample, on the one hand, and alligators and ground birds in our sample, on the other, may not be due simply to the lack of data for allosaur or tyrannosaur chicks.

Large theropods show negative allometry between femur length (a proxy for overall body size) and the lengths of more distal hindlimb elements, such as the tibia and metatarsals (Holtz 1994; Currie 2003*a*; Bybee et al. 2006; Foster and Chure 2006; McCrea et al. 2014*a*; Persons and Currie 2016; fig. 4.13 here), and overall leg length becomes proportionally shorter with respect to body torso, at least in *Tyrannosaurus* (Hutchinson et al. 2011). The bigger the theropod, the relatively shorter its lower hindlimb elements, perhaps indicating a reduced capacity for rapid locomotion among the largest forms (cf. Coombs 1978; Alexander 1985; Thulborn 1990; Christiansen 1999; Farlow et al. 2000; Hutchinson and Garcia 2002; Hutchinson 2004; Hutchinson et al. 2005, 2007, 2011; Sellers and Manning 2007; Gatesy et al. 2009; Bates et al. 2010, 2012; but see Paul 2008*a* for an alternative view).

Of greater interest for the present discussion is what size-related changes in limb proportions might tell us about the nature of routine walking in dinosaurs. Positive allometry in the digit III length as a function of leg length relationship in large theropods means negative allometry if the relationship is turned around, with leg length examined as a function of digit III length (McCrea et al. 2014*a*). The seemingly conflicting results about the digit length/limb length relationship observed for young alligators and birds, on the one hand, and large theropods, on the other, may possibly be reconciled by examining the limb length/digit III length relationship across the entire size range of non-avian dinosaurs and basal birds (fig. 4.14).

For non-avian theropods as a group, as well as basal birds and ornithischians, the ratio of overall hindleg length (femur + tibia + metatarsal III) to digit III length (excluding the ungual) rises quickly from smaller animals to those with total hindleg lengths in the 500–1500 mm range (femur lengths 200–400 mm), peaking for theropods among ornithomimosaurs and oviraptorosaurs. Beyond that peak, the ratio declines as theropods become increasingly bigger

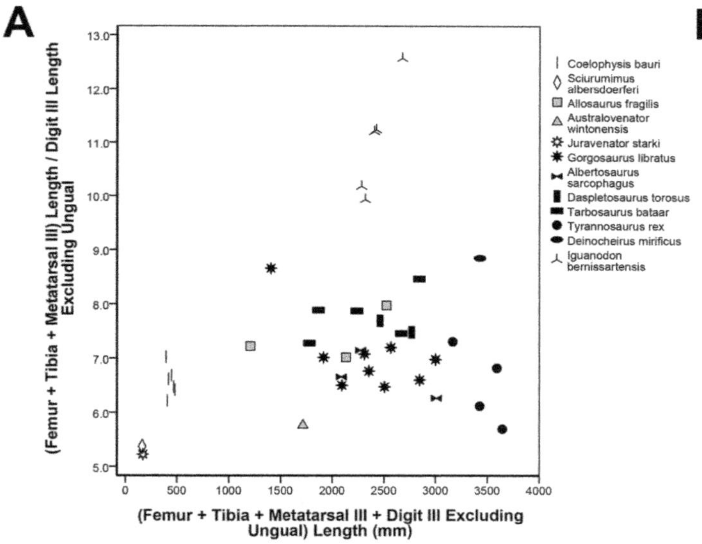

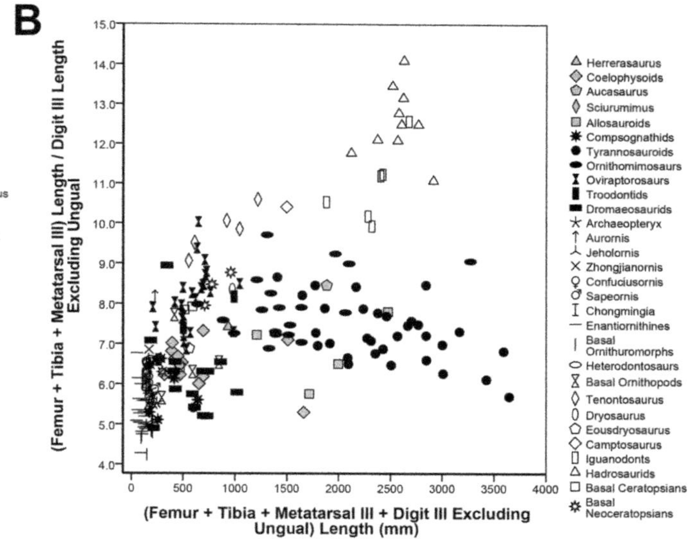

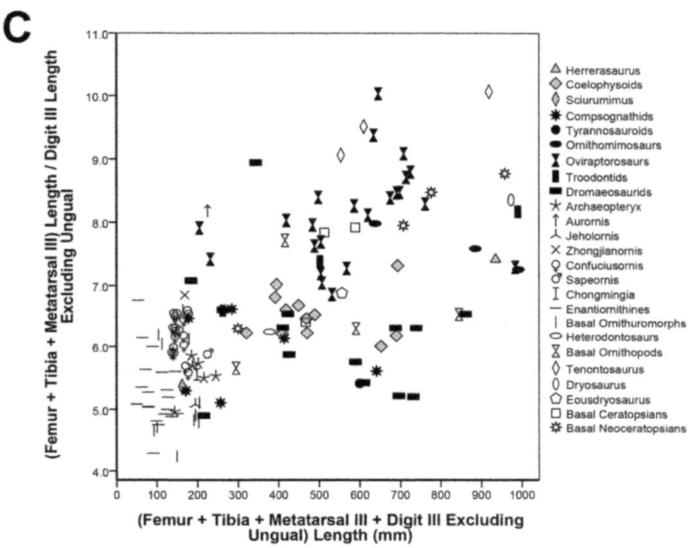

4.14. Ratio of a proxy for "overall" leg length (femur + tibia + metatarsal III) to digit III length excluding the ungual, as a function of a proxy for "total" limb length (femur + tibia + metatarsal III + digit III excluding the ungual): *A*, in selected dinosaur species, and *B*, *C*, across a wide range of dinosaur groups (including some basal birds to round out the picture for smaller-bodied forms). "Male" and "female" symbols are used in *B* and *C* as convenient labels only, not to identify the sex of the specimens so labeled. *C*, Detail of *B* for smaller forms (total limb length proxy 1 m or less). Overall leg length relative to digit III length increases across all groups (albeit with considerable scatter) in the sample for smaller forms (*C*) and continues to increase in large ornithopods (*A*, *B*). In contrast, for theropods with a total limb length proxy greater than 1.5 m, the ratio of overall leg length to digit III length decreases. The big coelophysoid *Dilophosaurus* and the megaraptorid *Australovenator* have unusually short overall limb length/digit III length ratios for dinosaurs of their size (*A*, *B*).

(allosaurs and tyrannosaurs). Two of the medium-sized to large theropods, however, the coelophysid *Dilophosaurus* and the megaraptorid *Australovenator*, have conspicuously shorter legs than expected from the overall relationship.

The increase in the overall hindleg length/digit III length ratio among smaller theropods (some of which are juveniles, and some of them adult animals) is consistent with the ontogenetic pattern seen in modern alligators and ground birds, all of which as adults would be comparable in body size to medium-sized theropods. That is, smaller theropods—both ontogenetically and across species—"grow into their toes," or lengthen their legs with respect to that portion of the foot that serves as a proxy for footprint length, as body size increases (cf. Houck et al. 1990; Brinkman et al. 1998). For theropods bigger than ornithomimosaurs and oviraptorosaurs, the trend reverses, and overall leg length becomes relatively shorter with respect to digit III length.

With the presently available data there is little or no difference between ontogenetic trends (e.g., in *Gorgosaurus libratus*) and those across species (cf. Persons and Currie 2016). The hypothesis that a size-related reversal in values of the overall limb length/digit III length ratio occurred during ontogeny of very large theropods could be tested should a fossil assemblage consisting of a large number of very small through adult animals, with complete limbs and feet, of a single species ever become available.

Although it is interesting that *Allosaurus* differs from tyrannosaurs in having a relatively shorter distal limb compared with femur length (figs. 4.8, 4.13), and relatively longer digits compared with metatarsal III lengths (figs. 4.9*B*, 4.10*C*, 4.10*D*), the relationship between overall limb length and digit III length does not greatly differ between the two groups of large theropods (fig. 4.11). While these differences in hindlimb and hindfoot proportions between *Allosaurus* and the tyrannosaurs might have affected limb kinematics in a way that would affect trackway pattern, what those differences might be is hard to say.

Among small-bodied dinosaurs, basal birds and ornithischians plot among non-avian theropods (fig. 4.14C) in the leg length/digit III length relationship. Unlike theropods, however, ornithopods do not reverse the overall leg/digit III length ratio trend in large forms. Instead, *Tenontosaurus*, iguanodonts, and hadrosaurs continue to lengthen femur + tibia + metatarsal III length relative to digit III length (fig. 4.14*A*, 4.14*B*).

Apart from the possible effects of size-related changes in the limb length/digit length ratio on relative stride length, there are few obvious, consistent differences among theropod groups in limb and foot proportions of the kind that might be expected to affect trackway patterns. Given their relatively long hindlimbs compared with the length of digit III (Fig. 4.14*B*), oviraptorosaurs might be expected to make trackways with relatively high stride/footprint length ratios. At the other extreme, the big coelophysoid *Dilophosaurus* and the megaraptorid *Australovenator* differ from other theropods of comparable total leg length in having an unusually short overall leg length/digit III length ratio (fig. 4.14*B*); the same also seems true of some dromaeosaurids. We might therefore predict trackways made by these dinosaurs to have shorter stride lengths for a given footprint length than trackways made by most allosaurs, tyrannosaurs, or *Aucasaurus*.

For theropods and other bipedal dinosaurs engaged in routine walking, we might expect very small individuals (e.g., hatchlings) to make trackways with low stride length/footprint length ratios. This ratio might be expected to increase with increasing footprint length up to trackways with footprints in the midsize range (say, 25–30 cm long), after which the stride might be expected to show a relative decrease with respect to footprint length.

This hypothesis can be tested by examining dinosaur trackways, to which we now turn.

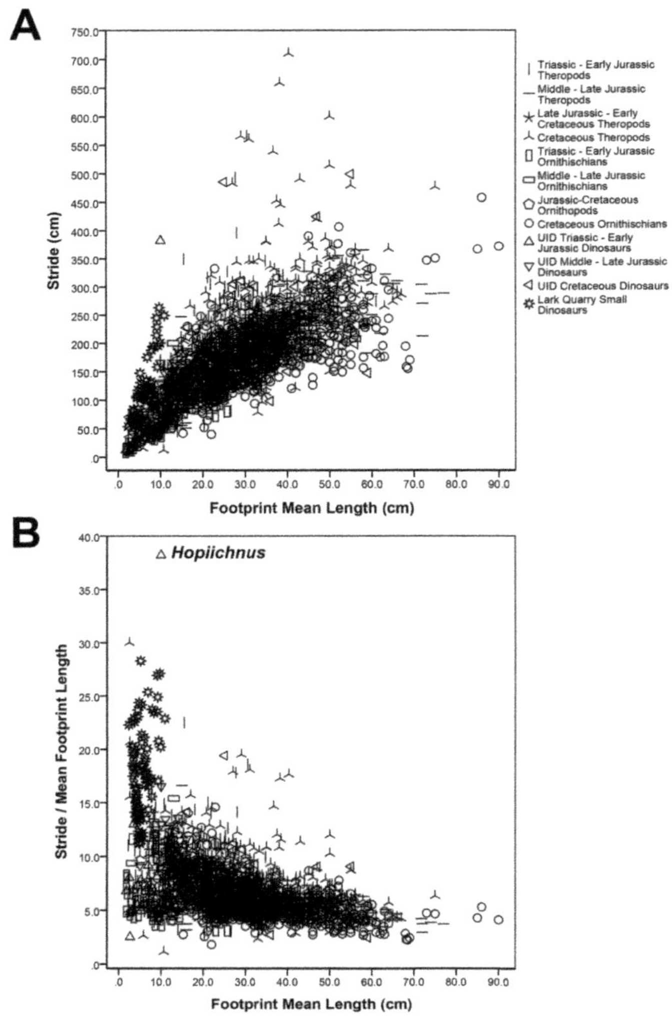

4.15. Stride and footprint length in bipedal and potentially bipedal dinosaurs; data from table A4.2. UID = unidentified dinosaur. *A*, Stride length vs. footprint length. There is a "main sequence" of points in the relationship curve that flattens out with increasing trackmaker size, indicating that bigger dinosaurs take relatively shorter steps (for a given footprint size) during normal walking. Points above this main sequence presumably indicate animals moving in a manner faster than walking, with the maximum stride length occurring among dinosaurs with a footprint length of about 40 cm. *B*, Stride/footprint length ratio as a function of increasing stride length. Data points in these graphs include the controversial cases of *Hopiichnus* and the Lark Quarry "stampeding" small dinosaurs.

TRACKWAYS

Trackway data must be treated with caution, because the preserved values of both footprint length and stride length can be affected by the manner of preservation (Manning 2008; Falkingham 2014). Even so, with a large enough sample size real trends should emerge from within the fog of extraneous extramorphological noise.

Unsurprisingly, there is an unmistakable tendency for bigger dinosaurs to take longer strides (fig. 4.15A), but the relationship is distinctly nonlinear. There is a blur of points defining a "main sequence" of cases, in which stride length increases with increasing footprint length, but the trend flattens out a bit among the biggest dinosaurs. This main sequence presumably shows the typical relationship between stride length and footprint length for animals walking in a routine manner (cf. Dalman and Weems 2013). Above the main sequence are points reflecting dinosaurs taking unusually long steps, presumably running (cf. Farlow 1981; Vierra and Torres 1995; Irby 1996*a*, 1996*b*). The maximum stride length increases from small dinosaurs to those with print lengths of about 40 cm (mostly identified as theropods), and then declines among the larger trackmakers (cf. Thulborn 1990; Casanovas et al. 1995*a*; Pérez-Lorente 1996; Cotton

Table 4.2. Stride/footprint length ratio as a function of footprint size class. Trackways are assigned to a size class on the basis of mean footprint length within the trackway (defined in the entries for all theropods). Values in parentheses for all theropods and all ornithischians exclude trackways in which the stride/footprint length ratio is greater than 10, thus excluding likely running dinosaurs, and restricting the median to values of likely walking animals; if no data are presented in parentheses, there were no running animals in the size sample. Data for the contested *Hopiichnus* and Lark Quarry trackways are not included in these tabulations.

Group	Size class	Minimum	Maximum	Median	Number of trackways
All theropods	1: 0–5.0 cm	5.0	30.0 (9.4)	8.1 (6.9)	22 (13)
	2: 5.1–10.0 cm	2.7	17.0 (10.0)	9.2 (7.2)	74 (43)
	3: 10.1–20.0 cm	1.2	22.5 (10.0)	8.7 (8.1)	431 (331)
	4: 20.1–30.0 cm	2.8	19.5 (10.0)	7.1 (6.9)	509 (467)
	5: 30.1–40.0 cm	2.4	18.6 (10.0)	6.3 (6.3)	459 (447)
	6: 40.1–50.0 cm	2.9	17.6 (8.9)	5.7 (5.7)	186 (182)
	7: >50.0 cm	2.9	8.7	5.1	89
Triassic–Early Jurassic theropods	0–5.0 cm	5.4	13.7	8.0	15
	5.1–10.0 cm	3.9	17.0	10.0	41
	10.1–20.0 cm	3.2	22.5	8.9	98
	20.1–30.0 cm	2.8	17.8	7.5	109
	30.1–40.0 cm	4.2	18.6	6.2	105
	40.1–50.0 cm	3.8	7.6	5.4	17
	>50.0 cm	3.6	4.5	4.1	2
Mid-Late Jurassic theropods	5.1–10.0 cm	5.0	14.1	9.2	15
	10.1–20.0 cm	3.2	16.6	8.1	74
	20.1–30.0 cm	3.5	9.5	7.1	64
	30.1–40.0 cm	4.8	9.9	6.3	96
	40.1–50.0 cm	4.2	7.3	5.6	54
	>50.0 cm	2.9	6.9	4.8	29
Cretaceous theropods	0–5.0 cm	5.0	30.0	10.1	7
	5.1–10.0 cm	2.7	11.2	8.3	18
	10.1–20.0 cm	1.2	15.7	8.8	246
	20.1–30.0 cm	4.0	19.5	7.0	335
	30.1–40.0 cm	2.4	18.1	6.4	257
	40.1–50.0 cm	2.9	17.6	5.7	112
	>50.0 cm	3.8	8.7	5.3	57
All ornithischians	0–5.0 cm	4.4	11.6 (9.4)	6.2 (5.9)	16 (15)
	5.1–10.0 cm	3.7	13.0 (9.9)	5.7 (5.3)	45 (32)
	10.1–20.0 cm	2.7	15.4 (9.5)	5.9 (5.8)	78 (72)
	20.1–30.0 cm	1.8	14.6 (9.5)	5.3 (5.3)	199 (195)
	30.1–40.0 cm	2.9	7.8	4.9	169
	40.1–50.0 cm	2.7	8.7	4.5	76
	>50.0 cm	2.3	7.8	4.1	73
Triassic–Early Jurassic ornithischians	0–5.0 cm	4.4	11.6	5.9	15
	5.1–10.0 cm	3.7	13.0	5.7	39
	10.1–20.0 cm	3.6	9.2	5.8	23
	20.1–30.0 cm	3.1	11.3	4.4	11
	30.1–40.0 cm	4.1	5.1	4.6	2
Mid-Late Jurassic ornithischians	0–5.0 cm	9.4			1
	5.1–10.0 cm	4.7	6.8	6.1	4
	10.1–20.0 cm	4.8	15.4	7.1	17
	20.1–30.0 cm	4.4	8.7	5.8	6
	30.1–40.0 cm	4.3			1
	40.1–50.0 cm	4.1			1
	>50.0 cm	3.7			1
Cretaceous ornithischians	5.1–10.0 cm	5.3	7.9	6.6	2
	10.1–20.0 cm	2.7	14.0	5.8	34
	20.1–30.0 cm	1.8	14.6	5.4	175
	30.1–40.0 cm	2.9	7.8	4.9	166
	40.1–50.0 cm	2.7	8.7	4.5	74
	>50.0 cm	2.3	7.8	4.1	72

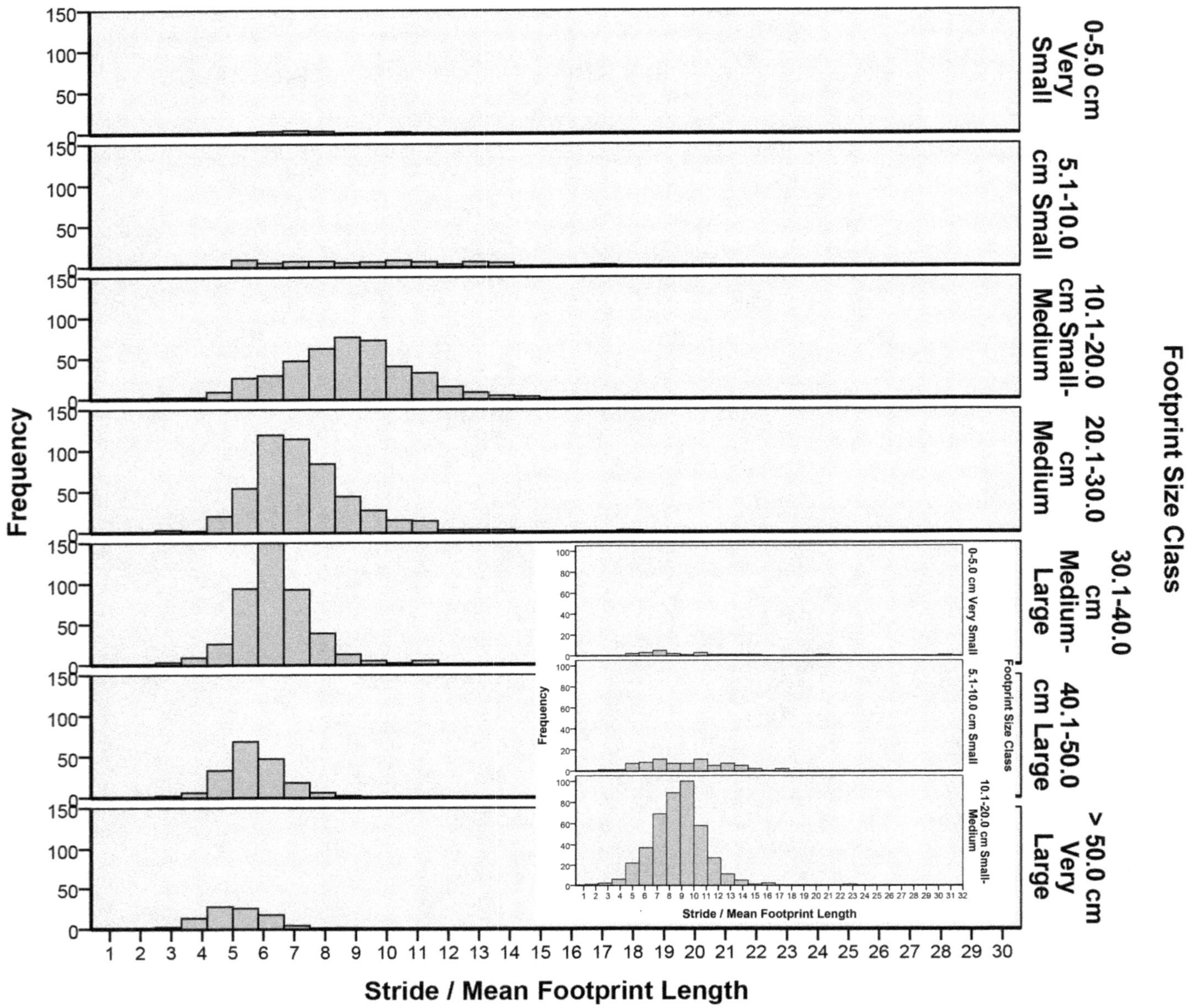

4.16. Distribution of the stride/footprint ratio in trackways attributed to theropod dinosaurs. *Inset* expands distribution for the three smallest size classes. Data are pooled across Mesozoic tracksites of all ages; see table 4.2 for a breakdown on the basis of different stratigraphic intervals.

et al. 1998; Pérez-Lorente and Romero Molina 2001; Pérez-Lorente 2015). The relationship between the ratio of stride length to footprint length, plotted against footprint length, is shaped like a right triangle, with points for running dinosaurs defining the hypotenuse (fig. 4.15*B*); bigger dinosaurs take relatively shorter steps.

For the smaller dinosaurs, examining the relationship in terms of the stride/footprint length ratio (fig. 4.15*B*) has one early Mesozoic data case, plotted as an unidentified trackmaker, with a relative stride length well above all other cases. This is the infamous *Hopiichnus* (Welles 1971), a poorly preserved dinosaur trackway (interpreted by Lockley and Gierliński 2006 as a variant of *Anomoepus*) that is suspected to have missing footprints (Thulborn 1990), in which case its stride would be less astonishing.

Another set of controversial small dinosaur trackways with particularly long stride/footprint length ratios are those from the famous Lark Quarry site in Queensland (Thulborn and Wade 1984), originally described as having been made by running small ornithopods (ichnogenus *Wintonopus*) and theropods (ichnogenus *Skartopus*) panicked by the approach of a large theropod (Romilio and Salisbury [2011, 2014] reinterpreted this large dinosaur as an ornithopod, an interpretation contested by Thulborn [2013]; this data case is treated in this study as an unidentified Cretaceous trackmaker, although Farlow leans toward the idea that it was a theropod). Romilio et al. (2013) argued that *Skartopus* is a preservational variant of *Wintonopus*, and so all of the little trackmakers were ornithopods, and that they may have been swimming or half-floating in water, rather than running flat out across the substrate; Thulborn (2016) vigorously disputed these interpretations. In any case, even if *Hopiichnus* and the contested Lark Quarry trackmakers are excluded, the pattern of the relative stride length vs. footprint length ratio remains much the same. What is of greater interest for us at the moment, however, is not the maximum stride lengths of running animals, but rather the stride lengths of normally walking dinosaurs. Unfortunately, any such patterns are buried in the main sequence blur of points in figure 4.15.

Consequently we need to look at the data a different way. Dinosaur trackways are assigned to a series of size classes on the basis of mean footprint length, after which we can examine the minimum, maximum, and median stride/footprint length for each combination of size class, trackmaker type, and stratigraphic age (table 4.2; figs. 4.16–4.18). The median value for each size/type/age category is used, rather than the mean, because the median is less affected by extreme cases (in this case, running dinosaurs), but the medians actually differ very little from the means.

For trackways attributed to theropod dinosaurs from all intervals of the Mesozoic Era (table 4.2, fig. 4.16), the distribution of values of the stride/footprint length ratio rather nicely matches expectations based on the limb length/digit length ratio (fig. 4.14). The median value of the stride/footprint ratio increases from the value seen in the smallest theropods (size class 1: footprint length 5 cm or less) to peak in size class 2 (footprint length 5–10 cm) or 3 (footprint length 10–20 cm), depending on whether trackways with a stride/footprint length ratio greater than 10 (animals that most likely were running) are included in calculation of the median. These would be animals roughly comparable in size to many oviraptorosaurs and ornithomimosaurs (table 4.3). The median values (and also the maximum values) of the stride/footprint length ratio decline as the dinosaurs get bigger than size class 3.

Among trackways attributed to bipedal or potentially bipedal ornithischians (table 4.2, fig. 4.17), probably mostly or entirely ornithopods (particularly for data cases of middle Jurassic age and younger), the pattern is a bit different. There is little difference in the median stride/footprint length ratio among size classes 1–3, but the ratio decreases from there to the biggest trackmakers. For each size class, the median stride/footprint length ratio is less than for trackways attributed to theropods (cf. Lockley et al. 2009). This is an interesting result, assuming that a lower stride/footprint length ratio wasn't the primary reason that the ornithischian trackways were identified as such in the first place.

Table 4.3. Correspondence of dinosaur size with the estimated footprint size class to which the dinosaurs belonged, and the ratio of (femur + tibia + metatarsal III) length to digit III length (excluding the ungual), in non-avian dinosaur groups. The footprint size class to which a dinosaur belonged was estimated on the assumption that the total length of digit III (including the ungual) approximates the length of the footprint that the dinosaur would have made.

Group	Range of footprint size classes of the group	(Femur + tibia + metatarsal III) length/digit III length excluding ungual		
		Median	Range	N
Herrerasaurus	3	7.4		1
Coelophysoids	2–4	6.5	5.3–7.3	14
Aucasaurus	4	8.5		1
Sciurumimus	1	5.4		1
Allosauroids	3–5	6.9	5.8–7.8	4
Compsognathids	1–2	6.1	5.1–6.6	5
Tyrannosauroids	3–7	7.2	5.4–8.7	32
Ornithomimosaurs	2–6	7.9	6.9–9.7	20
Oviraptorosaurs	1–3	8.2	6.9–10.0	24
Troodontids	1–3	7.4	6.6–8.2	3
Dromaeosaurids	1–4	6.3	4.9–8.9	15
Heterodontosaurs	2	6.3		1
Basal ornithopods	2–3	6.4	5.7–7.7	4
Tenontosaurus	2–3	9.9	9.1–10.6	5
Dryosaurus	3	8.4		1
Eousdryosaurus	3	6.9		1
Camptosaurus	4	10.4		1
Iguanodonts	4–5	10.9	9.9–12.6	6
Hadrosaurids	4–5	12.5	11.1–14.1	10
Basal ceratopsians	2	7.8	6.4–7.9	2
Basal neoceratopsians	2–3	8.2	6.3–8.8	4

There is another interesting feature in the ornithischian data, that being bimodality of relative stride length in size class 2 (footprint length 5.1–10.0 cm), which mainly consists of dinosaurs of Triassic and Early Jurassic age. The animals with a relatively high stride/footprint length ratio are from Weems' (1987) Late Triassic Culpeper Basin, Virginia site, footprints that he originally named *Gregaripus*, while the trackways associated with the lower value of the ratio are mainly typical *Anomoepus*. Why the *Gregaripus*-makers were so frisky is uncertain; perhaps they were particularly tasty morsels that had a lot to run away from.

Summing theropod and ornithischian trackway data for size classes across the entire Mesozoic era might be misleading, however. If some time intervals or trackmaker clades are represented by considerably more data cases than others—as is definitely the case (table 4.2)—this could result in biased values of the median stride/footprint length ratio for particular size classes. Bias could also be introduced because some of the trackways come from sites with large numbers of trackmakers of the same kind that were possibly moving through the area at the same time as a herd. If so,

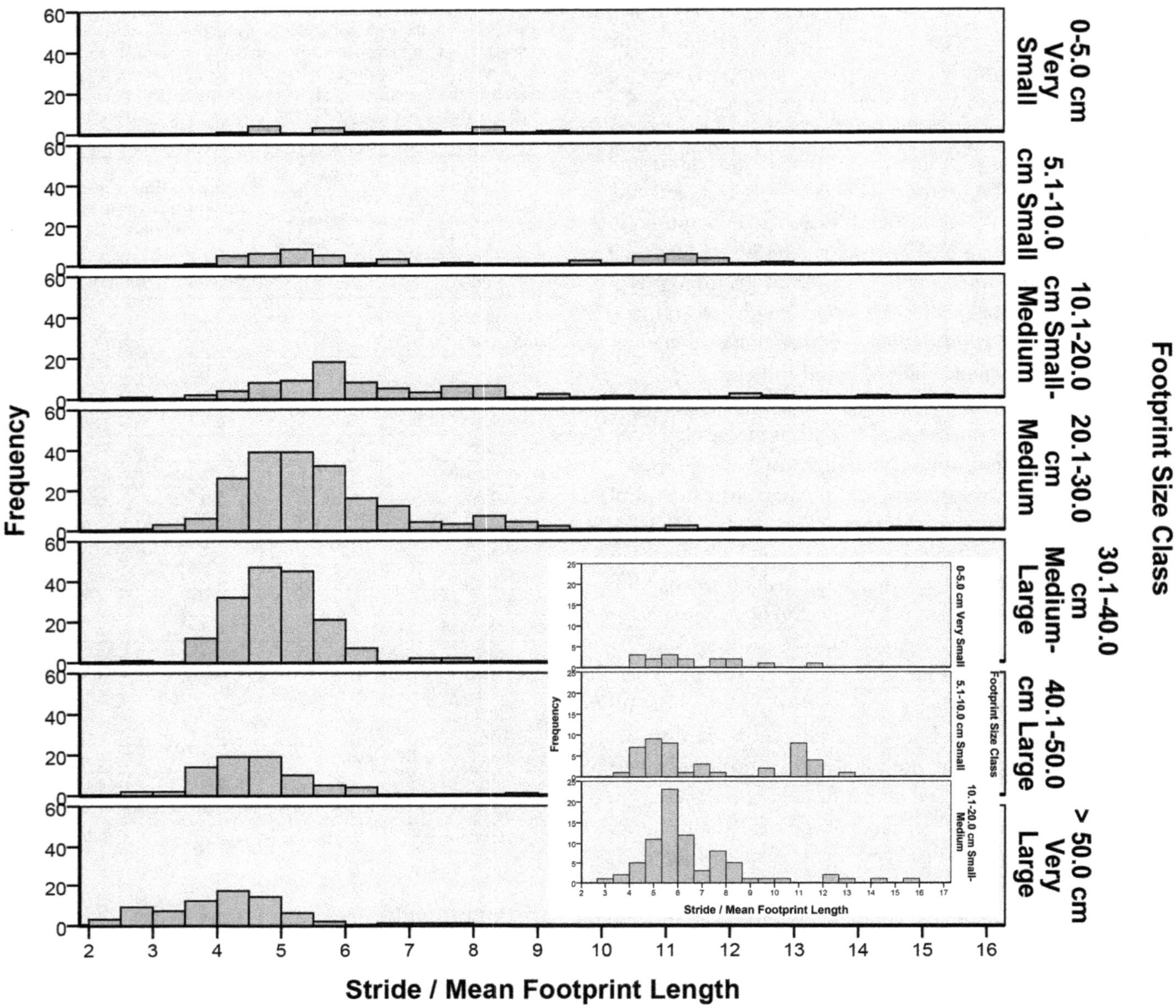

4.17. Distribution of the stride/footprint ratio in trackways attributed to ornithischian (mainly ornithopod) dinosaurs. *Inset* expands distribution for the three smallest size classes. Data are pooled across Mesozoic tracksites of all ages; see table 4.2 for a breakdown on the basis of different stratigraphic intervals.

the locomotion of individual animals (as expressed in each animal's stride/footprint length ratio) might have been affected by its desire to keep up with its neighbors, in which case the ratio values of trackways would not be independent points. The possibility of such biases can be evaluated by breaking the sample down into ichnofaunas from different stratigraphic intervals and geographic regions, restricting the plotted size and ichnofauna categories into those with at least 10 trackways (fig. 4.18).

This treatment shows that there is indeed some variability in values of the median stride/footprint ratio across ichnofaunas, and also severely underrepresented stratigraphic intervals. The smallest size class (1) is represented by 10 or more trackways only for the Triassic–Early Jurassic interval (fig. 4.18A). For any size class that is represented by several ichnofauna/trackmaker categories with at least 10 trackways, there is a fair degree of scatter in values of the median stride/footprint length ratio (e.g., size class 3 for theropods, and size class 4 for ornithischians; fig. 4.18*B*, 4.18*C*). Even so, the trend toward decreasing median values of the stride/footprint length ratio for animals from size class 2 upward seen in the pooled theropod and ornithischian samples (figs. 4.16, 4.17) seems to hold up. For trackways attributed to theropods, at least, there is no trackway evidence for increasing the ratio over the time interval from the early Mesozoic through the Cretaceous (fig. 4.18*B*; cf. Farlow et al. 2000).

4.18. Median values of the stride/footprint length ratio as a function of trackmaker size class (table 4.2) in trackways attributed to theropods and ornithischians of different stratigraphic intervals and faunas. To be included in these graphs, each trackway age/fauna category had to be represented by at least 10 trackways. *A*, Trackways grouped by trackmaker category and stratigraphic interval. Note that ornithischian medians consistently plot below theropod medians. All trackmaker/stratigraphic interval categories show the same pattern of declining median values of the stride/footprint length ratio from size class 3 to size class 7. Size class 1, the smallest dinosaurs, is represented here only by early Mesozoic dinosaurs. *B*, Trackways attributed to theropods from different ichnofaunas. *C*, Trackways attributed to ornithischians from different ichnofaunas. Although there is scatter in the values of the median stride/footprint length ratios for dinosaurs in the same size class among different ichnofaunas, both theropod and ornithischian patterns show a decline in the median value for animals bigger than size class 2, and possibly an increase in the median value from size class 1 to size class 2 (*A*).

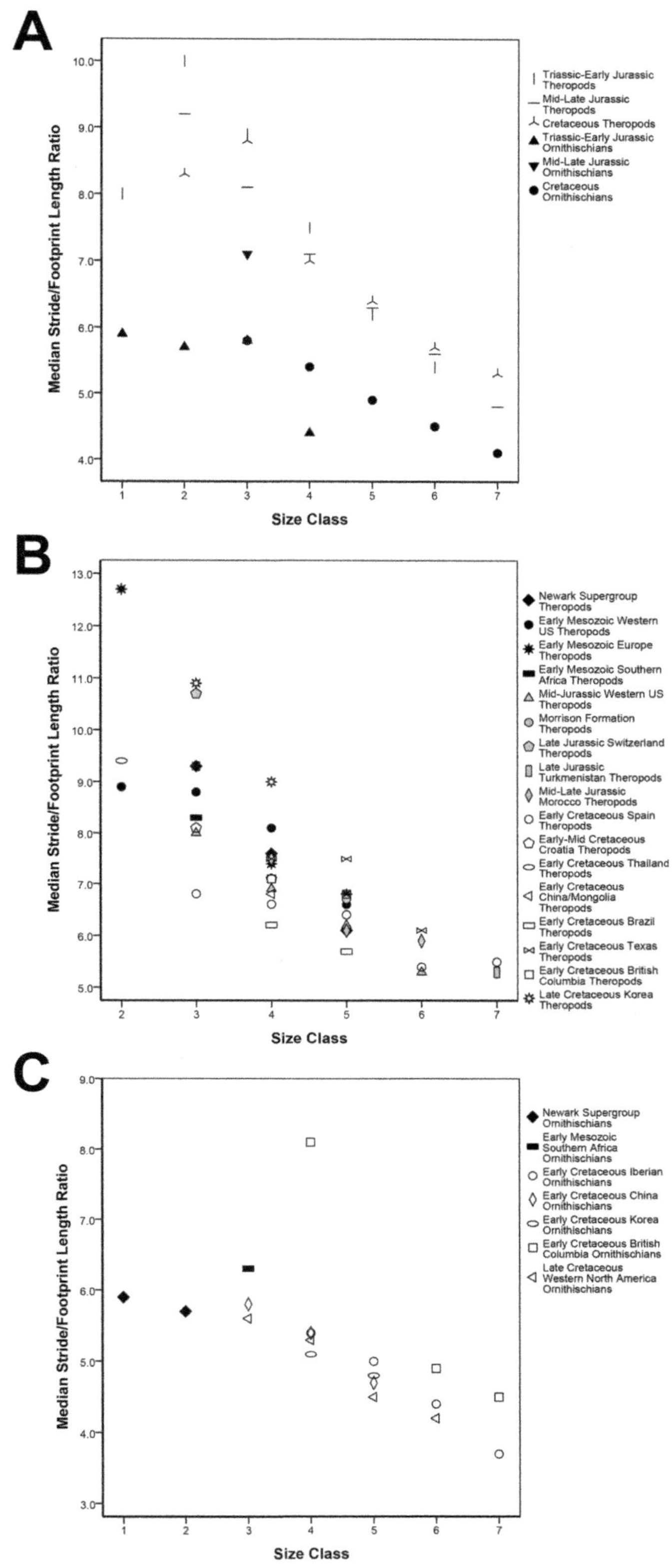

For theropods the trackway data are at least partly consistent with trends inferred from skeletal data (fig. 4.14). It is possible, but not certain (due to the paucity of data for post–Early Jurassic trackways), that the stride/footprint length ratio of walking animals would indeed have increased from trackways made by very small bipedal dinosaurs (footprint lengths 5 cm or less) to trackways with footprint lengths somewhere in the range of 10–20 cm. For walking dinosaurs bigger than that, the ratio would be expected to decrease (fig.4.14*B*), and the trackway data unambiguously support this conclusion (figs. 4.16, 4.18A, 4.18*B*). Judging from the skeletal data (fig. 4.14), this pattern of initial increase and subsequent decrease in relative stride length might occur both intraspecifically and interspecifically.

Although the match between expectations based on theropod limb proportions and trackway data tempts one to say that the former explains the latter, the ornithischian data caution against jumping to that conclusion. Anatomy seems not to be destiny in large ornithopods, because—all other things being equal—the increase in leg length relative to digit III length in big ornithopods, with values of the ratio substantially higher than those for large theropods (fig. 4.14A, 4.14*B*), might be expected to translate into higher stride length/footprint length ratios in big ornithopods than in big theropods. That is clearly not the case; in fact, the situation is just the opposite, with trackways attributed to large theropods showing as large or larger stride/footprint length ratios than those attributed to large ornithopods (cf. Thulborn 1990; Farlow et al. 2000). Conceivably this shortening of the step length is related to quadrupedalism, rather than bipedalism, being the norm in large ornithopods (Maidment and Barrett 2014; Barrett and Maidment 2017). Alternatively, if big ornithopods showed greater development of a "heel" pad behind the digital portion of the foot (Thulborn 1990), this would have exaggerated soft-tissue foot and footprint length, decreasing the stride/footprint length ratio. Or perhaps the decline in median stride/footprint length with increasing

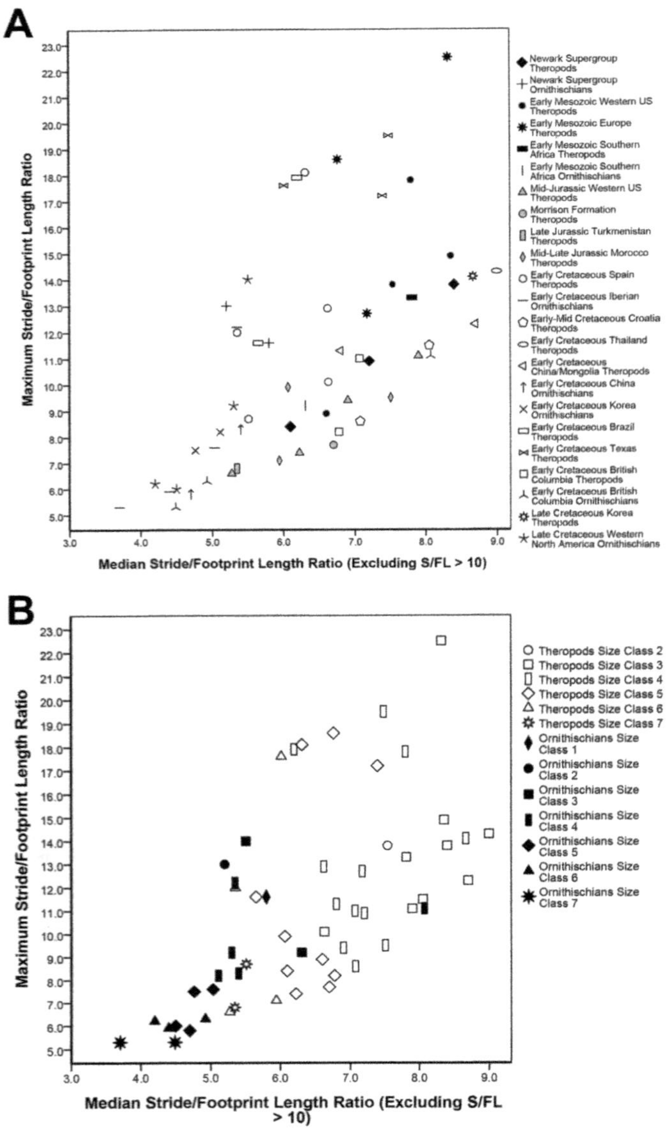

4.19. Relationship between the median and maximum stride/footprint length ratio in categories of bipedal and potentially bipedal non-avian dinosaurs. To reduce the effects of extremely high values of the stride/footprint length ratio on the median values for trackways made by presumed walking dinosaurs, medians are calculated without data cases in which the stride/footprint length ratio is greater than 10. To be included in these graphs, each ichnofaunal sample had to be represented by at least 10 trackways. *A*, Data broken down by ichnofaunal sample. *B*, Data broken down by trackmaker type and footprint size class.

size among both large theropods and ornithopods is related more to other anatomical constraints (cf. Hutchinson et al. 2011), or greater caution during locomotion, on the part of bigger than of smaller animals, than to size-related changes in limb proportions.

As previously noted, the overall relationship between the stride/footprint ratio and trackmaker size (with footprint length as the proxy of the latter) is shaped like a right triangle (fig. 4.15*B*): The lowest values of the ratio do not change with increasing footprint length, but the maximum ratios consistently become smaller as the animals get bigger. There is, furthermore, a clear positive correlation between median and maximum values of the stride/footprint length ratio for all trackmaker categories (table 4.2, fig. 4.19), with smaller maximum and median values seen in the larger size classes. To some extent median values could be expected to increase automatically as maximum values get bigger, given the definition of a median, but the opposite does not have to happen; median values could increase or decrease among sets of data cases with the same maximum value, depending on the distribution of values less than the maximum value. Consequently the positive correlation between median and maximum values of the stride/footprint length ratio reinforces the conclusion drawn from the overall relationship between maximum values of the ratio and footprint size: that small to medium-sized dinosaurs were bouncier than their bigger kin. Unfortunately, if these size changes apply both to ontogenetic, intraspecific sequences and also across species to comparisons of adult animals of different body sizes, this means that they cannot be used to distinguish trackways of young individuals of large-bodied species from trackways of adults of small-bodied species. It also means that one cannot assume, on the basis of relative stride length alone, that a big trackmaker with a relatively low stride/footprint length ratio was a different kind of animal than a much smaller trackmaker with a relatively high stride/footprint length ratio.

Looking ahead a few chapters, one specimen has long struck Farlow as particularly matching expectations for what the trackway of a baby dinosaur should look like (fig. 8.13*D*). This is a lovely little specimen in the collection of Yale University's Peabody Museum of Natural History, from the Early Jurassic of Massachusetts, catalogued under the ichnospecies name *Arachnichnus dehiscens*. The footprint length is about 2.7 cm. The trackmaker had a very pigeon-toed gait and a short stride (mean 6.8 cm). To Farlow's eye, this trackway could have been made by a juvenile of the kind of animal that would grow up to be an *Anomoepus*-maker. Perhaps it would have made a nice pet.

In these first few chapters we have examined the implications of proportions of foot and/or limb skeletons of non-avian dinosaurs, ground birds, and even crocodylians for interpreting dinosaur ichnology, looking at data in several different ways. But except, perhaps, in grade-B Hollywood horror movies, skeletons do not make footprints. Intact feet, with the enveloping tissues surrounding the bones, do. So now we must begin looking at intact feet.

Intraspecific Variability in Pedal Size and Shape in *Alligator mississippiensis*

5

INTRODUCTION

In previous chapters we considered intraspecific variability in skeletal digital and phalangeal proportions in crocodylians, birds, and non-avian dinosaurs. We now turn our attention to dimensions of "intact" feet, with soft tissues and horny claws investing the foot skeleton. We begin, in this chapter, with a quantitative analysis of intraspecific variability in pedal size and proportions in the American alligator. Dodson (2003: 889) made the wry observation that "It has become almost (but not quite) a rite of passage for paleontologists to study crocodylians early in their careers." I am as happily guilty of this charge as anybody.

STUDY AREAS AND SPECIMENS

Most of the alligators measured during this study came from three places: (1) Rockefeller Wildlife Refuge (RWR; Grand Chenier, LA); (2) Par Pond, Savannah River Site (New Ellenton, SC); (3) Lake Placid–Lake Okeechobee area, central Florida (table 5.1).

Alligators from RWR have been intensively studied from the standpoints of ecology, physiology, functional morphology, morphometrics, osteohistology, and ichnology (e.g., Chabreck 1966; Joanen 1969; Joanen and McNease 1970, 1972; Coulson et al. 1973; Chabreck and Joanen 1979; Coulson and Hernandez 1983; Reilly and Elias 1998; Farlow and Britton 2000; Davis et al. 2001; Seebacher et al. 2003; Farlow and Elsey 2004, 2010; Farlow et al. 2005; Bonnan et al. 2008; Livingston et al. 2009; Woodward et al. 2011, 2014, 2015). Most of the RWR alligators in the present study were animals that had hatched from wild eggs in the early 1970s and been kept in outdoor pens as part of a study of breeding biology. These reptiles had been routinely fed by Louisiana Department of Wildlife and Fisheries (LDWF) personnel, but also were able to supplement their diets by wild prey that unwisely strayed

Table 5.1. Breakdown of alligators measured during this study, on the basis of geographic region, sex, and body size. Total number of alligators across all regions = 96. Sex unknown/uncertain refers to alligators that could not be sexed, whose sex could not be determined with certainty, or that were not sexed for whatever reason. Parenthetic numbers in the table (*N* of alligators) indicate the total number of animals in a sample when it was impossible to measure the total lengths of all animals; the number preceding the parentheses is the count of alligators whose total length could be measured. See Farlow and Britton (2000) for details about particular animals (other than those from Florida).

Geographic region	Group	Total length (mm)			*N* of alligators
		Minimum	Maximum	Mean	
South Carolina	Total sample	518	3,450	1,306.0	12 (13)
	Females	1,050	2,246	1,402.0	4
	Males	3,230	3,450	3,340.0	2 (3)
	Sex unknown/uncertain	518	594	564.0	6
Louisiana	Total sample	440	3,873	2,494.3	43
	Females	1,550	2,845	2,431.6	26
	Males	1,613	3,873	2,855.1	15
	Sex unknown/uncertain	440	766	603.0	2
Florida	Total sample	1,549	3,470	2,609.8	24
	Females	—			0
	Males	3,130	3,470	3,306.9	7
	Sex unknown/uncertain	1,549	2,972	2,322.8	17
South Carolina/Alabama cross	Sex unknown/uncertain = total sample	375	478	447.3	4 (7)
Unknown	Total sample	546	1,924	899.2	9
	Females	1,583	1,924	1,753.5	2
	Males	—			0
	Sex unknown/uncertain	546	697	655.1	7

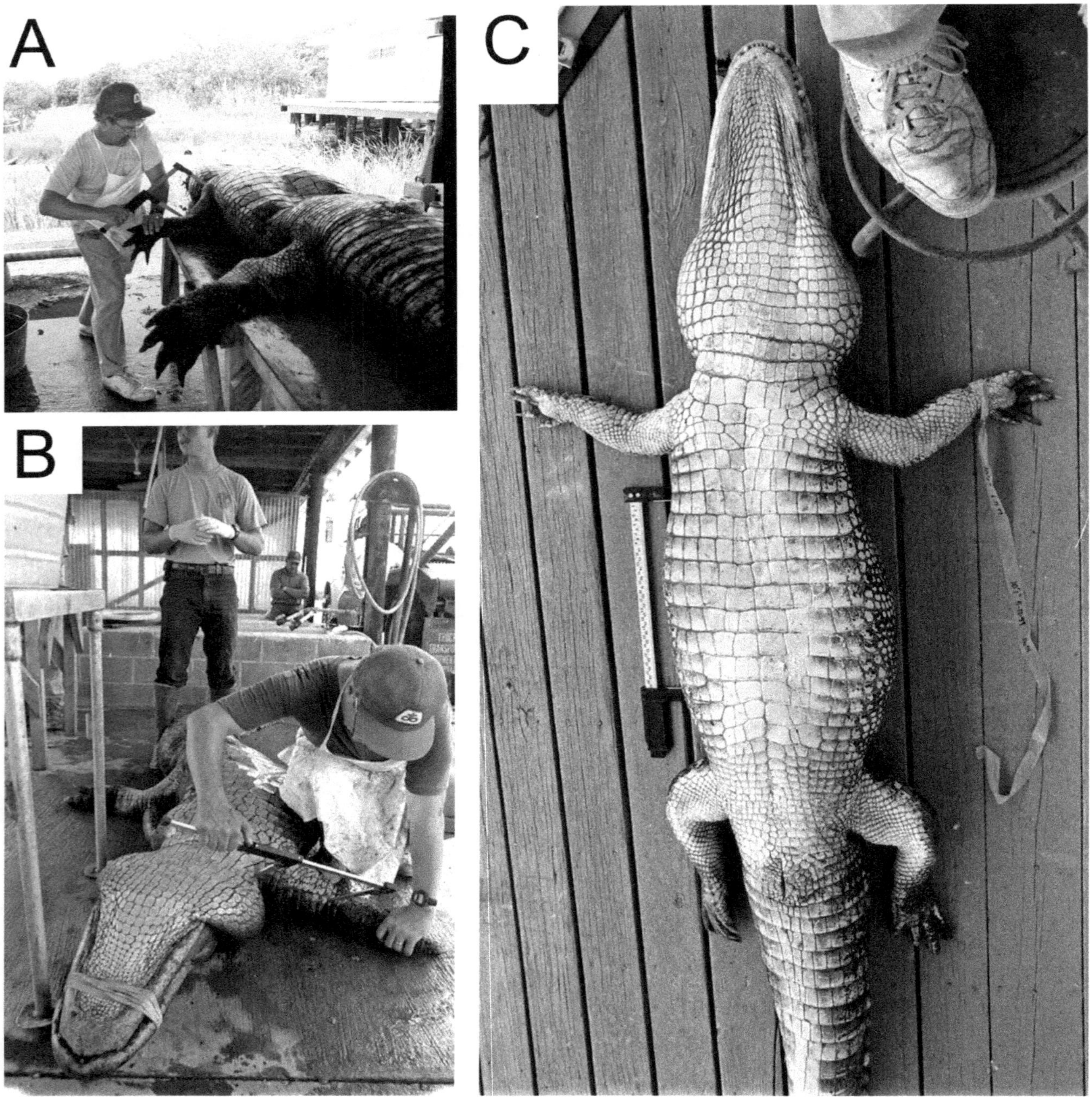

Above, **5.1.** Alligator carcasses. *A, B,* The author measuring alligator carcasses, Rockefeller Wildlife Refuge, Grand Chenier, Louisiana. *C,* Rockefeller Wildlife Refuge roadkill male 2; total length = 3073 mm; snout-vent length (measured to anterior edge of cloaca) = 1542 mm.

Facing, **5.2.** Hindfeet of alligators seen in ventral (plantar) view. Claws are present on digits I–III; digit IV (the outermost toe) lacks a claw. Feet are identified as left (L) or right (R). Specimens are arranged in order of increasing total length (TL) of the intact alligator. *A,* Par Pond (Savannah River Site, South Carolina) wild juvenile 3 (L); TL = 58 cm; digit III free length (IIIFree) = 27 mm. *B,* Par Pond wild juvenile 2 (L), TL = 59 cm, IIIFree = 30 mm. *C,* CITES 0054237 (R), wild Florida alligator, sex not determined, TL = 155 cm, IIIFree = 58 mm. *D,* RWR 19 (L), wild Louisiana male, TL = 177 cm, IIIFree = 76 mm. *E,* RWR 24 (L), wild Louisiana female, TL = 198 cm, IIIFree = 91 mm. *F,* RWR 23 (R), wild Louisiana male, TL = 213 cm, IIIFree = 92 mm. *G,* Par Pond EEB (R), wild South Carolina female, TL = 225 cm, IIIFree = 100 mm. *H,* RWR 17 (L), wild Louisiana male, TL = 234 cm, IIIFree = 98 mm. *I,* RWR 26 (R), wild Louisiana male, TL = 244 cm, IIIFree = 106 mm. *J,* RWR 20 (R), captive Louisiana female, TL = 267 cm, IIIFree = 114 mm. *K,* RWR 15 (R), captive Louisiana female, TL = 268 cm, IIIFree = 120 mm. *L,* CITES 0043361 (L), wild Florida alligator, sex not determined, TL = 272 cm, IIIFree = 117 mm. *M,* RWR 11 (L), captive Louisiana female, TL = 272 cm, IIIFree = 119 mm. *N,* RWR 28 (L), captive Louisiana female, TL = 277 cm. IIIFree = 125 mm. *O,* CITES 0056495 (L), wild Florida alligator, sex not determined, TL = 279 cm, IIIFree = 115 mm. Note significant wear on the claws of digits I and II. *P,* RWR 7 (R), captive Louisiana female, TL = 281 cm, IIIFree = 121 mm. Note significant wear on the claw of digit I. *Q,* CITES 0041015 (R), wild Florida alligator, sex not determined, TL = 282 cm, IIIFree = 129 mm. *R,* RWR wild roadkill male 2 (R), TL = 307 cm, IIIFree = 120 mm. *S,* CITES 0040187 (R), wild Florida male, TL = 325 cm, IIIFree = 143 mm. *T,* CITES 0041602 (R), wild Florida male, TL = 340 cm, TL = 140 mm. *U,* Par Pond 00AC (L), wild South Carolina male, TL = 345 cm, IIIFree = 128 mm. *V,* RWR 27 (R), captive Louisiana male, TL = 361 cm, IIIFree = 147 mm. *W,* RWR 31 (L), captive Louisiana male, TL = 361 cm, IIIFree = 146 mm. *X,* RWR 13 (R), captive Louisiana male, TL = 376 cm, IIIFree = 124 mm. *Y,* RWR 22 (L), captive Louisiana male, TL = 381 cm, IIIFree = 149 mm.

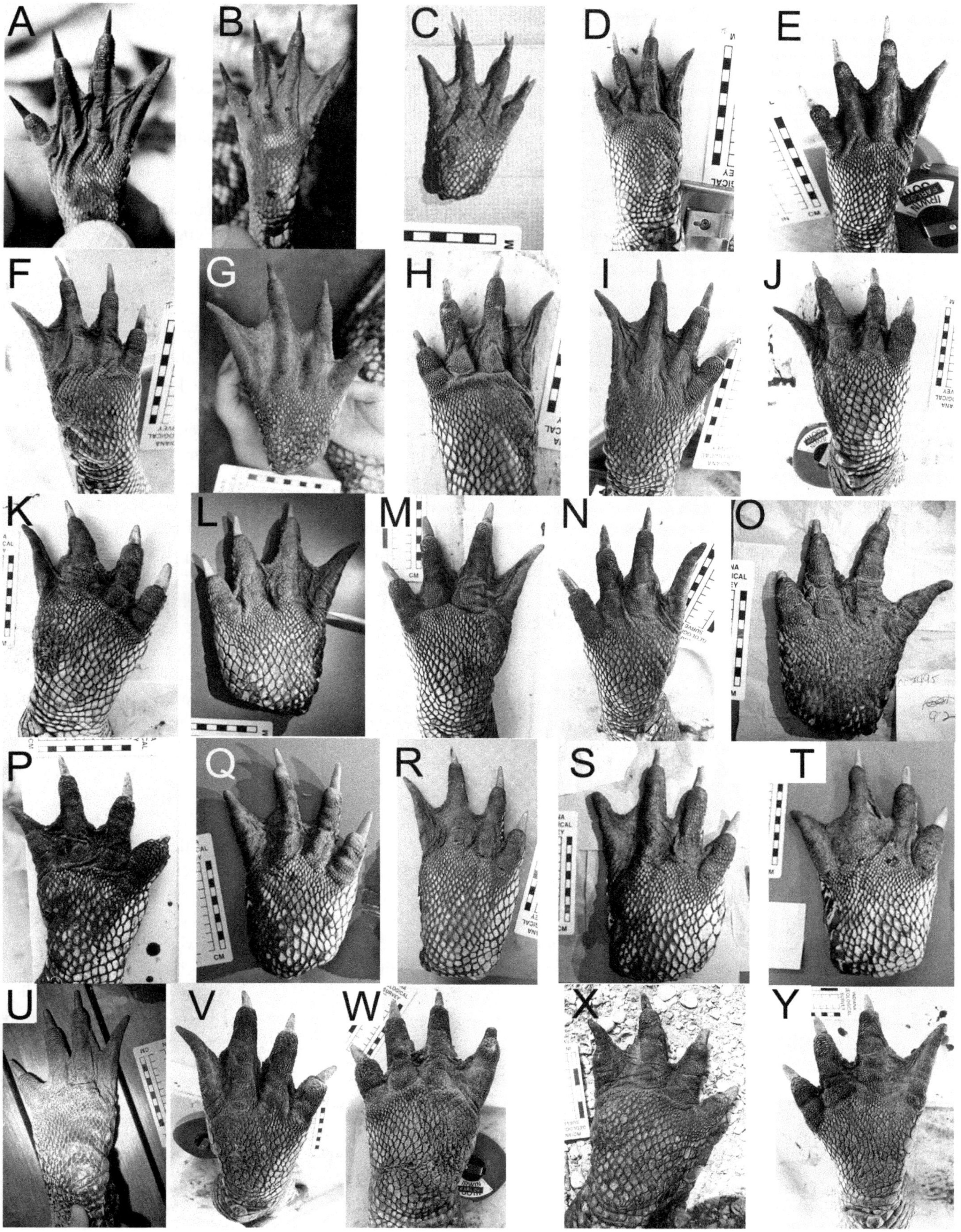
A
B
C
D
E
F
G
H
I
J
K
L
M
N
O
P
Q
R
S
T
U
V
W
X
Y

into their pens. Because these alligators were well nourished and protected from hunters, they had grown larger than wild alligators in the same area. In 1999 budgetary constraints forced LDWF personnel to discontinue the study, and the alligators had to be sacrificed. In addition to these alligators, some of the RWR animals were very young captives or wild-caught individuals (fig. 5.1). A few were animals originally captured at RWR that had been kept as captives in the Brisatron (see below) for many years.

Par Pond is a large (c. 1,100-ha) artificial reservoir on the Savannah River Site whose alligator population has been studied by scientists of the Savannah River Ecology Laboratory (SREL; University of Georgia) for many years (Murphy 1981; Brandt 1991*a*, 1991*b*; Brisbin et al. 1996). I measured both adult and young alligators that were captured in Par Pond and released after measurement.

The central Florida alligators were animals killed during the annual legal hunt and brought to a slaughterhouse for processing. Each animal received a CITES (Convention on International Trade in Endangered Species) tag number, which I used to label that specimen in this study.

Numerous alligators were kept in captivity at SREL. Several medium-sized individuals were maintained in the Brisatron, an indoor aquatic research facility. Some of these animals were originally from RWR, but others are of unknown provenance. Many small alligators are kept at SREL in aquaria and small artificial ponds. Some of these are offspring of "nuisance" parents from Alabama and South Carolina that bred in captivity at SREL, but many are of unknown origin.

Where possible, alligators were sexed by cloacal probing (Chabreck, 1963). This technique is generally used for animals whose total length is at least 350 mm (Murphy 1981; Brandt 1991*b*). Because I was a novice at this rather intimate procedure, I only used sex determinations that were confirmed by a more experienced gator-groper than I.

Most of the alligators in my Florida sample had been butchered without my being present, with their feet frozen until my arrival, making cloacal examination impossible. However, male alligators grow to considerably larger sizes than females, the latter generally reaching maximum total lengths of about 2.8 m (McIlhenny 1935; Chabreck and Joanen 1979; Murphy 1981; Joanen and McNease 1987; Coulson and Hernandez 1983), but occasionally attaining lengths of as much as 3.1 m (Woodward et al. 1995). I identified any Florida alligator with a total length of at least 3 m as a male; smaller animals were not sexed. Of the seven Florida specimens that I identified as males on the basis of total length, two were about 3.1 m in length, and so could have been extremely large females. It seems unlikely, however, that there should have been two females that happened to be of nearly record size in my small sample.

Most alligators of 1.83 m total length are sexually mature (Joanen and McNease 1980), although Wilkinson (1983) reported that females in coastal South Carolina do not become sexually mature until they reach a length of 2 m (also see Wilkinson and Rhodes 1997). Although I used 1.83 m as my criterion for maturity, in actuality the smallest alligator assumed to be sexually mature in the present study was about 1.92 m long. Only one of my South Carolina females was identified as mature, and her total length was well above 2 m.

Because capture of most alligators in this study was fortuitous, I had no control over the number of animals from each geographic region or of each sex or size. This precluded "balanced" evaluation of the effects of geographic origin, sex, and size on hindfoot shape (with equal numbers of animals in each combination of geography vs. sex vs. size). However, I tried to isolate the effects of each potentially important variable as best I could.

One contrast that I was not able to evaluate quantitatively was the possible effect of living in captivity as opposed to living in the wild on foot shape. However, I saw no indication that the feet of captive individuals differed from wild alligators in any of the parameters that I measured.

THE ALLIGATOR PES

The alligator hindfoot is plantigrade (figs. 5.2, 5.3). Scales on the underside of the foot are finer beneath the toes than beneath more proximal parts of the foot. The pes has four digits (the manus has five). Digits I–III bear well-developed claws, that of I being the stoutest, while digit IV lacks claws and terminates bluntly in skin. Digits I–III project increasingly farther from the proximal end of the foot, and digit IV projects about as far as II. The toes are conspicuously webbed, with the webbing between pairs of toes extending

Facing, **5.3.** Features of the alligator foot. *A–C,* Relation between skeletal structure and external features of the right pes of RWR wild male 2 (TL = 307 cm); the foot is shown in surface plantar view in Fig. 5.2*R*. *A,* Computed tomography (CT) scan, and *B,* conventional x-ray study of the foot. Bright pinpoints of light in *B* correspond to the centers of the digital pads of digit I. *C,* Longitudinal CT scan of digit I. *A–C* show that digit free lengths, as measured on alligators in this study, include the distal ends of the metatarsals. The first digital pad of digit I clearly coincides with the metatarsophalangeal joint, and the second digital pad is associated with the joint between phalanges I1 and I2 (the latter best seen in *C*). The horny sheaths covering the unguals extend very little beyond the distal ends of the unguals. *D, E,* Cast of a right footprint of RWR 27, a large captive Louisiana male; figure 5.2*V* shows the foot that made this print. *D,* View from directly overhead; the digital claws dug deeply into the substrate (the clawtip of digit III is directed straight upward in this view), and left downward drag marks during emplacement. Digit IV is less shallowly impressed than the other three digits; the alligator put more weight on the inside than on the outside of its foot. *E,* Oblique view from the rear of the print. Note obvious presence of interdigital webs in the footprint.

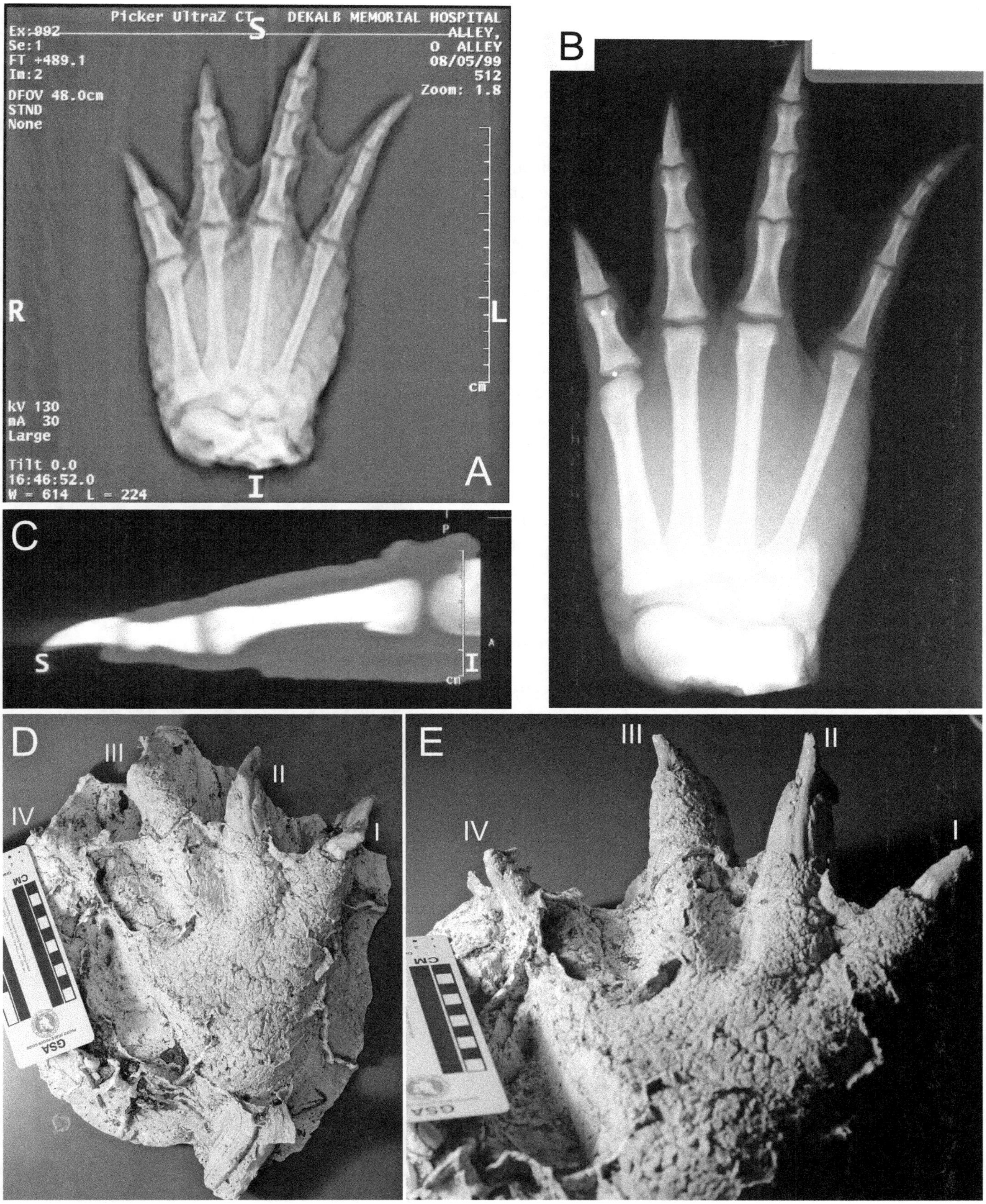
Picker UltraZ CT
DEKALB MEMORIAL HOSPITAL
Ex:992
Se:1
FT +489.1
Im:2
DFOV 48.0cm
STND
None
ALLEY,
O ALLEY
08/05/99
512
Zoom: 1.8
S
R
L
cm
kV 130
mA 30
Large
Tilt 0.0
16:46:52.0
W = 614 L = 224
I
A
B
C
P
S
I
cm
D
III
II
IV
I
CM
GSA
E
III
II
IV
I
CM
GSA

increasingly distally from the medial to the lateral side of the foot; the distal edge of the web between III and IV is thus much farther from the proximal end of the foot than the edge of the web between I and II. In plantar view, the proximal ends of the toes are sometimes, but not always, marked by creases that separate the toes from the "palm" of the foot.

Apart from digit I, pedal digits are not clearly subdivided by creases into digital pads (digital pads are much better developed on the fingers of the manus, but even here pad development seems rather haphazard compared with the clearly delimited pedal digital pads of many ground birds and [as seen in well-preserved footprints] bipedal non-avian dinosaurs). The fleshy part of pedal digit I is usually subdivided into two digital pads. Sometimes these pads abut one another, with only a narrow crease separating them. Sometimes, however, there is a clear gap between the two pads. In many cases the gap is simply due to nearly vertical or distinctly concave-downward (with the plantar surface of the foot directed downward) margins of the distal end of the first (most proximal) and the proximal end of the second pad. In other instances, however, there are distinct creases delimiting an interpad space, reminiscent of that seen in the toes of galliform birds (Lucas and Stettenheim 1972), between the two digital pads. The different kinds of boundaries between the two pads of digit I are not discrete categories, but rather grade into each other, and deciding whether to recognize an interpad space was often difficult.

Some alligator feet have very worn claws (fig. 5.2*O*, 5.2*P*). Of 54 alligators that had a total length of at least 1,830 mm (see below) and measurable digit I claws, 6 animals (11%) had digit I claws so heavily worn that this markedly affected the claw length. Because the horny claw does not extend much beyond the terminal limit of the underlying ungual (fig. 5.3*A*, 5.3*C*; table A5.1), wear was not restricted to the horny claw sheath, but sometimes even blunted the tip of the ungual. Digit lengths and claw lengths affected in this matter by wear were not used in most morphometric analyses.

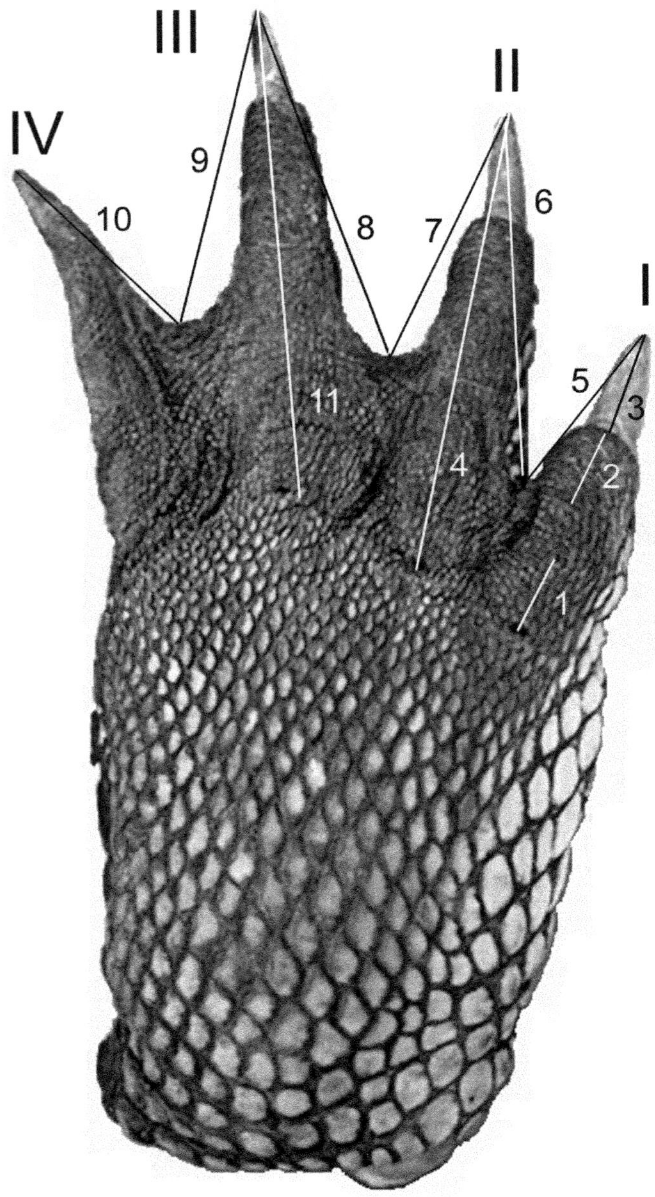

5.4. Underside of an alligator right hindfoot, illustrating conventions for measuring digital parameters. Not every measurement is illustrated, but at least one example of each kind of measurement is shown. *Key to measurements*: 1, Digit I first pad length. An interpad space was recognized on this toe between the first and the second digital pad. 2, Digit I second pad length. 3, Digit I claw length. 4, Digit II free length. 5, Digit I lateral hypex length. 6, Digit II medial hypex length. 7, Digit II lateral hypex length. 8, Digit III medial hypex length. 9, Digit III lateral hypex length. 10, Digit IV medial hypex length. 11, Digit III free length.

MEASUREMENTS AND METHODS

Several measurements of overall body proportions were taken (table A5.2; Farlow and Britton 2000):

Total length: Measured from the snout to the tip of the tail.

Snout-vent length: Measured from the snout to the anterior (SVLant) or posterior (SVL) end of the cloaca.

Tail length: This parameter was not directly measured, but instead was estimated (a slight underestimate) by subtracting SVLant from total length.

Total leg length: Measured as the sum of upper leg length (measured along the anterior edge of the hindlimb, from the lateral edge of the body to the middle of the knee) and lower leg length (measured along the anterior edge of the hindlimb, from the middle of the knee to the crook of the ankle).

Pes length: Measured in two ways: For most analyses (particularly comparisons with animal body size and leg length), overall pes length was measured along the dorsal side of the foot, from the crook of the ankle to the tip of the claw of digit III, with digit III held straight out. This was the easiest way to measure foot length in live, restrained alligators. For some comparisons, though, a "pes print proxy" was measured as the "non-toe palm" length of the foot (described immediately below) plus the "free" length of digit III (also described below).

Length of the toe region of the foot: The "non-toe palm" length of the foot was measured as the distance from the heel of the foot (identified as the most convex part of the back of the foot when it was dorsiflexed) to a line cutting across the bases of digits I and IV, in ventral view. The length of the toe region of the foot was approximated by subtracting the length of the non-toe palm length from overall pes length. As discussed below, however, the length of the toe region of the foot as defined this way includes the distal ends of the metatarsals.

In addition, several measurements of pes (and sometimes hindfoot print) size were made (fig. 5.4). Measurements less than 10 mm in length were taken to the nearest 0.5 mm. Measurements of at least 10 mm were taken to the nearest millimeter, except in a few cases (involving smaller alligators) in which a measurement was so close to the half-way point between two integer millimeter lengths that I recorded it to the 0.5 mm.

Digit I length: Measured from the crease at the proximal end of the toe to the tip of the claw. In this and other toe lengths, measurements were made with the toe held in a straight line, but not stretched.

Lengths of the first (proximal) and second (distal) pads of digit I: Measured parallel to the long axis of the toe. I took care not to deform the pads while putting calipers on them. In two alligators there was no crease subdividing the fleshy part of digit I into pads. For analytical purposes I assumed that the length of the first digital pad of these two specimens was the same as the length of the total fleshy part of the toe, and assigned a length of zero to the second digital pad. This obviously affected analytical procedures involving log-transformed variables, due to the impossibility of calculating the logarithm of zero; such cases had to be excluded from the analyses.

Free lengths of digits II and III: Even though webbing connects the toes over half or more of their lengths, it is still possible to see the individual toes as discrete units proximal to the webs. Leonardi (1987) used the term "free length" to designate the length of a digit distal to a line connecting the hypexes on either side of the digit. Using the human hand as an analogy, a digit's free length is how far it projects as a discrete entity beyond the palm. In the alligator pes, one could with equal justification describe free lengths relative to the interdigital webs or the bases of the toes proximal to the webs. I chose the latter here (fig. 5.4), but used the webs to define hypex lengths (below). Sometimes there were clear creases at the bases of the toes, in which case the free length was measured from the crease to the claw tip. Where such proximal creases did not occur, I measured a digit's free length from an imaginary line connecting the non-web hypexes on either side of the digit to the tip of the claw. Because of concerns that I may not have measured free lengths the same way in 1998 as I did in later years, for some analyses I analyzed measurements involving the free lengths of digits II and III both with, and without, 1998 free length measurements (the alligators measured in 1998 were all captives [with a large proportion of fairly small individuals], animals from Louisiana and South Carolina, small individuals bred from South Carolina and Alabama parents, and captives of unknown origin). Recognizing the proximal end of digit IV in a consistent manner was so difficult that I did not consider free lengths of this digit in my analyses. Manipulations and x-ray studies (fig. 5.3) of toes indicated that digit free lengths measured as described here include the distal-most ends of the metatarsals, thus exaggerating digit lengths as compared with osteological measurements.

Claw lengths: I measured claw lengths of digits I–III on their undersurfaces, from the terminal end of the fleshy part of the toe to the claw tip. The length of digit I excluding the claw, and the free lengths of II and III without their claws, were computed by subtracting the claw length from the overall length of the toe (free lengths in the cases of digits II and III).

Hypex lengths: Measured from a digit's claw tip to the web connecting the toe to its neighboring toe. Hypex lengths were measured with the toes extended and separated from each other, but not to the point that the web was tautly stretched. Where the web between two toes had an acute indentation, this was the spot on the web used in making the measurement, even if it was closer to one toe than to its partner. Where there was no such indentation, the midpoint on the web between the two toes was used in making the measurement. Digits II and III each had both a medial and a lateral hypex length. Digit I only had a

Table 5.2. Summary measurements of total length and hindfoot length (measured for individuals with unworn claws on digit III) of alligators examined during this study. All measurements are in millimeters. "Mature" alligators are defined as those with a total length of at least 1,830 mm. Sexed animals are those mature individuals whose sex was determined with certainty. Data for pedal parameters of mature alligators are in table 5.3.

Sample group	Parameter	Minimum	Maximum	Mean	Maximum/minimum ratio	*N*
All alligators	Total length	375	3,873	2,124.4	10.3	92
	Pes length	32	252	125.7	7.87	65
	Digit III free length (all data)	22	154	87.0	7.00	88
	Digit III free length (excl. 1998)	22	154	101.2	7.00	62
Mature alligators	Total length	1,924	3,873	2,802.0	2.01	57
Mature females	Total length	1,924	2,845	2,613.9	1.48	22
Mature males	Total length	2,134	3,873	3,296.1	1.81	20

lateral hypex length, and digit IV only a medial hypex length.

Pes print length: Measured from the back of heel impression of the sole of the foot to the tip of digit III. However, I also took hindfoot print lengths from the literature, in which the ways the lengths were measured were not necessarily identical to the way I measured print lengths.

For some analyses I created a simple proxy of overall hindfoot size by taking the mean of the log-transformed values of digit I length, digit II and III free lengths, and the hypex lengths (medial, lateral, or both as appropriate) of digits I through IV. Each of the log-transformed digit length measurements could then be scaled by subtracting the simple size proxy from each log-transformed digit length measurement. A discriminant analysis using such scaled digit length parameters was one of the methods used to evaluate sexual dimorphism in alligator foot shape.

Both bivariate and multivariate analyses of allometry were done. Variables were log transformed prior to analysis, and allometry was assumed if the 95% confidence interval (CI) of the slope of the relationship did not include 1 (Rayner 1985; Leduc 1987). Some relationships met this criterion of allometry, but only barely, with the lower or upper limit of the 95% CI being very close to one. Reduced major axis (RMA) relationships between variables were considered to be "only barely" allometric if the upper CI limit of RMA slope was between 0.95 and 1.00, or if the lower CI limit was between 1.00 and 1.05.

For multivariate analyses of allometry, I compared each of the log-transformed digit length variables used to calculate the simple hindfoot size proxy with a modification of the latter, in which the size proxy in each analysis was calculated as the mean of all the log-transformed digit length parameters except the one being analyzed. The reduced major axis of the slope of this relationship was then compared with a multivariate allometric coefficient for the same digit length variable calculated in a principal components analysis (using a covariance matrix). The multivariate allometric coefficient was calculated by dividing the loading of each log-transformed variable on the first principal component (PC1) by the mean PC1 loading of all the variables (cf. Kowalewski et al. 1997; Hammer and Harper 2006).

RESULTS

Size Range

Alligators measured during this study ranged from about 37 to 387 cm in total length (table 5.1). Hatchling alligators are generally about 20–30 cm in total length (Kellogg 1929 McIlhenny 1935; Goodwin and Marion 1978; Fuller 1981; Wilkinson 1983; Brandt 1991*a*). Adult males as much as 18 feet (549 cm) long are said to have been killed in the past (e.g., McIlhenny 1935), but Woodward et al. (1995) expressed doubt about these claims, and suggested that the maximum total length of alligators is about 460 cm (15 feet). I would give a great deal to be able to measure an alligator that big! In any case, my sample spanned most of the size range of the species, even if it didn't include the size extremes at either end.

The range of hindfoot measurements is proportionally less (as expressed as the ratio of maximum/minimum values) than the range of total lengths (table 5.2), reflecting the allometric relationship between pes length and overall body size (see below).

Variability of Pedal Measurements in Mature Alligators

The variability of hindfoot length measurements in adult alligators was analyzed in two ways: first, on a parameter-by-parameter basis, with variable sample sizes for particular parameters (table 5.3), and second, by using data for only those animals for which all measurements could be made,

Table 5.3. Variability of pedal measurements in mature alligators (total length at least 1,830 mm). All measurements are in millimeters. CV = coefficient of variation; *N* = number of alligators. Treatment: "all" = all mature alligators; "unworn" = alligators with unworn pedal claws; male and female breakdowns are for animals of known sex with unworn claws (for those parameters of which claw length is a component). Data for 1998 are included in this tabulation except where otherwise indicated.

Parameter	Treatment	Minimum	Maximum	Mean (CV)	Maximum/minimum ratio	*N*
Pes length	All	142	252	193.6 (15.9)	1.77	33
	Unworn	142	252	193.5 (16.3)	1.77	28
	Females	142	203	181.6 (10.2)	1.43	15
	Males	150	252	212.3 (16.5)	1.68	12
Length of toe region of foot	All	78	147	106.7 (18.3)	1.88	26
	Unworn	78	147	107.6 (18.1)	1.88	25
	Females	78	128	101.7 (14.9)	1.64	15
	Males	84	147	116.3 (19.5)	1.75	10
Digit I length	All	53	104	71.7 (17.4)	1.96	54
	Unworn	53	100	72.0 (16.2)	1.89	48
	Females	54	81	71.6 (11.3)	1.50	19
	Males	53	100	78.3 (18.2)	1.89	17
Digit I lateral hypex length	All	33	61	49.5 (15.5)	1.85	54
	Unworn	33	61	50.1 (14.8)	1.85	48
	Females	35	61	49.7 (14.2)	1.74	19
	Males	33	61	52.6 (16.3)	1.85	17
Digit I first pad length	All	11	43	20.6 (31.5)	3.91	54
	Females	13	37	20.9 (26.9)	2.85	21
	Males	15	43	22.9 (35.2)	2.87	19
Digit I second pad length	All	0	40	23.6 (31.6)	—[a]	54
	Females	0	32	24.2 (31.8)	—[a]	21
	Males	16	40	25.9 (29.0)	2.5	19
Digit I length excluding claw	All	31	75	47.2 (22.1)	2.42	54
	Females	31	55	46.0 (14.9)	1.77	21
	Males	32	75	53.9 (23.4)	2.34	19
Digit I claw length	All	14	32	23.8 (15.2)	2.29	54
	Unworn	18	32	24.5 (12.9)	1.78	48
	Females	19	32	24.9 (13.8)	1.68	19
	Males	18	29	24.8 (12.5)	1.61	17
Digit II free length	All	74	127	100.8 (13.3)	1.72	53
	All (excl. 1998)	74	127	100.9 (13.4)	1.72	52
	Unworn	74	127	101.2 (13.5)	1.72	50
	Females	79	111	99.2 (8.2)	1.41	19
	Males	77	127	108.6 (13.9)	1.65	20
Digit II medial hypex length	All	60	113	84.8 (15.3)	1.88	53
	Unworn	60	113	84.9 (15.6)	1.88	50
	Females	61	92	81.8 (11.2)	1.51	19
	Males	67	113	94.5 (14.1)	1.69	18
Digit II lateral hypex length	All	46	84	65.3 (15.3)	1.83	53
	Unworn	46	84	65.2 (15.8)	1.83	50
	Females	47	74	63.0 (13.0)	1.57	19
	Males	48	84	71.0 (15.2)	1.75	18
Digit II free length excluding claw	All	56	103	77.8 (16.0)	1.84	52
	All (excl. 1998)	56	103	77.9 (16.1)	1.84	51
	Females	60	87	75.8 (9.8)	1.45	19
	Males	59	103	84.7 (16.3)	1.75	21
Digit II claw length	All	14	29	23.2 (12.4)	2.07	52
	Unworn	17	29	23.4 (11.1)	1.71	50
	Females	17	29	23.4 (11.2)	1.71	19
	Males	18	27	23.6 (10.1)	1.50	18

Table 5.3. *continued*

Table 5.3. *continued*

Parameter	Treatment	Minimum	Maximum	Mean (CV)	Maximum/minimum ratio	*N*
Digit III free length	All	84	154	116.2 (14.4)	1.83	54
	All (excl. 1998)	84	154	116.4 (14.5)	1.83	53
	Unworn	84	154	116.5 (14.7)	1.83	49
	Females	91	127	113.6 (9.6)	1.40	18
	Males	92	154	126.9 (14.2)	1.67	20
Digit III medial hypex length	All	53	110	80.6 (16.9)	2.08	54
	Unworn	53	110	80.7 (17.4)	2.08	49
	Females	59	95	79.9 (12.4)	1.61	18
	Males	63	110	90.1 (15.9)	1.75	18
Digit III lateral hypex length	All	50	108	78.1 (17.8)	2.16	54
	Unworn	50	108	78.7 (18.0)	2.16	49
	Females	50	100	77.3 (15.9)	2.00	18
	Males	57	108	86.0 (16.4)	1.89	18
Digit III free length excluding claw	All	67	131	95.9 (16.1)	1.96	53
	All (excl. 1998)	67	131	96.0 (16.2)	1.96	52
	Females	72	103	92.5 (10.1)	1.43	20
	Males	72	131	105.4 (15.9)	1.82	21
Digit III claw length	All	5	26	20.5 (16.9)	5.2	53
	Unworn	14	26	20.9 (12.2)	1.86	49
	Females	17	26	21.2 (11.9)	1.53	18
	Males	17	24	21.9 (9.3)	1.41	18
Digit IV medial hypex length	All	30	74	49.7 (23.5)	2.47	54
	Females	32	71	49.5 (20.1)	2.22	21
	Males	33	74	55.4 (22.5)	2.24	19

[a] Parameter cannot be calculated because doing so would require division by zero.

Table 5.4. Comparison of variability of total length and pedal measurements in mature alligators with unworn claws. Data are for those alligators for which all measurements could be made and include 1998 data. Measurements are in millimeters. *N* = 22.

Parameter	Minimum	Maximum	Mean (CV)	Maximum/minimum ratio
Total length	1,924	3,873	2,843.3 (22)	2.01
Pes length	142	252	193.9 (17)	1.77
Length of toe region of foot	78	147	107.5 (19)	1.88
Digit I length	53	100	74.4 (20)	1.89
Digit I lateral hypex length	33	61	49.7 (19)	1.85
Digit I first pad length	13	35	21.5 (31)	2.69
Digit I second pad length	14	40	25.9 (30)	2.86
Digit I length excluding claw	31	75	49.1 (26)	2.42
Digit I claw length	18	31	24.6 (15)	1.72
Digit II free length	77	127	102.6 (15)	1.65
Digit II medial hypex length	61	113	86.9 (18)	1.85
Digit II lateral hypex length	47	84	65.3 (18)	1.79
Digit II free length excluding claw	59	103	79.5 (17)	1.75
Digit II claw length	18	27	23.1 (9.5)	1.50
Digit III free length	91	149	119.3 (14)	1.64
Digit III medial hypex length	59	110	83.3 (19)	1.86
Digit III lateral hypex length	50	108	79.2 (20)	2.16
Digit III free length excluding claw	72	128	97.7 (16)	1.78
Digit III claw length	17	26	21.6 (11)	1.53
Digit IV medial hypex length	32	74	52.0 (26)	2.31

Table 5.5. Approximate variability in total length, and predicted variability in pes length, in mature alligators from wild populations. The data sources reported alligator lengths in terms of size classes. If the size classes were described as single numbers (e.g., the 5-foot size class), I assumed that all the animals in that size class were of that length. If size classes were reported as a range of numbers (e.g., 2.5–3 m), I assumed that the length of all animals in that size class was the midpoint between the two numbers defining the size class. Total lengths were converted to millimeters prior to all calculations. Predicted pes lengths were calculated using the regression equation linking pes length and total length in table 5.6. CV = coefficient of variation.

Population	Parameter	Mean (CV)	Source of total length data
Par Pond, Savannah River Site, South Carolina, 1974–1976	Total length	2,754 (20)	Murphy (1981)
	Pes length	189 (17)	
Par Pond, Savannah River Site, South Carolina, 1986–1988	Total length	2,613 (23)	Brandt (1991b)
	Pes length	181 (21)	
Rockefeller Wildlife Refuge, Louisiana, 1966	Total length	2,122 (18)	Chabreck (1966)
	Pes length	151 (15)	
Louisiana, 1983	Total length	2,133 (19)	Taylor and Neal (1984)
	Pes length	151 (16)	
Welder Wildlife Refuge, Texas, 1998	Total length	2,145 (24)	Altrichter and Sherman (1999)
	Pes length	152 (21)	
Gainesville, Florida area, 1981–1983	Total length	2,504 (21)	Delany and Abercrombie (1986)
	Pes length	174 (18)	
L-39 Transect, Everglades, Florida, May 1967	Total length	2,197 (19)	Hines et al. (1968)
	Pes length	155 (16)	
Shark River Valley Loop Road Transect, Everglades, Florida, April 1967	Total length	2,302 (22)	Hines et al. (1968)
	Pes length	162 (19)	

Table 5.6. Bivariate relationships between parameters of hindfoot size and body size in alligators. For parameters in which claw length is a component of foot size, only data for animals with unworn claws were used. Data were log transformed prior to analysis; all measurements are in millimeters. CI, confidence interval; RMA = reduced major axis; SEE = standard error of estimate (regression model of relationship). Note that all RMA slopes were significantly different than 1, indicating allometry.

Independent variable	Dependent variable	r^2	SEE	RMA slope	95% CI of RMA slope	*N*
Total length	Pes length[a]	0.992	0.0243	0.877	0.857–0.898	61
	Digit III free length	0.979	0.0351	0.822	0.796–0.849	84
	Digit III free length (excluding 1998 data)	0.984	0.0276	0.868	0.840–0.897	62
	Digit III free length excluding claw	0.971	0.0414	0.836	0.805–0.867	88
	Digit III free length excluding claw (excl. 1998 data)	0.981	0.0301	0.902	0.871–0.934	66
	Simple foot size proxy	0.991	0.0251	0.874	0.855–0.893	75
Snout-vent length (anterior)	Pes length	0.991	0.0269	0.876	0.855–0.898	62
	Simple foot size proxy	0.992	0.0255	0.871	0.850–0.892	58
Shoulder-hip length	Pes length	0.992	0.0253	0.854	0.835–0.874	65
	Digit III free length	0.982	0.0354	0.799	0.772–0.827	65
	Digit III free length (excluding 1998 data)	0.988	0.0275	0.821	0.791–0.851	39
	Digit III free length excluding claw	0.975	0.0418	0.804	0.773–0.837	68
	Digit III free length excluding claw (excluding 1998 data)	0.987	0.0296	0.846	0.816–0.878	42
	Simple foot size proxy	0.992	0.0255	0.849	0.829–0.869	60
Leg length excluding foot	Pes length	0.987	0.0322	0.908	0.882–0.934	65
	Digit III free length	0.974	0.0429	0.849	0.815–0.885	65
	Digit III free length (excluding 1998 data)	0.986	0.0305	0.917[b]	0.882–0.955	39
	Digit III free length excluding claw	0.965	0.0497	0.853	0.814–0.894	68
	Digit III free length excluding claw (excluding 1998 data)	0.986	0.0307	0.944[b]	0.909–0.981	42
	Simple foot size proxy	0.986	0.0336	0.901	0.874–0.930	60

[a] Regression parameters: slope = 0.853; *y*-intercept = −0.789.
[b] Only barely allometric (upper CI limit of RMA slope between 0.95 and 1.00, or lower CI limit between 1.00 and 1.05).

Table 5.7. Bivariate relationships of log-transformed parameters of hindfoot size in alligators, broken down by sex and geographic origin; total length is the "independent" variable in all cases. Where claw length is a component of the parameter analyzed, data cases are animals with unworn claws. RMA slopes significantly different than 1 are indicated in **bold**. CI = confidence interval; RMA = reduced major axis; SEE = standard error of estimate.

Parameter	Treatment	r^2	SEE	RMA slope	95% CI of RMA slope	*N*
Pes length[a]	Females	0.962	0.0231	**0.844**	0.774–0.920	25
	Males	0.971	0.0228	**0.899**[b]	0.814–0.993	16
	South Carolina alligators	0.991	0.0270	**0.888**[b]	0.831–0.950	12
	Louisiana alligators	0.985	0.0269	**0.918**[b]	0.878–0.960	34
Digit III free length	Females	0.935	0.0269	**0.762**	0.685–0.848	28
	Females (excl. 1998 data)	0.843	0.0239	0.898	0.712–1.134	18
	Males	0.944	0.0246	**0.792**	0.711–0.882	24
	Males (excl. 1998 data)	0.937	0.0257	**0.815**	0.719–0.923	21
	South Carolina alligators	0.982	0.0375	**0.865**[b]	0.786–0.952	12
	South Carolina alligators (excl. 1998 data)	0.995	0.0225	**0.851**	0.798–0.906	9
	Louisiana alligators	0.962	0.0311	**0.831**	0.777–0.887	39
	Louisiana alligators (excl. 1998 data)	0.967	0.0303	**0.849**	0.794–0.909	33
	Florida alligators	0.937	0.0246	0.957	0.841–1.089	20
Digit III free length excluding claw	Females	0.910	0.0307	**0.753**	0.666–0.851	30
	Females (excl. 1998 data)	0.848	0.0244	0.986	0.797–1.220	20
	Louisiana females	0.866	0.0285	**0.778**	0.653–0.928	24
	Louisiana females (excl. 1998 data)	0.843	0.0251	0.993	0.793–1.243	19
	Males	0.915	0.0335	**0.866**[b]	0.754–0.995	23
	Males (excl. 1998 data)	0.921	0.0316	0.910	0.793–1.044	22
	Louisiana males	0.923	0.0375	0.851	0.714–1.013	15
	Louisiana males (excl. 1998 data)	0.934	0.0349	0.893	0.754–1.057	14
	South Carolina alligators	0.975	0.0460	0.894	0.798–1.001	12
	South Carolina alligators (excl. 1998 data)	0.994	0.0264	**0.875**	0.816–0.938	9
	Louisiana alligators	0.950	0.0356	**0.851**	0.789–0.916	41
	Louisiana alligators (excl. 1998 data)	0.965	0.0314	**0.880**	0.822–0.941	35
	Florida alligators	0.929	0.0283	1.032	0.907–1.175	22

[a] This parameter could not be measured on most of the Florida sample of alligators, owing to the way specimens were prepared prior to my working on them.
[b] Only barely allometric (upper CI limit of RMA slope between 0.95 and 1.00, or lower CI limit between 1.00 and 1.05).

ensuring that variability comparisons among parameters involved the same individual alligators (table 5.4). Results were comparable in both data treatments.

The coefficient of variation (CV: standard deviation as a percentage of the mean) of pedal parameters in mature alligators generally ranges from 10% to 20%, and maximum values of any parameter are generally 1.5 to 2 times the minimum values. Lengths of the digit I pads are considerably more variable than other parameters, however. Most measurements of digit I (the lateral hypex length is an exception), and also the medial hypex length of IV, are more variable than measurements of digits II and III. Somewhat surprisingly, claw lengths are among the least variable parameters. Male alligators usually have more variable pedal measurements than females, reflecting the greater size range of males than females, but female claw lengths are more variable than those of males. Overall pes length is less variable than alligator total length (tables 5.2–5.4).

Of course, my alligators did not come from a single wild population, and so the variability of pedal parameter lengths reported here may not be typical of natural populations. I did a quick-and-dirty check of this possibility by estimating the variability of total length in wild populations (table 5.5), using a regression of pes length on total length (table 5.6: footnote a) to predict the pes lengths in those populations. For my sample of mature alligators, CV of total length is 19%, and CV of pes lengths (table 5.3) is 16%. These values are close to those estimated for wild populations, suggesting that my data in fact provide a good estimate for the variability in hindfoot length that one would expect to see among mature alligators in wild populations.

Hindfoot Length and Alligator Size

Parameters of hindfoot length are negatively allometric with alligator total length, shoulder-hip length, and hindleg length (table 5.6; figs. 4.2, 4.3, 5.5; cf. tables A1.2, 4.1), as previously reported by Farlow and Britton (2000). The length of the "non-toe palm" of the foot seems to become relatively longer, compared with the free length

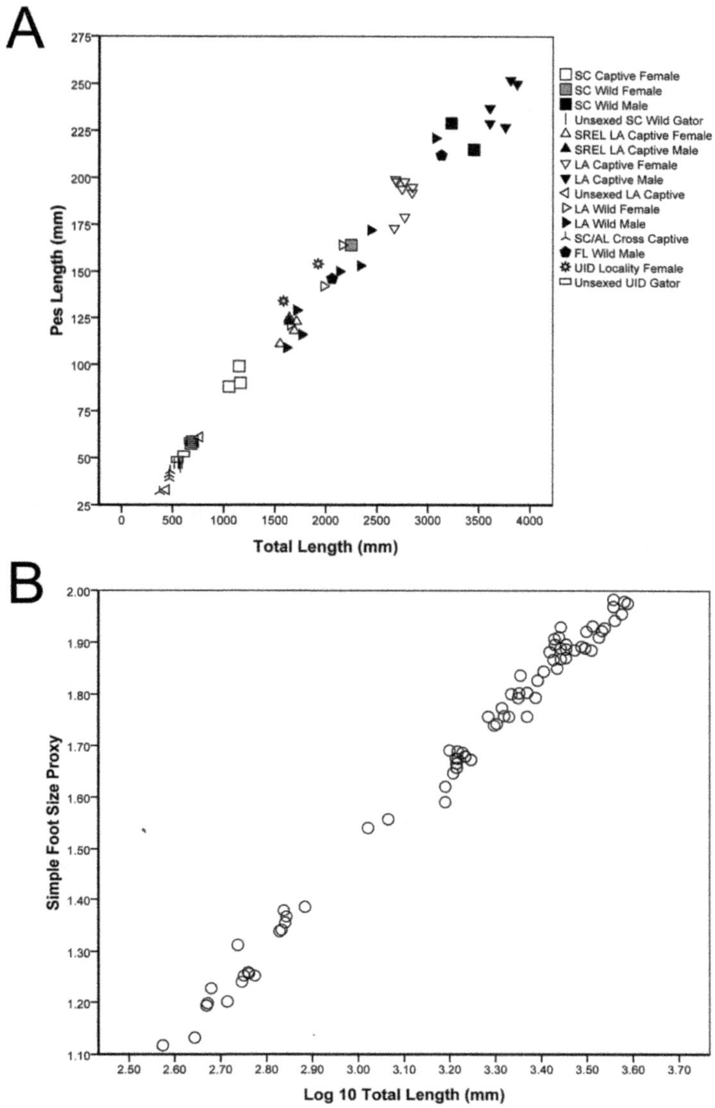

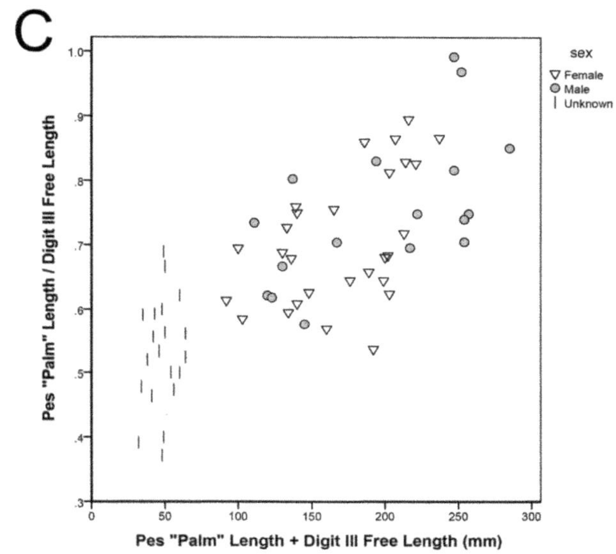

5.5. *A*, *B*, Alligator size and hindfoot size. *A*, Pes length plotted as a function of total length. *Key to specimen identification*: AL = Alabama, FL = Florida, LA = Louisiana, SC = South Carolina, SREL = Savannah River Ecology Laboratory, UID = unidentified. *B*, Relationship between log-transformed total length and a proxy for overall foot size, the latter calculated as the mean of the log-transformed pedal hypex lengths, digit I total length, and the free lengths of digits II and III. *C*, Ratio of the length of the "non-toe palm" of the hindfoot (roughly, but not exactly, equivalent to metatarsal III length) to digit III free length in alligators with unworn claws on digit III plotted as a function of overall foot length (in this case, measured as the "pes print proxy"). Note the relative increase in length of the "palm" with increasing foot length (cf. fig. 5.2).

of digit III, as alligators become larger (figs. 4.2*D*, 5.2, 5.5*C*).

If the data are broken down by sex or geographic region (table 5.7), most RMA slopes still indicate negative allometry (although sometimes just barely), and in some treatments the 95% confidence interval doesn't exclude 1. However, most of the treatments in table 5.7 do not include small juveniles; most ontogenetic changes in alligator body proportions occur in smaller individuals, with less change between half-grown and fully grown animals (Farlow and Britton 2000). Such a difference between small and large alligators is nicely shown by the relationship between the simple foot size proxy and log-transformed total length (fig. 5.5*B*). Even with the log transformations, there is a kink in the plotted relationship, with animals with a total length of less than about 795 mm (log value = 2.90) following a different, steeper trend than larger alligators.

Wilkinson and Rice (1996) measured body size and hindfoot length (the latter "measured from the first single extended scute posterior to the heel to the anterior end of the middle toe, not including the nail" [p. 431]) of 248 alligators from the Santee River delta of coastal South Carolina. For animals with a total length of 120 cm or more, they found a linear relationship between total length or snout-vent length and hindfoot length, but including juveniles required nonlinear relationships. For alligators with a total length less than 120 cm, total length was about 10.4 times hindfoot length, and for alligators of at least 120 cm total length, total length was 11.5 times hindfoot length. Their results are consistent with mine.

Analysis of covariance (ANCOVA; with body size as the covariate) of pedal size parameters against sex (table 5.8) suggests that for a given total length, females have larger feet than males; this result applies whether data are pooled across all regions, or data for Louisiana alone (the geographic

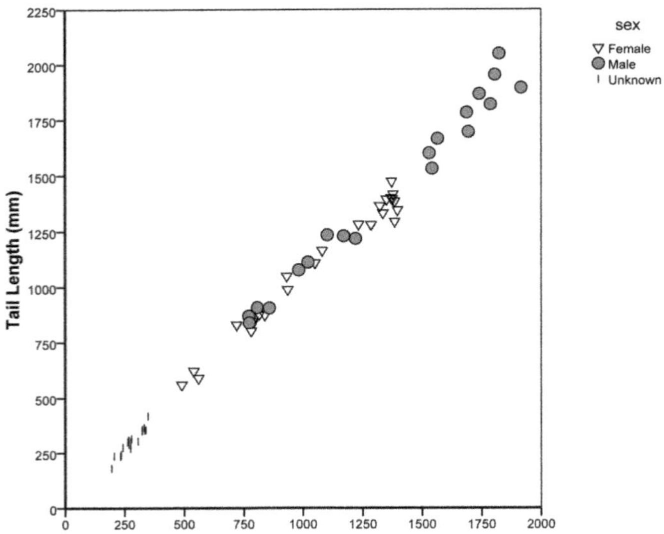

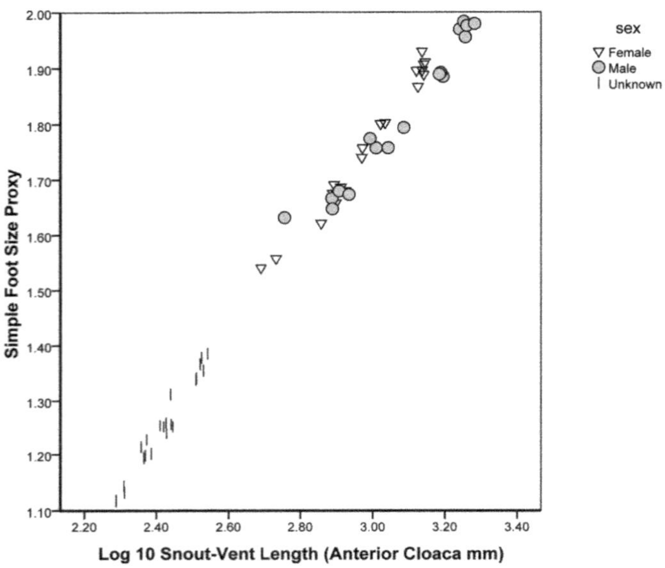

Top, **5.6.** Relationship between tail length and snout-vent length (SVL) (measured to the anterior edge of the cloaca) in alligators. For a given SVL, large males seem to have slightly longer tails than those of females.

Bottom, **5.7.** Relationship between the simple foot size proxy and log-transformed snout-vent length. Note kink in the relationship, with a different trend for small alligators that were not sexed than for large individuals that were (cf. fig. 5.5*B*). Among large alligators, females tend to have relatively larger feet than males.

region with the largest sample and the largest size range of alligators) are used. However, this result may be somewhat misleading from an ichnological standpoint, because males seem to have relatively longer tails than do females (table 5.8). The difference between the sexes isn't great, only amounting to a few centimeters in large animals (fig. 5.6; Webb and Messel [1978] observed the same kind of sexual difference in *Crocodylus porosus*). This difference seems to account for some of the sexual difference in the foot length/total length relationship. If pes length is compared with SVLant, a parameter of overall body size that largely excludes the tail, ANCOVA reveals no difference between the sexes (table 5.8). On the other hand, if the simple foot size proxy is compared with SVLant, females continue to show relatively larger feet than do males (fig. 5.7—and note once again the difference in the trends for large and small alligators; cf. fig. 5.5*B*).

There is no significant difference in the hindfoot size/body size relationship across geographic regions (table 5.8), with the possible exception of digit III free length, which approaches statistical significance. However, the sampling of large as opposed to small alligators across geographic regions was uneven; the South Carolina sample was dominated by fairly young animals, and the Louisiana and Florida samples by larger individuals. Consequently the ANCOVA results in table 5.8 might be suspect. I checked this possibility by doing comparable ANCOVAs across geographic regions for mature alligators only; results were similar to those in table 5.8. There is thus little evidence for geographic differences in foot size/body size relationships.

Alligator Size and Pes Shape

Bivariate RMA relationships between various parameters of pes shape are summarized in table 5.9, and bivariate plots of these relationships are shown in figures 5.8–5.14. Several of these comparisons indicate allometry. The free lengths of digits II and III (whether including or excluding the claw) are negatively allometric with respect to the length of digit I, while the medial hypex length of digit IV is positively allometric with respect to the length of digit I. The medial hypex length of digit IV is also positively allometric with respect to the free lengths of digits II and III. The free lengths of digits II and III are, in contrast, isometric with respect to each other. Both the medial and lateral hypex lengths of III (and possibly II) are positively allometric with respect to the free length of that digit, but the lateral hypex length of digit I is isometric with respect to the length of digit I. The length of the second digital pad of digit I looks isometric with respect to the length of the first digital pad of digit I, but interpretation of the relationship is complicated by the haphazard way in which the crease separating the two pads develops (see below). Claw lengths are negatively allometric with respect to the lengths of the fleshy portions of the digits, but the lengths of the claws on digits I–III are isometric with respect to each other.

Restating and summarizing these results more qualitatively, as the alligator foot gets bigger, the first and fourth toes become proportionately longer than the second and third toes, the fourth toe becomes proportionately longer than

Table 5.8. Analysis of covariance (ANCOVA) of pedal size parameters against sex and geographic region in alligators. Results are reported for analyses in which the assumptions of ANCOVA were met (*P* values of Levene's test of equality of error variances and lack-of-fit test of at least .05). Where claw length is a component of a parameter, analyses were done using animals with unworn claws. F = females; FL = Florida; LA = Louisiana; M = males; SC = South Carolina; SVLant = snout-vent length measured from the snout to the anterior edge of the cloaca.

Covariate	Parameter	Test factor	Treatment	*P* of *F*-test	*N* of alligators
Total length	Pes length	Sex	All sites	.032[a]	25 F: 16 M
			LA	.049[a]	19 F: 13 M
	Digit III free length	Sex	All sites	.007[a]	28 F: 22 M
			All sites (excl. 1998)	.016[a]	18 F: 21 M
			LA	.006[a]	22 F: 15 M
			LA (excl. 1998)	.009[a]	17 F: 14 M
	Digit III free length excluding claw	Sex	All sites	.032[a]	30 F: 23 M
			All sites (excl. 1998)	.039[a]	20 F: 22 M
			LA	.001[e]	24 F: 15 M
			LA (exc. 1998)	.033[a]	19 F: 14 M
	Simple foot size proxy	Sex	All sites	<.001[a]	23F: 21 M
			LA	<.001[a]	18 F: 14 M
SVLant	Pes length	Sex	All sites	.470	23 F: 16 M
			LA	.273	17 F: 13 M
	Tail length	Sex	All sites	.017[b]	26 F: 17 M
			LA	.036[b]	20 F: 13 M
	Simple foot size proxy	Sex	All sites	.001[a]	20 F: 16 M
			LA	<.001[a]	15 F: 13 M
Total leg length	Pes length	Sex	All sites	.585	25 F: 17 M
			LA	.422	19 F: 13 M
	Simple foot size proxy	Sex	All sites	.013[a]	21 F: 17 M
			LA	.009[a]	16 F: 13 M
Total length	Pes length	Geographic region	SC and LA[c]	.744	12 SC: 34 LA
	Digit III free length	Geographic region	All sites	.089	12 SC: 39 LA: 20 FL
			All sites (excl. 1998)	.053	9 SC: 33 LA: 20 FL
	Digit III free length excluding claw	Geographic region	All sites	.173	12 SC: 41 LA: 20 FL
			All sites (excl. 1998)	.118	9 SC: 35 LA: 22 FL
	Simple foot size proxy[d]	Geographic region	All sites	.877	11 SC: 34 LA: 18 FL
Total leg length	Pes length	Geographic region	SC and LA[c]	.824	13 SC: 34 LA
	Digit III free length	Geographic region	SC and LA[c]	.103	13 SC: 34 LA
			SC and LA[c] (excl. 1998)	.109	9 SC: 28 LA
	Digit III free length excluding claw	Geographic region	SC and LA[c]	.109	13 SC: 36 LA
			SC and LA[c] (excl. 1998)	.163	9 SC: 30 LA
	Simple foot size proxy[d]	Geographic region	SC and LA[c]	.905	12 SC: 31 LA

[a] Difference between estimated marginal means of sexes differs at .05 level (females larger than males) by the Bonferroni test.
[b] Difference between estimated marginal means of sexes differs at .05 level (males larger than females) by the Bonferroni test.
[c] Florida excluded owing to small sample size.
[d] For this analysis the log of the covariate had to be used in order to pass the lack-of-fit test and/or Levene's test of equality of error variances ($P > .05$).
[e] For this analysis the log of digit III free length excluding the claw had to be used in order to pass Levene's test of equality of error variances ($P > .05$). The difference between estimated marginal means of sexes differs at the .05 level (females larger than males) by the Bonferroni test.

the first toe, and the second and third toes remain constant in their relative lengths. The indentations separating digit III from its neighboring toes become relatively deeper. The claws become relatively shorter than the rest of their respective toes. Even so, most of these shape changes are rather subtle (cf. fig. 5.2).

Inspection of figures 5.8–5.14 reveals interesting details of these cases of minor allometric shape changes. In many cases the relationships look nearly linear for alligators of most sizes, but with departures from the overall trend mainly affecting the very biggest animals, nearly all of which are males (particularly figs. 5.8*A*, 5.8*E*, 5.8*F*, 5.9*A*, 5.9*B*, 5.10*A*, 5.11*A*–5.11*C*). Generalizing from these relationships, both digit I and IV seem to become relatively longer than digits II and III in the biggest alligators, and digit IV becomes relatively longer than digit I; these trends hold both for the entire lengths of the digits, and for the fleshy portions of the digits (i.e., excluding the claws) only. In addition, the claw lengths seem to become

Table 5.9. Bivariate relationships between parameters of hindfoot size in alligators. For parameters in which claw length is a component of foot size, only data for animals with unworn claws were used. Data were log transformed prior to analysis; all measurements are in millimeters. For measurements involving the free lengths of digits II–III, analyses were run both including (first line) and excluding (second line) data collected during 1998. RMA slopes significantly different than 1 indicated in **bold**. CI = confidence interval; RMA = reduced major axis; SEE = standard error of estimate.

Independent variable	Dependent variable	r^2	SEE	RMA slope	95% CI of RMA slope	*N*
Digit I length	Digit II free length	0.978	0.0370	**0.924**[b]	0.894–0.955	85
		0.982	0.0296	**0.927**[b]	0.894–0.961	59
	Digit III free length	0.978	0.0375	**0.917**	0.887–0.947	83
		0.977	0.0333	**0.919**[b]	0.882–0.958	57
	Digit IV medial hypex length	0.963	0.0570	**1.092**[b]	1.046–1.139	86
Digit II free length	Digit III free length	0.988	0.0275	0.999	0.975–1.023	85
		0.992	0.0191	1.000	0.977–1.024	59
	Digit IV medial hypex length	0.952	0.0646	**1.185**	1.130–1.244	88
		0.938	0.0618	**1.162**	1.088–1.241	63
Digit III free length	Digit IV medial hypex length	0.948	0.0672	**1.197**	1.138–1.259	87
		0.935	0.0642	**1.170**	1.093–1.253	62
Digit I length excluding claw	Digit II free length excluding claw	0.958	0.0524	**0.903**	0.864–0.944	91
		0.962	0.0437	**0.916**[b]	0.871–0.963	65
	Digit III free length excluding claw	0.964	0.0471	**0.886**	0.851–0.923	92
		0.959	0.0442	**0.899**	0.854–0.947	66
Digit II free length excluding claw	Digit III free length excluding claw	0.980	0.0351	0.985	0.956–1.015	90
		0.985	0.0269	0.984	0.954–1.016	64
Digit I length	Digit I lateral hypex length	0.985	0.0340	1.002	0.975–1.029	86
Digit II free length	Digit II medial hypex length	0.974	0.0427	**1.055**[b]	1.019–1.092	89
		0.987	0.0240	0.994	0.965–1.023	63
	Digit II lateral hypex length	0.979	0.0375	**1.039**[b]	1.007–1.071	89
		0.982	0.0291	1.009	0.975–1.045	63
Digit III free length	Digit III medial hypex length	0.979	0.0391	**1.078**[b]	1.045–1.112	88
		0.977	0.0337	1.030	0.990–1.071	62
	Digit III lateral hypex length	0.976	0.0416	**1.087**	1.051–1.125	87
		0.976	0.0354	**1.064**[b]	1.022–1.108	62
Digit I length excluding claw	Digit I claw length	0.929	0.0673	**0.895**[b]	0.843–0.950	87
Digit II free length excluding claw	Digit II claw length	0.889	0.0799	0.939	0.871–1.012	89
		0.886	0.0631	**0.826**	0.753–0.905	63
Digit III free length excluding claw	Digit III claw length	0.912	0.0732	0.981	0.918–1.049	88
		0.906	0.0576	**0.844**	0.777–0.917	62
Digit I first pad length	Digit I second pad length[a]	0.845	0.117	1.079	0.986–1.181	91
	Digit I claw length	0.856	0.0960	0.927	0.848–1.013	87
Digit I second pad lengtha	Digit I claw length	0.839	0.102	**0.852**	0.775–0.938	86
Digit I claw length	Digit II claw length	0.956	0.0509	0.959	0.915–1.005	85
	Digit III claw length	0.967	0.0455	0.984	0.945–1.025	83
Digit II claw length	Digit III claw length	0.972	0.0419	1.030	0.992–1.069	85

[a] Excludes cases in which digit I second pad length = 0 (no crease separates the fleshy portion of digit I into separate pads), owing to the impossibility of taking the logarithm of 0.

[b] Only barely allometric (upper CI limit of RMA slope between 0.95 and 1.00, or lower CI limit between 1.00 and 1.05).

Facing, **5.8.** Bivariate comparisons of overall digit lengths in alligators. When a claw length is a component of an overall digit length, only data for animals with unworn claws are plotted. In all plots, data points are broken down by sex. Data for all years are plotted in these graphs. *A,* Digit I length vs. digit II free length. The relationship looks slightly curvilinear, with digit II lengths becoming relatively shorter in the largest alligators, all of which are males. *B,* Digit I length vs. digit III free length. Once again the relationship looks slightly curvilinear. There is increasing scatter about the trend for the big males, but the curve for the relationship between digit III free length and digit I length does not appear to flatten out (that is, digit III free length does not become proportionately shorter with respect to digit I length) among the biggest animals as much as digit II free length does (*A*). *C,* Digit I length vs. digit IV medial hypex length. Again, the amount of scatter with this relationship is greater among the biggest animals. Here there is a suggestion that the digit IV medial hypex length becomes relatively longer, compared with digit I length, in large individuals. *D,* Digit II free length vs. digit III free length. The graph suggests that digit III free length may become proportionally longer than digit II free length among the biggest males, but this could instead merely be increasing scatter of points for the largest individuals. Otherwise there is no suggestion of allometry in the plot. *E,* Digit II free length vs. digit IV medial hypex length. Once again there is considerable scatter of points for the biggest animals, and as with the relationship between digit IV medial hypex length and digit I length (*C*), digit IV may be proportionally longer than digit II free length in the largest alligators. *F,* Digit III free length vs. digit IV medial hypex length. As before, there is considerable scatter of points for the biggest alligators, with a suggestion that the digit IV medial hypex length becomes proportionally longer.

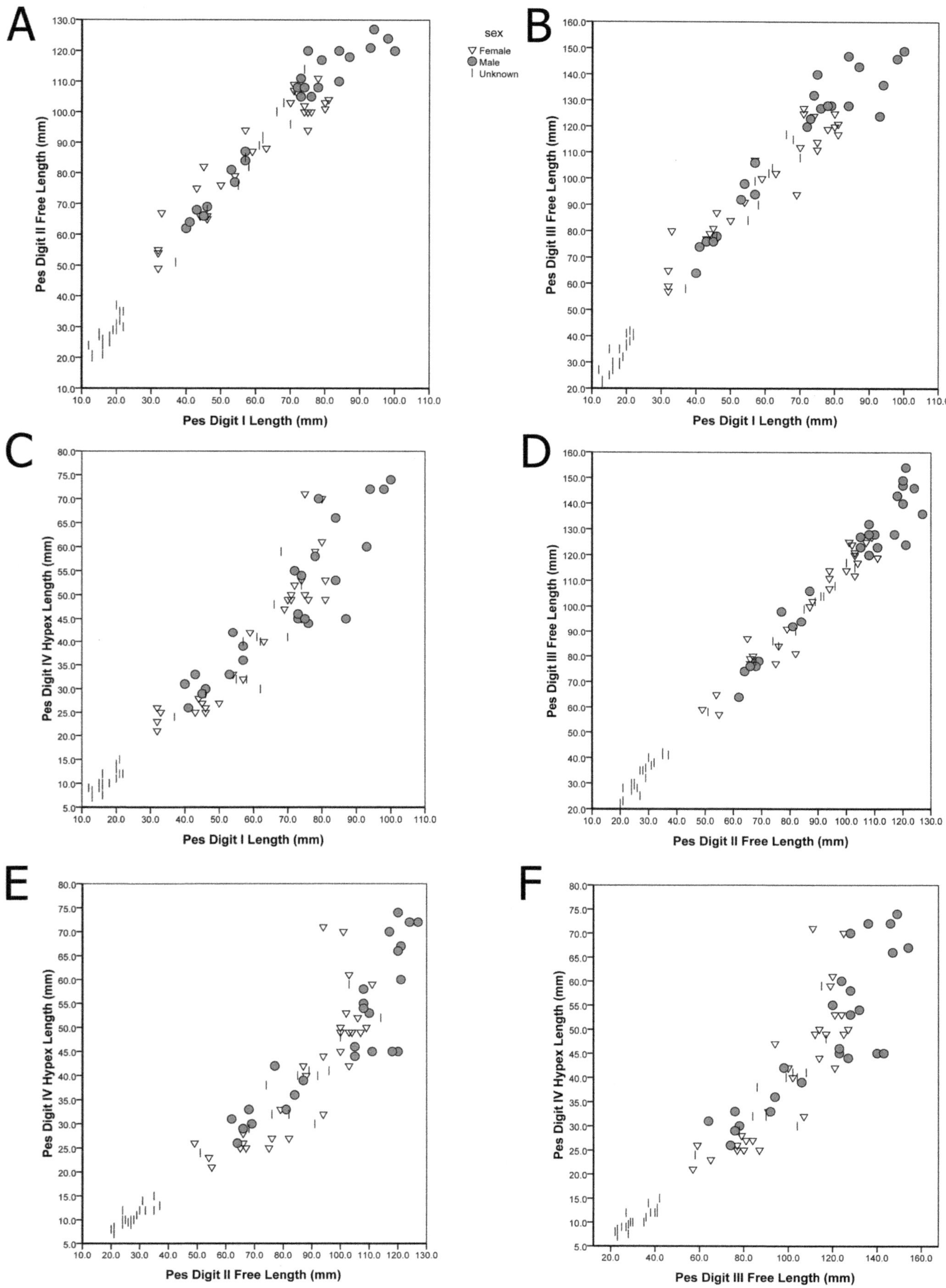
A
B
C
D
E
F
sex
Female
Male
Unknown
Pes Digit I Length (mm)
Pes Digit II Free Length (mm)
Pes Digit III Free Length (mm)
Pes Digit IV Hypex Length (mm)
Pes Digit I Length (mm)
Pes Digit III Free Length (mm)
Pes Digit I Length (mm)
Pes Digit IV Hypex Length (mm)
Pes Digit II Free Length (mm)
Pes Digit III Free Length (mm)
Pes Digit II Free Length (mm)
Pes Digit IV Hypex Length (mm)
Pes Digit III Free Length (mm)
Pes Digit IV Hypex Length (mm)

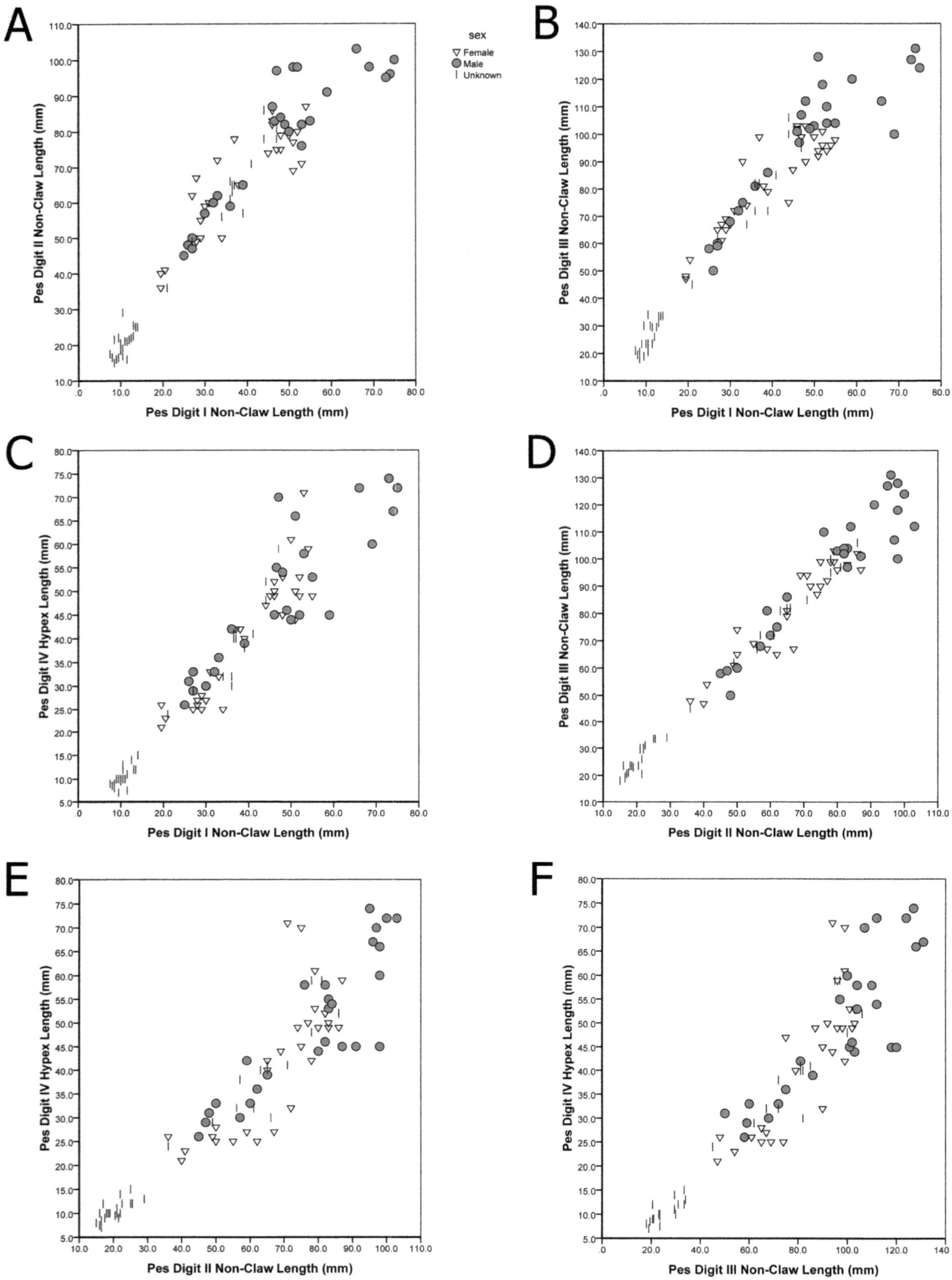
A
Pes Digit II Non-Claw Length (mm)
Pes Digit I Non-Claw Length (mm)
sex
Female
Male
Unknown
B
Pes Digit III Non-Claw Length (mm)
Pes Digit I Non-Claw Length (mm)
C
Pes Digit IV Hypex Length (mm)
Pes Digit I Non-Claw Length (mm)
D
Pes Digit III Non-Claw Length (mm)
Pes Digit II Non-Claw Length (mm)
E
Pes Digit IV Hypex Length (mm)
Pes Digit II Non-Claw Length (mm)
F
Pes Digit IV Hypex Length (mm)
Pes Digit III Non-Claw Length (mm)

Facing, **5.9.** Bivariate comparisons of "fleshy" digit lengths (excluding claws) in alligators. In all plots, data points are broken down by sex. Data for all years are plotted in these graphs. *A,* Digit I length vs. digit II free length. The relationship looks curvilinear, as with the corresponding relationship for total digit lengths (fig. 5.8*A*), indicating that the latter reflects not just claw lengths but also the lengths of the fleshy parts of the digits. Digit II free length seems to become proportionally shorter than digit I length among the large males. *B,* Digit I length vs. digit III free length. As with the relationship for total digit lengths (fig. 5.8*B*), there is considerable scatter among data points for the big males, and digit III free length becomes proportionally shorter than digit I length. *C,* Digit I length vs. digit IV medial hypex length. Again the relationship looks very much like its counterpart for total digit lengths (fig. 5.8*C*). *D,* Digit II free length vs. digit III free length. The relationship looks linear or nearly so, with considerable scatter of points for large males. *E,* Digit II free length vs. digit IV medial hypex length. Again the relationship is similar to its counterpart for total digit lengths (fig. 5.8*E*). *F,* Digit III free length vs. digit IV medial hypex length. Yet again the relationship is similar to its counterpart for total digit lengths (fig. 5.8*F*).

Table 5.10. Principal component analysis (using a covariance matrix) of log-transformed measurements of alligator pedal digit lengths. Number of specimens = 79 (1998 data included in the analysis).

Parameter	Component 1 loading (raw [rescaled])
Digit I length	0.272 (0.996)
Digit I lateral hypex length	0.270 (0.995)
Digit II free length	0.251 (0.994)
Digit II medial hypex length	0.266 (0.995)
Digit II lateral hypex length	0.260 (0.996)
Digit III free length	0.249 (0.994)
Digit III medial hypex length	0.269 (0.997)
Digit III lateral hypex length	0.271 (0.997)
Digit IV medial hypex length	0.296 (0.988)
Eigenvalue (% of variance explained)	0.644 (98.887)

Kaiser-Meyer-Olkin measure of sampling adequacy = 0.951; Bartlett's test of sphericity: χ^2 = 2,590.634, $P < .001$.

disproportionately shorter than the fleshy portions of the toes in the biggest males.

As previously noted, the boundary separating the first and second digital pads of the alligator digit I is not as clearly defined as the creases separating digital pads in typical bird and non-avian tridactyl dinosaur footprints. In a few alligators there was no subdivision of the fleshy portion of digit I into separate digit pads at all; for the purposes of this study, the entire fleshy part of the toe was assigned to the first digital pad. In these animals the first digital pad in consequence was disproportionately long (fig. 5.13A, 5.13*B*), and the second digital pad was assigned a length of zero (fig. 5.13*E*, 5.13*F*).

The second, more distal digital pad of digit I is often a little longer than the first (fig. 5.12A, 5.14). If the lengths of the two digital pads are expressed as percentages of the total length of the fleshy portion of the toe, the mean and median

Table 5.11. Multivariate relationships between parameters of hindfoot size in alligators. Data were log transformed prior to analysis. Each log-transformed parameter was examined as a dependent variable in an RMA analysis with respect to a foot size factor, the latter calculated as the mean of all log-transformed parameters except the one being examined. Only data for animals with unworn claws were used. Analyses were run including measurements made in 1998 (first line), and also excluding 1998 data (second line). RMA slopes significantly different than 1 indicated in **bold**. In addition to the RMA analysis, a multivariate allometric coefficient for each log-transformed variable against overall size was calculated by dividing the loading of that variable on the first component [PC1] of a principal component analysis by the mean of the loadings of all the log-transformed variables on the first component. In both versions of this latter analysis (excluding and including 1998 measurements of digit II–III free lengths), PC1 accounted for nearly 99% of the data variance. CI = confidence interval; PCA = principal component analysis; RMA = reduced major axis; SEE = standard error of estimate.

Dependent variable	r^2	SEE	RMA slope	95% CI of RMA slope	PCA allometric coefficient	*N*
Digit I length	0.990	0.0275	**1.025**[a]	1.001–1.048	1.019	79
	0.989	0.0257	**1.046**[a]	1.016–1.077	1.038	55
Digit I lateral hypex length	0.988	0.0303	1.019	0.994–1.045	1.011	79
	0.985	0.0299	1.033	0.999–1.069	1.025	55
Digit II free length	0.985	0.0308	**0.937**[a]	0.911–0.963	0.940	79
	0.992	0.0201	**0.960**[a]	0.937–0.984	0.960	55
Digit II medial hypex length	0.987	0.0308	0.999	0.994–1.026	0.996	79
	0.989	0.0234	**0.958**[a]	0.930–0.986	0.960	55
Digit II lateral hypex length	0.990	0.0259	**0.974**[a]	0.952–0.997	0.974	79
	0.986	0.0265	**0.965**[a]	0.934–0.998	0.964	55
Digit III free length	0.986	0.0301	**0.931**[a]	0.906–0.957	0.933	79
	0.989	0.0235	**0.956**[a]	0.928–0.984	0.955	55
Digit III medial hypex length	0.992	0.0248	1.013	0.993–1.034	1.007	79
	0.989	0.0236	0.983	0.955–1.012	0.981	55
Digit III lateral hypex length	0.991	0.0257	1.018	0.996–1.040	1.015	79
	0.988	0.0257	1.021	0.990–1.052	1.016	55
Digit IV medial hypex length	0.968	0.0540	**1.135**	1.089–1.183	1.109	79
	0.952	0.0575	**1.137**	1.069–1.210	1.102	55

[a] Only barely allometric (upper CI limit of RMA slope between 0.95 and 1.00, or lower CI limit between 1.00 and 1.05).

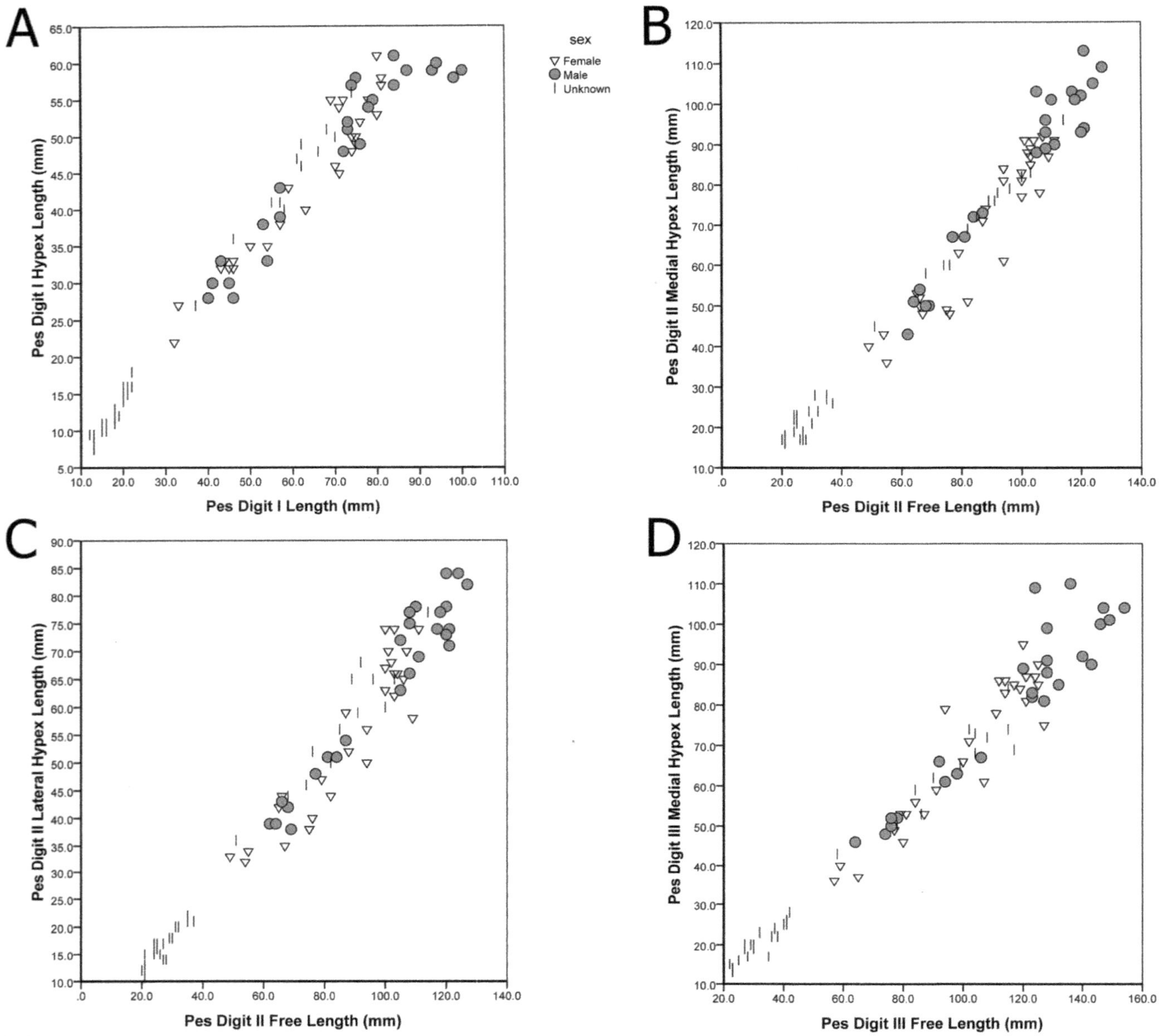

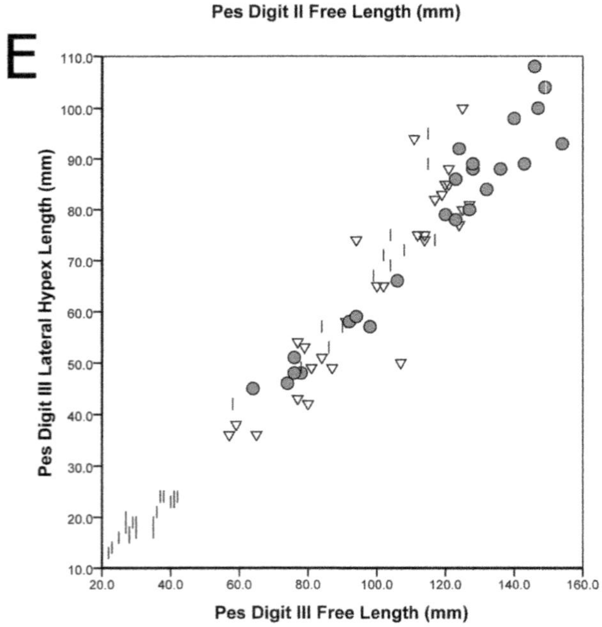

5.10. Bivariate comparisons of overall digit lengths and hypex lengths. Only data for animals with unworn claws are plotted. In all plots, data points are broken down by sex. Data for all years are plotted in these graphs. *A*, Digit I length vs. digit I lateral hypex length. The relationship looks linear for all but the largest animals (all males), for which the curve sharply flattens out, such that the hypex length becomes considerably shorter with respect to digit I length. *B*, *C*, Digit II free length vs. digit II medial (*B*) and lateral (*C*) hypex lengths. Both plots suggest a very slight (and highly questionable) proportional increase in the hypex lengths relative to the digit free length in the biggest alligators. *D*, *E*, Digit III free length vs. digit III medial (*D*) and lateral (*E*) hypex lengths. Both relationships look nearly linear, with increasing scatter among points for the larger animals, but the scatter around the relationship drifts mainly upward, suggesting that the hypex lengths may become proportionally longer than the digit free length in the largest alligators.

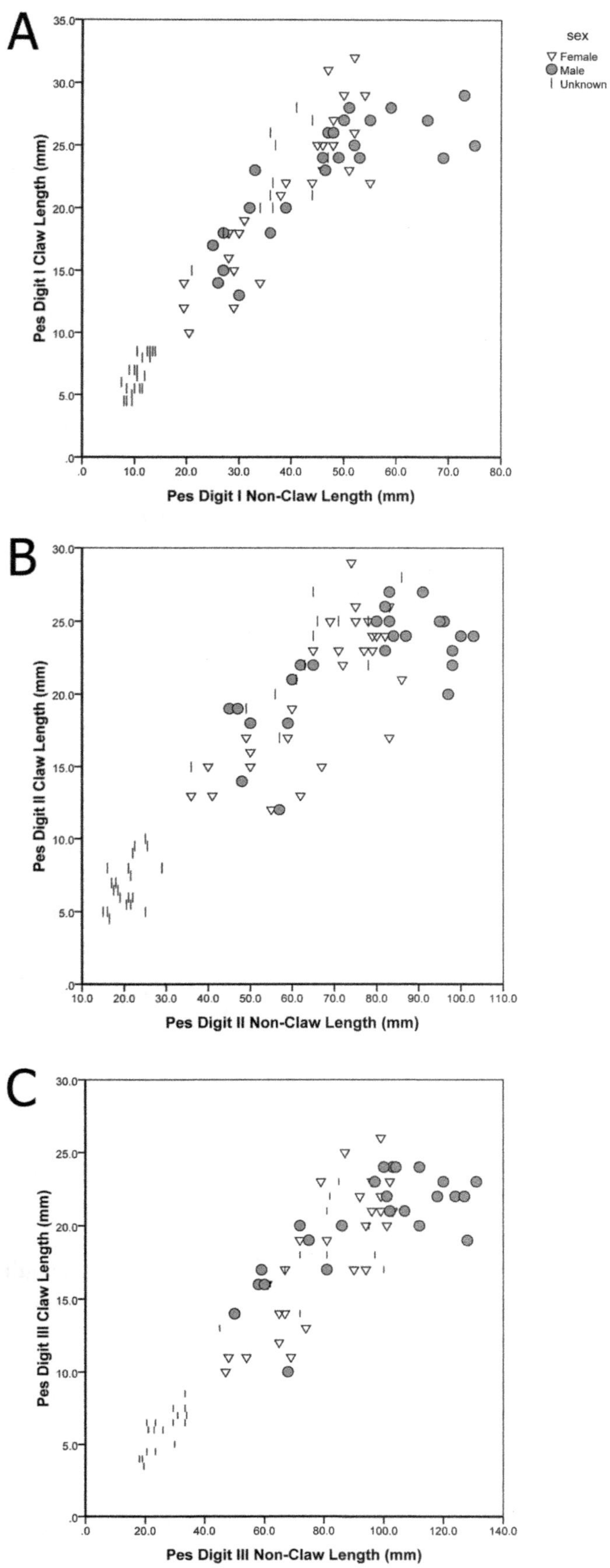

5.11. Bivariate comparisons of claw lengths with respect to lengths of the fleshy portions of toes. Only data for animals with unworn claws are plotted. In all plots, data points are broken down by sex. Data for all years are plotted in these graphs. *A*, Digit I length vs. digit I claw length. The relationship generally looks linear, except among the biggest males, for which the claw length becomes considerably shorter with respect to the length of the fleshy part of the digit. *B*, Digit II free length vs. digit II claw length. There is increasing scatter among points for larger alligators, and the biggest males show relatively shorter claw lengths. *C*, Digit III free length vs. digit III claw length. The relationship looks very similar to that for digit I, with big males having noticeably relatively shorter claws compared with the length of the fleshy part of the free length.

contributions of the first digital pad to the length of the fleshy part of the toe are 45% and 44%, respectively, while the mean and median percentage contributions of the second digital pad to length of the fleshy length of the toe are 50% and 52%, respectively. Neither size nor sex seems to affect the relative size of either digital pad (fig. 5.13*B*, 5.13*F*).

If there was a sufficiently large gap separating the two digital pads of digit I, this was considered to be an interpad space. About a quarter of the alligators in my sample were identified as having an interpad space (figs. 5.13–5.15). Where an interpad space was recognized, its proportional contribution to the length of the fleshy part of digit I was not obviously affected by animal size or sex (fig. 5.13*D*). In those alligators that did show an interpad space, its length came at the expense of both of the digital pads (fig. 5.15).

Multivariate allometric analyses (tables 5.10, 5.11) yielded results consistent with the bivariate analyses. The RMA slopes of individual log-transformed foot shape variables against a footprint size factor that included all variables except the one of interest were very similar to multivariate principal component analysis allometric coefficients, and most were very close to 1. However, some of the 95% confidence intervals around the slopes excluded 1 (albeit just barely in several comparisons). The strongest suggestion of allometry involves the medial hypex length of digit IV, which becomes relatively longer with respect to the rest of the foot.

Examination of figure 5.2 suggests a possible ontogenetic change in alligator pes shape in parameters that I did not measure: the relative stoutness of toes and their claws. Very small alligators have relatively skinny toes with needle-like claws, while large alligators have fleshier toes and claws whose basal transverse widths seem to be broader, compared with the lengths of the claws. The relative stoutness of the toes of large alligators in my sample may have been inflated by the fact that so many of my big animals were well-fed captives, but even some of the wild large alligators (fig. 5.2*O*, 5.2*S*, 5.2*T*) seem to have rather plump toes. Recall that the possibility of stouter toes in larger alligators was also seen in some limited skeletal data (fig. 4.4).

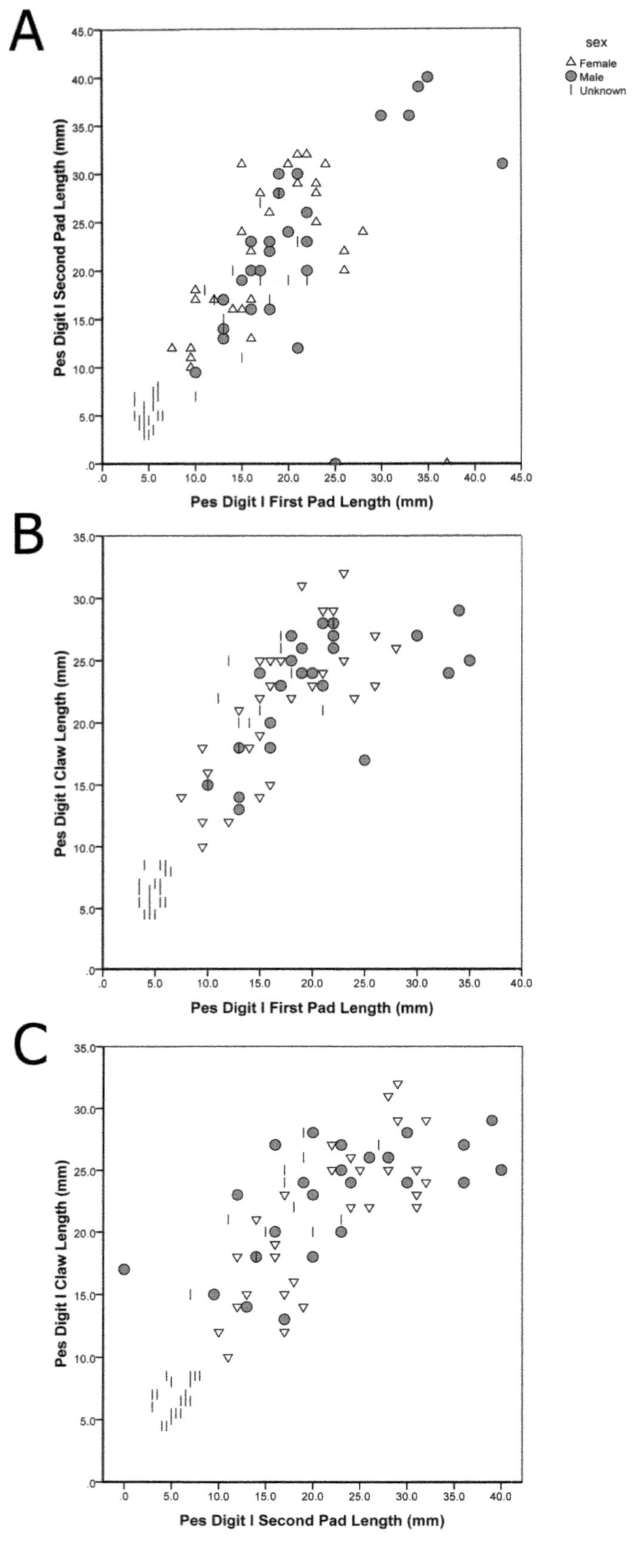

Left, **5.12.** Bivariate comparisons of pad lengths and claw lengths of digit I. In comparisons involving claws, only data for animals with unworn claws are plotted. In all plots, data points are broken down by sex. Data for all years are plotted in these graphs. *A,* Length of the first (more proximal) digital pad vs. the length of the second (more distal) digital pad. For two animals, there was no crease subdividing the fleshy portion of the toe into separate digital pads; for these animals the entire pad length was assigned to the first digital pad, and the second pad was assigned a length of zero. A third conspicuous outlier point is that for the male alligator with the longest first digital pad. Apart from these outliers, the relationship may be linear. *B,* Length of the first digital pad vs. claw length. There is an obvious increase in scatter in the relationship for the larger animals, but note that in the biggest males the claw is relatively short compared with the length of the pad. *C,* Length of the second digital pad vs. claw length. As in the comparison between the lengths of the first digital pad and the claw, the larger animals have a relatively short claw compared with the pad length. Because this relative shortening of the claw in the biggest animals occurs in comparison with the lengths of both digital pads, it is probably not an artifact of the way the pads were measured (in which lengths of undivided digital pads were all assigned to the first digital pad).

Facing, **5.13.** Length of subdivisions of the fleshy portion of digit I plotted as a function of alligator total length. *A, B,* First digital pad length. For two of the alligators, the fleshy portion of digit I was not subdivided into digital pads, so 100% of the fleshy portion length was assigned to the first digital pad. *C, D,* Length of the interpad space. In most alligators no distinct interpad space was recognized. *E, F,* Length of the second digital pad. In the two alligators with no subdivision of the fleshy portion of the digit, the length of the second digital pad was by definition zero. There was no obvious change in the proportional lengths of the two digital pads or the interpad space as a function of alligator size or sex.

Sexual Comparisons of Pedal Shape

ANCOVAs of hindfoot length parameters against sex (table 5.12) indicated only one statistically significant contrast. For alligators from Louisiana alone, females have a relatively longer digit I lateral hypex length, as compared with digit I length, than do males; the difference is trivial, however, amounting on average to no more than a couple millimeters in mature animals. Furthermore, this sexual difference does not hold up if the sexes are compared across all sites. A few other comparisons approached significance ($.10 > P > .05$), but in none of these was this true for the "all sites" as well as the "Louisiana alone" comparison.

As a further test of sexual differences in hindfoot shape, I performed a discriminant analysis using the same parameters of digit length considered in table 5.11. Each digit length parameter was scaled by subtracting the simple footprint size proxy (mean of all the log-transformed digit length parameters) from that digit length parameter, with the scaled digit length parameters then employed in the discriminant analysis. Data for 23 females and 22 males were used in the analysis. For none of the scaled digit length parameters did the Wilks's λ test indicate a significant difference ($P < .05$) between the sexes. It was therefore impossible to do a stepwise discriminant analysis. If I forced the discriminant analysis to consider all of the scaled digit length parameters

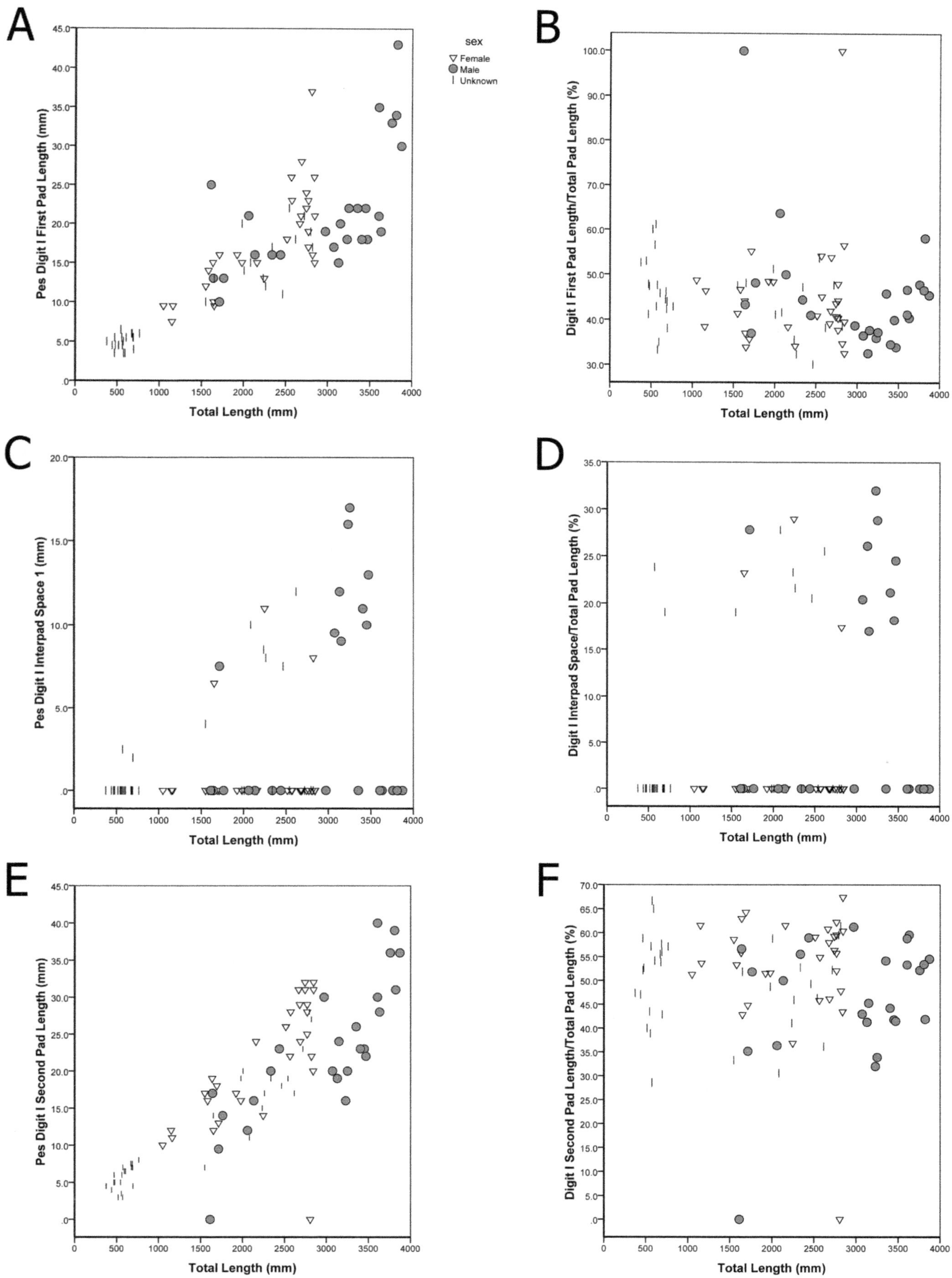
A
Pes Digit I First Pad Length (mm)
Total Length (mm)
sex
Female
Male
Unknown
B
Digit I First Pad Length/Total Pad Length (%)
Total Length (mm)
C
Pes Digit I Interpad Space 1 (mm)
Total Length (mm)
D
Digit I Interpad Space/Total Pad Length (%)
Total Length (mm)
E
Pes Digit I Second Pad Length (mm)
Total Length (mm)
F
Digit I Second Pad Length/Total Pad Length (%)
Total Length (mm)

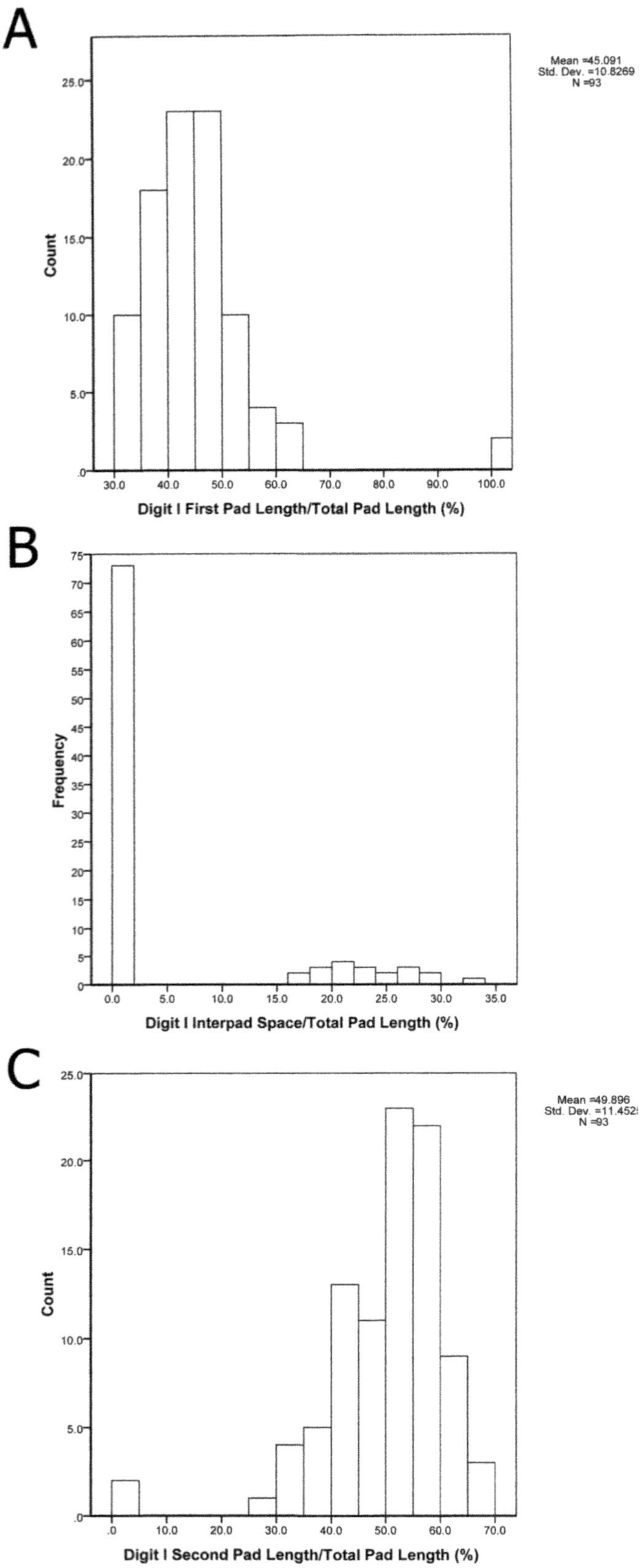

Left, **5.14.** Frequency distributions of lengths of subdivisions of the fleshy portion of digit I, expressed as a percentage of the total length of the fleshy portion of the digit. *A,* First digital pad length. *B,* Length of the interpad space. *C,* Second digital pad length.

Below, **5.15.** Effect of the presence of an interpad space separating the two digital pads of digit I on the relative lengths of the two digital pads. A comparison is made of digital pad length as a function of alligator total length in alligators with and without an interpad space. *A,* Length of the first digital pad. *B,* Length of the second digital pad. Both pad lengths are relatively shorter in alligators that have an interpad space, indicating that space for the latter has been "carved out" at the expense of both digital pads.

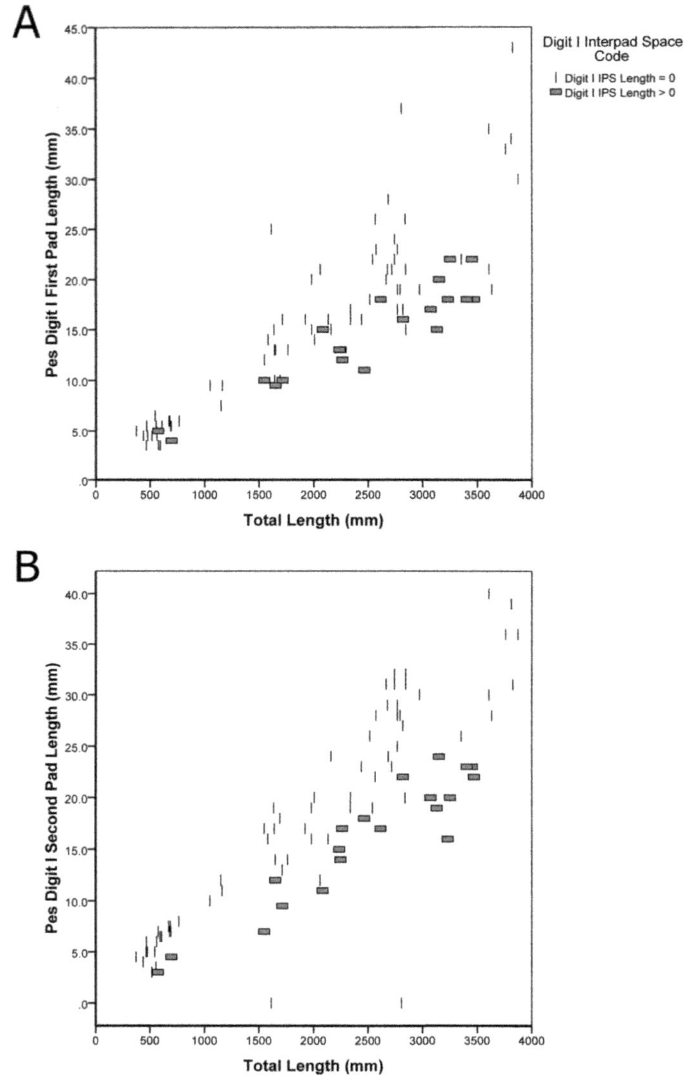

Table 5.12. Analysis of covariance (ANCOVA) of hindfoot parameters against sex in alligators of known sex. Results are reported for analyses in which the assumptions of ANCOVA were met (*P* values of Levene's test of equality of error variances and lack-of-fit test of at least .05); where necessary, variables were log transformed prior to analysis in an attempt to match the assumptions of ANCOVA. Analyses were done for alligators from all geographic areas combined, and separately for Louisiana alligators (the region for which my sample size was largest). Where claw length was a component of a parameter, analyses were done using animals with unworn claws.

Covariate	Parameter	Treatment	*P* of *F*-test	*N* of alligators (females: males)
Digit I length	Digit II free length	All sites	.194	17: 20
		Louisiana[a]	.337	16: 13
	Digit III free length	All sites	.070	16: 20
	Digit IV medial hypex length	All sites	.927	29: 22
		Louisiana	.234	23: 14
Digit II free length	Digit IV medial hypex length	All sites	.591	19: 21
		Louisiana	.466	18: 14
Digit III free length	Digit IV medial hypex length	All sites	.357	18: 21
		Louisiana	.753	17: 14
Digit I length	Digit I lateral hypex length	All sites	.246	28: 22
		Louisiana[b]	.029	23: 14
Digit II free length	Digit II medial hypex length	All sites	.343	19: 21
		Louisiana	.516	18: 14
	Digit II lateral hypex length	All sites	.986	19: 21
		Louisiana	.793	18: 14
Digit III free length	Digit III medial hypex length	All sites	.515	18: 21
	Digit III lateral hypex length	All sites	.094	18: 21
		Louisiana	.139	17: 14
Digit I length excluding claw	Digit I claw length	All sites	.185	29: 22
		Louisiana	.101	23: 14
Digit II free length excluding Claw	Digit II claw length	All sites	.254	19: 21
		Louisiana	.059	18: 14
Digit III free length excluding claw	Digit III claw length	All sites	.214	18: 21
		Louisiana	.093	17: 14
Digit I first pad length	Digit I second pad length	All sites[a]	.172	30: 23
		Louisiana[a]	.281	24: 14
	Digit I claw length	All sites	.591	29: 22
		Louisiana	.090	23: 14
Digit I second pad length	Digit I claw length	All sites	.382	29: 22
		Louisiana	.533	23: 14
Digit I claw length	Digit II claw length	All sites	.399	27: 22
		Louisiana	.701	21: 14
	Digit III claw length	All sites	.151	26: 22
		Louisiana	.535	20: 14
Digit II claw length	Digit III claw length	All sites	.291	26: 23
		Louisiana	.635	20: 15

[a] Variables were log transformed prior to analysis.
[b] Difference between estimated marginal means of sexes differs at .05 level (females larger than males) by Bonferroni test.

simultaneously, the analysis correctly assigned 62% of the alligator cases to the correct case; 53% of cases were correctly assigned to sex in a cross-validation (leave-one-out) check. This is little better than flipping a coin. As with the ANCOVA analyses, discriminant analysis provides no compelling evidence, then, for sexual differences in pes shape.

However, as described in the previous section, several bivariate comparisons of pedal proportions suggest that the main departures from linear trends occur in the biggest alligators, most of which are males. Because there are no females in this size range, it is hard to say whether these reflect size alone, or whether there could be a sexual component as well.

Regional Comparisons of Pedal Shape

Marked differences in the distribution of body sizes of alligators in the samples from Louisiana, South Carolina, and Florida (table 5.12) made ANCOVA comparisons across regions tricky, due to the confounding effects of possible size allometry. To mitigate this, I restricted most of my

Table 5.13. ANCOVAs of alligator hindfoot parameters against geographic origin (FL or LA) for mature animals (total length at least 1,830 mm). Except where otherwise indicated, results are reported for analyses in which the assumptions of ANCOVA were met (*P* values of Levene's test of equality of error variances, and lack-of-fit test, of at least .05). When claw length was a component of a parameter, analyses were done using animals with unworn claws. All = all cases; ANCOVA = analysis of covariance; FL = Florida; LA = Louisiana; overlap = large Louisiana alligators bigger than the largest Florida alligators excluded.

Covariate	Parameter	Treatment	*P* of *F*-test	*N* of alligators (LA: FL)
Digit I length	Digit I lateral hypex length	Overlap[a]	.003[b]	32: 19
	Digit IV medial hypex length	All	.039[c]	27: 17
		Overlap	.042[c]	23:17
	Digit I lateral hypex length	All[d]	<.001[b, e]	27: 17
		Overlap[a]	<.001[b]	23: 17
Digit II free length	Digit III free length[d]	All	.429	25: 17
	Digit IV medial hypex length	All	.005[c]	28: 18
		Overlap	.009[c]	24: 18
Digit III free length	Digit IV medial hypex length	All	.010[c]	27: 18
		Overlap	.017[c]	23: 18
Digit I length excluding claw	Digit II free length excluding claw	All	.855	28: 20
	Digit III free length excluding claw	All	.232	29: 20
Digit II free length excluding claw	Digit III free length excluding claw	All	.579	27: 20
Digit II free length	Digit II medial hypex length	All	.403	28: 18
	Digit II lateral hypex length	All	.110	28: 18
Digit III free length	Digit III medial hypex length	All	.003[c]	27: 18
		Overlap	.003[c]	23: 18
	Digit III lateral hypex length	All	.862	27: 18
Digit I length excluding claw	Digit I claw length	All	.714	27: 17
Digit II free length excluding claw	Digit II claw length	All	.332	28: 18
Digit III free length excluding claw	Digit III claw length	All	.137	27: 18
Digit I first pad length	Digit I second pad length	All[d]	.006[c]	29: 20
		Overlap	.002[c]	21: 20
	Digit I claw length	All	.757	27: 17
Digit I second pad length	Digit I claw length	All	.305	27: 17
Digit I claw length	Digit II claw length	All	.055	25: 17
	Digit III claw length	All	.374	24: 16
Digit II claw length	Digit III claw length	All	.006[c]	25: 17
Digit II lateral hypex length[d, e]	Digit III medial hypex length	All	<.001[c]	25: 17
Digit II lateral hypex length	Digit IV medial hypex length	All	<.001[c]	28: 18

[a] Louisiana alligators smaller than alligators in the Florida sample also excluded.
[b] Difference in estimated marginal means significant at .05 level by Bonferroni test; Florida larger than Louisiana.
[c] Difference in estimated marginal means significant at .05 level by Bonferroni test; Louisiana larger than Florida.
[d] Both variables were log transformed prior to analysis.
[e] Fails Levene's test of equality of error variances; $P < .05$.

comparisons to mature animals (total length at least 1,830 mm) from Louisiana and Florida, usually omitting the South Carolina sample (which was dominated by smaller alligators).

Some of the contrasts between Louisiana and Florida alligators appear to be significant ($P < .05$; table 5.13; fig. 5.16). Although all of the biggest alligators in my sample came from Louisiana, in those comparisons where the ANCOVAs indicate significant regional differences in relative digit lengths, this result applies even if the largest Louisiana alligators are removed from the comparison. Alligators from Louisiana seem to have a relatively longer digit IV medial hypex length than their Florida kin. Whether this indicates a longer

Facing, **5.16.** Possible differences in hindfoot shape in mature (total length at least 1830 mm) alligators from Louisiana and Florida; also see table 5.13. *A–C,* Note the tendency for Louisiana alligators to have a longer digit IV medial hypex length, compared with the lengths of digits I–III, than those of Florida alligators. *D,* Louisiana alligators also tend to have longer medial hypex lengths of digit III, compared with the free length of digit III. *E,* Louisiana alligators may have a relatively long second digital pad of digit I, compared with the length of the first digital pad, than those of Florida alligators. Note, though, that one of the Louisiana alligators had a digit I pad that was not subdivided; as measured in this study, this was plotted as a very long digit I first pad length, and a digit I second pad of zero length. *F,* Louisiana alligators also may have relative longer claws on digit III, compared with the length of the claw on digit II, than those of Florida alligators, although there is considerable overlap in the two sets of points (note that in several instances Florida points are plotted on top of, and mostly obscuring, Louisiana points). Although these regional differences in alligator pes shape are suggestive, they are not completely convincing, because all of the very largest alligators in this study were from Louisiana only.

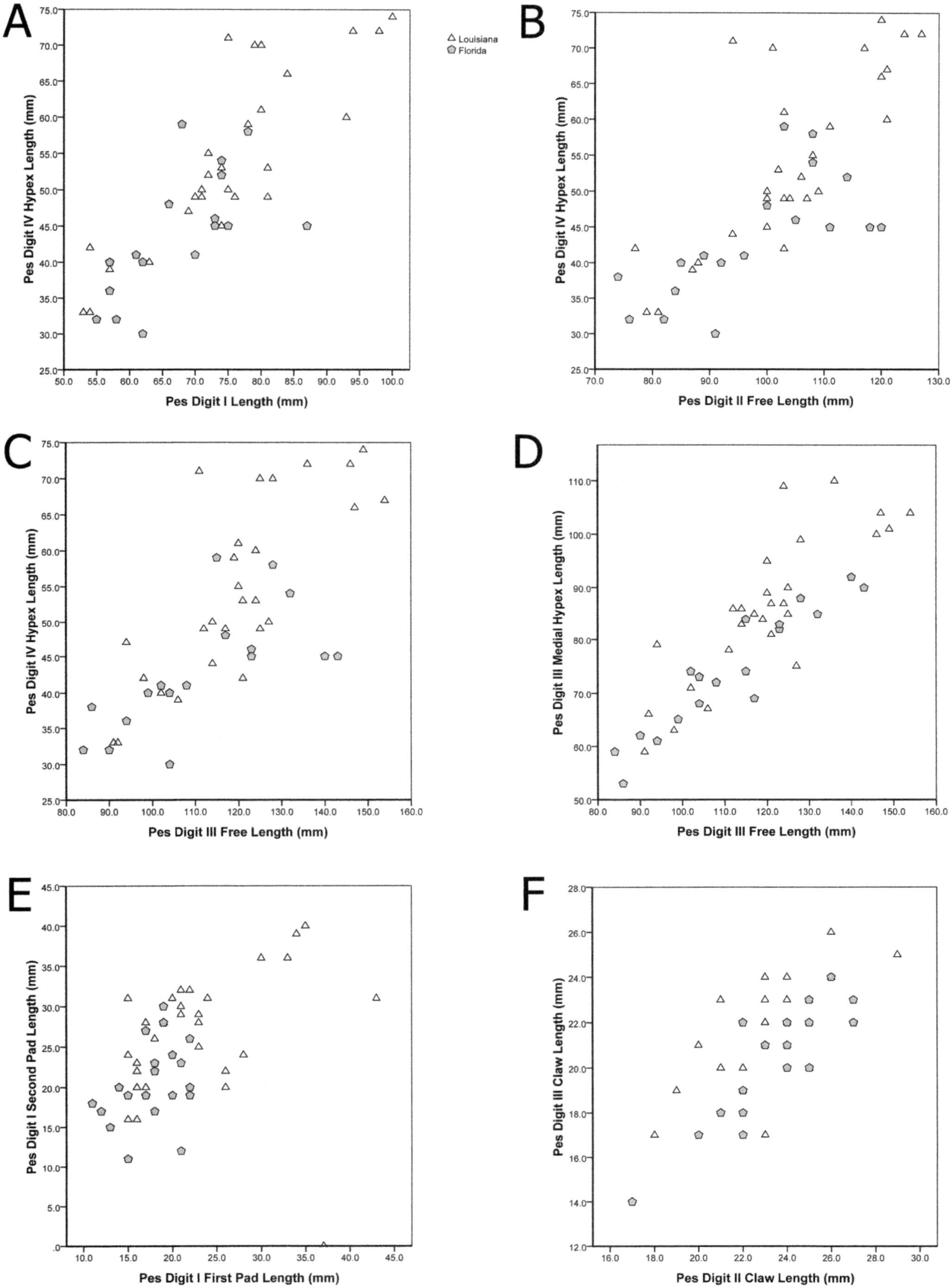
A
B
C
D
E
F
Louisiana
Florida
Pes Digit IV Hypex Length (mm)
Pes Digit I Length (mm)
Pes Digit II Free Length (mm)
Pes Digit III Free Length (mm)
Pes Digit III Medial Hypex Length (mm)
Pes Digit I Second Pad Length (mm)
Pes Digit I First Pad Length (mm)
Pes Digit III Claw Length (mm)
Pes Digit II Claw Length (mm)

Table 5.14. Comparison of occurrences of an interpad space separating the two digital pads of pedal digit I in alligators from South Carolina, Louisiana, and Florida.

Region	Number of alligators Without an interpad space	With an interpad space	Total
South Carolina	9	4	13
Louisiana	38	4	42
Florida	11	11	22
Total	58	19	77

$\chi^2 = 13.038$; two-sided $P = .001$.

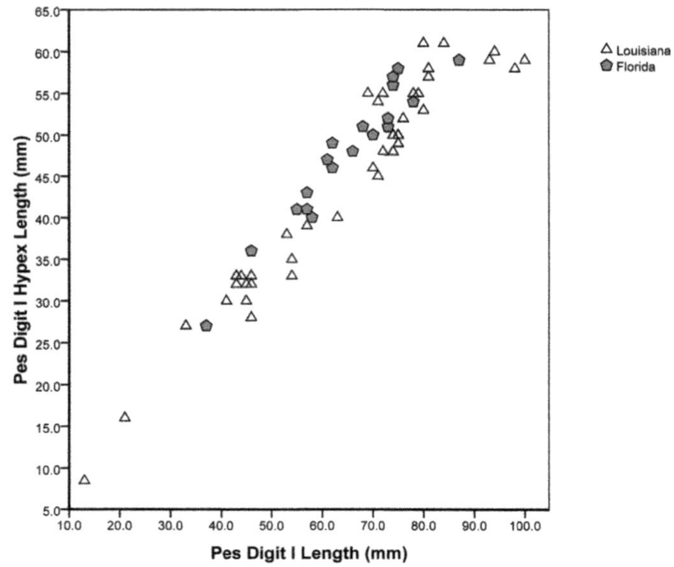

5.17. Comparison of the relationship between the lateral hypex length of digit I and digit I length in alligators with unworn claws from Florida and Louisiana. Florida alligators tend to have a longer hypex length.

projection of digit IV beyond the interdigital web it shares with digit III, or simply a relatively longer digit IV more generally, is unknown, because the medial hypex length was the only length measurement of digit IV I made (see above). In any case, the length difference isn't necessarily trivial. For example, the estimated mean difference in digit IV medial hypex length between Louisiana and Florida alligators, with digit II free length as the covariate, is 7 mm for a digit II free length of 101 mm.

Louisiana alligators also may have a slightly longer digit III medial hypex length, compared with digit III free length, than do Florida alligators. For a digit III free length of about 117 mm, the estimated mean difference in digit III medial hypex length between the two regions is about 6 mm. Digit III therefore may project a bit further beyond the web that joins it with digit II in alligators from Louisiana than in Florida alligators.

Another category of regional differences involves the relative size and proportions of digit I. Louisiana alligators seem to have a slightly longer second pad of digit I, compared with the first pad length, than do Florida alligators. For an average first pad length of 19.7 mm, the estimated mean second pad length is 25.6 mm and 21.1 mm in Louisiana and Florida alligators, respectively—about a 21% difference between the two estimated second pad lengths.

A distinct difference in the feet of alligators across regions seems to be the incidence of animals in which there is a big enough gap separating the two digital pads of digit I to warrant recognizing an interpad space between them (table 5.14). Fewer than half of the alligators from South Carolina showed an interpad space, and only about 10% of the Louisiana alligators did so. Half of the Florida alligators, in contrast, had an interpad space. These differences are significant by the χ^2 test (table 5.14).

ANCOVA of digit I lateral hypex length between Louisiana and Florida, with digit I length as the covariate, is problematic because of unequal error variances in the two populations, even after log transformation of variables (table 5.13). Inspection of the relationship between the two variables, contrasted between the two regions (fig. 5.17), suggests that Florida alligators tend to have a relatively longer digit I hypex length than do Louisiana alligators, compared with digit I length. Interpretation of the difference is complicated by the fact that very large males depart quite markedly from the trend of other alligators of both sexes, suggesting that there is some maximum digit I hypex length that isn't exceeded even if digit I length continues to get larger. If the comparison between regions is restricted to mature alligators of overlapping size (digit I lengths roughly between 37 and 87 mm; fig. 5.17), Florida alligators continue to have a proportionately longer digit I hypex length than Louisiana alligators for a given digit I length, and now the difference is unambiguously statistically significant (table 5.13).

As a further test of regional differences in hindfoot shape, I performed a stepwise discriminant analysis using the same parameters of digit length considered in table 5.11, scaled as in the discriminant analysis that searched for sexual differences in pedal digit lengths. Comparison was made between mature Florida and Louisiana alligators with unworn claws. The Wilks's λ test indicated significant ($P < .05$) regional differences in scaled digit I lateral hypex length (longer in Florida than Louisiana alligators), digit II lateral hypex length (Florida longer than Louisiana), digit III medial hypex length (Louisiana longer than Florida), and digit IV medial hypex length (Louisiana longer than Florida); and scaled digit II free length was nearly significant ($P = .59$; Florida longer than Louisiana), results mostly consistent with the above-reported bivariate comparisons. Curiously, the stepwise procedure selected scaled digit I lateral hypex

Table 5.15. Forward stepwise discriminant analysis of log-transformed digit lengths in mature alligators with unworn claws from Florida and Louisiana). Digit lengths scaled by subtracting the mean of all the log-transformed variables employed in the analysis from the value of each log-transformed variable. The digit length variables examined in the analysis were the same as those in table 5.11. Function 1 eigenvalue = 1.797; canonical correlation = 0.802.

Variable	Standardized canonical discriminant function coefficients (correlations between individual variables and discriminant function)		
Scaled digit II lateral hypex length	−0.602 (−0.564)		
Scaled digit III medial hypex length	0.878 (0.374)		
Scaled digit IV medial hypex length	0.669 (0.496)		
Classification results			
Geographic origin	**Predicted geographic origin (original grouped cases [cross-validated grouped cases])**		
	Louisiana	**Florida**	**Total**
Louisiana	19 (19)	3 (3)	22 (22)
Florida	0 (0)	16 (16)	16 (16)

92.1% of both original and cross-validated grouped cases correctly classified.

length in the first step of the analysis, before discarding it in the final step; the final discriminant function became a contrast between scaled digit II lateral hypex length, on the one hand, and scaled digit III and digit IV medial hypex lengths, on the other (table 5.15). In contrast to the poor performance of stepwise discriminant analysis in detecting differences between the sexes, in the comparison of alligators from the two regions the procedure did a very good job; only 3 out of 22 Louisiana alligators were mistakenly classified as Floridians, and all of the Florida alligators were correctly classified.

Focusing on the contrast between Florida and Louisiana alligators identified by the discriminant analysis (fig. 5.18), the relative length of the digit III medial hypex length, compared with the length of the digit II lateral hypex length, seems to be especially effective at distinguishing mature alligators from the two regions (fig. 5.18A). ANCOVA gives the same result, albeit with the caveat that the comparison failed Levene's test of the equality of error variances, even when the two variables are log transformed (table 5.13). Even though there is more overlap between alligators of the two regions than in the preceding comparison, Louisiana alligators also have a relatively longer digit IV medial hypex length for a given digit II lateral hypex length than do Florida alligators, and ANCOVA shows the contrast to be significantly different (table 5.13).

In summary, my data suggest that alligators may show regional differences in some pedal proportions, and in the

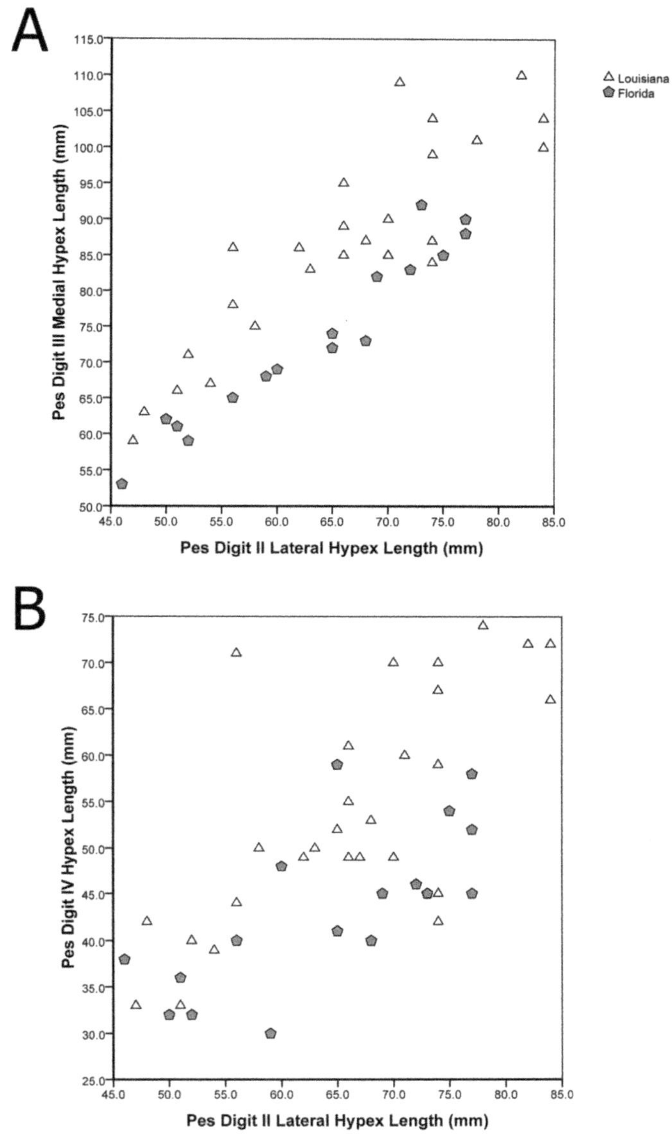

5.18. Comparisons of the relationship between the lateral hypex length of digit II and *A*, the medial hypex length of digit III, and *B*, the medial hypex length of digit IV, of mature alligators from Florida and Louisiana. In both comparisons, Florida alligators have a relatively long digit II lateral hypex, compared with the other hypex length.

extent to which an interpad space splits the nonclaw portion of digit I, although it would be nice to see these results confirmed with larger sample sizes. In this context it is interesting to note that there is genetic differentiation of American alligator populations, with a Louisiana/Texas western group of populations being distinctly different from a Florida/southern Georgia eastern set of populations (Davis et al. 2002). Most of the regional differences between Louisiana and Florida alligators in pedal proportions are subtle, involving population differences of around 5%–10% in predicted pedal parameter lengths, given specified values of other pedal parameters; the biggest regional difference seems to

Table 5.16. Values of the coefficient of relative dispersion (D_d) about the reduced major axis for bivariate comparisons of alligator pedal measurements. Values are calculated for raw measurements, and also for log-transformed data (parentheses). Data are for alligators of all sizes, from all geographic regions. For some comparisons, involving parameters of which claw lengths are components, D_d is calculated two ways, either including or excluding animals with heavily worn claws. Including animals with worn claws slightly increases the value of D_d; increases of comparable magnitude occur in other such comparisons, but to save space are not reported here.

Variables			D_d	N
Digit I length	Digit II free length	All feet	8.60 (1.98)	66
		Excluding worn claws	7.18 (1.67)	59
	Digit III free length	All feet	9.67 (2.19)	67
		Excluding worn claws	8.28 (1.83)	57
	Digit IV medial hypex length	All feet	14.9 (3.61)	92
		Excluding worn claws	14.2 (3.51)	86
Digit II free length	Digit III free length	Excluding worn claws	5.13 (1.01)	59
	Digit IV medial hypex length	Excluding worn claws	14.6 (3.28)	63
Digit III free length	Digit IV medial hypex length	Excluding worn claws	14.9 (3.34)	62
Digit I length excluding claw	Digit II free length excluding claw		12.2 (2.70)	65
	Digit III free length excluding claw		12.2 (2.71)	66
Digit II free length excluding claw	Digit III free length excluding claw		7.18 (1.43)	64
Digit I length	Digit I lateral hypex length	Excluding worn claws	8.58 (2.18)	86
Digit II free length	Digit II medial hypex length	Excluding worn claws	5.62 (1.24)	63
	Digit II lateral hypex length	Excluding worn claws	7.19 (1.58)	63
Digit III free length	Digit III medial hypex length	Excluding worn claws	8.57 (1.78)	62
	Digit III lateral hypex length	Excluding worn claws	8.46 (1.81)	62
Digit I length excluding claw	Digit I claw length	Excluding worn claws	20.0 (5.38)	87
Digit II free length excluding claw	Digit II claw length	Excluding worn claws	18.7 (4.54)	63
Digit III free length excluding claw	Digit III claw length	Excluding worn claws	17.3 (3.99)	62
Digit I first pad length	Digit I second pad length	All feet	40.7 (10.06)	93 (91)[a]
	Digit I claw length	Excluding worn claws	26.2 (8.84)	87
Digit I second pad length	Digit I claw length	Excluding worn claws	28.4 (9.61)	87 (86)[a]
Digit III medial hypex length	Digit I lateral hypex length	All feet	10.9 (2.65)	92
	II medial hypex length	All feet	8.03 (1.92)	92
	II lateral hypex length	All feet	9.43 (2.09)	92
	III lateral hypex length	All feet	10.7 (2.16)	92
	IV medial hypex length	All feet	14.7 (3.36)	92
Digit I claw length	Digit II claw length	Excluding worn claws	11.4 (4.34)	85
	Digit III claw length	Excluding worn claws	12.4 (3.99)	83
Digit II claw length	Digit III claw length	Excluding worn claws	9.02 (3.52)	85

[a] Difference in number of cases between raw and untransformed data is due to loss of cases when the second pad is regarded as missing (due to the absence of a crease subdividing the fleshy part of the toe), with length defined as zero (which cannot be log transformed).

be in the relative sizes of the digital pads of digit I. Most of the regional differences in alligator pedal proportions (table 5.13), however, do not seem to be statistically significant.

Overall Intraspecific Variability in Pedal Shape

To have a basis for comparing alligator hindfoot shape variability with shape variability of the feet and footprints of other species, I calculated three sets of quantitative variability indices in my pooled alligator sample. First, I determined the coefficient of relative dispersion about the reduced major axis (D_d), a two-dimensional equivalent of the coefficient of variation (Imbrie 1956), of bivariate relationships, using both raw and log transformed data (table 5.16). For the most part, only specimens with unworn claws were used in these calculations. Next, I scaled pedal measurements of alligators of all sizes to a common value of a specified parameter, after which I calculated the ratio of the maximum to the minimum value of each scaled measurement (table 5.17). I did two variations of this treatment; in one, the length of digit I without the claw was the parameter to which all other measurements were scaled, and in the other the free length of digit III without the claw was the scaling parameter. The particular values of these two scaling parameters selected (49 mm for digit I length excluding the claw, and 98 mm for digit III free length without the claw) were means for the alligators in my sample. Separate determinations were done for all alligator feet, and also for alligator feet with unworn

Table 5.17. Variability of pedal measurements scaled to a common dimension in alligator feet of all sizes. Data were scaled in two ways: to a digit I length (excluding claw) of 49 mm (scale 1), and to a digit III free length (excluding claw) of 98 mm (scale 2). When claw length is a component of a parameter, results are reported for all feet, and also for feet with unworn claws only.

Parameter	Treatment	Minimum (mm)	Maximum (mm)	Maximum/minimum ratio	*N*
Digit I length	Scale 2; all feet	53.5	91.1	1.71	66
	Scale 2; excluding worn claws	56.0	91.1	1.63	60
Digit I lateral hypex length	Scale 1; all feet	37.9	74.7	1.97	92
	Scale 1; excluding worn claws	37.9	74.7	1.97	86
	Scale 2; all feet	39.4	71.9	1.82	66
	Scale 2; excluding worn claws	39.9	71.9	1.80	60
Digit I first pad length	Scale 1: all feet	14.8	49.0	3.32	93
	Scale 2; all feet	13.3	42.2	3.17	66
Digit I second pad length	Scale 1; all feet	0	33.0	—[a]	93
	Scale 2; all feet	0	35.3	—[a]	66
Digit I length excluding claw	Scale 2: all feet	35.0	67.6	1.93	66
Digit I claw length	Scale 1; all feet	13.9	39.7	2.85	93
	Scale 1; excluding worn claws	16.3	39.7	2.43	87
	Scale 2: all feet	14.1	32.7	2.31	66
	Scale 2; excluding worn claws	19.8	32.7	1.65	60
Digit II free length	Scale 1; all feet	80.1	156.8	1.96	66
	Scale 1; excluding worn claws	80.1	156.8	1.96	63
	Scale 2; all feet	86.4	118.9	1.38	65
	Scale 2; excluding worn claws	90.5	118.6	1.31	62
Digit II medial hypex length	Scale 1; all feet	62.2	125.2	2.01	92
	Scale 1; excluding worn claws	62.2	125.2	2.01	89
	Scale 2; all feet	70.3	110.7	1.57	65
	Scale 2; excluding worn claws	70.3	110.7	1.57	62
Digit II lateral hypex length	Scale 1; all feet	49.0	98.0	2.00	89
	Scale 1; excluding worn claws	49.0	98.0	2.00	89
	Scale 2; all feet	55.2	86.2	1.56	65
	Scale 2; excluding worn claws	55.2	81.3	1.47	62
Digit II free length excluding claw	Scale 1; all feet	63.6	114.3	1.80	65
	Scale 2; all feet	66.7	96.0	1.44	64
Digit II claw length	Scale 1; all feet	14.6	43.6	2.98	91
	Scale 1; excluding worn claws	15.1	43.6	2.88	89
	Scale 2; all feet	14.1	33.5	2.37	64
	Scale 2; excluding worn claws	16.8	33.5	1.99	62
Digit III free length	Scale 1: all feet	88.1	176.4	2.00	67
	Scale 1; excluding worn claws	88.1	176.4	2.00	62
Digit III medial hypex length	Scale 1; all feet	63.3	124.1	1.96	93
	Scale 1: excluding worn claws	65.3	124.1	1.90	88
	Scale 2; all feet	67.6	106.8	1.58	66
	Scale 2; excluding worn claws	67.6	106.8	1.58	62
Digit III lateral hypex length	Scale 1; all feet	53.5	117.6	2.20	92
	Scale 1: excluding worn claws	61.6	117.6	1.91	87
	Scale 2; all feet	60.0	99.0	1.65	66
	Scale 2; excluding worn claws	69.0	99.0	1.44	62
Digit III free length excluding claw	Scale 1; all feet	71.0	137.2	1.93	66
Digit III claw length	Scale 1; all feet	4.5	39.2	8.81	92
	Scale 1; excluding worn claws	14.4	39.2	2.73	88
	Scale 2; all feet	5.0	31.1	6.21	66
	Scale 2; excluding worn claws	14.5	31.1	2.14	62
Digit IV medial hypex length	Scale 1: all feet	32.0	73.0	2.28	92
	Scale 2; all feet	35.9	74.0	2.06	66

[a] Parameter cannot be calculated because of the impossibility of division by zero.

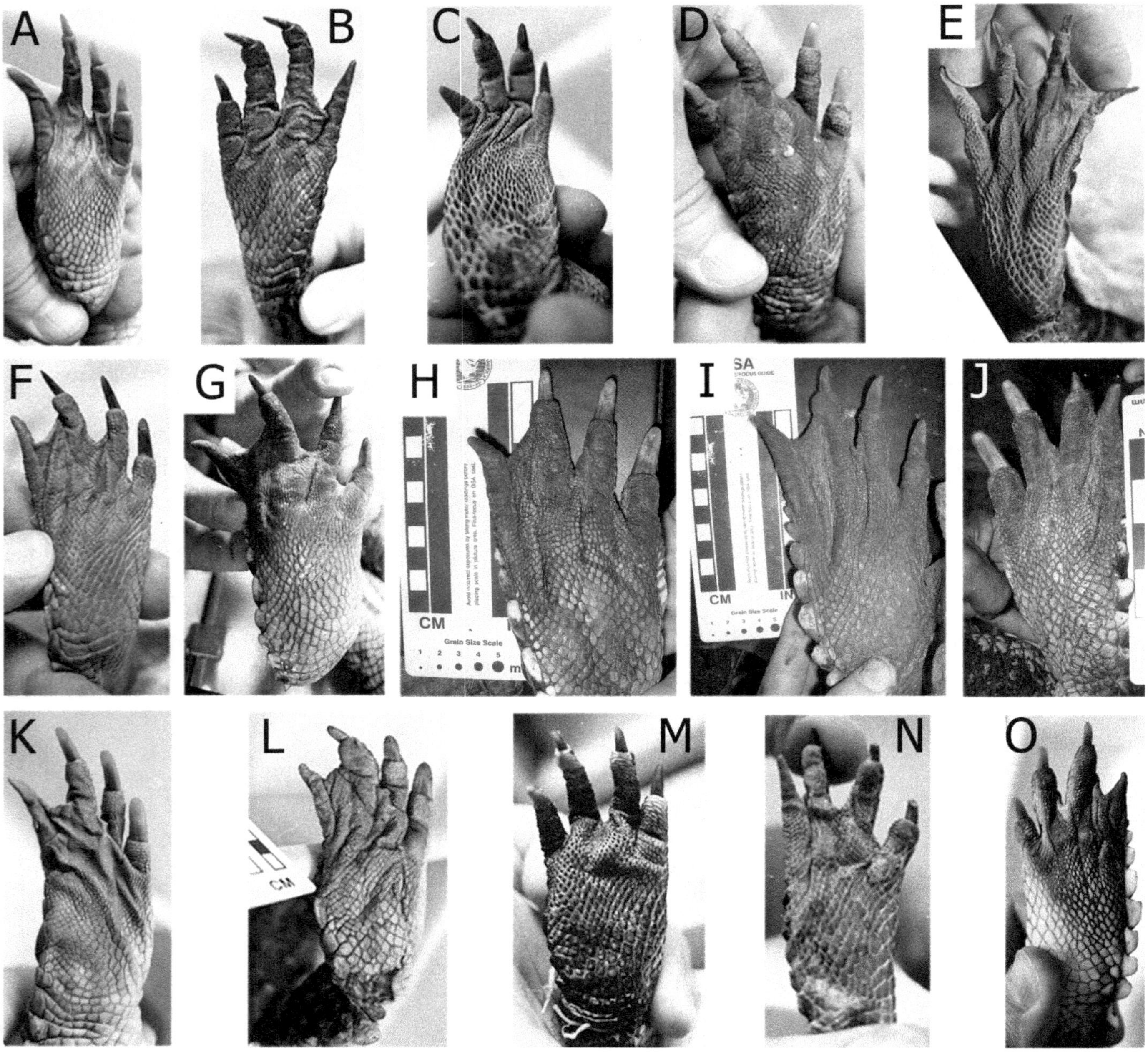

5.19. Comparison of plantar surfaces of the hindfeet of extant crocodylian species. Most of these are spirit-preserved specimens (making them hard to photograph in a consistent orientation) that are small individuals of their species. *A*, FMNH 229981, *Caiman crocodilus* (formerly identified as *C. sclerops*). *B*, FMNH 5777, *Caiman latirostris*. *C*, FMNH 5778, *Melanosuchus niger*. *D*, FMNH 3721, *Mecistops cataphractus*. *E*, FMNH 115355, *Crocodylus moreletii*. *F–J*, *Crocodylus acutus*. *F*, FMNH 15230. *G*, USNM 3172676. *H–J*, Live females (photographs courtesy of Cory Kumagai), Costa Rica. *H*, Animal with a total length of 199.5 cm. *I*, Animal with a total length of 212 cm. *J*, Animal with a total length of 243 cm. *K*, FMNH 63739, *Crocodylus palustris*. *L*, FMNH 39168, *Crocodylus niloticus*. *M*, USNM 193289, *Osteolaemus tetraspis*. *N*, USNM 11086, *Tomistoma schlegelii*. *O*, USNM 229860, *Gavialis gangeticus*.

claws. Finally, the standard error of the estimate (SEE) in bivariate regression analyses of log-transformed parameters (tables 5.9, 5.11) provided a third measure of shape variability.

Most D_d values are 15% or less for raw data, and 5% or less with log-transformed data (table 5.16). Higher values are seen in bivariate comparisons involving the pads of digit I (the high variability of digit I pad lengths in alligators has previously been noted in this chapter) and comparisons of claws against the fleshy parts of toes. Including animals with heavily worn claws increases values of D_d slightly, an unsurprising result.

Maximum/minimum ratios of scaled measurements (table 5.17) commonly take values of 2 or less. Higher ratios are associated with scaled values of digit I pad lengths, some hypex lengths, and claw lengths. Maximum/minimum ratios of scaled claw lengths were, unsurprisingly, higher if animals with heavily worn claws were included than if they were not included; the difference is particularly striking for the scaled lengths of the digit III claw.

Note that including alligators with heavily worn claws has a greater impact on maximum/minimum ratios than on D_d. This observation can be generalized to all estimates of parameter variability in terms of D_d and ratios of scaled variables. The latter will show the effects of extreme, outlier cases more than the former does.

SEE values of bivariate log-transformed digit length parameters (table 5.9) range from about 0.02 to 0.1. Once again the higher values are associated with digit pad lengths. SEE values of multivariate relationships (table 5.11) range from 0.02 to 0.06, with the highest values associated with the digit IV medial hypex length.

COMPARISONS WITH OTHER CROCODYLIAN SPECIES

Although a thorough examination of hindfoot shape across species of extant crocodylians is beyond the scope of the present study, I did examine spirit-preserved or live specimens of the feet of a least a few individuals of most of the living crocodylian species, some of which are illustrated in figure 5.19 (compare with fig. 5.2 for American alligators). Most of the museum "pickles" were small individuals of their respective species, due to the difficulty of preserving and storing intact carcasses of large crocodylians. Consequently many of the observations I will now report should be regarded as provisional.

In all extant crocodylians (Milàn and Hedegaard 2010; Grigg and Kirshner 2015) and some of their fossil relatives as well (Nesbitt, 2011; Nesbitt et al. 2013*b* there are four toes on the pes. The lengths of the toes and their respective metatarsals are such that digits I through III project progressively farther forward, although the extent of the difference varies among species (compare fig. 5.19*A* and 5.19*H*; also see Milàn and Hedegaard 2010), but then digit IV reverses the trend and does not extend as far forward as III. Webbing connects the toes of these aquatic reptiles, and how far the connecting webbing extends forward (distally) between toes increases from I–II, to II–III, to III–IV. There are nearly always two distinct digital pads on digit I. In all species digits I–III bear claws (with the length and/or stoutness of the claws decreasing from I–III), while IV does not. Small individuals seem to have more needle-like claws than older individuals. It is not unusual to see significant wear truncating the claws and leaving a sharp, beveled surface on the underside of the claws

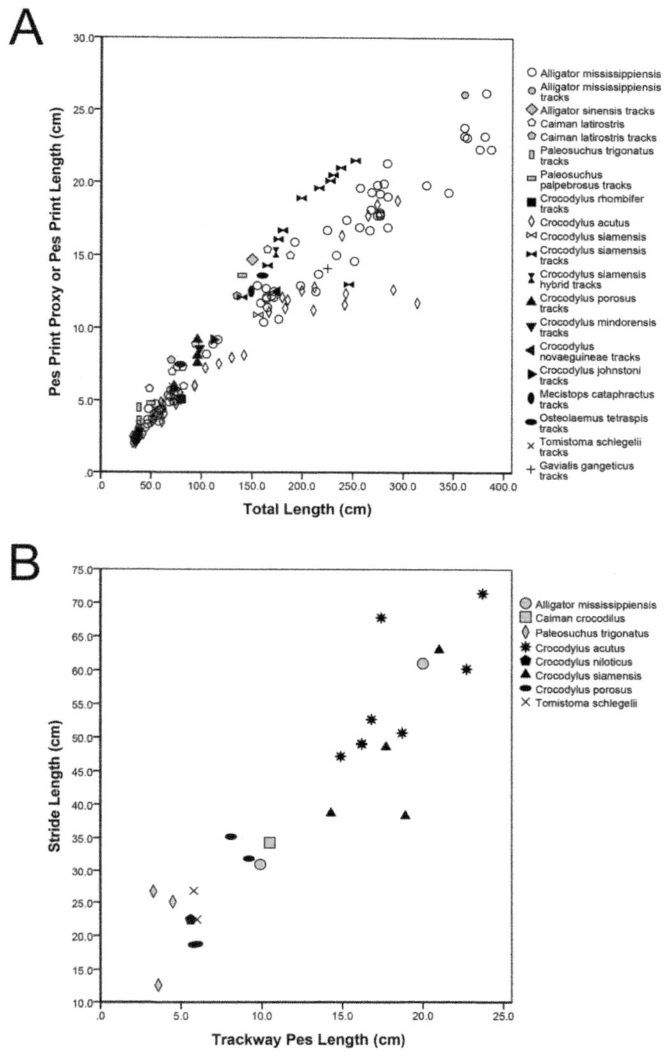

5.20. *A*, Pes or pes print length plotted as a function of animal total length across crocodylian species. *B*, Stride length vs. pes print length in crocodylian trackways. *Data from from Leonardi (1984), Padian and Olsen (1984), Kubo (2008), Carpenter (2009), Kumagai (2010), Milàn and Hedegaard (2010), Rajkumar et al. (2015), B. Simpson (personal communication), and this study.*

(table A5.1). When such wear occurs, it is heaviest on digit I, and becomes less severe outward across digits II and III. In general, then, extant crocodylians are fairly uniform in pedal morphology.

Nonetheless there are differences among species (or within species, or between the left and right feet of the same animal; also see Milàn and Hedegaard 2010). The forward extent of the webbing between toes varies, most notably between digits III and IV, where the webbing connects a markedly greater portion of the lengths of these toes in *Crocodylus* than in alligators, caimans, *Osteolaemus*, and *Tomistoma*. The extent to which the fleshy portions of digits II–IV are subdivided by creases into digital pads is variable. One often

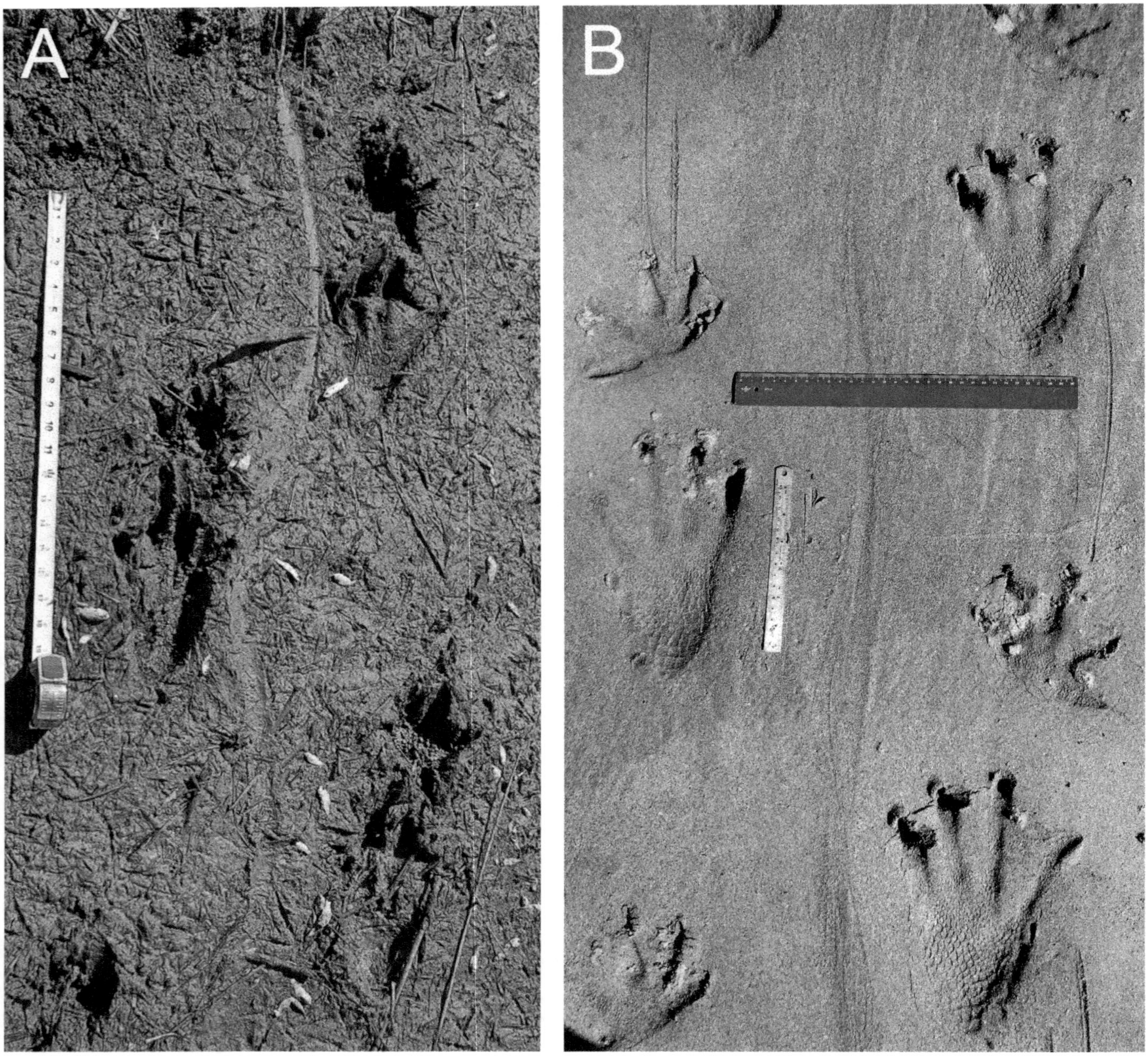

5.21. Trackways of extant crocodylians. *A*, *Alligator mississippiensis*, Rockefeller Wildlife Refuge, Louisiana (Farlow and Elsey 2010); 19 inches (48.3 cm) of measuring tape are exposed. *B*, *Crocodylus acutus*, northwest Costa Rica (composite image created from digital photographs by Nathan Robinson and Frank Paladino; Farlow et al. 2018); the larger scale is marked off in 30 1-cm increments.

sees two or three digital pads on II and III, but sometimes there are no clearly delimited digital pads, or only one, and when the latter situation occurs, the clear pad is at the distal end of the fleshy portion of the toe, adjacent to the claw. Digit III is sometimes divided into as many as four digital pads. Digit IV is the most variable of the four toes in the extent to which digital pads can be recognized. Sometimes there is no subdivision of the toe into digital pads, or only a single clear, basal pad, with the distal remainder of the toe undivided; and sometimes there are as many as five digital pads. Crocodiles and gharials have very large scales or scutes lining the outer edge of the metatarsal portion of the foot, a feature absent from alligators and caimans (Milàn and Hedegaard 2010; Grigg and Kirshner 2015).

Footprints and trackways of crocodylians and their close kin, both modern and ancient, have received occasional but increasing attention from paleontologists over the years (von Huene 1913; Reineck and Howard 1978; Padian and Olsen 1984; Bennett 1992; Moratalla et al. 1995; Foster and Lockley 1997; Fuentes Vidarte and Meijide Calvo 2001; Mazin et al.

2003; Lockley and Meyer 2004; McCrea et al. 2004; Erickson 2005; Pascual Arribas et al. 2005; Avanzini et al. 2007; Kubo 2008; Carpenter 2009; Milàn et al. 2010 [and many references therein]; Martin 2013; Contessi and Fanti 2012; Abbassi et al. 2015; Hadri et al. 2015; Rajkumar et al. 2015; Vila et al. 2015; Segura et al. 2016; Mateus et al. 2017; Farlow et al. 2018; Klein et al. 2018). Measurements of dimensions of pedal prints made by crocodylians of known size are obviously unavailable for fossil trackways, but some data are available for extant crocodylians (fig. 5.20A), although measurements are scant for large adults of most species. As with American alligators (tables 4.1, 5.6, 5.7) the available data for pes and pes print length across crocodylian species seem to show a curvilinear relationship with animal total length, with hindfoot length becoming relatively shorter as the animals get bigger. Put another way, the data suggest that crocodylians generally, and not just alligators, grow into their feet. There is a lot of scatter in the relationship, but the extent to which this reflects real differences among species, as opposed to different measurement protocols, is uncertain.

Although data are once again limited, trackway features also seem fairly uniform across crocodylian species (figs. 5.20*B*, 5.21). Differences across species are likely to be quantitative rather than qualitative (Farlow et al. 2018).

The results presented in this chapter have provided an indication of how much shape variability one could expect to see among hindfeet of one archosaurian species and some limited information about the degree of variability across species in pedal shape. However, before applying these results to interpreting dinosaur footprints, we must first see if they are consistent with what is seen in other kinds of archosaurs. In particular, we must examine intraspecific variability in foot shape in the other major branch of archosaurs, the branch that includes birds. If the degree of foot shape variability in the feet and footprints of birds is equivalent to that seen in alligators, we can more confidently apply such data to considerations of intraspecific vs. interspecific variability in dinosaur footprints. So we now turn our attention to birds.

6 Footprints of the Emu (*Dromaius novaehollandiae*) and Other Ground Birds

INTRODUCTION

In previous chapters we have considered intraspecific and interspecific variability in foot skeleton shape in dinosaurs, birds, and crocodylians, and intact foot shape in alligators. We now turn our attention from feet to footprints—specifically those of large ground birds, and one species of ground bird in particular. In so doing we will examine within-taxon and across-taxon variability not just in anatomical features (feet), but also in products of the interaction between those anatomical features and a sedimentary substrate (footprints). We will concern ourselves not just with features that are directly related to pedal form (such as digital lengths), but also with features that are less tightly constrained by anatomy (such as interdigital angles). We will also examine factors that affect footprint quality, and explore whether footprint quality is correlated with quantitative measurements of footprint shape.

In the present chapter we will concern ourselves with intraspecific variability in footprints made by emus (*Dromaius novaehollandiae*), addressing the question of how different in shape footmarks can be, and yet have been made by the same species of trackmaker. Why emus? As noted by ichnologist Tony Martin (2014: 310), "nearly every non-avian theropod and most ornithopod trackways bear similar basic patterns or features like those made by equivalently sized modern birds . . . it is no wonder that dinosaur ichnologists still turn to birds—especially large flightless ones—as their default models for how bipedal dinosaurs moved and made tracks."

More specifically, emus were selected for this study because they are the largest extant tridactyl, archosaurian bipeds; although the kinematics of emu locomotion may not be identical to that of bipedal non-avian dinosaurs (Abourachid and Renous 2000; Farlow et al. 2000), these birds probably provide as close an approximation to the trackmaking activities of their Mesozoic predecessors as one can find in the modern world. Furthermore, emus were readily available for study at the Fort Wayne (Indiana) Children's Zoo—where we received substantial logistical support for our work.

We will then compare intraspecific variability of emu prints with foot and footprint variability in other extant ground bird species, and with alligators. Finally, we will consider the extent to which footprints made by different clades of ground birds can be distinguished from one another, and how well foot and footprint shape track (if you will pardon the expression) the phylogenetic relationships of ground birds. In the course of this analysis we will find it necessary to compare not just bird footprints, but also the feet of ground birds.

FOOTPRINT COLLECTING METHODS

Emus

We (my students and I) collected measurements and casts (negative copies) of emu footprints from 1988 through 1995 (fig. 6.1); throughout this chapter, I will use the plural pronoun (we) to distinguish operations done by both my students and myself, and the singular pronoun (I) to identify actions and ideas for which I alone am responsible (or to blame).

During each episode of footprint collecting (a "run") a single (occasionally more than one) bird was allowed or forced to move across a prepared substrate (table 6.1). Each footprint of each bird was assigned its own unique identifying code.

Our most intensive footprint collecting took place during the summer of 1988. Three adult emus were run repeatedly across the prepared soil surface of a large (about 1½ acre) animal holding pen adjacent to the African Veldt exhibit of the Fort Wayne Children's Zoo. Prior to a run, the soil was turned over and smoothed with a tractor, and wetted with sprinklers. An emu was released from a pen inside a zoo barn, and driven into the Veldt pen. It then walked and trotted (no bird probably ever ran at its top speed) across the soil, after which it was eventually chased from the pen back into the barn. We then measured trackways, collected soil samples adjacent to footprints for laboratory analysis, and made plaster casts of prints. Large numbers of footprints were collected during these runs, 30 or more at a time.

During the same summer we also collected footprints from emu chicks. These birds were housed in a different part of the zoo. Prior to a run, a layer of pottery clay ("red clay") was spread across an 8 foot × 4 foot fiberwood sheet, smoothed, and wetted. The board was placed in a confined area, so that a chick could not easily get around it. The young bird was encouraged to walk across the clay surface—sometimes the birds seemed to find it slippery and difficult to walk on—and was then removed from the room. I didn't want to risk having chicks destroy good prints by treading on them, and so far fewer footprints were made during these runs with juveniles. As with the adults, sediment samples adjacent to each footprint were collected, trackways measured, and footprints cast.

We ran the juveniles repeatedly over the course of the summer, collecting footprints of the same bird at different sizes. The baby birds were fed a mixture of chick starter feed, rabbit pellets, and commercial ratite diet. Initially they were fed *ad libitum*, but one of the chicks developed a condition—not uncommon in captive ratites—in which the limbs became too weak for the bird's weight, and it had to be euthanized. Thinking that this was due to too-rapid growth,

Table 6.1. Description of emu footprint-collecting sessions ("runs") made during the course of this study. Runs marked with an asterisk (*) were used in "emu means" analyses (see text) from among the runs of juveniles that grew enough over the course of study for detectable increases in footprint dimensions to occur. If no runs for a particular bird are marked, all runs were used in calculating mean values for the bird.

Bird	Run	Date	Substrate (substrate code)
1988 Juvenile 1: captive-hatched bird run at the Fort Wayne Zoo; about 6 weeks old; bird mass = 5.08 kg	1	23 May 1988	"Red clay" (1): wetted potter's clay spread out on a 8 foot by 4 foot fiberboard sheet
1988 Juvenile 2: captive-hatched bird run at the Fort Wayne Zoo; bird was 4 days old on first run. Unfortunately, this bird developed a common malady of captive ratites in which it grew in weight too quickly for the strength of its legs, and had to be euthanized in mid-June 1988	2	27 May 1988; bird mass on this date = 0.34 kg	Red clay (1)
	6*	7 June 1988; bird mass on this date = 0.71 kg	Red clay (1)
1988 Adult 3: captive-hatched bird run at the Fort Wayne Zoo; bird mass = 35 kg	3	31 May 1988	"Veldt holding pen" (2): a roughly 1½-acre dirt-floored pen adjacent to the African Veldt exhibit of the Fort Wayne Zoo.
	5	3 June 1988	Veldt holding pen (2)
	17	15 July 1988	"Kangaroo yard" (3): an area of mixed sand and crushed brick in the middle of the Australian Exhibit, Fort Wayne Zoo
1988 Juvenile 4: captive-hatched bird run at the Fort Wayne Zoo; about 2½–3 weeks old on first run	4	2 June 1988; bird mass on this date = 0.89 kg	Red clay (1)
	10	16 June 1988; bird mass on this date = 1.84 kg	Red clay (1)
	13	28 June 1988; bird mass on this date = 2.24 kg	Red clay (1)
	18*	19 July 1988; bird mass = 4.5 kg	Red clay (1)
1988 Adult 5: captive-hatched bird run at the Fort Wayne Zoo; bird mass = 38 kg	7	8 June 1988	Veldt holding pen (2)
	8	10 June 1988	Veldt holding pen (2)
	14	7 July 1988	Kangaroo yard (3)
1988 Adult 6: captive-hatched bird run at the Fort Wayne Zoo; bird mass = 40 kg	9	15 June 1988	Veldt holding pen (2)
	11	17 June 1988	Veldt holding pen (2)
	12	21 June 1988	Kangaroo yard (3)
	15	7 July 1988	Kangaroo yard (3)
1988 Juvenile 7: captive-hatched bird run at the Fort Wayne Zoo; about 2 weeks old; bird mass = 1.16 kg	16	14 July 1988	Red clay (1)
Unidentified 1988 juvenile	19	11 October 1988	Kangaroo yard (3)
1989 Juvenile A: captive-hatched bird run at the Fort Wayne Zoo; about 2 months old on first run; bird mass = 2.75 kg	20	15 June 1989	"Damp" sand (4): sand was collected from outdoors. It had rained the previous night, and so the sand was damp, but we added no more water to it
	24	16 June 1989	"Wet" sand (6): water was added to the sand
	26	19 June 1989	"Very wet" sand (7): so much water was added that standing water covered the sand surface
	31	22 June 1989	Red clay (1)

Table 6.1. *continued*

Bird	Run	Date	Substrate (substrate code)
1989 Adult B: captive-hatched bird run at the Fort Wayne Zoo	21	15 June 1989	Damp sand (4)
1989 Juvenile C: captive-hatched bird run at the Fort Wayne Zoo; about 2½ months old on first run; bird mass = 5.84 kg	22	16 June 1989	Damp sand (4)
	23	16 June 1989	Wet sand (6)
	25	19 June 1989	Very wet sand (7)
	27	20 June 1989	Pond sediment (8): mixture of mud and sand as dug from a pond, with no additional water
	28	20 June 1989	"Wetted" pond sediment (9): pond sediment to which some water was added
	29	20 June 1989	"Very wet" pond sediment (10): enough water was added to the pond sediment to make it quite wet, but without standing water covering it
	30	20 June 1989	"Very very wet" pond sediment (11): still more water was added to the pond sediment, such that standing water covered most of the substrate
	32	22 June 1989	Red clay (1): in runs 32–34, we attempted to make the clay progressively wetter
	33	22 June 1989	Red clay (1)
	34	22 June 1989	Red clay (1)
Nearly grown juvenile (bird 12 or 13): captive-hatched bird run at the Fort Wayne Zoo	37	7 August 1990	Unknown
Nearly grown juvenile (bird 12): captive-hatched bird run at the Fort Wayne Zoo	35	4 September 1990	Kangaroo yard (3)
Nearly grown juvenile (bird 13): captive-hatched bird run at the Fort Wayne Zoo	36	7 September 1990	Kangaroo yard (3)
Glen Rose emu: adult bird run at Fossil Rim Wildlife Park, Glen Rose, Texas	38	13 October 1990	Available soil in bird's pen
Juvenile emu, probably tag number 00-0021-26ED; captive-hatched bird run at the Fort Wayne Zoo	39	13 May 1992	"Australian Aviary" exhibit (13), Fort Wayne Zoo: available substrate, mixture of soil and rotting wood chips and plant fibers
Juvenile 00-0025-5B48	40	13 May 1992	Australian Aviary (13)
Juvenile 00-0021-6A52	41	13 May 1992	Australian Aviary (13)
Juvenile 00-0021-26ED	42*	13 May 1992	Australian Aviary (13)
	46	19 August 1993	Bird run as an adult in a barn at the Earl Wells farm, using available substrate (mainly reindeer dung!)
Juvenile 00-001D-8823; captive-hatched bird run at the Fort Wayne Zoo	43	11 June 1993	Australian Aviary (13)
	47*	5 October 1993	Australian Aviary (13)
Juvenile 00-0021-35E4: captive-hatched bird run at the Fort Wayne Zoo	44*	11 June 1993	Australian Aviary (13)
	47	5 October 1993	Australian Aviary (13)
Juvenile 00-0025-D8A4; captive-hatched bird run at the Fort Wayne Zoo	45	11 June 1993	Australian Aviary (13)

zookeepers cut the ration of young birds back to 20 minutes of *ad libitum* feeding three times a day.

Soil or sediment samples were collected around the periphery of footprints. Because the juveniles didn't sink deeply into the red clay substrate, we only scraped shallowly around the prints in collecting sediment samples. Adults in the Veldt pen made deeper tracks; we sampled soil at and several centimeters below the soil surface at four points around the periphery of Veldt footprints.

In the laboratory, soil and clay samples were weighed wet, and then dried overnight at 110° C in an oven, allowed to cool in a desiccator, and weighed again. The soil and clay water content at the time of footprint formation was then expressed as a percentage of the sediment dry weight: 100 × (wet weight – dry weight)/dry weight.

We measured sand and clay ratios (as percentage of soil dry weight) in our soil samples using a simplified version of the method described by Bouyoucos (1936). We discovered, however, that our method was consistently underestimating the percentage of sand, and overestimating the percentage of clay, in our samples. We collected 20 additional soil samples from the Veldt pen and split each sample, with half being analyzed by our method, and half by a soils lab using protocol ASTM D422-63 (ASTM 1972). From this we derived empirical corrections for bringing our sand and clay proportional measurements into line with standard methods.

Bird	Run	Date	Substrate (substrate code)
Juvenile D59B1; captive-hatched bird run at the Fort Wayne Zoo	48*	5 May 1994	Australian Aviary (13)
	69	21 October 1994	Australian Aviary (13)
Juvenile 1E95BD; captive-hatched bird run at the Fort Wayne Zoo	49	5 May 1994	Australian Aviary (13)
	52*	26 May 1994	Australian Aviary (13)
	68	21 October 1994	Australian Aviary (13)
Juvenile D4264; captive-hatched bird run at the Fort Wayne Zoo. Subsequent to this run, the bird developed the bad limb condition and had to be put down	50	5 May 1994	Australian Aviary (13)
Juvenile 1E95BD or D59B1	51	5 May 1994	Australian Aviary (13)
Wild emu trail 26, Winton-Jundah Road, Winton Shire, Queensland, Australia	53	Observed 9 June 1994	Soil along roadside
Wild emu trail 27, road to Lark Quarry dinosaur tracksite, Winton Shire, Queensland, Australia	54	Observed 9 June 1994	Soil along roadside
Wild emu trail 28, Lark Quarry road	55	Observed 11 June 1994	Soil along roadside
Wild emu trail 29, Lark Quarry road	56	Observed 11 June 1994	Soil along roadside
Wild emu trail 42, Lark Quarry road	70	Observed 11 June 1994	Soil along roadside
Wild emu trail 30, Idalia National Park, Queensland, Australia	57	Observed 14 June 1994	Soil along roadside
Wild emu trail 31, Idalia National Park	58	Observed 14 June 1994	Soil along roadside
Wild emu trail A, Bribie Island, Queensland, Australia	59	Observed 20 June 1994	Sandy ditch along roadside
Wild emu trail B, Bribie Island	60	Observed 20 June 1994	Not recorded
Wild emu trail C, Bribie Island	61	Observed 20 June 1994	Not recorded
Wild emu trail D, Bribie Island	62	Observed 20 June 1994	Not recorded
Wild emu trail E, Bribie Island	63	Observed 20 June 1994	Not recorded
Wild emu trail F, Bribie Island	64	Observed 20 June 1994	Sandy dry creek bed
Wild emu trail G, Bribie Island	65	Observed 20 June 1994	Sandy soil
Two-year-old captive female caught in northwestern New South Wales, Australia, and kept at Outback Emu Farm, Queensland	71	17 June 1994	Available soil in pen
Three-year-old first-generation captive male (parents from western Queensland), Outback Emu Farm	71	17 June 1994	Available soil
Footprints from unspecified 2-year-old and 3-year-old captive birds from Queensland and Western Australia, Outback Emu Farm	71	17 June 1994	Available soil
Captive-hatched adult male, Klaus Radandt farm, run at Fort Wayne Zoo	66	June 1991	Veldt pen (2)
1995 Subadult A: captive-hatched bird run at Fort Wayne Zoo	75	4 October 1995	Kangaroo yard (3)
1995 Subadult B: captive-hatched bird run at Fort Wayne Zoo	76	4 October 1995	Kangaroo yard (3)

In the analyses that follow, we compare various footprint parameters with the sand and clay content of the soil, but not the silt content. We made this choice because silt values were not directly measured by the method we used, but only calculated from what was left over of a soil sample's dry weight after sand and clay values were determined. One can see the influence of the silt content in our analyses by considering the sand content of a sample to be inversely related to the total content of fine particles—that is, silt and clay particles.

For some analyses involving the three large 1988 adult emus, I wanted an overall assessment of footprint quality. I devised an ordinal scale (table 6.2) that expressed footprint quality in four defined classes, with the possibility of

Table 6.2. Emu footprint preservation classes. In addition to the described preservation classes, intermediate classes (1.5, 2.5, and 3.5) were used for borderline footprints whose quality of preservation didn't clearly fall into one of the defined preservation classes.

Class	Description
1 (Poor)	Toemarks poorly discernible or unrecognizable
2 (Fair)	Toemarks faint, blurred, or distorted, but recognizable
3 (Good)	Toemarks fairly clear and sharp, at least over substantial portions; some digital pads recognizable
4 (Excellent)	All digit impressions nearly or completely sharp and clear; scale impressions may be present

6.1. Collecting emu (*Dromaius novaehollandiae*) footprints at the Fort Wayne, Indiana Children's Zoo and in Australia. *A–C*, Emus and emu footprints made in the African Veldt exhibit holding pen. *A*, *B*, Emus making footprints on a freshly prepared surface. *C*, The author kneeling beside a sequence of footprints made during a run. The plastic flags were used to mark individual footprints, plaster casts of which could then later be labeled. *D*, The author inspecting a freshly created sequence of footprints made by a juvenile emu in the Fort Wayne Zoo Australian Aviary exhibit. *E*, Trackway of a wild emu, Lark Quarry Road, Queensland, Australia (Run 54, emu trail 27).

recognizing intermediate classes between the defined classes. This scale was subsequently modified for application to non-avian dinosaur footprints (Belvedere and Farlow 2016).

After our intensive runs with the three large emus, the birds were released into the kangaroo yard of the Australian wildlife exhibit of the Fort Wayne Zoo. In the center of this walk-through exhibit was an area of bare sand and crushed brick. We collected additional footprints of the three adults from this area in 1988, as well as prints of other emus in subsequent years.

In 1989 we worked mainly with two juveniles, running them repeatedly over different substrates over the course of a week. We ran the birds on sand, sediment from the bottom of a nearby pond, and red clay, to which we added progressively more water in an attempt to vary sediment water content. As in the 1988 juvenile runs, the 1989 birds were run in a confined space, which limited the number of footprints collected during each run.

After 1989 our footprint collecting procedure was less systematic. Because the Fort Wayne Zoo traded their grown emus to local animal breeders for chicks each year, we obtained a new set of juveniles every year. We generally ran each new set of birds at least once after the zoo opened for the season. Most of these runs were done in the Australian Aviary exhibit, in the emus' pen. The substrate here was not the best, and consisted of soil and rotting wood chips and plant fibers. Some runs were done in the Kangaroo Yard and the African Veldt pen, though. In some instances we were able to collect prints of birds run both as juveniles and as subadults or adults.

In 1994 I spent three weeks in Queensland, Australia, collecting measurements and casts of footprints of wild or captive emus. Several casts were taken in Winton Shire, near the famous Lark Quarry dinosaur footprint site (Thulborn and Wade 1984). Another set of prints was collected on Bribie Island, along the Queensland coast, and a third at a commercial emu farm. Additional observations on emu trails were made, but no casts taken, at Idalia National Park. Trails were measured the same way as with those of zoo birds. In my data analysis I assume that different trackways from a given area were made by different birds, but there is obviously no way to be sure of this, except where footprints in different trails are of markedly different size.

In many analyses the data cases were individual footprints, with no concern about which bird made which track. Thus one of the 1988 adults can be represented by tens of footprints, while a wild Australian emu may be represented by a single print. This seemed a reasonable approach, given that in dinosaur tracksites one may have no idea as to the number of individual animals responsible for the footprints. However, this unequal representation of individual birds might conceivably bias our results, and so we often did parallel analyses ("emu or bird means" treatment) in which the data cases constituted mean values of the footprint parameters for a given bird (or trackway, in the case of data collected in the field in Australia). If a bird was represented by but a single footprint, its measurements took the place of mean values. Thus the data for a single Australian footprint were given equal weight to average values for numerous prints of one of the zoo birds. This analytical approach will also be used with non-avian dinosaur footprints in chapter 8, and designated the "mean or single" treatment.

However, generating mean values for individual birds created a new problem for those captive juveniles that grew enough over the course of our study to be represented by runs in which footprint sizes differed markedly. One of those runs was selected for each such growing emu for use in our "emu or bird means" analyses, generally the run with the largest number of prints. Runs selected in this manner are indicated by an asterisk in table 6.1.

Although in our discussion I use the terms "juvenile," "subadult," and "adult" to characterize particular birds, these were only loose descriptions based on our impression of the overall sizes of birds. In most instances we did not know the reproductive status of larger birds. We therefore created size classifications of the footprints themselves. For analyses in which individual footprints constituted the data cases, I divided the length range between the smallest and the largest footprints into thirds, and assigned footprints in any given run to size class on the basis of mean footprint lengths in that run. Size class 1 comprises footprints with a field length less than 9.31 cm; size class 2 includes footprints between 9.32 and 15.11 cm in field length, and size class 3 includes footprints greater than 15.11 cm in field length. For the emu means treatment, in which data cases were mean values for each bird, we divided the emus into two somewhat arbitrary age/size classes: "immature birds" (IIILGL < 17 cm) and "adults" (IIILGL ≥ 17 cm). This cut-off point was selected because it seemed to be a break in the size distribution of birds in our sample, and also because it made for roughly comparable numbers of birds between the two size classes. (I will use a slightly different cut-off point between small and large emu prints in the emu means treatment in chapter 8.)

Milàn (2006) independently studied the effects of substrate and gait on emu footprint morphology, working with captive birds in Denmark. As appropriate, his results will be compared with mine throughout this study.

Other Bird Species

In order for this study to be completed within the lifetimes of the persons conducting it, it was impossible to devote the same effort to the collection and analysis of footprints of every bird species considered as we did to emu prints. A compromise had to be made between intensive and extensive coverage of species.

We obtained casts of footprints of at least 45 species of birds other than emus. Most of these were ground-living birds, but we also collected prints of some arboreal and water-bird species (table 6.3). Most of our birds were zoo or farm captives (fig. 6.2), but some prints were taken from free-living animals. Some of the latter were footprints of extinct species (moa, dromornithids, gastornithids; uncertainty about the number of species represented by these footprints accounts for our statement about the minimum number of species represented in our prints), which we obtained from museum collections.

No attempt was made to vary the substrates associated with footprints for birds other than emus. For the most part, when working with captive birds, we simply used dirt surfaces available in their pens. We wetted and thoroughly mixed the substrate into as smooth a consistency as possible, and then retreated from a bird's pen to allow the animal to walk across the prepared surface. By positioning this surface across a path customarily trod by the bird, it was usually possible to get footprints with a minimum of fuss and disturbance to the animal.

Most species are represented by one or only a few individual birds, and for most species only a few footprints were collected (for some species but a single print). We therefore could not analyze intraspecific variability in footprint shape as thoroughly as we did for emus.

Table 6.3. Bird species other than emus for which footprints were collected during this study. Terminology generally follows Sibley and Monroe (1990, 1993) Dickinson (2003), or Clements (2007). Museum acronyms: PU = Princeton University; MNZ = Museum of New Zealand Te Papa Tongarewa; WWU = Western Washington University.

Family	Species	Individual(s) and locality
Struthionidae	*Struthio camelus* (ostrich)	Adult male and female, Fort Wayne Zoo
Casuariidae	*Casuarius casuarius* (southern cassowary)	Adult male, San Antonio Zoo
		Adult, Brookfield (Illinois) Zoo
Rheidae	*Rhea americana* (greater rhea)	Adult female, Cleveland Zoo
	Pterocnemia (or *Rhea*) *pennata*[a] (lesser rhea)	Adult male, Oklahoma City Zoo
Apterygidae	*Apteryx australis* (southern brown kiwi)	Adult, San Diego Zoo
		Adult male, Brookfield Zoo
Dinornithiforms	Unidentified moa species, Pleistocene, New Zealand	YPM VPPU 019443
		NMNZ S.024314
		NMNZ S.000461
		NMNZ S.000462
		NMNZ TMP.0101493 (formerly 55-1296)
		NMNZ S.024309
		NMNZ S.000986
		NMNZ S.024310 (two prints on same slab)
		NMNZ S.024311
Tinamidae	*Eudromia elegans* (elegant crested tinamou)	Undetermined number of individuals, Evansville (Indiana) Zoo
Anseranatidae	*Anseranas semipalmata* (magpie goose)	Adult male, Cincinnati Zoo
Uncertain	Unidentified dromornithid species, mid-Tertiary, Tasmania	Several prints, Queen Victoria Museum and Art Gallery, Tasmania
Gastornithid prints	*Rivavipes giganteus* (Mustoe et al. 2012), Chuckanut Formation, Lower Eocene, Whatcom County, Washington	WWU-TR-059B
		WWU-TR-066
Megapodiidae	*Alectura lathami* (Australian brush-turkey)	Adult, Fort Wayne Zoo
Cracidae	*Crax daubentoni*, (yellow-knobbed currasow)	Adult male and female, San Diego Zoo
Phasianidae	*Gallus* (domestic chicken[b])	Adult female, Fort Wayne Zoo
		Adult female Plymouth Barred Rock, Jed Freels farm, Auburn, Indiana
		Cornish Cross chick, Freels farm
	Meleagris (domestic turkey[b])	Adult female, Fort Wayne Zoo
	Alectoris philbyi (Philby's partridge)	Adult male and female, San Diego Zoo
	Polyplectron napoleonis (Palawan peacock-pheasant)	Adult, San Diego Zoo
	Polyplectron malacense (Malayan peacock-pheasant)	Adult male, Brookfield Zoo
	Polyplectron bicalcaratum (gray peacock-pheasant)	Adult male, Cincinnati Zoo
	Afropavo congensis (Congo peafowl)	Adult female, Cincinnati Zoo
	Pavo cristatus (Indian peafowl)	Adult female, Fort Wayne Zoo
	Pavo muticus (green peafowl)	Adult male, Cincinnati Zoo
	Lophura swinhoii (Swinhoe's pheasant)	Adult male, Cleveland Zoo

Whenever possible, we made the same measurements that we took on emu prints. Unfortunately, for many species, particularly small-bodied forms, the substrate was unable to capture as much detail of the foot surface as we were able to see in emu prints. Consequently we have measurements of individual digital pads of only a few species. As will be seen, we attempted to compensate for this shortcoming by examining the feet of study skins of extant and extinct ground bird species.

Many of the species examined in this part of our project have a rear toe (digit I), which is missing in emus. Whenever possible we measured this toemark. However, the measurement we were usually able to take was not the actual length of the digit I mark, but only the distance from the toetip of I to the nearest point on the metatarsal pad; to avoid confusion, we term this the digit I "span" rather than "length."

Footprints of moa and dromornithids presented special difficulties. Known specimens of these footmarks are very nondescript, lacking well-defined digital pads; digits II and IV may differ very little in length. All or nearly all of our prints of these extinct species were isolated singletons, which meant that we could not identify them as rights or lefts from trackway patterns. Identifications of footprint symmetry

Family	Species	Individual(s) and locality
	Lophura ignita (crested fireback)	Adult male, Cincinnati Zoo
	Lophura nycthemera (silver pheasant)	Adult male, Bob Cassell farm, Larwill, Indiana
	Crossoptilon auritum (blue eared pheasant)	Adult male and female, San Diego Zoo
	Chrysolophus pictus (golden pheasant)	Male, Toledo Zoo
	Phasianus colchicus (common or ring-necked pheasant)	Adult male, Akron (Ohio) Zoo
		Undetermined number of birds (probably mostly hens), Bob Cassell farm
	Tragopan temminckii (Temminck's tragopan)	Adult male, San Diego Zoo
	Ammoperdix heyi (sand partridge)	Adult male and female, San Diego Zoo
	Lophophorus impejanus (Himalayan monal)	Adult male and female, San Diego Zoo
	Rollulus rouloul (crested partridge or roulroul)	Adult female, Brookfield Zoo
Odontophoridae	*Callipepla squamata* (scaled quail)	Adult, Fort Wayne Zoo
Accipitridae	*Sagittarius serpentarius* (secretarybird)	Adult female, Toledo Zoo
Cariamidae	*Cariama cristata* (red-legged seriema)	Adult, San Diego Zoo
	Chunga burmeisteri (black-legged seriema)	Adult, Cleveland Zoo
Bucorvidae	*Bucorvus abyssinicus* (northern or Abyssinian ground-hornbill)	Adult, San Diego Zoo
Burhinidae	*Burhinus bistriatus* (double-striped thick-knee)	Adult, Cleveland Zoo
Charadriidae	*Charadrius vociferus* (killdeer)	Footprints probably made by this species, banks of St. Joseph River, Fort Wayne, Indiana; footprints of at least two birds
Gruidae	*Grus antigone* (sarus crane)	Juvenile, Louisville Zoo
	Grus canadensis (sandhill crane)	Adult female, Brookfield Zoo
Psophiidae	*Psophia crepitans* (gray-winged trumpeter)	Adult, San Diego Zoo
Rallidae	*Gallirallus australis* (weka)	Adult, San Diego Zoo
Otididae	*Ardeotis kori* (Kori bustard)	Adult male, Toledo Zoo
	Eupodotis gindiana[c] (buff-crested bustard)	Adult, San Diego Zoo
Ardeidae	*Ardea herodias* (great blue heron)	Dried pond, Camp Alexander Mack, Milford, Indiana
		Banks of St. Joseph River, Fort Wayne, Indiana
		Freels farm, Auburn, Indiana
Columbidae	*Otidiphaps nobilis* (pheasant pigeon)	Adult, San Diego Zoo
		Adult female, Cincinnati Zoo

[a]Dickinson (2003) and Clements (2007) assign this species to the genus *Pterocnemia*, but Sibley and Monroe (1990, 1993) placed it in *Rhea*.

[b]Lucas and Stettenheim (1972: 2) argue against using scientific names for domestic chickens and turkeys because "species names of the presumed ancestors of domestic descendants do not adequately identify" different breeds of domestic fowl.

[c]Dickinson (2003) assigns this species to the genus *Lophotis*.

made in this study were my best guesses, based on comparison with foot skeletons, and many of them could well be wrong. Interpretations of footprint morphology in moa and dromornithids that depend on the identification of digits II and IV should therefore be regarded with caution.

Gastornithid footprints also require special comment. I measured casts (negative copies) of two footprints given the name *Rivavipes giganteus* (Mustoe et al. 2012; cf. Buffetaut 2004). These were received long after the digitizing equipment used to generate measurements of other ground bird footprints was no longer operable. Consequently measurements of the gastornithid prints were made directly from the casts and from tracings thereof using digital calipers. Because I obtained the gastornithid prints after most of the data analyses of the interspecific bird footprint sample had been completed, and because I only had two gastornithid prints, I did not include the gastornithid prints in most of the interspecific analyses.

Study Skins

In addition to measuring footprints of ground birds, I also measured some features of the feet of study skins of ground birds in several museum collections. This allowed me to

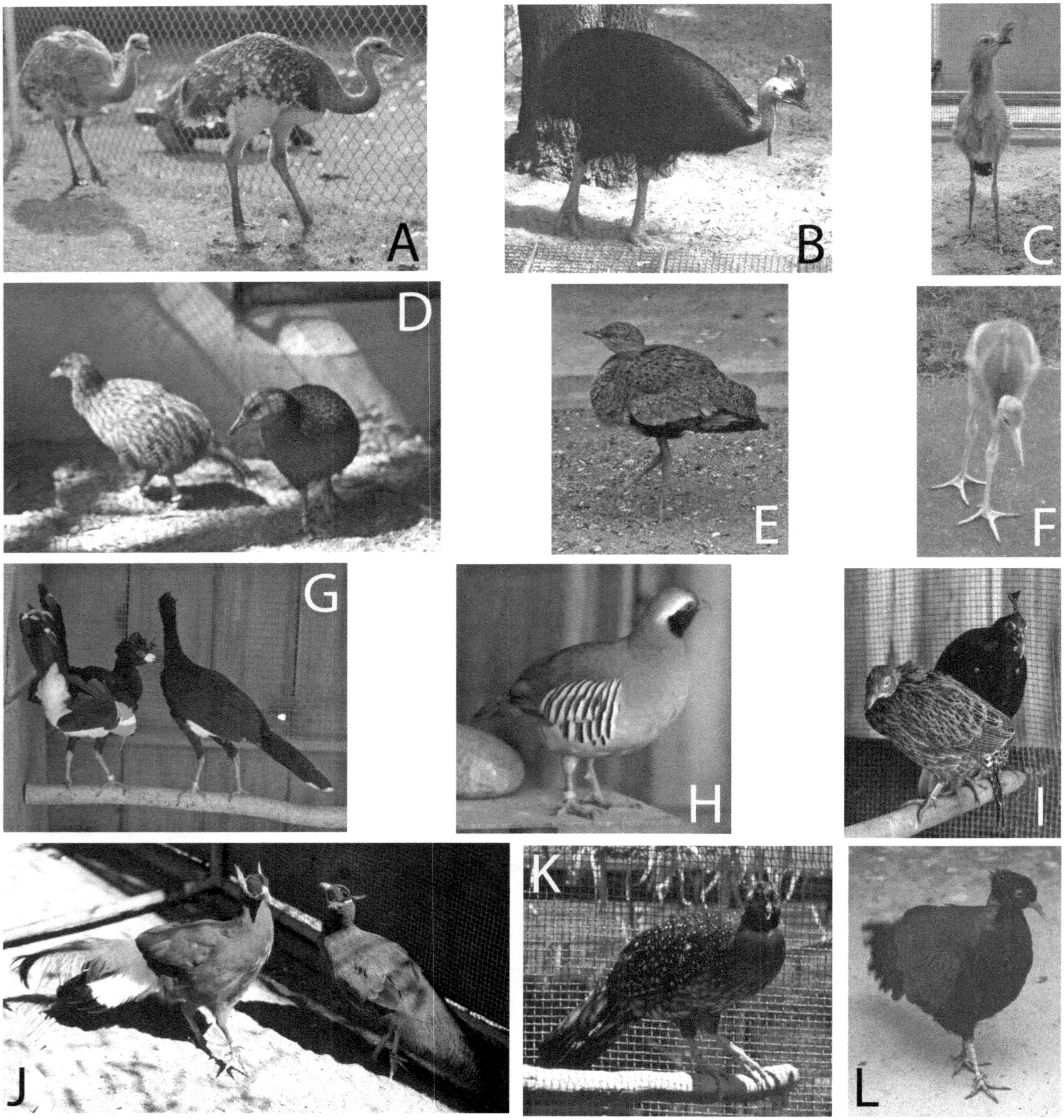

6.2. Samples of the ground birds other than emus for which footprints were collected in the study described in the text. *A*, Lesser rhea (*Pterocnemia pennata*), Oklahoma City Zoo. *B*, Southern cassowary (*Casuarius casuarius*), San Antonio Zoo. *C*, Red-legged seriema (*Cariama cristata*), San Diego Zoo. *D*, Weka (*Gallirallus australis*), San Diego Zoo. *E*, Buff-crested bustard (*Eupodotis gindiana*), San Diego Zoo. *F*, Juvenile sarus crane (*Grus antigone*), Louisville Zoo. *G*, Yellow-knobbed currasow (*Crax daubentoni*), San Diego Zoo. *H*, Philby's partridge (*Alectoris philbyi*), San Diego Zoo. *I*, Himalayan monal (*Lophophorus impejanus*) San Diego Zoo. *J*, Blue eared pheasant (*Crossoptilon auritum*), San Diego Zoo. *K*, Temminck's tragopan (*Tragopan temminckii*), San Diego Zoo. *L*, Pheasant pigeon (*Otidiphaps nobilis*), San Diego Zoo.

obtain additional data for comparing foot and footprint shape variability in emus, and also provided comparative data for other species. Furthermore, data from study skins are more nearly comparable to the data for alligator feet (chapter 5) than are footprints, in that each data case represents a single animal. In contrast, data cases in our footprint study are individual prints, or mean values of prints for a particular bird.

Overall foot dimensions (lengths, widths) could not readily be measured on the stiff, dried feet of study skins, and neither could metatarsal pad dimensions or hypex lengths. I limited myself to measurements of toe lengths, and components of toe lengths (pad and claw lengths). Parameters were measured in a manner similar to the way their counterparts were measured in footprints, with the exception of the way certain atypical digital pads were measured (see below).

However, the feet of study skins are essentially mummified, and so the lengths of digits and digital pads will differ somewhat from those of fresh material. Consequently I did not use study skins in analyses of the coefficient of relative dispersion about the reduced major axis (see below), but limited myself to the cruder parameters of maximum/minimum ratios of scaled parameters.

The final sources of comparative data provided relative toe lengths measured on bird study skins of digits I, II, and IV, excluding the claw (as percentages of the length of digit III without the claw), for many species of western Eurasian, North African, Australian, New Zealand, and Antarctic birds (Cramp 1977, 1980, 1983; Marchant and Higgins 1990*a*, 1990*b*, 1993; Higgins and Davies 1996). These well-known ornithological compendia are often abbreviated as BWP (for *Birds of the Western Palaearctic*) and HANZAB (*Handbook of Australian, New Zealand, and Antarctic Birds*). I selected data for ground and water birds, but did not employ data for small arboreal forms.

FOOTPRINT AND TRACKWAY MEASUREMENTS

In order to understand the measurements we made on emu footprints, a brief discussion of the topography of an emu's footprint is needed (fig. 6.3). In emus and other ground birds, the proximal (rear) edge of a footprint is made by a large pad of soft tissues that surrounds the trochlea of the digit III portion of the distal end of the tarsometatarsus. Impressions of the individual toes radiate outward from this metatarsal pad (terminology of Lucas and Stettenheim 1972) in a symmetrical fashion. The metatarsal pad is usually one of the deeper parts of a footprint.

For each of the three digits (II, III, and IV), there is a shallow region of the footprint that separates the metatarsal pad from the proximal end of the first (most proximal) digital pad. These regions are interpad spaces, in the terminology of Lucas and Stettenheim (1972). The interpad space on digit III that separates the metatarsal pad from the remainder of the toe's impression is subdivided in two by a crease, rather like that seen in the domestic chicken (Lucas and Stettenheim 1972). In this study I modify the terminology of Lucas and Stettenheim (1972) by using the term "pad gap" to describe these spaces between the metatarsal pad and the proximal ends of digital pads. I will restrict the term "interpad space" to significant gaps between digital pads within digit impressions in those bird species where such gaps occur (most notably galliforms).

Examination of osteological materials and x-ray studies (fig. 6.3) of intact emu feet indicates that the proximal phalanx of each digit spans its respective pad gap; the distal end of each proximal phalanx coincides with the toe's first (most proximal) digital pad (terminology of Lucas and Stettenheim 1972). Beyond the pad gap, the soft (non-claw) portion of digit II is usually split by a crease into two abutting digital pads (cf. Milàn 2006). The soft portion of digit III is usually separated by creases into three digital pads. The soft portion of IV can be subdivided into as many as three pads, but these are often faint and frequently cannot be discerned in footprints; consequently in most analyses we treated the non-claw portion of digital IV as a single unit. Unlike the situation in domestic chickens (Lucas and Stettenheim 1972), the digital pads of any toe of the emu are not themselves separated by additional, large, distinct interpad spaces. Milàn (2006: 407) reported that digit III of the emu foot "bears three prominent digital pads that are clearly separated by interpad spaces," but this is more a matter of definition than a real difference between his observations and ours. As will be considered later in this study, the strongly delimited interpad spaces between the digital pads of toes of galliform birds are quite different from the creases that separate the pads of emus.

Emus lack a rear toe (digit I). There is little webbing between toes; using the terminology applied to bird feet (cf. Lucas and Stettenheim 1972 [and references therein]; Elbroch and Marks 2001) the emu foot would be described as semipalmate (at best); the degree of interdigital webbing is notably less than in galliform birds.

Claws on digits II, III, and IV are well developed. Leonardi (1987: 47, item 52) states that "A nail is a blunt structure terminating a digit; a claw is a sharply pointed structure." The horny structure capping the ungual in emus is hard to characterize in these terms; it can be sharply pointed, although not to the extent seen in volant raptors. However, the clawmarks on footprints of a secretarybird (*Sagittarius serpentarius*) that we collected appear no sharper than the toetip marks on some of our emu footprints. Further complicating things, though, is the fact that the claws on emu (and

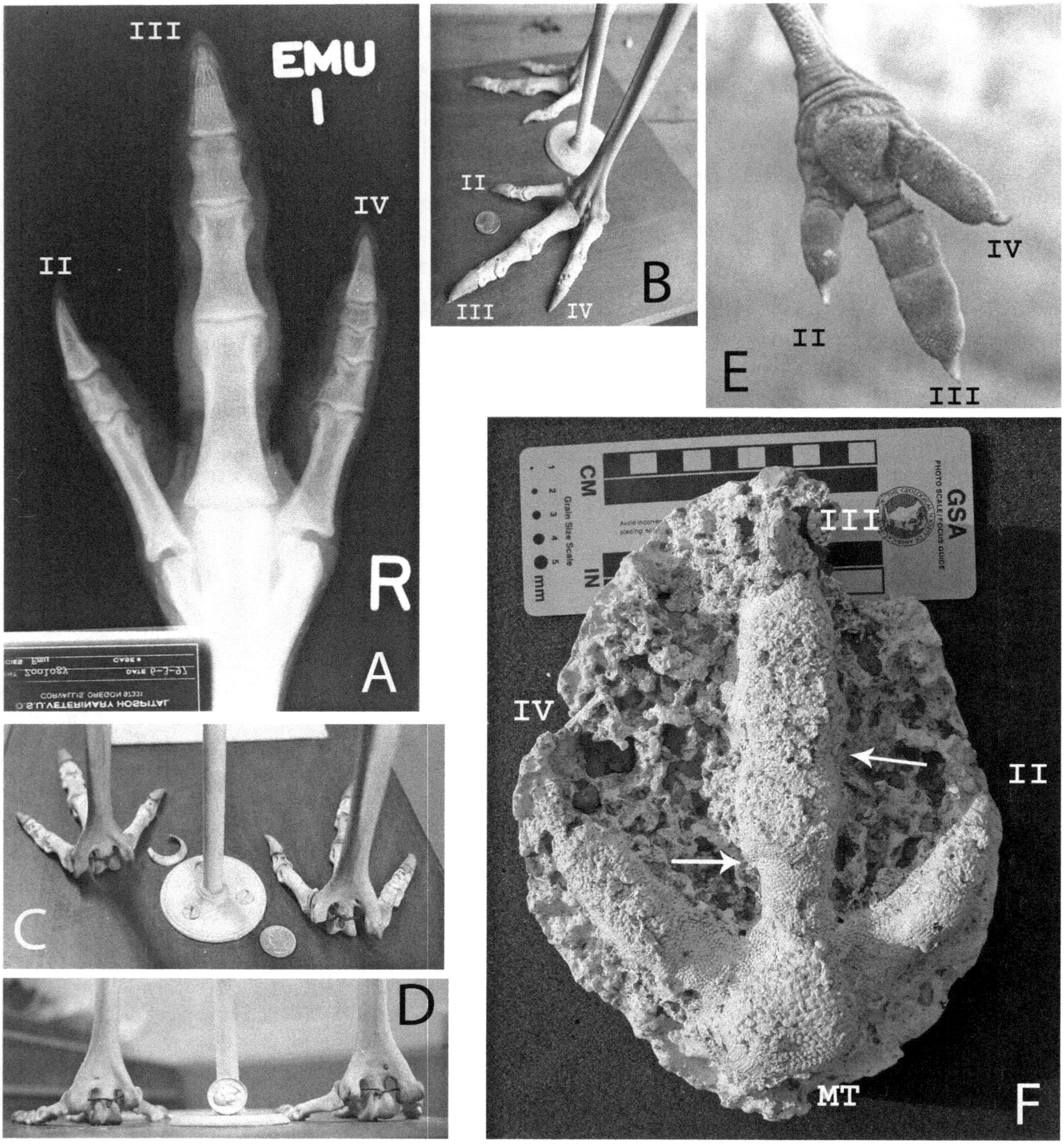

6.3. Feet and a footprint of the emu (*Dromaius novaehollandiae*). *A*, X-ray image of an intact foot, Oregon State University. The phalanges of digits II–IV and the joints between them are clearly visible. A ghostly outline of the soft tissues surrounds the foot skeleton. *B–D*, Three views of the feet of YPM ORN 102523 (formerly 02126). Note the large trochlea at the distal end of digit III of the tarsometatarsus. *E*, Ventral view of the right foot of juvenile emu 00-0025-5B58 (photograph taken May 13, 1992). Note the well-developed metatarsal pad, the conspicuous creases separating digital pads of digit III, and the fainter crease subdividing the non-claw portion of digit II. *F*, Cast of a very well-preserved right footprint, 06038830 of adult emu 3, Fort Wayne Zoo, made June 3, 1988. The metatarsal pad (MT) is clearly visible, along with the two creases (*arrows*) that delimit the first digital pad of digit III. Because this is a plaster cast of the original footprint, left-right symmetry is reversed with respect to the actual footprint. The same symmetry reversal is true for all footprint casts: Left footprints appear to the viewer as rights, and vice versa.

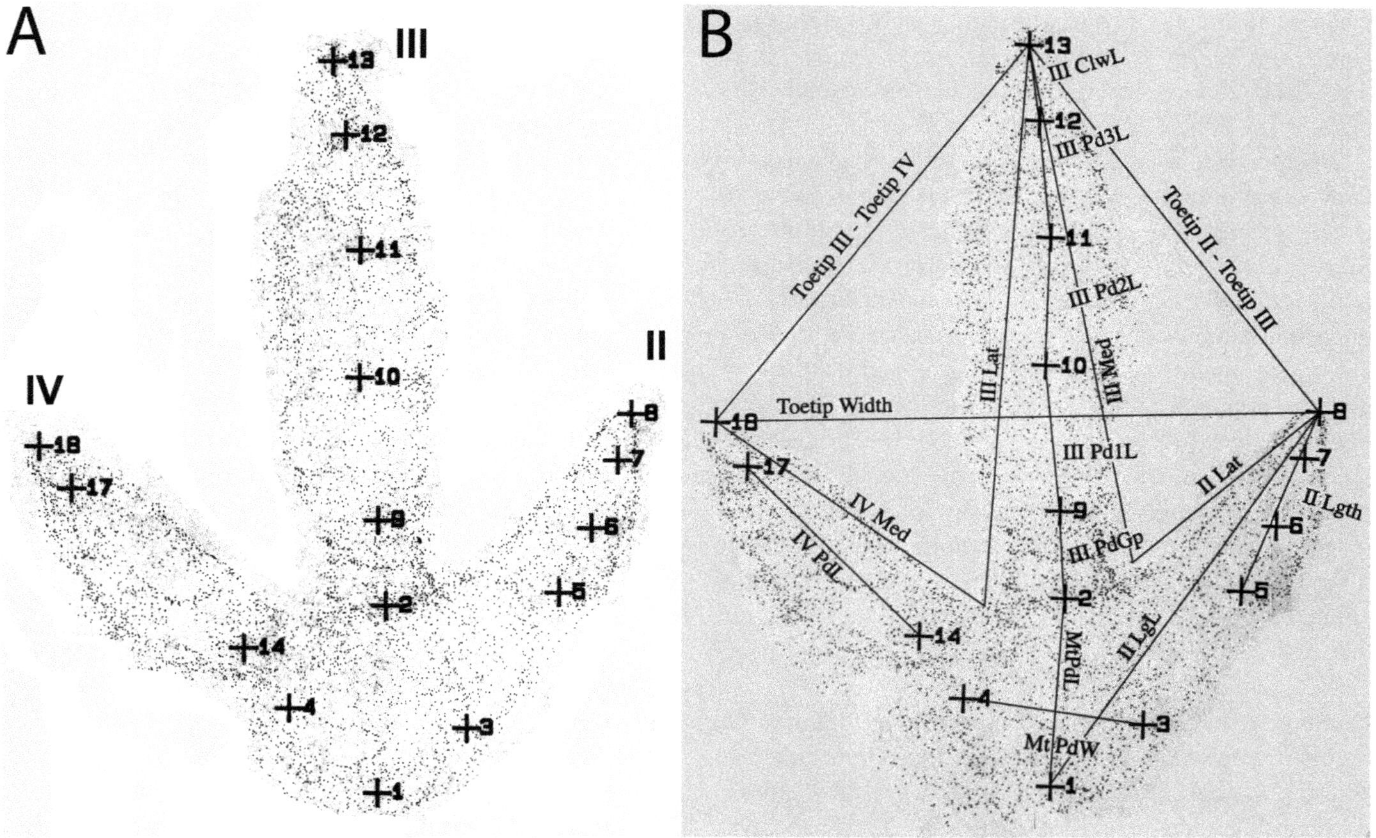

6.4. Reference points on emu right footprint 06038830 (compare with fig. 6.3*F*). *A*, Numbered "crosses" indicate the positions of the reference points: *1*, proximal end of metatarsal pad; *2*, distal end of metatarsal pad; *3*, medial edge of metatarsal pad; *4*, lateral edge of metatarsal pad; *5*, proximal end of first digital pad of digit II; *6*, crease between first and second digital pads of II; *7*, base of claw of digit II; *8*, tip of claw of digit II; *9*, proximal end of first digital pad of digit III; *10*, crease between pads one and two of digit III; *11*, crease between pads two and three of III; *12*, base of claw of digit III; *13*, tip of claw of digit III; *14*, proximal end of first digital pad of digit IV; *15*, crease between pads one and two of digit IV (not recognized in this print); *16*, crease between pads two and three of digit IV (not recognized in this print); *17*, base of claw of digit IV; *18*, tip of claw of digit IV. B, Some of the linear dimensions measured from the reference points in *A* and other footprint features.

alligator; chapter 5; table A5.1) toes often become blunted by abrasion (something that I suspect could also have happened to the claws of large carnivorous dinosaurs). I will use the term "claw" to characterize emu toetips.

Eighteen anatomical reference points were digitized on footprints, where possible (fig. 6.4):

1–4. Metatarsal pad: I digitized four points on the metatarsal pad: (1) The most proximal point on the pad; (2) the most distal point; (3) the most medial (toward the inside of the foot) point; (4) the most lateral (toward the outside of the foot) point.

5, 9, and 14. Proximal ends of the first digital pad of digits II, III, and IV.

6, 10, 11, 15, and 16. Interpad creases: These are the creases separating the digital pads of particular toes. For example, interpad crease 1–2 of digit II separates the two digital pads of that toe. The interpad creases are usually clearly defined on digits II and III of well-formed footprints, but are often faint or absent on digit IV.

7, 12, and 17. Clawbases of digits II, III, and IV: The proximal end of each digit claw was defined as the point where the base of the claw meets the distal-most point of the last (most distal) digital pad.

8, 13, and 18. Clawtip (or terminus) of each digit: In most footprints the most distal end of the toemark was made by the claw. In some poorly formed footprints, however, it was impossible to be certain that this was so. I therefore used a more neutral term, the "terminus," to designate the distal ends of such poorly made clawmarks.

Most of the measurements of emu footprints analyzed during this study were generated by computer from the digitized reference points. The three-dimensional spatial coordinates of each reference point were digitized with a Polhemus Navigation Sciences 3Space Isotrak system. Data were recorded in a computer, and distances between digitized points, and angles defined by points, were calculated from those data. Additional footprint and trail measurements were made in the field, directly from the trackways.

In contrast to what is done in most other chapters of this book, linear dimensions of bird footprints in this chapter are usually expressed in centimeters, rather than millimeters. Most published reports of dinosaur footprints report linear measurements in centimeters, and so I thought it appropriate to do the same for our bird prints. Furthermore, the computer program used to generate lengths and widths from digitized reference points expressed these linear dimensions in centimeters, and so I used the same units in all measurements made manually. However, in some analyses of footprint shape variability it was necessary to convert footprint measurements from centimeters to millimeters in order to make the results comparable with other data sets. The units employed in any given analysis are identified in the relevant data table.

Parameters related to overall footprint length and width: Some dimensions were measured in more than one way, and so it was necessary to create different variable names to distinguish among them. Most of these names are abbreviations, and some of them are admittedly uneuphonious (but if you think these names are ugly, wait till you see the beauties in chapter 8).

Footprint field length: A "quick and dirty" measurement of print length. For some footprints this was taken in the field before a cast of the footprint was made, but for others it was made in the laboratory from the cast. Footprint field length was measured from the rear edge of the metatarsal pad impression to the tip of the impression of digit III (cf. Elbroch and Marks 2001).

IIILGL: Measured in the same way as the field length, but always taken in the laboratory off a cast of the footprints, and calculated as the distance between the digitized positions of the appropriate reference points (points 1 and 13 in fig. 6.4). Unsurprisingly, IIILGL values are usually very similar to field lengths, but field lengths are sometimes slightly greater than IIILGL. The abbreviation used to designate this parameter was chosen to be consistent with two other parameters defined below (IILGL and IVLGL). Don't try to read too much into the suffix "LGL"; it is merely a convenient label for "length."

Thulborn length: Named for Tony Thulborn, who measured footprint lengths in this manner in a classic study (Thulborn and Wade 1984) of the Lark Quarry (Winton Formation, mid-Cretaceous, Queensland) dinosaur footprints. Thulborn length was taken as the maximum length of the footprint parallel to the long axis of digit III. Measuring the footprint in this manner follows the recommendation of Leonardi (1987: 46, item 33) that "the footprint long axis should always correspond with the axis of digit III." Like IIILGL, the Thulborn length will usually be very similar to the field length.

Thulborn width: Also measured as described by Thulborn and Wade (1984). This is the maximum width across the footprint, perpendicular to the Thulborn length. This measurement follows the recommendation made by Leonardi (1987: 46, item 34) about taking footprint width. I am somewhat skeptical that it is a meaningful parameter, because where the greatest width of a footprint occurs will not necessarily be the same anatomical region among different prints. In many cases a print will be widest at the toetips, but in other cases the print will be widest proximal to the toetips. I defer to tradition by taking this measurement—but I don't do much with it.

Toetip width (2Tp4): Distance across the clawtips of digits II and IV, or at least the terminal ends of digits II and IV if it isn't clear that that these terminal ends are the clawtips (cf. Elbroch and Marks 2001). Wright (1996), following some older literature, calls this parameter the "span."

Footprint size index: Following Thulborn and Wade (1984), this is the square root of the quantity (Thulborn length × Thulborn width).

Parameters related to the dimensions of components of footprints (fig. 6.4):

Metatarsal pad length and width: I measured metatarsal pad length (MtPdL) from the most proximal to the most distal points of the pad, as described above. Metatarsal pad width (MtPdW) was taken as the distance across the most medial and lateral points of the pad, as described above. Measured this way, the pad width was close to being, but generally was not exactly, perpendicular to the pad length.

Metatarsal pad-digit gaps (IIPdGp, IIIPdGp, IVPdGp): I designated the distance across a toe's proximal interpad space (as per Lucas and Stettenheim 1972), from a specified point on the metatarsal pad to the proximal end of the toe's first digital pad, as that digit's pad gap. For digit II, the specified point on the metatarsal pad is the medial edge as digitized. For digit III, it is the metatarsal pad's distal edge. For digit IV, it is the lateral edge. These points were not necessarily the closest points on the metatarsal pad to the proximal end of the relevant digit's first digital pad.

Digital pad lengths: For descriptive purposes I designated the most proximal digital pad of a toe as the first pad, the pad after that (next most distal), as the second pad, and so on. Thus IIPd1L is the length of the first digital pad of digit II, and IIIPd3L is the length of the third (most distal) digital pad of digit III. Pad lengths were computed as the distances between their defining creases (for pads surrounded on both sides by other pads), or the distance from a defining crease to the proximal (first pad of a toe) or distal (final pad of a toe) edge of the pad. Because of the way the computer calculated digital pad lengths, if the defining creases for a pad were not present, the pad length could not be measured. Consequently such unmeasurable pads were not included in analyses of pad length variability. For measurements made on the feet of study skins of emus and other ground birds, if the crease separating what would otherwise have been two distinct digital pads was absent, the pad was treated as a single pad. Furthermore, as was done with alligators (chapter 5), all of the length was treated as belonging to what would usually have been the more proximal digital pad, and what would usually have been the more distal pad was assigned a value of zero. This procedure automatically increases the variability of digital pad lengths as compared with the variability of computer-generated lengths.

Claw lengths (IIClwL, IIIClwL, IVClwL): These were measured from the clawbases to the clawtips, as defined above.

Parameters related to overall digit lengths: Inspection of the ichnological literature reveals variation in the ways footprint toe lengths are measured (cf. Rich and Green 1974; Covacevich and Rich 1982; Gierliński and Potemska 1987; Leonardi 1987; Moratalla et al. 1988*a*; Thulborn 1990; Wright 1996; Dalla Vecchia et al. 2001; Romilio and Salisbury 2011; Castanera et al. 2013*a*, 2016; Dalman and Weems 2013; Foster 2015; Hübner 2016; Buckley et al. 2016; Lallensack et al. 2016; Martin 2016; Richter and Böhme 2016; Falk et al. 2017; Razzolini et al. 2017; Sciscio et al. 2017), no doubt occasioned by the exigencies of preservation. One of my goals in this study was to provide ichnologists with a thorough analysis of footprint shape variability in a single species. I wanted my results to be comparable with previously published studies of dinosaur footprints. I therefore utilized several different ways of measuring overall digit lengths. In a (perhaps hopeless) attempt to minimize confusion, I created different abbreviations to characterize these different length measurements:

Digit length from the proximal end of the first digital pad to the clawbase (IIPdL, IIIPdL, and IVPdL); "PdL" connotes the length of the "pad" portion of the toe, the toe length without the claw. Obviously these measurements could be made only on those prints with clear clawbases.

Digit length from the proximal end of the first digital pad to the toetip (claw or terminus; IILgth, IIILgth, IVLgth); "Lgth" is meant to convey that this is one of the common ways of measuring digit length.

Distance from the medial edge of the metatarsal pad to the clawbase of digit II (IIFRNC), distance from the distal edge of the metatarsal pad to the clawbase of digit III (IIIFRNC), and distance from the lateral edge of the metatarsal pad to the clawbase of digit IV (IVFRNC): These measurements could only be made when a distinct clawbase was recognizable in a toe impression. The "FRNC" doesn't stand for anything, alas; by this point we were running out of clever abbreviations for variable names, and were just combining letters to label different parameters.

Distance from the medial edge of the metatarsal pad to the toetip (claw or terminus) of digit II (IIFRL), distance from the distal edge of the metatarsal pad to the toetip of digit III (IIIFRL), and distance from the lateral edge of the metatarsal pad to the toetip of digit IV (IVFRL): comparable to the FRNC measurements, except that the present measurements include the claw. The "FRL" suffix, as you might have guessed by now, doesn't stand for anything, but was just another convenient distinguishing label.

Distance from the proximal edge of the metatarsal pad to the clawbase (IILGNC, IIILGNC, IVLGNC): Again, these are measurements that could only be made if a recognizable clawbase was present. For once I used a meaningful variable abbreviation; "LGLNC" essentially means "length—no claw."

Distance from the proximal edge of the metatarsal pad to the toetip (IILGL, IIILGL, IVLGL): These measurements are analogous to the way that some toe lengths are measured in dinosaur footprints (e.g., Moratalla et al. 1988*a*; Lallensack et al. 2016; Martin 2016). IIILGL, as already described, is a measurement comparable to footprint field length.

Hypex lengths: Congleton (1990) measured toe lengths of tridactyl dinosaur footprints from the junction between a toe and its neighbor (hypex—Leonardi 1987) to the tip of the toe. When two toes meet at an angle, identification of the hypex is unambiguous. Frequently, however, the junction between two toes is a smooth curve. When that happened, I located the hypex midway between the two toes. Hypex lengths are particularly useful when digital pads cannot be discerned in toemarks, such that PdL and Lgth measurements cannot be taken. Hypex lengths might potentially be used to characterize the extent of interdigital webbing between toes if the hypex is taken as situated at the distal edge of the webbing between two toes (as was done with alligator feet in chapter 5); small values of hypex lengths compared with other toe length measurements would then indicate that webbing extends well out along the length of the toe. I measured four hypex lengths: digit II lateral length (length of II measured from its hypex with digit III to its tip), digit III medial length (length of III measured from its hypex with II to its tip), digit III lateral length (length of III measured from its hypex with IV to its tip), and digit IV medial length (length of IV measured from its hypex with III to its tip).

Parameters related to widths across different regions of the footprint:

Distance from the proximal end of digit II to the proximal end of digit III (2Px3): This is the distance from the proximal end of the first digital pad of II to the proximal end of the first digital pad of III. In like manner, I also measured distance from the proximal end of digit III to the proximal end of digit IV (3Px4) and distance from the proximal end of digit II to the proximal end of digit IV (2Px4).

Other footprint width measurements:

Distance from the clawbase of digit II to the clawbase of digit III (2Bs3)
Distance from the clawbase of digit III to the clawbase of digit IV (3Bs4)
Distance from the clawbase of digit II to the clawbase of digit IV (2Bs4).
Distance from the toetip of digit II to the toetip of digit III (2Tp3)
Distance from the toetip of digit III to the toetip of digit IV (3Tp4)
Toetip width (2Tp4) has already been described.

Measurements across digits II and IV (2Px4, 2Bs4, and 2Tp4) provide a useful proxy for interdigital angles (to be discussed next) in some data analyses. Large values of footprint widths, relative to footprint lengths, are associated with large interdigital angles.

Interdigital angles (IDAs; "divarication of digits" of Leonardi [1987]): Although angles between toemarks of dinosaur footprints are commonly measured and reported in the ichnological literature, paleontologists often express doubts about their value. It is commonly believed, for example, that interdigital angles are so variable, even among footprints made by the same species, that they will have little value in discriminating among ichnotaxa and trackmakers (cf. Thulborn 1990). Furthermore, the published literature shows little consistency among authors in the way that interdigital angles are measured. IDAs are generally measured from "best-fit" lines drawn through the long axes of toemarks. Some workers require that such lines run from toetips to meet at a common point along the rear margin of the footprint. Others permit best-fit lines through toemarks to meet at different points. Additional ambiguity about the placement of best-fit lines arises if toemarks are curved, rather than straight. Because of the inherent subjectivity in the way best-fit lines are drawn through toemarks, it is likely that there would be differences among workers in interdigital angles measured from the same footprint.

For this reason we measured IDAs of emu prints in several different ways and compared the resulting values. Some of our IDA measurements were of the traditional best-fit variety. However, we also measured IDAs in a very different manner, based on the digitized reference points of footprints.

Best-fit interdigital angles: In all cases, best-fit IDAs were made from images of a footprint projected onto a plane. Measurements were made at the deepest part of the footprint; as Milàn (2006) reported, the interdigital angle of deeply impressed prints may differ from the shallowest to the deepest parts of such prints. Different abbreviations were chosen to identify the slightly different ways that different persons made these projections. Certain common mensuration conventions were used in most of these measurements (cf. Falk et al. 2011, 2017): (A) In most best-fit measurements, we did not require that best-fit lines meet at a common point along the rear edge of a footprint. (B) If a toemark curved continuously over its length, greater weight was placed on the proximal than on the distal end of the toemark in deciding where to draw the best-fit line through it. (C) An exception to the preceding convention was followed where there was a distinct "dog-leg" angle between the main part of the toemark comprising the proximal end of the first digital pad through the toetip, and the interpad space separating this part of the toe from the metatarsal pad impression. Where that happened,

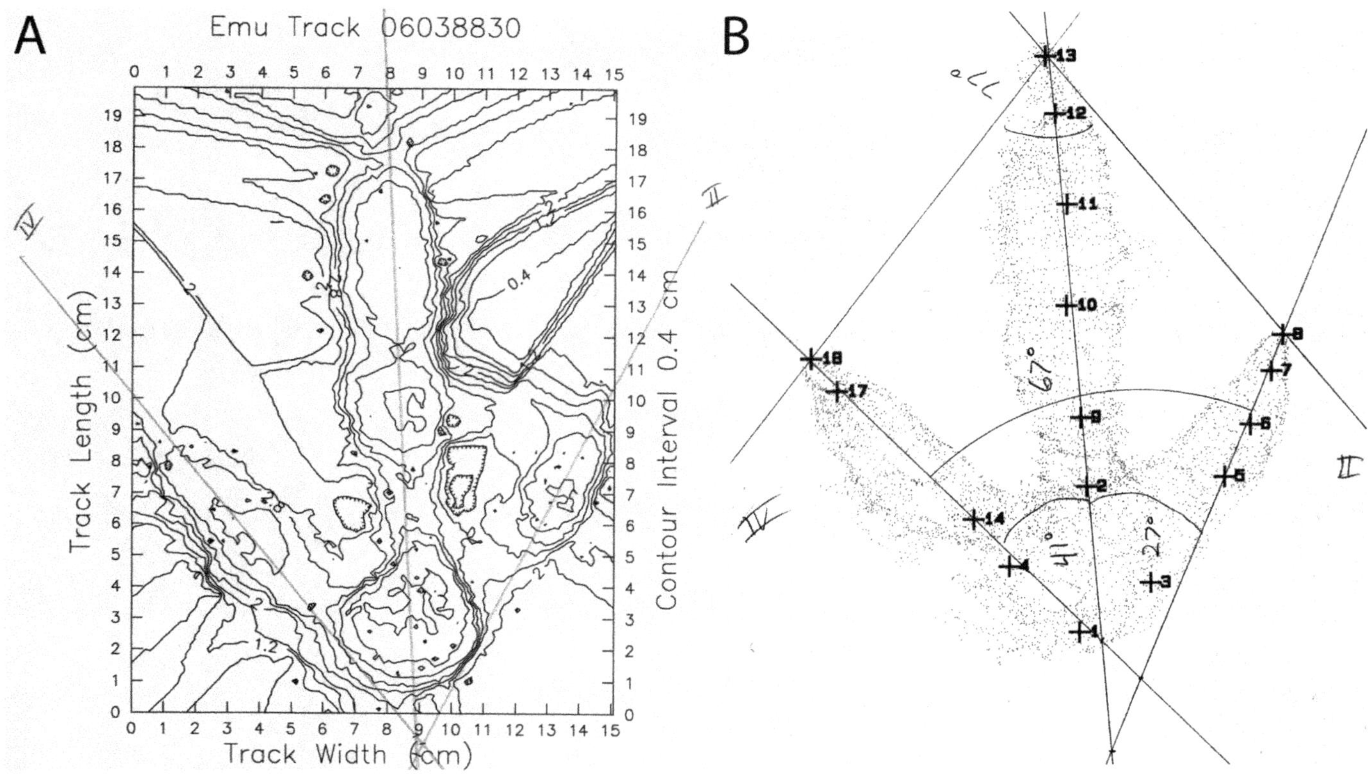

6.5. Actual data sheets (with written notes about the print in *B*) showing how best-fit lines for measuring interdigital angles (IDAs) were drawn through emu right footprint 060388. *A*, TOPO IDAs. *B*, JOF IDAs. See text for difference between TOPO and JOF IDAs.

the best-fit line was drawn disregarding the interpad space. (D) Clawmarks were given less weight than the fleshy part of a toe in locating the best-fit line through the long axis of the toemark. If a clawmark diverged sharply from rest of the toemark, the claw was disregarded in locating the best-fit line through the toe. (E) IDA II–III designates the interdigital angle between digits II and III, IDA III–IV that between digits III and IV, and IDA II–IV that between digits II and IV.

MCC IDAs: Using an opaque projector, the image of a footprint cast was projected onto a sheet of paper and its outline drawn by student research assistant Jana Benson (née McClain, hence "MCC" used to label these measurements). She then drew best-fit lines through toemarks and measured IDAs. MCC IDAs were measured only on emu prints collected during 1988.

TOPO IDAs: The entire surface of well-formed footprint casts was digitized in three dimensions, after which data were gridded using Golden Software Inc.'s program Surfer, and a topographic map of the footprint was generated (fig. 6.5A). Best-fit lines were then drawn through the toemarks and IDAs measured by me.

TRACE IDAs: For selected footprints a rigid sheet of transparent plastic was placed atop a footprint cast, and the outline of the print then traced by me onto a sheet of plastic lying on the rigid plastic. Best-fit lines and IDAs were then measured as before.

JOF IDAs: A computer program was written that plotted the digitized anatomical reference points onto a sheet of paper (fig. 6.5*B*). I drew best-fit lines through these points. Where possible, the conventions regarding proximal vs. distal portions of toemarks where they curved, and disregarding clawmarks where they diverged from the remainder of the toe, were followed. This was not always possible, however. Sometimes the only reference points that could be recognized on a toemark were the proximal end of the first digital pad and the toetip. When this happened the best-fit line could only connect these two points.

"Heel" spread angle: Because emus are digitigrade animals, the rear edge of the metatarsal pad does not constitute the anatomical heel of an emu's foot. However, the term "heel" in quotation marks is a useful shorthand for describing the rear margin of the print. Using the drawings of footprint outlines employed in measuring MCC IDAs, Benson also measured interdigital angles (here designated spread angles [SA]) based on lines constrained to run from the toetips of II and IV to meet at a common point at the middle of the proximal end of the metatarsal pad.

Reference point–based IDAs: Using calculated differences between digitized reference points on emu footprints, we calculated angles from the distances among triads of such points. In describing these IDAs below, the following format is used: first reference point–reference point forming the vertex of the angle (v)–second reference point.

Two of these reference point–based IDAs are similar in concept to the "heel" spread angle:

Clawbase spread angle (SA): The angle formed by the clawbase of digit II–proximal end of the metatarsal pad (v)–clawbase of digit IV.

Toetip spread angle (SA): Formed by the toetip of digit II–proximal end of the metatarsal pad (v)–toetip of digit IV. Values of the toetip spread angle should be very close to those of the "heel" spread angle; they are essentially two separate measurements of the same parameter, made in different ways.

The remaining reference point–based IDAs, in contrast to the IDAs described so far, have their vertices pointed toward the distal end of the footprint:

Toebase IDA: The angle formed by the proximal end of the first digital pad of digit II–proximal end of the first digital pad of digit III (v)–proximal end of the first digital pad of digit IV.

Clawbase IDA: Formed by clawbase II–clawbase III (v)–clawbase IV.

Toetip IDA: Formed by toetip II–toetip III (v)–toetip IV.

Olsen/Weems plots: Olsen (1980) pioneered an intriguing way of looking at the shapes of tridactyl dinosaur footprints by plotting the length of the digit III impression that extends distally beyond the tips of digits II and IV against the length of the footprint proximal to the tips of II and IV (here called the "backfoot") for footprints of the Newark Supergroup (fig. I.3). He showed that in such a scatterplot the dinosaur ichnotaxa *Grallator, Anchisauripus*, and *Eubrontes* "form a continuum similar to what would be expected of tracks of individuals of different ages perhaps representing a single dinosaur species" (Olsen 1980: 370; an opinion reiterated, albeit with reservations, by Olsen et al. 1998), but went on to caution that this did not necessarily mean that they were indeed made by a single species. A modification of Olsen's approach was systematically applied by Weems (1992) in a consideration of the ichnotaxonomy of Newark Supergroup footprints.

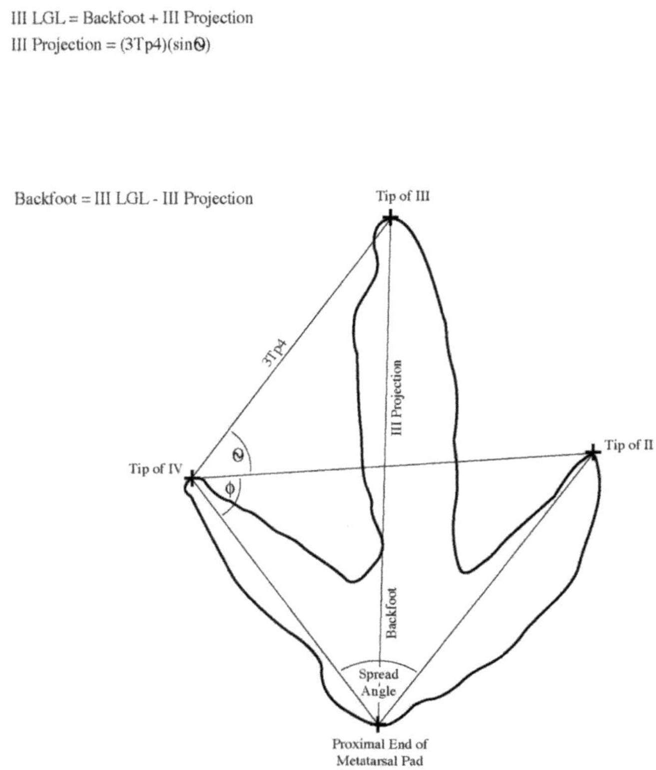

6.6. Calculation of the projection of digit III beyond the distal ends of digits II and IV. "Backfoot" designates the length of the footprint proximal to a line connecting the tips of II and IV. Backfoot and the digit III projection were not measured directly but were approximated using a geometric argument: It is assumed that the digit III projection is perpendicular to a line segment connecting toetips II and IV (variable 2Tp4); this assumption is not completely valid, but gives results that are close enough. The cosine of angle θ formed by 2Tp4 and 3Tp4 (toetips III and IV) is calculated using the law of cosines and the values of 2Tp3 (toetips II and III), 3Tp4, and 2Tp4. The digit III projection is then approximated as 3Tp4 multiplied by the sine of angle θ. Backfoot length is finally approximated by subtracting the digit III projection from IIILGL (i.e., the distance from the proximal end of the metatarsal pad to the toetip of digit III).

Because of the possibility that such plots might record ontogenetic changes in footprint shape, I prepared Olsen/Weems plots for our emu footprints. The digit III projection was calculated as the portion of footprint length that projected beyond a line connecting the toetips of II and IV (in actuality, what I approximated was the height of a triangle whose base ran between the toetips of II and IV, and whose apex was the toetip of III). This parameter was subtracted from overall footprint length to determine the length of the footprint proximal to the tips of II and IV (fig. 6.6). However, because in bird footprints the "heel" of the track is formed by the distal end of the digit III portion of the tarsometatarsus, and not the metatarsophalangeal joint of digit IV (the latter

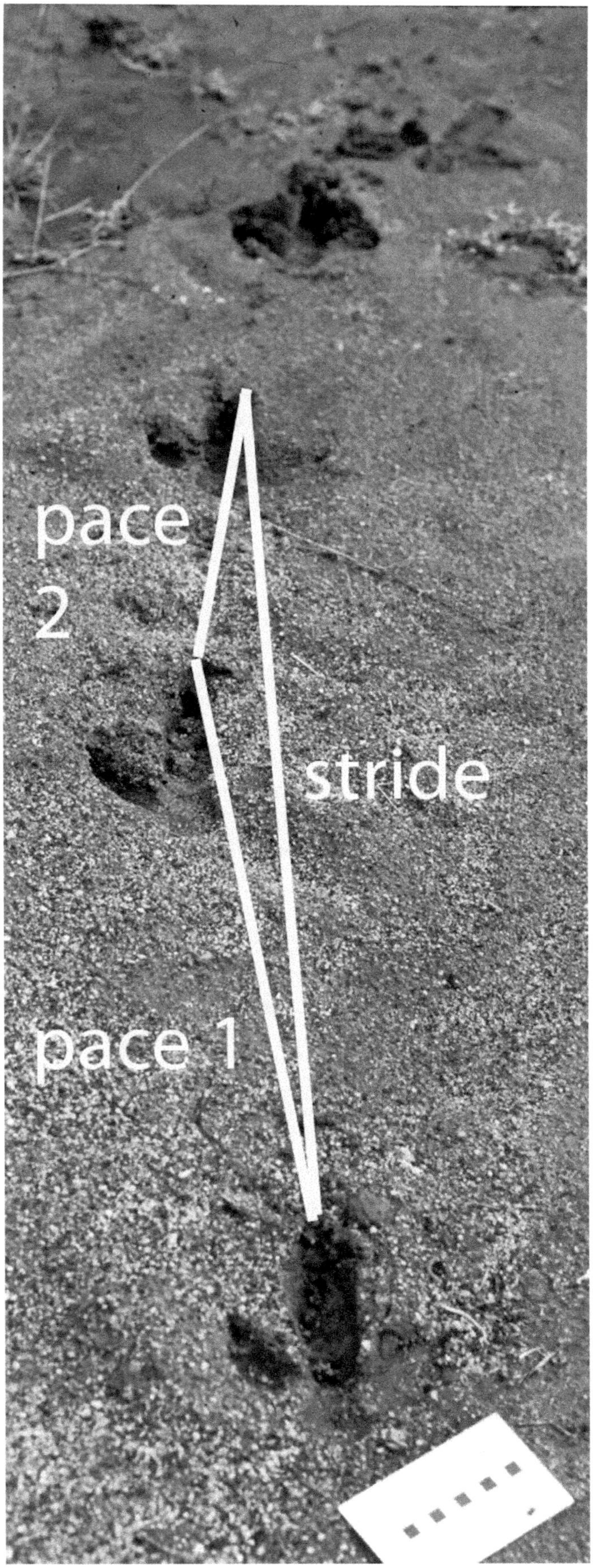

6.7. Wild emu trackway 30, Idalia National Park, Queensland, Australia, illustrating how paces and strides were measured. The photograph shows the trackway in oblique view, with the bird having been moving away from the viewer, rather than as seen from directly overhead. Scale bar marked in centimeters.

as in most tridactyl dinosaur footprints; Thulborn 1990; Farlow et al., 2000), my Olsen/Weems plots are only analogous to those for non-avian dinosaurs.

For many emu footprints we were able to measure associated trackway parameters:

Pace: The pace is the distance from a given point on a footprint to the same point on the next footprint of the opposite foot (fig. 6.7). We used the tip of digit III as our reference point for measuring paces (and strides). Any given footprint is formed at the end of one pace, and the beginning of another, and it is conceivable that the trackmaker's movements while making either pace could affect the shape of the footprint. I therefore distinguished between the pace ending in a footprint, and the pace beginning with that footprint.

Opposite stride: The stride is the distance from a given point on one footprint to the same point on the next footprint made by the same foot (fig. 6.7). I used the term "opposite" stride in reference to any given footprint, as the stride made as the opposite foot swings forward while the footprint of concern is being created.

Pace angulation: This is the angle defined by a pace ending with a footprint, the pace that begins with the same footprint, and the associated opposite stride (fig. 6.7). For trackways of large emus, I calculated the pace angulation from these three parameters using the law of cosines. However, this method has limitations. The value of the pace angulation is vulnerable to slight errors of measurement of paces and strides as the pace angulation approaches 180 degrees. For example, if the measured value of the opposite stride is slightly larger than the sum of the measured values of the two paces—a situation that frequently occurs with nearly linear trackways in the field—the pace angulation cannot be calculated. In such cases I assumed that the true value was in fact 180 degrees. If the trackmaker moves in a straight line, this is the maximum value of the pace angulation. However, if the trackmaker abruptly changes direction, pivoting away from the trackway midline on one foot as the opposite foot is brought forward, the pace angulation can have a value greater than 180 degrees. For trackways made by large emus, we did not knowingly measure pace angulations when such sharp changes of direction were made. For some juvenile emus, however, we made casts of entire trackway segments and brought them back to the laboratory. In some of these casts with direction changes, pace angulations greater than 180 degrees were measured.

Footprint rotation (divarication of foot from midline of Leonardi [1987]): This is the angle formed by the long axis of the footprint with respect to the trackmaker's direction of travel. Following Leonardi (1987), if the footprint angles outward with respect to the animal's direction of travel, the rotation is characterized as positive, and negative if the footprint angles inward with respect to the trackmaker's direction of travel. I measured footprint rotation several different ways.

One set of measurements was based on the compass bearings of footprints (sighted from the rear of the metatarsal pad to the toetip of digit III). The mean of the bearings of two successive footprints defines the overall forward motion of the trackmaker's body over the course of that step. If the bearing of the left footprint is less than this mean, and the bearing of the right footprint greater than this mean (for northbound trails, 360 degrees must be added to values east of 0 degrees [due north] during this calculation), then the two footprints angle outward. If the bearing of the left footprint is greater than this mean, and the bearing of the right footprint is less than this mean, the two footprints angle inward.

However, because any given footprint is situated at both the end and the beginning of a pace, that footprint's bearing can be compared with the mean of its bearing with that of the previous footprint or with that of the following footprint. I made both comparisons, using the abbreviations ROTPREV and ROTFOLL, respectively. Footprint rotations measured

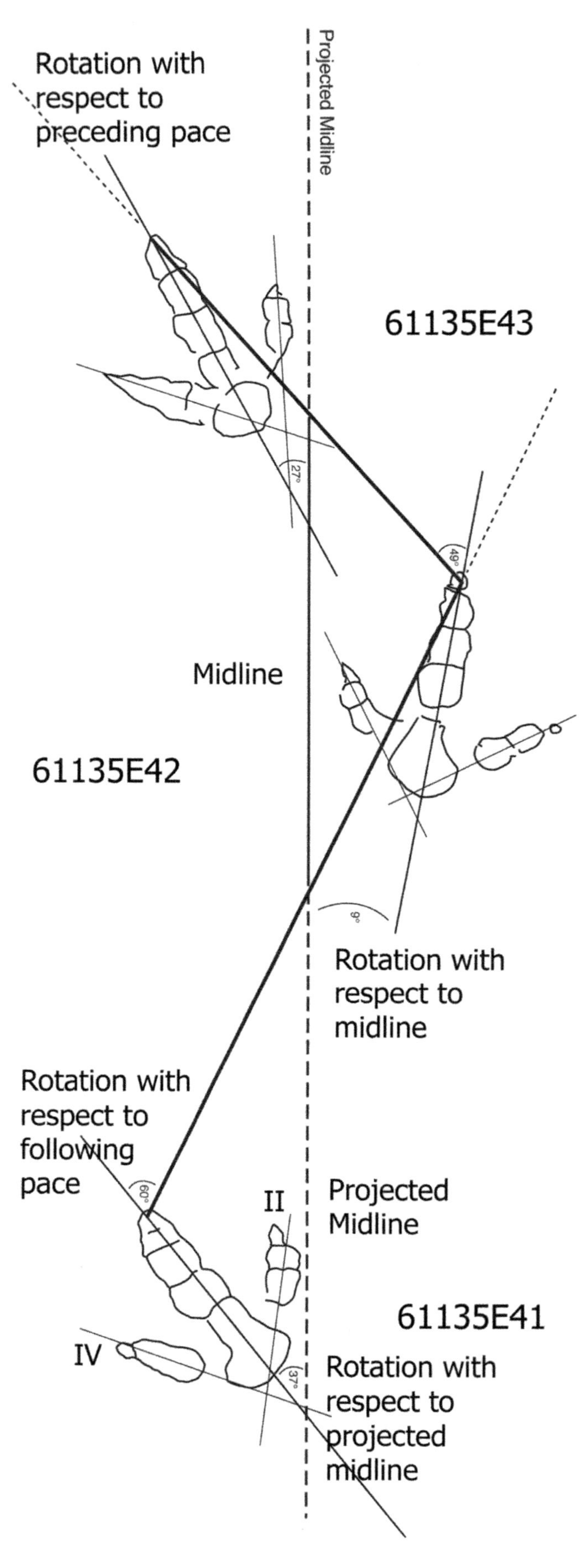

6.8. Tracing of trackway sequence made by juvenile emu 00-0021-35E4 at the Fort Wayne Zoo, June 11, 1993 (Run 43). Two paces (footprints 61135E41 to 61135E42 and prints 61135E42 to 61135E43) and a stride (prints 61135E41 to 61135E43) were preserved in this slab. The midline of the trackway was established opposite print 61135E42 as a line segment connecting the midpoints of the two successive paces that ended in, and began with, that footprint. The midline was then projected both behind and forward from the trackway midline. Long axes were drawn through the midlines of digit III of the three footprints. Footprint rotation could then be measured with respect to the trackway midline (print 61135E42) or the projected midline (footprints 61135E41 and 61135E43). Footprint rotation also could be measured with respect to the pace that begins with that footprint (rotation with respect to following pace) for prints 61135E41 and 61135E42, and the pace that ends with that footprint for prints (rotation with respect to preceding pace) 61135E42 and 61135E43. Footprint rotation has a positive value if the long axis of digit III turns more outward (away from the trackway midline) than whatever feature the angle is being measured against, and has a negative value if the long axis of digit III turns more inward (toward the trackway midline) than whatever feature the angle is being measured against. For emu trackways, footprint rotation with respect to the trackway midline, projected midline, and following pace is usually positive, and footprint rotation with respect to the preceding pace is usually negative.

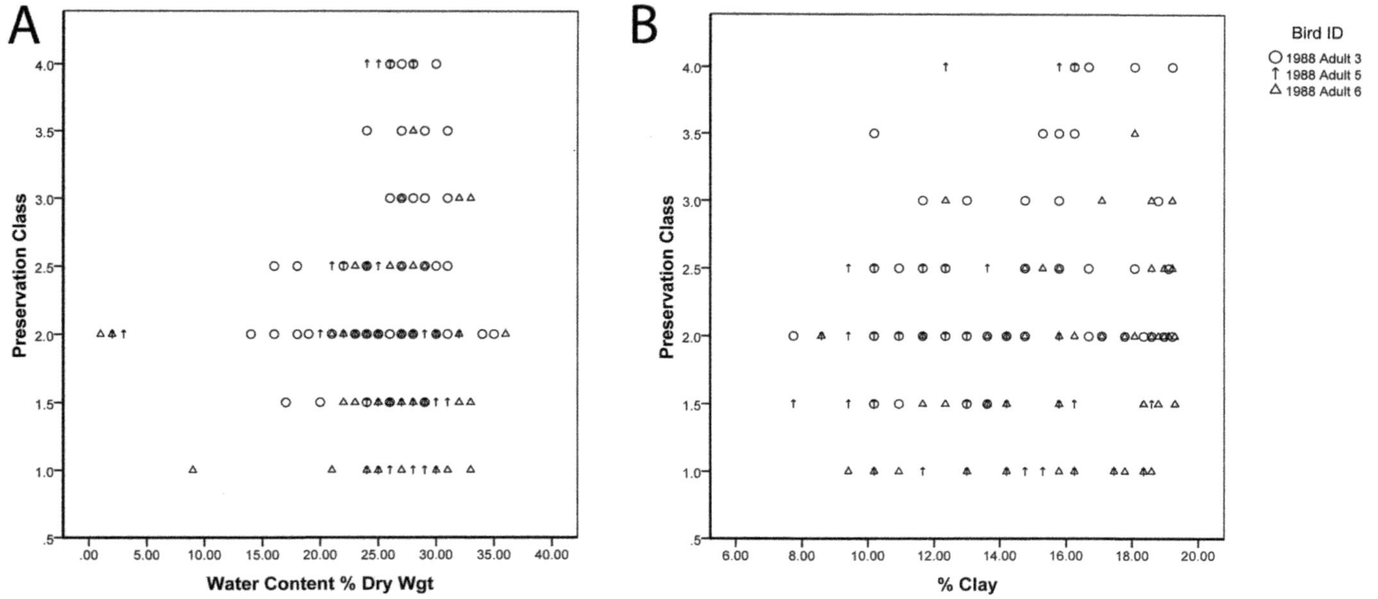

6.9. Footprint preservation class (1 = poor quality; 4 = highest quality; table 6.2) as a function of water (*A*) and clay (*B*) content of soil in the Veldt holding pen of the Fort Wayne Zoo. Data points are for adult emus 3, 5, and 6.

this way were determined for trackways of both large and small emus.

In those cases in which entire trackway segments of young emus were cast as slabs and returned to the laboratory, footprint rotations could be measured in additional ways (fig. 6.8). Some of these expressed rotations relative to the trackway midline. This was determined by locating the midpoints of two successive paces, and then connecting these midpoints with a line segment. The rotation of a footprint was then measured relative to the midline determined by the pace ending in the footprint and the pace beginning with the footprint. For the first and last footprint in a collected trail segment the trackway midline obviously could not be measured this way. In such cases I measured the footprint's rotation with respect to a projection from the trail midline associated with the following footprint (in the case of the first print in a series) or the previous footprint (for the final footprint in a series). Finally, for collected trail segments we also measured footprint rotation with respect to the orientation of the pace ending in the footprint and the pace beginning with the footprint.

Given the number and complexity of my data analyses, rather than discuss all of the kinds of statistical and other numerical analyses that I carried out in our study of emu and other bird footprints here, I will indicate the analyses I did for each aspect of our study as we come to that part of the study. As elsewhere in this book, tables of data summarizing results of analyses done for this chapter will appear in two places. Results that are most immediately relevant to the point being made will appear as tables in this chapter; results that are less immediately relevant, or that appear in very long tables, will be relegated to the Appendix.

RESULTS

Factors Affecting Footprint Quality and Measurements

Footprint Quality. Over the range of values observed in the Fort Wayne Zoo's Veldt holding pen, soil conditions were only loosely correlated with footprint quality (fig. 6.9). Excessively dry soil resulted in poor quality footprints, as also observed by Milàn (2006) for emus, and Falk et al. (2017) for chickens. The best-formed footprints (preservation classes 3–4) were associated with a soil water content of at least 20%, but this level was no guarantee of high-quality prints; water content was not significantly correlated with footprint preservation class (neither Kendall's τb nor Spearman's ρ had statistically significant values). The best footprints were associated with a clay content of 10% or better (and a sand content of 40% or less; again cf. Milàn [2006] and Falk et al. [2017]), but this, too, did not necessarily result in high-quality tracks. Footprint preservation class showed a weak but significant negative correlation with sand content (Kendall's $\tau b = -0.128$, two-tailed $P = .034$, $n = 161$; Spearman's $\rho = -0.170$, two-tailed $P = .031$, $n = 161$), and just missed a significant positive correlation with clay content (Kendall's $\tau b = 0.108$, two-tailed $P = .069$, $n = 161$; Spearman's $\rho = 0.145$, two-tailed

Table 6.4. Comparison of substrate parameters of low-quality (preservation class 1) and high-quality (preservation class 4) emu footprints from the Veldt holding pen.

Parameter	Preservation class	Minimum	Maximum	Mean	Number of footprints
Soil water content (% dry weight)	1	9	33	26.2	26
	4	24	30	26.4	10
Sand content (% dry weight)	1	35.6	47.8	38.4	26
	4	34.6	39.2	36.5	10
Clay content (% dry weight)	1	9.4	18.6	14.2	26
	4	12.3	19.2	16.3	10

$P = .067$, $n = 161$; however, it could be argued that because we might expect better footprints with a higher clay content, a one-tailed probability should be used, in which case the relationship between clay content and preservation class would be statistically significant).

Another way of looking at differences between low- and high-quality footprints is to compare substrate parameters for the very worst and best tracks (table 6.4). Mean values for preservation classes 1 and 4 did not differ significantly in water content, but were significantly different (*t*-test; $P < .05$) for sand and clay content, although these differences amounted to no more than a couple percentage points.

Dramatic differences in footprint appearance nonetheless occurred at the extremes of substrate conditions. For example, compare footprint 06038830, made by adult emu 3 (fig. 6.3*F*), with footprint 06108821, made by bird 5 (fig. 6.10). Track 06108821 (preservation class 2) was associated with a soil water content of 3%, a sand content of 39.2%, and a clay content of 9.4%. As with this footprint, Milàn (2006: 408), in his study of footprint formation by emus, noted that "collapsing of . . . [dry] sand caused the digits to appear significantly broader and more rounded than they are." Falk et al. (2017) observed much the same in chicken footprints. In contrast, print 06038830 (preservation class 4) was associated with a soil water content of 28%, a sand content of 34.6%, and a clay content of 16.7%.

Intriguingly enough, the differences in appearance between these two footprints are reminiscent of those thought to distinguish footprints of non-avian theropods from those of large ornithopods (chapter 2), suggesting that under the right substrate conditions a theropod foot might make an ornithopod-like footprint. The appropriate conditions might involve sand in either a very dry or a water-saturated (perhaps underwater) setting. Whether footprints made under those conditions would likely be preserved is, of course, another matter.

6.10. Cast of footprint 06108821 (a left), made by adult emu 5. II = terminus of impression of digit II.

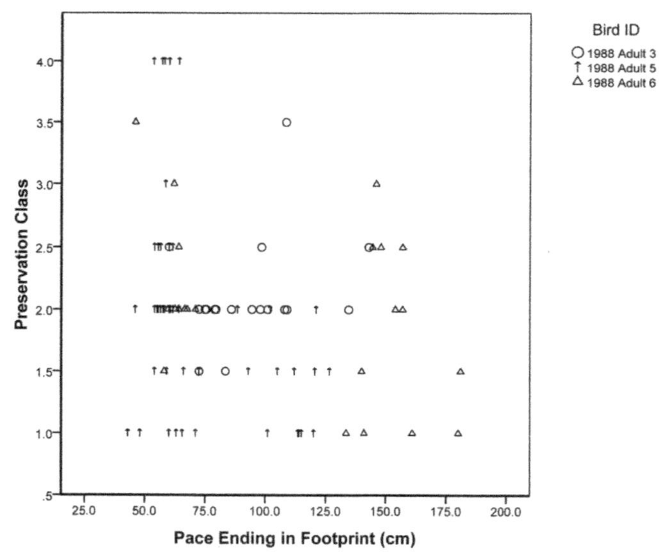

6.11. Footprint quality as a function of ending pace in the 1988 adult emus.

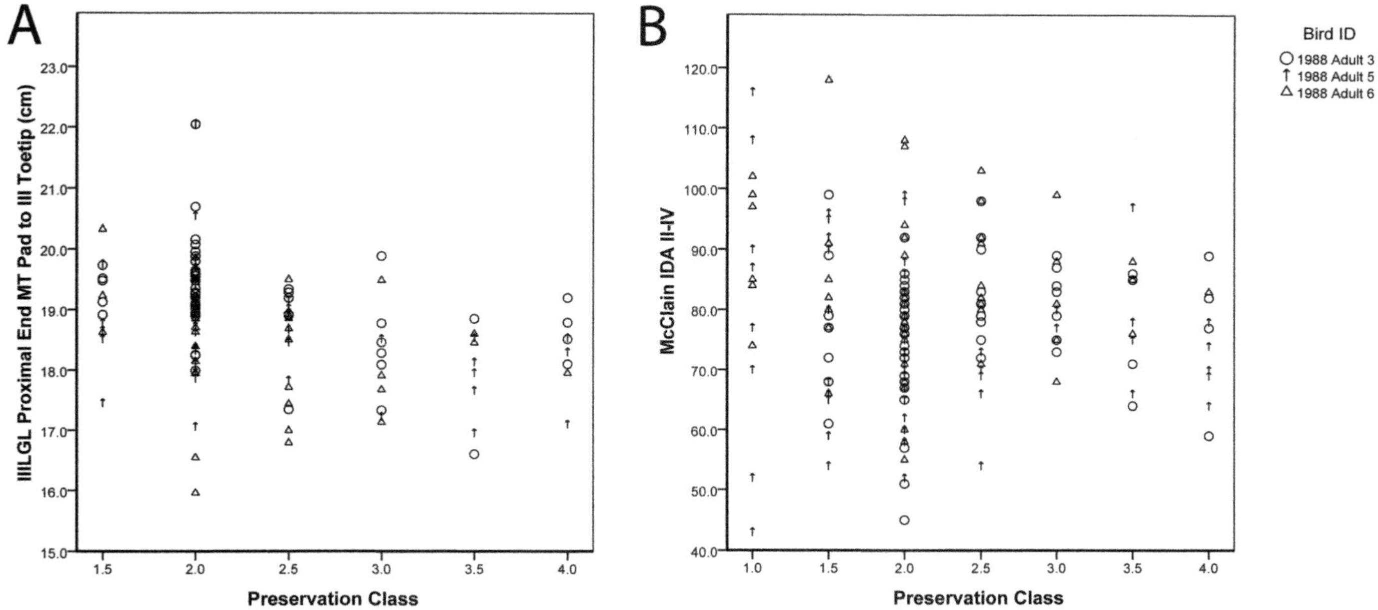

6.12. *A*, Footprint length (IIILGL) as a function of print preservation class in the 1988 adult emus. *B*, McClain IDA II–IV as a function of footprint preservation class in the 1988 adult emus. IDA = interdigital angle; MT = metatarsal pad.

Print quality may bear some relation to bird gait. Although short step lengths did not guarantee well-formed prints, high-quality footprints were only located at the ends of short paces (fig. 6.11); a similar result was obtained in comparing the length of the pace that begins with a footprint with that print's preservation class. However, the association of high-quality prints with short steps should be taken with caution. All of the footprints in preservation class 4 in figure 6.11 were made by the same bird, and were made one immediately after another in a single trackway sequence. Without these preservation class 4 prints, the pattern is less striking.

Footprint preservation class showed no correlation with pace angulation or footprint rotation.

Footprint Quality, Substrate Conditions, and Footprint and Trackway Measurements. High-quality footprints tended to be slightly shorter than poor-quality prints (fig. 6.12A). For each of the 1988 adult birds, footprint length IIILGL was significantly negatively correlated with preservation class (bird 3: Kendall's $\tau b = -0.380$, two-tailed $P = .001$; Spearman's $\rho = -0.508$, two-tailed $P < .001$; $n = 48$; adult 5: Kendall's $\tau b = -0.263$, two-tailed $P = 0.022$; Spearman's $\rho = -0.351$, two-tailed $P = .018$; $n = 45$; adult 6: Kendall's $\tau b = -0.279$, two-tailed $P = .033$; Spearman's $\rho = -0.379$, two-tailed $P = .025$; $n = 35$). The magnitude of the effect can be several millimeters. For example, the mean value of IIILGL for footprints of bird 3 in preservation class 1 is 19.4 cm ($n = 5$), while the mean value for prints in preservation class 4 is 18.6 cm ($n = 4$). The aggregate length of digit III (including the claw) showed a similar tendency to decrease with increasing foot quality—unsurprisingly, because digit III length constitutes a significant part of overall footprint length. Other than that, linear dimensions of footprints showed no consistent relationship with footprint quality among the three adult emus.

Bird 6 showed significant negative correlations between IIILGL and water or sand content (water: $r = -0.619$, $P < .001$, $n = 32$; sand: $r = -0.415$, $P = .039$, $n = 25$), and bird 5 came close, having a significant (but weak) negative correlation between IIILGL and water content ($r = -0.282$, $P = .091$, $n = 37$). None of the birds had a significant correlation between IIILGL and clay content, and bird 3 showed no significant correlations of IIILGL with any of these substrate variables.

The maximum values of best-fit IDAs may also decline with increasing footprint quality (fig. 6.12*B*). However, none of the three 1988 adult emus by themselves showed a statistically significant relationship between best-fit or single-point IDAs and preservation class. Pooling data for all three 1988 adult emus (a reasonable procedure because of the similar values of IDAs among these birds) resulted in a statistically significant negative correlation between TOPO IDA II–IV and preservation class (Kendall's $\tau b = -0.451$, $P < .001$; Spearman's $\rho = -0.573$, $P < .001$; $n = 43$), but not between other IDAs and preservation class. If the association between IDA and footprint quality is real, it appears to be weak (and note that, despite their differences in appearance, footprints 06038830 [fig. 6.3*F*] and 06108821 [fig. 6.10] have very similar

Table 6.5. Correlations between substrate parameters and interdigital angles using pooled data for the 1988 adult emus. For independent variables, the numbers in each cell are Pearson's *r*, two-tailed *P*, and the number of footprints. See text for descriptions of abbreviations used in dependent variables.

Dependent variable	Independent variable		
	% Water	*% Sand*	*% Clay*
MCC IDA II–IV	0.055; .544; 152	−0.061; .478; 137	0.105; .220; 137
JOF IDA II–IV	−0.024; .839; 76	−0.039; .760; 63	0.009; .944; 63
TOPO IDA II–IV	−0.042; .821; 32	0.062; .741; 31	0.064; .733; 31
TRACE IDA II–IV	0.619; .008; 17	−0.217; .403; 17	0.313; .222; 17
"Heel" spread angle	0.229; .004; 154	−0.110; .198; 138	0.205; .016; 138
Toebase IDA	0.173; .127; 79	−0.081; .519; 65	0.032; .798; 65
Toetip IDA	−0.016; .879; 93	−0.137; .224; 81	0.164; .144; 81
Toetip spread angle	0.158; .163; 80	0.153; .202; 71	−0.025; .836; 71

IDAs). There was no consistent correlation between substrate parameters and IDAs for the three 1988 adults (table 6.5).

Data for the 1989 juvenile emus are considerably fewer than for the 1988 adults, for the reasons previously described. However, there was no obvious effect of substrate type on footprint length or interdigital angle for either bird (fig. 6.13).

Only one potentially convincing candidate for an interaction between a trackway parameter and substrate conditions was observed: Footprint rotation showed an apparent decline with increasing water content in 1988 adult 5. Even here, however, there are grounds for skepticism. The footprints with the most positive and the most negative rotations were both made when the bird was changing directions, such that the manner by which footprint rotation is calculated would result in atypical values. If these two extreme points are disregarded, there is little suggestion of a relationship between footprint rotation and soil water content.

Bird Gait and Footprint Measurements. With one exception, there were no significant correlations between footprint length and pace length, pace angulation, and footprint rotation in the three 1988 adult emus. The exception was bird 6, which showed a significant negative correlation between IIILGL and ending ($r = -0.649$, $P < .001$, $n = 26$) and beginning ($r = -0.672$, $P < .001$, $n = 29$) pace (fig. 6.14A). The four short footprints responsible for this result were all made during a single trackway sequence when the bird was trotting, and sliding on its feet in a manner that caused the footprints to be slightly curved, thus reducing overall length.

Rather perversely, 1989 juvenile A showed a positive correlation between ending pace and IIILGL ($r = 0.524$, $P = .015$, $n = 21$) (fig. 6.14B), while 1989 juvenile C had negative correlations between IIILGL and both ending and beginning pace ($r = -0.479$ and -0.480, respectively; $P = .015$ and $n = 25$ in both cases). Juvenile A also had a significant positive

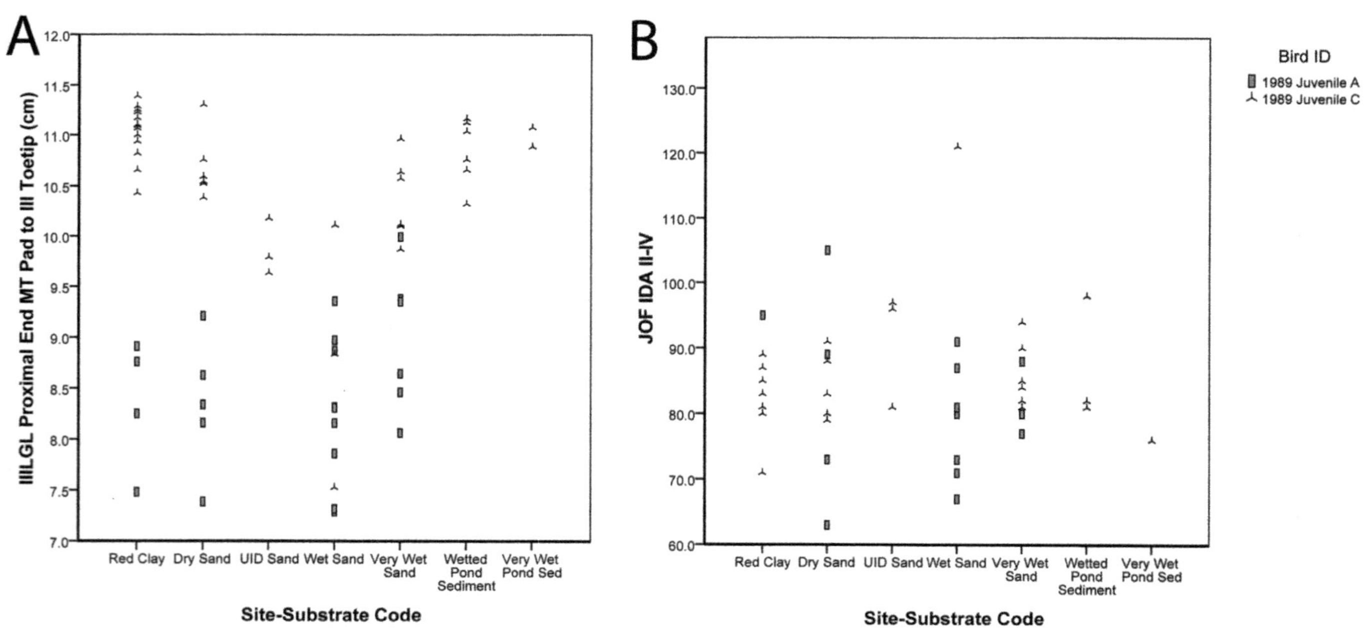

6.13. *A*, Footprint length (IIILGL) and substrate type in the 1989 juvenile emus. *B*, Interdigital angle and substrate type in the 1989 juvenile emus.

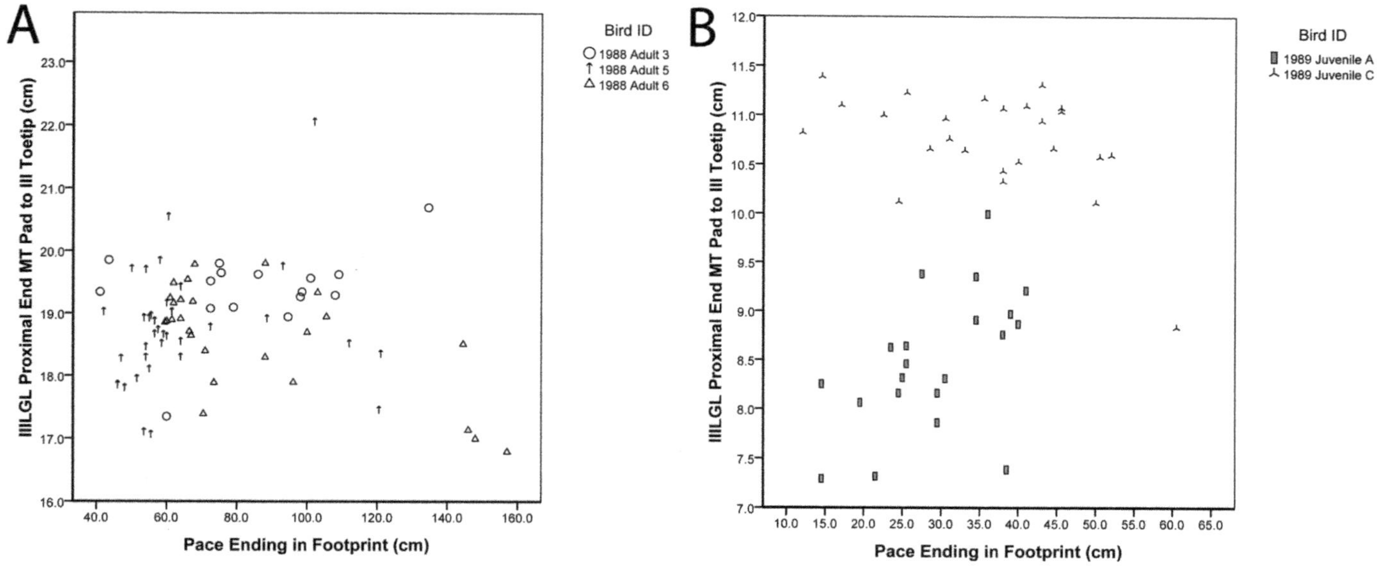

6.14. Footprint length (IIILGL) as a function of ending pace in *A*, the three 1988 adult emus, and *B*, the 1989 juvenile emus. MT = metatarsal pad.

correlation between IIILGL and pace angulation ($r = 0.621$, $P = .008$, $n = 17$). Otherwise neither juvenile showed significant correlations between footprint length and gait parameters.

For juvenile C, the negative correlation between IIILGL and pace length was due mainly to the effects of two footprints, both of which were made in wet sand as the bird was moving quickly. The distal end of digit III dug into the substrate, curving over its length, and this caused IIILGL to be considerably shorter than usual. Without these two prints the negative correlation between step length and footprint length largely vanishes. The positive correlation between step length and footprint length in juvenile A is harder to explain (fig. 6.15); it involves several footprints, and is not associated with any particular substrate.

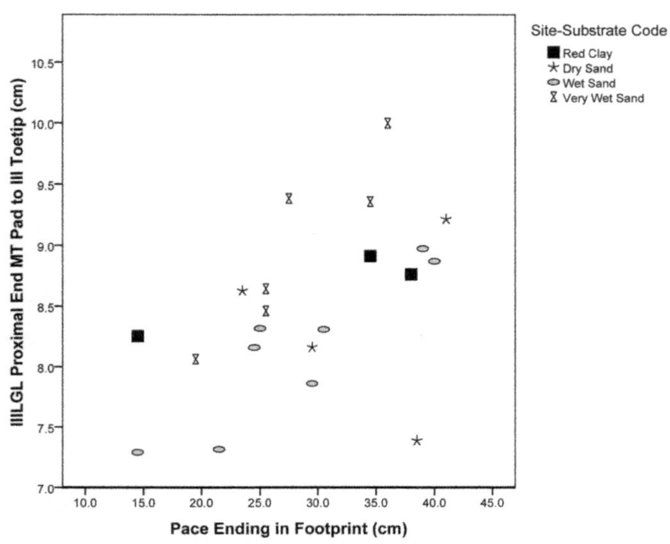

6.15. Footprint length (IIILGL) as a function of ending pace and substrate in 1989 juvenile emu A. MT = metatarsal pad.

If data for the 1988 adult emus are considered alone, there is a weak positive correlation between interdigital angle and pace length (fig. 6.16). However, this correlation is lost if data are pooled for all large (size class 3) footprints (fig. 6.17; data for only size class 3 footprints are pooled because of the negative relationship between IDA and footprint size described below). Similarly, there is no suggestion of a relationship between interdigital angle and step length for other emu footprint size classes.

Possible Joint Effects of Substrate and Gait on Footprint Parameters. We have thus far considered possible effects of substrate and bird gait variables on footprint measurements one at a time. However, it is conceivable that these variables exert their influence in tandem. I investigated this possibility using multiple regression and data for the three 1988 adult birds. I considered two dependent variables, IIILGL and MCC IDA II–IV. To increase the number of data points in each analysis, I pooled data for birds whenever that was reasonable. Thus for IIILGL I combined data for birds 5 and 6, because these two birds had very similar footprint lengths. For MCC IDA II–IV I pooled data for all three 1988 adult emus, because they had similar interdigital angles.

As independent variables I used soil water content, soil clay content, ending pace, pace angulation, and footprint rotation (ROTPREV). In each analysis, variables were added in a stepwise fashion, with the probability of F to enter a variable in each step being $P \le .05$, and the probability of F to remove a variable in each step being $P \ge .10$. In the first

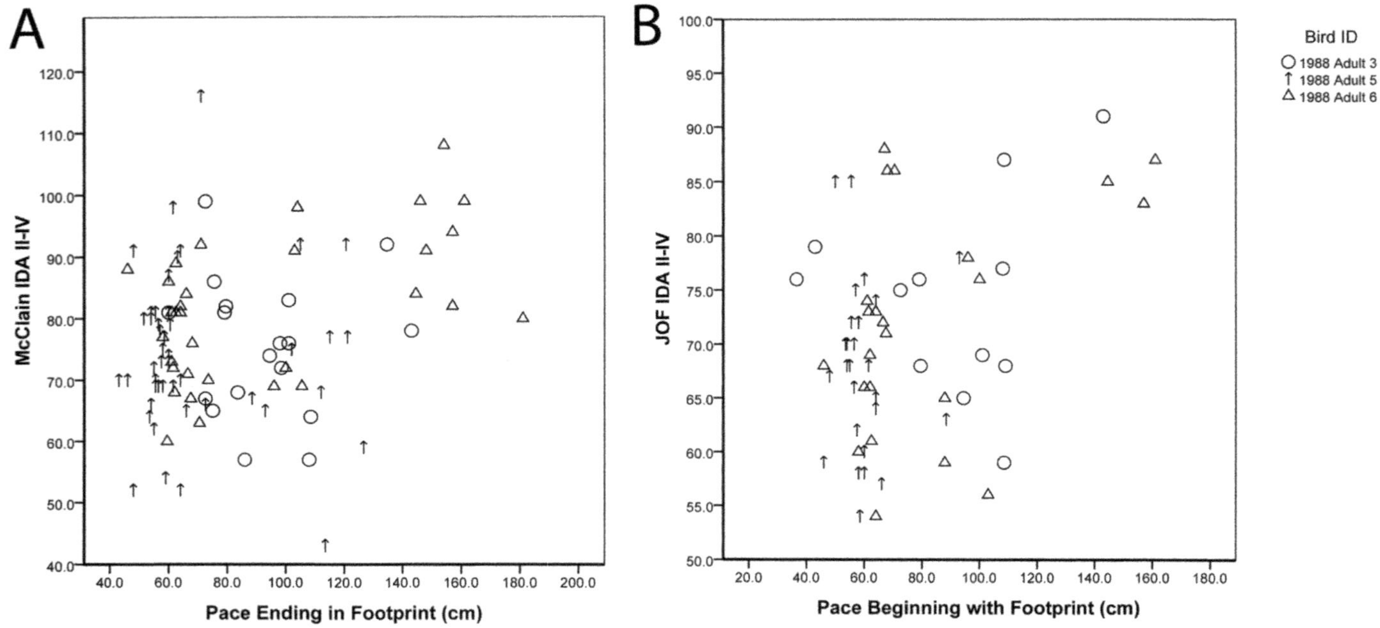

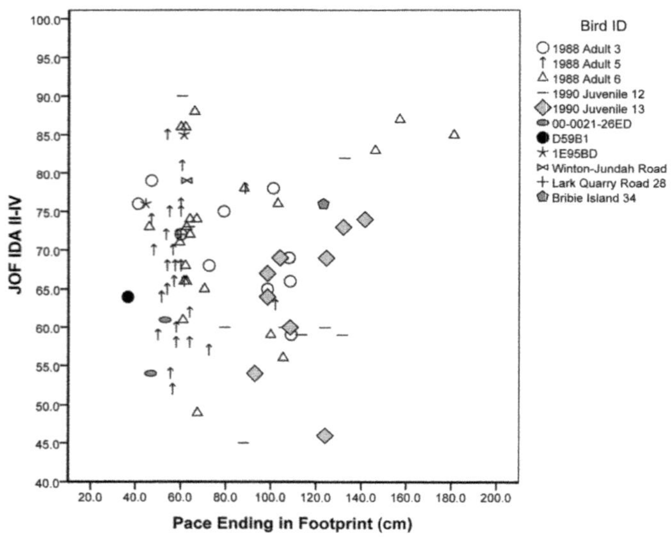

Above, **6.16.** *A,* McClain (MCC) IDA II–IV as a function of ending pace in the 1988 adult emus. The correlation is weak but significantly positive (r = 0.260, P =.010, n = 97); a comparable result is observed if beginning pace is substituted for ending pace. *B,* If JOF IDA II–IV is substituted for MCC IDA II–IV, there is a significant positive correlation with beginning pace (r = 0.354, P = .005, n = 61), but not with ending pace. IDA = interdigital angle.

Left, **6.17.** JOF IDA II–IV as a function of pace length in large (size class 3) emu footprints. IDA = interdigital angle.

step, the independent variable that has the highest correlation with the dependent variable is examined, and an F-test is performed to see if the correlation is significantly different from zero (no correlation); if the probability is .05 or less, the first variable is added to the regression equation. In the next step, the remaining independent variables were considered. Ordinarily the independent variable with the highest partial correlation with the dependent variable would be entered into the equation if its probability of F to enter the equation met the selection criteria, and the continued presence of the first independent variable in the regression equation reconsidered on the basis of the probability of F to remove the variable from the equation.

In the present analyses, however, only one variable qualified for inclusion in each analysis. In the case of IIILGL, the independent variable selected was water content (r = 0.442, P < .027, n = 25); for MCC IDA II–IV, it was ending pace (r = 0.366, P = .017, n = 42). Thus joint action of substrate and bird gait variables on footprint parameters could not be demonstrated with my data.

Footprint Quality, Substrate Features, and Bird Gait: Conclusions. It would be rash, however, to claim that there is no effect of substrate conditions or the way an animal moves on the quality and shape of the footprints that are made; comparing footprints 06038830 (fig. 6.3F) and 06108821 (fig. 6.10) is enough to make the point that such effects exist. However, my analyses found few correlations between footprint parameters and independent substrate or gait variables that were consistent among birds and across different ways of

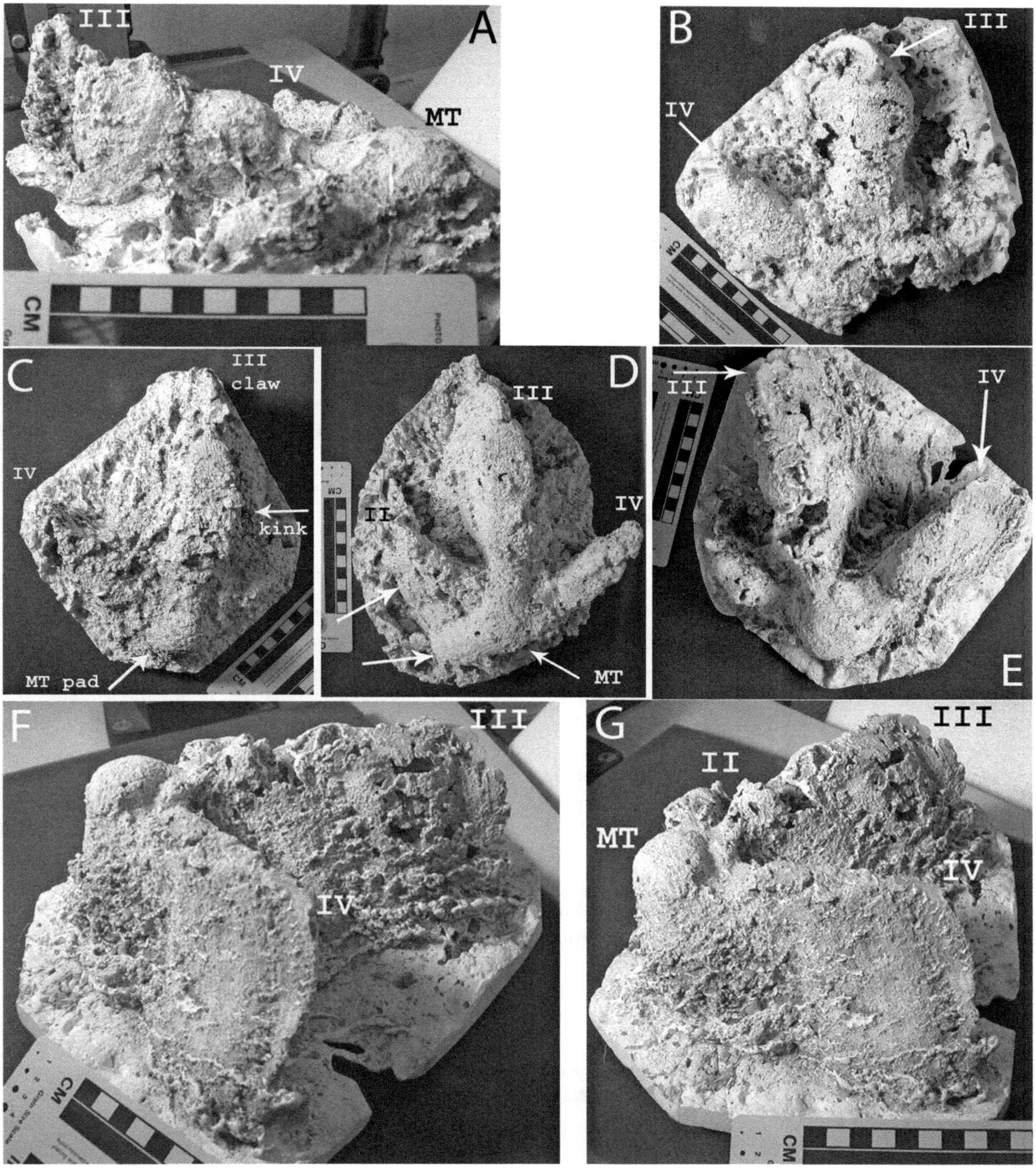

6.18. Casts of unusual footprints of the 1988 adult birds, showing conspicuous slide marks and/or deep penetration of the sediment by the toes. Individual toes are labeled for purposes of orientation. MT = metatarsal pad. Scale bar is marked in centimeters in all photographs. *A*, Footprint 06088824, made by bird 5; note significantly deeper penetration of the sediment by the distal end of digit III than by the metatarsal pad. *B*, Footprint 06178803 (bird 6); note curvilinear mark made by the tip of digit III during emplacement. *C*, Footprint 05318815 (bird 3); note conspicuous kink about midway along the length of the digit III impression. *D*, Footprint 06178808 (bird 6); note slide marks of the metatarsal pad and the medial side of the digit II impression. *E–G*, Three views of footprint 05318824 (bird 3); note curved slide mark made by the tip of digit IV during emplacement, as well as the slide mark along the lateral edge of the digit III impression, made as the toe shifted medially during emplacement. See Milàn (2006) for emu footprints showing similar slide marks and grooves made during emplacement.

Table 6.6. Comparison of interdigital angles (measured in degrees) of plunging/sliding footprints with those of more typical emu footprints. Data are pooled values for the 1988 adult emus. In each cell, the numbers reported are mean (range), and number of footprints. Starred (*) values of the mean indicate a statistically significant difference (two-tailed *t*-test; $P < .05$) between "typical" and plunging/sliding prints for that particular interdigital angle (IDA).

	McClain IDA II–III	JOF IDA II–III	McClain IDA III–IV	JOF IDA III–IV	McClain IDA II–IV	JOF IDA II–IV
"Typical" prints	32.8* (2–61), 142	28.3* (0–43), 78	43.6 (15–76), 144	41.6 (23–62), 85	76.8* (43–118), 149	70.2* (49–93), 74
Plunging/sliding prints	38.3* (9–59), 29	32.7* (21–47), 19	46.2 (33–61), 29	43.3 (33–56), 16	84.6* (57–108), 29	77.2* (66–90), 14

Table 6.7. Correlations between different length and width measurements of emu footprints. In each cell, the numbers are Pearson's *r*, two-tailed *P*, and the number of footprints.

	Field length	Thulborn length	IIILGL	Thulborn width	Toetip width	Clawbase width
Field length		1.000; $P < .001$; 263	0.995; $P < .001$; 489			
Thulborn length			0.992; $P < .001$; 263			
Thulborn width					0.980; $P < .001$; 179	0.982; $P < .001$; 89
Toetip width						0.997; $P < .001$; 174

measuring the same parameter; most of the correlations that we observed were rather weak. Although it is axiomatic that footprint shape should reflect the way the trackmaker's foot interacts with the substrate, the substrate and gait parameters that we measured were unable to capture the undoubted complexity of the interaction.

This point can be emphasized by considering the effects on footprint shape of an interaction between emu feet and the substrate that is hard to quantify. Several footprints of the 1988 adult emus were unusual in showing conspicuous slide marks, and/or atypically deep penetration of the substrate by the toes (fig. 6.18). These prints were made by birds that were moving at faster speeds than usual, as indicated by longer ending and beginning paces. Such "plunging/sliding" prints had larger interdigital angles than more typical footprints made by the same three adult birds (table 6.6), but as already described, longer paces didn't invariably translate into larger interdigital angles.

Whether my results can be generalized and applied to interpreting dinosaur footprints, or are specific to where and how we studied emu footprint formation, is a question I cannot answer. For the purposes of our present study, it is enough to say that, with a few exceptions to be considered later, the effects of substrate conditions and bird movements on footprint measurements are small enough that they can be considered random noise and disregarded in analyzing variability in footprint measurements.

Variability of Footprint Measurements

Recall that one of the key questions of interest in this study can be summarized as follows: How different can two footprints be, and yet have been made by the same species of trackmaker? As stated, the question is succinct, but addressing it requires complexity. Different in what way? Size, shape, or both?

We will consider variability in both size and shape in emu footprints. We begin by comparing variability in footprint parameters among footprints made by the same individual bird, using the coefficient of variation (CV), expressed as a percentage: (standard deviation/mean) × 100. To save space, I will report only those determinations of CV for each bird that are based on 10 or more footprints. In some cases, juvenile birds grew enough over the course of our study that there were detectable differences in print dimensions from one run to the next; in those cases we report values broken down by run.

Tables A6.1–A6.7 summarize variability in linear footprint dimensions and interdigital angles in our emus. Unsurprisingly, the various measurements of footprint length or width are very highly correlated (table 6.7). CV values for overall footprint size (table A6.1) are about 1%–10%. Footprint width measurements that reflect the positions of the toetips (toetip width and [often] Thulborn width) are more variable than length measurements. Clawbase width is considerably less variable than toetip width, roughly comparable in CVs to footprint lengths. For very young emus, toetip widths are only a bit less than lengths, but for adults toetip widths are noticeably less than footprint lengths.

The footprint size index, which Thulborn and Wade (1984: 417) found to be "remarkably consistent within each trackway" of the dinosaur trails they studied, and less variable than measurements of footprint length and width, failed to accomplish the same task with emu prints. The size index is less variable than toetip-based footprint widths, but has equivalent or larger CVs than footprint length.

Measurements of the individual components of footprint length or toe length generally have CVs somewhere between

Table 6.8. Comparison of variability among footprints in different regions of digit III of the 1988 adult emus. In this analysis, each footprint had to have measurements for all variables considered. This greatly reduced the number of footprints for each bird, but facilitates comparison of variability among the measurements. All measurements are in centimeters. See text for explanations of abbreviations used here.

Bird code	Parameter	Minimum	Maximum	Mean (CV)	Number of footprints
	IIIPdGp	1.6	3.0	2.44 (18)	10
	IIIPd1L	3.4	4.0	3.72 (3.9)	10
1988 adult 3	IIIPd2L	2.9	3.6	3.15 (6.3)	10
	IIIPd3L	2.2	3.5	2.85 (14)	10
	IIIClwL	1.5	2.7	2.28 (16)	10
	IIIPdGp	1.4	2.9	2.41 (15)	15
	IIIPd1L	3.0	3.9	3.53 (7.7)	15
1988 adult 5	IIIPd2L	3.0	4.0	3.34 (8.7)	15
	IIIPd3L	2.0	3.5	2.64 (16)	15
	IIIClwL	1.5	2.6	1.92 (16)	15
	IIIPdGp	1.3	2.5	2.01 (16)	14
	IIIPd1L	3.5	4.1	3.83 (5.5)	14
1988 adult 6	IIIPd2L	2.9	3.8	3.34 (10)	14
	IIIPd3L	1.9	3.9	2.79 (16)	14
	IIIClwL	1.4	2.6	1.89 (17)	14

5% and 30%, and are usually larger than CVs of overall footprint or toe length (tables A6.2–A6.5), as would be expected for theoretical reasons (Lande 1977). (The same is true for the corresponding measurements of the feet of emu study skins, as will be discussed later.) Measurements of the gap between the metatarsal pad and the proximal end of a toe, and of claw lengths, are particularly variable. This observation is especially evident if CVs are calculated for different portions of a toe for only those footprints in which all of the relevant measurements can be made (table 6.8). Thus for digit III of each of the 1988 adult emus, the metatarsal pad gap and claw lengths are the most variable. The length of the first digital pad is the least variable, and variability increases with the second and third digital pads.

Given the preceding observations, it isn't surprising that overall toe length measurements (table A6.5) that include distances from the metatarsal pad, or the claw length, as part of their length tend to be more variable than toe length measurements based only on the digital pad portion of the toe. Consequently Lgth toe lengths are sometimes more variable than PdL lengths, and FRNC lengths sometimes more variable than PdL measurements, despite that fact that PdL lengths are included in Lgth and FRNC lengths. Occasional very high CVs (e.g., those of IIPdL and IILgth of 1988 juvenile 7) are due to unusual interactions between the foot and the substrate during footprint formation (fig. 6.19).

Individual juvenile and adult emus are comparable in variability of the same length measurement.

The various spread angles and best-fit IDAs are positively correlated, but not always strongly (table 6.9). Because they measure IDAs in a very different manner, there is no reason to expect strong correlations between toebase, clawbase, or toetip IDAs and best-fit IDAs.

IDA II–III is usually more variable than IDA III–IV. IDA II–IV is often less variable than IDA III–IV (tables A6.6, 6.10). IDA II–III shows a modest positive correlation with footprint rotation relative to trail midline and preceding pace (table 6.11), but not other parameters of footprint rotation. That such correlations might exist makes sense. Negative and less positive values of print rotation mean that the footprint toes more nearly inward with respect to the bird's direction of

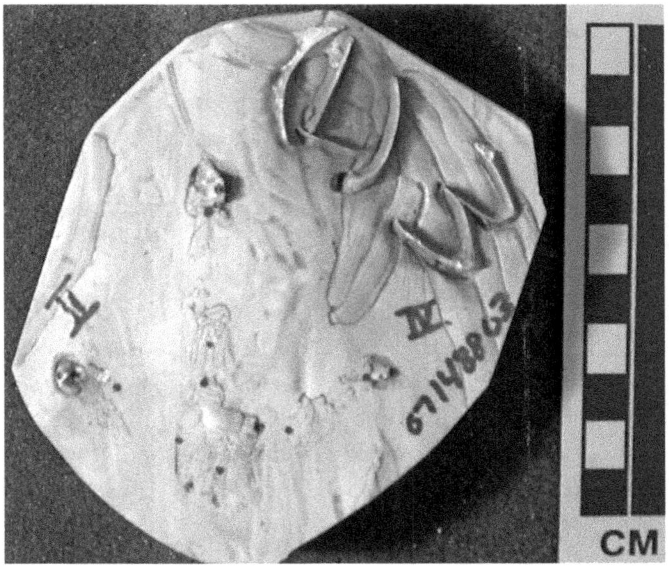

6.19. Cast of footprint 07148803, made by 1988 juvenile 7. Note the unusually short impression of digit II (left side of the footprint, as shown here).

Table 6.9. Correlations between selected interdigital angle (IDA) and spread angle (SA) measurements. In each cell, the numbers are Pearson's *r*, two-tailed *P*, and the number of footprints. See text for explanations of abbreviations.

	MCC IDA II–IV	JOF IDA II–IV	TOPO IDA II–IV	TRACE IDA II–IV	"Heel" SA	Clawbase SA	Toetip SA
MCC IDA II–IV		0.795; $P < .001$; 149	0.592; $P < .001$; 42	0.553; .006; 23	0.713; $P < .001$; 256	0.755; $P < .001$; 87	0.716; $P < .001$; 152
JOF IDA II–IV			0.777; $P < .001$; 79	0.772; $P < .001$; 25	0.830; $P < .001$; 150	0.854; $P < .001$; 189	0.881; $P < .001$; 357
TOPO IDA II–IV				0.680; $P < .001$; 29	0.437; .004; 42	0.588; $P < .001$; 40	0.634; $P < .001$; 74
TRACE IDA II–IV					0.357; .094; 23	0.617; .011; 16	0.775; $P < .001$; 21
"Heel" SA						0.914; $P < .001$; 87	0.915; $P < .001$; 153
Clawbase SA							0.964; $P < .001$; 167

travel, and this might press digits II and III together. However, it is surprising that only two parameters of footprint rotation suggest such an effect. IDA III–IV values show little effect of footprint rotation. IDA II–IV shows a modest positive correlation with footprint rotation relative to trail midline—unsurprising, given the similar correlation between IDA II–III and the same footprint rotation parameter.

Inspection of tables A6.6 and A6.7 suggests that reference-point IDAs tend to be less variable than best-fit IDAs. This impression is confirmed by table 6.10, in which the two kinds of IDAs are compared for the 1988 adult emus; all of the different IDAs had to be measurable in order for a footprint to be used in compiling the data in table 6.10.

An unexpected result emerges in comparisons of a parameter of overall footprint length (table 6.12), the length of a footprint subunit (table 6.13), and an interdigital angle (table 6.14) among the three 1988 adults and all adult emus. The range of values, means, and CVs for the three individual 1988 adults approximate those for all footprints of adult emus. Furthermore, the "bird means" treatment (in which data cases are the mean values of parameters for each bird) is similar in these features to the aggregate of all adult emu prints (in which individual birds are represented by wildly varying numbers of footprints). This suggests an intriguing possibility: In an assemblage of footprints made by animals at or near their full size, it may not be necessary to know how many individual trackmakers made the prints in order to get a reasonable approximation of the distribution of values of footprint parameters for the population of adult animals as a whole (Gand [1976] suggested that a sample of as few as 10 footprints might suffice for this; see below). However, this is not true if the track assemblage includes prints made by younger, smaller animals.

Table 6.10. Comparison of variability of selected interdigital angle (IDA) measurements in the 1988 adult emus. In this analysis, all of the selected IDAs had to have been measured in order for the print to be included in the analysis. This reduces the number of footprints for which IDAs were measured individually, but facilitates comparison of variability among IDAs. All angles are measured in degrees. Each of the three birds was represented in this analysis by 16 footprints. See text for descriptions of abbreviations.

Bird code	Parameter	Minimum	Maximum	Mean (CV)
1988 adult 3	JOF IDA II–III	22	40	31.3 (18)
	MCC IDA II–III	10	52	33.0 (30)
	JOF IDA III–IV	31	62	42.7 (20)
	MCC IDA III–IV	31	61	43.8 (20)
	JOF IDA II–IV	56	93	74.1 (13)
	MCC IDA II–IV	45	98	76.8 (16)
	"Heel" SA	66	87	79.3 (6.5)
	Toetip SA	72	92	82.7 (6.5)
	Toebase IDA	101	130	112.0 (7.8)
	Toetip IDA	70	89	77.1 (8.1)
1988 adult 5	JOF IDA II–III	16	38	26.3 (24)
	MCC IDA II–III	20	47	30.7 (28)
	JOF IDA III–IV	32	52	43.7 (13)
	MCC IDA III–IV	34	53	44.9 (11)
	JOF IDA II–IV	54	89	69.9 (12)
	MCC IDA II–IV	64	97	75.7 (12)
	"Heel" SA	76	93	79.9 (5.7)
	Toetip SA	74	93	80.3 (5.9)
	Toebase IDA	98	115	106.0 (4.0)
	Toetip IDA	76	87	80.8 (4.0)
1988 adult 6	JOF IDA II–III	15	49	29.8 (33)
	MCC IDA II–III	22	59	34.3 (27)
	JOF IDA III–IV	33	54	39.9 (13)
	MCC IDA III–IV	35	56	44.9 (13)
	JOF IDA II–IV	49	90	69.7 (15)
	MCC IDA II–IV	63	103	79.1 (13)
	"Heel" SA	70	89	79.6 (7.8)
	Toetip SA	67	89	77.2 (8.0)
	Toebase IDA	95	119	110.3 (7.1)
	Toetip IDA	72	87	81.8 (5.1)

Table 6.11. Correlations between best-fit interdigital angles (IDAs) and footprint rotation parameters. In each cell, numbers are Pearson's *r*, two-tailed *P*, and number of footprints. See text for descriptions of abbreviations.

Footprint rotation parameter	JOF IDA II–III	JOF IDA III–IV	JOF IDA II–IV
Rotation relative to previous print	0.031; .687; 173	0.055; .471; 176	0.077; .318; 170
Rotation relative to following print	0.171; .023; 178	0.150; .042; 183	0.191; .011; 175
Rotation relative to trail midline	0.496; .005; 31	0.192; .319; 29	0.553; .002; 29
Rotation relative to preceding pace	0.500; .001; 45	−0.134; .396; 42	0.267; .087; 42
Rotation relative to following pace	0.077; .620; 44	0.150; .343; 42	0.135; .393; 42

Table 6.12. Comparison of footprint lengths (IIILGL) in the 1988 adult emus and various groupings of adult emus. All measurements are in centimeters. In all but the "emu means" treatment, cases are individual footprints. In the "emu means" treatment, cases are mean values for each individual bird; these data include adult measurements for birds that were initially run as juveniles.

Bird category	Minimum	Maximum	Mean (CV)	Number of cases
1988 adult 3	16.6	22.0	19.13 (4.6)	50
1988 adult 5	17.0	22.1	18.63 (4.8)	54
1988 adult 6	16.0	20.3	18.58 (4.8)	52
1988 adults: pooled data	16.0	22.1	18.77 (4.9)	156
All adult emu footprints	16.0	22.1	18.79 (4.9)	201
Emu means adults	17.7	20.4	18.96 (3.8)	16

Table 6.13. Comparison of the length of the first digital pad of digit III in the 1988 adult emus and various groupings of adult emus. All measurements are in centimeters. Other data reporting conventions as in table 6.12.

Bird category	Minimum	Maximum	Mean (CV)	Number of cases
1988 adult 3	2.9	4.8	3.85 (11)	37
1988 adult 5	3.0	4.1	3.53 (7.2)	42
1988 adult 6	3.4	4.3	3.88 (5.8)	55
1988 adults: pooled data	2.9	4.8	3.76 8.9)	134
All adult emu footprints	2.9	5.3	3.80 (9.4)	186
Emu means adults	3.3	5.0	3.86 (9.9)	17

Table 6.14. Comparisons of JOF interdigital angle II–IV in the 1988 adult emus and various groupings of adult emus. All measurements are in degrees. Other data reporting conventions as in table 6.12.

Bird category	Minimum	Maximum	Mean (CV)	Number of cases
1988 adult 3	56	93	74.4 (13)	32
1988 adult 5	51	89	68.6 (13)	39
1988 adult 6	49	90	72.0 (14)	38
1988 adults: pooled data	49	93	71.5 (14)	109
All adult emu footprints	49	104	73.1 (14)	154
Emu means adults	62	90	75.6 (10)	15

Ontogenetic Changes in Footprint Shape

One of the key questions in this study is whether emus show changes during ontogeny in footprint proportions (a possibility suggested by Olsen [1980] and Olsen et al. [1998] for non-avian dinosaurs), or whether footprints of baby emus will be miniature copies of those of their elders. I considered this matter with both bivariate and multivariate analyses. Because dinosaur footprints are often highly variable in quality of preservation, it isn't always possible to measure all of the same footprint parameters among a large number of prints at a site. The fewer variables being considered, the greater the number of prints on which those parameters are likely to be measurable. Consequently bivariate analyses will often be useful in quantitative palichnology. On the other hand, bivariate analyses are necessarily less informative about overall footprint shape than are multivariate analyses, and so both approaches are necessary.

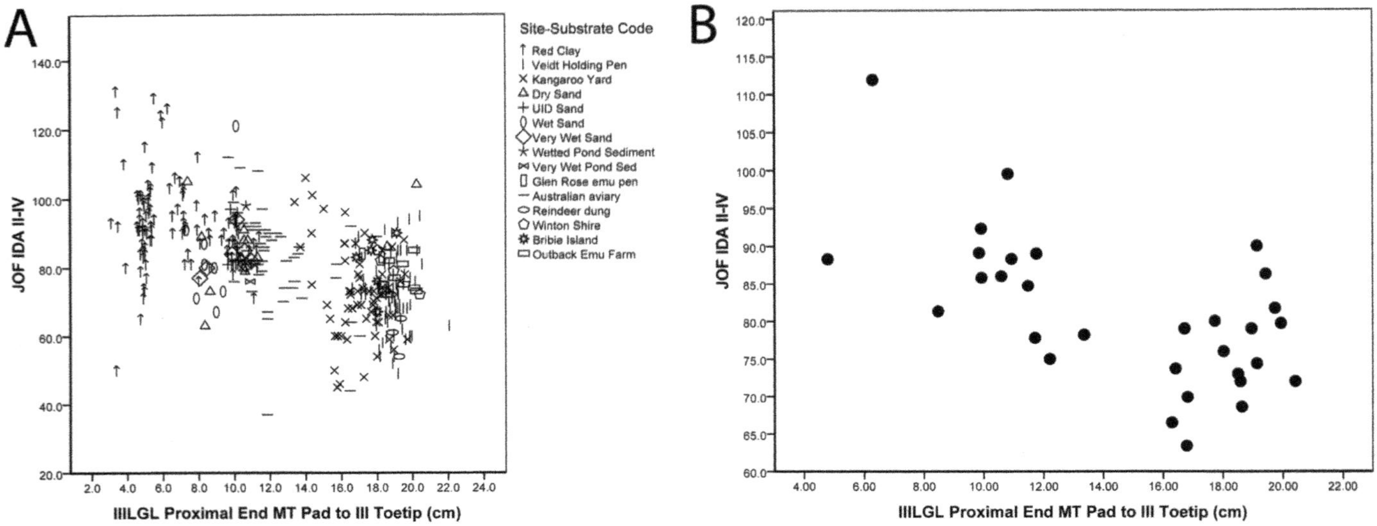

6.20. Interdigital angle II–IV (JOF IDA II–IV) as a function of footprint length (IIILGL) in emus. *A*, Data points are individual footprints, with each bird represented by a variable number of footprints; site-substrate code identifies footprints in terms of where they were made (including the nature of the substrate). *B*, Data points are mean values for each emu, with each bird represented by a single point. MT = metatarsal pad.

Bivariate Comparisons. I first analyzed the relationships between combinations of log-transformed variables (table A6.8). If the slope of the line between two variables differs significantly from 1 (which value would indicate isometry in the relationship between log-transformed variables), this reveals an allometric relationship between the variables.

Data were analyzed in several different ways. To begin with, I did analyses in which individual footprints were treated as data cases ("all footprints" treatment). In these analyses, particular birds were represented by different numbers of prints; thus the three 1988 adults were represented by numerous prints, while other adult emus contributed far fewer measurements to the data. I also did analyses ("emu means" treatment) in which mean values of measurements of footprints made by a particular bird (captives) or in a particular trackway (wild emus) constituted the data cases; in these analyses, each bird or trackway would be represented by a single "average" footprint. This treatment sometimes required choosing among runs made by juvenile emus as they grew, so that the means would be for footprints of about the same size. When this was necessary, I chose the run that had the most footprints; details about which runs were used in the "emu means" are provided in table 6.1.

Three kinds of bivariate analysis were performed. In one of these, the length of a subunit of the footprint (first digital pad of digit III [IIIPd1L]) was selected as the independent variable, and the lengths of other subunits of the print, and also measurements of footprint width, were compared with it. In the second analysis, the independent variable was the overall length of digit III (IIILgth), with which other footprint lengths and widths were compared. In the third analysis, miscellaneous bivariate comparisons of the kind done with alligator feet (chapter 5) as well as comparisons prompted by our multivariate analyses (see below) were done. Data were analyzed by "model II" regression (Sokal and Rohlf 1995), employing reduced major axis (RMA) analysis of log-transformed data (Imbrie 1956; Rayner 1985; Leduc 1987).

Unsurprisingly, the correlation coefficient and coefficient of determination (r^2) are respectably high for most equations (tables A6.8, A6.9). Most length measurements of footprint subunits (metatarsal pad, pad gaps, digital pads) have slopes against IIIPd1L and IIILgth that are somewhat less than 1, but claw lengths have slopes somewhat greater than 1 (table A6.8). The same is true for the length of the digit claws relative to the fleshy portions of the toes, especially for digits III and IV (table A6.9). The slope of the length of digit IV against the lengths of digits II and III is positive, both including and excluding claw lengths; and the slope of the length of the claw of digit IV against the lengths of the claws of digits II and III is positive (table A6.9). In many, but not all, of these relationships the slope is significantly greater or less than 1; in some cases whether the slope significantly differs from 1 depends on whether one looks at the "all footprints" or "bird means" treatment. It should be noted, too, that in some relationships with slopes significantly different than 1 they are only "barely" so by the standards used in this book (upper confidence interval [CI] limit of the RMA slope between 0.95 and 1.00, or lower CI limit of the RMA slope between 1.00 and 1.05).

In most cases in which a measurement was made on emu study skins that is comparable to a measurement made on footprints, the slopes of the relationship in the two data sets

Table 6.15. Changes in interdigital angle with increasing footprint size during the growth of individual emus. See table 6.1 for the dates of runs. In each of the interdigital angle cells, the numbers reported are mean (range), and number of footprints. Interdigital angles are in degrees.

Bird	Run	Mean footprint length (IIILGL; cm)	JOF IDA II–III	JOF IDA III–IV	JOF IDA II–IV
1988 juvenile 2	2	3.6	24.4 (0–37), 9	71.6 (40–94), 7	97.3 (50–131), 7
	6	4.8	28.8 (13–41), 36	58.6 (25–75), 35	88.3 (65–101), 31
1988 juvenile 4	4	5.3	42.7 (35–51), 17	55.5 (43–70), 15	97.9 (88–115), 18
	10	7.1	34.8 (23–46), 13	60.7 (54–65), 12	94.5 (81–105), 12
	13	8.1	36.8 (22–49), 10	52.8 (39–63), 10	89.7 (76–112), 10
	18	9.9	35.0 (25–41), 19	50.3 (42–66), 19	85.8 (77–101), 18
00-0021-26ED	42	11.5	25.3 (20–32), 3	59.3 (58–61), 3	84.7 (78–91), 3
	46	19.1	15.0 (8–21), 4	47.5 (45–51), 4	61.8 (54–72), 5
00-001D-8823	43	13.3	21.7 (18–25), 3	56.3 (53–58), 3	77.7 (74–83), 3
	47	16.8	28.5 (19–38), 2	47.0 (41–53), 2	76.0 (60–92), 2
00-0021-35E4	44	12.2	25.4 (20–31), 5	49.8 (43–61), 5	75.0 (65–91), 5
	47	16.0	22.3 (11–45), 4	39.5 (32–60), 4	61.7 (44–81), 4
D59B1	48	10.9	41.5 (31–62), 6	49.2 (45–58), 6	88.3 (81–109), 6
	69	17.3	30.0 (22–38), 2	46.5 (42–51), 2	76.5 (64–89), 2
1E95BD	49	12.0	32.5 (28–37), 2	55.0 (52–58), 2	87.0 (85–89), 3
	52	13.4	38.2 (34–50), 6	42.4 (35–50), 5	78.2 (71–86), 5
	68	18.5	31.0 (29–33), 2	49.5 (43–56), 2	75.3 (65–85), 3

were very similar. Thus both footprints and study skins suggest ontogenetic negative allometry in the total length of digit II with respect to the length of digit III (table A6.8), near isometry or very slight positive allometry of the length of digit IV with respect to digit III (table A6.8), slight positive allometry of the length of digit IV with respect to digit II (table A6.9), positive allometry of claw lengths with respect to the lengths of the fleshy portion of digits II and especially IV (table A6.9), and positive allometry of the length of the claw of digit IV with respect to the length of the claws of both digits II and III (table A6.9). However, the study skin and footprint data give conflicting results about allometry in the length of the claw of digit III with respect to the fleshy part of the toe, and the relative lengths of the claws of digits II and III (table A6.9).

There is rather strong evidence for allometric relationships between footprint width and footprint length. In all equations relating width across digits II–III, III–IV, and II–IV, the slope of the width against a footprint length parameter is less than 1, and in most of these equations the upper limit of the 95% CI is well below 1 (tables A6.8, A6.9). With increasing size, emu footprints become relatively narrower, with digits II and IV relatively closer together—a rather unexpected result.

This would be expected to affect interdigital angles, and such is indeed the case: Best-fit IDAs are negatively correlated with footprint size, regardless of substrate (fig. 6.20). For JOF measurements of IDAs, the correlations with IIILGL, using individual footprints as the data cases, are as follows: (a) JOF IDA II–III: $r = -0.289$, $P < .001$, $n = 407$; (b) JOF IDA III–IV: $r = -0.564$, $P < .001$, $n = 407$; (c) JOF IDA II–IV: $r = -0.580$, $P < .001$, $n = 380$. For the bird means treatment, the correlations between IIILGL and interdigital angles are: (d) JOF IDA II–III: $r = -0.414$, $P = .021$, $n = 31$; (e) JOF IDA III–IV: $r = -0.590$, $P < .001$, $n = 33$; (f) JOF IDA II–IV: $r = -0.617$, $P < .001$, $n = 31$. IDA III–IV shows a stronger negative correlation with footprint size than does IDA II–III. (In contrast, reference point–based IDAs show weak or no correlation with footprint size.) Decreasing values of IDAs with increasing footprint size are seen not only among birds of different size, but also for individual birds as they grew (table 6.15).

Multivariate Comparisons. In addition to simple bivariate comparisons of footprint parameters, I also examined possible ontogenetic changes in footprint shape involving multiple variables in the same analysis. For these analyses only the "emu means" treatment of variables was employed (tables 6.16–6.23).

Two versions of this kind of analysis were done. In the first set of analyses I performed a principal component analysis (PCA; table 6.16), using the log-transformed total lengths of digits II, III, and IV as the parameters of digit length, as well as dimensions of the metatarsal pad, digit hypex lengths, and measurements across toebases and toetips. I also calculated the RMA slope of each of the preceding log-transformed footprint variables against a footprint scaling factor calculated as the mean of all other log-transformed footprint variables except the one being analyzed (table 6.17). In a more complex version of the same kind of analysis (tables 6.18–6.19) digits II–IV were subdivided into their component digital pad and claw lengths; dimensions of the metatarsal pads,

Table 6.16. Principal component analysis (using a covariance matrix) of log-transformed measurements of emu footprints ("emu means" treatment). Number of birds = 22. Variable abbreviations as described in text.

Parameter	Component 1 loading (raw [rescaled])
MtPdL	0.088 (0.934)
MtPdW	0.101 (0.948)
IIPdGp	0.097 (0.914)
IIIPdGp	0.123 (0.899)
IVPdGp	0.114 (0.946)
IILgth	0.110 (0.983)
IIILgth	0.128 (0.988)
IVLgth	0.126 (0.985)
IILat	0.104 (0.967)
IIIMed	0.134 (0.986)
IIILat	0.129 (0.992)
IVMed	0.123 (0.977)
Proximal end II to proximal end III	0.101 (0.969)
Proximal end III to proximal end IV	0.102 (0.974)
Proximal end II to proximal end IV	0.098 (0.979)
Toetip II to toetip III	0.115 (0.980)
Toetip III to toetip IV	0.112 (0.973)
Toetip II to toetip IV	0.103 (0.982)
Eigenvalue (% of variance explained)	0.227 (93.389)

Kaiser-Meyer-Olkin measure of sampling adequacy = 0.796; Bartlett's test of sphericity: $\chi^2 = 955.764$, $P < .001$.

Table 6.18. Principal component analysis (using a covariance matrix) of log-transformed measurements of emu footprints ("emu means" treatment). Number of birds = 13. Variable abbreviations as described in text.

Parameter	Component 1 loading (raw [rescaled])
MtPdL	0.106 (0.970)
MtPdW	0.122 (0.967)
IIPdGp	0.122 (0.955)
IIPd1L	0.109 (0.949)
IIPd2L	0.105 (0.931)
IIClwL	0.130 (0.944)
IILat	0.120 (0.980)
IIIPdGp	0.136 (0.914)
IIIPd1L	0.129 (0.990)
IIIPd2L	0.128 (0.970)
IIIPd3L	0.147 (0.975)
IIIClwL	0.191 (0.969)
IIIMed	0.152 (0.996)
IIILat	0.146 (0.994)
IVPdGp	0.131 (0.968)
IVPdL	0.133 (0.989)
IVClwL	0.156 (0.924)
IVMed	0.139 (0.993)
Proximal end II to proximal end III	0.120 (0.986)
Proximal end III to proximal end IV	0.121 (0.986)
Proximal end II to proximal end IV	0.122 (0.993)
Clawbase II to clawbase III	0.126 (0.987)
Clawbase III to clawbase IV	0.105 (0.980)
Clawbase II to clawbase IV	0.113 (0.992)
Toetip II to toetip III	0.132 (0.987)
Toetip III to toetip IV	0.125 (0.981)
Toetip II to toetip IV	0.121 (0.990)
Eigenvalue (% of variance explained)	0.459 (94.382)

Table 6.17. Multivariate relationships between parameters of footprint size in emus. Data ("emu means" treatment) were log transformed prior to analysis. Each log-transformed parameter was examined as a dependent variable in a reduced major axis (RMA) analysis with respect to a footprint size factor, the latter calculated as the mean of all log-transformed parameters except the one being examined. In addition to the RMA analyses, a multivariate allometric coefficient for each log-transformed variable was calculated by dividing the loading of that variable on the first component (PC1) of a principal component analysis (PCA) by the mean of the loadings of all the log-transformed variables on the first component. PC1 accounted for more than 93% of the data variance (table 6.16). Number of cases in all analyses = 22. RMA slopes significantly different than 1 are indicated in **bold**. Variable abbreviations as described in text. CI = confidence interval; SEE = standard error of estimate.

Dependent variable	r^2	SEE	RMA slope	95% CI of slope	PCA allometric coefficient
MtPdL	0.866	0.0354	0.835	0.694–1.005	0.789
MtPdW	0.892	0.0359	0.951	0.807–1.120	0.905
IIPdGp	0.824	0.0455	0.943	0.758–1.174	0.869
IIIPdGp	0.780	0.0656	1.231	0.956–1.585	1.102
IVPdGp	0.882	0.0425	1.082	0.911–1.285	1.022
IILgth	0.962	0.0225	1.004	0.915–1.102	0.986
IIILgth	0.970	0.0227	**1.168**	1.075–1.268	1.147
IVLgth	0.965	0.0246	**1.152**	1.054–1.260	1.129
IILat	0.928	0.0294	0.960	0.842–1.094	0.932
IIIMed	0.965	0.0260	**1.238**	1.133–1.353	1.201
IIILat	0.979	0.0190	**1.173**	1.095–1.256	1.156
IVMed	0.946	0.0301	**1.140**[a]	1.019–1.275	1.102
Proximal end II to proximal end III	0.936	0.0271	0.930	0.822–1.051	0.905
Proximal end III to proximal end IV	0.946	0.0248	0.932	0.833–1.043	0.914
Proximal end II to proximal end IV	0.958	0.0211	**0.892**[a]	0.809–0.984	0.878
Toetip II to toetip III	0.954	0.0257	1.056	0.953–1.170	1.030
Toetip III to toetip IV	0.939	0.0290	1.029	0.913–1.159	1.004
Toetip II to toetip IV	0.961	0.0212	0.933	0.849–1.026	0.923

[a]Only barely allometric (upper CI limit of RMA slope between 0.95 and 1.00, or lower CI limit between 1.00 and 1.05).

Table 6.19. Slopes of reduced major axis (RMA) relationships between various log-transformed linear dimensions of emu footprints (using data from the "emu means" treatment") and a footprint scaling factor. The footprint scaling factor with which each log-transformed dependent variable was compared was the mean of all the log-transformed variables other than that dependent variable. Number of cases in all analyses = 13. RMA slopes significantly different than 1 are indicated in **bold**. Variable abbreviations as described in text. CI = confidence interval; PCA = principal component analysis; SEE = standard error of estimate.

Dependent variable	r^2	SEE	RMA slope	95% CI of slope	PCA allometric coefficient
MtPdL	0.938	0.0283	**0.839**[a]	0.706–0.997	0.821
MtPdW	0.931	0.0345	0.972	0.809–1.166	0.945
IIPdGp	0.907	0.0406	0.985	0.794–1.222	0.945
IIPd1L	0.892	0.0394	0.883	0.698–1.117	0.844
IIPd2L	0.861	0.0440	0.869	0.661–1.142	0.813
IIClwL	0.882	0.0494	1.067	0.833–1.367	1.007
IILat	0.959	0.0259	0.943	0.822–1.083	0.930
IIIPdGp	0.818	0.0661	1.150	0.832–1.590	1.053
IIIPd1L	0.977	0.0206	1.012	0.914–1.121	0.999
IIIPd2L	0.940	0.0339	1.022	0.863–1.210	0.991
IIIPd3L	0.947	0.0360	1.171	1.000–1.372	1.139
IIIClwL	0.929	0.0549	**1.558**	1.294–1.875	1.479
IIIMed	0.990	0.0160	**1.193**	1.116–1.275	1.177
IIILat	0.987	0.0174	**1.144**	1.060–1.235	1.131
IVPdGp	0.933	0.0366	1.046	0.874–1.252	1.015
IVPdL	0.976	0.0218	1.042	0.939–1.157	1.030
IVClwL	0.837	0.0714	1.320	0.976–1.785	1.208
IVMed	0.985	0.0181	**1.088**[a]	1.002–1.181	1.077
Proximal end II to proximal end III	0.972	0.0213	0.942	0.842–1.055	0.930
Proximal end III to proximal end IV	0.972	0.0216	0.949	0.848–1.063	0.937
Proximal end II to proximal end IV	0.986	0.0154	0.947	0.875–1.025	0.945
Clawbase II to clawbase III	0.972	0.0222	0.987	0.882–1.105	0.976
Clawbase III to clawbase IV	0.958	0.0229	**0.828**[a]	0.720–0.952	0.813
Clawbase II to clawbase IV	0.984	0.0152	**0.878**[a]	0.806–0.955	0.875
Toetip II to toetip III	0.971	0.0236	1.036	0.923–1.162	1.022
Toetip III to toetip IV	0.960	0.0267	0.983	0.858–1.127	0.968
Toetip II to toetip IV	0.979	0.0184	0.945	0.858–1.042	0.937

[a]Only barely allometric (upper CI limit of RMA slope between 0.95 and 1.00, or lower CI limit between 1.00 and 1.05).

hypex lengths, and measurements between toebases, clawbases, and toetips were also considered.

In a second set of analyses, I scaled footprint measurements in a slightly different manner. I first created a footprint scaling factor that was calculated as the mean of all the log-transformed measurements used in the analysis. This scaling factor was then subtracted from each of the log-transformed measurements. As done in the analyses described in the preceding paragraphs, two versions of this kind of analysis were done, a "simpler" version in which toe lengths were expressed as their total lengths, and a more "complex" version in which the lengths of digits II–IV were subdivided into their component digital pad and claw lengths. I also examined the correlations between the scaled footprint parameters and both the scaling factor and the log-transformed value of IIILGL (tables 6.20, 6.21).

As previously described, birds in the "emu means" data set were subdivided into two size/age groups, "immature" birds (IIILGL ≤17 cm) and "adults" (IIILGL >17 cm). The scaled footprint measurements were then used in a forward stepwise discriminant analysis. The analyses were done in such a way as to minimize the value of Wilks's λ, therefore creating discriminating functions that yielded the greatest separation of the two emu size classes. The success of the discriminating functions was evaluated both by using all of the data cases and by using a cross-validation (leave-one-out procedure) in which each case was classified by functions derived from all cases other than itself. Because the number of cases in the two size classes was unequal, I did discriminant analyses in which the prior probability that a case belonged to either of the size classes was the same, and also in which the prior probability was adjusted according to the number of cases in that size class.

In the RMA analyses (tables 6.17, 6.19), parameters associated with the lengths of digits III and IV (especially IIILgth, IIIClwL, IVLgth, IIIMed, IIILat, and IVMed) showed

Table 6.20. Significant bivariate correlations between "simple" scaled footprint parameters and footprint size (data from the "emu means" treatment). This analysis was "simple" in the sense that digit lengths were expressed as total lengths (IILgth, IIILgth, and IVLgth). Each parameter was log-transformed, and a footprint scaling factor was calculated as the mean of all the log-transformed footprint variables. This scaling factor was then subtracted from each log-transformed variable to create the variable's scaled value. Variables considered in this analysis were the same as those in table 6.16. Number of cases = 22.

Parameter	Parameter of footprint size	Pearson's *r*	Two-tailed *P*
MtPdL	Scaling factor	−0.575	.005
	Log IIILGL	−0.542	.009
IIILgth	Scaling factor	0.608	.003
	Log IIILGL	0.670	.001
IVLgth	Scaling factor	0.529	.011
	Log IIILGL	0.567	.006
IIIMed	Scaling factor	0.694	<.001
	Log IIILGL	0.715	<.001
IIILat	Scaling factor	0.698	<.001
	log IIILGL	0.714	<.001
Proximal end II to proximal end III	Log IIILGL	−0.424	.049
Proximal end II to proximal end IV	Scaling factor	−0.561	.007
	Log IIILGL	−0.612	.002
Toetip II to toetip IV	Log IIILGL	−0.444	.038

Table 6.21. Significant bivariate correlations between "complex" scaled footprint parameters and footprint size (data from the "emu means" treatment). This analysis was "complex" in the sense that digit lengths were subdivided into digital pad and claw lengths. Each parameter was log transformed, and a footprint scaling factor was calculated as the mean of all the log-transformed footprint variables. This scaling factor was then subtracted from each log-transformed variable to create the variable's scaled value. Variables considered in this analysis were the same as those in table 6.18. Number of cases = 13.

Parameter	Parameter of footprint size	Pearson's *r*	Two-tailed *P*
MtPdL	Scaling factor	−0.662	.014
	Log IIILGL	−0.630	.021
IIIClwL	Scaling factor	0.780	.002
	Log IIILGL	0.806	.001
IIIMed	Scaling factor	0.845	<.001
	Log IIILGL	0.856	<.001
IIILat	Scaling factor	0.728	.005
	Log IIILGL	0.736	.004
Clawbase III to clawbase IV	Scaling factor	−0.743	.004
	Log IIILGL	−0.733	.004
Clawbase II to clawbase IV	Scaling factor	−0.753	.003
	Log IIILGL	−0.780	.002

positive allometry with respect to overall footprint size, as expressed in a multivariate footprint scaling factor that excluded the parameter being examined, while parameters associated with footprint width (especially the widths across clawbases II and IV) usually showed negative allometry. Note that in both sets of RMA analyses the allometric coefficient derived from the PCA is generally similar to the slope of the corresponding RMA relationship.

Many of the scaled variables showed significant correlations with a scaling factor that did include the variable of interest in its calculation, or with log-transformed IIILGL (tables 6.20, 6.21). Because the scaling factor in these analyses included the variable of interest in its calculation, these analyses are somewhat suspect, in that there could be problems of autocorrelation. That said, parameters associated with the length of digit III (IIILgth, IIIMed, IIILat, and IIIClwL) once again suggested a relative increase in size with increasing footprint size (tables 6.20, 6.21; fig. 6.21), while the length of the metatarsal pad, and parameters associated with the transverse width of the footprint, suggested relative decreases in size with increasing footprint size.

Both versions of the discriminant analysis were remarkably effective in distinguishing between footprints of birds of the two size classes. In the "simpler" version of the discriminant analysis, scaled IIILgth was the primary variable that distinguished between immature and adult emus, being relatively longer in the adults (table 6.22). The standardized canonical discriminant function coefficient of scaled IILgth had the opposite sign from that of scaled IIILgth, consistent with the possible relative decrease of IILgth in its bivariate relationship with IIILgth (table A6.8), but its correlation with the discriminant function was trivially low. In the "complex" version of the discriminant analysis (table 6.23), the scaled lengths of the claws of digits III and IV had positive standardized canonical discriminant function coefficients, but the correlation of scaled digit IV claw length with the function was much lower than that of scaled digit III claw length. Scaled metatarsal pad width had a negative standardized canonical discriminant function coefficient, and its correlation with the discriminant function was even less than that of scaled digit IV claw length. Thus once again parameters associated with the relative size of digit III seem to be the main actors in distinguishing footprints of adult emus from those of immature birds.

These multivariate results are consistent with several of the "emu means" bivariate relationships previously examined (tables A6.8, A6.9). In particular, several measurements of print transverse dimensions showed negative allometry with IIIPd1L, IIILgth, and other parameters related to the size of digit III. This suggests that what causes parameters associated with digit III length to be relatively larger in adult birds, and to be good discriminators of adult from immature emus, in the multivariate analyses is the fact that they are good proxies

Table 6.22. Forward stepwise discriminant analysis of scaled log-transformed linear dimensions (the same parameters as in table 6.16) of emu footprints (using data from the "emu means" treatment"). Parameters were scaled by subtracting the mean of all the log-transformed variables from the value of each log-transformed variable. Birds were assigned to two groups, "immatures" (IIILGL ≤ 17 cm) and "adults" (IIILGL > 17 cm). Number of cases = 22. Only those variables that were included in the final discriminant function are shown, and they are entered in the table in the order in which they were entered in the forward stepwise procedure. Box's *M* test of the equality of group covariances matrices = 2.371 ($P = .550$). Function 1 Eigenvalue = 1.446; canonical correlation = 0.769; Wilks's λ = 0.409 ($P < .001$). Analyses were carried out both under the assumption that the prior probability that a case belonged to either size group (immatures or adults) was equal across groups, and with the prior probability of group membership adjusted according to the number of cases in each group; classification results were the same. The success of the discriminating function was evaluated both using all of the original data cases and using a cross-validation (leave-one-out procedure) in which each case was classified by functions derived from all cases other than itself. Regardless of whether or not the leave-one-out procedure was used, the discriminating function was 91.9% effective in assigning cases to their true size class. All immature birds were correctly classified as such, but two adult birds (1985 adult 5 and the Radandt adult male) were incorrectly classified as immatures.

Variable	Standardized canonical discriminant function coefficients (correlations between individual variables and discriminant function)
Digit III length (IIILgth)	1.165 (0.825)
Digit II length (IILgth)	−0.659 (−0.059)

Table 6.23. Forward stepwise discriminant analysis of scaled log-transformed linear dimensions (the same parameters as in table 6.18) of emu footprints (using data from the "emu means" treatment). Parameters were scaled by subtracting the mean of all the log-transformed variables from the value of each log-transformed variable. Birds were assigned to two groups, "immatures" (IIILGL ≤ 17 cm) and "adults" (IIILGL > 17 cm). Number of cases = 13. Only those variables that were included in the final discriminant function are shown, and they are entered in the table in the order in which they were entered in the forward stepwise procedure. Box's *M* test of the equality of group covariances matrices = 10.642 ($P = .321$). Function 1 Eigenvalue = 8.609; canonical correlation = 0.947; Wilks's λ = 0.104 ($P < .001$). Results were the same regardless of whether the prior probability that a case belonged to either size group (immatures or adults) was equal across groups, or whether the prior probability of group membership was adjusted according to the number of cases in each group. The success of the discriminating function was evaluated both using all of the original data cases and using a cross-validation (leave-one-out procedure) in which each case was classified by functions derived from all cases other than itself. Regardless of which prior probability procedure was used, and regardless of whether or not the leave-one-out procedure was used, the discriminating function was 100% effective in assigning cases to their true size class.

Variable	Standardized canonical discriminant function coefficients (correlations between individual variables and discriminant function)
Digit III claw length (IIIClwL)	1.266 (0.610)
Digit IV claw length (IVClwL)	0.926 (0.164)
Metatarsal pad width (MtPdW)	−0.816 (−0.092)

of footprint length that have an underlying contrast with the parameters associated with footprint width, when both are used to calculate the footprint scaling factors. That is, the relative narrowing of footprints with increasing footprint size causes the footprint scaling factor to take smaller values for bigger emus than would be the case if the footprint scaling factor were calculated without including footprint transverse dimensions. The multivariate results are therefore also consistent with the ontogenetic decrease in interdigital angles (table 6.15; fig. 6.20).

Olsen/Weems Plots. Olsen (1980) observed that, with increasing footprint size across the classic Newark Supergroup tridactyl ichogenera *Grallator*, *Anchisauripus*, and *Eubrontes*, the projection of digit III beyond the toetips of digits II and IV becomes relatively shorter in comparison with the length of the print proximal to the toetips of II and IV (here designated the "backfoot"). Although he later (Olsen 1995; Olsen et al. 1998) concluded that this could in part be an interspecific trend, in his 1980 paper Olsen speculated that this kind of pattern might also characterize ontogenetic changes in footprint shape in a single dinosaur species.

My emu data provide an opportunity to consider whether such an ontogenetic allometric change in footprint shape does in fact occur. As in Olsen's dinosaur footprint sample, the emu data indicate a relative decline in the projection of digit III beyond the other two digits with increasing footprint size (fig. 6.22*A*, 6.22*B*). The RMA slope of the relationship (log-transformed data; digit III projection the dependent variable) is 0.856 (95% CI = 0.802–0.920, $r^2 = 0.893$, $n = 392$), confirming negative allometry.

However, the projection of digit III beyond the other two digits is a composite parameter. It depends on the relative lengths of the three toes, but also on the interdigital angle between II and IV. For any specified values of digital lengths, the larger the interdigital angle, the greater the projection of digit III beyond II and IV will be.

The decrease in IDAs with increasing footprint size in emus markedly contrasts with Olsen's non-avian dinosaur footprints, in which IDAs increased with footprint length, from *Grallator* through *Anchisauripus* to *Eubrontes*. Consequently the similar declines in the digit III projection relative to the length of the backfoot in emus and dinosaurs may have very different causes.

Such considerations led Olsen et al. (1998) to modify Olsen's earlier (1980) argument by "correcting" backfoot length to eliminate the effects of varying interdigital angles (fig. 2.22*C*). Applying the "cosine correction" to my emu data dramatically changes the relationship between the digit III projection and backfoot length (fig. 6.22*D*, 6.22*E*): Again using log-transformed data, the RMA slope relating "corrected digit III projection" (dependent variable) to "corrected" backfoot length now becomes 1.088 (95% CI = 1.024–1.156, r^2 =

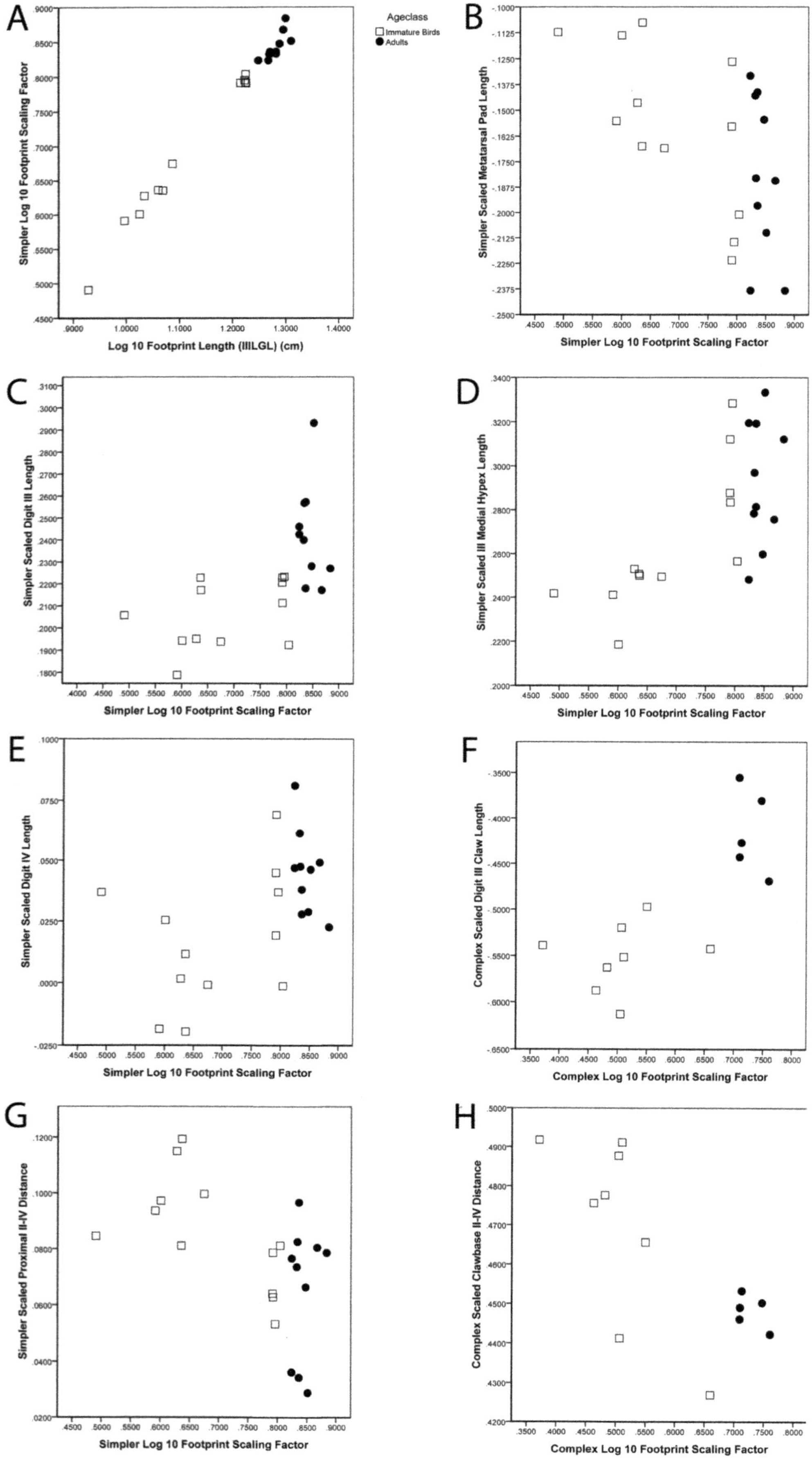

6.21. Emu footprint dimensions scaled against a multivariate proxy for overall footprint size. Using the "emu means" data set, for each bird both a "simpler" and a "complex" footprint scaling factor were created by calculating the mean of all the log-transformed footprint measurements. The simpler scaling factor was based on the footprint parameters in table 6.16, and the complex scaling factor was based on the footprint parameters in table 6.18. Each footprint parameter was then scaled by subtracting the footprint scaling factor from the log-transformed value of the parameter. The emus were placed into arbitrary size/age classes ("immature birds" vs. "adults" on the basis of whether footprint length was 17 cm or less or was greater than 17 cm. *A*, Plotting the simpler scaling factor against overall footprint size (log IIILGTH) shows that the scaling factor is indeed a good proxy for footprint size; the same is true for the complex scaling factor. *B–H*, Relationships between scaled footprint parameters and the appropriate scaling factor. The length of the metatarsal pad (*B*) and two measurements of footprint width (*G*, *H*) become relatively smaller with increasing footprint size, whereas three aspects of the length of digit III (*C*, *D*, *F*), and digit IV length (*E*) become relatively larger.

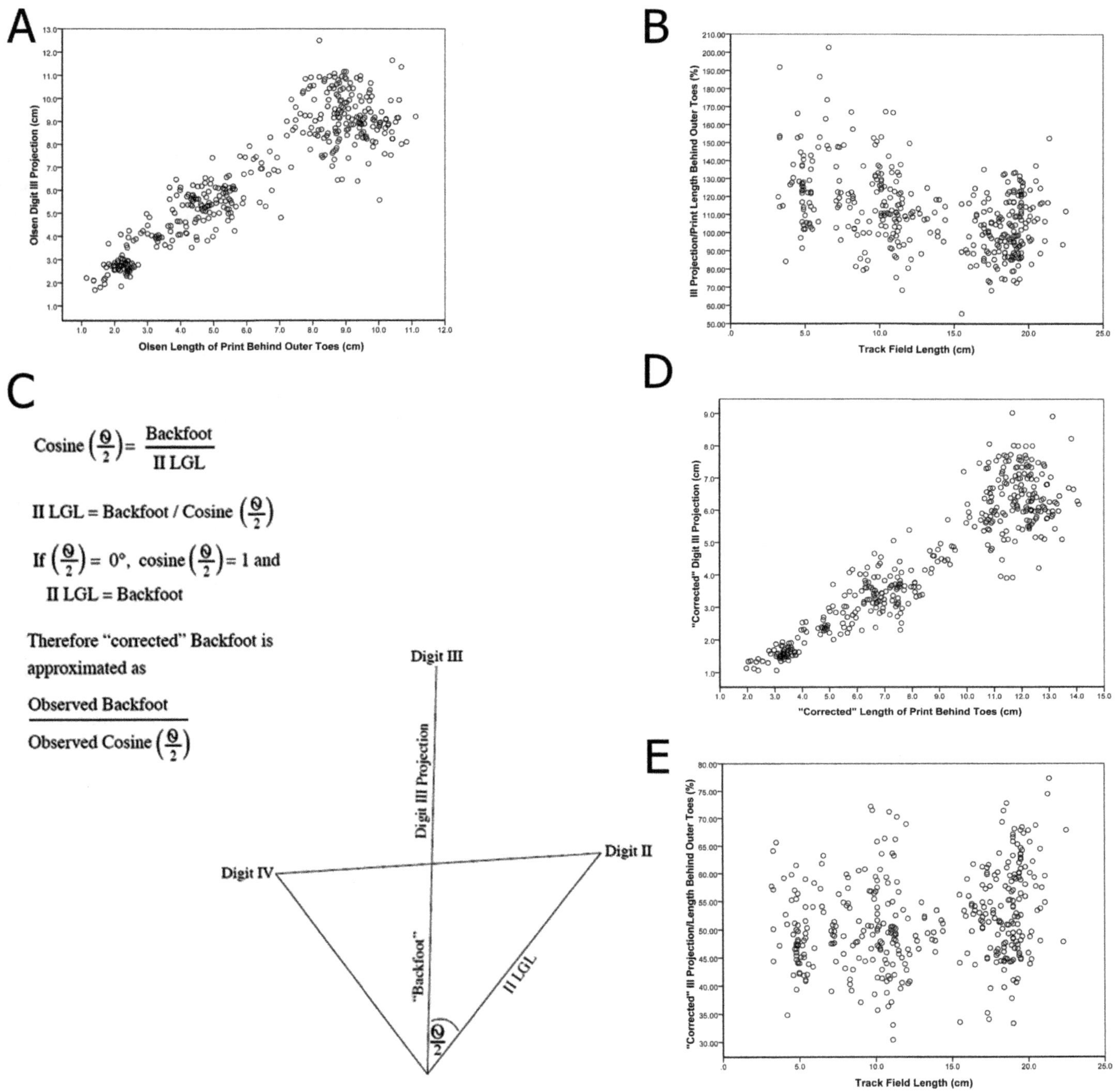

6.22. Length of the projection of digit III beyond the distal limits of digits II and IV in emu footprints as a function of the length of the footprint behind (proximal to) to toetips of II and IV in emus ("backfoot"). *A*, Just as Olsen (1980) observed for an array of tridactyl footprints attributed to theropods of increasing size from the Early Jurassic Newark Supergroup, with increasing footprint size, the digit III projection becomes relatively shorter. *B*, This is more clearly seen if the ratio of the digit III projection to backfoot length is plotted against total footprint length. *C*, However, Olsen et al. (1980) recognized that backfoot length depends in part on interdigital angle II–IV, so they introduced a "cosine correction" that approximates the value of backfoot length if this angle were 0 degrees, with digits II and IV parallel to digit III. To simplify the argument, the interdigital angle used here is the spread angle (θ) between digits II and IV, and these two toes are assumed to be roughly equal in length. It is further assumed that IIILGL bisects the spread angle into two equal angles, each of whose value is θ/2. One of these angles will be a right angle whose hypotenuse is IILGL. The cosine of (θ/2) will be equal to backfoot/IILGL. This in turn means that IILGL = backfoot/(cosine [θ/2]). If (θ/2) = 0 degrees, cosine (θ/2) = 1, and backfoot = IILGL. Thus one can determine what value backfoot would take if (θ/2) were 0 degrees by calculating the length of IILGL ("corrected" backfoot) from the actual values of backfoot and (θ/2) by dividing the observed value of backfoot by the observed value of (θ/2). *D*, *E*, If the cosine correction is applied to emu footprints, the degree to which digit III projects beyond the limits of digits II and IV no longer decreases with increasing footprint size. Consequently the relative decrease in the digit III projection with increasing footprint size in emu prints is due entirely to the ontogenetic decrease in interdigital angle II–IV.

0.914). If anything, this is slight positive allometry, rather than the negative allometry observed in Olsen's size series of dinosaur footprints. Unless non-avian dinosaur feet grew differently than emu feet, this suggests that the trend Olsen observed might be interspecific rather than intraspecific in nature. We will return to this topic in chapter 9.

The putative negative allometry of digit III projection vs. backfoot length is further compromised when doing the analysis with the bird means treatment rather than the individual footprint treatment. Using log-transformed data, the RMA slope of (uncorrected) digit III projection vs. (uncorrected) backfoot length (the latter the independent variable) is once again less than 1 (0.871), again suggesting negative allometry. However, the 95% CI is 0.669–1.132 ($r^2 = 0.894$, $n = 31$), and so we cannot reject a hypothesis of isometry.

Overall Intraspecific Variability in Footprint Shape

As with alligator feet (chapter 5), I calculated three sets of overall quantitative variability in footprint shape in our emu print sample: (1) the coefficient of relative dispersion about the reduced major axis (D_d; Imbrie 1956), using both raw and log-transformed data (tables A6.10, 6.11); (2) the standard error of the estimate (SEE) of regression equations involving log-transformed variables (tables A6.8, A6.9, 6.17, 6.19); and (3) maximum/minimum ratios of footprint parameters scaled to a common parameter (table A6.12). Separate calculations were done using all of the emu footprints in our sample, and also on a more restricted set of well-preserved footmarks. The latter included prints of captive adult birds with footprint preservation indices (table 6.2) of 3 or better, and also high-quality footprints of other captive and wild emus, both juveniles and adults, that were not used in the study of factors affecting footprint quality. Footprints of juvenile birds made in red clay were excluded from the high-quality sample.

For raw data, D_d values involving smaller subunits of footprints (metatarsal pad dimensions, metatarsal pad–digit pad gaps, digital pad lengths, and claw lengths) compared against the length of the first pad of digit II or III are generally around 15%–25% (table A6.10). Values based on log-transformed data are generally less than 10%. The coefficient of relative dispersion is a bit higher if parameters are compared with the length of the first pad of digit II than if compared with the length of the first pad of digit III. D_d is, unsurprisingly, a bit less when calculated for high-quality prints than for the entire emu footprint sample (most notably for comparisons involving claw lengths). Comparisons involving metatarsal pad–digital pad gap lengths and claw lengths are the most variable. Comparisons involving the smaller, more distal digital pads of a toe are more variable than those involving the larger, more proximal digital pads.

Comparisons in which parameters are compared with entire digits have D_d values of 15% or less (raw data; 5% or less with log-transformed data).

In the preceding discussion, individual footprints are the data cases, and particular birds are represented by varying numbers of prints. This may provide a realistic comparison for dinosaur footprint assemblages, in which the number of individual animals responsible for the prints is often unknown. However, at many tracksites it is possible to group most of the footprints into trackways, and it seems likely that most of the trackways will have been made by different animals (but this obviously cannot be known with certainty, unless there is considerable difference in footprint size among trackways).

Consequently it is also interesting to examine shape variability of emu prints according to the "emu means" treatment, in which data cases are means for prints made by the same bird (table A6.11). This treatment also provides numbers more directly comparable to the way alligator pedal shape was analyzed (chapter 5). Unsurprisingly, D_d values are usually somewhat less in the bird means treatment than when individual footprints constitute the data cases. Restricting the bird means treatment to high-quality prints generally (but not invariably) results in a slightly lower D_d than if the calculations are based on all of the footprints of each bird. Once again, comparisons involving claw lengths are the most variable.

SEE values of log-transformed bivariate relationships are unsurprisingly higher for the "all footprint" than for the emu means treatment (tables A6.8, A6.9). With log IIIPd1L as the independent variable, SEE takes values between 0.03 and 0.14 (all footprints treatment) or 0.02 and 0.11 (bird means treatment); with log IIILgth as the independent variable, SEE takes values between 0.03 and 0.14 (all footprints treatment) or 0.02 and 0.08 (bird means treatment) (table A6.8). High values are associated with claw lengths, pad gaps, and sometimes metatarsal pad lengths and widths and digital pad lengths; low values are associated with measurements across the proximal ends of digital impressions, clawbases, and toetips.

For miscellaneous bivariate relationships involving overall digit lengths and transverse widths, fleshy toe lengths vs. claw lengths, or comparing one claw length with another (table A6.9), "emu means" SEEs are again lower than "all footprint" SEEs, and the highest SEEs again are those involving claw lengths (values of 0.10 or more for the "all footprints" treatment, and 0.07 or more for the "emu means" treatment). Comparisons involving overall digit lengths, lengths of the fleshy portions of toes, hypex lengths, and transverse dimensions across prints take SEE values of 0.07 or less ("all footprints") or 0.04 or less ("emu means").

Multivariate SEE values (all computed with "emu means" treatment; tables 6.17, 6.19) again show higher values (0.04 or more) for pad gaps and claw lengths, and lower values (0.03 or less) for overall digit lengths, lengths of digits excluding claws, hypex lengths, and transverse dimensions. For digits II and III, SEE values increase distally from proximal pads to more distal pads to claws (table 6.19).

Maximum/minimum ratios of scaled footprint parameters (table A6.12) generally take values of 4 or less; if we restrict ourselves to high-quality footprints, most maximum/minimum ratios have values of about 2 or less. However, scaled claw lengths can be considerably more variable than that, with maximum/minimum ratios of 4–8 even with high-quality footprints. The lengths of gaps between the metatarsal pad and the most proximal ends of digits are also among the more variable scaled footprint parameters, although not to the same degree as claw lengths. Parameters scaled against a measure of overall digit length (e.g., IIIPdL) are a bit less variable than when scaled against a single digital pad (e.g., IIPd1L, IIIPd1L).

Maximum/minimum ratios of scaled parameters made on emu footprints can be compared with equivalent data for the feet of emu study skins (table A6.13). Some of the study skin birds did not subdivide the digital pads of digits II and III in the usual fashion, however, and this inflated the values of some maximum/minimum ratios, and made others impossible to compute. For "normal" emu feet, however, maximum/minimum ratios made on study skins are only a bit less variable than their counterparts from high-quality emu footprints (table A6.12). As with footprints, scaled claw lengths of study skin feet are the most variable parameters.

Comparisons With Other Species

Intraspecific Variability of Footprint and Foot Size. As already noted, given the small number of footprints and individuals of sample bird species other than emus in our collection, statements about intraspecific variability in print measurements can only be tentative. However, in those species for which an individual bird (or a group of birds of similar size) was represented by at least 10 footprints, variability (as measured by the coefficient of variation, CV) is comparable to what we observed for individual emus (table A6.14). Overall footprint lengths and toetip widths have CVs less than 10%. Individual toe lengths usually have CVs less than 15%, and often less than 10%. Best-fit IDAs are characterized by CVs that are generally less than 30%, and reference point–based IDAs have CVs less than 10%. Comparisons of relative variability of different parts of footprints are generally similar to what we described for emus.

Gand (1976) collected selected measurements from 93 footprints made by domestic chickens in a French farmyard. He provided no information on the number of individual birds represented in his footprint sample. Judging from the range of values for particular parameters, Gand's birds included no young chicks. Consequently his chicken data (table A6.15) are best compared with pooled data for adult emus, the latter with individual footprints constituting data cases (table A6.16). The CV values for the parameters Gand measured in his chicken footprints were less than 10%. Gand's chicken footprints were therefore somewhat less variable in parameters of footprint size than the pooled adult emu sample. Intriguingly, Gand's CV values were comparable to those for the single chicken hen whose prints we collected at the Fort Wayne Zoo (table A6.14), duplicating my results for emus, in which the variability of parameters of individual adults matched that of the entire adult emu sample (table 6.12). Of equal interest, Gand reported that a sample of 10 chicken footprints was sufficient to show the same variability as all 93 of his footprints combined.

Comparing adult emus with the mature alligators (table 5.3) is a little tricky. Emu feet and alligator feet differ considerably in overall pes shape, particularly in the number of digits. Furthermore, as noted in chapter 5, digital pads are hard to recognize in the alligator pes (except in digit I), making it impossible to measure some parameters the same way in alligators and emus. Thus the lengths of digits II–IV in the alligator study were free lengths, rather than the aggregate lengths of the digital pads and claws examined in emu footprints. Finally, the data cases in the alligator study were individual feet, with one foot measured per alligator, while the data cases for adult emus are individual footprints, with wildly varying numbers of footprints per bird.

Even so, it is instructive to examine intraspecific variability of those parameters of adult emu footprints that are as close to corresponding parameters in alligator feet as possible (table A6.17). Even if the parameters are not identical, they may be similar enough to make meaningful comparisons of intraspecific variability between the crocodylian and the ratite. Data are presented for a treatment in which individual footprints are the data cases, and also an "emu means" treatment. In addition, I analyzed intraspecific variability in the feet of study skins of adult or nearly adult emus (table A6.18), a treatment that, as in the alligator study, is based on one foot per bird.

Unsurprisingly, the "bird means" treatment usually shows less variability than the treatment in which footprints are the data cases (table A6.16), whether one considers coefficients of variation or maximum/minimum ratios. Variability of parameters in the "emu means" treatment is generally similar to that for feet of emu study skins (tables A6.17, A6.18).

Feet of mature alligators are about as variable as footprints of adult emus for some, but not all, parameters, if the individual footprints themselves are the data cases. There is an element of an "apples-and-oranges" situation in this comparison, however, because some of the variability in the emu footprint assemblage is extramorphological in origin, representing the exigencies of interaction between emu feet and substrates. Parameter variability would in this case be expected to be larger than if measurements were strictly morphological in character, having been made just on emu feet. Indeed, if the "emu means" footprint treatment and the emu study skin data are compared with the alligator data, the emu data are generally much less variable than the alligator numbers, except for claw lengths, in which emu prints and feet are about as variable as alligator feet (table A6.17).

How much weight should be placed on these results is uncertain. Like the alligator sample, the emu footprint and study skin samples are not representative samples from natural populations. I did not know the reproductive status of any of the emus I regarded as adults or near-adults. Consequently I cannot say whether the degree of variability in our emu footprint sample matches what would be observed in the wild.

However, it is at least plausible that sexually mature alligators would show a larger degree of intraspecific variability in pedal size parameters than emus do. Although—contrary to what was once thought—alligators, like emus, show determinate growth (Woodward et al. 2011; Wilkinson et al. 2016; cf. Lee et al. 2013), alligators reach sexual maturity well before growing to their final size, very unlike emus and nearly all other birds (Lee et al. 2013; Padian and Lamm 2013).

In both emus and alligators the variability of individual digital pad lengths (digit I of alligators and digits II and III of emus) is greater than that of entire digits. In both species claw lengths are likewise among the most variable pedal parameters.

Generalizing from my data on emu footprints and study skins, alligator feet, Gand's chicken footprint data, and my limited footprint data for ground birds other than chickens, we can make predictions about the variability of linear measurements of footprints made by conspecific, adult, tridactyl dinosaurs. Footprint lengths and digit lengths should have CVs from 10% (or less) to as much as 20%, and maximum/minimum parameter ratios up to about 1.5–2. Because non-avian dinosaurs, like alligators, reached sexual maturity well before, and at a smaller size than, somatic maturity (Horner et al. 1999, 2000; Sander 2000; Lee and Werning 2008; Lee et al. 2013), it is likely that the values would have been closer to the larger end of the range. Measurements of clawmarks might have CVs of 20%–30%, and maximum/minimum ratios around 2–3. Digital pad impressions would have CVs of 10%–20% and maximum/minimum ratios of 1.5–2.5 (judging from ground birds), or CVs of 20%–30% and maximum/minimum ratios of 2.5–4 (judging from alligators). However, my subjective impression, based on examination of numerous tridactyl dinosaur footprints, is that non-avian dinosaurs are more likely to have been like ground birds in digital pad development, and less haphazard in pad morphology than alligators.

It should be emphasized that the predictions of intraspecific variability in footprint dimensions of the preceding paragraph apply only to adult dinosaurs. Given the enormous size difference between hatchlings and adults of large-bodied dinosaur species (nearly 20-fold; cf. Carpenter 1999: Table 11.3), intraspecific variability in footprint measurements for dinosaur species can be expected to be considerably greater than the values in the preceding paragraph (cf. table 5.2 for alligators). On the other hand, if individuals of any dinosaur species had a tendency to segregate and associate on the basis of age class (cf. Varricchio et al. 2008; Myers and Fiorillo 2009), this would reduce intraspecific variability of footprint measurements at sites where the prints were made by groups of conspecific trackmakers.

Given all of these uncertainties, limits of intraspecific footprint size variability may be less interesting and useful than limits of intraspecific shape variability.

Intraspecific Variability of Footprint and Foot Shape. Values of the coefficient of relative dispersion about the reduced major axis (D_d) are a little higher for emu footprints (table A6.10) than for equivalent bivariate comparisons of alligator feet (table 5.16). This is not surprising, given that shape variability in the bird prints includes an extramorphological component, reflecting foot-sediment interactions, that will be absent in measurements of feet alone. The effects of this extramorphological component are reduced by examining only high-quality emu prints (table A6.10, values in parentheses), and/or by computing D_d values for means of footprint measurements in trackways, and not on individual prints (table A6.11), making alligator and emu coefficients more alike. As with emus, D_d values for alligators are higher for comparisons involving individual digital pads and claws than for comparisons involving entire digit lengths.

However, given the differences in foot shape between emus and alligators, and in the variables that can be measured in feet or footprints of the two species, many of the comparisons of the two species summarized thus far do not involve the same parameters. For example, the only digital pads on alligator feet that could be measured were on digit I, a toe entirely missing in emus.

If alligator and emu comparison are restricted to just those parameters that are the same, or nearly the same, in the two species (table A6.19), D_d values are usually (but not always) higher for emu footprints than alligator feet, if individual emu footprints of whatever quality are the data cases. The relationship between digit III medial and lateral hypex lengths is the most notable exception to the preceding statement; the much lower value for emus may possibly reflect the considerably greater projection of digit III beyond the interdigital webbing of emu feet than alligator feet. This presumably causes hypex lengths in emus to be affected more by the length of the digit than by the extent of digital webbing between toes, thereby introducing a major degree of autocorrelation in comparisons of hypex lengths involving the same digit. The same thing may also account for the fairly low values of D_d of the comparison of digit III length and its medial and lateral hypex lengths in emus.

Once again, the "emu means" and/or "high-quality" footprint treatments generally reduce D_d values for emu prints, and make comparisons with alligator feet more the same kind of thing. D_d values of untransformed data for both species are then usually 10% or less, except for comparisons involving claw lengths or (in alligators) hypex lengths.

Comparing the parameters that Gand (1976) measured on his chicken footprints (digital lengths and footprint width; table A6.15) gives D_d values of about 7%–9%. Emu footprints (table A6.10) are rather more variable (10%–15%) in comparisons of digit lengths than chicken prints are, if the entire sample of emu prints (of whatever quality) is the basis for the comparison. If we restrict the comparison to high-quality emu prints, the difference between the two bird species is lessened, but not eliminated. Comparisons of footprint width with digit III length between emus and chickens yield very similar D_d values.

Maximum/minimum ratios of scaled foot and footprint parameters (tables 5.17, A6.12, A6.13, A6.20) show similar patterns in alligator feet, the feet of emu study skins, and emu footprints. Scaled digital pad and claw lengths are more variable (maximum/minimum ratios commonly 2 or more) than scaled overall digit lengths (maximum/minimum ratios generally less than 2). Some of the variability of scaled pad lengths in alligator feet and emu study skins is a function of the way pad lengths were measured in study skins; if creases did not subdivide what would ordinarily be two pads, the entire length of the combined pad was assigned to the more proximal pad, and the more distal pad was assigned a value of zero. However, this procedure was not followed with emu footprints, which nonetheless are more variable in maximum/minimum ratios of individual digital pads than of overall digit lengths.

Scaled foot and footprint parameters for domestic chickens, greater rheas, and two species of bustard (tables A6.15, A6.21–A6.23) have values consistent with our observations on emus and alligators. Most scaled parameters have values of 2 or less, with higher values associated with claws or unusual configurations of digital pads.

Summing up, analyses of the intraspecific shape variability of alligator feet and the feet or footprints of five ground bird species (emus, greater rheas, chickens, kori bustards, and great bustards) yield results that are consistent enough to suggest that they may be applicable to the footprints of tridactyl non-avian dinosaurs. Values of the coefficient of relative dispersion about the reduced major axis should be 5%–25% (for raw data; the values will be less for log-transformed data) for footprints made by the same trackmaker species, and maximum/minimum ratios of scaled parameter lengths should usually be 5 or less. Values at the higher end of the range will be associated with measurements of claws and/or digital pads, and lower values with measurements of overall toe lengths.

We might also expect values at the higher end of the variability range where footprints were made under a wide range of substrate conditions, resulting in both good and poor print preservation, as in our "all emu prints" samples. Footprints preserved under the best of conditions, with limited extramorphological distortion of impression shape, will likely have D_d values and maximum/minimum scaled parameter ratios toward the lower ends of the ranges summarized in the preceding paragraph, as in our "high-quality" emu prints.

Trackway Parameters

Step Lengths. Larger emus obviously are able to take longer steps than smaller emus (fig. 6.23A). Captive birds in our zoo runs often appeared to be agitated at least part of the time, and this is reflected in the distribution of step lengths of captive as opposed to wild birds. Figure 6.23A plots pace length as a function of footprint length; note that for larger birds (footprint length at least 15 cm) there were many more paces of 75 cm or more for captive birds than were measured in trackways of wild Australian emus. If we restrict ourselves to such larger birds, we can compare the frequency distribution of step lengths of captive as opposed to wild birds with a larger number of pace measurements (fig. 6.23*B*, 6.23*C*): Trackways of both captive and wild large emus have a modal pace length of ca. 60–65 cm, suggesting that this represents the normal step length of a reasonably content big bird. Once again the greater number of long steps of captive than of wild large birds is apparent.

A curious observation that I made on captive emus serves as a note of caution about interpreting the absolute speeds of bipedal dinosaurs from trackway measurements of footprint length and stride length. On more than one occasion I saw a walking emu stop in midstride, with both feet on the ground

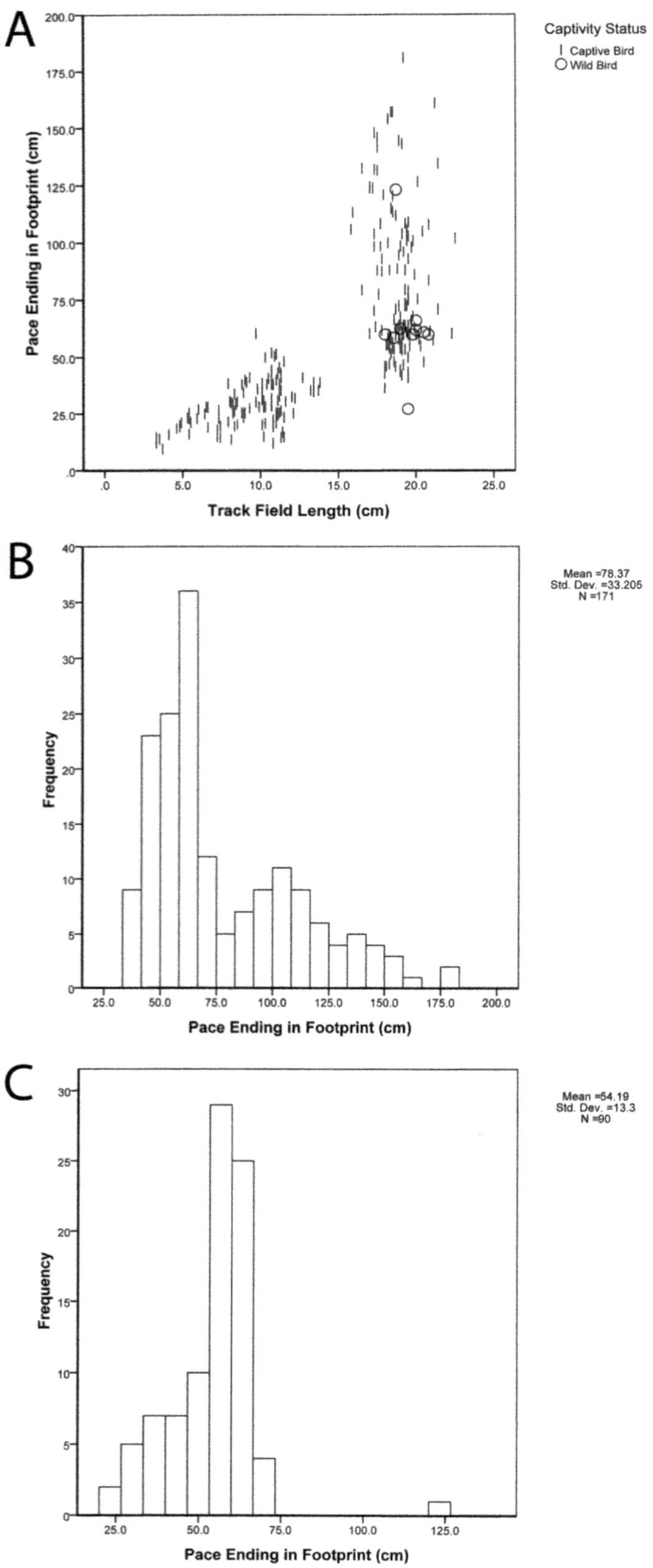

Left, **6.23.** Emu pace lengths. Cases represent individual measurements of paces and footprint lengths, so some birds are represented by many footprints. *A*, Pace length as a function of the length of the footprint in which it ends (a proxy for bird size). Large emus obviously take longer steps than smaller, younger birds. *B*, *C*, Frequency distribution for larger emus (size class 3; footprint length is 15 cm or less). These plots represent many more measurements of pace lengths than in *A*, because the lengths of all the individual footprints were not always measured for birds of known size. *B*, Captive birds. *C*, Wild emus in Australia.

Below, **6.24.** Pace angulation as a function of stride length in *A*, smaller (footprint length less than about 15 cm), and *B*, larger (footprint length at least 15 cm) emus. The smaller emus were all zoo birds; the larger emus included both captive and wild birds. In birds of all sizes, pace angulation increased rapidly with stride length until successive left and right footprints were aligned (pace angulation = 180 degrees), after which there was no change in pace angulation with increasing stride. In some trackway segments made by smaller emus, the pace angulation was greater than 180 degrees owing to a sudden change in direction on the part of the bird, in which a footprint crossed the trackway midline.

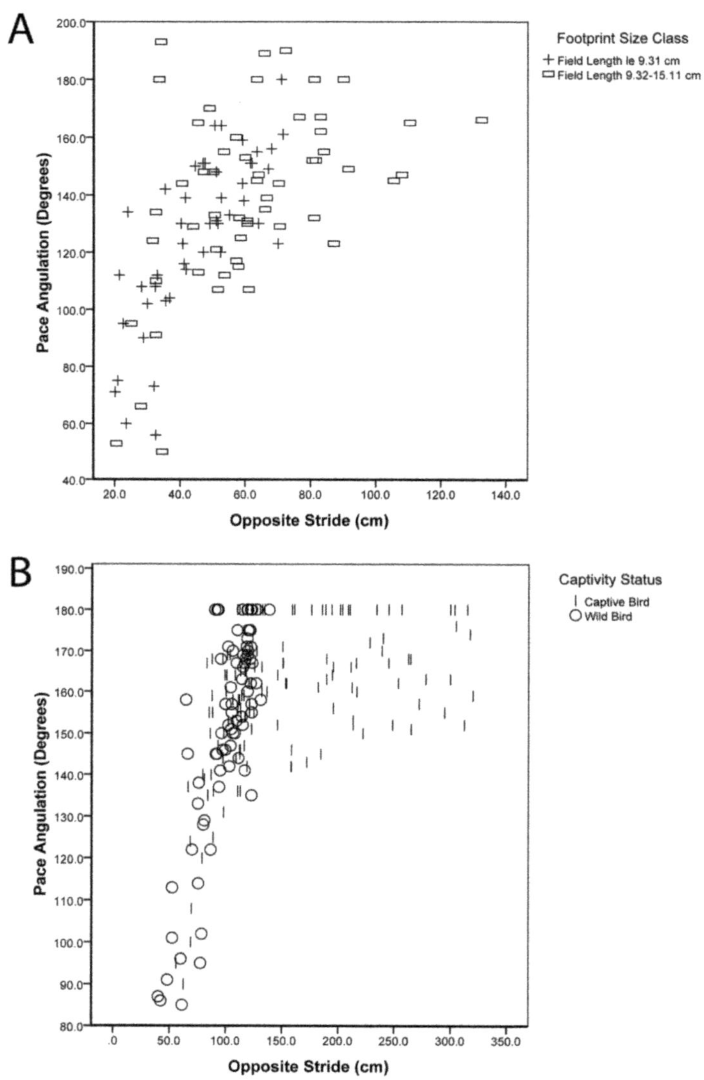

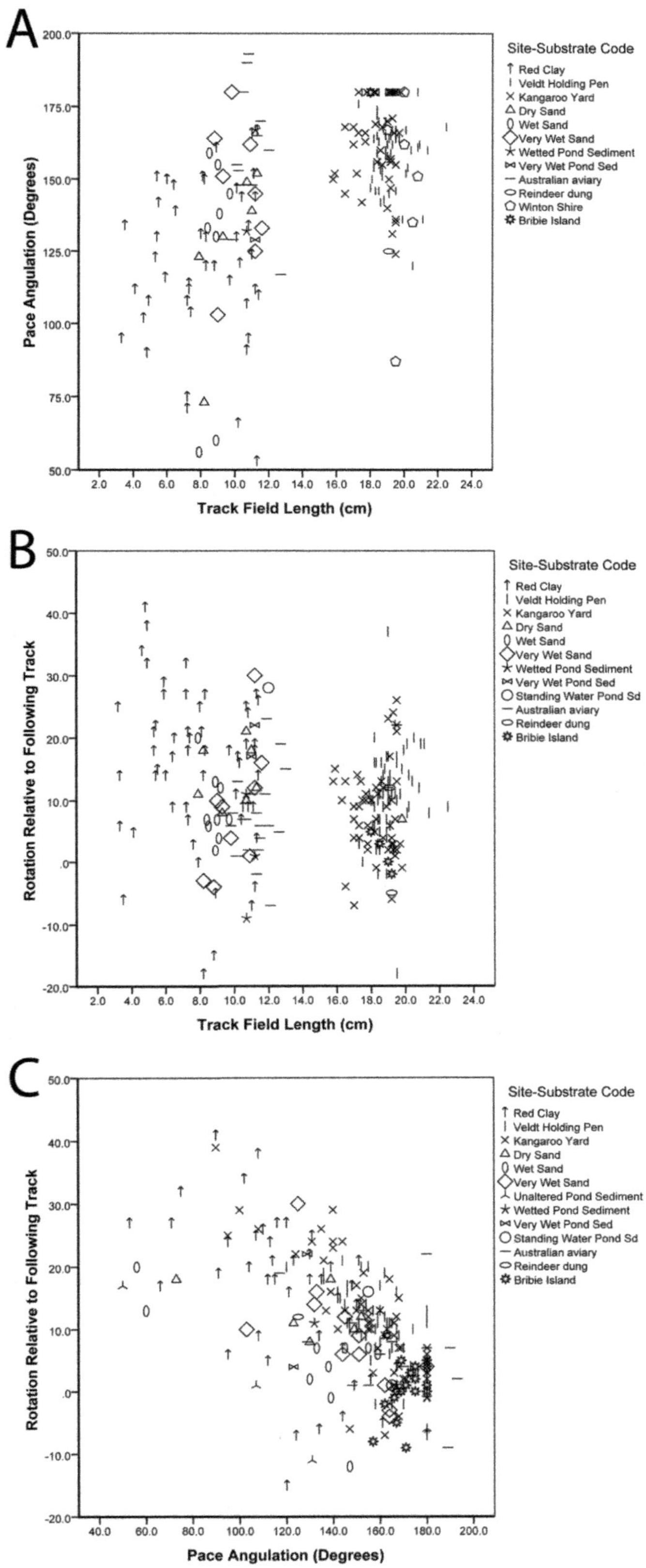

6.25. Interactions among emu size (as indicated by track field length), pace angulation, and footprint rotation. *A*, Pace angulation vs. track field length. Large emus tend to show larger values than those for small emus, but there is much scatter in the relationship, particularly for small emus. *B*, Footprint rotation vs. track field length. Small emus tend to show greater outward (more positive) footprint rotation than that for large emus, but again there is much scatter in the relationship. *C*, Footprint rotation vs. pace angulation. With increasing pace angulation, footprints go from being turned strongly outward (highly positive values) to being directed nearly straight ahead (rotation = 0) to even being turned inward (negative values). Among small emus there is an unexplained tendency for prints made in wet sand to have lower values of footprint rotation for a given value of pace angulation.

in a typical pace, and hold that pose for several minutes. The bird then resumed walking. One would never have known that it had stopped from the trackway pattern. Falk et al. (2017) observed the same trackmaking behavior in chickens, and named it "start-stop walking."

Pace Angulation. In both small and large emus, pace angulation rapidly increased with increasing stride until footprints were essentially in a straight line (pace angulation 180 degrees or close to it), after which there was no change in pace angulation with increasing stride (fig. 6.24). For large emus (footprint length at least 15 cm), captive and wild birds showed no difference in the relationship, except that, as already described, more of our captive birds took long steps. My data suggest that trackways of large emus may tend to have slightly higher pace angulations than those of smaller emus (fig. 6.25A; pace angulation vs. track field length: $r = 0.560$, $P < .001$, $n = 192$, for single measurements of pace angulations, rather than averages for individual birds). Because many of our footprints of young emus were made on slippery red clay, the relationship between pace angulation and emu size could be an artifact of substrate conditions. However, low pace angulations of young birds were also observed in tracks made in sand (fig. 6.25A). In general, there was a broader spread of values of pace angulations for juvenile as opposed to adult emus.

Footprint Rotation. Most of the various ways of measuring footprint rotation showed at least slightly positive correlations with each other (table 6.24). The exception was rotation measured with respect to the preceding pace, which showed a significant correlation only with rotation measured with respect to the trackway midline. All of the measures of footprint rotation except rotation relative to the preceding pace were negatively correlated (but not always significantly so) with both track field length and pace angulation (table 6.24; fig. 6.25*B*, 6.25*C*). Small emus tended to have greater (more positive) outward rotation than large emus, although there was considerable scatter of values for birds of all sizes. Substrate did not obviously affect these relationships, except that there was an unexplained tendency for footprints made

Table 6.24. Correlations between different measurements of rotation of emu footprints, and between measurements of footprint rotation, and footprint size, and pace angulation. In each cell, the numbers are Pearson's *r*, two-tailed *P*, and the number of footprints. ROTPREV = rotation with respect to previous footprint; ROTFOLL = rotation with respect to following footprint; MIDLINE = rotation with respect to trackway midline; PROJMID = rotation with respect to projected midline; PACEFOLL = rotation with respect to following pace; PACEPREV = rotation with respect to preceding pace.

	ROTPREV	ROTFOLL	MIDLINE	PROJMID	PACEFOLL	PACEPREV
ROTPREV		0.357; <.001; 241	0.493; .005, 31	0.819; .001; 12	0.601; <.001; 31	−0.043; .779; 45
ROTFOLL			0.855; <.001; 32	0.742; .009; 11	0.871; <.001; 45	0.207; .255; 32
MIDLINE				———[a]	0.881; <.001; 32	0.454; .009; 32
PROJMID					0.811; .002; 11	−0.151; .640; 12
PACEFOLL						0.047; .797; 32
Track field length	−0.203; .002; 236	−0.198; .002; 238	−0.206; .266; 31	−0.297; .168; 23	−0.502; <.001; 44	0.251; .096; 45
Pace angulation	−0.222; .001; 242	−0.582; <.001; 243	−0.594; <.001; 31	———[a]	−0.890; <.001; 31	0.402; .025; 31

[a]Could not be calculated because—by the definition of the projected midline—there were no footprints for which the trackway midline could be calculated.

Table 6.25. Comparison of footprint rotation (degrees) of wild and captive large emus (footprint length at least 15 cm).

Bird	Rotation measured relative to following track		Number of footprints
	Mean (median)	Range	
Captive birds			
1988 adult 3	17.0 (15.5)	7–39	34
1988 adult 5	10.9 (11.0)	−18–37	58
1988 adult 6	6.2 (4.0)	−6–22	40
1989 adult B	7	———	1
1990 juvenile 12	7.0 (10.0)	−7–15	9
00-0021-26ED	3.5 (3.5)	−5–12	2
D59B1	12	———	1
1E95BD	22.0 (22.0)	22–22	2
All captives	10.8 (11.0)	−18–39	156
Wild birds			
Bribie Island 33	0	———	1
Bribie Island 34	1	———	1
Bribie Island 36	4.5 (4.5)	4–5	2
Bribie Island 37	2.9 (3.0)	−2–9	7
Bribie Island 38	0.1 (1.0)	−8–4	13
All wild birds	1.3 (1.0)	−8–9	24

in wet sand to have lower values of footprint rotation for a specified value of the pace angulation (fig. 6.25C). Among adult or nearly adult emus, captive emus showed greater outward (positive) footprint rotation than wild birds (table 6.25).

Interspecific Comparisons of Footprint Shape: Can We Tell Them Apart?

We have now explored, in what has probably felt like excruciating detail, variability in size and shape of footprints made by a single species of trackmaker, the emu. We now ask how well we can distinguish emu prints from those of other large ground birds, and how well footprint shape serves as a proxy for the phylogenetic relationships of ground birds.

The higher level phylogenetic relationships of modern birds (Neornithes) are a contentious issue (cf. Cooper et al. 2001; Edwards et al. 2002; García-Moreno et al. 2003; Mayr and Clarke 2003; Zelenitsky and Modesto 2003; Chubb 2004; Cracraft et al. 2004; Dyke and Van Tuinen 2004; Fain and Houde 2004; Harrison et al. 2004; Poe and Chubb 2004; Kaiser 2007; Livezey and Zusi 2007; Slack et al. 2007; Hackett et al. 2008; Harshman et al. 2008; Mayr 2008, 2012, 2017; Morgan-Richards et al. 2008; Phillips et al. 2010; Allentoft et al. 2012; Haddrath and Baker 2012; Naish 2012; Worthy and Scofield 2012; Baker et al. 2014; Mitchell et al. 2014; Prum et al. 2015; Nesbitt and Clarke 2016; Angst and Buffetaut 2017; Grealy et al. 2017; Yonezawa et al. 2017; and many other studies cited in the preceding publications). Figure 6.26 is a generalized diagram of hypothesized phylogenetic relationships among the birds whose footprints were examined in the present study, primarily based on the topology recognized in the Tree of Life project (see caption to fig. 6.26 for details).

The sample of birds whose footprints were collected (table 6.3; figs. 6.27–6.33) is but a small sample of the diversity of extant or recently extinct birds. It is already a rather selective list, in that for the most part it consists of species with feet that are as similar to those of tridactyl, bipedal, non-avian dinosaurs as possible. For most of the bird species we had data for only one trackway, and in some cases, owing to the exigencies of how footprint-making runs went, or what prints were available as fossils, only a single footprint. Consequently our conclusions can only be tentative.

It quickly became apparent, as we collected prints of birds—particularly small birds—that the footprints we collected often did not record details of the plantar surface of the foot (e.g., lengths of digital pads and interpad spaces). My interspecific analyses of footprint shape consequently did not examine such details of digital subdivisions, but only considered overall digit lengths.

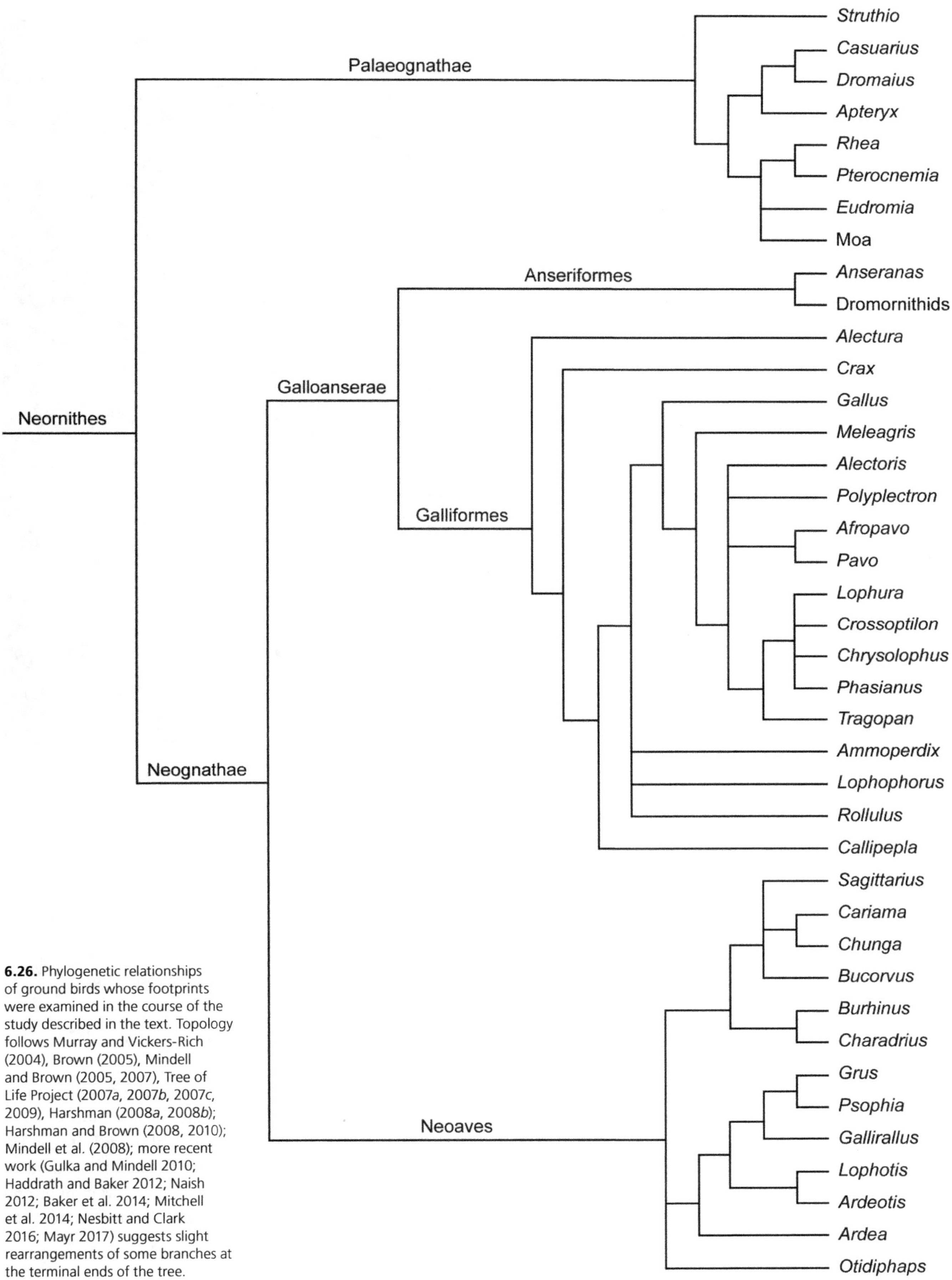

6.26. Phylogenetic relationships of ground birds whose footprints were examined in the course of the study described in the text. Topology follows Murray and Vickers-Rich (2004), Brown (2005), Mindell and Brown (2005, 2007), Tree of Life Project (2007*a*, 2007*b*, 2007*c*, 2009), Harshman (2008*a*, 2008*b*); Harshman and Brown (2008, 2010); Mindell et al. (2008); more recent work (Gulka and Mindell 2010; Haddrath and Baker 2012; Naish 2012; Baker et al. 2014; Mitchell et al. 2014; Nesbitt and Clark 2016; Mayr 2017) suggests slight rearrangements of some branches at the terminal ends of the tree.

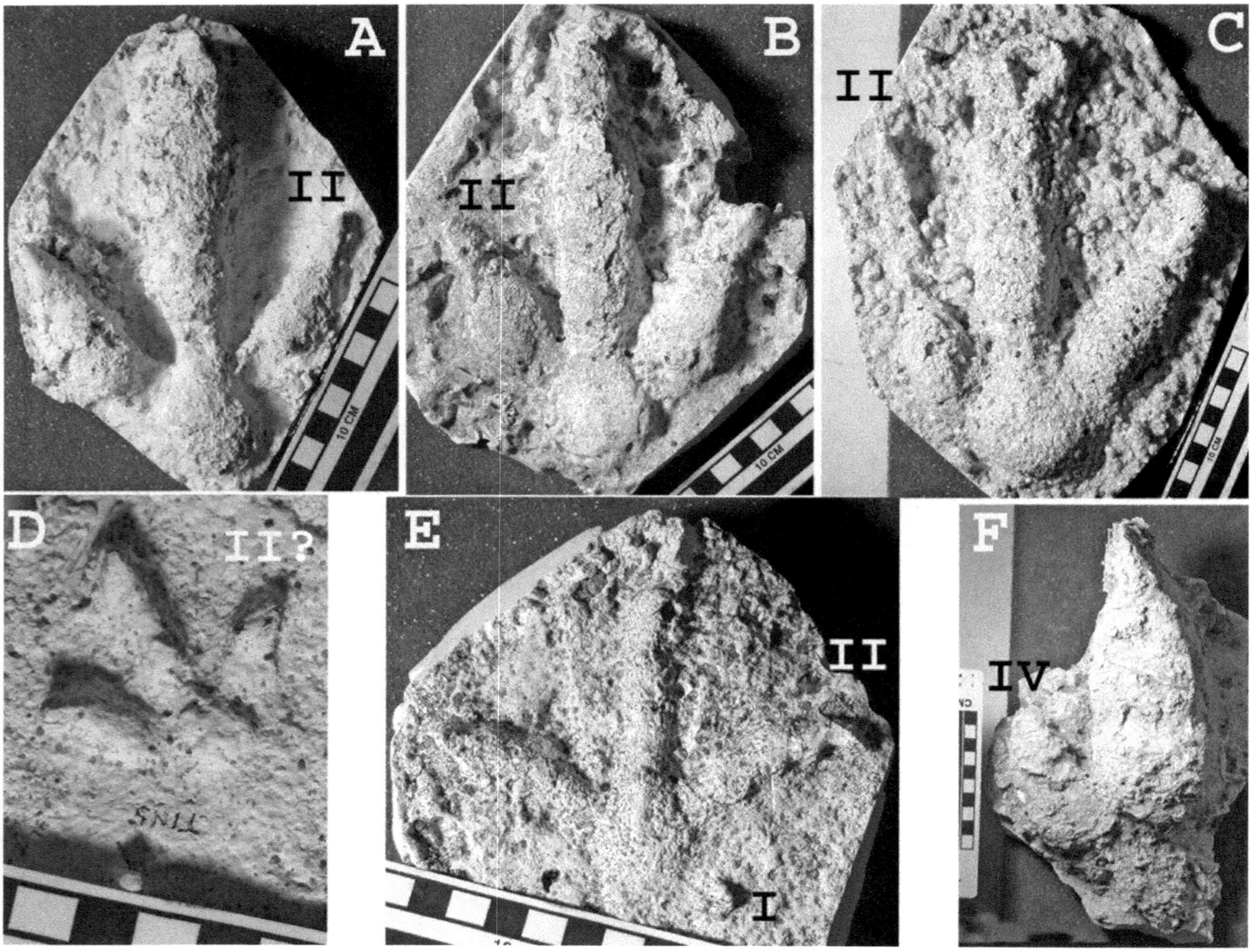

6.27. Footprints of ratites and tinamous. In this and the following series of photographs of footprints, scale bars are marked off in centimeters. All of the illustrated specimens are casts (negative copies) of the actual prints, so left-right symmetry is reversed. Impression of digit II (or IV) is labeled to assist interpretation. *A*, Right print of female greater rhea (*Rhea americana*), Cleveland Zoo. *B*, Left print of male lesser rhea (*Pterocnemia pennata*), Oklahoma City Zoo. *C*, Left print of male southern cassowary (*Casuarius casuarius*), San Antonio Zoo. *D*, Right (?) print of elegant crested tinamou (*Eudromia elegans*), Evansville Zoo. *E*, Right print of southern brown kiwi (*Apteryx australis*), San Diego Zoo. Impression of digit I is labeled. *F*, Right footprint of male ostrich (*Struthio camelus*), Fort Wayne Zoo.

Facing, **6.28.** Casts of footprints of moa (cf. Tennyson 1998; Worthy and Holdaway 2002; Lockley et al. 2007*a*). Left-right symmetry of these prints is uncertain; my own best guess of which toe impression was made by digit II is labeled. On some of the casts an informal number used to identify the cast in our lab is written on the cast, but this is not the original cast's institutional catalog number, which itself was modified from what is written on the casts in the photographs by the museum after the cast was labeled. *A*, NMNZ S.024314. *B*, NMNZ S.000462. *C*, NMNZ TMP.0101493 (formerly 55-1296). *D*, NMNZ S.024309. *E*, NMNZ S.000986. *F*, NMNZ S.024311. *G*, NMNZ S.024310, two footprints on the same slab (not necessarily made by the same bird).

Even so, there were interesting differences in features of the plantar surfaces of the digits across species that are of considerable interest for this study. In emus (fig. 6.3) the digital pads on the underside of a toe abut against their neighbors, with little space separating them. In contrast, in domestic chickens (fig. 6.34) there can be interpad spaces separating the digital pads that can be as long as the digital pads themselves. These differences in plantar surface form led me to examine the plantar surfaces of the feet of both study skin and alcohol-preserved bird specimens of several different species (figs. 6.35–6.40, plus many specimens not illustrated).

In addition to the bird species whose footprints, study skins, and spirit-preserved feet we were able to examine, I also considered data on the relative lengths of digits I, II, and IV, as percentages of the length of digit III, from two extensive collections of species accounts: HANZAB (Marchant and Higgins 1990*a*, 1990*b*, 1993; Higgins and Davies 1996), and BWP (Cramp 1977, 1980, 1983).

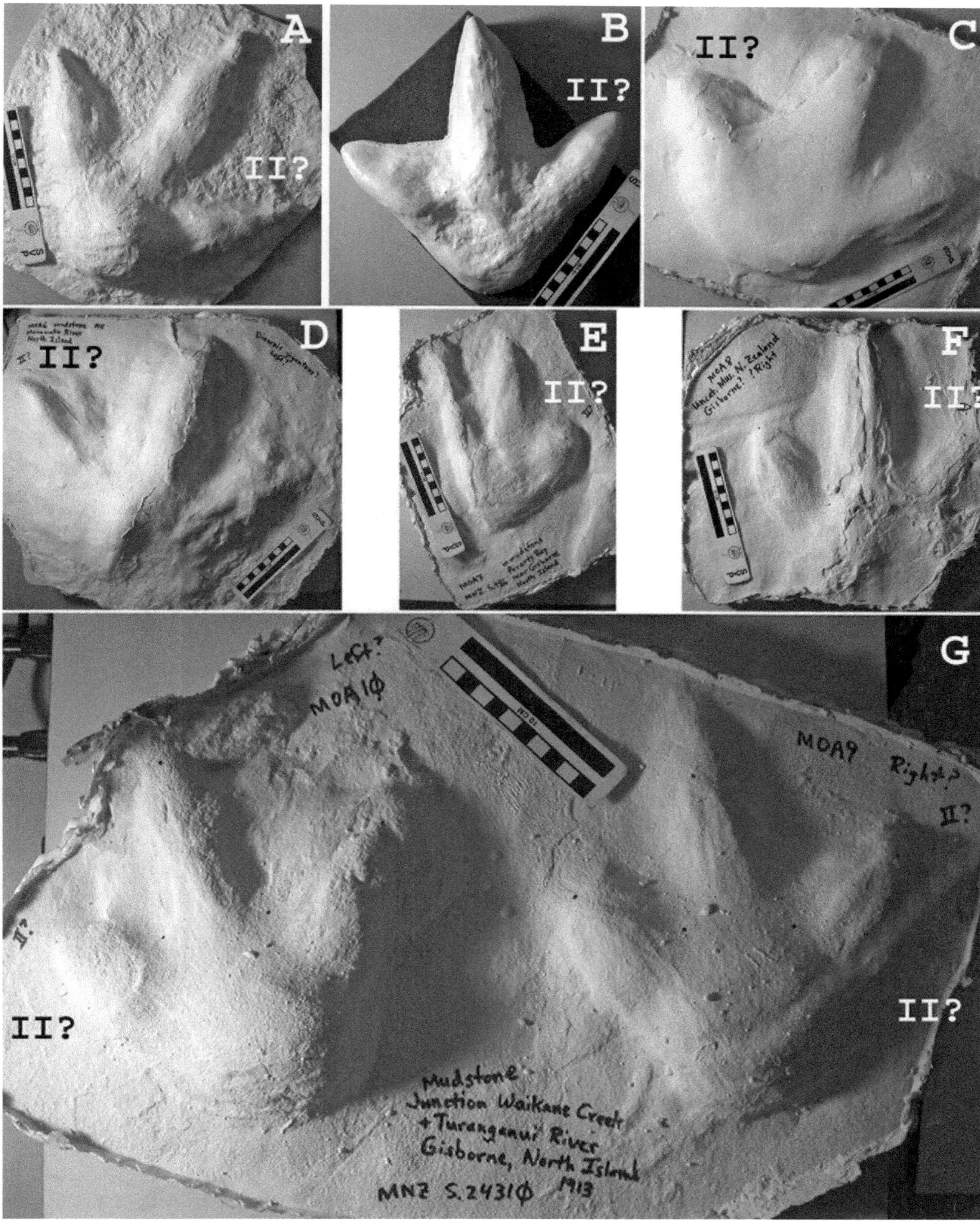

A
II?
B
II?
C
II?
D
II?
E
II?
F
II?
MOA8
Uncat. Mus. N. Zealand
Gisborne? Right
G
Left?
MOA10
MOA9
Right?
II?
II?
II?
Mudstone
Junction Waikane Creek
+ Turanganui River
Gisborne, North Island
MNZ S.24310
1913

6.29. Casts of footprints of dromornithids (mihirungs; cf. Rich and Green 1974; Rich and Gill 1976; Rich 1979; Rich and Molnar 1996; Murray and Vickers-Rich 2004) from Tasmania; the original specimens are held in the collection of the Queen Victoria Museum and Art Gallery (Launceston, Tasmania). Symmetry of footprints is uncertain, but my own best guess is indicated by labeling the impression of digit II. *A*, QVM:1991:GFV:35. *B*, QVM:1972:GFV:8. *C*, QVM:1972:GFV:9. *D*, QVM:1972:GFV:5. *E*, QVM:1972:GFV:4.

Comparison of Interspecific vs. Intraspecific Variability of Footprint Dimensions. I first considered whether the kinds of linear measures of footprint shape used in this study are even capable of recognizing when footprints of more than one bird species are included in a track sample. The analyses were done in three ways, using mean values of footprint parameters for each individual bird—i.e., the bird means treatment.

First, I examined bivariate relationships within and across bird species, employing parameters of variability (D_d and SEE) previously used to express intraspecific variability in emu footprint shape. Two versions of each analysis were done. In one of these, each data case was an individual bird, and so there were far more emus in the sample than any other bird species. Footprints of ostriches, moa, and mihirungs were included, if the bivariate comparison of interest could be made. In the second approach, only one individual per species was used in the analysis (henceforth referred to as the "one per species" treatment), and footprints of ostriches, moa, and mihirungs were excluded. My third analysis was more "quick and dirty" than the bivariate comparisons. All of the footprint size parameters were first scaled to a common value of footprint length (IIILGL = 15 cm). I then compared maximum and minimum values for a selected set of scaled footprint parameters for emus, some of the other categories of birds for which footprints were collected, and across all of our birds.

In all bivariate comparisons (table A6.24), the two versions of interspecific analysis indicate greater variability than the corresponding intraspecific analysis for emus, so it is possible to detect interspecific variability using the kinds of linear measurements of footprint shape employed in this study. In the simple comparison of maximum and minimum values of scaled footprint parameters (table A6.25), the range of values for all bird footprints combined exceeds that for emus alone in nearly all comparisons (cf. chapter 2 for skeletal proportion measures). Furthermore, despite the smaller number of footprints for categories of birds other than emus in our sample, several of the scaled parameters of birds other than emus are either less than the minimum scaled values for emus, greater than the maximum scaled values for emus, or both.

As encouraging as these results are, it is probably a mistake to make too much of them. They do not guarantee that such comparisons will always be able to distinguish interspecific vs. intraspecific variability in all possible combinations of particular bird species and numbers of individual footprints per bird species. Note, for example, that the scaled values of several footprint parameters of rheas and bustards in our sample fall within the range of scaled values for emus.

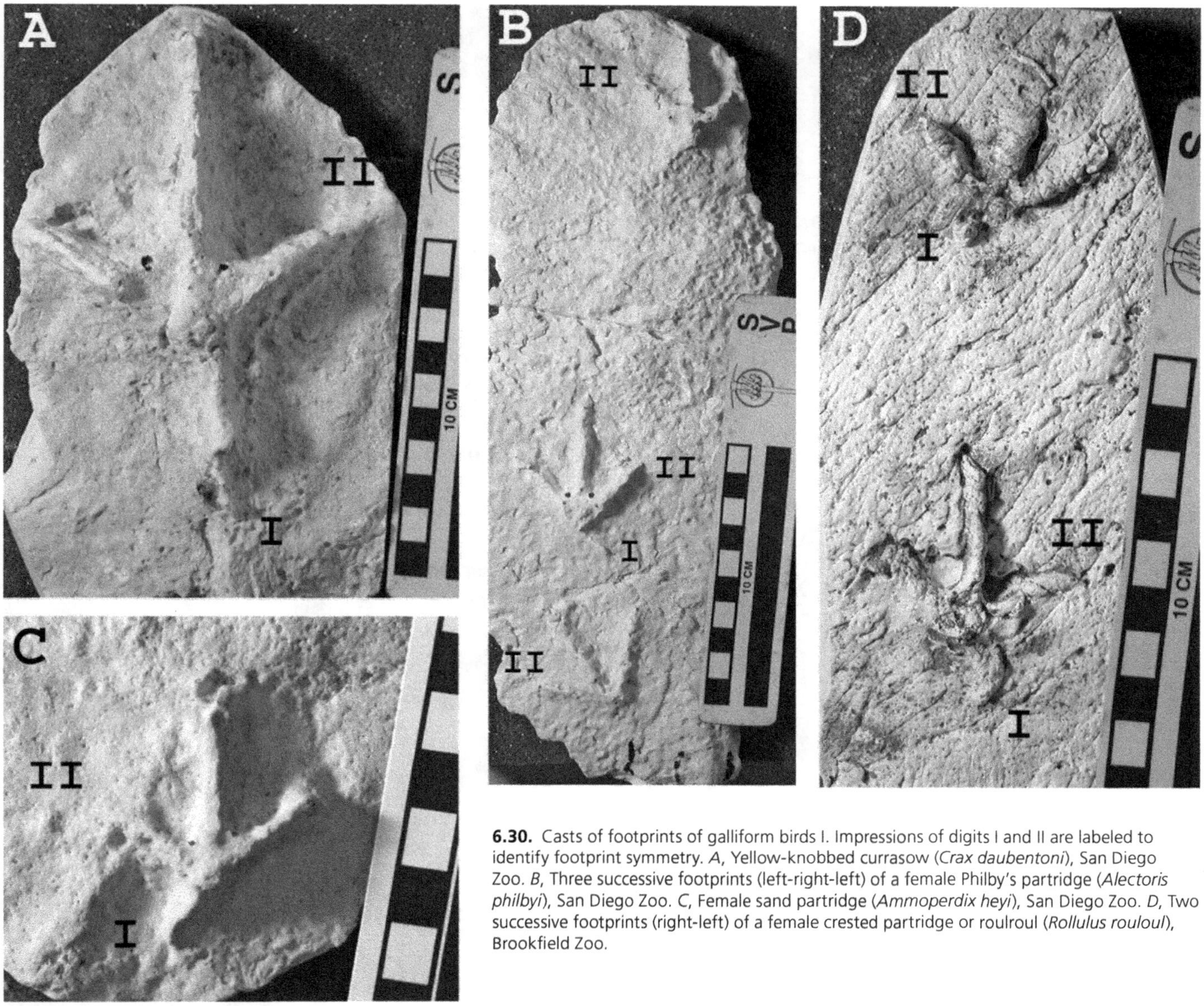

6.30. Casts of footprints of galliform birds I. Impressions of digits I and II are labeled to identify footprint symmetry. *A*, Yellow-knobbed currasow (*Crax daubentoni*), San Diego Zoo. *B*, Three successive footprints (left-right-left) of a female Philby's partridge (*Alectoris philbyi*), San Diego Zoo. *C*, Female sand partridge (*Ammoperdix heyi*), San Diego Zoo. *D*, Two successive footprints (right-left) of a female crested partridge or roulroul (*Rollulus rouloul*), Brookfield Zoo.

On the other hand, if in a tridactyl dinosaur footprint assemblage all measures of footprint shape variability were larger than their counterparts in the emu footprints described in this study, and if it was reasonably certain that this greater variability was not extramorphological in origin, then one could plausibly hypothesize that more than one species of trackmaker was responsible for the footprint assemblage.

To What Extent Does "Gross" Footprint Shape Serve as a Proxy for Phylogenetic Relationships? Because we were generally unable to capture the details of digital pad and other "fine" features in our casts of footprints of smaller ground birds, my examination of footprint shapes of different species and groups of birds was necessarily limited to the "gross" parameters of footprint shape considered in my study of emu footprints. These gross features include the length and width of the metatarsal pad; the lengths of the pad gaps separating the metatarsal pads from the proximal ends of digit impressions; overall lengths of the impressions of digits II–IV; digit hypex lengths; distances between the proximal ends of digits II, III, and IV; and distances between the tips of digits II, III, and IV. We also considered the digit I span, but I will discuss this parameter separately. Most of the following analyses were done using the bird means treatment.

As in some of my analyses of emu footprints, for many analyses I scaled footprint parameters by subtracting from each log-transformed variable the mean of all log-transformed variables used in that particular analysis. These scaled parameters were then used in both canonical variate and cluster analyses. Some of the analyses were carried out before I received casts of the probable gastornithid footprints, and so those footprints were not used in those analyses.

A
II
B
II
I
CM
C
10 CM
II
II
I
I
II
II
I
D
10 CM
II
I
E
II
I
F
II
I
10 CM
G
II
I
H
II
I
J
II
I
I
I
II
I
10 CM
K
II
I
L
II
IV
I
CM
M
II
I

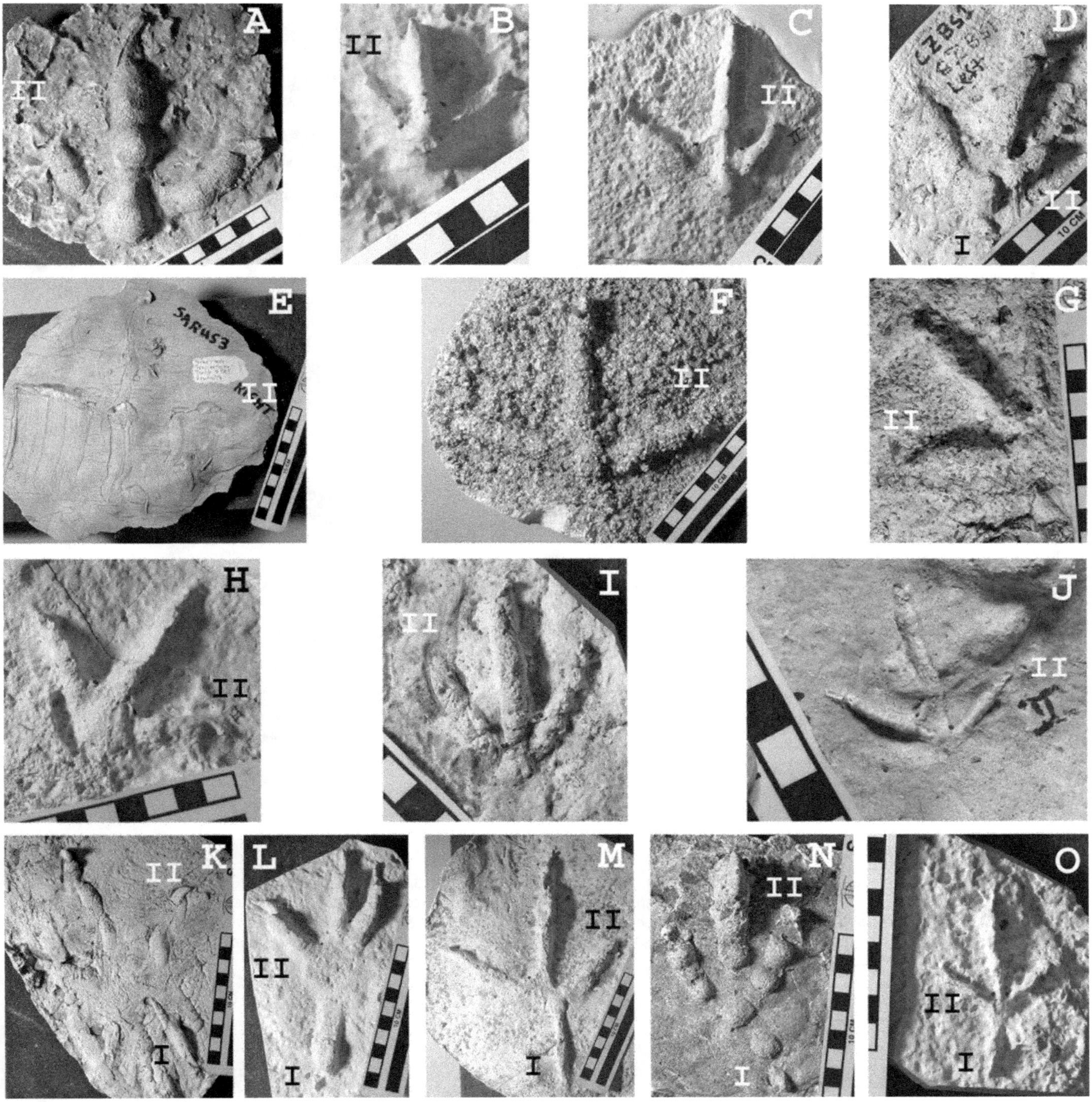

Facing, **6.31.** Casts of footprints of galliform birds II. Impressions of digits I and II are labeled to identify footprint symmetry. *A*, Female Himalayan monal (*Lophophorus impejanus*), San Diego Zoo. *B*, Cornish cross domestic chick (*Gallus*), Jed Freels farm, Auburn, Indiana. *C*, Sequence of four successive footprints (left-right-left-right) with very short pace lengths between them, domestic chicken hen, Fort Wayne Zoo. *D*, Male Swinhoe's pheasant (*Lophura swinhoii*), Cincinnati Zoo. *E*, Blue eared pheasant (*Crossoptilon auritum*), San Diego Zoo. *F*, Male crested fireback (*Lophura ignita*), Cincinnati Zoo. *G*, Common or ring-necked pheasant (*Phasianus colchicus*), Bob Cassell farm, Larwill, Indiana. *H*, Palawan peacock-pheasant (*Polyplectron napoleonis*), San Diego Zoo. *I*, Male gray peacock-pheasant (*Polyplectron bicalcaratum*), Cincinnati Zoo. *J*, Congo peahen (*Afropavo congensis*), Cincinnati Zoo. *K*, Male green peacock (*Pavo muticus*), Cincinnati Zoo. *L*, Indian peahen (*Pavo cristatus*), Fort Wayne Zoo. *M*, Domestic turkey hen (*Meleagris*), Fort Wayne Zoo.

Above, **6.32.** Casts of footprints of a variety of ground birds. Impressions of digits I and II are labeled to identify footprint symmetry. *A*, Male kori bustard (*Ardeotis kori*), Toledo Zoo. *B*, Buff-crested bustard (*Eupodotis* [or *Lophotis*] *gindiana*), San Diego Zoo. *C*, Red-legged seriema (*Cariama cristata*), San Diego Zoo. *D*, Black-legged seriema (*Chunga burmeisteri*), Cleveland Zoo. *E*, Juvenile sarus crane (*Grus antigone*), Louisville Zoo. *F*, Female sandhill crane (*Grus canadensis*), Brookfield Zoo. *G*, Weka (*Gallirallus australis*), San Diego Zoo. *H*, Gray-winged trumpeter (*Psophia crepitans*), San Diego Zoo. *I*, Double-striped thick-knee (*Burhinus bistriatus*), Cleveland Zoo. *J*, Killdeer (*Charadrius vociferus*), St. Joseph River, Fort Wayne, Indiana. *K*, Male magpie goose (*Anseranus semipalmata*), Cincinnati Zoo. *L*, Northern or Abyssinian ground-hornbill (*Bucorvus abyssinicus*), San Diego Zoo. *M*, Great blue heron (*Ardea herodias*), St. Joseph River, Fort Wayne, Indiana. *N*, Female secretarybird (*Sagittarius serpentarius*), Toledo Zoo. *O*, Pheasant pigeon (*Otidiphaps nobilis*), San Diego Zoo.

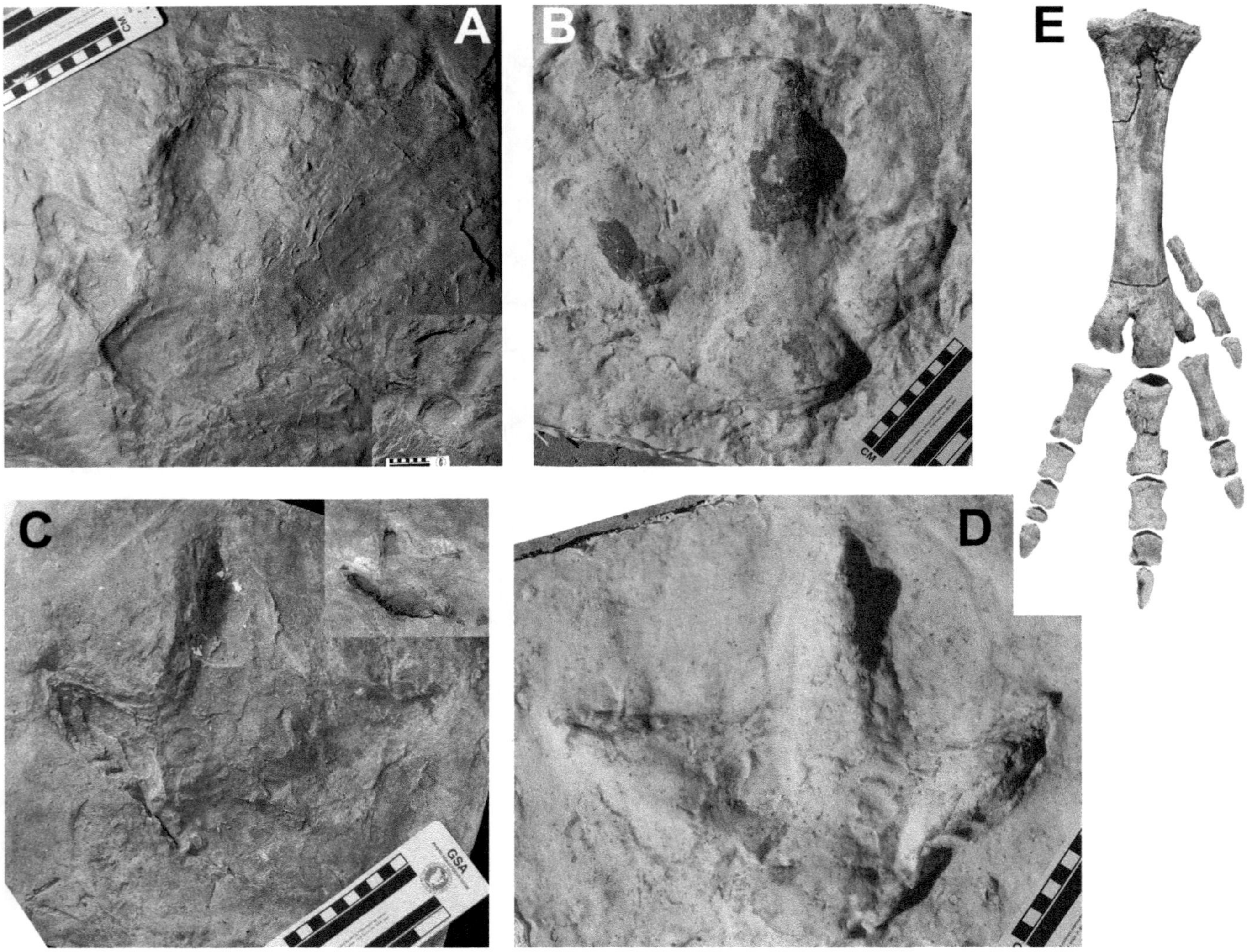

6.33. Gastornithiform footprints and foot. *A–D,* Casts of large bird footprints (*Rivavipes giganteus*) from the Lower Eocene Chuckanut Formation of Washington State (Mustoe et al. 2012). *A, B,* WWU-TR-059B, right footprint; *A* is a positive relief copy of the original print, and *B* is a negative relief copy. *Inset* in *A* shows the actual footprint. *C, D,* WWU-TR-066, another right footprint; *C* is a positive relief copy, and *D* a negative relief copy. *Inset* in *C* shows the actual footprint. Photographs of the actual footprints, and casts, supplied by George E. Mustoe. *E,* Right foot of *Gastornis* (or *Diatryma*) *gigantea* (AMNH 6169) (image from Matthew and Granger 1917, plate 32).

We will examine the extent to which footprint shape serves as a proxy for phylogenetic propinquity by beginning with very simple bivariate comparisons, and proceeding to more complicated multivariate analyses. We begin with the relationship between two parameters of footprint shape nearly always reported in studies of tridactyl dinosaur footprints, the ratio of which is often examined in attempts to identify or discriminate among ichnotaxa and trackmakers: footprint length and width (for an extensive, but by no means exhaustive, list, see Agirrezabala et al. 1985; Lockley et al. 1986, 1996, 1998*a*, 1998*d*, 2001*b*, 2006*b*, 2006*c*, 2006*e*, 2008*b*; 2011*a*, 2012*b*, 2013, 2014*a*, 2014*c*; Leonardi et al. 1987*b*; Lockley 1987; Casanovas et al. 1989*a*, 1989*b*, 1990*a*, 1990*b*, 1991*a*–1991*c*, 1992, 1993*a*–1993*c*, 1995*a*–1995*k*, 1997, 1998; Thulborn 1990; Calvo 1991; Platt and Meyer 1991; Leonardi and Spezzamonte 1992; Moratalla et al. 1992, 1994; Weems 1992; Bengoechea et al. 1993; Parkes 1993; Pérez-Lorente 1993, 1996, 2003, 2013; Carvalho 1995; Matsukawa et al. 1995, 1999, 2005; Lee 1997; Pérez-Lorente et al. 1997, 2000; Clark and Barco Rodríguez 1998; Dalla Vecchia and Mietto 1998; Fuentes Vidarte and Meijide Calvo 1998; Sarjeant et al. 1998; Nicosia et al. 1999; Álvarez et al. 2000; Blanco et al. 2000; Dalla Vecchia and Tarlao 2000; Lange-Bandré and Lafon 2000; Mietto et al. 2000; Pascual-Arribas and Sanz Pérez 2000; Dalla Vecchia et al. 2001; Pérez-Lorente and Romero Molina 2001; Gierliński and Niedźwiedzki 2002; Le Loeuff et al. 2002; Moreno and Pino 2002; Rogers 2002; Lingham-Soliar et al. 2003; Meyer and Thüring 2003; Pérez-Lorente and Ortega 2003; Romano

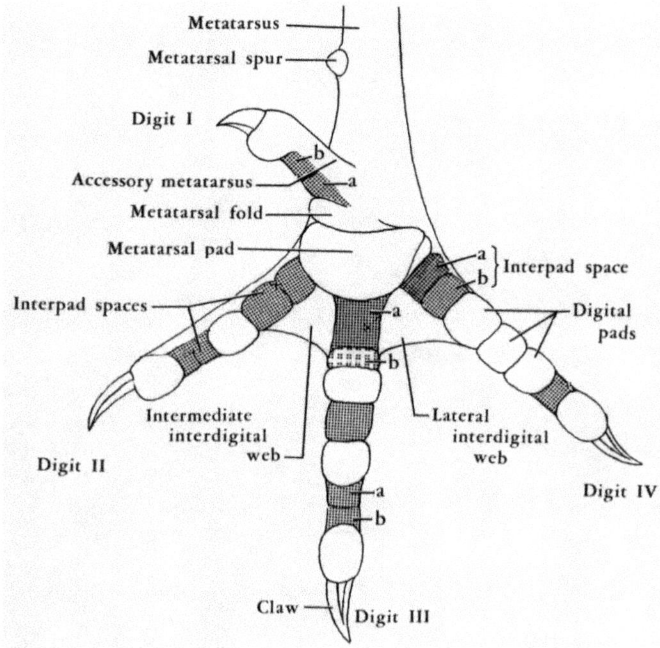

6.34. Features of the plantar surface of the foot of the domestic chicken (image reproduced from Lucas and Stettenheim [1972: Fig. 43A]). Note the presence of both digital pads and interpad spaces.

and Whyte 2003; Romero Molina et al. 2003*a*, 2003*b*; Day et al. 2004; Leonardi and dos Santos 2004; Moreno et al. 2004, 2012; Barco et al. 2005, 2006; Clark et al. 2005; Gand and Demathieu 2005; Pascual-Arribas et al. 2005, 2009; Piubelli et al. 2005; Azuma et al. 2006; Huh et al. 2006; Li et al. 2006; Milner et al. 2006, 2009; Zhang et al. 2006; Jiménez Vela and Pérez-Lorente 2006–2007; Requeta Loza et al. 2006–2007; Gand et al. 2007; Hadri et al. 2007; Rodríguez de la Rosa 2007; Wings et al. 2007; Ansorena et al. 2007–2008; Fujita et al. 2008; Petti et al. 2008, 2011; Rubilar-Rogers et al. 2008; Schulp et al. 2008; Boutakiout et al. 2009*a*; Díaz-Martínez et al. 2009; Gierliński et al. 2009; Ishigaki et al. 2009; Kim et al. 2009; Lockley and Gierliński 2009; Sullivan et al. 2009; Xing et al. 2009*a*, 2009*b*, 2011*a*, 2013*c*–2013*e*, 2014*a*–2014*e*, 2014*g*–*k*; Belvedere et al. 2010; Ishigaki 2010; Pascual-Arribas and Hernández-Medrano 2010, 2011; Wagensommer et al. 2010, 2012; Melero Rubio and Pérez-Lorente 2011; Nouri et al. 2011; Ősi et al. 2011; Contessi and Fanti 2012; Dalman 2012; Fiorillo and Adams 2012; Fujita et al. 2012; Kellner et al. 2012; J. Li et al. 2012; Martin et al. 2012; Moreau et al. 2012*b*, 2014; Rodríguez de la Rosa et al. 2012; Schulp and Al-Wosabi 2012; Álvarez Bermejo et al. 2013; Castanera et al 2013*a*; Dalman and Weems 2013; Fanti et al. 2013*a*; Shell and Boss 2013; Cobos et al. 2014; Fiorillo et al. 2014*a*; Lockwood et al. 2014; Marzola and Dalla Vecchia 2014; McCrea et al. 2014*a*; Pond et al. 2014; Razzolini et al. 2014; R. Li et al. 2015; Martin 2016; Razzolini et al. 2016*b*; Salisbury et al. 2016; Sciscio et al. 2016—Whew! Had enough?). Footprints of non-avian large theropods are generally considered to be different from those of large ornithopods in having a relatively smaller width/length ratio.

I will characterize width as the distance across the tips of digits II and IV (fig. 6.41). Some forms plot either well or just above (cranes [*Grus*], currasow [*Crax*], pheasants [Phasianidae]) or below (cassowaries [*Casuarius*], rheas [*Rhea*, *Pterocnemia*]) the interspecific "main sequence," but there is considerable nearness or even overlap of points for species that are not close relatives (emus and kori bustard [*Ardeotis*]; kiwi [*Apteryx*], pheasants, and many neoavians). Whatever its merits for distinguishing between large non-avian theropods and large ornithopods (for which it probably does work), this simple bivariate relationship is not particularly impressive in discriminating among major ground bird groups. So if two sets of footprints plot well away from each other, their makers probably were different kinds of animals, but if a series of footprints all fall on the same trend, they were not necessarily made by the same kind of trackmaker.

As already discussed, another set of comparisons frequently used in analyses of the shape of tridactyl footprints of non-avian dinosaurs (Olsen 1980; Weems 1992; Gierliński 1996*b*; Olsen et al. 1998; cf. Dalla Vecchia et al. 2001; Piubelli et al. 2005; Lockley 2009; Sullivan et al. 2009; Nouri et al. 2011; Petti et al. 2011; Xing et al. 2011*b*, 2014*b*, 2014*c*; Dalman 2012; Moreau et al. 2012*a*, 2012*b*, 2014; Castanera et al. 2013*a*; Dalman and Weems 2013; Cobos et al. 2014; Razzolini et al. 2016*a*; cf. Salisbury et al. 2016) is comparison of the projection of digit III beyond the tips of digits II and IV with the length of the footprint ("backfoot") proximal to a line connecting the tips of digits II and IV, and also the relative proportions of the three parameters (toetip II–toetip III distance, toetip III–toetip IV distance, and toetip II–toetip IV distance), of the "anterior triangle," either by themselves and along with the backfoot length.

If we compare the projection of digit III beyond the limits of digits II and IV against the backfoot length (fig. 6.42A), the relationship is curvilinear, as is the corresponding intraspecific plot for emus alone (fig. 6.22A), with the digit III projection becoming proportionately shorter with increasing footprint size. Most of the birds in our footprint fall along an avian "main sequence," although there are some forms—quite literally odd birds (*Grus*, *Bucorvus*, *Casuarius*, one moa)—that plot well away from it.

In emus the proportional drop in the digit III projection with increasing footprint size was related to a decrease in IDA II–IV with increasing print size (fig. 6. 20). The same seems to be true more generally for the ground birds in our sample (fig. 6.42*B*): small birds, both juvenile emus and adults of small-bodied species, often—but not invariably—have a

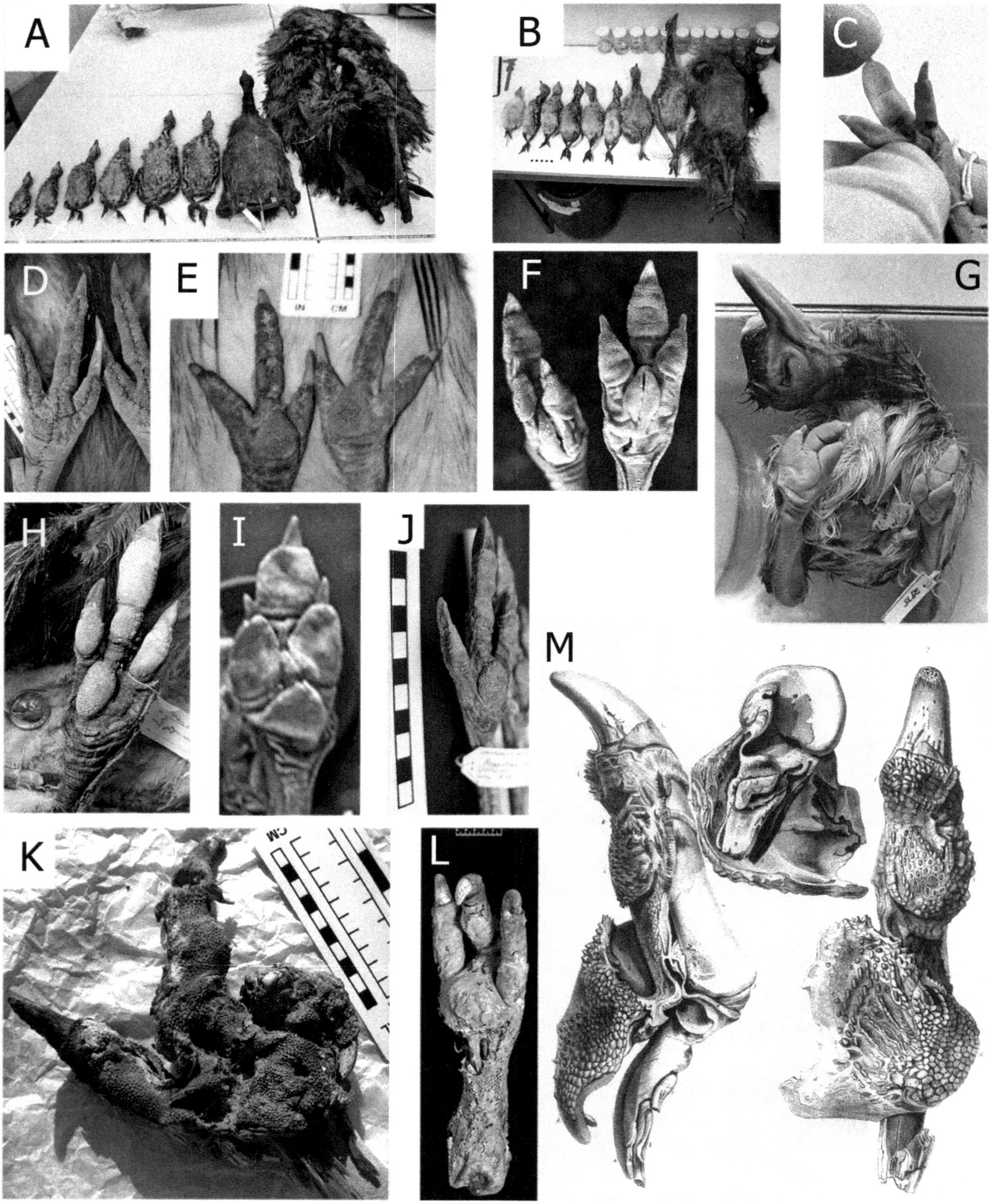
A
B
C
D
E
IN
CM
F
G
H
I
J
K
L
M

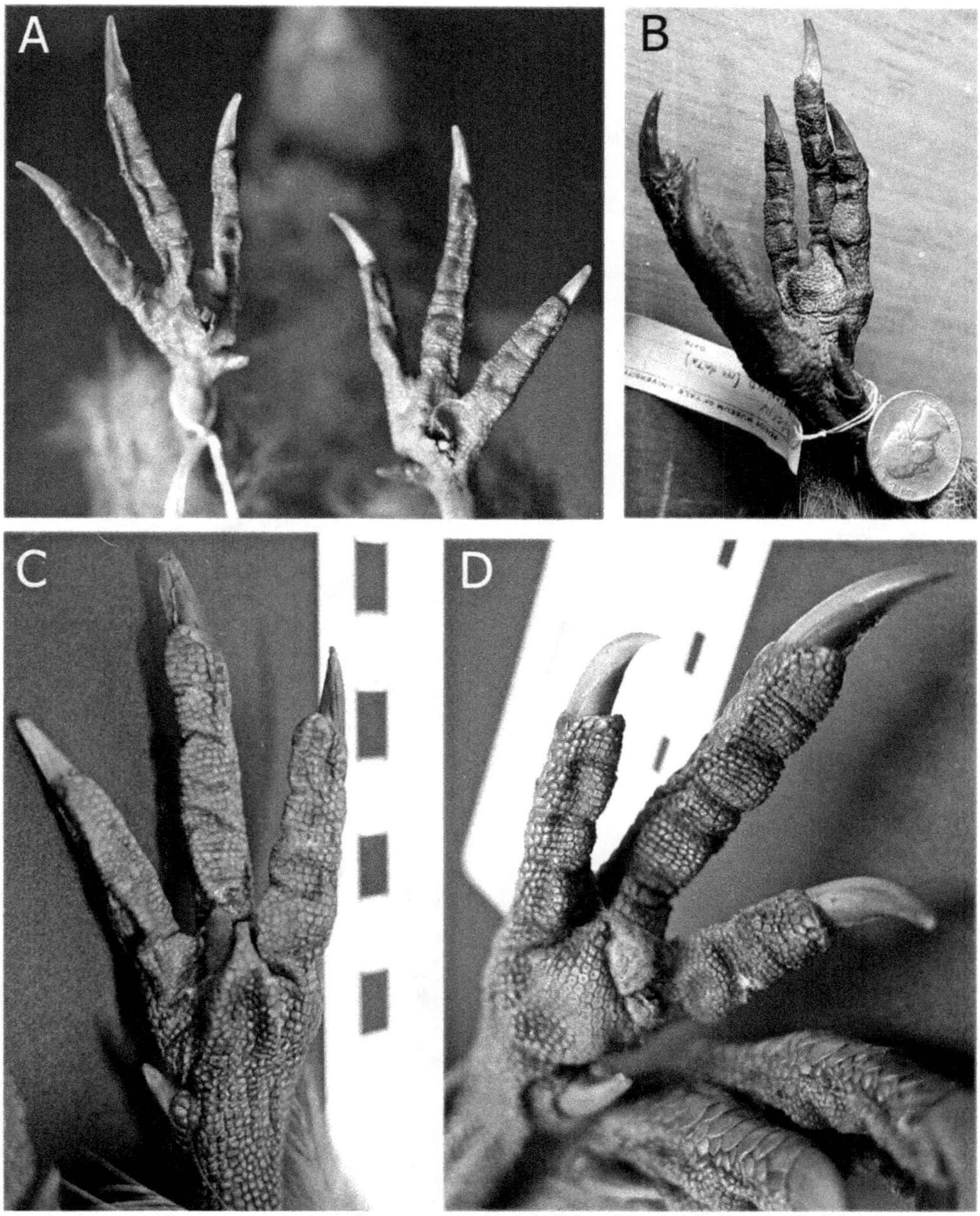

6.36. Feet of kiwi. *A, B,* Little spotted kiwi (*Apteryx owenii*). *A,* YPM ORN 100228 (formerly 7060). Both feet are about 62 mm long from the back of the metatarsal pad to the clawtip of digit III. The right foot is on the left in the photograph. *B,* YPM ORN 022887, *Apteryx owenii.* Both feet appear in the photograph, but the underside of the right foot (right side of photograph) is best seen. American quarter (diameter 24 mm) is shown for scale. *C,* FMNH 4616, male North Island brown kiwi (*A. mantelli*), left foot. *D,* FMNH 88262, male southern brown kiwi (*A. australis*), right foot. The kiwi foot is rather reminiscent of that of galliform birds, showing both digital pads and interpad spaces. Scale bars in *C* and *D* are marked off in centimeters.

Facing, **6.35.** More-or-less intact feet of living and extinct ratites. Black and white scale bars are marked off in centimeters. *A–C,* Emus (*Dromaius novaehollandiae*). Growth series of emus in the collections of *A,* the University of Michigan Museum of Zoology (Ann Arbor), and *B,* the Biodiversity Program of the Queensland Museum (Brisbane, Australia). Note the increase in the length of the portion of the foot corresponding to the tarsometatarsus relative to the length of digit III on progression from small to large birds (cf. fig. 4.7*D*). *C,* UMMZ Birds 2970, larger of two unhatched emu embryos, right foot. Digit III length (from base of most proximal digital pad to tip of claw) = 17 mm. Digits II and IV are not subdivided into digital pads, and digit III is subdivided by a single crease into two digital pads, the more proximal of which presumably corresponds to the first digital pad of more mature birds (fig. 6.3). *D,* FMNH 87322, female *Casuarius bennetti,* right foot. Digits II and IV are undivided into digital pads. Digit III is divided into three pads, separated by interpad spaces. *E,* FMNH 87321, female *Casuarius* sp., both feet. F–H) Greater rheas (*Rhea americana*). *F,* UMMZ Birds 216665, juvenile rhea, right foot on right. Total length of digit III is about 18 mm. *G,* UMMZ Birds 2975, rhea hatchling. Total length of digit III is about 22 mm. *H,* YPM ORN 022814, female rhea, left foot. American quarter (diameter 24 mm) is shown for scale. Note that in this bird, like the two very young individuals (*F* and *G*), the pads of digits II and IV are undivided (in some rhea specimens, however, the pad of digit II is subdivided into two digital pads, separated by a faint crease), but the pad of digit III is subdivided into a short proximal and longer distal digital pads. The configuration of digit III seen in these three specimens occurs in many specimens of greater rheas, but in some specimens the longer, more distal digital pad of digit III is itself further subdivided into two pads. *I, J,* Lesser rhea, *Pterocnemia pennata. I,* UMMZ Birds 134715, right foot of juvenile; Digit II length is about 15 mm. *J,* FMNH 370474, immature individual, right foot. Feet of the lesser rhea are very similar to those of the greater rhea. *K–M,* Mummified feet of moa. *K, L, Megalapteryx didinus.* K) NMNZ S.023808, left foot. Digit IV is bent back upon itself. The fleshy part of digit II (at left in the photograph) appears to be faintly subdivided into two digital pads, and digit III possibly into three digital pads. *L,* NHMUK A16, left foot. Note cm scale at top of photograph. There may be three or four creases on the underside of digit II, but this would be a rather unusual configuration for digit II. There may also be two or three creases on the underside of digit III, but again these are faint. The overall impression for this species is that if the fleshy undersides of toes were indeed subdivided into digital pads, these were not as well developed as in many other large ground birds. *M,* The Tiger Hill specimen of *Dinornis robustus* (formerly *D. novaezealandiae*). Digit II is very well preserved in this specimen and clearly subdivided by a crease into two digital pads (drawing from Owen 1879, plate 71).

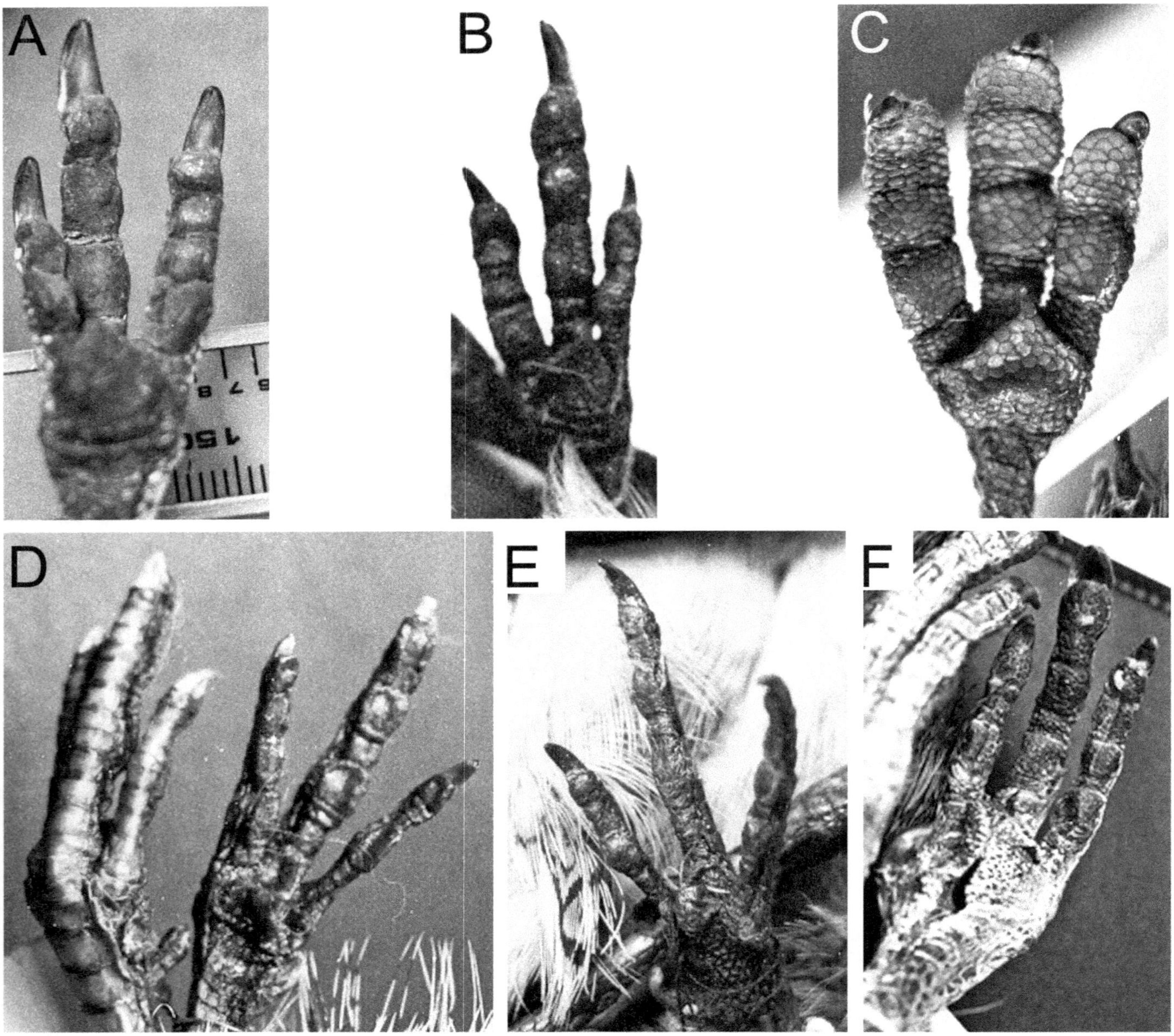

6.37. Feet of tinamous (Tinamidae). *A*, FMNH 22374, female elegant crested tinamou (*Eudromia elegans*), left foot. Total length of digit III is about 27 mm. *B*, FMNH 17183, male quebracho crested tinamou (*E. formosa*), right foot. Total length of digit III is about 26 mm. *C*, FMNH 5287, female Patagonian tinamou (*Tinamotis ingoufi*), right foot. Total length of digit III is about 26 mm. The feet of these species are similar to those of emus and bustards in lacking digit I, and also in having interdigital pads that closely abut each other in a toe, with little or no space separating them. *D*, FMNH 25683, male Darwin's nothura (*Nothura darwinii*). The right foot is at the right of the photograph. Total length of digit III is about 15 mm. *E*, FMNH 8926, female spotted nothura (*N. maculosa*), left foot. Total length of digit III is about 22 mm. *F*, FMNH 19383, male gray tinamou (*Tinamus tao*), right foot. Total length of digit III is about 28 mm. In contrast with the tinamous illustrated in *A–C*, the birds in *D–F* do have a digit I. In addition, there are distinct interpad spaces separating digital pads in digits II and III. In these features the feet of these tinamous resemble those of galliform birds (figs. 6.34, 6.38).

substantially larger IDA II–IV than do big ground birds. If we correct the backfoot–digit III projection relationship for this size-related drop in IDA II–IV (fig. 6.42C), most of the bird footprints in our sample now fall along the same linear trend as our ontogenetic sample of emus; indeed, across their ontogenetic size range emus show overlap with numerous other bird species. However, footprints of moa, dromornithids, cassowaries, cranes, and ground hornbills plot well away from the main trend of ground birds.

Because across most ground bird species the relationship between the digit III projection and backfoot length is similar to that which Olsen (1980) observed in the *Grallator-Anchisauripus-Eubrontes* series, this might support my earlier suggestion that this is more likely to be an interspecific

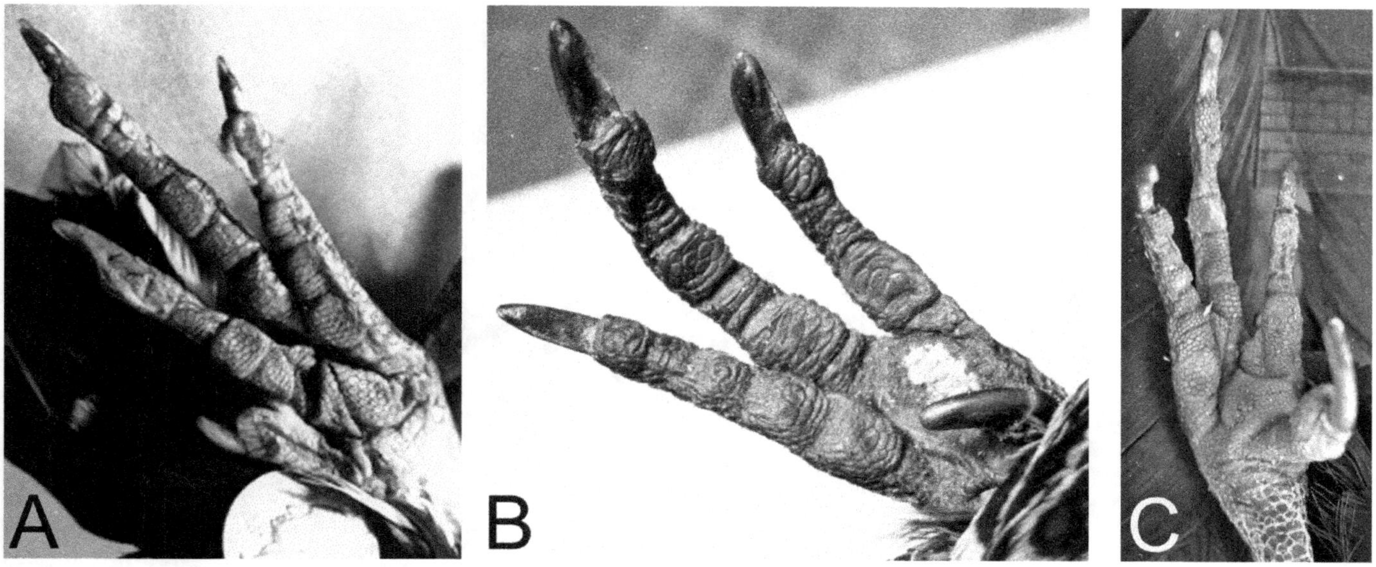

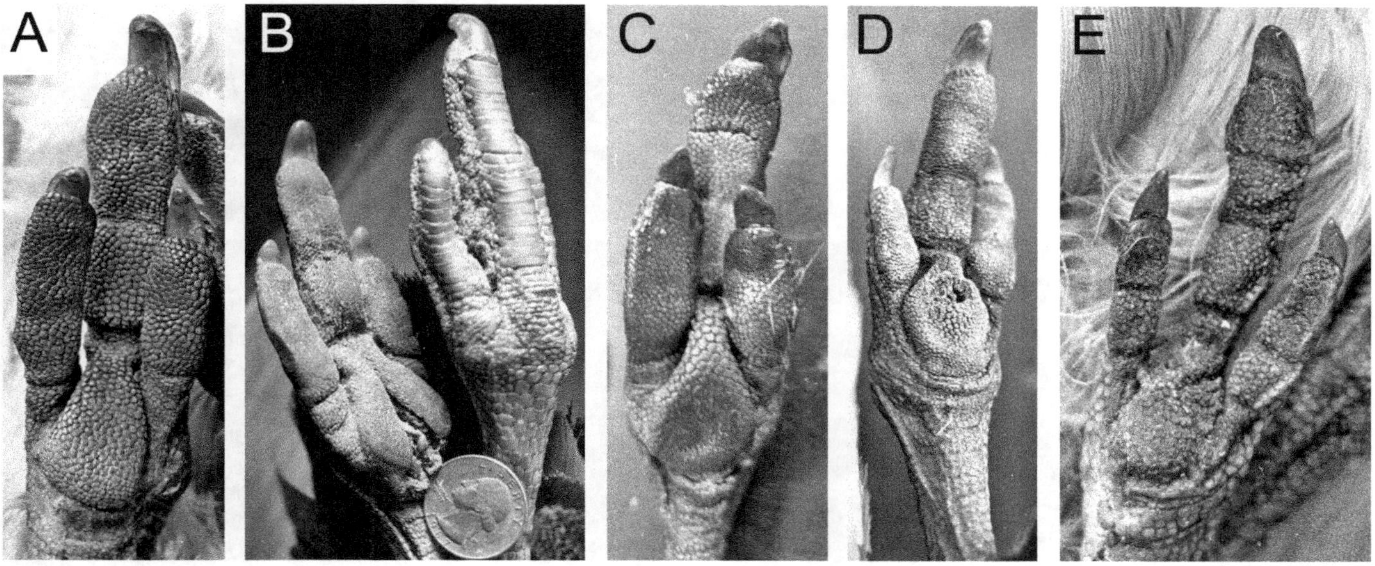

Top, **6.38.** Feet of galliform birds; scales not available in these photographs. *A, B,* Phasianids. *A,* YPM ORN 006753, left foot of a domestic chicken, with American quarter (diameter 24 mm) shown for scale. *B,* USNM 390110, Himalayan monal (*Lophophorus impejanus*), right foot. *C,* MSU OR.2608, crested guan (*Penelope purpurascens*; Cracidae), right foot. These birds show the typical galliform foot morphology, with both digital pads and interpad spaces on the undersides of toes.

Above, **6.39.** Feet of bustards (Otididae). *A, B,* Kori bustard (*Ardeotis kori*). *A,* UMMZ Birds 97916, male, right foot. Length of digit III is about 66 mm. *B,* YPM ORN 048917; right foot on left. American quarter (diameter 24 mm) provides scale. *C, D,* Great bustard (*Otis tarda*). *C,* UMMZ Birds 57476, right foot. Digit III length is about 50 mm. *D,* UMMZ Birds 114965, left foot. Length of digit III not completely measured, but length of digit II is about 25 mm. *E,* UMMZ Birds 97918, male white-bellied bustard (*Eupodotis senegalensis*), left foot. Length of digit III is about 25 mm. Bustard feet look remarkably like those of small emus (fig. 6.3).

than an intraspecific relationship. Unfortunately, things are not that simple. Both within emus (fig. 6.20) and, at least in our sample of footprints, across species of ground birds (fig. 6.42*B*), the interdigital angle is smaller in big than in little birds—just the opposite of what occurs in the *Grallator-Anchisauripus-Eubrontes* series. Consequently we cannot, by analogy with our bird data, conclude with certainty from Olsen plots alone that the non-avian dinosaur trend is in fact interspecific or intraspecific. What we can say, however, is that such plots do not serve to distinguish among any but the more oddball footprint shapes among ground birds.

Weems (1992) modified Olsen's digit III projection/back-foot length plots by dividing both of those parameters by footprint width, and used such ratios to separate out different

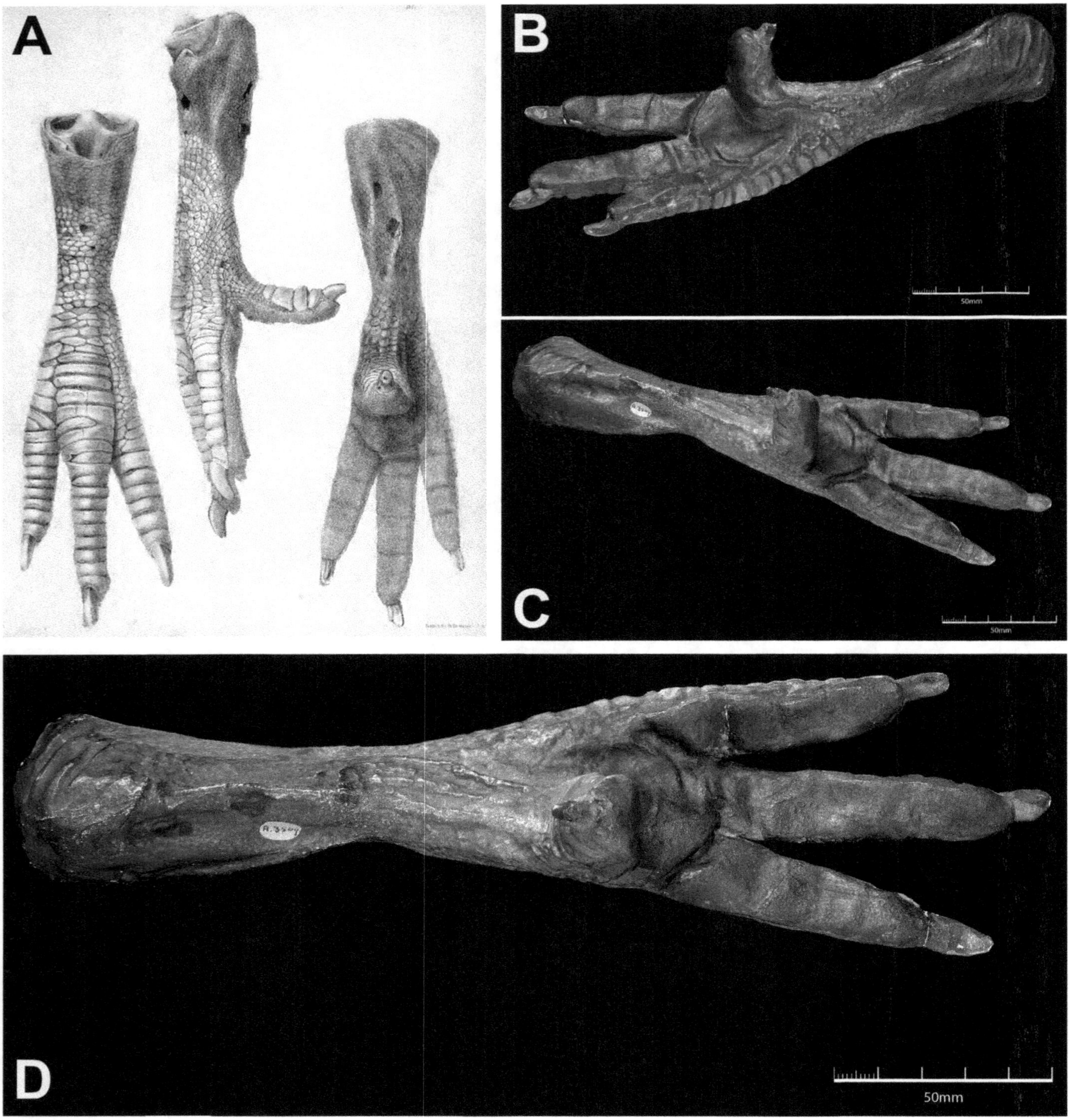

6.40. Right foot of the extinct dodo (*Raphus cucullatus*). *A*, Dorsal, medial, and plantar views of the British Museum foot, now lost (cf. Parish 2013; Claessens et al. 2015); image from Strickland and Melville (1849, plate 6). *B–D*, Three views of the plantar surface of a cast of the former British Museum foot (now Natural History Museum [NHMUK] A3509). Note that digit II is longer than digit IV (cf. fig. 1.7); a large number of digital pads and/or interpad spaces can be seen on the underside of digit II. *Photographs courtesy Image Resources, Natural History Museum, London; copyright NHM, London.*

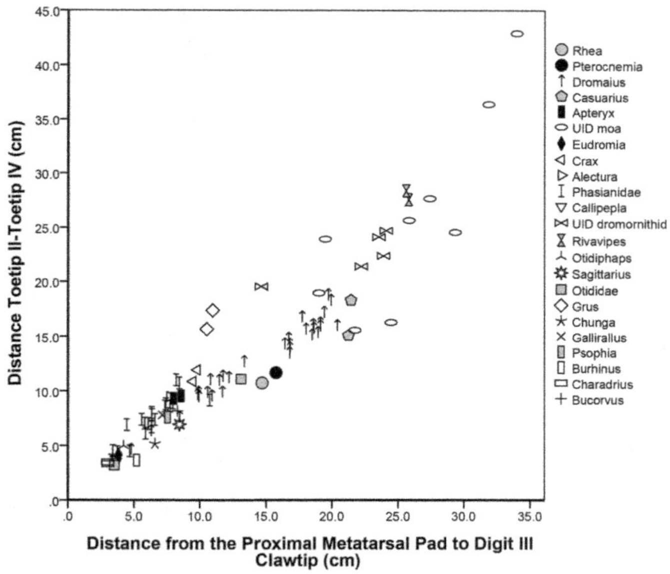

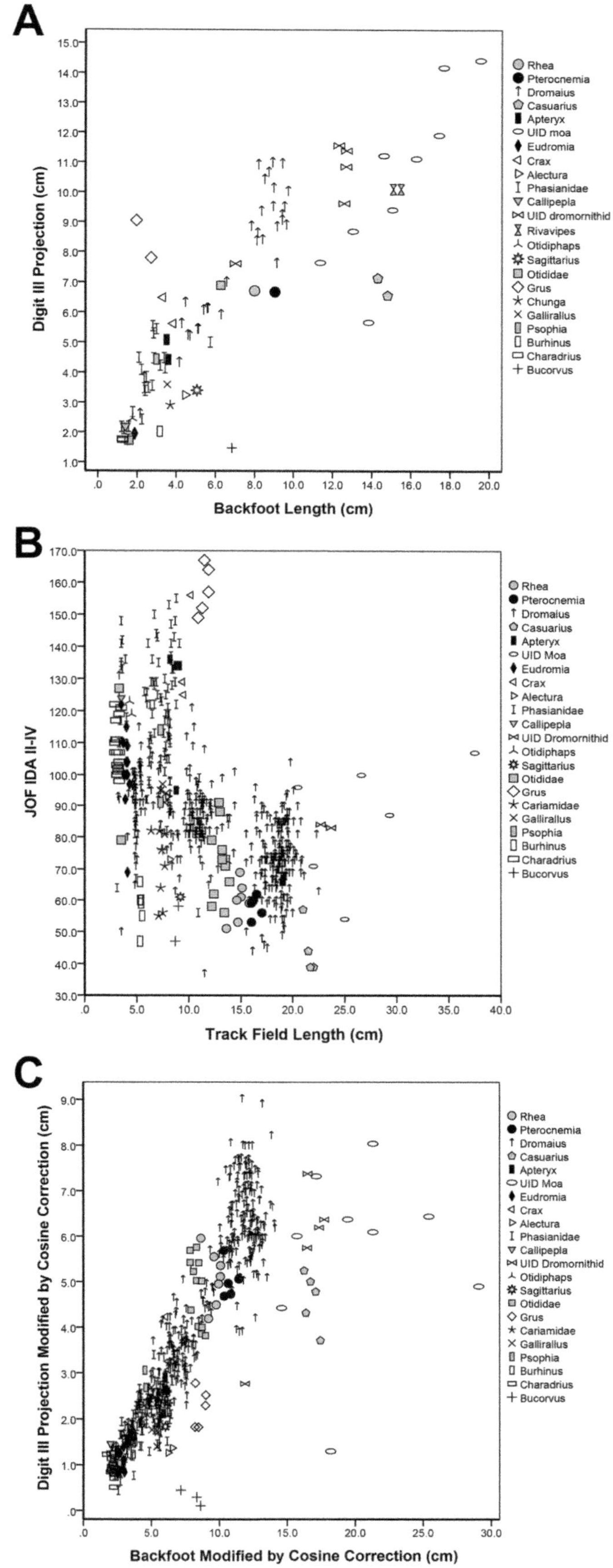

Above, **6.41.** Length (IIILGL) vs. toetip width in footprints of ground birds. Although some species and groups plot above (phasianids, cranes) or below (rheas, cassowaries) the overall trend, there is considerable overlap of points across groups.

Right, **6.42.** Projection of digit III beyond the limits of digits II and IV plotted as a function of backfoot length across species of ground birds. Both parameters were approximated as described in figure 6.6. *A*, Raw data ("bird means" treatment; *Rivavipes* included). Note the tendency of the graphed curve to flatten out with increasing footprint size, indicating a relative decrease in the digit III projection. *B*, Across the ground bird footprints in the study sample, however, small birds often have substantially larger interdigital angles (IDAs) than do big birds. (In *B* and *C*, plotted points are for individual footprints, rather than for means across all footprints of the same bird.) *C*, Correcting the digit III projection–backfoot length relationship for differences in interdigital angles among bird footprints causes points for most ground birds to fall along a linear trend. Data points for some birds (moa, dromornithids, *Casuarius*, *Buvcorvus*, *Grus*), however, continue to fall away from the avian "main sequence."

ichnotaxa of Newark Supergroup tridactyl saurischian footprints. A Weems plot (fig. 6.43) does indeed result in some groupings and separations of bird footprints; the emu prints form a nice group, for example. In addition to separating out the more extreme morphologies (moa, cassowaries, hornbill, cranes, dromornithids) from more standard ground bird footprint shapes, the plot also separates rheas and emus, on the basis of a relatively long backfoot/footprint width ratio, from galliforms. However, kiwi plot among the galliforms, and bustards among the emus. The various members of Neoaves do not plot together, away from ratites and galliforms, but instead are all over the graph. Weems plots thus seem to have some utility in discriminating among some bird species, and even more inclusive groups, but do not consistently group species together in a manner that reflects their phylogenetic relationships.

We now consider a related feature, the anterior triangle, created by joining the toetips of digits II, III, and IV (cf.

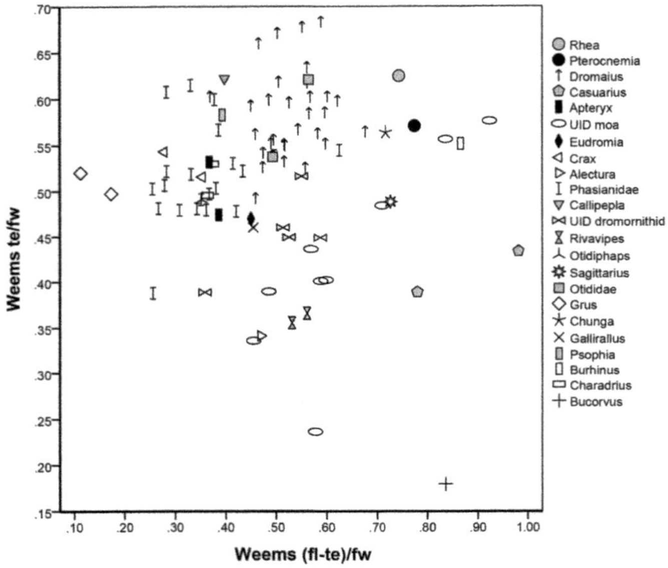

Above, **6.43.** Weems plot of digit III projection (or toetip extension [te]) plotted as a function of backfoot length ("footprint length – toetip extension" [fl – te]), with both parameters divided by footprint width (fw) ("bird means" treatment; *Rivavipes* included) in footprints of ground birds. The more extreme morphologies (moa, *Casuarius*, dromornithids, *Grus*) are separated from that of more typical birds by this plot. Emus and rheas tend to have a relatively larger backfoot/footprint width ratio than do galliforms.

Right, **6.44.** Cluster analysis of scaled toetip distances (II–III, III–IV, and II–IV) across footprints of ground birds. This "anterior triangle" of a footprint does not yield any obvious phylogenetic structure to the groupings. Although many emu cases group together, those for some emus and other palaeognaths, galliforms, and neoavian birds seem scattered willy-nilly among the clusters.

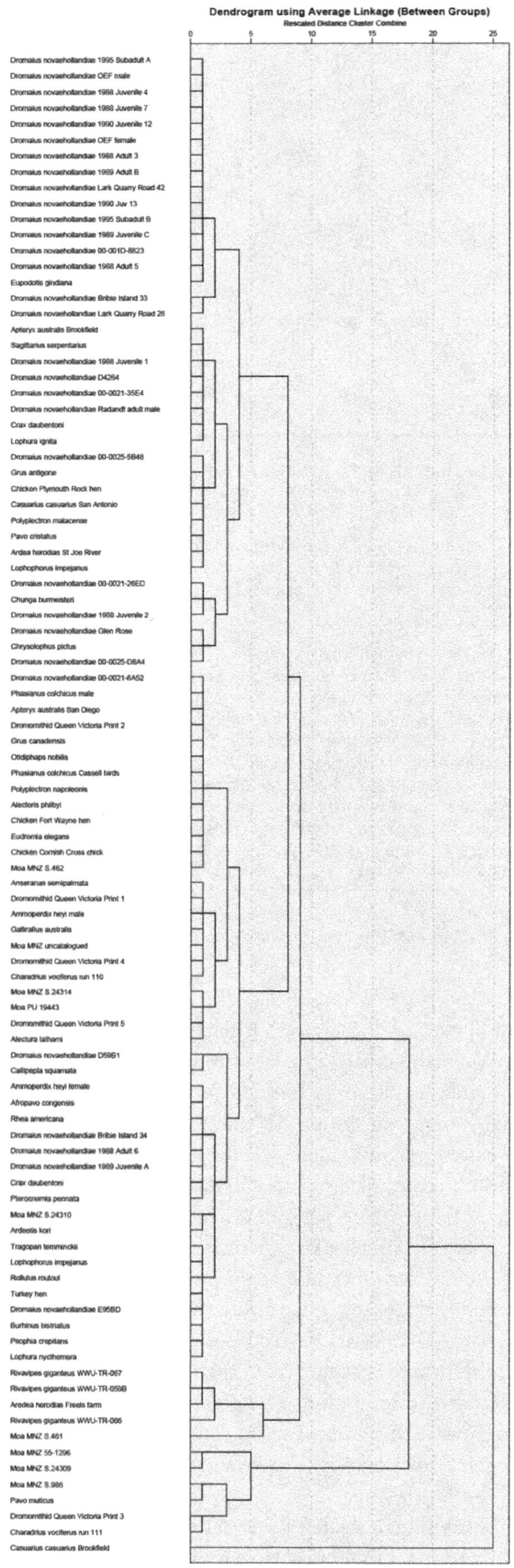

Lockley 2009). Because the footprints of the ground-hornbill (*Bucorvus*) are so obviously different from other birds, this species will be excluded from this and the remaining analyses, except where otherwise indicated. To remove size as a confounding variable, I scale the three toetip distances as already described.

I then compare the scaled toetip distances among our birds by cluster analysis. Clusters in this and other such analyses are created using the between-groups linkage method, based on the squared Euclidean distance. Distances between clusters are expressed in terms of values between 0 and 25. Data cases are from the bird means treatment.

Because only three measurements have to be made on one or more footprints in order for a bird to be included in the analysis, in this, like the analysis of digit III projection vs. backfoot, there are quite a few data cases (fig. 6.44). Unfortunately, there is very little structure to the resulting clusters. Although more than half of the emu cases cluster together, or come close to clustering together, the remaining emu cases are well separated from their conspecifics. Furthermore, emu cases cluster together with those of galliform, anseriform, and neoavian birds, and the galliforms, anseriforms, and neoavians themselves do no better at clustering with their close relatives than with birds of the other major groups. The anterior triangle by itself seems to be of little use in assigning footprints of birds to supraspecific groups in anything approximating what is thought to be their true phylogenetic relationships (cf. fig. 6.26).

We now turn from the tips of the toes to the lengths thereof. In their individual species accounts, both HANZAB and BWP report the relative lengths of pedal digits I, II, and IV as percentages of the length of digit III. This large set of data makes it possible to consider the extent to which relative digit lengths alone can serve to discriminate among bird feet, and thus potentially among footprints made by those feet. We also consider the extent to which relative digit lengths discriminate among the ground bird footprints in our sample.

Among our footprints (fig. 6.45*A*), those of *Struthio* stand out, unsurprisingly, as very different from those of all other birds, due to the absence of digit II and a proportionately very short impression of digit IV compared with digit III. *Casuarius* also stands out in having a proportionately long digit II (reflecting the size of the claw on this toe). *Dromaius* and *Rhea* have less extreme foot shapes, but tend to have proportionately shorter digits II and IV than do galliforms; put another way, digit III is relatively larger in emus and rheas than it is in galliforms. However, *Apteryx*, *Eudromia*, and some moa are more like galliforms than rheas and emus. Members of Neoaves can be found in the same regions as ratites or galliforms. Bustards plot close to the emus and rheas.

Relative lengths of digits II and IV, compared with III, all measured excluding the claws, from HANZAB and BWP are plotted for several families and other groups of extant birds (fig. 6.45*B*). Most of the bird species fall along another avian "main sequence," in that species with a relatively long or short digit II also tend to have a correspondingly long or short digit IV. But this isn't invariably true, and so in this comparison there are again some odd birds. Some herons (Ardeidae) and the magpie goose (Anseranatidae) have a relatively longer digit II than digit IV, while plovers, some oystercatchers (Charadrii), and ibises (Threskiornithidae) have a relatively longer IV than II. Among "main sequence" birds, emu, cassowary, bustards (Otididae), sandgrouse (Pteroclididae), and coursers/pratincoles (Glareolidae) define the lower end (relatively short digits II and IV) of the sequence, while sandpipers (Scolopacidae) and rails (Rallidae) characterize the upper end. As in our footprint sample, galliform birds plot higher along the main sequence than do bustards and the emu, and the latter two groups are close together. So even as limited a morphometric comparison as relative lengths of digits II–IV has some use in discriminating among feet and footprints of birds. However, there is so much overlap among bird families and other groups that are not closely related, and separation of families that are near relatives (e.g., Glareolidae vs. Scolopacidae), that the utility of relative digit lengths in grouping birds into higher phylogenetic categories is rather questionable. Relative toe lengths can tell some groups apart, but cannot readily group them together.

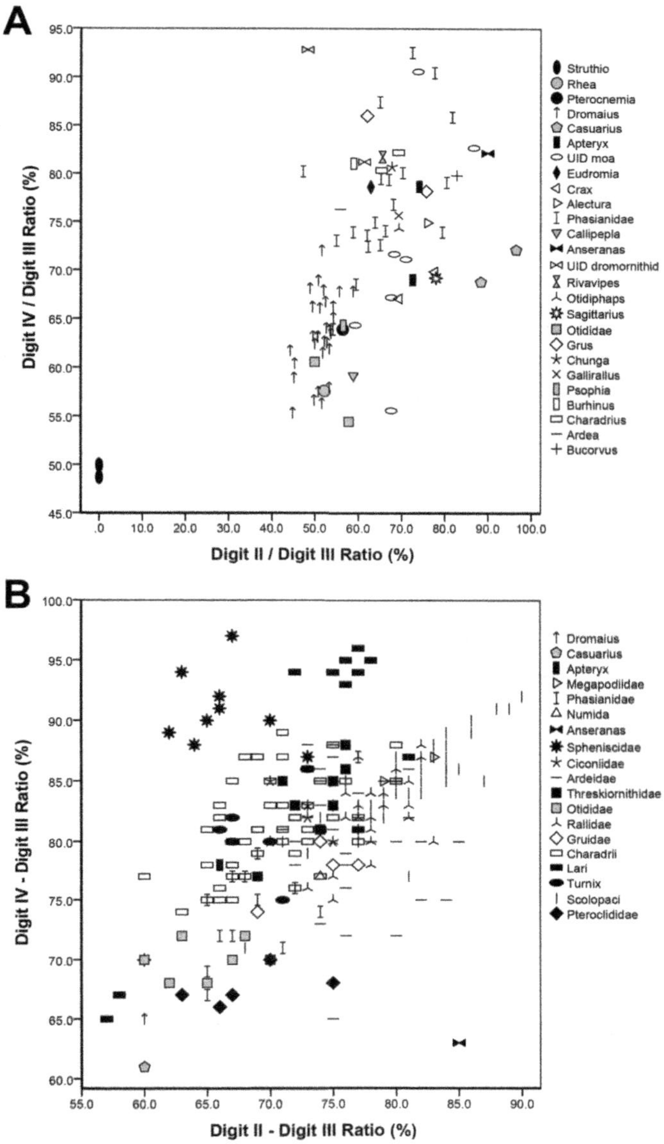

6.45. Relative digit lengths as discriminators among footprints and feet of various bird groups. The lengths of digits II and IV are expressed as percentages of the lengths of digit III. *A*, Footprints ("bird means" treatment). *B*, Toe lengths of bird study skins (excluding the claw; data from HANZAB and BWP: Cramp 1977, 1980, 1983; Marchant and Higgins 1990*a*, 1990*b*, 1993; Higgins and Davies 1996); phylogenetic categories in *B* after Naish (2012). In both comparisons, relative toe lengths separate birds with unusual pedal proportions from more typical birds. There is also some utility in distinguishing among the pedal proportions of birds with more typical feet. For example, galliform birds tend to have proportionately longer digits II and IV than do emus and bustards. However, both plots show overlap or close proximity between groups of birds that are not closely related (e.g., emus and bustards) and separation of groups that are closely related (such as emus and cassowaries in *A*).

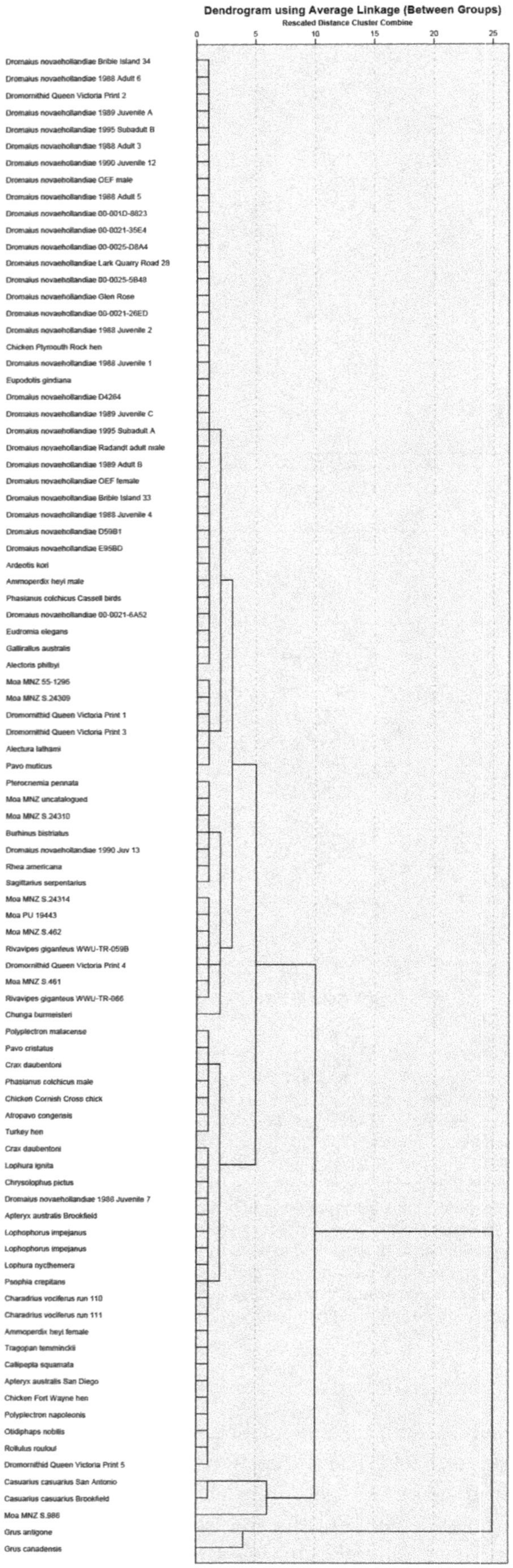

Left, **6.46.** Cluster analysis of scaled toetip distances (II–III, III–IV, and II–IV) and backfoot length across footprints of ground birds. Most of the emu cases now group together, as do the two cassowary and the two crane cases, and most of the galliforms. However, the emu cluster also includes bustards, galliforms, and neoavians, and the galliform cluster also includes kiwi, one emu, trumpeter, and pheasant pigeon.

Facing, **6.47.** Cluster analysis of 18 scaled footprint dimensions (metatarsal pad lengths and widths; gaps between the metatarsal pad and the proximal ends of digits II–IV; lengths of the impressions of digits II–IV; hypex lengths of digits II–IV; distances between the proximal ends of digits II, III, and IV; and distances between the tips of digits II, III, and IV) were log transformed. Each log-transformed measurement was then scaled by subtracting from it the mean of all the 18 log-transformed measurements. Although the cluster analysis generally separates footprints of palaeognaths from those of galliforms, there are nonetheless some odd pairings; see text for details.

We must therefore add variables to the analyses. The hope is that using more footprint shape parameters will make it more likely that ground bird footprints will group themselves in meaningful patterns. However, adding new parameters comes with a cost: the more parameters that must be measurable, the fewer the number of footprints in which it will be possible to make all of the necessary measurements.

We first add backfoot length to the three distance measurements of the anterior triangle, once again scaling the four measurements to remove footprint size (fig. 6.46). Adding backfoot length generates substantially more structure to the clusters than obtained from considering the anterior triangle alone. Nearly all of the emu cases now group together. The two cassowary (*Casuarius*) cases now link up, as do the two crane (*Grus*) cases. Most of the galliforms now join up in two very closely linked, similar clusters.

However, oddities remain in the clusters. The emu cluster also includes bustards (*Eupodotis, Ardeotis*), as well as tinamou (*Eudromia*), dromornithids, brush-turkey (*Alectura*), and weka (*Gallirallus*), and some galliforms (*Alectoris, Ammoperdix, Pavo,* and a domestic chicken). The mainly galliform grouping also includes kiwi (*Apteryx*), trumpeter (*Psophia*), killdeer (*Charadrius*), dromornithid, and pheasant pigeon (*Otidiphaps*). Most of the moa footprints fall in a cluster with cases for rheas (*Rhea, Pterocnemia*), thick-knee (*Burhinus*), secretarybird (*Sagittarius*), seriema (*Chunga*), and *Rivavipes*. So we have made some phylogenetic progress, but still have a ways to go.

Adding scaled lengths of the digit III projection and the distances from the proximal edge of the metatarsal pad to the tips of digits II (IILGL) and IV (IVLGL) to the scaled lengths of the backfoot and the toetip distances in a cluster analysis results in very little change from the clusters created in the preceding analysis, and so it will not be considered further.

For my most complex cluster analysis I employed the scaled values of 18 parameters: length and width of the metatarsal pad; the lengths of the pad gaps separating the

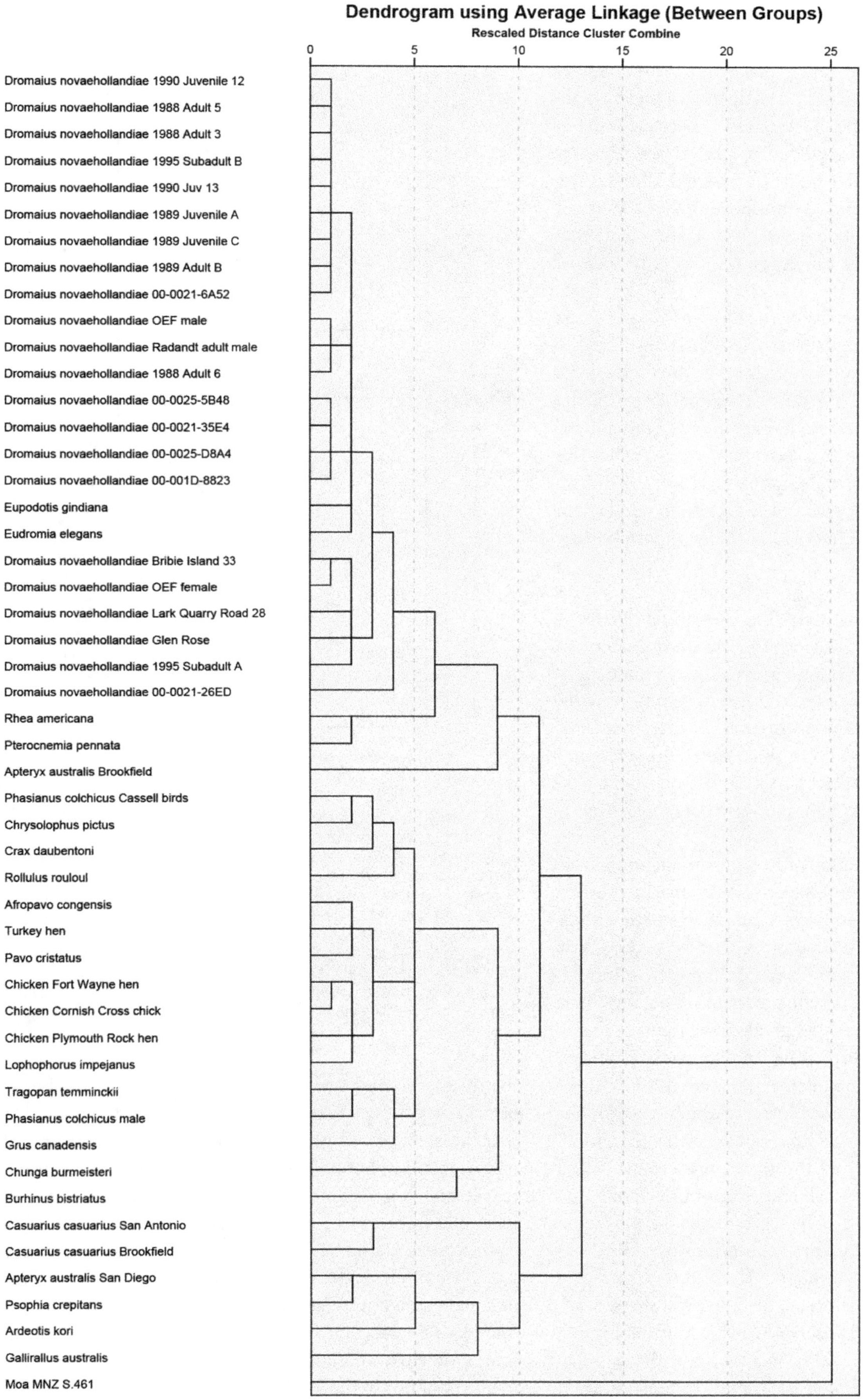
Dendrogram using Average Linkage (Between Groups)
Rescaled Distance Cluster Combine
0
5
10
15
20
25
Dromaius novaehollandiae 1990 Juvenile 12
Dromaius novaehollandiae 1988 Adult 5
Dromaius novaehollandiae 1988 Adult 3
Dromaius novaehollandiae 1995 Subadult B
Dromaius novaehollandiae 1990 Juv 13
Dromaius novaehollandiae 1989 Juvenile A
Dromaius novaehollandiae 1989 Juvenile C
Dromaius novaehollandiae 1989 Adult B
Dromaius novaehollandiae 00-0021-6A52
Dromaius novaehollandiae OEF male
Dromaius novaehollandiae Radandt adult male
Dromaius novaehollandiae 1988 Adult 6
Dromaius novaehollandiae 00-0025-5B48
Dromaius novaehollandiae 00-0021-35E4
Dromaius novaehollandiae 00-0025-D8A4
Dromaius novaehollandiae 00-001D-8823
Eupodotis gindiana
Eudromia elegans
Dromaius novaehollandiae Bribie Island 33
Dromaius novaehollandiae OEF female
Dromaius novaehollandiae Lark Quarry Road 28
Dromaius novaehollandiae Glen Rose
Dromaius novaehollandiae 1995 Subadult A
Dromaius novaehollandiae 00-0021-26ED
Rhea americana
Pterocnemia pennata
Apteryx australis Brookfield
Phasianus colchicus Cassell birds
Chrysolophus pictus
Crax daubentoni
Rollulus rouloul
Afropavo congensis
Turkey hen
Pavo cristatus
Chicken Fort Wayne hen
Chicken Cornish Cross chick
Chicken Plymouth Rock hen
Lophophorus impejanus
Tragopan temminckii
Phasianus colchicus male
Grus canadensis
Chunga burmeisteri
Burhinus bistriatus
Casuarius casuarius San Antonio
Casuarius casuarius Brookfield
Apteryx australis San Diego
Psophia crepitans
Ardeotis kori
Gallirallus australis
Moa MNZ S.461

metatarsal pads from the proximal ends of digit impressions; overall lengths of the impressions of digits II–IV from their proximal ends to their tips; digit hypex lengths; distances between the proximal ends of digits II, III, and IV; and distances between the tips of digits II, III, and IV. Because some of these parameters (e.g., pad gap lengths) take very small values for footprints of the smaller birds, I converted all measurements to millimeters before log transforming them, so as not to have to deal with negative values for measurements that in centimeters would have values less than 1. It just seemed like a tidier thing to do.

As previously noted, increasing the number of parameters that must be measurable in data cases dramatically drops the number of data cases in the analysis, but we nonetheless have a fair number of birds left (fig. 6.47). This analysis yields the most structure yet. All of the emus that remain in the analysis now group together. The two cassowary cases remain together. The two rhea species also group together, and link up with the emus, one of the kiwi cases, and *Eudromia* (which hides among the emus). There is a cluster composed nearly entirely of galliforms.

Nonetheless, phylogenetic infelicities remain. *Eupodotis* continues to crash the emu party. The remaining *Grus* case joins with the galliforms. Cassowaries are most like one of the kiwi, trumpeter, kori bustard, and weka, and far away in the dendrogram from their close relatives, the emus. Finally, one of the moa is very different from all of the other birds (unless this is simply because I misinterpreted the symmetry of the print, which is a distinct possibility). Even so, the phylogenetic cup constructed from footprint data looks slightly better than half-full.

The 18 scaled parameters of gross footprint shape can also be used in canonical variate analyses (CVAs), in which we examine footprints of birds in three large phylogenetic groups (fig. 6.26): palaeognaths, Galloanserae, and Neoaves. CVAs were carried out in a forward stepwise manner, minimizing the value of Wilks's λ, creating discriminating functions using the variables that yielded greatest separation of taxa. The success of the discriminating functions was evaluated both using all of the original data cases, and also in a cross-validation (leave-one-out procedure) in which each case is classified by functions derived from all cases other than itself. Because the sample sizes of the three bird groups varied considerably, I did CVAs in which the prior probability that a case belonged to any particular group (taxon) was equal across groups, and also in which the prior probability was adjusted according to the number of cases in each group.

Of the 18 scaled parameters of gross footprint shape available for our analyses, stepwise CVA picks four (table 6.26; fig. 6.48): metatarsal pad width, toetip width, the pad gap separating the proximal end of the digit II impression from the metatarsal pad, and the width across the proximal ends of the impressions of digits II and IV. (If it should seem odd that the stepwise CVA selected three scaled parameters of footprint width, but none of footprint length, keep in mind that several length parameters were used in creating the scaling factor [mean of all log-transformed parameters], and so parameters of length play a role in the values of the four scaled parameters selected by the stepwise CVA.) These four parameters do a not bad job of assigning bird footprints to the major phylogenetic categories of palaeognaths, galloanserans, or neoavians, getting it right at least 84% of the time, in all variations of the CVA carried out.

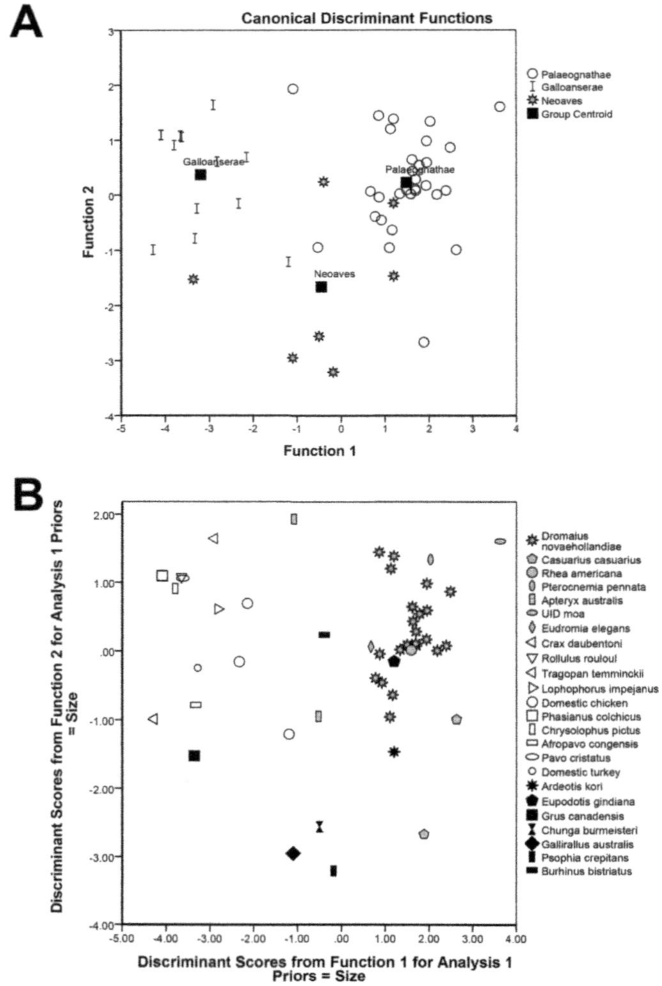

6.48. Discriminant scores from a canonical variate analysis employing the same 18 scaled gross footprint shape parameters used in the cluster analysis in figure 6.47. *A*, Bird data cases identified to major group, with the centroids for each group indicated. *B*, Bird data cases identified to species or other group. Palaeognaths separate fairly well from Galloanserae along Discriminant Function 1, whereas Neoaves generally fall between these two major groups. In contrast, palaeognaths differ little from galloanseraeans in their position along Discriminant Function 2, but neoavians tend to plot lower (more negatively) along Discriminant Function 2 than the other two bird groups. Even so, neoavians tend not to form a tight group, with some forms being more like birds in other groups than like other neoavians.

Table 6.26. Forward stepwise discriminant (canonical variate) analyses of footprint linear dimensions across three major bird groups (Palaeognaths, Galloanserae, and Neoaves). Metatarsal pad lengths and widths, metatarsal pad–digit pad gaps, digit lengths, hypex lengths, and widths across the proximal and distal ends of digit lengths were all log transformed prior to analysis. Each log-transformed parameter was then scaled by subtracting the mean of all log-transformed variables employed in the analysis from it. For the four scaled parameters selected by the stepwise procedure, the *P* value of the Wilks's λ test was <.001, indicating significant differences in the means of these scaled variables across the three bird groups. Box's *M* test of the equality of group covariance matrices = 39.046; *F*-test = 1.553; *P* = .057.

Variable	**Function** 1 (*Eigenvalue = 4.271; 89.9% of variance; canonical correlation = 0.900*)	2 (*Eigenvalue = 0.481; 10.1% of variance; canonical correlation = 0.570*)
	Canonical discriminant function coefficients (correlations between individual variables and discriminant function)	
Metatarsal pad width	10.427 (0.630)	2.956 (0.087)
Distance from toetip II to toetip IV	−18.511 (−0.526)	12.323 (0.034)
Metatarsal–digit II pad gap	−2.490 (0.249)	15.208 (0.864)
Distance from proximal end of digit II to proximal end of digit IV	10.790 (0.450)	−7.069 (0.263)

Classification results (errors in classification in **bold**): Classification was done in two ways. In the first method, the prior probability that a case belonged to a particular group was assumed to be equal for all groups. In the second method, the prior probability was weighted according to the observed proportions of cases in each group, which varied considerably across the three bird groups. Results for the second classification are reported only where they differ from those of the equal prior probability run.

Group	**Predicted group membership: original grouped cases (cross-validated grouped cases)** Palaeognaths	Galliforms	Neoaves	**Total**
Paleognaths	27 (27)	**1 (1)**	**2 (2)**	30
	28 (27)	**1 (1)**	**1 (2)**	30
Galloanserae	0 (0)	12 (12)	**1 (1)**	13
Neoaves	**2 (3)**	**1 (1)**	4 (3)	7
	3 (3)	**1 (1)**	3 (3)	7

Prior probabilities equal 86.0% of original grouped cases correctly classified; 84.0% of cross-validated grouped cases correctly classified.
Prior probabilities weighted according to observed number of cases in each group: 86.0% of original grouped cases correctly classified; 84.0% of cross-validated cases correctly classified.

These results can be visualized by plotting the two discriminant function scores (table 6.26) for each of the cases against each other (fig. 6.48). More positive values of discriminant function (DF) 1 are associated with a relatively broad metatarsal pad and a relatively larger distance across the proximal ends of digits II and IV, while more negative values are associated with relatively greater values of toetip width across digits II and IV. Palaeognaths tend to have positive scores along DF 1, while members of Galloanserae tend to have negative scores. Members of Neoaves tend to fall between palaeognaths and galloanserans along DF 1, but with considerable scatter. DF 2 most is most strongly associated with the digit II gap from the metatarsal pad, with more positive values indicating a relatively larger pad gap. Along DF 2, there is very little difference between palaeognaths and galloanserans, but neoavians tend to be more negative (but again with considerable scatter).

The errors of classification are equally interesting, however. One of the two kiwi (*Apteryx*) whose footprints are represented in our sample is always misclassified as belonging to Galloanserae, and the other is always assigned to Neoaves. The buff-crested bustard (*Eupodotis*) is always assigned to palaeognaths, and the kori bustard (*Ardeotis*) usually is. The sandhill crane (*Grus*) is always misidentified as a galloanseran. One of the domestic chickens is always assigned to Neoaves, and one of the cassowaries (*Casuarius*) usually is. The thick-knee (*Burhinus*) is always identified as a palaeognath. (Regrettably, the gastornithid prints [*Rivavipes*] were not included in the CVA, because it was not possible to measure all of the parameters used in the CVA in any of those three prints.)

The classification errors are readily seen in the discriminant function plots (fig. 6.48). The buff-crested bustard (*Eupodotis*) plots squarely among the emus, just as it did in the cluster analysis based on the same scaled footprint shape parameters (fig. 6.47), and the kori bustard (*Ardeotis*) is very nearby. The thick-knee (*Burhinus*) is also near the palaeognaths. The two kiwi points (*Apteryx*) are between the palaeognaths and the galloanserans along DF 1, with one plotting high along DF 2 (with a score as positive as that of any galloanseran) and the other lower (closer to the neoavians). The sandhill crane (*Grus*) is close to the galloanserans, and one of the *Casuarius* points takes a fairly negative, neoavian-like, score of DF 2.

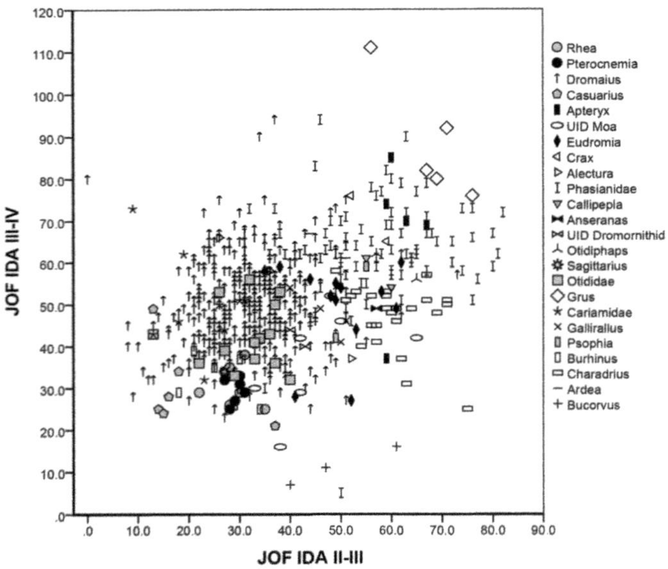

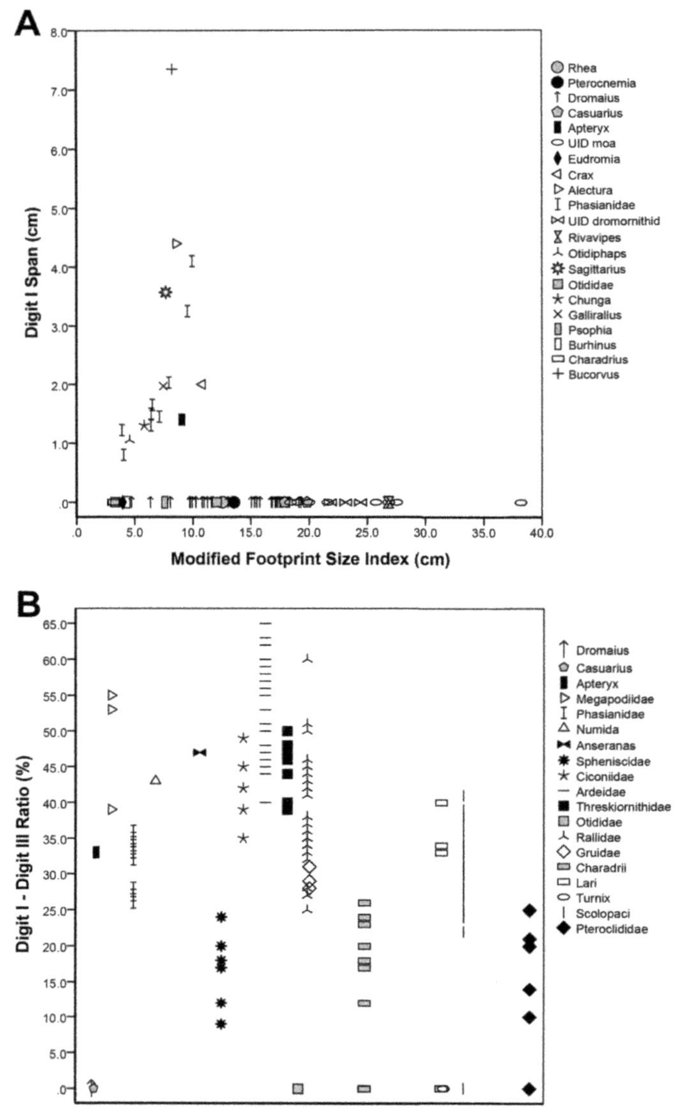

Above, **6.49.** Interdigital angle (IDA) II–III vs. IDA III–IV in footprints of ground birds other than ostrich; data cases are individual footprints.

Right, **6.50.** Digit I length in bird footprints and feet. *A*, Digit I span in footprints of ground birds (bird means treatment) plotted as a function of the footprint size index (square root of [footprint length × toetip width]). Digit I span is assigned a value of zero for birds that lack this toe, or for footprints in which there is no impression of digit I. Footprints of ratites other than kiwi lack a digit I impression, and kiwi cases plot near those of phasianids. Note that data points for bustards plot among those for the ratites. *B*, Digit I length plotted as a percentage of digit III length in several bird groups (data from HANZAB and BWP; the horizontal axis has no units but is simply a way of spacing out the various bird groups in the graph). Relative digit I length is highly variable within and across bird families, with much overlap across groups, and several groups have species in which digit I is absent.

Interdigital Angles. We have already noted that IDA II–IV can be (but isn't always) larger in small than in large ground birds (fig. 6.42*B*). Plotting IDA II–III vs. IDA III–IV (fig. 6.49) creates a pattern like a shotgun blast, although there may be a very weak tendency for the two interdigital angles to be positively correlated (Pearson's $r = 0.351$, $P < .001$, $n = 698$). Emus alone plot over much of the bivariate space, although they tend not to occur at the highest values of either interdigital angle. Phasianids and *Grus* tend to have relatively large values of both angles. Killdeer (*Charadrius*) footprints tend to have higher values of IDA II–III than IDA III–IV, while seriema (Cariamidae) prints do the opposite. Once again, kiwi plot among the galliforms, and bustards and thick-knees among the emus and rheas. Thus comparing interdigital angles has some value in characterizing individual groups of ground birds (cf. Falk et al. 2011), but does little to indicate the phylogenetic relationships among those groups.

Digit I Length. Digit I generally does not register in footprints of tridactyl non-avian dinosaurs, and so it is of less interest for our purposes than digits II–IV. However, for completeness sake I will briefly describe its occurrence in tracks and feet of ground birds (fig. 6.50). The relative length of digit I is highly variable, both within and among bird groups. Ostriches, rheas, emus, and cassowaries lack a digit I, but digit I is present in moa and kiwi, and in kiwi the first digit is about as long, relative to overall footprint size or the length of digit III, as it is in many galliforms. Tinamous either have a very small digit I, or lose it altogether (Davies 2002). Digit I ranges from reasonably long to absent in plovers, sandpipers, and sandgrouse (fig. 6.50*B*). Bustards are once again very like emus in losing digit I. Thus the proportional length of digit I seems not to be very useful in distinguishing among groups of ground birds. The substantial variability of digit I length in ground bird feet and footprints may have implications for interpreting the presence or absence of a digit I impression in footprints of non-avian dinosaurs (chapter 8).

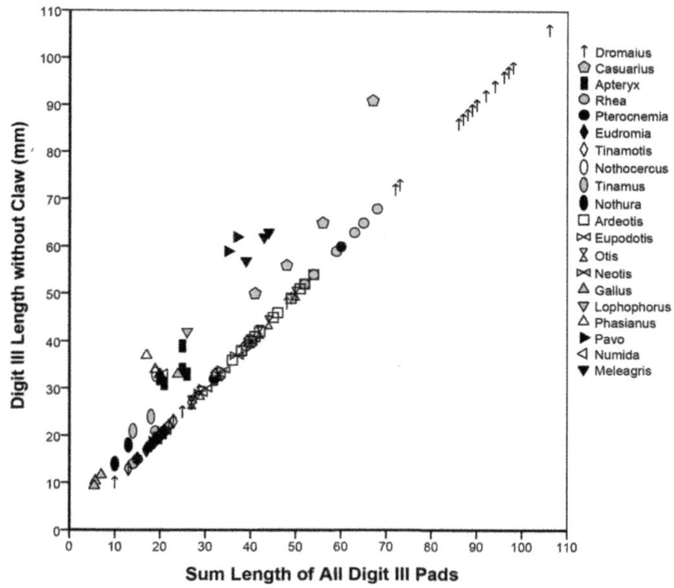

6.51. Extent of development of interpad spaces on the plantar surface of digit III in study skins of ground birds. The vertical axis is the total length of the fleshy portion of digit III, from the proximal end of the first digital pad to the base of the claw, whereas the horizontal axis sums the lengths of the digital pads (but not the interpad spaces) of digit III. For birds with little or no development of interpad spaces, the two variables will be identical. The greater the development of interpad spaces, the more the total length of the fleshy part of digit III will exceed the aggregate length of the digital pads. Although measurements made on dried study skins should be interpreted cautiously, the graph shows the difference between birds with an emu-like toe plantar configuration (little or no development of interpad spaces), and those with a galliform-like configuration (well-developed interpad spaces; fig. 6.34).

Subdivisions of Toes. As previously noted, the preservation of footprints of smaller species of our ground birds was often not good enough to allow comparison of the development of digital pads and claw lengths. However, some observations could be made, and I supplemented my observations of bird footprints (figs. 6.27–6.33) with observations on bird study skins (figs. 6.34–6.40). Among the bird groups I examined, there is a general tendency for the plantar surfaces of toes to be either like those of emus (fig. 6.3*E*, 6.3*F*) in showing digital pads that abut their neighbors with very little gap between them, or like those of galliforms (figs. 6.34, 6.38), in which substantial interpad spaces separate the digital pads (fig. 6.51). Remarkably, kiwi feet (figs. 6.36, 6.51) resemble those of galliforms in this feature, in addition to the other similarities to galliform feet and footprints already discussed. To lesser extent, cassowaries (fig. 6.35*D*, 6.35*E*) also seem somewhat galliform-like in the development of interpad spaces. In contrast, bustard feet (fig. 6.39) are again like those of emus in showing digital pads without interpad spaces. Tinamous (fig. 6.37) are particularly interesting, in that some forms (*Eudromia*, *Tinamotis*) have an emu-like plantar pattern, with little or no development of interpad spaces, while others (*Nothura*, *Tinamus*) are more like galliforms in their toe plantar patterns; equally interesting, tinamous that have an emu-like plantar pattern are also like emus in lacking digit I, while tinamous with a galliform-like plantar pattern are also like galliforms in having digit I. Rhea toes (fig. 6.35*F*–6.35*J*) tend to fuse digital pads that are separate in emus. Although preservation of the toes in moa mummies leaves much to be desired (fig. 6.35*K*–6.35*M*), the plantar surfaces appear to have been similar to those of emus.

The recently and regrettably extinct dodo (fig. 6.40; cf. Parish 2013; Claessens et al. 2015) had a very distinctive foot. Digit II was longer than digit IV, and digit II shows a greater number of digital pads than expected for this toe—perhaps some of them were actually interpad spaces.

The development of digital pads and interpad spaces seems largely to parallel overall footprint dimensions in the way that footprints of one ground bird group are either similar to or different from those of other bird groups. Determining why this might be so is beyond the scope of the present study, but brief speculation might be forgiven. One possibility is that ground birds with an emu-like foot are more cursorial than those with a more galliform-like foot. Or perhaps the galliform-like foot is better constructed for scratch-digging, or perching?

Trackway Parameters. At this point I should reiterate the warning with which our discussion of bird footprints other than those of emus began: Given the small number of individual birds for which trackway measurements were made, and in most cases the small number of trackway measurements for each bird, my discussion of differences in trackway pattern among different species and larger groups of ground birds can only be tentative. My observations will mainly consider the extent to which the trackway pattern of emus is similar to, or different from, that of other birds. In all of the graphs presented in this section, data points are for individual measurements, and not "bird means" summaries.

Unsurprisingly, bigger birds generally take longer steps than do smaller birds, and there is considerable overlap in this relationship between emus and other bird species (fig. 6.52A). The three secretarybird (*Sagittarius*) pace lengths are well above the main smear of points. At short stride lengths (both in absolute terms [fig. 6.52*B*] or relative to footprint length [fig. 6.52*C*]), increases in stride length result in rapid increases in the pace angulation. Although most birds plot among the emus in the stride length/pace angulation relationship, phasianids seem to have higher pace angulations at shorter stride lengths than do emus. Eventually the relationship flattens out at a pace angulation near 180 degrees,

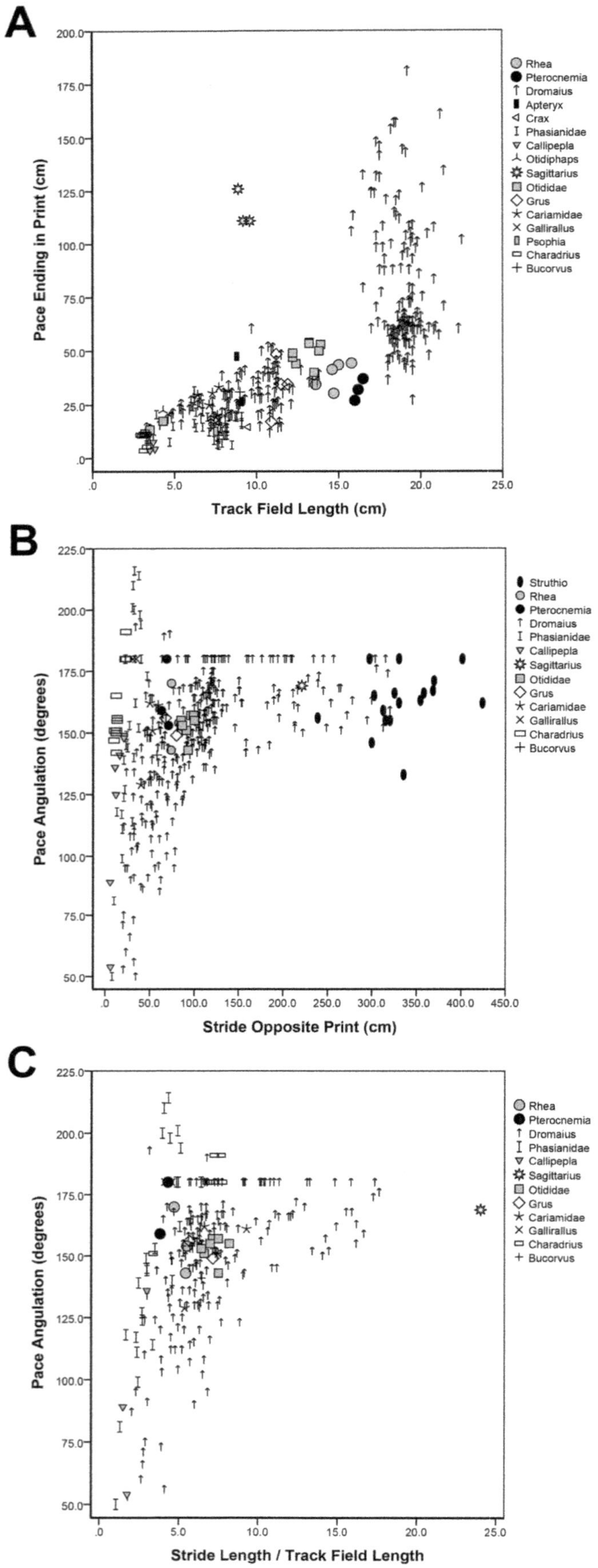

Left, **6.52.** Features of trackways of ground birds; data cases are measurements of individual steps and pace angulations (i.e., not the bird means treatment). *A*, Pace length vs. footprint length. Except for agitated birds that were trotting or even running, most bird trackways show a common and not very surprising trend of increasing pace length with increasing animal size (with footprint length as a proxy for the latter). Pace angulation increases with increasing stride length, both as absolute stride (*B*) and the stride/footprint length ratio (*C*), until the trackway becomes linear, with a pace angulation of 180 degrees. Values greater than 180 degrees indicate bird movements with change of direction. Ostrich data (*Struthio*) are not included in *C* because track field length as measured on tracks of other birds could not be measured on footprints of this species.

Facing, **6.53.** Additional features of trackways of ground birds. *A*, Across all of the ground birds in the study sample, there is no relationship between footprint length and footprint rotation (unlike the situation for emus alone, in which footprint rotation becomes less positive [i.e., decreasing tendency for footprints to turn outward with respect to the bird's direction of travel] with increasing emu size). Killdeer (*Charadrius*) and pheasant (phasianid) trackways show markedly greater inward (negative) rotation of footprints than do emu prints. *B*, Unsurprisingly, two different ways of measuring rotation of the same footprint are positively correlated. Once again phasianid and killdeer trackways differ from those of emus in showing negative rotation of footprints. *C*, Pace angulation vs. footprint size. Smaller birds (killdeer being an exception) more commonly show lower values of the pace angulation than do larger birds. *D*, As pace angulation increases, footprint rotation calculated by comparison with the following footprint decreases. Although most bird points plot among the emus, for any given value of the pace angulation pheasants have less positive rotation than do emus.

when prints of the left and right side are nearly aligned, except during sudden changes of direction on the part of the trackmaker.

Across the bird species in our sample, neither footprint rotation nor pace angulation shows obvious change with footprint (and thus bird) size (fig. 6.53*A*, 6.53*C*). Unsurprisingly, footprint rotation with respect to the previous print is positively correlated with footprint rotation with respect to the following print (fig. 6.53*B*). Footprint rotation seems to become less positive with increasing pace angulation (fig. 6.53*D*). Phasianid and *Charadrius* trackways show more inward (negative) rotation of footprints with respect to the direction of travel than do emu trackways (fig. 6.53*A*), while trackways of most other birds in our sample plot among emu points. Our limited data thus suggest that trackway measurements are only somewhat useful in discriminating among different kinds of ground bird trackmakers.

CONCLUSIONS

Like an emu on the move, in this chapter we have covered a lot of ground, so a brief summary of our results is in order; a more detailed summary will appear in chapter 7. Substrate conditions and bird gait were not found to have consistent effects on emu footprint shape. Measurements of the individual components of emu prints (particularly the gaps between the metatarsal pad and the most proximal digital

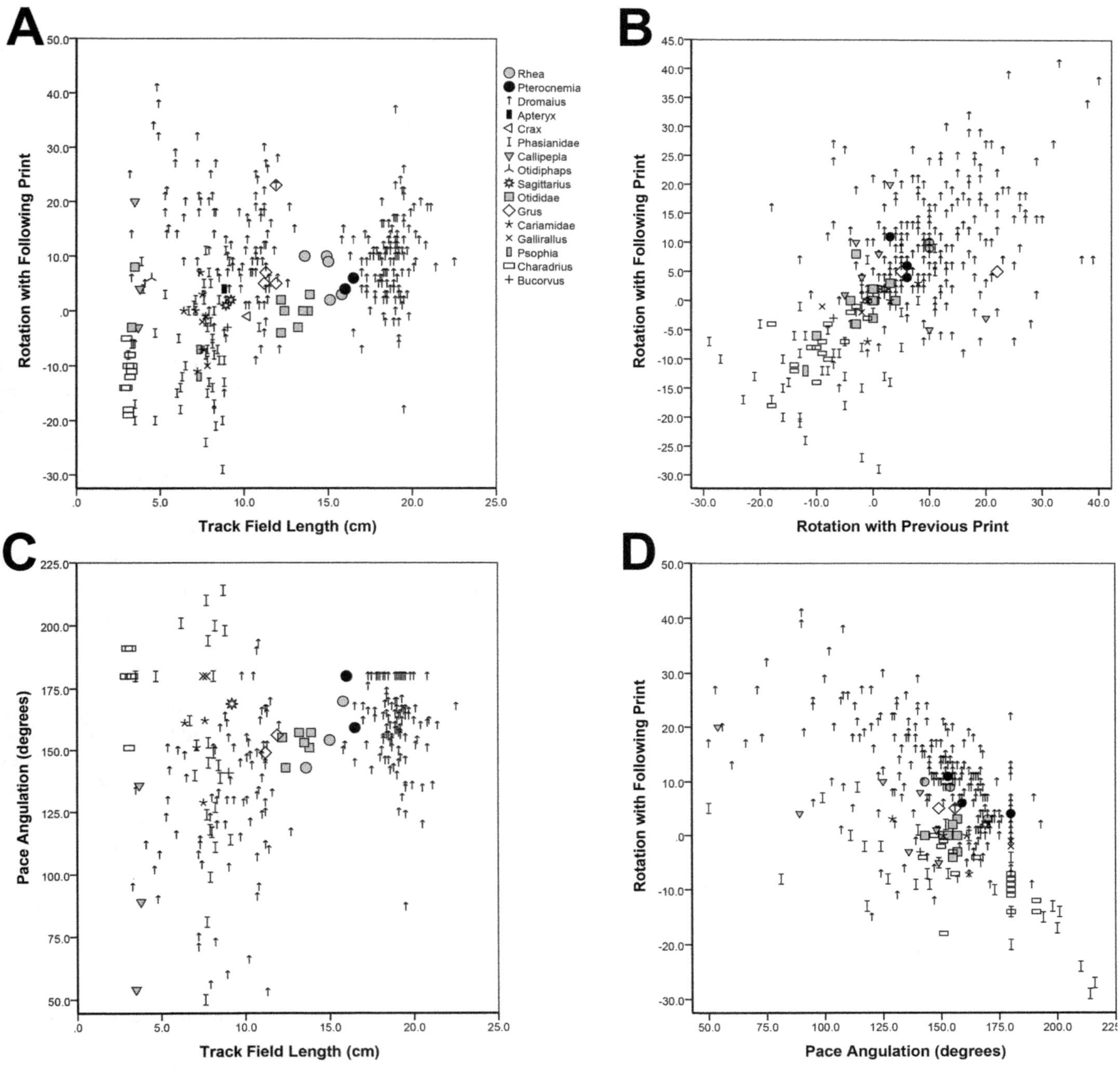

pads, and the clawmarks) tend to be more variable in shape than measures of overall digit or footprint length. The extent of shape variability in footprints of grown or nearly grown emus is about the same as the corresponding measures of the feet of emu study skins.

As emus grow from hatchlings to adults, footprint width becomes less, relative to footprint length, as interdigital angle II–IV decreases. Claws become relatively longer as emus get bigger, and digit IV becomes a bit longer with respect to the other toes.

Footprint samples that include tracks made by more than one ground bird species are generally more variable in shape than samples of emu footprints only, as one would hope. Unfortunately, footprint shape is only partially successful at indicating the phylogenetic relationships of trackmakers.

Although we are getting close to turning our attention to footprints of non-avian dinosaurs, we are not finished with footprints of emus and other ground birds. We will look at bird tracks once again in chapter 8, when we compare bird footprint shape variability with footprint shape variability in non-avian dinosaurs.

7 Summing Up the Comparative Analyses

"The time has come," the Walrus said, "To talk of many things."

Lewis Carroll, *Through the Looking-Glass and What Alice Found There*

In the preceding chapters, we have looked at many kinds of data in many different ways. Before we consider how this information might apply to interpreting the makers of non-avian dinosaur footprints, it seems appropriate to pause and review the main conclusions drawn from our comparative studies of foot skeletons, intact feet, and footprints. To make it easier for the reader to refer back to the relevant material, I will summarize the results by chapter.

Chapter 1. We started by considering whether foot skeletons of closely related archosaurs (crocodylians, ground birds, or non-avian dinosaurs) differ enough across species that it is possible to say that a collection of foot skeletons represents one, or more than one, species. The idea here was that if we cannot tell the foot skeletons of extinct forms apart, it is likely that we won't be able to tell their footprints apart.

We examined a series of measures of foot shape variability as applied to phalangeal and digital proportions. The first set of measures were essentially bivariate in character. The simplest of these was a ratio, that of the maximum to the minimum value of the phalangeal or digital dimension of interest scaled to a common value of another phalangeal or digital dimension (usually phalanx III1 length or digit III length), as applied to foot skeletons of specimens known or thought to belong to the same species, and also to multispecies samples. In addition to this very crude, quick and dirty, measure of shape variability, we also considered the coefficient of variation of the dimensions scaled to the common value. A third bivariate measure of shape variability examined was the coefficient of relative dispersion around the reduced major axis (D_d), using log-transformed versions of the phalanx and digit lengths in the calculations.

More complicated measures of shape variability took more than two phalangeal or digital dimensions into account. Phalangeal and digital dimensions were "GM scaled" by first log-transforming them, and then subtracting the mean of all the log-transformed dimensions from each log-transformed dimension. Multivariate scaling was also examined by using the coefficient of relative dispersion around the reduced major axis once again, but this time with each log-transformed variable being compared against the mean of all other log-transformed variables in the analysis besides itself.

An issue of concern in comparing intraspecific with interspecific samples was whether a "bogus" apparent increase in variability might arise simply from larger sample sizes that resulted from pooling of specimens from more than one species, without there being any real difference in shape among the species in the analysis. This was evaluated in two ways, using the maximum/minimum ratio of scaled parameters. The first approach involved random selection, repetitively sampling smaller numbers of specimens than the actual number of specimens for a single species (the moa *Dinornis robustus*) with a decently large sample size (24) of data cases, to see if the maximum/minimum ratio increased as the number of specimens selected increased. Each time the specified number of cases was randomly selected from among all of the *D. robustus* data cases, there was a maximum and a minimum value of the scaled parameter of interest, and thus a single value for the maximum/minimum ratio. After this sampling was done for the specified number of replicates (15), the median value of the maximum/minimum ratio across the replicates was calculated. This median value was calculated across the replicates for trials in which the specified number of randomly selected cases increased: 5 cases, 10 cases, and 15 cases randomly and repeatedly selected from among the 24 *D. robustus* cases in total.

The second approach was similar: randomly, repetitively sampling a smaller number of specimens than the actual number of specimens in pooled interspecific samples, to see if these smaller, multispecies samples still showed greater shape variability than the same number of intraspecific data cases. The median value of the maximum/minimum ratio of the scaled parameter across the repetitive trials was again calculated as the measure of variability in all these trials, the idea being that the higher the value of the median for a series of trials, the greater the difference between the largest and the smallest values of the scaled parameter in the selected data cases.

For a group of particular interest to this study, the moa of New Zealand, the question of whether phalangeal and digital proportion differences occurred across species was also considered by analyses of covariance (ANCOVAs).

Yet another kind of analysis also used the reduced major axis (RMA) between log-transformed variables. This analysis looked for allometry, both bivariate and multivariate, in both intraspecific and interspecific samples by determining whether the RMA slope took a value significantly different than 1.

So what did we find?

As expected, the median of the maximum/minimum ratio of scaled parameters across replicates did increase with an increasing number of data cases in the *D. robustus* sample. Even so, interspecific samples of data cases of both crocodylians and moa generally showed higher values of the maximum/minimum ratio than did intraspecific samples, even when the number of multispecies cases was the same as the number of intraspecific cases. For moa this was confirmed by the ANCOVA, with at least some moa species having significantly different phalangeal and digital proportions than others.

We then looked at the maximum/minimum ratio of scaled parameters for many individual species and interspecific groups of crocodylians, large ground birds, and non-avian dinosaurs. In most taxa, whether these were single species or interspecific samples, maximum/minimum ratios of scaled parameters were less for the bigger and more proximal phalanges than for the smaller, more distal phalanges. Often, but with many exceptions, multispecies groups showed higher values of the maximum/minimum ratio of scaled phalanx and digit lengths than did individual species.

For the large ("big seven") individual phalanges (II1–2, III2–3, IV1–2, all scaled to a common length of III1), a rough cut-off for thinking that more than one species is present in the sample is if the maximum/minimum ratio of scaled proximal phalanx lengths takes values of 1.5 or more, and for scaled distal phalanx lengths values of 2 or more, although interpreting how many more than one species are present is difficult, and maybe even impossible. For the overall lengths of digits II and IV scaled to a common length of digit III, values of the ratio of 1.3 or more likely indicate that more than one species is included in the sample.

The maximum/minimum ratio of scaled parameters admittedly is a crude measure of shape variability. It nonetheless is positively correlated with the more sophisticated shape variability measures described earlier, and these in turn are correlated with each other. They all tell the same story, albeit not always with the same clarity.

In *Alligator mississippiensis*, skeletal measures of overall pes size showed negative allometry with measures of overall alligator size, and skeletal measures of digit length showed negative allometry with metatarsal III length. Alligators seem to grow into their feet, and especially their toes. Some of the more distal non-ungual phalanges showed negative allometry with respect to some of the more proximal phalanges of the digits. The length of digit II shows negative allometry relative to the length of digit III, while digit IV looks isometric relative to digit III (but see the results from chapter 5 for what seem more definitive results about allometry of digit lengths). Most of the size-related changes in pedal proportions seen in alligators also seem to be true across crocodylian species; contrary to my expectations, interspecific relationships do not show stronger allometry than intraspecific relationships.

In moa, there were very few data for foot skeletons of very young birds, and so I was unable to say anything definite about intraspecific allometry of phalangeal or digital length proportions for the two species for which I had respectable sample sizes (*D. robustus* and *Anomalopteryx didiformis*). Across moa species, in contrast, the length of the digital portion of the foot compared with tarsometatarsus length shows negative allometry, a result similar to that seen in crocodylians. Across moa species, phalanx width does not show allometry with digit III length.

Chapter 2. As useful as it is to establish that variability in phalangeal and digital proportions is generally greater across than within species of potential trackmakers, this relationship is only part of the story. The pattern of similarity and dissimilarity of pedal shape is also important. Do potential trackmakers that are close relatives generally have foot skeletons more similar in shape to those of their close relatives than to forms that are much more distant relatives? This question was examined using the same data sets of specimens employed in chapter 1.

In a cluster analysis of GM-scaled lengths of non-ungual phalanges (data were not available for phalanx widths and ungual lengths) of crocodylians, *Alligator mississippiensis* tended to plot together, apart from other crocodylian species, as would be expected from the results reported in chapter 1, in which various shape measures were more variable across crocodylian species than within *A. mississippiensis*. There was a suggestion that conspecific specimens of other species also tended to be more like each other than like specimens of other crocodylian species. However, the cluster analysis showed no tendency for pedal skeleton shapes of different crocodylian species to link up in a hierarchical pattern consistent with their known phylogenetic relationships.

Similar analyses (using considerably more data than I had for crocodylians, including measurements of ungual lengths and phalanx widths) were carried out for digits II–IV of moa, extant ratites, and large ground birds across many clades. Principal component analyses (PCAs) and cluster analyses tended to group foot skeletons of some species (*Anomalopteryx didiformis, Pachyornis elephantopus, Megalapteryx didinus*) together, apart from other moa species. The two species of *Dinornis* also tended to group together, but members of each of those two species were not necessarily more like each other than like the other *Dinornis* species. The two stout-toed moa (*Emeus crassus* and *Euryapteryx curtus*) grouped together, but didn't form consistent subclusters comprising the two species). At higher levels in the dendrogram, however, the way the various moa clusters joined together did not reflect what are thought to be the phylogenetic relationships of the species.

The same is true for the corresponding analysis for extant ratites. Emus (*Dromaius novaehollandiae*) and each of the two rhea species (*Pterocnemia pennata* and *Rhea americana*) have phalangeal proportions more like members of their own species than like other ratite species. Cassowaries (*Casuarius* spp.) and kiwi (*Apteryx* spp.) are more like congeneric ratites than like specimens of other genera, and the two rheas are more like each other than like other ratites as well. Beyond that, the pattern becomes messy. Emus are more like rheas than like their close relatives, the cassowaries. Kiwi as a cluster link up with a supercluster composed of emus, rheas, and cassowaries. And we must not forget ostriches (*Struthio camelus*), whose didactyl foot is so different from those of any of the tridactyl ratites.

Some phylogenetic pattern persists if we look at big ground birds of all clades. Within moa, the clusters composed mainly of *Dinornis* spp., *A. didiformis*, *P. elephantopus*, and *E. crassus* + *E. curtus* continue to form, and link up to create a moa supercluster. The *Dromaius*, *Casuarius*, *Apteryx*, and *Rhea* + *Pterocnemia* clusters likewise persist, as does the more inclusive emu + rhea cluster. But untidiness is present at both low and high hierarchical levels. Emus actually join bustards (*Ardeotis*) before both of them join with rheas. The moa *M. didinus* joins the goose *Cnemiornis*, and then kiwi, and finally seriemas (*Cariama*), forming a cluster that then joins the main moa-cassowary-rhea-emu-bustard supercluster, after which all of these groups together form a final link with *Genyornis*.

Turning at last to digits II–IV of non-avian dinosaurs, PCAs and cluster analysis cause theropods other than dromaeosaurids to tend to group together (albeit along with some small ornithopods!), as do dromaeosaurids, many long-clawed, medium-sized ornithischians (along with, interestingly enough, *Plateosaurus*), and big ornithopods. Among large theropods, there is no obvious tendency for members of the same clade (e.g., allosaurs and tyrannosaurs) to be more like each other than like other large theropods. Within large ornithopods, foot skeletons of *Iguanodon bernissartensis* and to a lesser extent *Edmontosaurus* spp. tend to be more like others of the same kind than like other kinds of big ornithopods.

The bottom line in all this seems to be that phalangeal and digital proportions can tell low taxonomic categories apart—that is, conspecific, and often congeneric, forms tend to have foot skeletons whose proportions are more like each other than like those of more distantly related forms. There is also some suggestion of a hierarchical phylogenetic signal in pedal proportions, particularly at lower hierarchical links in dendrograms, but there are enough exceptions to ring many alarm bells. Similarities in pedal skeletal proportions do not necessarily reflect patterns of phylogenetic closeness.

An interesting trend is seen across clades of ground birds, non-avian theropods, and ornithischians: the pedal phalanges become relatively broader as one goes from small-bodied to large-bodied forms. However, this does not necessarily occur within individual clades (e.g., moa).

For interpreting tridactyl dinosaur footprints, we would ideally like to do more than recognize that there are (however many) different kinds of trackmakers represented in a track assemblage. We would also like to say something about what kinds of trackmakers there were. Theropods tend to have relatively longer non-ungual phalanges, shorter unguals, and narrower toes than do ornithischians.

Chapter 3. For ground birds and non-avian dinosaurs, the influence of phalangeal proportions on overall toe shape, as it might be expressed in the toe impressions of footprints, was examined in yet another different way than in previous chapters. We tested a hypothesis that the overall shape of toe impressions of theropods would take a V-shape, while the overall shape of toe impressions of ornithopods would take a U-shape.

Toe-tapering profiles (TTPs) were created by examining the transverse distal widths of pedal phalanges as a function of the cumulative lengths of the individual phalanges. The effects of size were factored out by expressing lengths and widths as percentages of the length of the first phalanx of digit III. For the more complete TTPs, toe-tapering curves (TTCs) were mathematically fit, using mean values for species represented by more than one specimen.

The hypothesized difference in overall digit shape between theropods and ornithopods did not show up in the TTPs and TTCs constructed from measurements of foot skeletons. Instead, we saw that larger dinosaurs had steeper TTPs and TTCs than smaller dinosaurs, whether they were non-avian theropods, birds, or ornithischians. As seen in previous chapters, ornithischians tend to have stouter toes, and large ornithopods much steeper TTPs, than other kinds of dinosaurs. Some species or genera within clades have markedly different TTPs and TTCs than members of the same clade, and some species have TTPs and TTCs more like those of species in other clades than like those of members of the same clade. Thus the examination of TTPs reinforces conclusions from previous chapters about the extent to which pedal proportions discriminate among potential groups of trackmakers, and the degree to which pedal proportions are consilient with phylogenetic propinquity.

Chapter 4. As previously seen in chapter 1, American alligators show unambiguous negative allometry of skeletal measures of foot length with overall body size and leg length. The length of metatarsal III is negatively allometric with respect to total alligator length (snout to tailtip) and femur length. The ratio of metatarsal III length to digit III length increases with respect to overall foot length (metatarsal III + digit III length) as very small alligators grow, but it flattens out among larger individuals. Alligators appear to grow into their feet and their toes, and the limited data suggest that the same is also true for other crocodylian species. Digit width may be positively allometric with respect to digit length in alligators, bigger individuals being stouter-toed.

Domestic chickens may show positive allometry of tarsometatarsus length with respect to the combined length of the femur and tibiotarsus, a trend that contrasts with the metatarsal length vs. femur length relationship in alligators. Like alligators, however, chickens show negative allometry of digit III length with respect to femur length and tarsometatarsus length, and also relative to the combined lengths of femur and tibiotarsus. The ratio of femur + tibiotarsus + tarsometarsus length to digit III length is positively correlated with respect to a linear measure of overall bird size, the cube root of body mass. Like alligators, chickens seem to grow into the digital portions of their feet.

Study skin measurements of both emus and greater rheas ontogenetically show negative allometry of the length of digit III relative to the length of a rough measure of the length of the tarsometatarsus (as measured on study skins). The ratio of rough tarsometatarsus length to digit III length, plotted against the combined length of the tarsometatarsus and digit III, shows a pattern similar to that seen in alligators, in which the ratio increases very rapidly with increasing size among young birds, but then flattens out among large birds. Thus the ground bird data indicate that, as with alligators, the birds grow into their toes, consistent with the skeletal results in chapter 1.

Data for non-avian theropods are much more limited than for large ground birds. In *Gorgosaurus libratus*, however, the length of metatarsal III shows negative allometry against femur length, and also femur + tibia length, as with alligators, but unlike chickens. *Tarbosaurus bataar* likewise shows negative allometry of metatarsal III length with femur length. Unlike alligators, emus, rheas, and chickens, *Coelophysis bauri* and *Gorgosaurus libratus* show positive allometry of digit III length relative to the length of metatarsal III, at least over the size range for which data are available. The same seems to be true across species of large non-avian theropods.

Like theropods, ornithopods across species shorten the length of the metatarsus relative to femur length as they get larger. Small ornithopods overlap small non-avian theropods in the relationship between metatarsal III length and femur length, and medium-sized and bigger ornithopods similarly are close to basal theropods and allosaurs in this relationship. In contrast, medium-sized and large ornithopods have shorter metatarsals for any given femur length than do coelurosaurs. For a given femur length, ornithopods have a shorter digit III than do non-avian theropods. On the other hand, ornithopods fall along the same metatarsus length/digit III length relationship as do theropods. Combining all of these relationships, medium-sized and large ornithopods show a shorter digit III relative to the combined lengths of femur + tibia + metatarsal III than do theropods.

The seemingly conflicting results about the digit III length/limb length relationship seen in alligators and ground birds, on the one hand, and large theropods on the other, are possibly reconciled by examining the way the ratio of limb length (femur + tibia + metatarsal III) to digit III length changes from very small to very large non-avian theropods, with "total" limb length (femur + tibia + metatarsal III + digit III) serving as the proxy for overall dinosaur size. The ratio increases from small non-avian theropods to forms with a total limb length of about a meter or so, consistent with what is seen in the alligators and the ground birds. Above this size, however, the trend reverses, and the ratio decreases. Unlike theropods, ornithopods show a constant increase in the limb length/digit III length relationship from small to very large forms.

On the assumption that the limb length/digit III length relationship should affect the relative step lengths of dinosaurs, we then examined the relationship between stride length and footprint length for considerably more than 2,000 non-avian dinosaur trackways. There were too few trackways for very small non-avian dinosaurs for us confidently to say whether the stride/footprint length ratio increases from very small dinosaurs to animals of moderate size, but the ratio unambiguously declines from moderate-sized to very large dinosaurs in trackways attributed to both theropods and ornithopods. This is exactly as one would predict from the limb length/digit III length relationship seen in theropods, but not what one would predict for ornithopods. In any case, what the trackway data indicate is that differences in relative step length between trackways that differ in footprint length do not by themselves necessarily indicate different kinds of trackmaker; step length has to be scaled against footprint size for meaningful comparisons.

Chapter 5. Turning from foot skeletons to intact feet, we first looked at intraspecific variability in pes size and shape in *Alligator mississippiensis*. Measures of hindfoot size (overall pes length, digit III free length, a simple size proxy created as the mean of several log-transformed pedal measurements) showed negative allometry with measures of overall alligator size (total length, snout-vent length, shoulder-hip length, leg length excluding the autopodium). Alligators grow into their feet, as previously seen in the chapters dealing with skeletal data. By some measures, female alligators have relatively larger hindfeet than do males, but males have relatively longer tails. There was no sign of regional variation in the alligator size/hindfoot size relationship in animals from Louisiana, South Carolina, and Florida.

As the alligator foot gets bigger, the first and fourth toes become relatively longer with respect to the second and third toes, the fourth toe (measured as the medial hypex length) becomes proportionately longer than the first toe, and the second and third toes remain constant in their relative lengths. (Some of these results differ from what we saw with alligator foot skeletons, but because there were many more measurements made on intact feet than on skeletons, I am more inclined to trust the results derived from the intact feet.) As alligators get bigger, the hypex indentations separating digit III from its neighboring toes become relatively deeper. The claws become relatively shorter than the rest of their respective toes. Most of these shape changes are, however, rather subtle. In many cases bivariate relationships look nearly linear for alligators of most sizes, with departures from the overall trend mainly affecting the very biggest animals, nearly all of which are males. Although this was not analyzed quantitatively, the toes of alligators visually become relatively stouter with increasing animal size, and the claws change from needle-like in shape to being relatively broader. In large alligators, claws are sometimes blunted or even beveled by wear. About one-fourth of the alligators in my sample had an interpad space separating the first and second digital pad of digit I, but such spaces are not as clearly defined as they are in the toes of some bird feet (e.g., galliforms). No differences in foot shape were seen between the sexes, but there were possible regional differences.

Bivariate and multivariate measures of pedal shape variability comparable to those examined for foot skeletons of crocodylians, ground birds, and non-avian dinosaurs were analyzed for intact alligator feet. Maximum/minimum ratios of scaled measurements commonly take values of 2 or less. The higher values of variability measures are seen in bivariate comparisons involving the lengths of the pads of digit I, in comparisons of the relative lengths of the claws and the fleshy parts of toes, and in some comparisons involving hypex lengths.

Chapter 6. After analyses concerned mainly with feet, in this chapter we finally turned our focus to footprints, starting with an analysis of footprint shape in the emu (*Dromaius novaehollandiae*). The highest quality footprints made by adult emus had a soil water content of at least 20% of soil dry weight, and a clay content 10% or better of soil dry weight. There were no consistent effects of substrate conditions or bird gait on measurements of footprint dimensions.

For individual emus, the coefficient of variation (CV) for footprint measurements is usually 10% or less, and measurements of footprint width are more variable than measurements of footprint length. Measurements of the individual components of footprint length or toe length generally show CVs somewhere between 5% and 30%, and are usually larger than CVs of overall footprint or toe length; measurements of the gap between the metatarsal pad and the proximal end of a toe mark, and of claw lengths, are particularly variable. Interdigital angle (IDA) II–III is usually more variable than IDA III–IV. Interestingly, the range of values, means, and CVs observed individually for the three adult emus represented by the largest number of footprints in our sample are about the same as for all footprints of adult emus combined.

Both bivariate and multivariate approaches were used in investigating the extent of ontogenetic shape change in emu footprints. In most cases in which a measurement could be made on emu study skins that is equivalent to a measurement made on footprints, the relative variability of different footprint parameters was similar to that seen in footprints.

The length of digit IV is positively allometric with respect to the lengths of digits II and III (as with alligators), and claws appear to become relatively longer as emus get bigger (unlike alligators). A very consistent result is that measures of footprint width are negatively allometric with respect to footprint size. Related to this, the value of IDA II–IV decreases with increasing print length; decreasing values of IDAs with increasing footprint size occur not only among emus of different size, but also for individual birds as they grow. With increasing size, emu footprints become relatively narrower, with digits II and IV relatively closer together. With increasing footprint size, emu prints show a relative decline in the projection of digit III beyond the other two digits, but this is mainly due to the decrease in IDA II–IV with increasing bird size.

Several measures of intraspecific footprint shape variability in emus were examined. Some of these were simple bivariate measures, and others more complex multivariate measures. Bivariate measures, as in earlier chapters, included the coefficient of relative dispersion about the reduced major axis (D_d), using both raw and log-transformed data, the standard error of the estimate (SEE) of regression equations involving log-transformed variables, and maximum/minimum ratios of footprint parameters scaled to a common parameter. Separate calculations were done using all of the emu footprints in our sample, and also on a more restricted set of well-preserved footmarks.

For raw data, D_d values in bivariate comparisons involving smaller subunits of footprints (metatarsal pad dimensions, metatarsal pad–digit pad gaps, digital pad lengths, and claw lengths), compared against the length of the first pad of digit II or III, are generally around 15%–25%. Values based on log-transformed data are generally less than 10%. D_d values are, unsurprisingly, a bit less when calculated for high-quality prints than for the entire emu footprint sample (most notably for comparisons involving claw lengths). Comparisons involving metatarsal pad–digital pad gap lengths and claw lengths are the most variable. Comparisons involving the smaller, more distal digital pads of a toe are more variable than those involving the larger, more proximal digital pads. Bivariate comparisons in which parameters are compared with entire digits have D_d values of 15% or less (raw data; 5% or less with log-transformed data). D_d values are usually somewhat less in the bird means treatment than when individual footprints constitute the data cases.

Maximum/minimum ratios of footprint dimensions scaled to a common value (e.g., the length of the first digital pad of digit III) generally take values of 4 or less; for high-quality footprints, most maximum/minimum ratios have values of about 2 or less. However, scaled claw lengths can be considerably more variable, with maximum/minimum ratios of 4–8, even with high-quality footprints. The lengths of gaps between the metatarsal pad and the most proximal ends of digits again are among the more variable scaled footprint parameters, although not to the same degree as claw lengths. For feet of emu study skins, maximum/minimum ratios of scaled parameters are only a bit less variable than their counterparts from high-quality emu footprints. As with footprints, scaled claw lengths of study skin feet are the most variable parameters.

Multivariate SEE relationships in which log-transformed footprint dimensions are compared against the mean of all other log-transformed footprint dimensions (all computed using the "emu means" treatment) show higher values for pad gaps and claw lengths, and lower values for overall digit lengths, lengths of digits excluding claws, hypex lengths, and transverse dimensions. For digits II and III, SEE values increase distally from proximal pads to more distal pads to claws.

Our sample of footprints made by birds other than emus was limited, but for those ground bird species in which an individual bird (or a group of birds of similar size) was represented by at least 10 footprints, variability (as measured by the coefficient of variation, CV) is comparable to what we observed for individual emus.

Analyses of the intraspecific shape variability of alligator feet and the feet (of study skins) or footprints of five ground bird species (emus, greater rheas, chickens, kori bustards, and great bustards) yield results that are consistent enough to suggest that they may be applicable to the footprints of tridactyl non-avian dinosaurs. Values of the coefficient of relative dispersion about the reduced major axis in bivariate relationships should be 5%–25% (for raw data; the values will be less for log-transformed data) for footprints made by the same trackmaker species, and maximum/minimum ratios of scaled parameter lengths should usually be 5 or less. Values at the higher end of the range will be associated with measurements of claws and/or digital pads, and lower values with measurements of overall toe lengths.

As was done with skeletal foot proportions (chapter 1), we compared the variability of intraspecific and interspecific footprint proportions, to see if the kinds of linear measures of footprint shape used in this study are even capable of recognizing when footprints of more than one bird species are included in a track sample. The analyses were done using mean values of footprint parameters for each individual bird ("bird means" treatment). We employed the usual measures of D_d and SEE, and also the range of values of footprint dimensions scaled to a common value, as measures of variability. In nearly all bivariate comparisons interspecific samples showed greater variability than for emus only, so it is possible in at least some cases to detect interspecific variability using the kinds of linear measurements of footprint shape employed in this study.

We then analyzed the extent to which footprint shape serves as a proxy for the phylogenetic relationships of ground birds, examining measures often used to characterize footprint shape in studies of non-avian tridactyl dinosaur footprints. Simple plots of footprint width against footprint length are not very effective at discriminating among trackmakers, except for those species with very distinctive pedal proportions; most of the footprints of different bird species in our sample plot close to a common width/length relationship.

The same is true in plots of the digit III projection against backfoot length. The footprints of most ground bird species in our sample fell along a common trend. Our ontogenetic series of emu prints plotted right among prints of bird species of different sizes, indicating that this kind of plot does not distinguish intraspecific ontogenetic size-related series of footprints from samples of footprints made by species of different body size.

Relative lengths of digits also do not work very well as proxies for phylogeny. They serve to distinguish some bird groups from others, but do not group them together in a manner reflecting the relationships of those groups. The absence or presence of digit I, or its relative length when it is present, is especially feckless in this.

The relative size of interdigital angles II–III and III–IV has some value in distinguishing individual groups of ground birds, but does little to indicate the phylogenetic relationships among those groups. Similarly, the dimensions of the "anterior triangle" (the relative lengths of line segments connecting toetips II–III, III–IV, and II–IV) seem of little value in assigning footprints of birds to supraspecific groups in anything approximating what is thought to be their true phylogenetic relationships.

The closest approximation of a topology of footprint shape relationships among ground bird species to their evolutionary relationships was achieved using GM scaling of 18 footprint parameters: length and width of the metatarsal pad; the lengths of the pad gaps separating the metatarsal pads from the proximal ends of digit impressions; overall lengths of the impressions of digits II–IV from their proximal ends to their tips; digit hypex lengths; distances between the proximal ends of digits II, III, and IV; and distances between the tips of digits II, III, and IV. A cluster analysis of these GM-scaled variables generally groups footprints of birds of the same species together, or nearby. It also commonly (but not invariably) groups galliform birds together. However, there remain some glaring discrepancies between tree topology based on footprint proportions and phylogenetic relationships. The overall message seems to be that with a sufficiently large number of characters being examined, footprint shape has some utility in recognizing the major groups to which trackmakers belong, but that one should be prepared for nasty surprises, in which ground birds (and non-avian tridactyl dinosaurs?) that were not at all closely related would make footprints that are very similar in shape. This is, not surprisingly, much the same result we saw in chapter 2 in comparisons of pedal proportions of foot skeletons.

And now we are finally ready to turn our attention to dinosaur footprints and trackways.

Noah's Ravens: Interpreting the Makers of Tridactyl Dinosaur Footprints of the Newark Supergroup, Early Jurassic, Eastern North America

8

What a wonderful menagerie! . . . At first men supposed that the strange and gigantic races which I had described, were mere creatures of imagination, like the Gorgons and Chimaeras of the ancient poets. But now that hundreds of their footprints, as fresh and distinct as if yesterday impressed upon the mud, arrest the attention of the sceptic on the ample slabs of our cabinets, he might as reasonably doubt his own corporeal existence as that of these enormous and peculiar races.

Edward Hitchcock (1858: 190)

Pliny Moody, bless his heart,
Gave Ed Hitchcock quite a start.
"They're bird tracks!" poor old Eddie swore—
Alas! They were dinosaur.

Geological clerihew by J. O. Farlow;
from Sarjeant (1983–1984)

INTRODUCTION

After considering foot skeletons of non-avian dinosaurs, ground birds, and crocodylians; the fleshed-out feet of birds and alligators; and footprints of ground birds, in previous chapters, we are finally ready to tackle the matter that prompted my interest in such things in the first place: interpreting how many kinds of dinosaurs, and what kinds of dinosaurs, were responsible for an ichnofauna of tridactyl footprints. The footprint fauna we will examine is the classic assemblage of beautifully preserved three-toed dinosaur footprints from the early Mesozoic Newark Supergroup of eastern North America; I will specifically be looking at the classic ichnogenera *Grallator*, *Anchisauripus*, *Eubrontes*, and *Anomoepus*.

These are, of course, the tracks that were studied by Edward Hitchcock (e.g., 1858, 1863) and subsequent workers (e.g., Thorpe 1929; Lull 1953; Baird 1957; Olsen 1980, 1995; Weems 1987, 1992, 2003, 2006*a*; Demathieu 1990; Weems and Kimmel 1993; Olsen et al. 1998; Farlow and Galton 2003; Olsen and Rainforth 2003; Rainforth 2003, 2005, 2007; Smith and Farlow 2003; Getty 2005; Dalman 2012; Dalman and Weems 2013; Getty et al. 2012, 2015, 2017). The same ichnogenera have been interpreted as present in other ichnofaunas around the world, most often from early Mesozoic sites, but also even from Cretaceous localities (for an extensive, but by no means exhaustive list: Demathieu 1990; Gierliński 1991, 1995, 1996*b*; Leonardi 1994; Thulborn 1994, 1998; Lockley et al. 1996, 2008*a*, 2011*a*, 2011*b*, 2014*b*, 2014*g*, 2015; Gierliński and Pieńkowski 1999; Avanzini 2000; Demathieu et al. 2002; Haubold and Klein 2000, 2002; Lockley and Meyer 2000; Clark et al. 2004, 2005; 2009; Niedźwiedzki and Pieńkowski 2004, 2016; Rodríguez-de la Rosa et al. 2004; Gierliński and Niedźwiedzki 2005; Avanzini et al. 2006, 2012; Azuma et al. 2006; Hamblin et al. 2006; Klein et al. 2006, 2011; Lockley and Eisenberg 2006; Lockley and Gierliński 2006, 2009; Lucas et al. 2006; Matsukawa et al. 2006; Milner et al. 2006; Gand et al. 2007; Hadri et al. 2007; Hunt and Lucas 2007; Ishigaki et al. 2009; Sullivan et al. 2009; Wilson et al. 2009; Cook et al. 2010; Ishigaki 2010; Li et al. 2010; De Valais 2011; Diedrich 2011; Lockley 2011; Niedźwiedzki 2011; Lagnaoui et al. 2012; Moreau et al. 2012*a*, 2012*b*, 2014, Xing et al. 2014*f*, 2014*i*, 2014*k*, 2015*c*, 2015*e*, 2015*g*, 2015*h*; Xing et al 2016*a*–2016*e*, 2016*g*, 2016*i*, 2016*k*, 2016*m*, 2017*a*, 2017*b*; Fiorillo et al. 2014*a*; Marzola and Dalla Vecchia 2014; Castanera et al. 2015, 2016; Li et al 2015; Pieńkowski et al. 2015; Scisio et al. 2016; Wang et al. 2016; Xing and Lockley 2016; Wagensommer et al. 2016; Bordy et al. 2017).

A generation ago, the great French ichnologist Georges Demathieu (1990) did a brief study of three of the then-recognized Newark Supergroup tridactyl dinosaur ichnospecies: *Anomoepus intermedius*, *Eubrontes giganteus*, and *Grallator sillimani*. Demathieu examined several of the same specimens that I will consider in this chapter, and made some of the same, or at least similar, measurements of them that I made. Demathieu (1990: 108) concluded that the "three ichnogenera are well separated statistically; this is particularly true of *Eubrontes* and *Grallator*, which cannot be synonymized." As will be seen, Demathieu was quite prescient. Using more footprint characters than Demathieu did, measured on more specimens than Demathieu measured, and analyzed differently than by Demathieu, I came to conclusions about the makers of small *Eubrontes* (Demathieu's *Grallator*), large *Eubrontes*, and *Anomoepus* that are remarkably consistent with his interpretation.

In addition to *Grallator, Anchisauripus, Eubrontes,* and *Anomoepus*, I give some attention to a fifth ichnogenus. *Kayentapus* was initially described by Welles (1971) from the Early Jurassic of the American Southwest, and an ichnospecies of this form was later recognized by Weems (1992) from the Late Triassic of Virginia. This ichnotaxon has also been recognized at several other sites around the world (Mietto et al. 2000; Gierliński and Niedźwiedzki 2005; Piubelli et al. 2005; Lockley et al. 2006*a*; Avanzini and Petti 2008; Petti et al. 2008, 2011; Li et al. 2010; Wagensommer et al. 2012; Xing et al. 2013*c*; 2014*b*, 2014*d*, 2014*f*, 2014*h*; 2015*c*; Wang et al. 2016; Sciscio et al. 2017).

Let me be emphatic about what the present chapter is not: It is not a systematic revision of the ichnofauna of the Newark Supergroup. That would be a career-long project in itself (see Rainforth [2005] for an inkling of the magnitude of such an undertaking). As noted by Lockley et al. (2014: 174), "There has been little consistency in the morphometric methodology adopted by various workers for differentiating theropod tracks and little attempt to undertake systematic revisions based on the results of such morphometric analyses."

Well. I am not trying to do a study exactly of the kind, the absence of which was noted by Lockley and his colleagues, but rather something parallel to it. My goals are far more modest: using the quantitative approaches applied in previous chapters, and the results of those analyses, to make a minimum estimate of how many different kinds of dinosaurs, and what groups of dinosaurs, made the classic footprints to which such names as *Grallator, Anchisauripus, Eubrontes, Kayentapus,* and *Anomoepus* have been applied. As suggested by Hornung et al. 2016: 222), "numerical analysis may expose a higher diversity of trackmakers than is obvious from qualitative analysis of the track record."

I don't claim that what I have done is anything more than a first pass at the problem. In the time I have had, it has not been possible for me to deal with all of the potentially measurable specimens of interest. I am also acutely aware of the fact that the data used in this study in most cases are taken from a small number of specimens in each trackway—often a single footprint. As was undoubtedly apparent in my analyses of alligator feet and emu footprints, I am much happier with sample sizes in the many-tens or even hundreds. I nonetheless hope that my data constitute a representative sample of the relevant ichnological diversity, but the conclusions that I reach about the identities and number of kinds of trackmakers are best regarded as hypotheses for further testing.

Particular emphasis in this study is placed on *Grallator, Anchisauripus,* and *Eubrontes*, which Rainforth (2005) interpreted to be various nominal species within *Eubrontes*. I consider two questions about these footprints. Is the degree of variability seen among specimens assigned to *Eubrontes* comparable to, less than, or greater than that seen in known intraspecific foot and footprint samples? Are the kinds of footprint shape changes seen with increasing size across *Eubrontes* consistent with the kinds of changes seen with increasing size in foot and footprint samples known to be intraspecific growth series, or are they more like the kinds of changes seen from individuals of different size across species?

We will need some "measuring sticks" that allow us to assess whether the amount of shape variability seen in a sample of footprints is consistent with their having been made by a single species of trackmaker, or whether it is more likely that more than one species was responsible for the prints. The emu footprints collected in this study (chapter 6) unambiguously represent a single trackmaker species, and so shape variability of footprints within emus will serve as one criterion for evaluating whether the shape variability of a sample of non-avian dinosaur footprints is equivalent to, less than, or greater than the amount of shape variability seen among footprints known to have been made by a single species.

This is obviously not an infallible criterion. It is too much to suppose that intraspecific footprint shape variability would be uniform among all species of ground birds or non-avian dinosaurs. Furthermore, unlike the emus, the dinosaurs whose trackways will be compared with the emus—even those dinosaurs whose footprints are interpreted as potentially representing homogeneous groups (that is, coming from a single kind of trackmaker)—did not come from individuals living at close to a single instant in time. But the emus provide one standard of comparison, which is one more than we previously had. Comparisons will also be made with intact alligator feet; crocodylian, non-avian dinosaur, and bird foot skeletons; and bird study skins. All of these were examined in their own right in earlier chapters, but now they will be considered again in the specific context of what light they shed about the biodiversity reflected in the Newark Supergroup dinosaur footprints.

As in previous chapters, the approach taken here will of necessity require numerous tables of data and graphs. In an attempt once again to make the analysis more user-friendly, so that the reader will not be wandering about lost in statistical thickets, the data tables presented in the main text will be those most germane to the discussion. Other tables—particularly those that are very long, but that nonetheless provide important information—will be relegated to the Appendix.

Table 8.1. Tentative assignment of Newark Supergroup tridactyl dinosaur footprint specimens to ichnospecies or other categories; these are specimens used in some way during this study, for trackway data and/or footprint shape analysis. Where I could find an explicit assignment of a specimen to an ichnotaxon in Olsen and Rainforth (2003) or Rainforth (2005), I simply followed that assignment. Where I didn't find such an explicit assignment, I sometimes used Rainforth's (2005) account of old and current names to convert from the name on the specimen label to her terminology. Of course, this assumes that the old names were correct identifications to begin with, which may be a generous assumption. In other cases I made my own interpretation of what the tracks should be called. Locality names refer to sites in Massachusetts, unless otherwise indicated. I have tried to group ichnotaxa of the same kind together, except when more than one form occurs on the same catalogued slab. Fm = formation; UID = unidentified. For some trackways there may actually be more than the number of footprints described in the "Comments" column; my count is of the prints that clearly belonged to that trackway and/or were in some way measurable.

Catalog number and locality	Old or traditional name (often name on specimen label)	Current ichnotaxon or other name used here	Comments
AC 4/1a; Pliny Moody home, South Hadley	*Grallator cursorius*	*Eubrontes cursorius*	Trackway of four footprints (figs. 8.1, 8.15*A*)
AC 12/3; Turners Falls	*Grallator tenuis*	*Eubrontes tenuis*	Two trackways of three footprints each
	Anomoepus isp.	*Anomoepus scambus*	Sequence of five footprints
AC 30/1; Lily Pond	*Grallator cuneatus*	*Eubrontes josephbarratti*	Several small footprints (figs. 8.6, 8.15*D*, 8.15*E*)
	Anchisauripus tuberosus	*Eubrontes loxonyx*	My prints B, G, I, L, and O in nice trackway (figs. 8.6, 9.15*K*)
AC 60/33; Turners Falls	*Grallator*?	*Eubrontes* isp.	Fig. 8.4*G*
AC 64/2; Turners Falls	*Grallator* or *Anchisauripus*	*Eubrontes* isp.	Sequence of three small prints
YPM VP 003314; Horse race, Montague	*Anchisauripus tuberosus*	*Eubrontes loxonyx*	
YPM VP 003323; Middlesex County, Connecticut	*Anchisauripus sillimani*	*Eubrontes sillimani*	Figs. 8.4*A*, 8.15*F*
AC 2/3; Smith's Ferry	*Anchisauripus exsertus*	*Eubrontes exsertus*	Sequence of three prints
AC 24/1; Lily Pond	*Anchisauripus exsertus*	*Eubrontes exsertus*	Sequence of three prints
AC 20/6; Lily Pond, Gill	*Anchisauripus exsertus*	*Eubrontes exsertus*	Sequence of four prints
	Apatichnus circumagens	*Anomoepus scambus*	Sequence of 10 prints
YPM VP 003303; Franklin County	*Anchisauripus exsertus*	*Eubrontes exsertus*	Figs. 8.4*B*, 8.15*L*
YPM VP 000361; North Branford, Connecticut	*Anchisauripus exsertus*	*Eubrontes exsertus*	Fig. 8.4*C*
YPM VP 001210; North Branford, Connecticut	*Anchisauripus tuberosus*	*Eubrontes loxonyx*	Fig. 8.4*D*
YPM VP 003348; Franklin County	*Anchisauripus hitchcocki*	*Eubrontes hitchcocki*	
YPM VP 002054C; Franklin County	*Anchisauripus sillimani*	*Eubrontes sillimani*	Fig. 8.4*E*
AC 56/1; Lily Pond	*Anchisauripus minusculus*	*Eubrontes minusculus*	Big print (figs. 8.7*I*, 8.15*M*)
	Anchisauripus hitchcocki	*Eubrontes hitchcocki*	Small print (figs. 8.7*I*, 8.15*B*)
AC 14/4; Turners Falls	*Anchisauripus divaricatus*	*Eubrontes divaricatus*	Sequence of three prints (fig. 8.7*E*)
ACM.ICH.196; locality unknown	*Anchisauripus*?	*Eubrontes* isp.	Sequence of three prints
ACM.ICH.1928-0; locality unknown	*Anchisauripus*?	*Eubrontes* isp.	Sequence of five prints
AC 10/1; Below Smith's Ferry, Northampton	*Anchisauripus minusculus*	*Eubrontes exsertus*	Sequence of three prints (fig. 8.7*F*)
AC 16/1; Lily Pond	*Anchisauripus minusculus*	*Eubrontes minusculus*	Sequence of two large prints (figs. 8.7*K*, 8.15*N*)
	Anchisauripus isp.	*Eubrontes* isp.	Small print
AC 22/1; Horse race, Gill	*Anchisauripus minusculus*	*Eubrontes tuberatus*	Sequence of three prints (fig. 8.7*G*, 8.7*H*)
AC 13/4; Orchard Quarry, Turners Falls	*Anchisauripus*?	*Eubrontes* isp.; I label some of the better small prints in this slab; Hitchcock (1863) assigns them to more than one species, but they all look alike to me	My print A
	Anchisauripus?		My print C
	Anchisauripus?		My print G
	Anchisauripus?		My print H
	Anchisauripus?		My print J
	Anchisauripus?		My print K (figs. 8.10*A*, 8.15*J*)
	Eubrontes platypus	*Eubrontes platypus*	My prints B1 and B2 in trackway (fig. 8.10*A*, 8.10*B*)
YPM VP 003315B; Franklin County	*Anchisauripus tuberosus*	*Eubrontes loxonyx*	Fig. 8.4*F*
AC 9/14; Middlefield, Connecticut	*Anchisauripus sillimani*	*Eubrontes sillimani*	Numerous footprints from multiple individuals (figs. 8.2, 8.3, 8.15*G*, 8.15*H*)
AC 45/2; Turners Falls, probably Lily Pond	*Anchisauripus minusculus*	*Eubrontes minusculus*	Sequence of three prints (figs. 8.7*A*–8.7*D*, 8.15*O*)
AC 52/15; Ferry	*Anchisauripus minusculus*	*Eubrontes minusculus*	Sequence of two prints
YPM VP 056559; Turners Falls	*Anchisauripus minusculus*	*Eubrontes minusculus*	Fig. 8.7*J*

Table 8.1. *continued*

Catalog number and locality	Old or traditional name (often name on specimen label)	Current ichnotaxon or other name used here	Comments
AC 49/1; Turners Falls, south side	*Anchisauripus sillimani* or *tuberosus*	*Eubrontes sillimani*	My "up" (B) trail (three prints; fig. 8.5*B* left side)
	Anchisauripus sillimani or *tuberosus*	*Eubrontes sillimani*	My "down" (A) trail (four prints; figs. 8.5*B* right side, 8.15*I*)
AC 43/4; Turners Falls, probably Lily Pond	*Anchisauripus tuberosus*	*Eubrontes loxonyx*	Sequence of five footprints (fig. 8.5*A*)
YPM VP 003337A; Franklin County	*Anchisauripus hitchcocki*	*Eubrontes hitchcocki*	Figs. 8.4*H*, 8.15*C*
AC 15/3: Mt. Tom, east side, near Northampton	*Eubrontes giganteus*	*Eubrontes giganteus*	Figs. 8.8
AC 45/8; Smith's Ferry	*Eubrontes giganteus*	*Eubrontes giganteus*	
AC 58/1; locality unknown	*Eubrontes divaricatus*	*Eubrontes divaricatus*	Sequence of four prints
AC 7/1; below Smith's Ferry	*Eubrontes giganteus*	*Eubrontes giganteus*	Sequence of three prints (fig. 8.9*F*)
YPM VP 002098; Franklin County	*Eubrontes giganteus*	*Eubrontes giganteus*	Sequence of four prints (fig. 8.9*A*–8.9*E*)
USNM 481087; Aldie, VA "Rainy Day" slab	*Anchisauripus?*	*Eubrontes* isp.	Bigger of two prints on slab
AC 8/1; Smith's Ferry	*Eubrontes giganteus*	*Eubrontes giganteus*	Sequence of seven prints (fig. 8.9*H*, 8.9*I*)
AC 45/1; Turner's Falls, probably Lily Pond	*Eubrontes giganteus*	*Eubrontes giganteus*	"Early" trail (three prints; fig. 8.9*M*; trackway E)
	Eubrontes giganteus	*Eubrontes giganteus*	"Late" trail (three prints; figs. 8.9*M*, 8.9*N*; trackway L)
AC 43/2; Northampton	*Eubrontes giganteus*	*Eubrontes giganteus*	Sequence of two prints (fig. 8.9*G*)
AC 9/9 and 9/10; Lily Pond, Turners Falls	*Gigandipus caudatus*	*Eubrontes caudatus*	Sequence of seven prints (fig. 8.9*J*–8.9*L*)
AMNH FR 2046; Franklin County	*Eubrontes* isp.	*Eubrontes* isp.	Sequence of six prints; measurements made from photograph by Carl Mehling
Gary Gaulin Site, East Berlin Fm, Holyoke	Large *Eubrontes*	*Eubrontes* isp.	Sequence of 10 prints
Nash Dinosaur Tracksite, Granby	Large *Eubrontes*	*Eubrontes* isp.	Sequence of six prints
AC 16/2; South Hadley	*Anomoepus* isp.	*Anomoepus scambus*	Sequence of four prints
YPM VP 003397A; Massachusetts	*Anomoepus intermedius*	*Anomoepus scambus*	Fig. 8.11*F*
YPM VP 003384	*Anomoepus* cf. *curvatus*	*Anomoepus scambus*	
YPM VP 003398; Franklin County	*Anomoepus intermedius*	*Anomoepus scambus*	Fig. 8.13*G*
YPM VP 007732; Rockland County, New York	*Anchisauripus* isp.	UID bipedal dinosaur	Grallatorid or *Anomoepus*? (fig. 8.13*F*)
AC 52/1; Turners Falls	*Anomoepus* isp.	*Anomoepus scambus*	Sequence of five prints
AC 52/10; Lily Pond, Turners Falls	*Anomoepus curvatus*	*Anomoepus scambus*	Right trail (fig. 8.11*A*–8.11*E*); sequence of five prints
	Anomoepus curvatus	*Anomoepus scambus*	Left trail (fig. 8.11*A*); sequence of three prints
AC 52/14; Lily Pond	*Anomoepus curvatus*	*Anomoepus scambus*	Trackway of four very shallow prints
AC 53/7; Lily Pond	*Anomoepus curvatus*	*Anomoepus scambus*	Three nice trackways (two of four prints, one of three)
AC 26/10; Field's orchard, Turners Falls	*Anomoepus intermedius*	*Anomoepus scambus*	Left trail (prints A–D; fig. 8.12*C*, 8.12*D*)
	Anomoepus intermedius	*Anomoepus scambus*	"Diagonal" trail (prints 1–5; fig. 8.12*C*)
AC 26/21; Turners Falls	*Tridentipes (Steropoides) elegans*	*Anomoepus scambus*	Sequence of four "so-so" footprints
AC 44/2; Turners Falls (probably Lily Pond)	*Anomoepus curvatus*	*Anomoepus scambus*	Sequence of five small prints
AC 48/1; Ferry, Turners Falls	*Anomoepus intermedius*	*Anomoepus scambus*	Right trail (11 prints; fig. 8.12*A*–8.12*B*)
	Anomoepus intermedius	*Anomoepus scambus*	Left (nine prints; P or sitting) trail (fig. 8.12*A*)
AC 19/8; Turners Falls	*Tridentipes uncus?*	*Anomoepus scambus?*	Sequence of seven small, poorly preserved prints
AC 19/18; Turners Falls	*Plesiornis quadrupes*	*Anomoepus scambus*	Sequence of three footprints
ACM.ICH.8; locality unknown	*Anomoepus?*	*Anomoepus scambus*	Sequence of three footprints
YPM VPPU 021238; Essex County, New Jersey (fig. 10.13*B*)	*Anomoepus* isp.	*Anomoepus* isp.	My A trail (four prints)
			My print C
			My print D
YPM VP 002045; Franklin County	*Anomoepus gracillimus*	*Anomoepus scambus*	Sequence of three prints (fig. 8.11*G*)

Catalog number and locality	Old or traditional name (often name on specimen label)	Current ichnotaxon or other name used here	Comments
YPMVPPU 021722; Essex County, New Jersey	*Anomoepus* isp.	*Anomoepus* isp.	Sequence of four prints; baby dinosaur? (fig. 8.13*C*)
YPM VPPU 023665; Essex County, New Jersey	*Anomoepus* isp.	*Anomoepus scambus*	Sequence of two prints
YPM VPPU 002019; no locality data	cf. *Anomoepus minimus*	*Anomoepus scambus*	Sequence of six prints; baby dinosaur? (fig. 8.13*E*)
YPM VPPU 002021	*Anomoepus curvatus*	*Anomoepus scambus*	Sequence of seven prints
YPM VP 003473; Franklin County	*Arachnichnus dehiscens*	*Arachnichnus dehiscens*	Sequence of six prints; baby dinosaur? (fig. 8.13*D*)
YPM VP 002095B / 002102; Franklin County (fig. 10.13*A*)	*Anchisauripus hitchcocki*	There are three trackways on this big slab. I'm not certain which trail is supposed to be which; none of them looks identifiable to me	My trail A (seven prints); UID bipedal dinosaur?
	Anomoepus intermedius		My trail B (eight prints); UID bipedal dinosaur?
	Anomoepus isp.?		My trail C (five prints); UID bipedal dinosaur?
YPM VP 002030; Franklin County	*Argoides minimus*	*Argoides minimus*	My numbered trail (17 prints); UID bipedal dinosaur
AC 1/4; Marsh's quarry, Montague	*Tridentipes insignis*	*Steropoides loripes*	Sequence of three prints; preservational variant of *Anomoepus*?
YPM VP 002090; Franklin County	*Steropoides diversus*	*Eubrontes diversus*	My lettered trail; sequence of five prints of UID bipedal dinosaur
YPM VP 002031; Franklin County	*Argoides minimus*	*Argoides minimus*	Final five footprints in slab; UID bipedal dinosaur
YPM VP 002132; Franklin County	*Argoides minimus*	*Argoides minimus*	Sequence of five prints of UID bipedal dinosaur
AC 11/1; Removed from sidewalk in Greenfield; quarried from Turners Falls?	*Platypterna varica*	*Platypterna concamerata*	Sequence of 10 poor-quality footprints; some kind of bipedal dinosaur
AC 42/2; Turners Falls	*Plectropterna lineans*	*Plectropterna lineans*	Sequence of five prints of small dinosaur
AC 6/1; Lily Pond, Gill	*Eubrontes* or *Anchisauripus* isp.?	UID bipedal dinosaur	Sequence of two prints of larger animal
	Tridentipes uncus	*Anomoepus scambus*?	Sequence of seven or eight prints of smaller animal
ANSP 9735; Connecticut		UID bipedal dinosaur	Sequence of three prints
ANSP 10021; Connecticut		UID bipedal dinosaur	Sequence of three prints
ANSP 15641; Turners Falls?	*Anomoepus* isp.	*Anomoepus* isp.	Sequence of two prints
YPM VPPU 021713	*Atreipus acadianus*	*Atreipus acadianus*	Sequence of two pes prints of quadruped
YPM VPPU 021808; Wolfville Fm, Paddy Island, Nova Scotia	*Atreipus acadianus*	*Atreipus acadianus*	Sequence of two pes prints of quadruped
YPM VPPU 018581; Brunswick Fm, Hunterdon Co., New Jersey	*Atreipus milfordensis*	*Atreipus milfordensis*	Sequence of two pes prints of quadruped

MATERIALS AND METHODS

Measurements were made on Newark Supergroup tridactyl dinosaur footprints in several museum collections and in the field (table 8.1; figs. 8.1–8.15). It is quite possible that some of the ichnotaxonomic names that I have applied to footprints in table 8.1 are incorrect, but for the present study this isn't a problem. For my purposes, I just needed some label by which to identify particular footprints and trackways. Similarly, I will also use Weems's (1992) *Kayentapus minor* as a convenient label for the footprints he described, without offering an opinion about whether this is the best moniker for them. Keep in mind that my main goal is not to sort out the ichnotaxonomy of the footprints (that is a project for others), but rather to make an interpretation of the number and kinds of trackmakers that were responsible for them.

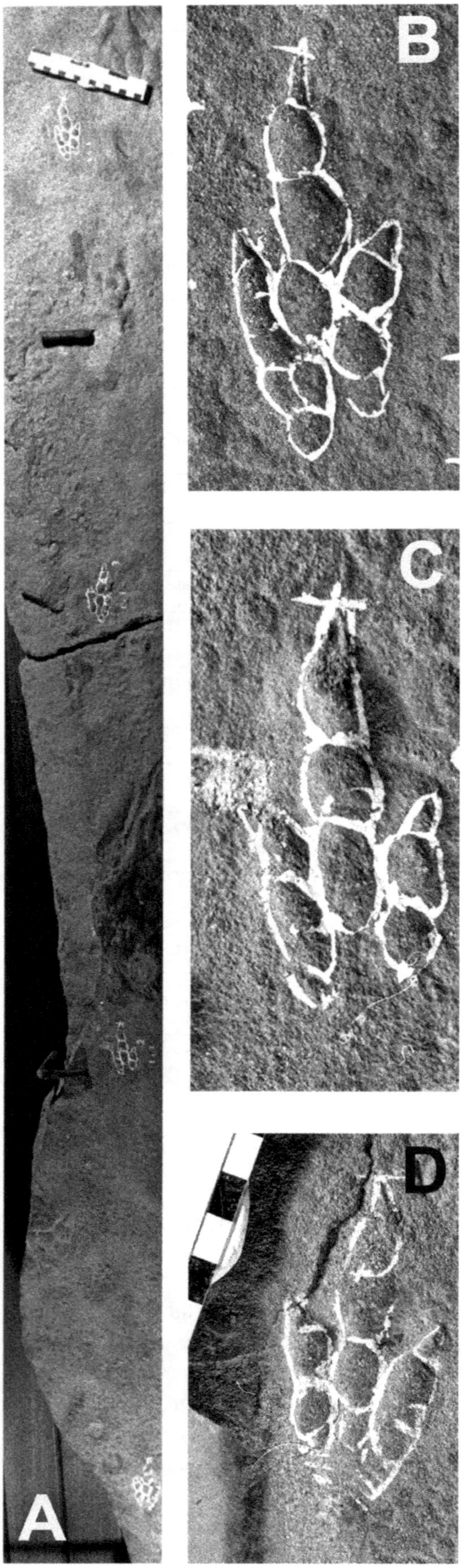

Left, **8.1.** Beneski Museum of Natural History, Amherst College AC 4/1a, trackway of four successive footprints, assigned to *Eubrontes cursorius*. The trackway is preserved as a natural cast (convex hyporelief), so left-right symmetry is reversed from that of the actual footprints. Scale bar near the top of the photograph in *A* is marked off in centimeters. Outlines around footprints in these photographs were made by persons other than the author. *A*, Entire trackway. *B*, Fourth footprint in the trackway (a right). *C*, Second footprint (right). *D*, First footprint (left). *This and other photographs of specimens from the collections of the Beneski Museum of Natural History used courtesy of the Trustees of Amherst College.*

Below, **8.2.** AC 9/14, a spectacular slab containing numerous footprints in several trackways, assigned to the medium-sized ichnotaxon *Eubrontes sillimani*. The prints are preserved as natural casts, so left-right symmetry is reversed from that of the actual footprints. Arbitrary number and letter labels were assigned to the various footprints. Line segments connect footprints thought (some perhaps questionably) to be in trackways. Scale bar to the right side of the slab is marked off in centimeters. UID = unidentified.

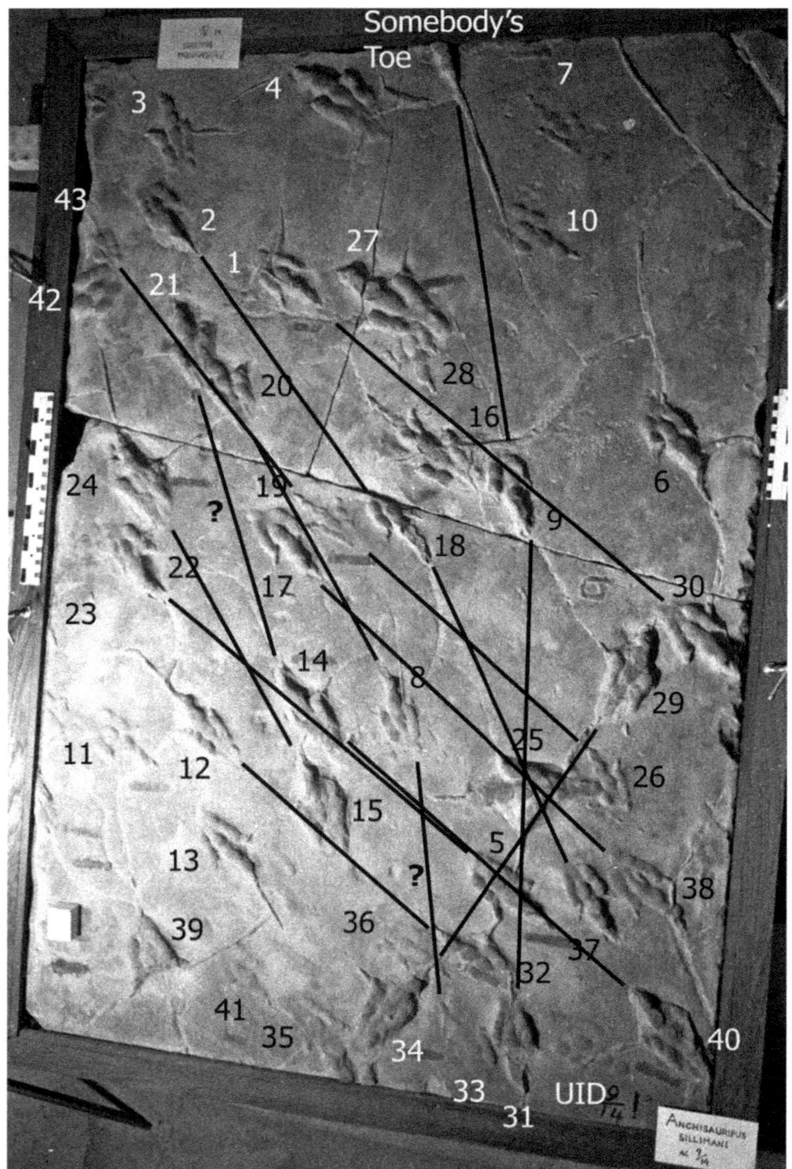

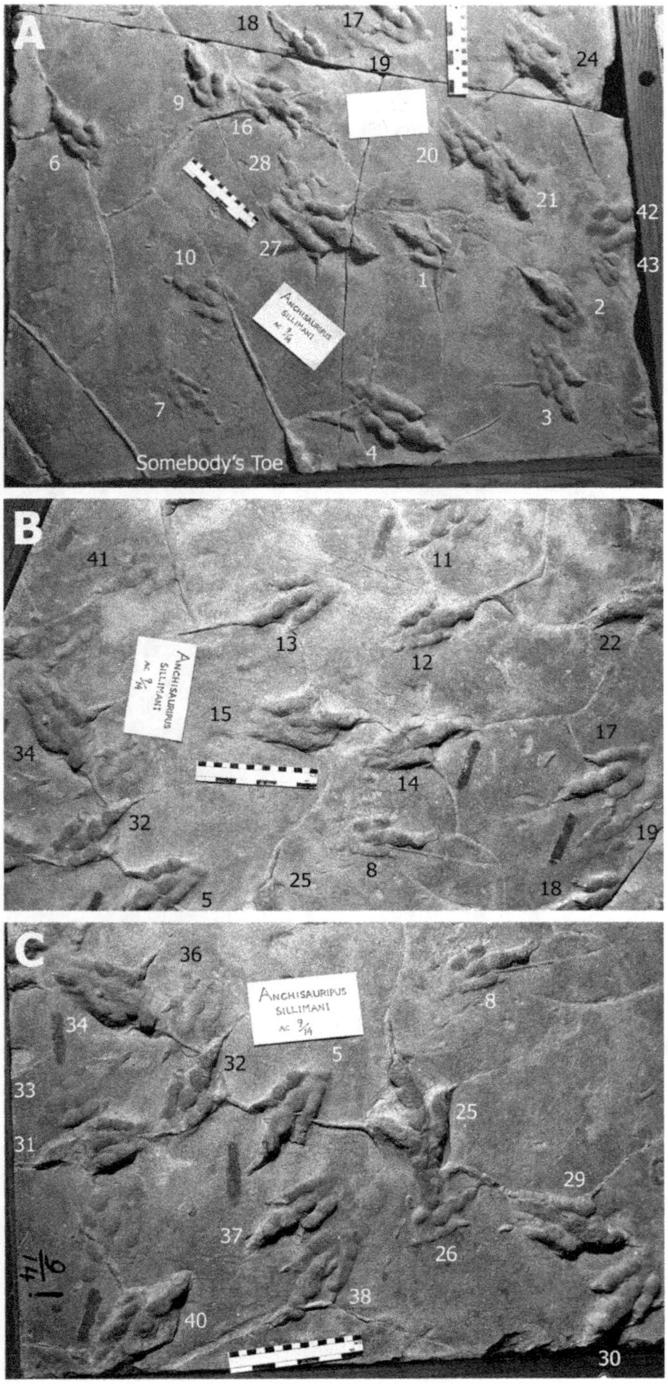

8.3. Details from AC 9/14, illustrating the morphology of individual footprints; see figure 8.2 for how the three details relate to the complete slab. Scale bars are marked off in centimeters and inches. *A*, Detail showing the top portion of the slab (as depicted in fig. 8.2), but here turned 180 degrees relative to the way it appears in figure 8.2. *B*, Detail showing much of the lower left-hand portion of the slab (as depicted in fig. 8.2). *C*, Detail showing much of the lower right-hand portion of the slab (as depicted in fig. 8.2).

In many cases, a museum catalog number has been assigned to a rock slab containing many individual footprints. In such instances I applied my own labels to the individual prints; these appear on the figures illustrating the slabs, and/or figure captions.

Measurements (fig. 8.16; table A8.1) were variously made from the footprints and trackways themselves, casts of the footprints, and tracings of footprint outlines on plastic sheets. To reduce errors due to extramorphological variability, measurements of footprint dimensions were generally made on well-preserved prints (cf. Belvedere and Farlow 2016), although trackway measurements were often made on less well preserved specimens. To reduce errors due to individual interpretation, most measurements and nearly all tracings were made by a single person (myself). Some of the measurements illustrated in figure 8.16 did not ultimately figure in the analysis that follows, either because they could be made on too few tracks, or they were found not to be of much use in discriminating among different footprint shapes.

An unavoidable problem that we face in working with dinosaur footprints relates to the vagaries of footprint preservation. It simply doesn't work out that all of the measurements we would like to make can be made on all of the footprints in a trackway or a tracksite assemblage. There may be only one or a few well-preserved prints in any particular trackway, for example, and some trackmakers will often be represented by a single isolated footprint. These limitations impose two conditions on shape analysis of our dinosaur footprints that must be recognized.

The first of these is that, whenever possible, data cases will consist of mean values of measurements among the individual footprints in a trackway. It is often not possible to make all of the measurements on each of the prints in the trackway, but taking the mean values of all the measurable parameters across all of the prints in the trackway will increase the chances of getting a complete set of measurements of a composite footprint for that particular trackmaker. Of course, this also means that those parameters that are most often preserved and measurable in footprints will have the best chances of being accurately represented by mean values. For well-preserved footprints, we can hope that the mean values of those parameters will have a reasonable chance of approximating the shapes of the parts of the trackmaker's foot that impressed those features in the recording substrate. In contrast, those parameters that can be measured in only one or a few footprints in a trackway will run a greater risk of failing accurately to record the true shape of the part of the trackmaker's foot responsible for those features. Similarly, those trackmakers represented by a single footprint run the same risk of having their dimensions of footprint shape being inaccurate because of extramorphological variability. There

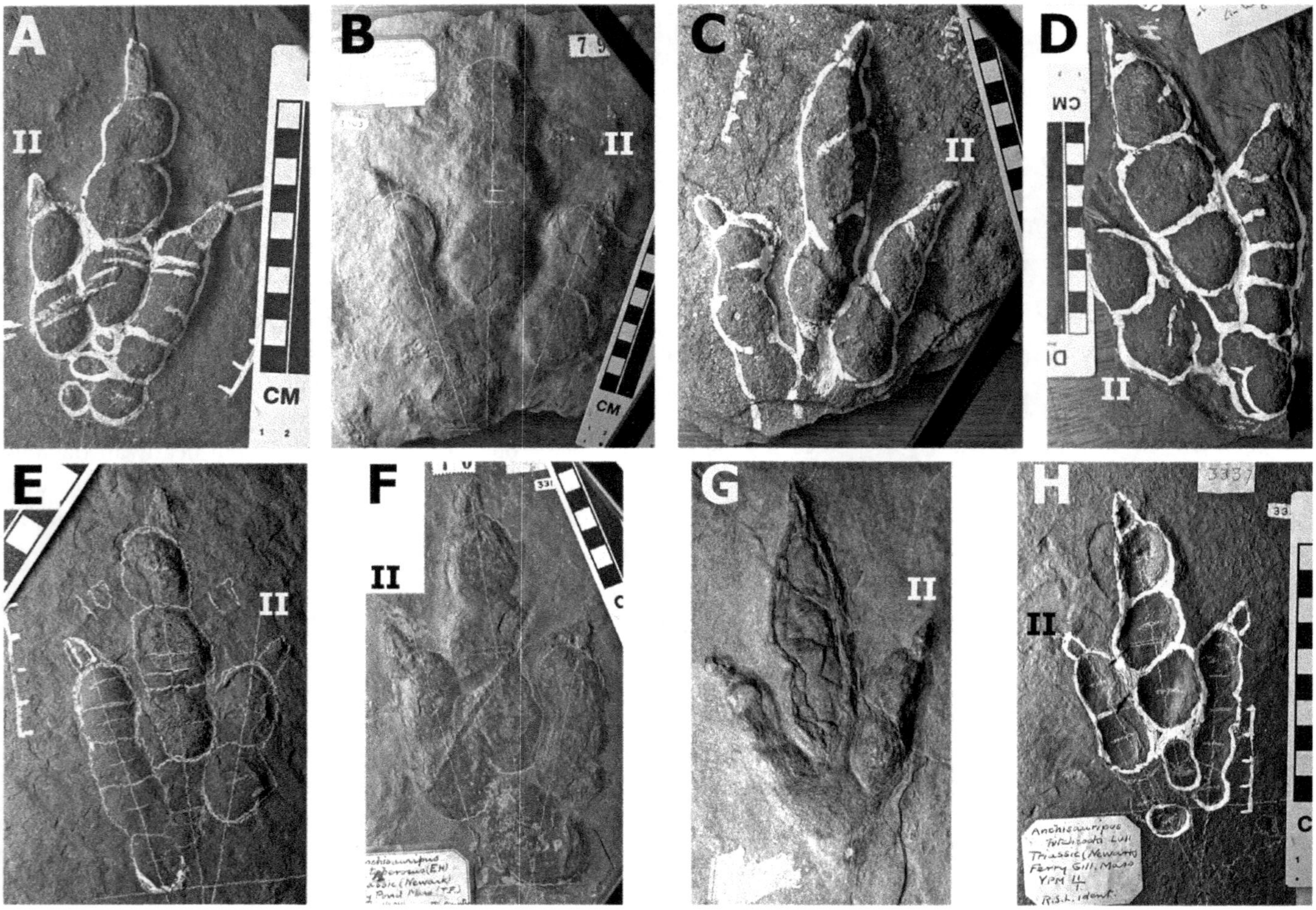

8.4. Medium-sized forms of *Eubrontes*. *A*, YPM VP 003323, *E. sillimani*. *B*, YPM VP 003303, *E. exsertus*. *C*, YPM VP 000361, *E. exsertus*. *D*, YPM VP 001210, *E. loxonyx*. *E*, YPM VP 002054C, *E. sillimani*. *F*, YPM VP 003315B, *E. loxonyx*. *G*, AC 60/33, *Eubrontes* sp. *H*, YPM VP 003337A, *E. hitchcocki*. Symmetry is indicated by labeling digit II on all footprints; scale bars are marked off in centimeters.

is not much that can be done about this problem, except to keep it in mind as a source of data "noise" that could potentially mask the true footprint/foot shape signal.

Thus the data cases used in the dinosaur footprint shape analyses in this chapter consist of parameter means across footprints in trackways with more than one print in them (or a single measurement, if that parameter can only be measured on one print in a trackway), and parameter values of singleton footprints when trackmakers are represented by just that one print. This approach will be designated the "mean-or-single" treatment.

The second restriction imposed by the limits of footprint preservation on our analyses is that the particular set of parameters used in an analysis is often an uneasy compromise between the number of footprint shape parameter measurements that we would like to use and the sample size of data cases for which those parameters can be measured. Neither a large number of shape parameters that can be measured for only a few data cases nor a very small number of measurable parameters obtained from a large number of data cases is ideal. The exact compromise between sample size and number of parameters will depend on the available data.

A set of measurements that were made on footprints of ground birds (chapter 6), and that I would have liked to measure on the dinosaur footprints here, are the various digit hypex lengths (digit II lateral hypex length, digit III medial and lateral hypex lengths, and digit IV medial hypex lengths). Unfortunately, the interdigital angles of many of the small and medium-sized *Eubrontes* are so small that the toemarks are squashed together, making recognition of the hypex points difficult or impossible. Although I was unable to employ hypex lengths in this study, in most tridactyl dinosaur footprints from other ichnofaunas these parameters can be measured. I recommend including them whenever possible, at least on a trial basis, but recognize that they can be particularly variable parameters, even among footprints in the same trackway (Farlow et al. 2006; Lallensack et al. 2016; Wings et al. 2016).

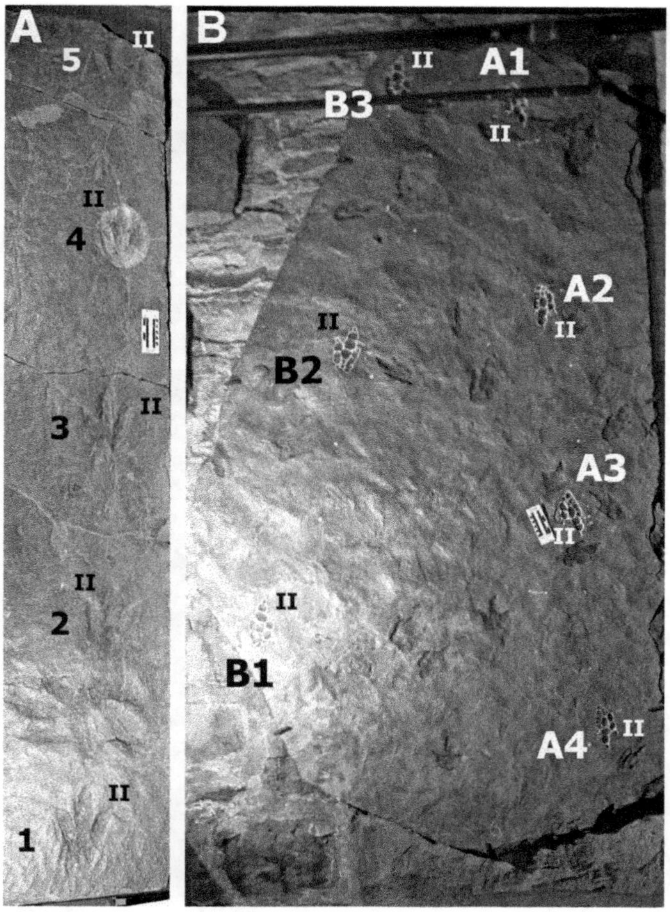

8.5. Trackways of medium-sized *Eubrontes*. Digit II is labeled to indicate footprint symmetry. *A*, AC 43/4, *E. loxonyx*. *B*, AC 49/1, two trackways (headed in opposite direction, labeled A and B), assigned to *E. sillimani*.

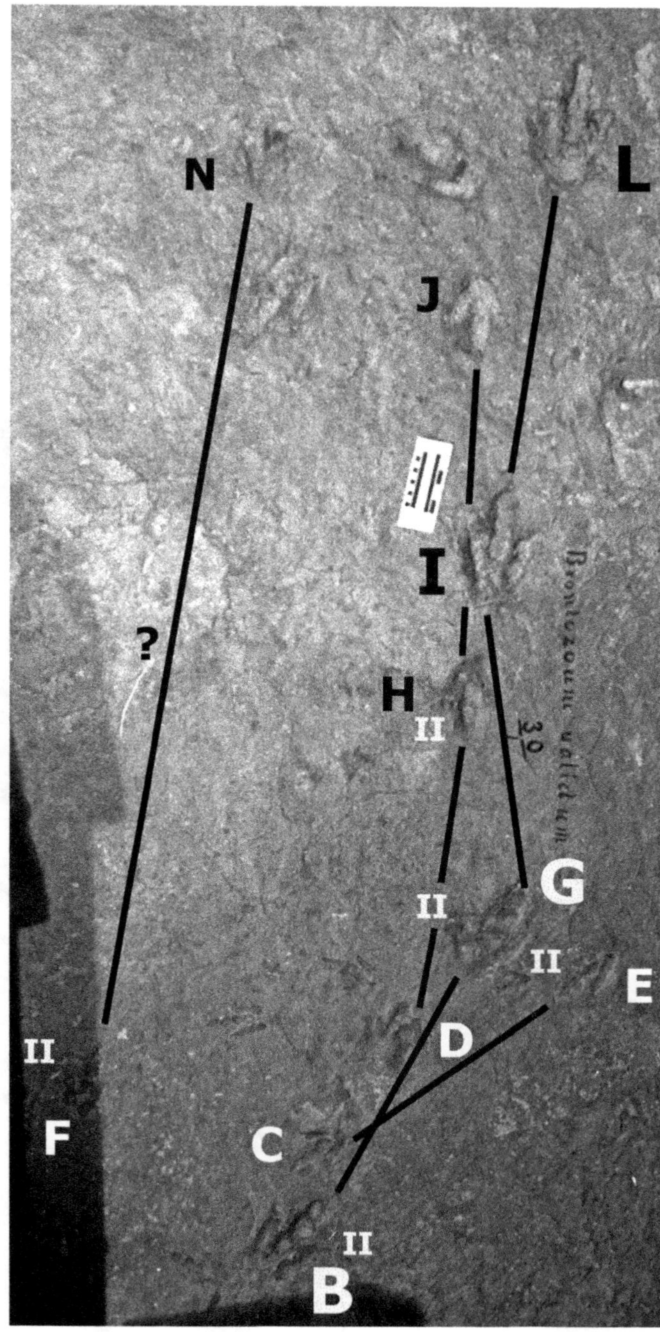

8.6. AC 30/1, large slab with several footprints of small to medium-sized *Eubrontes*. Scale bar is marked in centimeters and inches. Line segments connect prints assigned to the same trackway. Digit II of at least one print in each trackway is labeled to establish footprint symmetry. Prints B–G–I–L are assigned to *E. loxonyx*. Prints J–H–D, E–C and possible sequence F–N are assigned to *E. josephbarratti*.

In many ichnological studies of dinosaur tracks, lengths of individual toemarks are measured from the back of the footprint (the "heel"—keeping in mind that this is not the true, anatomical heel) to the tip of the digit (analogous to the way the digit length parameters IILGL, IIILGL, and IVLGL were measured on emu footprints; chapter 6). Digit lengths can be measured this way even when the individual digital pads are not clearly delimited, allowing a larger sample size of measurable footprints. However, because the Newark Supergroup footprints are often very well preserved, it is frequently possible with these tracks also to measure digit lengths from the proximal end of the first digital pad (distal to the metatarsophalangeal pad) to the tip of the toe (analogous to the IILgth, IIILgth, and IVLgth digit length measurements of emu footprints). The lengths of digits II and III were measured in this way, but the length of digit IV includes what was likely the metatarsophalangeal pad of this digit (cf. Baird 1957; Farlow et al. 2000).

I created toeprint-tapering profiles analogous to the toe-tapering profiles made from skeletal measurements (chapter 3) for digits II and III of footprints; there were too few footprints in which digit IV preserved enough digital pads clearly enough to make this possible. The cumulative lengths of impressions of digital pads and claws, and the widths of digital pads, were scaled as percentages of the length of the first digital pad of digit III.

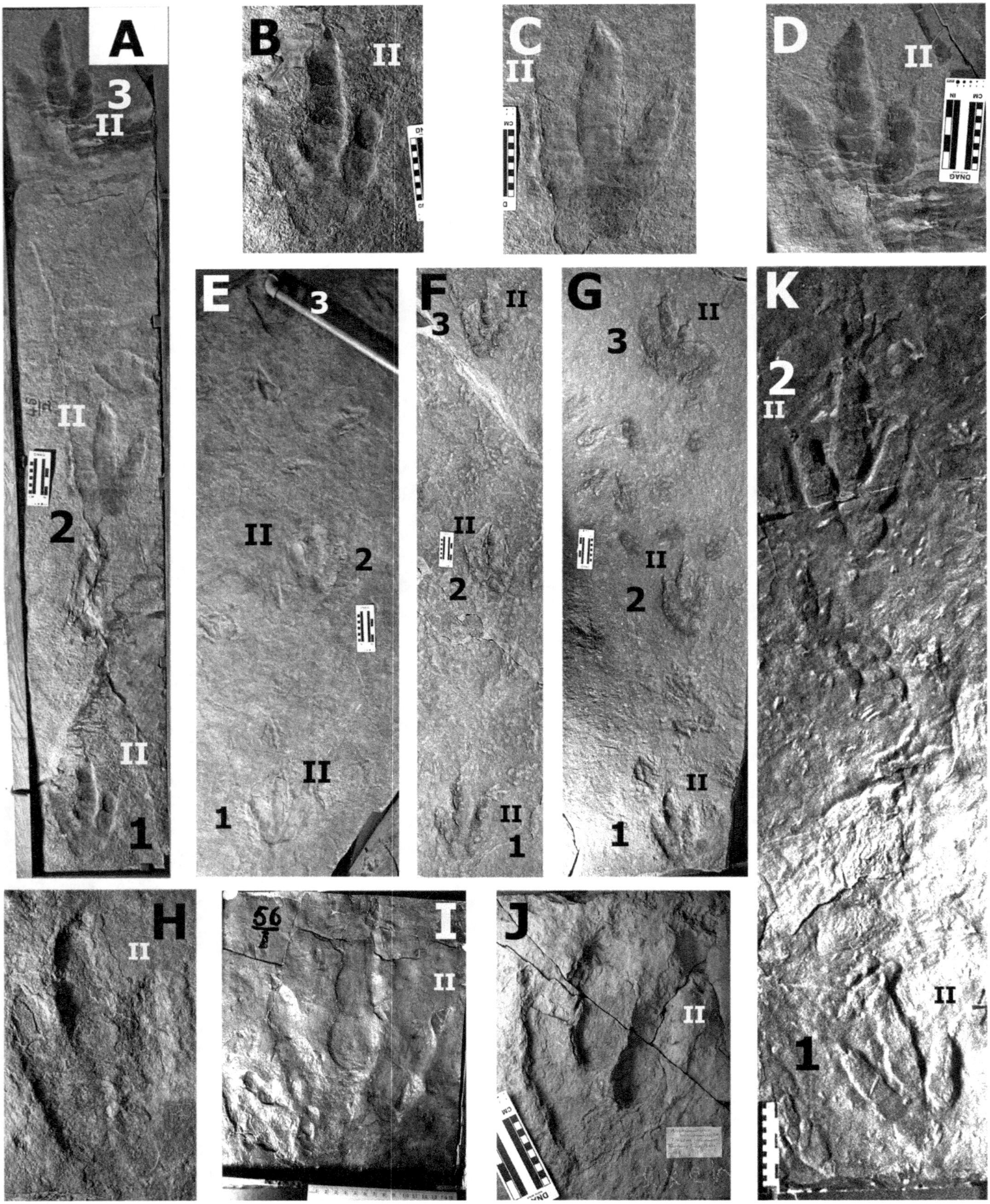
A
B
C
D
E
F
G
H
I
J
K
56

Facing, **8.7.** Medium-sized to large forms of *Eubrontes*. Scale bars are marked off in centimeters and sometimes also inches. Digit II is labeled on prints to indicate symmetry. *A–D*, AC 45/2, *E. minusculus*. *A*, Sequence of three prints. *B–D*, Individual views of the first, second, and third prints, respectively, of AC 45/2. *E*, AC 14/4, *E. divaricatus*, sequence of three prints. *F*, AC 10/1, *E. exsertus*, sequence of three prints. *G*, *H*, AC 22/1, *E. tuberatus*, sequence of three prints. *H*, First print of AC 22/1. *I*, AC 56/1, two prints. A large print, directed upward, is assigned to *E. minusculus*. A second, smaller print (hard to see in this photograph), *E. hitchcocki*, is in the lower left-hand corner of the photograph, directed in nearly the opposite direction from that in the larger print. *J*, YPM VP 056559, *E. minusculus*. *K*, AC 16/1, sequence of two prints of *E. minusculus*.

Because the data analyses in this chapter are long and often complicated, I beg the reader's indulgence (as I did in chapter 6 with emu footprints) for not describing all of the kinds of data analyses here in the methods section of the chapter, thus separating such descriptions from the results thereof, as is usually done in technical papers. Instead I will describe the specific analyses employed as they are used, along with the results of those analyses.

I will also frequently attempt to remove absolute size of footprints as a confounding variable in analyses by a variety of different scaling methods. To keep these straight in my mind, I created numerous abbreviation names for the scaling parameters. These are often long and admittedly less than euphonious. I will explain what each of these ghastly abbreviations means as I use it.

8.8. AC 15/3, the type specimen of *Eubrontes giganteus*. Natural cast (negative copy) of a right footprint.

Table 8.2. Principal component (PC) analysis (using a covariance matrix) of log-transformed measurements of Newark Supergroup dinosaur trackways or singleton footprints (mean or single treatment; number of trackways = 24).

Parameter	Component loadings (raw [rescaled])		
	PC1	PC2	PC3
Digit II first pad length	0.191 (0.976)	−0.003 (−0.014)	−0.011 (−0.057)
Digit II second pad length	0.180 (0.969)	−0.015 (−0.080)	0.025 (0.137)
Digit II claw length	0.119 (0.813)	0.068 (0.464)	0.039 (0.269)
Digit III first pad length	0.174 (0.978)	−0.006 (−0.032)	0.020 (0.112)
Digit III second pad length	0.153 (0.941)	−0.018 (−0.112)	0.033 (0.200)
Digit III third pad length	0.166 (0.963)	−0.013 (−0.075)	0.017 (0.098)
Digit III claw length	0.112 (0.828)	0.035 (0.259)	0.022 (0.163)
Digit IV length	0.168 (0.980)	0.011 (0.062)	−0.008 (−0.046)
Digit II first pad width	0.201 (0.985)	−0.006 (−0.029)	−0.016 (−0.080)
Digit II second pad width	0.190 (0.987)	−0.010 (−0.050)	−0.010 (−0.054)
Digit III first pad width	0.205 (0.988)	−0.019 (−0.090)	−0.019 (−0.090)
Digit III second pad width	0.193 (0.959)	−0.046 (−0.230)	0.013 (0.067)
Digit III third pad width	0.214 (0.962)	−0.049 (−0.220)	0.005 (0.023)
Proximal end of digit II–proximal end of digit III	0.194 (0.971)	0.009 (0.046)	−0.026 (−0.131)
Proximal end of digit III–proximal end of digit IV	0.163 (0.963)	0.023 (0.134)	−0.014 (−0.080)
Proximal end of digit II–proximal end of digit IV	0.153 (0.918)	0.031 (0.185)	0.022 (0.133)
Toetip II–toetip III	0.155 (0.960)	0.000 (0.001)	−0.004 (−0.027)
Toetip III–toetip IV	0.157 (0.957)	0.027 (0.165)	−0.017 (−0.101)
Toetip II–toetip IV	0.185 (0.946)	0.038 (0.197)	−0.037 (−0.191)
Cumulative variance explained	91.574	94.210	95.574

Kaiser-Meyer-Olkin measure of sampling adequacy = 0.833; Bartlett's test of sphericity: $\chi^2 = 953.249$, $P < .001$.

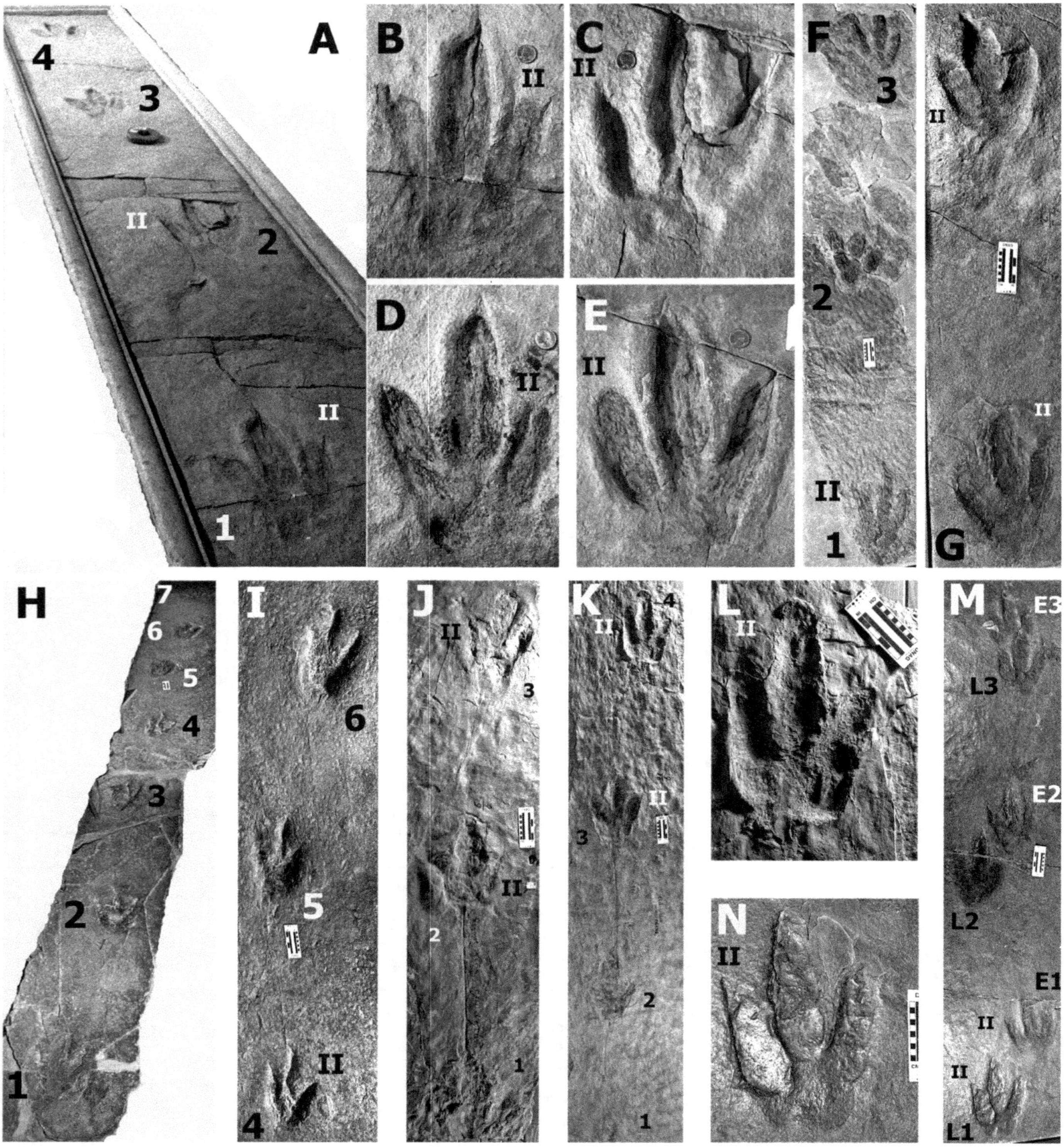

8.9. Additional specimens of footprints of large forms of *Eubrontes*. Footprints in *A–I*, *M*, and *N* are those of *E. giganteus*; *J–L* are from *E. caudatus*. Scale bars are marked off in centimeters and sometimes also inches. Digit II is labeled to indicate symmetry. *A–E*, YPM VP 002098, oblique view of sequence of four prints. *B–E*, Prints 1 through 4, respectively, of YPM VP 002098. *F*, AC 7/1, sequence of three prints. *G*, AC 43/2, sequence of two prints. *H*, *I*, AC 8/1, sequence of seven prints; *H* = oblique view of the trackway; *I* = nearly overhead view of prints 4 through 6 of the trackway. *J*, AC 9/9, and *K*, AC 9/10, two segments of a single trackway. *L*, Footprint 4 of AC 9/10. *M*, *N*, AC 45/1. *M* shows two trackways (L1–L3, E1–E3) on the same slab. *N*, Footprint L1 of AC 45/1.

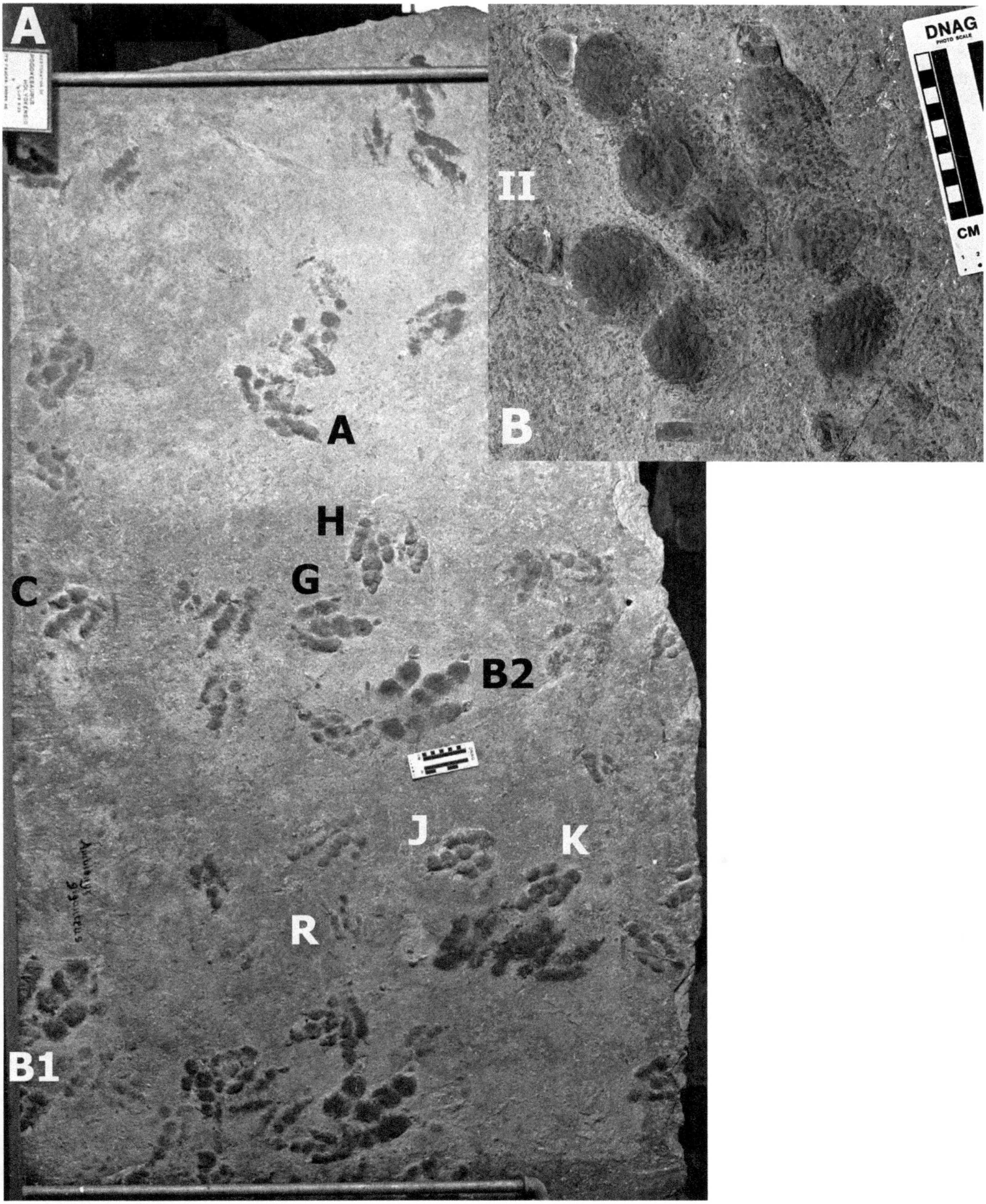

8.10. *A*, AC 13/4, large and small forms of *Eubrontes*. Scale bars are marked off in centimeters and sometimes also inches. Prints B1 and B2 constitute a sequence of prints assigned to *E. platypus*. The other labeled footprints are here treated simply as *Eubrontes* sp. *B*, Print B2.

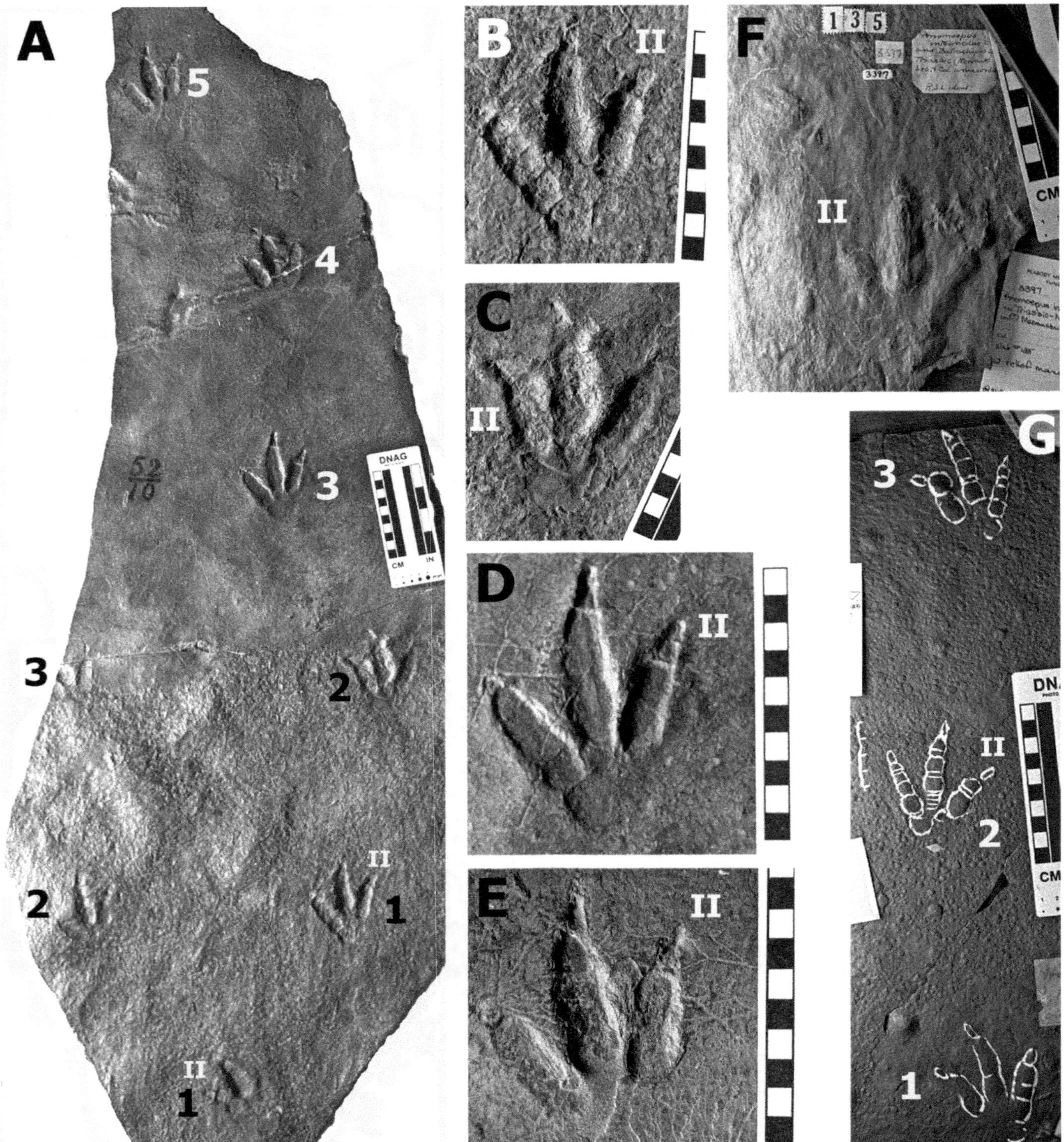

Above, **8.11.** *Anomoepus scambus*: small tridactyl footprints attributed to ornithischian dinosaurs. Scale bars are marked off in centimeters and sometimes also inches. Digit II is labeled on some prints to indicate symmetry. *A–E*, AC 52/10, two trackways on the same slab. *B–E*, Prints 1, 2, 3, and 5, respectively, of the right trail of AC 52/10. *F*, YPM VP 003397A. *G*, YPM VP 002045.

Facing, **8.12.** Additional *Anomoepus scambus* footprints. Digit II is labeled on some prints to indicate symmetry. Scale bars are marked off in centimeters and inches. *A*, *B*, AC 48/1. Two trackway sequences are labeled. *B*, Detail of prints 5–9 of the right-hand trail of AC 48/1. *C*, *D*, AC 26/10. Two trackway sequences are labeled. *D*, Detail showing the left-side trackway sequence of AC 26/10.

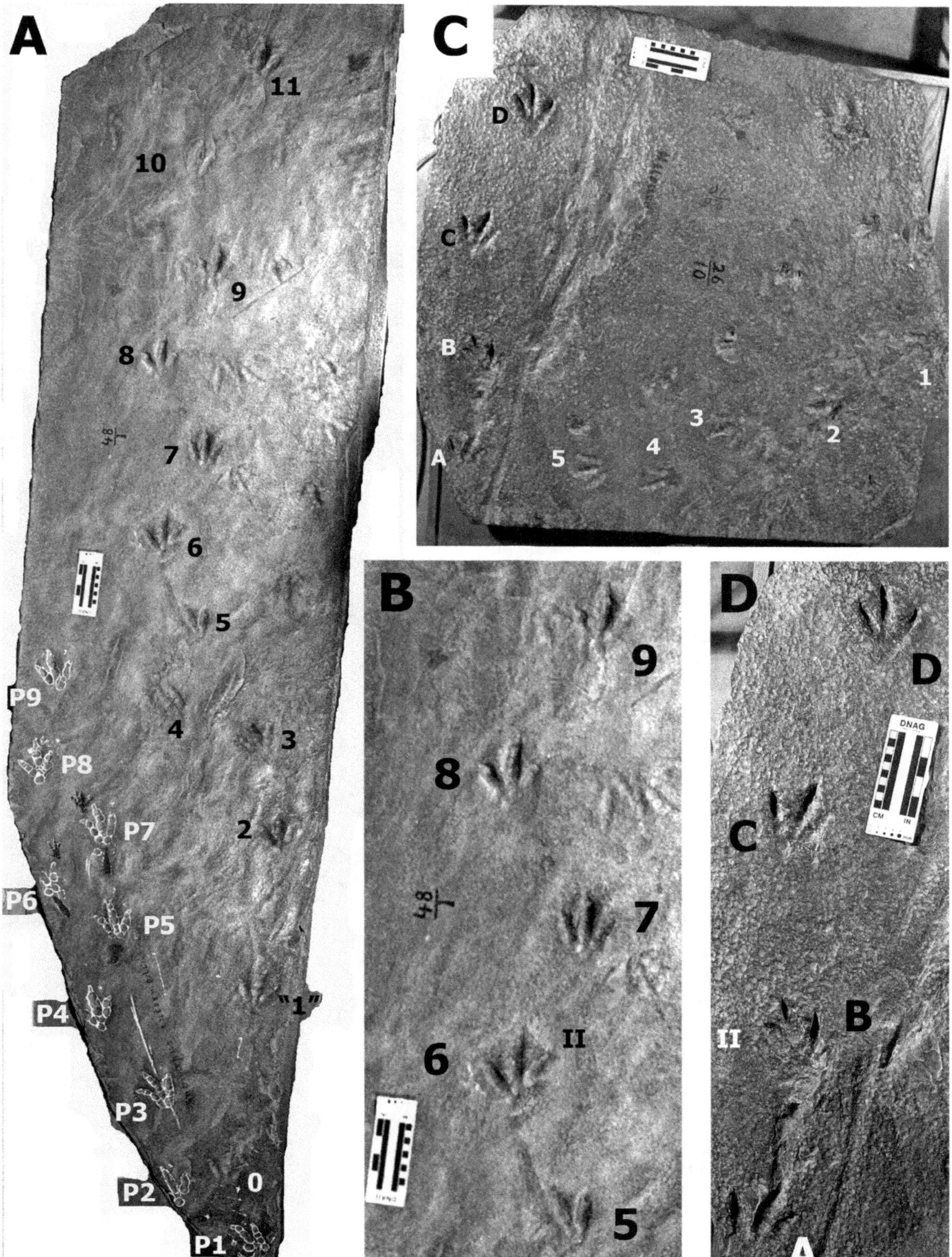
A
11
10
9
8
7
6
5
4
3
2
"1"
0
P9
P8
P7
P6
P5
P4
P3
P2
P1
C
D
C
B
A
5
4
3
2
1
B
9
8
7
6
II
5
D
D
C
II
B
A
DNAG
CM
IN

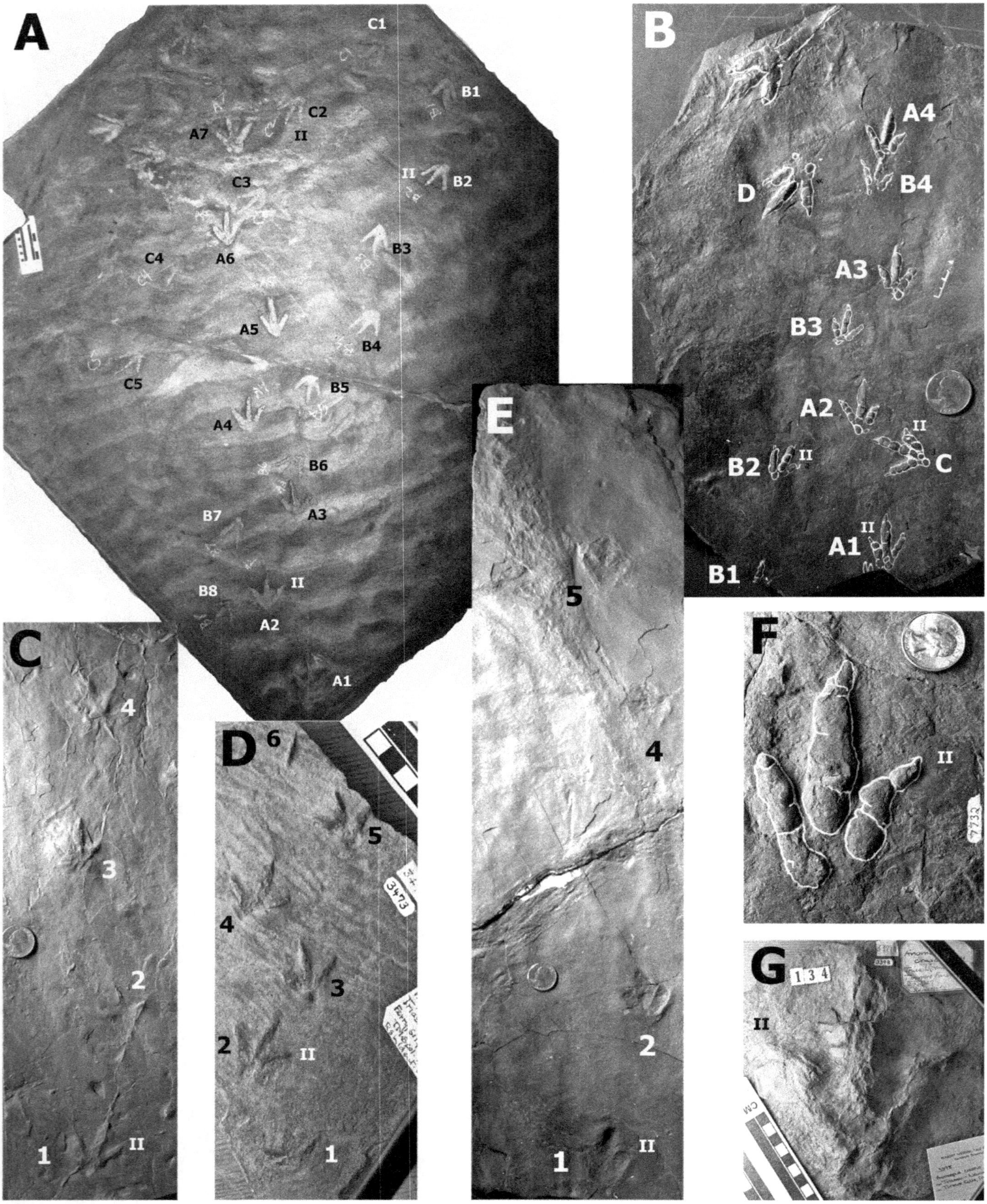
A
C1
B1
C2
A7
II
C3
II
B2
A6
C4
B3
A5
B4
C5
B5
A4
B6
A3
B7
B8
II
A2
A1
B
A4
D
B4
A3
B3
A2
II
B2
II
C
II
A1
B1
C
4
3
2
1
II
D
6
5
4
3
2
II
1
3473
E
5
4
2
1
II
F
II
G
1 3 4
II
CM

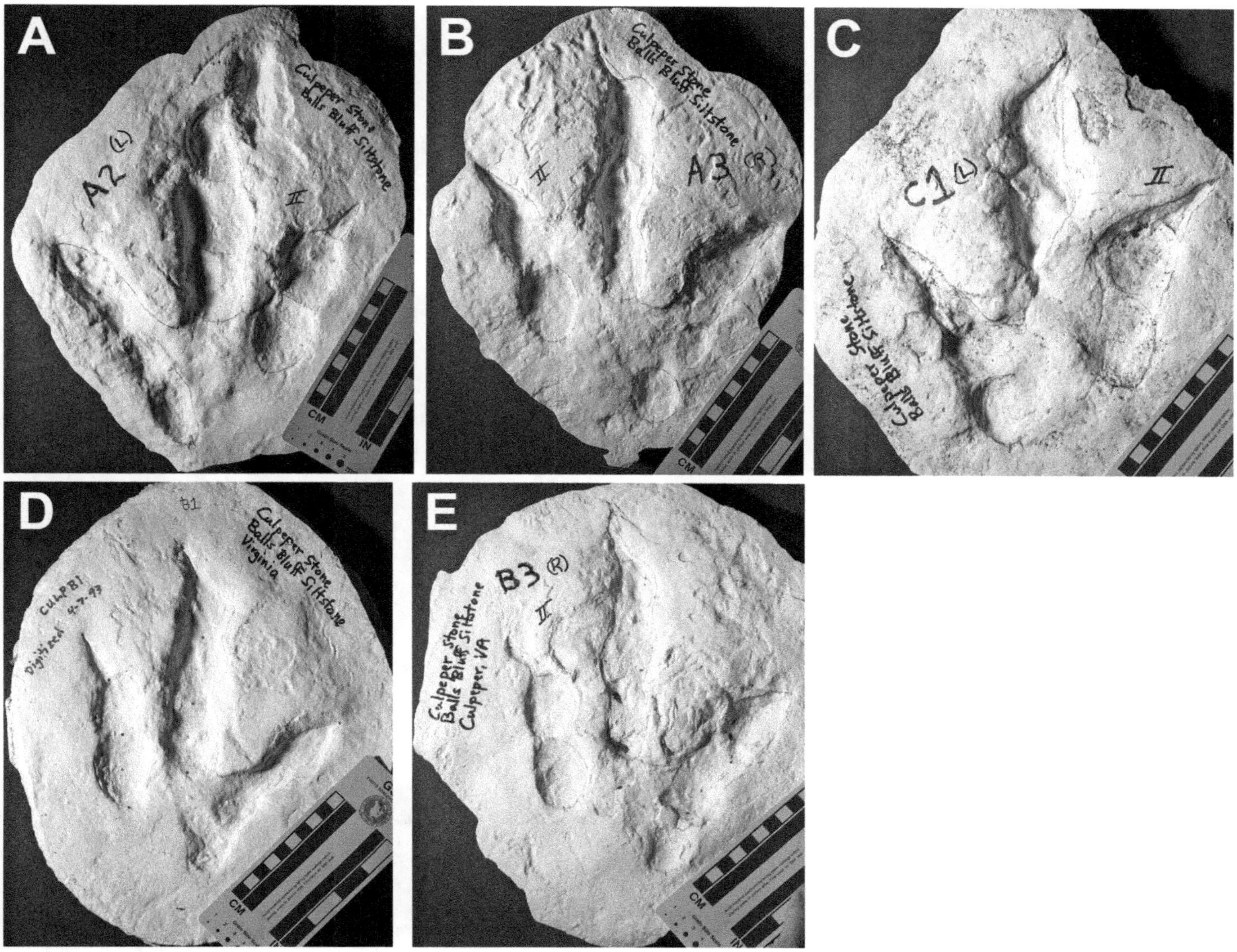

Facing, **8.13.** Miscellaneous small tridactyl footprints from the Newark Supergroup. Scale bars are marked off in centimeters; American quarter coin (diameter about 24 mm) is used for scale in some photographs. *A,* Large slab with footprints in at least three trackways. Two of the trackways are assigned catalog numbers (YPM VP 002102, *Anomoepus,* and YPM VP 002095B, *Anchisauripus,* but all are here regarded merely as tracks of otherwise unidentified bipedal dinosaurs. *B,* YPM VPPU 021238, slab with prints in at least four trackways (individually labeled A–D on the photograph), tentatively identified as *Anomoepus scambus. C,* YPM VPPU 021722, trackway of a small dinosaur, tentatively ascribed to *A. scambus. D,* YPM VP 003473, trackway catalogued as *Arachnichnus dehiscens,* but here identified simply as that of a small bipedal dinosaur—possibly a young juvenile. *E,* YPM VPPU 002019, trackway tentatively assigned to *A. scambus. F,* YPM VP 007732, catalogued as *Anchisauipus* isp., here identified simply as a small bipedal dinosaur. *G,* YPM VP 003398, footprint tentatively assigned to *A. scambus.*

Above, **8.14.** Casts of *Kayentapus minor* prints, Balls Bluff Siltstone, Virginia (Weems 1992). These casts were made from molds taken in the field by Glen Kuban. Scale bar is marked off in centimeters. Prints are from trackways labeled A, B, and C; different identifying labels were attached to the same trackways by Weems, but which of Weems's trails correspond to trackways labeled A–C here is unknown. *A, B,* Two footprints from trackway A. *C,* Footprint from trackway C. *D, E,* Two footprints from trackway B.

HOW MANY KINDS OF NEWARK SUPERGROUP TRACKMAKER?

Identifying Possible Trackmaker Groups

Multivariate Comparisons. After several rounds of trial and error, searching for the best compromise between number of variables and number of data cases in my data set, I selected a set of 19 footprint shape variables (table 8.2) that could be measured on all singleton footprints, or at least one footprint in trackways (i.e., mean-or-single data treatment), of 24 individual dinosaurs (on the assumption that each trackway or singleton print was made by a different animal). Analyses based on these 19 variables are here designated "maximum information" analyses because they are based on this best compromise between number of detailed variables of footprint size and shape (lengths and widths of individual digital pads, and lengths of clawmarks) and number of data cases.

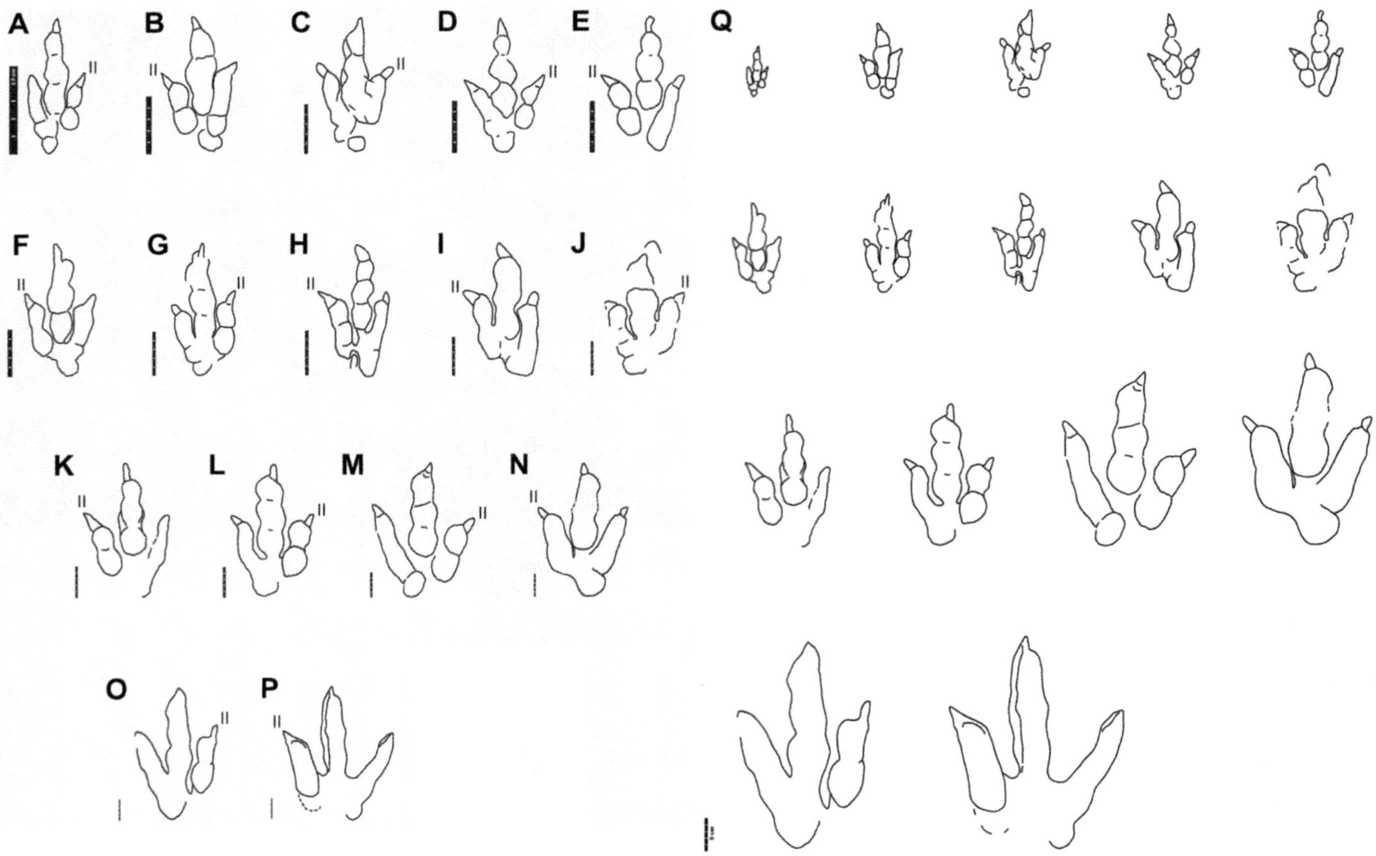

8.15. Interpretive drawings of *Eubrontes* prints from the Newark Supergroup. *A–Q*, Footprints are drawn to the same size; scale bar = 5 cm in all cases. *A*, AC 4/1a, *E. cursorius*, footprint 4 (fig. 8.1*B*). *B*, AC 56/1, *E. hitchcocki* (fig. 8.7*I*, smaller footprint). *C*, YPM VP 003337A, *E. hitchcocki* (fig. 8.4*H*). *D*, AC 30/1, footprint E (my label), *E. josephbarratti* (fig. 8.6). *E*, AC 30/1, footprint F (my label), *E. josephbarratti* (fig. 8.6). *F*, YPM VP 003323, *E. sillimani* (fig. 8.4*A*). *G*, AC 9/14, footprint 38 (my label), *E. sillimani* (figs. 8.2, 8.3). *H*, AC 9/14, footprint 3 (my label), *E. sillimani* (figs. 8.2, 8.3). *I*, AC 49/1, *E. sillimani*, second footprint of my A ("down") trackway (fig. 8.5*B*). *J*, AC 13/4, footprint K (my label), *Eubrontes* isp. (fig. 8.10*A*). *K*, AC 30/1, footprint G (my label), *E. loxonyx* (fig. 8.6). *L*, YPM VP 003303, *E. exsertus* (fig. 8.4*B*). *M*, AC 56/1, *E. minusculus* (fig. 8.7*I*, large print). *N*, AC 16/1, *E. minusculus*, second print in trackway (fig. 8.7*K*). *O*, AC 45/2, *E. minusculus*, third print in trackway (fig. 8.7*D*). *P*, AC 58/1, *E. divaricatus*, print 3. *Q*, The same set of footprints as in *A–P*, but shown in their actual relative sizes; scale bar again = 5 cm.

As typical with data sets of the kind employed here, a principal component analysis (PCA) using the log-transformed variables generated a first component (PC1) that accounted for more than 90% of the data variance (table 8.2). As usual, this first component is mainly associated with size, spreading the tracks out from little *Eubrontes* (or *Grallator*) *cursorius* and the somewhat larger *Anomoepus scambus* at the small end, through several nominal ichnospecies of *Eubrontes* (formerly *Anchisauripus*) in the midsize range, to *Kayentapus minor* and *Eubrontes* (formerly *Anchisauripus*) *minusculus* and *E. giganteus* at the large end (fig. 8.17A).

PC2 accounts for only just under 2% of the data variance, but separates the tracks in an interesting manner. Positive values of PC2 are associated with relatively long clawmarks of digits II and III, and also relatively large values of measures that are to varying degrees associated with footprint width (toebase and toetip distances between digits III and IV, and II and IV), and a long digit IV. Negative values of PC2 are associated with relatively long and broad digital pads of digits II and III.

PC3 accounts for less than 1% of the sample variance, but also separates the tracks in an interesting way. Positive values of PC3 are mainly associated with a relatively long distal (second) digital pad and a long clawmark of digit II, with relatively long digital pads and the clawmark of digit III, and a relatively large distance between the bases (proximal ends) of the impressions of digits II and IV. Negative values of PC3 are mainly associated with relatively large values of the distances between toetips II and IV and toetips III and IV, and relatively large values of the distances between toebases II and III and III and IV.

Because PC2 and PC3 both result in some separation of ichnotaxa, plotting PC3 against PC2 segregates the various Newark Supergroup ichnotaxa into distinct regions of morphospace (fig. 8.17*B*). More positive values of PC2 separate the bigger forms of *Eubrontes* (*minusculus* and *giganteus*),

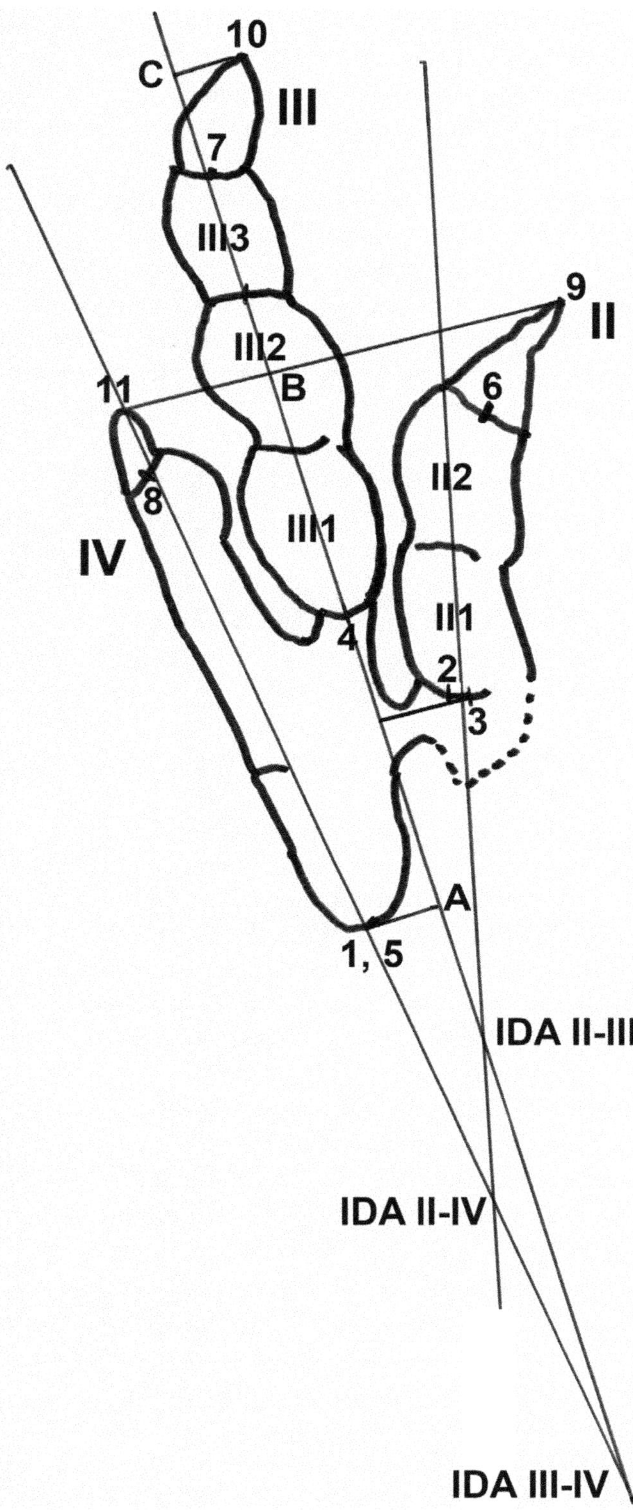

8.16. Anatomical landmarks on tridactyl dinosaur footprints (modified from Farlow et al. [2006: Fig. 36]; cf. fig. 6.4 for emu footprints, and also Belvedere et al. [2010]; Moreau et al. [2012*b*]). Digits II, III, and IV, and the individual digital pads of digits II and III, are labeled (II1, II2, III1–III3). Anatomical landmarks: *1*, footprint "heel"; *2*, proximal end of the digit II impression (may coincide with landmark *3*, but need not do so); *3*, digit II proximal bend point (location where the medial margin of the footprint undergoes a sharp change of "slope" as it turns from the medial side of the proximal edge of digit II and heads toward the digit III midline, the proximal end of digit IV, and the "heel"); *4*, proximal end of the digit III impression; *5*, proximal end of the digit IV impression (in this specimen, this landmark coincides with the "heel," but that will not always be the case)—note that digit IV length in this study includes what is likely the metatarsophalangeal pad of this digit; *6*, digit II clawbase; *7*, digit III clawbase; *8*, digit IV clawbase; *9*, digit II clawtip; *10*, digit III clawtip; *11*, digit IV clawtip. Footprint field length is the distance between the "heel" (landmark *1*) and the terminal end of digit III (landmark *10*), measured on the footprints themselves; the same measurement is termed "heel" to toetip III when measured from tracings of the footprint midline. Line segment A–B–C is drawn through the middle of the long axis of the digit III impression (putting greater emphasis on the fleshy portion of the digit impression than on the clawmark). Landmark A marks where a line segment perpendicular to the digit III long axis intersects the "heel"; landmark B is where a line segment connecting the clawtips of digits II and IV crosses the digit III long axis; landmark C is where a line segment perpendicular to the digit III long axis intersects the clawtip of digit III. Line segment A–B is the backfoot length, and line segment B–C is the projection of digit III beyond the tips of digits II and IV. An unlabeled line segment indicates the distance between the digit II proximal bend point (landmark *3*) and the digit III long axis. Interdigital angles (IDAs) II–III, III–IV, and II–IV are defined by the intersection of lines passing through the long axes of the digit impressions (with greater emphasis placed on the fleshy portions of the toe impressions than on the clawmarks); note that these lines are not constrained to come together at the footprint "heel."

and especially *Anomoepus* and *Kayentapus*, from many (but not all) specimens of the small to medium-sized forms of *Eubrontes*. PC3 separates *Anomoepus* and big *Eubrontes* (more negative values) from *Kayentapus* and most specimens of small to medium-sized *Eubrontes* (more positive values).

The 24 mean-or-single data cases used in the PCA were also compared in a cluster analysis in which each of the 19 log-transformed footprint measurements (table 8.2) of each data case was scaled by subtracting the mean of all log-transformed measurements of that data case from each of the data case's individual log-transformed measurements. The clustering procedure creates a dendrogram with three major groups (fig. 8.18), each of which could arguably be further subdivided into smaller groups (if we had more data).

One of these major groups (figs. 8.18, 8.19) includes many of the smaller and medium-sized specimens of *Eubrontes* (*E. cursorius* [fig. 8.1], *E. sillimani* [e.g., figs. 8.2, 8.3, 8.4A, 8.4*E*, 8.5*B* (trackway A)], *E. hitchcocki* [e.g., fig. 8.7*I* (small print)], and *E. loxonyx* [fig. 8.6 (trackway B-G-I-L-O)]). A second major group includes *Anomoepus* (figs. 8.11A–8.11*E* [right trackway], 8.12A, 8.12*B* [right trackway]) and, interestingly enough, *Kayentapus* (fig. 8.14). The third major group includes large specimens of *Eubrontes* (e.g., figs. 8.7*I* [large print], 8.7*J*, 8.7*K*, 8.9*A*–8.9*E*), as well as some medium-sized specimens of the same ichnogenus (e.g., figs. 8.4*B*, 8.5*B* [trackway B], 8.6 [trackway F-N?]). For convenience sake, and as a nod to R. S. Lull's (1953) traditional ichnogenera, I will designate these groups the "*Anchisauripus*," *Anomoepus*, and "*Eubrontes*" clusters, respectively.

I put quotation marks around the names "*Anchisauripus*" and "*Eubrontes*" in the names of the maximum information clusters because I am using these names in an informal

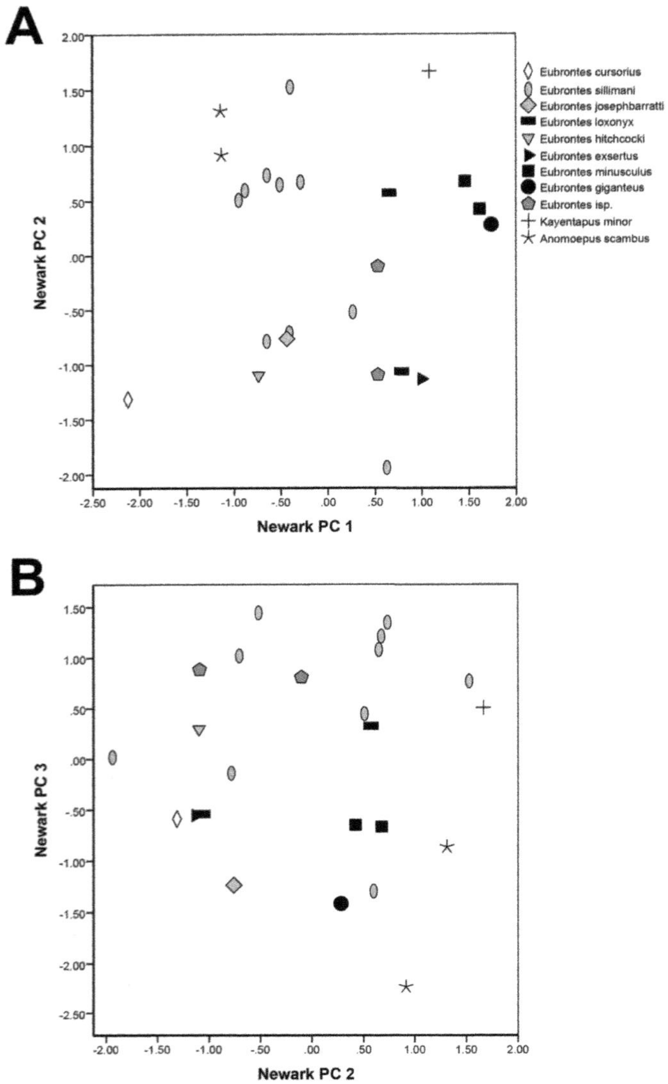

8.17. Principal component analysis (PCA) of log-transformed measurements of shape of Newark Supergroup tridactyl dinosaur footprints (table 8.2). *A*, PC1 vs. PC2. PC1 is mainly associated with overall footprint size; PC2 separates forms with relatively long clawmarks, a long digit IV, and relatively broad measures of footprint width (positive values) from forms with relatively long and broad digital pads of digits II and III (negative values). *B*, PC2 vs. PC3. PC3 separates forms with relatively long digital pads and claws of digit III, relatively long distal (second) digital pads and claws of digit II, and a broad toebase II–IV width (positive values) from forms with broad toebase II–III and III–IV widths, and broad toetip widths (negative values).

sense, and not as "official" ichnogeneric names. This is because the specimens in the *"Anchisauripus"* and the *"Eubrontes"* clusters include some, but not all, of the nominal species of *Eubrontes*. When I use the name *Eubrontes* without quotation marks, in contrast, I am using the name in an official sense. The name *Anomoepus* is used in official, ichnogeneric sense throughout, including the name of the maximum information cluster to which *Anomoepus* specimens belong. *Kayentapus* is also a member of the *Anomoepus* maximum information cluster, but is so much bigger than *Anomoepus* (and also comes from a different ichnofauna in eastern North America than the others), that in some analyses I will separate it from the *Anomoepus* cluster into one of its own, or consider *Anomoepus* without including *Kayentapus*. The results of the cluster analysis yield groupings of ichnotaxa that seem broadly consistent with those obtained from the PCA.

The features that separate the three clusters can be viewed another way by examining histograms of the same 19 scaled parameters employed in the cluster analysis (figs. 8.20–8.22; excluding *Kayentapus* cases). Visual examination of the histograms suggests the following differences among the three clusters:

Members of the *"Anchisauripus"* cluster are characterized by: a relatively short first (most proximal) digital pad on digit II, and relatively long second and possibly third digital pads on digit III; relatively narrow digital pads on digits II and III; relatively long claws on digits II and III; relatively narrow width across the proximal ends (toebases) of digits II and III, and a relatively long distance between the bases of digits II and IV; a relatively narrow toetip width (distance between the tips of digits II and IV).

Members of the *"Eubrontes"* cluster are characterized by: a relatively long first digital pad of digit II, and a relatively short second digital pad on digit III; relatively broad digital pads on digits II and III; relatively short claws on digits II and III; a relatively broad toebase II–III width, and a relatively short distance between the bases of digits II and IV; a relatively broad toetip width.

Because there are only two cases in the *Anomoepus* cluster (excluding *Kayentapus*), generalizations can only be tentative, but members of this cluster possibly are characterized by: a relatively long first digital pad on digit II, and relatively short or very short digital pads on digit III; relatively narrow digital pads; relatively long claws on digits II and III; a relatively long digit IV; a relatively long distance between the bases of digits II and IV; a very long toetip III–toetip IV distance; a very broad toetip width.

These results are broadly congruent with those of the PCA.

A forward stepwise canonical variate analysis (CVA) does a remarkable job of discriminating among the three clusters (table 8.3). Of the 19 scaled variables available for consideration, the CVA chooses 7: the length and width of the first (most proximal) digital pad of digit III; the distances between toebases III and IV, and II and IV; and all three toetip distances. For all seven of these variables, the *P* value of the *F*-statistic of the Wilks's λ test was $< .001$, indicating significant

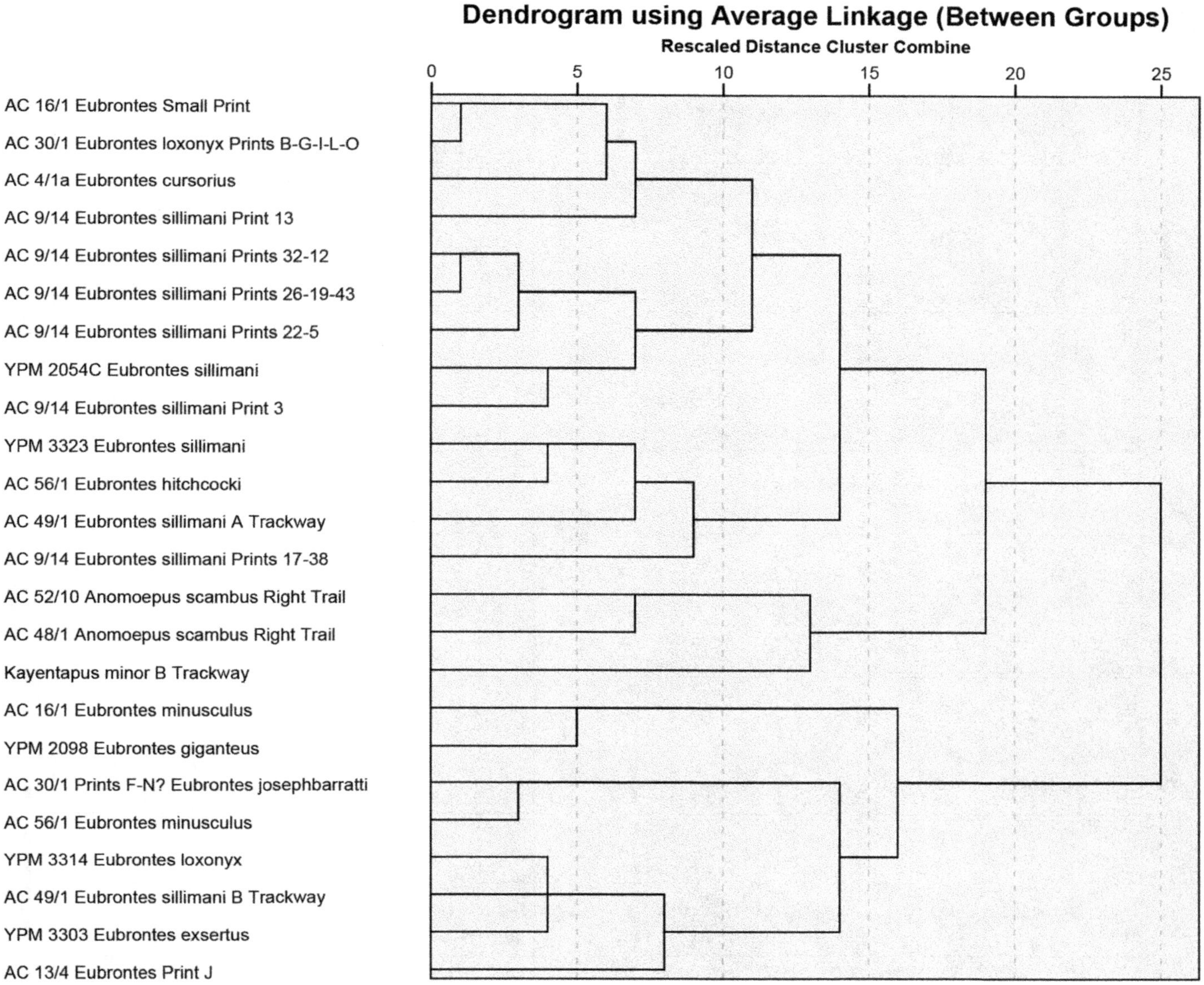

Above, **8.18.** "Maximum information" cluster analysis of Newark Supergroup tridactyl dinosaur footprints. The analysis is based on the same 24 individual specimens (mean-or-single data treatment) and the same 19 footprint measurements used in the principal component analysis (PCA) (table 8.2, fig. 8.17). It is designated the "maximum information" cluster analysis because it, like the PCA, is based on the most detailed parameters of footprint size and shape that could be used in this study. All measurements for each data case were log transformed, and each measurement scaled by subtracting from that log-scaled measurement the mean value of all 19 log-transformed measurements for that data case. The 24 individual dinosaur trackmakers fall into at least three distinct clusters: one containing most of the small and medium-sized specimens of *Eubrontes* ("*Anchisauripus*" cluster), one with the larger and some medium-sized specimens of *Eubrontes* ("*Eubrontes*" cluster), and one with *Anomoepus* and possibly *Kayentapus* specimens (*Anomoepus* cluster).

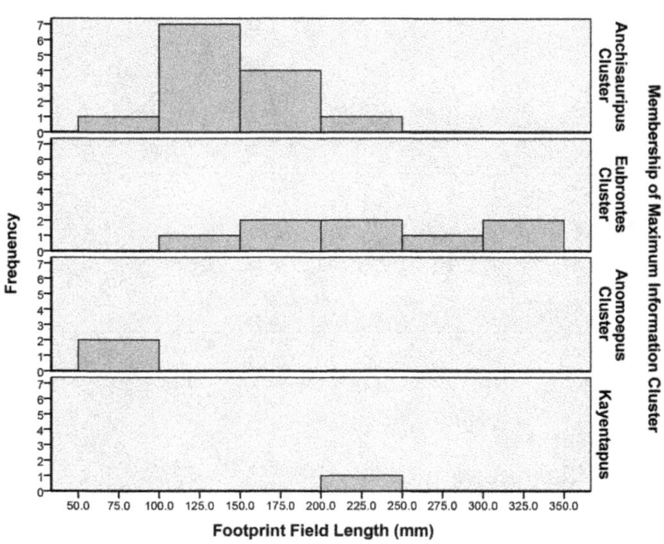

Right, **8.19.** Size-frequency distribution of the 24 trackways and singleton prints in the groups recognized by the maximum information cluster analysis (fig. 8.18). The informal names applied to the clusters reflect in part the traditional nomenclature of R. S. Lull (1953).

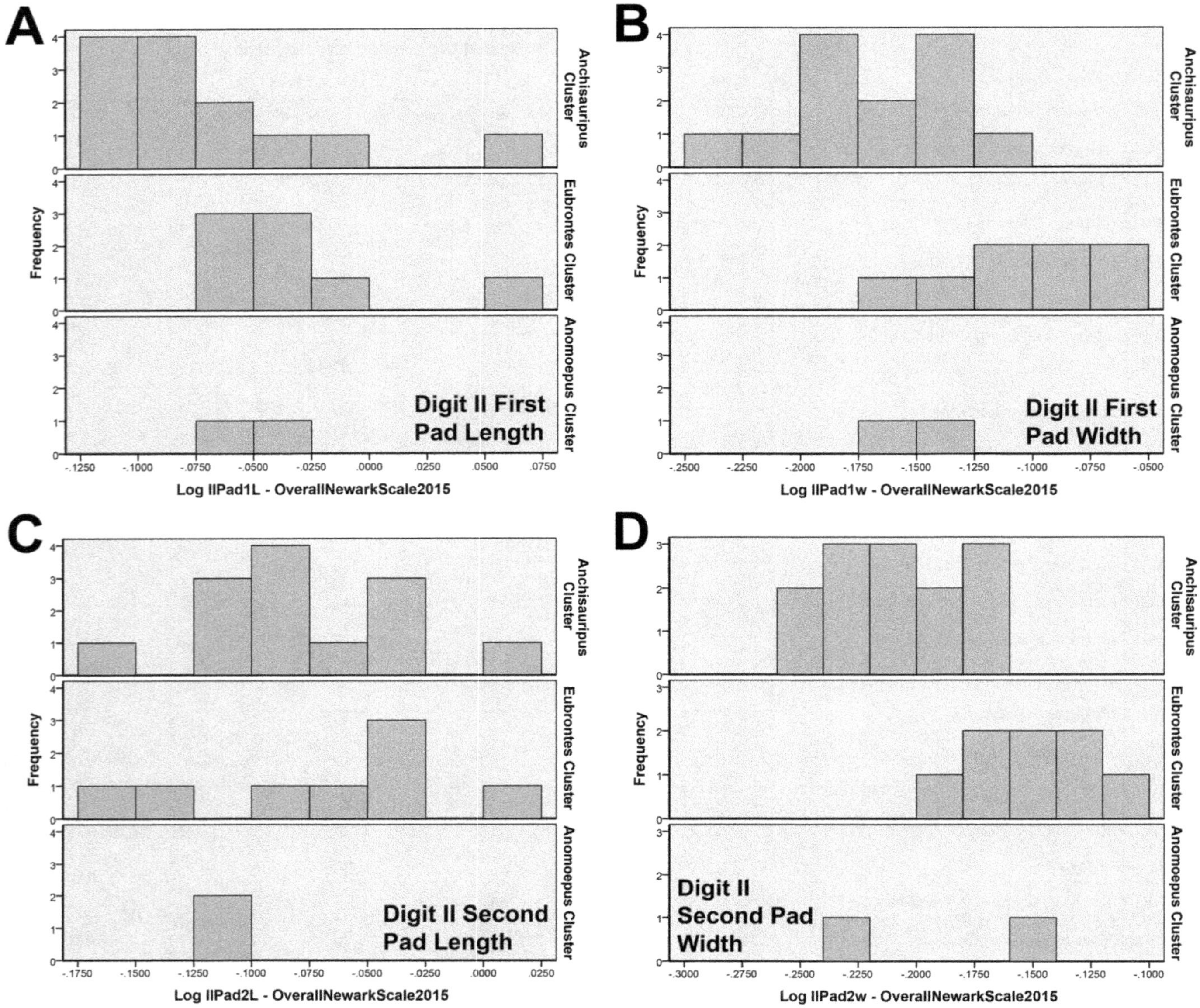

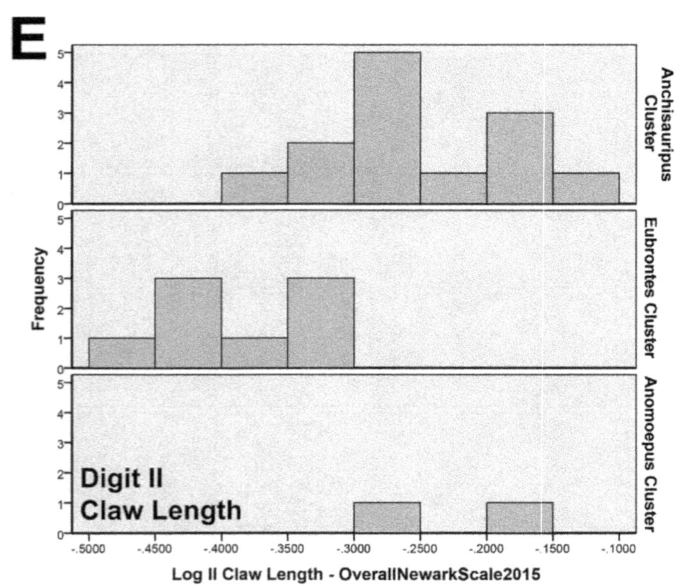

8.20. Size-frequency distributions of digit II pad lengths and widths, and digit II claw length, of the three "maximum information" clusters (excluding *Kayentapus* from the *Anomoepus* cluster). All measurements were "GM-scaled" by subtracting the mean of all 19 log-transformed footprint parameters (table 8.2; labeled "OverallNewarkScale2015" in the graphs) from each log-transformed footprint parameter. *A*, First pad length. *B*, First pad width. *C*, Second pad length. *D*, Second pad width. *E*, Claw length. Members of the *"Anchisauripus"* cluster seem to have a relatively shorter first digital pad, relatively narrower digital pads, and a relatively longer claw than members of the *"Eubrontes"* cluster. Although there are only two *Anomoepus* cases, they may be characterized by a relatively longer first digital pad, and a relatively shorter second digital pad, than members of the *"Anchisauripus"* cluster, and by a relatively shorter second digital pad, relatively narrower digital pads, and longer claws than members of the *"Eubrontes"* cluster. GM = geometric mean.

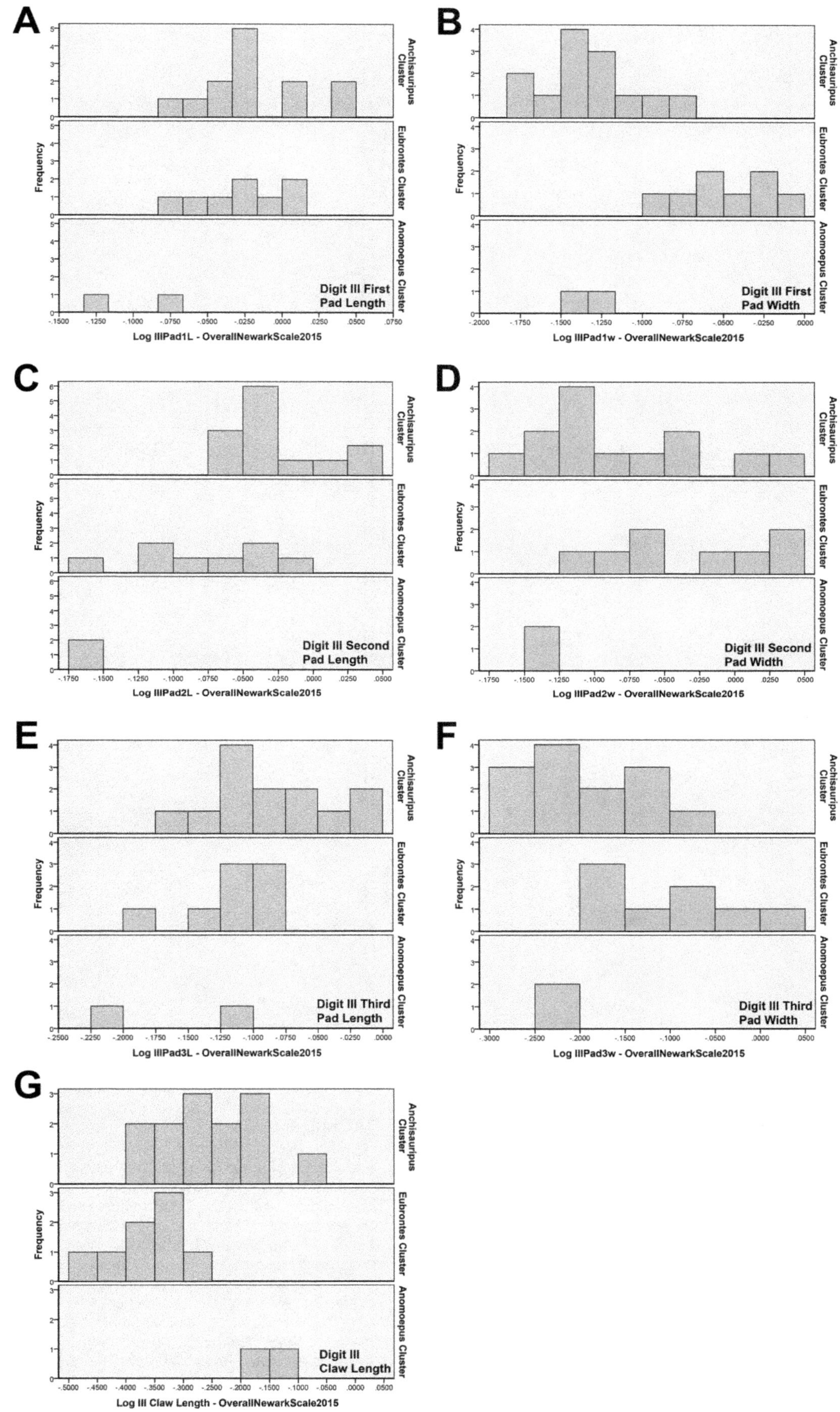

8.21. Size-frequency distributions of digit III pad lengths and widths, and digit III claw length, of the three "maximum information" clusters (excluding *Kayentapus* from the *Anomoepus* cluster). All measurements were "GM-scaled" by subtracting the mean of all 19 log-transformed footprint parameters (table 8.2; labeled "OverallNewarkScale2015" in the graphs) from each log-transformed footprint parameter. *A*, First pad length. *B*, First pad width. *C*, Second pad length. *D*, Second pad width. *E*, Third pad length. *F*, Third pad width. *G*, Claw length. Members of the "*Anchisauripus*" cluster seem to have a relatively longer second digital pad, and possibly a relatively longer third digital pad, narrower pad widths, and a relatively longer claw than members of the "*Eubrontes*" cluster. The two *Anomoepus* cases have relatively shorter digital pads than members of both the "*Anchisauripus*" and "*Eubrontes*" clusters, narrower digital pads than members of at least the "*Eubrontes*" cluster, and relatively longer claws than members of at least the "*Eubrontes*" cluster. GM = geometric mean.

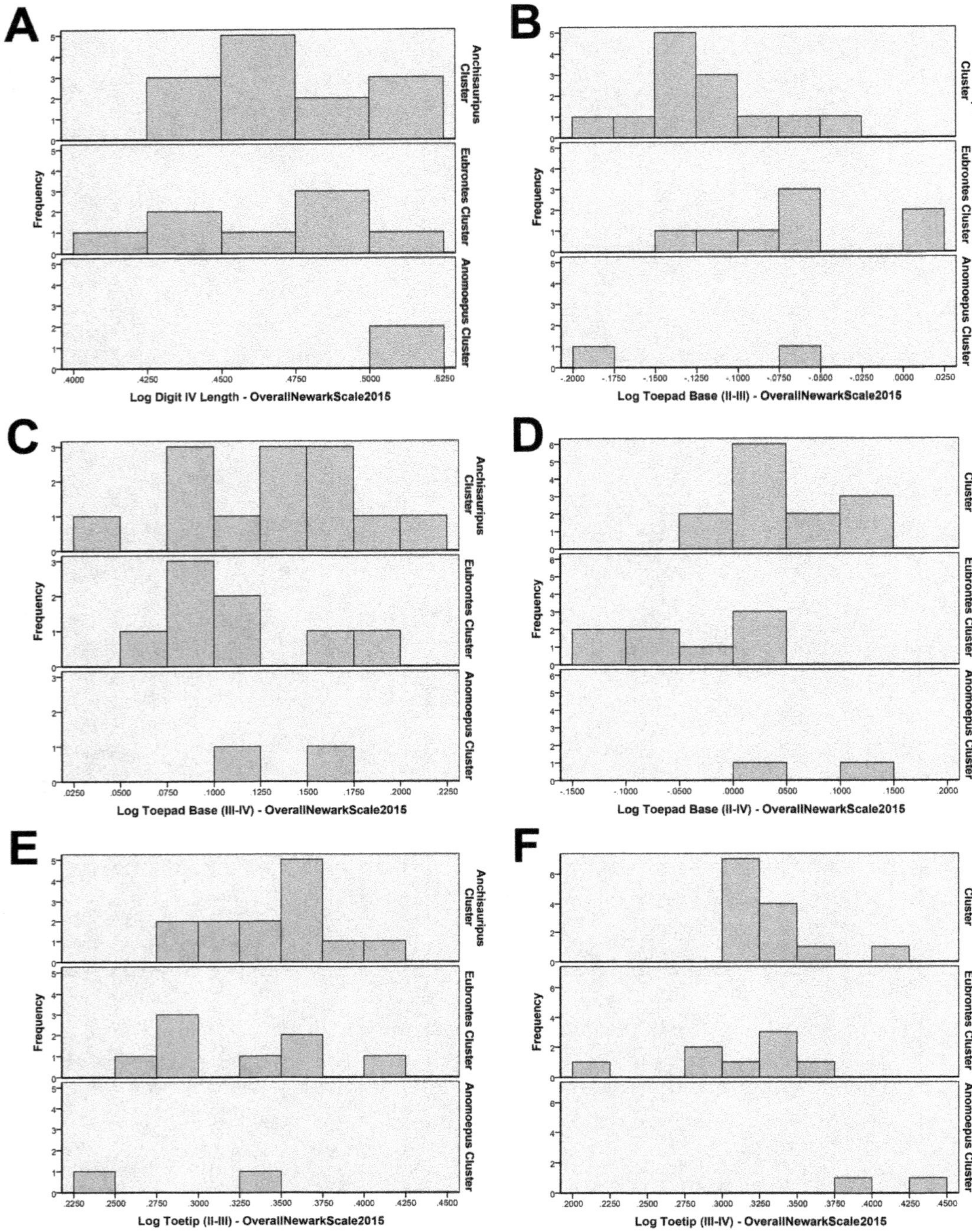

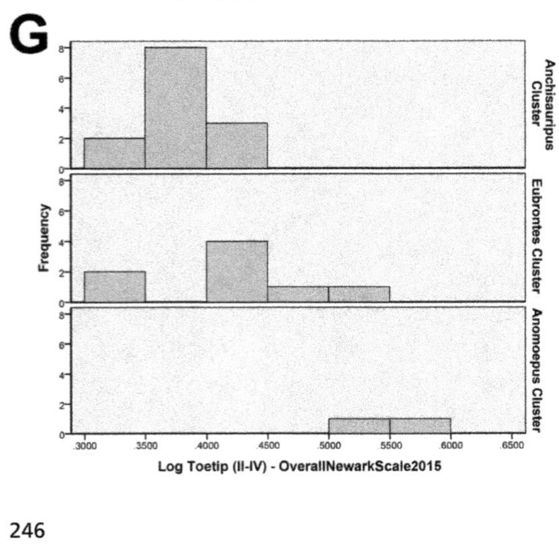

8.22. Size-frequency distributions of miscellaneous footprint parameters of the three "maximum information" clusters (excluding *Kayentapus* from the *Anomoepus* cluster). All measurements were "GM-scaled" by subtracting the mean of all 19 log-transformed footprint parameters (table 8.2; labeled "OverallNewarkScale2015" in the graphs) from each log-transformed footprint parameter. *A*, Digit IV length. *B*, Distance between the proximal ends (toepad bases) of digits II and III. *C*, Distance between the proximal ends of digits III and IV. *D*, Distance between the proximal ends of digits II and IV. *E*, Distance between toetips II and III. *F*, Distance between toetips III and IV. *G*, Distance between toetips II and IV (toetip width). Members of the *"Anchisauripus"* cluster have a narrower toebase II–III, a longer toebase II–IV, possibly a longer toetip II–III and III–IV distances, and a narrower toetip width (toetip II–IV) than members of the *"Eubrontes"* cluster. The two *Anomoepus* cases have a relatively longer digit IV than members of both the *"Anchisauripus"* and *"Eubrontes"* cluster, a relatively longer toebase II–IV than members of the *"Eubrontes"* cluster, a shorter toetip II–III distance than members of the *"Anchisauripus"* and possibly also the *"Eubrontes"* clusters, and a relatively longer toetip III–IV distance and toetip width than members of both the *"Eubrontes"* and *"Anchisauripus"* clusters. GM = geometric mean.

Table 8.3. Forward stepwise discriminant (canonical variate) analysis of log-transformed Newark Supergroup tridactyl dinosaur footprint measurements (the same set of measurements used in the principal component analysis [table 8.2]), all scaled by subtracting the mean of all the log-transformed variables from each log-transformed measurement. Groups compared are the "maximum information" clusters. Results are the same whether or not the prior probability of group membership is adjusted for the number of cases in each group. Classification results: 100% of original grouped cases and 100% of cross-validated grouped cases were correctly classified. Asterisk (*) indicates largest absolute correlation between each variable and either discriminant function.

Variable	Standardized canonical discrimination function coefficients (correlations between individual variables and discriminant functions)	
	Function 1	Function 2
Digit III first pad length	1.156 (0.075)	0.029 (0.284)*
Digit III first pad width	1.330 (0.243)	−0.282 (−0.516)*
Distance from toebase III to toebase IV	1.412 (−0.026)	−0.002 (0.104)*
Distance from toebase II to toebase IV	−1.901 (−0.163)	0.381 (0.228)*
Distance from toetip II to toetip III	0.212 (0.048)	0.831 (0.287)*
Distance from toetip III to toetip IV	−1.385 (−0.179)	0.063 (−0.191)*
Distance from toetip II to toetip IV (toetip width)	−0.536 (−0.112)	−0.895 (−0.596)*
Eigenvalue (% of variance)	28.999 (90.3)	3.119 (9.7)
Canonical correlation	0.983	0.870

differences in the means of all variables among the three groups. The eigenvalues of the canonical discriminant functions and the canonical correlations (especially of the first function) are quite large, indicating that the functions do a splendid job of telling the three clusters apart.

Positive values of discriminant function 1 are associated with the scaled length and width of the first pad of digit III, the distance between the bases (proximal ends) of digits III and IV, and the distance between the tips of digits II and III; negative values are associated with the distance between the bases of digits II and IV and the distances between toetips III and IV and between toetips II and IV. The *Anomoepus* and *Kayentapus* cases have strongly negative discriminant function 1 scores, members of the "*Eubrontes*" cluster (including both large and a few medium-sized forms) have strongly positive scores, and members of the "*Anchisauripus*" cluster have neutral to slightly negative scores (fig. 8.23).

Positive values of discriminant function 2 are especially associated with the scaled distances between toebases II and IV, and between toetips II and III, while negative values are associated with the width of the first pad of digit III and the distance between toetips II and IV. The larger forms of *Eubrontes*, as well as *Kayentapus* and *Anomoepus*, have strongly negative scores for discriminant function 2, while the smaller versions of *Eubrontes* have neutral or positive scores (fig. 8.23).

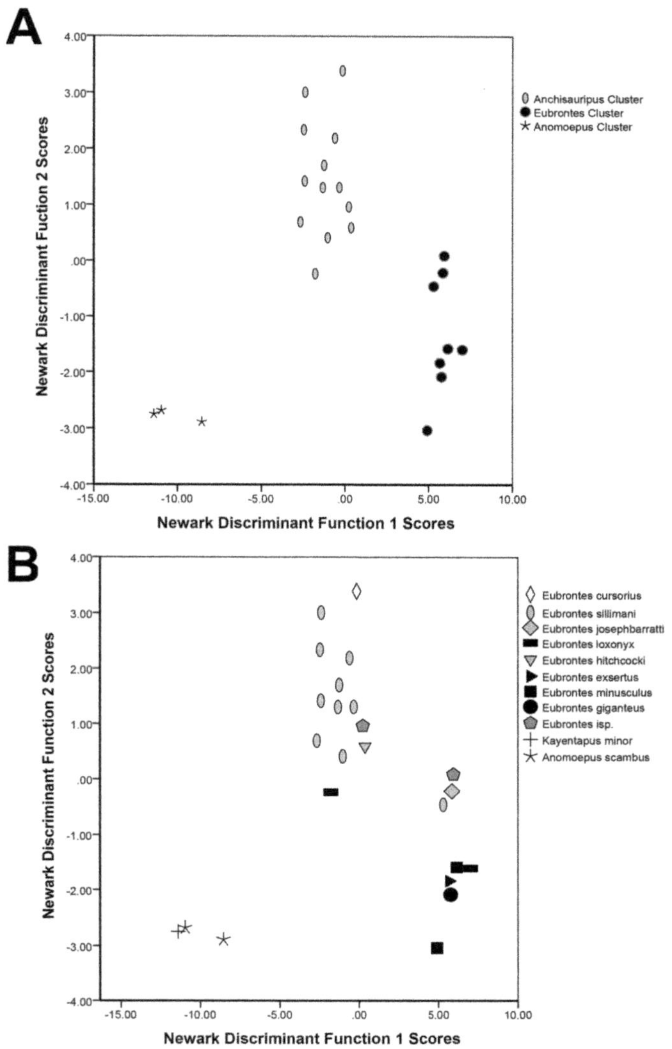

8.23. Canonical variate (discriminant) analysis of the 24 trackways and singleton prints in the groups recognized by the maximum information cluster analysis (fig. 8.18). All measurements for each data case were log transformed, and each measurement was scaled by subtracting from that log-scaled measurement the mean value of all 19 log-transformed measurements for that data case. Discriminant Function 1 separates forms with relatively long scaled lengths and widths of the first pad of digit III, distances between the proximal ends of digits III and IV, distances between the tips of digits II and III (positive scores) from forms with relatively large distances between the bases of digits II and IV, between the tips of digits III and IV, and between the tips of digits II and IV (negative scores). Discriminant Function 2 separates forms with a relatively long first digital pad of digit III, and especially with relatively long distances between the proximal ends of digits II and IV, as well as between the tips of digits II and III (positive scores) from forms with a relatively broad first pad of digit III and a relatively long distance between the tips of digits II and IV (negative scores). Plotting Discriminant Function 2 against Discriminant Function 1 results in dramatic separation of the groups recognized by the maximum information cluster analysis. *A*, Data cases identified by membership in the cluster groups. *B*, Data cases identified by ichnotaxon.

Remarkably, a test of the ability of discriminant functions to identify which cluster data cases belong to shows 100% accuracy (table 8.3). This holds true both when the functions test the predicted group membership of all the data cases

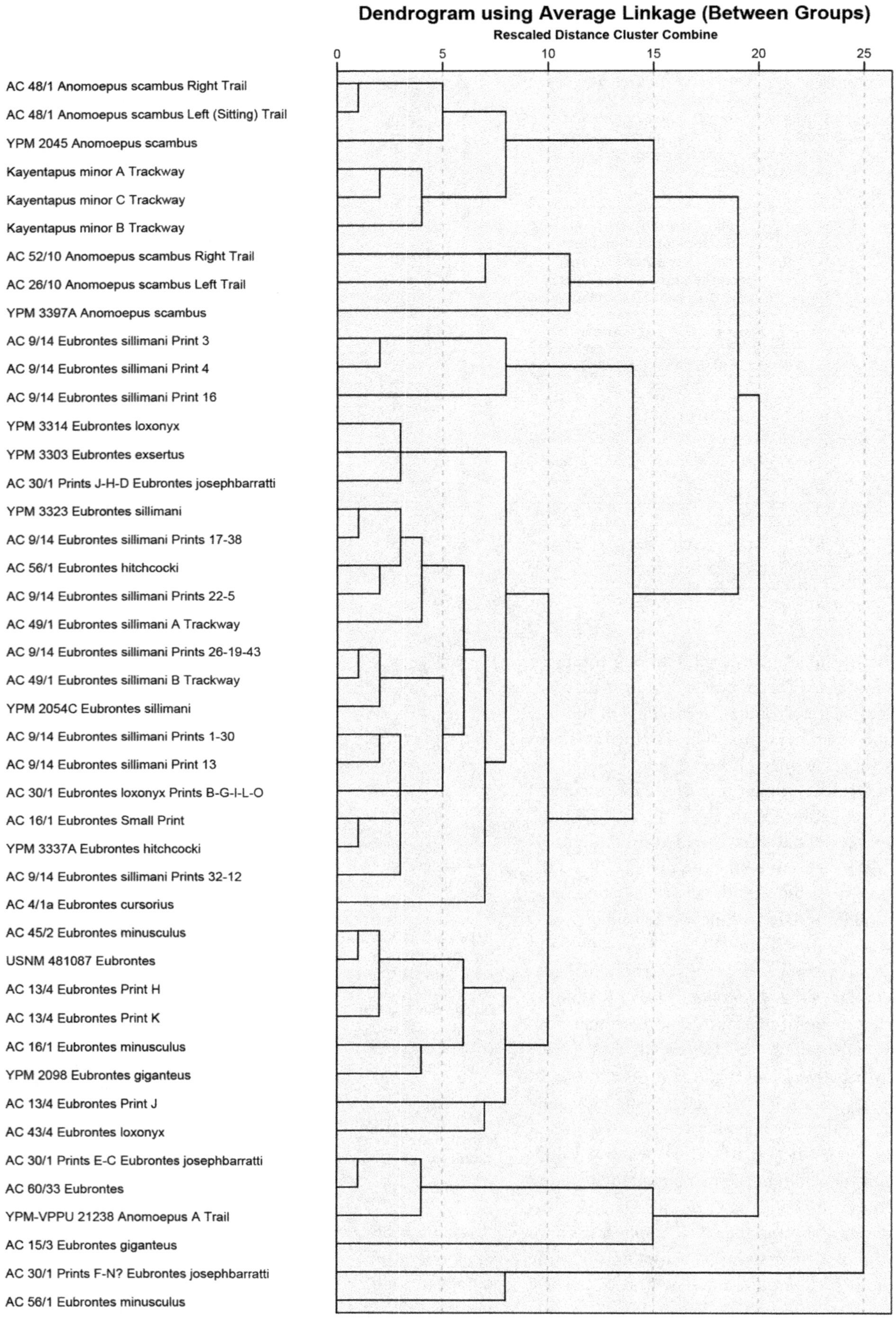
Dendrogram using Average Linkage (Between Groups)
Rescaled Distance Cluster Combine
0
5
10
15
20
25
AC 48/1 Anomoepus scambus Right Trail
AC 48/1 Anomoepus scambus Left (Sitting) Trail
YPM 2045 Anomoepus scambus
Kayentapus minor A Trackway
Kayentapus minor C Trackway
Kayentapus minor B Trackway
AC 52/10 Anomoepus scambus Right Trail
AC 26/10 Anomoepus scambus Left Trail
YPM 3397A Anomoepus scambus
AC 9/14 Eubrontes sillimani Print 3
AC 9/14 Eubrontes sillimani Print 4
AC 9/14 Eubrontes sillimani Print 16
YPM 3314 Eubrontes loxonyx
YPM 3303 Eubrontes exsertus
AC 30/1 Prints J-H-D Eubrontes josephbarratti
YPM 3323 Eubrontes sillimani
AC 9/14 Eubrontes sillimani Prints 17-38
AC 56/1 Eubrontes hitchcocki
AC 9/14 Eubrontes sillimani Prints 22-5
AC 49/1 Eubrontes sillimani A Trackway
AC 9/14 Eubrontes sillimani Prints 26-19-43
AC 49/1 Eubrontes sillimani B Trackway
YPM 2054C Eubrontes sillimani
AC 9/14 Eubrontes sillimani Prints 1-30
AC 9/14 Eubrontes sillimani Print 13
AC 30/1 Eubrontes loxonyx Prints B-G-I-L-O
AC 16/1 Eubrontes Small Print
YPM 3337A Eubrontes hitchcocki
AC 9/14 Eubrontes sillimani Prints 32-12
AC 4/1a Eubrontes cursorius
AC 45/2 Eubrontes minusculus
USNM 481087 Eubrontes
AC 13/4 Eubrontes Print H
AC 13/4 Eubrontes Print K
AC 16/1 Eubrontes minusculus
YPM 2098 Eubrontes giganteus
AC 13/4 Eubrontes Print J
AC 43/4 Eubrontes loxonyx
AC 30/1 Prints E-C Eubrontes josephbarratti
AC 60/33 Eubrontes
YPM-VPPU 21238 Anomoepus A Trail
AC 15/3 Eubrontes giganteus
AC 30/1 Prints F-N? Eubrontes josephbarratti
AC 56/1 Eubrontes minusculus

Facing, **8.24.** Cluster analysis of Newark Supergroup tridactyl dinosaur footprints. To increase the sample size of animals represented in the analysis (mean-or-single data treatment), the clusters are based on nine scaled, log-transformed variables: the lengths of digits II–IV from the proximal ends of their first digital pads to their toetips, and the toebase and toetip distances of digits II–IV. Number of data cases = 44.

Right, **8.25.** Cluster analysis of Newark Supergroup tridactyl dinosaur footprints. To increase the sample size of animals represented in the analysis (mean-or-single data treatment), the clusters are based on 10 scaled, log-transformed variables: backfoot length, digit III projection, toetip distances of digits II–IV, the distances from the footprint "heel" to toetips II and IV, the distances from the bending point along the medial edge of digit II, near its proximal end, to toetip II, to the long axis of digit III, and to the footprint "heel." Number of data cases = 50.

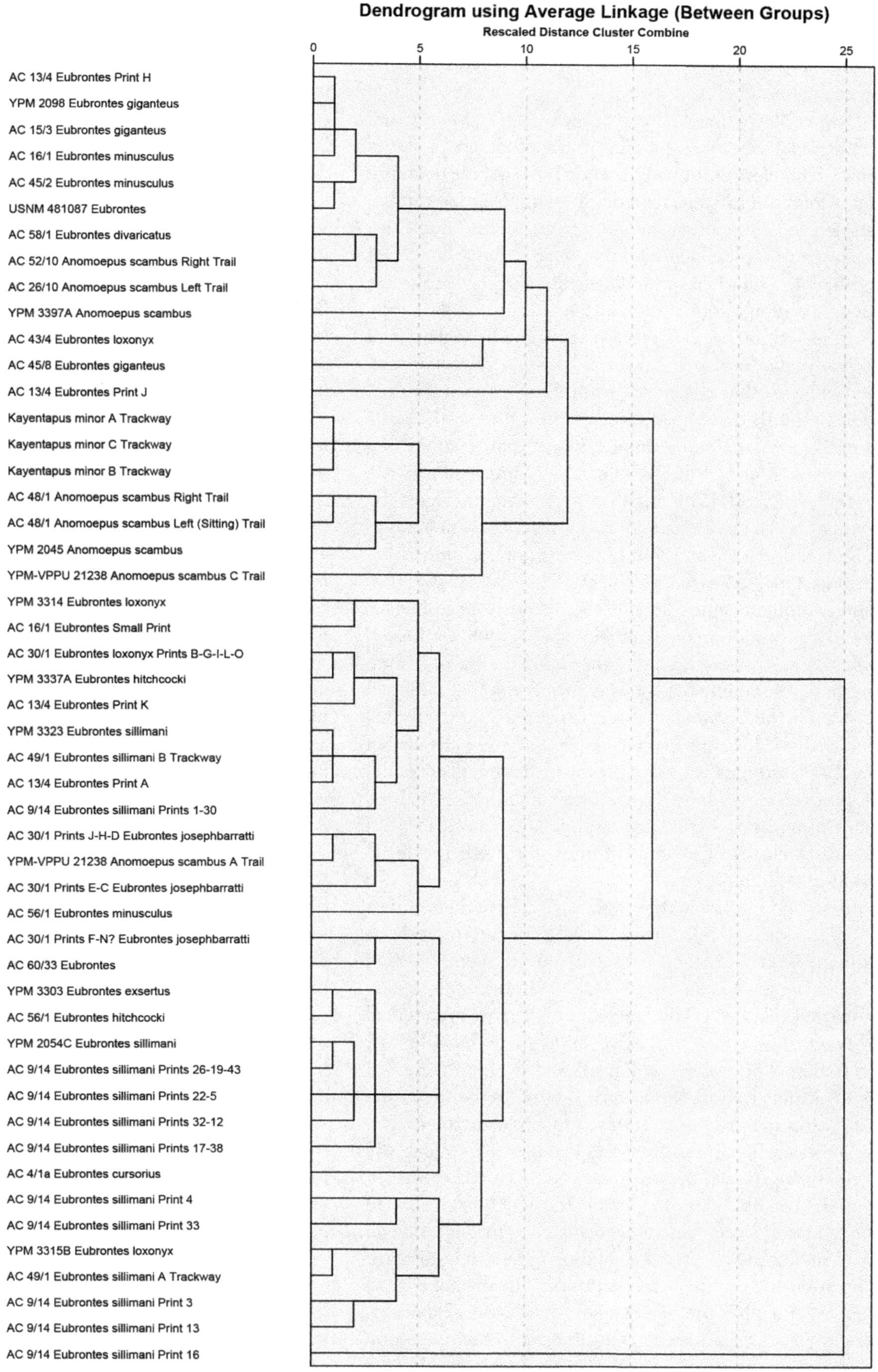

against their known group membership, and also when the test uses the leave-one-out procedure, in which each case is classified on the basis of discriminant functions created using all data cases other than itself.

Regrettably, a sample size of 24 mean-or-single data cases, presumably representing 24 individual dinosaurs, is rather small. I therefore examined the effects of reducing the number of footprint variables, and of using variables that capture less detailed information about footprints, on the sample size of dinosaurs represented in cluster analyses, and the extent to which less informative variables still manage to recognize the "maximum information" clusters.

One such analysis (fig. 8.24) uses the same toebase and toetip distance measurements as did the maximum information analyses, but eliminates measurements of individual digit II and III digital pad lengths and widths, and digit II and III clawmark lengths, instead using overall lengths (from the proximal end of the first digital pad impression to the toetip) of digits II–IV. Using these 9 variables increases the number of data cases to 44. The results are, up to a point, mostly consistent with those of the maximum information analysis. Once again most of the small to medium-sized *Eubrontes* group together. Most of the data cases identified as *Anomoepus* again group together, together with the *Kayentapus* cases. However, some of the *Eubrontes* cases, including both large and small forms, specimens that clumped together in the "*Eubrontes*" cluster in the maximum information analysis, now link up more closely with the equivalent of the "*Anchisauripus*" cluster than with other members of the "*Eubrontes*" cluster from the maximum information analysis. More annoyingly, two *Eubrontes*, one large and one medium-sized, cluster together apart from all the other tridactyl prints in the sample.

A second cluster analysis (fig. 8.25) uses some even less detailed characters. Instead of lengths of the impressions of digits II–IV from their proximal ends to the tips of their clawmarks, some anatomically less exact measurements are employed (fig. 8.16). The length of digit III is represented by two measurements, the backfoot length and the digit III projection. The lengths of digits II and IV are represented by the distances from the footprint "heel" to the tips of the clawmarks of those digits. Instead of toebase distances, two distances are measured from the bending point along the medial margin near the proximal end of the impression of digit II. One of the distances is from the digit II bend point to the footprint "heel," and the second is from the digit II bend point to the midline long axis of the impression of digit III. The distance from the bend point along the medial edge of digit II to the tip of the clawmark of digit II is also employed. Toetip distances are once again used, as in the maximum information analysis. This set of 10 parameters results in a sample size of 50 animals, more than twice the number in the maximum information cluster.

Most of the small and medium-sized *Eubrontes* still group together, but otherwise the clusters identified in the maximum information analysis now start to look a bit scrambled. Some of the *Anomoepus* still join with *Kayentapus*, but other *Anomoepus* cases link first with several large *Eubrontes* before joining the *Anomoepus-Kayentapus* cluster. One of the smaller *Eubrontes* links up with all the other dinosaur prints in the sample.

Bivariate Comparisons. We can increase sample sizes still further by doing bivariate comparisons of the variables that the PCA and CVA identified as most important in segregating the data cases into distinct groups, as well as other comparisons of potential interest. Once again we will be using the mean-or-single data treatment. We start by comparing the relative widths of the toemarks (figs. 8.26–8.28). Although there is a great deal of overlap in the distribution of points, large as well as some medium-sized *Eubrontes* tend to have a relatively broad first digital pad of digit III, compared with the length of that pad, while *Anomoepus*, *Kayentapus* and small to medium-sized *Eubrontes* tend to have a relatively narrower first pad of digit III (fig. 8.26A).

Things look generally similar if we compare the widths of the second digital pad of digits II and III (located roughly midway out along the lengths of those toes) against the overall lengths of digits II and III, without (fig. 8.27) or including (fig. 8.28A, 8.28*B*) the impressions of the clawmarks. There is again considerable overlap among ichnotaxa in the trends. *Anomoepus* again tends to be aligned with the trend for smaller *Eubrontes*. *Kayentapus* is narrow-toed in all comparisons. For digit III, medium-sized *Eubrontes* look stouter-toed than smaller *Eubrontes*, perhaps with a tendency for their relative stoutness to become greater with increasing size. This seems consistent with the tendency for digits to become broader with increasing size seen in dinosaur foot skeletons (figs. 2.20–2.23, 4.12). It is surprising, then, that in comparisons of digital pad width against the length of digit III, the biggest forms of *Eubrontes* have relatively narrower toes than do many medium-sized *Eubrontes* (figs. 8.27*B*, 8.28*B*).

There are too few data points to say much about the relative width of digit IV (fig. 8.28C), except to note that *Kayentapus* again looks rather narrow-toed compared to the other ichnotaxa, and there is a suggestion that medium-sized *Eubrontes* may be broader-toed than smaller forms.

Another set of comparisons involves measures related to footprint breadth. Given the architecture of the footprint, distances across the proximal ends of the digit impressions (fig. 8.16) can be as much or more dimensions parallel to, as opposed to dimensions across, the footprint long axis, but

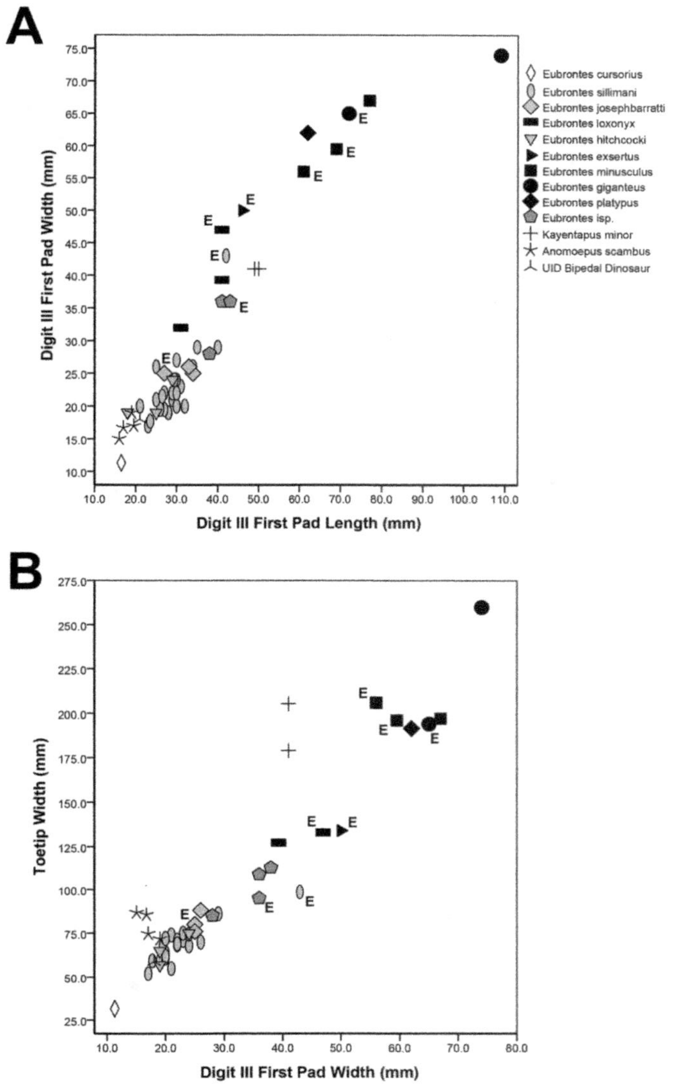

8.26. Bivariate comparisons involving the width of the first (most proximal) digital pad of digit III. In this and subsequent figures, specimens labeled "E" are members of the "*Eubrontes*" maximum information cluster (fig. 8.18); unlabeled large *Eubrontes* could not be classified in the cluster analysis because one or more of the footprint parameters used to define the cluster could not be measured. Many (but not all) of the smaller and mid-size range *Eubrontes* prints would be members of the *"Anchisauripus"* cluster but are not labeled, because a large number of them plot so close together. *A*, Length vs. width of the first digital pad. Smaller forms of *Eubrontes*, as well as *Kayentapus* and *Anomoepus*, have relatively narrow pads compared with their lengths, while the medium-sized and larger forms of *Eubrontes* have relatively broader pads. *B*, Toetip width (distance between the tips of digits II and IV) vs. the width of the first digital pad of digit III. *Anomoepus* and *Kayentapus* have a relatively broad toetip width compared with *Eubrontes*.

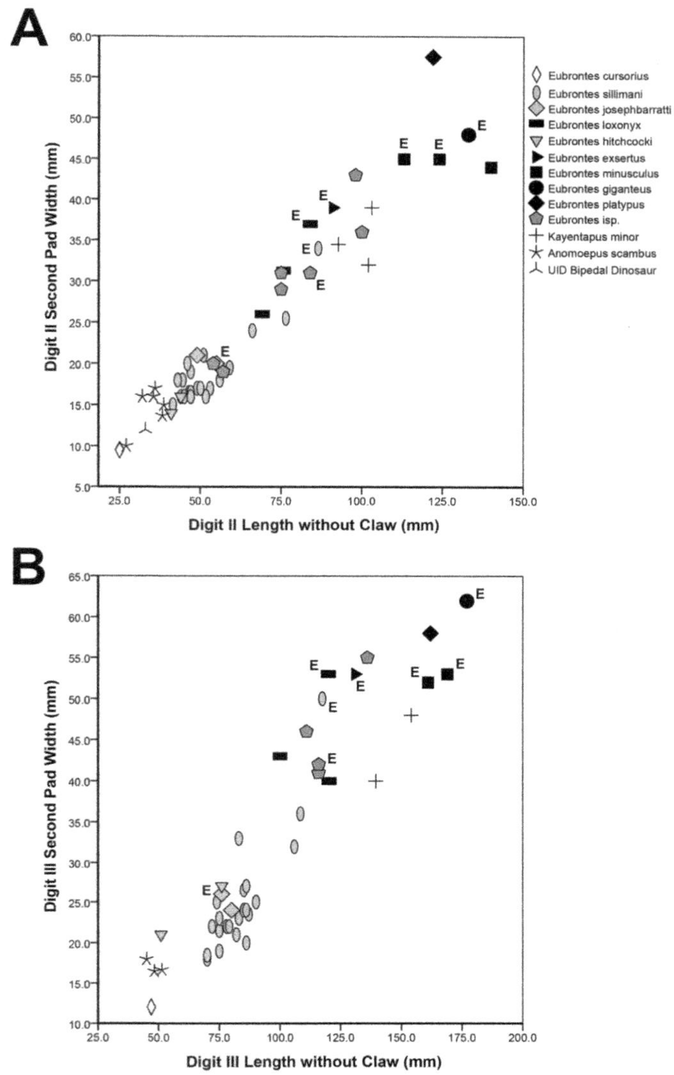

8.27. Overall length of digits II and III (excluding the clawmark) compared with the width of the second digital pad (a measure of the width of the toemark about halfway out along its length). *A*, Digit II. *B*, Digit III. In both comparisons, *Anomoepus* and the smaller *Eubrontes* fall along the same trend. In both digits II and III, medium-sized *Eubrontes* plot rather higher, suggesting that these forms are relatively stouter-toed. *Kayentapus* and most specimens of large *Eubrontes* seem to have relatively less stout toes than do medium-sized *Eubrontes*, close to expectations from extrapolation of the trend for small *Eubrontes* to large sizes.

they nonetheless partake of the latter. There is little difference among the ichnotaxa in the relationship between footprint length and the distance between the proximal ends of digits II–III, and III–IV (fig. 8.29A, 8.29*B*), but for a given footprint length *Kayentapus* and possibly *Anomoepus* tend to have a larger distance between the proximal ends of digits II–IV than does *Eubrontes* (fig. 8.26C).

Anomoepus and *Kayentapus* have a relatively larger toetip (digit II–digit IV) width (figs. 8.26*B*, 8.30) than do most *Eubrontes*, in all comparisons of this parameter with other footprint dimensions, and *Kayentapus* shows a large distance between the tips of digits III–IV compared with the distance between the tips of digits II–III (fig. 8.31).

As was discussed in chapter 6 (and will be considered again in this chapter), the lengths of claw impressions in my bird footprint sample are particularly variable among tracks made by the same species, which predisposes one to treat the relative lengths of claw impressions in the Newark

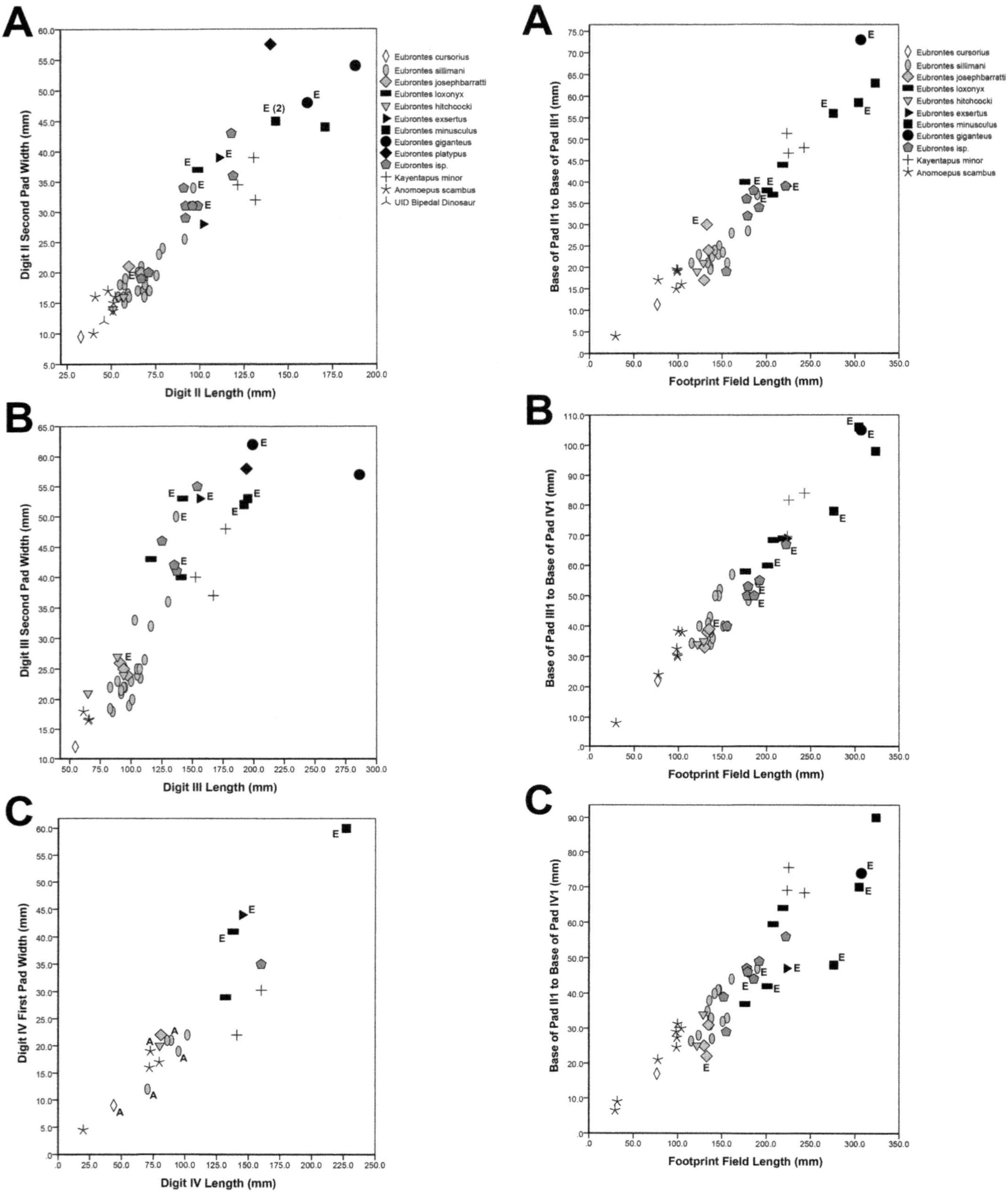
A
Digit II Second Pad Width (mm)
Digit II Length (mm)
Eubrontes cursorius
Eubrontes sillimani
Eubrontes josephbarratti
Eubrontes loxonyx
Eubrontes hitchcocki
Eubrontes exsertus
Eubrontes minusculus
Eubrontes giganteus
Eubrontes platypus
Eubrontes isp.
Kayentapus minor
Anomoepus scambus
UID Bipedal Dinosaur
A
Base of Pad II1 to Base of Pad III1 (mm)
Footprint Field Length (mm)
Eubrontes cursorius
Eubrontes sillimani
Eubrontes josephbarratti
Eubrontes loxonyx
Eubrontes hitchcocki
Eubrontes exsertus
Eubrontes minusculus
Eubrontes giganteus
Eubrontes isp.
Kayentapus minor
Anomoepus scambus
B
Digit III Second Pad Width (mm)
Digit III Length (mm)
B
Base of Pad III1 to Base of Pad IV1 (mm)
Footprint Field Length (mm)
C
Digit IV First Pad Width (mm)
Digit IV Length (mm)
C
Base of Pad II1 to Base of Pad IV1 (mm)
Footprint Field Length (mm)

Facing left, **8.28.** Overall lengths (including the clawmark) and relative widths of digits. *A*, *B*, Lengths of digits II and III compared with the width of the second digital pad. *A*, Digit II. The number 2 in parentheses associated with the label E indicates two data cases that plotted too close together to be labeled separately. *B*, Digit III. The results and interpretation for *A* and *B* are mostly the same as for the comparable plots excluding the clawmarks (fig. 8.27). *C*, Digit IV length vs. the width of its first digital pad (interpreted as coinciding with the metatarsophalangeal joint of this digit [Baird 1957; Farlow et al. 2000]). In this and subsequent figures, specimens labeled "A" are members of the *"Anchisauripus"* maximum information cluster (fig. 8.18) Compared with *Anomoepus* and small to medium-sized forms of *Eubrontes*, *Kayentapus* has a very narrow digit IV, whereas larger forms of *Eubrontes* have a broader digit.

Facing right, **8.29.** Comparisons of the distances between the proximal ends (toebases) of digits II–IV with footprint length. *A*, Toebase II–III. *B*, Toebase III–IV. *C*, Toebase II–IV. *Kayentapus* and possibly *Anomoepus* have a noticeably greater distance between the proximal ends of digits II and IV than does *Eubrontes*, but there is less, if any difference among ichnotaxa in toebase distances in the other comparisons.

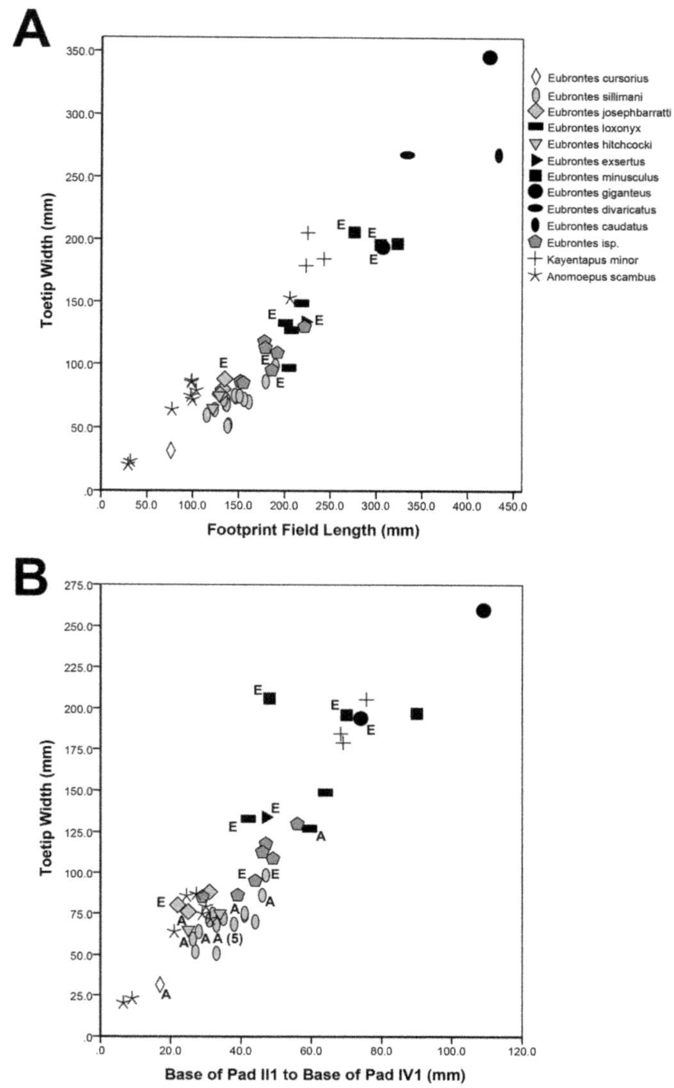

Right, **8.30.** Toetip width (distance between the tips of digits II and IV). *A*, Toetip width vs. footprint length. *Anomoepus* and *Kayentapus* footprints tend to be broader than those of *Eubrontes*. *B*, Toetip width vs. the width across the proximal ends (bases) of digits II and IV. This comparison shows much the same pattern as in *A*, except that there is a more distinct tendency for large *Eubrontes*, and smaller *Eubrontes* that fall in the *"Eubrontes"* maximum information cluster, to have broader toetip widths than the smaller *Eubrontes*, particularly those cases that plot in the *"Anchisauripus"* maximum information cluster.

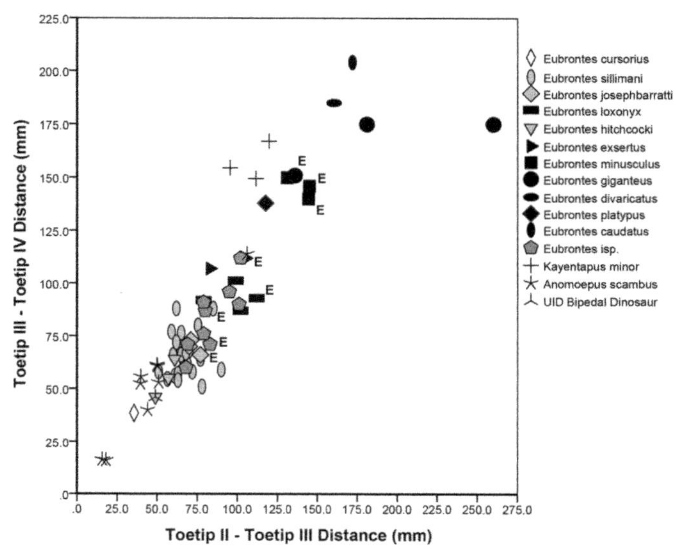

Bottom right, **8.31.** Toetip II–III vs. toetip III–IV. *Anomoepus* measurements fall along the same trend as for *Eubrontes*, but *Kayentapus* has a relatively wide toetip III–IV distance.

Supergroup dinosaur footprints with caution. Furthermore, during the actual process of measuring footprint parameters, I frequently found it difficult to be certain where a clawmark ended and extramorphological features like mud cracks began (I have observed [as has Martin 2014] in modern bird footprints made in drying mud that toenail impressions act as zones of weakness in the mud from which cracks propagate; cf. fig. 8.2). Caveats duly noted, *Kayentapus* has relatively long clawmarks on digit II. The bigger forms of *Eubrontes* seem to have relatively shorter clawmarks in impressions of this toe than do many, but not all, smaller representatives of the same ichnogenus (fig. 8.32*A*). *Anomoepus* seems to have a relatively long clawmark in impressions of digit III (fig. 8.32*B*).

Measures of overall digit lengths (figs. 8.33, 8.34) show only subtle differences among the ichnotaxa in the length of digit II relative to digit III (figs. 8.33*A*, 8.33*B*, 8.34), although *Anomoepus*, *Kayentapus*, and possibly larger *Eubrontes* show a slight tendency to have a relatively longer digit II than do small to medium-sized *Eubrontes*. The relative length of digit IV, compared with the lengths of both digits III and II, in contrast, yields more distinct separation of ichnotaxa (figs. 8.33*C*–8.33*F*, 8.34, 8.35). Although there is, as usual, much overlap among the groups, *Anomoepus* and large *Eubrontes* generally have a relatively longer digit IV than do small to

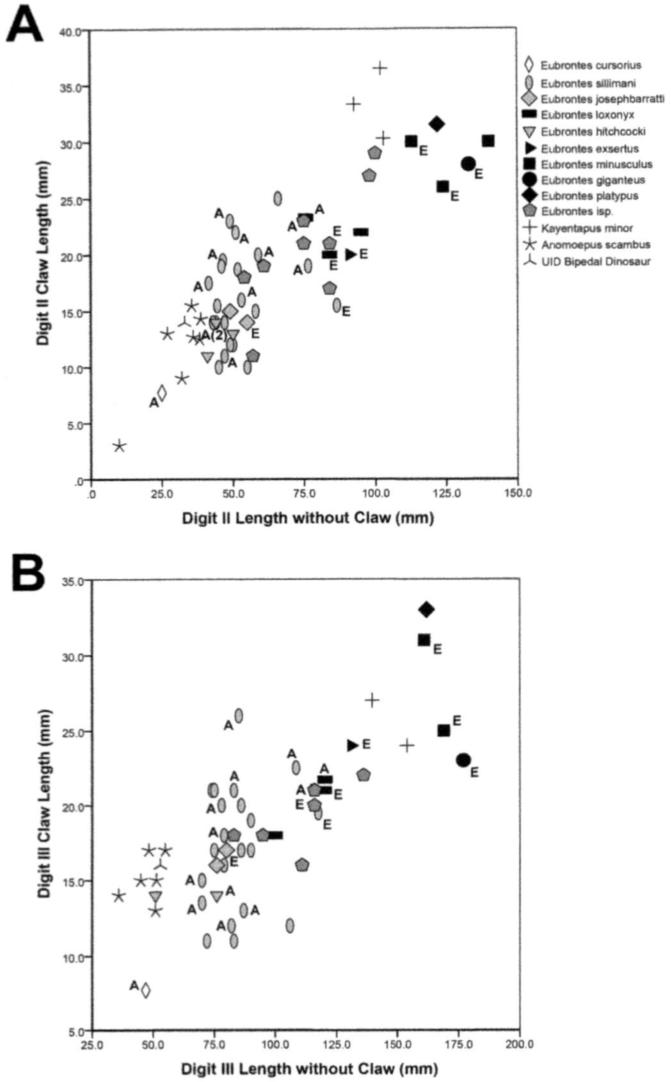

Above, **8.32.** Lengths of clawmark impressions compared with the length of the digit impression excluding the claw. *A*, Digit II. *Kayentapus* has relatively long clawmarks, whereas the bigger *Eubrontes* have relatively short clawmarks. *B*, Digit III. *Anomoepus* tends to have relatively long clawmarks. Data for digit IV were not plotted owing to a scarcity of large footprints in which the clawmarks could be distinguished from the rest of the digit impression.

Facing, **8.33.** Bivariate comparisons of overall digit lengths. In *A*, *C*, and *E*, digit lengths are measured from the proximal end of the digit impression to its tip (for digit IV, the length measured in this manner also includes the metatarsophalangeal pad). In *B*, *D*, and *F*, digit lengths are measured from the "heel" of the footprint to the tip of the toe impression. *A*, *B*, Digit II length vs. digit III length. *Anomoepus* and *Kayentapus* may have a slightly longer digit II length, relative to the length of digit III, than do most *Eubrontes*. *C*, *D*, Digit IV length vs. digit III length. *Anomoepus* and large *Eubrontes* have a relatively longer digit IV than do *Kayentapus* and small to mid-sized *Eubrontes*, and *Kayentapus* has a particularly short digit IV (relative to digit III). *E*, *F*, Digit IV length vs. digit II length. Large *Eubrontes* may have a relatively long digit IV, whereas *Kayentapus* has a relatively short digit IV.

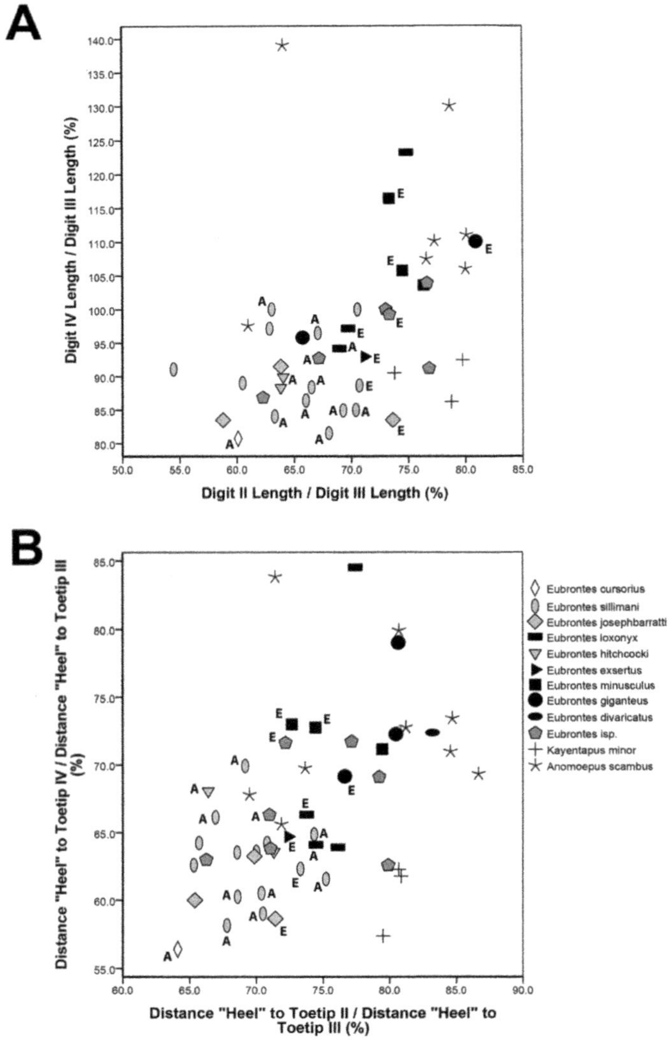

8.34. Digit length ratios: digit II and IV lengths as percentages of digit III length (cf. fig. 8.33). *A*, Digit lengths measured from the proximal end of the first digital pad to the tip of the toe. *B*, Digit lengths measured from the "heel" to the tip of the toe. *Anomoepus* and *Kayentapus* mostly have a relatively long digit II compared with digit III, and the larger forms of *Eubrontes* may tend to have a relatively longer digit II than do smaller *Eubrontes*. *Anomoepus* and large *Eubrontes* have a relatively longer digit IV than do smaller *Eubrontes*, whereas *Kayentapus* has a relatively short digit IV.

mid-sized *Eubrontes*. *Kayentapus* has a relatively shorter digit IV, compared with the length of digit III, than do *Anomoepus* and large *Eubrontes*, and a particularly short digit IV compared with the length of digit II (figs. 8.33*E*, 8.33*F*, 8.34, 8.35).

The distance from the bend point along the medial margin of the digit II impression to the midline of the digit III impression, compared with the length of the footprint (fig. 8.36A), another measure of relative footprint width at the proximal end of the print, tends to be larger in *Anomoepus* and *Kayentapus* than in *Eubrontes*. Within *Eubrontes*, larger forms may have a relatively greater distance from the digit

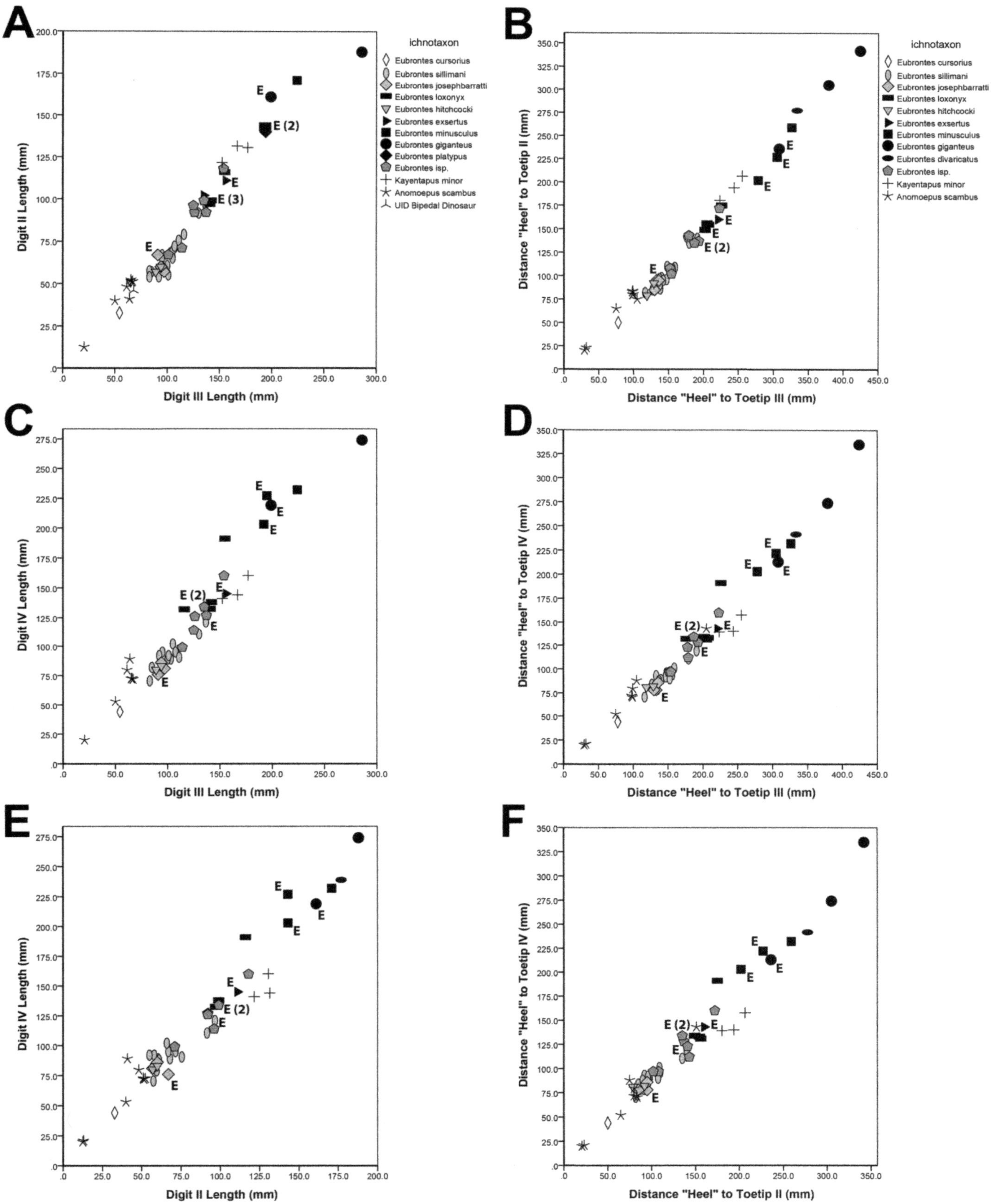
A
B
C
D
E
F
ichnotaxon
Eubrontes cursorius
Eubrontes sillimani
Eubrontes josephbarratti
Eubrontes loxonyx
Eubrontes hitchcocki
Eubrontes exsertus
Eubrontes minusculus
Eubrontes giganteus
Eubrontes platypus
Eubrontes divaricatus
Eubrontes isp.
Kayentapus minor
Anomoepus scambus
UID Bipedal Dinosaur
Digit II Length (mm)
Digit III Length (mm)
Digit IV Length (mm)
Distance "Heel" to Toetip II (mm)
Distance "Heel" to Toetip III (mm)
Distance "Heel" to Toetip IV (mm)
E
E (2)
E (3)

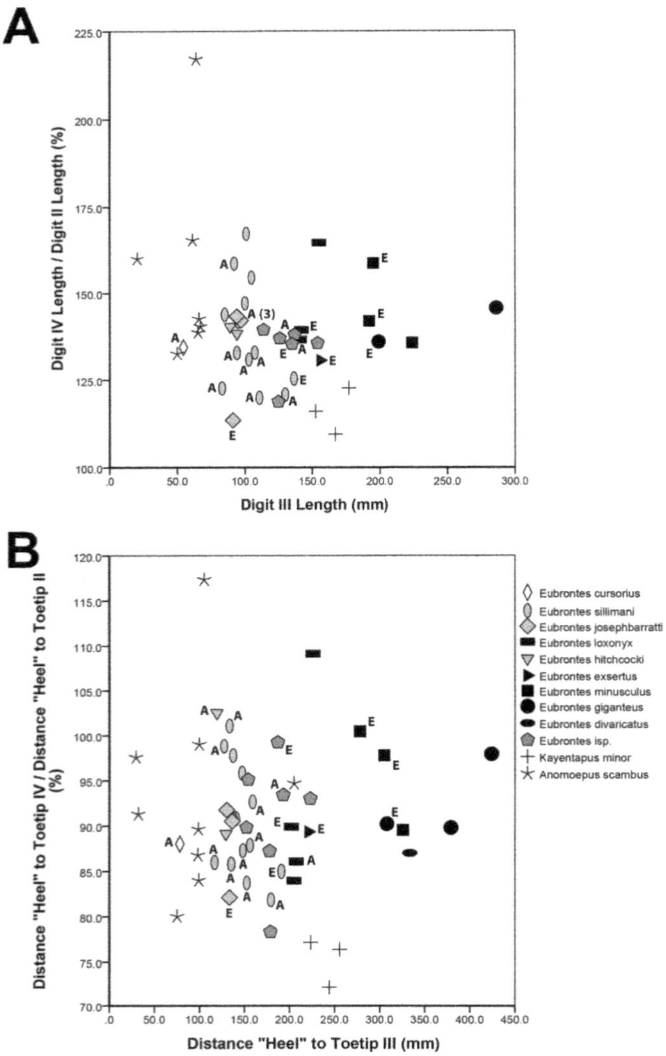

8.35. Measures of the relative lengths of digits II and IV, compared with overall footprint size. *A*, Digit IV and II lengths measured from the proximal end of the first digital pad to the tip of the toe. *B*, Digit IV and II lengths measured from the "heel" to the tip of the toe. *Kayentapus* has a distinctly short digit IV relative length (cf. figs. 8.33, 8.34).

II bend point to the digit III midline than do smaller forms. *Anomoepus* and *Kayentapus* also have a longer relative distance from the digit II bend point to the "heel" than does *Eubrontes* (fig. 8.36*B*).

Previous analyses of Newark Supergroup tridactyl dinosaur footprint shape have emphasized the projection of digit III beyond the tips of digits II and IV (also called "toetip extension" [TE]) and the portion of footprint length proximal to the digit III projection (in this book termed "backfoot") (Olsen 1980; Olsen et al. 1998; Weems 1992; also see chapter 6 of this book). It is therefore useful to compare several footprint dimensions against backfoot length among the different footprint categories (fig. 8.37). Compared with backfoot length, *Anomoepus* and *Kayentapus* show relatively

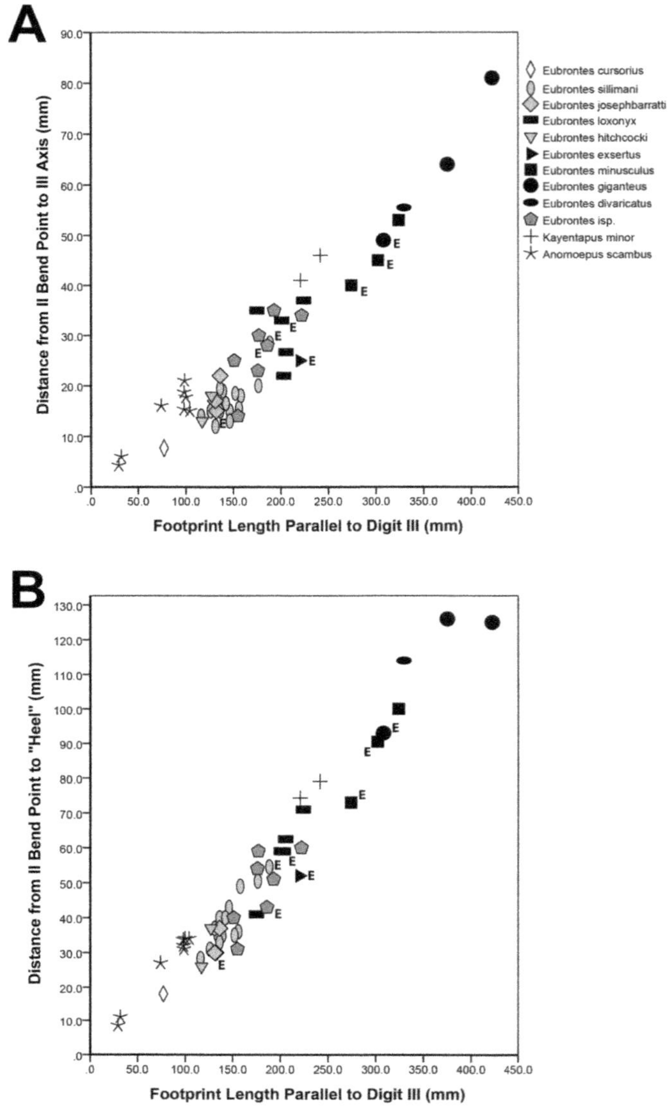

Above, **8.36.** Bivariate comparisons involving the bend point along the medial margin of the proximal end of the digit II impression. *A*, Distance from the bend point to the footprint midline, compared with overall footprint length. *Anomoepus* and *Kayentapus*, and possibly the largest and some medium-sized *Eubrontes,* are relatively broad footprints in this comparison, while small and most mid-sized *Eubrontes* are relatively narrow. *B*, Distance from the digit II bend point to the "heel" as a function of overall footprint length. *Anomoepus* and *Kayentapus* may have a somewhat longer bend point–to–"heel" distance than that in *Eubrontes*.

Facing, **8.37.** Footprint dimensions compared with backfoot length (footprint length [FL]–toetip extension [TE]: Olsen 1980; Olsen et al. 1998; Weems 1992). It was not possible to label all of the points in the "*Eubrontes*" cluster in all of the graphs, because of the density of overlap. *A*, Toebase II–toebase IV distance. *Anomoepus* and *Kayentapus* have a relatively longer toebase distance than most *Eubrontes* (especially the larger forms). *B*, Toetip II–toetip III distance. *Anomoepus* has a relatively shorter distance than most of the other forms. *C*, Toetip III–toetip IV distance. *Kayentapus* has a relatively greater distance than the other forms. *D*, Toetip width (toetip II–toetip IV distance). *Anomoepus*, *Kayentapus*, and perhaps the larger *Eubrontes* have a relatively broader footprint than do smaller *Eubrontes*. *E*, Digit III projection (i.e., TE). *Anomoepus* and large *Eubrontes* have a relativly shorter toetip extension than the other forms. *F*, Distance from the proximal bend point of digit II to the long axis of digit III. *Anomoepus* and *Kayentapus* are broader footprints than *Eubrontes* in this comparison.

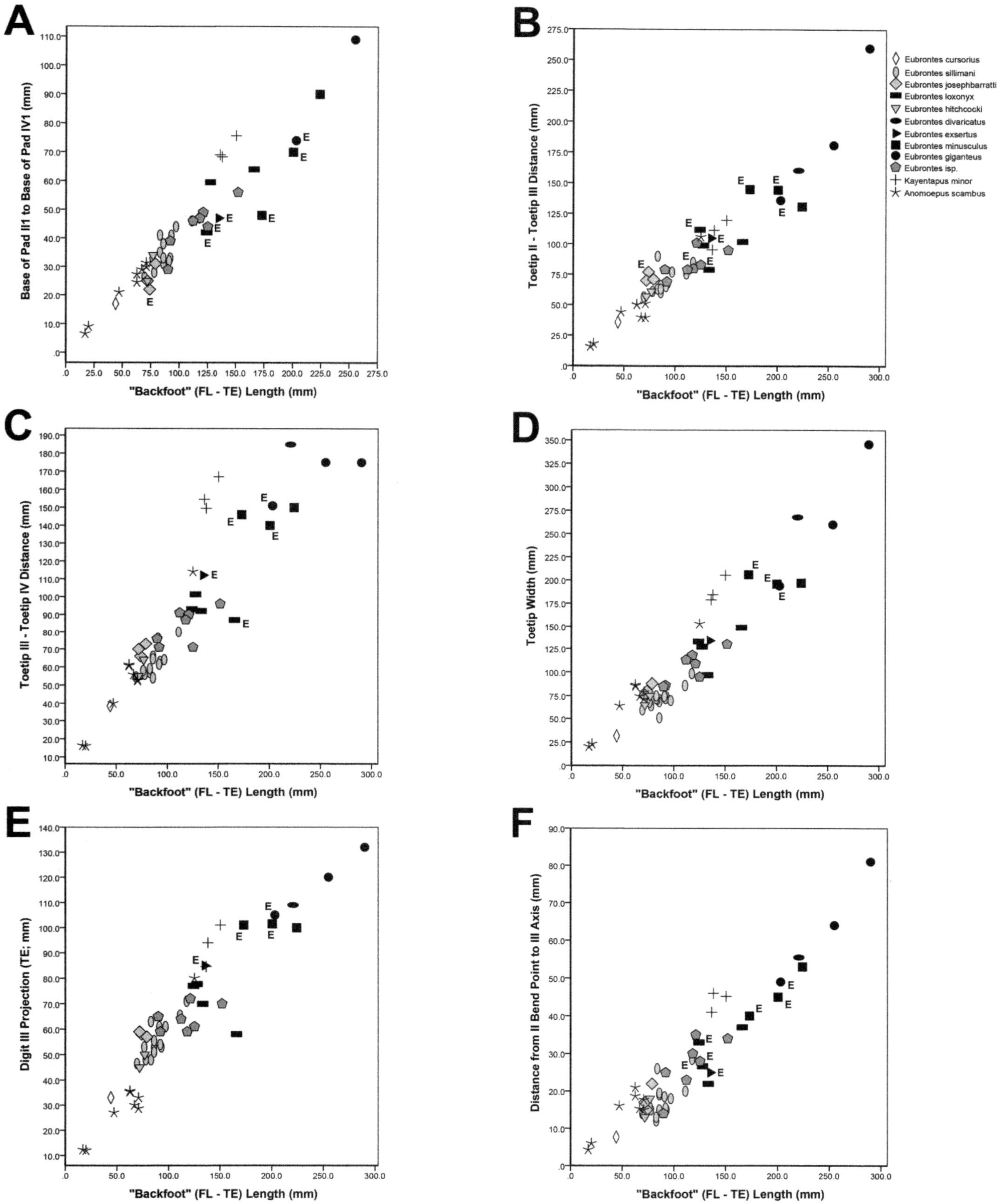
A
Base of Pad II1 to Base of Pad IV1 (mm)
"Backfoot" (FL - TE) Length (mm)
B
Toetip II - Toetip III Distance (mm)
"Backfoot" (FL - TE) Length (mm)
Eubrontes cursorius
Eubrontes sillimani
Eubrontes josephbarratti
Eubrontes loxonyx
Eubrontes hitchcocki
Eubrontes divaricatus
Eubrontes exsertus
Eubrontes minusculus
Eubrontes giganteus
Eubrontes isp.
Kayentapus minor
Anomoepus scambus
C
Toetip III - Toetip IV Distance (mm)
"Backfoot" (FL - TE) Length (mm)
D
Toetip Width (mm)
"Backfoot" (FL - TE) Length (mm)
E
Digit III Projection (TE; mm)
"Backfoot" (FL - TE) Length (mm)
F
Distance from II Bend Point to III Axis (mm)
"Backfoot" (FL - TE) Length (mm)

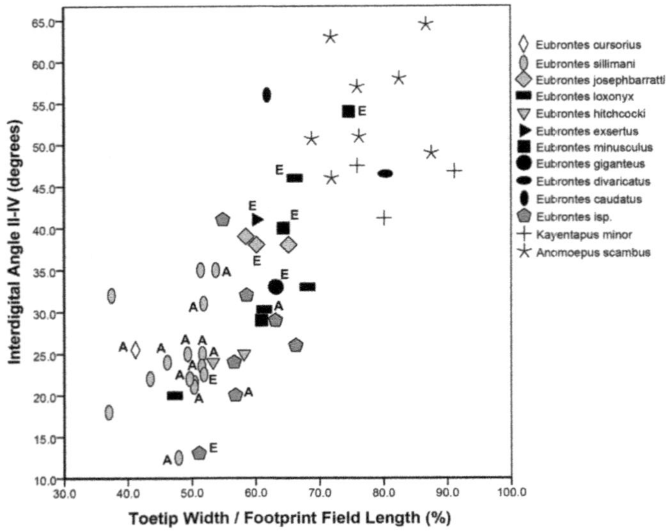

8.38. Interdigital angle II–IV plotted as a function of the ratio of toetip width (distance between the distal ends of digits II and IV) to footprint length. Relatively wider footprints are, unsurprisingly, correlated with a larger interdigital angle (for *Eubrontes*, Kendall's τb = 0.500, $P < .001$; Spearman's $\rho = 0.692$, $P < .001$; $N = 38$).

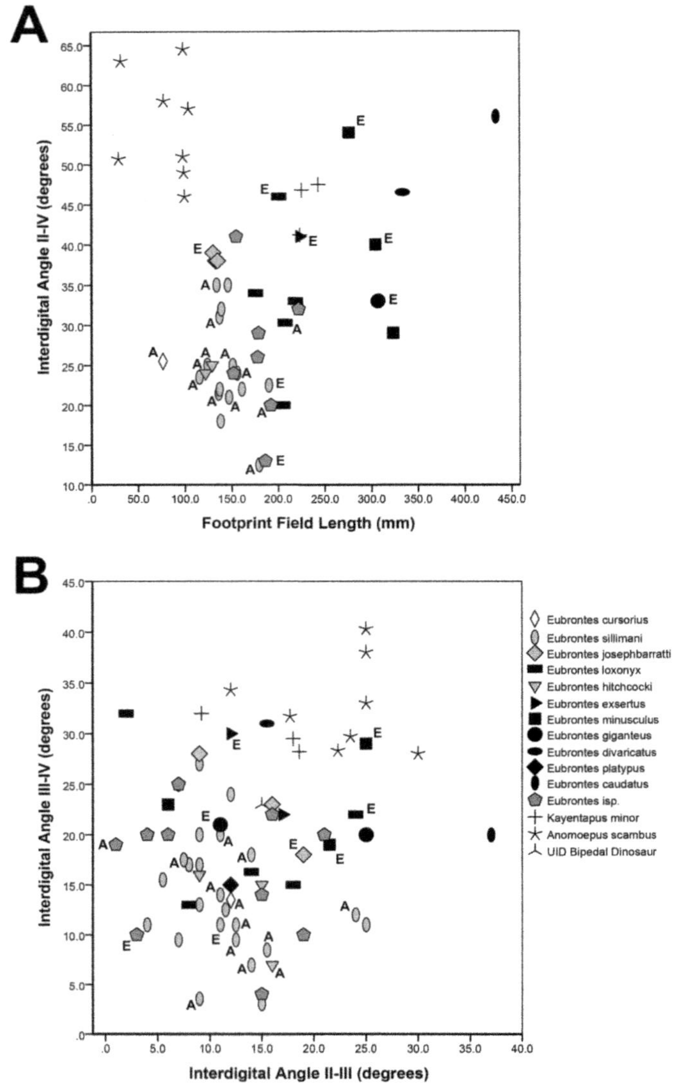

8.39. Interdigital angles. *A*, Interdigital angle II–IV vs. footprint length. Across *Eubrontes*, the angle increases with increasing footprint size ($r = 0.533$, $P < .001$, $N = 39$). *Kayentapus* tends to have a large interdigital angle for its size, compared with most *Eubrontes*, and *Anomoepus* has a very large interdigital angle, relative to footprint length, that puts it well above the *Eubrontes* size-related trend. *B*, Interdigital angle II–III vs. interdigital angle III–IV. There is no obvious pattern in the relationship among the various forms of *Eubrontes*. *Anomoepus* shows a larger interdigital angle II–III than most *Eubrontes*, and *Kayentapus* and especially *Anomoepus* have a very wide interdigital angle III–IV compared with interdigital angle II–III.

large measures related to footprint breadth (toebase II–IV distance, toetip III–IV distance [*Kayentapus*, but not *Anomoepus*], toetip width, and distance from the digit II proximal bend point to the long axis of digit III). *Anomoepus* and large *Eubrontes* have a relatively shorter digit III projection than *Kayentapus* and most small and medium-sized *Eubrontes*.

A relatively broader footprint is strongly correlated with a larger interdigital angle (IDA) II–IV (fig. 8.38). Across *Eubrontes*, IDA II–IV increases with increasing footprint size (fig. 8.39*A*), the opposite of what was observed in the ontogenetic series of emu footprints (fig. 6.20). *Anomoepus* has a very wide IDA II–IV for its size. *Eubrontes* shows no pattern among putative ichnotaxa in the relationship between IDAs II–III and III–IV (fig. 8.39*B*). *Kayentapus* and *Anomoepus* consistently have a wider IDA III–IV than IDA II–III, giving them wider IDAs III–IV for specified values of IDA II–III than in *Eubrontes*.

Are the Groups Significantly Different? Our multivariate and bivariate analyses definitely show at least two distinct morphotypes among the Newark Supergroup footprint assemblage: *Anomoepus* (and also *Kayentapus*) vs. *Eubrontes*. Beyond that, the data suggest the possibility of at least two disparate morphotypes within *Eubrontes*: small and some medium-sized *Eubrontes* (including the "*Anchisauripus*" cluster) vs. other medium-sized, and also large, *Eubrontes* (including the "*Eubrontes*" cluster). The question then becomes whether these groups really do differ enough from each other to warrant assigning them to different kinds of trackmaker.

I examined this matter in two ways. First, I did analyses of covariance (ANCOVAs) across the three main footprint shape categories: *Anomoepus* (for now excluding *Kayentapus*), "*Anchisauripus*," and "*Eubrontes*." I made three different, but related, kinds of comparisons, mainly looking at the parameters that the CVA (table 8.3) picked as best separating the groups, using either backfoot length or the length of the first digital pad of digit III as the covariate (table 8.4).

One of the group comparisons was among the "*Anchisauripus*," "*Eubrontes*," and *Anomoepus* maximum information

Table 8.4. Analyses of covariance (ANCOVA) of selected footprint size parameters against trackmaker group. In most analyses backfoot length is the covariate; where digit III first digital pad width is the test parameter, the length of the first digital pad of digit III is the covariate. Data cases reflect the "mean or single" data treatment. Unless otherwise indicated, the assumptions of ANCOVA were met (*P* values of Levene's test of equality of error variances and lack-of-fit test of at least .05). Anch = *"Anchisauripus"* cluster; AC = prints on AC 9/14; Ano = *Anomoepus* cluster plus two other Amherst College *Anomoepus* specimens (but excluding *Kayentapus*); AA = all *Anomoepus*; E = *"Eubrontes"* cluster; Big E = *Eubrontes* with footprint field length ≥ 25 cm.

Parameter	Group comparison	P of F-test	Result of comparison (Bonferroni test)	Number of cases
Distance between proximal ends of first digital pads of digits III–IV (toebase III–IV)	Anch: E: Ano	.634	No significant differences between groups	13 Anch: 8 E: 4 Ano
	AC: E: AA	.259	No significant differences between groups	10 AC: 8 E: 7 AA
	AC: Big E: AA[a]	.489	No significant differences between groups	10 AC: 4 Big E: 7 AA
Distance between proximal ends of first digital pads of digits II–IV (toebase II–IV)	Anch: E: Ano	.030	Anch significantly different than E (Anch > E)	13 Anch: 8 E: 4 Ano
	AC: E: AA	.028	AC significantly different than E (AC > E)	10 AC: 8 E: 8 AA
	AC: Big E: AA	.006	AC significantly different than Big E (AC > E); Big E significantly different than AA (AA > Big E)	10 AC: 4 Big E: 8 AA
Distance between toetip II–toetip III	Anch: E: Ano[b]	.073	No significant differences between groups	13 Anch: 8 E: 4 Ano
	AC: E: AA	.296	No significant differences between groups	10 AC: 8 E: 9 AA
	AC: Big E: AA[a]	.444	No significant differences between groups	10 AC: 6 Big E: 9 AA
Distance between toetip III–toetip IV	Anch: E: Ano[a]	.142	No significant differences between groups	13 Anch: 8 E: 4 Ano
	AC: E: AA[a]	.027	AC significantly different than AA (AA > AC)	10 AC: 8 E: 9 AA
	AC: Big E: AA[a]	.023	AC significantly different than AA (AA > AC)	10 AC: 6 Big E: 9 AA
Distance between toetip II–toetip IV (toetip width)	Anch: E: Ano[c]	<.001	Anch significantly different than Ano (Ano > Anch); E significantly different than Ano (Ano > E)	13 Anch: 8 E: 4 Ano
	AC: E: AA[a]	<.001	AC significantly different than E (E > AC); AC significantly different than AA (AA > AC); E significantly different than AA (AA > E)	11 AC : 8 E: 9 AA
	AC: Big E: AA[a]	<.001	AC significantly different than Big E (Big E > AC); AC significantly different than AA (AA > AC)	11 AC: 6 Big E: 9 AA
Digit III projection	Anch: E: Ano[c]	<.001	Anch significantly different than Ano (Anch > Ano); E significantly different than Ano (E > Ano)	13 Anch: 8 E: 4 Ano
	AC: E: AA	.015	AC significantly different than AA (AC > AA)	10 AC: 8 E: 9 AA
	AC: Big E: AA	.013	AC significantly different than AA (AC > AA)	10 AC: 6 Big E: 9 AA
Distance from digit II proximal bend point to long axis of digit III	Anch: E: Ano	.018	Anch significantly different than Ano (Ano > Anch)	13 Anch: 8 E: 4 Ano
	AC: E: AA	.012	AC significantly different than AA (AA > AC)	11 AC: 8 E: 8 AA
	AC: Big E: AA	.003	AC significantly different than AA (AA > AC)	11 AC: 6 Big E: 8 AA
Digit III first pad width	Anch: E: Ano	.003	Anch significantly different than E (E > Anch)	13 Anch: 8 E: 4 Ano
	AC: E: AA[a]	<.001	AC significantly different than E (E > AC)	22 AC: 8 E: 4 AA
	AC: Big E: AA	<.001	No significant differences between groups, but AC comes close to being significantly different than Big E (*P* = .064), with Big E > AC	22 AC: 4 Big E: 4 AA

[a]Test parameter and covariate log transformed prior to analysis to equalize error variances or to pass lack-of-fit test.
[b]*P* of lack-of-fit test < .05, but log transforming variables results in *P* of Levene's test of equality of error variances being < .05.
[c]*P* of Levene's test of equality of error variances < .05, even after log transformation of variables.

clusters (fig. 8.18). To increase the sample size, two Amherst College trackways of *Anomoepus* that did not yield a complete enough set of measurements to be included in the cluster analyses were added to the *Anomoepus* sample in the present comparison.

A second comparison was made that again used the *"Eubrontes"* cluster, but instead of comparing it with the *"Anchisauripus"* cluster, this second analysis substituted all of the trackways on slab AC 9/14 (fig. 8.2), making what seems the plausible assumption that these could all represent a homogeneous grouping (members of the same biological species?). The third group used in this second comparison constituted all the trackways identified as *Anomoepus*, again on the assumption that these constitute a homogeneous group. This almost doubled the number of *Anomoepus* cases.

The third comparison again used the AC 9/14 and "all *Anomoepus*" sample, but instead of comparing them with the *"Eubrontes"* maximum information cluster, the comparison was made with all *Eubrontes* with a footprint length of at least 25 cm. This eliminated the medium-sized prints in the *"Eubrontes"* maximum information cluster, but added some large *Eubrontes* that were not well enough preserved

Table 8.5. Independent-samples *t*-test for equality of means of the 19 GM-scaled footprint parameters used to create the maximum information cluster. Comparisons are made between the *"Anchisauripus"* (Anch) cluster (*N* = 13) and the *"Eubrontes"* (E) cluster (*N* = 8). In all comparisons the *P* value of Levene's test for equality of variances is >.05. Significant differences in mean values between the two cluster groups are indicated in **bold**. Relative sizes of the mean values for the two cluster groups are reported for comparisons in which the difference is statistically significant, or approaches statistical significance.

Parameter	*P* of *t*-test	Comparison of mean values
Digit II first pad length	**0.041**	E > Anch
Digit II first pad width	**<0.001**	E > Anch
Digit II second pad length	0.669	——
Digit II second pad width	**<0.001**	E > Anch
Digit II claw length	**<0.001**	Anch > E
Digit III first pad length	0.540	——
Digit III first pad width	**<0.001**	E > Anch
Digit III second pad length	**0.007**	Anch > E
Digit III second pad width	0.075	E > Anch
Digit III third pad length	0.157	——
Digit III third pad width	**0.005**	E > Anch
Digit III claw length	**0.005**	Anch > E
Digit IV length	0.660	——
Distance from base of digit II to base of digit III	**0.005**	E > Anch
Distance from base of digit III to base of digit IV	0.315	——
Distance from base of digit II to base of digit IV	**0.001**	Anch > E
Toetip II to toetip III	0.286	——
Toetip III to toetip IV	0.129	——
Toetip II to toetip IV	0.065	E > Anch

to allow measurement of all the parameters used in creating the maximum information clusters.

For *Eubrontes* alone, I did *t*-tests of the 19 GM-scaled parameters used in creating the maximum information clusters, with the *"Anchisauripus"* and *"Eubrontes"* clusters comprising the groups compared.

There seem to be significant differences between/among at least some of the various small to medium-sized *Eubrontes* (*"Anchisauripus"* cluster and combined AC 9/14 sample), the *"Eubrontes"* cluster, and the *Anomoepus* samples in one or more of the proportions involving the distance across the proximal ends of digits II and IV, the distance between the proximal bend point of digit II and the long axis of digit III, the distance between toetips III and IV, toetip width, digit III projection, and the width of the first digital pad of digit III (table 8.4). For the *"Anchisauripus"* and *"Eubrontes"* clusters, *t*-tests of mean values of GM-scaled footprint shape parameters show significant differences between the two clusters for some of the digital pad lengths and widths, claw lengths, and distances between clawbases (table 8.5). These results are consistent with what was found in the multivariate and bivariate comparisons.

There are, then, significant differences among the *"Eubrontes," "Anchisauripus,"* and *Anomoepus* (and probably also *Kayentapus*) morphologies (table 8.6). The question then becomes: What kind(s) of differences are these? Variation within a single species, or variation across species, or some mixture of both? We will consider this question using several different approaches.

BACK TO THE BIRDS AND THE ALLIGATORS: COMPARISONS OF DINOSAUR FOOTPRINT SHAPE VARIABILITY WITH KNOWN INTRASPECIFIC AND INTERSPECIFIC VARIABILITY OF FOOT AND FOOTPRINT SHAPE

Comparisons of Within-Trackway Footprint Shape Variability

A few of the dinosaur trackways in my sample consisted of enough measurable footprints to allow rough comparisons of within-trackway variability with that seen in my ground bird trackways, using our old quick-and-dirty standby, the maximum/minimum ratio of scaled parameters (tables A8.2–A8.6; fig. 8.40). Within-trackway variability in these terms is comparable for the bird and the dinosaur trackways. Indeed, the dinosaur trackways fall at the low end of values for trackways

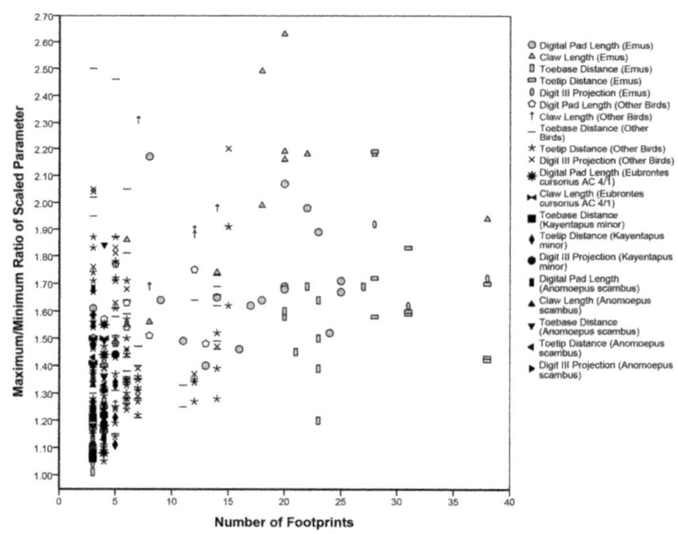

8.40. Comparison of shape variability (maximum/minimum ratio of scaled parameters) among footprints made by the same individual bird, with shape variability of footprints within trackways of Newark Supergroup non-avian dinosaurs. Individual pad lengths and widths, and claw lengths are scaled as percentages of the length of the first digital pad of digit III; toebase and toetip distances, and digit III projection are scaled as percentages of backfoot length. Each data case represents a value for a single bird, or a single dinosaur trackway (tables A8.2–A8.6). Note that dinosaur trackways show variability at the low end of the maximum/minimum ratio of scaled parameters seen in modern bird trackways.

Table 8.6. Summary of hypothesized proportional differences in footprint (and foot) shape across the three kinds of trackmakers inferred on the basis of footprint morphotypes recognized in this study. Anch = "*Anchisauripus*" (most small and medium-sized *Eubrontes* isp.) cluster; AC = prints on AC 9/14 (fig. 8.2); Ano = *Anomoepus* cluster (excluding *Kayentapus*); AA = all *Anomoepus*; E = "*Eubrontes*" cluster (large and some medium-sized *Eubrontes* isp.); Big E = *Eubrontes* with footprint field length ≥ 25 cm. Relative size in the Anch: E: Ano group comparison is based on GM-scaling of log-transformed parameters. Relative size in the AC: E: AA and AC: Big E: AA group comparisons is based on parameters scaled as percentages of either the length of the first digital pad of digit III (digital pad lengths and widths, claw lengths) or backfoot length (digit IV length, toebase and toetip distances, digit III projection, and distance from the digit II proximal bend point to the long axis of digit III). If groups are combined within parentheses, such as (E and Anch) and (E and Ano), in the comparison of relative size, the groups therein are inferred to be similar in the relative size of that parameter, unless otherwise stated. Figures relevant to the comparison are cited; the prefix "cf." before a cited figure indicates that the figure cited does not show exactly the same proportional comparison made in the cell, but a related comparison, in which the parameter of interest is compared against a different footprint parameter than that used to rank relative sizes of the scaled parameter across groups in that cell (described in the preceding sentences in this table).

Scaled parameter	Group comparison	Rank of relative size of scaled parameter
Digit II		
First digital pad relative length	Anch: E: Ano	(E and Ano) > Anch (fig. 8.20*A*)
	AC: E: AA	(E and AA) > AC
	AC: Big E: AA	(Big E and AA) > AC
	Conclusion: The "*Eubrontes*"-maker and the *Anomoepus*-maker had a relatively longer first digital pad of digit II than the "*Anchisauripus*"-maker	
First digital pad relative width	Anch: E: Ano	E > (Anch and Ano); Ano possibly > Anch (fig. 8.20*B*)
	AC: E: AA	(E and AA) > AC
	AC: Big E: AA	AA > Big E > AC
	Conclusion: The "*Eubrontes*"-maker had a relatively broader first digital pad of digit II than the "*Anchisauripus*"-maker, and the *Anomoepus*-maker may have had a broader pad than the "*Anchisauripus*"-maker	
Second digital pad relative length	Anch: E: Ano	(Anch and E) > Ano (fig. 8.20*C*)
	AC: E: AA	E > Anch; AA relative size of AA uncertain
	AC: Big E: AA	AA > (AC and Big E)
	Conclusion: Nothing consistent	
Second digital pad relative width	Anch: E: Ano	E > Anch; relative size of Ano uncertain (fig. 8.20*D*)
	AC: E: AA	AA > E > AC (cf. figs. 8.27*B*, 8.28*A*)
	AC: Big E: AA	AA > (Big E and AC)
	Conclusion: The "*Eubrontes*"-maker may have had a wider second digital pad of digit II than the "*Anchisauripus*"-maker, and the *Anomoepus*-maker may have had a relatively broader second digital pad of digit II than both the "*Anchisauripus*"-maker and the "*Eubrontes*"-maker	
Claw length	Anch: E: Ano	(Anch and Ano) > E (fig. 8.20*E*)
	AC: E: AA	(AC and AA) > E (cf. fig. 8.32*A*)
	AC: Big E: AA	(AC and AA) > Big E
	Conclusion: The "*Anchisauripus*"-maker and the *Anomoepus*-maker both had a relatively longer digit II claw than the "*Eubrontes*"-maker	
Digit III		
First digital pad length	Anch: E: Ano	(Anch and E) > Ano (fig. 8.21*A*)
First digital pad width	Anch: E: Ano	E > (Anch and Ano) (fig. 8.21*B*)
	AC: E: AA	(E and AA) > AC (fig. 8.26*A*)
	AC: Big E: AA	(Big E and AA) > AC
	Conclusion: The "*Eubrontes*"-maker had a relatively broader first digital pad of digit III than the "*Anchisauripus*"-maker, and the *Anomoepus*-maker may have had a broader pad than the "*Anchisauripus*"-maker	
Second digital pad length	Anch: E: Ano	Anch > E > Ano (fig. 8.21*C*)
	AC: E: AA	AC > (E and AA) (cf. figs. 8.27*B*, 8.28*B*)
	AC: Big E: AA	AC > (Big E and AA)
	Conclusion: The "*Anchisauripus*"-maker had a relatively longer second digital pad of digit III than both the "*Eubrontes*"-maker and the *Anomoepus*-maker	
Second digital pad width	Anch: E: Ano	E > Anch > Ano (fig. 8.21*D*)
	AC: E: AA	E > (AC and AA); AA possibly > AC
	AC: Big E: AA	(Big E and AA) > AC
	Conclusion: The "*Eubrontes*"-maker had a relatively broader second digital pad of digit III than the "*Anchisauripus*"-maker, and possibly also the *Anomoepus*-maker	
Third digital pad length	Anch: E: Ano	Anch > E > Ano (fig. 8.21*E*)
	AC: E: AA	AC possibly > (E and AA)
	AC: Big E: AA	AC possibly > Big E; relative size of AA uncertain
	Conclusion: The "*Anchisauripus*"-maker may have had a relatively longer third digital pad of digit III than the "*Eubrontes*"-maker	

Table 8.6. *continued*

Scaled parameter	Group comparison	Rank of relative size of scaled parameter
Digit III		
Third digital pad width	Anch: E: Ano	E > Anch > Ano (fig. 8.21*F*)
	AC: E: AA	(E and AA) > AC
	AC: Big E: AA	(Big E and AA) > AC
	Conclusion: The *"Eubrontes"*-maker had a relatively broader third digital pad of digit III than the *"Anchisauripus"*-maker	
Claw length	Anch: E: Ano	Ano > Anch > E (fig. 8.21*G*)
	AC: E: AA	AA > AC > E (cf. fig. 8.32*B*)
	AC: Big E: AA	AA > AC > Big E
	Conclusion: The *Anomoepus*-maker had a relatively longer claw than both the *"Anchisauripus"*-maker and the *"Eubrontes"*-maker, and the *"Anchisauripus"*-maker had a relatively longer claw than the *"Eubrontes"*-maker	
Digit IV length	Anch: E: Ano	Ano > (Anch and E) (fig. 8.22*A*)
	AC: E: AA	AA > E > AC (cf. figs. 8.33*C*–8.33*F*, 8.34, 8.35)
	AC: Big E: AA	AA > Big E > AC
	Conclusion: The *Anomoepus*-maker had a relatively longer digit IV than both the *"Anchisauripus"*-maker and the *"Eubrontes"*-maker, and the *"Eubrontes"*-maker may have had a relatively longer digit IV than the *"Anchisauripus"*-maker	
Distances across proximal ends of digital pads		
II–III	Anch: E: Ano	E possibly > (Anch and Ano) (fig. 8.22*B*)
	AC: E: AA	E possibly > (AC and AA) (cf. fig. 8.29*A*)
	AC: Big E: AA	Big E > AC
	Conclusion: The *"Eubrontes"*-maker probably had a relatively larger distance between the proximal ends of the digit II and III impressions than either the *"Anchisauripus"*-maker or the *Anomoepus*-maker	
III–IV	Anch: E: Ano	No difference among groups (fig. 8.22*C*)
	AC: E: AA	No difference among groups (cf. fig. 8.29*B*)
	AC: Big E: AA	No difference among groups
II–IV	Anch: E: Ano	(Anch and Ano) > E (fig. 8.22*D*)
	AC: E: AA	(AC and AA) > E (fig. 8.37*A*) (cf. fig. 8.29*C*)
	AC: Big E: AA	(AC and AA) > Big E
	Conclusion: Both the *"Anchisauripus"*-maker and the *Anomoepus*-maker had a relatively longer distance between the proximal ends of the digit II and IV impressions than the *"Eubrontes"*-maker	
Toetip distances		
II–III	Anch: E: Ano	Anch > Ano, and Anch possibly > E as well; E possibly > Ano (fig. 8.22*E*)
	AC: E: AA	No difference among groups (fig. 8.37*B*)
	AC: Big E: AA	AC possibly > Big E
	Conclusion: The *"Anchisauripus"*-maker may have had a relatively greater toetip II–III distance than the *"Eubrontes"*-maker	
III–IV	Anch: E: Ano	Ano > (E and Anch) (fig. 8.22*F*)
	AC: E: AA	AA possibly > (E and AC) (fig. 8.37*C*; cf. fig. 8.31)
	AC: Big E: AA	AA > (AC and Big E)
	Conclusion: The *Anomoepus*-maker had a relatively greater toetip III–IV distance than either the *"Anchisauripus"*-maker or the *"Eubrontes"*-maker	
II–IV	Anch: E: Ano	Ano > (E and Anch); E possibly > Anch (fig. 8.22*G*)
	AC: E: AA	AA > E > AC (fig. 8.37*D*; cf. fig. 8.30)
	AC: Big E: AA	AA > Big E > AC
	Conclusion: The *Anoemopus*-maker had a relatively greater toetip width than the *"Eubrontes"*-maker, which in turn had a relatively greater toetip width than the *"Anchisauripus"*-maker	
Digit III projection	AC: E: AA	(AC and E) > AA; AC possibly > A
	AC: Big E: AA	AC > Big E; AC possibly > AA; AA probably > Big E (fig. 8.37*E*)
	Conclusion: The *"Anchisauripus"*-maker probably had a relatively longer digit III projection than the *Anomoepus*-maker	
Distance from the digit II proximal bend point to the long axis of digit III	AC: E: AA	AA > AC (figs. 8.36A, 8.37*F*)
	AC: Big E: AA	AA > AC
	Conclusion: The *Anomoepus*-maker had a relatively larger distance from the digit II proximal bend point to the long axis of digit III than the *"Anchisauripus"*-maker	

in the modern bird sample. This isn't that surprising, given that the parameters of interest could only be measured on reasonably well preserved dinosaur footprints.

Within-Group and Across-Group Shape Variability

Dinosaurs. Of greater interest, of course, is whether there is more shape variability across than within potentially homogeneous groups of dinosaur tracks like the maximum information clusters, and how this variability compares with that of known intraspecific and interspecific samples of ground bird footprints. Tables A8.7–A8.12 summarize such comparisons for several scaled footprint shape parameters; *Kayentapus* is not included in these analyses because of very small sample sizes. Two kinds of measures of variability are employed.

One of these measures once again consists of footprint parameters scaled as percentages of either the length of the first digital pad of digit III (individual digital pad lengths and widths, and claw lengths) or backfoot length (toebase and toetip distances, and digit III projection). Variability of parameters scaled in this manner is expressed as the usual maximum/minimum ratio (for groups represented by three or more data cases) and the coefficient of variation of the scaled parameter (for groups represented by six or more data cases).

The second measure of variability employs the "GM-scaled" parameters from the PCA (table 8.2) and the maximum information cluster analysis (fig. 8.18), in which the mean of the log-transformed parameter values is subtracted from each of the log-transformed parameter values. Variability is then expressed as the range of the GM-scaled parameters and the standard deviation of the GM-scaled parameters.

The maximum information clusters potentially constitute "homogeneous" footprint groups, but they are not the only candidates for that status. Other plausible contenders, as already discussed, include all of the prints on slab AC 9/14, all of the large *Eubrontes*, and all *Anomoepus*. The footprint shape variability within these potentially homogeneous groups can be compared with that of larger, increasingly more inclusive combinations of these groups. In most comparisons (tables A8.7–A8.12), the more inclusive group does indeed show greater variability by one or more measures than at least one of the homogeneous groups included within it.

However, this result could be misleading. As we considered in our discussion of shape variability of pedal skeletons within and across species of moa (chapter 1), greater variability in samples of more inclusive groups than in homogeneous groups might be due merely to a larger number of specimens, and not to real differences among the homogeneous groups within those larger samples.

For a selected set of variables (mostly those involved in the CVA [table 8.3]), this possibility was evaluated in the same manner as was done for moa foot skeletons. For inclusive groups composed of two homogeneous groups, 15 trials were performed in which a number of cases, corresponding to the number of cases in the smaller homogeneous group, were randomly selected from the pooled sample of data cases in the more inclusive group. In each trial, the maximum/minimum ratio (where the number of data cases in the smaller homogeneous group was at least three) and the coefficient of variation (where the number of data cases in the smaller homogeneous group was at least six) of the scaled parameter were calculated. In addition, where the number of data cases warranted, the range and standard deviation of the GM-scaled parameters were calculated.

The median value from the 15 trials for each measure of variability for each of the selected scaled test variables was then compared with the values for the individual homogeneous groups. If the median of the 15 trials for a measure of variability of a scaled parameter in the more inclusive sample is larger than the value of the measure of variability of that scaled parameter for the individual homogeneous group(s), this indicates that more than half of the 15 trials of randomly selected cases resulted in measures of variability of the scaled parameter greater than the same measure for the individual homogeneous group(s). This would indicate that the greater variability of scaled parameters in the more inclusive group than in the homogeneous group(s) is not simply a matter of more data cases in the more inclusive group.

And here's the payoff: For all of the scaled parameters examined (table A8.12), at least two of the four measures of variability had median values in the 15 trials of data cases selected from the more inclusive group that were greater than the values for at least one of the homogeneous groups within that more inclusive group. This result is not surprising, given the results of the ANCOVAs (table 8.4) and *t*-tests (table 8.5).

Multiple measures of variability, then, suggest that at least some of the potentially homogeneous groups of Newark Supergroup dinosaur footprints really do differ from one another. Then the question becomes whether they are different enough to merit recognizing them as proxies for different kinds (species?) of trackmakers. To address this question, we must haul out the measuring sticks created in this and earlier chapters of this book from our examination of modern bird footprints, bird study skins, and alligator hindfeet.

Maximum/Minimum Ratios of Scaled Parameters of Ground Bird Footprints: Effects of Sample Size and Differences Among Parameters. The first measuring sticks that we will use to compare bird with non-avian dinosaur footprints are, yet again, the simple ratios of the maximum to

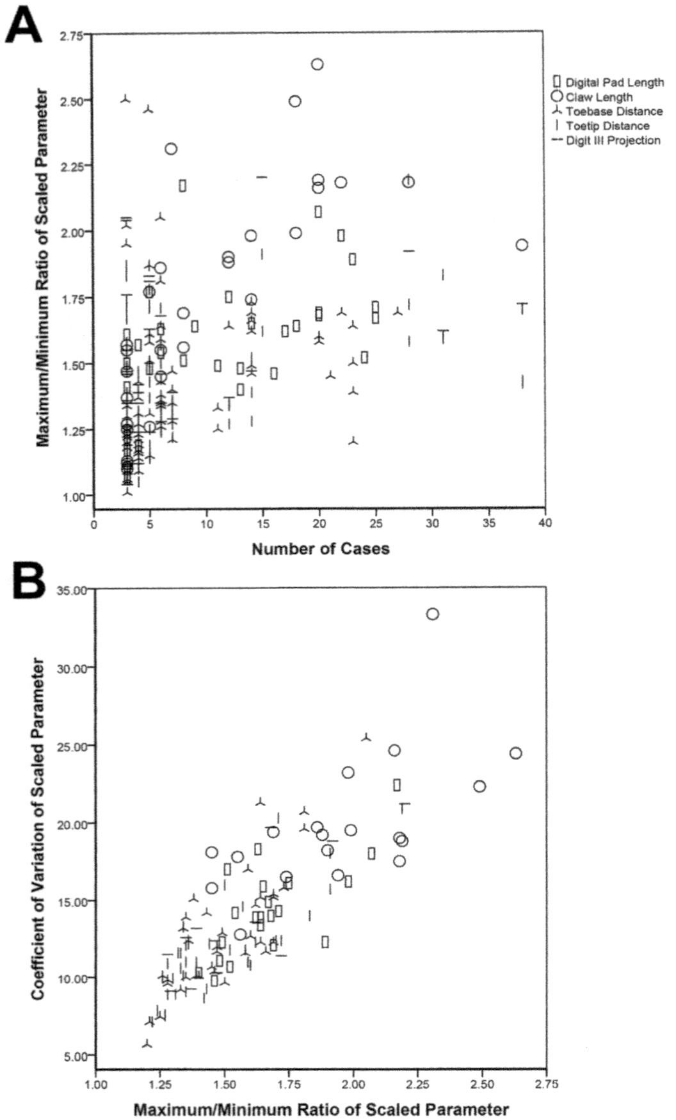

8.41. Variability of footprint shape parameters in prints of ground birds. Each data case represents the maximum/minimum ratio of the scaled shape parameter among footprints made by an individual bird, of each of the species for which prints were collected (and the parameter could be measured) during this study (chapter 6). Digital pad lengths and claw lengths are scaled as percentages of the length of the first digital pad of digit III; distances between the bases (proximal ends), and also the tips, of digits II, III, and IV, and the projection of digit III beyond the tips of digits II and IV, are scaled as percentages of backfoot length. *A*, Maximum/minimum ratio of the scaled parameter as a function of the number of individual footprints (number of cases) collected for each bird. Data are reported for birds in which at least 3 measurements of the scaled parameter of interest were available. Note that the minimum value of the ratio creeps upward with an increasing number of footprints for which the parameter could be measured for that particular bird. Scaled claw lengths tend to be more variable than other scaled parameters. *B*, Relationship between the maximum/minimum value of the scaled parameter and the coefficient of variation of the scaled parameter (reported when at least 6 measurements of the scaled parameter of interest were available). The two measures of footprint shape variability are nicely correlated for birds represented by 6 or more footprint measurements.

the minimum values of scaled parameters. As has repeatedly been stressed in this and in earlier chapters, these measuring sticks are canons that must be adjusted for sample size—number of individual footprints or number of individual trackmakers, as the case may be.

The footprint shape parameters that we will initially examine in bird footprints are, to the extent this is possible, the same as those we have already considered in the dinosaur footprint sample: relative lengths of subunits of digital impressions (digit pad lengths, claw lengths); relative lengths of distances between the ends of digital impressions (distances between the proximal ends [toebases], and between the distal ends [toetips], of digital impressions); and measures of overall print length (backfoot length, projection of digit III beyond the tips of digits II and IV). The lengths of the impressions of the first and second digital pads of digit II, the second and third digital pads of digit III, and the first digital pad of digit IV (the more distal digital pads of digit IV too often fail to be delimited by interpad creases to be used here), and also the lengths of individual claws, will be scaled as percentages of the length of the impression of the first digital pad of digit III. The distances between the proximal ends of the impressions of digits II–III, III–IV, and II–IV; the distances between the toetips of digits II–III, III–IV, and II–IV; and the projection of digit III beyond the tips of digits II and IV will be scaled as percentages of backfoot length.

Maximum/minimum ratios of scaled parameters will be reported for individual birds, and for pooled samples of individual birds (particularly adult or nearly adult emus), for those parameters represented by at least three footprints for that bird or group of birds. For those samples consisting of six or more footprints, the coefficient of variation of the scaled parameter will also be reported. For the adult or nearly adult emu sample, results will be reported for treatments in which individual footprints comprise data cases, and also for "emu means" treatments in which data cases represent the mean values for each individual emu.

Results are reported in tables A8.2–A8.6 and depicted graphically in figures 8.40–8.45. Generalizing from all of these data, several conclusions seem warranted. The first is that, while samples consisting of only a few footprints sometimes show just as much variability in the maximum/minimum ratio of a scaled parameter as do samples consisting of large numbers of footprints, the converse is not true. That is, the observed lowest values of the maximum/minimum ratio increase with the number of footprints, and the number of individual birds, in the sample. If a bird is represented by only a few footprints, which are all reasonably well preserved, the maximum/minimum ratio of the sample can take values very close to 1, but as the number of footprints in the sample

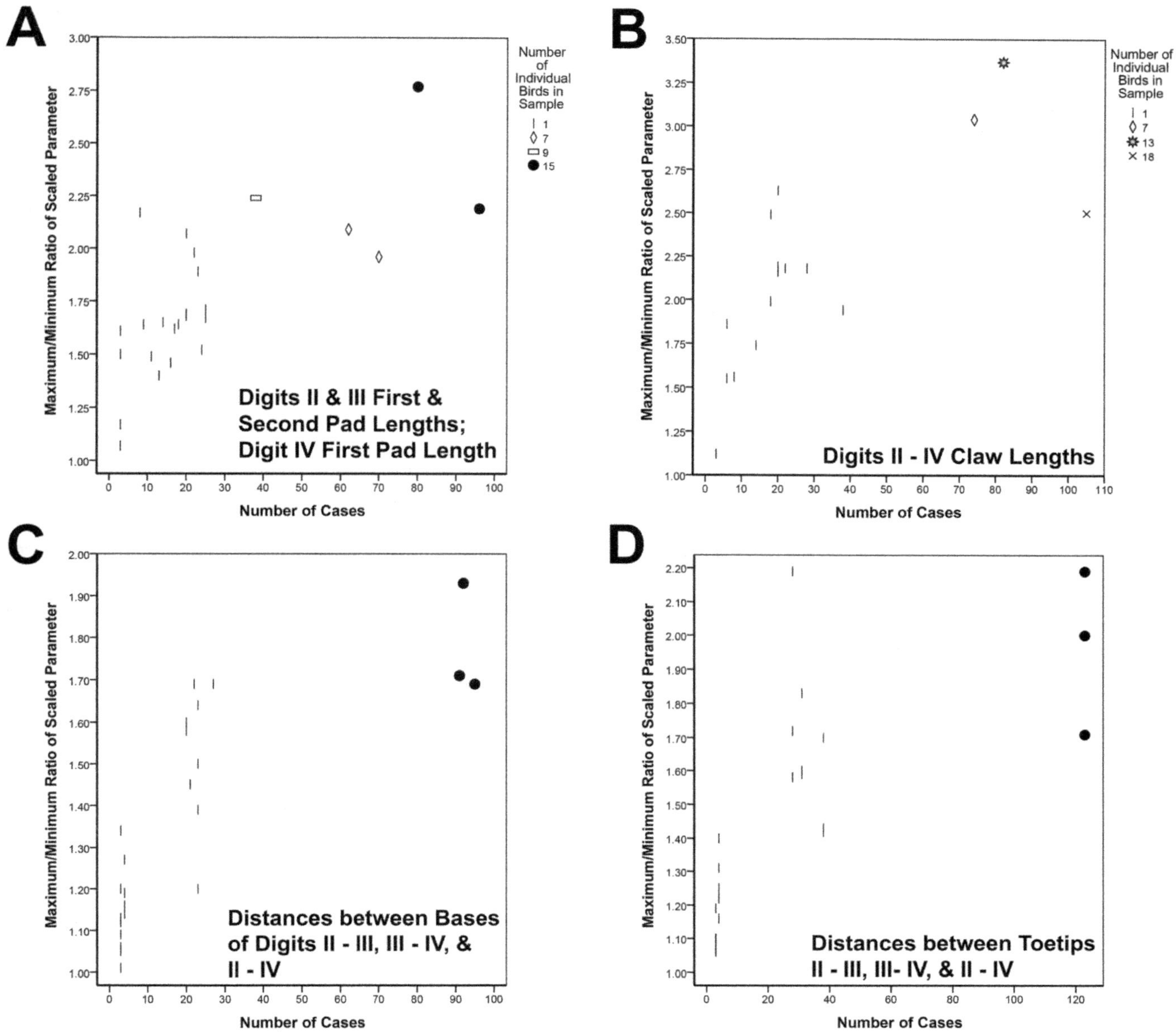

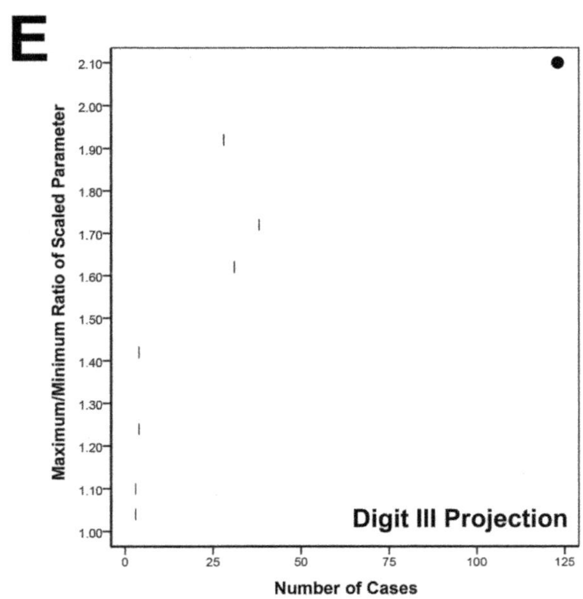

8.42. Maximum/minimum ratios of scaled parameters as a function of the number of footprints (labeled "number of cases" on the *x*-axis), and the number of individual birds, in adult and nearly adult emus. Where points for the number of individual birds labeled on the graphs = 1, points represent values for single birds; numbers >1 represent pooled samples of footprints for adult or nearly adult birds. *A*, *B*, Footprint length parameters scaled as a percentage of the length of the first digital pad of digit III. *A*, Scaled digital pad lengths. *B*, Scaled claw lengths. *C–E*, Footprint length parameters scaled as a percentage of backfoot length. *C*, Distances between the proximal ends of the first digital pads of digits II, III, and IV. *D*, Distances between the tips of the toes of digits II, III, and IV. *E*, Projection of digit III beyond the tips of digits II and IV. Note the tendency for the maximum/minimum ratio to increase in what appears to be an asymptotic fashion with an increase in the number of footprints (data cases) and the number of birds for all scaled footprint parameters. For scaled pad and claw lengths, the asymptotic maximum/minimum ratio looks to take a value of 3–4; for toebase and toetip distances, and digit III projection, about 2 or a bit more.

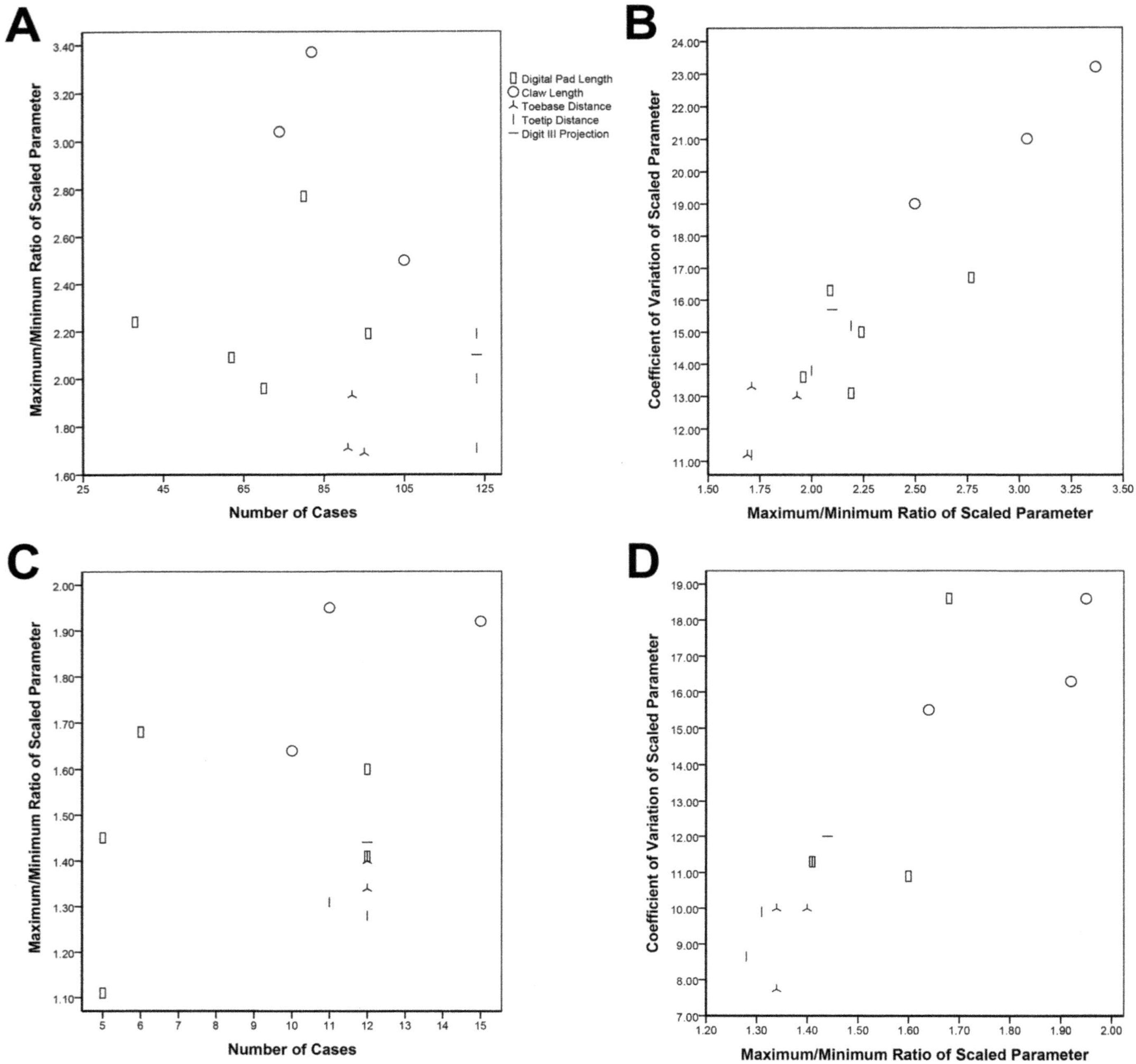

8.43. Comparison of two treatments of intraspecific shape variability in footprints of adult and nearly adult emus. *A*, *B*, Individual footprint treatment: Pooled footprint samples of birds, without regard to the number of individual birds responsible for those prints (individual emus are represented by a variable number of footprints, with prints of 1988 adult emus 3, 5, and 6 dominating the sample). *A*, Maximum/minimum ratio of scaled parameters as a function of the number of footprints (data cases). With this treatment, because of the large number of footprints of even the plotted point represented by the smallest number of footprints, there is no suggestion of an increase in the ratio with an increase in the number of footprints in the sample. Note that scaled claw lengths are more variable than other scaled parameters. *B*, The maximum/minimum ratio of each scaled parameter is tightly correlated with the coefficient of variation of the same scaled parameter. *C*, *D*, Bird means treatment: For each emu, the value of each parameter is the mean of all values of that parameter for that bird. Consequently each emu is represented by one value (the mean) of each measurable footprint parameter, and the number of cases represents the number of individual birds in the sample comparison. *C*, With the smaller number of data cases in this treatment (the maximum number of data cases [individual emus] = 15), there is once again a suggestion that the maximum/minimum ratio of scaled parameters increases with the number of cases. As always, scaled claw lengths are the most variable parameters. Note that the bird means treatment reduces variability of all parameters as compared with the individual footprint treatment, because the effects of extreme values (presumably due to extramorphological variability) are reduced in the process of calculating mean values for each parameter. *D*, The maximum/minimum ratio of scaled parameters is again nicely correlated with the coefficient of variation of the same scaled parameters.

increases to 10 or more, the ratio for even the less variable scaled parameters takes values of at least 1.3 or 1.4. This is, of course, not very surprising: The more prints there are in a sample, the greater the opportunity for extreme specimens that will depart from the other prints in the sample. However, the relationship seems not to be linear; judging from the emu sample (fig. 8.42), the value of the maximum/minimum ratio seems to flatten out asymptotically as the number of footprints in the sample approaches 100.

The second generalization is that if the sample size of footprints for a bird or group of birds is at least six, the maximum/minimum ratio of the scaled parameter nicely correlates with the coefficient of variation of the same scaled parameter (fig. 8.43), as was also observed (fig. 1.16) for skeletal proportions of phalanges and digits of ground birds and non-avian dinosaurs. Consequently the maximum/minimum ratio of scaled parameters does seem to provide a useful quick-and-dirty measure of footprint shape variability, as long as the sample size is taken into account.

The third generalization is that the bird means treatment reduces the variability of footprint shape, as measured by the maximum/minimum ratio of scaled parameters, over the treatment in which individual footprints constitute the data cases (fig. 8.43; cf. table A6.16).

Finally, as was observed for emu footprints (chapter 6) using other measures of variability, or other parameters against which to scale footprint shape parameters than those used in this chapter, claw lengths are the most variable shape parameters of bird footprints. The maximum/minimum ratio of scaled claw lengths for pooled samples of adult and nearly adult emus can take values of 3 or more if the data cases are individual footprints, and up to about 2 if the data cases are mean values for the individual birds (fig. 8.43). Scaled digital pad lengths take maximum/minimum ratios for the adult/nearly adult emu sample of 2–2.8 (data cases = individual footprints) and 1.1–1.7 (bird means treatment). Scaled toebase and toetip distances, and the scaled digit III projection, of the pooled emu sample appear to be less variable still, taking values of 1.6–2.2 (data cases = individual footprints) and 1.3–1.4 (bird means treatment). The numbers look even better (lower values of the maximum/minimum ratio) if one restricts the footprints examined to the very best preserved specimens (preservation class 3 or more for 1988 adult emus 3, 5, and 6; tables A8.2–A8.6; cf. table A6.12). Interestingly, as noted in chapter 6, the values of the maximum/minimum ratio of scaled parameters of the emu footprints look to be in the same ballpark as values of their anatomical analogs as measured on study skins of emus, greater rheas, and bustards (tables A6.13, A6.21–A6.23).

I will cautiously use these results as some of the criteria for assessing whether the degree of variability in footprint shapes in the Newark Supergroup dinosaur track sample suggests a single or more than one species of trackmaker.

Scaled Parameters: Comparisons of Emus With Dinosaurs. Because values of the maximum/minimum ratio of scaled parameters increase with the number of footprints or trackways in the sample (figs. 8.40–8.45), comparing variability among the groups of dinosaur trackways with what is seen

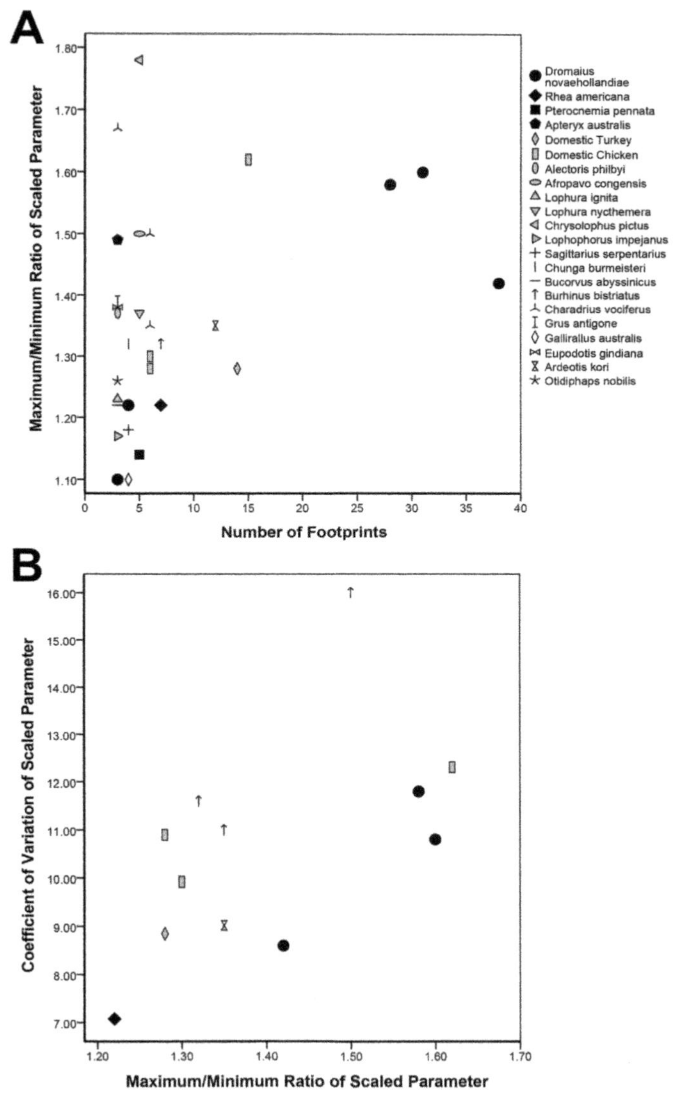

8.44. Variability of toetip width (distance between the tips of digits II and IV) scaled as a percentage of backfoot length, broken down by individual birds of different species. *A*, Maximum/minimum ratio of scaled toetip width as a function of the number of footprints for which the parameter could be measured. Although individual birds represented by a small number of footprints can have high values of the scaled parameter, the minimum value of the ratio increases with the number of footprints by which a particular bird is represented. *B*, Relationship between the maximum/minimum value of the scaled parameter and the coefficient of variation of the scaled parameter (reported where at least 6 measurements of the scaled parameter of interest were available). The two measures of footprint shape variability are correlated for birds represented by 6 or more footprint measurements (cf. fig. 8.41).

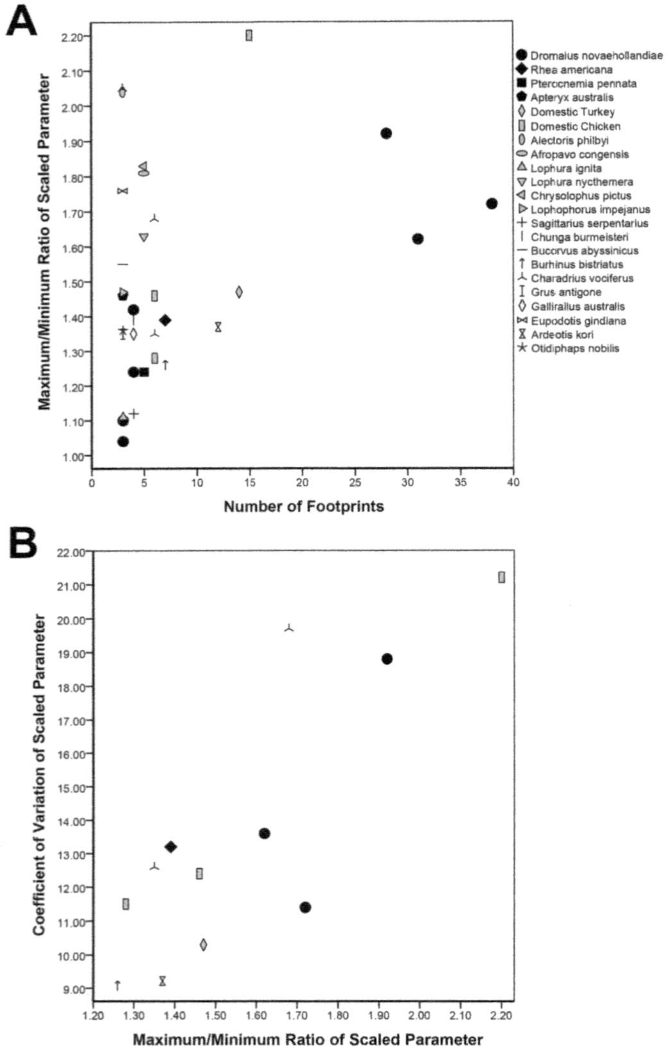

8.45. Variability of the projection of digit III beyond the tips of digits II and IV scaled as a percentage of backfoot length, broken down by individual birds of different species. *A*, Maximum/minimum ratio of scaled digit III projection as a function of the number of footprints for which the parameter could be measured. Although individual birds represented by a small number of footprints can have high values of the scaled parameter, the minimum value of the ratio increases with the number of footprints by which a particular bird is represented. *B*, Relationship between the maximum/minimum value of the scaled parameter and the coefficient of variation of the scaled parameter (reported where at least 6 measurements of the scaled parameter of interest were available). The two measures of footprint shape variability are correlated for birds represented by 6 or more footprint measurements (cf. fig. 8.41).

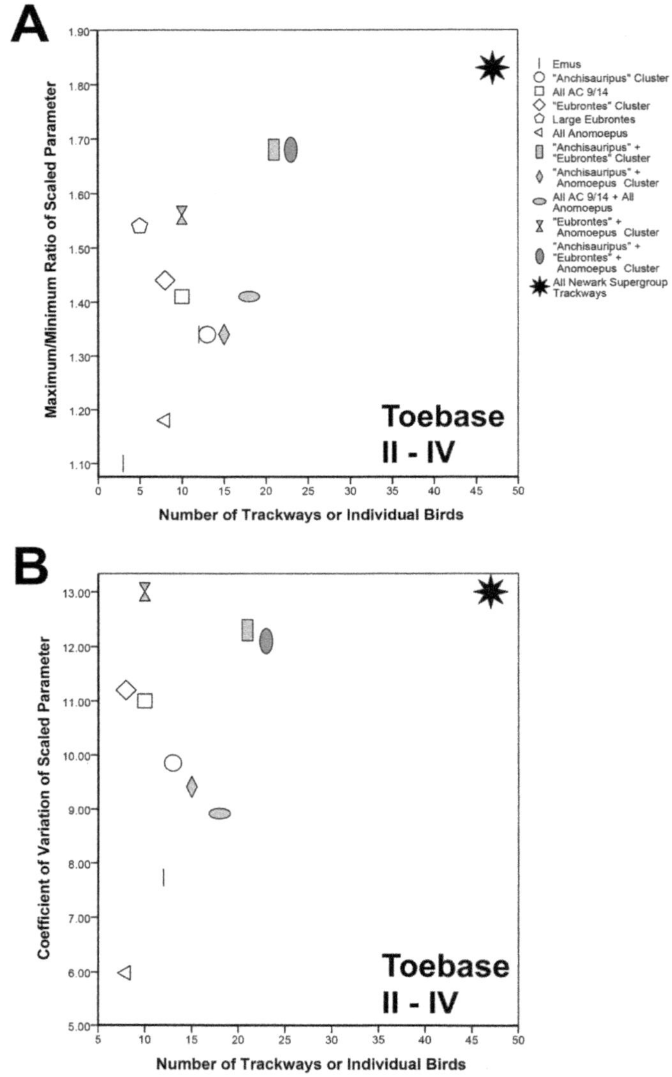

8.46. Comparison of variability of the scaled distance across the proximal ends of the impressions of digits II and IV (toebase distance II–IV) in footprints of emus, with both within-group and across-group variability in trackways of Newark Supergroup non-avian dinosaurs. Data cases are based on means for individual emus (all of the footprints made by each bird), and means for individual dinosaur trackways. The two points for emus in *A* are for good-quality (preservation class of 3 or higher) prints of the three 1988 Fort Wayne Zoo birds, and for all adult emus (*N* = 12). Groups of dinosaur trackways represent (with various degrees of likelihood) potentially homogeneous (single kind of trackmaker) groups (*"Anchisauripus," "Eubrontes,"* and *Anomoepus* clusters, all footprints on the slab AC 9/14, all large *Eubrontes* trackways, and all trackways identified as *Anomoepus*: plotted with *open* symbols), as well as more inclusive combinations of those potentially homogeneous groups (plotted with *shaded* symbols). This comparison of emus with non-avian dinosaurs should be taken with a grain of salt, however, because the distance between the proximal end of digits II and IV is not strictly comparable in bird (fig. 6.4) and non-avian dinosaur footprints (fig. 8.16). *A*, Maximum/minimum ratio of toebase II–IV distance scaled as a percentage of backfoot length. *B*, Coefficient of variation of toebase II–IV scaled as a percentage of backfoot length.

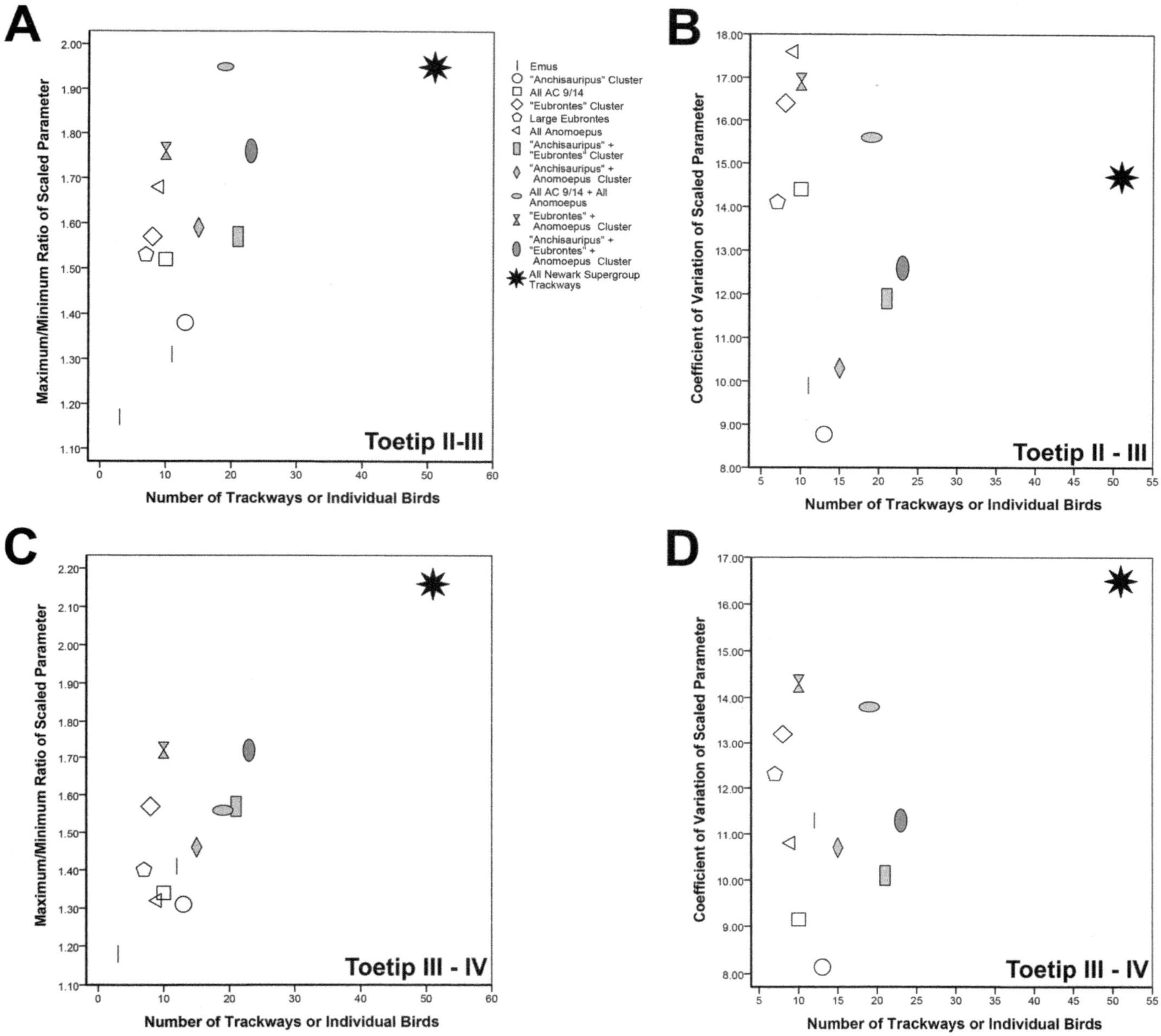

8.47. Comparison of variability of the scaled distances across the toetips of digits II and III, and III and IV, in footprints of emus, with both within-group and across-group variability in trackways of Newark Supergroup non-avian dinosaurs. Data cases and plotting conventions are the same as those used in figure 8.46. *A*, *B*, Toetip II–III distance scaled as a percentage of backfoot length. *A*, Maximum/minimum ratio of scaled distance. *B*, Coefficient of variation of scaled distance. *C*, *D*, Toetip III–IV distance scaled as a percentage of backfoot length. *C*, Maximum/minimum ratio of scaled distance. *D*, Coefficient of variation of scaled distance.

in the emu trackways requires taking the number of trackways used in the comparisons into account. Data cases used in the comparisons will be mean values; for the emus, these will be mean values for all of the footprints made by a single bird, and for the dinosaurs these will be mean values for all of the footprints in a trackway, on the assumption (possibly unwarranted for some trackway slabs) that each trackway was made by a different individual dinosaur.

For the emus, two data samples will be used: (1) "emu means" values of well-preserved (preservation class ≥ 3) footprints of the three 1988 adult emus; and (2) "emu means" values for all of the emus treated as adults (chapter 6). The reason for restricting the emus selected for the present comparisons with dinosaurs to adult or nearly adult birds is that emus show ontogenetic changes in interdigital angles likely to affect the shape parameters considered here (chapter 6), and most or all of the dinosaurs in our sample are plausibly interpreted as having been adults. That is, given their size, I doubt that many of the footprints in my dinosaur sample were made by hatchlings or other very young individuals.

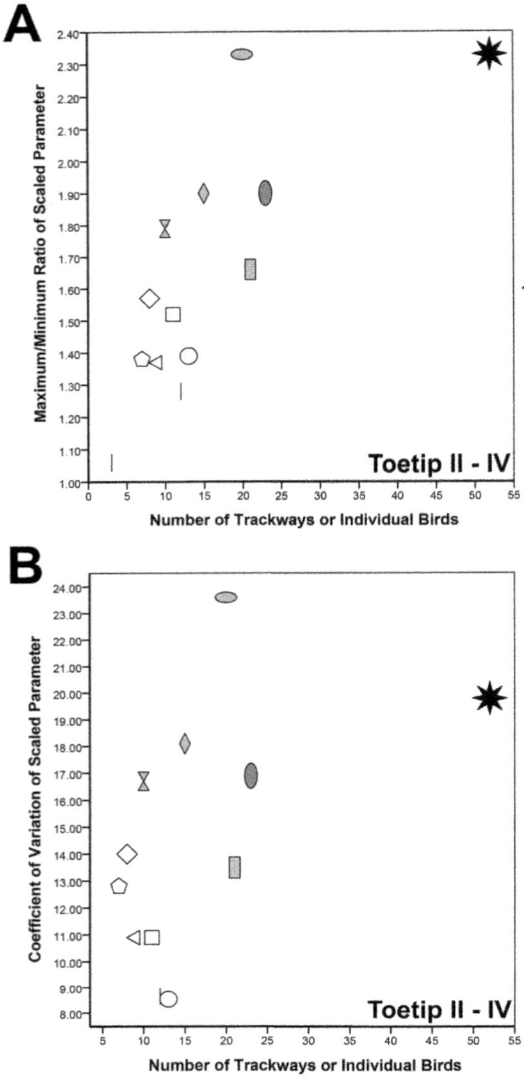

8.48. Comparison of variability of the scaled distances across the toetips of digits II and IV (toetip width), in footprints of emus, with both within-group and across-group variability in trackways of Newark Supergroup non-avian dinosaurs; toetip width scaled as a percentage of backfoot length. Data cases and plotting conventions are the same as those used in figure 8.46. *A*, Maximum/minimum ratio of scaled distance. *B*, Coefficient of variation of scaled distance.

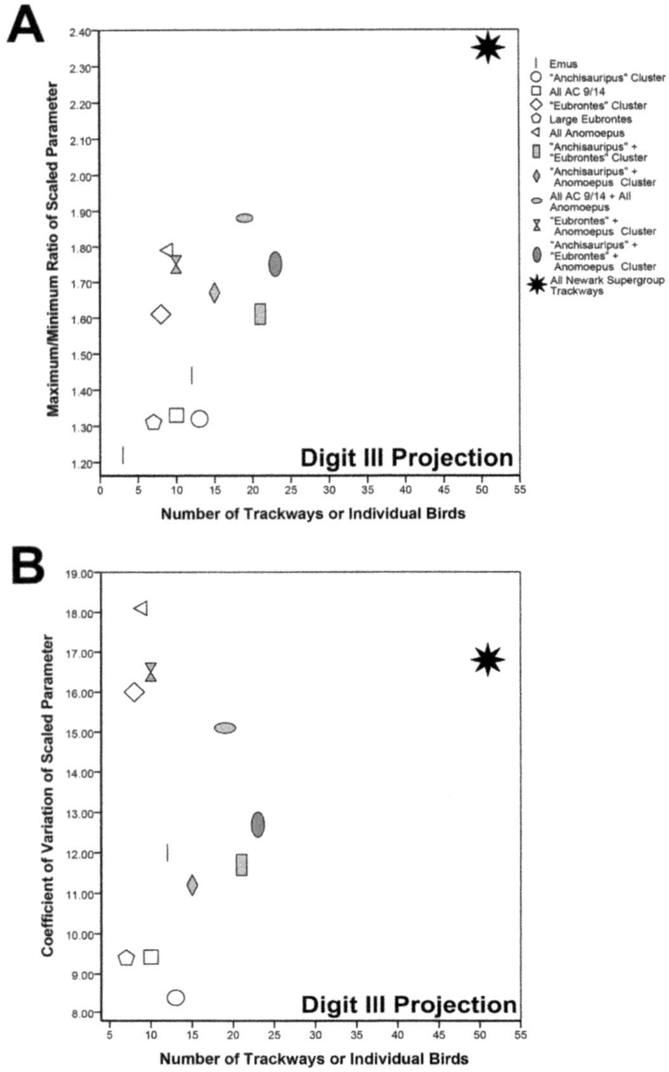

8.49. Comparison of variability of the projection of digit III beyond the tips of digits II and IV, in footprints of emus, with both within-group and across-group variability in trackways of Newark Supergroup non-avian dinosaurs; digit III projection scaled as a percentage of backfoot length. Data cases and plotting conventions are the same as those used in figure 8.46. *A*, Maximum/minimum ratio of scaled distance. *B*, Coefficient of variation of scaled distance.

Groups of dinosaur trackways compared with the emus will consist of the same potentially homogeneous (single kind of trackmaker) groups, as well as more inclusive combinations of those potentially homogeneous groups.

The footprint shape variables that I will first compare between emus and dinosaurs are some of the more interesting (more potentially useful in discriminating among groups) of those that we have already examined for the dinosaur footprints. These include toebase and toetip distances, and the projection of digit III beyond the tips of digits II and IV, either scaled as a percentage of backfoot length, or examined in bivariate relationships with backfoot length. As in tables A8.7–A8.11, data will be compared between emus and dinosaurs for groups represented by at least three (maximum/minimum ratios of scaled parameters) or six (coefficients of variation of scaled parameters) measurements of that scaled parameter.

As already seen with bird footprints (figs. 8.40–8.46), the variability of scaled parameters increases as the number of data cases increases from less to more inclusive categories of dinosaur tracks, whether expressed in terms of the maximum/minimum ratio or the coefficient of variation of the scaled parameter (figs. 8.46–8.49). For comparisons based on the maximum/minimum ratio of the scaled parameter, dinosaur track samples composed of two or more potentially homogeneous groups show greater variability than the equivalent ratios calculated for emus (which take maximum values of 1.3–1.5), even when the number of dinosaur trackways in the more inclusive sample is close to the number of individual birds in the emu sample. However, some of the putatively homogeneous dinosaur track samples also show greater variability than do the emu samples, although usually not as much as the more inclusive dinosaur track samples. In any case, values of the maximum/minimum ratio of scaled parameters of 1.7 or more are well beyond any of the emu values, and also greater than nearly all the values of the ratio for the potentially homogeneous dinosaur groups.

The coefficient of variation of scaled parameters tells a similar story. For the scaled parameters examined, emus take values of about 7–12. Coefficients of variation for the scaled toetip II–III and toetip III–IV distances, and the scaled digit III projection, do not show very "clean" separation of the potentially homogeneous as opposed to the more inclusive dinosaur groups, but except for the scaled toetip II–III distance and the scaled digit III projection, values of the coefficient of variation of 14 or more are only seen in dinosaur groups composed of two or more potentially homogeneous groups.

There are, of course, different parameters against which to scale test parameters than the ones employed thus far in our comparison of shape variability of emu as opposed to non-avian dinosaur footprints. Previous chapters have examined several of these in considerable detail for foot skeletons, intact alligator feet, and emu footprints, and for completeness sake they will briefly be reconsidered here. In the comparisons that follow, the emu sample will include both young and adult birds; in many comparisons this will cause the emu sample to be more variable than it would be if the sample had been restricted to adult birds (the latter as in figs. 8.46–8.49).

Tables A8.13–A8.16 present "simple" and "complex" comparisons of footprint shape parameters. The "simple" comparisons are so named because they involve parameters other than the lengths of individual digital pads and claws: toebase and toetip distances, lengths of digits II and IV, backfoot length, and digit III projection; all of these parameters are scaled to a single common value of footprint length. The "complex" comparisons also involve toebase distances, but also digital pad lengths of digits II and III, the length of digit IV without the claw, clawbase distances, and the distance from the back edge ("heel") of the footprint to the clawbase of digit II; these parameters are scaled to a single common value of digit III length without the claw.

Variability of comparisons of parameters scaled in this manner are compared using, once again, the maximum/minimum ratio of the scaled parameters (tables A8.13, A8.14), but also two other measures of variability (tables A8.15, A8.16). For these latter comparisons, each parameter is "GM-scaled" by first log-transforming all of the variables in the "simple" or "complex" comparison, then taking the mean of all of the log-transformed variables in the comparison, and finally subtracting that mean from each of the log-transformed parameters of interest. We then examine both the range and the standard deviation of the GM-scaled parameters.

Comparisons are variously made among the entire *Eubrontes* footprint sample (that is, all of the footprints assigned to this ichnogenus), the "potentially homogeneous" AC 9/14 and "all *Anomoepus*" samples, emu (both juveniles and adults) footprints, pooled footprints of the various species of phasianids, and footprints of all the birds represented in this study (chapter 6). Because there are many more individuals of some bird species (e.g., obviously, emus) than others, the "all bird" comparisons are made in two ways: one in which each bird is included in the analysis, regardless of how many conspecifics also appear in the analysis, and one in which only one individual per bird species is used in the analysis.

In most of the tables, footprint shape variability in the comparisons is ranked across the groups, with the most variable group being ranked 1, the second most variable group being ranked 2, and so on, allowing for ties. Examination of the tables is rather tedious (which is why, of course, they are relegated to the Appendix), and so summary tables of the results are presented in tables A8.17 and A8.18 (and, yes, I'll admit that things may be getting a bit out of hand when one starts providing tables that summarize the results of other tables).

In most comparisons of scaled variables, the sample consisting of all *Eubrontes* is more variable than the AC 9/14 and *Anomoepus* samples (table A8.17)—not at all surprising, given the results reported earlier in this chapter. More interestingly, in most comparisons the *Eubrontes* sample is also more variable than the emu footprint sample, suggesting that the *Eubrontes* sample could well represent more than one species of trackmaker. In contrast, the AC 9/14 and *Anomoepus* samples are about as often less as more variable than the emu sample (table A8.18), suggesting that each of these could indeed represent single-species homogeneous assemblages. The "all bird" and phasianid interspecific footprint samples are usually more variable than any of the non-avian dinosaur footprint samples in these comparisons (tables A8.13–A8.15).

Other Measures of Shape Variability: Comparisons of Feet and Footprints of Non-avian Dinosaurs, Birds, and Alligators. In previous chapters, numerous bivariate comparisons of shape variability were made using the standard error of estimate (SEE) of the regression equation and the coefficient of dispersion around the reduced major axis (D_d). The results of such analyses for emu and other bird footprints, and intact

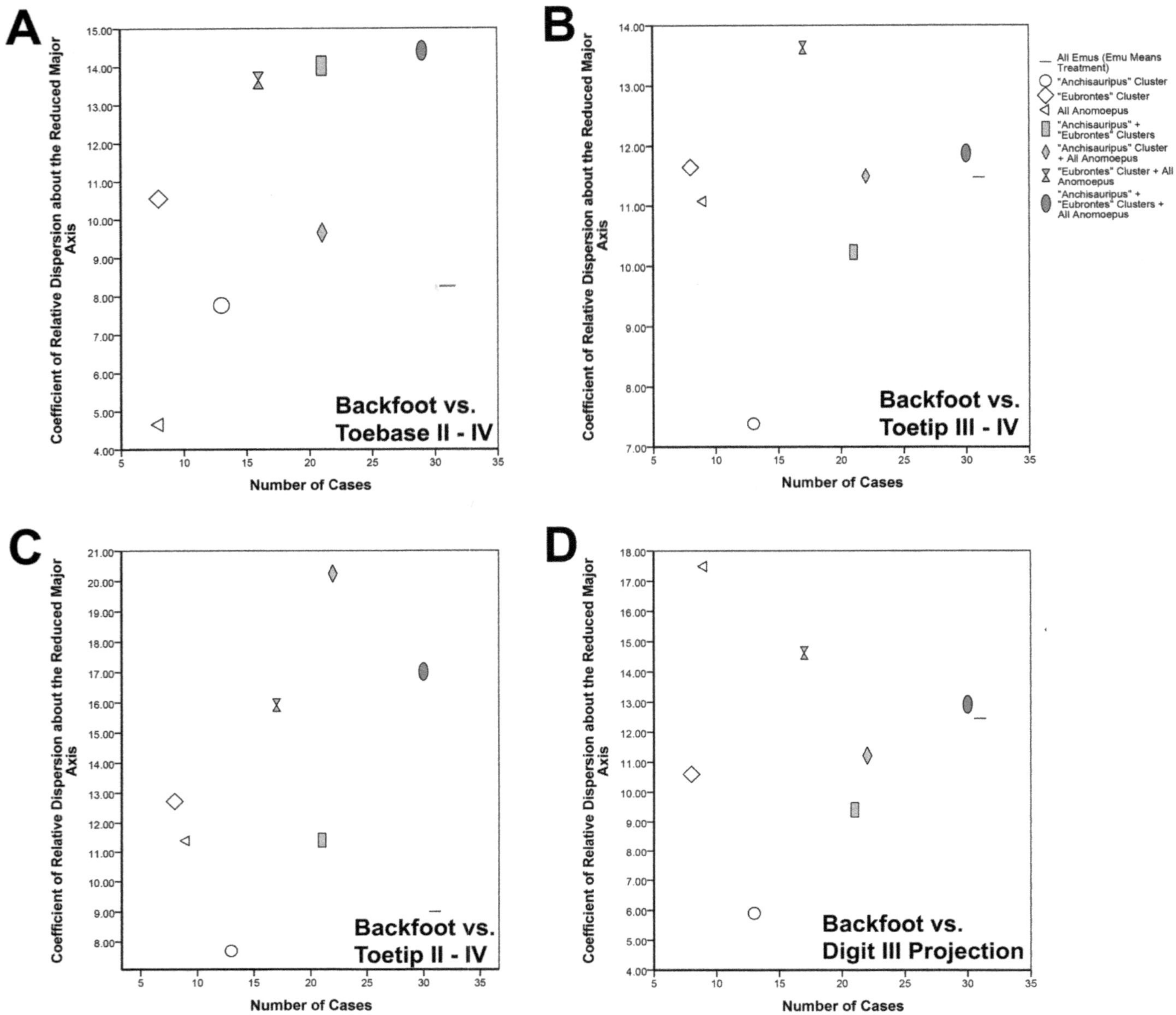

8.50. Variability of footprint shape, as expressed in bivariate shape comparisons using the coefficient of relative dispersion around the reduced major axis (D_d), in footprints of emus and both potentially homogeneous and more inclusive groups of Newark Supergroup non-avian dinosaurs. Data cases and plotting conventions are the same as those used in figure 8.46; emu cases are means for all birds, both young and adult. D_d is calculated using untransformed measurements in millimeters. All footprint parameters are analyzed relative to backfoot length. *A*, Distance between the proximal ends (toebases) of digits II and IV. *B*, Distance between the toetips of digits III and IV. *C*, Distance between the toetips of digits II and IV (toetip width). *D*, Digit III projection.

alligator feet, are extracted from earlier appendix tables and compared with the same or analogous relationships for the sample of all *Eubrontes* in tables A8.19 and A8.20, and summarized in table A8.21. The *Eubrontes* sample is once again more variable than the emu footprint sample in nearly all comparisons, and is also more variable than the alligator intact pes sample in most of the comparisons that can be made. The *Eubrontes* sample is less variable than the "all bird" footprint sample in most comparisons.

Comparisons in which emu footprints can be more variable than the entire *Eubrontes* sample include those involving backfoot length (table A8.19), prompting a further look at this parameter (fig. 8.50). Even though including juvenile as well as adult emus results in the number of birds being greater than the number of dinosaur trackways, the D_d values for emus are comparable to the low end of values for potentially homogeneous dinosaur groups for bivariate comparisons between backfoot length and the distance across the proximal ends of digits II and IV (fig. 8.50A—but keep in mind that these are not strictly comparable measurements in emus and dinosaurs) and the distance across toetips II and IV (fig. 8.50C). In contrast, emu D_d values for comparisons involving the toetip III–IV distance and the digit III projection are comparable to those for the most inclusive dinosaur group in this set of graphs, the combined *"Anchisauripus"* and *"Eubrontes"* clusters and the all *Anomoepus* sample (fig. 8.50*B*, 8.50*D*). For these last two comparisons, the pooled dinosaur trackway sample does not exceed expectations for a single species, based on the emu sample. However, bear in mind that the emu sample used in the D_d calculations includes footprints of both very young/small and large/adult animals, which may not be true of the pooled non-avian dinosaur footprint sample.

The distances between the proximal ends of digits II and IV, between toetips III and IV, and between toetips II and IV (toetip width), compared against backfoot length, take larger D_d values for most combinations of potentially homogeneous groups than for the individual groups, although in some comparisons the D_d value for the combined *"Anchisauripus"* and *"Eubrontes"* clusters is less than that of the *"Eubrontes"* cluster alone. D_d values for digit III projection vs. backfoot length are nearly always larger for combinations of groups than for at least one of the potentially homogeneous groups in that combination, but not all of them, with the all *Anomoepus* group being particularly variable.

Shape Variability in Footprints of Dinosaurs vs. Emus: Conclusions

The various measures of shape variability generally do show greater variability for the more inclusive groups of dinosaur tracks than for emus, suggesting that the former really do represent samples from more than one kind of trackmaker. However, the same is also true for some measures of footprint shape in some categories of potentially homogeneous dinosaur footprint types, suggesting that these tracks were either made by species with greater footprint shape variability than in emus, or that they actually are samples of prints from more than one kind of trackmaker—i.e., that they are not so homogeneous as I might have wished.

PATTERNS OF SIZE-RELATED DIFFERENCES IN FOOTPRINT AND FOOT SHAPE OF EMUS, ALLIGATORS AND DINOSAURS

Overall Size-Related Differences in Footprint Shape in Emus

Given that there do seem to be shape differences between members of the *"Anchisauripus"* and *"Eubrontes"* clusters, and that these two clusters are also separated largely (but not entirely) by size, the question arises whether the proportional differences between the two clusters are of the kind we would expect to see in an intraspecific, ontogenetic sequence—and only in such an intraspecific sequence. Although the Newark Supergroup dinosaurs obviously were not emus, our emu footprint sample nonetheless provides a known ontogenetic sequence with which the dinosaur sample can be compared. Using the "emu means" treatment, I will compare the shapes of prints in two size classes, using a slightly different cutoff between small and large emu prints than that used in chapter 6.

Much as was done with the Newark Supergroup tridactyl dinosaur footprint sample, I sought the best compromise between the number of footprint dimensions, and the number of birds for which those dimensions could be measured, using the "emu means" data set. The set of footprint parameters selected (lengths of digital pads II1, II2, III1–III3; claw lengths of digits II and III; digit IV length; distances between the proximal ends [toebases] and toetips of digits II–IV) was comparable to that used in creating the maximum information clusters of dinosaur tracks (table 8.2), except that no measurements of digital pad widths were available for the emu prints. I then used those log-transformed parameters in a PCA (fig. 8.51*A*), and after subtracting from each log-transformed measurement the mean of all of the log-transformed measurements ("OverallScale2015"), a cluster analysis, of the data cases (fig. 8.51*B*).

Neither the cluster analysis nor the PCA shows separation of the smaller from larger emu footprints (and so a PCA table for the emu prints comparable to table 8.2 is not presented here). Although there are differences between small and large emu prints, particularly in parameters related to relative footprint width (chapter 6), these seem to be swamped by the lack of much difference in other footprint proportions. The smaller emu footprints do not separate from the large emu prints as neatly as the *"Anchisauripus"* cluster separates from the *"Eubrontes"* cluster (figs. 8.17, 8.18).

There are other ways of looking at the emu data, however. A stepwise discriminant analysis using the same set of scaled parameters as the cluster analysis (table 8.7; fig. 8.51*C*) does a pretty good job of distinguishing small from large emu prints, selecting the scaled lengths of digit IV and the clawmark of digit III as the most important parameters by which prints of different size differ (cf. table 2.32). The *t*-tests of the means of scaled parameters (table 8.8) show some differences between small and large prints: Emu prints in the smaller size class have relatively broader footprints and a relatively longer second digital pad of digit II than prints in the larger class, while prints in the larger class have a relatively longer digit IV and digit III clawmark (figs. 8.52, 8.53). Seven of 14 reduced major axis analyses in which the log-transformed footprint variable is compared against the mean of the other 13 log-transformed variables besides itself show allometry:

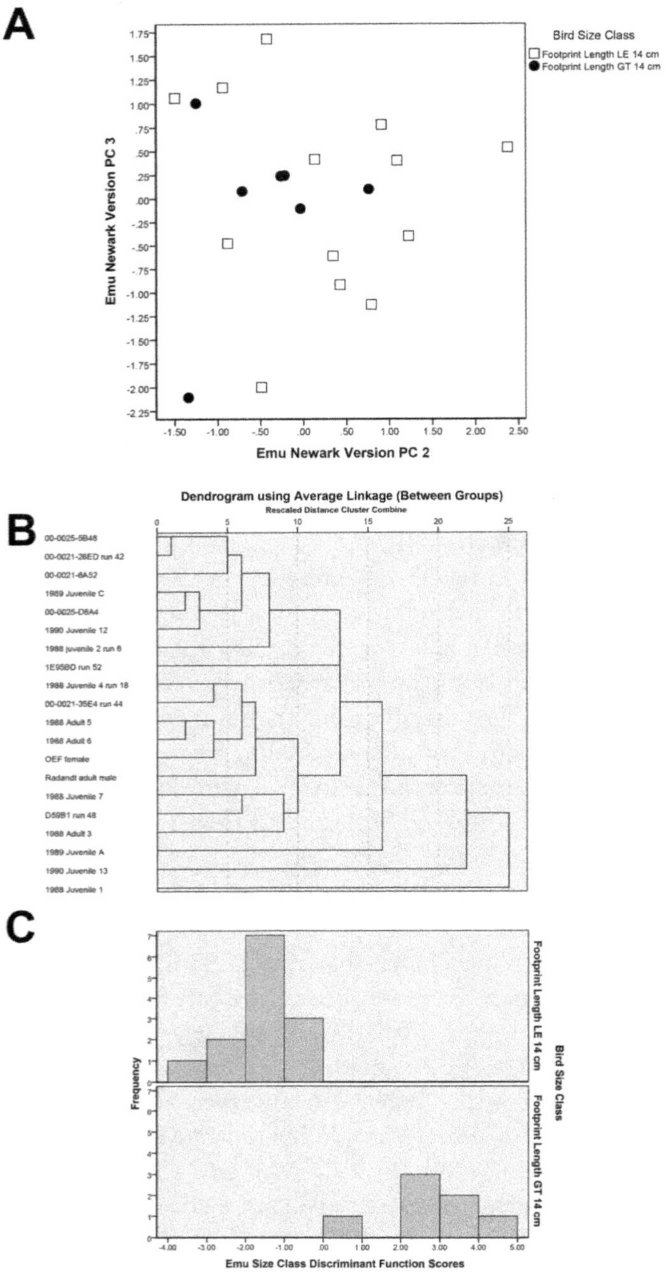

8.51. Overall shape comparison of small (footprint length ≤14 cm) and large (footprint length >14 cm) emu footprints (emu means treatment). Analyses are based on log-transformed values of the lengths of digital pads II1, II2, III1–III3; claw lengths of digits II and III; digit IV length; distances between the proximal ends (toebases) of digits II–III, III–IV, and II–IV; distances between the toetips of digits II–III, III–IV, and II–IV. *A*, Principal components 2 and 3 (PC 2, PC 3) of a principal component analysis. *B*, Dendrogram of cases "GM-scaled" by subtracting the mean of the log-transformed variables from each log-transformed variable. Seven of the birds constitute the larger size class: 1988 adults 3, 5, and 6, OEF female, Radandt adult male, and 1990 juveniles 12 and 13; see chapter 6 for details. Neither analysis shows obvious separation between footprints of the two size classes. *C*, Results of stepwise discriminant analysis (table 8.7) of "GM-scaled" parameters. Digit IV length and digit III claw length are the two main scaled parameters upon which the discriminant function is based. GM = geometric mean.

Table 8.7. Forward stepwise discriminant analysis of log-transformed emu footprint measurements (emu means treatment), all scaled by subtracting the mean of all of the log-transformed variables from each log-transformed measurement. Parameters used in the analysis include the lengths of the digital pads and claws of digits II and III, digit IV length, and the distances between the three toebases and toetips. Footprints of small (size class 1: footprint length ≤ 14 cm) or large (size class 2: footprint length > 14 cm) groups are compared. Results are the same whether or not the prior probability of group membership is adjusted for the number of cases in each group. Classification results: 95% of original grouped cases and 95% of cross-validated cases were correctly classified.

Variable	Standardized canonical discrimination function coefficients (correlations between individual variables and discriminant function)
Digit IV length	1.555 (0.490)
Digit III claw length	0.747 (0.404)
Digit III third pad length	1.118 (–0.057)
Eigenvalue	4.120
Canonical correlation	0.897

Table 8.8. Independent-samples *t*-test for equality of means of the 14 GM-scaled footprint parameters used in the discriminant analysis of emu footprints (table 8.7; emu means treatment). Comparisons are made between small (size class [SC] 1: footprint lengths ≤ 14 cm; *N* = 13) and large (size class [SC] 2: footprint lengths > 14 cm; *N* = 7) prints. In most comparisons the *P* value of Levene's test for equality of variances is >.05; where this condition is not met, the test is made without the assumption of equal variances. Significant differences in mean values between the two cluster groups are indicated in **bold**. Relative sizes of the mean values for the two cluster groups are reported for comparisons in which the difference is statistically significant, or approaches statistical significance.

Parameter	*P* of *t*-test	Comparison of mean values
Digit II first pad length	.827	———
Digit II second pad length	**.013**	SC1 > SC2
Digit II claw length	.578	———
Digit III first pad length	.070	SC2 > SC1
Digit III second pad length	.334	———
Digit III third pad length	.628	———
Digit III claw length	**.003**	SC2 > SC1
Digit IV length	**.001**	SC2 > SC1
Distance from base of digit II to base of digit III	**.020**	SC1 > SC2
Distance from base of digit III to base of digit IV	.252	———
Distance from base of digit II to base of digit IV	**.005**	SC1 > SC2
Toetip II to toetip III	.521	———
Toetip III to toetip IV	.073	SC1 > SC2
Toetip II to toetip IV	**.007**	SC1 > SC2

8.52. Size-frequency distributions of digits II–III pad and claw lengths, and digit length, in small and large emu footprints. All measurements were "GM-scaled" by subtracting the mean (labeled "OverallScale2015mm" in the graphs) of all 14 log-transformed footprint parameters (table 8.8) from each log-transformed footprint parameter; cf. figures 8.20–8.22 showing analogous distributions for the "maximum information" dinosaur footprint clusters. *A*, Digit II first pad length. *B*, Digit II second pad length. *C*, Digit II claw length. *D*, Digit III first pad length. *E*, Digit III second pad length. *F*, Digit III third pad length. *G*, Digit III claw length. *H*, Digit IV length. Large emu prints show a relatively longer digit III first pad length, digit III claw length, and digit IV length, than small emu prints, whereas small emu prints show a relatively longer digit II second pad length. GM = geometric mean.

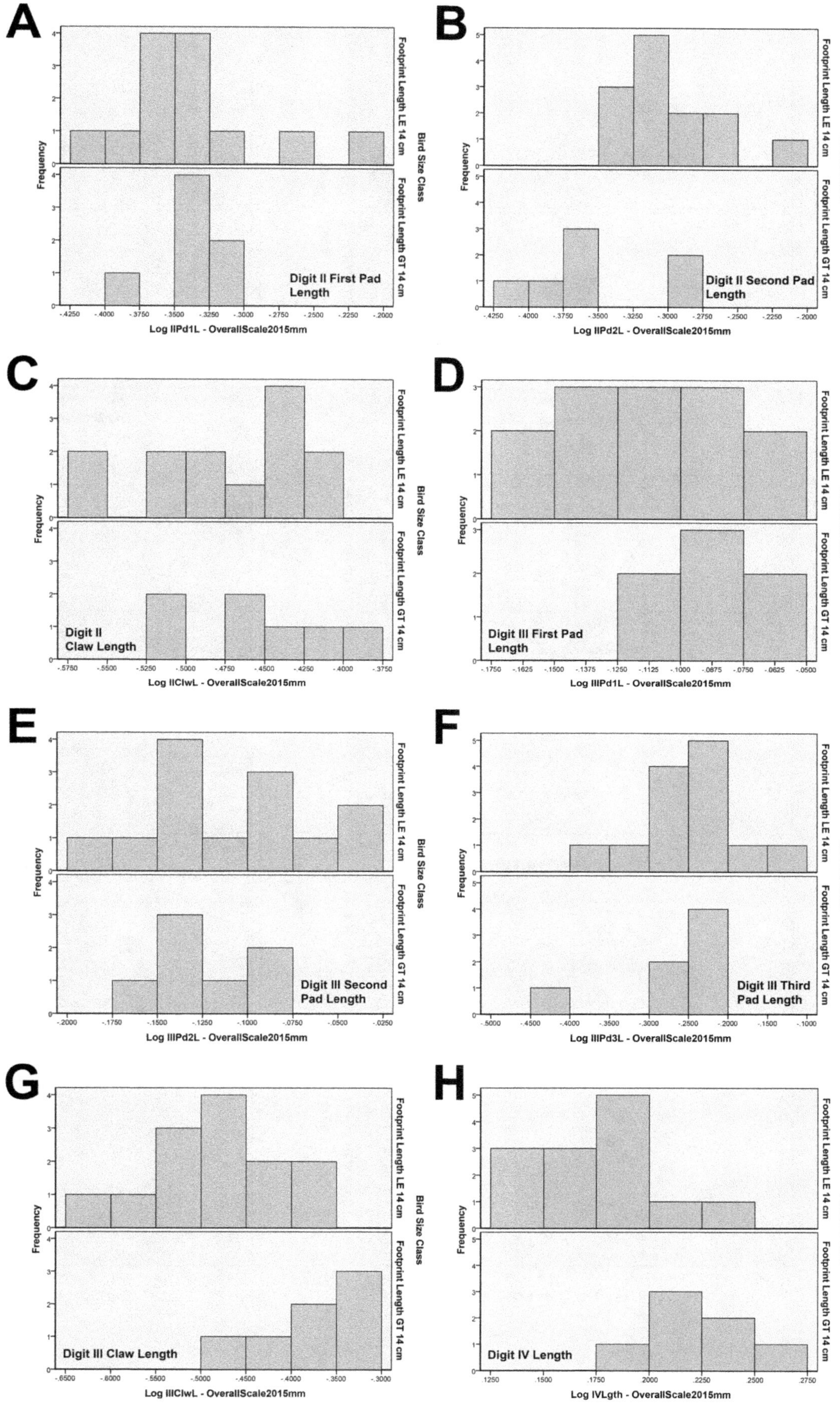

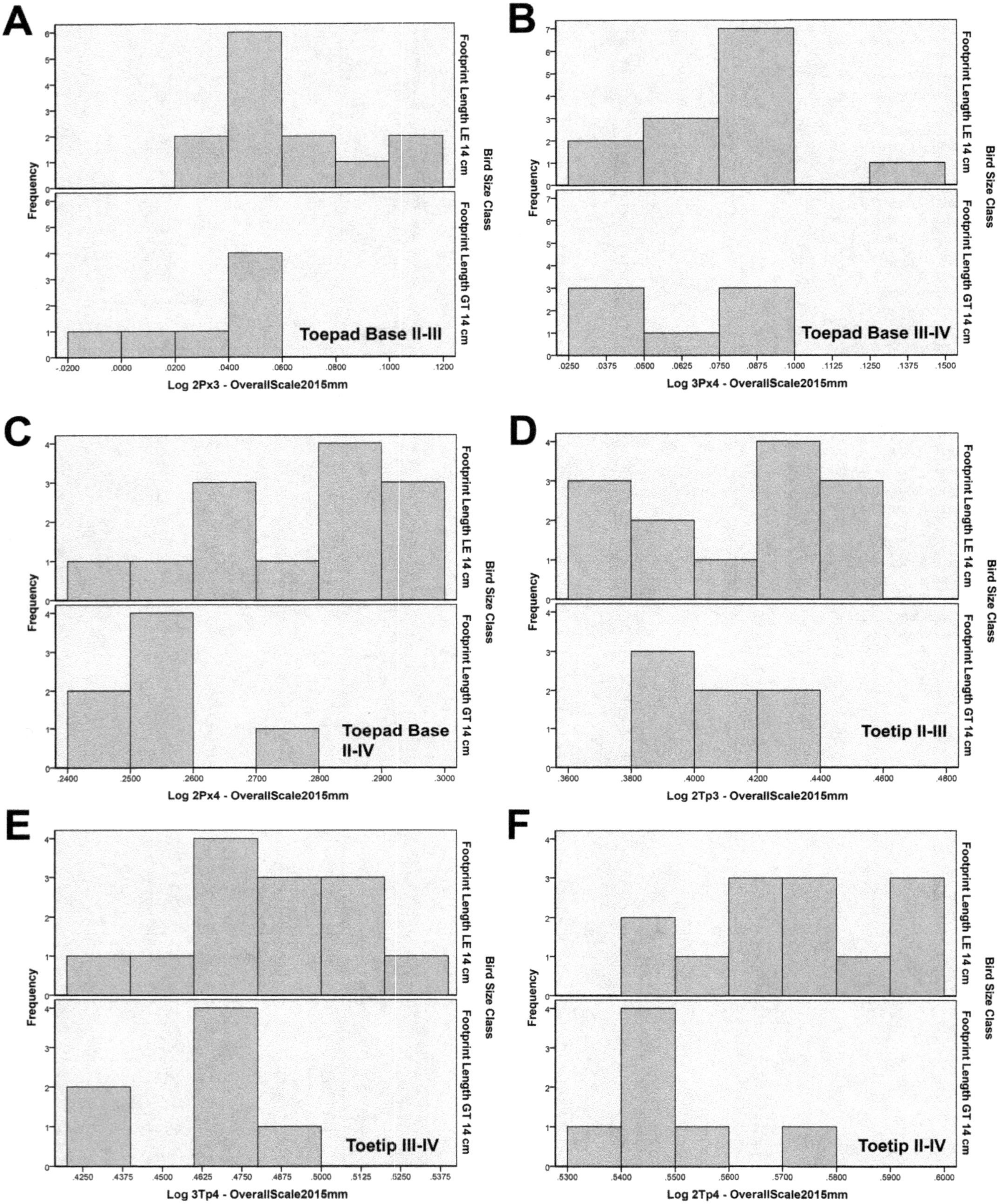
A
Frequency
Bird Size Class
Footprint Length LE 14 cm
Footprint Length GT 14 cm
Toepad Base II-III
-.0200 .0000 .0200 .0400 .0600 .0800 .1000 .1200
Log 2Px3 - OverallScale2015mm
B
Frequency
Bird Size Class
Footprint Length LE 14 cm
Footprint Length GT 14 cm
Toepad Base III-IV
.0250 .0375 .0500 .0625 .0750 .0875 .1000 .1125 .1250 .1375 .1500
Log 3Px4 - OverallScale2015mm
C
Frequency
Bird Size Class
Footprint Length LE 14 cm
Footprint Length GT 14 cm
Toepad Base II-IV
.2400 .2500 .2600 .2700 .2800 .2900 .3000
Log 2Px4 - OverallScale2015mm
D
Frequency
Bird Size Class
Footprint Length LE 14 cm
Footprint Length GT 14 cm
Toetip II-III
.3600 .3800 .4000 .4200 .4400 .4600 .4800
Log 2Tp3 - OverallScale2015mm
E
Frequency
Bird Size Class
Footprint Length LE 14 cm
Footprint Length GT 14 cm
Toetip III-IV
.4250 .4375 .4500 .4625 .4750 .4875 .5000 .5125 .5250 .5375
Log 3Tp4 - OverallScale2015mm
F
Frequency
Bird Size Class
Footprint Length LE 14 cm
Footprint Length GT 14 cm
Toetip II-IV
.5300 .5400 .5500 .5600 .5700 .5800 .5900 .6000
Log 2Tp4 - OverallScale2015mm

Facing, **8.53.** Size-frequency distributions of parameters associated with footprint width in small and large emu prints. All measurements were "GM-scaled" by subtracting the mean (labeled "OverallScale2015mm" in the graphs) of all 14 log-transformed footprint parameters (table 8.8) from each log-transformed footprint parameter; cf. figure 8.22 showing analogous distributions for the "maximum information" dinosaur footprint clusters. *A*, Distance between the proximal ends (toepad bases) of digits II and III. *B*, Distance between the proximal ends of digits III and IV. *C*, Distance between the proximal ends of digits II and IV. *D*, Distance between toetips II and III. *E*, Distance between toetips III and IV. *F*, Distance between toetips II and IV (toetip width). Large emu prints show a relatively narrower toebase II–IV, and narrower toetips III–IV and II–IV, than those in small emu prints. GM = geometric mean.

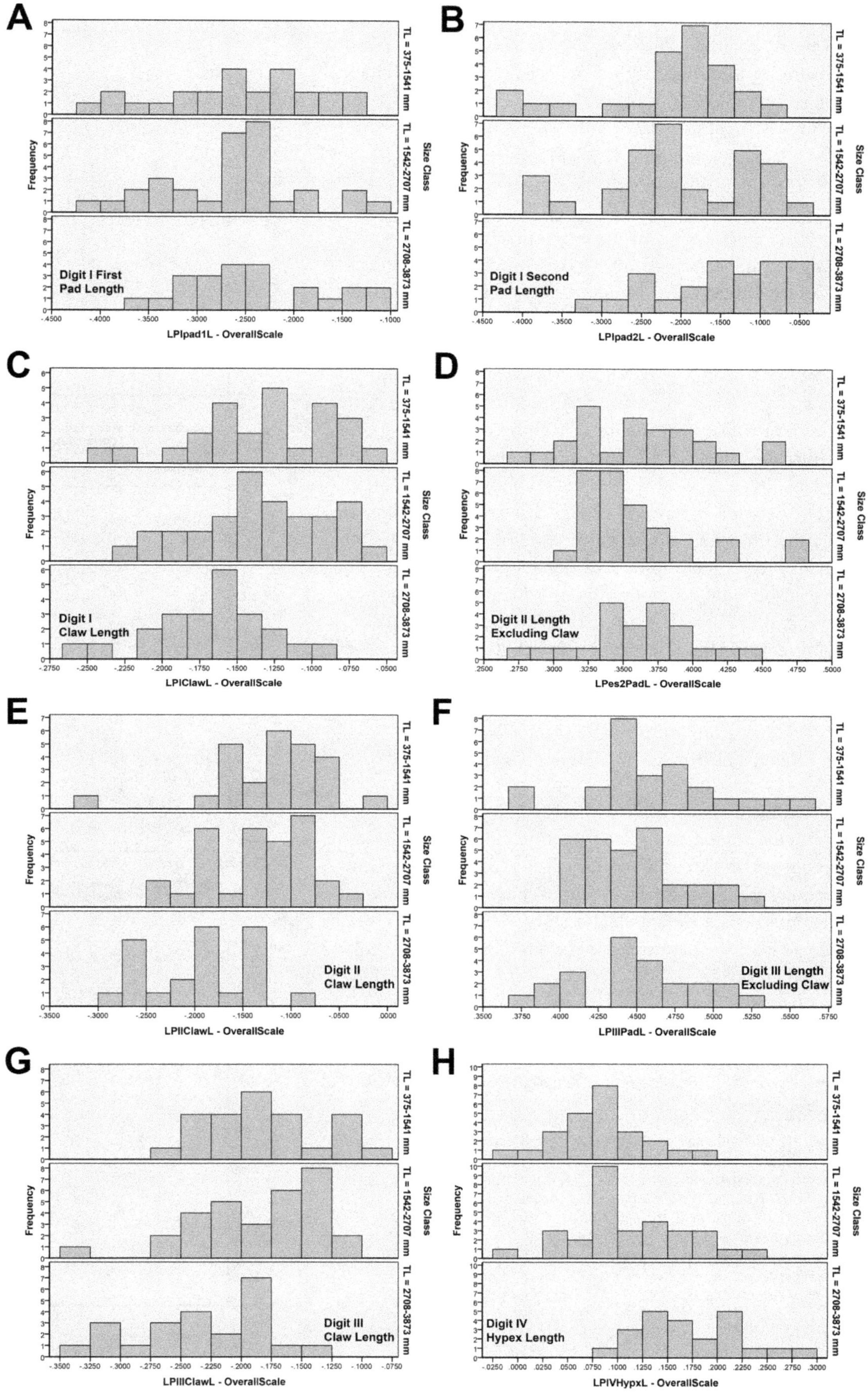

Right, **8.54.** Size-frequency distributions of parameters of hindfoot shape in alligators of different size classes; TL = alligator total length. All measurements were "GM-scaled" by subtracting the mean (labeled "OverallScale" in the graphs) of the eight log-transformed footprint parameters examined here from each log-transformed footprint parameter. *A*, Digit I first pad length. *B*, Digit I second pad length. *C*, Digit I claw length. *D*, Digit II length excluding claw. *E*, Digit II Claw length. *F*, Digit III length excluding claw. *G*, Digit III claw length. *H*, Digit IV hypex length. Note the decrease in relative claw lengths, and the increase in the digit IV hypex length, on progression from small to large alligators. GM = geometric mean.

The lengths of the first digital pad of digit III, the digit III clawmark, and digit IV show positive allometry, while most measures of footprint width show negative allometry (table A8.23).

Overall Size-Related Differences in Foot Shape in Alligators

The other species of extant archosaur that has figured prominently throughout this study is the American alligator, for which we have a set of hindfoot measurements from individuals spanning a considerable size range (chapter 5). These data can be examined the same way as for emu and *Eubrontes* prints, with the parameters GM-scaled and compared across alligator size class (fig. 8.54)—although note that the measurements are not identical to those of either the emu or dinosaur prints (due to inclusion of data for digit I, and the inability to recognize clear digital pads on digits II–IV).

The second digital pad of digit I is relatively longer in large than in small alligators (fig. 8.54*B*). Unlike emu footprints, alligator feet show a clear tendency for the lengths of claws to become relatively shorter with increasing animal size (fig. 8.54*C*, 8.54*E*, 8.54*G*; cf. table 5.9). Like emus, alligator feet show an increase in the length of a measure of digit IV length with increasing size (fig. 8.54*H*; cf. table 5.9).

Comparing Intraspecific vs. Interspecific Size-Related Differences in Foot and Footprint Shape

Ground Birds. Some of the size-related changes seen in footprint shape among emus can be compared with size-related changes across my entire sample of ground bird prints. However, because it was generally impossible to measure digital pad dimensions in the smaller bird tracks (chapter 6), it is necessary to scale the footprint dimensions of interest differently

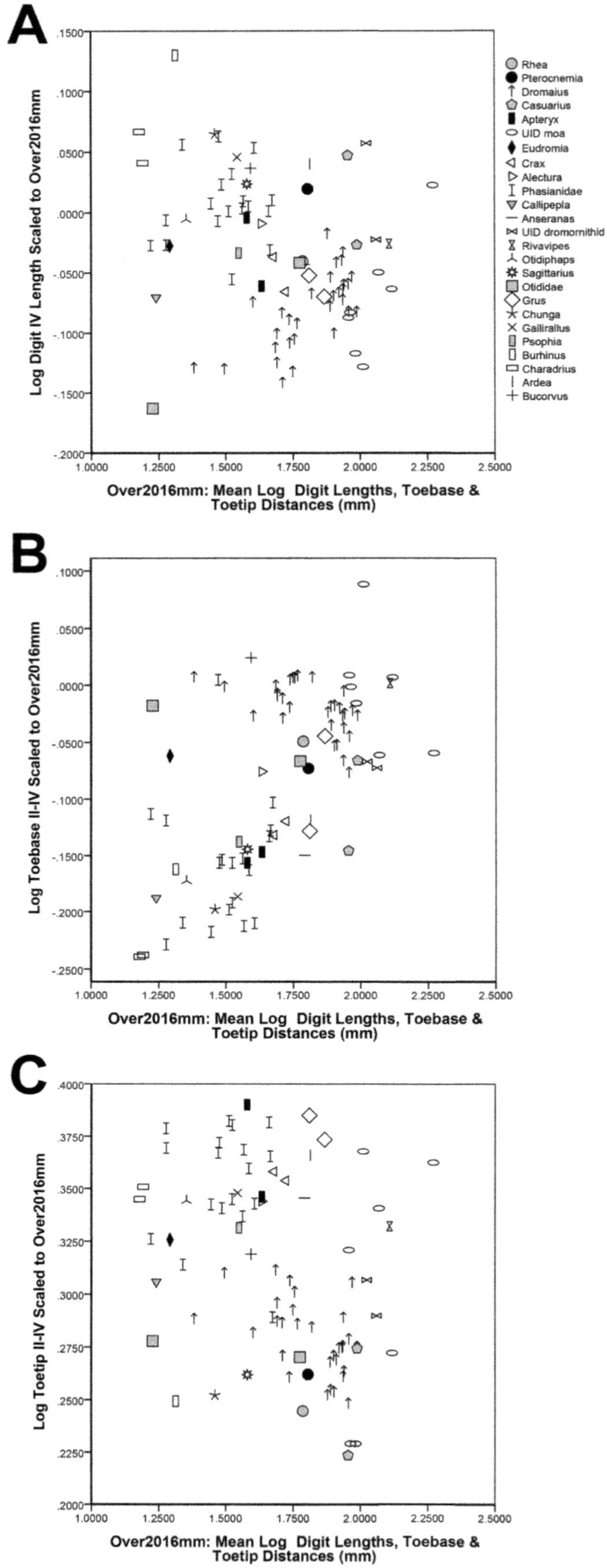

8.55. Comparing intraspecific and interspecific size-related changes in ground bird footprint shape (bird means data treatment). Footprints were scaled by taking the mean of the log-transformed lengths of digits II–III, and the toebase and toetip distances among digits II–III, and subtracting that mean (labeled "Over2016mm" in the graphs) from each of the three log-transformed parameters of interest. In all three graphs the number of emus = 30, the total number of birds is 84, and the number of birds in a "one individual per species" analysis is 37. *A*, Digit IV length. Within emu data, this parameter increases relative to overall footprint size as birds get bigger: $r = 0.700$, $P < .001$; cf. fig. 8.52*H*). For the entire bird footprint sample, on the other hand, the correlation is either negative (all birds: $r = -0.288$, $P = .008$) or nonexistent (one per species: $r = -0.096$, $P = .572$). *B*, Distance between the proximal ends (toebases) of digits II and IV. The correlation in emus is negative ($r = -0.575$, $P = .001$; cf. fig. 8.53*C*). For the entire bird sample, by contrast, the correlation is positive (all birds: $r = 0.579$, $P < .001$; one per species: $r = 0.405$, $P = .013$). *C*, Toetip width. The correlation in emus is again negative ($r = -0.504$, $P = .005$; cf. fig. 8.53*F*). For the entire bird sample, the correlation is also negative ($r = -0.364$, $P = .001$), but nonexistent with the one-per-species treatment ($r = -0.110$, $P = .516$).

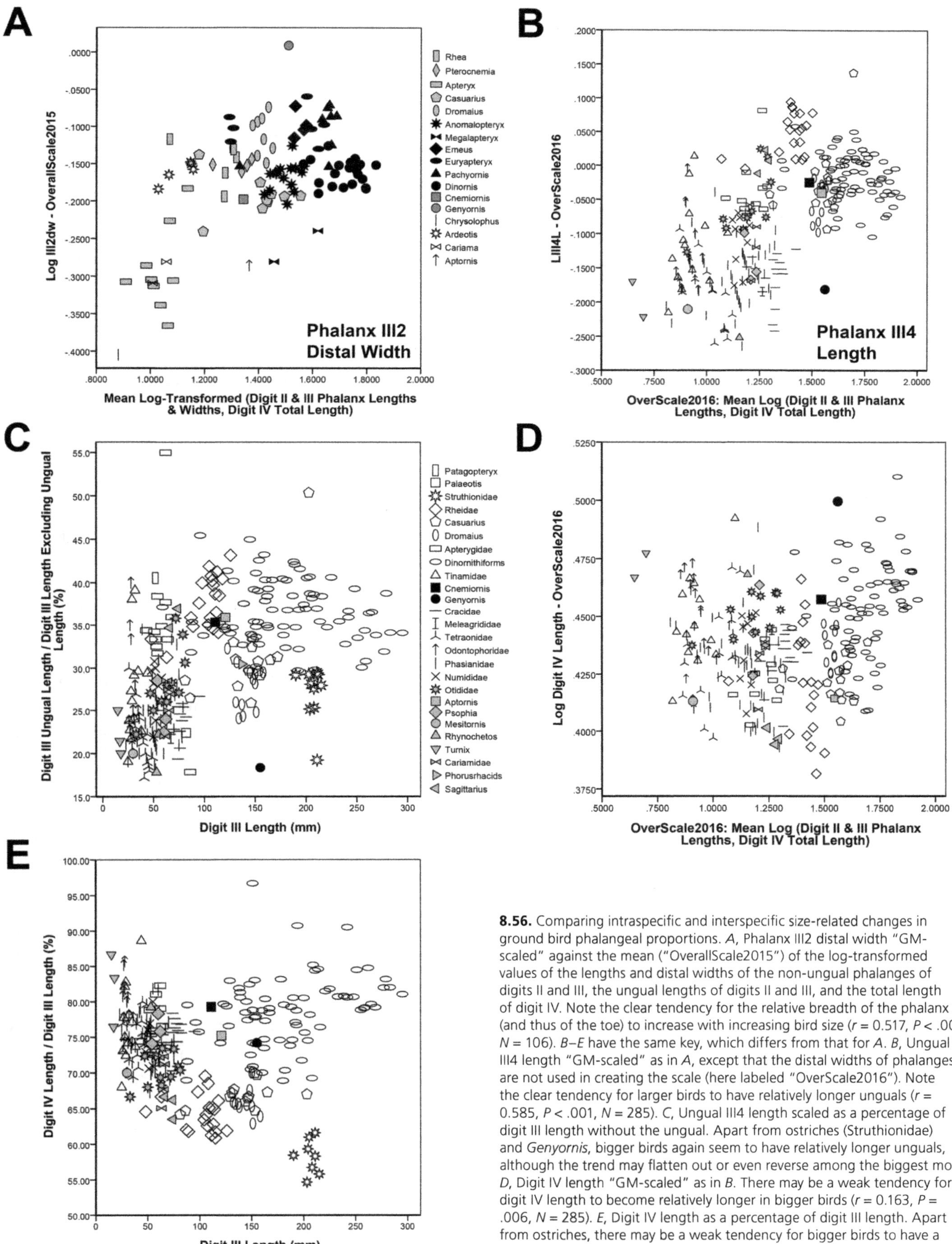

8.56. Comparing intraspecific and interspecific size-related changes in ground bird phalangeal proportions. *A*, Phalanx III2 distal width "GM-scaled" against the mean ("OverallScale2015") of the log-transformed values of the lengths and distal widths of the non-ungual phalanges of digits II and III, the ungual lengths of digits II and III, and the total length of digit IV. Note the clear tendency for the relative breadth of the phalanx (and thus of the toe) to increase with increasing bird size ($r = 0.517$, $P < .001$, $N = 106$). *B–E* have the same key, which differs from that for *A*. *B*, Ungual III4 length "GM-scaled" as in *A*, except that the distal widths of phalanges are not used in creating the scale (here labeled "OverScale2016"). Note the clear tendency for larger birds to have relatively longer unguals ($r = 0.585$, $P < .001$, $N = 285$). *C*, Ungual III4 length scaled as a percentage of digit III length without the ungual. Apart from ostriches (Struthionidae) and *Genyornis*, bigger birds again seem to have relatively longer unguals, although the trend may flatten out or even reverse among the biggest moa. *D*, Digit IV length "GM-scaled" as in *B*. There may be a weak tendency for digit IV length to become relatively longer in bigger birds ($r = 0.163$, $P = .006$, $N = 285$). *E*, Digit IV length as a percentage of digit III length. Apart from ostriches, there may be a weak tendency for bigger birds to have a relatively longer digit IV. GM = geometric mean.

to make the comparisons, using overall digit lengths of digits II and III instead of the lengths of the individual digital pads and claws. Digit IV in the emus is again seen to become relatively longer with increasing bird size (fig. 8.55A; cf. fig. 8.52*H*). If we do a quick-and-dirty correlation analysis that disregards phylogeny or the number of individual birds represented by each species (I can just imagine the screaming from some readers), the footprint sample among bird species as a whole shows a weak negative correlation between digit IV length and overall size, or no correlation at all (fig. 8.55A). The toebase and toetip II–IV widths (fig. 8.55*B*, 8.55C) again show bigger emus to have relatively narrower prints (cf. fig. 8.53*C*, 8.53*F*). However, for the entire bird footprint sample, toebase II–IV width seems to increase with increasing print size, and the toetip width shows at best a weak (and rather unconvincing) negative correlation with print size.

Data from bird foot skeletons allow an independent set of comparisons of cross-species size-related changes in pedal proportions. Bigger birds tend to have relatively broader phalanges (fig. 8.56A; cf. fig. 2.12). Across species ungual III4 becomes relatively longer with increasing bird size (fig. 8.56*B*, 8.56C), as seen intraspecifically in emu footprints (fig. 8.52G). In contrast to what was seen across our interspecific bird footprint sample (fig. 8.55A), the skeletal sample suggests that digit IV may become relatively longer across species in bigger birds (fig. 8.56*D*, 8.56*E*)—with ostriches a conspicuous exception—as seen more convincingly in the emu ontogenetic footprint sample (fig. 8.52*H*).

Crocodylians. Skeletal data indicate that there could be an interspecific increase in the relative length of digit IV with increasing size (fig. 8.57), as seen ontogenetically in American alligators (fig. 8.54*H*). However, because the skeletal sample, particularly of larger specimens, is dominated by alligators, this conclusion must be regarded with a grain of salt.

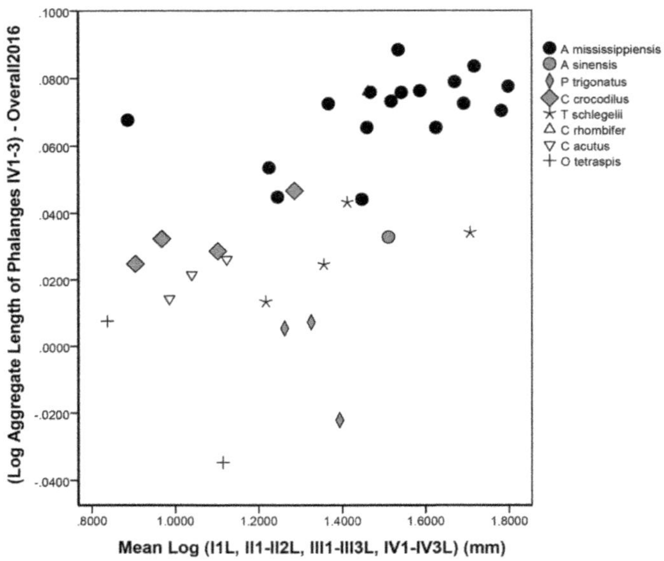

8.57. Relative length of digit IV across foot skeletons of different crocodylian species. Because of problems of measuring the more distal phalanges (particularly unguals) in osteological specimens of crocodylians (chapter 1), only non-ungual phalanges of digits I–III and the first three phalanges of digit IV were used in the comparison. The log-transformed aggregate length of the first three phalanges of digit IV was "GM-scaled" against the mean ("Overall2016") of this log-transformed digit IV length and the log-transformed lengths of the individual phalanges of digits I–III. The data suggest a possible increase in the relative length of digit IV with increasing foot size across crocodylian species ($r = 0.588$, $P < .001$, $N = 35$), but note that nearly all of the large specimens are American alligators. GM = geometric mean.

Comparing Size-Related Changes in Non-avian Dinosaur Footprints With Size-Related Changes in Bird and Alligator Feet, and Bird Footprints

We are now ready to see if the differences in footprint proportions between the "*Anchisauripus*" and "*Eubrontes*" maximum information clusters, and across the size range of the nominal species of *Eubrontes*, are consistent with expectations for an intraspecific or interspecific trend, or both. We will compare the non-avian dinosaur footprint sample with the results already described for birds and alligators, and even throw in a new data set (bird study skins) different from those already examined. In the process we will also again compare *Anomoepus* and *Kayentapus* with the *Eubrontes* sample.

The "*Eubrontes*" maximum information cluster differs from the "*Anchisauripus*" maximum information cluster (tables 8.4–8.6, figs. 8.20–8.22) in having a relatively longer first digital pad of digit II, relatively shorter clawmarks, and relatively broader measurements related to footprint width. Across the size range of *Eubrontes* (tables A8.19, A8.22, A8.23), the first digital pad of digit II, some measures of footprint width, and digital pad widths show positive allometry with various measures of footprint size, and digits II and IV become relatively longer than digit III with increasing size. These shape changes can now be examined in more detail vis-à-vis the above-described intraspecific and interspecific trends for birds and crocodylians.

Length of the First Digital Pad of Digit II (figs. 8.20A, 8.52A, 8.58). As previously noted, this digital pad becomes relatively longer as *Eubrontes* gets bigger, and is relatively longer in the "*Eubrontes*" than the "*Anchisauripus*" maximum information cluster. Figure 8.58 examines the relative length of this digital pad another way, as a percentage of the length of the first digital pad of digit III. Scaled this way, there is once again a significant difference between members of the "*Anchisauripus*" and "*Eubrontes*" maximum

Table 8.9. Analyses of covariance (ANCOVA) of selected footprint size parameters against emus of different size. Size class 1 = birds with IIILGL footprint length ≤ 14 cm; size class 2 = birds with IIILGL footprint length > 14 cm (emu means treatment). Unless otherwise indicated, the assumptions of ANCOVA were met (*P* values of Levene's test of equality of error variances and lack-of-fit test of at least .05). LSD = least significant difference; SC = size class.

Test parameter	Covariate	*P* of *F*-test	Result of comparison (LSD test)	Number of cases (SC 1: SC 2)
Digit II first pad length		0.346	No difference	13: 7
Digit II second pad length[a]		0.214	No difference	13: 7
Digit II claw length[a]	Digit III first pad length	0.648	No difference	14: 13
Digit III second pad length		0.876	No difference	14: 15
Digit III third pad length		0.934	No difference	14: 14
Digit III claw length[a]		0.010	SC2 > SC1	14: 17
Digit IV length from proximal end of metatarsal pad to clawtip[a, b, c]		0.449	No difference	14: 17
Toebase II to toebase III		0.960	No difference	14: 17
Toebase III to toebase IV[a, b]		0.074	No difference	14: 17
Toebase II to toebase IV[b]	Backfoot length	0.548	No difference	14: 17
Toetip II to toetip III[c]		0.381	No difference	14: 16
Toetip III to toetip IV[a]		0.097	No difference	14: 17
Toetip II to toetip IV[a]		0.720	No difference	14: 17
Digit III projection[a]		0.068	No difference	14: 17

[a]Test parameter and covariate were log transformed prior to analysis to equalize error variances.
[b]The way the parameter was measured in emu footprints is only analogous to the way it is measured in non-avian dinosaur footprints.
[c]*P* of lack-of-fit test < .05.

information clusters, and the relative length of the first pad of digit II again increases with increasing size across *Eubrontes* (fig. 8.58A), although the relationship may flatten out in the biggest specimens.

In emus, a known intraspecific assemblage, the relative length of the first pad of digit II decreases with increasing footprint size as examined here (fig. 8.58*B*), although an analysis of covariance of the length of this pad, using the length of the first pad of digit III as the covariate, detects no difference between small and large emu footprints (table 8.9). Neither does GM-scaled pad length show any difference between small and large emu footprints (table 8.8; fig. 8.52A).

Even though the desiccated nature of the digital pads of bird study skins means that quantitative comparisons of their dimensions should be taken with caution, the relative size of the first pad of digit II seems to decrease with increasing foot size in emus, and possibly large ground birds more generally (fig. 8.58C). Thus the size-related trend seen in the relative size of the first pad of digit II in emus (and maybe across big bird species) is just the opposite of what is seen in *Eubrontes*.

Distinct digital pads could not be recognized in digit II of the pes of alligators, but the first digital pad of digit I shows no difference in relative length between big and little alligators (fig. 8.54A).

The relative increase in the length of the first digital pad of digit II between the "*Anchisauripus*" and "*Eubrontes*" maximum information clusters, and with increasing size across *Eubrontes* more generally, is thus not consistent with what is seen ontogenetically in either of the two extant species of large archosaurs examined. Neither is it consistent with what is seen across species in my bird study skin sample (fig. 8.58C).

Anomoepus has high values of the scaled length of the first pad of digit II, despite being small footprints (table 8.6; figs. 8.20A, 8.58A), while *Kayentapus* falls among the *Eubrontes* cases.

Length of the Second Digital Pad of Digit III (figs. 8.21C, 8.52E, 8.59). The "GM-scaled" length of the second pad of digit III is shorter in the "*Eubrontes*" maximum information cluster than in the "*Anchisauripus*" maximum information cluster (table 8.5; fig. 8.21C). As scaled in figure 8.59A, fewer parameters have to be measurable than in the creation of figure 8.21C, and so more data cases are available. The relative length of the second digital pad of digit III as scaled here does not differ significantly between the "*Anchisauripus*" and "*Eubrontes*" maximum information clusters. Although it looks like the ratio could decrease with increasing footprint size across all *Eubrontes*, the parameters are not significantly correlated; there is, however, significant negative allometry between the "GM-scaled" length of the second pad of digit III and overall print size (table A8.23), and also between the length of the second and first digital pads of digit III (table A8.19).

As with *Eubrontes*, emus may show a negative correlation between the two variables examined here (fig. 8.59*B*), but

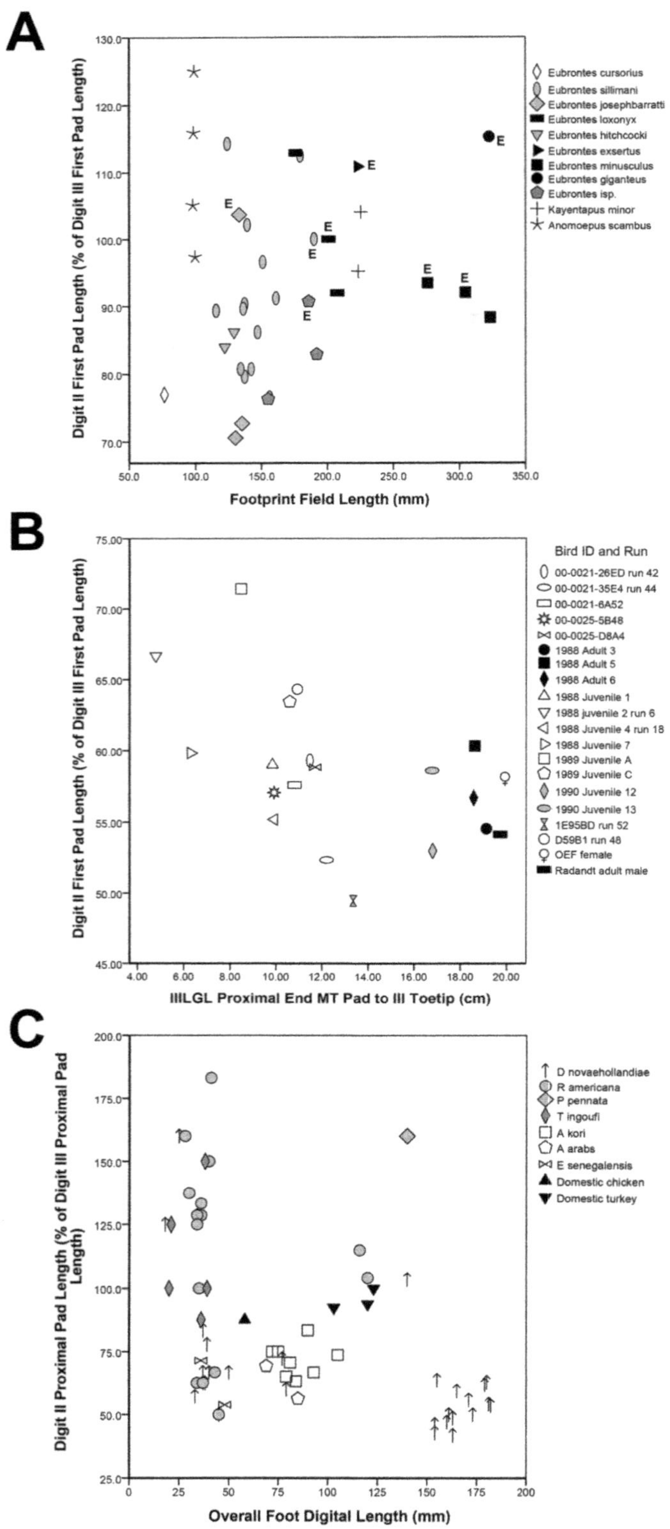

8.58. Comparison of the length of the first digital pad of digit II (scaled as a percentage of the length of the first digital pad of digit III) in non-avian dinosaur and bird footprints, and bird study skins, of different size. *A*, Non-avian dinosaur footprints. The relative length of the first digital pad of digit II as scaled here is significantly different between members of the *"Anchisauripus"* and *"Eubrontes"* maximum information clusters (*P* of Mann-Whitney U test = .013), and is positively correlated with increasing footprint size across all *Eubrontes* (Kendall's $\tau b = 0.279$, $P = .028$; Spearman's $\rho = 0.405$, $P = .024$; $N = 31$; cf. table A8.23). *Anomoepus* shows high values for the scaled length of the first pad of digit II, despite being small footprints, whereas *Kayentapus* falls among the *Eubrontes* cases. *B*, Emu footprints (emu means treatment); see table 6.1 for identification of track-making "runs." In contrast with *Eubrontes*, in emu prints the length of the first pad of digit II decreases relative to the length of the first pad of digit III with increasing print size (here expressed as parameter IIILGL, the length of the footprint from the proximal end of the metatarsal pad to the tip of digit III) (Kendall's $\tau b = -0.379$, $P = .019$; Spearman's $\rho = -0.537$, $P = .015$; $N = 20$). *C*, Bird study skins. Given the results for emu footprints, it isn't surprising to see a decrease in the relative length of the first (proximal) pad of digit II as the length of the digital portion of the foot increases with increasing emu foot size (Kendall's $\tau b = -0.405$, $P = .002$; Spearman's $\rho = -0.602$, $P = .001$; $N = 29$). Although there is a lot of scatter in the data, the same may also be true across species (Kendall's $\tau b = -0.367$, $P < .001$; Spearman's $\rho = -0.552$, $P < .001$; $N = 66$).

the correlation again is not significant (cf. tables A8.19 and A8.23, the latter of which shows no suggestion of allometry between "GM-scaled" length of the digital pad and overall footprint size [cf. table 8.8; fig. 8.52*E*]). In contrast to the footprints, study skins (fig. 8.59*C*) provide stronger support for a negative correlation between the scaled length of the second digital pad and foot length in emus. For the entire sample of bird feet there may also be a negative correlation, but it isn't particularly convincing.

Distinct digital pads could not be recognized in digits II or III of the pes of alligators, but it is worth noting that, as already described, the second digital pad of digit I becomes relatively longer with increasing alligator size when GM scaled (table A8.24; fig. 8.54*B*), which is opposite the trend seen in the second digital pad of digit III of *Eubrontes*. However, the length of the second digital pad of digit I does not show significant positive allometry against the length of the first digital pad in a simple bivariate relationship (table 5.9).

Taking everything into consideration, then, the possible relative decrease in the length of the second digital pad of digit III from the "*Anchisauripus*" to the "*Eubrontes*" maximum information clusters, and with increasing size across *Eubrontes*, is at best weakly consistent with (emus), or possibly inconsistent with (alligators), what is seen ontogenetically in the two extant species of large archosaurs examined. Neither is it particularly consistent with what is seen across species in my bird study skin sample (fig. 8.59C).

Anomoepus departs from the *Eubrontes* trend in having a relatively short second pad of digit III despite being small prints (table 8.6; fig. 8.21C, 8.59A).

Claw Lengths (figs. 8.20E, 8.21G, 8.32, 8.52C, 8.52G, 8.54C, 8.54E, 8.54G, 8.56B, 8.56C, 8.60; 8.61A). The "GM-scaled" claw lengths of these digits seem to be shorter in the "*Eubrontes*" maximum information cluster than in the "*Anchisauripus*" maximum information cluster (table 8.5; figs. 8.20*E*, 8.21G). As scaled in figure 8.60 (8.60A, 8.60*B*) the relative claw lengths continue to be significantly different

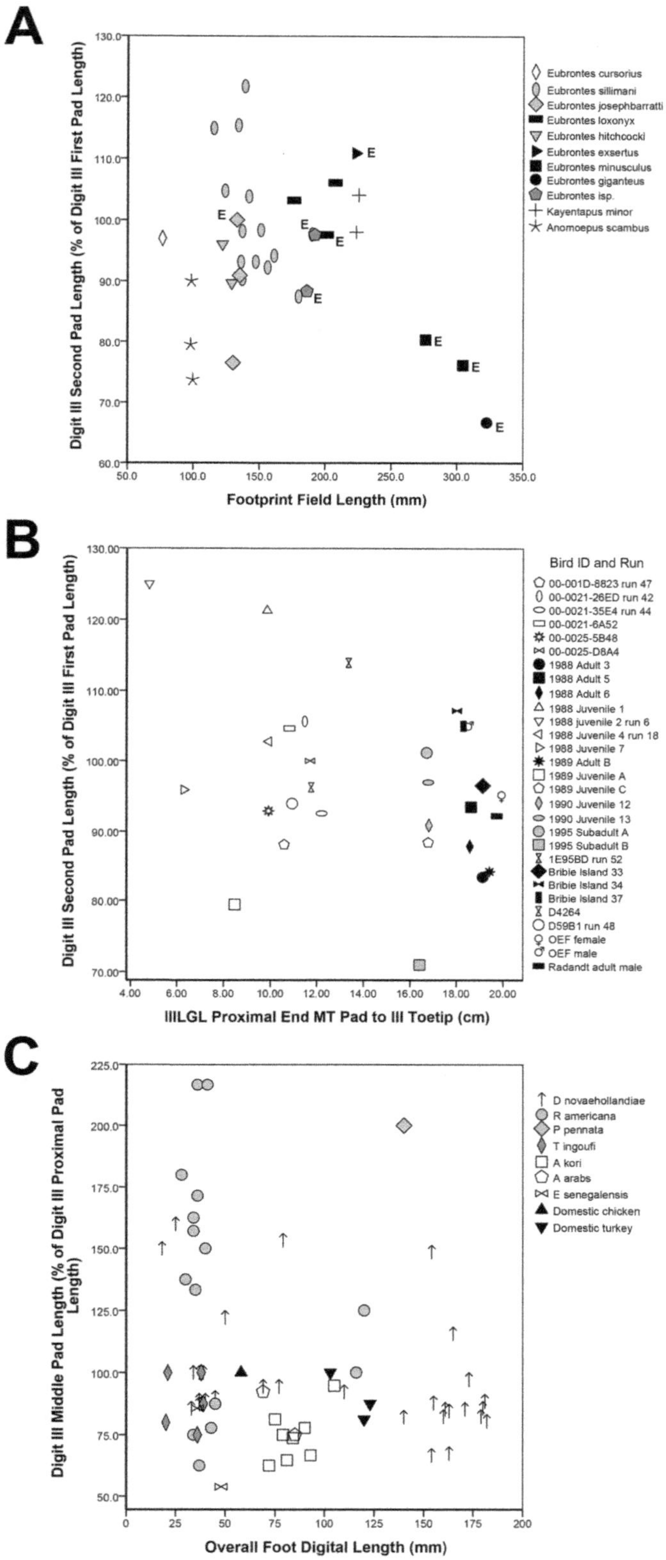

8.59. Comparison of the length of the second digital pad of digit III (scaled as a percentage of the length of the first digital pad of digit III) in non-avian dinosaur and bird footprints, and bird study skins, of different size. *A*, Non-avian dinosaur footprints. The relative length of the second digital pad of digit III as scaled here does not differ between members of the *"Anchisauripus"* and *"Eubrontes"* maximum information cluster (*P* of Mann-Whitney U test = .210). The scaled pad length looks as though it could decrease with increasing footprint size across all *Eubrontes*, but the parameters are not significantly correlated (Kendall's $\tau b = -0.166$, $P = .209$; Spearman's $\rho = -0.238$, $P = .214$; $N = 29$). Note the relatively short second pad of digit III of *Anomoepus*. *B*, Emu prints (emu means treatment). As with *Eubrontes*, there could be negative allometry between the two variables, but the correlation is not significant (Kendall's $\tau b = -0.193$, $P = .143$; Spearman's $\rho = -0.266$, $P = .163$; $N = 29$). *C*, Bird study skins provide stronger support for a negative correlation between the scaled length of the second digital pad and foot length in emus (Kendall's $\tau b = -0.317$, $P = .014$; Spearman's $\rho = -0.494$, $P = .005$; $N = 31$). For the entire sample of bird feet there may also be a negative correlation (Kendall's $\tau b = -0.196$, $P = .020$; Spearman's $\rho = -0.310$, $P = .010$; $N = 68$), but the relationship isn't very impressive.

between the two maximum information clusters. However, the claw lengths do not show significant negative allometry with overall footprint size across *Eubrontes* (table A8.23; cf. table A8.19), and the correlations between relative claw length (as presented in fig. 8.60A, 8.60*B*) and footprint length, while negative, are not statistically significant.

The "GM-scaled" claw lengths of emu footprints possibly (digit II) and definitely (digit III) increase with increasing print size (table A8.23; fig. 8.52*C*, 8.52*G*; cf. tables 8.8, 8.9; table A8.19). As scaled in figure 8.60 (8.60*C*, 8.60*D*), however, there is little or no relationship between footprint size and scaled digit II claw length, but there is a weak positive correlation between footprint size and scaled digit III claw length. In the corresponding comparison involving bird study skins, both for emus alone (fig. 8.60*E*) and across all the birds in my sample (fig. 8.60*F*), there is no significant relationship between scaled claw length and overall foot digit length (digit III length as measured on the study skin). On the other hand, skeletal data (fig. 8.56*B*, 8.56*C*) suggest an increase in the relative length of ungual III4 with increasing bird size, at least when the parameters are GM scaled.

For alligators, the GM-scaled lengths of the claws of digits I–III show a consistent decline in relative size with increasing alligator size (table A8.24; fig. 8.54*C*, 8.54*E*, 8.54*G*). The claw length of digit I shows significant negative allometry with respect to the length of digit I excluding the claw (but not with the length of the first digital pad of digit I; cf. fig. 8.61A), and the same may be true between the lengths of the non-ungual portions of digits II and III and their claws, although this is less certain (table 5.9).

All things considered, the decline in relative claw length between the "*Anchisauripus*" and "*Eubrontes*" maximum information clusters, and possibly (but less convincingly) across *Eubrontes* more generally, is the opposite of what is seen ontogenetically in emu footprints, but consistent with what is seen in the hindfeet from small to large alligators. If there is a decrease in relative claw length with increasing size across *Eubrontes*, this would not match what is seen across species in the bird skeletal samples.

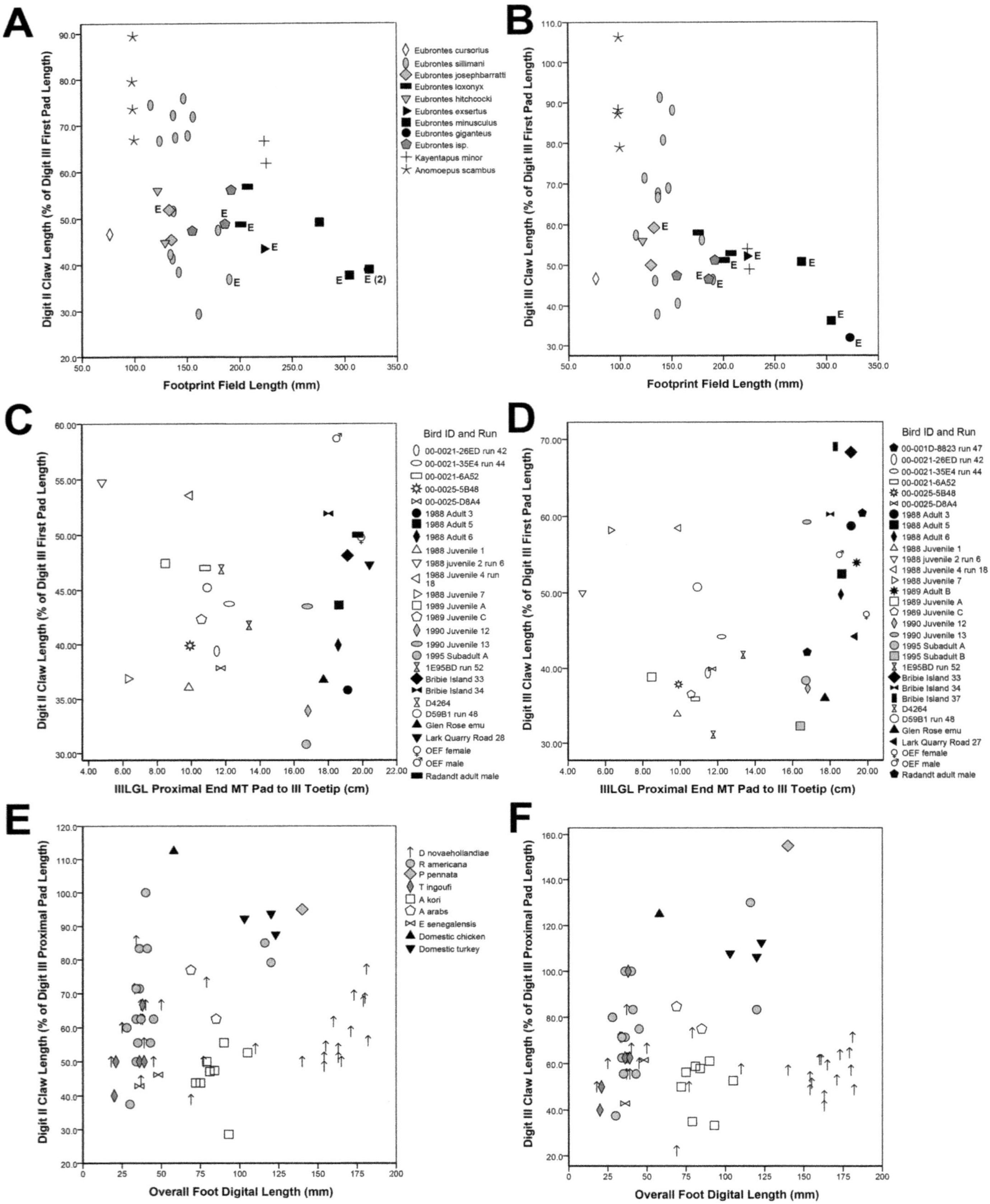
A
Digit II Claw Length (% of Digit III First Pad Length)
Footprint Field Length (mm)
Eubrontes cursorius
Eubrontes sillimani
Eubrontes josephbarratti
Eubrontes loxonyx
Eubrontes hitchcocki
Eubrontes exsertus
Eubrontes minusculus
Eubrontes giganteus
Eubrontes isp.
Kayentapus minor
Anomoepus scambus
B
Digit III Claw Length (% of Digit III First Pad Length)
Footprint Field Length (mm)
C
Digit II Claw Length (% of Digit III First Pad Length)
IIILGL Proximal End MT Pad to III Toetip (cm)
Bird ID and Run
00-0021-26ED run 42
00-0021-35E4 run 44
00-0021-6A52
00-0025-5B48
00-0025-D8A4
1988 Adult 3
1988 Adult 5
1988 Adult 6
1988 Juvenile 1
1988 juvenile 2 run 6
1988 Juvenile 4 run 18
1988 Juvenile 7
1989 Juvenile A
1989 Juvenile C
1990 Juvenile 12
1990 Juvenile 13
1995 Subadult A
1E95BD run 52
Bribie Island 33
Bribie Island 34
D4264
D59B1 run 48
Glen Rose emu
Lark Quarry Road 28
OEF female
OEF male
Radandt adult male
D
Digit III Claw Length (% of Digit III First Pad Length)
IIILGL Proximal End MT Pad to III Toetip (cm)
Bird ID and Run
00-001D-8823 run 47
00-0021-26ED run 42
00-0021-35E4 run 44
00-0021-6A52
00-0025-5B48
00-0025-D8A4
1988 Adult 3
1988 Adult 5
1988 Adult 6
1988 Juvenile 1
1988 juvenile 2 run 6
1988 Juvenile 4 run 18
1988 Juvenile 7
1989 Adult B
1989 Juvenile A
1989 Juvenile C
1990 Juvenile 12
1990 Juvenile 13
1995 Subadult A
1995 Subadult B
1E95BD run 52
Bribie Island 33
Bribie Island 34
Bribie Island 37
D4264
D59B1 run 48
Glen Rose emu
Lark Quarry Road 27
OEF female
OEF male
Radandt adult male
E
Digit II Claw Length (% of Digit III Proximal Pad Length)
Overall Foot Digital Length (mm)
D novaehollandiae
R americana
P pennata
T ingoufi
A kori
A arabs
E senegalensis
Domestic chicken
Domestic turkey
F
Digit III Claw Length (% of Digit III Proximal Pad Length)
Overall Foot Digital Length (mm)

Facing, **8.60.** Comparison of the lengths of the claws of digits II and III (scaled as a percentage of the length of the first digital pad of digit III) in non-avian dinosaur and bird footprints, and bird study skins, of different sizes. *A, B,* Digits II and III claw lengths in non-avian dinosaur footprints. As scaled here, the relative claw lengths are significantly different between the *"Anchisauripus"* and *"Eubrontes"* maximum information clusters (P values for Mann-Whitney U test = .005 [digit II claw length] and .030 [digit III claw length]). The correlations between relative claw length (as presented here) and footprint length across *Eubrontes*, however, while negative, are not statistically significant (for neither digit claw is the P value of the Kendall's τb or Spearman's $\rho < .05$). *Anomoepus* has particularly long claws on both digits II and III, and *Kayentapus*, relatively long claws on digit II. *C, D,* Digits II and III claw lengths in emu footprints (IIILGL = length of the footprint from the proximal end of the metatarsal [MT] pad to the tip of digit III). The length of the digit II claw length as scaled here does not show a significant correlation with footprint size, but the scaled length of the digit III claw length shows a weak positive correlation (Kendall's $\tau b = 0.295$, $P = .020$; Spearman's $\rho = 0.442$, $P = .013$; $N = 31$). *E, F,* Bird study skins. Neither emus alone nor all birds combined show significant relationships (P value of the Kendall's τb or Spearman's $\rho < .05$) for the relationship between digit III length (= "overall foot digital length" as measured on the study skin) and either claw length as scaled here.

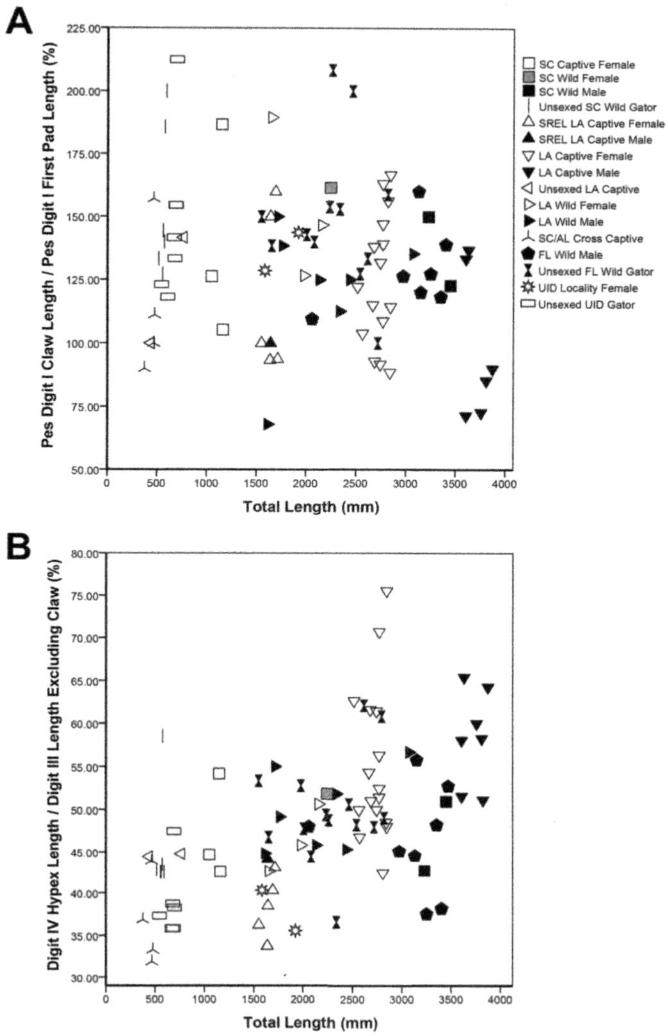

8.61. Scaled hindfoot proportions in alligators. *A,* Length of the claw of digit I (scaled as a percentage of the first digital pad of digit I) as a function of alligator total length. Neither Kendall's τb nor Spearman's ρ shows a significant correlation between claw length as scaled here and alligator size. *B,* Length of the medial hypex of digit IV scaled as a percentage of digit III length (excluding the claw) as a function of alligator total length; the correlation is positive (Kendall's $\tau b = 0.439$, $P < .001$; Spearman's $\rho = 0.603$, $P < .001$; $N = 81$).

Anomoepus has particularly long claws on both digits II and III, essentially an extrapolation of the trend seen across *Eubrontes* (table 8.6; figs. 8.20*E*, 8.21*G*, 8.60*A*, 8.60*B*). *Kayentapus* has relatively long claws on digit II, but not digit III.

Digit IV Length (figs. 8.22*A*, 8.33*C*–8.33*F*, 8.34, 8.35, 8.52*H*, 8.54*H*, 8.56*D*, 8.56*E*, 8.61*B*, 8.62). The "GM-scaled" digit IV length does not differ significantly between the "*Eubrontes*" and the "*Anchisauripus*" maximum information clusters (table 8.5); scaled as a percentage of the length of digit III, digit IV length does not quite show a statistically significant difference between the two clusters. Across *Eubrontes*, the "GM-scaled" digit IV length does show significant positive allometry with at least some "GM-scaled" measures of overall footprint size (tables A8.22, A8.23), and digit IV length shows positive allometry with digit III length in a simple bivariate relationship (table A8.19). Intriguingly, there is a strong positive correlation between "GM-scaled" digit IV length and overall footprint size within the "*Eubrontes*" maximum information cluster (fig. 8.62A).

Emu footprints, like *Eubrontes* (especially the "*Eubrontes*" maximum information cluster), show a clear tendency for multivariate-scaled digit IV to become relatively longer with increasing footprint size (tables 8.8, A8.23; figs. 8.52*H*, 8.62*C*). Digit IV length scaled as a percentage of digit III length, however, does not show a positive correlation with overall footprint length (fig. 8.62*D*), nor is there allometry between digit IV length and digit III length in a simple bivariate relationship (table A8.19). In contrast, study skins perversely suggest a possible negative correlation between digit IV length (scaled as a percentage of digit III length), both within emus and across my ground bird sample (fig. 8.62*E*). Against that, the much more extensive skeletal data (fig. 8.56*D*, 8.56*E*) suggest an increase in the relative length of digit IV across bird species (the emu skeletal data span too small a size range to say much about what happens within this species), at least if one excludes ostriches from consideration.

Alligators show a clear tendency for digit IV hypex length to become relatively longer from little to big animals (tables 5.9, A8.24; figs. 8.54*H*, 8.57), and crocodylians may show the same trend across species.

Bearing in mind the caveat that in this study digit IV length was not measured the same way in non-avian dinosaur and bird footprints, and alligator feet, the increase in the relative length of digit IV with increasing animal size is one of the stronger, more consistent signals seen (albeit not without exception in some comparisons), both within

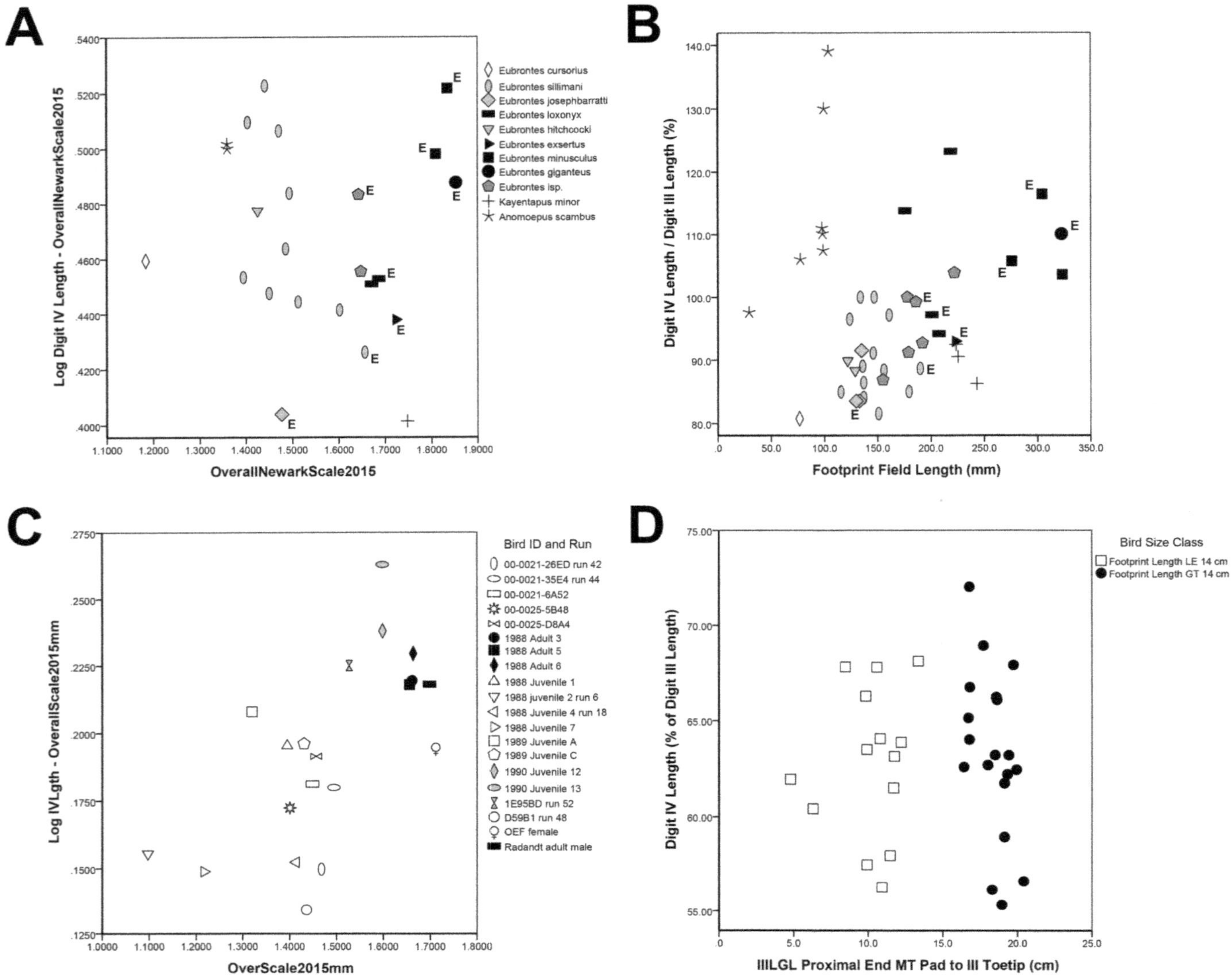

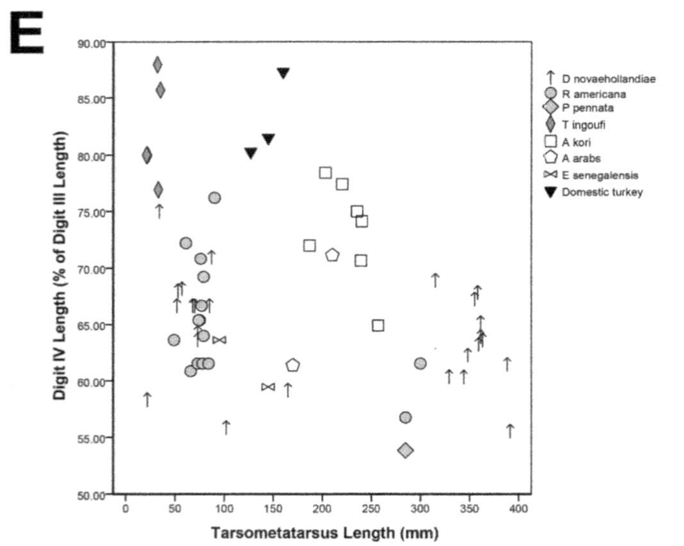

8.62. Scaled digit IV length in non-avian dinosaur and bird footprints, and bird study skins. *A*, *B*, Relative digit IV length of dinosaur prints. *A*, Digit IV length "GM-scaled" against footprint size ("OverallNewarkScale2015") as in figure 8.22*A*. Across *Eubrontes* the relationship shows no correlation, but there is a strong positive correlation between scaled digit IV length and footprint size among cases in the "*Eubrontes*" maximum information cluster ($r = 0.820$, $P = .013$, $N = 8$). *B*, Digit IV length (scaled as a percentage of digit III length) as a function of footprint length. The ratio comes close to being significantly different between members of the "*Anchisauripus*" and "*Eubrontes*" maximum information cluster (*P* of Mann-Whitney U test = .053). The two variables are positively correlated across *Eubrontes* (Kendall's $\tau b = 0.464$, $P < .001$; Spearman's $\rho = 0.636$, $P < .001$; $N = 33$). In both *A* and *B*, *Anomoepus* shows a relatively long digit IV, whereas *Kayentapus* has a relatively short digit IV. *C*, *D*, Relative digit IV length in emu footprints. *C*, Digit IV (IVLgth) "GM-scaled" against footprint size ("OverScale2015mm") as in figure 8.52*H*. The two variables are positively correlated ($r = 0.642$, $P = .002$, $N = 20$). *D*, Digit IV length as a percentage of digit III length shows no change with increasing footprint length (IIILGL). *E*, Bird study skins. Emus come close to having a weak but significant negative relationship between digit IV length (as a percentage of digit III length) and a proxy for bird size, tarsometatarsus length (Kendall's $\tau b = -0.261$, $P = .071$; Spearman's $\rho = -0.375$, $P = .065$, $N = 25$). Across all the bird species in the sample, the negative correlation remains weak but is significant (Kendall's $\tau b = -0.225$, $P = .012$; Spearman's $\rho = -0.323$, $P = .012$; $n = 60$). GM = geometric mean.

and in some comparisons across species. The strong positive correlation between "GM-scaled" digit IV length and overall footprint size in the "*Eubrontes*" maximum information cluster is especially striking. Could this be an intraspecific size-related trend within the cluster?

Anomoepus is strikingly different from *Eubrontes* in having a relatively long digit IV, despite being a small footprint (table 8.6; figs. 8.22A, 8.62A, 8.62B). *Kayentapus*, in contrast, has a relatively short digit IV.

Distances Between the Proximal Ends of Digit Impressions (figs. 8.22B–8.22D, 8.29, 8.37A, 8.53A–8.53C, 8.55B, 8.63). The distances between the proximal ends of the digit impressions (especially III–IV and II–IV) in non-avian dinosaur footprints are as much or more measures of footprint length as footprint width (fig. 8.16), but do contribute to the breadth across prints. The "GM-scaled" distance between the proximal ends of digits II and III (toebase II–III) is significantly different in the "*Eubrontes*" than the "*Anchisauripus*" maximum information cluster, taking larger values in the former (table 8.5); scaled as a percentage of backfoot length, a significant difference in the relative toebase II–III distance between the two clusters continues to hold true (fig. 8.63A). The toebase II–III distance becomes relatively longer with increasing footprint size across all *Eubrontes* (tables A8.19, A8.22, A8.23).

There is no difference in the "GM-scaled" toebase III–IV distance between the "*Anchisauripus*" and "*Eubrontes*" maximum information clusters, nor is there any relative change in this distance across *Eubrontes* (see preceding list of tables).

The "GM-scaled" toebase II–IV distance is longer in the "*Anchisauripus*" than the "*Eubrontes*" maximum information cluster (table 8.5), and the toebase II–IV distance differs significantly between the two clusters when examined in an ANCOVA with backfoot length as the covariate (table 8.4), or when the toebase II–IV distance is scaled as a percentage of backfoot length (fig. 8.63*B*). Whether and how the relative length of the toebase II–IV distance changes with increasing size across *Eubrontes* depends on how the parameter is scaled (again, same set of tables and figures). Keep in mind here that the toebase II–IV distance in the non-avian dinosaur prints is more a dimension of footprint length than width (fig. 8.16).

The intraspecific trend in emu footprints is strikingly different from what is seen in *Eubrontes*. Toebase distances II–III and II–IV—which, as measured in emu footprints, are more nearly associated with footprint breadth than is true of *Eubrontes* (fig. 6.4)—are relatively greater in footprints of small than large emus in most, although not all, ways of scaling the distances, and show negative correlations between toebase distance and footprint size; the same is true for some measures of the toebase III–IV distance (tables 8.8, 8.9, A8.19, A8.23; fig. 8.63C). The relative decrease in the toebase II–IV distance with increasing size in emus differs from what is seen among all the bird footprints examined in this study, in which the toebase II–IV distance increases with increasing print size (fig. 8.55*B*, 8.63*D*).

Thus the differences between the "*Anchisauripus*" and "*Eubrontes*" maximum information clusters are emphatically not consistent with what is seen intraspecifically in emu footprints: the proximal ends of emu footprints become relatively narrower with increasing size, while those of *Eubrontes* become relatively broader. The *Eubrontes* pattern is more like what occurs across bird species.

Anomoepus combines a relatively narrow toebase II–III distance with relatively long toebase II–IV distance (figs. 8.22*B*, 8.22*D*, 8.63*A*, 8.63*B*). *Kayentapus* has relatively long toebase II–III and III–IV distances.

Undoubtedly contributing to the proximal widths of footprints are the widths of the most proximal digital pads. Members of the "*Eubrontes*" maximum information cluster have relatively broader digital pads than members of the "*Anchisauripus*" maximum information cluster (tables 8.4, 8.5; figs. 8.26A, 8.27, 8.28), and across *Eubrontes* toe pads become stouter with increasing size (table A8.23). Although I didn't measure digital pad widths in emu footprints, it is likely that the relative width of toe impressions in footprints would increase with increasing trackmaker size, as it seems to do in alligators, and also both within and across bird and non-avian dinosaur species (figs. 2.12, 2.13, 2.20–2.24, 4.4, 4.12, 5.2, 8.56A), and so this is not a useful consideration in deciding if a size-related increase in toe impression width is an intraspecific vs. an interspecific trend.

Toetip Distances (figs. 8.22E–8.22G, 8.26B, 8.30, 8.31, 8.37B–8.37D, 8.53D–8.53F, 8.55C, 8.64). "GM-scaled" toetip distances do not differ significantly between the "*Anchisauripus*" and "*Eubrontes*" maximum clusters, although the toetip II–IV width comes close to being significant, with "*Eubrontes*" having relatively broader toetip widths than "*Anchisauripus*" (table 8.5). There is strong positive allometry between footprint size and toetip II–IV width across *Eubrontes* as a whole (tables A8.19, A8.22, A8.23). Scaled against backfoot length (table 8.4; fig. 8.64A), "*Eubrontes*" shows a significant difference from "*Anchisauripus*" only in the toetip II–IV width, with "*Eubrontes*" being relatively broader than "*Anchisauripus*."

Because IDA II–IV is correlated with relative footprint width (fig. 8.38), the angle increases with increasing footprint size across *Eubrontes* (fig. 8.39A).

The "GM-scaled" toetip II–IV width is less in large than in small emu footprints (table 8.8), and nearly all toetip distances show negative allometry with measures of footprint

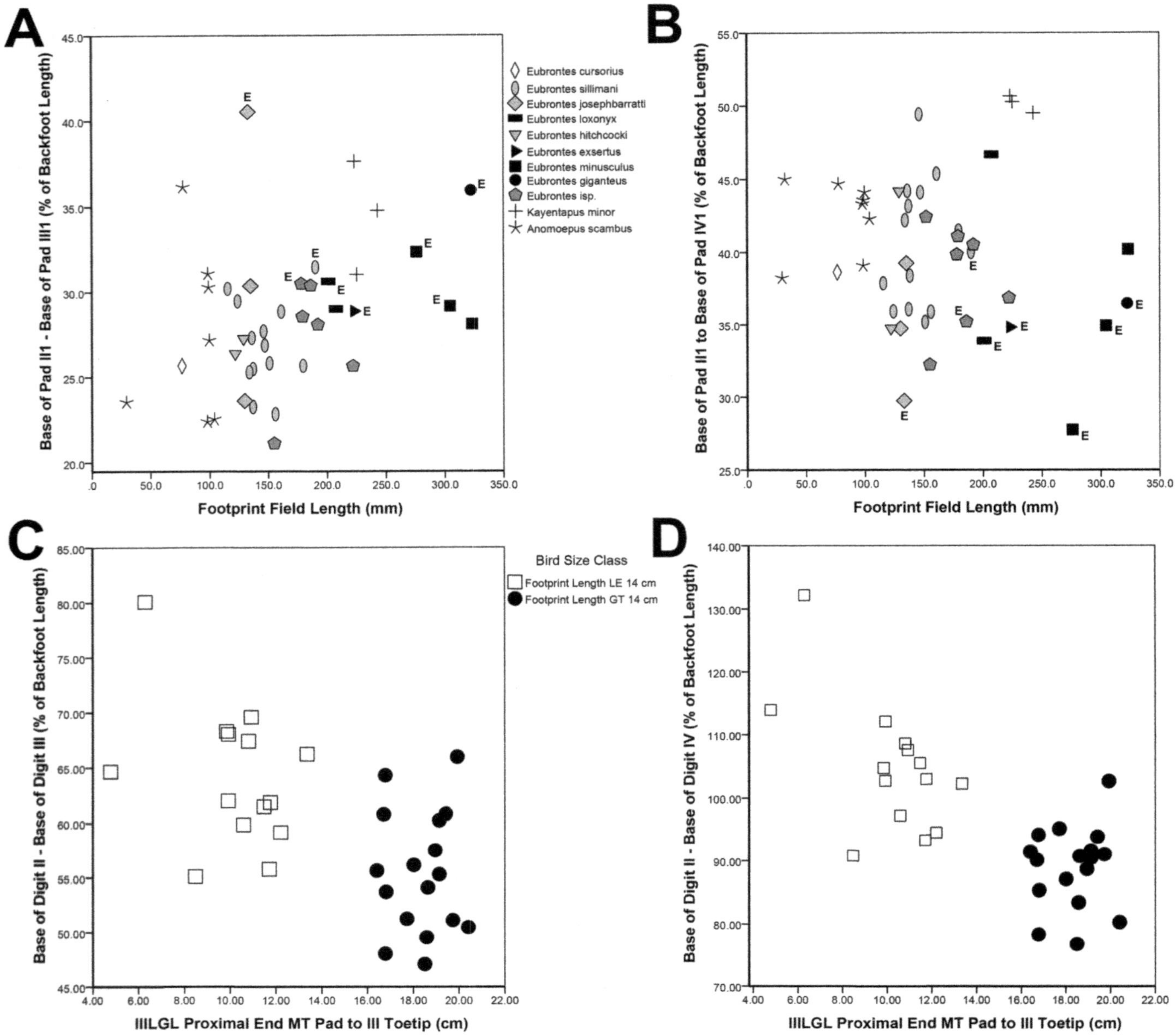

8.63. Distance between the proximal ends of digit impressions (toebase distances), scaled as percentages of backfoot length, as a function of size in non-avian dinosaur and emu footprints. *A*, Toebase II–III distance in non-avian dinosaur prints. The scaled distance is significantly different between members of the *"Anchisauripus"* and *"Eubrontes"* maximum information cluster (*P* of Mann-Whitney U test < .001). There may be a weak positive correlation across *Eubrontes*, but the relationship is not quite statistically significant (Kendall's $\tau b = 0.214$, $P = .080$; Spearman's $\rho = 0.305$, $P = .08$, $N = 33$). *Kayentapus* and some *Anomoepus* have a relatively long toebase II–III distance compared with *Eubrontes*. *B*, Toebase II–IV distance in non-avian dinosaur prints. The scaled distance is significantly different between members of the *"Anchisauripus"* and *"Eubrontes"* maximum information cluster (*P* of Mann-Whitney U test = .008). *Eubrontes* otherwise shows no size-related trend in this comparison. Both *Anomoepus* and *Kayentapus* have a relatively long toebase II–IV distance. *C*, Toebase II–III distance in emu prints shows a decrease in relative size with increasing footprint length (IIILGL) (Kendall's $\tau b = -0.390$, $P = .002$; Spearman's $\rho = -0.553$, $P = .001$; $N = 31$). D, Toebase II–IV distance in emu prints shows an even stronger negative correlation with footprint length (Kendall's $\tau b = -0.476$, $P < .001$; Spearman's $\rho = -0.657$, $P < .001$; $N = 25$).

length (tables A8.19, A8.23). Scaled against backfoot length, the toetip II–IV width is negatively correlated with footprint length (fig. 8.64*B*). As emu prints get bigger, they become proportionally narrower, and IDA II–IV becomes smaller (fig. 6.20). Thus as seen with breadth across the proximal end of the footprint, the toetip width:footprint size relationship seen intraspecifically in emus is opposite the pattern seen in *Eubrontes*. In contrast to the moderately strong positive correlation between "GM-scaled" toebase II–IV width and print size across bird species (fig. 8.55*B*), the "GM-scaled" toetip width shows a visually unimpressive but possibly negative correlation with footprint size (fig. 8.55C).

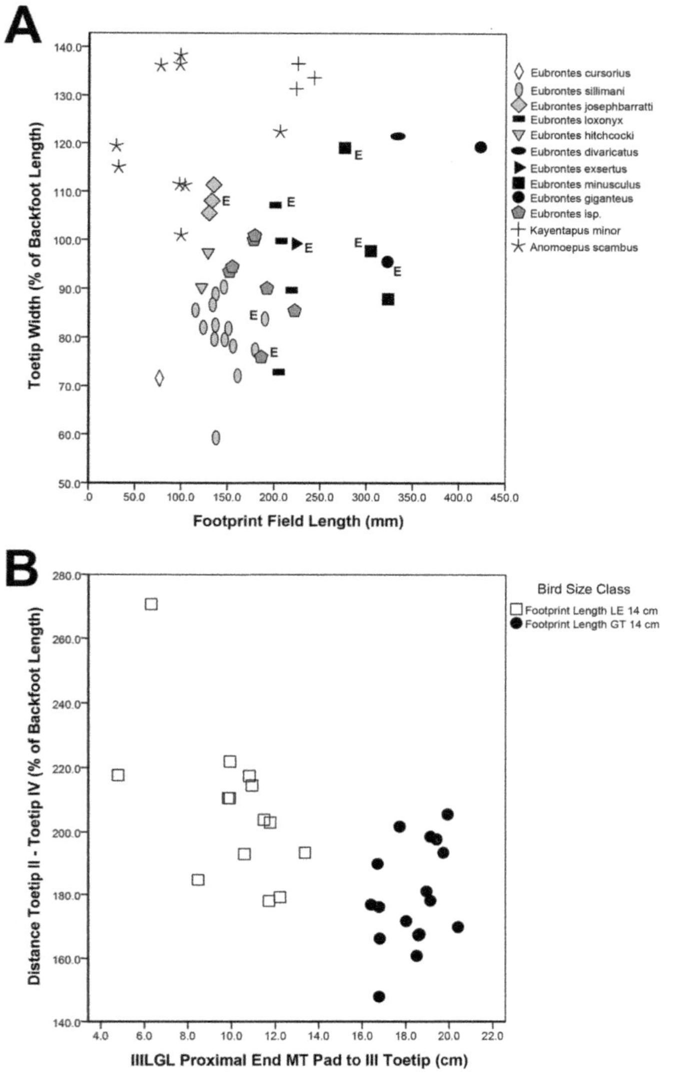

8.64. Toetip width (scaled as a percentage of backfoot length) as a function of footprint length in non-avian dinosaur and emu footprints. *A*, Dinosaur footprints. The scaled distance is significantly different between members of the *"Anchisauripus"* and *"Eubrontes"* maximum information clusters (*P* of Mann-Whitney U test = .025). Across *Eubrontes* a weak positive correlation between scaled toetip width and footprint length comes close to being significant (Kendall's $\tau b = 0.194$, $P = .087$; Spearman's $\rho = 0.286$, $P = .081$; $N = 38$). Both *Anomoepus* and *Kayentapus* have a relatively broad toetip width. *B*, Emu footprints show a significant negative correlation between scaled toetip width and footprint size (Kendall's $\tau b = -0.377$, $P = .003$; Spearman's $\rho = -0.523$, $P = .003$; $N = 31$).

Anomoepus and *Kayentapus* both show a very broad toetip II–IV width and a large IDA II–IV (figs. 8.38, 8.39, 8.64*A*). This is particularly noteworthy given the small size of *Anomoepus*, in contrast to what is seen in small *Eubrontes*.

Digit III Projection (figs. 6.42, 8.37*E*, 8.65). How far digit III projects beyond the tips of digits II and IV is a complex parameter; it is a function of the relative lengths of the metatarsals, the relative lengths of digits II–IV, and IDA II–IV. Both *Eubrontes* and emu prints show significant negative allometry between digit III projection and backfoot length (table A8.19; fig. 8.65), with the slope for *Eubrontes* differing more from isometry than that of emu prints. Despite this, with backfoot as the covariate, digit III projection does not differ significantly between members of the *"Anchisauripus"* and *"Eubrontes"* maximum information clusters (table 8.4) or between the small and large emu footprint size classes (table 8.9—although the difference here comes close to being significant).

Based on what is observed in footprints of ground birds, the relationship between digit III projection and footprint size generally does not seem to be a good discriminator between intraspecific and interspecific samples, because so many species plot along the same trend (figs. 6.42*A*, 6.42*C*, 8.65*C*, 8.65*D*). However, the relationship does separate "oddball" species from the main sequence of bird prints, and it is interesting to note that the largest bird footprints (some moa, dromornithids, and *Rivavipes*) have distinctly lower relative values of the digit III projection than one would extrapolate from the emu trend. In like manner, the larger forms of *Eubrontes* do not continue the trend defined by smaller forms of *Eubrontes* (fig. 8.65*B*). Consequently the relationship across *Eubrontes* looks more like an interspecific than an intraspecific pattern.

Anomoepus has a relatively shorter digit III projection than *Eubrontes*, but *Kayentapus* looks similar to *Eubrontes* in that regard.

Toe-Tapering Profiles (fig. 8.66). Plots of the cumulative lengths of phalanges against the distal widths of those phalanges were shown to have utility in distinguishing among foot skeletons of some major groups of dinosaurs, albeit with only limited success within those groups (chapter 3). Analogous toe-tapering profiles can be constructed for well-preserved dinosaur footprints. The lengths and widths of the digital pads, and lengths of the claws, in the profiles are scaled as percentages of the length of the first digital pad of digit III (which itself is, of course, fixed at a value of 100), but profiles are presented only for the impressions of digits II and III, because the impression of digit IV does not always show distinct boundaries between digital pads.

Within *Eubrontes*, for digit II there is a tendency for members of the *"Eubrontes"* maximum information cluster to have relatively greater lengths of the first digital pad, as well as the cumulative lengths of the two digital pads together, than members of the *"Anchisauripus"* maximum information cluster (fig. 8.66*B*), but the difference between the two clusters is lost when the length of the clawmark is added to the total. This is consistent with the relative lengths of the two pads and the claws when GM-scaled (fig. 8.20A, 8.20C, 8.20*E*). There is also a tendency for members of the

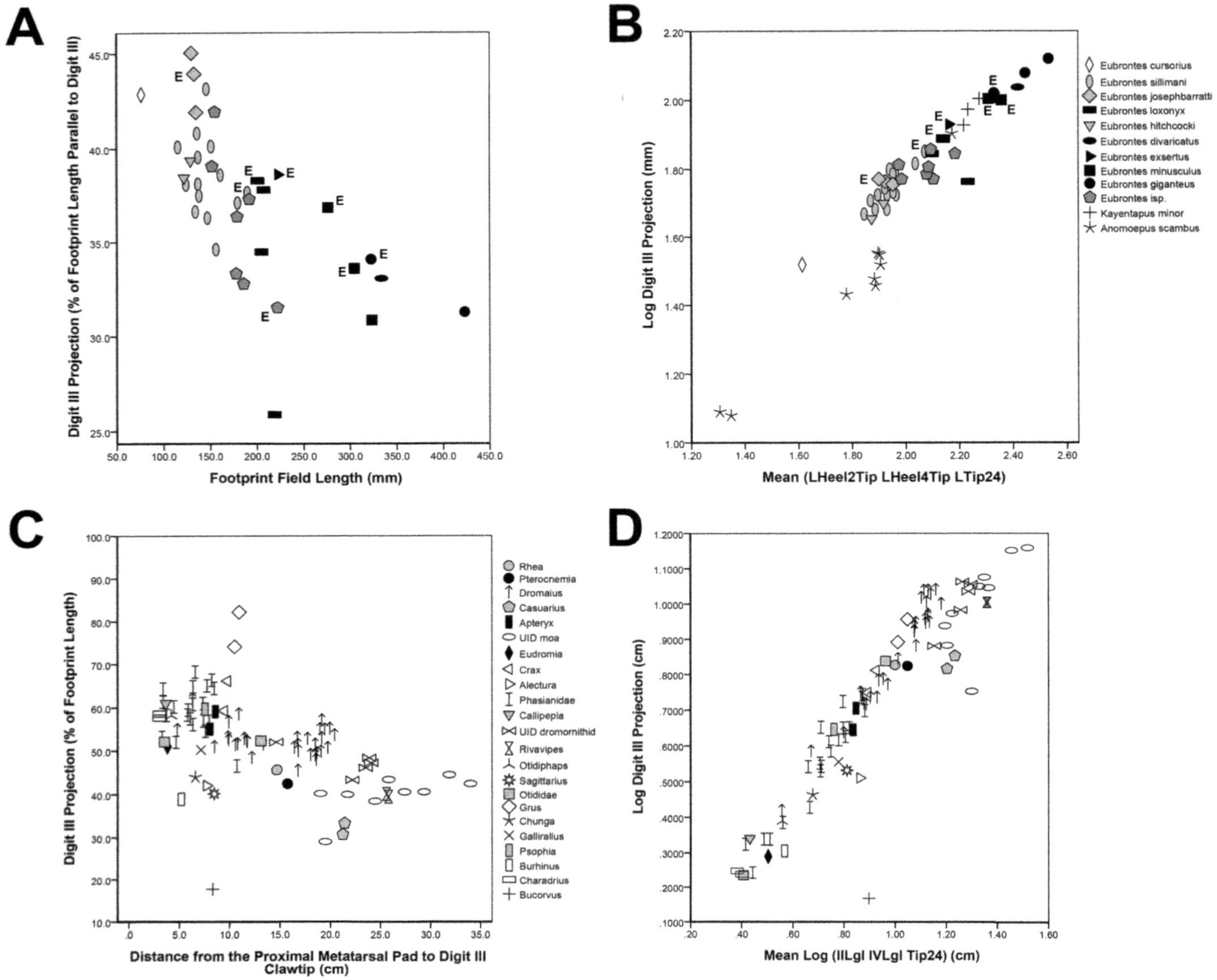

8.65. Digit III projection in footprints of non-avian dinosaurs (*A*, *B*) and ground birds (*C*, *D*). In *A* and *C*, digit III projection is expressed as a percentage of overall footprint length. In *B* and *D*, the log-transformed digit III projection is plotted against a proxy of overall footprint size, the mean of the log-transformed values of the distances from the proximal end of the metatarsal pad to the tips of digits II (LHeel2Tip in dinosaurs, IILgl in birds) and IV (LHeel4Tip in dinosaurs, IVLgl in birds), and the log-transformed distance between the tips of digits II and IV (toetip width). *A*, *Eubrontes*. Note the tendency for larger prints (including some footprints assigned to the "*Eubrontes*" maximum information cluster) to have relatively lower values of the digit III projection than smaller *Eubrontes*. *B*, In absolute terms, digit III projection increases as footprint size increases across *Eubrontes*, but note the tendency for the slope of the relationship to become less with increasing print size (the break point in the relationship is positioned at mean values of the proxy of footprint size of about 2). *Anomoepus* generally has a relatively low digit III projection, but *Kayentapus* falls among the *Eubrontes* cases. *C*, Apart from obvious outlier forms like cranes (*Grus*), cassowaries (*Casuarius*), and ground hornbill (*Bucorvus*), the various ground bird species show a tendency for the relative digit III projection to decrease with increasing print size. The intraspecific trend for emus looks rather flatter than the interspecific trend. *D*, Again apart from *Bucorvus*, most of the bird prints fall along a common trend. The emu trend is linear in this double-logarithmic plot. Note that the larger bird prints (cassowaries, moa, dromornithids, *Rivavipes*) fall away from the emu trend, indicating a relatively shorter digit III projection, much as was seen in large as opposed to smaller *Eubrontes* specimens (*B*).

"*Eubrontes*" cluster to have broader toe pads than members of the "*Anchisauripus*" cluster, again as was seen with GM scaling (fig. 8.20 *B*, 8.20*D*). However, the beauty of this pattern is spoiled by the behavior of one of the points for *Eubrontes giganteus* (naturally, it is the type specimen of the ichnospecies, AC 15/3; fig. 8.8), which looks more like a member of the "*Anchisauripus*" cluster (but not in other comparisons: figs. 8.24, 8.25). The toe-tapering profile for digit III shows less separation of digital pad relative lengths between members of the "*Anchisauripus*" and "*Eubrontes*" clusters than does digit II, but there remains a tendency for the first and third digital pads to be relatively broader for "*Eubrontes*" than "*Anchisauripus*" (cf. fig. 8.21*B*, 8.21*F*). Unfortunately, as with digit II, AC 15/3 has relatively narrow digital pads on digit III.

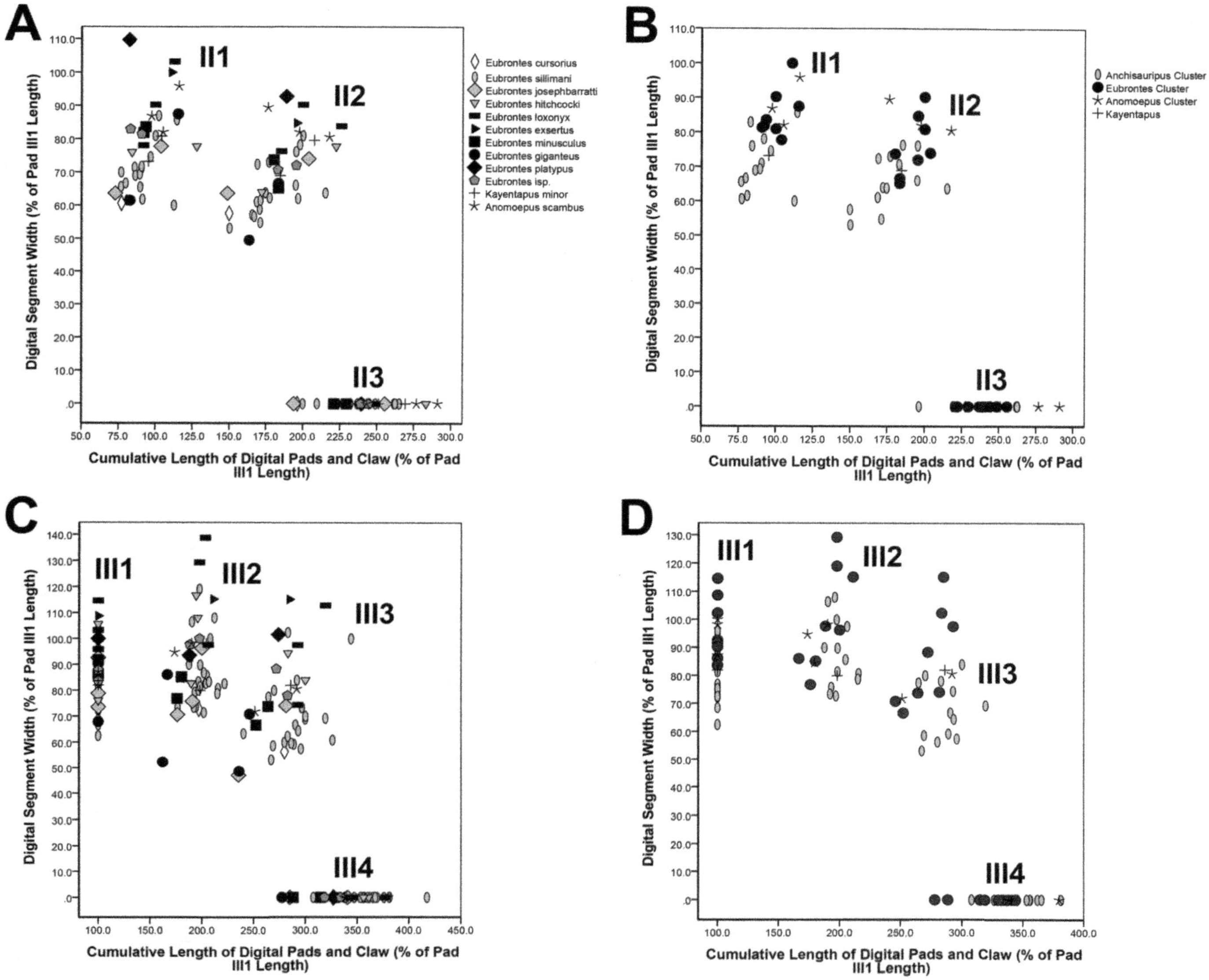

8.66. Toeprint tapering profiles of Newark Supergroup dinosaur footprints for digits II (*A*, *B*) and III (*C*, *D*), broken down by nominal ichnotaxon (*A*, *C*) and membership in the maximum information clusters (*B*, *D*).

Points for *Anomoepus* and *Kayentapus* plot among those of the "*Eubrontes*" cluster for digit II (cf. fig. 8.20), although the relatively long claws of *Anomoepus* give it an especially long total cumulative scaled length. *Anomoepus* and *Kayentapus* are not so much like the "*Eubrontes*" as opposed to the "*Anchisauripus*" cluster in the digit III toe-tapering profile (fig. 8.66*D*).

Trackway Parameters

Before examining trackway patterns, some limitations of my data have to be identified. Because it wasn't possible to determine which of the *Kayentapus* prints in my sample (fig. 8.14) went with which set of trackway measurements in Weems's (1992) description of these footprints, they will not be included in the numerical analyses that follow. Because of the way I usually measured footprint rotation with respect to the trackmaker's direction of travel (using compass bearings of prints; see discussion in chapter 6), it was not possible to make measurements of footprint rotation for several of the dinosaur trackways in museum collections.

Step Lengths (fig. 8.67). Big dinosaurs generally took longer steps than smaller dinosaurs—no surprise there. The pattern of relative step length (pace or stride divided by footprint length) shows the same pattern seen in dinosaur trackways more generally (fig. 4.15), with values generally declining with increasing footprint size at footprint lengths greater than about 10 cm. *Anomoepus* shows particularly short steps for its footprint length, not only in comparison with small forms

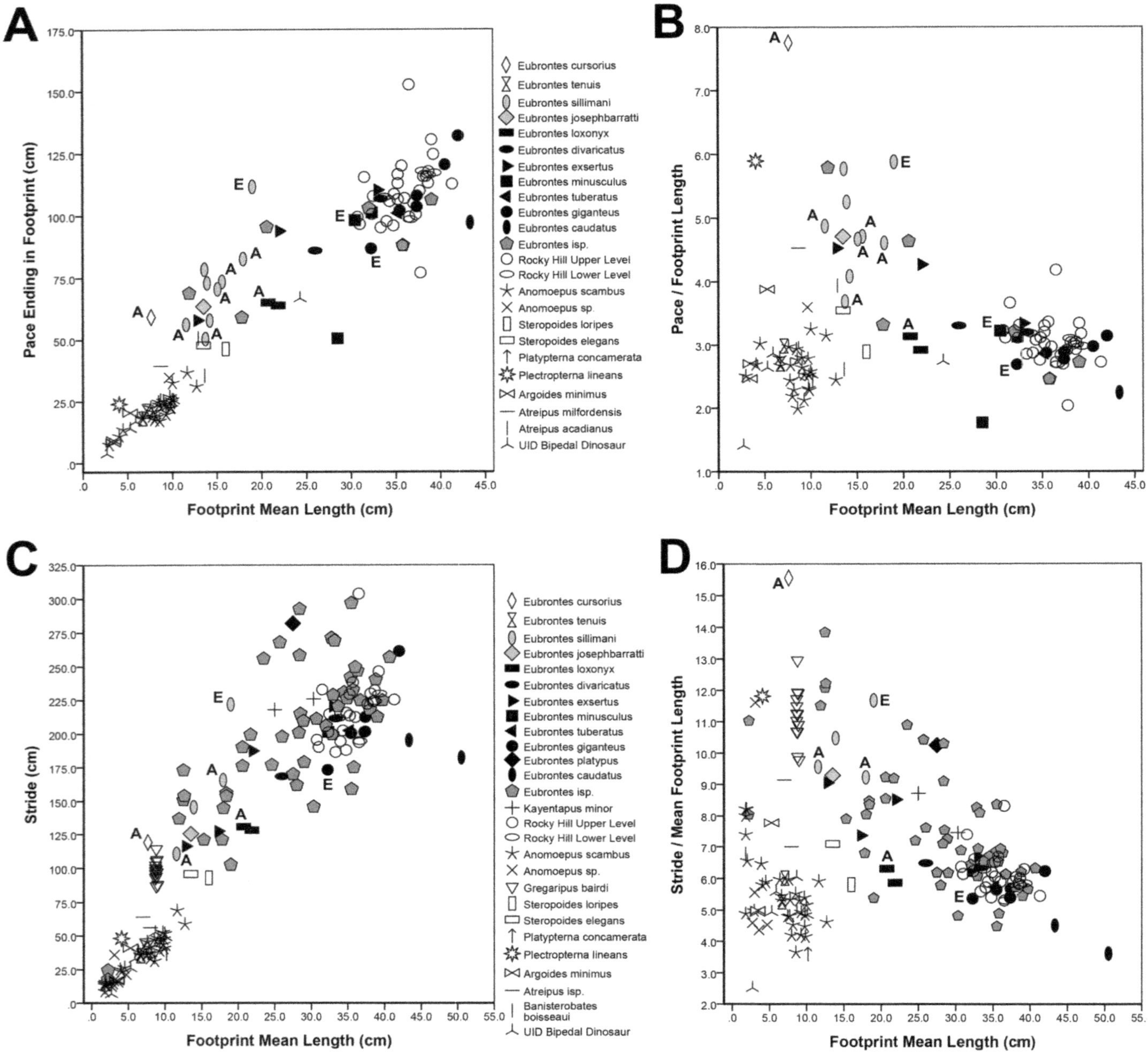

8.67. Mean step length as a function of mean footprint length of Newark Supergroup tridactyl dinosaur trackways (table A4.2). Weems's (1987) trackways of *Grallator* and *Eubrontes* are plotted here as *Eubrontes* isp. *A*, Pace length. *B*, Relative pace length. *C*, Stride length. *D*, Relative stride length.

of *Eubrontes*, but also compared with *Gregaripus*, which like *Anomoepus* is attributed to an ornithischian (Weems 1987—and note that Weems later [2006*a*] reinterpreted *Gregaripus* as a preservational variant of *Anomoepus*).

Pace Angulation (fig. 8.68A). Bipedal dinosaur trackways usually show nearly linear trackways, with little "zigzaggedness," and so have pace angulations approaching 180 degrees. Most of the *Eubrontes* trackways have pace angulations of 155 degrees or more. Values for *Anomoepus* overlap those of *Eubrontes*, but some *Anomoepus* trackways take lower values.

Footprint Rotation (fig. 8.68B). Large *Eubrontes* tend to show positive (outward) rotation of footprints with respect to the trackmaker's direction of travel, while *Anomoepus* generally shows negative (inward) rotation. Although I had no quantitative data for smaller *Eubrontes* trackways, inspection of photographs of such trackways (figs. 8.2, 8.5) suggests that they tend to show positive rotation.

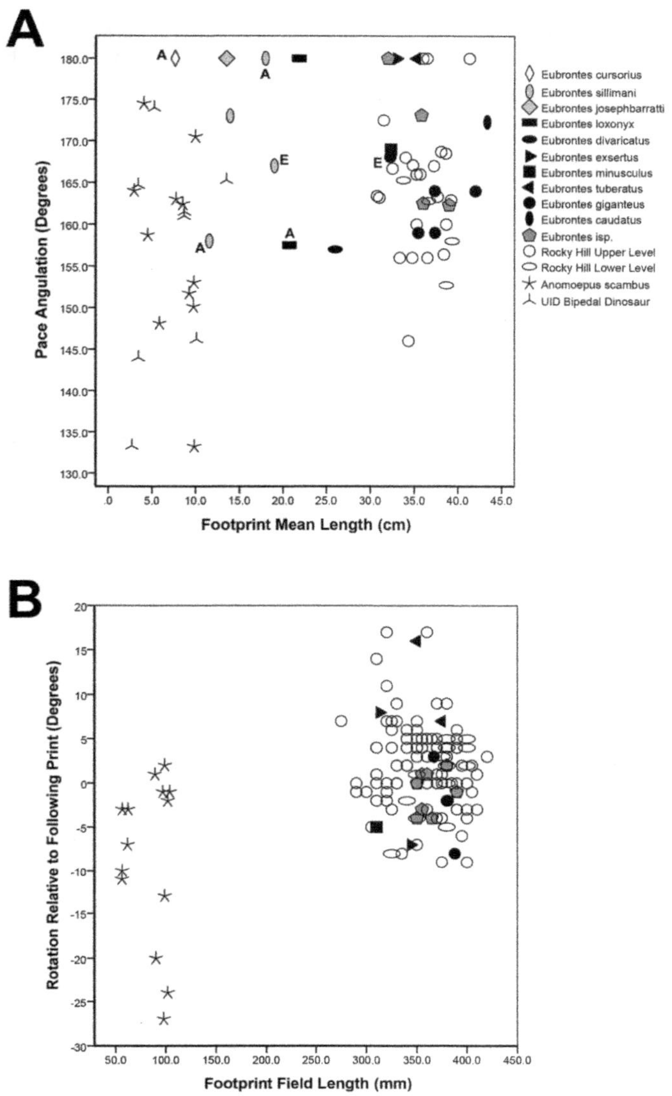

8.68. Trackway angles. Parameters are plotted against footprint lengths for convenience in illustration, not because of any inferred relationship between the variables. *A*, Pace angulation. *Anomoepus* shows a tendency to have slightly lower values than does *Eubrontes*. *B*, Footprint rotation. Unlike in *A*, data cases here are measurements for individual prints, not means for trackways. *Anomoepus* shows a strong tendency toward inward (negative) rotation of footprints with respect to the trackmaker's direction of travel, whereas the other tracks (all of them likely to be those of some species of *Eubrontes*) show a tendency for outward (positive) rotation of footprints.

How Many Kinds of Newark Trackmakers? Conclusions

We have now looked at the footprints and trackways in our sample in many different ways, and the time has come for me to go out on a limb, stick my neck out, or whatever other metaphor one prefers, about the number of kinds of dinosaurs represented. First the easy ones: *Anomoepus* and *Kayentapus* look different enough from *Eubrontes* that it is almost certain that their makers differed from the makers of *Eubrontes*, and very likely from each other as well. This conclusion should surprise no one, because it is consistent with what everybody has previously said, from Edward Hitchcock's time to the present. Dalman and Weems (2013), interestingly enough, think that there may be more than one valid ichnospecies of *Anomoepus*, and presumably more than one species of trackmaker.

The other matter, of course, is what to make of *Eubrontes*. Multiple measures of shape variability suggest that the amount of shape variability in footprints across this ichnogenus is greater than one would expect to see among feet and/or footprints of a single species, and some of the patterns of shape change with size across *Eubrontes* are not consistent with what is seen in ontogenetic series of feet or footprints of modern archosaurs, and/or are more like what is seen across species. The reasonably consistent differences in proportions between the "*Anchisauripus*" and "*Eubrontes*" maximum information clusters lead me to think that there were at least two different kinds of trackmakers represented by *Eubrontes*, one of them responsible for the "*Anchisauripus*" cluster, and the other for the "*Eubrontes*" cluster. The features in which they differ are described in some detail in table 8.6, but can be summarized here: the "*Eubrontes*"-maker had a relatively broad print, relatively stout toes with short claws, and a relatively long first digital pad of digit II, while the "*Anchisauripus*"-maker had relatively slim toes with long claws and relatively long distal digital pads on digit III. The two trackmaker kinds likely differed in adult size, with the "*Eubrontes*"-maker being bigger than the "*Anchisauripus*"-maker.

As mentioned at the outset of this chapter, my conclusions about there being at least three distinct kinds of tridactyl dinosaur trackmaker in the Newark Supergroup ichnofauna are consistent with what Demathieu (1990) concluded about the three then-recognized ichnogenera *Grallator*, *Eubrontes*, and *Anomoepus*. The characters that Demathieu employed in his study included overall lengths of the print, lengths of digits II–III, digit III projection, footprint width, and interdigital angle. He analyzed the linear dimensions as ratios (III length/II length, III length/IV length, IV length/II length, III length/digit III projection, footprint length/width, footprint length/digit III length, footprint length/digit III projection), calculating the means, standard deviations, and coefficients of variation of the various ratios. On this basis, Demathieu concluded that what were then called *Anomoepus intermedius*, *Eubrontes giganteus*, and *Grallator sillimani* were distinctly different from each other. *Anomoepus* had a wide IDA II–IV, rather broad toemarks of triangular shape, and rather similar lengths of the impressions of digits II–IV. *Eubrontes* was characterized by relatively stout toes, rather broad prints, digit impressions of rather similar length, and relatively short clawmarks. *Grallator* had slender toemarks, low IDA II–IV, relatively long clawmarks, and a relatively long digit III projection. Demathieu's interpretations are consistent with my more detailed analyses. More recently, Wagensommer et al. (2016) used Weems plots (Weems 1992) to compare small tridactyl footprints from the Lower Jurassic Etjo Formation of Namibia with Newark Supergroup *Grallator, Anchisauripus*, and *Eubrontes*. Their results led them to conclude that *Grallator* and *Eubrontes* are distinctly different from each other, and that *Anchisauripus* might be a variant of *Grallator*. These conclusions also seem consistent with my results.

Castanera et al. (2015) used geometric morphometrics to analyze shapes of tridactyl dinosaur footprints from the Early Cretaceous Huérteles Formation of Spain, and to compare those prints with a few specimens of classic *Grallator, Anchisauripus*, and *Eubrontes* (in the traditional usage of names) from the Newark Supergroup. Landmarks used in this analysis comprised the toetips, "heel," and locations related to the widths across the first digital pads of digits II and IV, and the second digital pad of digit II. Interestingly, these authors found that one of the two *Anchisauripus* prints (one of the prints in AC 9/14) in their sample could not be distinguished from *Grallator* (AC 1/4a), and that two large *Eubrontes* (AC 45/1 and AC 15/3) were quite different from *Grallator* and *Anchisauripus*. These results, too, seem consistent with my findings.

Lockley and Hunt (1995) presented a further reason for thinking that large (*Eubrontes* in their usage) and small (*Grallator* in their usage) versions of *Eubrontes* (in the broad sense used here) from the western U.S. might represent different taxa of trackmaker: "In the western United States . . . no tracksites reveal obvious intermediate tracks between *Grallator* and *Eubrontes* . . . what is termed *Anchisauripus* in the East is missing, or at least has not been identified in the West. We therefore conclude that *Grallator* and *Eubrontes* probably do not belong to the same species" (Lockley and Hunt 1995: 120).

I offer the conclusion that the makers of big and little *Eubrontes* were different kinds of dinosaur as an hypothesis for further testing rather than a firm conclusion. If future measurements of well-preserved specimens continue to support the distinction between the two *Eubrontes* morphotypes associated with my maximum information clusters, this will bolster confidence that the differences between them are not artifacts of extramorphological variability. And while we are wishing for new discoveries of beautifully preserved footprints, we might as well go all out and also wish for discoveries of well-preserved skeletons. What would really support my hypothesis is discovery of some bipedal, functionally tridactyl Early Jurassic dinosaurs from eastern North America that differ in size and skeletal morphology from each other, that have foot skeletons that match the proportions of the "*Anchisauripus*" and "*Eubrontes*" morphotypes, and that through unambiguous indications of skeletal maturity (Padian and Lamm 2013) show that the smaller form was nearly or fully grown. This is a good place to mourn the loss of *Podokesaurus* (Lull 1953).

WHAT KINDS OF DINOSAURS MADE THE FOOTPRINTS?

Differences in pedal digital and phalangeal proportions of foot skeletons among groups of potential dinosaurian makers of tridactyl footprints were discussed in some detail in chapters 2–4. To apply this information to footprints requires a bit of data massage, however. The asymmetrical shape of the "heel" end of typical tridactyl dinosaur prints suggested to Baird (1957: 459) that the most proximal phalanges of digits II–IV of theropod dinosaurs did not routinely impress the substrate to the same degree. In most species of "*Grallator* and in *Anchisauripus* and some species of *Eubrontes* . . . the metatarso-phalangeal pad of digit II impresses less deeply [than that of digit IV] or not at all." Thulborn (1990: 115) likewise observed that the metatarsophalangeal pads of digits II and III impressed "less commonly" than that of digit IV. Baird and Thulborn agreed with some earlier workers that the digital pads (nodes) of footprints likely corresponded with the joints between phalanges, or between phalanges and metatarsals.

Farlow (2001: 413) accordingly proposed that "the impression of digit II in a theropod footprint can be approximated by adding half the length of phalanx II1 to the combined lengths of phalanges II2 and II3. In like manner . . . the toemark of digit III would consist of half the length of phalanx III1 plus the combined lengths of III2-III4." The length of digit IV in a footprint would consist of the combined lengths of all five phalanges of that digit, plus a bit more associated with the distal end of metatarsal IV.

These lengths would, of course, be underestimates to the extent that the horny toe claws extended beyond the terminal ends of their underlying unguals, which seems to be quite variable among species of extant crocodylians and ground birds (table A5.1; figs. 1.4*H*, 1.4*I*, 1.5, 5.3, 6.3*A*). They would also be underestimates to the extent that they exclude soft tissues at the joints between pedal phalanges. However, judging from x-ray studies of intact feet (figs. 1.5, 5.3*A*, 5.3*B*, 6.3*A*) this is likely a minor source of error. Furthermore, my observations of phalangeal foot skeletons of non-avian dinosaurs preserved in articulation suggest very little gap between the distal and proximal ends of adjoining phalanges involved in toe joints.

Proxies for the lengths of the fleshy (non-claw) toemarks in footprints can be estimated, on the assumption that the final digital pad in the toemark underlay the joint between the last non-ungual phalanx and the ungual phalanx, as half the length of the first phalanx plus the combined lengths of the remaining non-ungual phalanges, and half the length of the ungual phalanx. The proxy for the claw length in the footprint would then be half the length of the ungual.

Digit I is a little tricky. It usually did not register in tridactyl dinosaur footprints, but as we shall see, its relative length does have some bearing on identifying likely trackmakers. I will therefore simply use the aggregate length of phalanges I1 and I2 in my comparisons.

The only skeletal proxies for measures of overall footprint width, or the widths of the individual toemarks, are derived from the transverse widths of phalanges (cf. figs. 1.5, 5.3, 6.3). The greatest widths of toemarks will usually be the widths of the first (most proximal) digital pads. For digits II and III, the skeletal proxy for this width is the distal transverse width of the first phalanx of each of those digits. For digit IV the appropriate skeletal proxy for the width of the first digital pad would be either the distal transverse width of the fourth metatarsal (which I didn't measure, alas) or the proximal transverse width of the first phalanx (which I didn't include for all specimens in the database). Lacking those transverse measurements, the skeletal proxy for digit IV width that I will use is the distal transverse width of phalanx IV1, which—in those footprints in which creases separating the digital pads of digit IV are nice and distinct—would likely be a proxy for the width of the second digital pad of digit IV.

The skeletal proxy for footprint width that I will use is a proxy for the distance from the bend point on the inner (medial) edge of the proximal end of the digit II impression (fig. 8.16, item 3) to the midline of the digit III impression. This will be the distal transverse width of phalanx II1 plus half the distal transverse width of phalanx III1. I grant that this is not a completely satisfactory proxy for footprint width—it is, more accurately, a proxy for about half of the width of the proximal end of the print. But this seems to be the best skeletal proxy for footprint width to be had from my measurements of pedal phalanges.

Summarizing, then, the skeletal proxies for footprint measurements (apart from individual digit impression widths):

- Digit II first pad length proxy: ½ phalanx II1 length + ½ phalanx II2 length
- Digit III first pad length proxy: ½ phalanx III1 length + ½ phalanx III2 length
- Digit III second pad length proxy: ½ phalanx III2 length + ½ phalanx III3 length
- Digit II length proxy: ½ phalanx II1 length + phalanx II2 length + phalanx II3 length
- Digit II length proxy excluding claw: ½ phalanx II1 length + phalanx II2 length + ½ phalanx II3 length
- Digit II claw length proxy: ½ phalanx II3 length
- Digit III length proxy: ½ phalanx III1 length + phalanx III2 length + phalanx III3 length + phalanx III4 length
- Digit III length proxy excluding claw: ½ phalanx III1 length + phalanx III2 length + phalanx III3 length + ½ phalanx III4 length
- Digit III claw length proxy: ½ phalanx III4 length
- Proxy for distance from digit II bend point to digit III midline: phalanx II1 distal width + ½ phalanx III1 distal width

Obviously these proxies do not exactly approximate the dimensions of the footprint measures for which they serve as stand-ins. Digital pads may have been arthral in location, but their lengths would not necessarily be exactly equal to the sum of half of the lengths of the phalanges which they incorporated in the intact foot. But these are, in my opinion, the best—maybe the only—way to make comparisons between foot skeletons and footprints. Interestingly, Dalman and Weems (2013) did rather similar calculations, but in the opposite direction, estimating phalangeal and ungual proportions from measurements of footprints.

I will examine these various footprint dimension proxies among candidates for the makers of the Newark Supergroup footprints. There are limitations in the available data that must be acknowledged up front. The most glaring of these, of course, is that most of the dinosaurs whose feet I have been able to measure are of later Jurassic and Cretaceous age. However, the data that I do have for early Mesozoic forms seem consistent with what we see in post–Early Jurassic dinosaurs, and so I don't think this a huge problem.

And speaking of huge, many of the dinosaurs for which I have foot skeletal measurements were much bigger animals than their Newark Supergroup predecessors (even if the bigger *Eubrontes*-makers were carnivorous dinosaurs, many tyrannosaurs could easily have had them for lunch, had they ever met). In my comparisons I will therefore put greatest emphasis on dinosaurs whose foot skeletons show a footprint digit III length proxy of 350 mm or less.

The final limitation to my data is the limited data that I have for the feet of basal sauropodomorphs. In addition to the possible basal sauropodomorph *Eoraptor*, I have measured two prosauropod feet (and those are casts of the actual

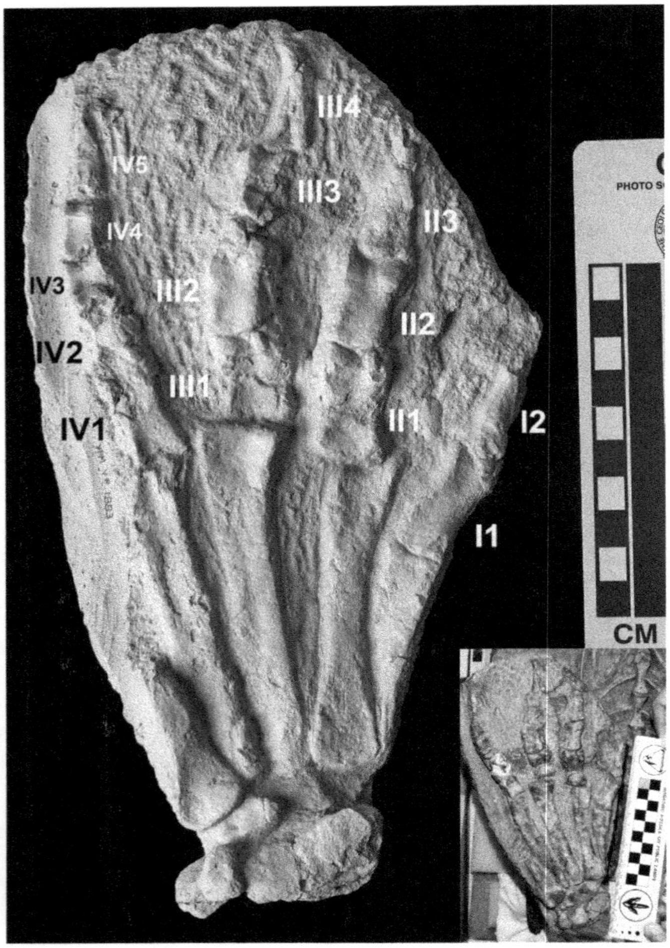

Above, **8.69.** Underside of the skeleton of the right foot of YPM VP 1883, *Anchisaurus polyzelus*. The foot is preserved in situ in surrounding matrix. The larger image is a cast of the foot; the *inset* shows the actual foot itself. The individual bones are variably preserved. The distal two thirds of phalanx III1, all of III2, much of the left side of phalanx III3 (as seen here), and most of IV2 are reconstructed (Galton 1976: Figs. 11, 22E).

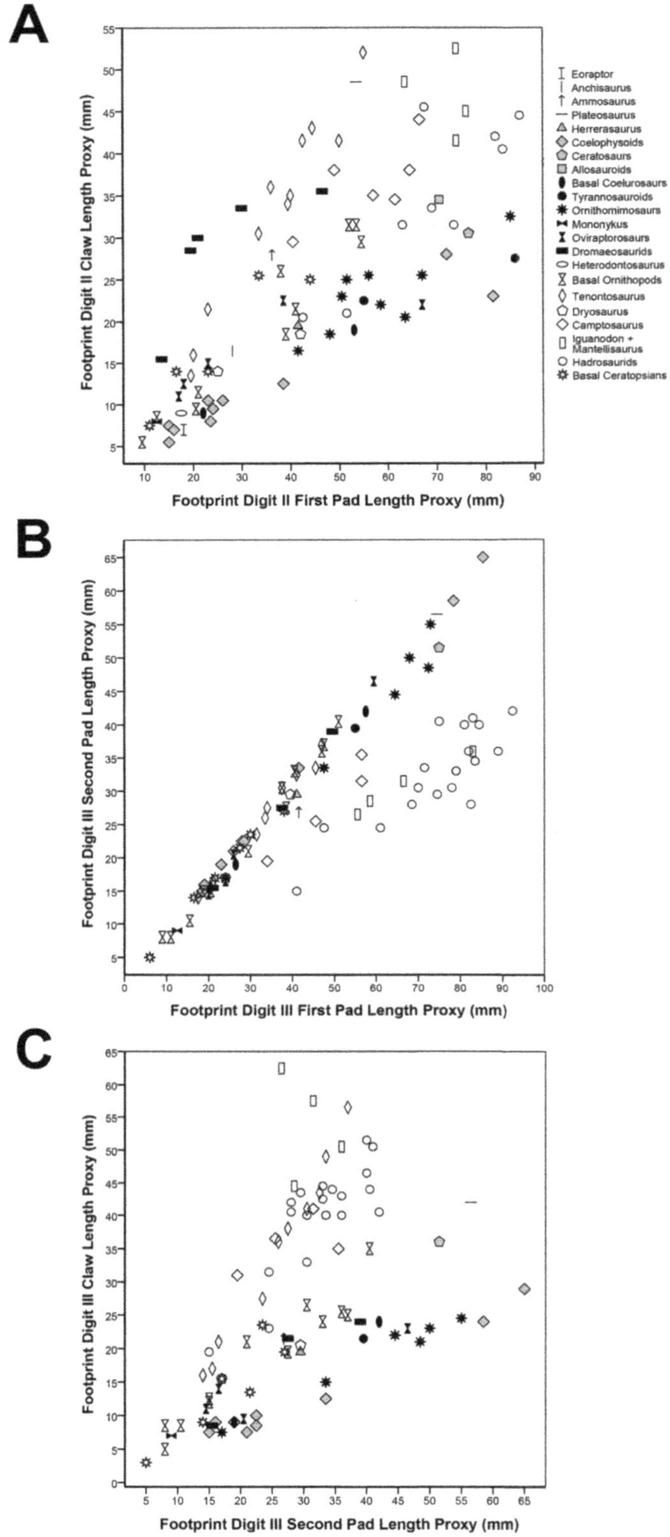

Right, **8.70.** Skeletal proxies for footprint digital pad and claw proportions. Calculation of pad and claw length proxies is as described in the text; plotted data are limited to forms roughly within the size range of the potential makers of Newark Supergroup tridactyl dinosaur footprints (digit III length proxy up to about 250 mm). *A*, Digit II claw length proxy vs. digit II first pad length proxy. Although there is overlap, *Eoraptor* and theropods other than dromaeosaurids tend to have relatively shorter claws than most ornithischians, *Anchisaurus*, *Ammosaurus*, and *Plateosaurus*. *B*, Digit III second pad length proxy vs. first pad length proxy. Data points for dinosaurs other than large ornithopods fall along the same trend, with very little scatter. *C*, Digit III claw length proxy vs. second pad length proxy. Although there is overlap between smaller theropods and ornithischians, large theropods, *Ammosaurus*, and *Plateosaurus* tend to have relatively shorter claws than large ornithopods.

specimens), one each of *Plateosaurus* (table A1.1; fig. 1.10C) and *Anchisaurus* (fig. 8.69). A third prosauropod foot (*Ammosaurus*, but see below) was measured by D. D. Gillette at my request, and following my specifications (table A1.1). Consequently my comments about the feet of these dinosaurs are only provisional. To increase the sample size of basal sauropodomorphs, in some comparisons I will supplement my own data with measurements from the published literature. The latter should be regarded with caution, because there is no guarantee that the authors whose numbers I used measured pedal bones the same way I did.

It should also be noted that Yates (2004) regards the genus *Ammosaurus* as a junior synonym of *Anchisaurus*, and considers *Anchisaurus* to be a basal sauropod. The Arizona specimen that will be discussed under the name *Ammosaurus* in the material that follows is regarded by Yates (2004: 6) as "an indeterminate primitive sauropodomorph." I will continue to use the name *Ammosaurus* for this specimen as a convenient historical label rather than any conviction about its identity on my part.

Differences Among Dinosaur Groups in Skeletal Proxies of Footprint Dimensions

In nearly all comparisons of footprint dimension proxies (figs. 8.70–8.72), small ornithischians and small theropods show considerable overlap, suggesting that it would be difficult to distinguish footprints of little dinosaurs of these two broad groups on the basis of most of the characters that will be examined here, especially for early Mesozoic forms (cf. Baron et al. 2017*a*, 2017*b*). The clearest differences among theropods and ornithischians show up at larger sizes, in comparisons of animals like allosaurs and tyrannosaurs with animals like hadrosaurs and *Iguanodon*—creatures that obviously have little relevance to interpreting the makers of Early Jurassic footprints.

It will be helpful, in the comparisons that follow, to have some notion of how the sizes of the various skeletal proxies of footprint parameters relate to the dimensions of the footprint parameters themselves. Summary data for the latter are provided in table 8.10.

Table 8.10. Dimensions (mm) of selected Newark Supergroup tridactyl dinosaur footprint parameters (mean or single data treatment).

Ichnotaxon	Parameter	Mean (range)	*N*
Eubrontes cursorius	Digit II first pad length	12.7	1
	Digit II length excluding claw	25	1
	Digit II length	32.7	1
	Digit III first pad length	16.5	1
	Digit III second pad length	16	1
	Digit III length excluding claw	47	1
	Digit III length	54.5	1
	Digit IV length	44	1
Eubrontes sillimani	Digit II first pad length	26.5 (21–45)	27
	Digit II length excluding claw	52.4 (41.5–86.5)	26
	Digit II length	66.7 (54–96.5)	26
	Digit III first pad length	29.2 (21–42)	27
	Digit III second pad length	28.3 (22–41)	26
	Digit III length excluding claw	83.9 (70–117.5)	25
	Digit III length	101.1 (83–136.5)	24
	Digit IV length	92.1 (70.5–121)	15
Eubrontes minusculus	Digit II first pad length	62.8 (57–68)	3
	Digit II length excluding claw	125.7 (113–140)	3
	Digit II length	152.3 (143–171)	3
	Digit III first pad length	69.0 (61–77)	3
	Digit III second pad length	50.7 (49–52.5)	2
	Digit III length excluding claw	165.0 (161–169)	2
	Digit III length	203.7 (192–224)	3
	Digit IV length	220.7 (203–232)	3
Eubrontes giganteus	Digit II first pad length	86.5 (83–90)	2
	Digit II length excluding claw	133	1
	Digit II length	174.5 (161–188)	2
	Digit III first pad length	90.5 (72–109)	2
	Digit III second pad length	58.0 (48–68)	2
	Digit III length excluding claw	177	1
	Digit III length	242.5 (199–286)	2
	Digit IV length	246.5 (219–274)	2
Anomoepus scambus	Digit II first pad length	18.8 (14–20.5)	6
	Digit II length excluding claw	31.1 (10–38.7)	7
	Digit II length	38.7 (12.5–52.3)	8
	Digit III first pad length	17.9 (16–19.5)	4
	Digit III second pad length	14.9 (14–15.5)	3
	Digit III length excluding claw	47.8 (36–55)	6
	Digit III length	56.3 (20.5–67)	7
	Digit IV length	60.0 (20–89)	8

Digits II and III Relative Pad and Claw Lengths (figs. 8.32, 8.70A, 8.70C, 8.71A, 8.71B, 8.73A, 8.73B). Recall that the "*Anchisauripus*"-maker had a relatively short first digital pad of digit II, while the "*Eubrontes*"-maker and the *Anomoepus*-maker had a relatively long such pad (fig. 8.20A); the "*Eubrontes*"-maker had a relatively short digit II clawmark, while the "*Anchisauripus*"-maker and the *Anomoepus*-maker had relatively long clawmarks on digit II (fig. 8.20E). The little dinosaur responsible for *Eubrontes cursorius* had a first pad length whose size might make it hard to say, on the basis of the pad length–claw length relationship, whether the trackmaker was a theropod or an ornithischian (table 8.10; fig. 8.70A), although the clawmark looks short enough compared with the length of the pad (fig. 8.73A) to make me favor the theropod identification (it should be kept in mind, throughout this discussion, that the true clawmark length would have been added to by the impression of the horny covering around the ungual). The same is true if one compares the length of the clawmark against the aggregate length of the fleshy part of the digit II impression (figs. 8.32A, 8.71A). It is worth noting that the relevant skeletal proxies suggest that relative clawmark lengths for coelophysoids would be particularly short.

The medium-sized forms of *Eubrontes* show quite a bit of scatter in the digit II first pad length to claw length, or overall digit II length (without claw) to claw length, comparison (figs. 8.32A, 8.73A), making comparisons with the skeletal proxy (fig. 8.70A) problematic. A typical *E. sillimani* (I will use this form throughout this discussion as a convenient stand-in for medium-sized *Eubrontes* more generally) has a first digital pad length of about 27 mm (table 8.10) and

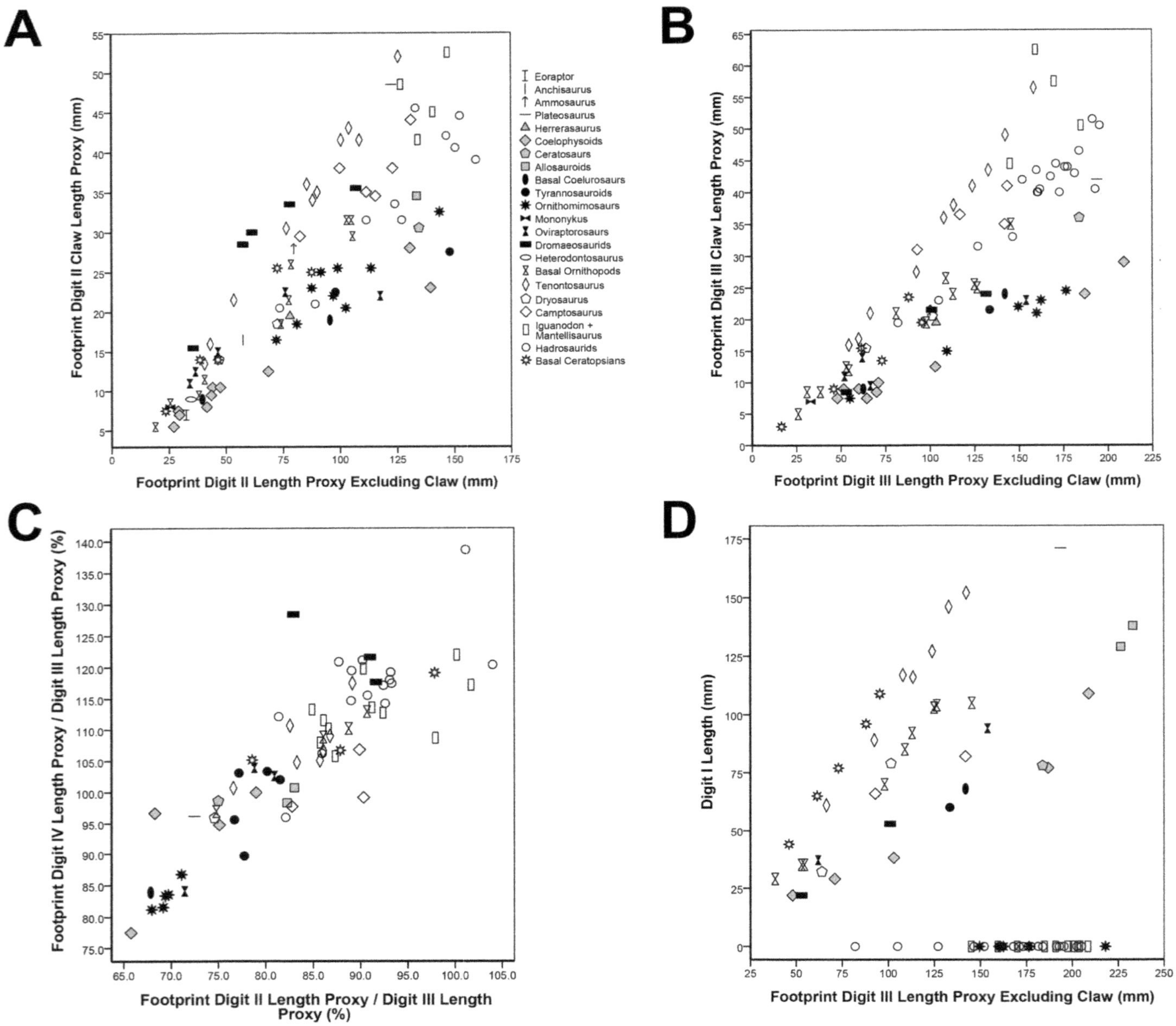

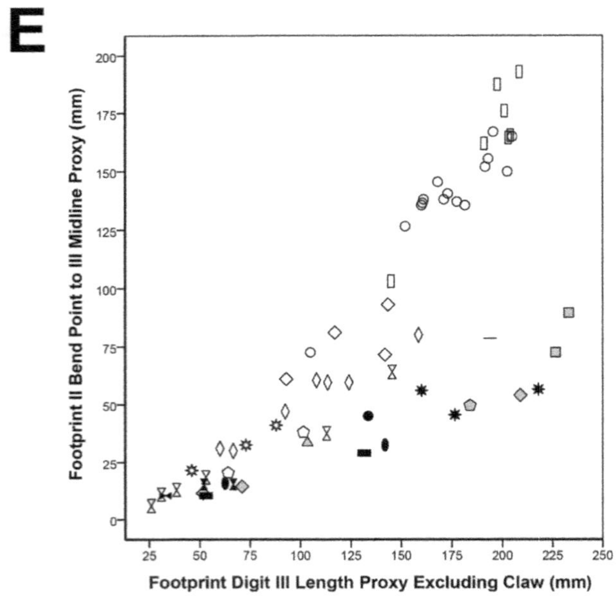

8.71. Skeletal proxies for footprint proportions. In *A*, *B*, *D*, and *E*, plotted data are limited to forms roughly within the size range of the potential makers of Newark Supergroup tridactyl dinosaur footprints (digit III length proxy up to about 250 mm). Calculation of digital proxies is as described in the text. *A*, Digit II length proxy excluding claw vs. claw length proxy. Large ornithopods, *Ammosaurus*, and *Plateosaurus* tend to have relatively longer claws than large theropods, but there is much overlap among theropods, *Eoraptor*, *Anchisaurus*, and ornithischians at small sizes. *B*, Digit III length proxy excluding claw vs. claw length proxy. Large ornithopods and *Plateosaurus* again tend to have relatively longer claws than do most large theropods (other than *Spinosaurus*; fig. 1.9*A*). *C*, Relative digit lengths. *Plateosaurus* and theropods other than dromaeosaurids tend to have relatively short digits II and IV compared with the length of digit III. *D*, Digit I length vs. digit III length proxy. Ornithomimosaurs and big ornithopods have lost digit I (length = 0). Otherwise theropods (other than *Spinosaurus*—again, fig. 1.9*A*) have a relatively shorter digit I than *Plateosaurus* and the ornithischians. *E*, Proxy for footprint width from the digit II bend point to the midline of digit III vs. digit III length proxy. Most ornithischians have a relatively broader foot than *Plateosaurus* and the theropods.

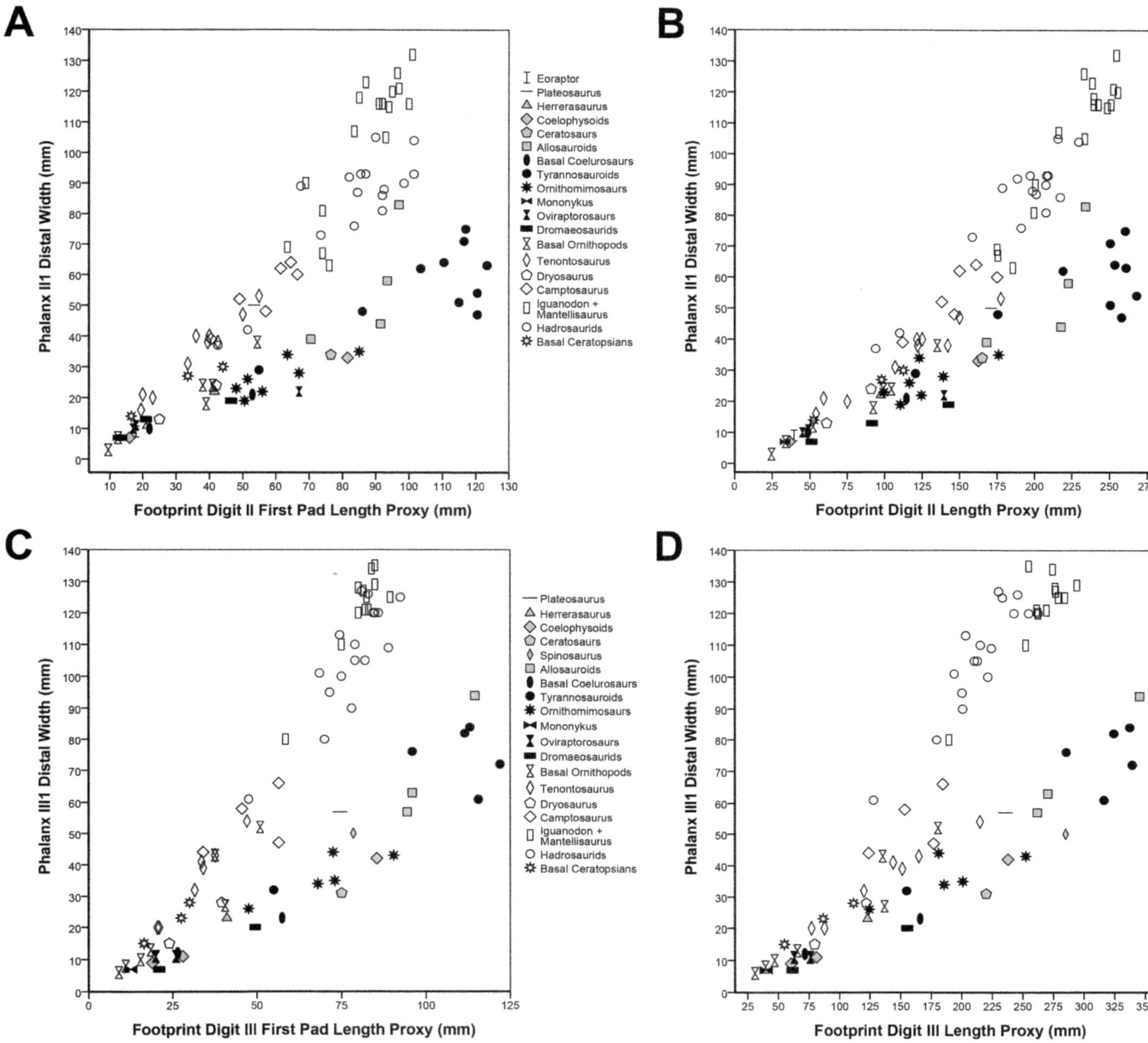

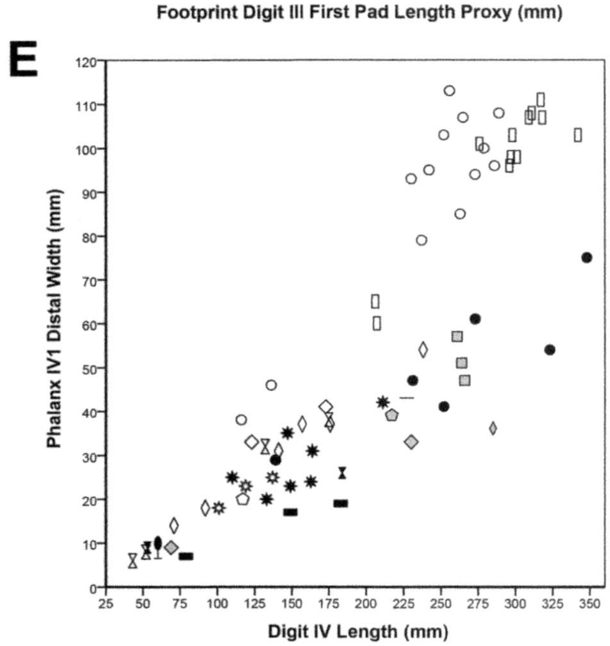

8.72. Skeletal proxies for phalanx and digit relative widths; plotted data are limited to forms roughly within the size range of the potential makers of Newark Supergroup tridactyl dinosaur footprints and somewhat larger (digit III length proxy 350 mm or less). *A*, Digit II first pad length proxy vs. its corresponding width proxy. *Plateosaurus* and all but the smaller ornithischians have a relatively broader pad width proxy than do *Eoraptor* and theropods. *B*, Digit II length proxy vs. digit II first pad width proxy. Once again large ornithopods have a relatively broader toe than do large theropods, but there is overlap between small theropods, *Eoraptor*, and small ornithischians in the relationship. *Plateosaurus* plots at the boundary between theropod and ornithopod points. *C*, Digit III first pad length proxy vs. its corresponding width proxy. As in *A*, all but the smaller ornithischians clearly have a broader first digital pad than do theropods. *Plateosaurus*, however, here is more like theropods than ornithopods. *D*, Digit III length proxy vs. digit III first pad width proxy; interpretation is similar to that for *C*. *E*, Digit IV length proxy vs. digit IV second pad width proxy. As usual, small theropods and *Eoraptor* overlap small ornithopods in this relationship, but large theropods and *Plateosaurus* have a relatively narrower digit IV than that of large ornithopods.

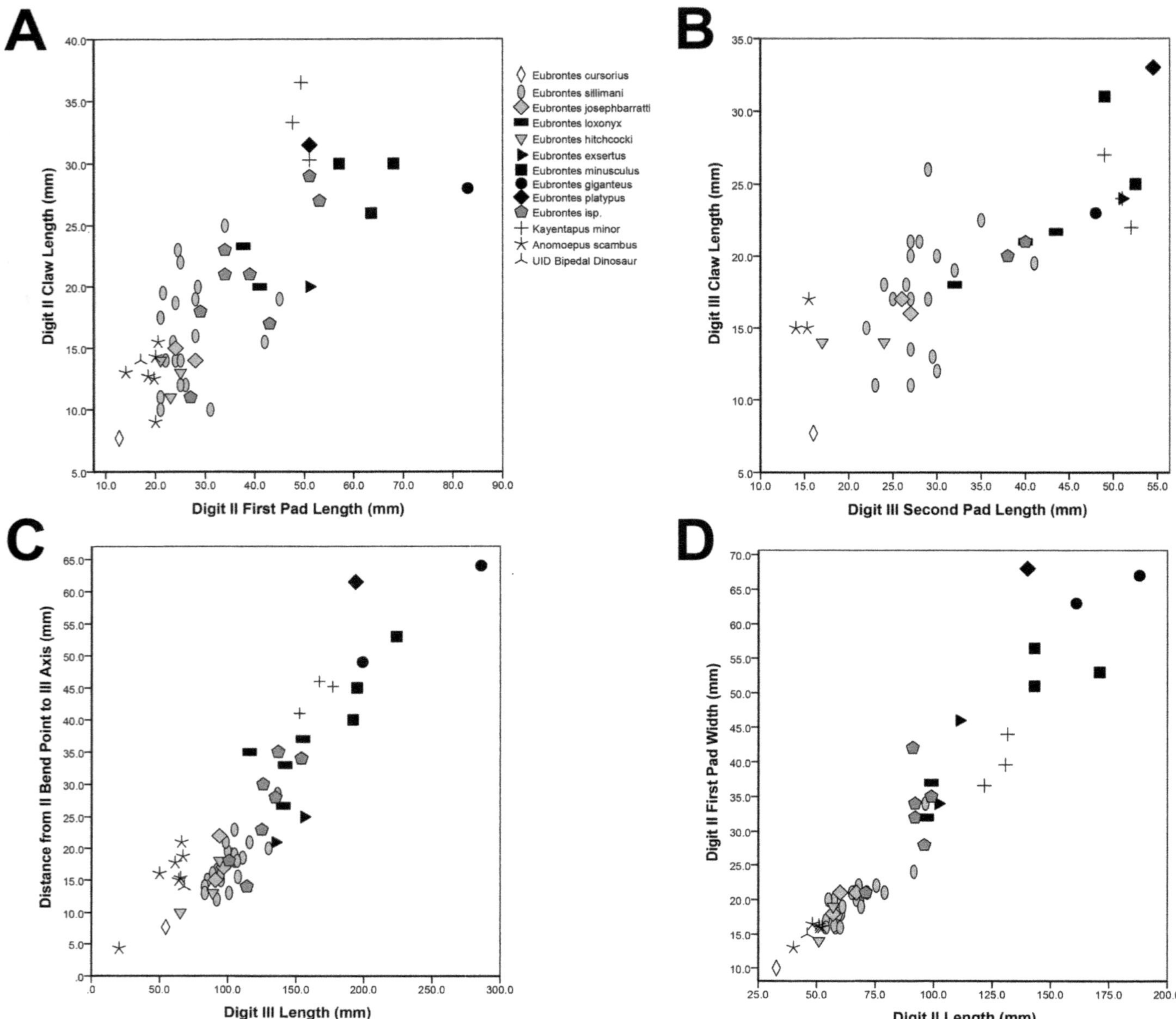

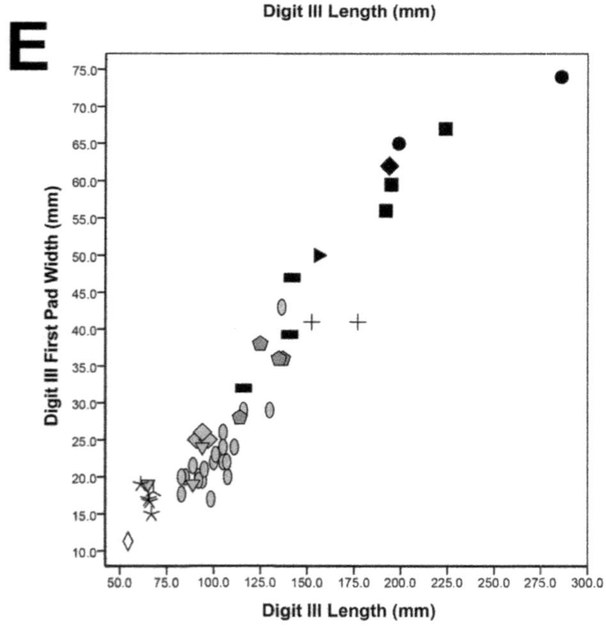

8.73. Proportions of Newark Supergroup tridactyl dinosaur footprints for comparison with skeletal proxies. *A*, Digit II first (most proximal) pad impression length vs. clawmark length. *B*, Digit III second pad impression length vs. clawmark length. *C*, Digit III length impression vs. the distance from the medial proximal bend point of digit II to the long axis of the digit III impression. *D*, Digit II length impression vs. the width of the first digital pad impression. *E*, Digit III length impression vs. the width of the first digital pad impression.

a clawmark length anywhere from around 10 to 25 mm (fig. 8.73A); this is a broad enough range that the skeletal proxy could be consistent with theropods, *Anchisaurus*, or smaller ornithischians (fig. 8.70A); the same holds true for a comparison of the fleshy part of the overall digit II impression against the claw length (figs. 8.32A, 8.71A). Recall, though, that for many medium-sized *Eubrontes* it was difficult to be certain that the terminal end of the clawmark impression was correctly distinguished from extramorphological features of the footprint.

The bigger *Eubrontes*, however, are less ambiguous. A typical *E. minusculus* has a digit II first pad impression about 63 mm long, and a typical *E. giganteus* a pad about 87 mm long (table 8.10, fig. 8.73A); lengths of the corresponding clawmark impressions are in the 25–30 mm range. This puts them among theropods other than dromaeosaurids on the basis of claw pad and length skeletal proxies (fig. 8.70A), and well below values for *Plateosaurus* and nearly all ornithischians. The length of the fleshy portion of the digit II impression of both *E. minusculus* and *E. giganteus* is about 130 mm (table 8.10); this again puts them among theropods in the corresponding skeletal proxies for the relative lengths of the fleshy part of the toe and the claw (fig. 8.71A). (Interestingly, some very unusually preserved large tridactyl prints from the Early Cretaceous of Spain [Platt and Meyer 1991; Huerta et al. 2012] likewise indicate relatively short claws compared with length of the fleshy part of the toe, thereby also indicating that the trackmakers were likely theropods.)

Kayentapus and most *Anomoepus*, in contrast, have relatively long clawmarks on digit II. A typical *Anomoepus* has a first digital pad length of about 19 mm (table 8.10) and a clawmark length of about 13–16 mm (fig. 8.73A). The pad length puts *Anomoepus* right among a scrum of small-dinosaur points in the relevant skeletal proxy plot (fig. 8.70A), but perhaps more like ornithischians than theropods (particularly coelophysoids). The length of the digit II impression apart from the claw of *Anomoepus* averages about 31 mm (table 8.10; fig. 8.32A), a size relative to the clawmark length which is again more like ornithischians than most theropods (fig. 8.71A). *Kayentapus* has a digit II first pad length of about 50 mm, and a clawmark length of 30–37 mm (fig. 8.73A), which puts it more among points for ornithischians than most theropods, but well below *Plateosaurus* (fig. 8.70A). The length of the fleshy part of digit II of *Kayentapus* is about 100 mm (fig. 8.32A), which gives it an ornithopod-like or perhaps prosauropod-like fleshy vs. claw length relationship (fig. 8.71A).

The "*Anchisauripus*," "*Eubrontes*," and *Anomoepus* maximum information clusters show marked differences in the relative lengths of the first and second digital pads, and the clawmark, of digit III (fig. 8.21A, 8.21C, 8.21G). The relative sizes of proxies for lengths of the first and second digital pads show remarkably little variation among dinosaur groups, apart from the clear inference that larger ornithopods should have a relatively short second pad (fig. 8.70*B*). Interesting as this is, it is of little use for interpreting Newark Supergroup prints.

More useful is the comparison of relative sizes of the second digital pad proxy and the clawmark proxy of digit III (fig. 8.70C). Although there is again overlap between the smallest theropods and ornithischians, for dinosaurs with a second pad proxy length of 20 mm or more there is a marked tendency for *Ammosaurus* and *Plateosaurus* (this comparison could not be made, alas, for *Anchisaurus*: fig. 8.69), and theropods, to show a relatively shorter clawmark length than ornithopods, with coelophysoids having particularly short relative clawmark proxy lengths. Although *E. cursorius* is small enough (table 8.10) that its maker's foot skeleton could be expected to plot in the pile-up of overlapping points for small theropods and ornithischians, it is interesting that its point plots quite low in the corresponding footprint bivariate relationship (fig. 8.73*B*), a theropod-like feature. Medium-sized *E. sillimani* (table 8.10) once again show a rather broad range of clawmark lengths relative to the length of the second pad of digit III, anywhere from around 10–25 mm for a second pad length of about 30 mm (fig. 8.73*B*), which would put them into a region between theropod points only at the lower end, and an area of overlap between theropods with relatively long clawmarks, and ornithopods with relatively short clawmarks at the upper end (fig. 8.70C). The biggest *Eubrontes* (digit III second pad lengths of 50 mm or more) have clawmarks of 20–35 mm length (fig. 8.73*B*), which would put them well below expected ornithopod territory (fig. 8.70C), among points for *Plateosaurus* and theropods. In contrast to its behavior in the relationship between the lengths of the first digital pad and the clawmark of digit II, *Kayentapus* plots like a theropod in the comparison of the relative lengths of the second digital pad and clawmark lengths of digit III (fig. 8.73*B*). *Anomoepus*, in contrast, has a relatively long digit III clawmark, just as with the relative length of its digit II clawmark, which puts it nicely into the ornithischian region of the proxy plot (fig. 8.70C). Without going into detail, the same conclusions about the identities of the makers of *Eubrontes*, *Anomoepus*, and *Kayentapus* made on the basis of comparison of the relative lengths of the digit III second pad and clawmark can be drawn from comparison of the overall length of the fleshy part of the digit III impression and the length of the clawmark (figs. 8.32*B*, 8.71*B*).

Overall Digit Lengths (figs. 2.18A, 8.34A, 8.71C). Although there is a zone of overlap, *Plateosaurus* and theropods other than dromaeosaurids tend to have relatively shorter toemark length proxies of digits II and IV, compared with the length

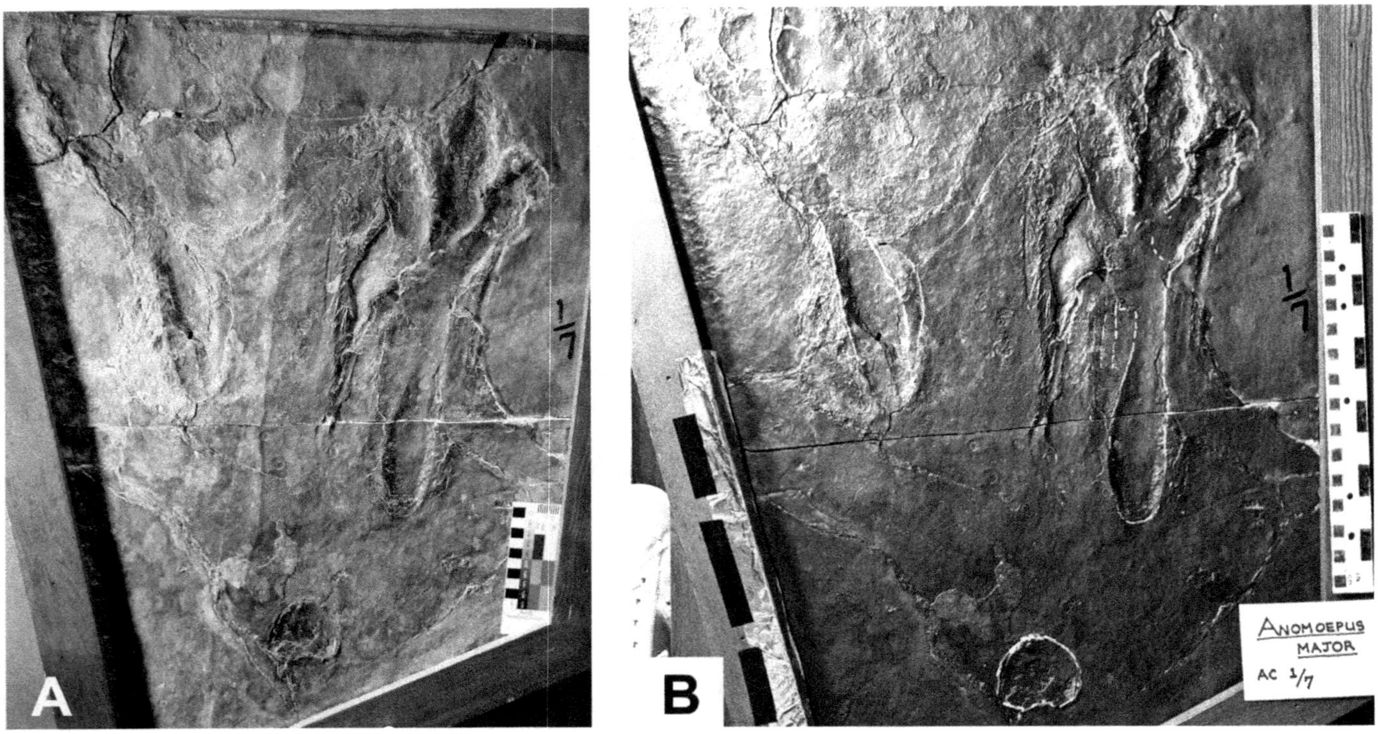

8.74. Two views of AC 1/7, impressions made by a sitting dinosaur, under different lighting conditions. *A*, Photographed in slightly oblique view. *B*, Photographed more nearly flat on, with stronger raking light, and with parts of the traces outlined. The specimen shows left and right footprints with metatarsal impressions, and a probable impression of the tissues covering the ischium below and between the two footprints. Preservation of the footprints isn't great, but there is no clear indication of a digit I impression, which would have been expected to touch the substrate in such a crouch had this toe been present and well developed.

of the toemark III length proxy, than do ornithischians. *Kayentapus* and most *Eubrontes* look more like expectations for saurischian than ornithischian prints in this comparison. *Anomoepus* differs from the other ichnogenera in having relatively long impressions of digits II and IV, and would fall near the border between saurischians and ornithischians in the proxy graph, perhaps somewhat more like the latter group.

Digit I Length (fig. 8.71D). As previously discussed (fig. 2.25), in those non-avian dinosaur groups that have not completely lost digit I, this toe is relatively longer (compared with the length of digit III) in basal sauropodomorphs (prosauropods) and ornithischians than in theropods, albeit—as usual—with considerable overlap among smaller-bodied forms. The same conclusion applies if we plot the length of digit I against the skeletal proxy for the length of that part of digit III likely to register in footprints (fig. 8.71*D*). Digit I left no impression in any of the Newark Supergroup tridactyl prints in my database of measurements (which did not include *Gigandipus*), although it could be expected to register in prints made by dinosaurs that were either sitting (fig. 8.74; cf. Gierliński 1994; Lockley et al. 2003*a*; Milàn et al. 2008; Milner et al. 2009; Romano and Citton 2017) or otherwise causing their metatarsals to press against the substrate (cf. Farlow et al. 2015 and numerous references therein). A crouching theropod the size of a typical *E. sillimani* (digit III impression ca. 100 mm long) might leave a digit I impression somewhere around 25–50 mm long, while a sitting ornithischian of comparable size might leave a digit I impression with a length of 50 mm or more (fig. 8.71*D*). For a little trackmaker with a digit III impression of the size of *E. cursorius* or *Anomoepus*, in contrast, there would be little difference in the expected length of a digit I impression made by theropods or ornithischians.

If the dinosaur was walking in the normal manner, digit I would probably not register at all in most theropods (fig. 1.8), and possibly not the smallest ornithischians (cf. Baron et al. 2017*b*, but might do so in prints of most basal ornithopods, and in prosauropods (figs. 1.10C, 1.11, 8.69, 8.71*D*). Consequently the absence of a digit I impression in the medium-sized and larger forms of *Eubrontes* and *Kayentapus* may mean that their makers are unlikely to have been prosauropods or ornithopods. We will revisit this question again below.

Relative Widths of Digital Pads and Toes (figs. 2.20–2.24, 8.72). Members of the "*Eubrontes*" maximum information cluster tend to have broader digital pads than members of the "*Anchisauripus*" maximum information cluster and members of the *Anomoepus* maximum information cluster (table 8.6; figs. 8.20, 8.21). Across all dinosaur groups in our skeletal sample, there is a clear tendency for toes to become relatively stouter with increasing size. Excluding once again the smallest forms, at any given size ornithischians tend to have relatively fatter toes than do *Plateosaurus* (except for the distal width of the first phalanx of digit II compared with the length of the proxy for the length of the first digital pad of digit II) and theropods, whether one is comparing distal widths with pad proxy lengths or with overall toemark proxy lengths (fig. 8.72).

Judging from some modern ratites (figs. 1.5, 6.3*A*), the width of the fleshy part of the toe at the level of the distal end of the first phalanx of digits II and III would be roughly 1.4–1.6 times the width of the distal end of the underlying first phalanx itself; the value may be more like 1.7× for the alligator (fig. 5.3*A*). I will use 1.5× as a rough estimate of how much broader the first digital pad of digits II and III of a dinosaur print would be expected to be than the width of the distal end of the first phalanx of each toe.

Eubrontes cursorius and *Anomoepus* are small enough that I am reluctant to say much about whether they are more theropod-like or ornithischian-like in the relative widths of their toemarks. It is worth noting, however, that *Anomoepus* seems to have relatively broader first digital pads of digits II and especially III than does *E. cursorius* (fig. 8.73*D*, 8.73*E*), which could suggest more likely ornithischian and theropod affinities, respectively. *Eubrontes sillimani* is getting big enough that it might be possible to say more about the nature of its maker from its pad widths, except that I have rather few data for foot skeletons of the expected size for the maker of this form (typical digit II impression length of 67 mm, and a digit III impression length of about 100 mm: table 8.10; fig. 8.72*B*, 8.72*D*).

The bigger forms of *Eubrontes* have typical digit II impression lengths of 150–200 mm, and digit III impression lengths of 200–300 mm (table 8.10; fig. 8.73*D*, 8.73*E*). If the makers of such prints had been theropods, the distal widths of phalanges II1 and III1 would be expected to be perhaps 20–50 mm, and 30–70 mm, respectively (fig. 8.72*B*, 8.72*D*). Using the 1.5× conversion to soft-tissue width, these values become 30–75 mm for the width of the first pad of the digit II impression, and 45–105 mm for the width of the first pad of the digit III impression. The single *Plateosaurus* specimen for which I have data has distal widths of phalanges II1 and III1 of 50 mm and 57 mm, respectively, resulting in expected pad widths of 75 mm and 85 mm, respectively. The only ornithischians near the size range of big *Eubrontes* are iguanodonts and hadrosaurs, which have very fat toes, of course (fig. 8.72), but don't seem very relevant to the current discussion. In any case, the observed first pad widths of big *Eubrontes* are 50–70 mm for digit II, and 55–75 mm for digit III. These ranges are consistent with predictions for theropods and *Plateosaurus*, which is regrettably not very helpful in deciding between the two groups.

Kayentapus is somewhat smaller than big *Eubrontes*, with a digit II impression length of 120–130 mm and a digit III impression length of 150–175 mm (fig. 8.73*D*, 8.73*E*). If its maker was a theropod, it would have an expected phalanx II1 distal width of about 15–30 mm, and an expected phalanx III1 distal width of about 20–30 mm (fig. 8.72*B*, 8.72*D*). With the 1.5× conversion, these translate into expected first digital pad widths of 23–45 mm (digit II) and 30–45 mm (digit III), which ranges include the observed digital pad widths of *Kayentapus* (about 35–45 mm for digit II, and 40 mm for digit III; fig. 8.73*D*, 8.73*E*). Ornithischians of the same size would be expected to have a phalanx II1 distal width of about 25–40 mm (expected first digital pad width of 37–60 mm), and a phalanx III1 distal width of about 40–60 mm (expected first digital pad width of 60–90 mm). The lower end of the estimated digit II expected first pad widths of ornithischians overlaps the observed digital pad widths of *Kayentapus*, but the same is not true for the first digital pad of digit III. On balance *Kayentapus* seems more theropod-like (or possibly prosauropod-like) than ornithischian-like in relative toe widths.

Distance from Digit II Bend Point to Digit III Midline (figs. 8.36A, 8.71E, 8.73C). *Anomoepus* and *Kayentapus* are both broader than most *Eubrontes* in this measure of footprint basal width. Ornithischians tend to be relatively broader than *Plateosaurus* and most theropods in the skeletal proxy for this footprint parameter. The distance from the digit II bend point to the digit III midline will be affected by more than simply the widths of the digital pads (and their underlying skeletal proxies). The value of IDA II–III will also contribute to this distance, and the value of this angle in turn presumably reflects, at least in part, the shape of the metatarsals and their distal articular surfaces. Consequently I won't do more than say that, based on the skeletal proxy examined here, *Anomoepus* and *Kayentapus* could be more ornithischian-like than theropod-like or *Plateosaurus*-like, while *Eubrontes* may be more theropod-like or *Plateosaurus*-like.

Toe-Tapering Profiles (figs. 3.23, 8.66). Inspection of the TTPs of digits II and III of non-avian dinosaurs suggests that *Plateosaurus* and ornithischians would be expected to have proportionally broader digital pads along the lengths of both toes than would theropods, although with some overlap between fatter-toed theropods and skinnier-toed ornithischians. Ornithischians and *Plateosaurus* would also be expected to

Table 8.11. Summary of comparisons of Newark Supergroup tridactyl dinosaur footprints with skeletal proxies of footprint features of theropods (T), ornithischians (O), and prosauropods (P). The letter in each cell indicates the group of trackmaker whose skeletal proxy seems most like what is seen in that footprint type; a question mark indicates uncertainty about which category of trackmaker candidate the footprint category is most like.

Footprint feature	*Eubrontes cursorius*	*Eubrontes sillimani*	*Eubrontes minusculus* and *Eubrontes giganteus*	*Anomoepus scambus*	*Kayentapus minor*
Digit II claw length vs. digit II first pad length	T?	?	T	O?	O
Digit II claw length vs. digit II length excluding clawmark	T?	?	T	O	O/P
Digit III claw length vs. digit III second pad length	T?	T/O	T/P	O	T
Digit III claw length vs. digit III length excluding clawmark	T?	T/O	T/P	O	T
Relative lengths of digits II, III, and IV	T/P	T/P	T/P	O?	T/P
Digit I length vs. digit III length	T/O?	T?	T?	T/O?	T?
Relative width of first digital pad of digit II	T?	?	T/P	O?	T/P?
Relative width of first digital pad of digit III	T?	?	T/P	O?	T/P
Distance from digit II proximal bend point to digit III long axis	T/P?	T/P?	T/P?	O?	O?

have relatively longer clawmarks than theropods, albeit again with some overlap.

Trackmaker Identification from Footprint Parameter Proxies: Conclusions. Interpretations of trackmaker identities based on comparisons of footprint features and skeletal proxies are summarized in table 8.11. On balance, little *Eubrontes cursorius* looks more theropod-like than anything else, with any ambiguities in this identification being due to its diminutive size. Somewhat surprisingly, *E. sillimani* appears more ambiguous, again probably due to its relatively small size, but I lean toward theropod. The bigger forms of *Eubrontes* have several features that could be consistent with either a theropod or a prosauropod interpretation, but the relatively short clawmark on the digit II impression, and the lack of any imprint of digit I, make me vote for the theropod candidate for the maker of these prints as well; we will revisit this question shortly. In nearly all features *Anomoepus* comes across as ornithischian-like, a conclusion that will surprise nobody (cf. Olsen and Rainforth 2003; Wilson et al. 2009; Dalman and Weems 2013; Becerra et al. 2016). *Kayentapus* is a bit puzzling. It shows features consistent with all three major groups of trackmakers, but perhaps more theropod-like characters than others (cf. Weems [2006*a*, 2006*b*] for additional features that may support a theropod identification for the trackmaker).

Complicating the identification of the groups to which early Mesozoic bipedal trackmakers belonged is the fact that the three major clades—theropods, sauropodomorphs, and ornithischians—dinosaurs that were quite different from each other in the later Mesozoic, were represented in the early Mesozoic by forms that weren't quite so different in body form, which has greatly complicated interpretation of their phylogenetic relationships (Baron et al. 2017*a*, 2017*a1*, 2017*b*; Langer et al. 2017). Thus *Eoraptor*, which has been treated (at least implicitly) throughout this book as a basal-most sauropodomorph (Sereno et al. 2013), has also been interpreted as a basal theropod (Baron et al. 2017*a*), and *Herrerasaurus*, which has been treated here as a theropod, may have been closer to sauropodomorphs (Baron et al. 2017*a*). All of these animals, along with basal ornithischians (Baron et al. 2017*a*, 2017*b*) were slimly built dinosaurs with feet that by many or most of the criteria discussed in this book would be rather theropod-like; many or most of them may have been at least omnivorous, and some of them were likely faunivorous (Cabreira et al. 2016; Baron et al. 2017*a*; Müller et al. in press).

Consequently it isn't inconceivable that some hypothetical Early Jurassic survivor of basal-most sauropodomorph stock (more basal than prosauropods, which will be discussed below) might have survived to make theropod-like footprints in eastern North America. Similarly, some of the Newark Supergroup footprints interpreted here as having been made by small theropods might in fact have been made by ornithischians. However, I would guess that any such "pseudo-theropods" would in life have been so much like small theropods in appearance and biology that distinguishing between them (at higher than, say, the species or genus level), and the footprints they made, would have been—had there been contemporary naturalists to consider such things—so difficult as to be almost meaningless. One inevitably thinks of the old saw about whether something that looks, walks, and quacks like a duck should be recognized as being something other than a duck.

A footnote (if you'll pardon the expression yet again) to identification of the makers of small to medium-sized forms of *Eubrontes*—and maybe even *Anomoepus*—arises from consideration of the Late Triassic bipedal pseudosuchian *Poposaurus* (Gauthier et al. 2011; Schachner et al. 2011; Bates and Schachner 2012; Kubo and Kubo 2012. The foot of this reptile was dinosaur-like (figs. 3.23, 8.75), and so it

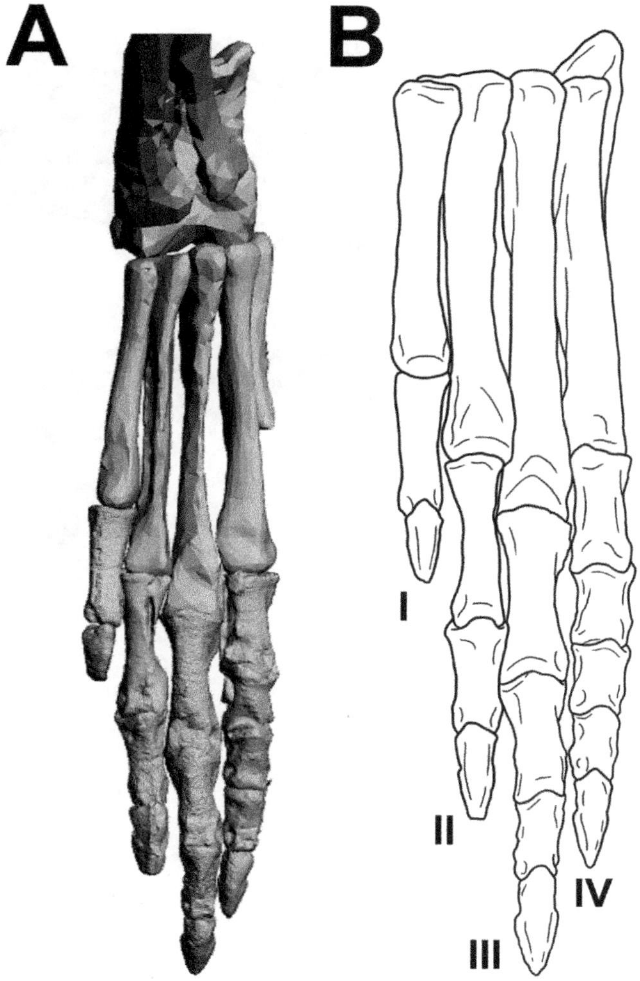

8.75. Left foot of *Poposaurus gracilis* (YPM VP.057100). *A*, Digital model of distal shank and foot (image created by Emma R. Schachner and John Cody Sarrazin). *B*, Interpretative drawing of foot (image by Emma R. Schachner).

is possible that some early Mesozoic tridactyl prints were made by such "dinocrocs" instead of dinosaurs (Farlow et al. 2014). Whether such a case of mistaken identity would in fact arise depends in part on whether the rather long digit I of *Poposaurus* routinely touched the ground—an issue that will arise again when we reconsider the possibility that basal sauropodomorphs (prosauropods) might have made some of the larger *Eubrontes*. In any case, poposauroids are presently restricted to the Triassic (Nesbitt et al. 2013*a*), and so I consider it unlikely—but not impossible—that any of the Newark Supergroup smaller *Eubrontes* were made by such animals.

The Prosauropod Hypothesis for the Maker of Large *Eubrontes*. Given the above-described ambiguities over the nature of the dinosaurs responsible for large specimens of *Eubrontes*, it isn't surprising that there has been dissent from the interpretation favored here, that these are footprints of large (for the Early Jurassic) theropods. A strong minority opinion has championed basal sauropodomorphs (prosauropods) as the maker of many, most, or all large *Eubrontes*.

Bock (1952: 405) was an early proponent of this hypothesis: "the large tracks of the Connecticut Valley and Whitehall, New Jersey, representing *Eubrontes giganteus* E. Hitchcock, may well match up with those of *Plateosaurus* or the similar sized *Pachysaurus*, which is abundant in the German Knollenmergel. Von Huene . . . shows the left foot of *Pachysaurus wetzelianus*, whose size will fit well in the huge track of *Eubrontes giganteus*. . . . Not only are the dimensions, such as length, thickness and position of the phalanges, close, but the presence of the well-defined first toe rounds up a picture of close relationship to the likewise huge *Gigandipus caudatus*."

In describing large forms of *Eubrontes*, Lull (1953: 178) thought that their makers were "relatively huge, ponderous, bipedal forms, definitely dinosaurian and probably Theropodous although not perhaps strictly carnivorous in their habits for they lack the trenchant, raptorial type of claws generally associated with beasts of prey." More recently, Miller et al. (1989: 213) used similar reasoning to identify the maker of *Eubrontes* from the Early Jurassic of Utah as a plateosaurid prosauropod: "The very broad phalangeal pads strongly suggest a heavy animal, and the short, blunt claw marks do not seem to represent a carnivore."

The strongest case for prosauropods as the makers of large *Eubrontes* (including *Gigandipus*), however, has been presented by R. E. Weems (1987, 2003, 2006*a*), who has forcefully argued that the guilty trackmaker was most likely a bipedal sauropodomorph with a pedal structure something like that of *Plateosaurus*. Weems has presented both morphological and ecological reasons in support of his case. We will first consider his morphological arguments.

Plateosaurus had a somewhat long body trunk compared with its hindlimb length (cf. Galton 1976; Christian and Preuschoft 1996; van Heerden 1997; Galton and Upchurch 2004), and for this and other reasons has been interpreted as a facultative biped or even a quadruped (cf. Van Heerden 1979; Weishampel and Westphal 1986; Wellnhofer 1993*b*; Moser 2003). Although conceding the relatively long trunk of *Plateosaurus*, Weems (2006*a*) nonetheless argued that the animal's tail was long and massive enough to counteract the long trunk, enabling the dinosaur to be at least facultatively bipedal. Weems (2003) also noted that digit III of the pes is the longest toe, unlike basal sauropodomorphs that were almost certainly quadrupedal, but like the condition seen in typical theropods, which were clearly bipedal. Other workers have also concluded that *Plateosaurus* was mainly or even entirely bipedal on the basis of the construction of its forelimb and hand (Bonnan and Senter 2007; also see Cooper 1981; Reisz et al. 2005 for [adult] *Massospondylus*; Yates et al. 2010)

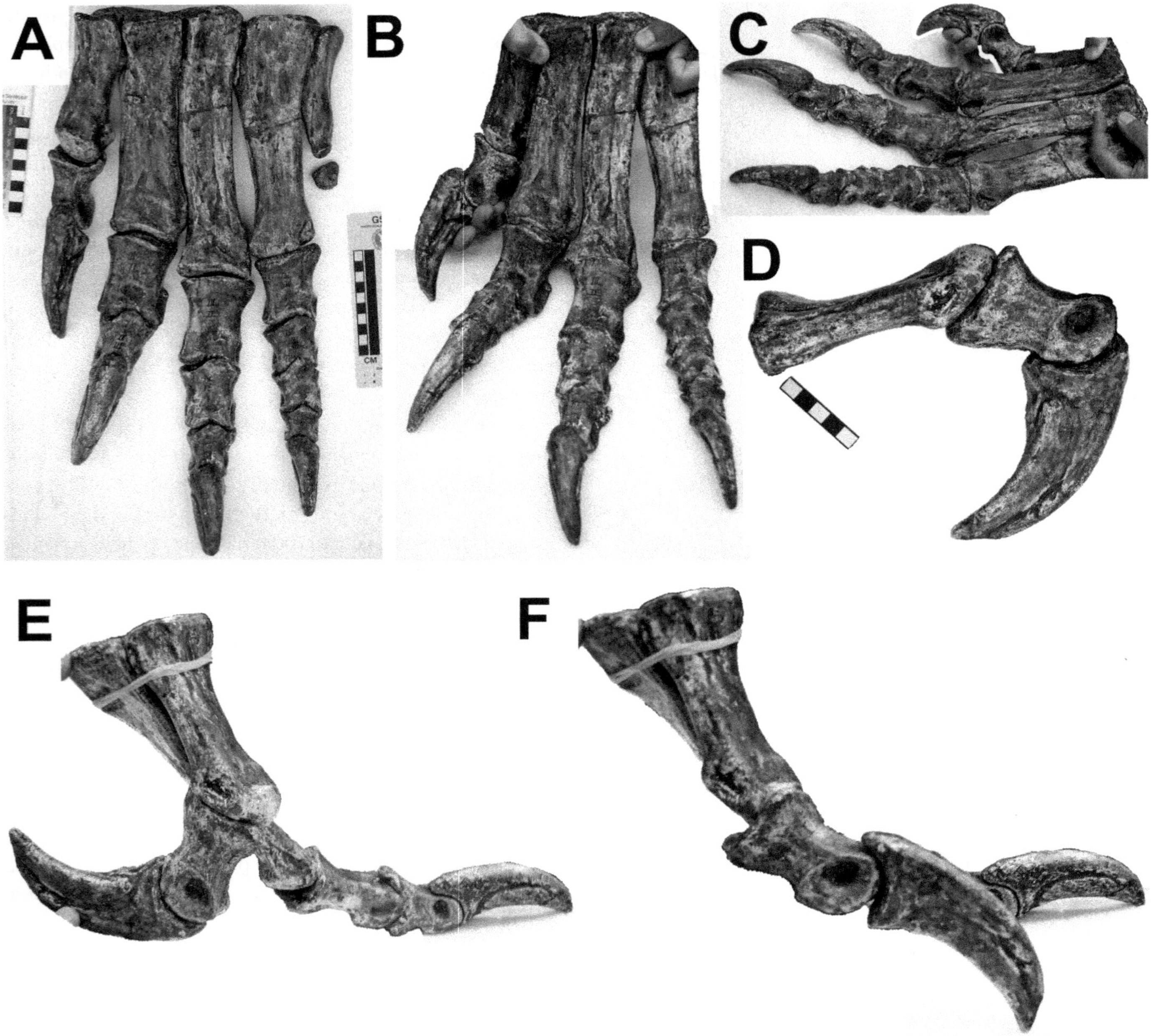

8.76. CM 11908, cast of a *Plateosaurus* foot from Germany. The foot was previously identified as a right (McIntosh 1981) but is a left. *A*, The foot laid out flat. *B–F*, Various manipulations of the foot skeleton, in some cases with a rubber band and human hands (the latter mostly screened out in the photographs) holding bones together. *B*, Anterior view of metatarsals and digits I–IV with the proximal metatarsus elevated and metatarsal and digit I slightly extended. *C*, Same arrangement of bones as *B*, but with the foot viewed obliquely from the lateral side. *D*, Medial view of metatarsal and digit I with the bones maximally ventroflexed at the joints. *E*, Medial view of metatarsals and digits I and II, with digit I ventroflexed as in *D*. *F*, Medial view of metatarsals and digits I and II, but with digit I pointed forward in what seems a "comfortable" position (the proximal end of phalanx I-1 slipped downward against the distal end of the metatarsal as the picture was taken; it was hard to have enough hands to hold all the bones in the desired positions). *All photographs copyright Carnegie Museum of Natural History.*

and the dynamics of motion inferred from computer modeling of joints in the skeleton (Mallison 2010*a*, 2010*b*, 2011).

So it seems that at least some prosauropods were bipeds. Weems (2003: 297) went on to propose that "the metatarsal that forms the origin of digit I [of *Plateosaurus*] fits snugly against the other metatarsals at an acute angle, which indicates that the base of digit I was rotated inward at least 60° relative to the orientation of the other four digits." In addition, Weems (2003: Fig. 18.11) suggested that the long digit I of *Plateosaurus* could swing fore and aft along the distal end of metatarsal I through about 180 degrees. This would have allowed the dinosaur to walk with its first toe bent behind

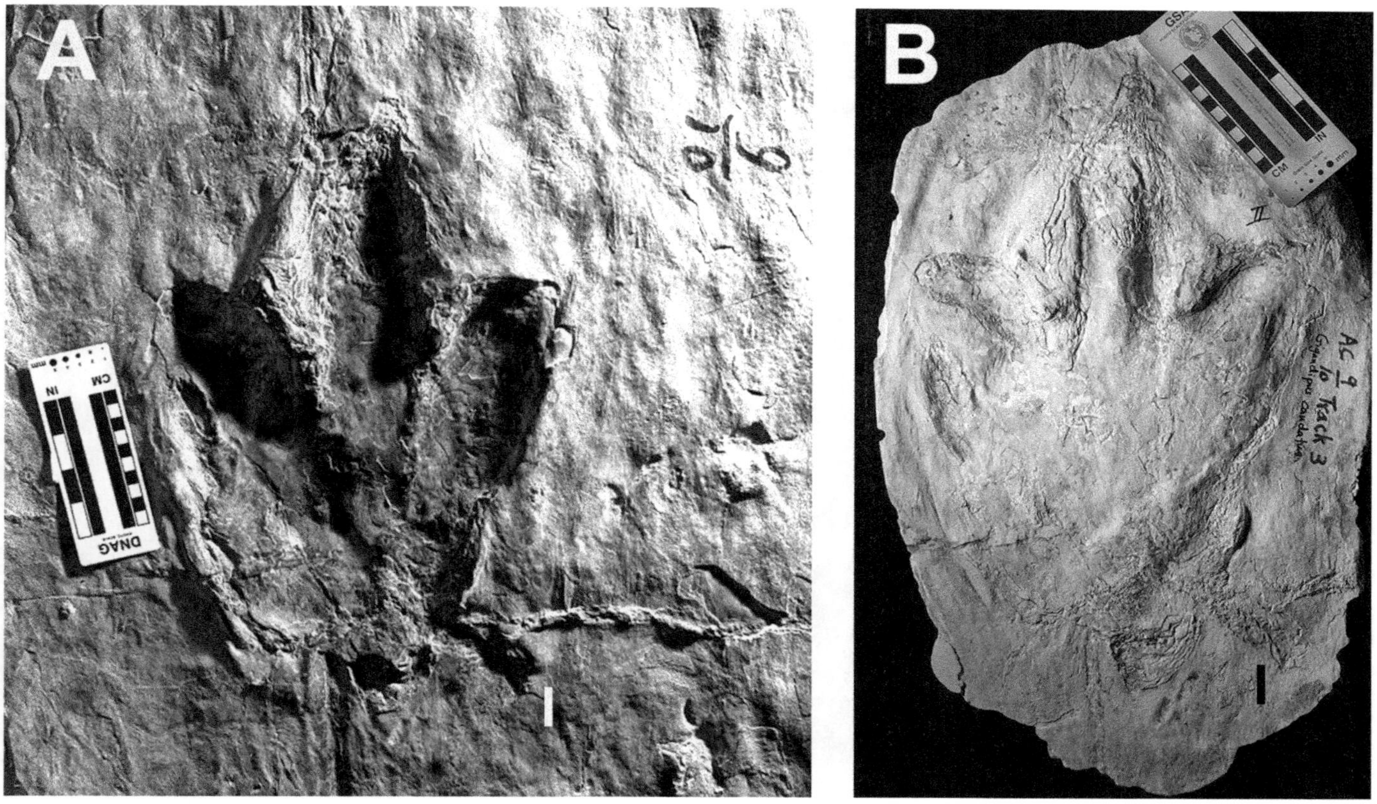

8.77. The third footprint in AC 9/10, *Eubrontes caudatus* (fig. 8.9*K*); digit I labeled in both panels. *A*, The footprint itself. *B*, A positive copy of the footprint. This is not a very good copy (it was not possible to reinforce the mold from which it was made), but it nicely shows the configuration of the digit I impression.

metatarsals II and III, and off the ground, and so *Plateosaurus* could have been functionally tridactyl—and capable of making three-toed footprints—during normal walking. Interestingly, Cooper [1981: 796–797] argued that the articulation of pedal digit I of *Massospondylus* with its metatarsal allowed the toe to be "held in a permanently extended position as in *Deinonychus* . . . digit I of *Massospondylus* . . . was analogous to the dew-claw of some modern mammals, and . . . its feet were functionally tridactylous"—just the opposite, that is, of the way Weems thought *Plateosaurus* habitually held its digit I.

My manipulations of the same *Plateosaurus* foot skeleton cast studied by Weems (2003) suggest that extreme ventroflexion of the kind hypothesized by Weems, and extreme dorsoextension of the kind proposed by Cooper, are both possible, based on the fit of the joints between the metatarsal and phalanges of digit I (fig. 8.76), but I have some doubt that either of these was the routine carriage of the digit. Mallison's (2010*b*: Fig. 6) digital reconstruction of the foot of *Plateosaurus* in "probable standing pose" shows digit I angling medially relative to digits I–IV, consistent with that aspect of Weems's interpretation. Mallison's interpretation of the range of motion of digit I relative to its metatarsus includes ventroflexion of the kind proposed by Weems, and dorsoextension as proposed by Cooper, but does not show digit I either flexed backward/downward, or cocked upward, in the foot's "probable standing pose," instead being roughly parallel to the other toes in medial view, with the ungual of digit I just clearing the substrate (cf. fig. 8.76*B*, 8.76*C*, 8.76*F* here).

Weems (2003) suggested that the maker of large *Eubrontes* was also responsible for the peculiar trackway known as *Eubrontes* (formerly *Gigandipus*) *caudatus* (figs. 8.9*J*–8.9*L*; 8.77). In addition to the highly unusual tail drag mark seen in this trackway, there are impressions of what very much looks like digit I associated with some of the prints. The digit I impression appears to curve medially and backward from the inside of the print, and Weems (2003: 302) suggested that "*Gigandipus* tracks were made by a dinosaur in which the clockwise rotation of digit I away from the plane of digits II to IV was more extreme than in the foot of *Plateosaurus*."

While I agree that it is possible that *E. caudatus* might well be a preservational rather than anatomical variant of standard large *Eubrontes*, I am suspicious of the appearance of that curved digit I impression. For one thing, it is very slim (figs. 8.9*J*–8.9*K*, 8.77), and not at all like the stout mark I would expect to see in a prosauropod digit I impression. Furthermore, it is suspiciously similar in shape to the kind

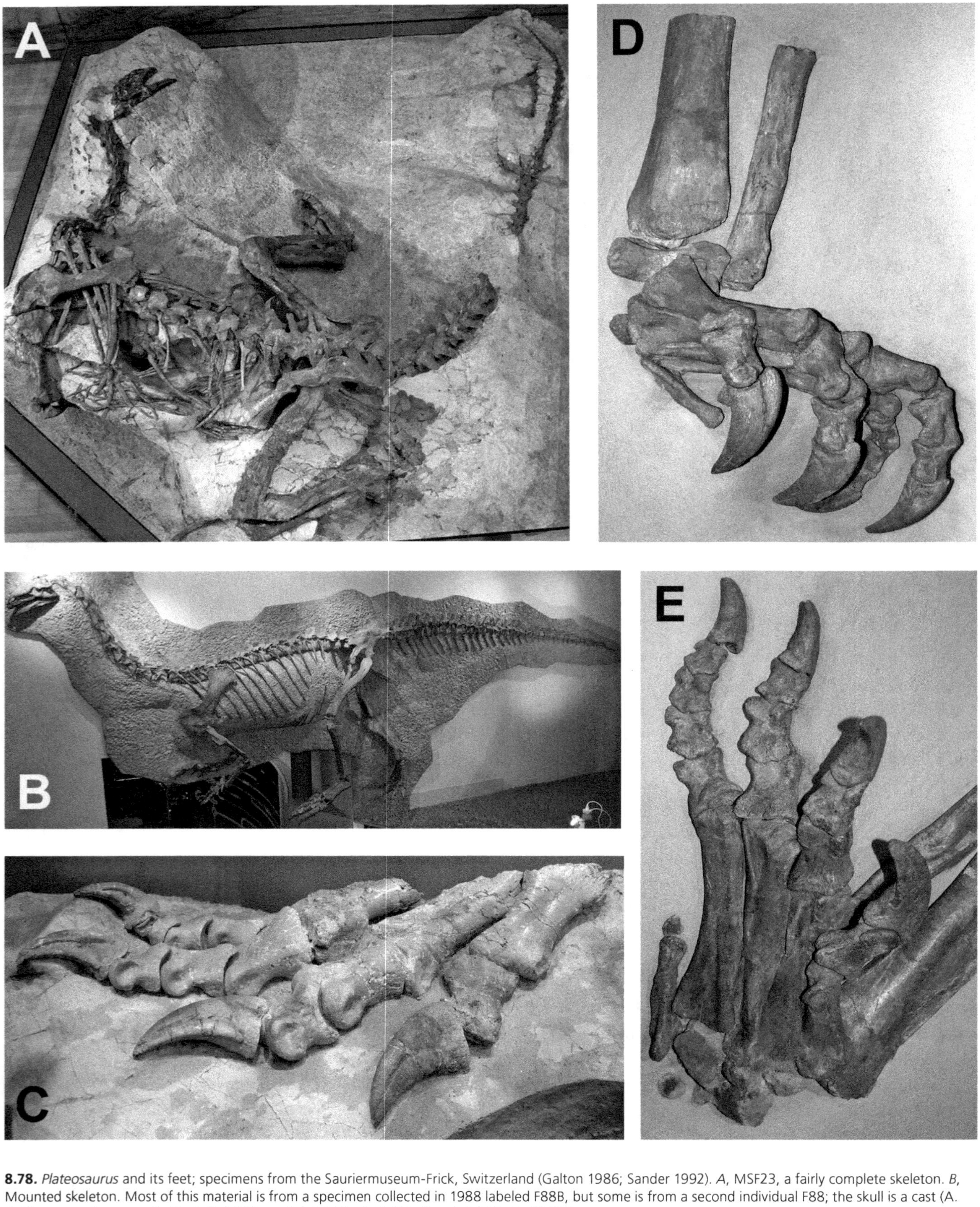

8.78. *Plateosaurus* and its feet; specimens from the Sauriermuseum-Frick, Switzerland (Galton 1986; Sander 1992). *A*, MSF23, a fairly complete skeleton. *B*, Mounted skeleton. Most of this material is from a specimen collected in 1988 labeled F88B, but some is from a second individual F88; the skull is a cast (A. Oettl, personal communication). *C*, MSF3, partial right foot. *D*, *E*, Left and right feet of MSF4.

of mark that Gatesy et al. (1999) described as an extramorphological creation of the penetration of soft sediment by a typically oriented theropod digit I (but I grant that the *Gigandipus* prints appear to be rather shallowly impressed, rather than deeply punched into the substrate). I don't think that this part of Weems's argument is very strong.

An additional reason for doubting that digit I was habitually held in the position advocated by Weems comes from the preservation of articulated foot skeletons of *Plateosaurus* from the Late Triassic Knollenmergel bonebeds of central Europe (Sander 1992). The unfortunate dinosaurs seem to have died slow and possibly gruesome deaths after becoming trapped in sticky mudflats. Their skeletons are preserved with "the legs spread apart in the 'frog-kick' pose" (Sander 1992: 261); "The leg is by no means preserved in one plane, however. The bones are actually sloping down into the sediment towards their distal end so that the foot comes to lie considerably deeper in the sediment than the acetabulum" (Sander 1992: 284).

That being the case, the position of the bones of the foot in these poor creatures may say something about how the toes were held in life, or at least at the end thereof. If *Plateosaurus* had routinely walked with digit I strongly ventroflexed and bent backward behind the rest of the metatarsals, I would have expected some of the trapped individuals in the Knollenmergel mudstones to have digit I preserved in this position. As the foot sank into the sediment, the first toe would have been pushed downward by the animal's weight, and been pinned down in the mud by the rest of the foot. Alternatively, if an animal *in extremis* had been trying to claw its way free from an entombing mire, I would expect all of its pedal digits to be ventroflexed to their maximum extent.

The way the Knollenmergel plateosaurs are actually preserved (Sander 1992: Figs. 3, 7, 12; fig. 8.78 here) does not seem consistent with the extreme ventroflexion of digit I hypothesized by Weems. In some specimens (fig. 8.78C) the first toe looks to be extended forward in a manner similar to Mallison's "probable standing pose." However, in at least one specimen (fig. 8.78*D*) the claws are curled downward as though the dinosaur had been trying to claw free, but metatarsal digit I is bent downward/backward no more than the other metatarsals. In the same individual, digit I of the opposite foot is actually bent upward in a manner approaching the carriage proposed by Cooper. Although it would be a mistake to push interpretations based on death poses too far, I think the way the feet of *Plateosaurus* are preserved in the Knollenmergel implies that during regular walking digit I is more likely to have been held in a position like Mallison's standing pose than the extreme ventroflexion of Weems or the extreme dorsiflexion of Cooper (noting, of course, that Cooper was talking about *Massospondylus* rather than *Plateosaurus*).

Keeping that in mind, and as previously noted (fig. 2.25), prosauropods have a relatively long digit I compared with the length of digit III. The same trend is seen if the length of digit I is plotted against the proxy for the length of the footprint digit III impression (figs. 8.71*D*, 8.79A). As with the pseudosuchian *Poposaurus*, the oddball theropod *Spinosaurus*, and a host of basal ornithopods and ceratopsians, digit I is relatively long across the board in prosauropods and basal sauropodiforms (von Huene 1907–1908, 1932; Broom 1911; van Hoepen 1920*a*; Young 1941*a*, 1941*b*; Galton 1971, 1976*a*, 1976*b*, 1984*a*, 1986, 1998; Bonaparte 1971; Galton and Cluver 1976; Cooper 1981 Galton and Van Heerden 1985; Moser 2003; Yates 2003; Galton and Kermack 2010; Sertich and Loewen 2010; Xing et al. 2016*j*; Y.-M. Wang et al. 2017), in contrast to the relatively short digit I in most theropods. Although it would depend on the thickness of the soft tissues underlying the pedal phalanges, and the length of the horny claw surrounding the ungual of the first toe, it seems likely to me that the long digit I of prosauropods would routinely have left some kind of mark in the substrate (cf. D'Orazi Porchetti and Nicosia [2007: Fig. 10])—unlike what is seen in most large *Eubrontes*.

Farlow and Galton (2003) described poorly preserved possible manus prints associated with a *Eubrontes* trackway at Dinosaur State Park (Rocky Hill, CT). The better-preserved of the two prints seemed to record the impressions of four fingers, but we did not think that the impression matched the expected skeletal morphology of any potential trackmaker. We noted that our manus print lacked a large clawmark for digit I, unlike the putative prosauropod trace fossil *Navahopus* (Baird 1980). Subsequent workers have argued about whether the large manual clawmark of *Navahopus* is in fact anatomical rather than extramorphological, and whether the *Navahopus*-maker was in fact a prosauropod (cf. Rainforth 2003; Irmis 2005; Hunt and Lucas 2006; and references therein). Weems (2006*a*) noted this controversy, and suggested that the Dinosaur State Park handprint morphology would better fit a prosauropod than a theropod manus, particularly if the trackmaker had extended the distal ends of its digits upward off the ground during locomotion. The argument is interesting, and I do not reject it out of hand, but it isn't clear to me that a prosauropod hand would necessarily have been more likely to leave a four-fingered or knobbed impression than would a theropod hand (cf. Gilmore 1920: Figs. 60, 62; Galton 1976*a*: Fig. 7; Welles 1984: Fig. 29; Xu et al. 2009; Carrano and Choiniere 2016).

Another of Weems's (2003, 2006*a*) morphological arguments against theropod and for prosauropod candidacy of

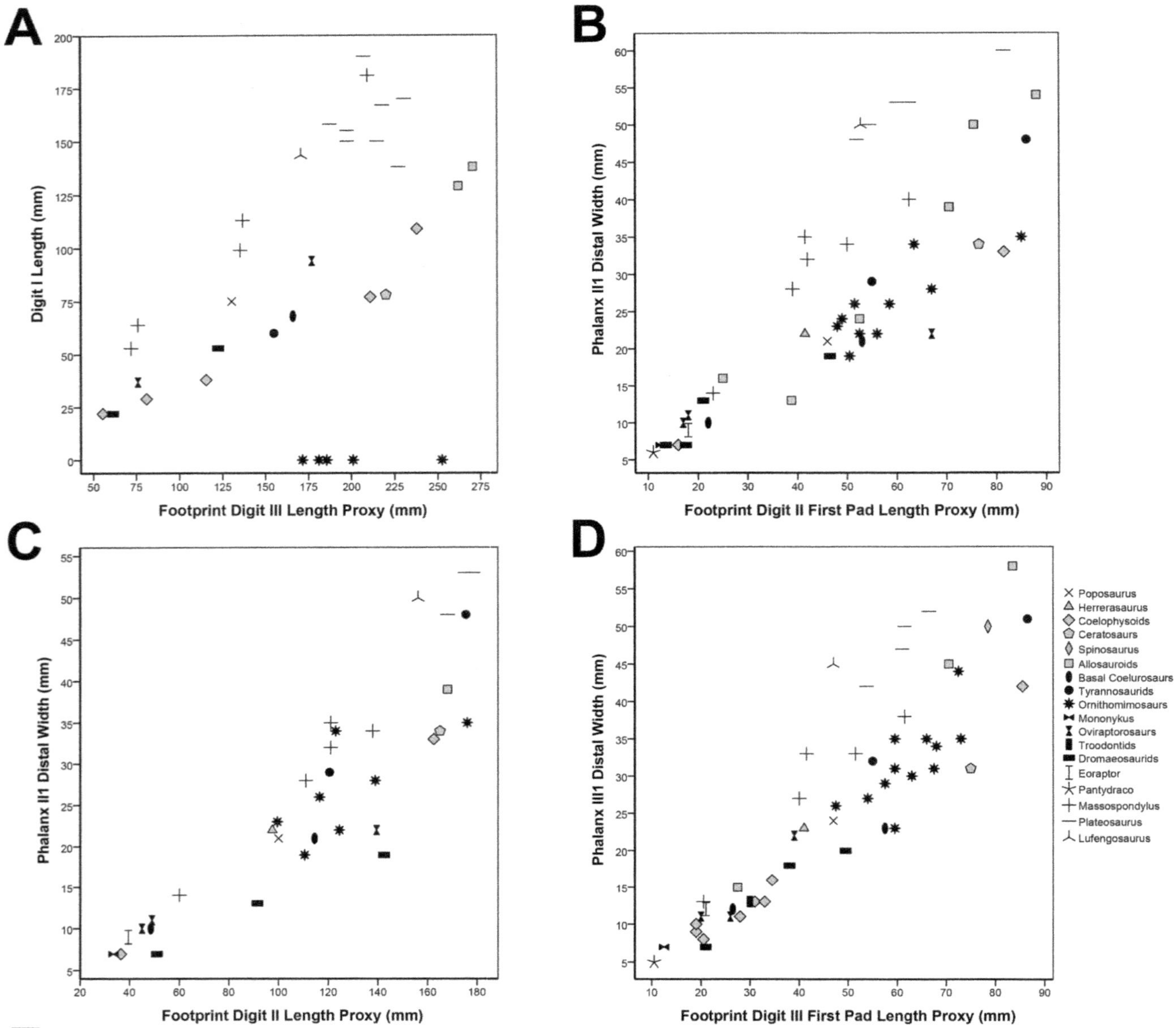

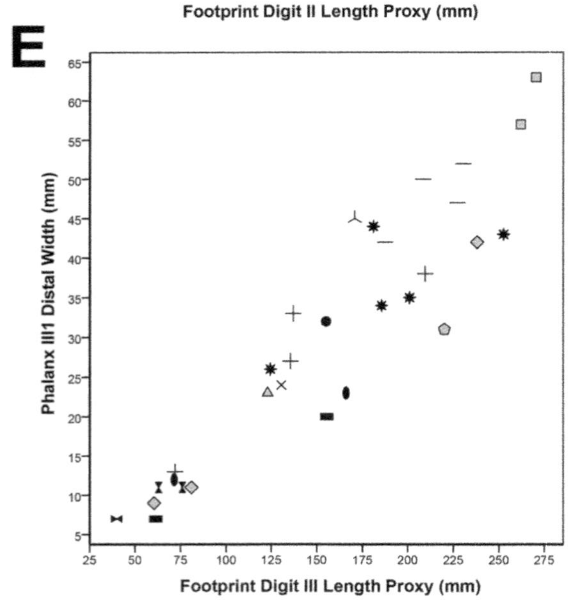

8.79. Detailed comparison of the proportions of basal sauropodomorph (mostly prosauropod) feet with those of other dinosaurs and the pseudosuchian *Poposaurus*. Data are limited to forms roughly within the size range of the potential makers of Newark Supergroup tridactyl dinosaur footprints (digit III length proxy 350 mm or less). Data for *Plateosaurus*, *Massospondylus*, *Pantydraco*, and *Lufengosaurus* in these plots are taken from the literature (Broom 1911; van Hoepen 1920*a*; von Huene 1907–1908, 1932; Young 1941*a*; Galton and Cluver 1976; Galton and Kermack 2010), so measurements may not have been made the same way as for the study described in the text; data for *Ammosaurus*, *Anchisaurus*, and *Eoraptor*, on the other hand, are taken from this study. *A*, Digit I length vs. the proxy for the length of the digit III impression in footprints (cf. fig. 8.71*D*). *Poposaurus* and the prosauropods have a relatively longer digit I than most theropods. *B*, Phalanx II1 distal width vs. the proxy for the length of the first digital pad of digit II (cf. fig. 8.72*A*). Large prosauropods have a relatively stouter toe than theropods (especially coelophysoids) in this comparison, although *Allosaurus* comes close to the prosauropod trend. *Poposaurus* plots among the theropods (as it does in *C*–*E*). *C*, Phalanx II1 distal width vs. the proxy for the overall length of the digit II impression (cf. fig. 8.72*B*). Separation of theropods from prosauropods is less marked than in *B*. *D*, Phalanx III1 distal width vs. the proxy for the length of the first digital pad of digit III (cf. fig. 8.72*C*). Once again prosauropods tend to be broader-toed than theropods (especially coelophysoids), although some bigger theropods come close to the prosauropod relationship. *E*, Phalanx III1 distal width vs. the proxy for the overall length of digit III (cf. fig. 8.72*D*). There is again overlap between the stouter-toed large theropods and prosauropods, but coelophysoids are skinnier-toed.

Table 8.12. Comparison of relative toe pad widths of well-preserved large *Eubrontes* (mean or single treatment) with toe pad width proxies from selected dinosaur foot skeletons. For foot skeletons, the value in parentheses increases the raw value of the phalanx distal width by a factor of 1.5 to account for soft tissues.

Specimen	Digit	First pad width (prints) or first phalanx distal width (skeletons) divided by first pad or proxy length (%)	First pad width (prints) or first phalanx distal width (skeletons) divided by digit impression or proxy length (%)
Footprints			
AC 56/1 *Eubrontes minusculus* (digit II length = 143 mm; digit III length = 192 mm)	II	89.5	35.7
	III	91.8	29.2
AC 16/1 *Eubrontes minusculus* (digit II length = 143 mm; digit III length = 195 mm)	II	89.0	39.5
	III	86.2	30.5
AC 45/2 *Eubrontes minusculus* (digit II length = 171 mm; digit III length = 224 mm)	II	77.9	31.0
	III	87.0	29.9
AC 15/3 *Eubrontes giganteus* (digit II length = 188 mm; digit III length = 286 mm)	II	74.4	35.6
	III	67.9	25.9
YPM 2098 *Eubrontes giganteus* (digit II length = 161 mm; digit III length = 199 mm)	II	75.9	39.1
	III	90.3	32.7
AC 13/4 B1-B2 *Eubrontes platypus* (digit II length = 140 mm; digit III length = 194 mm)	II	133.3	48.6
	III	100.0	32.0
Foot skeletons			
USNM 10924 *Plateosaurus longiceps* (digit II length proxy = 171 mm; digit III length proxy = 236 mm)	II	92.6 (138.9)	29.2 (43.9)
	III	76.0 (114.0)	24.2 (36.2)
UCMP 37302 *Dilophosaurus wetherilli* (digit II length proxy = 163 mm; digit III length proxy = 238 mm)	II	40.2 (60.4)	20.2 (30.4)
	III	48.8 (73.3)	17.6 (26.4)
MCF-PVPH-236 *Aucasaurus garridoi* (digit II length proxy = 165 mm; digit III length proxy = 220 mm)	II	44.2 (66.2)	20.6 (30.9)
	III	41.3 (62.0)	14.1 (21.1)
AMNH 6125 *Allosaurus* sp.	II	67.4 (101.1)	—
AMNH 324 *Allosaurus fragilis*	II	65.8 (98.7)	—
	III	84.0 (125.9)	—
DNM DINO 11541 *Allosaurus* sp.	II	65.8 (98.7)	—
	III	69.0 (103.6)	—
DNM Quarry Wall *Allosaurus fragilis* (digit II length proxy = 168 mm)	II	54.9 (82.4)	23.2 (34.8)
	III	63.4 (95.1)	—
UMNH VP C481 *Allosaurus* sp.	II	45.3 (67.9)	—
USNM 4734 *Allosaurus fragilis*	II	61.4 (92.0)	—
	III	78.5 (117.7)	—
USNM 8423 *Allosaurus fragilis*	III	74.5 (111.7)	—
MOR 693 *Allosaurus fragilis* (digit II length proxy = 218 mm; digit III length proxy = 262 mm)	II	47.8 (71.7)	20.2 (30.3)
	III	60.0 (90.0)	21.8 (32.6)
SMA 0005 *Allosaurus fragilis* (digit II length proxy = 223 mm; digit III length proxy = 271 mm)	II	61.7 (92.6)	26.0 (39.0)
	III	65.6 (98.4)	23.2 (34.9)
SDSM 30510 *Allosaurus fragilis*	II	33.3 (50.0)	—

the large *Eubrontes*-maker relates to digit stoutness: "the gracile pes elements of such theropods [as *Dilophosaurus*] seem undersized in comparison with the massive footpads of a *Eubrontes* or a *Gigandipus* footprint" (Weems 2003: 304). Indeed, prosauropod toes tend to be stouter than theropod toes (table 8.12; figs. 8.72, 8.79*B*–8.79*E*), and coelophysoids are especially slim-toed. However, if allowance is made for the thickness of enveloping soft tissues around the distal ends of proximal phalanges by using the 1.5× multiplication factor described earlier (table 8.12), and those "corrected" widths are then presented as percentages of the first digital pad length skeletal proxy or the digit impression length skeletal proxy, theropods like *Dilophosaurus* and *Aucasaurus* have relative first digital pad widths that approach the low end of values for large *Eubrontes*. *Allosaurus*, in fact, overlaps the range of large *Eubrontes* values (cf. Gierliński et al. 2001, 2004), and *Plateosaurus*, if anything, is a bit too stout-toed. Based as they are on inferred proxies for pad and digit impression lengths,

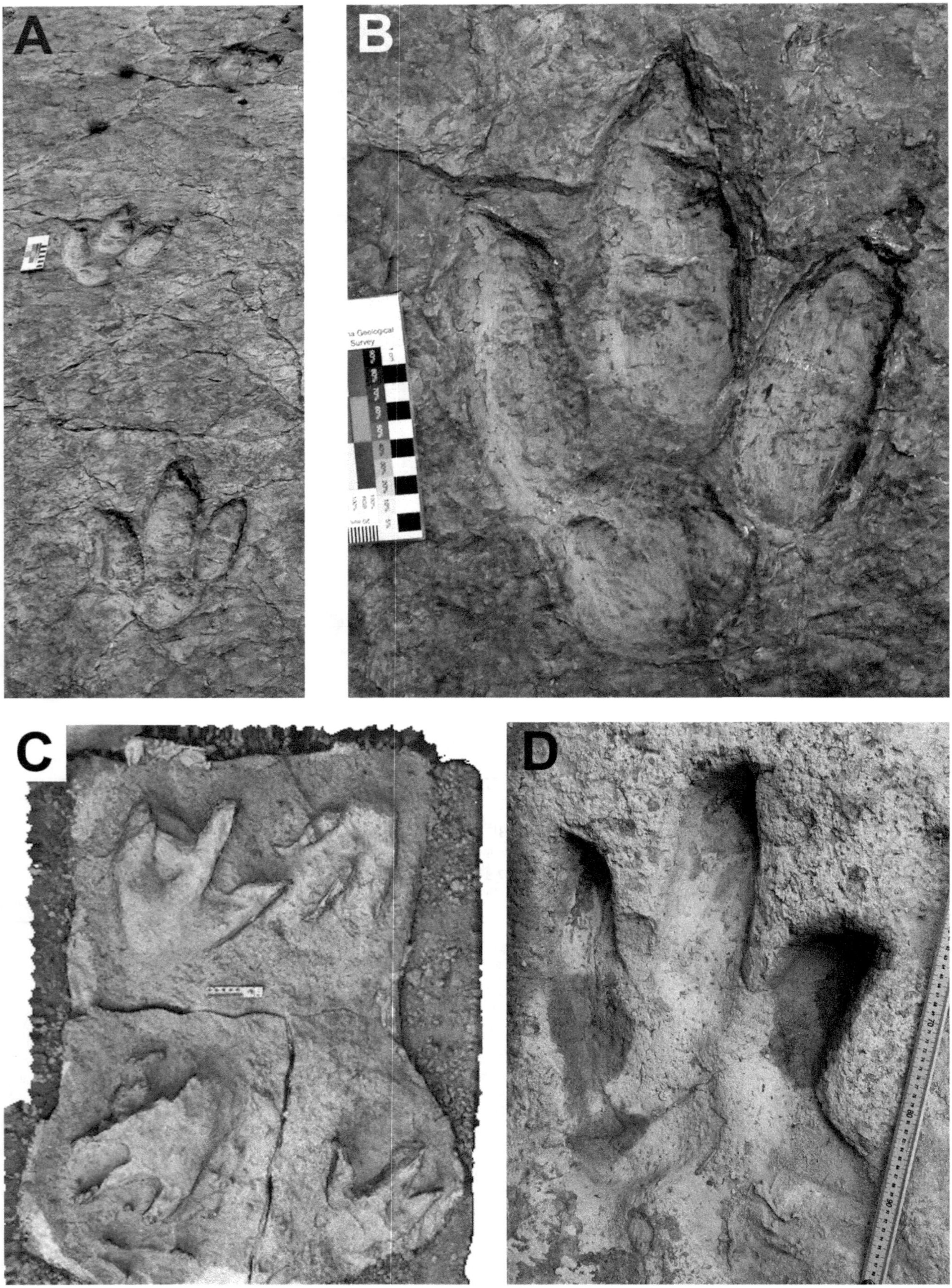
A
B
C
D

and digit pad widths, these numbers should be taken with entire evaporite pans of salt, but they suggest that the relative thickness of *Eubrontes* toemarks don't necessarily exclude theropods as candidates for their makers.

Weems (2006*a*: 373) also suggested that "digit divarication angles in [large] *Eubrontes* are relatively low. . . . By the Late Jurassic and Cretaceous, most theropod tracks show much wider digit divarication relative to earlier forms. . . . In contrast, the track maker of [large] *Eubrontes* showed no tendency toward developing the kind of divaricated digits that would be expected of a theropod that was evolving toward the later Mesozoic pes condition." This argument is a little tricky to evaluate, because when comparing measurements of divarication/interdigital angles (IDAs) in dinosaur footprints across studies, I am never certain that all authors made the measurements the same way (recall the discussion of the various ways of measuring interdigital angles in emu prints presented in chapter 6). However, I am confident that my own measurements of "best-fit" IDAs of footprints attributed to theropods from the Early Cretaceous Glen Rose Formation of Texas (e.g., Farlow et al. 2006: Table 5) were made the same way as the measurements of Newark Supergroup footprints in the present study. As previously discussed, IDA II–IV tends to increase with increasing footprint size across *Eubrontes* (fig. 8.39A). The larger forms have IDAs anywhere from 30 to 55 degrees. This range encompasses the values I obtained for Glen Rose Formation prints I attributed to theropods in the just-cited study (cf. fig. 8.80). Consequently I don't find this argument against theropods as makers of large *Eubrontes* convincing. If, however, it turns out that there really is a tendency for footprints thought to have been made by theropods of the later Mesozoic to have larger interdigital angles than typical early Mesozoic large *Eubrontes*, I suspect that this is simply a manifestation of the larger size of later Mesozoic big theropods.

Weems (2006*a*), citing Welles (1984), also suggested that metatarsal IV of *Dilophosaurus* distally angles more sharply away from the long axis of metatarsal III than does metatarsal II, such that a footprint made by *Dilophosaurus* would have a larger IDA III–IV than IDA II–III. Although Rainforth (2003) thought the distal divarication of metatarsal IV was an artifact of preservation, the same curvature occurs in metatarsal IV of *Allosaurus* (cf. fig. 1.8*B*), and so I don't rule out the possibility that a greater IDA III–IV than IDA II–III could characterize some large theropod footprints. While inspection of other feet in figure 1.8 suggests that this character doesn't necessarily hold for all large theropods, it is worth noting that in my measurements of Newark Supergroup *Eubrontes* (table A8.1), IDA III–IV often is larger than IDA II–III. So I don't think this is a strong argument against the theropod hypothesis for the large *Eubrontes*-maker.

I now turn to features of prosauropod feet that seem to me to militate against such dinosaurs as big *Eubrontes*-makers, and to favor large theropods as candidates. Chief among these is the relative lengths of the unguals (figs. 1.8, 1.10, 8.69, 8.70A, 8.70C, 8.71A, 8.71*B*, 8.76, 8.78, 8.81A–8.81*E*). Prosauropods, like many ornithischians, have whopping big unguals compared with the length of the rest of the toe skeleton. Keep in mind that in life there would have been a horny sheath surrounding the ungual which would have made the impression of the intact claw in sediment likely look proportionally even longer. Even if the claws became heavily abraded (as we saw with alligator feet), I would still expect prosauropod feet to leave relatively longer clawmarks than large theropod feet. In contrast, the clawmarks in large *Eubrontes* do not seem especially long (figs. 8.7–8.10), and their relative lengths look more like expectations for theropods than for prosauropods (compare fig. 8.73A, 8.73*B* with figs. 8.70A, 8.70*C*, 8.81A, 8.81C, 8.81*D* and fig. 8.32 with figs. 8.71A, 8.71*B*, 8.81*B*, 8.81*E*).

In addition, prosauropods might leave footprints with proportionally longer impressions of digits II and IV (compared with the length of the impression of digit III) than would most theropods other than dromaeosaurids (figs. 8.71C, 8.81*F*). The relative length of the digit II impression, at least (the digit IV/digit III ratio looks more ambiguous), of *Eubrontes* (fig. 8.34A) looks more like expectations for theropods than prosauropods.

To be fair to Weems's morphological argument, however, he did not propose that *Plateosaurus* itself, or any other known prosauropod, was the *Eubrontes*-maker. Rather, he hypothesized that a presently unknown "anatomically advanced bipedal plateosaurid could have made tracks of the *Eubrontes* type" (Weems 2003: 301–302; also see Weems 2006*a*: 373).

In addition to his morphological arguments, Weems (2003, 2006*a*) advanced some ecological arguments in favor of the prosauropod hypothesis. He noted that *Plateosaurus* is the most common large dinosaur found in the Late Triassic of central Europe, just as *Eubrontes* is the most common large dinosaur footprint type in the Early Jurassic of eastern North America. Weems thought that the occurrence of *Plateosaurus* in bonebeds was indicative of group behavior, and

Facing, **8.80.** Early Cretaceous footprints attributed to large theropods. *A*, *B*, Trackway PP146 from the Enciso Group, Peñaportillo, Spain (Pérez-Lorente 2015). *A*, Oblique view of the trackway. *B*, Second footprint in *A* (a left). *C*, *D*, Large theropod footprints from the Glen Rose Formation, Dinosaur Valley State Park, Glen Rose, Texas (Farlow et al. 2015). *C*, Cast of footprints from the Opossum Branch Site. *D*, Large footprint from the Blue Hole site. Compare the divarication of digits II–IV of these prints with those of large *Eubrontes* (figs. 8.9, 8.10).

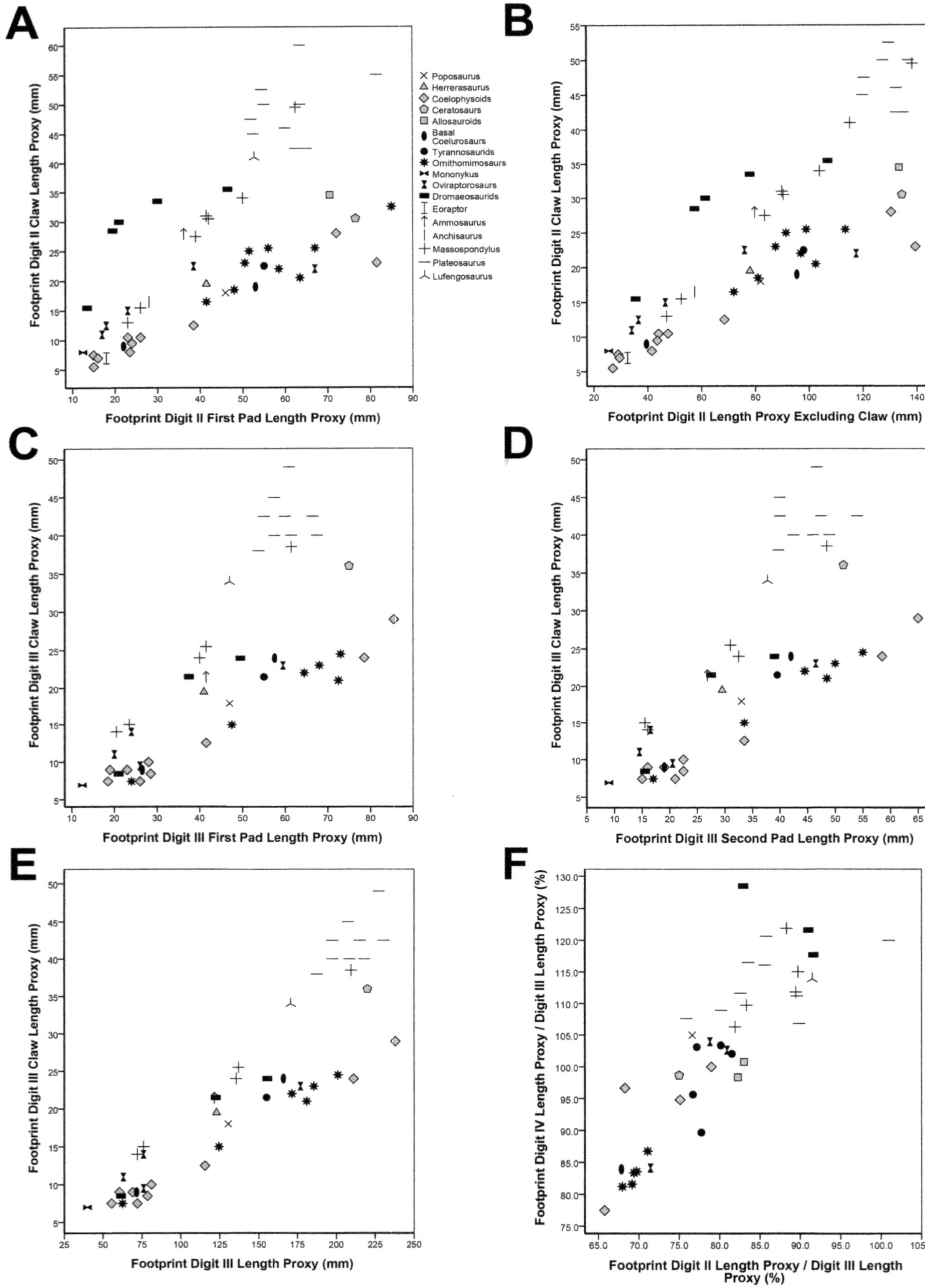
A
Footprint Digit II Claw Length Proxy (mm)
Footprint Digit II First Pad Length Proxy (mm)
Poposaurus
Herrerasaurus
Coelophysoids
Ceratosaurs
Allosauroids
Basal Coelurosaurs
Tyrannosaurids
Ornithomimosaurs
Mononykus
Oviraptorosaurs
Dromaeosaurids
Eoraptor
Ammosaurus
Anchisaurus
Massospondylus
Plateosaurus
Lufengosaurus
B
Footprint Digit II Claw Length Proxy (mm)
Footprint Digit II Length Proxy Excluding Claw (mm)
C
Footprint Digit III Claw Length Proxy (mm)
Footprint Digit III First Pad Length Proxy (mm)
D
Footprint Digit III Claw Length Proxy (mm)
Footprint Digit III Second Pad Length Proxy (mm)
E
Footprint Digit III Claw Length Proxy (mm)
Footprint Digit III Length Proxy (mm)
F
Footprint Digit IV Length Proxy / Digit III Length Proxy (%)
Footprint Digit II Length Proxy / Digit III Length Proxy (%)

Facing, **8.81.** Comparison of skeletal proxies for footprint impressions, in this case of relative claw lengths and overall digit lengths, in basal sauropodomorph (mostly prosauropod) feet with those of other dinosaurs and the pseudosuchian *Poposaurus*. Data are limited to forms roughly within the size range of the potential makers of Newark Supergroup tridactyl dinosaur footprints (digit III length proxy 350 mm or less). Data for *Plateosaurus*, *Massospondylus*, and *Lufengosaurus* in these plots are from the sources cited in figure 8.79. *A*, Digit II claw length vs. digit II first pad length (cf. fig. 8.70*A*). *B*, Digit II claw length vs. length of digit II proxy excluding the clawmark (cf. fig. 8.71*A*). *C*, Digit III claw length vs. digit III first pad length. *D*, Digit III claw length vs. digit III second pad length (cf. fig. 8.70*C*). *E*, Digit III claw length vs. length of digit III proxy excluding the clawmark (cf. fig. 8.71*B*). Although there is overlap, in all comparisons involving relative claw length prosauropods tend to have larger relative claw length proxies than do theropods (other than dromaeosaurids and the ceratosaur *Aucasaurus*) and *Poposaurus*. *F*, Overall digit proxy length ratios (cf. fig. 8.71*C*). Prosauropods tend to have relatively longer digit II and IV proxies than do theropods (other than dromaeosaurids) and *Poposaurus*.

that the occurrence of multiple parallel *Eubrontes* trackways likewise showed group behavior of a kind unlikely to have been present in large carnivorous dinosaurs.

All of these points have been called into question. Sander (1992) concluded that the Knollenmergel plateosaurs had been mired attritionally, one at a time, rather than catastrophically, in herds (although this does not necessarily mean that the beasts were not gregarious). Getty et al. (2012, 2015, 2017) have challenged the interpretation that the maker of large *Eubrontes* was a gregarious form. And finally, even if the makers of large *Eubrontes* were theropods, they may not necessarily (all?) have been carnivores (cf. Xu et al. 2009)—although speaking now strictly in terms of aesthetics rather than science, I'll be rather disappointed in them if they all turn out to have been herbivores!

But even if large *Eubrontes*-makers had been mainly or exclusively carnivores, the abundance of their tracks in footprint assemblages would not be unprecedented. In dinosaur ichnofaunas from the later Jurassic and the Cretaceous, the abundance of footprints attributed to large theropods, relative to those of large herbivorous dinosaurs, is often (but not always) much larger than expectations for the relative abundance of large carnivores and large herbivores in the living fauna—sometimes with tracks of theropods even outnumbering those of sauropods and large ornithischians (cf. Leonardi 1989; Foster and Lockley 2006; Petti et al. 2008; Belvedere et al. 2010, 2013; Moreno et al. 2012; Wagensommer et al. 2012; Abbassi and Madanipour 2014; Farlow et al. 2015; Pérez-Lorente 2015; Xing and Lockley 2016; D'Orazi Porchetti et al. 2016; Richter and Böhme 2016). Leonardi (1989) suggested that this was due to greater activity on the part of carnivorous than herbivorous dinosaurs, an interpretation supported by Farlow (2001) on the basis of what is known about the relative magnitudes of the daily movement distances and home range sizes of extant large mammalian carnivores and herbivores.

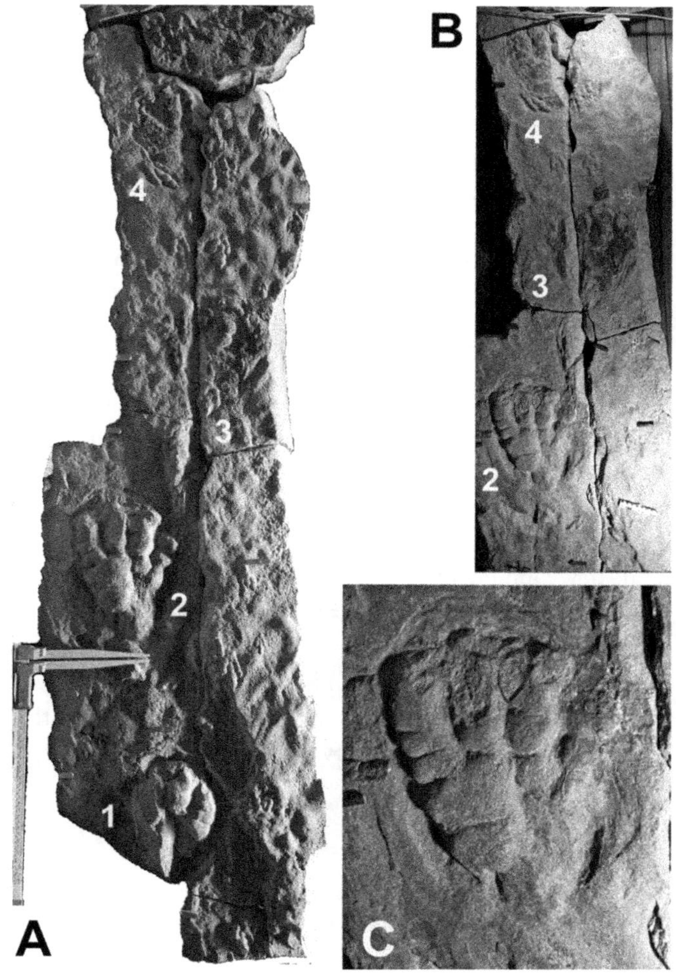

8.82. AC 4/1a, *Otozoum moodii* (cf. Rainforth 2003). *A*, Shot of the slab in rather harsh oblique light, emphasizing the topography of the slab, which is preserved as a negative copy (cast) of the original trackway. The four footprints—1 left, 2 right, 3 left, 4 right—are labeled. Large set of calipers has a total length of 86 cm. *B*, Portion of the same slab under different lighting. Note numerous other small tridactyl footprints, including the trackway of *Eubrontes cursorius* (fig. 8.1), running from top to bottom in the photograph. *C*, Detail of the second and best-preserved footprint in the trackway.

By the Late Jurassic and Cretaceous, prosauropods had long since "shuffled off this mortal coil" (when stealing a phrase, always steal from the best), and yet footprints rather like those of large Early Jurassic *Eubrontes* continued to be made by bipedal dinosaurs (fig. 8.80). Indeed, as noted at the beginning of this chapter, some workers even apply the name *Eubrontes* itself to tridactyl prints from the Early Cretaceous of China. Many or most of these tracks were surely made by theropods, in which case it seems more parsimonious to me to hypothesize that Newark Supergroup large *Eubrontes* were made by theropods—maybe not *Dilophosaurus* or other coelophysoids [cf. Gierliński et al. 2001, 2004]—than to attribute them to an unknown bipedal basal sauropodomorph with a foot structure different from that of known prosauropods, a foot structure with some of the same features seen in theropods.

But having said that, I must back off a bit. Earlier in this chapter (table 8.11) I noted that in some features large *Eubrontes* do not unambiguously point to theropods as opposed to prosauropods as their makers. Furthermore, a recurrent theme in this book has been the convergence of foot and footprint form among archosaurs that are not particularly close relatives.

If the moa *Megalapteryx* thinks it's a kiwi, if bustard feet and footprints can look like those of diminutive emus, if *Plateosaurus* feet look rather like *Tenontosaurus* feet, and if the suchian *Poposaurus* had a foot that might be mistaken for that of a dinosaur, it is certainly not inconceivable that a presently unknown large, bipedal prosauropod might have evolved a foot similar to that of large theropods. The rather labile development of digit I in non-avian dinosaurs (fig. 2.25) and ground birds (figs. 2.14, 6.50), for example, suggests that significant reduction or loss of digit I in some hypothetical bipedal basal sauropodomorph is not out of the question.

Keep in mind that the analyses presented earlier in this chapter could at best identify the minimum number of potential trackmakers. Consequently I am open to the possibility that, with more data, it might be possible to identify more than one kind of maker among the dinosaurs responsible for large *Eubrontes*. Although I would be astonished if all, or even most, early Mesozoic large *Eubrontes* had been made by prosauropods, I would be only mildly surprised if some of them had been.

Before we finish, it must be noted that there already is an uncommon ichnotaxon from the Newark Supergroup that nearly all workers would agree was made by prosauropods: *Otozoum* (fig. 8.82). Rainforth (2003) published a detailed account of the ichnotaxon, and noted that, apart from the size difference, the pes of *Otozoum* is rather similar to the pedal morphology of *Anchisaurus* (fig. 8.69)—which makes one wonder about the ontogenetic age of YPM VP 1883 at the time it died. The pes print of *Otozoum* has four rather stout digit impressions, which certainly matches expectations for a prosauropod pes. Otherwise I have not carefully examined footprints of this ichnogenus, and so I have nothing more to say about it. In any case, its maker was clearly not a tridactyl dinosaur, and thus not one of Noah's Ravens, and so will receive no further attention here.

Final Thoughts

9

"This is ridiculous," she fumed. "Further study is needed!" You could say that about anything. You could say that about . . . *paleontology*.

Christopher Buckley, *Boomsday*, 2007, p. 252 (italics in the original)

Figures often beguile me, particularly when I have the arranging of them myself; in which case the remark . . . would often apply with justice and force: "There are three kinds of lies: lies, damned lies and statistics."

Samuel L. Clemens, *Mark Twain's Own Autobiography: The Chapters from the North American Review*, 1906–1907

Well. I have used lots of statistics, and tried very hard not to lie.

To summarize: Multiple measures of within-taxon and across-taxon variability in the foot skeletons, intact feet, and/or footprints of crocodylians, non-avian dinosaurs, and ground birds, and of intraspecific vs. interspecific size-related trends in pedal and footprint proportions, lead me to conclude that at least three different kinds of dinosaurs were responsible for the Newark Supergroup footprints assigned to the classic Early Jurassic Connecticut Valley ichnogenera *Grallator*, *Anchisauripus*, *Eubrontes*, and *Anomoepus*. Comparisons with skeletal data indicate that the *Anomoepus*-maker was pretty certainly an ornithischian. At least one kind of dinosaur, most likely a theropod species of modest adult size, was responsible for *Grallator* and most *Anchisauripus*. A third kind of dinosaur, again probably a theropod, was the maker of some medium-sized and larger *Anchisauripus* and big examples of *Eubrontes*. The Late Triassic *Kayentapus*-maker would have been a fourth kind of dinosaur, yet again probably a theropod. With additional well-preserved specimens thrown into the database, it might be possible to subdivide these trackmaker categories into still more kinds of dinosaurs (cf. Dalman and Weems 2013).

None of these conclusions is particularly new or startling, in which case it might be objected that all my effort has amounted to an unnecessary waste of time. The response to this objection, it seems to me, is obvious. If increasing the number of measurable footprint characters, or at least looking at characters in different ways, generates results that support earlier hypotheses about the number and kinds of trackmakers, this will bolster the confidence one has in those hypotheses. This is similar to discovering that new characters and new specimens used in a phylogenetic analysis support the results of previous phylogenetic analyses that had been based—depending on the study—on fewer characters or specimens. Even so, while I hope that I have made a contribution to sorting out the biodiversity of Connecticut Valley trackmakers, I think that my conclusions remain only tentative. There is always more that could be done.

In this chapter I will tie up some miscellaneous loose ends and offer some suggestions, based on what I have learned over the course of this study, about ways to improve interpretations of the makers of tridactyl dinosaur prints in other ichnofaunas, and comparisons of footprints and trackmakers across footprint faunas. I will end by posing some questions about the limits of tridactyl footprints as proxies for the biodiversity of their makers.

LOOSE ENDS

Footprint Preservation and Extramorphological Variability. Although I have from time to time acknowledged (especially in chapter 6) variability in footprint shape that reflects the influences of sediment consistency, or the foot-substrate interaction, this has not been a major thrust of this study. Such influences have, however, become an important emphasis in vertebrate ichnology (Thulborn and Wade 1989; Allen 1989, 1997; Platt and Meyer 1991; Brand 1996; Gatesy et al. 1999; Gatesy 2001, 2003; Nadon 2001; Romano and Whyte 2003; Manning 2004; Milàn et al. 2004, 2005, 2006; Henderson 2006; Milàn 2006; Milàn and Bromley 2006, 2008; Graversen et al. 2007; Jackson et al. 2009, 2010; Wilson et al. 2009; Falkingham et al. 2010, 2011*a*, 2011*b*; Avanzini et al. 2012; Huerta et al. 2012; Platt et al. 2012; Carvalho et al. 2013; Schanz et al. 2013; Alcalá et al. 2014*b*; Cariou et al. 2014; Falkingham 2014, 2016; Falkingham and Gatesy 2014; Razzolini et al. 2014; Dai et al. 2015; Lockley and Xing 2015;

Pérez-Lorente 2015; Therrien et al. 2015; Cobos et al. 2016; Gatesy and Ellis 2016; Loope and Milàn 2016; Milàn and Falkingham 2016; Scisio et al. 2016; Xing et al. 2016*m*).

Among the issues raised by this research is the not-so-simple matter of deciding what the boundaries of a footprint are. "Defining where a track begins and ends is not trivial due to the continuous nature of track and surrounding substrate, particularly in fine-grained, cohesive substrates that exhibit large areas of deformation" (Falkingham 2016: 74). In this study, I have usually (but not always) worked with negative copies (casts) of footprints, because I find it easier to interpret a shape coming out of a plane toward me than a shape sinking into a plane away from me. In defining such features as the proximal ends of toemarks, and the "heel," I have probably used (unconsciously) "the inflexion point . . .the location at which the slope of the edge of the track impression changes, usually from curving upward to curving downward" (Falkingham 2016: 74). It undoubtedly helps that many of the footprints I have worked with, whether of emus or non-avian dinosaurs, have been clearly registered (preservation class 2 or 3 in the scheme of Belvedere and Farlow [2016]; cf. Hornung et al. [2016]). My emu prints have all been "true tracks" rather than "undertracks" (Marty et al. 2016), and I suspect that many of the Newark Group prints I examined were true tracks or close to being true tracks. I operate under the assumption (or at least the hope) that the vagaries of footprint creation collectively operate as a filter, through which, with enough data, especially when taken from high-quality prints, it is nonetheless possible to detect a real morphological signal despite preservational noise.

I offer the trackway data discussed in chapter 4 as a reason for hope. Manning (2008) and Falkingham (2014) showed that footprint length measurements are substantially affected by whether a print is a surface, true track, or an undertrack, and that this could affect estimates of trackmaker speed by as much as tenfold. That would be preservational noise with a vengeance! It is almost certain that some (many?) of the measured trackways in table A4.2 are based on undertracks, or if true tracks have dubious measurements due to other confounding extramorphological influences. Even so, the patterns of stride length relative to footprint length documented in figures 4.15–4.19 look pretty consistent and convincing, suggesting that preservational noise has not completely effaced the trackway signal.

Geometric Morphometrics. Several authors have examined footprint shape variability using geometric morphometrics (Rasskin-Gutman et al. 1997; Rodrigues and Santos 1999; Belvedere 2008; Clark and Brett-Surman 2008; Castanera et al. 2015, 2016; Hornung et al. 2016; Lallensack et al. 2016), and Wings et al. (2016: 69) suggest that this approach "may become a new standard in vertebrate ichnology." In this book I have used traditional morphometrics, partly because I felt more comfortable with such techniques. However, I also felt that traditional morphometric analyses of footprint shape would allow more direct comparison of the proportions of footprints (e.g., digit pad lengths and widths, and claw mark lengths) with those of pedal skeletal features (phalanx lengths and widths, and ungual lengths), and with measurements of intact feet. However, when it comes to data analysis, I am a "both/and" rather than an "either/or" kind of guy (as should be obvious from the many graphs and tables in this book that examine the same feature in different ways), and so I think that traditional and geometric morphometrics can both yield useful results. It would be very interesting simultaneously to use the kinds of detailed measurements described in figure 8.16 along with geometric shape analysis in a study of well-preserved dinosaur footprints. One hopes that they would tell a similar story.

Intraspecific and Interspecific Variability in Foot and Footprint Shape. In chapters 5 and 6, I examined intraspecific (including ontogenetic) variability in the size and shape of feet and footprints of alligators and emus, respectively. These two species sit on either side of the extant phylogenetic bracket around non-avian dinosaurs, and so comparing results of analyses of foot and footprint shape in these extant animals should provide useful data for interpreting intraspecific variability in the feet and footprints of non-avian dinosaurs. But these are only two species, of course. It would be very interesting and useful to have comparable data for other species of extant crocodylians and ground birds (the study skin data in chapter 6 were presented as rough approximations of the data potentially available for other bird species)—and so, yes, the contemptuous comment from the character in Christopher Buckley's comical novel notwithstanding, I do think that more study is needed. And not just of crocodylian and bird feet and footprints. It would also be nice to have more data on foot skeletons of non-avian dinosaurs. A dream come true for me would be for somebody to find a death assemblage of several articulated skeletons of a single species of allosaur or tyrannosaur spanning a wide range of sizes and ages, all with a complete set of bones from the hindlimbs and feet.

For me, the single most surprising result of this study was the decline in interdigital angle (IDA) II–IV and in relative footprint width from footprints of small to large emus (figs. 6.20, 6.21*G*, 6.21*H*). I was expecting to see just the opposite, based on what occurs across small to large *Eubrontes* (figs. 8.38, 8.39A). Is this a peculiarity of the emus I happened to work with, or with emus more generally? Or does it occur in other species of ratites, and in other clades of large ground birds, like bustards? Is the decline in IDA II–IV due to a morphological change in the foot itself, or is it related to differences in the foot/substrate interaction between smaller and bigger birds? At this point I just don't know; footprint measurements from youngsters to adults of more species of modern big ground birds would be very nice to have.

It would also be interesting to examine carefully the distal ends of the tarsometatarsus from hatchling to adult emus (cf. Falk et al. 2011; Buckley 2015; Buckley et al. 2016). Would there be any size-related change in the angulation between the distal articular ends of the two peripheral metatarsals and the tarsometatarsal shaft? If so, this might be expressed in changes in the value of IDA II–IV between little and big individuals of the birds.

What makes this matter of some interest is that one of the features thought to distinguish (albeit imperfectly) footprints of Mesozoic birds from those of non-avian dinosaurs is a larger IDA II–IV in bird prints (Lockley and Rainforth 2002; Wright 2004; but note that Falk et al. [2011] and Buckley et al. [2016] do not consider this to be the most useful discriminator). Weems and Kimmel (1993) described a small tridactyl footprint from the Upper Triassic Culpeper Basin of Virginia that they noted was similar in form to Edward Hitchcock's *Plesiornis* from the Lower Jurassic Portland Formation of Massachusetts. They noted the larger IDA II–IV of *Plesiornis* (ca. 70 degrees) compared with that of larger, more typical early Mesozoic footprints attributed to theropods (30–50 degrees), which led them to "view *Plesiornis pilulatus* as a primitive bird or near-bird" (Weems and Kimmel 1993: 396; cf. Gierliński 1996*a*, Gierliński et al. 2017).

While this interpretation is certainly plausible, if the decrease in IDA II–IV with increasing print size in emu prints occurs in big ground birds more generally, might it also have been true of non-avian, bipedal, tridactyl dinosaurs? Might the *Plesiornis*-maker have been a very young individual of a non-avian dinosaur species that would grow up to make footprints with lower values of IDA II–IV? For that matter, could some of the many Cretaceous footprints presently attributed to birds in fact be prints of baby non-avian dinosaurs?

If the Dinosaur Dream Wish List Fairy were to grant my above-expressed desire for a series of feet from individuals of different size/ontogenetic age of a single species of theropod of large adult size, one of the things I would examine carefully is the distal ends of metatarsals II–IV (cf. Farlow et al. 2013; Buckley 2015), as suggested above for emus. Would there be any size-related change in the angulation between the distal articular ends of the two peripheral metatarsals and their shafts?

APPLICATIONS

Site Description. Imagine that we have a newly discovered tracksite with lots of trackways of tridactyl bipedal dinosaurs. How might the results of this study be dragooned into helping to make an interpretation of the biodiversity of trackmakers responsible for the footprints?

We would need to start by considering the abundance and quality of the footprints. A quantitative analysis of footprint size and shape would be of limited value if there are only a few footprints at the site, while a site with some 10,000 individual footprints would probably require judicious sampling, unless the researcher were blessed with an army of assistants. If the footprints are, for whatever reason, of uniformly poor quality (cf. Belvedere and Farlow 2016), it probably wouldn't be worth the effort to make a lot of detailed measurements on them. It is worth remembering here that I selected the classic Newark Supergroup dinosaur footprints as the text case for the methods of this study in the first place because so many of them are so beautifully registered/preserved.

But assuming that the footprints are reasonably well preserved, and that there are enough of them to make quantitative analysis worthwhile, the next step would be carefully to examine the prints and the trackways in which they occur, and decide which of the measurements described in this study (or other measurements that I did not consider) can be made, and/or, if the researcher wants to employ geometric morphometrics, which landmarks can be defined. I suspect that these decisions will to some extent have to be made on an ad hoc basis from site to site, depending on how well the tracks are preserved. Can discrete digital pads and clawmarks be recognized, or are the toe impressions simple undivided troughs poking forward from the "heel"? This will determine whether digital pad lengths and widths, and claw lengths, are measured, or whether it is possible only to measure overall digit lengths. It will also determine whether the overall digit lengths are measured from the proximal end of the first digital pad (as in "Lgth" measures of emu prints; chapter 6), or whether overall digit lengths are measured from the "heel" to the toetip (analogous to "LGL" measures of emu prints). Because different persons might "see" the boundaries of footprints differently, it would be a good plan in documenting tracksites to have the same person make all the measurements on the footprints, or at least to be sure that all persons taking such data are following the same explicit measurement protocols.

Assuming that the individual footprints at the site can be assigned to trackways, the next step is to examine within-trackway variability of footprint parameters. Using many of the same parameters employed in the present study, Farlow et al. (2006: Table 3) did this for footprints at a tracksite from the Early Cretaceous Glen Rose Formation of Texas. The least variable within-trackway footprint parameters at the site included the distances between toetips II–III, III–IV, and II–IV, and the backfoot length. Somewhat more variable were the distances from the "heel" to hypex II–II, and from the "heel" to toetip II. The most variable within-trackway parameters were the digit III projection and the digit II lateral hypex length, the digit III medial hypex length, and the digit IV medial hypex length. One might want to put more emphasis on the least variable footprint parameters at the site, and treat the most variable footprint parameters with a bit of caution.

At this point I would not conclude that the parameters that were most or least variable at this Texas site would necessarily be the most or least variable within-trackway parameters at all sites, even within the Glen Rose Formation. I suspect that the ranking of within-trackway variability of parameters will have to be determined on a site-by-site basis.

After this the next step would be to select the parameters used for principal component analysis (PCA), cluster analysis, and discriminant or canonical variate analysis (CVA). As in chapter 8, it would be necessary to look for the best compromise between the number of trackways or singleton footprints (assuming that the analysis is done as a mean-or-single treatment), and the number of parameters that can be measured on the footprints. This, too, will likely have to be done on an ad hoc basis.

If the analyses suggest the existence of distinct morphological groupings within the footprint sample, these can be further investigated, as was done in chapter 8, with larger sample sizes using bivariate analyses, using the variables that the PCA and CVA identified as having the greatest impact in separating the groups. Shape variability can also be compared across the identified groups using analysis of covariance (ANCOVA) to see if the groups are indeed significantly different in proportions.

The shape variability within and across groups can also be compared with that observed in intact alligator feet, emu footprints, and bird study skins, as reported in chapters 5, 6, and 8. Is the variability within or across groups comparable to, less than, or greater than that observed in the modern species? Data for a variety of both simple (e.g., maximum/minimum ratios, and standard deviations or coefficients of variation of scaled parameters, and the coefficient of dispersion around the reduced major axis) and more complex measures of variability (GM-scaled parameters) within species are provided in data tables throughout this book. If the variability (however it is measured) between potential footprint morphotypes is consistently greater than that observed in the modern species, this would suggest—but, I emphasize again, not prove—that more than one kind of trackmaker is involved. If the morphotypes differ in size, are the differences between them consistent with what is observed in comparisons of footprints or feet from juveniles to adults of emus and alligators? If the former, it might not be possible to reject the hypothesis that we are seeing an ontogenetic sequence, but if the latter, we would suspect that we are looking at tracks of different kinds of animals. Keep in mind, too, that the methods developed in this study can probably yield but a minimum estimate of the biodiversity of a footprint sample.

I would not include trackway parameters (pace, stride, pace angulation, footprint rotation) in the characters used to search for distinct morphotypes. I would, rather, stick with measurements of the footprints themselves. This is because the way the animal places its feet on the ground is to a great degree under the animal's control, while the shape of its foot presumably is not. It seems to me, therefore, that the trackway pattern constitutes a different category of variables than individual footprint measurements, and should be treated separately. But once putative morphotypes are identified on the basis of print shape, it would then be appropriate to ask if there are recognizable differences in trackway pattern between the footprint morphotypes, as was done in chapter 8.

Once we have an interpretation of the minimum number of kinds of trackmakers, we can attempt to identify the kinds of dinosaurs responsible for the footprints. Chapters 2, 3, 4, and 8 examined this matter in some detail, noting characters of phalangeal and digital proportions that differ among different kinds of bipedal or potentially bipedal tridactyl dinosaurs. Some distinctions are quite dramatic, such as those between large theropods and large ornithopods in digit stoutness (figs. 2.17, 2.24, 3.24). Other distinctions, such as those between small ornithischians and small theropods, are rather trickier (unless, of course, there is a digit I impression).

Of particular concern to me, because the footprints that interest me most are the tridactyls from the Glen Rose Formation of Texas, is the question of the extent to which footprints of large theropods can be told apart on the basis of footprint morphology alone. The results are, regrettably, somewhat discouraging. The shapes of foot skeletons of large theropods (apart from *Spinosaurus*) don't seem to differ that much (figs. 2.16–2.21, 3.1–3.7; cf. Farlow 2001, Farlow et al. 2013, 2014; Romano and Citton 2017). Farlow et al. (2013) did note, however, that tyrannosaurs appear to have a relatively longer phalanx IV2 length, compared with the length of phalanx IV1, than do allosaurs, something that might manifest itself in the relative lengths of the basal-most digital pads of digit IV, if distinct digital were present in such prints. Farlow et al. (2013) also noted that tyrannosaurs tend to have a longer metatarsal IV length for a given metatarsal II length than do allosaurs (cf. Holtz 1994), and this might cause the impression of digit IV to stick further forward, relative to the digit II impression, in tyrannosaur than in allosaur footprints. Other than those features, though, most large theropod foot skeletons look pretty much alike, and so one would suspect that their prints would do the same.

I suggest that tridactyl dinosaur footprints should be identified as to their zoological affinities on the basis of distinctive features of footprint morphology alone. Put another way, the footprint should be identified only to the systematic level at which the print can be correlated with skeletal features of trackmaker candidates. If, for example, a footprint morphology can only be correlated with skeletal features of large theropods generally—and not specifically with coelophysoids, ceratosaurs, allosaurs, or tyrannosaurs to the exclusion of the other categories of large theropods—then the print should be labeled a large theropod track. I also suspect that many indifferently preserved prints should be identified as bipedal dinosaur prints, and leave it at that.

Now, it will undoubtedly be possible to say, on the basis of stratigraphic and geographic occurrence, what group of large theropods is responsible for the print identified as a large theropod print. A large theropod print from the Early Cretaceous of Texas is most likely to have been made by an allosaur (Langston 1974; Farlow 2001; Carpenter 2016), and a large theropod print from the Late Cretaceous of western North America is most likely to have been made by a tyrannosaur (Lockley and Hunt 1995; Fanti et al. 2013*a*; McCrea et al. 2014*a*). But I would not identify such prints specifically as allosaur or tyrannosaur tracks unless there are morphological features in them that can be correlated with allosaur or tyrannosaur pedal morphology to the exclusion of all other large theropods. Making such an identification would run the risk that should a footprint of very similar morphology be found in a formation from another time and place, the conclusion would be drawn that this indicates the presence of allosaurs or tyrannosaurs at that new location, when in fact all it could tell you is that some kind of large theropod was stomping around the landscape.

Comparisons Across Sites and Footprint Faunas. The discussion so far has focused on interpreting the number and kinds of trackmakers responsible for the tridactyl footprints of a particular locality. Obviously we would like to be able to do more than that, and make comparisons across sites, both within the same stratigraphic unit and across stratigraphic units and geographic areas. How well will the approaches described in this book work for this?

I must first acknowledge that making such comparisons may not be easy. For one thing, as already emphasized, the greater the number of footprint characters that can be measured, the better the chances of discriminating among footprint morphotypes in a meaningful way. Once again it should be noted that such quantitative analyses are best made on well-preserved prints. Even so, the circumstances of preservation across sites and formations conceivably may be such that it will be difficult to find a large number of the same characters that can be measured on footprints across sites. If this turns out to be the case, then from the standpoint of the methods of this book, all ichnology (like politics, according to onetime U.S. Congressman Tip O'Neill) is local.

But I doubt (or at least I hope) that the situation is that bad, and certainly the published ichnological literature is replete with studies that have attempted just such cross-faunal comparisons. In this book we have explored numerous ways of measuring lengths and other linear dimensions, and also interdigital angles, of tridactyl footprints, to the point of sometimes creating ghastly names to distinguish different versions of a particular kind of measurement. Assuming that it is possible to measure the same characters in tracks across sites, whatever characters are available for comparison should be measured using the same protocols. Otherwise we run the risk of concluding, on the basis of some parameter(s), that two sets of prints from different sites were morphologically different, when the difference was simply due to the fact that the parameter(s) in question was (were) measured in different ways between sites. I have often wondered, when reading published studies of dinosaur footprints, just how the authors measured their track parameters. With this in mind, I urge persons who report measurements of dinosaur footprints to be very explicit about how their measurements are made.

This brings us at last to a matter that I have deliberately downplayed throughout this book: ichnotaxonomy. What are we to make of the plethora of names applied to footprints of tridactyl dinosaurs over the years?

In an ideal ichnological world, every species of tridactyl trackmaker would have a pedal skeletal and soft-tissue morphology that would be uniquely its own, such that well-preserved footprints would be clearly recognizable as having been made by that species, and only that species. In addition, feet and footprints of one species would be more like those of its close relatives than like those of other, more distantly related species.

But as this book has made abundantly clear, we don't live in such a world. Some groups (such as crocodylians and large theropods) show considerable conservatism in pedal morphology across species, in some groups closely related species have rather dissimilar feet (emus and cassowaries), and some species that are distantly related at best have very similar feet (emus and bustards).

Suppose that it is indeed possible to make a large set of the same measurements, using the same protocols, on all the big tridactyl dinosaur prints that have been assigned to the ichnogenus *Eubrontes* from Mesozoic tracksites of different ages across the world. Would all of these prints turn out to be more like each other than like prints assigned to *Megalosauripus*, *Irenesauripus*, *Kayentapus*, or other tridactyl ichnotaxa? I can't even guess, but it would be an exercise certainly worth doing; Castanera et al. (2016) have taken an interesting first step in this kind of analysis in a comparison of named ichnospecies of *Eubrontes*, *Grallator*, *Kayentapus*, and selected other theropod ichnogenera.

One approach worth trying might be to extend the rationale of Belvedere and Farlow (2016) to create a numerical scale to indicate the strength of the similarity between two footprint morphotypes or ichnotaxa. How many distinct characters do they have in common, and what kinds of characters are those? I would think, for example, that characters that measure the dimensions of the marks of digital pads and claws would be more useful than characters based on gross shape alone, such as overall digit impression lengths. I visualize a numerical (ordinal?) scale of similarity, with each reasonably independent (one wouldn't want to count as separate characters parameters that include each other, such as the lengths of individual digital pads along with aggregate length of the same digit) shape parameter shared in common being assigned a value of one. Parameters that are deemed more compelling (such as lengths of individual digital pads) might be assigned a higher value than those thought to be less useful (such as total length from the "heel" to the toetip of a digit impression)—perhaps a score of two rather than one? The higher the similarity score between two footprints, the greater the chance would be that their similarity in shape is meaningful.

But suppose, as in the example above, all big *Eubrontes* did in fact turn out to be more like each other than like other named ichnotaxa. Could we then conclude that the makers of all large *Eubrontes* were more closely related to each other than to the makers of other ichnotaxa? The results reported in this book suggest that, in the words of a rather raffish character from the Gershwin opera *Porgy and Bess*, it ain't necessarily so. Recall one last time the feet and footprints of emus and kori bustards. If footprints of these birds were to be found as fossils, I am pretty sure that they would be assigned to the same ichnogenus, although possibly different ichnospecies within that ichnogenus. And yet these birds are almost as distantly related as two modern birds can be.

I don't want to leave the wrong impression here, and suggest that I consider ichnotaxonomy to be of no value—because I don't. I suspect—or at least hope—that in most cases similarities in footprint shape do in fact reflect—at some systematic level—similarities in their makers. I do, however, think that we need to be aware of the limitations of such inferences, at least when we are talking about the ichnotaxonomy of tridactyl dinosaur footprints. Such prints do preserve a record of dinosaur biodiversity, even if it is seen in a mirror, dimly (and I promise that that is the final literary allusion I will make in this book). I simply warn that using named ichnotaxa as proxies for dinosaur diversity is something that must be done with caution, and with awareness of possible pitfalls. Knowledge of limitations is, after all, useful knowledge.

Appendix

A

Table A1.1. Measurements of foot skeletons of crocodylians, nonavian dinosaurs, and ground birds. Metatarsal (MT) III (or tarsometatarsus [TMT]) and phalanx lengths (L) and widths (pw = proximal transverse; dw = distal transverse); FL = femur length; ID = identity; SVL = snout-vent length (whether to anterior or posterior end of vent unspecified); SVLant = snout-vent length measured to anterior edge of vent; SVLpost = snout-vent length measured to posterior end of vent; TL = total length. All measurements are in millimeters. Empty cells indicate that the bone was missing, the bone couldn't be measured, or the bone was not measured (for whatever reason). "Absent" means the bone does not occur in that species (e.g., the bones of digit II of *Struthio camelus*), instead of not being preserved in that particular specimen. Museum abbreviations with specimen numbers are identified in table I.1. The institutional catalog numbers of several specimens have been slightly or drastically changed since I originally measured them. I have tried to keep track of such changes in the entries. This is important because some of the figures in this book were drafted using the older numbers. In some entries there have also been changes in the systematic nomenclature of the specimens; these changes are noted where I was aware of them.

Omit taxon and specimen	**Measurements**				**Comments**
Crocodylians					
Alligator mississippiensis USNM 211228	MT III L = 75				Left foot
	I1: 24L	I2: 24L			
	II1: 26L	II2: 18L	II3: 20L		
	III1: 28L	III2: 19L	III3: 14L	III4: 18L	
	IV1: 24L	IV2: 15L	IV3: 10L	IV4: 5.5	
Alligator mississippiensis USNM 211235	II1: 38L	II2: 22L			
		III2: 25L	III3: 18L		
Alligator mississippiensis USNM 216198	SVL = 280; FL = 42; MT III L = 23				Left foot
	I1: 7L				
	II1: 8L	II2: 5.5L			
	III1: 8.5L	III2: 6L			
	IV1: 7.5L				
Alligator mississippiensis USNM 216200	MT III L = 10				Right foot
	I1: 4.5L				
	II1: 5L	II2: 4L			
	III1: 4.5L	III2: 4L	III3: 3L		
	IV1: 3.5L				
Alligator mississippiensis USNM 216208	SVL = 280; FL = 43; MT III L = 24				Left foot
	I1: 7.5L				
	II1: 9L	II2: 6L			
	III1: 9L				
	IV1: 7.5L				
Alligator mississippiensis USNM 291916	TL = 1,445; SVLant = 695; FL = 99; MT III L = 52				Right foot of female
	I1: 16L				
	II1: 19L	II2: 12L			
	III1: 19	III2: 13L	III3: 9.5L		
	IV1: 17L	IV2: 10L	IV3: 6.5L		
Alligator mississippiensis USNM 300659	TL = 620; SVLant = 293; SVLpost = 306; FL = 44; MT III L = 24				Left foot
	I1: 7.5L				
	II1: 9L	II2: 5.5L			
	III1: 9L				
Alligator mississippiensis USNM 312673	TL = 2,685; FL = 185; MT III L = 90				Measurements based on both feet of female
	I1: 28L				
	III1: 33L	III2: 21L			

Table A1.1. *continued*

Omit taxon and specimen	Measurements				Comments
Alligator mississippiensis USNM 313406	MT III L = 10				Hatchling
	I1: 3L				
	II1: 3.5L				
	III1: 3.5L				
	IV1: 3L				
Alligator mississippiensis USNM 313409	TL = 610; FL = 56; MT III L = 28				Right foot of male
	I1: 9.5L	I2: 10L			
	II1: 11L	II2: 7L	II3: 8.5L		
	III1: 11L	III2: 7.5L	III3: 6L	III4: 7.5L	
	IV1: 9L	IV2: 6L	IV3: 4L	IV4: 2.5L	
Alligator mississippiensis USNM 313410	FL = 20; MT III L = 11				Right foot
	I1: 3.5L				
	II1: 4L				
	III1: 4L				
	IV1: 3.5L				
Alligator mississippiensis UFHerpetology 115605	MT III L = 32				Right foot
	I1: 9.5L	I2: 12L			
	II1: 12L	II2: 7.5L	II3: 11L		
	III1: 12L	III2: 8.5L	III3: 6L	III4: 10L	
	IV1: 9L	IV2: 6.5L	IV3: 4L		
Alligator mississippiensis CITES 0027580	TL = 2,007; MT III L = 63				Right foot
	I1: 19L				
	II1: 22L	II2: 14L			
	III1: 23L	III2: 15L, 8.1pw	III3: 10L		
	IV1: 21L	IV2:13L	IV3: 8L	IV4: 3.5	
Alligator mississippiensis CITES 0040187	TL = 3,251; MT III L = 113				Right foot
	I1: 34L				
	II1: 40L	II2: 24L			
	III1: 41L	III2: 26L, 16.7pw	III3: 19L		
	IV1: 36L	IV2: 21L	IV3: 14L	IV4: 8L	
Alligator mississippiensis CITES 0041015	TL = 2,819; MT III L = 91				Right foot
	I1:27L				
	II1: 34L	II2: 21L			
	III1: 37L	III2: 24L, 13.6pw	III3: 17L		
	IV1: 32L	IV2: 18L	IV3: 13L	IV4: 8.5L	
Alligator mississippiensis CITES 0041062	TL = 3,404; MT III L = 107				Right foot
	I1: 31L				
	II1: 36L	II2: 23L			
	III1: 39L	III2: 25L, 16.7pw	III3: 19L		
	IV1: 34L	IV2: 19L			
Alligator mississippiensis CITES 0042178	TL = 3,150; MT III L = 97				Right foot
	I1: 29				
	II1: 34L	II2: 20L			
	III1: 36L	III2: 23L, 15.4pw	III3: 17L		
	IV1: 31L	IV2: 18L	IV3: 7.5		
Alligator mississippiensis CITES 0043361	TL = 2,718; MT III L = 88				Left foot
	I1: 26L				
	II1: 30L	II2: 18L			
	III1: 32L	III2: 21L, 11.5pw	III3: 15L		
	IV1: 28L	IV2: 17L	IV3: 11L	IV4: 7L	

Omit taxon and specimen	Measurements				Comments
Alligator mississippiensis CITES 0046863	TL = 1,981; MT III L = 67				Right foot
	I1: 19L				
	II1: 22L	II2: 14L			
	III1: 24L	III2: 16L, 8.8pw	III3: 12L		
	IV1: 20L	IV2: 13L	IV3: 8.5L	IV4: 5L	
Alligator mississippiensis CITES 0054237	TL = 1,549; MT III L = 44				Right foot
	I1: 13L				
	II1: 15L	II2: 9.5L			
	III1: 16L	III2: 10L, 5.8pw	III3: 7.5L		
	IV1: 14L	IV2: 8.5L	IV3: 5L	IV4: 2L	
Alligator mississippiensis CITES 0056493	TL = 2,083; MT III L = 71				Left foot
	I1: 21L				
	II1: 25L	II2: 16L			
	III1: 26L	III2: 17L, 9.1pw	III3: 13L		
	IV1: 24L	IV2: 13L	IV3: 9L	IV4: 5L	
Alligator mississippiensis SMM Z79.1.286	MT III L = 50				Measurements based on both feet
	I1: 16L	I2: 13L			
	II1: 19L	II2: 12L	II3: 14L		
	III1: 19L	III2: 13L	III3: 11L	III4: 12L	
	IV1: 17L	IV2: 11L	IV3: 7L	IV4: 3.5L	
Alligator mississippiensis CM Herps S 9102	FL = 129; MT III L = 63				Right foot
	I1: 18L				
	II1: 21L	II2: 14L			
	III1: 22L	III2: 15L	III3: 11L		
	IV1: 19L	IV2: 12L	IV3: 8L	IV4: 4.5L	
Alligator mississippiensis hatchling CM Herps 33944	FL = 24; MT III L = 13				Measurements based on both feet; the smaller phalanges were at the limit of my ability to measure
	I1: 4L				
	II1: 5L	II2: 3.5L			
	III1: 5L	III2: 3.5L	III3: 3L		
	IV1: 4L	IV2: 3L	IV3: 2L	IV4: 1L	
Alligator mississippiensis FMNH 8205	MT III L = 93				Right foot
	I1: 27L				
	II1: 32L	II2: 20L			
	III1: 35L	III2: 22L	III3: 17L		
	IV1: 30L	IV2: 17L	IV3: 11L		
Alligator mississippiensis FMNH 22027	MT III L = 97				Left foot
	I1: 26L	I2: 29L			
	II1: 32L	II2: 20L	II3: 29L		
	III1: 32L	III2: 22L	III3: 15L	III4: 25L	
	IV1: 29L	IV2: 18L	IV3: 11L	IV4: 6.5L	
Alligator mississippiensis FMNH 22029	MT III L = 109				Left foot
	I1: 34L				
	II1: 41L	II2: 25L			
	III1: 44L	III2:28L	III3: 20L		
	IV1: 38L	IV2: 22L	IV3: 15L	IV4: 8L	
Alligator mississippiensis FMNH 31321	MT III L = 49				Right foot
	I1: 16L	I2: 17L			
	II1: 19L	II2: 12L	II3: 15L		
	III1: 19L	III2: 11L	III3: 9L	III4: 11L	
	IV1: 17L	IV2: 8.5L	IV3: 5.5L		
Alligator sinensis FMNH 197946	MT III L = 63				Right foot
	I1: 20L				
	II1: 21L	II2: 15L			
	III1: 21L	III2: 13L	III3: 10L		
	IV1: 18L	IV2: 10L	IV3: 7L		

Table A1.1. *continued*

Omit taxon and specimen	Measurements				Comments
Caiman crocodilus UMMZ Herps 130547	FL = 66; MT III L = 34				Measurements based on both feet
	I1: 11L	I2: 11L			
	II1: 12L	II2: 9L	II3: 11L		
	III1: 12L	III2: 8.5L	III3: 7.5	III4: 8.5L	
	IV1: 11L	IV2: 6.5L	IV3: 4L	IV4: 2.5L	
Caiman crocodilus UMMZ Herps 134962	FL = 42; MT III L = 23				Left foot
	I1: 7.5L	I2: 8L			
	II1: 8L	II2: 6L			
	III1: 7.5L	III2: 5.5L			
	IV1: 6.5L	IV2: 4L	IV3: 2.5L		
Caiman crocodilus UFHerpetology 99067	MT III L = 29				Measurements based on both feet
	I1: 9L	I2: 8.5L			
	II1: 10L	II2: 7.5L			
	III1: 10L	III2: 7L	III3: 6L	III4: 6L	
	IV1: 9L	IV2: 5.5L			
Caiman crocodilus UFHerpetology 14361	MT III L = 13				Measurements based on both feet
	I1: 4.5L	I2: 3.5L			
	II1: 5L	II2: 3.5L			
	III1: 5L	III2: 3.5	III3: 3L		
	IV1: 4L	IV2: 2.5			
Caiman crocodilus FMNH 98960	MT III L = 21				Right foot
	I1: 7.5L	I2: 7.5L			
	II1: 8L	II2: 6L	II3: 7.5L		
	III1: 8L	III2: 5.5L	III3: 4.5L		
	IV1: 7L	IV2: 4L	IV3: 2.5L		
Caiman crocodilus (formerly *C. sclerops*) CM Herps 33977	FL = 26; MT III L = 15L				Right foot
	I1: 5L				
	II1: 5L	II2: 3.5L			
	III1: 5L	III2: 3.5L	III3: 3L		
	IV1: 4.5L	IV2: 2.5L	IV3: 1.5L	IV4: 1L	
Caiman crocodilus (formerly *C. sclerops*) CM Herps 91419	FL = 28; MT III L = 16				Right foot
	I1: 5L				
	II1:5.5L				
	III1: 5.5L				
	IV1: 4.5L				
Caiman crocodilus CM Herps 111946	FL = 32; MT III L = 18				Measurements based on both feet
	I1: 6L	I2: 5.5L			
	II1: 6L	II2: 4.5L	II3: 4.5L		
	III1: 6L	III2: 6L	III3: 3.5L	III4: 4.5L	
	IV1: 5.5L				
Caiman crocodilus CM Herps 114437	FL = 80; MT III L = 38				Measurements based on both feet
	I1: 13L				
	II1: 14L				
	III1: 14L	III2: 9.5L	III3: 8L		
	IV1: 11L	IV2: 7L			
Caiman crocodilus CM Herps 145669 or 145670	MT III L = 17				Two specimens are in the same box, with no indication of which goes with which catalog number. I measured the left foot of one of them
	I1: 5.5L				
	II1: 5.5L	II2: 4.5L			
	III1: 6L	III2: 4L	III3: 3.5L		
	IV1: 5.5L	IV2: 3L	IV3: 1.5L	IV4: 1L	
Caiman yacare USNM 297784	SVL = 178; FL = 26; MT III L = 15				Right foot
	I1: 5L				
	II1: 5.5L				
	III1: 5.5L				
	IV1: 4.5L	IV2: 2.5L	IV3: 1.5L	IV4: 1L	

Omit taxon and specimen	Measurements				Comments
Caiman sp. UMMZ Herps 129398	MT III L = 48				Right foot
	I1: 15L				
	II1: 16L				
	IV1: 14L	IV2: 7.5L			
Paleosuchus palpebrosus UMMZ Herps 219018	SVL = 625; FL = 83; MT III L = 43				Left foot of female
	I1: 13L				
	II1: 14L	II2: 11L			
	III1: 14L	III2: 10L			
	IV1: 11L				
Paleosuchus trigonatus UMMZ Herps 46113	FL = 85; MT III L = 42				Measurements based on both feet
	I1: 12L				
	II1: 14L	II2: 11L			
	III1: 13L	III2: 9.5L	III3: 8.5L		
	IV1: 11L	IV2: 6.5L	IV3: 4L	IV4: 2.5L	
Paleosuchus trigonatus UMMZ Herps 129764	FL = 73; MT III L = 35				Left foot
	I1: 12L	I2: 11L			
	II1: 11L	II2: 9.5L	II3: 10L		
	III1: 10L	III2: 8L	III3: 6.5L	III4: 10L	
	IV1: 9.5L	IV2: 5.5L	IV3: 3.5L	IV4: 2L	
Paleosuchus trigonatus USNM 300660	FL = 89; MT III L = 42				Right foot of male
	I1: 12L				
	III1: 12L	III2: 9L			
	IV1: 10L				
Paleosuchus trigonatus USNM 302052	SVL = 783; FL = 112; MT III L = 47				Left foot of male
	I1: 16L	I2: 13L			
	II1: 16L	II2: 12L	II3: 9L		
	III1: 15L	III2: 11L	III3: 9.5L	III4: 8L	
	IV1: 12L	IV2: 7L	IV3: 4.5L		
Paleosuchus trigonatus UFHerpetology 56316	TL = 450; SVL = 240; FL = 36; MT III L = 18				Measurements based on both feet
	I1: 6L				
	II1: 6.5L	II2: 5L	II3: 5L		
	III1: 6L	III2: 4.5L			
	IV1: 5L	IV2: 2.5L			
Wannaganosuchus brachymanus SMM P.72.34.214	MT III L = 42				Right foot
	I1: 13L	I2: 9.5L			
	II1: 14L	II2: 9L	II3: 10L		
	III1: 13L	III2: 9L	III3: 7L	III4: 8L	
	IV1: 11L	IV2: 6.5L			
?*Procaimanoidea* sp. SMM P89.12.1c	TL = 800; FL = 51; MT III L = 29				Cast of skeleton
	II1: 10L	II2: 8L			
	III1: 10L	III2: 8.5L	III3: 7L	III4: 5.5L	
	IV1: 8L	IV2: 4.5L			
Crocodylus acutus UFHerpetology 99202	TL = 433; SVL = 220; MT III L = 15				Left foot
	I1: 5.5L	I2: 6L			
	II1: 6L	II2: 5L	II3: 4.5L		
	III1: 6L	III2: 4.5L	III3: 4L	III4: 4L	
	IV1: 5L	IV2: 3L	IV3: 2L	IV4: 1L	
Crocodylus acutus FMNH 22028	MT III L = 81				Right foot
	I1: 31L	I2: 31L			
	II1: 30L	II2: 20L	II3: 25L		
	III1: 32L	III2: 19L	III3: 15L	III4: 19L	
	IV1: 24L	IV2: 14L	IV3: 9L	IV4: 7L	

Table A1.1. *continued*

Omit taxon and specimen	Measurements				Comments
Crocodylus acutus FMNH 98937	MT III L = 19				Measurements based on both feet
	I1: 7L				
	II1: 7L	II2: 4.5L	II3: 5L		
	III1: 7.5L	III2: 4.5L	III3: 3.5L	III4: 4L	
	IV1: 5.5L	IV2: 3.5L	IV3: 2.5L	IV4: 1.5L	
Crocodylus acutus FMNH 98938	MT III L = 22				Left foot
	I1: 8L	I2: 7L			
	II1: 8.5L	II2: 6L	II3: 7L		
	III1: 9L	III2: 5.5L	III3: 4.5L	III4: 5.5L	
	IV1: 7L	IV2: 4L	IV3: 3L	IV4: 1.5L	
Crocodylus acutus CM Herps 6450	FL = 210; MT III L = 95				Right foot
	I1: 34L	I2: 33L			
	II1: 35L	II2: 24L	II3: 27L		
	III1: 36L	III2: 22L	III3: 17L	III4: 24L	
Crocodylus rhombifer FMNH 34677	MT III L = 55				Left foot of female
	I1: 17L				
	II1: 18L	II2: 11L			
	III1: 19L	III2: 12L	III3: 9L		
	IV1: 17L	IV2: 10L	IV3: 7.5L		
Crocodylus mindorensis USNM 252669	SVL = 460; FL = 61; MT III L = 30				Measurements based on both feet
	I1: 11L	I2: 10L			
	II1: 11L	II2: 7.5L			
	III1: 11L	III2: 7.5L			
	IV1: 8.5L				
Crocodylus mindorensis USNM 252670	FL = 62; MT III L = 29				Right foot of female
	I1: 11L	I2: 12L			
	II1: 11L	II2: 8L	II3: 9.5L		
	III1: 11L	III2: 7.5L	III3: 6L	III4: 7.5L	
	IV1: 9L	IV2: 5.5L			
Crocodylus siamensis UMMZ Herps 129718	FL = 72; MT III L = 37				Left foot
	II1: 12L	II2: 9L			
	III1: 13L	III2: 8.5L	III3: 7.5L		
	IV1: 11L	IV2: 6.5L	IV3: 4.5L	IV4: 2L	
Crocodylus johnstoni USNM 233976	FL = 37L; MT III L = 18L				Left foot
	I1: 6L	I2: 5L			
	II1: 6.5L	II2: 4.5L	II3: 4.5L		
	III1: 7L	III2: 4L			
	IV1: 6L	IV2: 3.5L	IV3: 2L		
Osteolaemus tetraspis USNM 19448	FL = 81; MT III L = 36				Left foot
	I1: 10L				
	II1: 12L	II2: 8.5L			
	III1: 11L	III2: 7.5L			
	IV1: 8.5L	IV2: 5.5L	IV3: 4L	IV4: 3L	
Osteolaemus tetraspis UFHerpetology 99204	TL = 242; SVL = 160; FL = 27; MT III L = 13				
	I1: 4L				
	II1: 4.5L	II2: 3.5L			
	III1: 4.5L	III2: 3L	III3: 2.5L		
	IV1: 3.5L	IV2: 2L	IV3: 1.5L	IV4: 1L	
Osteolaemus tetraspis FMNH 98936	MT III L = 23				Left foot
	I1: 8L	I2: 7L			
	II1: 9L	II2: 5.5L	II3: 6.5L		
	III1: 8.5L	III2: 5.5L	III3: 6.5L	III4: 5L	
	IV1: 6.5L	IV2: 3L	IV3: 2.5L		

Omit taxon and specimen	Measurements					Comments
Osteolaemus tetraspis UNSM ZM-15431	TL = 1,320; FL = 110; MT III L = 48					Measurements based on both feet
	I1: 19L					
	II1: 18L					
	III1: 18L	III2: 11L	III3: 8.5L			
	IV1: 14L	IV2: 7.5L	IV3: 5L	IV4: 3.5L		
Tomistoma schlegelii UMMZ Herps 128552	FL = 81; MT III L = 41					Left foot
	I1: 14L	I2: 17L				
	II1: 15L	II2: 10L	II3: 15L			
	III1: 15L	III2: 9L	III3: 7.5	III4: 11L		
	IV1: 12L	IV2: 7L	IV3: 5L	IV4: 3L		
Tomistoma schlegelii UMMZ Herps 129397	FL = 93; MT III L = 42					Right foot
	I1: 16L					
	II1: 17L	II2: 10L				
	III1: 18L	III2: 10L	III3: 8L			
	IV1: 14L	IV2: 8L	IV3: 6.5L	IV4: 2L		
Tomistoma schlegelii UMMZ Herps 174416	FL = 59; MT III L = 30					Right foot
	I1: 10L	I2: 13L				
	II1: 11L	II2: 7.5L	II3: 11L			
	III1: 11L	III2: 7L	III3: 5.5L	III4: 8.5L		
	IV1: 8.5L	IV2: 5L	IV3: 3.5L	IV4: 2L		
Tomistoma schlegelii USNM 52972	FL = 209; MT III L = 91					Measurements based on both feet
	I1: 30L	I2: 41L				
	II1: 36L	II2: 21L	II3: 34L			
	III1: 34L	III2: 21L	III3: 16L	III4: 23L		
	IV1: 27L	IV2: 17L	IV3: 11L	IV4: 5L		
Tomistoma schlegelii FMNH 206755	MT III L = 87					Measurements based on both feet
	I1: 33L					
	II1: 35L	II2: 22L				
	III1: 35L	III2: 22L	III3: 17L			
	IV1: 26L	IV2: 17L				
Gavialis gangeticus FMNH 22025	MT III L = 105					Right foot
	I1: 27L	I2: 30L				
	II1: 29L	II2: 20L	II3: 27L			
	III1: 31L	III2: 21L	III3: 17L	III4: 20L		
	IV1: 26L	IV2: 15L	IV3: 11L			
Gavialis gangeticus FMNH 82681	MT III L = 42					Right foot
	I1: 13L					
	II1: 13L	II2: 9L				
	III1: 13L	III2: 8.5L	III3: 7L			
	IV1: 11L	IV2: 6.5L				
Basal theropods						
Herrerasaurus ischigualastensis PVSJ 373	MT III L = 156					Measured cast FMNH PR 1806; measurements based on both feet
	I1: 40L					
	II1: 49L, 24pw, 22dw	II2: 34L, 17dw	II3: 39L			
	III1: 50L, 27pw, 23dw	III2: 32L, 19dw	III3: 27L, 18dw	III4: 39L		
	IV1: 41L, 25pw, 21dw		IV3: 18dw	IV4: 20L, 15dw	IV5: 33L	
Coelophysoids						
Segisaurus halli YPM VP 055662	II1: 25L	II2: 21L	II3: 21L			Cast of UCMP specimen; left foot, which is difficult to measure because phalanges are rotated about their long axes in an inconsistent manner; lengths measured along dorsal surface of bones
	III1: 26L	III2: 20L	III3: 18L	III4: 18L		
	IV1: 14L	IV2: 14L	IV3: 11L	IV4: 14L	IV5: 16L	

Table A1.1. *continued*

Omit taxon and specimen	Measurements					Comments
Coelophysis bauri MNA V3318	I1: 12L	I2: 10L				Measurements based on both feet
	II1: 17L	II2: 13L	II3: 15			
	III1: 21L	III2: 16L, 7dw	III3: 14L	III4: 15L		
	IV1: 10L	IV2: 7L	IV3: 6L	IV4: 5L	IV5: 15L	
Coelophysis bauri MNA V3320	MT III L = 128					Left foot
	I1: 18L	I2: 17L				
	II1: 30L, 10pw, 10dw	II2: 8dw	II3: 23L			
	III1: 36L, 13pw, 13dw	III2: 26L, 10dw	III3: 23L, 8dw			
	IV1: 19L, 9pw, 9dw	IV2: 14L, 8dw	IV3: 11L, 7dw	IV4: 10L, 7dw		
Coelophysis bauri MNA uncatalogued right foot A		I2: 12L				
	II1: 30L	II2: 22L	II3: 21L			
Coelophysis bauri MNA uncatalogued right foot B	II1: 27L	II2: 20L	II3: 16L			
	III1: 30L					
Coelophysis bauri Ghost Ranch right foot	II1: 19L, 7dw	II2: 13L, 5dw	II3: 14L			
	III1: 21L, 9pw, 9dw	III2: 17L, 7dw	III3: 15L	III4: 18L		
	IV1: 12L, 6dw	IV2: 8L, 5dw	IV3: 7L, 5dw	IV4: 5L		
Coelophysis bauri SMP VP-630	I1: 15L					Left foot of specimen in Harrisburg block (sediment matrix); lengths of II2 and IV2 may be a bit off
	II1: 25L	II2: 21L				
	III1: 30L	III2: 22L	III3: 20L	III4: 15L		
	IV1: 14L	IV2: 11L	IV3: 9L			
Coelophysis bauri SMP uncatalogued "*Syntarsus* morph"	I1: 11L	I2: 7L				Measurements based on both feet of specimen in Harrisburg block
	II1: 17L	II2: 13L	II3: 11L			
	III1: 20L	III2: 17L				
	IV1: 12L	IV2: 7L	IV3: 5L	IV4: 5L	IV5: 10L	
Coelophysis bauri SMP uncatalogued "*Coelophysis / Riorribasaurus*" morph"	II1: 28L	II2: 20L	II3: 19L			Left foot of specimen in Harrisburg block
	III1: 31L	III2: 24L				
Coelophysis bauri CMNH 10971	I1: 16L					Left foot of specimen lying on right side in block
			II3: 22L			
	III1: 33L	III2: 24L	III3: 21L	III4: 17L		
	IV1: 19L	IV2: 13L	IV3: 10L	IV4: 9L	IV5: 18L	
Coelophysis bauri CMNH 10971		II2: 27L	II3: 25L			Large left foot
	III1: 39L, 16dw	III2: 30L, 13dw	III3: 26L			
	IV1: 23L, 14pw, 12dw					
Coelophysis bauri CMNH 10971	II1: 18L		II3: 12L			Small right foot
	III1: 20L		III3: 13L			
	IV1: 11L	IV2: 7L	IV3: 6L	IV4: 4L		
Coelophysis bauri CMNH 10971		II2: 23L, 7dw				Right foot removed from block
	III1: 38L, 15pw, 13dw	III2: 28L				
	IV1: 20L, 10pw, 10dw	IV2: 15L, 9dw	IV3: 12L, 8dw			
Coelophysis bauri CMNH 10971	I1: 17L	I2: 12L				Measurements based on left and right feet; removed from block and put on display. Foot bones glued to the base
	II1: 28L, 9dw	II2: 7dw	II3: 19L			
	III1: 32L, 13pw, 11dw	III2: 24L, 9dw	III3: 21L, 7dw	III4: 20L		
	IV1: 19L, 9dw	IV2: 13L, 7dw	IV3: 10L, 6dw	IV4: 8L, 6dw	IV5: 19L	
Coelophysis bauri USNM block specimen	II1: 21L, 6pw	II2: 13L				
	III1: 23L, 8dw	III2: 18L	III3: 14L, 5dw			
	IV1: 12L	IV2: 9L	IV3: 6L, 6dw		IV5: 11L	

Omit taxon and specimen	Measurements					Comments
Dilophosaurids						
Dilophosaurus wetherilli UCMP 37302	MT III L = 300					Measured right foot of cast (RTMP 83.26.2)
	I1: 62L	I2: 47L				
	II1: 93L, 40pw, 33dw	II2: 70L, 27dw	II3: 46L			
	III1: 100L, 49pw, 42dw	III2: 71L, 33dw	III3: 59L, 27dw	III4: 58L		
	IV1: 71L, 37pw, 33dw	IV2: 47L, 29dw	IV3: 36L, 26dw	IV4: 29L, 21dw	IV5: 47L	
Dilophosaurus wetherilli TMM 43646-61	MT III L = 298					Right foot
	I1: 46L	I2: 31L				
	II1: 83L	II2: 61L	II3: 56L			
	III1: 92L	III2: 65L	III3: 52L	III4: 48L		
	IV1: 61L	IV2: 43L	IV3: 33L	IV4: 22L	IV5: 41L	
Ceratosaurs						
Velocisaurus unicus MUCPv 41	II1: 20L, 8pw	II2: 12L				Right foot; measured cast (RTMP 95.36.5)
	III1: 21L, 10pw, 10dw	III2: 17L, 8dw	III3: 14L			
	IV1: 14L	IV2: 7L	IV3: 5L	IV4: 4L		
Deltadromeus agilis SGM Din 2	II1: 143L, 50pw, 37dw					Measurements based on both feet
	III1: 137L, 63pw, 52dw					
	IV1: 104L, 39pw, 38dw		IV3: 43L, 30dw	IV4: 32L, 24dw	IV5: 65L	
Aucasaurus garridoi MCF-PVPH-236	I1: 46L	I2: 32L				Left foot; data from Cecilia Succar
	II1: 98L, 34dw	II2: 55L, 26dw	II3: 61L			
	III1: 90L, 31dw	III2: 60L, 29dw	III3: 43L, 23dw	III4: 72L		
	IV1: 54L, 39dw	IV2: 40L, 32dw	IV3: 32L, 27dw	IV4: 30L, 21dw	IV5: 61L	
Noasaur Museum National de Niger	MT III L = ca. 139					Right foot; specimen lies on left side in block
		I2: 13L				
			II3: 21L			
	III1: 36L	III2: 26L	III3: 22L			
	IV1: 22L	IV2: 11L	IV3: 8L	IV4: 7L	IV5: 22L	
Megalosauroids						
Spinosaurus aegyptiacus Sereno specimen C	I1: 117L	I2: 90L				Right foot; phalanx IV1 is lost, but a cast of it was made before it was lost
	II1: 100L, 50pw, 49dw		II3: 115L			
	III1: 94L, 56pw, 50dw	III2: 63L, 58pw, 49dw	III3: 56L, 46pw, 48dw	III4: 119L		
	IV1: 75L, 47pw, 36dw	IV2: 51L, 45pw, 38dw	IV3: 40L, 43pw, 39dw	IV4: 33L, 40pw, 40dw	IV5: 86L	
Allosauroids						
Allosaurus fragilis USNM V 4734	MT III L = 327					Left foot; quite hard to distinguish real bone from reconstruction
	II1: 108L, 54dw	II2: 68L, 40dw				
	III1: 105L, 73dw	III2: 80L, 49dw	III3: 57L, 40dw			
		IV2: 37L, 40dw	IV3: 25L, 33dw			
Allosaurus fragilis USNM V 8423	MT III L = 355					Left foot
	II1: 108L, 63pw, 61dw					
	III1: 107L, 81pw, 70dw	III2: 81L, 53dw				
		IV2: 48L, 42dw			IV5: 79L	

Table A1.1. *continued*

Omit taxon and specimen	Measurements					Comments
Allosaurus fragilis MOR 693 ("Big Al")	MT III L = 357					Measurements based on both feet, mainly from casts; what I identified as II3 may be III4, and vice versa
	I1: 63L	I2: 66L				
	II1: 109L, 49pw, 44dw	II2: 74L, 40	II3: 89L			
	III1: 110L, 66pw, 57dw	III2: 79L, 46dw	III3: 57L, 32dw	III4: 71L		
	IV1: 84L, 55pw, 51dw	IV2: 49L, 43dw	IV3: 38L, 36dw	IV4: 27L, 30dw	IV5: 66L	
Allosaurus fragilis SMA 0005 ("Big Al II")	MT III L = 353					Mostly or entirely right foot. Ungual II3 could be III4 and vice versa; unguals were not found articulated with rest of digits but were associated
	I1: 66L	I2: 72L				
	II1: 113L, 66pw, 58dw	II2: 74L, 45dw	II3: 92L			
	III1: 115L, 79pw, 63dw	III2: 77L, 54dw	III3: 61L, 40dw	III4: 75L		
	IV1: 75L, 57pw, 47dw	IV2: 49L, 45dw	IV3: 43L, 41dw	IV4: 29L, 36dw	IV5: 70L	
Allosaurus fragilis CM 11844	II1: 112L	II2: 80L	II3: 104			Jack McIntosh thought these bones, at least, are real
	III1: 116L	III2: 102L	III3: 77L			
Allosaurus fragilis AMNH 324	II1: 139L, 75dw	II2: 89L, 54dw				Left (?) foot. Phalanx III2 may be pathologically short. There is another foot (right?). With the same catalog number, but it seems to have more phalanges than should be present in a complete foot
	III1: 131L, 104pw, 89dw	III2: 80L, 64dw	III3: 119L			
	IV1: 92L, 79pw, 68dw				IV5: 87L	
Allosaurus fragilis SCMG0177		I2: 69L				Measured cast of left foot; ID of some bones uncertain
	III1: 103L, 67pw	III2: 73L, 46dw	III3: 56L, 37dw			
			IV3: 33L			
Allosaurus sp. AMNH 6125	II1: 115L, 69pw, 64dw	II2: 74L, 48dw				
Allosaurus sp. AMNH 6128	IV1: 77L, 62pw, 57dw	IV2: 51L, 48dw	IV3: 39L, 42dw	IV4: 28L, 34dw	IV5: 66L	Not entirely certain that the ungual belongs to digit IV, but it fits; the ungual tip is missing a bit of material, so the value used here is a minimum length estimate
Allosaurus sp. MWC 363	IV1: 82L, 60dw	IV2: 58L, 52dw	IV3: 56L, 46dw	IV4:31L		Right foot; toe is preserved in situ in rock
Allosaurus sp. DNM DINO 3987-3996	MT III L = 277					Right foot; specimen in situ in quarry face
	II1: 84L, 45pw, 39dw	II2: 57L, 29dw	II3: 69L			
	III1: 88L, 45dw	III2: 53L	III3: 39L			
Allosaurus sp. DNM DINO 11541	MT III L = 307					Right foot: what I interpreted as II1 was identified by DNM as IV1, and vice versa
	II1: 91L, 54pw, 50dw	II2: 60L, 33dw				
	III1: 95L, 75pw, 58dw	III2: 72L, 37dw				
	IV1: pw = 50	IV2: 39L, 40dw	IV3: 30L, 35dw	IV4: 19L, 27dw		
Allosaurus sp. UMNH VP C481	MT III L = 221					Left foot
	I1: 38L	I2: 31L				
	II1: 63L, 28pw, 24dw	II2: 42L, 17dw				
	III1: 71L, 33pw, 27dw		III3: 36L, 20dw			
	IV1: 42L, 25pw, 24dw	IV2: 26L, 22dw		IV4: 15L, 15dw	IV5: 40L	

Omit taxon and specimen	Measurements					Comments
Saurophaganax maximus OMNH several catalog numbers	MT III L = 427					Probably a composite of two individuals of comparable size Each bone is assigned its own catalog number. Ungual II3 could be III4, and vice versa. Phalanx II1 looks somewhat pathological
	I1: 83L					
	II1: 93L, 89pw, 79dw		II3: 93L			
	III1: 137L, 114pw, 94dw	III2: 92L, 74dw	III3: 63L, 52dw	III4: 122L		
	IV1: 89L, 77pw, 68dw	IV2: 57L, 63dw				
Acrocanthosaurus atokensis NCSM 14345	MT III L = 460					Measurements based on BHI cast of right foot
	I1: 68L	I2: 59L				
	II1: 138L, 93pw, 83dw	II2: 84L, 60dw				
	III1: 143L, 111pw, 97dw	III2: 93L, 78dw				
	IV1: 115L, 91pw, 85dw	IV2: 53L, 64dw	IV3: 43L, 53dw			
"Camptosaurus amplus" YPM 1879	I1: 71L	I2: 51L				Cast of right foot; identified as large theropod (possibly *Allosaurus*) following Galton (2015)
	II1: 114L, 87pw, 83dw	II2: 80L, 72dw	II3: 97L			
	III1: 125L, 103pw, 80dw	III2: 81L, 69dw	III3: 53L, 58dw			
	IV1: 90L, 70pw, 64dw	IV2: 63L, 58dw	IV3: 41L, 52dw	IV4: 29L, 46dw		
Basal coelurosaurs						
Nedcolbertia justinhoffmani CEUM 5071	II1: 27L, 10dw	II2: 17L, 7dw	II3: 18L			Cast of right foot; Brownstein (2017) interprets this species as a basal ornithomimosaur
	III1: 31L, 14pw, 12dw	III2: 22L, 10dw	III3: 16L, 7dw	III4: 18L		
	IV1: 19L, 10dw	IV2: 11L, 9dw	IV3: 8L, 8dw	IV4: 5L, 7dw	IV5: 17L	
Tanycolagreus topwilsoni TPII 2000-09-29	I1: 41L	I2: 27L				Cast of right foot
	II1: 59L, 23pw, 21dw	II2: 47L, 13dw	II3: 38L			
	III1: 68L, 23dw	III2: 47L, 19dw	III3: 37L, 14dw	III4: 48L		
	IV1: 49L, 18dw	IV2: 32L, 20dw	IV3: 24L, 14dw	IV4: 19L, 13dw		
Tyrannosauroids						
Alectrosaurus olseni AMNH 6554	I1: 54L					
	II1: 103L, 51pw, 48dw	II2: 69L, 38dw	II3: 55L			
	III3: 106L, 65pw, 51dw	III2: 67L, 43dw	III3: 57L, 35dw			
	IV1: 70L, 43pw, 47dw	IV2: 49L, 43dw	IV3: 38L, 39dw	IV4: 25L, 30dw	IV5: 49L	
Gorgosaurus libratus AMNH 5423	II1: 39dw	II2: 30dw	II3: 69L			
		III2: 74L, 38dw	III3: 59L, 31dw			
	IV1: 76L, 41dw	IV2: 55L, 35dw	IV3: 39L, 31dw	IV4: 27L, 24dw	IV5: 55L	
Gorgosaurus libratus USNM V 12814	I1: 90L					Measurements based on both feet of panel mount; measurements hard to take, and difficult to tell real bone from fabrication
	II1: 146L, 54dw	II2: 95L, 35dw	II3: 100L			
	III1: 151L	III2: 97L	III3: 75L	III4: 94L		
	IV1: 101L, 56dw	IV2: 69L	IV3: 50L, 44dw		IV5: 77L	
Gorgosaurus libratus CMN 2120	MT III L = 620					Right foot; skeleton is mounted in plaster base, making some measurements hard to get
	II1: 170L, 69dw	II2: 115L, 56dw				
	III1: 167L, 91pw, 81dw	III2: 118L, 65dw	III3: 90L, 62dw			
	IV1: 126L, 78dw	IV2: 88L, 66dw	IV3: 62L, 64dw	IV4: 40L, 48dw	IV5: 105L	

Table A1.1. *continued*

Omit taxon and specimen	Measurements					Comments
Gorgosaurus libratus CMN 11593	I1: 99L					Measurements made on several loose phalanges; may come from both feet
		II2: 53dw				
	III1: 166L, 84pw, 80dw					
		IV2: 83L, 60dw	IV3: 57L, 51dw	IV4: 39L, 43dw	IV5: 91L	
Gorgosaurus libratus FMNH PR2211 ("Elmer")	MT III L = 335					Right foot; a few mm missing from tip of II3
	I1: 34L	I2: 26L				
	II1: 69L, 33pw, 29dw	II2: 41L, 22dw	II3: 45L			
	III1: 66L, 35pw, 32dw	III2: 44L, 27dw	III3: 35L, 21dw	III4: 43L		
	IV1: 42L, 28pw, 29dw	IV2: 28L, 26dw	IV3: 18L, 22dw	IV4: 13L, 18dw	IV5: 38L	
Gorgosaurus libratus ROM 1247	MT III L = 539					Digit I phalanges from right foot, but other bones from left foot; some material missing from tip of III4
	I1: 90L	I2: 69L				
	II1: 148L, 47dw	II2: 93L, 38dw	II3: 91L			
	III1: 137L, 57pw, 61dw	III2: 94L, 49dw	III3: 74L, 39dw	III4: 80L		
	IV1: 98L, 55pw, 54dw	IV2: 70L, 48dw	IV3: 49L, 40dw	IV4: 34L, 33dw	IV5: 72L	
Gorgosaurus libratus RTMP 73.30.1	III1: 127L, 55pw, 51dw	III2: 60L, 40dw	III3: 39L, 36dw			
	IV1: 82L, 40pw, 34dw					
Gorgosaurus sp., Two Medicine Fm Indianapolis Children's Museum	MT III L = 535					Right foot
	I1: 82L					
	II1: 150L, 75pw, 63dw	II2: 97L, 49dw	II3: 89L			
	III1: 145L, 84pw, 73dw	III2: 97L, 59dw	III3: 76L, 46dw			
	IV1: 101L, 73pw, 65dw			IV4: 34L, 39dw		
Gorgosaurus sp., Two Medicine Fm, Linster specimen		II2: 84L, 43dw				Left foot
		III2: 88L, 54dw	III3: 67L, 43dw	III4: 80+L		
	IV1: 88L, 64pw, 58dw	IV2: 65L, 50dw	IV3: 46L, 44dw		IV5: 71L	
Albertosaurus sarcophagus ROM 807	III1: 197L, 102pw, 92dw	III2: 122L, 68dw	III3: 57dw			
Albertosaurus sp. CMN 11315	I1: 73L					Measurements made on several loose phalanges; may come from both feet
	II1: 46dw	II2: 82L, 34dw				
	III1: 129L, 68pw, 60dw	III2: 81L, 46dw				
	IV1: 81L, 51pw, 53dw	IV2: 60L, 46dw				
Albertosaurus sp. MOR 657	MT III L = 525					Left foot: many uncertainties about this foot, particularly about the ID of the unguals, and of the phalanges of digit IV; what is identified as ungual IV5 may be missing as much as 10 mm from tip
	II1: 143L, 85pw, 71dw	II2: 90L, 59dw	II3: 89L			
	III1: 137L, 93pw, 82dw	III2: 86L, 68dw	III3: 67L, 54dw	III4: 103L		
		IV2: 86L, 70dw	IV3: 70L, 54dw	IV4: 45L, 57dw		
	89+ L					
Albertosaurus sp. USNM V 10754	II1: 95L, 60dw		II3: 65L			Measurements based on both feet; ID of some bones uncertain Actual length of II2 2–3 mm less than value used here
	III1: 73pw	III2: 90L, 45dw	III3: 59L, 41dw			
	IV1: 76L, 48dw		IV3: 38L, 40dw		IV5: 64L	

Omit taxon and specimen	Measurements					Comments
Daspletosaurus torosus CMN 350	MT III L = 554					Left foot
	II1: 145L, 87pw, 70dw					
	III1: 142L, 89pw, 86dw	III2: 118L, 63dw				
		IV2: 61L, 65dw				
Daspletosaurus torosus RTMP 85.62.1	I1: 69L	I2: 89L				Measurements based on both feet
	II1: 141L, 71dw	II2: 83L, 51dw				
	III1: 133L, 83pw, 82dw	III2: 86L, 67dw	III3: 67L, 44dw			
	IV1: 100L, 70dw	IV2: 63L	IV3: 41L			
Daspletosaurus sp. MOR 590	MT III L = 501					
	II1: 130L, 78 pw, 62dw	II2: 77L, 62 pw, 49dw	II3: 77L			
	III1: 117L, 84pw, 76dw	III2: 75L, 63dw	III3: 61L, 57pw, 47dw	III4: 91L		
	IV1: 84L, 67pw, 61dw	IV2: 58L, 66pw, 57dw	IV3: 39L, 50dw	IV4: 30L, 44pw, 43dw	IV5: 62L	
Daspletosaurus sp. RMDRC 06-005		I2: 72L				Left foot; measurements from Anthony Maltese
	II1: 133L, 64dw	II2: 88L, 56dw	II3: 99L			
	III1: 137L, 80dw	III2: 87L, 71dw	III3: 70L, 54dw			
		IV2: 71L, 66dw	IV3: 50L, 58dw	IV4: 35L, 50dw	IV5: 90L	
Tarbosaurus bataar	I1: 69L	I2: 50L				Cast obtained from Gaston Design; several mm missing from tip of I2
	II1: 141L, 83pw, 75dw	II2: 93L, 57dw	II3: 97L			
	III1: 137L, 90pw, 84dw	III2: 89L	III3: 75L, 56dw	III4: 105L		
	IV1: 101L, 74pw, 75dw	IV2: 70L, 66dw	IV3: 46L, 58L	IV4: 33L, 48dw	IV5: 98L	
Tyrannosaurus rex (or *Nanotyrannus*) BNHM 2002.4.1 ("Jane")	MT III L = 550					Measurements based on both feet
	I1: 87L					
	II1: 139L, 57pw, 51dw	II2: 91L, 40dw	II3: 90L			
	III1: 139L, 67pw, 58dw	III2: 93L, 50dw				
	IV1: 91L, 52pw, 51dw	IV2: 69L, 44dw	IV3: 47L, 38dw	IV4: 37L, 37dw		
Tyrannosaurus rex (or *Nanotyrannus*) BHI 6437	MT III L = 539					Left foot
	I1: 77L					
	II1: 136L	II2: 85L				
	III1: 135L, 54dw	III2: 87L, 43dw	III3: 68L, 33dw			
	IV1: 94L, 51dw	IV2: 80L				
Tyrannosaurus rex BMRP 2006.4.4 ("Petey")		II2: 103L, 41dw				IDs of phalanges III3 and III4 are questionable. There are two additional phalanges, which I could not identify–possibly another animal?
		III2: 151L, 57dw	III3: 54L, 40dw			
	IV1: 104L, 56pw, 55dw			IV4: 36L	IV5: 86L	
Tyrannosaurus rex Indianapolis Children's Museum ("Bucky")	II1: 169L, 107pw, 101dw	II2: 123L, 76dw				Measurements based on both feet
			III3: 103L, 65dw			
	IV1: 119L, 107dw	IV2: 96L, 95dw		IV4: 46L, 68dw		

Table A1.1. *continued*

Omit taxon and specimen	Measurements					Comments
Tyrannosaurus rex FMNH PR2081 ("Sue")	MT III L = 677					
		I2: 113L				
	II1: 201L, 126pw, 117dw	II2: 134L, 94dw	II3: 167L			
	III1: 189L, 145pw, 137dw	III2: 131L, 114dw	III3: 117L, 88dw	III4: 148L		
	IV1: 144L, 114pw, 126dw	IV2: 106L, 105dw	IV3: 75L, 95dw			
Tyrannosaurus rex LACM 23845 (*"Albertosaurus megagracilis"/ "Gorgosaurus lancensis"*)	I1: 79L					Right foot; lengths of phalanges II1 and II2 estimated from Molnar (1980: text fig. 7)
	II1: 152L	II2: 103L				
	III1: 145L, 78pw, 72dw	III2: 99L, 60dw	III3: 84L	III4: 84L		
	IV1: 90L, 68dw	IV2: 72L, 61dw	IV3: 52L	IV4: 33L, 45dw		
Tyrannosaurus rex MOR 009	II1: 197L, 94dw					
		IV2: 97L, 90dw				
Tyrannosaurus rex MOR 555	MT III L = 650					Cast of left foot
	I1L 120L	I2: 106L				
	II1: 223L, 114pw, 95dw	II2: 147L, 92dw				
	III1: 213L, 141pw, 129dw	III2: 140L, 110dw	III3: 120L, 88dw	III4: 137L		
	IV1: 149L, 108pw, 110dw	IV2: 109L, 108dw	IV3: 73L, 95dw	IV4: 55L, 77dw	IV5: 145L	
Tyrannosaurus rex LACM 23844	II1: 195L, 113pw, 103dw	II2: 133L, 86dw				Measurements based on phalanges from both feet
	III1: 195L, 128pw, 117dw	III2: 133L, 101dw	III3: 117L, 83dw			
	IV1: 137L, 105pw, 111dw	IV2: 99L, 98dw	IV3: 69L, 85dw	IV4: 51L, 71dw		
Tyrannosaurus rex ("Peck's rex")	I1: 103L					Measurements made on casts of phalanges
	III1: 211L, 132pw, 121dw	III2: 143L, 100dw	III3: 120L, 83dw	III4: 152L		
	IV1: 146L, 109pw, 119dw		IV3: 71L, 94dw	IV4: 50L, 76dw		
Tyrannosaurus rex BHI 6230 ("Wyrex")	MT III L = 661					Measured cast of right foot
	I1: 99L	I2: 94L				
	II1: 177L, 112pw, 97dw	II2: 117L, 77dw	II3: 138L			
	III1: 175L, 122pw, 111dw	III2: 116L, 95dw	III3: 95L, 72dw	III4: 130L		
	IV1: 126L, 102pw, 104dw	IV2: 87L, 91dw	IV3: 57L, 76dw	IV4: 45L, 65dw	IV5: 128L	
Tyrannosaurus rex BHI "Stan"	I1: 118L	I2: 105L				Cast of left foot
	II1: 209L, 137pw, 107dw		II3: 177L			
	III1: 194L, 149pw, 129dw	III2: 141L, 105dw				
	IV1: 147L, 115pw, 117dw	IV2: 103L, 106dw	IV3: 68L, 90dw	IV4: 53L, 77dw	IV5: 149L	
Tyrannosaurus rex MOR N rex ("Nathan")	MT III L = 563+; part of proximal end missing					Right foot
	II1: 219L	II2: 125L	II3: 163L			
			III3: 103L	III4: 159L		
Ornithomimosaurs						
Sinornithomimus dongi (individual # 3)	II1: 36L	II2: 18L				Left foot
	III1: 34L	III2: 25L	III3: 18L			
		IV2: 12L	IV3: 10L			
Sinornithomimus dongi (individual # 4)	III1: 27L	III2: 21L	III3: 13L	III4: 15L		Left foot
	IV1: 15L	IV2: 9L	IV3: 4L	IV4: 6L	IV5: 14L	

Omit taxon and specimen	Measurements					Comments
Sinornithomimus dongi largest individual (#10)	MT III L = 257L					Measurements based on both feet
	I1: 15L					
	II1: 55L	II2: 28L	II3: 33L			
	III1: 55L, 26dw	III2: 40L, 21dw	III3: 27L, 17dw	III4: 30L		
	IV1: 35L	IV2: 19L	IV3: 12L	IV4: 12L	IV5: 30L	
Struthiomimus altus AMNH 5339	Digit I absent					Specimen measured was a cast (RTMP 85.8.3); measurements based on both feet
	II1: 84L	II2: 33L	II3: 44L			
	III1: 77L	III2: 52L	III3: 37L	III4: 44L		
	IV1: 44L, 30dw	IV2: 23L, 24dw	IV3: 17L, 22dw	IV4: 18L, 18dw	IV5: 41L	
Struthiomimus altus AMNH 5375	Digit I absent					
	II1: 83L, 31pw, 28dw					
	III1: 79L, 42pw, 35dw	III2: 53L, 28dw	III3: 21L, 24dw			
Struthiomimus altus CMN 930 P-0102	Digit I absent					Measurements based on both feet; possibly a few mm missing from tip of II3
	II1: 73L, 29pw, 26dw	II2: 30L, 21dw	II3: 50L			
	III1: 70L, 31dw	III2: 49L, 27dw	III3: 38L, 21dw			
	IV1: 42L, 27pw, 27dw	IV2: 21L, 26dw	IV3: 15L, 25dw	IV4: 15L, 21dw		
Struthiomimus altus ROM 1790	Digit I absent					Left foot; possibly some material missing from tip of II3
		II2: 28L, 18dw	II3: 43L			
	III1: 63L, 27dw	III2: 45L, 22dw	III3: 35L, 17dw			
	IV1: 23dw	IV2: 20L, 21dw	IV3: 14L, 20dw	IV4: 15L, 18dw	IV5: 38L	
Struthiomimus altus BHI 1266 ("*S. sedens*")	MT III L = 465					Cast of left foot
	Digit I absent					
	II1: 118L, 42pw, 35dw	II2: 52L, 29dw	II3: 65L			
	III1: 105L, 50pw, 43dw	III2: 76L, 35dw	III3: 55L, 27dw	III4: 69L		
	IV1: 60L, 39pw, 42dw	IV2: 36L, 35dw	IV3: 25L, 30dw	IV4: 27L, 29dw	IV5: 63L	
Ornithomimus edmontonicus ROM 851	MT III L = 316					Right foot
	Digit I absent					
	II1: 75L, 22dw	II2: 30L, 17dw				
	III1: 75L	III2: 54L	III3: 42L			
	IV1: 40L	IV2: 28L	IV3: 17L	IV4: 18L		
Ornithomimus edmontonicus CMN 8632 P-1602	Digit I absent					Left foot
	II1: 80L	II2: 31L, 20dw				
	III1: 73L, 30dw	III2: 53L, 23dw	III3: 23dw			
		IV2: 23L	IV3: 17L, 21dw	IV4: 18L, 19dw		
Ornithomimus sp. LACM 47520	Digit I absent					Right foot
		II2: 52L, 29dw				
		III2: 61L, 27dw		III4: 53L		
		IV2: 38L, 33dw	IV3: 25L, 29dw	IV4: 25L, 25dw	IV5: 57L	
Ornithomimus edmontontoncicus (formerly *Dromiceiomimus brevitertius*) ROM 797	Digit I absent					Right foot
	II1: 73L, 21pw, 19dw	II2: 28L, 16dw	II3: 46L			
	III1: 69L, 23dw	III2: 50L, 18dw	III3: 36L, 15dw			
	IV1: 41L, 20dw	IV2: 22L, 19dw	IV3: 15L, 18dw	IV4: 16L, 15dw	IV5: 39L	

Table A1.1. *continued*

Omit taxon and specimen	Measurements					Comments
"Ornithomimus velox" MNA Pl 1726A	MT III L = 385					Left foot; measured cast of specimen
	Digit I absent					
	II1: 77L, 27pw, 22dw	II2: 35L, 21dw	II3: 51L			
	III1: 67L, 38pw, 35dw	III2: 52L, 30dw	III3: 39L, 20dw			
	IV1: 47L, 23dw	IV2: 26L, 24dw	IV3: 18L, 23dw	IV4: 19L, 20dw	IV5: 39L	
Gallimimus bullatus Gaston Design cast	MT III L = 470					Left foot
	Digit I absent					
	II1: 90L, 36pw, 34dw	II2: 37L, 27dw	II3: 41L			
	III1: 84L, 55pw, 44dw	III2: 61L, 34dw	III3: 36L, 28dw	III4: 42L		
	IV1: 47L, 35dw	IV2: 28L, 30dw	IV3: 17L, 27dw	IV4: 16L, 23dw	IV5: 39L	
Ornithomimid; possibly part of CMN 0930	IV1: 41L, 28pw, 27dw	IV2: 25L, 24dw	IV3: 17L, 22dw	IV4: 18L, 19dw		Right toe
Ornithomimid MOR 450	Digit I absent					Right foot; tip of II3 missing; IV5 missing 4–5 mm from tip
	II1: 67L, 28pw, 23dw	II2: 29L, 18dw	II3: 37+ L			
	III1: 60L, 30pw, 26dw		III3: 29L, 17dw			
	IV1: 36L, 25dw	IV2: 17L, 21dw	IV3: 12L, 20dw	IV4: 14L, 17dw	IV5: 27+ L	
Ornithomimid RTMP 79.14.701	Digit I absent					Right foot
	II1: 69L, 27pw, 24dw	II2: 29L, 18dw				
		III2: 48L, 22dw	III3: 38L, 18dw			
		IV2: 21L	IV3: 14L, 22dw			
Ornithomimid RTMP 79.14.864	Digit I absent					
		III2: 48L, 22dw	III3: 37L, 18dw			
Ornithomimid RTMP 90.26.1	Digit I absent					Right foot
		II2: 35L, 21dw	II3: 47L			
	III1: 79L, 34dw	III2: 57L, 26dw	III3: 43L, 20dw	III4: 46L		
		IV2: 26L, 25dw	IV2: 18L, 23dw	IV3: 17L, 20dw		
Ornithomimid RTMP 96.61.1	Digit I absent					Measurements based on both feet
	II1: 71L, 23dw					
	III1: 67L, 31pw, 29dw	III2: 48L, 23dw	III3: 37L, 17dw			
	IV1: 38L, 23pw, 21dw	IV2: 33L	IV3: 16L, 18dw	IV4: 16L, 16dw		
Hell Creek Fm Fort Peck ornithomimid BHI	MT III L = 390					Right foot
	Digit I absent					
	II1: 92L, 33pw, 28dw	II2: 42L, 23dw	II3: 51L			
	III1: 84L, 43pw, 35dw	III2: 62L, 28dw	III3: 48L, 22dw	III4: 49L		
	IV1: 47L, 29pw, 31dw	IV2: 29L, 27dw	IV3: 20L, 24dw	IV4: 21L, 22dw	IV5: 47L	
"Coelosaurus antiquus" AMNH 2551	II1: 92L, 39pw, 31dw					Right foot
	III1: 83L, 51pw, 38dw					

Omit taxon and specimen	Measurements					Comments
"Arkansaurus fridayi" AU 74-16	MT III L = 389					Cast of right foot borrowed from J. Kirkland. What is identified here as III1 could be II1
	III1: 95L, 54pw, 43dw					
	IV1: 59L, 48pw, 42dw					
Alvarezsaurids						
Mononykus olecranus GI N107/6	I1:7L					Left foot; ungual III4 missing about 1 mm from tip; actual measured III4L as preserved = 13 mm
	II1: 15L, 9pw, 7dw	II2: 10L, 7dw	II3: 16L			
	III1: 16L, 9pw, 7dw	III2: 9L, 6dw	III3: 9L, 7dw	III4: 14L		
	IV1: 11L, 9pw, 8dw	IV2: 8L, 8dw	IV3: 7L, 7dw	IV4: 9L, 7dw		
Oviraptorosaurs						
Oviraptor philoceratops IGM 100/972	I1: 31L	I2: 33L				Measurements based on both feet
	II1: 47L	II2: 30L, 17dw	II3: 45L			
	II1: 49L, 25pw, 22dw	III2: 29L				
	IV1: 36L	IV2: 24L	IV3: 21L	IV4: 19L	IV5: 36L	
Chirostenotes pergracilis CMN 8538	MT III L = 230					Right foot; ungual IV5 possibly missing a few mm from tip
	I1: 58L	I2: 36L				
	II1: 77L, 25pw, 22dw	II2: 57L, 17dw	II3: 44L			
	III1: 76L, 35pw	III2: 43L, 22dw	III3: 50L, 17dw	III4: 46L		
	IV1: 58L, 26dw	IV2: 34L, 19dw	IV3: 26L, 18dw	IV4: 30L, 17dw	IV5: 36L	
Chirostenotes pergracilis RTMP 79.20.1	II1: 40L, 14pw, 11dw					Proximal end of III1 is crushed, so length could be a bit off
	III1: 53L, 19dw					
Elmisaurus elegans MOR 752	II1: 10dw	II2: 23L, 8dw	II3: 18L			Identification of unguals uncertain; II3 could actually be IV5
	III1: 32L, 11dw	III2: 20L, 10dw	III3: 21L, 8dw	III4: 19L		
	IV1: 21L, 12pw, 11dw	IV2: 12L, 9dw	IV3: 10L, 8dw	IV4: 12L		
Conchoraptor gracilis Gaston Design cast	II1: 22L, 10dw	II2: 12L, 8dw	II3: 22L			Measurements based on both feet, which are preserved in situ in block
	III1: 24L, 11dw	III2: 16L, 9dw	III3: 13L, 8dw	III4: 22L		
	IV1: 15L, 8pw, 9dw	IV2: 9L, 7dw	IV3: 7L, 6dw	IV4: 7L, 6dw	IV5: 15L	
Conchoraptor gracilis Gaston Design cast	I1: 15L	I2: 15L				Left foot
	II1: 24L, 11dw	II2: 12L, 10dw	II3: 25L			
	III1: 28L	III2: 16L	III3: 13L			
	IV1: 17L	IV2: 11L, 10dw	IV3: 8L, 9dw	IV4: 7L, 8dw	IV5: 19L	
Troodontids						
*Troodon formosus*MOR 748	I1: 19L					Measurements probably based on both feet; ungual II3 missing a few mm from tip
	II1: 37L		II3: 31L			
	III1: 42L		III3: 24L			
		IV2: 19L				
Troodon formosus MOR 563		II2: 13L, 8dw	II3: 27L			Measurements based on both feet; ungual II3 probably missing a few mm from tip
	III1: 36L, 15pw, 13dw	III2: 24L, 12dw	III3: 23L, 10dw			
	IV1: 20L, 12dw	IV2: 16L, 11dw		IV4: 15L, 9dw		
Dromaeosaurids						
Saurornitholesteslangstoni MOR 660	MT III L = 116					Measurements based on both feet
	II1: 21L, 14pw, 13dw	II2: 21L, 9dw	II3: 60L			
	III1: 45L, 15pw, 13dw		III3: 25L, 10dw			
	IV1: 37L, 13dw	IV2: 23L, 11dw				

Table A1.1. *continued*

Omit taxon and specimen	Measurements					Comments
Saurornitholesteslangstoni RTMP 88.121.39	I1: 21L					
	II1: 23L, 15pw, 14dw		II3: 64L			
	III1: 48L, 15pw, 14dw		III3: 27L, 9dw	III4: 35L		
	IV1: 39L, 13dw	IV2: 24L, 11dw	IV3: 19L, 10dw	IV4: 19L, 8dw		
Deinonychus antirrhopus MOR 747 (individual with complete foot)	MT III L = 147					Measurements mainly based on right foot; probably a few mm missing from tips of III4 and IV5
	I1: 23L	I2: 30L				
	II1: 31L	II2: 29L, 13dw	II3: 67L			
	III1: 49L, 23pw	III2: 26L, 16dw	III3: 29L, 13dw	III4: 43L		
	IV1: 42L, 18pw, 17dw	IV2: 29L, 16dw	IV3: 21L, 15dw	IV4: 21L, 11dw	IV5: 36L	
Deinonychus antirrhopus MOR 747 same number as preceding; individual with incomplete feet)	I1: 23L					Measurements based on both feet
	II1: 31L, 20pw, 19dw	II2: 13dw				
	III1: 46L, 23pw, 18dw	III2: 30L	III3: 27L, 13L			
	IV1: 40L, 17pw, 17dw	IV2: 21L, 15dw		IV4: 21L, 12dw		
Buitreraptor gonzalezorum MPCA 238	I1: 14L	I2: 9L				Right foot
	II1: 21L, 6pw, 7dw	II2: 14L, 5dw				
	IV1: 21L, 5pw					
Velociraptor mongoliensis	I1: 17L					Cast of left foot of Osmólska specimen from Matt Smith
	II1: 20L	II2: 19L	II3: 57L			
		III2: 21L	III3: 17L			
		IV2: 20L	IV3: 15L			
Velociraptor mongoliensis	I1: 17L					Right foot of Gaston Design cast
	II1: 22L	II2: 19L				
		IV2: 21L	IV3: 16L	IV4: 16L	IV5: 29L	
Bambiraptor feinbergi FIP 001	I1: 12L	I2: 10L				Measurements based on both feet; made from cast from D. Burnham
	II1: 14L, 7pw, 7dw	II2: 13L, 5dw	II3: 31L			
	III1: 27L, 8pw, 7dw	III2: 15L, 6dw	III3: 16L, 5dw	III4: 17L		
	IV1: 22L, 6pw, 7dw	IV2: 17L, 6dw	IV3: 11L, 5dw	IV4: 12L, 5dw	IV5: 17L	
Archaeopteryx siemensii HMN 1880/81 (Berlin specimen)	MT III L = 39					Measurements based on both feet. Phalanges not especially well preserved, and hard to measure
	I1: 6L	I2: 7L				
	II1: 9L	II2: 7L	II3: 8L			
	III1: 10L	III2: 9L	III3: 8L	III4: 8L		
	IV1: 6L	IV2: 6L	IV3: 5L	IV4: 6L	IV5: 6L	
Patagopteryx deferrariisi MACN-N 03	I1: 9L	I2: 13L				Right foot. Tip of III4 abraded or broken, missing 1–2 mm; length used here includes 2 mm added to actual measurement. IV5 possibly also a bit abraded, but actual measurement used here
	II1: 12L	II2: 8L	II3: 16L			
	III1: 15L	III2: 13L	III3: 9L	II4: 15L		
			IV3: 4L	IV4: 5L	IV5: 11L	
Patagopteryx deferrariisi	I1: 9L					Measurements made on "so-so" private cast
	II1: 12L					
	III1: 17L		III3: 9L	III4: 15L		
	IV1: 8L					
Palaeotis weigelti GM 4362	MT III L = 209					Measurements based on both feet
	Digit I absent					
	II1: 26L	II2: 11L	II3: 13L			
	III1: 32L	III2: 21L	III3: 14L	III4: 15L		
	IV1: 24L				IV5: 11L	

Omit taxon and specimen	Measurements				Comments
Struthionidae					
Struthio camelus MSU mounted skeleton	Digits I and II absent				Right foot
	III1: 86L	III2: 50L	III3: 27L	III4: 41L	
	IV1: 76L	IV2: 25L	IV3: 13L	IV4: 7	
Struthio camelus USNM uncatalogued	Digits I and II absent				Right foot
	III1: 85L				
	IV1: 78L				
Struthio camelus USNM 345220	Digits I and II absent				Left or right uncertain
	III1: 83L, 38pw, 36dw	III2: 51L, 35dw	III3: 30L, 31dw	III4: 48L	
	IV1: 76L, 21pw, 16dw	IV2: 25L, 19dw	IV3: 14L, 16dw		
Struthio camelus USNM 343621	Digits I and II absent				Measurements based on both feet of this male bird
	III1: 75L, 41pw, 40dw	III2: 45L, 38dw	III3: 27L, 34dw	III4: 43L	
	IV1: 67L, 24pw, 18dw	IV2: 23L, 20dw	IV3: 13L, 16dw	IV4: 8L, 10dw	
Struthio camelus USNM 224856	Digits I and II absent				Measurements based on both feet
	III1: 88L, 39pw, 36dw	III2: 51L, 36dw	III3: 29L, 33dw	III4: 47L	
	IV1: 73L, 21pw, 15dw	IV2: 25L, 19dw	IV3: 14L, 16dw	IV4: 8L, 7dw	
Struthio camelus USNM 346697	Digits I and II absent				Measurements based both feet
	III1: 83L, 40pw, 38dw	III2: 50L, 36dw	III3: 27L, 31dw	III4: 45L	
	IV1: 77L, 22pw, 16dw	IV2: 27L, 18dw	IV3: 13L, 14dw	IV4: 8L, 9dw	
Struthio camelus USNM 291160	Digits I and II absent				Measurements based both feet
	III1: 86L, 41pw, 40dw	III2: 51L, 37dw	III3: 29L, 32dw	III4: 42L	
	IV1: 75L, 23pw, 17dw	IV2: 24L, 18dw	IV3: 12L, 14dw	IV4: 7L, 9dw	
Struthio camelus Amherst College	TMT L = 485				
	Digits I and II absent				
	III1: 93L, 43pw, 40dw	III2: 54L, 39dw	III3: 30L, 36dw	III4: 34L	
	IV1: 78L, 25pw, 16dw	IV2: 27L, 21dw	IV3: 15L, 16dw	IV4: 10L, 9dw	
Struthio camelus FMNH 85690	TMT L = 447				Measurements based on both feet
	Digits I and II absent				
	III1: 82L, 38pw, 37dw	III2: 49L, 34dw	III3: 26L, 28dw	III4: 46L	
	IV1: 70L, 19pw, 15dw	IV2: 22L, 19dw	IV3: 11L, 14dw	IV4: 8L, 10dw	
Struthio camelus FMNH 105097	TMT L = 451				Measurements based on both feet of female bird
	Digits I and II absent				
	III1: 83L, 43pw, 43dw	III2: 50L, 39dw	III3: 30L, 35dw	III4: 48L	
	IV1: 75L, 22pw, 16dw	IV2: 25L, 20dw	IV3: 14L, 16dw	IV4: 9L, 10dw	
Struthio camelus FMNH 106776	TMT L = 463				
	Digits I and II absent				
	III1: 85L	III2: 50L	III3: 28L	III4: 45L	

Table A1.1. *continued*

Omit taxon and specimen	Measurements					Comments
Rheidae						
Rhea americana University of Illinois mount	Digit I absent					Left foot of mounted skeleton
	II1: 45L, 15pw, 12dw	II2: 8L, 10dw	II3: 21L			
	III1: 43L, 22pw, 18dw	III2: 26L, 15dw	III3: 12L, 15dw	III4: 28L		
	IV1: 35L, 17pw, 14dw	IV2: 6L, 13dw	IV3: 5L, 11dw	IV4: 4L, 10dw	IV5: 21L	
Rhea americana USNM 614471	Digit I absent					Juvenile male bird; measurements based on both feet
	II1: 22L	II2: 4L	II3: 10L			
	III1: 20L	III2: 11L	III3: 5L	III4: 12L		
	IV1: 16L	IV2: 3L	IV3: 1L	IV4: 2L	IV5: 9L	
Rhea americana USNM 614473	Digit I absent					Male chick; measurements based on both feet
	II1: 18L	II2: 3L				
	III1: 17L	III2: 10L	III3: 5L			
	IV1: 13L	IV2: 2L	IV3: 1L	IV4: 1L		
Rhea americana YPM ORN 109071 (formerly 11524)	Digit I absent					Measurements based on both feet; what I identified as II3 could be IV5, and vice versa
	II1: 45L, 15pw, 12dw	II2: 8L, 11dw	II3: 22L			
	III1: 41L, 21pw, 17dw	III2: 25L, 15dw	III3: 11L, 13dw	III4: 31L		
	IV1: 35L, 16pw, 14dw	IV2: 6L, 13dw	IV3: 5L, 13dw	IV4: 5L, 12dw	IV5: 23L	
Rhea americana FMNH 104061	MT III L = 306					Measurements based on both feet of female bird
	Digit I absent					
	II1: 42L, 11dw	II2: 8L	II3: 22L			
	III1: 38L, 15dw	III2: 22L	III3: 10L, 10dw	III4: 28L		
	IV1: 29L	IV2: 7L	IV3: 5L	IV4: 4L	IV5: 21L	
Rhea americana FMNH 105749	TMT L = 330					Measurements based on both feet
	Digit I absent					
	II1: 44L, 17pw, 11dw	II2: 7L	II3: 23L			
	III1: 42L, 16dw	III2: 26L, 14dw	III3: 10L, 13dw	III4: 31L		
	IV1: 33L	IV2: 6L	IV3: 5L	IV4: 4L	IV5: 23L	
Rhea americana FMNH 105754	TMT L = 329					Female bird
	Digit I absent					
	II1: 42L, 13pw, 10dw	II2: 8L	II3: 23L			
	III1: 41L, 20pw, 15dw	III2: 24L, 13dw	III3: 11L, 13dw	III4: 31L		
	IV1: 31L, 14pw, 12dw	IV2: 6L, 12dw	IV3: 4L	IV4: 4L	IV5: 22L	
Rhea americana FMNH 339615	TMT L = 308					Measurements based on both feet of female bird
	Digit I absent					
	II1: 40L, 12pw, 11dw	II2: 7L, 10dw	II3: 23L			
	III1: 39L, 18pw, 15dw	III2: 24L, 12dw	III3: 11L, 12dw	III4: 31L		
	IV1: 31L, 15pw, 13dw	IV2: 6L, 12dw	IV3: 5L, 11dw	IV4: 5L, 9dw	IV5: 26L	
Rhea americana FMNH 339616	Digit I absent					Measurements based on both feet of female bird
	II1: 43L, 12pw, 11dw	II2: 7L, 10dw	II3: 21L			
	III1: 41L, 18pw, 15dw	III2: 26L, 13dw	III3: 11L, 13dw	III4: 28L		
	IV1: 32L, 14pw, 13dw	IV2: 7L, 12dw	IV3: 4L, 11dw	IV4: 4L, 10dw	IV5: 21L	

Omit taxon and specimen	Measurements					Comments
Rhea americana LACM BIRD 93063	TMT L = 258 (+?); tarsal elements possibly not attached					Measurements made on both feet of immature male; bone articular ends quite porous
	Digit I absent					
	II1: 34L, 11pw, 9dw	II2: 6L, 8dw				
	III1: 34L, 16pw, 14dw	III2: 20L, 12dw	III3: 10L, 10dw	III4: 23L		
	IV1: 26L, 13pw, 10dw	IV2: 5L, 9dw	IV3: 3L, 8dw	IV4: 3L, 7dw	IV5: 17L	
Rhea americana LACM BIRD 93064	TMT L = 266 (+?); tarsal elements possibly not attached					Measurements made on both feet of immature female; bone articular ends quite porous
	Digit I absent					
	II1: 38L, 12pw, 9dw	II2: 7L, 8dw				
	III1: 37L, 16pw, 13dw	III2: 22L, 11dw	III3: 10L, 10dw	III4: 26L		
	IV1: 30L, 13pw, 10dw	IV2: 5L, 9dw	IV3: 3L, 9dw	IV4: 3L, 8dw		
Rhea americana LACM uncatalogued	TMT L = 283					Measurements based on both feet
	Digit I absent					
	II1: 41L, 13pw, 12dw	II2: 13L, 11dw	II3: 20L			
	III1: 40L, 18pw, 16dw	III2: 26L, 15dw	III3: 14L, 13dw	III4: 31L		
	IV1: 32L, 15pw, 13dw	IV2: 8L, 13dw	IV3: 5L, 12dw	IV4: 5L, 11dw	IV5: 18L	
Rhea americana juvenile carcass received from animal breeder	Digit I absent					Left foot; skeleton dissected from soft tissues
	II1: 27L	II2: 5L	II3: 12L			
	III1: 26L	III2: 15L	III3: 7L	III4: 15L		
	IV1: 21L	IV2: 3L	IV3: 3L	IV4: 3L	IV5: 12L	
Pterocnemia pennata USNM 28814	Digit I absent					Right foot
	II1: 45L, 14pw, 13dw	II2: 12L, 13dw	II3: 25L			
	III1: 41L, 20pw, 18dw	III2: 29L, 16dw	III3: 14L, 15dw	III4: 34L		
	IV1: 35L, 18pw, 14dw	IV2: 9L, 14dw	IV3: 6L, 13dw	IV4: 5L, 10dw	IV5: 23L	
Pterocnemia pennata IRSNB 5403	TMT L = 295					Left foot of male bird
	Digit I absent					
	II1: 42L	II2: 11L	II3: 23L			
	III1: 42L	III2: 27L	III3: 13L	III4: 30L		
	IV1: 32L	IV2: 9L	IV3: 6L	IV4: 5L	IV5: 21L	
Pterocnemia pennata IRSNB 48774	TMT L = 285					Left foot of female bird
	Digit I absent					
	II1: 41L	II2: 10L	II3: 25L			
	III1: 40L	III2: 27L	III3: 13L	III4: 33L		
	IV1: 32L	IV2: 8L	IV3: 5L	IV4: 5L	IV5: 21L	
Pterocnemia pennata IRSNB 48297	TMT L = 265					Left foot
	Digit I absent					
	II1: 39L	II2: 12L	II3: 22L			
	III1: 38L	III2: 26L	III3: 13L	III4: 27L		
	IV1: 31L	IV2: 8L	IV3: 4L	IV4: 4	IV5: 19L	
Pterocnemia pennata FMNH 104108	TMT L = 283					Left foot of male bird
	Digit I absent					
	II1: 41L	II2: 11L	II3: 23L			
	III1: 40L	III2: 28L	III3: 15L	III4: 32L		
	IV1: 33L	IV2: 8L	IV3: 5L	IV4: 4L	IV5: 21L	

Table A1.1. *continued*

Omit taxon and specimen	Measurements					Comments
Pterocnemia pennata FMNH 288197	TMT L = 301					Measurements based on both feet of female bird
	Digit I absent					
	II1: 44L, 14pw, 13dw	II2: 14L, 12dw	II3: 28L			
	III1: 42L, 21pw, 18dw	III2: 30L, 17dw	III3: 16L, 15dw	III4: 38L		
	IV1: 34L, 16pw, 14dw	IV2: 9L, 14dw	IV3: 5L, 13dw	IV4: 6L, 12dw	IV5: 24L	
Pterocnemia pennata FMNH 339617	Digit I absent					Measurements based on both feet of juvenile bird
	II1: 35L, 11pw, 11dw	II2: 9L, 9dw	II3: 17L			
	III1: 33L, 14pw, 14dw	III2: 22L, 12dw	III3: 11L, 11dw	III4: 23L		
	IV1: 26L, 12pw, 11dw	IV2: 5L, 11dw	IV3: 4L, 10dw	IV4: 4L, 8dw	IV5: 16L	
Pterocnemia pennata FMNH 339618	Digit I absent					Measurements based on both feet
	II1: 44L, 14pw, 12dw	II2: 13L, 11dw	II3: 25L			
	III1: 42L, 20pw, 17dw	III2: 29L, 15dw	III3: 14L, 14dw	III4: 30L		
	IV1: 33L, 15pw, 13dw	IV2: 8L, 13dw	IV3: 5L, 12dw	IV4: 5L, 10dw	IV5: 19L	
Dromaiidae						
Dromaius novaehollandiae YPM ORN 102523 (formerly 02126)	TMT L = 330					Left foot
	Digit I absent					
	II1: 46L, 15pw, 12dw	II2: 13L, 11dw	II3: 23L			
	III1: 54L, 25pw, 21dw	III2: 34L, 19dw	III3: 16L, 16dw	III4: 27L		
	IV1: 39L, 18pw, 14dw	IV2: 14L, 12dw	IV3: 9L, 11dw	IV4: 4L, 10dw	IV5: 22L	
Dromaius novaehollandiae YPM ORN 102525 (formerly 2128)	Digit I absent					Measurements based on both feet
	II1: 50L, 14pw, 12dw	II2: 14L, 11dw	II3: 26L			
	III1: 56L, 27pw, 21dw	III2: 36L, 18dw	III3: 19L, 16dw	III4: 27L		
	IV1: 40L, 17pw, 14dw	IV2: 15L, 13dw	IV3: 9L, 12dw	IV4: 5L, 10dw	IV5: 21L	
Dromaius novaehollandiae YPM ORN 102524 (formerly 2127)	Digit I absent					Measurements based on both feet
	II1: 55L, 16pw, 14dw	II2: 12L, 14dw	II3: 26L			
	III1: 63L, 29pw, 25dw	III2: 39L, 23dw	III3: 19L, 18dw	III4: 30L		
	IV1: 43L, 20pw, 16dw	IV2: 15L, 16dw	IV3: 9L, 14dw	IV4: 5L, 13dw	IV5: 23L	
Dromaius novaehollandiae QMO12657	Digit I absent					Left foot of male bird
	II1: 48L, 16pw, 14dw	II2: 12L, 12dw	II3: 29L			
	III1: 59L, 31pw, 26dw	III2: 38L, 23dw	III3: 20L, 19dw	III4: 33L		
	IV1: 41L, 21pw, 17dw	IV2: 16L, 15dw	IV3: 10L, 14dw	IV4: 6L, 14dw	IV5: 25L	
Dromaius novaehollandiae QM O11686	Digit I absent					Right foot
	II1: 47L, 15pw, 12dw	II2: 14L, 11dw	II3: 28L			
	III1: 57L, 29pw, 24dw	III2: 37L, 20dw	III3: 21L, 16dw	III4: 33L		
	IV1: 43L, 19pw, 15dw	IV2: 17L, 13dw	IV3: 9L, 13dw	IV4: 7L, 12dw	IV5: 24L	

Omit taxon and specimen	Measurements					Comments
Dromaius novaehollandiae USNM 345521	Digit I absent					Measurements based on both feet
	II1: 47L, 15pw, 12dw	II2: 15L, 12dw	II3: 27L			
	III1: 55L, 27pw, 24dw	III2: 36L, 21dw	III3: 18L, 18dw	III4: 28L		
	IV1: 39L, 19pw, 15dw	IV2: 14L, 14dw	IV3: 9L, 13dw	IV4: 5L, 12dw	IV5: 23L	
Dromaius novaehollandiae USNM 343393	TMT L = 351					Measurements based on both feet
	Digit I absent					
	II1: 44L, 15pw, 12dw	II2: 13L, 12dw	II3: 26L			
	III1: 50L, 27pw, 22dw	III2: 33L, 20dw	III3: 18L, 18dw	III4: 31L		
	IV1: 37L, 18pw, 14dw	IV2: 13L, 13dw	IV3: 8L, 12dw	IV4: 6L, 13dw	IV5: 23L	
Dromaius novaehollandiae USNM 500380	TMT L = 348					
	Digit I absent					
	II1: 42L	II2: 10L				
	III1: 53L	III2: 34L	III3: 16L			
	IV1: 38L	IV2: 14L	IV3: 9L	IV4: 6L		
Dromaius novaehollandiae UMA 3966	Digit I absent					Left foot of young bird; not completely defleshed, so some measurements hard to get
	II1: 40L					
	III1: 45L	III2: 29L	III3: 19L	III4: 30L		
	IV1: 33L	IV2: 12L	IV3: 7L			
Dromaius novaehollandiae Amherst College	TMT L = 355					Left foot
	Digit I absent					
	II1: 45L, 13pw, 11dw	II2: 12L, 11dw	II3: 24L			
	III1: 55L, 25pw, 21dw	III2: 34L	III3: 19L, 16dw	III4: 26L		
	IV1: 37L, 15pw, 14dw	IV2: 14L, 12dw	IV3: 9L, 11dw	IV4: 7L, 11dw	IV5: 21L	
Dromaius novaehollandiae FMNH 104536	Digit I absent					Measurements based on both feet of male bird
	II1: 50L, 17pw, 13dw	II2: 14L, 14dw	II3: 27L			
	III1: 57L, 29pw, 23dw	III2: 36L, 20dw	III3: 19L	III4: 33L		
	IV1: 41L	IV2: 15L, 15dw	IV3: 9L	IV4: 6L	IV5: 24L	
Dromaius novaehollandiae FMNH 313620	TMT L = 377					Measurements based on both feet
	Digit I absent					
	II1: 53L, 15pw, 15dw	II2: 15L, 13dw	II3: 29L			
	III1: 64L, 28pw, 24dw	III2: 37L, 21dw	III3: 21L, 18dw	III4: 36L		
	IV1: 44L, 17pw, 15dw	IV2: 15L, 14dw	IV3: 9L, 12dw	IV4: 6L, 13dw	IV5: 27L	
Dromaius novaehollandiae FMNH 339613	TMT L = 393					Measurements based on both feet of female bird
	Digit I absent					
	II1: 54L, 15pw, 13dw	II2: 15L, 12dw	II3: 29L			
	III1: 61L, 27pw, 22dw	III2: 38L, 19dw	III3: 17L, 17dw	III4: 34L		
	IV1: 45L, 18pw, 16dw	IV2: 17L, 13dw	IV3: 10L, 13dw	IV4: 6L, 12dw	IV5: 27L	

Table A1.1. *continued*

Omit taxon and specimen	Measurements					Comments
Casuariidae						
Casuarius casuarius AMNH SKEL 3870	Digit I absent					Right foot
	II1: 36L, 20pw, 17dw	II2: 14L, 13dw	II3: 55L			
	III1: 51L, 25pw, 19dw	III2: 35L, 17dw	III3: 25L, 15dw	III4: 33L		
	IV1: 35L, 18pw, 14dw	IV2: 14L, 13dw	IV3: 9L, 13dw	IV4: 8L, 12dw	IV5: 29L	
Casuarius casuarius AMNH SKEL 962	Digit I absent					Measurements based on both feet
	II1: 34L, 18pw, 16dw	II2: 11L, 14dw	II3: 55L			
	III1: 48L, 25pw, 21dw	III2: 33L, 17dw	III3: 19L, 14dw	III4: 31L		
	IV1: 34L, 18pw, 15dw	IV2: 13L, 15dw	IV3: 8L, 15dw	IV4: 6L, 12dw	IV5: 27L	
Casuarius casuarius AMNH SKEL 963	Digit I absent					Measurements based on both feet of male bird
	II1: 34L	II2: 10L	II3: 48L			
	III1: 46L	III2: 31L	III3: 20L	III4: 32L		
	IV1: 31L	IV2: 12L	IV3: 8L	IV4: 8L	IV5: 27L	
Casuarius casuarius USNM OR.023788	Digit I absent					Left foot
	II1: 39L, 19pw, 16dw	II2: 13L, 14dw	II3: 48L			
	III1: 52L, 26pw, 20dw	III2: 38L, 17dw	III3: 26L, 14dw	III4: 30L		
	IV1: 38L, 19pw, 15dw	IV2: 14L, 13dw	IV3: 10L, 14dw	IV4: 9L, 12dw	IV5: 26L	
Casuarius casuarius FMNH 314889	TMT L = 315					Measurements based on both feet
	Digit I absent					
	II1: 45L, 23pw, 19dw	II2: 15L, 17dw	II3: 61L			
	III1: 59L, 31pw, 23dw	III2: 39L, 20dw	III3: 28L, 16dw	III4: 39L		
	IV1: 42L, 22pw, 18dw	IV2: 17L, 15dw	IV3: 12L, 15dw	IV4: 10L, 14dw	IV5: 30L	
Casuarius casuarius LACM 102774	Digit I absent					Measurements based on both feet
	II1: 49L, 25pw, 19dw	II2: 15L, 21dw	II3: 67L			
	III1: 63L, 32pw, 24dw	III2: 41L, 21dw	III3: 27L, 18dw	III4: 41L		
	IV1: 46L, 26pw, 19dw	IV2: 18L, 17dw	IV3: 13L, 16dw		IV5: 35L	
Casuarius casuarius uncatalogued Berkeley juvenile female J.R. Hutchinson 002	Digit I absent					Right foot
	II1: 21L, 11pw, 10dw	II2: 6L, 11dw	II3: 26L			
	III1: 32L, 16pw, 12dw	III2: 22L, 9dw	III3: 14L, 9dw	III4: 18L		
	IV1: 21L, 12pw, 10dw	IV2: 8L, 8dw	IV3: 6L, 8dw	IV4: 4L, 7dw	IV5: 16L	
Casuarius casuarius uncatalogued Berkeley juvenile male J.R. Hutchinson 003	Digit I absent					Left foot
	II1: 22L, 10pw, 9dw	II2: 6L, 8dw	II3: 24L			
	III1: 29L, 16pw, 14dw	III2: 20L, 11dw	III3: 14L, 8L	III4: 18L		
	IV1: 20L, 10dw	IV2: 8L, 9dw	IV3: 5L, 8dw	IV4: 4L, 7dw	IV5: 15L	
Casuarius unappendiculatus YPM ORN 110931 (formerly 13894)	Digit I absent					Right foot
	II1: 40L	II2: 15L	II3: 56L			
	III1: 57L	III2: 38L	III3: 25L	III4: 35L		
	IV1: 40L	IV2: 17L	IV3: 11L	IV4: 9L	IV5: 31L	

Omit taxon and specimen	Measurements					Comments
Casuarius unappendiculatus USNM 320946	Digit I absent					Male bird; measurements approximate due to considerable tissue around joints
	II1: 42L	II2: 14L				
	III1: 54L	III2: 38L	III3: 25L			
	IV1: 39L	IV2: 16L	IV3: 12L	IV4: 8L		
Casuarius unappendiculatus FMNH 93274	TMT L = 303					Left foot of female bird
	Digit I absent					
	II1: 39L, 21pw, 16dw	II2: 12L, 16dw	II3: 46L			
	III1: 55L, 27pw, 21dw	III2: 38L, 18dw	III3: 24L, 17dw	III4: 36L		
	IV1: 37L, 21pw, 16dw	IV2: 15L, 14dw	IV3: 11L, 12dw	IV4: 7L, 13dw	IV5: 29L	
Casuarius unappendiculatus FMNH 104271	TMT L = 296					Right foot of male bird
	Digit I absent					
	II1: 39L	II2: 13L	II3: 59L			
	III1: 55L	III2: 39L	III3: 24L	III4: 35L		
	IV1: 39L	IV2: 16L	IV3: 11L, 14dw	IV4: 9L, 14dw	IV5: 32L	
Casuarius bennetti LACM 107030 FB	TMT L = 257					Right foot of male bird
	Digit I absent					
	II1: 34L, 18pw, 14dw	II2: 10L, 13dw	II3: 62L			
	III1: 50L, 23pw, 18dw	III2: 36L, 16dw	III3: 24L, 14dw	III4: 37L		
	IV1: 34L, 19pw, 15dw	IV2: 15L, 13dw	IV3: 10L, 13dw	IV4: 7L, 11dw	IV5: 32L	
Casuarius sp. YPM ORN 102974 (formerly 4351)	Digit I absent					
	II1: 40L	II2: 12L	II3: 59L			
	III1: 53L	III2: 36L	III3: 22L	III4: 36L		
	IV1: 36L	IV2: 15L	IV3: 9L	IV4: 8L	IV5: 30L	
Casuarius sp. YPM ORN 102520 (formerly 2123)	Digit I absent					Measurements based on both feet
	II1: 33L	II2: 10L	II3: 48L			
	III1: 51L	III2: 34L	III3: 19L	III4: 29L		
	IV1: 35L	IV2: 13L	IV3: 9L	IV4: 8L	IV5: 25L	
Casuarius sp. LACM uncatalogued	TMT L = 342					Measurements based on both feet
	Digit I absent					
	II1: 44L, 24pw, 19dw	II2: 14L, 21dw	II3: 82L			
	III1: 63L, 33pw, 26dw	III2: 44L, 23dw	III3: 28L, 19dw	III4: 68L		
	IV1: 46L, 25pw, 18dw	IV2: 18L, 18dw	IV3: 12L, 16dw	IV4: 9L, 17dw	IV5: 51L	
Apterygiidae						
Apteryx australis YPM ORN 102519 (formerly 2122)	II1: 28L	II2: 22L	II3: 16L			Left foot of mounted skeleton; the ungual on digit IV as mounted is the fourth phalanx in the mount
	III1: 28L	III2: 24L	III3: 21L	III4: 13L		
	IV1: 19L				IV5: 10	
Apteryx australis AMNH SKEL 5372	II1: 21L, 12pw, 8dw	II2: 14L, 7dw	II3: 17L			Right foot
	III1: 22L, 11pw, 9dw	III2: 15L, 9dw	III3: 13L, 7dw	III4: 18L		
	IV1: 15L, 10pw, 8dw	IV2: 7L, 7dw	IV3: 7L, 7dw	IV4: 8L, 6dw	IV5: 13L	
Apteryx haastii NMNZ OR.000960 (formerly DM 960)	I1: 5L					Right foot; all unguals covered by sheaths
	II1: 20L, 8pw, 7dw	II2: 14L, 6dw				
	III1: 19L, 9pw, 7dw	III2: 14L, 6dw	III3: 12L, 6dw			
	IV1: 13L, 9pw, 6dw	IV2: 6L, 6dw	IV3: 6L, 5dw	IV4: 6L, 5dw		

Table A1.1. *continued*

Omit taxon and specimen	Measurements					Comments
Apteryx mantelli NMZN OR.000950 (formerly DM 950)	II1: 14L, 6pw, 4dw	II2: 9L, 4dw	II3: 10L			Measurements based on both feet
	III1: 13L, 6pw, 5dw	III2: 10L, 4dw	III3: 9L, 4dw	III4: 11L		
	IV1: 10L, 5pw, 4dw	IV2: 5L, 4dw	IV3: 5L, 4dw	IV4: 4L, 3dw	IV5: 8L	
Apteryx mantelli NMNZ S.000951 (formerlyDM 951)	I1: 3L	I2: 9L				Measurements based on both feet
	II1: 17L, 7pw, 6dw	II2: 12L, 5dw	II3: 12L			
	III1: 17L, 7pw, 6dw	III2: 13L, 5dw	III3: 11L, 5dw	III4: 13L		
	IV1: 12L, 7pw, 5dw	IV2: 6L, 5dw	IV3: 6L, 4dw	IV4: 6L, 4dw	IV5: 11L	
Apteryx mantelli AM LB7099		I2: 11L				Measurements based on both feet of immature female
	II1: 17L, 7pw, 6dw	II2: 11L, 5dw	II3: 13L			
	III1: 16L, 7pw, 6dw	III2: 12L, 5dw	III3: 11L, 4dw	III4: 15L		
	IV1: 12L, 6pw, 5dw	IV2: 6L, 5dw	IV3: 5L, 4dw	IV4: 6L, 4dw	IV5: 11L	
Apteryx mantelli AM LB7289	TMT L = 75					Measurements based on both feet of female bird
		I2: 13L				
	II1: 21L, 9pw, 7dw	II2: 14L, 6dw	II3: 16L			
	III1: 21L, 9pw, 7dw	III2: 15L, 6dw	III3: 12L, 5dw	III4: 16L		
	IV1: 15L, 9pw, 6dw	IV2: 8L, 6dw	IV3: 6L, 5dw	IV4: 7L, 5dw	IV5: 13L	
Apteryx mantelli FMNH 85778	TMT L = 64					Measurements based on both feet of female bird
	I1: 3L	I2: 10L				
	II1: 17L, 8pw, 6dw	II2: 11L	II3: 12L			
	III1: 17L	III2: 12L	III3: 11L	III4: 13L		
	IV1: 12L	IV2: 6L	IV3: 6L	IV4: 6L	IV5: 10L	
Apteryx mantelli. NMNZ OR.001152	I1: 5L	I2: 11L				Left foot; catalog number was written on foot, and didn't correspond to number in the catalog
	II1: 16L, 7pw, 5dw	II2: 11L, 5dw	II3:13L			
	III1: 16L, 7pw, 6dw	III2: 12L, 5dw	III3: 10L, 4dw	III4: 13L		
	IV1: 11L, 6pw, 5dw	IV2: 6L, 4dw	IV3: 5L, 5dw	IV4: 5L, 4dw	IV5: 10L	
Apteryx owenii USNM 18279	I1: 4L	I2: 11L				Right foot
	II1: 18L	II2: 12L	II3: 13L			
	III1: 16L	III2: 12L	III3: 11L	III4: 13L		
	IV1: 12L	IV2: 6L	IV3: 6L	IV4: 6L	IV5: 11L	
Apteryx owenii NMNZ OR.023036	I1: 4L	I2: 16L				Mainly left foot bones, but possibly some rights, of adult female; several measurements approximate
	II1: 19L, 7pw, 5dw	II2: 13L, 5dw	II3: 20L			
	III1: 16L, 8pw, 5dw	III2: 12L, 5dw	III3: 12L, 4dw	III4: 22L		
	IV1: 12L, 7pw, 5dw	IV2: 6L, 5dw	IV3: 6L, 5dw	IV4: 8L, 4dw	IV5: 17L	
Apteryx owenii NMNZOR.001145	II1: 11L	II2: 9L				Right foot of juvenile bird; measurements approximate
	III1: 11L	III2: 8L	III3: 7L			
	IV1: 8L	IV2: 3L	IV3: 4L	IV4: 4L		

Omit taxon and specimen	Measurements					Comments
Apteryx sp. University of Illinois	II1: 20L, 8pw, 7dw	II2: 13L, 6dw	II3: 14L			Left foot of mounted skeleton
	III1: 18L, 9pw, 8dw	III2: 13L, 7dw	III3: 12L, 5dw	III4: 15L		
	IV1: 13L, 9pw, 7dw	IV2: 7L, 6dw	IV3: 7L, 6dw	IV4: 8L, 5dw	IV5: 14L	
Apteryx sp. AMNH SKEL 3739	II1: 18L	II2: 12L				Right foot
	III1: 17L	III2: 13L	III3: 12L			
	IV1: 14L	IV2: 7L	IV3: 6L	IV4: 7L		
Apteryx sp. AMNH SKEL 1513	II1: 21L, 8pw, 6dw	II2: 13L, 5dw	II3: 17L			Right foot
	III1: 19L, 8pw, 6dw	III2: 14L, 5dw	III3: 12L, 6dw	III4: 17L		
	IV1: 15L, 7pw, 5dw	IV2: 8L, 5dw	IV3: 7L, 5dw	IV4: 7L, 5dw	IV5: 14L	
Dinornithiforms						
Megalapteryx didinus NMNZ S.023430	TMT L = 214					Measurements probably based on both feet
	I1: 33L	I2: 47				
	II1: 64L, 35pw, 28dw	II2: 39L, 24dw	II3: 59L			
	III1: 70L, 36pw, 29dw	III2: 43L, 24dw	III3: 24L, 23dw	III4: 57L		
	IV1: 46L, 33pw, 27dw	IV2: 26L, 24dw	IV3: 24L, 22dw	IV4: 24L, 23dw	IV5: 56L	
Megalapteryx didinus NMNZ S.023700	I1: 23L	I2: 38L				Measurements based on both feet
	II1: 51L, 28pw, 20dw	II2: 29L, 20dw	II3: 50L			
	III1: 51L, 26pw, 21dw	III2: 33L, 18dw	III3: 27L, 17dw	III4: 48L		
	IV1: 33L, 24pw, 20dw	IV2: 20L, 18dw	IV3: 20L, 17dw		IV5: 41L	
Megalapteryx didinus NMNZ S.025657	TMT L = 189					Measurements mainly based on right foot, but symmetry of digit I uncertain
	I1: 27L	I2: 38L				
	II1: 55L, 31pw, 25dw	II2: 30L	II3: 48L			
	III1: 58L, 31pw, 27dw	III2: 40L, 23dw	III3: 31L, 21dw	III4: 52L		
	IV1: 41L, 28pw, 22dw	IV2: 22L, 21dw	IV3: 19L, 20dw	IV4: 20L, 21dw	IV5: 42L	
Megalapteryx didinus NMNZ S.028206	TMT L = 140					Measurements probably based on both feet
	I1: 20L	I2: 28L				
	II1: 43L, 24pw, 16dw	II2: 25L, 16dw	II3: 44L			
	III1: 45L, 23pw, 17dw	III2: 30L, 15dw	III3: 25L, 16dw	III4: 45L		
	IV1: 31L, 22pw, 16dw	IV2: 17L, 14dw	IV3: 15L, 13dw	IV4: 16L, 14dw	IV5: 37L	
Anomalopteryx didiformis NHMUK (BMNH) A.3	TMT L = 146					*"Dinornis parvus"* right foot
	II1: 39L, 24pw, 18dw	II2: 19L, 17dw	II3: 36L			
	III1: 43L, 25pw, 19dw	III2: 25L, 17dw	III3: 19L, 16dw	III4: 36L		
	IV1: 32L, 23pw, 18dw	IV2: 14L, 17dw	IV3: 11L, 18dw	IV4: 12L, 16dw	IV5: 29L	

Table A1.1. *continued*

Omit taxon and specimen	Measurements					Comments
Anomalopteryx didiformis NMNZ S.023511	TMT L = 193					Measurements probably based on both feet
	II1: 50L, 33pw, 25dw					
	III1: 54L, 33pw, 26dw					
	IV1: 39L, 31pw, 24dw					
Anomalopteryx didiformis NMNZ S.035274	TMT L = 179					Measurements probably based on both feet
	II1: 46L, 28pw, 21dw	II2: 27L, 20dw	II3: 37L			
	III1: 52L, 30pw, 23dw	III2: 30L, 21dw	III3: 22L, 20dw	III4: 34L		
	IV1: 38L, 28pw, 20dw	IV2: 17L, 18dw	IV3: 15L, 18dw	IV4: 13L, 19dw	IV5: 30L	
Anomalopteryx didiformis AM LB5439		I2: 23L				Measurements probably based on both feet; ungual II3 could be III4, and vice versa
	II1: 49L, 32pw, 23dw	II2: 26L, 24dw	II3: 41L			
	III1: 54L, 31pw, 25dw	III2: 31L, 22dw	III3: 24L, 22dw	III4: 37L		
	IV1: 40L, 29pw, 22dw	IV2: 17L, 19dw	IV3: 11L, 20dw	IV4: 14L, 23dw	IV5: 31L	
Anomalopteryx didiformis AM LB5465	TMT L = 165					Measurements probably based on both feet; possibly a bit missing from tip of III4
	II1: 41L, 24pw, 19dw	II2: 20L, 17dw				
	III1: 46L, 24pw, 21dw	III2: 26L, 19dw	III3: 17L, 17dw	III4: 31L		
	IV1: 31L, 24pw, 19dw	IV2: 14L, 18dw	IV3: 10L, 18dw	IV4: 11L, 18dw	IV5: 23L	
Anomalopteryx didiformis AM LB5506	TMT L = 176					Measurements probably based on both feet
	I1: 16L					
	II1: 44L, 27pw, 21dw	II2: 21L, 21dw				
	III1: 52L, 30pw, 23dw	III2: 30L, 21dw	III3: 17L, 22dw			
	IV1: 40L, 29pw, 22dw	IV2: 17L, 20dw	IV3: 13L, 20dw	IV4: 13L, 19dw		
Anomalopteryx didiformis AM LB5510	TMT L = 197					Measurements based on both feet. Ungual II3 could be III4, and vice versa. Possibly a bit missing from tip of IV5
	II1: 51L, 29pw, 24dw	II2: 27L, 22dw	II3: 42L			
	III1: 55L, 32pw, 26dw	III2: 34L, 24dw	III3: 24L, 23dw	III4: 40L		
	IV1: 41L, 31pw, 24dw	IV2: 18L, 23dw	IV3: 15L, 22dw	IV4: 15L, 23dw	IV5: 30L	
Anomalopteryx didiformis AM LB5512	TMT L = 187					Measurements probably based on both feet. IDs of unguals uncertain, and are best guesses based on size and joint matching. Tip of II3 possibly abraded
	II1: 50L, 32pw, 23dw	II2: 26L, 22dw	II3: 30L			
	III1: 57L, 35pw, 24dw	III2: 33L, 20dw	III3: 21L, 22dw	III4: 37L		
	IV1: 43L, 32pw, 23dw	IV2: 20L, 21dw	IV3: 15L, 22dw	IV4: 14L, 20dw	IV5: 30L	
Anomalopteryx didiformis AM LB5545	TMT L = 158					Measurements probably based on both feet
		I2: 22L				
	II1: 38L, 25pw, 19dw	II2: 22L, 17dw	II3: 34L			
	III1: 44L, 27pw, 20dw	III2: 28L, 18dw	III3: 20L, 17dw	III4: 37L		
	IV1: 35L, 25pw, 18dw	IV2: 17L, 15dw	IV3: 13L, 16dw	IV4: 13L, 17dw	IV5: 29L	

Omit taxon and specimen	Measurements					Comments
Anomalopteryx didiformis AM LB5546	II1: 47L, 31pw, 24dw	II2: 26L, 21dw	II3: 35L			Measurements probably based on both feet; possibly a bit missing from tips of unguals II3 and IV5
	III1: 52L, 33pw, 24dw	III2: 33L, 23dw	III3: 23L, 21dw	III4: 46L		
	IV1: 37L, 30pw, 24dw	IV2: 18L, 21dw	IV3: 14L, 24dw	IV4: 14L, 23dw	IV5: 37L	
Anomalopteryx didiformis AM LB5548	TMT L = 174					Measurements probably based on both feet
	II1: 45L, 26pw, 20dw	II2: 21L, 19dw	II3: 28L			
	III1: 51L, 29pw, 22dw	III2: 30L, 20dw	III3: 20L, 19dw	III4: 38L		
	IV1: 38L, 26pw, 21dw	IV2: 17L, 19dw	IV3: 12L, 20dw	IV4: 13L, 19dw	IV5: 32L	
Anomalopteryx didiformis AM LB5550	TMT L = 173					Measurements probably based on both feet; ungual II3 could be III4, and vice versa
	II1: 43L, 27pw, 21dw	II2: 23L, 19dw	II3: 36L			
	III1: 49L, 29pw, 23dw	III2: 30L, 21dw	III3: 22L, 18dw	III4: 38L		
	IV1: 37L, 27pw, 20dw	IV2: 17L, 19dw	IV3: 14L, 19dw	IV4: 13L, 18dw	IV5: 32L	
Anomalopteryx didiformis AM LB5551	TMT L = 161					Measurements probably based on both feet; probably a bit missing from tip of IV5
	II1: 42L, 27pw, 20dw	II2: 22L, 18dw	II3: 29L			
	III1: 46L, 27pw, 22dw	III2: 28L, 19dw	III3: 21L, 16dw	III4: 33L		
	IV1: 34L, 25pw, 19dw	IV2: 15L, 19dw	IV3: 12L, 18dw	IV4: 12L, 19dw	IV5: 25L	
Anomalopteryx didiformis AM LB5796	II1: 50L, 30pw, 24dw	II2: 25L, 23dw	II3: 46L			
	III1: 57L, 34pw, 28dw	III2: 36L, 24dw	III3: 24L, 24dw			
	IV1: 40L, 32pw, 25dw	IV2: 18L, 23dw	IV3: 13L, 23dw	IV4: 14L, 25dw	IV5: 40L	
Anomalopteryx didiformis AM LB5820	TMT L = 83					Immature bird; bone ends very porous and not distinct, so some IDs could be incorrect. Ungual II3 could be III4, and vice versa
	II1: 22L, 12pw, 10dw	II2: 10L, 9dw	II3: 14L			
	III1: 22L, 13pw, 11dw	III2: 13L, 10dw	III3: 8L, 9dw	III4: 16L		
	IV1: 16L, 12pw, 10dw				IV5: 12L	
Anomalopteryx didiformis WO63.23 Ind A	TMT L = 169					There are some additional phalanges (including an ungual), possibly from digit IV
	II1: 39L, 28pw, 20dw	II2: 20L, 19dw				
	III1: 45L, 29pw, 22dw					
	IV1: 32L, 28pw, 20dw					
Anomalopteryx didiformis WO 153	TMT L = 189					Measurements probably based on both feet; phalanx II2 seems rather long. Ungual III4 could be II3
	II1: 48L, 32pw, 24dw	II2: 35L, 25dw				
	III1: 57L, 35pw, 28dw	III2: 37L, 28dw		III4: 45L		
	IV1: 41L, 32pw, 25dw	IV2: 24L, 24dw	IV3: 14L, 23dw		IV5: 40L	

Table A1.1. *continued*

Omit taxon and specimen	Measurements					Comments
Anomalopteryx didiformis WO 171	TMT L = 188					
	II1: 47L, 33pw, 25dw	II2: 25L, 21dw	II3: 44L			
	III1: 54L, 33pw, 25dw	III2: 32L, 22dw	III3: 24L, 23dw	III4: 43L		
	IV1: 41L, 32pw, 23dw	IV2: 19L, 21dw	IV3: 15L, 21dw	IV4: 15L, 22dw	IV5: 37L	Measurements probably based on both feet
Anomalopteryx didiformis WO 237.4		II2: 23L, 22dw	II3: 38L			
	III1: 51L, 32pw, 25dw	III2: 28L, 23dw	III3: 22L, 22dw	III4: 38L		
	IV1: 36L, 29pw, 23dw	IV2: 15L, 24dw			IV5: 34L	
Anomalopteryx didiformis WO 237.8	TMT L = 202					Right foot. Phalanx IV2 doesn't closely fit against IV1; from the other side, or possibly another bird? There is another phalanx with this catalog number that could be from another individual as well
	II1: 50L, 35pw, 27dw	II2: 28L, 25dw	II3: 43L			
	III1: 56L, 36pw, 28dw	III2: 34L, 25dw	III3: 27L, 24dw	III4: 45L		
	IV1: 42L, 34pw, 25dw	IV2: 19L, 24dw	IV3: 14L, 23dw	IV4: 14L, 24dw	IV5: 30L	
Anomalopteryx didiformis WO 258	TMT L = 195					There are phalanges of two birds under this catalog number. The smaller bones are darker, more poorly preserved, and less numerous. I assembled bones of the larger bird here
	II1: 49L, 33pw, 25dw	II2: 24L, 22dw	II3: 43L			
	III1: 56L, 34pw, 26dw	III2: 34L, 25dw	III3: 21L, 24dw	III4: 33L		
	IV1: 41L, 33pw, 25dw	IV2: 16L, 22dw	IV3: 13L, 23dw	IV4: 11L, 24dw	IV5: 32L	
Anomalopteryx didiformis Southland Museum E80.4	II1: 51L	II2: 27L	II3: 44L			Left foot of mummified specimen
	III1: 55L					
Euryapteryx curtus AM LB6424	II1: 34L, 19pw, 16dw	II2: 11L, 14dw	II3: 19L			Measurements probably based on both feet
	III1: 37L, 22pw, 18dw	III2: 18L, 16dw	III3: 10L, 14dw	III4: 23L		
	IV1: 29L, 19pw, 15dw	IV2: 10L, 14dw	IV3: 6L, 11dw	IV4: Absent	IV5: 18L	
Euryapteryx curtus AM LB6425	II1: 40L, 24pw, 20dw	II2: 13L, 17dw				Measurements probably based on both feet of female bird
	III1: 43L, 28pw, 23dw	III2: 21L, 19dw	III3: 10L, 20dw	III4: 25L		
	IV1: 33L, 25pw, 19dw	IV2: 12L, 20dw	IV3: 8L, 17dw	IV4: Absent	IV5: 22L	
Euryapteryx curtus AM LB6603.1	TMT L = 151					Measurements probably based on both feet
	II1: 41L, 24pw, 19dw	II2: 15L, 18dw	II3: 24L			
	III1: 43L, 28pw, 22dw	III2: 23L, 19dw	III3: 11L, 21dw	III4: 27L		
	IV1: 33L, 24pw, 18dw		IV3: 9L, 17dw	IV4: Absent	IV5: 21L	
Euryapteryx curtus AM LB6618	TMT L = 125					Measurements probably based on both feet
	II1: 35L, 20pw, 16dw	II2: 11L, 15dw	II3: 21L			
	III1: 37L, 22pw, 18dw	III2: 18L, 16dw	III3: 10L, 16dw	III4: 23L		
	IV1: 30L, 21pw, 16dw	IV2: 12L, 15dw	IV3: 9L, 14dw	IV4: Absent	IV5: 19L	

Omit taxon and specimen	Measurements					Comments
Euryapteryx curtus AM LB6637	TMT L = 128					Measurements probably based on both feet
	II1: 35L, 21pw, 16dw	II2: 12L, 15dw	II3: 18L			
	III1: 37L, 25pw, 19dw	III2: 18L, 15dw	III3: 11L, 15dw	III4: 21L		
	IV1: 27L, 21pw, 16dw	IV2: 9L, 16dw	IV3: 8L, 14dw	IV4: Absent	IV5: 16L	
Euryapteryx curtus NMNZ S.025656	TMT L = 223					Measurements based on both feet
	II1: 60L					
	III1: 61L, 49pw, 39dw	III2: 34L, 33dw	III3: 19L, 34L	III4: 46L		
	IV1: 51L, 45pw, 34dw	IV2: 20L, 32dw	IV3: 17L, 30dw	IV4: Absent	IV5: 38L	
Euryapteryx curtus NMNZ S.031552	TMT L = 203					Right foot; ungual III4 tip possibly a bit abraded
	II1: 58L, 41pw, 33dw	II2: 25L, 28dw	II3: 36L			
	III1: 62L, 40pw, 37dw	III2: 33L, 33dw	III3: 21L, 31dw	III4: 35L		
	IV1: 49L, 39pw, 32dw	IV2: 21L, 30dw	IV3: 17L, 28dw	IV4: Absent	IV5: 32L	
Euryapteryx curtus CM AV 8378	TMT L = 229					
	I1: 19L	I2: 30L				
	II1: 65L, 42pw, 33dw	II2: 25L, 29dw	II3: 42L			
	III1: 64L, 42pw, 35dw	III2: 33L, 31dw	III3: 20L, 30dw	III4: 46L		
	IV1: 55L, 42pw, 32dw	IV2: 22L, 30dw	IV3: 17L, 27dw	IV4: Absent	IV5: 37L	
Euryapteryx curtus CM AV 8472	II1: 72L, 46pw, 36dw	II2: 28L, 34dw	II3: 47L			Left foot; tip of ungual IV5 possibly a bit abraded
	III1: 72L, 49pw, 37dw	III2: 36L, 35dw	III3: 25L, 36dw	III4: 49L		
	IV1: 50L, 43pw, 32dw	IV2: 21L, 33dw	IV3: 20L, 31dw	IV4: Absent	IV5: 37L	
Euryapteryx curtus CM AV 8622	II1: 68L, 48pw, 34dw	II2: 26L, 32dw	II3: 45L			Measurements based on both feet
	III1: 73L, 47pw, 38dw	III2: 38L, 31dw	III3: 25L, 31dw	III4: 44L		
	IV1: 56L, 41pw, 33dw	IV2: 18L, 27dw	IV3: 13L, 27dw	IV4: Absent	IV5: 37L	
Euryapteryx curtus CM AV 15,034	TMT L = 212					Measurements based on both feet of "subadult" bird; phalangeal surfaces have porous, pitted appearance. Ungual II3 possibly missing a small bit of tip
	II1: 56L	II2: 21L	II3: 38L			
	III1: 61L	III2: 30L	III3: 16L	III4: 42L		
	IV1: 46L	IV2: 15L	IV3: 12L	IV4: Absent	IV5: 32L	
Emeus crassus NMNZ S.000469 (formerly DM 469)	TMT L = 202					Right foot
	II1: 55L, 33pw, 27dw	II2: 20L, 26dw	II3: 36L			
	III1: 60L, 41pw, 32dw	III2: 32L, 29dw	III3: 18L, 24dw	IIII4: 39L		
	IV1: 44L, 35pw, 27dw	IV2: 18L, 24dw	IV3: 14L, 24dw	IV4: Absent	IV5: 32L	
Emeus crassus NMNZ S.000470 (formerly DM 470)	TMT L = 225					Measurements probably based on both feet
		I2: 26L				
	II1: 60L, 36pw, 29dw	II2: 27L, 25dw	II3: 39L			
	III1: 65L, 44pw, 32dw	III2: 34L, 30dw	III3: 22L, 27dw	III4: 42L		
	IV1: 50L, 37pw, 29dw	IV2: 21L, 26dw	IV3: 19L, 24dw	IV4: Absent	IV5: 36L	

Table A1.1. *continued*

Omit taxon and specimen	Measurements					Comments
Emeus crassus AM LB6092	TMT L = 206					Right foot
	II1: 57L, 36pw, 27dw	II2: 26L, 26dw	II3: 37L			
	III1: 64L, 42pw, 30dw	III2: 34L, 29dw	III3: 22L, 28dw	III4: 42L		
	IV1: 49L, 35pw, 27dw	IV2: 21L, 25dw	IV3: 16L, 24dw	IV4: Absent	IV5: 35L	
Emeus crassus CM AV 8305	II1: 56L, 33pw, 25dw	II2: 24L, 22dw	II3: 36L			Right foot of "subadult" bird
	III1: 61L, 38pw, 30dw	III2: 31L, 26dw	III3: 20L, 23dw	III4: 39L		
	IV1: 46L, 31pw, 25dw	IV2: 18L, 22dw	IV3: 16L, 20dw	IV4: Absent	IV5: 32L	
Emeus crassus CM AV 8309		I2: 29L				Measurements probably based on both feet of female bird
	II1: 66L	II2: 30L	II3: 47L			
	III1: 71L	III2: 39L	III3: 25L	III4: 52L		
	IV1: 55L	IV2: 22L	IV3: 19L	IV4: Absent	IV5: 42L	
Emeus crassus CM AV 8332		I2: 26L				Measurements probably based on both feet
	II1: 55L	II2: 26L	II3: 33L			
	III1: 58L	III2: 33L	III3: 22L	III4: 38L		
	IV1: 44L	IV2: 18L	IV3: 16L	IV4: Absent	IV5: 29L	
Emeus crassus CM AV 8342		I2: 23L				Measurements probably based on both feet
	II1: 54L	II2: 26L	II3: 37L			
	III1: 61L	III2: 34L	III3: 21L	III4: 38L		
	IV1: 46L	IV2: 19L	IV3: 15L	IV4: Absent	IV5: 34L	
Emeus crassus CM AV 8359	II1: 56L	II2: 25L	II3: 39L			Left foot
	III1: 61L	III2: 33L	III3: 22L	III4: 44L		
	IV1: 45L	IV2: 18L	IV3: 15L	IV4: Absent	IV5: 34L	
Pachyornis australis NMNZ S.027896	TMT L = 218					Measurements probably based on both feet
	I1: 22L	I2: 38L				
	II1: 75L, 43pw, 33dw	II2: 30L, 31dw	II3: 53L			
	III1: 78L, 43pw, 35dw	III2: 41L, 32dw	III3: 30L, 31dw	III4: 57L		
	IV1: 58L, 42pw, 33dw	IV2: 22L, 30dw	IV3: 18L, 31dw	IV4: 19L, 29dw	IV5: 44L	
Pachyornis geranoides AM LB6020	TMT L = 123					Measurements probably based on both feet of male bird
	II1: 34L, 21pw, 15dw	II2: 15L, 15dw	II3: 25L			
	III1: 36L, 23pw, 17dw	III2: 17L, 15dw	III3: 13L, 15dw	III4: 30L		
	IV1: 26L, 21pw, 16dw	IV2: 11L, 14dw	IV3: 8L, 13dw	IV4: 8L, 13dw	IV5: 21L	
Pachyornis geranoides AM LB6064	TMT L = 154					Measurements based on both feet. Phalanges were glued together in toes. IDs of bones in assembled toes mostly look OK, except that III3 is missing. One of the assembled digits IV seems to consist of phalanges 1, 2, 4, and 5; while the other seems to consist of 1, 3, and 4
	II1: 45L, 20dw	II2: 18L, 17dw	II3: 30L			
	III1: 45L, 25pw, 22dw	III2: 25L, 19dw		III4: 33L		
	IV1: 35L, 24pw, 20dw	IV2: 13L, 18dw	IV3: 10L, 15dw	IV4: 10L, 16dw	IV5: 24L	
Pachyornis geranoides WO 82.39	TMT L = 132					Ungual II3 could be III4, and vice versa. In addition to the measured phalanges, the box included two phalanges of a smaller bird, which are darker in color than the measured bones
	II1: 41L, 25pw, 21dw	II2: 20L, 19dw	II3: 37L			
	III1: 45L, 30pw, 24dw	III2: 27L, 21dw	III3: 18L	III4: 31L		
					IV5: 31L	

Omit taxon and specimen	**Measurements**					**Comments**
Pachyornis geranoides WO mounted skeleton	II1: 35L, 22pw, 16dw	II2: 16L, 14dw	II3: 25L			Left foot of mount. The mount is a composite of two specimens, but the foot comes from only one of them
	III1: 37L, 23pw, 18dw	III2: 19L, 15dw	III3: 12L, 12dw			
	IV1: 27L, 22pw, 16dw	IV2: 12L, 14dw	IV3: 10L, 13dw	IV4: 9L, 13dw	IV5: 19L	
Pachyornis elephantopus NHMUK (BMNH) unnumbered mount	TMT L = 215					Left foot of skeleton as mounted
	II1: 66L, 54pw, 41dw	II2: 22L, 36dw	II3: 47L,			
	III1: 79L, 55pw, 43dw	III2: 39L, 38dw	III3: 28L, 32dw	III4: 47L		
	IV1: 59L, 48pw, 37dw	IV2: 21L, 35dw	IV3: 15L, 30dw	IV4: 13L, 31dw	IV5: 44L	
Pachyornis elephantopus AMNH FR 7305	II1: 71L, 46pw, 36dw	II2: 32L, 31dw	III2: 55L			Measurements probably based on both feet
	III1: 72L, 47pw, 39dw	III2: 41L, 35dw	III3: 32L, 31dw	III4: 59L		
	IV1: 55L, 45pw, 36dw	IV2: 21L, 34dw	IV3: 15L, 30dw	IV4: 18L, 29dw	IV5: 43L	
Pachyornis elephantopus AMNH FR 7307	II1: 69L, 44pw, 36dw	II2: 30L, 32dw	II3: 49L			Measurements probably based on both feet
	III1: 77L, 47pw, 40dw	III2: 45L, 38dw	III3: 28L, 32dw	III4: 58L		
	IV1: 54L, 42pw, 34dw	IV2: 20L, 29dw	IV3: 13L, 27dw	IV4: 15L, 27dw	IV5: 42L	
Pachyornis elephantopus NMNZ S.025868	TMT L = 256					
	II1: 80L, 49pw, 40dw	II2: 50L, 40dw				
	III1: 88L, 51pw, 41dw					
	IV1: 64L, 52pw, 41dw					
Pachyornis elephantopus NMNZ S.0326721	TMT L = 219					Measurements probably based on both feet. Phalanx III2 may actually be II2, and III4 may be II3
	III1: 74L, 50pw, 42dw	III2: 39L, 41dw	III3: 28L, 35dw	III4: 52L		
	IV1: 57L, 46pw, 37dw	IV2: 20L, 33dw	IV3: 16L, 30dw	IV4: 11L, 30dw	IV5: 44L	
Pachyornis elephantopus AM LB5946	TMT L = 226					
	II1: 73L, 50pw, 37dw	III2: 28L, 32dw	III3: 51L			
	III1: 78L, 56pw, 44dw	III2: 40L, 38dw	III3: 29L, 35dw	III4: 57L		
	IV1: 60L, 46pw, 36dw	IV2: 17L, 35dw	IV3: 14L, 32dw		IV5: 44L	
Pachyornis elephantopus CM AV 8382		I2: 31L				Measurements probably based on both feet. Ungual II3 could be III4, and vice versa
	II1: 63L	II2: 28L	II3: 48L			
	III1: 67L	III2: 40L	III3: 25L	III4: 56L		
	IV1: 54L	IV2: 20L	IV3: 15L	IV4: 17L	IV5: 43L	
Pachyornis elephantopus CM AV 8383	TMT L = 220					Left foot. In this and other specimens of this species, phalanges IV2 and IV3 are markedly asymmetrical in length, being much shorter on the medial than the lateral side
	I1: 23L	I2: 35L				
	II1: 73L, 47pw, 36dw	II2: 32L, 32dw	II3: 52L			
	III1: 80L, 48pw, 41dw	III2: 45L, 40dw	III3: 31L, 35dw	III4: 60L		
	IV1: 58L, 45pw, 39dw	IV2: 25L, 36dw	IV3: 17L, 34dw	IV4: 20L, 32dw	IV5: 45L	
Pachyornis elephantopus CM AV 8385		I2: 30L				Measurements probably based on both feet
	II1: 65L	II2: 24L	II3: 52L			
	III1: 73L	III2: 39L	III3: 21L	III4: 56L		
	IV1: 53L	IV2: 20L	IV3: 13L	IV4: 14L	IV5: 41L	

Table A1.1. *continued*

Omit taxon and specimen	Measurements					Comments
Pachyornis elephantopus CM AV 8388	TMT L = 213					Measurements probably based on both feet
		I2: 31L				
	II1: 66L, 46pw, 35dw	II2: 30L, 31dw	II3: 49L			
	III1: 74L, 49pw, 42dw	III2: 41L, 39dw	III3: 29L, 35dw	III4: 54L		
	IV1: 53L, 41pw, 36dw	IV2: 16L, 33dw	IV3: 14L, 30dw	IV4: 17L, 28dw	IV5: 42L	
Pachyornis elephantopus CM AV 8389	TMT L = 214					Measurements probably based on both feet. Several of my bone IDs differ from labels on the bones; my IDs make better matches between joints
		I2: 26L				
	II1: 69L, 48pw, 37dw	II2: 25L, 31dw	II3: 51L			
	III1: 69L, 51pw, 41dw	III2: 36L, 35dw	III3: 21L, 35dw	III4: 51L		
	IV1: 51L, 44pw, 34dw	IV2: 16L, 30dw	IV3: 13L, 30dw	IV4: 14L, 30dw	IV5: 38L	
Pachyornis elephantopus CM AV15,029	I1: 15L	I2: 30L				Measurements probably based on both feet
	II1: 71L	II2: 29L	II3: 52L			
	III1: 72L	III2: 43L	III3: 28L	III4: 58L		
	IV1: 56L	IV2: 20L	IV3: 15L	IV4: 18L	IV5: 43L	
?Pachyornis elephantopus CM AV 8384	I1: 16L	I2: 21L				Measurements probably based on both feet of immature bird. Articular ends of phalanges quite pitted. Phalanges IV2 and IV3 don't show the marked medial vs. lateral length asymmetry seen in adults of this species. Ungual II3 could be III4, and vice versa
	II1: 45L, 29pw, 23dw	II2: 22L, 18dw	II3: 31L			
	III1: 49L, 32pw, 26dw	III2: 26L, 20dw	III3: 19L, 20dw	III4: 32L		
	IV1: 34L, 28pw, 22dw	IV2: 16L, 19dw	IV3: 12L, 17dw	IV4: 12L, 17dw	IV5: 26L	
Dinornis robustus NHMUK (BMNH) A.32039-42	II1: 88L, 53pw, 44dw	II2: 40L, 39dw	II3: 69L			Right foot of "*Dinornis maximus*"; bones are attached to base, making measurement difficult
	III1: 99L, 56pw, 49dw	III2: 46L, 41dw	III3: 33L, 39dw	III4: 70L		
	IV1: 69L, 51pw, 42dw	IV2: 30L, 36dw	IV3: 23L, 33dw	IV4: 19L, 34dw	IV5: 59L	
Dinornis robustus NHMUK (BMNH) 32040	TMT L = 269					Right foot
	II1: 55L		II3: 44L			
	III1: 62L, 37pw, 29dw	III2: 24L, 23dw	III3: 21L, 20dw	III4: 44L		
	IV1: 45L, 27dw	IV2: 29L, 23dw	IV3: 15L, 20dw	IV4: 15L, 19dw	IV5: 42L	
Dinornis robustus YORYM:204.20.a	TMT L = 345					Right foot; a few mm missing from tip of II3
	II1: 81L, 49pw, 38dw	II2: 40L, 29dw	II3: 58L			
	III1: 86L, 45pw, 35dw	III2: 42L, 33dw	III3: 33L, 30dw	III4: 48L		
	IV1: 61L, 43pw, 36dw	IV2: 28L, 33dw	IV3: 23L, 27dw	IV4: 19L, 31dw	IV5: 47L	
Dinornis robustus AMNH FR 7301	I1: 29L	I2: 37L				
	II1: 79L, 47pw, 36dw	II2: 31L, 33dw	II3: 53L			
	III1: 88L, 52pw, 41dw	III2: 46L, 34dw		III4: 51L		
	IV1: 60L, 43pw, 39dw	IV2: 28L, 33dw	IV3: 21L, 31dw	IV4: 23L, 29dw	IV5: 45L	
Dinornis robustus AMNH FR 7303	I1: 31L	I2: 47L				Measurements probably based on both feet
	II1: 87L, 53pw, 38dw	II2: 37L, 37dw	II3: 63L			
	III1: 95L, 57pw, 43dw	III2: 50L, 38dw	III3: 32L, 36dw	III4: 65L		
	IV1: 70L, 54pw, 40dw	IV2: 44L, 39dw	IV3: 23L, 37dw	IV4: 27L, 37dw	IV5: 55L	

Omit taxon and specimen	Measurements					Comments
Dinornis robustus UCMP 77209	TMT L = 485					Formerly CM AV 10,899; left foot
	I1: 35L	I2: 45L				
	II1: 93L, 53pw, 43dw	II2: 44L, 38dw	II3: 66L			
	III1: 103L, 55pw, 51dw	III2: 53L, 41dw	III3: 36L, 38dw	III4: 66L		
	IV1: 71L, 52pw, 43dw	IV2: 31L, 39dw	IV3: 25L, 36dw	IV4: 24L, 36dw	IV5: 53L	
Dinornis robustus NMNZ S.028225	TMT L = 254					Right foot of male bird. Ungual II3 possibly missing 3–4 mm from tip, and ungual III4 looks a bit abraded
	I1: 25L	I2: 35L				
	II1: 58L, 34pw, 26dw	II2: 28L, 24dw	II3: 41L			
	III1: 64L, 37pw, 28dw	III2: 34L, 26dw	III3: 24L, 24dw	III4: 45L		
	IV1: 43L, 35pw, 27dw	IV2: 21L, 25dw	IV3: 14L, 22dw	IV4: 14L, 21dw	IV5: 39L	
Dinornis robustus NMNZ S.032667	TMT L = 327					Measurments from both feet of female bird
	I1: 26L	I2: 50L				
	II1: 75L, 41pw, 33dw	II2: 36L, 30dw	II3: 61L			
	III1: 83L, 42pw, 36dw	III2: 44L, 31dw	III3: 27L, 30dw	III4: 57L		
	IV1: 59L, 42pw, 32dw	IV2: 28L, 27dw	IV3: 20L, 28dw	IV4: 19L, 27dw	IV5: 52L	
Dinornis robustus NMNZ S.034088	TMT L = 487					Left foot
	I1: 35L	I2: 52L				
	II1: 98L, 61pw, 45dw	II2: 47L, 46dw	II3: 66L			
	III1: 110L, 61pw, 50dw	III2: 59L, 41dw	III3: 41L, 42dw	III4: 67L		
	IV1: 77L, 62pw, 46dw	IV2: 37L, 43dw	IV3: 29L, 44dw	IV4: 27L, 42dw	IV5: 57L	
Dinornis robustus NMNZ S.000211 (DM 211)	TMT L = 301					Measurements from both feet of female bird; ID of unguals possibly incorrect; ungual identified as II3 a bit abraded at tip
	II1: 68L, 39pw, 29dw	II2: 33L, 27dw	II3: 46L			
	III1: 73L, 39pw, 32dw	III2: 39L, 28dw	III3: 29L, 28dw	III4: 49L		
	IV1: 53L, 38pw, 31dw	IV2: 24L, 27dw	IV3: 17L, 28dw	IV4: 16L, 25dw	IV5: 43L	
Dinornis robustus NMNZ S.000411 (formerly DM 411)	TMT L = 331					Measurements based on both feet
	II1: 71L, 44pw, 31dw					
	III1: 79L, 42L, 35dw					
	IV1: 57L, 42pw, 33dw					
Dinornis robustus AM LB7111	I1: 34L	I2: 54L				Formerly CM AV 8485 69C, moa 460
	II1: 93L, 57pw, 43dw	II2: 44L, 39dw	II3: 70L			
	III1: 102L, 58pw, 48dw	III2: 54L, 42dw	III3: 36L, 39dw	III4: 70L		
	IV1: 77L, 58pw, 44dw	IV2: 35L, 40dw	IV3: 25L, 40dw	IV4: 24L, 35dw	IV5: 61L	
Dinornis robustus CM AV 8418	I1: 24L	I2: 45L				Right foot of female bird
	II1: 79L	II2: 35L	II3: 64L			
	III1: 87L	III2: 46L	III3: 27L	III4:65L		
	IV1: 63L	IV2: 28L	IV3: 20L	IV4: 20L	IV5: 53L	

Table A1.1. *continued*

Omit taxon and specimen	Measurements					Comments
Dinornis robustus CM AV 8423	I1: 37L	I2: 44L				Left foot
	II1: 96L, 52pw, 42dw	II2: 44L, 37dw	II3: 56L			
	III1: 105L, 54pw, 47dw	III2: 54L, 41dw	III3: 37L, 38dw	III4: 59L		
	IV1: 74L, 55pw, 44dw	IV2: 32L, 38dw	IV3: 25L, 36dw	IV4: 24L, 34dw	IV5: 46L	
Dinornis robustus CM AV 8466	I1: 34L	I2: 47L				Right foot
	II1: 91L, 55pw, 40dw	II2: 40L, 37dw	II3: 62L			
	III1: 101L, 58pw, 45dw	III2: 54L, 37dw	III3: 33L, 37dw	III4: 63L		
	IV1: 75L, 53pw, 43dw	IV2: 36L, 36dw	IV3: 25L, 37dw	IV4: 23L, 37dw	IV5: 53L	
Dinornis robustus CM AV 8473		I2: 47L				Right foot
	II1: 86L, 54pw, 39dw	II2: 39L, 38dw				
	III1: 93L, 54pw, 43dw	III2: 48L, 37dw	III3: 33L, 37dw			
	IV1: 66L, 55pw, 41dw	IV2: 31L, 36dw	IV3: 23L, 37dw	IV4: 24L, 37dw	IV5: 55L	
Dinornis robustus CM AV 8476	I1: 39L	I2: 57L				Mostly from right foot
	II1: 98L, 59pw, 46dw	II2: 47L, 42dw	II3: 80L			
	III1: 110L, 59pw, 47dw	III2: 53L, 41dw	III3: 38L, 41dw	III4: 78L		
	IV1: 77L, 57pw, 46dw	IV2: 34L, 41dw	IV3: 27L, 44dw	IV4: 27L, 37dw	IV5: 67L	
Dinornis robustus CM AV 8477	I1: 33L	I2: 46L				
	II1: 95L, 55pw, 41dw	II2: 45L, 35dw	II3: 63L			
	III1: 107L, 59pw, 49dw	III2: 55L, 40dw	III3: 36L, 38dw	III4: 61L		
	IV1: 73L, 54pw, 44dw	IV2: 33L, 39dw	IV3: 21L, 38dw	IV4: 24L, 34dw	IV5: 54L	
Dinornis robustus CM AV 8479	I1: 34L	I2: 52L				Left foot
	II1: 101L	II2: 45L	II3: 73L			
	III1: 111L	III2: 60L	III3: 39L	III4: 72L		
	IV1: 80L	IV2: 37L	IV3: 27L	IV4: 27L	IV5: 61L	
Dinornis robustus CM AV 8480	TMT L = 410					Measurements based on both feet
	I1: 28L	I2: 44L				
	II1: 84L, 49pw, 36dw	II2: 38L, 34dw	II3: 60L			
	III1: 91L, 50pw, 41dw	III2: 48L, 34dw	III3: 33L, 35dw	III4: 61L		
	IV1: 67L, 48pw, 40dw	IV2: 28L, 34dw	IV3: 21L, 32dw	IV4: 21L, 34dw	IV5: 51L	
Dinornis robustus CM AV 8481	I1: 30L	I2: 44L				Right foot of "immature" bird; articular ends of bones very porous
	II1: 83L, 49pw, 38dw	II2: 35L, 34dw	II3: 58L			
	III1: 93L, 51pw, 43dw	III2: 47L, 35dw	III3: 24L, 32dw	III4: 57L		
	IV1: 65L, 48pw, 38dw	IV2: 27L, 34dw	IV3: 20L, 31dw	IV4: 19L, 31dw	IV5: 49L	

Omit taxon and specimen	Measurements					Comments
Dinornis robustus CM AV 8484	TMT L = 470					Left foot
	I1: 33L	I2: 54L				
	II1: 109L, 68pw, 49dw	II2: 52L, 47dw	II3: 76L			
	III1: 115L, 67pw, 56dw	III2: 64L, 48dw	III3: 41L, 47dw	III4: 75L		
	IV1: 82L, 63pw, 51dw	IV2: 40L, 43dw	IV3: 24L, 44dw	IV4: 31L, 42dw	IV5: 62L	
Dinornis robustus CM AV 8492	I1: 32L	I2: 48L				Left foot of female bird
	II1: 89L, 52pw, 41dw	II2: 41L, 36dw	II3: 61L			
	III1: 99L, 54pw, 46dw	III2: 49L, 39dw	III3: 30L, 38dw	III4: 64L		
	IV1: 72L, 47pw, 42dw	IV2: 29L, 37dw	IV3: 21L, 35dw	IV4: 20L, 32dw	IV5: 53L	
Dinornis robustus CM AV 8493	I1: 27L	I2: 58L				Right foot
	II1: 92L, 57pw, 42dw	II2: 42L, 37dw	II3: 67L			
	III1: 105L, 58pw, 49dw	III2: 56L, 40dw	III3: 33L, 40dw	III4: 71L		
	IV1: 73L, 56pw, 45dw	IV2: 33L, 39dw	IV3: 24L, 37dw	IV4: 23L, 37dw	IV5: 58L	
Dinornis robustus CM AV 8494	II1: 89L	II2: 38L	II3: 60L			Measurements based on both feet of "immature" bird
	III1: 97L	III2: 51L	III3: 32L	III4: 59L		
	IV1: 68L	IV2: 31L	IV3: 22L	IV4: 22L	IV5: 50L	
Dinornis robustus CM AV 12,589	I1: 31L					Measurements based on both feet; tip of II3 a bit abraded
	II1: 82L	II2: 41L	II3: 54L			
	III1: 85L	III2: 46L	III3: 27L	III4: 56L		
	IV1: 61L	IV2: 31L	IV3: 20L	IV4: 21L	IV5: 48L	
Dinornis robustus CM AV 14,451	I1: 32L	I2: 52L				Right foot
	II1: 104L, 60pw, 45dw	II2: 46L, 41dw	II3: 66L			
	III1: 111L, 63pw, 50dw	III2: 54L, 41dw	III3: 36L, 41dw	III4: 68L		
	IV1: 77L, 58pw, 46dw	IV2: 33L, 39dw	IV3: 24L, 38dw	IV4: 23L, 38dw	IV5: 56L	
Dinornis novaezealandiae AM LB6292	TMT L = 310					Measurements based on both feet of male bird
	I1: 19L	I2: 27L				
	II1: 67L, 37pw, 27dw	II2: 29L, 25dw	II3: 40L			
	III1: 72L, 39pw, 32dw	III2: 38L, 28dw	III3: 23L, 25dw	III4: 45L		
	IV1: 52L, 36pw, 29dw	IV2: 24L, 25dw	IV3: 19L, 24dw	IV4: 17L, 21dw	IV5: 36L	
Dinornis novaezealandiae AM LB6401	TMT L = 365					Measurements based on both feet
	II1: 76L, 42pw, 33dw	II2: 34L, 31dw				
	III1: 80L, 42pw, 35dw	III2: 43L, 31dw	III3: 27L, 32dw			
	IV1: 59L, 43pw, 33dw	IV2: 28L, 29dw	IV3: 21L, 31dw	IV4: 19L, 29dw		
Dinornis novaezealandiae AM LB6403	II1: 65L, 37pw, 31dw	II2: 30L, 28dw	II3: 48L			Measurements based on both feet; ungual IV5 possibly missing a bit from the tip
	III1: 68L, 41pw, 34dw	III2: 37L, 30dw	III3: 27L, 28dw	III4: 46L		
	IV1: 49L, 37pw, 30dw	IV2: 23L, 27dw			IV5: 38L	

Table A1.1. *continued*

Omit taxon and specimen	Measurements					Comments
Dinornis novaezealandiae AM LB6873	TMT L = 410					Measurements based on both feet
	II1: 71L, 40pw, 33dw					
	III1: 76L, 42pw, 35dw					
	IV1: 55L, 42pw, 33dw					
Dinornis novaezealandiae WO 145	TMT L = 395					Measurements based on both feet
	I1: 31L	I2: 44L				
	II1: 82L, 49pw, 38dw	II2: 38L, 33dw	II3: 54L			
	III1: 86L, 50pw, 39dw	III2: 48L, 34dw	III3: 32L, 33dw			
	IV1: 62L, 48pw, 36dw	IV2: 29L, 32pw	IV3: 21L, 34dw	IV4: 22L, 31dw		
Dinornis novaezealandiae WO 51	TMT L = 360					All measurements except II3 from left foot
	II1: 71L, 43pw, 34dw	II2: 34L, 30dw	II3: 45L			
	III1: 75L, 45pw, 38dw	III2: 42L, 34dw	III3: 29L, 31dw	III4: 54L		
	IV1: 54L, 42pw, 35dw	IV2: 27L, 30dw	IV3: 22L, 30dw	IV4: 20L, 29dw	IV5: 38L	
Dinornis sp. AM LB7870	II1: 72L, 40pw, 29dw	II2: 35L, 25dw	II3: 45L			Measurements based on both feet; ungual identified as II3 could be III4, and vice versa
	III1: 77L, 40pw, 34dw	III2: 42L, 27dw	III3: 29L, 26L	III4: 44L		
	IV1: 55L, 42pw, 31dw	IV2: 24L, 26dw	IV3: 18L, 23dw	IV4: 14L, 22dw	IV5: 37L	
Tinamidae						
Tinamus solitarius YPM ORN 102482 (formerly 2085)	II1: 14L	II2: 10L	II3: 7L			Bones measured along dorsal surfaces
	III1: 13L	III2: 11L	III3: 8L	III4: 8L		
	IV1: 11L	IV2: 7L	IV3: 5L	IV4: 5L	IV5: 6L	
Crypturellus soui USNM 347331	II1: 10L	II2: 6L	II3: 3L			Right foot of male bird
	III1: 10L	III2: 7L	III3: 6L	III4: 5L		
	IV1: 7L	IV2: 5L	IV3: 4L	IV4: 2L	IV5: 3L	
Crypturellus soui USNM 344067	II1: 9L	II2: 6L	II3: 3L			Right foot of male bird
	III1: 9L	III2: 7L	III3: 5L	III4: 5L		
	IV1: 7L	IV2: 4L	IV3: 3L	IV4: 2L	IV5: 3L	
Crypturellus soui USNM 345745	I1: 2L	I2: 2L				Right foot of female bird
	II1: 9L	II2: 6L	II3: 3L			
	III1: 9L	III2: 7L	III3: 5L	III4: 4L		
	IV1: 6L	IV2: 4L	IV3: 3L	IV4: 2L	IV5: 2L	
Crypturellus undulatus USNM 345741	I1: 3L	I2: 4L				Male bird: measurements based on both feet
	II1: 11L	II2: 7L	II3: 5L			
	III1: 11L	III2: 8L	III3: 6L	III4: 6L		
	IV1: 8L	IV2: 5L	IV3: 3L	IV4: 3L	IV5: 4L	
Rhynchotus rufescens USNM 19000	I1: 6L	I2: 4L				Measurements based on both feet
	II1: 14L	II2: 8L	II3: 6L			
	III1: 14L	III2: 10L	III3: 7L	III4: 7L		
	IV1: 9L	IV2: 6L	IV3: 4L	IV4: 4L	IV5: 5L	
Rhynchotus rufescens USNM 612017	I1: 8L	I2: 7L				Measurements based on both feet
	II1: 15L	II2: 10L	II3: 8L			
	III1: 15L	III2: 11L	III3: 8L	III4: 10L		
	IV1: 12L	IV2: 8L	IV3: 6L	IV4: 5L	IV5: 8L	
Nothoprocta perdicaria USNM 321770	I1: 4L	I2: 4L				
	II1: 11L	II2: 8L	II3: 7L			
	III1: 12L	III2: 9L	III3: 6L	III4: 8L		
	IV1: 8L	IV2: 5L	IV3: 4L	IV4: 4L	IV5: 6L	

Omit taxon and specimen	Measurements					Comments
Nothura maculosa USNM 347608	I1: 4L	I2: 3L				Male bird
	II1: 10L	II2: 6L	II3: 4L			
	III1: 10L	III2: 7L	III3: 6L	III4: 6L		
	IV1: 7L	IV2: 4L	IV3: 3L	IV4:3L	IV5: 4L	
Nothura maculosa USNM 345020	I1: 5L	I2: 4L				Male bird
	II1: 10L	II2: 6L	II3: 5L			
	III1: 10L	III2: 7L	III3: 5L	III4: 5L		
	IV1: 7L	IV2: 4L	IV3: 3L	IV4: 3L	IV5: 4L	
Nothura maculosa USNM 347605	I1: 5L	I2: 2L				Male bird; I2 of both feet looks unusually short
	II1: 10L	II2: 6L	II3: 5L			
	III1: 10L	III2: 7L	III3: 5L	III4: 6		
	IV1: 8L	IV2: 5L	IV3: 3L	IV4: 3L	IV5: 4L	
Eudromia elegans USNM 345018	Digit I absent					Left foot of captive male
	II1: 10L	II2: 5L	II3: 6L			
	III1: 11L	III2: 8L	III3: 5L	III4: 7L		
	IV1: 8L	IV2: 4L	IV3: 3L	IV4: 3L	IV5: 4L	
Eudromia elegans USNM 345052	Digit I absent					Left foot of captive female
	II1: 10L	II2: 5L	II3: 7L			
	III1: 9L	III2: 7L	III3: 5L	III4: 8L		
	IV1: 8L	IV2: 4L	IV3: 3L	IV4: 3L	IV5: 6L	
Eudromia elegans USNM 344966	Digit I absent					Left foot of captive female
	II1: 11L	II2: 5L	II3: 7L			
	III1: 11L	III2: 7L	III3: 5L	III4: 9L		
	IV1: 9L	IV2: 4L	IV3: 3L	IV4: 3L	IV5: 6L	
Anatidae						
Cnemiornis calcitrans NMNZ S.035266	II1: 34L, 16pw, 12dw	II2: 16L, 11dw	II3: 34L			Left foot
	III1: 40L, 19pw, 17dw	III2: 26L, 14dw	III3: 16L, 12dw	III4: 29L		
	IV1: 29L, 16pw, 12dw	IV2: 15L, 10dw	IV3: 10L, 9dw	IV4: 10L, 9dw	IV5: 24L	
Gastornithidae						
Gastornis gigantea LACM 31752	I1: 46L					Measurements mostly or entirely from right foot
	II1: 93L, 35pw, 27dw	II2: 34L, 28dw				
	III1: 92L, 47pw, 37dw	III2: 47L, 35dw	III3: 32L, 32dw			
	IV1: 73L, 37pw, 31dw	IV2: 34L, 25dw	IV3: 21L, 24dw	IV4: 11L, 21dw		
Dromornithidae						
Genyornis newtoni NHMUK A.660	Digit I absent					Left foot
	II1: 80L, 22pw, 19dw	II2: 18L, 19dw	II3: 18L			
	III1: 77L, 42pw, 37dw	III2: 40L, 33dw	III3: 14L, 31dw	III4: 24L		
	IV1: 64L, 35pw, 27dw	IV2: 25L, 25dw	IV3: 11L, 21dw	IV4: absent	IV5: 15L	
Cracidae						
Penelope superciliaris USNM 345798	I1: 17L	I2: 10L				Measurements based on both feet of male bird
	II2: 17L	II2: 12L	II3: 10L			
	III1: 19L	III2: 15L	III3: 12L	III4: 11L		
	IV1: 12L	IV2: 8L	IV3: 7L	IV4: 8L	IV5: 8L	
Penelope superciliaris USNM 345796	I1: 18L	I2: 11L				Measurements based on both feet of male bird
	II1: 18L	II2: 12L	II3: 11L			
	III1: 18L	III2: 16L	III3: 14L	III4: 12L		
	IV1: 12L	IV2: 8L	IV3: 7L	IV4: 8L	IV5: 9L	

Table A1.1. *continued*

Omit taxon and specimen	Measurements					Comments
Penelope superciliaris USNM 345797	I1: 22L	I2: 12L				Measurements based on both feet of female bird
	II1: 21L	II2: 15L	II3: 12L			
	III1: 23L	III2: 18L	III3: 16L	III4: 13L		
	IV1: 14L	IV2: 9L	IV3: 9L	IV4: 10L	IV5: 10L	
Penelope purpurascens USNM 288718	I1: 25L	I2: 16L				Measurements based on both feet of male bird
	II1: 25L	II2: 17L	II3: 16L			
	III1: 25L	III2: 20L	III3: 18L	III4: 16L		
	IV1: 16L	IV2: 11L	IV3: 10L	IV4: 10L	IV5: 13L	
Penelope purpurascens USNM 347315	I1: 21L	I2: 13L				Left foot of male bird
	II1: 20L	II2: 16L	II3: 12L			
	III1: 21L	III2: 17L	III3: 16L	III4: 12L		
	IV1: 14L	IV2: 9L	IV3: 8L	IV4: 9L	IV5: 9L	
Penelope purpurascens USNM 322976	I1: 25L	I2: 15L				Left foot
	II1: 23L	II2: 17L	II3: 15L			
	III1: 25L	III2: 20L	III3: 17L	III4: 16L		
	IV1: 16L	IV2: 10L	IV3: 9L	IV4: 11L	IV5: 12L	
Aburria pipile (*Pipile cumanensis*) USNM 345800	I1: 20L	I2: 11L				Measurements based on both feet of female bird
	II1: 19L	II2: 14L	II3: 11L			
	III1: 20L	III2: 16L	III3: 14L	III4: 12L		
	IV1: 12L	IV2: 8L	IV3: 7L	IV4: 8L	IV5: 9L	
Nothocrax urumutum USNM 430759	I1: 17L	I2: 10L				Measurements based on both feet
	II1: 20L	II2: 14L	II3: 10L			
	III1: 20L	III2: 16L	III3: 13L	III4: 11L		
	IV1: 13L	IV2: 8L	IV3: 7L	IV4: 7L		
Crax tomentosa (*Mitu tomentosum*) USNM 346789	I1: 24L	I2: 13L				Measurements based on both feet
	II1: 25L	II2: 17L	II3: 14L			
	III1: 25L	III2: 19L	III3: 17L	III4: 15L		
	IV1: 17L	IV2: 11L	IV3: 9L	IV4: 9L	IV5: 12L	
Crax mitu (*Mitu mitu*) USNM 292879	I1: 25L	I2: 11L				Measurements based on phalanges of both feet of male bird
	II1: 25L	II2: 17L	II3: 13L			
	III1: 25L	III2: 20L	III3: 17L	III4: 12L		
	IV1: 18L	IV2: 11L	IV3: 9L	IV4: 10L	IV5: 8L	
Crax mitu (*Mitu mitu*) USNM 320690	I1: 25L	I2: 14L				Measurements based on both feet of male bird
	II1: 25L	II2: 18L	II3: 14L			
	III1: 25L	III2: 20L	III3: 17L	III4: 12L		
	IV1: 17L	IV2: 11L	IV3: 10L	IV4: 10L	IV5: 10L	
Crax mitu (*Mitu mitu*) USNM 345792	I1: 27L	I2: 15L				Measurements based on both feet of female bird
	II1: 27L	II2: 17L	II3: 15L			
	III1: 28L	III2: 20L	III3: 18L	III4: 16L		
	IV1: 18L	IV2: 12L	IV3: 10L	IV4: 11L	IV5: 12L	
Crax mitu (*Mitu mitu*) MSU 4439	I1: 25L	I2: 17L				Left foot of male bird
	II1: 26L					
	III1: 26L	III2: 20L	III3: 17L			
	IV1: 17L	IV2: 11L	IV3: 9L	IV4: 11L		
Crax rubra USNM 322977	I1: 28L	I2: 18L				Left foot
	II1: 30L	II2: 21L	II3: 18L			
	III1: 31L	III2: 24L	III3: 20L	III4: 20L		
	IV1: 21L	IV2: 13L	IV3: 11L	IV4: 12L	IV5: 15L	
Crax rubra USNM 321840	I1: 25L	I2: 16L				Left foot of female bird
	II1: 27L	II2: 17L	II3: 17L			
	III1: 27L	III2: 21L	III3: 18L	III4: 16L		
	IV1: 18L	IV2: 12L	IV3: 10L	IV4: 10L	IV5: 12L	
Crax rubra USNM 344394	I1: 25L	I2: 14L				Right foot of female bird
	II1: 26L	II2: 17L	II3: 13L			
	III1: 26L	III2: 20L	III3: 16L	III4: 14L		
	IV1: 18L	IV2: 11L	IV3: 9L	IV4: 10L	IV5: 11L	

Omit taxon and specimen	Measurements					Comments
Crax fasciolata USNM 346718	I1: 23L	I2: 13L				Right foot of male bird
	II1: 24L	II2: 17L	II3: 14L			
	III1: 24L	III2: 19L	III3: 16L	III4: 15L		
	IV1: 15L	IV2: 10L	IV3: 7L	IV4: 9L	IV5: 11L	
Crax alector USNM 321496	I1: 23L	I2: 14L				Left foot
	II1: 25L	II2: 17L	II3: 14L			
	III1: 26L	III2: 20L	III3: 17L	III4: 15L		
	IV1: 17L	IV2: 11L	IV3: 9L	IV4: 10L	IV5: 13L	
Crax alberti USNM 289739	I1: 24L	I2: 13L				Left foot of female bird
	II1: 25L	II2: 18L	II3: 13L			
	III1: 25L	III2: 21L	III3: 17L	III4: 13L		
	IV1: 16L	IV2: 11L	IV3: 10L	IV4: 10L	IV5: 11L	
Meleagrididae						
Meleagris gallopavo YPM ORN 102058 (formerly 365)	II1: 26L	II2: 18L	II3: 15L			Mounted skeleton. Phalanx III1 is broken all around; with a little dowel holding the two pieces together; this dowel could add as much as 1 mm to apparent length of the bone
		III2: 22L	III3: 17L	III4: 16L		
	IV1: 18L	IV2: 11L	IV3: 9L	IV4: 11L	IV5: 13L	
Meleagris gallopavo MSU OR.7136	I1: 15L					Left foot of female bird; all unguals covered by sheaths
	II1: 23L	II2: 15L				
	III1: 26L	III2: 17L	III3: 14L			
	IV1: 17L	IV2: 10L	IV3: 8L	IV4: 9L		
Meleagris gallopavo MSU OR.4809	I1: 16L					Right foot of female bird; all unguals covered by sheath
	II1: 25L	II2: 16L				
	III1: 27L	III2: 19L	III3: 14L			
	IV1: 20L	IV2: 10L	IV3: 8L	IV4: 8L		
Agriocharis ocellata (*Meleagris ocellata*) USNM 347792	I1: 14L	I2: 9L				Measurements based on both feet of female bird
	II1: 22L	II2: 15L	II3: 11L			
	III1: 23L	III2: 18L	III3: 14L	III4: 12L		
	IV1: 17L	IV2: 10L	IV3: 7L	IV4: 9L	IV5: 8L	
Tetraonidae						
Dendragapus canadensis (*Falcipennis canadensis*) USNM 288088	I1: 8L	I2: 4L				Measurements based on both feet of female bird
	II1: 12L	II2: 9L	II3: 6L			
	III1: 13L	III2: 11L	III3: 9L	III4: 7L		
	IV1: 9L	IV2: 5L	IV3: 4L	IV4: 5L	IV5: 6L	
Lagopus lagopus USNM 322639	I1: 6L	I2: 4L				Right foot of male bird
	II1: 12L	II2: 8L	II3: 7L			
	III1: 12L	III2: 9L	III3: 8L	III4: 7L		
	IV1: 7L	IV2: 4L	IV3: 3L	IV4: 5L	IV5: 6L	
Lagopus lagopus USNM 320684	I1: 4L	I2: 5L				Measurements based on both feet of male bird
	II1: 10L	II2: 7L	II3: 7L			
	III1: 11L	III2: 8L	III3: 7L	III4: 7L		
	IV1: 7L	IV2: 3L	IV3: 3L	IV4: 4L	IV5; 6L	
Lagopus lagopus USNM 320685	I1: 5L	I2: 4L				Measurements based on both feet of female bird
	II1: 10L	II2: 7L	II3: 6L			
	III1: 11L	III2: 8L	III3: 7L	III4: 6L		
	IV1: 7L	IV2: 4L	IV3: 3L	IV4: 4L	IV5: 5L	
Lagopus muta USNM 320683	I1: 3L	I2: 4L				Measurements based on both feet of male bird
	II1: 8L	II2: 6L	II3: 6L			
	III1: 9L	III2: 6L	III3: 5L	III4: 6L		
	IV1: 6L	IV2: 3L	IV3: 3L	IV4: 3L	IV5: 5L	
Lagopus muta USNM 289444	I1: 3L	I2: 3L				Measurements based on both feet of female bird
	II1: 9L	II2: 6L	II3: 5L			
	III1: 9L	III2: 7L	III3: 5L	III4: 5L		
	IV1: 6L	IV2: 3L	IV3: 2L	IV4: 3L	IV5: 5L	

Table A1.1. *continued*

Omit taxon and specimen	Measurements					Comments
Lagopus leucura USNM 289726	I1: 4L					Left foot of immature bird
	II1: 9L	II2: 6L	II3: 5L			
	III1: 10L	III2: 7L	III3: 6L	III4: 5L		
	IV1: 7L	IV2: 4L	IV3: 2L	IV4: 3L	IV5: 5L	
Bonasa umbellus MSU OR.2440	I1: 8L					Right foot of female bird; all unguals covered with sheaths
	II1: 12L	II2: 9L				
	III1: 13L	III2: 10L	III3: 9L			
	IV1: 9L	IV2: 6L	IV3: 5L	IV4: 6L		
Bonasa umbellus USNM 289428	I1: 9L	I12: 5L				Measurements based on both feet of male bird
	II1: 14L	II2: 11L	II3: 6L			
	III1: 16L	III2: 12L	III3: 11L	III4: 7L		
	IV1: 11L	IV2: 6L	IV3: 5L	IV4: 6L	IV5: 6L	
Bonasa umbellus UMA 1502	I1: 9L					Adult male bird
	II1: 13L	III2: 10L				
	III1: 14L	III2: 11L	III3: 10L	III4: 7L		
	IV1: 10L	IV2: 6L	IV3: 5L	IV4: 6L		
Bonasa umbellus UMA 3189	I1: 9L					Right foot
	II1: 13L	II2: 10L				
	III1: 13L	III2: 11L	III3: 10L			
	IV1: 9L	IV2: 6L	IV3: 5L	IV4: 5L		
Bonasa umbellus UMA 3928	I1: 8L					Right foot
	II1: 12L	II2: 9L				
	III1: 13L	III2: 11L	III3: 9L			
	IV1: 9L	IV2: 5L	IV3: 4L	IV4: 5L	IV5: 4L	
Bonasa umbellus UMA 3543	I1: 9L					Left foot
	II1: 14L	II2: 10L				
	III1: 14L	III2: 12L	III3: 10L			
	IV1: 10L	IV2: 6L	IV3: 4L	IV4: 6L		
Bonasa umbellus UMA 1127	I1: 9L					Right foot; measurements "iffier" than most, due to large amount of dried tissue around joints
	II1: 12L	II2: 9L				
	III1: 12L	III2: 10L	III3: 9L			
	IV1: 9L	IV2: 5L	IV3: 5L	IV4: 5L		
Centrocercus urophasianus USNM 17974	I1: 10L					Measurements based on both feet
	II1: 18L	II2: 12L	II3: 10L			
	III1: 19L	III2: 14L	III3: 11L			
	IV1: 13L	IV2: 8L	IV3: 6L	IV4: 7L		
Centrocercus urophasianus USNM 17968	I1: 10L					Measurements based on both feet of male bird
	II1: 17L	II2: 12L				
	III1: 20L	III2: 15L	III3: 13L			
	IV1: 13L	IV2: 8L	IV3: 6L	IV4: 7L		
Centrocercus urophasianus USNM 17985	I1: 9L					Left foot
	II1: 15L	II2: 10L	II3: 7L			
	III1: 16L	III2: 12L	III3: 10L			
	IV1: 10L	IV2: 6L	IV3: 5L	IV4: 6L		
Tympanuchus phasianellus USNM 17958	I1:9L	I2: 5L				Measurements based on both feet of male bird
	II1: 16L	II2: 12L	II3: 7L			
	III1: 16L	III2: 12L	III3: 10L	III4: 7L		
	IV1: 11L	IV2: 6L	IV3: 5L	IV4: 6L	IV5: 6L	
Tympanuchus phasianellus USNM 17964	I1: 9L	I2: 5L				Measurements based on both feet
	II1: 13L	II2: 10L	II3: 6L			
	III1: 15L	III2: 11L	III3: 9L	III4: 6L		
	IV1: 9L	IV2: 5L	IV3: 4L	IV4: 6L	IV5: 5L	
Tympanuchus phasianellus USNM 498713	I1: 10L	I2: 6L				Measurements based on both feet
	II1: 16L	II2: 11L	II3: 7L			
	III1: 16L	III2: 12L	III3: 9L	III4: 7L		
	IV1: 11L	IV2: 6L	IV3: 5L	IV4: 5L	IV5: 6L	

Omit taxon and specimen	Measurements					Comments
Tympanuchus cupido USNM 289576	I1: 10L	I2: 6L				Measurements based on both feet of male bird
	II1: 16L	II2: 11L	II3: 7L			
	III1: 17L	III2: 13L	III3: 11L	III4: 8L		
	IV1: 12L	IV2: 7L	IV3: 6L	IV4: 7L	IV5: 7L	
Tympanuchus cupido USNM 289376	I1: 10L	I2: 5L				Measurements based on both feet of male bird
	II1: 15L	II2: 11L	II3: 6L			
	III1: 16L	III2: 12L	III3: 10L	III4: 7L		
	IV1: 11L	IV2: 6L	IV3: 6L	IV4: 7L	IV5: 6L	
Tympanuchus pallidicinctus USNM 291884	I1: 10L	I2: 6L				Measurements based on both feet of male bird
	II1: 15L	II2: 11L	II3: 7L			
	III1: 15L	III2: 11L	III3: 10L	III4: 8L		
	IV1: 11L	IV2: 6L	IV3: 5L	IV4: 6L	IV5: 6L	
Tympanuchus pallidicinctus USNM 291885	I1: 8L	I2: 5L				Measurements based on both feet of female bird
	II1: 13L	II2: 9L	II3: 6L			
	III1: 14L	III2: 10L	III3: 8L	III4: 7L		
	IV1: 10L	IV2: 6L	IV3: 5L	IV4: 4L	IV5: 5L	
Odontophoridae						
Oreortyx pictus USNM 320874	I1: 6L	I2: 4L				Right foot of male bird
	II1: 10L	II2: 8L	II3: 6L			
	III1: 11L	III2: 9L	III3: 8L	III4: 7L		
	IV1: 7L	IV2: 5L	IV3: 4L	IV4: 5L	IV5: 5L	
Oreortyx pictus USNM 320054	I1: 6L	I2: 4L				Measurements based on both feet of female bird
	II1: 10L	II2: 8L	II3: 5L			
	III1: 11L	III2: 9L	III3: 8L	III4: 6L		
	IV1: 7L	IV2: 4L	IV3: 4L	IV4: 5L	IV5: 5L	
Oreortyx pictus USNM 226363	I1: 7L	I2: 4L				Measurements based on both feet of female bird
	II1: 10L	II2: 8L	II3: 6L			
	III1: 11L	III2: 9L	III3: 8L	III4: 7L		
	IV1: 7L	IV2: 5L	IV3: 4L	IV4: 5L	IV5: 5L	
Callipepla squamata USNM 18984	I1: 5L	I2: 3L				Left foot
	II1: 8L	II2: 7L	II3: 5L			
	III1: 9L	III2: 7L	III3: 6L	III4: 5L		
	IV1: 6L	IV2: 4L	IV3: 3L	IV4: 4L	IV5: 4L	
Callipepla squamata USNM 18985	I1: 5L	I2: 3L				Measurements based on both feet
	II1: 9L	II2: 7L	II3: 5L			
	III1: 9L	III2: 8L	III3: 7L	III4: 6L		
	IV1: 6L	IV2: 4L	IV3: 3L	IV4: 4L	IV5: 4L	
Callipepla squamata USNM 499260	I1: 6L	I2: 4L				Left foot of male bird
	II1: 9L	II2: 8L	II3: 5L			
	III1: 10L	III2: 8L	III3: 7L	III4: 6L		
	IV1: 7L	IV2: 4L	IV3: 3L	IV4: 5L	IV5: 4L	
Callipepla douglasii AMNH SKEL 1567	I1: 6L	I2: 4L				Right foot; unguals probably covered by sheaths, but sheaths are so thin that unguals can be seen beneath them, and unguals are about same length as sheaths. Joints of digit IV are too hard to see to allow measurement
	II1: 9L	II2: 7L	II3: 5L			
	III1: 9L	III2: 8L	III3: 7L	III4L 5L		
Callipepla californica USNM 492600	I1: 7L	I2: 4L				Right foot of male bird
	II1: 10L	II2: 8L	II3: 5L			
	III1: 11L	III2: 9L	III3: 8L	III4: 6L		
	IV1: 7L	IV2: 5L	IV3: 4L	IV4: 5L	IV5: 4L	

Table A1.1. *continued*

Omit taxon and specimen	Measurements					Comments
Callipepla californica USNM 346357	I1: 6L	I2: 4L				Left foot
	II1: 9L	II2: 7L	II3: 5L			
	III1: 10L	III2: 8L	III3: 7L	III4: 5L		
	IV1: 6L	IV2: 4L	IV3: 4L	IV4: 4L	IV5: 3L	
Callipepla californica USNM 321110	I1: 5L	I2: 3L				Measurements of both feet of female bird
	II1: 8L	II2: 7L	II3: 5L			
	III1: 9L	III2: 8L	III3: 7L	III4: 6L		
	IV1: 6L	IV2: 4L	IV3: 4L	IV4: 5L	IV5: 5L	
Callipepla gambelii USNM 320776	I1: 6L	I2: 4L				Left foot of male bird
	II1: 9L	II2: 7L	II3: 6L			
	III1: 10L	III2: 8L	III3: 7L	III4: 7L		
	IV1: 7L	IV2: 4L	IV3: 4L	IV4: 5L	IV5: 5L	
Callipepla gambelii USNM 18589	I1: 6L	I2: 4L				Measurements of both feet of female bird
	II1: 9L	II2: 7L	II3: 5L			
	III1: 9L	III2: 8L	III3: 7L	III4: 6L		
	IV1: 6L	IV2: 4L	IV3: 4L	IV4: 5L	IV5: 5L	
Callipepla gambelii USNM 225776	I1: 6L	I2: 4L				Left foot of male bird; measurements difficult due to tissue obscuring joints, especially on digit IV
	II1: 9L	II2: 7L	II3: 5L			
	III1: 10L	III2: 8L	III3: 7L	III4: 6L		
	IV1: 7L					
Callipepla gambelii USNM 346496	I1: 6L	I2: 4L				Right foot of female bird
	II1: 9L	II2: 8L	II3: 6L			
	III1: 10L	III2: 8L	III3: 7L	III4: 6L		
	IV1: 7L	IV2: 4L	IV3: 4L	IV4: 5L	IV5: 5L	
Philortyx fasciatus USNM 346717	I1: 6L	I2: 4L				Left foot
	II1: 9L	II2: 7L	II3: 5L			
	III1: 10L	III2: 7L	III3: 7L	III4: 6L		
	IV1: 6L	IV2: 4L	IV3: 4L	IV4: 5L	IV5: 4L	
Colinus nigrogularis USNM 288735	I1: 6L	I2: 3L				Left foot of female bird
	II1: 8L	II2: 6L	II3: 4L			
	III1: 8L	III2: 7L	III3: 6L	III4: 5L		
	IV1: 6L	IV2: 4L	IV3: 3L	IV4: 4L	IV5: 4L	
Colinus nigrogularis USNM 288737	I1: 6L	I2: 3L				Right foot of male bird
	II1: 8L	II2: 7L	II3: 4L			
	III1: 9L	III2: 7L	III3: 6L	III4: 5L		
	IV1: 6L	IV2: 4L	IV3: 4L	IV4: 4L	IV5: 4L	
Colinus nigrogularis USNM 346846	I1: 6L	I2: 3L				Right foot of female bird
	II1: 8L	II2: 7L	II3: 5L			
	III1: 9L	III2: 7L	III3: 6L	III4: 5L		
	IV1: 6L	IV2: 3L	IV3: 3L	IV4: 4L	IV5: 4L	
Odontophorus stellatus USNM 345804	I1: 9L	I2: 6L				Measurements of both feet of female bird
	II1: 11L	II2: 10L	II3: 7L			
	III1: 11L	III2: 9L	III3: 9L	III4: 7L		
	IV1: 9L	IV2: 4L	IV3: 4L	IV4: 6L	IV5: 5L	
Cyrtonyx montezumae USNM 346625	I1: 5L	I2: 5L				Right foot of male bird
	II1: 7L	II2: 6L	II3: 8L			
	III1: 8L	III2: 5L	III3: 7L	III4: 7L		
	IV1: 6L	IV2: 3L	IV3: 3L	IV4: 5L	IV5: 6L	
Cyrtonyx montezumae USNM 346453	I1: 5L					Measurements of both feet of male bird; ungual I2 abnormal
	II1: 6L	II2: 7L	II3: 8L			
	III1: 8L	III2: 5L	III3: 7L	III4: 8L		
	IV1: 6L	IV2: 3L	IV3: 3L	IV4: 5L	IV5: 7L	
Cyrtonyx montezumae USNM 429820	I1: 5L	I2: 3L				Left foot
	II1: 6L	II2: 7L	II3: 7L			
	III1: 8L	III2: 6L	III3: 7L	III4: 7L		
	IV1: 6L	IV2: 3L	IV3: 3L	IV4L	IV5: 6L	

Omit taxon and specimen	Measurements					Comments
Phasianidae						
Tetraogallus himalayensis AMNH SKEL 16360	I1: 10L					Right foot of male bird
	II1: 19L	II2: 15L				
	III1: 19L	III2: 17L	III3: 13L			
	IV1: 14L	IV2: 7L	IV3: 6L	IV4: 7L		
Alectoris graeca MSU OR.270	I1: 7L					Measurements based on both feet; unguals covered by sheaths
	I1: 12L	II2: 9L				
	III1: 12L	III2: 10L	III3: 8L			
	IV1: 8L	IV2: 5L	IV3: 4L	IV4: 5L		
Alectoris graeca MSU OR.3561	I1: 7L					Measurements based on both feet of male bird
	II1: 13L	II2: 11L				
	III1: 13L	III2: 12L	III3: 9L	III4: 8L		
	IV1: 9L	IV2: 6L	IV3: 5L	IV4: 6L	IV5: 7L	
Alectoris chukar USNM 343411	I1: 6L	I2: 4L				Right foot of female bird
	II1: 11L	II2: 9L	II3: 6L			
	III1: 12L	III2: 10L	III3: 8L	III4: 6L		
	IV1: 8L	IV2: 5L	IV3: 4L	IV4: 5L	IV5: 6L	
Alectoris chukar USNM 430377	I1: 5L	I2: 4L				Measurements based on both feet of male bird
	II1: 11L	II2: 8L	II3: 6L			
	III1: 12L	III2: 9L	III3: 8L	III4: 7L		
	IV1: 8L	IV2: 4L	IV3: 4L	IV4: 4L	IV5: 5L	
Alectoris chukar USNM 345098	I1: 8L					Left foot of male bird
	II1: 14L	II2: 11L	II3: 8L			
	III1: 14L	III2: 12L	III3: 10L	III4: 9L		
	IV1: 10L	IV2: 6L	IV3: 5L	IV4: 6L	IV5: 7L	
Alectoris chukar USNM 345048	I1: 7L	I2: 5L				Measurements based on both feet of male bird
	II1: 13L	II2: 10L	II3: 8L			
	III1: 14L	III2: 11L	III3: 9L	III4: 8L		
	IV1: 9L	IV2: 6L	IV3: 5L	IV4: 6L	IV5: 7L	
Alectoris barbara USNM 527	I1: 8L	I2: 5L				Left foot
	II1: 13L	II2: 11L	II3: 8L			
	III1: 14L	III2: 12L	III3: 10L	III4: 7L		
	IV1: 9L	IV2: 6L	IV3: 5L	IV4: 7L	IV5: 7L	
Alectoris rufa USNM 488551	I1: 7L	I2: 5L				Male bird
	II1: 12L					
	III1: 13L	III2: 10L	III3: 9L	III4L: 8L		
	IV1: 9L	IV2: 6L	IV3: 4L	IV4: 6L	IV5: 6L	
Ammoperdix griseogularis AMNH 3687	I1: 5L	I2: 7L				Left foot
	II1: 8L	II2: 6L				
	III1: 9L	III2: 8L	III3: 6L			
	IV1: 6L	IV2: 4L	IV3: 3L	IV4: 3L		
Francolinus pintadeanus USNM 343201	I1: 5L	I2: 4L				Measurements based on both feet of male bird
	II1: 9L	II2: 7L	II3: 5L			
	III1: 11L	III2: 8L	III3: 6L	III4: 7L		
	IV1: 7L	IV2: 4L	IV3: 3L	IV4: 4L	IV5: 5L	
Francolinus pintadeanus USNM 343199	I1: 6L	I2: 4L				Right foot of male bird
	II1: 9L	II2: 7L	II3: 6L			
	III1: 10L	III2: 8L	III3: 7L	III4: 7L		
	IV1: 7L	IV2: 4L	IV3: 4L	IV4: 4L	IV5: 5L	
Francolinus afer AMNH SKEL 5050	I1: 7L	I2: 6L				Right foot of female bird; claw sheaths present, but transparent, and ungual length ~ sheath length
	II1: 11L	II2: 8L	II3: 9L			
	III1: 12L	III2: 9L	III3: 8L	III4: 10L		
	IV1: 8L	IV2: 5L	IV3: 4L	IV4: 5L	IV5: 8L	

Table A1.1. *continued*

Omit taxon and specimen	Measurements					Comments
Perdix perdix USNM 226159	I1: 4L	I2: 4L				Left foot of male bird
	II1: 11L	II2: 8L	II3: 5L			
	III1: 12L	III2: 9L	III3: 7L	III4: 6L		
	IV1: 8L	IV2: 4L	IV3: 4L	IV4: 4L	IV5: 5L	
Melanoperdix niger USNM 320866	II1: 10L	II2: 8L	II3: 5L			Right foot of female bird; digit I deformed
	III1: 10L	III2: 9L	III3: 8L	III4: 4L		
	IV1: 7L	IV2: 5L	IV3: 4L	IV4: 4L	IV5:4L	
Coturnix coturnix USNM 345029	I1: 5L	I2: 3L				Male bird
	II1: 8L	II2: 6L	II3: 4L			
	III1: 9L	III2: 7L	III3: 5L	III4: 4L		
	IV1: 6L	IV2: 4L	IV3: 3L	IV4: 3L	IV5: 3L	
Coturnix pectoralis USNM 612659	I1: 5L	I2: 3L				Male bird
	II1: 7L	II2: 6L	II3: 4L			
	III1: 8L	III2: 6L	III3: 5L	III4: 4L		
	IV1: 5L	IV2: 4L	IV3: 3L	IV4: 3L	IV5: 3L	
Rollulus rouloul USNM 18764	I1: 7L					Left foot; ungual I2 deformed
	II1: 11L	II2: 8L	II3: 6L			
	III1: 11L	III2: 9L	III3: 8L	III4: 7L		
	IV1: 7L	IV2: 4L	IV3: 3L	IV4: 5L	IV5: 6L	
Ithaginis cruentus AMNH SKEL 21986	I1: 9L					Right foot of male bird
	II1: 18L	II2: 14L				
	III1: 17L	III2: 15L	III3: 14L			
	IV1: 11L	IV2: 6L	IV3: 6L	IV4: 9L		
Ithaginis cruentus AMNH SKEL 21985	I1: 9L					Left foot of male bird
	II1: 16L	II2: 13L				
	III1: 15L	III2: 14L	III3: 12L			
	IV1: 10L	IV2: 6L	IV3: 5L	IV4: 9L		
Tragopan satyra AMNH SKEL 13747	I1: 13L					Right foot of female bird
	II1: 18L	II2: 14L				
	III1: 18L	III2: 15L	III3: 14L			
	IV1: 12L	IV2: 7L	IV3: 6L	IV4: 9L		
Tragopan satyra AMNH SKEL 1318	I1: 15L	I2: 9L				Left foot
	II1: 21L	II2: 17L	II3: 12L			
	III1: 21L	III2: 17L	III3: 15L	III4: 14L		
	IV1: 15L	IV2: 8L	IV3: 8L	IV4L: 10L	IV5: 10L	
Tragopan caboti AMNH SKEL 3925	I1: 13L					Left foot of male bird
	II1: 19L	II2: 14L				
	III1: 18L	III2: 15L	III3: 14L			
	IV1: 12L	IV2: 7L	IV3: 6L	IV4: 9L		
Pucrasia macrolopha AMNH SKEL 4051	I1: 13L					Measurements based on both feet
	II1: 19L	II2: 15L				
	III1: 19L	III2: 15L	III3: 14L			
	IV1: 13L	IV2: 7L	IV3: 6L	IV4: 9L		
Lophophorus impejanus YPM ORN 102071 (formerly 378)	I1: 11L	I2: 10L				Right foot
	II1: 18L	II2: 15L	II3: 13L			
	III1: 18L	III2: 16L	III3: 15L	III4: 13L		
	IV1: 12L	IV2: 8L	IV3: 7L	IV4: 9L	IV5: 11L	
Gallus gallus USNM 318501	I1: 13L	I2: 7L				Measurements based on both feet of male bird
	II1: 17L	II2: 12L	II3: 11L			
	III1: 17L	III2: 13L	III3: 11L	III4: 11L		
	IV1: 11L	IV2: 7L	IV3: 6L	IV4: 7L	IV5: 9L	
Gallus gallus USNM 344674	I1: 12L	I2: 7L				Measurements based on both feet of male bird
	II1: 16L	II2: 12L	II3: 10L			
	III1: 17L	III2: 13L	III3: 12L	III4: 11L		
	IV1: 12L	IV2: 7L	IV3: 6L	IV4: 7L	IV5: 8L	

Omit taxon and specimen	Measurements					Comments
Gallus gallus USNM 18959	I1: 16L	I2: 9L				Measurements based on both feet of female domestic chicken
	II1: 19L	II2: 16L	II3: 11L			
	III1: 19L	III2: 15L	III3: 15L	III4: 12L		
	IV1: 13L	IV2: 8L	IV3: 8L	IV4: 9L	IV5: 10L	
Gallus gallus USNM 19183	I1: 20L	I2: 10L				Domestic chicken: IV5 abnormally short?
	II1: 25L	II2: 19L	II3: 13L			
	III1: 26L	III2: 19L	III3: 19L	III4: 14L		
	IV1: 17L	IV2: 11L	IV3: 9L	IV4: 12L		
Gallus varius AMNH SKEL 3926	I1: 11L					Measurements based on both feet
	II1: 17L	II2: 12L				
	III1: 17L	III2: 13L	III3: 11L			
	IV1: 11L	IV2: 7L	IV3: 6L	IV4: 8L		
Lophura leucomelanos USNM 343610	I1: 11L	I2: 7L				Left foot of male bird
	II1: 17L	II2: 13L	II3: 9L			
	III1: 18L	III2: 15L	III3: 13L	III4: 11L		
	IV1: 12L	IV2: 7L	IV3: 6L	IV4: 8L	IV5: 8L	
Lophura leucomelanos USNM 428600	I1: 10L	I2: 7L				Right foot of female bird
	II1: 15L	II2: 12L	II3: 9L			
	III1: 14L	III2: 12L	III3: 12L	III4: 11L		
	IV1: 10L	IV2: 6L	IV3: 6L	IV4: 7L	IV5: 8L	
Lophura leucomelanos USNM 429331	I1: 11L	I2: 7L				Right foot of male bird
	II1: 16L	II2: 13L	II3: 11L			
	III1: 17L	III2: 14L	III3: 12L	III4: 11L		
	IV1: 11L	IV2: 7L	IV3: 6L	IV4: 7L	IV5: 8L	
Lophura swinhoii YPM ORN 102984 (formerly 4361)	I1: 12L	I2: 7L				Right foot
	II1: 18L	II2: 13L	II3: 9L			
	III1: 17L	III2: 14L	III3: 12L	III4: 10L		
	IV1: 12L	IV2: 7L	IV3: 5L	IV4: 6L	IV5: 7L	
Lophura swinhoii USNM 320294	I1: 13L					Right foot of male bird: I2 looks abnormal
	II1: 20L	II2: 15L	II3: 10L			
	III1: 19L	III2: 16L	III3: 14L	III4: 12L		
	IV1: 13L	IV2: 8L	IV3: 7L	IV4: 8L	IV5: 8L	
Lophura swinhoii USNM 344676	I1: 11L	I2: 7L				Female bird;ungual III4 looks too short
	II1: 15L	II2: 11L	II3: 9L			
	III1: 15L	III2: 13L	III3: 11L			
	IV1: 11L	IV2: 7L	IV3: 6L	IV4: 6L	IV5: 7L	
Lophura swinhoii USNM 322024	II1: 19L	II2: 15L	II3: 11L			Measurements based on both feet of male bird; digit I looks deformed on both feet
	III1: 18L	III2: 15L	III3: 14L	III4: 13L		
	IV1: 13L	IV2: 7L	IV3: 7L	IV4: 7L	IV5: 9L	
Lophura nycthemera USNM 346404	I1: 13L	I2: 8L				Right foot of male bird
	II1: 17L	II2: 13L	II3: 10L			
	III1: 17L	III2: 13L	III3: 13L	III4: 11L		
	IV1: 11L	IV2: 7L	IV3: 5L	IV4: 7L	IV5: 8L	
Lophura nycthemera USNM 346204	I1: 12L	I2: 6L				Left foot of male bird
	II1: 16L	II2: 12L	II3: 7L			
	III1: 17L	III2: 14L	III3: 12L	III4: 8L		
	IV1: 12L	IV2: 7L	IV3: 6L	IV4: 7L	IV5: 7L	
Lophura nycthemera USNM 321103	I1: 13L	I2: 7L				Right foot of male bird
	II1: 18L	II2: 14L	II3: 10L			
	III1: 18L	III2: 15L	III3: 13L	III4: 11L		
	IV1: 12L	IV2: 7L	IV3: 6L	IV4: 7L	IV5: 8L	
Lophura erythrophthalma AMNH SKEL 3677	I1: 10L					Left foot of female bird
	II1: 16L	II2: 12L				
	III1: 16L	III2: 12L	III3: 11L			
	IV1: 11L	IV2: 7L	IV3: 5L	IV4: 7L		

Table A1.1. *continued*

Omit taxon and specimen	Measurements					Comments
Lophura erythrophthalma USNM 19730	I1: 9L	I2: 5L				Measurements based on both feet
	II1: 12L	II2: 9L	II3: 7L			
	III1: 13L	III2: 10L	III3: 8L	III4: 7L		
	IV1: 9L	IV2: 5L	IV3: 4L	IV4: 5L	IV5: 5L	
Lophura ignita USNM 318065	I1: 15L	I2: 10L				Right foot of male bird
	II1: 21L	II2: 15L	II3: 11L			
	III1: 21L	III2: 16L	III3: 14L	III4: 12L		
	IV1: 14L	IV2: 9L	IV3: 7L	IV4: 8L	IV5: 9L	
Lophura ignita USNM 18760	I1: 11L	I2: 7L				Left foot
	II1: 17L	II2: 12L	II3: 10L			
	III1: 18L	III2: 13L	III3: 11L	III4: 11L		
	IV1: 12L	IV2: 7L	IV3: 6L	IV4: 6L	IV5: 8L	
Lophura ignita USNM 19933	I1: 13L	I2: 8L				Left foot of male bird
	II1: 19L	II2: 14L	II3: 10L			
	III1: 19L	III2: 15L	III3: 14L	III4: 12L		
	IV1: 12L	IV2: 8L	IV3: 6L	IV4: 8L	IV5: 9L	
Lophura bulweri USNM 491472	I1: 14L	I2: 12L				Mostly from left foot of male bird; I2 could be from right foot
	II1: 18L	II2: 13L	II3: 8L			
	III1: 20L	III2: 16L	III3: 14L	III4: 10L		
	IV1: 14L	IV2: 13L	IV3: 8L	IV4: 6L	IV5: 8L	
Crossoptilon crossoptilon USNM 319196	I1: 14L	I2: 13L				Right foot of male bird
	II1: 21L	II2: 17L	II3: 16L			
	III1: 22L	III2: 18L	III3: 15L	III4: 18L		
	IV1: 15L	IV2: 9L	IV3: 7L	IV4: 9L	IV5: 14L	
Crossoptilon crossoptilon USNM 319195	I1: 13L	I2: 12L				Left foot of male bird
	II1: 22L	II2: 18L	II3: 15L			
	III1: 23L	III2: 18L	III3: 17L	III4: 17L		
	IV1: 15L	IV2: 9L	IV3: 7L	IV4: 11L	IV5: 14L	
Crossoptilon crossoptilon AMNH SKEL 11360	I1: 15L					Left foot of male bird
	II1: 22L	II2: 17L				
	III1: 23L	III2: 19L	III3: 15L	IV1: 16L		
	IV1: 16L	IV2: 10L	IV3: 8L	IV4: 9L		
Crossoptilon mantchuricum YPM ORN 102983 (formerly 4360)	I1: 14L	I2: 9L				Left foot; phalanges hard to measure because surrounded by dried tissue
	II1: 21L	II2: 16L	II3: 13L			
	III1: 21L	III2: 18L	III3: 15L	III4: 13L		
	IV1: 14L	IV2: 8L	IV3: 7L	IV4: 7L	IV5: 11L	
Crossoptilon mantchuricum USNM 344941	I1: 14L	I2: 10L				Left foot of male bird
	II1: 22L	II2: 17L	II3: 14L			
	III1: 23L	III2: 18L	III3: 15L	III4: 15L		
	IV1: 14L	IV2: 8L	IV3: 7L	IV4: 8L	IV5: 12L	
Crossoptilon mantchuricum AMNH SKEL 16348	I1: 13L					Right foot of male bird
	II1: 21L	II2: 16L				
	III1: 21L	III2: 16L	III3: 14L			
	IV1: 15L	IV2: 8L	IV3: 7L	IV4: 8L		
Crossoptilon mantchuricum AMNH SKEL 1992	I1: 12L					Right foot of male bird
	II1: 20L	II2L 15L				
	III1: 21L	III2: 16L	III3: 14L			
	IV1: 14L	IV2: 8L	IV3: 6L	IV4: 7L		
Crossoptilon auritum YPM ORN 102045 (formerly 352)	I1: 13L					Left foot; most unguals surrounded by sheaths
	II1: 20L	II2: 16L	II3: 14L			
	III1: 21L	III2: 16	III3: 14			
	IV1: 13L	IV2: 8L	IV3: 6L	IV4: 8L		
Catreus wallichi USNM 347598	I1: 10L	I2: 7L				Measurements based on both feet of female bird
	II1: 15L	II2: 13L	II3: 11L			
	III1: 16L	III2: 13L	III3: 12L	III4: 11L		
	IV1: 10L	IV2: 6L	IV3: 6L	IV4: 7L	IV5: 9L	

Omit taxon and specimen	Measurements					Comments
Catreus wallichi USNM 346852	I1: 10L	I2: 8L				Measurements based on both feet of female bird
	II1: 16L	II2: 13L	II3: 12L			
	III1: 17L	III2: 13L	III3: 12L	III4: 13L		
	IV1: 10L	IV2: 7L	IV3: 5L	IV4: 7L	IV5: 10L	
Catreus wallichi AMNH SKEL 5194	I1: 12L					Right foot of male bird
	II1: 18L	II2: 14L				
	III1: 19L	III2: 15L	III3: 13L			
	IV1: 12L	IV2: 7L	IV3: 6L	IV4: 8L		
Syrmaticus mikado AMNH SKEL 11424	I1: 11L					Measurements based on both feet of female bird
	II1: 16L	II2: 13L				
	III1: 16L	III2: 13L	III3: 13L			
	IV1: 11L	IV2: 7L	IV3: 6L	IV4: 8L		
Syrmaticus soemmerringii AMNH SKEL 27149	I1: 10L	I2: 6L				Left foot for all phalanges except IV5
	II1: 16L	II2: 13L	II3: 9L			
	III1: 17L	III2: 13L	III3: 12L	II4: 9L		
	IV1: 11L	IV2: 7L	IV3: 6L	IV4: 7L	IV5: 8L	
Syrmaticus reevesii AMNH SKEL 639	I1: 11L	I2: 8L				Left foot of male bird: ungual II3 looks abnormally short
	II1: 17L	II2: 14L				
	III1: 19L	III2: 15L	III3: 14L	III4: 12L		
	IV1: 12L	IV2: 7L	IV3: 6L	IV4: 9L	IV5: 9L	
Syrmaticus reevesii AMNH SKEL 3678	I1: 10L					Right foot
	II1: 17L	II2: 13L				
	III1: 18L	III2: 14L	III3: 12L			
	IV1: 12L	IV2: 6L	IV3: 5L	IV4: 8L		
Phasianus colchicus YPM ORN 102498 (formerly 2101)	II1: 14L	II2: 12	II3: 11L			Mounted skeleton; left foot
	III1: 16L	III2: 14L	III3: 12L	III4: 13L		
	IV1: 11L	IV2: 8L	IV3: 7L	IV4: 7L	IV5: 12L	
Phasianus colchicus YPM ORN 111007 (formerly 14346)	I1: 8L					Right foot; unguals I1, II3, and II4 covered by sheaths
	II1: 17L	II2: 13L				
	III1: 17L	III2: 14L	III3: 12L			
	IV1: 12L	IV2: 7	IV3: 6	IV4: 7	IV5: 11	
Phasianus colchicus UMMZ Birds 99579	I1: 9L	I2: 6L				Left foot of female bird
	II1: 17L	II2: 13L	II3: 8L			
	III1: 17L	III2: 13L	III3: 12L	III4: 9L		
	IV1: 12L	IV2: 7L	IV3: 6L	IV4: 7L		
Phasianus colchicus MSU OR.3129	I1: 8L					Right foot; all unguals covered by sheaths
	II1: 15L	II2: 11L				
	III1: 15L	III2: 12L	III3: 10L			
	IV1: 11L	IV2L: 7L	IV3: 5L	IV4: 6L		
Phasianus colchicus UMA 1519	I1: 9					Right foot, but digit I possibly left
	II1: 16L	II2: 12L				
	III1: 17L	III2: 13L	III3: 11L			
	IV1: 12L	IV2: 7L	IV3: 6L	IV4: 7L		
Phasianus colchicus versicolor USNM 346561	I1: 9L	I2: 6L				Right foot of male bird
	II1: 17L	II2: 13L	II3: 8L			
	III1: 18L	III2: 14L	III3: 12L	III4: 9L		
	IV1: 12L	IV2: 8L	IV3: 7L	IV4: 7L	IV5: 7L	
Chrysolophus pictus YPM ORN 102491 (formerly 2094)	I1: 9L	I2: 5L				Measurements based on both feet
	II1: 16L, 3pw, 3dw	II2: 12L, 3dw	II3: 8L			
	III1: 16L, 4pw, 3dw	III2: 13L, 3dw	III3: 11L, 3dw	III4: 9L		
	IV1: 11L, 4pw, 3dw	IV2: 7L, 2dw	IV3: 6L, 2dw	IV4: 7L, 2dw	IV5: 7L	

Table A1.1. *continued*

Omit taxon and specimen	Measurements					Comments
Chrysolophus pictus YPM ORN 109062 (formerly 11515)	I1: 9L	I2: 5L				Right foot
	II1: 16L	II2: 12L	II3: 8L			
	III1: 17L	III2: 14L	III3: 12L	III4: 9L		
	IV1: 11L	IV2: 7L	IV3: 5L	IV4: 6L	IV5: 6L	
Chrysolophus pictus AMNH SKEL 898	I1: 10L	I2: 5L				Left foot of male bird
	II1: 17L	II2: 13L	II3: 8L			
	III1: 17L	III2: 14L	III3: 12L	III4: 9L		
	IV1: 12L	IV2: 7L	IV3: 6L	IV4: 7L	IV5: 7L	
Chrysolophus pictus AMNH SKEL 1566	I1: 9L	I2: 4L				Left foot
	II1: 16L	II2: 12L				
	III1: 16L	III2: 13L	III3: 11L	III4: 8L		
	IV1: 11L	IV2: 7L	IV3: 5L	IV4: 6L		
Chrysolophus amherstiae USNM 289574	I1: 9L	I2: 5L				Right foot of female bird
	II1: 15L	II2: 11L	II3: 8L			
	III1: 15L	III2: 12L	III3: 11L	III4: 9L		
	IV1: 10L	IV2: 6L	IV3: 5L	IV4: 7L	IV5: 7L	
Chrysolophus amherstiae USNM 322022	I1: 10L	I2: 7L				Male bird
	II1: 17L	II2: 13L	II3: 10L			
	III1: 17L	III2: 14L	III3: 13L	III4: 8L		
	IV1: 12L	IV2: 7L	IV3: 6L	IV4: 8L	IV5: 7L	
Chrysolophus amherstiae AMNH SKEL 3439	I1: 10L					Left foot
	II1: 14L	II2: 12L	II3: 9L			
	III1: 15L	III2: 12L	III3: 11L	III4: 9L		
	IV1: 10L	IV2: 6L	IV3: 5L	IV4: 6L	IV5: 7	
Polyplectron chalcurum AMNH SKEL 4970	I1: 10L					Left foot of male bird
	II1: 15L	II2: 12L				
	III1: 15L	III2: 12L	III3: 12L			
	IV1: 10L	IV2: 7L	IV3: 6L	IV4: 9L		
Polyplectron bicalcaratum YPM ORN 102506 (formerly 2109)	I1: 10L					Measurements based on both feet
	II1: 16L	II2: 13L	II3: 8			
	III1: 15L	III2: 14L	III3: 12L	III4: 9L		
	IV1: 11L	IV2: 7L	IV3: 6L	IV4: 7L	IV5: 7L	
Polyplectron bicalcaratum AMNH SKEL 2908	I1: 12L					Left foot of male bird
	II1: 16L	II2: 13L				
	III1: 16L	III2: 13L	III3: 12L			
	IV1: 11L	IV2: 7L	IV3: 7L	IV4: 7L		
Polyplectron bicalcaratum AMNH SKEL 1904	I1: 10L					Measurements based on both feet
	II1: 15L	II2: 12L				
	III1: 15L	III2: 13L	III3: 11L			
	IV1: 10L	IV2: 7L	IV3: 6L	IV4: 7L		
Rheinardia ocellata AMNH SKEL 5865	I1: 16L					Left foot
	II1: 22L	II2: 16L				
	III1: 22L	III2: 17L	III3: 16L			
	IV1: 16L	IV2: 9L	IV3: 7L	IV4: 10L		
Rheinardia ocellata AMNH SKEL 4029	I1: 15L					Right foot
	II1: 19L	II2: 14L				
	III1: 20L	III2: 15L	III3: 14L			
	IV1: 13L	IV2: 8L	IV3: 6L	IV4: 8L		
Rheinardia ocellata AMNH SKEL 6046	I1: 17L					Left foot
	II1: 23L	II2: 16L				
	III1: 22L	III2: 17L	III3: 15L			
	IV1: 16L	IV2: 9L	IV3: 7L	IV4: 10L		

Omit taxon and specimen	Measurements					Comments
Argusianus argus USNM 19470	I1: 18L	I2: 9L				Measurements based on both feet of male bird
	II1: 24L	II2: 17L	II3: 12L			
	III1: 24L	III2: 18L	III3: 15L	III4: 13L		
	IV1: 17L	IV2: 10L	IV3: 8L	IV4: 10L	IV5: 9L	
Argusianus argus AMNH SKEL 13708	I1: 15L					Right foot of male bird
	II1: 21L	II2: 15L				
	III1: 22L	III2: 17L	III3: 15L			
	IV1: 16L	IV2: 8L	IV3: 7L	IV4: 9L		
Argusianus argus AMNH SKEL 4969	I1: 17L					Measurements based on both feet of female bird
	II1: 22L	II2: 16L				
	III1: 23L	III2: 17L	III3: 15L			
	IV1: 16L	IV2: 9L	IV3: 7L	IV4: 9L		
Pavo cristatus YPM ORN 102063 (formerly 370)	II1: 24L	II2: 20L				Left foot; bones measured along dorsal surfaces; unguals covered by horny sheath
	III1: 28L	III2: 23L	III3: 21L			
	IV1: 19L	IV2: 12L	IV3: 9L	IV4: 13		
Pavo cristatus USNM 320870	I1: 19L	I2: 12L				Right foot of male bird
	II1: 23L	II2: 17L	II3: 12L			
	III1: 26L	III2: 20L	III3: 17L	III4: 16L		
	IV1: 17L	IV2: 10L	IV3: 8L	IV4: 10L	IV5: 11L	
Pavo cristatus USNM 19917	I1: 18L	I2: 10L				Right foot
	II1: 22L	II2: 17L	II3: 12L			
	III1: 26L	III2: 19L	III3: 17L	III4: 13L		
	IV1: 17L	IV2: 10L	IV3: 8L	IV4: 10L	IV5: 9L	
Pavo muticus USNM 320990	I1: 19L					Left foot of female bird; I2 looks deformed
	II1: 24L	II2: 17L	II3: 12L			
	III1: 26L	III2: 20L	III3: 17L			
	IV1: 18L	IV2: 10L	IV3: 8L	IV4: 11L	IV5: 10L	
Afropavo congensis AMNH SKEL 22691	I1: 14L					Right foot of female bird
	II1: 18L	II2: 13L				
	III1: 18L	III2: 15L	III3: 14L			
	IV1: 12L	IV2: 7L	IV3: 6L	IV4: 8L		
Numididae						
Agelastes meleagrides USNM 19689	I1: 10L	I2: 6L				Right foot of male bird
	II1: 18L	II2: 13L	II3: 10L			
	III1: 18L	III2: 14L	III3: 11L	III4: 12L		
	IV1: 13L	IV2: 8L	IV3: 7L	IV4: 7L	IV5: 9L	
Phasidus niger (*Agelastes niger*) USNM 292396	I1: 10L	I2: 7L				Measurements based on both feet of female bird
	II1: 15L	II2: 11L	II3: 9L			
	III1: 16L	III2: 13L	III3: 11L	III4: 10L		
	IV1: 11L	IV2: 6L	IV3: 6L	IV4: 7L	IV5: 7L	
Phasidus niger (*Agelastes niger*) AMNH SKEL 4147	I1: 10L	I2: 7L				Measurements based on both feet of male bird
	II1: 15L	II2: 10L	II3: 9L			
	III1: 16L	III2: 12L	III3: 11L	III4: 11L		
	IV1: 11L	IV2: 6L	IV3: 6L	IV4: 7L	IV5: 8L	
Phasidus niger (*Agelastes niger*) AMNH SKEL 6044	I1: 9L					Right foot
	II1: 15L	II2: 8L				
	III1: 16L	III2: 12L	III3: 10L			
	IV1: 11L	IV2: 6L	IV3: 5L	IV4: 7L		
Numida meleagris USNM 291435	I1: 11L	I2: 8L				Left foot of male bird
	II1: 15L	II2: 11L	II3: 10L			
	III1: 17L	III2: 13L	III3: 10L	III4: 12L		
	IV1: 12L	IV2: 7L	IV3: 6L	IV4: 6L	IV5: 9L	

Table A1.1. *continued*

Omit taxon and specimen	Measurements					Comments
Numida meleagris USNM 292503	I1: 12L	I2: 8L				Female bird
	II1: 14L	II2: 10L	II3: 10L			
	III1: 16L	III2: 12L	III3: 9L	III4: 11L		
	IV1: 11L	IV2: 5L	IV3: 5L	IV4: 6L	IV5: 8L	
Numida meleagris AMNH SKEL 5124	I1: 11L	I2: 6L				Left foot
	II1: 16L	II2: 11L	II3: 10L			
	III1: 17L	III2: 12L	III3: 10L	III4: 12L		
	IV1: 11L	IV2: 7L	IV3: 6L	IV4: 6L	IV5: 6L	
Guttera plumifera AMNH SKEL 4258	I1: 12L	I2: 6L				Right foot
	II1: 17L	II2: 11L	II3: 7L			
	III1: 17L	III2: 13L	III3: 11L	III4: 9L		
	IV1: 12L	IV2: 7L	IV3: 6L	IV4: 7	IV5: 6L	
Guttera plumifera AMNH SKEL 4204	I1: 12L	I2: 6L				Right foot of female bird (digit I possibly from left foot)
	II1: 17L	II2: 12L				
	III1: 18L	III2: 13L	III3: 11L	III4: 10L		
	IV1: 13L	IV2: 7L	IV3: 6L	IV4: 7L		
Guttera plumifera USNM 428654	I1: 13L					Measurements based on both feet of female bird; ungual I2 deformed
	II1: 18L	II2: 13L	II3: 8L			
	III1: 19L	III2: 14L	III3: 12L	III4: 10L		
	IV1: 13L	IV2: 7L	IV3: 6L	IV4: 7L	IV5: 6L	
Guttera edouardi (*G. pucherani*) USNM 322592	I1: 14L	I2: 7L				Measurements based on both feet of male bird
	II1: 18L	II2: 13L	II3: 10L			
	III1: 19L	III2: 15L	III3: 11L	III4: 13L		
	IV1: 13L	IV2: 8L	IV3: 6L	IV4: 8L	IV5: 8L	
Guttera edouardi (*G. pucherani*) AMNH SKEL 4973	I1: 15L					Left foot
	II1: 18L	II2: 13L				
	III1: 20L	III2: 15L	III3: 12L			
	IV1: 14L	IV2: 7L	IV3: 6L	IV4: 8L		
Guttera edouardi (*G. pucherani*) AMNH SKEL 6045	I1: 13L					Measurements based on both feet
	II1: 18L	II2: 13L				
	III1: 20L	III2: 15L	III3: 12L			
	IV1: 14L	IV2: 8L	IV3: 7L	IV4: 7L		
Acryllium vulturinum USNM 346192	I1: 15L	I2: 7L				Female bird; ungual IV5 deformed on both feet
	II1: 19L	II2: 14L	II3: 9L			
	III1: 21L	III2: 15L	III3: 12L	III4: 11L		
	IV1: 14L	IV2: 8L	IV3: 7L	IV4: 7L		
Acryllium vulturinum AMNH SKEL 2567	I1: 15L					Left foot
	II1: 20L	II2: 14L				
	III1: 22L	III2: 16L	III3: 12L			
	IV1: 15L	IV2: 9L	IV3: 7L	IV4: 8L	IV5: 8L	
Acryllium vulturinum AMNH SKEL 11341	I1: 16L					Left foot of male bird
	II1: 21L	II2: 15L				
	III1: 22L	III2: 17L	III3: 13L			
	IV1: 15L	IV2: 9L	IV3: 7L	IV4: 8L		
Sagittariidae						
Sagittarius serpentarius USNM 321578	I1: 15L	I2: 17				Right foot
	II1: 16L	II2: 13L	II3: 17L			
	III1: 27L	III2: 11L	III3: 11L	III4: 17L		
	IV1: 15L	IV2: 4L	IV3: 4L	IV4: 7L	IV5: 14L	
Sagittarius serpentarius USNM 346684	I1: 14L	I2: 18L				Measurements based on both feet
	II1: 17L	II2: 13L	II3: 18L			
	III1: 29L	III2: 14L	III3: 11L	III4: 20L		
	IV1: 15L	IV2: 4L	IV3: 5L	IV4: 8L	IV5: 15L	

Omit taxon and specimen	Measurements					Comments
Sagittarius serpentarius USNM 431491	I1: 16L	I2: 20L				Measurements based on both feet of male bird
	II1: 17L	II2: 15L	II3: 19L			
	III1: 28L	III2: 14L	III3: 12L	III4: 20L		
	IV1: 15L	IV2: 5L	IV3: 5L	IV4: 8L	IV5: 16L	
Psophiidae						
Psophia crepitans YPM ORN 100228 (formerly 2130)	II1: 16L	II2: 16L	II3: 11L			Mounted skeleton; measured dorsal lengths of phalanges
	III1: 18L	III2: 15L	III3: 16L	III4: 9L		
	IV1: 13L	IV2: 8L	IV3: 7L	IV4: 9L	IV5: 10L	
Psophia crepitans YPM ORN 110540 (formerly 13138)	I1: 10L					Left foot
	II1: 18L	II2: 15L	II3: 14L			
	III1: 20L	III2: 15L	II3: 16L			
	IV1: 13L	IV2: 9L	IV3: 7L	IV4: 11L		
Psophia crepitans USNM 320971		I2: 6L				Right foot of captive female
	II1: 15L	II2: 14L	II3: 10L			
	III1: 19L	III2: 15L	III3: 15L	III4: 11L		
	IV1: 13L	IV2: 8L	IV3: 7L	IV4: 10L	IV5:9L	
Psophia viridis USNM 491334	I1: 11L	I2: 7L				Right foot
	II1: 18L	II2: 15L	II3: 11L			
	III1: 20L	III2: 15L	III3: 15L	III4: 12L		
	IV1: 13L	IV2: 9L	IV3: 7L	IV4: 9L	IV5: 9L	
Psophia leucoptera USNM 288609	I1: 9L	I2: 7L				Male bird; measurements based on both feet
	II1: 16L	II2: 13L	II3: 10L			
	III1: 17L	III2: 12L	III3: 13L	III4: 12L		
	IV1: 11L	IV2: 7L	IV3: 5L	IV4: 8L	IV5: 9L	
Mesitornithidae						
Mesitornis variegatus USNM 345128	I1: 12L	I2: 6L				Measurements based on both feet
	II1: 10L	II2: 8L	II3: 4L			
	III1: 10L	III2: 8L	III3: 7L	III4: 5L		
	IV1: 5L	IV2: 4L	IV3: 4L	IV4: 4L	IV5: 4L	
Rhynochetidae						
Rhynochetos jubatus USNM 322770	I1: 10L	I2: 7L				Measurements based on both feet of captive male
	II1: 19L	II2: 12L	II3: 7L			
	III1: 18L	III2: 15L	III3: 12L	III4: 8L		
	IV1: 13L	IV2: 9L	IV3: 7L	IV4: 7L	IV5: 6L	
Rhynochetos jubatus USNM 018994	I1: 9L	I2: 6L				Measurements based on both feet
	II1: 20L	II2: 12L	II3: 7L			
	III1: 18L	III2: 15L	III3: 12L	III4: 10L		
	IV1: 13L	IV2: 8L	IV3: 7L	IV4: 7L		
Cariamidae						
Cariama cristata University of Illinois	II1: 15L, 7pw, 5dw	II2: 14L, 6dw	II3: 13L			Left foot of mounted skeleton
	III1: 26L, 9pw, 7dw	III2: 15L, 6dw	III3: 11L, 6dw	III4: 14L		
	IV1: 19L, 7pw, 5dw	IV2: 7L, 5dw	IV3: 6L, 4dw	IV4: 5L, 4dw	IV5: 10L	
Cariama cristata USNM 19941	I1: 5L					Male bird; measurements based on both feet
	II1: 13L	II2: 13L				
	III1: 25L	III2: 15L	III3: 12L			
	IV1: 17L	IV2: 6L	IV3: 5L	IV4: 6L	IV5: 12L	
Cariama cristata AMNH SKEL 3998	I1: 7L	I2: 11L				Left foot
	II1: 13L	II2: 13L	II3: 15L			
	III1: 23L	III2: 16L	III3: 13L	III4: 13L		
	IV1: 16L	IV2: 7L	IV3: 5L	IV4: 8L	IV5: 11L	

Table A1.1. *continued*

Omit taxon and specimen	Measurements					Comments
Cariama cristata AMNH SKEL 1722	II1: 13L	II2: 13L				Female bird; measurements based on both feet
	III1: 25L	III2: 14L	III3: 12L	III4:14L		
	IV1: 18L	IV2: 6L	IV3: 4L	IV4: 6L	IV5: 10L	
Cariama cristata uncatalogued Berkeley J.R. Hutchinson specimen	I1: 4L	I2: 9L				Female bird
	II1: 13L, 6pw, 5dw	II2: 11L, 4dw	II3: 14L			
	III1: 24L, 8pw, 6dw	III2: 14L, 5dw	III3: 11L, 5dw	III4: 14L		
	IV1: 16L, 6pw, 4dw	IV2: 6L, 4dw	IV3: 4L, 4dw	IV4: 5L, 4dw	IV5: 10L	
Chunga burmeisteri AMNH SKEL 2601	I1: 5L					Left foot of female bird
	II1: 12L	II2: 12L				
	III1: 20L	III2: 14L	III3: 11L			
	IV1: 15L	IV2: 6L	IV3: 5L	IV4: 6L		
Chunga burmeisteri AMNH SKEL 4250	I1: 6L	I2: 10L				Right foot
	II1: 14L	II2: 13L	II3: 13L			
	III1: 23L	III2: 15L	III3: 12L	III4: 11L		
	IV1: 15L	IV2: 6L	IV3: 5L	IV4: 7L	IV5: 11L	
Phorusrhacids						
Procariama simplex FMNH P14525	II1: 21L, 9pw, 8dw	II2: 23L, 8dw				Measurements based on both feet
	III1: 39L, 16pw, 13dw	III2: 11dw		III4: 27L		
	IV1: 21L, 11pw, 8dw	IV2: 9L, 8dw	IV3: 8L, 7dw		IV5: 17L	
Prophororhacus incertus FMNH P14422	I1: 13L					Measurements probably based on both feet
	II1: 25L, 16pw, 13dw					
	III1: 45L, 22pw, 22dw	III2: 18dw	III3: 23L, 15L			
Palaeociconia cristata FMNH P13213	II1: 26L, 12pw, 11dw	II2: 23L, 9dw				Measurements based on both feet
	III1: 51L, 22pw, 17dw	III2: 32L	III3: 22L, 15dw	III4: 32L		
	IV1: 31L, 15pw, 12dw	IV2: 11L, 10dw				
Llallawavis scagliai MMP 5050	I1: 13L	I2: 19L				Measurements made for me on both feet by F. Degrange; cf. Degrange et al. (2015)
	II1: 22L, 10dw	II2: 23L, 9dw				
	III1: 42L, 15dw	III2: 25L, 14dw				
	IV1: 22L, 9dw	IV2: 10L, 9dw	IV3: 9L, 9dw	IV4: 9L, 7dw	IV5: 21L	
Rallidae						
Gallirallus australis USNM 226168	I1: 13L					Right foot of captive male; sheaths surround unguals
	II1: 24L	II2: 18L				
	III1: 23L	III2: 17L	III3: 15L			
	IV1: 16L	IV2: 10L	IV3: 8L	IV4: 10L		
Gallirallus australis USNM292885	II1: 23L	II2: 17L	II3: 12L			Right foot: digit III phalanges were loose in box, and I was not confident I could correctly identify them
	IV1: 15L	IV2: 9L	IV3: 8L	IV4: 10L	IV5: 9L	
Aptornithidae						
Aptornis otidiformis NMZ S.023040	II1: 40L, 19pw, 13dw					
	III1: 39L, 20pw, 15dw	III2: 26L, 13dw				
	IV1: 27L, 17pw, 12dw					

Omit taxon and specimen	Measurements					Comments
Aptornis defossor CM AV 6016	I1: 18L	I2: 19L				Measurements probably based on both feet
	II1: 39L, 16pw, 11dw	II2: 26L, 11dw	II3: 32L			
	III1: 39L, 18pw, 14dw	III2: 27L, 12dw	III3: 23L, 12dw	III4: 32L		
	IV1: 25L, 15pw, 13dw	IV2: 15L, 11dw	IV3: 12L, 10dw	IV4: 14L, 10dw	IV5: 25L	
Otididae						
Otis tarda USNM 289732	Digit I absent					Measurements based on both feet
	II1: 21L	II2: 8L	II3: 11L			
	III1: 27L	III2: 15L	III3: 8L	III4: 13L		
	IV1: 19L	IV2: 7L	IV3: 5L	IV4: 4L	IV5: 8L	
Otis tarda AMNH SKEL 1706	Digit I absent					Left foot
	II1: 18L	II2: 7L	II3: 8L			
	III1: 22L	III2: 12L	III3: 6L	III4: 10L		
	IV1: 16L	IV2: 5L	IV3: 4L	IV4: 3L	IV5: 6L	
Otis tarda AMNH SKEL 1705	Digit I absent					Right foot
	II1: 22L	II2: 8L	II3: 9L			
	III1: 28L	III2: 15L	III3: 7L	III4: 12L		
	IV1: 19L	IV2: 7L	IV3: 5L	IV4: 4L	IV5: 8L	
Ardeotis australis (*Choriotis australis*) USNM 347651	Digit I absent					Male bird; measurements based on both feet
	II1: 24L, 8pw, 6dw	II2: 9L, 5dw	II3: 11L			
	III1: 27L, 11pw, 9dw	III2: 15L, 8dw	III3: 8L, 7dw	III4: 14L		
	IV1: 21L, 9pw, 6dw	IV2: 7L, 6dw	IV3: 5L	IV4: 4L	IV5: 10L	
Ardeotis australis (*Choriotis australis*) LACM SN 89966	Digit I absent					Measurements based on both feet
	II1: 22L, 8pw, 5dw	II2: 9L, 5dw	II3: 10L			
	III1: 25L, 10pw, 8dw	III2: 14L, 7dw	III3: 7L, 7dw	III4: 12L		
	IV1: 20L, 8pw, 5dw	IV2: 6L, 4dw	IV3: 4L, 4dw	IV4: 4L, 3dw	IV5: 8L	
Ardeotis kori AMNH SKEL 5420	Digit I absent					Left foot
	II1: 25L	II2: 9L	II3: 14L			
	III1: 30L	III2: 15L	III3: 8L	III4: 19L		
	IV1: 22L	IV2: 7L	IV3: 4L	IV4: 5L	IV5: 12L	
Ardeotis kori AMNH SKEL 6246	Digit I absent					Measurements mainly from right foot. But IV4 and IV5 possibly from left
	II1: 27L	II2: 10L	II3: 11L			
	III1: 29L	III2: 15L	III3: 9L	III4: 15L		
	IV1: 22L	IV2: 8L	IV3: 5L	IV4: 5L		
Ardeotis kori FMNH 338448	TMT L = 233					Measurements based on both feet of male bird
	Digit I absent					
	II1: 28L, 10pw, 7dw	II2: 9L, 6dw	II3: 15L			
	III1: 33L, 14pw, 11dw	III2: 18L, 10dw	III3: 8L, 9dw	III4: 20L		
	IV1: 25L, 11pw, 8dw	IV2: 9L, 6dw	IV3: 6L, 6dw	IV4: 4L, 5dw	IV5: 12L	

Table A1.1. *continued*

Omit taxon and specimen	Measurements					Comments
Ardeotis kori FMNH 338449	TMT L = 238					Measurements based on both feet of male bird
	Digit I absent					
	II1: 30L, 11pw, 8dw	II2: 9L, 6dw	II3: 16L			
	III1: 35L, 14pw, 11dw	III2: 18L, 10dw	III3: 9L, 8dw	III4: 19L		
	IV1: 27L, 11pw, 8dw	IV2: 8L, 6dw	IV3: 5L, 6dw	IV4: 5L, 5dw	IV5: 12L	
Ardeotis kori FMNH 338450	TMT L = 240					Measurements based on both feet
	Digit I absent					
	II1: 30L, 11pw, 8dw	II2: 10L, 6dw	II3: 13L			
	III1: 33L, 14pw, 12dw	III2: 17L, 10dw	III3: 9L, 9dw	III4: 16L		
	IV1: 26L, 11pw, 8dw	IV2: 9L, 7dw	IV3: 5L, 6dw	IV4: 5L, 5dw	IV5: 10L	
Neotis denhami (*Neotis cafra*) USNM 321970	Digit I absent					Female
	II1: 20L	II2: 6L	II3: 9L			
	III1: 21L	III2: 11L	III3: 5L	III4: 10L		
	IV1: 16L	IV2: 5L	IV3: 3L	IV4: 3L	IV5: 7L	
Neotis denhami (*Neotis cafra*) AMNH SKEL 4267	Digit I absent					Measurements based on both feet
	II1: 21L	II2: 7L	II3: 10L			
	III1: 23L	III2: 12L	III3: 6L			
	IV1: 18L	IV2: 6L	IV3: 3L	IV4: 3L		
Eupodotis senegalensis AMNH SKEL 1710	Digit I absent					Left foot
	II1: 11L	II2: 5L	II3: 5L			
	III1: 15L	III2: 8L	III3: 4L	III4: 6L		
	IV1: 9L	IV2: 4L	IV3: 2L	IV4: 2L	IV5: 5L	
Turnicidae						
Turnix tanki AMNH SKEL 1581	Digit I absent					Measurements based on both feet
	II1: 5L	II2: 4L	II3: 2L			
	III1: 6L	III2: 4L	III3: 4L	IV4: 3L		
	IV1: 4L	IV2: 3L	IV3: 2L	IV4: 2L	IV5: 2L	
Turnix suscitator USNM 343204	Digit I absent					Right foot of female bird
	II1: 6L	II2: 4L				
	III1: 6L	III2: 5L	III3: 4L			
	IV1: 5L	IV2: 3L	IV3: 2L	IV4: 2L		
Turnix suscitator USNM 343205	Digit I absent					Right foot of female bird
	II1: 6L	II2: 4L				
	III1: 6L	III2: 5L	III3: 4L			
	IV1: 5L	IV2: 3L	III3: 2L	IV4: 2L		
Turnix suscitator USNM 347288	Digit I absent					Right foot of captive male
	II1: 6L	II2: 4L	II3: 3L			
	III1: 6L	III2: 5L	III3: 4L	III4: 3L		
	IV1: 5L	IV2: 4L	IV3: 2L	IV4: 2L	IV5: 2L	
Turnix velox AMNH SKEL 9653	Digit I absent					Left foot of female bird
	II1: 5L	II2: 3L				
	III1: 5L	III2: 4L	III3: 3L	III4: 3L		
	IV1: 4L	IV2: 3L	IV3: 2L	IV4: 2L	IV5: 2L	
Raphidae						
Raphus cucullatus USNM 16954	MT III L = 131					Cast of Oxford University left foot
	II1: 40L, 15pw, 10dw	II2: 21L, 8dw				
	III1: 36L, 16pw, 11dw	III2: 24L, 9dw	III3: 17L, 8dw			
	IV1: 26L, 13pw, 11dw	IV2: 14L, 9dw	IV3: 11L, 7dw	IV4: 10L, 7dw	IV5: 14L	

Omit taxon and specimen	Measurements					Comments
Sauropodomorphs						
Eoraptor lunensis PVSJ 512	MT III L = 81					Measurements based on right foot. Most measurements made by Farlow on actual material or casts, but some received from P. Sereno
	I1: 18L	I2: 14L				
	II1: 21L, 9dw	II2: 15L, 9dw	II3: 14L			
	III1: 12L, 13pw, 12dw	III2: 17L, 9dw				
	IV1: 16L, 8dw	IV2: 12L, 8dw	IV3: 10L, 8dw	IV4: 8L, 7dw	IV5: 14L	
Ammosaurus major MNA G2 7233 V744 and V745	II1: 41L	II2: 31L	II3: 56L			Measurements made on both feet by D. Gillette following my instructions; Yates (2004) considers this specimen to be an indeterminate basal sauropodomorph
	III1: 50L	III2: 33L	III3: 21L	III4: 43L		
				IV4: 28L	IV5: 33L	
Anchisaurus polyzelus YPM VP 1883	MT III L = 99					Cast of right foot. Measurements approximate; specimen is prepared in situ in rock, and seen from underside
	I1: 30L	I2:27L				
	II1: 30L	II2: 26L	II3: 33L			
				III4: 26L		
				IV4: 13L	IV5: 22L	
Plateosaurus longiceps USNMV10924	I1: 65L	I2: 106L				Cast of left foot from Halberstadt, Germany
	II1: 66L, 61pw, 50dw	II2: 41L, 37dw	II3: 97L			
	III1: 78L, 67pw, 57dw	III2: 71L, 60dw	III3: 42L, 43dw	III4: 84L		
	IV1: 48L, 54pw, 43dw	IV2: 41L, 37dw	IV3: 36L, 38dw	IV4: 29L, 31dw	IV5: 73L	
Ornithischians						
Basal ornithopods						
Heterodontosaurus tucki MCZ 4188	I1: 17L	I2: 15L				Right foot; I measured cast (RTMP 84.172.1)
	II1: 19L	II2: 16L	II3: 18L			
	III1: 21L	III2: 15L	III3: 13L			
	IV1: 17L	IV2: 12L	IV3: 11L	IV4: 9L	IV5: 13L	
Othnielosaurus consors BYU 631/163	MT III L = 79					Measured cast of left foot
	I1: 21L	I2: 15L				
	II1: 25L	II2: 16L	II3: 20L			
	III1: 26L, 15pw	III2: 17L	III3: 14L	III4: 22L		
	IV1: 20L	IV2: 12L	IV3: 10L	IV4: 8L	IV5: 17L	
Othnielosaurus consors Sauriermuseum Aathal specimen	I1: 43L	I2: 27L				Left foot; measured from underside
	II1: 45L, 18dw	II2: 33L, 17dw	II3: 37L			
	III1: 47L	III2: 30L, 20dw	III3: 25L, 20dw	III4: 39L		
	IV1: 31L, 20pw, 17dw	IV2: 17L	IV3: 21L, 15dw			
Othnielosaurus consors Witherell Quarry, Moffat County, Colorado			II3: 30L			Cast from Gaston Design
	III1: 36L, 22pw, 19dw	III2: 28L		III4: 32L		
					IV5: 25L	
Proctor Lake ornithopod SMU 73170/73171	MT III L = 49					Measurements based on both feet
	I1: 13L					
	II1: 16L, 7pw,7dw	II2: 9L, 7dw	II3: 17 L			
	III1: 13L, 9pw, 8dw	III2: 9L, 7dw	III3: 7L, 7dw	III4: 17L		
	IV1: 11L, 7pw, 6dw	IV2: 7L, 6dw	IV3: 6L, 5dw	IV4: 7L, 5dw	IV5: 12L	
Proctor Lake ornithopod SMU 73181	II1: 45L	II2: 24L				Left foot
	III1: 36L	III2: 23L	III3: 19L	III4: 42L		
	IV1: 29L		IV3: 15L	IV4: 13L	IV5: 33L	

Table A1.1. *continued*

Omit taxon and specimen	Measurements					Comments
"Laosaurus minimus" NMC P 2301/9483 Gilmore 1924	I1: 16					Measured a cast (RTMP 83.272.1)
	II1: 20L, 9pw, 7dw	II2: 12L, 7dw				
	III1: 18L, 11pw, 9dw	III2: 12L, 8dw	III3: 11L, 8dw			
	IV1: 14L, 6pw, 8dw	IV2: 9L, 7dw	IV3: 7L, 6dw	IV4: 6L, 7dw	IV5: 16L	
Orodromeus makelai MOR 331 juvenile	I1: 9L					Right foot; glued upside down on cardboard, so difficult to measure; phalanx lengths measured as articulated lengths in ventral view. II3 possibly missing 1–2 mm from tip; IV5 possibly missing much of tip
	II1: 11L, 4pw, 3dw	II2: 8L, 4dw	II3: 10+ L			
	III1: 10L, 5pw, 6dw	III2: 8L, 5dw	III3: 8L, 4dw	III4: 10L		
	IV1: 6L, 5pw, 4dw	IV2: 5L, 4dw	IV3: 5L, 3dw		IV5: 4+ L	
Willow Creek Anticline hypsilophodontid YPM VPPU 023246	I1: 17L	I2: 12L				Phalanges glued onto a cardboard lid; left foot as mounted. Unguals I2 and III4 are missing their tips, and so lengths are somewhat estimates
	II1: 19L, 9pw, 8dw	II2: 12L, 8dw				
	III1: 18L, 12pw, 10dw	III2: 13L, 9dw	III3: 8L, 8dw	III4: 17L		
	IV1: 13L, 10pw, 8dw	IV2: 9L, 8dw		IV4: 7L, 6dw		
Haya griva IGM 100/2013	MT III L = 83					Measured FMNH PR 3053, cast of the articulated right foot. Ungual II3 length may be a bit low; the tip looks bent a bit dorsally
	I1: 19L	I2: 16L				
	II1: 26L, 12dw	II2: 16L, 11dw	II3: 23L			
	III1: 21L, 18pw, 13dw	III2: 16L, 12dw	III3: 14L, 11dw	III4: 25L		
	IV1: 17L, 14pw, 11dw	IV2: 11L, 12dw	IV3: 10L, 10dw	IV4: 9L, 10dw		
Parksosaurus warreni ROM 804	MT III L = 141					Measurements based on both feet, which are rather banged up: crushed, shattered, heavily restored. A few mm missing from tip of III4
	I1: 49L	I2: 43L				
	II1: 47L, 24dw	II2: 29L, 23dw	II3: 52L			
	III1: 46L, 27dw	III2: 35L, 23dw	III3: 31L	III4: 48L		
	IV1: 35L, 23dw	IV2: 28L	IV3: 22L	IV4: 26L, 22dw		
Thescelosaurus neglectus MOR 797	I1: 56L	I2: 47L				Specimen partly prepared in situ in block; measurements based on both feet
	II1: 67L	II2: 40L	II3: 63L			
	III1: 55L	III2: 39L	III3: 33L	III4: 51L		
	IV1: 47L	IV2: 31L	IV3: 23L	IV4: 19L	IV5: 50L	
Thescelosaurus neglectus MOR 2932	MT III L = 140					Right foot
	I1: 48L	I2: 37L				
		II2: 31L, 29dw	II3: 47L			
	II1: 43L, 43dw	II2: 32L, 38L	III3: 29L	III4: 53L		
	IV1: 37L, 31pw, 32dw	IV2: 23L, 28dw	IV3: 18L, 26dw	IV4: 16L, 23dw	IV5: 38L	
Thescelosaurus neglectus USNM V 7577	I1: 46L					Right foot; measurements made on cast (AMNH 5745)
	II1: 52L, 24dw	II2: 30L, 23dw	II3: 43L			
		III2: 32L, 32dw	III3: 28L, 26dw	III4: 51L		
	IV1: 35L	IV2: 22L	IV3: 17L, 22dw	IV4: 15L, 20dw	IV5: 37L	
Thescelosaurus neglectus LACM 33542	I1: 63L	I2: 42L				Left foot
	II1: 66L, 41pw, 38dw	II2: 43L, 34dw	II3: 59L			
	III1: 59L, 54pw, 52dw	III2: 43L, 43dw	III3: 38L, 35dw	III4: 70L		
	IV1: 47L, 40pw, 38dw	IV2: 32L, 34dw	IV3: 22L, 31L	IV4: 22L, 27dw	IV5: 52L	

Omit taxon and specimen	Measurements					Comments
Tenontosaurus tilletti YPM VP 005460	II1: 73L, 59pw, 53dw	II2: 37L, 49dw	II3: 104L			Right foot. What I interpreted as II1 and III1 were reversed in the identification on the specimen. Unguals III4 and IV5 are both missing about 10 mm from tip
	III1: 56L, 65pw, 54dw	III2: 38L, 53dw	III3: 36L, 48dw	III4: 106+ L		
	IV1: 53L, 55pw, 54dw	IV2: 33L, 51dw	IV3: 27L, 46dw	IV4: 27L, 41dw	IV5: 85+ L	
Tenontosaurus tilletti YPM VP 005459	I1: 55L	I2: 56+ L				Right foot. Specimen (especially II2) has suffered some crushing or breakage. I2 may be missing 5 mm from tip. What I identify as IV4 could be IV3
	II1: 45L, 42pw, 40dw	II2: 27L, 35dw	II3: 72L			
	III1: 41L, 39dw	III2: 27L, 37dw	III3: 28L, 35dw	III4: 76L		
	IV1: 37L, 38pw, 37dw	IV2: 23L, 33dw		IV4: 21L, 27dw	IV5: 59 L	
Tenontosaurus tilletti AMNH 3014		II2: 23L	II3: 51L			Left foot; restoration work on this specimen is so thoroughly blended with the real bone that recognizing the latter is difficult
	III1: 32dw	III2: 23L	III3: 23L, 29dw			
		IV2: 21L	IV3: 20L			
Tenontosaurus tilletti AMNH 3022	I1: 30L					Right foot
	II1: 26L, 21dw	II2: 14L, 18dw	II3: 32L			
	III1: 24L, 25pw, 20dw	III2: 17L, 18dw	III3: 14L, 17dw	IV4: 34L		
Tenontosaurus tilletti VPPU 016338	MT III L = 136					Cast of right foot
	I1: 48L	I2: 41L				
	II1: 42L, 35pw, 31dw	II2: 25L, 30dw	II3: 61L			
	III1: 36L, 39pw, 32dw	III2: 27L, 29dw	III3: 20L, 29dw	III4: 55L		
	IV1: 33L, 32pw, 31dw	IV2: 24L, 28dw	IV3: 17L, 28dw	IV4: 17L, 25dw	IV5: 50L	
Tenontosaurus tilletti MOR 678	MT III L = 94					Right foot; I2 possibly missing 5 mm or more; II3 possibly missing 3–4 mm; III4 probably missing 5 mm; IV5 possibly missing 3–4 mm
	I1: 31L	I2: 25+ L				
	II1: 28L, 21pw, 20dw	II2: 18L, 20dw	II3: 40+ L			
	III1: 25L, 25pw, 20dw	III2: 17L, 19dw	III3: 16L, 18dw	III4: 37+ L		
	IV1: 22L, 19pw, 18dw	IV2: 14L, 17dw	IV3: 12L, 16dw	IV4: 12L, 15dw	IV5: 29+ L	
Tenontosaurus tilletti MOR 787	MT III L = 184+(?); possibly a bit missing at proximal end					Left foot; metatarsals and phalanges look dorsally crushed in some cases
	I1: 88L					
	II1: 56L		II3: 67L			
	II1: 44L		III3: 27L, 34dw			
	IV1: 33L	IV2: 21L	IV3: 19L	IV4: 19L	IV5: 58L	
Tenontosaurus tilletti OMNH 34783	II1: 50L, 38dw	II2: 29L, 36dw	II3: 68L			Left foot; phalanges of digit IV in articulation show a *Pachyornis*-like medial curvature
	III1: 40L	III2: 29L				
	IV1: 35L, 41pw, 38dw	IV2: 25L, 35dw	IV3: 20L, 33dw	IV4: 21L, 30dw		
Tenontosaurus tilletti OMNH 34784	I1: 50L	I2: 59L				Right foot
	II1: 43L, 35pw, 35dw	II2: 26L, 34dw				
	III1: 35L, 39pw, 35dw	III2: 25L, 34dw	III3: 24L, 32dw			
	IV1: 34L, 34dw	IV2: 21L, 30dw		IV4: 18L, 25dw		

Table A1.1. *continued*

Omit taxon and specimen	Measurements					Comments
Tenontosaurus tilletti OMNH 34785	MT III L = 81					Right foot; possibly a few mm missing from tips of II3 and IV5. In articulation, digit IV shows slight medial curvature along its length
	I1: 27L					
	II1: 24L, 18pw, 16dw	II2: 15L, 14dw	II3: 27			
	III1: 21L, 20pw	III2: 14L, 15dw	III3: 14L, 14dw	III4: 32L		
	IV1: 18L, 15pw, 14dw	IV2: 12L, 13dw	IV3: 9L, 13dw	IV4: 9L, 12dw	IV5: 23L	
Tenontosaurus tilletti OMNH 50270	MT III L = 217					Right foot; much of foot adhered tightly to its cradle, and couldn't be removed for easy measuring. Small bit of II3 tip missing
	I1: 79L	I2: 67L				
	II1: 66L, 47dw	II2: 34L, 48dw	II3: 83+ L			
	III1: 49L, 46dw	III2: 36L, 46dw	III3: 32L			
	IV1: 46L					
Tenontosaurus cf. *tilletti* OMNH 58340	MT III L = 178					Right foot; small portion of tip of III4 missing
	I1: 60L	I2: 67L				
	II1: 52L, 46+ pw, 38dw	II2: 33L, 39dw	II3: 83L			
	III1: 44L, 50pw, 43dw	III2: 31L, 41dw	III3: 30L, 38dw	III4: 82+ L		
	IV1: 38L, 37dw	IV2: 25L, 36dw	IV3: 23L, 32dw	IV4: 23L, 31dw	IV5: 67L	
Tenontosaurus sp. OMNH 02926	MT III L = 165					Measurements based on both feet
	I1: 57L	I2: 60L				
	II1: 50L, 43pw, 40dw	II2: 30L, 34dw	II3: 70L			
	III1: 40L, 41dw	III2: 27L, 38dw	III3: 25L, 35dw	III4: 72L		
	IV1: 32L, 39pw, 37dw	IV2: 23L, 33dw	IV3: 20L, 31dw	IV4: 22L, 29dw	IV5: 60L	
Tenontosaurustilletti OU 10132	I1: 73L	I2: 73L				Left foot
	II1: 56L	II2: 33L	II3: 86L			
	III1: 49L	III2: 33L	III3: 32L	III4: 87L		
	IV1: 41L	IV2: 26L	IV3: 23L	IV4: 23L	IV5: 72L	
Tenontosaurus sp. OMNH 04165	MT III L = 148					Left foot; several bones look dorsoventrally squashed. Tips of IV5 and especially III4 missing material, and extensively reconstructed
	I1: 53					
	II1: 45L	II2: 30				
	III1: 41L	III2: 27L	III3: 29L	III4: 50++ L		
	IV1: 34L		IV3: 20L	IV4: 19L	IV5: 57+ L	
Tenontosaurus dossi FWMSH 89C-2	I1: 81L	I2: 71 + L				Right foot; phalanges preserved in articulation. Material from tips of I2, III4, and especially II3 missing
	II1: 62L	II2: 40L	II3: 76++ L			
	III1: 53L	III2: 38L	III3: 29L	III4: 98+ L		
	IV1: 59L	IV2: 36L	IV3: 34L	IV4: 26L	IV5: 100L	
Dryosaurus altus CM 21786	II1: 61L, 27pw, 23dw					Measurements based on both feet. Actual L of III4 = 41; 1–2 mm likely missing from tip
	III1: 49L, 35pw, 28dw	III2: 31L, 25dw		III4: 43L		
	IV1: 36L, 21pw, 21dw	IV2: 23L, 22dw	IV3: 21L, 20dw	IV4: 14L, 17dw		
Dryosaurus sp. DNM (DMNH) specimen	I1: 20L	I2: 12L				II3 possibly would be 28 mm L if all of tip present. Tip of III4 also missing a bit.
	II1: 34L, 13pw, 13dw	II2: 16L, 11dw	II3: 26+ L			
	III1: 29L, 18pw, 15dw	III2: 19L, 14dw	III3: 15L, 13dw	III4: 31+ L		
				IV4: 8L	IV5: 21L	

Omit taxon and specimen	Measurements					Comments
Dryosaurus altus ACM 12370	MTIIIL = 150					Measured right hind limb
	I1: 50L	I2: 29L				
	II1: 60L, 24dw	II2: 24L, 22pw, 21dw	II3: 37L			
	III1: 44L, 35pw, 28dw	III2: 35L, 27pw, 25dw	III3: 24L, 23pw, 20dw	III4: 41L		
	IV1: 34L, 20dw	IV2: 21L, 22pw, 19dw	IV3: 16L, 18pw, 17dw	IV4: 14L, 15pw, 16dw	IV5: 32L	
Camptosaurus dispar YPM VP 001877	I1: 51L	I2: 49L				Cast (USNM 6014) of left foot
	II1: 84L, 62dw	II2: 39L, 52dw	II3: 69L			
	III1: 69L, 84pw, 69dw	III2: 33L, 60dw	III3: 29L, 51dw			
	IV1: 59L, 49dw	IV2: 25L, 48dw	IV3: 20L, 42dw	IV4: 19L, 38dw		
Camptosaurus dispar USNMV 4277	MT III L = 220					Right foot
	I1: 57L					
	II1: 88L, 73pw, 64dw	II2: 41L, 62dw	II3: 76L			
	III1: 76L, 80pw, 59dw	III2: 32L, 57dw	III3: 28L, 53dw			
	IV1: 57L, 55pw, 52dw	IV2: 24L, 45dw	IV3: 18L, 41dw	IV4: 19L, 35dw		
Camptosaurus sp. Western Paleo Labs "Campto B"	MT III L = 187					Measurements based on both feet
	II1: 72L, 54pw, 52dw	II2: 26L, 51dw	II3: 76L			
	III1: 59L, 69pw, 58dw	III2: 32L, 46dw	III3: 19L, 44dw	III4: 73L		
	IV1: 48L	IV2: 21L	IV3: 15L, 37dw	IV4: 20L, 39dw	IV5: 60L	
Camptosaurus sp. Western Paleo Labs "Campto C"	MT III L = 163					Right foot
	II1: 64L		II3: 60L			
	III1: 46dw	III2: 42dw	III3: 26L			
	IV1: 39L, 39pw, 36dw	IV2: 15L, 32dw		IV4: 12L, 27dw	IV5: 49L	
Camptosaurus sp. Bob Simon's "Wally"	MT III L = 291					Left foot
	I1: 49L	I2: 33L				
	II1: 75L, 58pw, 48dw	II2: 39L, 41dw	II3: 70L			
	III1: 72L, 63pw, 47dw	III2: 41L, 41dw	III3: 30L, 40dw	III4: 70L		
	IV1: 54L, 45pw, 41dw	IV2: 26L, 42dw	IV3: 21L, 33dw	IV4: 17L, 28dw	IV5: 55L	
Camptosaurus sp. CEUM 52457	MT III L = 145					Left foot; possibly a few mm missing from tip of I2
	I1: 37L	I2: 29+ L				
	II1: 56L, 44pw, 39dw	II2: 25L, 34dw	II3: 59L			
	III1: 46L, 53pw, 44dw	III2: 22L, 39dw	III3: 17L, 38dw	III4: 62L		
	IV1: 35L, 44pw, 33dw	IV2: 15L, 32dw	IV3: 13L, 29dw	IV4: 12L	IV5: 48L	
Iguanodon bernissartensis IRSNB R56 (formerly 22 L/1561/EFR 56)	FL = 990; MT III L = 345					Left foot
	Digit I absent					
	III1: 143L					
	IV1: 105L, 122pw, 98dw	IV2: 33L, 93dw	IV3: 25L, 80dw	IV4: 31L, 71dw	IV5: 103+ L	

Table A1.1. *continued*

Omit taxon and specimen	Measurements					Comments
Iguanodon bernissartensis IRSNB Vert-5144-1657 (formerly 23 D2/1657)	MT III L = 345					Measurements based on both feet
	Digit I absent					
	II1: 173L, 140pw, 133dw	II1: 37L, 107dw				
	III1: 125L, 168pw, 139dw	III2: 40L, 141dw	III3: 40L, 117dw			
	IV1: 107L, 145pw, 124dw	IV2: 52L, 99dw	IV3: 25L	IV4: 24L, 81dw		
Iguanodon bernissartensis IRSNB Vert-5144-1714 (formerly 25 G/1714)	FL = 1,010; MT III L = 373					Left foot
	Digit I absent					
	II1: 148L, 116dw	II2: 52L, 100dw	II3: 125L			
	III1: 166pw, 128dw	III2: 45L, 130dw	III3: 44L, 114dw			
	IV1: 120pw, 97dw	IV2: 37L, 111dw	IV3: 28L, 101dw	IV4: 33L, 77dw	IV5: 115L	
Iguanodon bernissartensis IRSNB R 51 (formerly 26 Q/1534/HT R51)	FL = 970; MT III L = 345					Measurements based on both feet; some of tip of IV5 missing
	Digit I absent					
	II1: 139L, 122pw, 115dw	II2: 49L, 97dw	II3: 130L			
	III1: 125L, 153pw, 125dw	III2: 40L, 121dw	III3: 33L, 103dw	III4: 149L		
	IV1: 107L, 124pw, 98dw	IV2: 37L, 105dw	IV3: 25L, 87dw	IV4: 29L, 73dw	IV5: 103+ L	
Iguanodon bernissartensis IRSNB R 52 (formerly 1536)	FL = 880; MT III L = 344					Measurements based on both feet
	Digit I absent					
	II1: 142L, 121pw, 126dw	II2: 51L, 94dw	II3: 111L			
	III1: 123L, 145pw, 134dw	III2: 45L, 124dw	III3: 36L, 102dw	III4: 132L		
	IV1: 109L, 121pw, 108dw	IV2: 36L, 108dw	IV3: 27L, 92dw	IV4: 29L, 75dw	IV5: 110L	
Iguanodon bernissartensis IRSNB Vert-5144-1715 (formerly 30 G2/1715)	FL = 925; MT III L = 321					Left foot
	Digit I absent					
	II1: 153L, 93pw, 96dw	II2:46L, 94dw				
		III2: 39L, 138dw	III3: 35L, 112dw			
		IV2: 31L, 94dw	IV3: 24L, 83dw	IV4: 27L, 68dw		
Iguanodon bernissartensis IRSNB Vert-5144-1562 (formerly 31 E2/1562)	FL = 950 (Left); MT III L = 333					Measurements based on both feet
	Digit I absent					
	II1: 149L, 119pw, 132dw	II2: 53L, 98dw	II3: 127L			
	III1: 138L, 148pw, 125dw	III2: 41L, 132dw	III3: 33L, 111dw	III4: 136L		
	IV1: 117L, 122pw, 111dw	IV2: 37L, 103dw	IV3: 27L, 89dw	IV4: 27L, 74dw	IV5: 109L	
Iguanodon bernissartensis IRSNB Vert-5144-1639 (formerly 28 R/1639)	FL = 960 (Left); MT III L = 347					Measurements based on both feet
	Digit I absent					
	II1: 125L, 120pw, 123dw	II2: 49L, 103dw	II3: 127L			
	III1: 124L, 121pw, 127dw	III2: 39L, 133dw	III3: 35L, 110dw	III4: 141L		
	IV1: 104L, 128pw, 107dw	IV2: 36L, 100dw	IV3: 27L, 84dw	IV4: 31L, 65dw	IV5: 111L	

Omit taxon and specimen	Measurements					Comments
Iguanodon bernissartensis IRSNB Vert-5144-1723 (formerly 10 V/1723)	MT III L = 325					Measurements based on both feet
	Digit I absent					
	II1: 135L, 114pw, 116dw	II2: 49L, 93dw	II3: 123L			
	III1: 131L, 144pw, 128dw	III2: 42L, 123dw	III3: 37L, 112dw			
	IV1: 100L, 125pw, 103dw	IV2: 33L, 99dw	IV3: 29L, 81dw	IV4: 27L, 68dw	IV5: 109L	
Iguanodon bernissartensis IRSNB Vert-5144-1710 (formerly 6/1710)	MT III L = 360					Right foot
	Digit I absent					
	III1: 115L					
	IV1: 128L, 103dw	IV2: 32L, 113dw	IV3: 27L	IV4: 31L	IV5: 124L	
Iguanodon bernissartensis IRSNB Vert-5144-1724 (formerly 16 Y/1724)	MT III L = 335					Right foot
	Digit I absent					
	II1: 135L, 124pw, 105dw	II2: 51L, 87dw	II3: 115L			
	III1: 125L, 142dw, 121dw	III2: 41L, 117dw	III3: 35L, 94DW	III4: 131L		
	IV1: 98L, 107pw, 98dw	IV2: 37L, 97dw	IV3: 27L, 85dw	IV4: 28L, 70dw	IV5: 107L	
Iguanodon bernissartensis IRSNB Vert-5144-1725 (formerly 20 Q/1725)	FL = 1,045; MT III L = 362					Right foot
	Digit I absent					
	II1: 129L, 118dw	II2: 41L, 96dw	II3: 134L			
	III1: 125L, 152pw	III2: 41L, 133dw	III3: 31L, 114dw	III4: 101L		
	IV1: 78L, 101dw	IV2: 34L, 99dw	IV3: 23L, 89dw	IV4: 22L	IV5: 119L	
Iguanodon bernissartensis IRSNB Vert-5144-1726 (formerly 19 U/1726)	Digit I absent					Right foot
	II1dw: 107L, 81dw	II2: 41L,74dw	II3: 105L			
	III1: 99L	III2: 34L	III3: 29L, 77dw	III4: 115L		
	IV1: 72L	IV2: 39L	IV3: 19L	IV4: 21L		
Iguanodon bernissartensis IRSNB R 53 (formerly 3 B/1727/EFR 53/slab skeleton 13)	MT III L = 340					Measurements based on both feet
	Digit I absent					
	II1: 136L, 121pw, 117dw	II2: 49L, 96dw				
	III1: 117L, 157pw, 128dw	III2: 43L, 128dw	III3: 39L, 105dw	III4: 136L		
	IV1: 136L	IV2: 37L, 103L	IV3: 27L, 93dw	IV4: 30L, 74dw	IV5: 113L	
Iguanodon bernissartensis IRSNB Vert-5144-1730 (formerly 2 F/1730/slab skeleton 3)	MI III L = 275					Measurements based on both feet
	Digit I absent					
	II1: 101L, 90dw	II2: 37L, 74dw	II3: 113L			
	III1: 88L	III2: 23L, 93dw	III3: 30L, 94dw	III4: 125L		
	IV1: 83L	IV2: 31L	IV3: 19L	IV4: 21L	IV5: 112L	
Iguanodon bernissartensis IRSNB R 54 (formerly F/1731/ EFR 54/slab skeleton 4)	FL = 990 (left); MT III L = 337					Left foot
	Digit I absent					
	II1: 133L, 123pw, 116dw	II2: 49L, 96dw	II3: 127L			
	III1: 115L, 140pw, 120dw	III2: 45L, 116dw	III3: 37L, 92dw	III4: 123L		
	IV1: 99L, 108pw, 96dw	IV2: 39L, 96dw	IV3: 27L, 84dw	IV4: 29L, 68dw	IV5: 102L	

Table A1.1. *continued*

Omit taxon and specimen	Measurements					Comments
Iguanodon bernissartensis IRSNB Vert-5144-1712 (formerly 1712/slab skeleton 17)	Digit I absent					Measurements based on both feet
	II1: 135L, 124pw, 120dw	II2: 55L, 99dw	II3: 133L			
	III1: 130L, 135dw	III2: 40L, 129dw	III3: 35L, 116dw	III4: 115L		
	IV1: 114L, 125pw, 108dw	IV2: 30L, 106dw	IV3: 27L, 90dw	IV4: 30L, 73dw	IV5: 110L	
Iguanodon bernissartensis IRSNB Vert-5144-1729 (formerly 2 M/1729/slab skeleton 5)	FL = 830					Left foot
	Digit I absent					
	II1: 116L, 111pw, 107dw	II2: 51L, 82dw	II3: 107L			
	III1: 111L, 139pw, 110dw	III2: 39L, 111dw	III3: 35L, 94dw	III4: 123L		
	IV1: 91L, 110pw, 95dw	IV2: 33L, 89dw	IV3: 25L, 75dw	IV4: 28L, 60dw		
Iguanodon bernissartensis IRSNB Vert-5144-1716 (formerly A/1716/skeleton 24)	MT III L = 337					Left foot
	Digit I absent					
	II1: 119pw, 104dw	II2: 52L, 91dw	II3: 117L			
	III1: 119L, 121dw	III2: 45L	III3: 40L, 107dw	IV4: 117L		
	IV1: 106dw	IV2: 36L, 98dw	IV3: 29L, 90dw	IV4: 33L, 74dw	IV5: 103L	
Iguanodon bernissartensis NHMUK R.2506	MT III L = 376					Measurements based mainly on left foot; unguals II3 and IV5 are missing up to several mm of material from tip
	Digit I absent					
	II1: 147L, 124pw, 121dw	II2: 47L, 93dw	II3: 132L			
	III1: 125L, 149pw, 129dw	III2: 45L, 121dw	III3: 36L, 101dw	III4: 151L		
	IV1: 103L, 131pw, 107dw	IV2: 37L, 100dw	IV3: 33L	IV4: 27L, 89dw	IV5: 118	
Iguanodon sp. NHMUK R.1863	II1: 98L, 75pw, 69dw	II2: 29L, 61dw	II3: 97L			There are several sets of phalanges with this catalog number; those associated in a digit belong together, but the digits collectively do not belong to the same animal
Iguanodon sp. NHMUK R.1863	III1: 111L, 126pw, 107dw		III3: 34L, 103dw	III4: 111L		
Iguanodon sp. NHMUK R.1863	IV1: 135L, 102pw, 96dw	IV2: 36L, 93dw				
Mantellisaurus atherfieldensis IRSNB R 57 (formerly 29 T/1551/Pt R57)	FL = 733 (right); MT III L = 269					Left foot
	Digit I absent					
	II1: 113L, 74pw, 63dw	II2: 39L, 52dw	II3: 83L			
	III1: 87L, 87pw, 80dw	III2: 30L, 78dw	III3: 27L, 61dw	III4: 89L		
	IV1: 81L, 65pw, 65dw	IV2: 21L, 55dw	IV3: 16L, 59dw	IV4: 19L, 40dw	IV5: 69L	
Mantellisaurus atherfieldensis NHMUK R.1829	MT III L = 286					Left foot
	Digit I absent					
	II1: 111L, 69pw, 67dw	II2: 37L, 56dw	II3: 83L			
	III1: 107L, 94pw, 79dw	III2: 33L, 74dw	III3: 26L, 61dw			
	IV1: 73L, 72pw, 60dw	IV2: 22L, 56dw	IV3: 16L	IV4: 18L, 43dw	IV5: 78L	

Omit taxon and specimen	Measurements					Comments
Hadrosauridae						
?*Saurolophus* AMNH 5271	Digit I absent					Right foot
	II1: 103L, 74pw, 73dw	II2: 44L, 66dw	II3: 63L			
	III1: 94dw	III2: 37L, 99dw	III3: 24L, 83dw			
		IV2: 21L, 72dw	IV3: 18L, 64dw	IV4: 19L, 59dw		
Juvenile lambeosaurine MOR 471	MT III L = 227					Right foot
	Digit I absent					
	II1: 59L, 29pw, 37dw	II2: 21L, 34dw	II3: 41L			
	III1: 65L	III2: 17L, 54dw	III3: 13L, 44dw	III4: 39L		
	IV1: 50L, 42pw, 38dw	III2: 12, 39dw	III3: 9L, 36dw	III4: 9L, 29dw	IV5: 40L	
Corythosaurus casuarius USNM V 15493	Digit I absent					Left foot; bones fixed in panel mount, so hard to measure
	II1: 143L	II2: 52L	II3: 99L			
	III1: 126L	III2: 41L	III3: 28L	III4: 88L		
	IV1: 107L	IV2: 40L	IV3: 33L	IV4: 28L	IV5: 97L	
Corythosaurus casuarius RTMP 80.40.1	MT III L = 404					Slab mount; measurements based on both feet. Digit IV looks heavily restored
	Digit I absent					
	II1: 142L	II2: 52L				
	III1: 122L, 116dw	III2: 43L	III3: 31L			
Corythosaurus casuarius LACM 3743/126137	MT III L = 288					Measurements based on both feet; specimen a panel mount, making measurement difficult
	Digit I absent					
	II1: 92L	II2: 34L	II3: 63L			
	III1: 93L	III2: 29L	III3: 20L	III4: 63L		
	IV1: 79L	IV2: 21L	IV3: 14L	IV4: 15L	IV5: 63L	
Corythosaurus casuarius ROM 845	MT III L = 365					Right foot
	Digit I absent					
	II1: 132L, 93dw	II2: 42L, 80dw	II3: 89L			
	III1: 119L, 105dw	III2: 39L, 100dw	III3: 27L, 88dw	III4: 85L		
	IV1: 112L, 92dw	IV2: 32L, 82dw	IV3: 19L, 71dw	IV4: 17L, 59dw		
Lambeosaurus lambei ROM 1218	MT III L = 370					Measurements based on both feet
	Digit I absent					
	II1: 123L, 83dw	II2: 46L, 73dw				
	III1: 121L, 110dw	III2: 37L, 108dw	III3: 29L, 92dw	III4: 89L		
	IV1: 96L, 93dw	IV2: 31L, 85dw	IV3: 19L, 76dw	IV4: 19L, 65dw		
Lambeosaurus lambei FMNH PR 380	Digit I absent					Measurements based on both feet
	II1: 119L, 92dw	II2: 45L, 72dw	II3: 84L			
	III1: 131L	III2: 34L, 106dw	III3: 22L, 88dw	III4: 81L		
	IV1: 109L	IV2: 23L	IV3: 15L	IV4: 17L	IV5: 75L	
Lambeosaurus lambei RTMP 82.38.1	Digit I absent					Left foot
	II1: 137L, 81dw	II2: 47L, 69dw	II3: 92L			
	III1: 133L, 138pw, 109dw	III2: 45L, 104dw	III3: 27L, 91dw	III4: 86L		
	IV1: 107L, 85dw	IV2: 26L, 85dw	IV3: 17L	IV4: 21L, 59dw	IV5: 92L	

Table A1.1. *continued*

Omit taxon and specimen	Measurements					Comments
Hypacrosaurus altispinus AMNH 5357	Digit I absent					Measurements based on both feet
	II1: 95L, 85pw	II2: 43L, 53dw	II3: 67L			
	III1: 105L, 80dw	III2: 35L, 87dw	III3: 26L, 74dw	III4: 66L		
	IV1: 85L	IV2: 28L, 64dw	IV3: 20L, 56dw	IV4: 19L, 49dw	IV5: 65L	
Hypacrosaurus altispinus CMN FV 8501	MT III L = 380					Right foot; bones hard to measure in the mounted skeleton
	Digit I absent					
	II1: 95L, 88pw, 89dw	II2: 40L, 78dw	II3: 91L			
	III1: 107L, 124pw, 95dw	III2: 36L, 94dw	III3: 31L, 82dw	III4: 80L		
	IV1: 90L, 110pw, 93dw	IV2: 23L, 89dw	IV3: 18L, 72dw	IV4: 15L, 65dw	IV5: 84L	
Hypacrosaurus altispinus RTMP 82.10.1	Digit I absent					Left foot
	II1: 115L, 72dw	II2: 48L, 66dw				
		III2: 34L, 81dw				
	IV1: 81L, 60dw	IV2: 16L, 60dw	IV3: 15L, 55dw	IV4: 15L		
Prosaurolophus maximus ROM 787	MT III L = 317					Right foot; possibly a bit of the tip of III4 is buried in plaster in the mount
	Digit I absent					
	II1: 122L, 93dw	II2: 49L, 72dw	II3: 99L			
	III1: 120L, 90dw	III2: 36L, 84dw	III3: 25L, 78L	III4: 80L		
	IV1: 90L, 95dw	IV2: 21L, 68dw	IV3: 18L, 78dw	IV4: 25L, 61dw	IV5: 88L	
Brachylophosaurus canadensis MOR 794	MT III L = 450					Right foot
	Digit I absent					
	II1: 144L, 111pw, 93dw	II2: 59L, 73dw	II3: 77L			
	III1: 137L, 145pw, 125dw	III2: 48L, 120dw	III3: 36L, 84dw	III4: 81L		
	IV1: 116L, 126pw, 100dw	IV2: 35L, 89dw	IV3: 26L, 78dw	IV4: 25L, 64dw	IV5: 77L	
Brachylophosaurus canadensis MOR 1071	MT III L = 298					Right foot
	Digit I absent					
	II1: 110L, 65dw	II2: 42L, 59dw				
	III1: 106L, 107pw	III2: 38L, 70dw	III3: 26L, 69dw			
	IV1: 80L, 67dw					
Gryposaurus incurvimanus ROM 764	MT III L = 355					Measurements based on both feet
	Digit I absent					
	II1: 122L, 79dw	II2: 44L, 73dw				
	III1: 115L, 113dw	III2: 34L, 110dw	III3: 25L, 94dw	III4: 87L		
	IV1: 96L, 79dw	IV2: 23L, 73dw	IV3: 18L	IV4: 17L, 61dw	IV5: 83L	
Edmontosaurus regalis Drumheller Dinosaur Museum	Digit I absent					Left foot. Skeleton is in a panel mount, making bones hard to measure. Specimen is composite of two individuals, but foot comes from one of them
	II1: 114L, 87dw	II2: 55L, 88dw	II3: 89L			
	III1: 105L, 95pw, 100dw	III2: 45L, 97dw	III3: 36L, 80dw	III4: 88L		
	IV1: 101L, 113dw	IV2: 36L, 83dw	IV3: 23L, 78dw	IV4: 21L, 74dw	IV5: 75L	

Omit taxon and specimen	Measurements					Comments
Edmontosaurus annectens DMNH 1493	MT III L = 325					Measurements based mainly on left foot
	Digit I absent					
	II1: 135L, 96pw, 90dw	II2: 62L, 85dw	II3: 78L			
	III1: 118L, 142pw, 120dw	III2: 54L, 125dw	III3: 37L, 111L	III4: 105L		
	IV1: 97L, 114pw, 96dw	IV2: 35L, 89dw	IV3: 34L, 83dw	IV4: 31L	IV5: 89L	
Edmontosaurus annectens LACM 23502	MT III L = 360					Left foot
	Digit I absent					
	II1: 137L, 117pw, 104dw	II2: 66L, 85dw	II3: 95L			
	III1: 126L, 138pw, 126dw	III2: 40L, 109dw	III3: 42L, 103dw	III4: 101L		
	IV1: 101L, 121pw, 108dw	IV2: 35L, 96dw	IV3: 29L, 91dw	IV4: 31L, 78dw	IV5: 93L	
Edmontosaurus annectens LACM 23504	MT III L = 200					Measurements based on both feet
	Digit I absent					
	II1: 70L, 50pw, 42dw	II2: 33L, 37dw	II3: 42L			
	III1: 66L, 72pw, 61dw	III2: 29L, 61dw	III3: 20L, 55dw	III4: 46L		
	IV1: 51L, 58pw, 46dw	IV2: 16L, 45dw	IV3: 15L, 42dw	IV4: 15L, 34dw	IV5: 39L	
Edmontosaurus annectens USNM 2414	Digit I absent					Left foot
	II1: 127L, 105dw	II2: 53L, 59dw	II3: 99L			
	III1: 119L, 120dw	III2: 51L, 123dw	III3: 37L, 102dw	III4: 115L		
	IV1: 91L, 103dw	IV2: 29L, 86dw	IV3: 21L, 62dw	IV4: 29L, 60dw	IV5: 82L	
Edmontosaurus annectens ROM 801	MT III L = 430					Measurements based on both feet
	Digit I absent					
	II1: 151L, 106dw	II2: 63L, 92dw				
	III1: 147L, 151dw	III2: 57L, 137dw	III3: 38L			
	IV1: 123L	IV2: 39L	IV3: 32L, 89dw	IV4: 27L, 84dw	IV5: 83L	
Edmontosaurus annectens ROM 867	MT III L = 363					Left foot
	Digit I absent					
	II1: 129L, 88dw	II2: 56L, 79dw	II3: 78L			
	III1: 122L, 105dw	III2: 42L, 118dw	III3: 30L, 94dw	III4: 80L		
	IV1: 103L	IV2: 27L, 75dw	IV3: 22L, 72dw	IV4: 25L, 63dw	IV5: 77L	
Edmontosaurus annectens Senckenberg museum mummy	MT III L = 329					Cast of right foot
	Digit I absent					
	II1: 139L, 92dw	II2: 57L, 82dw				
	III1: 120L, 120dw	III2: 49L, 123dw	III3: 31L, 108dw	III4: 103L		
	IV1: 102L, 94dw	IV2: 32L, 88dw	IV3: 26L, 80dw	IV4: 26L, 68dw	IV5: 87L	

Table A1.1. *continued*

Omit taxon and specimen	Measurements					Comments
Edmontosaurus sp. USNM V 5742	Digit I absent					
		II2: 56L, 92dw				
	III1: 115L, 139pw, 127dw	III2: 47L, 131dw	III3: 33L, 112dw	III4: 93L		
	IV1: 95L, 96pw, 107dw	IV2: 26L, 100dw	IV3: 24L, 87dw	IV4: 26L, 75dw	IV5: 94L	
Unidentified hadrosaur AMNH 3971	Digit I absent					
	II1: 130L, 86dw	II2: 54L, 77dw	II3: 98L			
		III2: 45L, 99dw	III3: 35L	III4: 97L		
		IV2: 35L, 79dw	IV3: 24L, 72dw	IV4: 24L, 62dw	IV5: 90L	
Unidentified hadrosaur AMNH 5899	Digit I absent					Some phalanges, especially on digit III, could have been misidentified; there were 4 nonungual phalanges in addition to the ungual in the specimen drawer
	II1: 114L, 78pw, 76dw	II2: 53L, 69dw	II3: 81L			
	III1: 108L, 101dw	III2: 29L, 78dw	III3: 27L, 73dw	III4: 84L		
Unidentified hadrosaur UCMP 125819	MT III L = 331					Right foot
	Digit I absent					
	II1: 114pw					
	III1: 125L, 147pw, 131dw					
	IV1: 100L, 121pw, 115dw	IV2: 35L, 102dw	IV3: 33L	IV4: 31L, 75dw		
Basal ceratopsians						
Psittacosaurus mongoliensis AMNH 6536	MT III L = 22					Right foot of juvenile; length of II1 difficult to measure due to way specimen is embedded in rock and plaster
	II1: 5L	II2: 4L, 3dw				
	III1: 7L	III2: 5L, 3dw	III3: 5L, 3dw	III4: 6L		
	IV1: 4L	IV2: 4L				
Psittacosaurusmeileyingensis Gaston Design cast	I1: 25L	I2: 19L				Left foot
	II1: 17L, 14dw	II2: 16L, 13dw	II3: 28L			
	III1: 18L, 15dw	III2: 15L, 13dw	III3: 13L, 11dw	III4: 18L		
	IV1: 18L, 13dw	IV2: 15L, 12dw	IV3: 13L, 11dw	IV4: 11L, 9dw		
Psittacosaurus major LHPV1	MT III L = 86					Measurements based on both feet, both of which look squashed dorsoventrally, so only lengths taken
	I1: 33L	I2: 32L				
	II1: 27L	II2: 19L	II3: 28L			
	III1: 24L	III2: 19L	III3: 15L	III4: 31L		
	IV1: 25L	IV2: 16L	IV3: 11L	IV4: 11L	IV5: 18L	
Psittacosaurus sp. Wyoming Dinosaur Center	MT III L = 41 (Left foot)					Right foot; measured from underside
	II1: 12L	II2: 10L	II3: 15L			
		III2: 9L	III3: 8L	III4: 12L		
	IV1: 10L	IV2: 10L		IV4: 6L	IV5: 12L	
Cerasinops hodgskissi MOR 300	I1: 60L	I2: 49L				
	II1: 51L, 30dw	II2: 37L, 27dw	II3: 50L			
	III1: 44L	III2: 32L	III3: 22L	III4: 39L		
	IV1: 38L, 25dw	IV2: 30L	IV3: 22L	IV4: 18L	IV5: 29L	
Montanoceratops sp. MOR 542	I1: 44L	I2: 33L				
	II1: 37L, 21dw	II2: 26L, 20dw				
	III1: 33L, 23dw	III2: 22L, 21dw	III3: 21L, 19dw	III4: 27L		
	IV1: 24L, 18dw	IV2: 17L, 17dw	IV3: 16L, 16dw	IV4: 15L, 16dw	IV5: 29L	

Omit taxon and specimen	Measurements					Comments
Leptoceratops gracilis CMN 8889	MT III L = 137					Left foot; proximal ends of III1 and IV1 possibly a bit squashed dorsoventrally
	I1: 49L	I2: 47L				
	II1: 40L, 29pw, 27dw	II2: 27L, 26dw	II3: 51L			
	III1: 35L, 35pw, 28dw	III2: 25L, 25dw	III3: 22L, 25dw	III4: 47L		
	IV1: 28L, 30pw, 23dw	IV2: 20L, 22dw	IV3: 16L, 21dw	IV4: 15L, 20dw	IV5: 40L	

Table A1.2. Reduced major axis relationships between log-transformed variables of animal size vs. metatarsal, phalangeal, and digital lengths in *Alligator mississippiensis* and other extant crocodylian species. To indicate the size range of specimens employed, the range of untransformed values (in millimeters) of the independent variable for all alligators in the data base is reported, but this will not necessarily be the range of values of the independent variable in all bivariate relationships (due to missing data). The "big seven" mean is the average of log-transformed values of the lengths of phalanges II1, II2, III1, III2, III3, IV1, and IV2, and serves as a proxy for overall size of the digital portion of the foot; for consideration of multivariate allometry, the log-transformed value of each of the lengths of the big seven phalanges is compared with the mean of the remaining six big seven phalanx log-transformed lengths. CI = confidence interval; RMA = reduced major axis. Comparisons in which the 95% CI of the RMA slope excludes 1 are indicated in **bold**. For some RMA relationships, in addition to intraspecific relationships of alligators, two interspecific comparisons are made. One is for all crocodylian specimens for which data were available, without regard to the number of specimens for each species. The second is based on one specimen per species, generally the biggest individual for which complete measurements could be made

Independent variable (range in alligators)	Dependent variable	Treatment	r^2	RMA slope	95% CI of slope	*N*
Total length (610–3,404)	Metatarsal III length	Alligators	0.972	**0.861**[a]	0.769–0.965	13
	"Big seven" mean	Alligators	0.929	**0.799**[a]	0.647–0.988	11
	Digit III length excluding ungual	Alligators	0.922	0.809	0.648–1.012	11
Femur length (20–244)	Metatarsal III length	Alligators	0.997	**0.911**[a]	0.874–0.950	11
		Crocodylians	0.990	**0.902**	0.873–0.931	42
		1 per species	0.985	0.917	0.841–1.001	12
	"Big seven" mean	Alligators	0.995	**0.862**[a]	0.781–0.951	6
		Crocodylians	0.980	**0.907**[a]	0.851–0.966	24
		1 per species	0.969	0.922	0.784–1.083	9
	Digit III length excluding ungual	Alligators	0.996	**0.868**	0.795–0.948	6
		Crocodylians	0.979	**0.890**	0.839–0.944	28
		1 per species	0.969	0.909	0.773–1.067	9
Metatarsal III length (10–113)	"Big seven" mean	Alligators	0.994	**0.948**[a]	0.911–0.986	19
		Crocodylians	0.991	0.972	0.943–1.001	43
		1 per species	0.985	0.944	0.860–1.037	11
	Digit III length excluding ungual	Alligators	0.989	**0.902**[a]	0.856–0.950	20
		Crocodylians	0.985	**0.947**[a]	0.913–0.983	48
III1L (3.5–44)	I1L	Alligators	0.996	**0.959**[a]	0.934–0.985	26
		Crocodylians	0.983	**0.954**[a]	0.922–0.987	62
		1 per species	0.990	0.996	0.932–1.065	13
	II1L	Alligators	0.998	**0.960**[a]	0.942–0.979	25
		Crocodylians	0.996	**0.974**[a]	0.958–0.991	61
		1 per species	0.996	1.012	0.973–1.054	14
	II2L	Alligators	0.990	**0.887**	0.847–0.928	23
		Crocodylians	0.972	**0.886**	0.846–0.929	55
		1 per species	0.978	**0.897**[a]	0.811–0.991	13
	III2L	Alligators	0.994	**0.915**	0.882–0.949	22
		Crocodylians	0.981	**0.939**[a]	0.904–0.975	57
		1 per species	0.985	0.972	0.895–1.055	13
	III3L	Alligators	0.988	**0.875**	0.828–0.924	20
		Crocodylians	0.974	**0.876**	0.834–0.919	48
		1 per species	0.974	**0.870**[a]	0.769–0.985	11
	IV1L	Alligators	0.997	**1.027**[a]	1.003–1.053	24
		Crocodylians	0.992	1.017	0.994–1.042	60
		1 per species	0.993	1.024	0.971–1.080	14
	IV2L	Alligators	0.988	**0.927**[a]	0.876–0.981	19
		Crocodylians	0.982	1.030	0.991–1.071	50
		1 per species	0.995	1.040	0.993–1.091	13

Independent variable (range in alligators)	**Dependent variable**	**Treatment**	***r*²**	**RMA slope**	**95% CI of slope**	***N***
Digit III length excluding ungual (11.5–92)	Digit II length excluding ungual	Alligators	0.998	**0.969**[a]	0.948–0.991	20
		Crocodylians	0.995	0.987	0.966–1.008	46
		1 per species	0.996	0.999	0.953–1.048	11
	Aggregate length of phalanges IV1–IV3	Alligators	0.996	1.033	0.998–1.070	17
		Crocodylians	0.990	**1.077**[a]	1.043–1.115	37
		1 per species	0.992	1.083	1.000–1.174	9
Digit II length excluding ungual (9–66)	Aggregate length of phalanges IV1–IV3	Alligators	0.995	**1.054**	1.014–1.096	17
		Crocodylians	0.983	**1.098**	1.051–1.147	39
		1 per species	0.984	1.063	0.957–1.179	10
"Big seven" mean excluding II1L	II1L	Alligators	0.997	1.003	0.975–1.032	19
		Crocodylians	0.994	1.019	0.994–1.044	43
		1 per species	0.994	1.056	0.995–1.119	11
"Big seven" mean excluding II2L	II2L	Alligators	0.996	**0.948**[a]	0.918–0.979	19
		Crocodylians	0.983	**0.909**	0.872–0.947	43
		1 per species	0.992	**0.885**	0.827–0.947	11
"Big seven" mean excluding III1L	III1L	Alligators	0.998	**1.055**[a]	1.031–1.079	19
		Crocodylians	0.993	**1.047**[a]	1.019–1.075	43
		1 per species	0.996	1.048	0.999–1.099	11
"Big seven" mean excluding III2L	III2L	Alligators	0.997	0.999	0.971–1.027	19
		Crocodylians	0.997	1.003	0.986–1.021	43
		1 per species	0.998	0.975	0.942–1.008	11
"Big seven" mean excluding III3L	III3L	Alligators	0.988	**0.934**[a]	0.883–0.989	19
		Crocodylians	0.978	**0.903**	0.861–0.947	43
		1 per species	0.982	**0.891**[a]	0.804–0.987	11
"Big seven" mean excluding IV1L	IV1L	Alligators	0.993	**1.115**	1.068–1.164	19
		Crocodylians	0.991	**1.079**[a]	1.047–1.112	43
		1 per species	0.990	**1.081**[a]	1.001–1.166	11
"Big seven" mean excluding IV2L	IV2L	Alligators	0.988	0.965	0.912–1.021	19
		Crocodylians	0.981	**1.075**[a]	1.029–1.123	43
		1 per species	0.993	**1.092**[a]	1.025–1.163	11

[a] Only barely allometric (upper CI limit of RMA slope between 0.95 and 1.00, or lower CI limit between 1.00 and 1.05).

Table A1.3. Reduced major axis relationships between log-transformed variables of animal size and phalangeal and digital lengths in moa. To indicate the size range of specimens employed, the range of untransformed values (in millimeters) of the independent variable for all specimens in the database is reported for selected variables, but this will not necessarily be the range of values of the independent variable in every one of the bivariate relationships (due to missing data). CI = confidence interval; RMA = reduced major axis. Comparisons in which the 95% CI of the RMA slope exclude 1 are indicated in **bold**. Intraspecific RMA relationships are reported for two species. In addition, three interspecific comparisons are made. One is for the pooled sample of the two species. A second is for all moa specimens in the sample, without regard to the number of specimens for each species. The third is based on one specimen per species, generally the biggest individual for which complete measurements could be made.

Independent variable	Dependent variable	Treatment (range of values [mm] of independent variable)	r^2	RMA slope	95% CI of slope	*N*
Tarsometatarsus length	"Big seven" mean	*Dinornis robustus* (254–487)	0.950	0.887	0.677–1.162	7
		Anomalopteryx didiformis (83–202)	0.873	1.072	0.839–1.369	14
		D. robustus + *A. didiformis* (83–487)	0.967	**0.736**	0.673–0.805	21
		All moa (83–487)	0.840	**0.794**	0.692–0.913	43
		1 per species (123–487)	0.779	0.801	0.477–1.344	9
	Digit III length excluding ungual	*Dinornis robustus*	0.925	1.011	0.755–1.355	8
		Anomalopteryx didiformis	0.986	**1.130**	1.052–1.214	15
		D. robustus + *A. didiformis*	0.928	**0.803**	0.707–0.912	23
		All moa	0.816	**0.831**[a]	0.722–0.956	49
		1 per species	0.751	0.817	0.463–1.443	9
	Digit III length	*Dinornis robustus*	0.935	0.943	0.720–1.235	8
		Anomalopteryx didiformis	0.967	1.123	0.999–1.262	14
		D. robustus + *A. didiformis*	0.924	**0.784**	0.685–0.897	22
		All moa	0.794	**0.850**[a]	0.727–0.994	46
		1 per species	0.722	0.760	0.407–1.421	9
III1L	II1L	*Dinornis robustus* (62–115)	0.980	1.007	0.949–1.068	27
		Anomalopteryx didiformis (22–57)	0.975	**0.901**[a]	0.836–0.971	22
		D. robustus + *A. didiformis* (22–115)	0.993	1.011	0.986–1.036	49
		All moa (22–115)	0.985	0.992	0.967–1.018	96
		1 per species (36–110)	0.982	0.975	0.864–1.101	9
	II2L	*Dinornis robustus*	0.786	1.026	0.816–1.290	25
		Anomalopteryx didiformis	0.860	1.165	0.963–1.409	22
		D. robustus + *A. didiformis*	0.937	**0.892**[a]	0.825–0.964	47
		All moa	0.835	1.071	0.976–1.176	92
		1 per species	0.853	1.055	0.714–1.557	9
	III2L	*Dinornis robustus*	0.887	**1.235**	1.061–1.437	26
		Anomalopteryx didiformis	0.961	1.047	0.947–1.157	20
		D. robustus + *A. didiformis*	0.940	**0.866**	0.802–0.935	46
		All moa	0.923	0.957	0.901–1.017	92
		1 per species	0.950	1.129	0.917–1.391	9
	III3L	*Dinornis robustus*	0.748	1.060	0.821–1.369	25
		Anomalopteryx didiformis	0.846	1.236	0.990–1.543	19
		D. robustus + *A. didiformis*	0.865	**0.819**	0.724–0.927	44
		All moa	0.773	1.043	0.929–1.172	89
		1 per species	0.860	1.074	0.736–1.566	9
	IV1L	*Dinornis robustus*	0.970	1.008	0.937–1.084	27
		Anomalopteryx didiformis	0.971	1.022	0.942–1.108	22
		D. robustus + *A. didiformis*	0.992	**0.964**[a]	0.939–0.990	49
		All moa	0.978	**0.969**[a]	0.939–0.999	96
		1 per species	0.942	0.983	0.785–1.232	9
	IV2L	*Dinornis robustus*	0.529	0.971	0.637–1.479	26
		Anomalopteryx didiformis	0.525	1.413	0.831–2.405	19
		D. robustus + *A. didiformis*	0.900	1.000	0.903–1.109	45
		All moa	0.833	1.090	0.990–1.199	89
		1 per species	0.934	1.081	0.848–1.377	9

Independent variable	Dependent variable	Treatment (range of values [mm] of independent variable)	r^2	RMA slope	95% CI of slope	*N*
Digit III length excluding ungual	Digit II length excluding ungual	*Dinornis robustus* (107–220)	0.959	0.980	0.894–1.074	24
		Anomalopteryx didiformis (43–119)	0.984	**0.905**[a]	0.846–0.969	18
		D. robustus + *A. didiformis* (43–220)	0.988	**1.063**[a]	1.026–1.101	42
		All moa (43–220)	0.982	1.023	0.993–1.054	85
		1 per species (66–210)	0.977	0.944	0.822–1.083	9
	Digit IV length excluding ungual	*Dinornis robustus*	0.896	0.941	0.812–1.091	25
		Anomalopteryx didiformis	0.784	0.985	0.731–1.326	17
		D. robustus + *A. didiformis*	0.970	1.045	0.988–1.105	42
		All moa	0.949	**1.090**[a]	1.035–1.147	82
		1 per species	0.975	1.031	0.893–1.191	9
Digit II length excluding ungual	Digit IV length excluding ungual	*Dinornis robustus*	0.886	1.081	0.925–1.263	25
		Anomalopteryx didiformis	0.722	0.961	0.673–1.372	17
		D. robustus + *A. didiformis*	0.973	0.955	0.905–1.007	42
		All moa	0.946	1.055	1.000–1.113	81
		1 per species	0.983	1.092	0.971–1.229	9
Digit III length	Digit II length	*Dinornis robustus* (151–295)	0.956	0.998	0.905–1.100	23
		Anomalopteryx didiformis (59–167)	0.958	0.984	0.867–1.116	15
		D. robustus + *A. didiformis* (59–295)	0.986	**1.059**[a]	1.017–1.103	38
		All moa(59–295)	0.984	**1.041**[a]	1.011–1.072	77
		1 per species (96–277)	0.979	0.988	0.866–1.127	9
	Digit IV length	*Dinornis robustus*	0.919	0.914	0.801–1.043	24
		Anomalopteryx didiformis	0.867	1.065	0.839–1.353	15
		D. robustus + *A. didiformis*	0.976	1.050	0.996–1.106	39
		All moa	0.964	**1.091**[a]	1.043–1.141	76
		1 per species	0.968	1.082	0.918–1.275	9
Digit II length	Digit IV length	*Dinornis robustus*	0.930	1.014	0.898–1.146	24
		Anomalopteryx didiformis	0.638	0.804	0.494–1.308	15
		D. robustus + *A. didiformis*	0.974	0.964	0.913–1.018	39
		All moa	0.964	1.034	0.989–1.082	75
		1 per species	0.992	**1.096**[a]	1.011–1.187	9
"Big seven" mean excluding II1L	II1L	*Dinornis robustus*	0.953	1.006	0.912–1.110	24
		Anomalopteryx didiformis	0.833	0.955	0.743–1.229	17
		D. robustus + *A. didiformis*	0.984	**1.177**	1.129–1.227	41
		All moa	0.933	1.021	0.962–1.083	82
		1 per species	0.940	0.953	0.757–1.199	9
"Big seven" mean excluding II2L	II2L	*Dinornis robustus*	0.853	0.980	0.814–1.180	24
		Anomalopteryx didiformis	0.806	1.285	0.975–1.695	17
		D. robustus + *A. didiformis*	0.966	0.960	0.904–1.021	41
		All moa	0.910	**1.113**[a]	1.037–1.193	82
		1 per species	0.882	1.039	0.740–1.459	9
"Big seven" mean excluding III1L	III1L	*Dinornis robustus*	0.931	1.033	0.915–1.166	24
		Anomalopteryx didiformis	0.826	0.941	0.727–1.218	17
		D. robustus + *A. didiformis*	0.983	**1.129**	1.082–1.178	41
		All moa	0.947	1.017	0.965–1.072	82
		1 per species	0.979	0.984	0.862–1.122	9

Table A1.3. *continued*

Independent variable	Dependent variable	Treatment (range of values [mm] of independent variable)	r^2	RMA slope	95% CI of slope	*N*
"Big seven" mean excluding III2L	III2L	*Dinornis robustus*	0.911	1.046	0.910–1.202	24
		Anomalopteryx didiformis	0.852	1.048	0.830–1.324	17
		D. robustus + A. didiformis	0.974	**0.882**	0.836–0.930	41
		All moa	0.966	0.965	0.925–1.006	82
		1 per species	0.979	1.135	0.995–1.295	9
"Big seven" mean excluding III3L	III3L	*Dinornis robustus*	0.709	1.111	0.830–1.487	24
		Anomalopteryx didiformis	0.576	1.467	0.878–2.450	17
		D. robustus + A. didiformis	0.865	**0.794**	0.699–0.903	41
		All moa	0.824	1.043	0.941–1.157	82
		1 per species	0.856	1.058	0.721–1.555	9
"Big seven" mean excluding IV1L	IV1L	*Dinornis robustus*	0.933	1.067	0.947–1.202	24
		Anomalopteryx didiformis	0.790	0.985	0.736–1.318	17
		D. robustus + A. didiformis	0.982	**1.061**[a]	1.015–1.108	41
		All moa	0.903	0.971	0.903–1.045	82
		1 per species	0.889	0.958	0.691–1.329	9
"Big seven" mean excluding IV2L	IV2L	*Dinornis robustus*	0.688	1.176	0.865–1.599	24
		Anomalopteryx didiformis	0.699	1.035	0.709–1.511	17
		D. robustus + A. didiformis	0.943	**1.122**[a]	1.036–1.215	41
		All moa	0.892	**1.111**[a]	1.028–1.201	82
		1 per species	0.949	1.075	0.871–1.327	9
Digit III length excluding ungual	Phalanx III2 distal width	*Dinornis robustus*	0.943	1.007	0.894–1.134	21
		Anomalopteryx didiformis	0.924	0.947	0.816–1.097	19
		D. robustus + A. didiformis	0.978	0.997	0.949–1.047	40
		All moa	0.859	1.025	0.934–1.126	77
		1 per species	0.825	0.982	0.634–1.520	9

[a] Only barely allometric (upper CI limit of RMA slope between 0.95 and 1.00, or lower CI limit between 1.00 and 1.05).

Table A1.4. Variability of listwise scaled parameters of phalangeal and digital pedal proportions in selected intraspecific and interspecific samples of crocodylians and ground birds. Phalangeal lengths were scaled against a common length of phalanx III1, and the lengths of digits II and IV (excluding the ungual) were scaled against the length of digit III (excluding the ungual). Shape variability for each scaled parameter is expressed as the ratio of the maximum value of the scaled parameter to the minimum value of the scaled parameter in each intraspecific or interspecific sample. Interspecific comparisons involve both two-species and multispecies comparisons. "Entire" samples indicate the maximum/minimum ratio of the scaled parameter for all specimens of that group for which data were available. For selected intraspecific and interspecific samples, the effects of sample size on the magnitude of the maximum/minimum ratio of the scaled parameter were examined by randomly selecting a specified number of cases from the sample (with either 15 or 30 such trials), with the median value of the 15 or 30 replicates being reported; higher values of the median indicate increased variability. For the species with the largest sample size, the moa *Dinornis robustus*, 5, 10, or 15 cases were randomly selected. For two-species or multispecies comparisons, the number of cases randomly selected was the number of cases of that scaled parameter for individual species with which the interspecific sample was compared. In addition, the final column reports the percentage of cases in the interspecific comparison for which the values of the maximum/minimum ratio of the scaled parameter calculated from the specified number of replicates were greater than the maximum/minimum value of the entire sample of the single species with which the interspecific sample is compared. *Ad* = *Anomalopteryx didiformis*, *Dr* = *Dinornis robustus*, and *Pe* = *Pachyornis elephantopus*.

Scaled parameter	Taxon	Sample or test	Number of cases (entire sample) or number of trials (analyses of effects of sample size on variability)	Maximum/minimum ratio (entire sample) or median value of maximum/minimum ratio (analyses of effects of sample size on variability)	Percentage of values of maximum/minimum ratio > intraspecific value
I1L	*Alligator mississippiensis*	Entire	19	1.18	
	All crocodylians	Entire	42	1.64	
		Random 19	30	1.51	100.0
II1L	*Alligator mississippiensis*	Entire	19	1.09	
	All crocodylians	Entire	42	1.20	
		Random 19	30	1.20	100.0
	Pachyornis elephantopus[a]	Entire	11	1.20	
	Anomalopteryx didiformis	Entire	17	1.10	
	Dinornis robustus	Entire	24	1.10	
		Random 5	15	1.06	
		Random 10	15	1.07	
		Random 15	15	1.09	
	A. didiformis + *D. robustus*	Entire	41	1.14	
		Random 17	30	1.11	73.3 (*Ad*) 60.0 (*Dr*)
	All moa	Entire	82	1.22	
		Random 11	15	1.13	0.0 (*Pe*) 93.3(*Ad*) 93.3 (*Dr*)
		Random 17	15	1.14	0.0 (*Pe*) 100.0 (*Ad*) 100.0 (*Dr*)
		Random 24	15	1.15	13.3 (*Pe*) 80.0 (*Ad*) 80.0 (*Dr*)
II2L	*Alligator mississippiensis*	Entire	19	1.26	
	All crocodylians	Entire	42	1.71	
		Random 19	30	1.71	100.0
	Pachyornis elephantopus	Entire	11	1.61	
	Anomalopteryx didiformis	Entire	17	1.29	
	Dinornis robustus	Entire	24	1.28	
		Random 5	15	1.13	
		Random 10	15	1.20	
		Random 15	15	1.24	
	A. didiformis + *D. robustus*	Entire	41	1.38	
		Random 17	30	1.30	53.3 (*Ad*) 73.3 (*Dr*)

Table A1.4. *continued*

Scaled parameter	Taxon	Sample or test	Number of cases (entire sample) or number of trials (analyses of effects of sample size on variability)	Maximum/minimum ratio (entire sample) or median value of maximum/minimum ratio (analyses of effects of sample size on variability)	Percentage of values of maximum/minimum ratio > intraspecific value
II2L	All moa	Entire	82	2.04	
		Random 11	15	1.49	20.0 (*Pe*) 100.0 (*Ad*) 100.0 (*Dr*)
		Random 17	15	1.65	60.0 (*Pe*) 86.7 (*Ad*) 93.3 (*Dr*)
		Random 24	15	1.86	100.0 (*Pe*) 100.0 (*Ad*) 100.0 (*Dr*)
III2L	*Alligator mississippiensis*	Entire	19	1.13	
	All crocodylians	Entire	42	1.44	
		Random 19	30	1.33	100.0
	Pachyornis elephantopus	Entire	11	1.21	
	Anomalopteryx didiformis	Entire	17	1.13	
	Dinornis robustus	Entire	24	1.20	
		Random 5	15	1.09	
		Random 10	15	1.16	
		Random 15	15	1.16	
	A. didiformis + D. robustus	Entire	41	1.37	
		Random 17	30	1.32	100.0 (*Ad*) 100.0 (*Dr*)
	All moa	Entire	82	1.48	
		Random 11	15	1.30	73.3 (*Pe*) 100.0 (*Ad*) 80.0 (*Dr*)
		Random 17	15	1.32	93.3 (*Pe*) 100.0 (*Ad*) 93.3 (*Dr*)
		Random 24	15	1.41	100.0 (*Pe*) 100.0 (*Ad*) 100.0 (*Dr*)
III3L	*Alligator mississippiensis*	Entire	19	1.38	
	All crocodylians	Entire	42	1.76	
		Random 19	30	1.50	96.7
	Pachyornis elephantopus	Entire	11	1.55	
	Anomalopteryx didiformis	Entire	17	1.50	
	Dinornis robustus	Entire	24	1.54	
		Random 5	15	1.21	
		Random 10	15	1.27	
		Random 15	15	1.49	
	A. didiformis + D. robustus	Entire	41	1.90	
		Random 17	30	1.62	83.3 (*Ad*) 80.0 (*Dr*)

Scaled parameter	Taxon	Sample or test	Number of cases (entire sample) or number of trials (analyses of effects of sample size on variability)	Maximum/minimum ratio (entire sample) or median value of maximum/minimum ratio (analyses of effects of sample size on variability)	Percentage of values of maximum/minimum ratio > intraspecific value
III3L	All moa	Entire	82	2.39	
		Random 11	15	1.70	73.3 (*Pe*) 80.0 (*Ad*) 80.0 (*Dr*)
		Random 17	15	1.83	80.0 (*Pe*) 86.7 (*Ad*) 80.0 (*Dr*)
		Random 24	15	1.95	100.0 (*Pe*) 100.0 (*Ad*) 100.0 (*Dr*)
IV1L	*Alligator mississippiensis*	Entire	19	1.23	
	All crocodylians	Entire	42	1.30	
		Random 19	30	1.26	86.7
	Pachyornis elephantopus	Entire	11	1.16	
	Anomalopteryx didiformis	Entire	17	1.18	
	Dinornis robustus	Entire	24	1.12	
		Random 5	15	1.08	
		Random 10	15	1.10	
		Random 15	15	1.11	
	A. didiformis + *D. robustus*	Entire	41	1.18	
		Random 17	30	1.14	20.0 (*Ad*) 70.0 (*Dr*)
	All moa	Entire	82	1.33	
		Random 11	15	1.16	33.3 (*Pe*) 26.7 (*Ad*) 86.7 (*Dr*)
		Random 17	15	1.18	80.0 (*Pe*) 46.7 (*Ad*) 93.3 (*Dr*)
		Random 24	15	1.19	80.0 (*Pe*) 60.0 (*Ad*) 100.0 (*Dr*)
IV2L	*Alligator mississippiensis*	Entire	19	1.34	
	All crocodylians	Entire	42	1.70	
		Random 19	30	1.35	53.3
	Pachyornis elephantopus	Entire	11	1.51	
	Anomalopteryx didiformis	Entire	17	1.35	
	Dinornis robustus	Entire	24	1.60	
		Random 5	15	1.15	
		Random 10	15	1.22	
		Random 15	15	1.56	
	A. didiformis + *D robustus*	Entire	41	1.62	
		Random 17	30	1.35	40.0 (*Ad*) 16.7 (*Dr*)

Table A1.4. *continued*

Scaled parameter	Taxon	Sample or test	Number of cases (entire sample) or number of trials (analyses of effects of sample size on variability)	Maximum/minimum ratio (entire sample) or median value of maximum/minimum ratio (analyses of effects of sample size on variability)	Percentage of values of maximum/minimum ratio > intraspecific value
	All moa	Entire	82	2.14	
		Random 11	15	1.35	40.0 (*Pe*) 53.3 (*Ad*) 13.3 (*Dr*)
		Random 17	15	1.59	60.0 (*Pe*) 93.3 (*Ad*) 46.7 (*Dr*)
		Random 24	15	1.75	93.3 (*Pe*) 100.0 (*Ad*) 86.7 (*Dr*)
Digit II length excluding ungual	*Dinornis robustus*	Entire	24	1.13	
		Random 5	15	1.04	
		Random 10	15	1.09	
		Random 15	15	1.13	
Digit IV length excluding ungual	*Dinornis robustus*	Entire	25	1.29	
		Random 5	15	1.07	
		Random 10	15	1.11	
		Random 15	15	1.19	

[a] The sample for this species in this table includes a small (presumably juvenile) individual identified as ?*P. elephantopus*, but excludes the specimen (CM AV 8622; previously identified as *Euryapteryx curtus*) that Allentoft et al. (2010) identified on molecular grounds as belonging to *P. elephantopus*.

Table A1.5. Analyses of covariance (ANCOVAs) of pedal and digital proportions across moa species. Ad = *Anomalopteryx didiformis*; Dn = *Dinornis novaezealandiae*; Dr = *Dinornis robustus*; Ecr = *Emeus crassus*; Ecu = *Euryapteryx curtus*; Pe = *Pachyornis elephantopus* (excluding both the small individual [CM AV 8384 identified as ?*P. elephantopus*] and CM AV 8622). In several tests both the test parameter and the covariate were log transformed prior to analysis in an effort (not always successful) more nearly to equalize error variances among species, and/or to improve performance in the lack-of-fit test. In all tests *P* of the *F*-test of the effects of species <.001, indicating significant differences among three or more species.

Test parameter	Results of comparison (Bonferroni test)	*N* of cases
Covariate = phalanx III1 length		
Phalanx II1 length[a, b]	Ad significantly different than Dn, Ecu, and Pe	27 Dr: 6 Dn: 8 Ecr: 11 Ecu: 22 Ad: 11 Pe
Phalanx II2 length[a]	Ad significantly different than Dr, Ecr, Ecu, and Pe	25 Dr: 5 Dn: 8 Ecr: 10 Ecu: 22 Ad: 11 Pe
Phalanx III2 length[a]	Ad significantly different than Dr, Dn, Ecr, Ecu, and Pe; Dr significantly different than Pe	26 Dr: 5 Dn: 8 Ecr: 11 Ecu: 20 Ad: 11 Pe
Phalanx III3 length	Ad significantly different than Dr, Ecr, and Ecu; Ecu significantly different than Pe	25 Dr: 5 Dn: 8 Ecr: 11 Ecu: 19 Ad: 11 Pe
Phalanx IV1 length[a, c]	Ad significantly different than Ecu; Dr significantly different than Ecr, Ecu, and Pe; Dn significantly different than Ecu	27 Dr: 6 Dn: 8 Ecr: 11 Ecu: 22 Ad: 12 Pe
Phalanx IV2 length[a, c]	Ad significantly different than Ecu and Pe; Dr significantly different than Pe; Dn significantly different than Ecu and Pe	26 Dr: 5 Dn: 8 Ecr: 10 Ecu: 19 Ad: 11 Pe
Covariate = digit III length excluding ungual		
Digit II length excluding ungual[d]	Ad significantly different than Dr, Dn, and Ecu; Dr significantly different than Pe; Dn significantly different than Pe	24 Dr: 5 Dn: 8 Ecr: 10 Ecu: 18 Ad: 10 Pe
Digit IV length excluding ungual	Ad significantly different than Dr, Dn, Ecr, and Ecu; Dr significantly different than Ecr, Ecu, and Pe; Dn significantly different than Ecr, Ecu, and Pe	25 Dr: 4 Dn: 8 Ecr: 10 Ecu: 17 Ad: 10 Pe

[a] Test parameter and covariate log transformed prior to analysis.
[b] *P* of Levene's test of equality of error variances, and *P* of lack-of-fit test, both <.05 even with log transformation of both variables.
[c] *P* of Levene's test of equality of error variances <.05, even with log transformation of both variables.
[d] *P* of lack-of-fit test <.05, even with log transformation of both variables, and so test parameter and covariate were not log transformed in the reported test.

Table A1.6. Maximum/minimum ratios of scaled lengths (scaled to a common length of phalanx III1) of the larger ("big seven") non-ungual phalanges of digits II–IV in crocodylians, birds, and non-avian dinosaurs. Numbers in each cell are: ratio (number of specimens). The range of values of III1 length (mm) is indicated for each species and for some genera. For *Iguanodon bernissartensis*, data are reported for "synthetic"/"blended" measurements (first row) and for medial phalangeal lengths only (second row). For some taxa, listwise values from table A1.4 are reported in a second row of data for that taxon. Individual species are compared with more inclusive groups; parameters for which the interspecific comparison results in larger values (greater variability) than for at least one of the species with which the comparison is made are indicated in **bold**; if the interspecific values are larger than all of the intraspecific values of the same parameter with which they are compared, the interspecific value is also *italicized.* Similarly, values for genera or more inclusive groups are compared with even more inclusive groups containing those genera or groups; parameters for which the more inclusive intergroup comparison results in larger values (greater variability) than for at least one of the less inclusive groups with which the comparison is made are indicated in **bold**; if the more inclusive group values are larger than all of the less inclusive group values of the same parameter, the more inclusive group value is also *italicized.*

Taxon (III1 length range)	II1L	II2L	III2L	III3L	IV1L	IV2L
Alligator mississippiensis (3.5–44)	1.21 (25) 1.09 (19)	1.60 (23) 1.26 (19)	1.42 (22) 1.13 (19)	1.53 (20) 1.38 (19)	1.23 (24) 1.23 (19)	1.34 (19) 1.34 (19)
Caiman crocodilus (5–14)	1.16 (10)	1.14 (10)	1.50 (9)	1.11 (8)	1.17 (10)	1.10 (8)
All extant crocodylians (*A. mississippiensis, A. sinensis, Paleosuchus trigonatus, P. palpebrosus, Caiman crocodilus, C. yacare, Tomistoma schlegelii, Crocodylus rhombifer, C. acutus, C. siamensis, C. johnstoni, C. mindorensis, Osteolaemus tetraspis, Gavialis gangeticus*)	**1.21** (61) **1.20** (42)	***1.71*** (55) **1.71** (42)	***1.80*** (57) **1.44** (42)	***1.76*** (48) **1.76** (42)	***1.30*** (60) **1.30** (42)	***1.70*** (50) **1.70** (42)
Casuarius casuarius (29–63)	1.15 (8)	1.42 (8)	1.12 (8)	1.26 (8)	1.08 (8)	1.12 (8)
Casuarius spp. (*C. casuarius, C. unappendiculatus, C. bennetti*)	**1.20** (16)	1.42 (16)	1.12 (16)	**1.34** (16)	**1.09** (16)	**1.18** (16)
Dromaius novaehollandiae (45–64)	1.13 (13)	1.45 (12)	1.14 (13)	1.52 (13)	1.12 (13)	1.27 (13)
D. novaehollandiae + *Casuarius* spp.	***1.38*** (29)	***1.46*** (28)	***1.26*** (29)	***1.79*** (29)	**1.12** (29)	***1.28*** (29)
D. novaehollandiae + *Ardeotis kori*	**1.17** (18)	**1.83** (17)	**1.32** (18)	**1.74** (18)	**1.17** (18)	**1.30** (18)
Rhea americana (17–43)	1.11 (13)	1.95 (13)	1.18 (13)	1.47 (13)	1.13 (13)	1.73 (13)
Pterocnemia pennata (33–44)	1.12 (9)	1.33 (9)	1.12 (9)	1.29 (9)	1.12 (9)	1.45 (9)
R. americana + *P. pennata*	***1.13*** (22)	***2.00*** (22)	***1.30*** (22)	***1.60*** (22)	**1.13** (22)	***1.90*** (22)
D. novaehollandiae + *R. americana* + *P. pennata*	***1.39*** (35)	***2.00*** (34)	***1.30*** (35)	***1.77*** (35)	***1.27*** (35)	***2.58*** (35)
Euryapteryx curtus (37–72)	1.11 (10)	1.36 (9)	1.15 (10)	1.49 (10)	1.24 (10)	1.41 (9)
Anomalopteryx didiformis (22–57)	1.19 (22) 1.10 (17)	1.52 (22) 1.29 (17)	1.18 (20) 1.13 (17)	1.50 (19) 1.50 (17)	1.18 (22) 1.18 (17)	1.47 (19) 1.35 (17)
E. curtus + *A. didiformis*	***1.21*** (32)	***2.07*** (31)	***1.33*** (30)	***2.11*** (29)	***1.28*** (32)	***1.73*** (28)
Pachyornis elephantopus[a] (49–88)	1.20 (12) 1.20 (11)	2.04 (12) 1.61 (11)	1.21 (12) 1.21 (11)	1.55 (12) 1.55 (11)	1.16 (13) 1.16 (11)	1.51 (12) 1.51 (11)
P. elephantopus[b] (67–88)	1.20 (11)	2.04 (11)	1.21 (11)	1.55 (11)	1.15 (12)	1.45 (11)
P. elephantopus[c] (49–88)	1.20 (13)	2.04 (13)	1.21 (13)	1.55 (13)	1.16 (14)	1.51 (13)
Pachyornis spp.[a] (*P. elephantopus, P. mappini, P. australis*)	1.20 (17)	2.04 (17)	**1.27** (17)	1.55 (16)	1.16 (17)	1.51 (16)
Dinornis robustus (62–115)	1.10 (27) 1.10 (24)	1.37 (25) 1.28 (24)	1.44 (26) 1.20 (24)	1.54 (25) 1.54 (24)	1.12 (27) 1.12 (24)	1.61 (26) 1.60 (24)
Dinornis spp. (*D. robustus, D. novaezelandiae*)	1.10 (34)	1.37 (31)	**1.45** (32)	1.54 (31)	1.12 (34)	1.61 (32)
D. robustus + *A. didiformis*	**1.19** (49) ***1.14*** (41)	***1.74*** (47) ***1.38*** (41)	***1.68*** (46) ***1.37*** (41)	***1.90*** (44) ***1.90*** (41)	**1.18** (49) **1.18** (41)	***1.64*** (45) ***1.62*** (41)
All moa spp. (*D. robustus, D. novaezealandiae, A. didiformis, Megalapteryx didinus, Emeus crassus, E. curtus, P. elephantopus, P. geranoides, P. australis*)	***1.22*** (96) ***1.22*** (82)	***2.21*** (92) ***2.04*** (82)	***1.78*** (92) ***1.48*** (82)	***2.39*** (89) ***2.39*** (82)	***1.33*** (96) ***1.33*** (82)	***2.16*** (89) ***2.14*** (82)

Taxon (III1 length range)	**II1L**	**II2L**	**III2L**	**III3L**	**IV1L**	**IV2L**
Large tridactyl ground birds (emu, cassowaries, rheas, moa, kori bustard)	***1.71*** (154)	***3.69*** (149)	***1.89*** (150)	***2.39*** (147)	***1.33*** (154)	***4.05*** (147)
Coelophysis bauri (20–39)	1.13 (10)	1.24 (9)	1.18 (11)	1.17 (9)	1.29 (11)	1.22 (10)
Allosaurus fragilis (88–131)	1.11 (7)	1.07 (6)	1.46 (7)	1.50 (5)	1.17 (3)	1.27 (4)
Allosaurus spp. (71–131)	**1.20** (9)	**1.17** (8)	1.46 (8)	1.50 (6)	**1.29** (4)	1.27 (6)
Tyrannosaurus rex (including *Nanotyrannus*; 135–213)	1.08 (8) 1.06 (6)	1.13 (7) 1.13 (6)	1.13 (9) 1.08 (6)	1.23 (7) 1.23 (6)	1.23 (9) 1.23 (6)	1.19 (8) 1.19 (6)
All tyrannosaurids (*T. rex* [including *Nanotyrannus lancensis*], *Tarbosaurus bataar, Daspletosaurus torosus, Daspletosaurus* sp., *Albertosaurus* sp., *Gorgosaurus libratus, Alectrosaurus olseni*, possibly other taxa)	**1.15** (19) **1.76** (23) **1.15** (14)	**1.14** (18) **2.02** (18) **1.14** (14)	 1.23 (20) **1.12** (14)	 **1.48** (20) **1.25** (14)	 1.23 (14)	 **1.40** (14)
Ornithomimosaurs (*Ornithomimus edmontonicus, O. velox, Struthiomimus altus, Dromiceiomimus brevetertius, Gallimimus bullatus, Sinornithomimus dongi*)	1.15 (16)	1.32 (15)	1.16 (17)	2.19 (17)	1.31 (13)	1.74 (17)
Tyrannosaurids + ornithomimosaurs	***1.19*** (35)	***1.78*** (33)	**1.76** (40)	***2.33*** (35)	***1.43*** (33)	***2.22*** (37)
Allosauroids (*Allosaurus fragilis, Allosaurus* sp., *Saurophaganax maximus, Acrocanthosaurus atokensis, Camptosaurus amplus*[d]) + tyrannosaurids	**1.64**(31)	**1.21**(28)	**1.86** (34)	**2.16** (26)	**1.36**(27)	**1.78** (29)
Allosaurus + *Tyrannosaurus*	***1.21*** (17)	***1.20*** (15)	**1.46** (17)	**1.50** (13)	**1.29** (13)	***1.68*** (14)
Large theropods (allosauroids, tyrannosaurids, *Dilophosaurus wetherilli, Aucasaurus garridoi*)	**1.64** (34)	**1.21** (30)	**1.86** (36)	**2.16** (28)	**1.36** (30)	**1.78**(31)
Tenontosaurus spp. (21–65)	1.24 (14)	1.29 (14)	1.14 (14)	1.29 (13)	1.39 (13)	1.28 (11)
Iguanodon bernissartensis (88–143)	1.37 (14) 1.26 (13)	1.55 (15) 1.73 (12)	1.50 (15) 1.40 (13)	1.43 (15) 1.42 (12)	1.86 (16) 1.54 (11)	1.80 (17) 1.93 (10)
All *Iguanodon* plus *Mantellisaurus atherfieldensis*	1.37 (17)	1.55 (18)	1.50 (18)	1.43 (19)	1.86 (19)	**2.02** (20)
Edmontosaurus spp. (*E. regalis, E. annectens)* (66–147)	1.13 (8)	1.23 (9)	1.44 (9)	1.39 (9)	1.26 (9)	1.55 (9)
All hadrosaurids (*Brachylophosaurus canadensis, Edmontosaurus* spp., *Gryposaurusincurvimanus, Prosaurolophusmaximus, Lambeosaurus lambei, Corythosaurus casuarius, Hypacrosaurus altispinus*)	**1.31** (23)	**1.63** (24)	**1.76** (24)	**2.04** (24)	**1.36** (23)	**1.96** (22)
Iguanodon and similar taxa + hadrosaurids	***1.56*** (40)	***1.77*** (42)	**1.76** (42)	**2.04** (43)	**1.86** (42)	***2.38*** (42)

[a] Includes a small (presumably juvenile) individual identified as ?*P. elephantopus*.
[b] Excludes the small (presumably juvenile) individual identified as ?*P. elephantopus*.
[c] Includes all specimens identified on morphological or molecular grounds as being *P. elephantopus*.
[d] Here regarded as the foot of a large theropod rather than an ornithopod (Galton and Powell 1980).

Table A1.7. Maximum/minimum ratios of scaled digital lengths (scaled to the length of digit III, with or without the ungual, as appropriate) in crocodylians, birds, and non-avian dinosaurs. Numbers in each cell are: ratio (number of specimens).The range of values of digit III length without the ungual (in millimeters) is indicated for each species and some genera. Individual species are compared with more inclusive groups; parameters for which the interspecific comparison results in larger values (greater variability) than for at least one of the species with which the comparison is made are indicated in **bold**; if the interspecific values are larger than all of the intraspecific values of the same parameter, the interspecific value is also *italicized.* Similarly, values for genera or more inclusive groups are compared with even more inclusive groups containing those genera or groups; parameters for which the more inclusive intergroup comparison results in larger values (greater variability) than for at least one of the less inclusive groups with which the comparison is made are indicated in **bold**; if the more inclusive group values are larger than all of the less inclusive group values of the same parameter, the more inclusive group value is also *italicized.*

Taxon	Digit II length excluding ungual	Digit II length	Digit IV length excluding ungual	Digit IV length
Alligator mississippiensis (11.5–92)	1.13 (20)	——	1.09 (13)[a]	——
Caiman crocodilus (11.5–31.5)	1.15 (7)	——	1.05 (3)[a]	——
All extant crocodylians	***1.24*** (46)	——	***1.25*** (28)[a]	——
Casuarius casuarius (63–131)	1.21 (8)	1.22 (8)	1.10 (7)	1.05 (7)
Casuarius spp.	**1.22** (16)	1.22 (15)	1.10 (15)	**1.09** (14)
Dromaius novaehollandiae (93–122)	1.18 (12)	1.10 (11)	1.13 (12)	1.11 (11)
D. novaehollandiae + *Casuarius* spp.	***1.49*** (28)	***1.29*** (26)	***1.16*** (27)	**1.11** (25)
D. novaehollandiae + *Ardeotis kori*	**1.38** (17)	**1.19** (16)	**1.28** (17)	**1.17** (15)
Rhea americana (32–81)	1.16 (13)	1.12 (10)	1.25 (13)	1.13 (11)
Pterocnemia pennata (66–88)	1.08 (9)	1.09 (9)	1.11 (9)	1.09 (9)
R. americana + *P. pennata*	**1.16** (22)	***1.15*** (19)	**1.25** (22)	***1.14*** (20)
D. novaehollandiae + *R. americana* + *P. pennata*	***1.43*** (34)	***1.26*** (30)	***1.27*** (34)	***1.15*** (31)
Euryapteryx curtus (65–133)	1.11 (9)	1.11 (8)	1.21 (9)	1.17 (9)
Anomalopteryx didiformis (43–119)	1.16 (18)	1.19 (15)	1.17 (17)	1.13 (15)
E. curtus + *A. didiformis*	***1.20*** (27)	***1.20*** (23)	***1.27*** (26)	***1.20*** (24)
Pachyornis elephantopus[b] (94–156)	1.24 (11)	1.17 (11)	1.18 (11)	1.15 (11)
All *Pachyornis* spp.	1.24 (15)	1.17 (14)	**1.25** (14)	1.15 (13)
Dinornis robustus (107–220)	1.13 (24)	1.13 (23)	1.29 (25)	1.23 (24)
Dinornis spp.	1.13 (30)	**1.14** (27)	**1.30** (30)	**1.25** (27)
D. robustus + *A. didiformis*	***1.21***(42)	***1.27*** (38)	***1.34*** (42)	***1.32*** (39)
All moa	***1.29*** (85)	***1.27*** (77)	***1.52*** (82)	***1.40*** (76)
Large tridactyl ground birds	***1.95*** (142)	***1.44*** (129)	***1.83*** (138)	***1.59*** (127)
All tyrannosaurids	1.11 (16)	1.06 (8)	1.21 (10)	1.18 (7)
Ornithomimosaurs	1.17 (11)	1.02 (5)	1.25 (12)	1.12 (6)
Tyrannosaurids + ornithomimosaurs	***1.33*** (27)	***1.14*** (13)	***1.51*** (22)	***1.38*** (12)
Allosauroids + tyrannosaurids	**1.26** (22)	**1.14** (10)	**1.24** (13)	1.18 (8)
Large theropods	**1.26** (24)	**1.19** (13)	1.21 (15)	1.18 (11)
Tenontosaurus spp. (49–130)	1.20 (13)	1.15 (9)	1.32 (8)	1.33 (8)
Iguanodon bernissartensis (141–212)	1.19 (14)	1.15 (11)	1.45 (14)	1.18 (11)
All *Iguanodon* and similar taxa	**1.22** (17)	**1.17** (13)	**1.49** (17)	1.18 (12)
Edmontosaurus spp. (115–242)	1.13 (8)	1.11 (6)	1.18 (9)	1.20 (8)
All hadrosaurids	**1.31** (23)	**1.20** (17)	**1.33** (21)	**1.38** (18)
Iguanodon and similar taxa + hadrosaurids	***1.36*** (40)	**1.20** (30)	**1.49** (38)	**1.38** (30)

[a] There is no ungual on digit IV of the crocodylian pes.
[b] Includes a small (presumably juvenile) individual identified as ?*P. elephantopus*, but excludes the specimen Allentoft et al. (2010) assigned to this species on molecular grounds.

Table A1.8. Coefficient of variation of scaled lengths (scaled to a common length of phalanx III1) of the larger non-ungual phalanges of digits II–IV in crocodylians, birds, and non-avian dinosaurs. The number of cases for each taxon, and the range of values of III1 length (in millimeters) for particular species, are the same as in table A1.6. Individual species are compared with more inclusive groups; parameters for which the interspecific comparison results in larger values (greater variability) than for at least one of the species with which the comparison is made are indicated in **bold**; if the interspecific values are larger than all of the intraspecific values of the same parameter, the interspecific value is also *italicized*. Similarly, values for genera or more inclusive groups are compared with even more inclusive groups containing those genera or groups; parameters for which the more inclusive intergroup comparison results in larger values (greater variability) than for at least one of the less inclusive groups with which the comparison is made are indicated in **bold**; if the more inclusive group values are larger than all of the less inclusive group values of the same parameter, the more inclusive group value is also *italicized*.

Taxon	II1L	II2L	III2L	III3L	IV1L	IV2L
Alligator mississippiensis	4.56	10.9	8.23	11.1	4.72	7.00
Caiman crocodilus	3.56	4.31	14.1	3.33	5.82	4.27
All extant crocodylians	***4.91***	***13.3***	**10.5**	***13.8***	***6.30***	***8.86***
Casuarius casuarius	4.68	11.5	3.73	7.84	3.21	3.89
Casuarius spp.	**5.35**	10.8	3.23	7.77	3.04	**5.09**
Dromaius novaehollandiae	3.93	11.6	3.35	11.1	3.41	6.22
D. novaehollandiae + *Casuarius* spp.	***9.60***	**11.0**	***5.32***	***16.9***	**3.33**	***6.31***
D. novaehollandiae + *Ardeotis kori*	**4.24**	**16.1**	**9.40**	**14.1**	**4.33**	**6.92**
Rhea americana	3.13	20.5	4.41	9.74	3.42	15.9
Pterocnemia pennata	3.37	9.59	4.01	8.17	3.52	10.2
R. americana + *P. pennata*	**3.24**	***24.4***	***7.59***	***13.4***	**3.43**	***19.7***
D. novaehollandiae + *R. americana* + *P. pennata*	***10.2***	***20.6***	***6.29***	***13.4***	***6.22***	***25.9***
Euryapteryx curtus	3.55	12.2	5.04	13.0	6.25	13.2
Anomalopteryx didiformis	3.78	9.42	4.40	10.4	3.52	9.20
E. curtus + *A. didiformis*	***4.83***	***16.7***	***8.89***	***20.2***	**5.30**	**11.7**
Pachyornis elephantopus[a]	5.16	17.8	6.18	11.1	4.39	13.0
All *Pachyornis* spp.[a]	4.78	15.0	**6.91**	10.3	4.07	12.2
Dinornis robustus	2.30	6.68	6.56	8.48	2.80	13.1
Dinornis spp.	**2.57**	6.46	6.36	**8.76**	2.61	11.9
D. robustus + *A. didiformis*	**3.14**	***10.1***	***9.45***	***14.4***	***3.55***	**11.5**
All moa	**3.97**	**14.1**	**8.98**	**16.7**	**4.86**	***13.4***
Large tridactyl ground birds	***9.77***	***29.7***	***11.7***	**18.7**	**5.66**	**22.5**
Coelophysis bauri	4.30	6.73	5.05	4.41	8.43	6.44
Allosaurus fragilis	3.72	2.96	13.6	14.7	7.91	10.7
Allosaurus spp.	**5.07**	**4.78**	12.7	13.5	**18.2**	**20.3**
Tyrannosaurus rex (including *Nanotyrannus*)	2.98	4.32	3.51	6.64	6.38	6.75
All tyrannosaurids	**3.57**	4.21	**9.12**	**12.4**	**6.43**	**9.80**
Ornithomimosaurs	4.01	9.33	3.89	14.6	7.37	13.7
Tyrannosaurids + ornithomimosaurs	***4.16***	***18.7***	**8.46**	**13.4**	***10.7***	***22.0***
Allosaurusfragilis + *Tyrannosaurusrex*	***4.78***	***4.88***	**9.51**	**10.2**	**7.62**	***14.7***
Allosauroids + tyrannosaurids	***7.75***	**4.74**	**10.3**	**12.9**	**7.45**	**13.3**
Large theropods	**7.71**	**4.88**	**10.0**	**12.1**	**7.74**	**13.0**
Tenontosaurus spp.	6.81	6.77	4.26	7.87	8.91	7.79
Iguanodon bernissartensis	8.60	10.4	10.2	10.9	15.7	16.8
All *Iguanodon* and similar taxa	**8.99**	**10.6**	9.63	10.7	15.3	**17.5**
Edmontosaurus spp.	4.10	7.43	11.3	11.7	6.92	14.4
All hadrosaurids	**7.23**	**14.4**	**16.4**	**17.6**	6.80	18.2
Iguanodon and similar taxa + hadrosaurids	**8.69**	**13.3**	**14.0**	**16.0**	**11.6**	***20.1***

[a] Includes a small (presumably juvenile) individual identified as ?*P. elephantopus*, but excludes a specimen assigned to this species on molecular grounds.

Table A1.9. Coefficient of variation of scaled digit lengths (scaled to a common length of digit III) in crocodylians, birds, and non-avian dinosaurs. The number of cases for each taxon, and the range of values of III1 length (in millimeters) for particular species, are the same as in table A1.7. Individual species are compared with more inclusive groups; parameters for which the interspecific comparison results in larger values (greater variability) than for at least one of the species with which the comparison is made are indicated in **bold**; if the interspecific values are larger than all of the intraspecific values of the same parameter, the interspecific value is also *italicized*. Similarly, values for genera or more inclusive groups are compared with even more inclusive groups containing those genera or groups; parameters for which the more inclusive intergroup comparison results in larger values (greater variability) than for at least one of the less inclusive groups with which the comparison is made are indicated in **bold**; if the more inclusive group values are larger than all of the less inclusive group values of the same parameter, the more inclusive group value is also *italicized*.

Taxon	Digit II length excluding ungual	Digit II length	Digit IV length excluding ungual	Digit IV length
Alligator mississippiensis	3.18	——	2.38[a]	——
Caiman crocodilus	4.22	——	2.63[a, b]	——
All extant crocodylians	***4.72***	——	***5.68***[a]	——
Dromaius novaehollandiae	4.98	3.34	3.52	2.84
D. novaehollandiae + *Casuarius* spp.	**12.1**	**8.11**	**3.58**	2.67
D. novaehollandiae + *Ardeotis kori*	**9.52**	**5.44**	**8.43**	**4.35**
Rhea americana	4.33	4.16	5.36	4.06
Pterocnemia pennata	2.57	2.58	3.17	2.89
R. americana + *P. pennata*	**3.72**	**3.44**	**4.52**	**3.84**
D. novaehollandiae + *R. americana* + *P. pennata*	***9.36***	***5.85***	**4.35**	**3.67**
Euryapteryx curtus	3.24	3.69	6.95	6.13
Anomalopteryx didiformis	3.60	5.01	4.62	3.78
E. curtus + *A. didiformis*	***4.91***	**4.71**	**6.42**	**5.32**
Pachyornis elephantopus[c]	5.37	4.47	4.81	4.15
All *Pachyornis* spp.[c]	**5.53**	**4.52**	**5.84**	3.92
Dinornis robustus	2.83	2.92	5.74	4.95
Dinornis spp.	2.56	**2.94**	5.54	4.86
D. robustus + *A. didiformis*	***4.44***	**4.49**	**5.60**	**4.95**
All moa	**4.40**	**4.24**	***7.45***	***6.38***
All tyrannosaurids	2.76	2.21	5.58	6.11
Tyrannosaurids + ornithomimosaurs	**7.10**	**4.73**	**14.4**	**11.7**
Large theropods	**4.91**	**4.85**	4.75	4.89
Tenontosaurus spp.	5.25	4.88	9.13	9.29
Iguanodon bernissartensis	4.58	4.86	8.46	5.17
All *Iguanodon* + similar taxa	**5.39**	**5.41**	**9.27**	5.14
All hadrosaurids	6.12	5.45	6.81	2.20
Iguanodon and similar taxa + hadrosaurids	**5.80**	**5.34**	**8.75**	***5.75***

[a] There is no ungual on digit IV of the crocodylian pes.
[b] There are only three specimens for this parameter.
[c] Includes a small (presumably juvenile) individual identified as ?*P. elephantopus*, but excludes a specimen assigned to this species on molecular grounds.

Table A1.10. Within-species and across-species variability of log-transformed values of the "big seven" phalanx lengths. Log-transformed phalanx lengths were "GM-scaled" (GM = geometric mean) by subtracting from each log-transformed length the mean of all seven of the log-transformed phalanx lengths. Variability in scaled phalanx lengths is expressed in terms of the range of values (top row) and the standard deviation of values (bottom row). Parameters for which the interspecific comparison results in larger values (greater variability) than for at least one of the species with which the comparison is made are indicated in **bold**; if the interspecific values are larger than all of the intraspecific values of the same parameter, the interspecific value is also *italicized*. Similarly, values for genera or more inclusive groups are compared with even more inclusive groups containing those genera or groups; parameters for which the more inclusive intergroup comparison results in larger values (greater variability) than for at least one of the less inclusive groups with which the comparison is made are indicated in **bold**; if the more inclusive group values are larger than all of the less inclusive group values of the same parameter, the more inclusive group value is also *italicized*.

Taxon (number of specimens)	II1L	II2L	III1L	III2L	III3L	IV1L	IV2L
Alligator mississippiensis (19)	0.0450	0.0561	0.0477	0.0438	0.1086	0.0906	0.0942
	0.0113	0.0155	0.0137	0.0099	0.0241	0.0274	0.0217
Caiman crocodilus (6)	0.0403	0.0347	0.0210	0.0237	0.0327	0.0566	0.0312
	0.0147	0.0137	0.0085	0.0083	0.0119	0.0189	0.0123
All extant crocodylians (43)	***0.0927***	***0.1453***	***0.1146***	***0.0623***	***0.2408***	***0.0935***	***0.2089***
	0.0189	***0.0366***	***0.0225***	***0.0126***	***0.0410***	***0.0290***	***0.0373***
Dromaius novaehollandiae (12)	0.0726	0.1292	0.0478	0.0501	0.1041	0.0385	0.0669
	0.0177	0.0402	0.0155	0.0136	0.0304	0.0114	0.0220
Rhea americana (13)	0.1055	0.2117	0.0890	0.0678	0.0913	0.0931	0.1694
	0.0268	0.0527	0.0232	0.0206	0.0290	0.0256	0.0503
Pterocnemia pennata (9)	0.0528	0.0874	0.0474	0.0222	0.0767	0.0357	0.1441
	0.0157	0.0298	0.0173	0.0084	0.0242	0.0116	0.0419
R. americana + *P. pennata* (22)	***0.1151***	**0.2117**	***0.1081***	**0.0678**	***0.1079***	***0.1070***	***0.2019***
	0.0413	***0.0689***	***0.0375***	**0.0169**	***0.0292***	***0.0355***	***0.0569***
D. novaehollandiae + *R. americana* + *P. pennata* (34)	***0.1934***	**0.2117**	***0.1081***	***0.0744***	***0.1111***	***0.1299***	***0.3301***
	0.0611	***0.0597***	***0.0320***	**0.0169**	***0.0312***	***0.0420***	***0.0988***
Euryapteryx curtus (8)	0.0583	0.0955	0.0630	0.0394	0.1311	0.0889	0.1119
	0.0184	0.0368	0.0257	0.0143	0.0412	0.0309	0.0411
Anomalopteryx didiformis (17)	0.0657	0.0851	0.0535	0.0499	0.1354	0.0746	0.0836
	0.0151	0.0207	0.0155	0.0147	0.0340	0.0172	0.0215
E. curtus + *A. didiformis* (25)	***0.1393***	***0.1793***	***0.1070***	***0.0768***	***0.2297***	***0.1296***	**0.1119**
	0.0436	***0.0456***	***0.0324***	***0.0184***	***0.0630***	***0.0415***	**0.0283**
Pachyornis elephantopus[a] (11)	0.0784	0.1537	0.0693	0.0638	0.1259	0.0857	0.1506
	0.0232	0.0414	0.0240	0.0188	0.0382	0.0271	0.0491
Dinornis robustus (24)	0.0443	0.0948	0.0550	0.0624	0.1330	0.0589	0.1621
	0.0111	0.0199	0.0138	0.0158	0.0307	0.0141	0.0333
D. robustus + *A. didiformis* (41)	***0.1050***	***0.1115***	***0.0908***	***0.1006***	***0.2008***	**0.0746**	***0.1981***
	0.0246	***0.0209***	***0.0208***	***0.0220***	***0.0453***	**0.0166**	**0.0311**
All moa (82)	***0.1480***	***0.2282***	***0.1248***	**0.1006**	***0.2534***	***0.1993***	***0.2952***
	0.0296	**0.0376**	**0.0262**	**0.0209**	**0.0495**	**0.0352**	**0.0409**
All tyrannosaurids (14)	0.0541	0.0483	0.0525	0.0381	0.0661	0.0937	0.1256
	0.0167	0.0120	0.0140	0.0116	0.0206	0.0217	0.0272
Ornithomimosaurs (10)	0.0647	0.0912	0.0650	0.0326	0.1063	0.0674	0.0919
	0.0208	0.0306	0.0191	0.0104	0.0318	0.0254	0.0259
Tyrannosaurids + ornithomimosaurs (24)	***0.1213***	***0.1849***	***0.1039***	***0.1259***	***0.1064***	***0.0951***	***0.2322***
	0.0380	***0.0578***	***0.0311***	***0.0476***	**0.0311**	***0.0257***	***0.0643***
Tenontosaurus spp. (10)	0.0609	0.0638	0.0345	0.0444	0.1251	0.1209	0.0795
	0.0202	0.0197	0.0115	0.0128	0.0409	0.0321	0.0263
Iguanodon bernissartensis (14)	0.1132	0.1997	0.0959	0.1680	0.1070	0.1743	0.2120
	0.0278	0.0476	0.0269	0.0405	0.0265	0.0456	0.0560
All hadrosaurids (20)	0.1151	0.1343	0.1632	0.1421	0.1792	0.1516	0.1975
	0.0337	0.0333	0.0429	0.0397	0.0490	0.0404	0.0501
Iguanodon and similar taxa + hadrosaurids (37)	***0.1434***	***0.2545***	**0.1632**	***0.2304***	***0.1851***	***0.1780***	***0.2843***
	0.0320	**0.0445**	**0.0382**	***0.0431***	**0.0447**	**0.0415**	***0.0644***

[a] Includes a small (presumably juvenile) individual identified as ?*P. elephantopus*, but excludes a specimen assigned to this species on molecular grounds.

Table A1.11. Within-species and across-species variability of log-transformed values of digit lengths (excluding the unguals). Log-transformed digit lengths were "GM-scaled" (GM = geometric mean) by subtracting from each length the mean of all three of the log-transformed digit lengths. Variability in scaled digit lengths is expressed in terms of the range of values (top row) and the standard deviation of scaled values (bottom row). Parameters for which the interspecific comparison results in larger values (greater variability) than for at least one of the species with which the comparison is made are indicated in **bold**; if the interspecific values are larger than all of the intraspecific values of the same parameter, the interspecific value is also *italicized.* Similarly, values for genera or more inclusive groups are compared with even more inclusive groups containing those genera or groups; parameters for which the more inclusive intergroup comparison results in larger values (greater variability) than for at least one of the less inclusive groups with which the comparison is made are indicated in **bold**; if the more inclusive group values are larger than all of the less inclusive group values of the same parameter, the more inclusive group value is also *italicized.*

Taxon (number of specimens)	Digit II length excluding ungual	Digit III length excluding ungual	Digit IV length excluding ungual
Alligator mississippiensis (17)	0.0369 0.0098	0.0278 0.0066	0.0474[a] 0.0115[a]
All extant crocodylians (36)	**0.0648** **0.0195**	**0.0667** **0.0116**	**0.1079[a]** **0.0249[a]**
Dromaius novaehollandiae (12)	0.0450 0.0154	0.0323 0.0089	0.0392 0.0124
Rhea americana (13)	0.0428 0.0117	0.0399 0.0119	0.0569 0.0146
Pterocnemia pennata (9)	0.0211 0.0083	0.0236 0.0062	0.0271 0.0094
R. americana + *P. pennata* (22)	**0.0428** **0.0106**	**0.0399** **0.0098**	**0.0569** **0.0128**
D. novaehollandiae + *R. americana* + *P. pennata* (34)	***0.1127*** ***0.0299***	***0.0523*** ***0.0141***	***0.0966*** ***0.0208***
Euryapteryx curtus (8)	0.0374 0.0115	0.0382 0.0127	0.0484 0.0191
Anomalopteryx didiformis (17)	0.0338 0.0101	0.0286 0.0074	0.0460 0.0140
E. curtus + *A. didiformis* (25)	***0.0722*** ***0.0195***	***0.0410*** **0.0092**	***0.0827*** ***0.0229***
Pachyornis elephantopus[b] (9)	0.0605 0.0165	0.0321 0.0127	0.0418 0.0148
Dinornis robustus (24)	0.0444 0.0099	0.0303 0.0073	0.0605 0.0124
D. robustus + *A. didiformis* (41)	***0.0523*** ***0.0126***	***0.0494*** ***0.0102***	***0.0613*** **0.0130**
All moa (79)	***0.0753*** **0.0150**	**0.0577** ***0.0132***	***0.1053*** **0.0205**
All tyrannosaurids (10)	0.0284 0.0100	0.0370 0.0106	0.0549 0.0160
Ornithomimosaurs (10)	0.0404 0.0127	0.0418 0.0120	0.0432 0.0124
Tyrannosaurids + ornithomimosaurs (20)	**0.0404** **0.0111**	***0.0887*** ***0.0307***	***0.0884*** ***0.0320***
Tenontosaurus spp. (8)	0.0360 0.0109	0.0506 0.0164	0.0706 0.0222
Iguanodon bernissartensis (14)	0.0470 0.0114	0.0646 0.0172	0.0968 0.0215
All hadrosaurids (20)	0.0550 0.0144	0.0715 0.0163	0.0672 0.0168
Iguanodon and similar taxa + hadrosaurids (37)	***0.0709*** ***0.0146***	***0.0725*** ***0.0184***	***0.1089*** ***0.0221***

[a] To increase the number of specimens, in crocodylians digit IV length is treated as the aggregate length of phalanges IV1–IV3.
[b] Includes a small (presumably juvenile) individual identified as ?*P. elephantopus*, but excludes a specimen assigned to this species on molecular grounds.

Table A1.12. Bivariate and multivariate comparisons of intraspecific and interspecific phalangeal and digital proportions in crocodylians, ground birds, and non-avian dinosaurs. See text for details about the composition of interspecific samples. With a few exceptions, data are reported for taxa represented by 10 or more specimens. Variability is expressed in terms of the coefficient of relative dispersion around the reduced major axis (D_d) of the log-transformed variables (for emus [*Dromaius novaehollandiae*] the value of D_d for the same relationship is also reported, using variables that have not been log transformed, in parentheses after the log-transformed relationship). The "big seven" phalanges are II1–II2, III1–III3, and IV1–IV2. Values of D_d for interspecific comparisons that are greater than at least one of the corresponding intraspecific comparisons are indicated in **bold**; if the interspecific comparison is greater than all of the corresponding intraspecific comparisons, the interspecific D_d value is also *italicized*. Similarly, values for genera or more inclusive groups are compared with even more inclusive groups containing those genera or groups; parameters for which the more inclusive intergroup comparison results in larger values (greater variability) than for at least one of the less inclusive groups with which the comparison is made are indicated in **bold**; if the more inclusive group values are larger than all of the less inclusive group values of the same parameter, the more inclusive group value is also *italicized*.

Independent variable	Dependent variable	Taxon	D_d	N
Total length	Metatarsal III length	*Alligator mississippiensis*	1.47	13
	"Big seven" mean	*Alligator mississippiensis*	2.11	11
	Digit III length excluding ungual	*Alligator mississippiensis*	2.11	11
Femur length	Metatarsal III length	*Alligator mississippiensis*	1.35	11
		Extant crocodylians	**1.65**	42
		Extant crocodylians one per species	**1.85**	12
Metatarsal III or tarsometatarsus length	Digit III length excluding ungual	*Alligator mississippiensis*	1.63	20
		Extant crocodylians	**2.14**	48
		Extant crocodylians one per species	**2.42**	11
		Anomalopteryx didiformis	0.56	15
		D. robustus + *A. didiformis*	**2.03**	23
		All moa	***2.90***	49
		All moa one per species	***3.50***	9
Metatarsal III or tarsometatarsus length	"Big seven" mean	*Alligator mississippiensis*	1.18	19
		Extant crocodylians	**1.82**	43
		Extant crocodylians one per species	**2.65**	11
		Anomalopteryx didiformis	0.83	14
		D. robustus + *A. didiformis*	**1.33**	21
		All moa	***2.79***	43
		All moa one per species	***3.61***	9
III1L	I1L	*Alligator mississippiensis*	1.76	26
		Caiman crocodilus	2.18	10
		A. mississippiensis + *C. crocodilus*	***2.32***	36
		All crocodylians	***3.41***	62
		All crocodylians one per species	***2.71***	13
III1L	II1L	*Alligator mississippiensis*	1.23	25
		Caiman crocodilus	1.66	10
		A. mississippiensis + *C. crocodilus*	**1.33**	35
		All crocodylians	***1.69***	61
		All crocodylians one per species	**1.64**	14
		Rhea americana	0.85	13
		R. americana + *P. pennata*	**0.86**	22
		Dromaius novaehollandiae	1.00 (3.85)	13
		Casuarius spp.[a]	1.39	16
		Euryapteryx curtus	0.71	10
		Pachyornis elephantopus[b]	1.25	11
			1.22	12
		Dinornis robustus	0.51	27
		Anomalopteryx didiformis	0.81	22
		E. curtus + *A. didiformis*	***1.22***	32
		D. robustus + *A. didiformis*	**0.78**	49
		All moa	**0.98**	96
		All moa one per species	**0.96**	9
		Coelophysis bauri	1.35	10
		Tyrannosaurids	0.71	19
		Ornithomimosaurs	0.87	16
		Tyrannosaurids + ornithomimosaurs	**0.83**	35
		Late Cretaceous Asiamerican theropods	***3.75***	42

Table A1.12. *continued*

Independent variable	Dependent variable	Taxon	D_d	*N*
		Tenontosaurus spp.[c]	1.45	15
		Iguanodon bernissartensis	1.65	14
		All hadrosaurs	1.46	23
III1L	II2L	*Alligator mississippiensis*	2.33	23
		A. mississippiensis + *C. crocodilus*	**2.78**	31
		All crocodylians	***4.30***	55
		All crocodylians one per species	***3.65***	13
		Rhea americana	5.59	13
		R. americana + *P. pennata*	**6.71**	22
		Dromaius novaehollandiae	3.45 (8.75)	12
		Casuarius spp.	2.52	16
		Pachyornis elephantopus	3.76	11
			4.30	12
		Dinornis robustus	1.62	25
		Anomalopteryx didiformis	2.38	22
		E. curtus + *A. didiformis*	**4.32**	31
		D. robustus + *A. didiformis*	**2.30**	47
		All moa	**3.72**	92
		All moa one per species	**3.18**	9
		Tyrannosaurids	0.63	18
		Ornithomimosaurs	2.37	15
		Tyrannosaurids + ornithomimosaurs	***2.53***	33
		Late Cretaceous Asiamerican theropods	***3.79***	39
		Tenontosaurus spp.	1.93	14
		Iguanodon bernissartensis	2.58	15
		All hadrosaurs	3.12	24
III1L	III2L	*Alligator mississippiensis*	1.70	22
		A. mississippiensis + *C. crocodilus*	**3.03**	31
		All crocodylians	***3.65***	57
		All crocodylians one per species	***3.06***	13
		Rhea americana	0.97	13
		R. americana + *P. pennata*	**1.78**	22
		Dromaius novaehollandiae	0.80 (3.33)	13
		Casuarius spp.	0.86	16
		Euryapteryx curtus	1.16	10
		Pachyornis elephantopus	1.58	11
			1.52	12
		Dinornis robustus	1.45	26
		Anomalopteryx didiformis	1.18	20
		E. curtus + *A. didiformis*	***2.38***	30
		D. robustus + *A. didiformis*	***2.12***	46
		All moa	**2.30**	92
		All moa one per species	**1.84**	9
		Coelophysis bauri	1.13	11
		Tyrannosaurids	1.86	23
		Ornithomimosaurs	0.88	17
		Tyrannosaurids + ornithomimosaurs	***1.90***	40
		Late Cretaceous Asiamerican theropods	***2.21***	44
		Tenontosaurus spp.	1.18	14
		Iguanodon bernissartensis	2.19	15
		All hadrosaurs	3.61	24
III1L	III3L	*Alligator mississippiensis*	2.42	20
		A. mississippiensis + *C. crocodilus*	**2.82**	28
		All crocodylians	***4.22***	48
		All crocodylians one per species	***4.17***	11
		Rhea americana	3.09	13

Independent variable	Dependent variable	Taxon	D_d	*N*
		R. americana + *P. pennata*	**3.91**	22
		Dromaius novaehollandiae	3.07 (10.6)	13
		Casuarius spp.	2.25	16
		Euryapteryx curtus	2.62	10
		Pachyornis elephantopus	2.69	11
			2.93	12
		Dinornis robustus	2.16	25
		Anomalopteryx didiformis	2.76	19
		E. curtus + *A. didiformis*	***5.50***	29
		D. robustus + *A. didiformis*	***3.36***	44
		All moa	**4.39**	89
		All moa one per species	**3.18**	9
		Tyrannosaurids	2.77	18
		Ornithomimosaurs	4.63	17
		Tyrannosaurids + ornithomimosaurs	**3.68**	35
		Late Cretaceous Asiamerican theropods	**4.02**	41
		Tenontosaurus spp.	2.27	14
		Iguanodon bernissartensis	2.62	15
		All hadrosaurs	4.11	24
III1L	IV1L	*Alligator mississippiensis*	1.30	24
		Caiman crocodilus	3.07	10
		A. mississippiensis + *C. crocodilus*	**1.99**	34
		All crocodylians	***2.50***	60
		All crocodylians one per species	***2.08***	14
		Rhea americana	0.99	13
		R. americana + *P. pennata*	0.97	22
		Dromaius novaehollandiae	0.86 (3.44)	13
		Casuarius spp.	0.66	16
		Euryapteryx curtus	1.63	10
		Pachyornis elephantopus	0.95	12
			1.00	13
		Dinornis robustus	0.64	27
		Anomalopteryx didiformis	0.95	22
		E. curtus + *A. didiformis*	**1.36**	32
		D. robustus + *A. didiformis*	**0.78**	49
		All moa	**1.15**	96
		All moa one per species	***1.81***	9
		Coelophysis bauri	2.80	11
		Tyrannosaurids	1.12	20
		Ornithomimosaurs	1.78	13
		Tyrannosaurids + ornithomimosaurs	**1.56**	33
		Late Cretaceous Asiamerican theropods	***2.73***	40
		Tenontosaurus spp.	2.43	14
		Iguanodon bernissartensis	3.14	16
		Edmontosaurus spp.	1.36	9
		All hadrosaurs	1.27	23
III1L	IV2L	*Alligator mississippiensis*	2.04	19
		A. mississippiensis + *C. crocodilus*	**2.47**	27
		All crocodylians	***3.81***	50
		All crocodylians one per species	1.98	13
		Rhea americana	4.33	13
		R. americana + *P. pennata*	**4.95**	22
		Dromaius novaehollandiae	1.58 (6.48)	13
		Casuarius spp.	1.29	16
		Pachyornis elephantopus	3.10	11
			3.62	12

Table A1.12. *continued*

Independent variable	Dependent variable	Taxon	D_d	*N*
III1L	IV2L	*Dinornis robustus*	2.89	26
		Anomalopteryx didiformis	2.40	19
		E. curtus + *A. didiformis*	**3.08**	28
		D. robustus + *A. didiformis*	**2.76**	45
		All moa	**3.56**	89
		All moa one per species	2.22	9
		Coelophysis bauri	1.50	10
		Tyrannosaurids	1.70	20
		Ornithomimosaurs	3.32	17
		Tyrannosaurids + ornithomimosaurs	**3.03**	37
		Late Cretaceous Asiamerican theropods	***4.68***	44
		Tenontosaurus spp.	2.74	12
		Iguanodon bernissartensis	3.84	17
		All hadrosaurs	4.02	22
Digit III length excluding ungual	Digit II length excluding ungual	*Alligator mississippiensis*	0.73	20
		A. mississippiensis + *C. crocodilus*	**0.92**	27
		All crocodylians	***1.49***	46
		All crocodylians one per species	***1.18***	11
		Rhea americana	1.05	13
		R. americana + *P. pennata*	0.88	22
		Dromaius novaehollandiae	1.05 (4.25)	12
		Casuarius spp.	1.17	16
		Pachyornis elephantopus	1.16	10
			1.12	11
		Dinornis robustus	0.56	24
		Anomalopteryx didiformis	0.63	18
		E. curtus + *A. didiformis*	**1.08**	27
		D. robustus + *A. didiformis*	***0.82***	42
		All moa	**0.93**	85
		All moa one per species	**0.96**	9
		Tyrannosaurids	0.51	16
		Ornithomimosaurs	0.99	11
		Tyrannosaurids + ornithomimosaurs	**0.97**	27
		Late Cretaceous Asiamerican theropods	***1.53***	29
		Tenontosaurus spp.	1.01	13
		Iguanodon bernissartensis	0.87	14
		All hadrosaurs	1.14	23
Digit III length excluding ungual	Aggregate length of phalanges IV1–IV3	*Alligator mississippiensis*	0.91	17
		A. mississippiensis + *C. crocodilus*	0.83	21
		All crocodylians	***1.85***	37
		All crocodylians one per species	***1.91***	9
Digit III length excluding ungual	Digit IV length excluding ungual	*Rhea americana*	1.13	13
		R. americana + *P. pennata*	0.95	22
		Dromaius novaehollandiae	0.78 (3.66)	12
		Casuarius spp.	0.51	15
		Pachyornis elephantopus	1.00	10
			0.98	11
		Dinornis robustus	1.06	25
		Anomalopteryx didiformis	1.02	17
		E. curtus + *A. didiformis*	**1.36**	26
		D. robustus + *A. didiformis*	***1.07***	42
		All moa	***1.49***	82
		All moa one per species	**1.04**	9
		Tyrannosaurids	0.78	10
		Ornithomimosaurs	1.11	12
		Tyrannosaurids + ornithomimosaurs	***1.50***	22

Independent variable	Dependent variable	Taxon	D_d	*N*
		Late Cretaceous Asiamerican theropods	***3.11***	25
		Iguanodon bernissartensis	1.62	14
		All hadrosaurs	1.11	21
Digit II length excluding ungual	Aggregate length of phalanges IV1–IV3	*Alligator mississippiensis*	1.15	17
		A. mississippiensis + *C. crocodilus*	**1.25**	22
		All crocodylians	***2.63***	39
		All crocodylians one per species	***2.90***	10
Digit II length excluding ungual	Digit IV length excluding ungual	*Rhea americana*	0.88	13
		R. americana + *P. pennata*	0.87	22
		Dromaius novaehollandiae	1.40 (5.64)	12
		Casuarius spp.	1.89	15
		Pachyornis elephantopus	1.43	9
			1.36	10
		Dinornis robustus	0.99	25
		Anomalopteryx didiformis	1.24	17
		E. curtus + *A. didiformis*	**2.18**	25
		D. robustus + *A. didiformis*	**1.13**	42
		All moa	**1.63**	81
		All moa one per species	0.85	9
		Tyrannosaurids	0.70	11
		Ornithomimosaurs	1.20	11
		Tyrannosaurids + ornithomimosaurs	**1.07**	22
		Late Cretaceous Asiamerican theropods	***3.91***	26
		Iguanodon bernissartensis	1.49	15
		All hadrosaurs	1.50	21
Digit III length	Digit II length	*Rhea americana*	0.69	10
		R. americana + *P. pennata*	0.64	19
		Dromaius novaehollandiae	0.70 (3.38)	11
		Casuarius spp.	1.07	15
		Pachyornis elephantopus	0.86	10
			0.82	11
		Dinornis robustus	0.53	23
		Anomalopteryx didiformis	1.05	15
		E. curtus + *A. didiformis*	1.00	23
		D. robustus + *A. didiformis*	**0.81**	38
		All moa	**0.80**	77
		All moa one per species	**0.85**	9
		Iguanodon bernissartensis	0.83	11
		All hadrosaurs	0.96	17
Digit III length	Digit IV length	*Rhea americana*	0.95	11
		R. americana + *P. pennata*	0.87	20
		Dromaius novaehollandiae	0.59 (2.96)	11
		Casuarius spp.	0.41	14
		Pachyornis elephantopus	0.79	10
			0.77	11
		Dinornis robustus	0.85	24
		Anomalopteryx didiformis	0.78	15
		E. curtus + *A. didiformis*	**1.05**	24
		D. robustus + *A. didiformis*	***0.88***	39
		All moa	***1.13***	76
		All moa one per species	***1.07***	9
		Iguanodon bernissartensis	0.87	10
		All hadrosaurs	1.07	18
Digit II length	Digit IV length	*Rhea americana*	1.03	10
		R. americana + *P. pennata*	1.02	19
		Dromaius novaehollandiae	0.80 (3.64)	11

Table A1.12. *continued*

Independent variable	Dependent variable	Taxon	D_d	*N*
Digit II length	Digit IV length	*Casuarius* spp.	1.02	14
		Pachyornis elephantopus	1.27	9
			1.22	10
		Dinornis robustus	0.70	24
		Anomalopteryx didiformis	1.29	15
		E. curtus + *A. didiformis*	**1.68**	22
		D. robustus + *A. didiformis*	**0.94**	39
		All moa	**1.20**	75
		All moa one per species	0.55	9
		Iguanodon bernissartensis	0.74	10
		All hadrosaurs	0.93	15
"Big seven" mean excluding II1L	II1L	*Alligator mississippiensis*	1.13	19
		A. mississippiensis + *C. crocodilus*	**1.15**	25
		All crocodylians	***1.94***	43
		All crocodylians one per species	***2.01***	11
		Rhea americana	2.06	13
		R. americana + *P. pennata*	**2.80**	22
		Dromaius novaehollandiae	1.22	12
		Casuarius spp.	1.37	16
		Pachyornis elephantopus	1.53	10
			1.59	11
		Dinornis robustus	0.72	24
		Anomalopteryx didiformis	1.12	17
		E. curtus + *A. didiformis*	**3.16**	25
		D. robustus + *A. didiformis*	**1.03**	41
		All moa	**2.05**	82
		All moa one per species	**1.95**	9
		Tyrannosaurs	0.77	14
		Ornithomimosaurs	1.17	10
		Tyrannosaurids + ornithomimosaurs	***1.45***	24
		Late Cretaceous Asiamerican theropods	***3.99***	27
		Tenontosaurus spp.	1.43	10
		Iguanodon bernissartensis	1.50	14
		All hadrosaurs	2.06	20
"Big seven" mean excluding II2L	II2L	*Alligator mississippiensis*	1.19	19
		A. mississippiensis + *C. crocodilus*	**2.06**	25
		All crocodylians	***3.46***	43
		All crocodylians one per species	***2.32***	11
		Rhea americana	5.61	13
		R. americana + *P. pennata*	**5.81**	22
		Dromaius novaehollandiae	3.28	12
		Casuarius spp.	2.39	16
		Pachyornis elephantopus	2.76	10
			3.16	11
		Dinornis robustus	1.39	24
		Anomalopteryx didiformis	1.49	17
		E. curtus + *A. didiformis*	**2.52**	25
		D. robustus + *A. didiformis*	***1.50***	41
		All moa	**2.71**	82
		All moa one per species	**3.09**	9
		Tyrannosaurids	0.63	14
		Ornithomimosaurs	2.09	10
		Tyrannosaurids + ornithomimosaurs	**2.04**	24
		Late Cretaceous Asiamerican theropods	***2.99***	27
		Tenontosaurus spp.	1.52	10
		Iguanodon bernissartensis	3.16	14

Independent variable	**Dependent variable**	**Taxon**	D_d	N
		All hadrosaurs	2.15	20
		Alligator mississippiensis	0.81	19
		A. mississippiensis + *C. crocodilus*	**1.16**	25
		All crocodylians	***1.95***	43
		All crocodylians one per species	***1.63***	11
		Rhea americana	1.83	13
		R. americana + *P. pennata*	**2.62**	22
		Dromaius novaehollandiae	1.13	12
		Casuarius spp.	1.03	16
		Pachyornis elephantopus	1.63	10
			1.61	11
		Dinornis robustus	0.87	24
"Big seven" mean excluding III1L	III1L	*Anomalopteryx didiformis*	1.13	17
		E. curtus + *A. didiformis*	**2.09**	25
		D. robustus + *A. didiformis*	**1.06**	41
		All moa	**1.81**	82
		All moa one per species	**1.11**	9
		Tyrannosaurids	0.58	14
		Ornithomimosaurs	1.27	10
		Tyrannosaurids + ornithomimosaurs	**1.08**	24
		Late Cretaceous Asiamerican theropods	***1.38***	27
		Tenontosaurus spp.	0.81	10
		Iguanodon bernissartensis	1.60	14
		All hadrosaurs	2.51	20
		Alligator mississippiensis	0.85	19
		A. mississippiensis + *C. crocodilus*	**1.22**	25
		All crocodylians	**1.19**	43
		All crocodylians one per species	**1.21**	11
		Rhea americana	1.92	13
		R. americana + *P. pennata*	1.54	22
		Dromaius novaehollandiae	0.99	12
		Casuarius spp.	1.29	16
		Pachyornis elephantopus	1.14	10
			1.18	11
		Dinornis robustus	1.07	24
"Big seven" mean excluding III2L	III2L	*Anomalopteryx didiformis*	1.15	17
		E. curtus + *A. didiformis*	**1.47**	25
		D. robustus + *A. didiformis*	***1.25***	41
		All moa	***1.51***	82
		All moa one per species	**1.32**	9
		Tyrannosaurids	0.61	14
		Ornithomimosaurs	0.70	10
		Tyrannosaurids + ornithomimosaurs	**1.79**	24
		Late Cretaceous Asiamerican theropods	***3.20***	27
		Tenontosaurus spp.	1.05	10
		Iguanodon bernissartensis	2.22	14
		All hadrosaurs	2.67	20

Table A1.12. *continued*

Independent variable	Dependent variable	Taxon	D_d	N
		Alligator mississippiensis	2.11	19
		A. mississippiensis + *C. crocodilus*	**2.45**	25
		All crocodylians	***3.97***	43
		All crocodylians one per species	***3.66***	11
		Rhea americana	3.06	13
		R. americana + *P. pennata*	2.91	22
		Dromaius novaehollandiae	2.51	12
		Casuarius spp.	2.11	16
		Pachyornis elephantopus	2.08	10
			2.61	11
		Dinornis robustus	2.15	24
"Big seven" mean excluding III3L	III3L	*Anomalopteryx didiformis*	2.52	17
		E. curtus + *A. didiformis*	**4.23**	25
		D. robustus + *A. didiformis*	***2.94***	41
		All moa	**3.82**	82
		All moa one per species	**3.52**	9
		Tyrannosaurids	1.20	14
		Ornithomimosaurs	2.22	10
		Tyrannosaurids + ornithomimosaurs	**1.87**	24
		Late Cretaceous Asiamerican theropods	**2.11**	27
		Tenontosaurus spp.	3.25	10
		Iguanodon bernissartensis	1.81	14
		All hadrosaurs	3.39	20
		Alligator mississippiensis	1.47	19
		A. mississippiensis + *C. crocodilus*	**2.07**	25
		All crocodylians	***2.35***	43
		All crocodylians one per species	***2.70***	11
		Rhea americana	2.25	13
		R. americana + *P. pennata*	**2.68**	22
		Dromaius novaehollandiae	0.87	12
		Casuarius spp.	1.02	16
		Pachyornis elephantopus	1.57	10
			1.77	11
		Dinornis robustus	0.90	24
"Big seven" mean excluding IV1L	IV1L	*Anomalopteryx didiformis*	1.32	17
		E. curtus + *A. didiformis*	**2.96**	25
		D. robustus + *A. didiformis*	**1.07**	41
		All moa	**2.52**	82
		All moa one per species	**2.73**	9
		Tyrannosaurids	1.22	14
		Ornithomimosaurs	1.52	10
		Tyrannosaurids + ornithomimosaurs	**1.48**	24
		Late Cretaceous Asiamerican theropods	***2.22***	27
		Tenontosaurus spp.	2.22	10
		Iguanodon bernissartensis	2.47	14
		All hadrosaurs	2.56	20

Independent variable	Dependent variable	Taxon	D_d	*N*
		Alligator mississippiensis	2.12	19
		A. mississippiensis + *C. crocodilus*	**2.56**	25
		All crocodylians	***3.85***	43
		All crocodylians one per species	**2.30**	11
		Rhea americana	4.29	13
		R. americana + *P. pennata*	3.92	22
		Dromaius novaehollandiae	1.85	12
		Casuarius spp.	1.37	16
		Pachyornis elephantopus	3.34	10
			3.89	11
		Dinornis robustus	2.31	24
"Big seven" mean excluding IV2L	IV2L	*Anomalopteryx didiformis*	1.80	17
		E. curtus + *A. didiformis*	**2.37**	25
		D. robustus + *A. didiformis*	**2.15**	41
		All moa	**3.08**	82
		All moa one per species	**2.10**	9
		Tyrannosaurids	1.38	14
		Ornithomimosaurs	1.87	10
		Tyrannosaurids + ornithomimosaurs	***2.13***	24
		Late Cretaceous Asiamerican theropods	***4.61***	27
		Tenontosaurus spp.	2.10	10
		Iguanodon bernissartensis	3.83	14
		All hadrosaurs	3.11	20

[a] Mostly = *Casuarius casuarius.*
[b] Specimens in this table exclude the bird identified as *P. elephantopus* on molecular grounds. Data in the first line exclude the juvenile specimen assigned to this species; data in the second line include it.
[c] Mostly = *Tenontosaurus tilletti.*

Table A1.13. Summary of selected bivariate and multivariate comparisons of intraspecific variablility in phalangeal proportions of crocodylians, ground birds, and non-avian dinosaurs. Variability is expressed in terms of the coefficient of relative dispersion around the reduced major axis (D_d) of the log-transformed variables. Results are reported for species for which data were available for at least 10 specimens; see table A1.12 for details. For all species, D_d is reported for the RMA (reduced major axis) relationship between the phalanx length of interest and the length of phalanx III1 (first row of data; data summarized from table A1.12). For some species, D_d is also reported for the RMA relationship between the phalanx length of interest and the mean of the "big seven" log-transformed variables excluding the phalanx length of interest (second row of data). Empty cells indicate comparisons that were either irrelevant (lengths of III1 scaled to itself were defined as taking a value of 1) or were not reported because of the small number of data cases.

Species	II1L	II2L	III1L	III2L	III3L	IV1L	IV2L
Alligator mississippiensis	1.23	2.33	——	1.70	2.42	1.30	2.04
	1.13	1.19	0.81	0.85	2.11	1.47	2.12
Caiman crocodilus	1.66	——	——	——	——	3.07	——
Rhea americana	0.85	5.59	——	0.97	3.09	0.99	4.33
	2.06	5.61	1.83	1.92	3.06	2.25	4.29
Dromaius novaehollandiae	1.00	3.45	——	0.80	3.07	0.86	1.58
	1.22	3.28	1.13	0.99	2.51	0.87	1.85
Euryapteryx curtus	0.71	——	——	1.16	2.62	1.63	——
Pachyornis elephantopus[a]	1.22	3.76	——	1.52	2.93	1.00	3.62
	1.59	3.16	1.61	1.18	2.61	1.77	3.89
Dinornis robustus	0.51	1.62	——	1.45	2.16	0.64	2.89
	0.72	1.39	0.87	1.07	2.15	0.90	2.31
Anomalopteryx didiformis	0.81	2.38	——	1.18	2.76	0.95	2.40
	1.12	1.49	1.13	1.15	2.52	1.32	1.80
Coelophysis bauri	1.35	——	——	1.13	——	2.80	1.50
Iguanodon bernissartensis	1.65	2.58	——	2.19	2.62	3.14	3.84
	1.50	3.16	1.60	2.22	1.81	2.47	3.83

[a] Specimens exclude the bird identified as *P. elephantopus* on molecular grounds, but include the juvenile specimen assigned to this species.

Table A1.14. Summary of selected bivariate and multivariate comparisons of interspecific variablility in phalangeal proportions of crocodylians, ground birds, and non-avian dinosaurs. Variability is expressed in terms of the coefficient of relative dispersion around the reduced major axis (D_d) of the log-transformed variables. Results are reported for comparisons for which data were available for at least 10 specimens; see table A1.12 for details. For all comparisons, D_d is reported for the RMA (reduced major axis) relationship between the phalanx length of interest and the length of phalanx III1 (first row of data), and also for the RMA relationship between the phalanx length of interest and the mean of the big-seven log-transformed variables excluding the phalanx length of interest (second row of data). Parameters for which the interspecific comparison results in larger values (greater variability) than for at least one of the species with which the comparison is made (table A1.13) are indicated in **bold**; if the interspecific values are larger than all of the intraspecific values of the same parameter, the interspecific value is also *italicized* (note, though, that in many comparisons there is only one species with which the interspecific sample can be compared, due to the number of cases in each of the species). Similarly, values for genera or more inclusive groups are compared with even more inclusive groups containing those genera or groups; parameters for which the more inclusive intergroup comparison results in larger values (greater variability) than for at least one of the less inclusive groups with which the comparison is made are indicated in **bold**; if the more inclusive group values are larger than all of the less inclusive group values of the same parameter, the more inclusive group value is also *italicized*.

Comparison	II1L	II2L	III1L	III2L	III3L	IV1L	IV2L
Two-species comparisons:							
A. mississippiensis + *C. crocodilus*	**1.33**	**2.78**	———	**3.03**	**2.82**	**1.99**	**2.47**
	1.15	**2.06**	**1.16**	**1.22**	**2.45**	**2.07**	**2.56**
R. americana + *P. pennata*	**0.86**	**6.71**	———	**1.78**	**3.91**	0.97	**4.95**
	2.80	**5.81**	**2.62**	1.54	2.91	**2.68**	3.92
E. curtus + *A. didiformis*	***1.22***	**4.32**	———	***2.38***	***5.50***	**1.36**	**3.08**
	3.16	**2.52**	**2.09**	**1.47**	**4.23**	**2.96**	**2.37**
D. robustus + *A. didiformis*	**0.78**	**2.30**	———	***2.12***	***3.36***	**0.78**	**2.76**
	1.03	***1.50***	**1.06**	***1.25***	***2.94***	**1.07**	**2.15**
Comparisons among three or more species:							
All crocodylians	**1.69**	**4.30**	———	**3.65**	**4.22**	**2.50**	**3.81**
	1.94	**3.46**	**1.95**	1.19	**3.97**	**2.35**	**3.85**
All moa	**0.98**	**3.72**	———	**2.30**	**4.39**	**1.15**	**3.56**
	2.05	***2.71***	**1.81**	***1.51***	**3.82**	**2.52**	***3.08***
Tyrannosaurids	0.71	0.63	———	1.86	2.77	1.12	1.70
	0.77	0.63	0.58	0.61	1.20	1.22	1.38
Ornithomimosaurs	0.87	2.37	———	0.88	4.63	1.78	3.32
	1.17	2.09	1.27	0.70	2.22	1.52	1.87
Tyrannosaurids + ornithomimosaurs	**0.83**	***2.53***	———	***1.90***	**3.68**	**1.56**	**3.03**
	1.45	**2.04**	**1.08**	***1.79***	**1.87**	**1.48**	***2.13***
Late K Asiamerican theropods	***3.75***	***3.79***	———	***2.21***	**4.02**	***2.73***	***4.68***
	3.99	***2.99***	***1.38***	***3.20***	**2.11**	***2.22***	***4.61***
Hadrosaurids	1.46	3.12	———	3.61	4.11	1.27	4.02
	2.06	2.15	2.51	2.67	3.39	2.56	3.11

Table A1.15. Summary of selected bivariate comparisons of intraspecific variability in digit proportions of crocodylians, ground birds, and non-avian dinosaurs. Variability is expressed in terms of the coefficient of relative dispersion around the reduced major axis (D_d) of the log-transformed variables. Results are reported for species for which data were available for at least 10 specimens; see table A1.12 for details. For all species, D_d is reported for the RMA (reduced major axis) relationship between the digit length of interest and the corresponding length of either digit III or digit IV. Empty cells (for alligators) indicate comparisons that could not be made due to the nature of my data.

Species	Digit III length excluding ungual compared with:		Digit II length excluding ungual compared with digit IV length excluding ungual	Digit III length compared with:		Digit II length compared with digit IV length
	Digit II length excluding ungual	Digit IV length excluding ungual		Digit II length	Digit IV length	
Alligator mississippiensis	0.73	0.91[a]	1.15[a]	——	——	——
Rhea americana	1.05	1.13	0.88	0.69	0.95	1.03
Dromaius novaehollandiae	1.05	0.78	1.40	0.70	0.59	0.80
Pachyornis elephantopus	1.12	0.98	1.36	0.82	0.77	1.22
Dinornis robustus	0.56	1.06	0.99	0.53	0.85	0.70
Anomalopteryx didiformis	0.63	1.02	1.24	1.05	0.78	1.29
Iguanodon bernissartensis	0.87	1.62	1.49	0.83	0.87	0.74

[a] Digit IV length without the ungual is approximated for crocodylians as the aggregate length of phalanges 1 through 3.
[b] Specimens exclude the bird identified as *P. elephantopus* on molecular grounds, but include the juvenile specimen assigned to this species.

Table A1.16. Summary of selected bivariate comparisons of interspecific variability in digit proportions of crocodylians, ground birds, and non-avian dinosaurs. Variability is expressed in terms of the coefficient of relative dispersion around the reduced major axis (D_d) of the log-transformed variables. Results are reported for comparisons for which data were available for at least 10 specimens; see table A1.12 for details. For all species, D_d is reported for the RMA (reduced major axis) relationship between the digit length of interest and the corresponding length of either digit III or digit IV. Empty cells indicate comparisons that could not be made, or for which the number of data cases was too small to justify reporting. Parameters for which the interspecific comparison results in larger values (greater variability) than for at least one of the species with which the comparison is made (table A1.15) are indicated in **bold**; if the interspecific values are larger than all of the intraspecific values of the same parameter, the interspecific value is also *italicized* (note, though, that in many comparisons there is only one species with which the interspecific sample can be compared, due to the number of cases in each of the species). Similarly, values for genera or more inclusive groups are compared with even more inclusive groups containing those genera or groups; parameters for which the more inclusive intergroup comparison results in larger values (greater variability) than for at least one of the less inclusive groups with which the comparison is made are indicated in **bold**; if the more inclusive group values are larger than all of the less inclusive group values of the same parameter, the more inclusive group value is also *italicized.*

Species	Digit III length excluding ungual compared with:		Digit II length excluding ungual compared with digit IV length excluding ungual	Digit III length compared with:		Digit II length compared with digit IV length
	Digit II length excluding ungual	Digit IV length excluding ungual		Digit II length	Digit IV length	
Two-species comparisons:						
A. mississippiensis + *C. crocodilus*	**0.92**	0.83[a]	**1.25**[a]	——	——	——
R. americana + *P. pennata*	0.88	0.95	0.87	0.64	0.87	1.02
E. curtus + *A. didiformis*	**1.08**	**1.36**	**2.18**	1.00	**1.05**	**1.68**
D. robustus + *A. didiformis*	***0.82***	***1.07***	**1.13**	**0.81**	***0.88***	**0.94**
Comparisons among three or more species:						
All crocodylians:	**1.49**	**1.85**[a]	**2.63**[a]	——	——	——
All moa	**0.93**	***1.49***	**1.63**	0.80	***1.13***	**1.20**
Tyrannosaurids	0.51	0.78	0.70	——	——	——
Ornithomimosaurs	0.99	1.11	1.20	——	——	——
Tyrannosaurids + ornithomimosaurs	**0.97**	***1.50***	**1.07**	——	——	——
Late K Asiamerican Theropods	***1.53***	***3.11***	***3.91***	——	——	——
Hadrosaurids	1.14	1.11	1.50	0.96	1.07	0.93

[a] Digit IV length without the ungual is approximated for crocodylians as the aggregate length of phalanges 1 through 3.

Table A1.17. Summary comparison of different measures of variability of phalangeal and digital proportions for comparisons involving one, two, or three or more species of birds and non-avian dinosaurs. Variability comparisons are summarized for (1) the first phalanges of digits II–IV (pooled samples of the lengths of II1 and IV1 scaled to a common value of the length of III1, and of log-transformed lengths of II1, III1, and IV1 "GM-scaled" against the mean of the log-transformed lengths all of the "big seven" phalanges); (2) the distal "big seven" non-ungual phalanges (lengths of II2, III2, III3, and IV2 scaled to a common length of III1; log-transformed lengths of the same distal phalanges "GM-scaled" against the mean of the log-transformed lengths of all the "big seven" phalanges); (3) digit lengths excluding the ungual (lengths of digit II and digit IV scaled to a common value of digit III length; log-transformed lengths of digits II, III, and IV "GM-scaled" against the mean of the log-transformed lengths of the three digits); and (4) digit lengths including the ungual (lengths of digit II and IV scaled to a common value of digit III length). Each row of data (for one, two, or three or more species) is split into two rows; the upper row indicates the median value for each measure of variability across all of the comparisons made involving the number of species and the particular parameters under consideration, while the lower row indicates the range of values of each measure of variability for those comparisons, and (in parentheses) the number of those comparisons that could be made. Empty cells indicate that too few comparisons could be made to justify reporting. GM = geometric mean; RMA = reduced major axis.

Number of species	Minimum/maximum ratio of scaled length	Coefficient of variation of scaled length	Range of GM-scaled length	Standard deviation of GM-scaled length	Coefficient of relative dispersion around RMA
(1) First phalanges (II1 and IV1 scaled to III1; GM-scaled II1–IV1)					
One	1.13	3.72	0.0736	0.0181	0.99
	1.08–1.86 (23)	2.30–15.7 (23)	0.0357–0.1743 (16)	0.0111–0.0456 (16)	0.51–3.14 (23)
Two	1.19	4.17	0.1111	0.0338	1.09
	1.10–1.86 (18)	2.57–18.2 (18)	0.0609–0.1393 (8)	0.0166–0.0436 (8)	0.78–2.43 (10)
Three or more	1.31	6.61	0.1256	0.0329	1.15
	1.09–1.86 (28)	3.04–11.6 (28)	0.0541–0.1993 (14)	0.0167–.0611 (14)	0.66–2.37 (13)
(2) Distal phalanges (II2, III2, III3, IV2)					
One	1.39	8.84	0.1055	0.0305	2.39
	1.07–2.04 (46)	2.96–20.5 (46)	0.0222–0.2120 (32)	0.0084–0.0560 (32)	0.80–5.59 (32)
Two	1.50	10.7	0.1117	0.0287	2.75
	1.14–2.11 (36)	4.26–24.4 (36)	0.0444–0.2297 (16)	0.0128–0.0689 (16)	1.18–6.71 (20)
Three or more	1.77	12.7	0.1382	0.0387	2.73
	1.12–4.05 (56)	3.23–29.7 (56)	0.0326–0.3301 (28)	0.0102–0.0988 (28)	0.63–4.63 (25)
(3) Digit lengths excluding unguals (digits II and IV scaled to digit III; GM-scaled digits II–IV)					
One	1.18	4.62	0.0423	0.0121	1.02
	1.08–1.45 (19)	2.57–8.46 (19)	0.0211–0.0968 (20)	0.0062–0.0215 (20)	0.45–1.62 (15)
Two	1.23	5.44	0.0523	0.0128	0.98
	1.13–1.49 (16)	2.56–9.52 (16)	0.0360–0.0827 (11)	0.0092–0.0229 (11)	0.82–1.58 (10)
Three or more	1.27	5.82	0.0691	0.0155	1.05
	1.10–1.95 (26)	2.76–14.4 (26)	0.0284–0.1127 (16)	0.0100–0.0320 (16)	0.51–1.50 (12)
(4) Digit lengths (digits II and IV scaled to digit III)					
One	1.13	4.11	——	——	——
	1.05–1.23 (18)	1.81–7.39 (18)	——	——	——
Two	1.19	4.79	——	——	——
	1.11–1.32 (14)	2.94–6.02 (14)	——	——	——
Three	1.19	5.39	——	——	——
	1.06–1.59 (24)	2.21–11.7 (24)	——	——	——

Table A4.1. Measurements of the feet of study skins of ground birds. All measurements are in millimeters. Pads are numbered with the most proximal labeled Pad1, and increasing in number more distally; values of zero on a more distal pad indicate that the more distal pad was not separated from the more proximal pad by a crease, and so the entire length was assigned to the more proximal pad. For birds with the "galliform" foot configuration, the length of the toe excluding the claw may be greater than the combined lengths of the digital pads because of the presence of interpad spaces. Because of the desiccated nature of study skins, all soft-tissue measurements of the foot should be regarded as approximate. Foot = overall length measurement of the toe region of the foot; L = length; TMT = rough tarsometatarsus length; see chapter 4 for a description. It was not possible to measure all parameters on all skins, and some parameters were not measured on some specimens.

Specimen	Measurements
Emus	
Dromaius novaehollandiae UMMZ Birds 2963	TMT: 70; foot: 51
Dromaius novaehollandiae UMMZ Birds 2964	TMT: 53; foot: 34 IIPad1L: 9 IIPad2L: 0 IIClawL: 6 IIIPad1L: 7 IIIPad2L: 7 IIIPad3L: 6 IIIClawL: 5 Digit II length excluding claw: 9; digit II length: 15; digit III length excluding claw: 20; digit III length: 25; digit IV length excluding claw: 13; digit IV length: 17
Dromaius novaehollandiae UMMZ Birds 2965	TMT: 73; foot: 50 IIPad1L: 6 IIPad2L: 5 IIClawL: 6 IIIPad1L: 9 IIIPad2L: 11 IIIPad3L: 7 IIIClawL: 6 Digit II length excluding claw: 11; digit II length: 17; digit III length excluding claw: 27; digit III length: 33; digit IV length excluding claw: 17; digit IV length: 21
Dromaius novaehollandiae UMMZ Birds 2967	TMT: 57; foot: 37 IIPad1L: 5 IIPad2L: 5 IIClawL: 4 IIIPad1L: 6 IIIPad2L: 6 IIIPad3L: 5 IIIClawL: 5 Digit II length excluding claw: 10; digit II length: 14; digit III length excluding claw: 17; digit III length: 22; digit IV length excluding claw: 12; digit IV length: 15
Dromaius novaehollandiae UMMZ Birds 2968	TMT: 165; foot: 79 IIPad1L: 9 IIPad2L: 7 IIClawL: 11 IIIPad1L: 15 IIIPad2L: 23 IIIPad3L: 0 IIIClawL: 11 Digit II length excluding claw: 16; digit II length: 27; digit III length excluding claw: 38; digit III length: 49; digit IV length excluding claw: 22; digit IV length: 29
Dromaius novaehollandiae UMMZ Birds 2970a unhatched chick	TMT: 34; foot: 25 IIPad1L: 8 IIPad2L: 0 IIClawL: 2 IIIPad1L: 5 IIIPad2L: 8 IIIPad3L: 0 IIIClawL: 3 Digit II length excluding claw:8; digit II length: 11; digit III length excluding claw: 13; digit III length: 16; digitIV length excluding claw: 10; digit IV length: 12
Dromaius novaehollandiae UMMZ Birds 2970b unhatched chick	TMT: 22; foot: 18 IIPad1L:5 IIPad2L: 0 IIClawL: 2 IIIPad1L: 4 IIIPad2L: 6 IIIPad3L:0 IIIClawL: 2 Digit II length excluding claw: 5; digit II length: 7; digit III length excluding claw: 10; digit III length: 12; digit IV length excluding claw: 6; digit IV length: 7
Dromaius novaehollandiae UMMZ Birds 119173	TMT: 315; foot: 155 IIPad1L: 21 IIPad2L: 20 IIClawL: 18 IIIPad1L: 33 IIIPad2L: 29 IIIPad3L: 24 IIIClawL: 17 Digit II length excluding claw: 41; digit II length: 59; digit III length excluding claw: 86; digit III length: 103; digit IV length excluding claw: 56; digit IV length: 71
Dromaius novaehollandiae MSU OR.8778 (accession number 1547)	TMT: 361; foot: 171 IIPad1L: 19 IIPad2L: 21 IIClawL: 20 IIIPad1L: 34 IIIPad2L: 29 IIIPad3L: 31 IIIClawL: 18 Digit II length excluding claw: 40; digit II length: 60; digit III length excluding claw: 94; digit III length: 112; digit IV length excluding claw: 59; digit IV length: 73
Dromaius novaehollandiae QM O4364	TMT: 102; foot: 45 IIPad1L: 5 IIPad2L: 5 IIClawL: 5 IIIPad1L: 10 IIIPad2L: 9 IIIPad3L: 9 IIIClawL: 6 Digit II length excluding claw: 10; digit II length: 15; digit III length excluding claw: 28; digit III length:34; digit IV length excluding claw: 15; digit IV length: 19
Dromaius novaehollandiae QM O4801	TMT: 87; foot: 40 IIPad1L: 6 IIPad2L: 6 IIClawL: 6 IIIPad1L: 9 IIIPad2L: 8 IIIPad3L: 8 IIIClawL: 6 Digit II length excluding claw: 12; digit II length: 18; digit III length excluding claw: 25; digit III length: 31; digit IV length excluding claw: 18; digit IV length: 22

Table A4.1. *continued*

Specimen	Measurements
Dromaius novaehollandiae QM O6159	TMT: 68; foot: 38 IIPad1L: 5 IIPad2L: 5 IIClawL: 5 IIIPad1L: 8 IIIPad2L: 7 IIIPad3L: 7 IIIClawL: 5 Digit II length excluding claw: 10; digit II length: 15; digit III length excluding claw: 22; digit III length: 27; digit IV length excluding claw: 15; digit IV length: 16
Dromaius novaehollandiae QM O6631	TMT: 280; foot: 110 IIPad1L: ——— IIPad2L: ——— IIClawL: 14 IIIPad1L: 26 IIIPad2L: 24 IIIPad3L: 22 IIIClawL: 15 Digit II length excluding claw: ———; digit II length: ———; digit III length excluding claw: 72; digit III length: 87; digit IV length excluding claw: ———; digit IV length: ———
Dromaius novaehollandiae QM O6160	TMT: 73; foot: 37 IIPad1L: 5 IIPad2L: 5 IIClawL: 5 IIIPad1L: 8 IIIPad2L: 7 IIIPad3L: 8 IIIClawL: 5 Digit II length excluding claw: 10; digit II length: 15; digit III length excluding claw: 23; digit III length: 28; digit IV length excluding claw: 15; digit IV length: 18
Dromaius novaehollandiae QM O6803	TMT: 400; foot: ——— IIPad1L: 22 IIPad2L: 19 IIClawL: 21 IIIPad1L: 34 IIIPad2L: 34 IIIPad3L: 28 IIIClawL: 19 Digit II length excluding claw: 41; digit II length: 62; digit III length excluding claw: 96; digit III length: 115; digit IV length excluding claw:——— ; digit IV length: ———
Dromaius novaehollandiae QM O10007	TMT: 85; foot: 39 IIPad1L: 7 IIPad2L: 7 IIClawL: 5 IIIPad1L: 9 IIIPad2L: 9 IIIPad3L: 7 IIIClawL: 5 Digit II length excluding claw: 14; digit II length: 19; digit III length excluding claw: 25; digit III length: 30; digit IV length excluding claw: 16; digit IV length: 20
Dromaius novaehollandiae QM O10008	TMT: 70; foot: 37 IIPad1L: 6 IIPad2L: 6 IIClawL: 4 IIIPad1L: 9 IIIPad2L: 8 IIIPad3L: 8 IIIClawL: 5 Digit II length excluding claw: 12; digit II length: 16; digit III length excluding claw: 25; digit III length: 30; digit IV length excluding claw: 17; digit IV length: 20
Dromaius novaehollandiae QM O11893	TMT: 182; foot: 90 IIPad1L:——— IIPad2L: 11 IIClawL: 7 IIIPad1L: 18 IIIPad2L: 17 IIIPad3L: 14 IIIClawL: 4 Digit II length excluding claw: ———; digit II length: ———; digit III length excluding claw: 49; digit III length: 53; digit IV length excluding claw: ———; digit IV length: ———
Dromaius novaehollandiae QM O16140	TMT: ———; foot: 154 IIPad1L: 18 IIPad2L: 15 IIClawL: 20 IIIPad1L: 39 IIIPad2L: 26 IIIPad3L: 25 IIIClawL: 19 Digit II length excluding claw: 33; digit II length: 53; digit III length excluding claw: 90; digit III length: 109; digit IV length excluding claw: 54; digit IV length: 71
Dromaius novaehollandiae QM O22509	TMT: 52; foot: 33 IIPad1L: 4 IIPad2L: 5 IIClawL: 5 IIIPad1L: 7 IIIPad2L: 6 IIIPad3L: 6 IIIClawL: 5 Digit II length excluding claw: 9; digit II length: 14; digit III length excluding claw: 19; digit III length: 24; digit IV length excluding claw: 12; digit IV length: 16
Dromaius novaehollandiae QM O29273	TMT: ———; foot: 77 IIPad1L: 13 IIPad2L: 9 IIClawL: 9 IIIPad1L: 18 IIIPad2L: 17 IIIPad3L: 13 IIIClawL: 9 Digit II length excluding claw: 22; digit II length: 31; digit III length excluding claw: 48; digit III length: 57; digit IV length excluding claw: 33; digit IV length: 40
Dromaius novaehollandiae Fort Wayne Zoo dead bird	TMT: ———; foot: 140 IIPad1L: 29 IIPad2L: 0 IIClawL: 14 IIIPad1L: 28 IIIPad2L: 23 IIIPad3L: 22 IIIClawL: 16 Digit II length excluding claw: 29; digit II length: 43; digit III length excluding claw: 73; digit III length: 89; digit IV length excluding claw: 40; digit IV length: 55
Dromaius novaehollandiae Oregon State University pickled 3-month-old bird	TMT: 186; foot: 91

Specimen	Measurements			
Dromaius novaehollandiae Oregon State bird 2	TMT: 355; foot: 161			
	IIPad1L: 17	IIPad2L: 19	IIClawL: ———	
	IIIPad1L: 34	IIIPad2L: 29	IIIPad3L: 26	IIIClawL: 21
	Digit II length excluding claw: 36; digit II length: ———; digit III length excluding claw: 89; digit III length: 110; digit IV length excluding claw: 54; digit IV length: 74			
Dromaius novaehollandiae Oregon State bird 3	TMT: 358; foot: 179			
	IIPad1L: 21	IIPad2L: 13	IIClawL: 23	
	IIIPad1L: 34	IIIPad2L: 28	IIIPad3L: 28	IIIClawL: 22
	Digit II length excluding claw: 34; digit II length: 57; digit III length excluding claw: 90; digit III length: 112; digit IV length excluding claw: 55; digit IV length: 76			
Dromaius novaehollandiae Oregon State bird 4	TMT: 359; foot: 173			
	IIPad1L: 18	IIPad2L: 17	IIClawL: 25	
	IIIPad1L: 36	IIIPad2L: 35	IIIPad3L: 23	IIIClawL: 23
	Digit II length excluding claw: 35; digit II length: 60; digit III length excluding claw: 94; digit III length: 117; digit IV length excluding claw: 57; digit IV length: 74			
Dromaius novaehollandiae Oregon State bird 5	TMT: 361; foot: 181			
	IIPad1L: 19	IIPad2L: 21	IIClawL: 27	
	IIIPad1L: 35	IIIPad2L: 31	IIIPad3L: 31	IIIClawL: 25
	Digit II length excluding claw: 40; digit II length: 67; digit III length excluding claw: 97; digit III length: 122; digit IV length excluding claw: 57; digit IV length: 78			
Dromaius novaehollandiae Oregon State bird 6	TMT: 329; foot: 154			
	IIPad1L: 15	IIPad2L: 16	IIClawL: 17	
	IIIPad1L: 35	IIIPad2L: 52	IIIPad3L: 0	IIIClawL: 19
	Digit II length excluding claw: 31; digit II length: 48; digit III length excluding claw: 87; digit III length: 106; digit IV length excluding claw: 50; digit IV length: 64			
Dromaius novaehollandiae Oregon State bird 7	TMT: 344; foot:			
	IIPad1L: 18	IIPad2L: 16	IIClawL: 22	
	IIIPad1L: 43	IIIPad2L: 29	IIIPad3L: 26	IIIClawL: 18
	Digit II length excluding claw: 34; digit II length: 56; digit III length excluding claw: 98; digit III length: 116; digit IV length excluding claw: 55; digit IV length: 70			
Dromaius novaehollandiae Oregon State bird 8	TMT: 363; foot: 160			
	IIPad1L: 16	IIPad2L: 16	IIClawL: 21	
	IIIPad1L: 34	IIIPad2L: 28	IIIPad3L: 27	IIIClawL: 21
	Digit II length excluding claw: 32; digit II length: 53; digit III length excluding claw: 89; digit III length: 110; digit IV length excluding claw: 50; digit IV length: 70			
Dromaius novaehollandiae Oregon State bird 9	TMT: 391; foot: 182			
	IIPad1L: 22	IIPad2L: 20	IIClawL: 23	
	IIIPad1L: 41	IIIPad2L: 33	IIIPad3L: 32	IIIClawL: 20
	Digit II length excluding claw: 42; digit II length: 65; digit III length excluding claw: 106; digit III length: 126; digit IV length excluding claw: 51; digit IV length: 70			
Dromaius novaehollandiae Oregon State bird 10	TMT: 388; foot: 165			
	IIPad1L: 19	IIPad2L: 19	IIClawL: 16	
	IIIPad1L: 32	IIIPad2L: 37	IIIPad3L: 21	IIIClawL: 19
	Digit II length excluding claw: 38; digit II length: 54; digit III length excluding claw: 90; digit III length: 109; digit IV length excluding claw: 51; digit IV length: 67			
Dromaius novaehollandiae Oregon State bird 11	TMT: 348; foot: 163			
	IIPad1L: 19	IIPad2L: 17	IIClawL: 21	
	IIIPad1L: 39	IIIPad2L: 33	IIIPad3L: 24	IIIClawL: 18
	Digit II length excluding claw: 36; digit II length: 57; digit III length excluding claw: 96; digit III length: 114; digit IV length excluding claw: 56; digit IV length: 71			
Dromaius novaehollandiae PU 9004	IIPad1L: 19	IIPad2L: 19	IIClawL: 20	
	IIIPad1L: 34	IIIPad2L: 28	IIIPad3L: 26	IIIClawL: 18
	Digit II length excluding claw: 38; digit II length: 58; digit III length excluding claw: 88; digit III length: 106; digit IV length excluding claw: 57; digit IV length: 75			

Table A4.1. *continued*

Specimen	Measurements
Dromaius novaehollandiae Torbert emu	TMT: ———; foot: 180 IIPad1L: 22 IIPad2L: 19 IIClawL: 24 IIIPad1L: 35 IIIPad2L: 30 IIIPad3L: 27 IIIClawL: 20 Digit II length excluding claw:41; digit II length: 65; digit III length excluding claw: 92; digit III length: 112; digit IV length excluding claw: 59; digit IV length: 76
Cassowaries	
Casuarius sp. uncataloged Berkeley juvenile specimen	IIPad1L: 23 IIPad2L: 0 IIClawL: 21 IIIPad1L: 14 IIIPad2L: 8 IIIPad3L: 19 IIIClawL: 12 Digit II length excluding claw: 23; digit II length: 44; digit III length excluding claw: 50; digit III length: 62; digit IV length excluding claw: 42; digit IV length: 51
Casuarius sp. uncataloged Berkeley juvenile specimen	IIPad1L: 23 IIPad2L: 0 IIClawL: 24 IIIPad1L: 17 IIIPad2L: 12 IIIPad3L: 19 IIIClawL: 13 Digit II length excluding claw: 23; digit II length: 47; digit III length excluding claw: 56; digit III length: 69; digit IV length excluding claw: 37; digit IV length: 48
Casuarius bennetti FMNH 87322	IIPad1L: 42 IIPad2L: 0 IIClawL: 52 IIIPad1L: 27 IIIPad2L: 19 IIIPad3L: 21 IIIClawL: 24 Digit II length excluding claw: 42; digit II length: 94; digit III length excluding claw: 91; digit III length: 115; digit IV length excluding claw: 57; digit IV length: 79
Kiwi	
Apteryx mantelli FMNH 4616	IIPad1L: 7 IIPad2L: 8 IIClawL: 8 IIIPad1L: 8 IIIPad2L: 6 IIIPad3L: 7 IIIClawL: 10 Digit II length excluding claw: 21; digit II length: 29; digit III length excluding claw: 31; digit III length: 41; digit IV length excluding claw: 27; digit IV length: 35
Apteryx mantelli FMNH 30379	IIPad1L: 8 IIPad2L: 9 IIClawL: 14 IIIPad1L: 9 IIIPad2L: 8 IIIPad3L: 8 IIIClawL: 17 Digit II length excluding claw:21; digit II length: 35; digit III length excluding claw: 34; digit III length: 51; digit IV length excluding claw: 26; digit IV length: 36
Apteryx australis FMNH 88262	IIPad1L: 9 IIPad2L: 8 IIClawL: 16 IIIPad1L: 9 IIIPad2L: 7 IIIPad3L: 9 IIIClawL: 18 Digit II length excluding claw: 25; digit II length: 41; digit III length excluding claw: 39; digit III length: 57; digit IV length excluding claw: 29; digit IV length: 43
Apteryx australis FMNH 88261	IIPad1L: 9 IIPad2L: 9 IIClawL: 17 IIIPad1L: 10 IIIPad2L: 7 IIIPad3L: 9 IIIClawL: 18 Digit II length excluding claw:23; digit II length: 40; digit III length excluding claw: 33; digit III length: 51; digit IV length excluding claw: 26; digit IV length: 39
Apteryx owenii FMNH 4669	IIPad1L: 7 IIPad2L: 7 IIClawL: 7 IIIPad1L: 6 IIIPad2L: 7 IIIPad3L: 7 IIIClawL: 12 Digit II length excluding claw: 21; digit II length: 28; digit III length excluding claw: 32; digit III length: 44; digit IV length excluding claw: 24; digit IV length: 32
Rheas	
Rhea americana UMMZ Birds 216665	TMT: 61; foot: 28 IIPad1L: 8 IIPad2L: 0 IIClawL: 3 IIIPad1L: 5 IIIPad2L: 9 IIIPad3L: 0 IIIClawL: 4 Digit II length excluding claw: 8; digit II length: 11; digit III length excluding claw: 14; digit III length: 18; digit IV length excluding claw: 10; digit IV length: 13
Rhea americana UMMZ Birds 110978	TMT: 79; foot: 36 IIPad1L: 8 IIPad2L: 0 IIClawL: 5 IIIPad1L: 6 IIIPad2L: 13 IIIPad3L: 0 IIIClawL: 6 Digit II length excluding claw: 8; digit II length: 13; digit III length excluding claw: 19; digit III length: 25; digit IV length excluding claw: 11; digit IV length: 16
Rhea americana UMMZ Birds 110976	TMT: 77; foot: 36 IIPad1L: 9 IIPad2L: 0 IIClawL: 5 IIIPad1L: 7 IIIPad2L: 12 IIIPad3L: 0 IIIClawL: 5 Digit II length excluding claw: 9; digit II length: 14; digit III length excluding claw: 19; digit III length: 24; digit IV length excluding claw: 11; digit IV length: 16

Specimen	Measurements
Rhea americana UMMZ Birds 110979	TMT: 66; foot: 34 IIPad1L: 9 IIPad2L: 0 IIClawL: 5 IIIPad1L: 7 IIIPad2L: 11 IIIPad3L: 0 IIIClawL: 5 Digit II length excluding claw: 9; digit II length: 14; digit III length excluding claw: 18; digit III length: 23; digit IV length excluding claw: 10; digit IV length: 14
Rhea americana UMMZ Birds 110980	TMT: 73; foot: 34 IIPad1L: 10 IIPad2L: 0 IIClawL: 4 IIIPad1L: 8 IIIPad2L: 13 IIIPad3L: 0 IIIClawL: 5 Digit II length excluding claw: 10; digit II length: 14; digit III length excluding claw: 21; digit III length: 26; digit IV length excluding claw: 12; digit IV length: 16
Rhea americana UMMZ Birds 2975	TMT: 49; foot: 30 IIPad1L: 11 IIPad2L: 0 IIClawL: 3 IIIPad1L: 8 IIIPad2L: 11 IIIPad3L: 0 IIIClawL: 3 Digit II length excluding claw: 11; digit II length: 14; digit III length excluding claw: 19; digit III length: 22; digit IV length excluding claw: 11; digit IV length: 14
Rhea americana UMMZ Birds 110977	TMT: 84; foot: 35 IIPad1L: 9 IIPad2L: 0 IIClawL: 5 IIIPad1L: 9 IIIPad2L: 12 IIIPad3L: 0 IIIClawL: 5 Digit II length excluding claw: 9; digit II length: 14; digit III length excluding claw: 21; digit III length: 26; digit IV length excluding claw: 11; digit IV length: 16
Rhea americana UMMZ Birds 100002	TMT: 300; foot: 116 IIPad1L: 23 IIPad2L: 0 IIClawL: 17 IIIPad1L: 20 IIIPad2L: 20 IIIPad3L: 12 IIIClawL: 26 Digit II length excluding claw: 23; digit II length: 40; digit III length excluding claw: 52; digit III length: 78; digit IV length excluding claw: 30; digit IV length: 48
Rhea americana UMMZ Birds 100003	TMT: 285; foot: 120 IIPad1L: 25 IIPad2L: 0 IIClawL: 19 IIIPad1L: 24 IIIPad2L: 30 IIIPad3L: 0 IIIClawL: 20 Digit II length excluding claw: 25; digit II length: 44; digit III length excluding claw: 54; digit III length: 74; digit IV length excluding claw: 25; digit IV length: 42
Rhea americana MSU OR.2794	TMT: 63; foot: 31
Rhea americana MSU OR.6345	TMT: 90; foot: 40 IIPad1L: 9 IIPad2L: 0 IIClawL: 6 IIIPad1L: 6 IIIPad2L: 9 IIIPad3L: 0 IIIClawL: 6 Digit II length excluding claw: 9; digit II length: 15; digit III length excluding claw: 15; digit III length: 21; digit IV length excluding claw: 11; digit IV length: 16
Rhea americana FMNH 101876	TMT: 75; foot: 41 IIPad1L: 11 IIPad2L: 0 IIClawL: 5 IIIPad1L: 6 IIIPad2L: 13 IIIPad3L: 0 IIIClawL: 5 Digit II length excluding claw: 11; digit II length: 16; digit III length excluding claw: 21 (*note:* this bird had a 2-mm interpad space between IIIPad1L and IIIPad2L); digit III length: 26: digit IV length excluding claw: 12; digit IV length: 17
Rhea americana FMNH 101877	IIPad1L: 5 IIPad2L: 5 IIClawL: 6 IIIPad1L: 7 IIIPad2L: 6 IIIPad3L: 5 IIIClawL: 6 Digit II length excluding claw: 10; digit II length: 16; digit III length excluding claw: 18; digit III length: 24; digit IV length excluding claw: 10; digit IV length: 16
Rhea americana FMNH 101880	TMT: 79; foot: 34 IIPad1L: 5 IIPad2L: 6 IIClawL: 5 IIIPad1L: 8 IIIPad2L: 6 IIIPad3L: 7 IIIClawL: 5 Digit II length excluding claw: 11; digit II length: 16; digit III length excluding claw: 21; digit III length: 26; digit IV length excluding claw: 13; digit IV length: 18
Rhea americana FMNH 101874	TMT: 76; foot: 37 IIPad1L: 5 IIPad2L: 5 IIClawL: 5 IIIPad1L: 8 IIIPad2L: 5 IIIPad3L: 6 IIIClawL: 5 Digit II length excluding claw: 10; digit II length: 15; digit III length excluding claw: 19; digit III length: 24; digit IV length excluding claw: 12; digit IV length: 17

Table A4.1. *continued*

Specimen	Measurements
Rhea americana FMNH 101875	TMT: 78; foot: 45 IIPad1L: 4 IIPad2L: 4 IIClawL: 5 IIIPad1L: 8 IIIPad2L: 7 IIIPad3L: 5 IIIClawL: 6 Digit II length excluding claw: 8; digit II length: 13; digit III length excluding claw: 20; digit III length: 26; digit IV length excluding claw: 11; digit IV length: 16
Rhea americana FMNH 101879	TMT: 74; foot: 43 IIPad1L: 6 IIPad2L: 4 IIClawL: 5 IIIPad1L: 9 IIIPad2L: 7 IIIPad3L: 5 IIIClawL: 5 Digit II length excluding claw: 10; digit II length: 15; digit III length excluding claw: 21; digit III length: 26; digit IV length excluding claw: 12; digit IV length: 17
Rhea americana FMNH 101753	IIPad1L: 28 IIPad2L: 0 IIClawL: 26 IIIPad1L: 22 IIIPad2L: 20 IIIPad3L: 17 IIIClawL: 30 Digit II length excluding claw: 28; digit II length: 54; digit III length excluding claw: 59; digit III length: 89; digit IV length excluding claw: 34; digit IV length: 59
Rhea americana FMNH 110481	IIPad1L: 29 IIPad2L: 0 IIClawL: 24 IIIPad1L:24 IIIPad2L: 21 IIIPad3L: 20 IIIClawL: 32 Digit II length excluding claw: 29; digit II length: 53; digit III length excluding claw: 65; digit III length: 97; digit IV length excluding claw: 33; digit IV length: 60
Rhea americana FMNH 110483	IIPad1L: 23 IIPad2L: 0 IIClawL: 20 IIIPad1L: 25 IIIPad2L: 34 IIIPad3L: 0 IIIClawL: 25 Digit II length excluding claw: 23; digit II length: 43; digit III length excluding claw: 59; digit III length: 84; digit IV length excluding claw: 26; digit IV length: 49
Rhea americana FMNH 81400	IIPad1L: 27 IIPad2L: 0 IIClawL: 21 IIIPad1L: 26 IIIPad2L: 25 IIIPad3L: 12 IIIClawL: 27 Digit II length excluding claw: 27; digit II length:48; digit III length excluding claw: 63; digit III length: 90; digit IV length excluding claw: 32; digit IV length: 55
Rhea americana FMNH 65376	IIPad1L: 30 IIPad2L: 0 IIClawL: 22 IIIPad1L: 26 IIIPad2L: 21 IIIPad3L: 18 IIIClawL: 25 Digit II length excluding claw: 30; digit II length: 52; digit III length excluding claw: 65; digit III length: 90; digit IV length excluding claw: 36; digit IV length: 56
Rhea americana FMNH 110484	IIPad1L: 27 IIPad2L: 0 IIClawL: 21 IIIPad1L: 27 IIIPad2L: 32 IIIPad3L: 0 IIIClawL: 24 Digit II length excluding claw: 27; digit II length: 48; digit III length excluding claw: 59; digit III length: 83; digit IV length excluding claw: 28; digit IV length: 49
Rhea americana FMNH 101870	IIPad1L: 30 IIPad2L: 0 IIClawL: 23 IIIPad1L: 27 IIIPad2L: 19 IIIPad3L: 22 IIIClawL: 29 Digit II length excluding claw: 30; digit II length: 53; digit III length excluding claw: 68; digit III length: 97; digit IV length excluding claw: 32; digit IV length: 54
Rhea americana dead bird from animal breeder	IIPad1L: 20 IIPad2L: 0 IIClawL: 12 IIIPad1L: 15 IIIPad2L: 24 IIIPad3L: 0 IIIClawL: 16 Digit II length excluding claw: 20; digit II length: 32; digit III length excluding claw: 39; digit III length: 55; digit IV length excluding claw: 24; digit IV length: 36
Pterocnemia pennata UMMZ Birds 134715	IIPad1L: 11 IIPad2L: 0 IIClawL: 4 IIIPad1L: ——— IIIPad2L: 10 IIIPad3L: 0 IIIClawL: 5 Digit II length excluding claw: 11; digit II length: 15; digit III length excluding claw: ———; digit III length: ———; digit IV length excluding claw: 11; digit IV length: 14
Pterocnemia pennata FMNH 279304	IIPad1L: 11 IIPad2L: 0 IIClawL: 4 IIIPad1L: 7 IIIPad2L: 5 IIIPad3L: 7 IIIClawL: 4 Digit II length excluding claw: 11; digit II length: 15; digit III length excluding claw: 19; digit III length: 23; digit IV length excluding claw: 13; digit IV length: 16
Pterocnemia pennata FMNH 279305	IIPad1L: 11 IIPad2L: 0 IIClawL: 4 IIIPad1L: 8 IIIPad2L: 12 IIIPad3L: 0 IIIClawL: 5 Digit II length excluding claw: 11; digit II length: 15; digit III length excluding claw: 20; digit III length: 25; digit IV length excluding claw: 13; digit IV length: 17
Pterocnemia pennata FMNH 370474	IIPad1L: 8 IIPad2L: 6 IIClawL: 10 IIIPad1L: 14 IIIPad2L: 11 IIIPad3L: 7 IIIClawL: 13 Digit II length excluding claw: 13; digit II length: 23; digit III length excluding claw: 32; digit III length: 45; digit IV length excluding claw: 15; digit IV length: 25

Specimen	Measurements
Pterocnemia pennata IRSNB 48297	TMT: 285; foot: 140 IIPad1L: 32 IIPad2L: 0 IIClawL: 19 IIIPad1L: 20 IIIPad2L: 40 IIIPad3L: 0 IIIClawL: 31 Digit II length excluding claw: 32; digit II length: 51; digit III length excluding claw: 60; digit III length: 91; digit IV length excluding claw: 32; digit IV length: 49
Tinamous	
Eudromia elegans FMNH Conover 780	IIPad1L: 6 IIPad2L: 7 IIClawL: 6 IIIPad1L:6 IIIPad2L: 6 IIIPad3L: 7 IIIClawL: 6 Digit II length excluding claw: 13; digit II length: 19; digit III length excluding claw: 19; digit III length: 25; digit IV length excluding claw: 16; digit IV length: 21
Eudromia elegans FMNH Conover 1937	IIPad1L: 6 IIPad2L: 4 IIClawL: 4 IIIPad1L: 5 IIIPad2L: 5 IIIPad3L: 5 IIIClawL: 6 Digit II length excluding claw: 10; digit II length: 14; digit III length excluding claw: 15; digit III length: 21; digit IV length excluding claw: 12; digit IV length: 15
Eudromia elegans FMNH Conover 2830	IIPad1L: 6 IIPad2L: 5 IIClawL: 5 IIIPad1L: 7 IIIPad2L: 7 IIIPad3L: 6 IIIClawL: 6 Digit II length excluding claw: 11; digit II length: 16; digit III length excluding claw: 20; digit III length: 26; digit IV length excluding claw: 17; digit IV length: 22
Eudromia elegans FMNH Conover 4788	IIPad1L: 7 IIPad2L: 5 IIClawL: 5 IIIPad1L: 7 IIIPad2L: 6 IIIPad3L: 6 IIIClawL: 7 Digit II length excluding claw: 12; digit II length: 17; digit III length excluding claw: 19; digit III length: 26; digit IV length excluding claw: 14; digit IV length: 19
Eudromia elegans FMNH Conover 22374	IIPad1L: 6 IIPad2L: 5 IIClawL: 5 IIIPad1L: 8 IIIPad2L: 7 IIIPad3L: 6 IIIClawL: 6 Digit II length excluding claw:11; digit II length: 16; digit III length excluding claw: 21; digit III length: 27; digit IV length excluding claw: 17; digit IV length: 23
Eudromia elegans FMNH Conover 62331	IIPad1L: 5 IIPad2L: 5 IIClawL: 4 IIIPad1L: 7 IIIPad2L: 6 IIIPad3L: 6 IIIClawL: 6 Digit II length excluding claw: 10; digit II length: 14; digit III length excluding claw: 19; digit III length: 25; digit IV length excluding claw:14; digit IV length: 17
Eudromia elegans PU 8816	IIPad1L: 5 IIPad2L: 5 IIClawL: 6 IIIPad1L: 7 IIIPad2L: 6 IIIPad3L: 6 IIIClawL: 7 Digit II length excluding claw: 10; digit II length: 16; digit III length excluding claw: 19; digit III length: 26; digit IV length excluding claw: 16; digit IV length: 21
Eudromia formosa FMNH Conover 13491	IIPad1L: 5 IIPad2L: 5 IIClawL: 5 IIIPad1L: 6 IIIPad2L: 6 IIIPad3L: 6 IIIClawL: 7 Digit II length excluding claw: 10; digit II length: 15; digit III length excluding claw: 18; digit III length: 25; digit IV length excluding claw: 15; digit IV length: 20
Eudromia formosa FMNH Conover 17185	IIPad1L: 5 IIPad2L: 5 IIClawL: 5 IIIPad1L: 6 IIIPad2L: 5 IIIPad3L: 6 IIIClawL: 6 Digit II length excluding claw: 10; digit II length: 15; digit III length excluding claw: 17; digit III length: 23; digit IV length excluding claw: 15; digit IV length: 20
Eudromia formosa FMNH Conover 17183	IIPad1L: 5 IIPad2L: 5 IIClawL: 5 IIIPad1L: 7 IIIPad2L: 7 IIIPad3L: 6 IIIClawL: 7 Digit II length excluding claw: 10; digit II length: 15; digit III length excluding claw: 20; digit III length: 27; digit IV length excluding claw: 15; digit IV length: 20
Eudromia elegans FMNH Conover 16499	IIPad1L: 4 IIPad2L: 5 IIClawL: 5 IIIPad1L: 7 IIIPad2L: 6 IIIPad3L: 5 IIIClawL: 6 Digit II length excluding claw: 9; digit II length: 14; digit III length excluding claw: 18; digit III length: 24; digit IV length excluding claw: 15; digit IV length: 19
Tinamotis pentlandii FMNH 62382	IIPad1L: 8 IIPad2L: 7 IIClawL: 5 IIIPad1L: 7 IIIPad2L: 7 IIIPad3L: 8 IIIClawL: 6 Digit II length excluding claw: 15; digit II length: 20; digit III length excluding claw: 22; digit III length: 28; digit IV length excluding claw: 20; digit IV length: 25

Table A4.1. *continued*

Specimen	Measurements
Tinamotis pentlandii FMNH Conover 2831	IIPad1L: 8 IIPad2L: 8 IIClawL: 3 IIIPad1L: 8 IIIPad2L: 7 IIIPad3L: 8 IIIClawL: 4 Digit II length excluding claw: 16; digit II length: 19; digit III length excluding claw: 23; digit III length: 27; digit IV length excluding claw: 19; digit IV length: 22
Tinamotis pentlandii FMNH 311856	IIPad1L: 10 IIPad2L: 8 IIClawL: 4 IIIPad1L: 9 IIIPad2L: 9 IIIPad3L: 9 IIIClawL: 4 Digit II length excluding claw: 18; digit II length: 22; digit III length excluding claw: 27; digit III length: 31; digit IV length excluding claw: 23; digit IV length: 27
Tinamotis ingoufi FMNH Conover5283	TMT: 22; foot: 21 IIPad1L: 5 IIPad2L: 4 IIClawL: 2 IIIPad1L: 4 IIIPad2L: 4 IIIPad3L: 5 IIIClawL: 2 Digit II length excluding claw: 9; digit II length: 11; digit III length excluding claw: 13; digit III length: 15; digit IV length excluding claw: 11; digit IV length: 12
Tinamotis ingoufi FMNH Conover 5284	TMT: 21; foot: 20 IIPad1L: 5 IIPad2L: 4 IIClawL: 2 IIIPad1L: 5 IIIPad2L: 4 IIIPad3L: 4 IIIClawL: 2 Digit II length excluding claw: 9; digit II length: 11; digit III length excluding claw: 13; digit III length: 15; digit IV length excluding claw: 11; digit IV length: 12
Tinamotis ingoufi FMNH Conover 5282	TMT: 32; foot: 38 IIPad1L: 9 IIPad2L: 7 IIClawL: 4 IIIPad1L: 6 IIIPad2L: 6 IIIPad3L: 7 IIIClawL: 6 Digit II length excluding claw: 16; digit II length: 20; digit III length excluding claw: 19; digit III length: 25; digit IV length excluding claw: 19; digit IV length: 22
Tinamotis ingoufi FMNH Conover 5281	TMT: 33; foot: 36 IIPad1L: 7 IIPad2L: 6 IIClawL: 4 IIIPad1L: 8 IIIPad2L: 6 IIIPad3L: 7 IIIClawL: 5 Digit II length excluding claw: 13; digit II length: 17; digit III length excluding claw: 21; digit III length: 26; digit IV length excluding claw: 16; digit IV length: 20
Tinamotis ingoufi FMNH Conover 15651	TMT: 35; foot: 39 IIPad1L: 8 IIPad2L: 7 IIClawL: 4 IIIPad1L: 8 IIIPad2L: 7 IIIPad3L: 8 IIIClawL: 5 Digit II length excluding claw: 15; digit II length: 19; digit III length excluding claw: 23; digit III length: 28; digit IV length excluding claw: 20; digit IV length: 24
Tinamotis ingoufi FMNH Conover 5287	IIPad1L: 8 IIPad2L: 7 IIClawL: 3 IIIPad1L: 6 IIIPad2L: 5 IIIPad3L: 8 IIIClawL: 5 Digit II length excluding claw: 15; digit II length: 18; digit III length excluding claw: 19; digit III length: 24; digit IV length excluding claw: 17; digit IV length: 20
Tinamotis ingoufi FMNH Conover 15652	IIPad1L: 7 IIPad2L: 6 IIClawL: 4 IIIPad1L: 6 IIIPad2L 5: IIIPad3L: 6 IIIClawL: 5 Digit II length excluding claw: 13; digit II length: 17; digit III length excluding claw: 17; digit III length: 22; digit IV length excluding claw: 16; digit IV length: 19
Nothocercus bonapartei FMNH Conover 8622	IIPad1L: 5 IIPad2L: 4 IIClawL: 7 IIIPad1L: 7 IIIPad2L: 6 IIIPad3L: 6 IIIClawL: 8 Digit II length excluding claw: 18; digit II length: 25; digit III length excluding claw: 33; digit III length: 41; digit IV length excluding claw: 26; digit IV length: 33
Tinamus major FMNH Conover 12370	IIPad1L: 5 IIPad2L: 4 IIClawL: 3 IIIPad1L: 5 IIIPad2L: 4 IIIPad3L: 4 IIIClawL: 4 Digit II length excluding claw: 13; digit II length: 16; digit III length excluding claw: 21; digit III length: 25; digit IV length excluding claw: 16; digit IV length: 19
Tinamus major FMNH Conover 19383	IIPad1L: 6 IIPad2L: 5 IIClawL: 3 IIIPad1L: 6 IIIPad2L: 6 IIIPad3L: 6 IIIClawL: 4 Digit II length excluding claw: 15; digit II length: 18; digit III length excluding claw: 24; digit III length: 28; digit IV length excluding claw: 20; digit IV length: 23
Nothura darwinii FMNH Conover 25683	IIPad1L: 4 IIPad2L: 2 IIClawL: 2 IIIPad1L: 4 IIIPad2L: 3 IIIPad3L: 3 IIIClawL: 1 Digit II length excluding claw: 8; digit II length: 10; digit III length excluding claw: 14; digit III length: 15; digit IV length excluding claw: 11; digit IV length: 12

Specimen	Measurements
Nothura maculosa FMNH Conover 8926	IIPad1L: 4 IIPad2L: 3 IIClawL: 4 IIIPad1L: 6 IIIPad2L: 4 IIIPad3L: 3 IIIClawL: 4 Digit II length excluding claw: 10; digit II length: 14; digit III length excluding claw: 18; digit III length: 22; digit IV length excluding claw: 16; digit IV length: 19
Bustards	
Ardeotis kori FMNH 5577	IIPad1L: 12 IIPad2L: 14 IIClawL: 8 IIIPad1L: 21 IIIPad2L: 26 IIIPad3L: 12 IIIClawL: 7 Digit II length excluding claw: 26; digit II length: 34; digit III length excluding claw: 49; digit III length: 56; digit IV length excluding claw: 37; digit IV length: 43
Ardeotis kori FMNH 6231	TMT: 220; foot: 105 IIPad1L: 14 IIPad2L: 12 IIClawL: 10 IIIPad1L: 19 IIIPad2L: 18 IIIPad3L: 15 IIIClawL: 10 Digit II length excluding claw: 26; digit II length: 36; digit III length excluding claw: 52; digit III length: 62; digit IV length excluding claw: 40; digit IV length: 48
Ardeotis kori FMNH 10588	TMT: 257; foot: 84 IIPad1L: 12 IIPad2L: 13 IIClawL: 9 IIIPad1L: 19 IIIPad2L: 14 IIIPad3L: 13 IIIClawL: 11 Digit II length excluding claw: 25; digit II length: 34; digit III length excluding claw: 46; digit III length: 57; digit IV length excluding claw: 30; digit IV length: 37
Ardeotis kori FMNH 192817	TMT: 187; foot: 75 IIPad1L: 12 IIPad2L: 11 IIClawL: 7 IIIPad1L: 16 IIIPad2L: 13 IIIPad3L: 12 IIIClawL: 9 Digit II length excluding claw: 23; digit II length: 30; digit III length excluding claw: 41; digit III length: 50; digit IV length excluding claw: 31; digit IV length: 36
Ardeotis kori FMNH 192818	TMT: 185; foot: 72 IIPad1L: 12 IIPad2L: 10 IIClawL: 7 IIIPad1L: 16 IIIPad2L: 10 IIIPad3L: 10 IIIClawL: 8 Digit II length excluding claw: 22; digit II length: 29; digit III length excluding claw: 36; digit III length: 44; digit IV length excluding claw: ———; digit IV length: ———
Ardeotis kori FMNH 32992	TMT: 235; foot: 90 IIPad1L: 15 IIPad2L: 13 IIClawL: 10 IIIPad1L: 18 IIIPad2L: 14 IIIPad3L: 13 IIIClawL: 11 Digit II length excluding claw: 28; digit II length: 38; digit III length excluding claw: 45; digit III length: 56; digit IV length excluding claw: 35; digit IV length: 42
Ardeotis kori FMNH 86079	IIPad1L: 10 IIPad2L: 11 IIClawL: 8 IIIPad1L: 15 IIIPad2L: 11 IIIPad3L: 12 IIIClawL: 9 Digit II length excluding claw: 21; digit II length: 29; digit III length excluding claw: 38; digit III length: 47; digit IV length excluding claw: 26; digit IV length: 32
Ardeotis kori FMNH 86081	IIPad1L: 11 IIPad2L: 10 IIClawL: 7 IIIPad1L: 17 IIIPad2L: 12 IIIPad3L: 13 IIIClawL: 9 Digit II length excluding claw: 21; digit II length: 28; digit III length excluding claw: 42; digit III length: 51; digit IV length excluding claw: 28; digit IV length: 33
Ardeotis kori FMNH 86082	TMT: 245; foot: 80
Ardeotis kori FMNH 86084	IIPad1L: 10 IIPad2L: 9 IIClawL: 8 IIIPad1L: 17 IIIPad2L: 24 IIIPad3L: 0 IIIClawL: 8 Digit II length excluding claw: 19; digit II length: 27; digit III length excluding claw: 41; digit III length: 49; digit IV length excluding claw: 25; digit IV length: 30
Ardeotis kori FMNH 192816	TMT: 240; foot: 93 IIPad1L: 14 IIPad2L: 16 IIClawL: 6 IIIPad1L: 21 IIIPad2L: 14 IIIPad3L: 16 IIIClawL: 7 Digit II length excluding claw: 30; digit II length: 36; digit III length excluding claw: 51; digit III length: 58; digit IV length excluding claw: 38; digit IV length: 43
Ardeotis kori FMNH 192819	TMT: 239; foot: 79 IIPad1L: 13 IIPad2L: 14 IIClawL: 10 IIIPad1L: 20 IIIPad2L: 15 IIIPad3L: 16 IIIClawL: 7 Digit II length excluding claw: 27; digit II length: 37; digit III length excluding claw: 51; digit III length: 58; digit IV length excluding claw: 34; digit IV length: 41

Table A4.1. *continued*

Specimen	Measurements
Ardeotis kori MSU.OR 6934	IIPad1L: 11 IIPad2L: 14 IIClawL: 11 IIIPad1L: 21 IIIPad2L: 13 IIIPad3L: 15 IIIClawL: 14 Digit II length excluding claw: 25; digit II length: 36; digit III length excluding claw: 49; digit III length: 63; digit IV length excluding claw: 34; digit IV length: 41
Ardeotis kori UMMZ Birds 97916	IIPad1L: 13 IIPad2L: 14 IIClawL: 9 IIIPad1L: 22 IIIPad2L: 16 IIIPad3L: 16 IIIClawL: 11 Digit II length excluding claw: 27; digit II length: 36; digit III length excluding claw: 54; digit III length: 65; digit IV length excluding claw: 36; digit IV length: 43
Ardeotis kori UMMZ Birds 97917	IIPad1L: 11 IIPad2L: 9 IIClawL: 6 IIIPad1L: 16 IIIPad2L: 12 IIIPad3L: 10 IIIClawL: ——— Digit II length excluding claw: 20; digit II length: 26; digit III length excluding claw: 38; digit III length: ———; digit IV length excluding claw: 25; digit IV length: 29
Ardeotis arabs FMNH 6247	TMT: 177; foot: 71
Ardeotis arabs FMNH 17009	TMT: 210; foot: 85 IIPad1L: 9 IIPad2L: 11 IIClawL: 10 IIIPad1L: 16 IIIPad2L: 12 IIIPad3L: 12 IIIClawL: 12 Digit II length excluding claw: 20; digit II length: 30; digit III length excluding claw: 40; digit III length: 52; digit IV length excluding claw: 29; digit IV length: 37
Ardeotis arabs FMNH 192820	TMT: 170; foot: 69 IIPad1L: 9 IIPad2L: 8 IIClawL: 10 IIIPad1L: 13 IIIPad2L: 12 IIIPad3L: 8 IIIClawL: 11 Digit II length excluding claw: 17; digit II length: 27; digit III length excluding claw: 33; digit III length: 44; digit IV length excluding claw: 19; digit IV length: 27
Ardeotis australis UMMZ Birds 214404	TMT: 201; foot: 80 IIPad1L: 11 IIPad2L: 12 IIClawL: 6 IIIPad1L: 16 IIIPad2L: 11 IIIPad3L: 11 IIIClawL: 7 Digit II length excluding claw: 23; digit II length: 29; digit III length excluding claw: 38; digit III length: 45; digit IV length excluding claw: 29; digit IV length: 34
Eupodotis senegalensis FMNH 5580	TMT: 145; foot: 48 IIPad1L: 7 IIPad2L: 8 IIClawL: 6 IIIPad1L: 13 IIIPad2L: 7 IIIPad3L: 9 IIIClawL: 8 Digit II length excluding claw: 15; digit II length: 21; digit III length excluding claw: 29; digit III length: 37; digit IV length excluding claw: 18; digit IV length: 22
Eupodotis senegalensis FMNH 5581	TMT: 95; foot: 36 IIPad1L: 5 IIPad2L: 5 IIClawL: 3 IIIPad1L: 7 IIIPad2L: 6 IIIPad3L: 6 IIIClawL: 3 Digit II length excluding claw: 10; digit II length: 13; digit III length excluding claw: 19; digit III length: 22; digit IV length excluding claw: 12; digit IV length: 14
Eupodotis senegalensis FMNH 6232	IIPad1L: 21 IIPad2L: 0 IIClawL: 6 IIIPad1L: 15 IIIPad2L: 22 IIIPad3L: 0 IIIClawL: 6 Digit II length excluding claw: 21; digit II length: 27; digit III length excluding claw: 37; digit III length: 43; digit IV length excluding claw: 26; digit IV length: 30
Eupodotis senegalensis FMNH 24492	IIPad1L: 10 IIPad2L: 6 IIClawL: 6 IIIPad1L: 14 IIIPad2L: 8 IIIPad3L: 8 IIIClawL: 7 Digit II length excluding claw: 16; digit II length: 22; digit III length excluding claw: 30; digit III length: 37; digit IV length excluding claw: 18; digit IV length: 23
Eupodotis senegalensis FMNH 192822	IIPad1L: 11 IIPad2L: 12 IIClawL: 6 IIIPad1L: 15 IIIPad2L: 11 IIIPad3L: 14 IIIClawL: 5 Digit II length excluding claw: 23; digit II length: 29; digit III length excluding claw: 40; digit III length: 45; digit IV length excluding claw: 29; digit IV length: 33
Eupodotis senegalensis MSU.OR 5406	IIPad1L: 5 IIPad2L: 5 IIClawL: 5 IIIPad1L: 8 IIIPad2L: 7 IIIPad3L: 6 IIIClawL: 7 Digit II length excluding claw: 10; digit II length: 15; digit III length excluding claw: 21; digit III length: 28; digit IV length excluding claw: 15; digit IV length: 19
Eupodotis senegalensis UMMZ Birds 97918	IIPad1L: 5 IIPad2L: 6 IIClawL: 5 IIIPad1L: 7 IIIPad2L: 6 IIIPad3L: 7 IIIClawL: 5 Digit II length excluding claw: 11; digit II length:16; digit III length excluding claw: 20; digit III length: 25; digit IV length excluding claw: 14; digit IV length: 18

Specimen	Measurements
Otis tarda FMNH 10174	IIPad1L: 13 IIPad2L: 10 IIClawL: 5 IIIPad1L: 16 IIIPad2L: 13 IIIPad3L: 11 IIIClawL: 7 Digit II length excluding claw: 23; digit II length: 28; digit III length excluding claw: 40; digit III length: 47; digit IV length excluding claw: 28; digit IV length: 34
Otis tarda FMNH 10175	IIPad1L: 8 IIPad2L: 7 IIClawL: 6 IIIPad1L: 13 IIIPad2L: 8 IIIPad3L: 8 IIIClawL: 7 Digit II length excluding claw: 15; digit II length: 21; digit III length excluding claw: 29; digit III length: 36; digit IV length excluding claw: 18; digit IV length: 23
Otis tarda FMNH 10210	IIPad1L: 8 IIPad2L: 6 IIClawL: 6 IIIPad1L: 12 IIIPad2L: 7 IIIPad3L: 8 IIIClawL: 7 Digit II length excluding claw: 14; digit II length: 20; digit III length excluding claw: 27; digit III length: 34; digit IV length excluding claw: 19; digit IV length: 23
Otis tarda FMNH 21007	IIPad1L: 11 IIPad2L: 12 IIClawL: 9 IIIPad1L: 20 IIIPad2L: 16 IIIPad3L: 14 IIIClawL: 11 Digit II length excluding claw: 23; digit II length: 32; digit III length excluding claw: 50; digit III length: 61; digit IV length excluding claw: 39; digit IV length: 47
Otis tarda FMNH 21008	IIPad1L: 13 IIPad2L: 11 IIClawL: 13 IIIPad1L: 18 IIIPad2L: 15 IIIPad3L: 11 IIIClawL: 20 Digit II length excluding claw: 24; digit II length: 37; digit III length excluding claw: 44; digit III length: 64; digit IV length excluding claw: 24; digit IV length: 35
Otis tarda FMNH 21014	IIPad1L: 10 IIPad2L: 9 IIClawL: 7 IIIPad1L: 13 IIIPad2L: 11 IIIPad3L: 9 IIIClawL: 8 Digit II length excluding claw: 19; digit II length: 26; digit III length excluding claw: 33; digit III length: 41; digit IV length excluding claw: 24; digit IV length: 29
Otis tarda FMNH 21015	IIPad1L: 11 IIPad2L: 8 IIClawL: 7 IIIPad1L: 13 IIIPad2L: 10 IIIPad3L: 10 IIIClawL: 8 Digit II length excluding claw: 19; digit II length: 26; digit III length excluding claw: 33; digit III length: 41; digit IV length excluding claw: 22; digit IV length: 28
Otis tarda FMNH 26277	IIPad1L: 10 IIPad2L: 6 IIClawL: 6 IIIPad1L: 11 IIIPad2L: 7 IIIPad3L: 9 IIIClawL: 8 Digit II length excluding claw: 16; digit II length: 22; digit III length excluding claw: 27; digit III length: 35; digit IV length excluding claw: 20; digit IV length: 24
Otis tarda FMNH 187598	IIPad1L: 10 IIPad2L: 8 IIClawL: 8 IIIPad1L: 14 IIIPad2L: 10 IIIPad3L: 8 IIIClawL: 9 Digit II length excluding claw: 18; digit II length: 26; digit III length excluding claw: 32; digit III length: 41; digit IV length excluding claw: 22; digit IV length: 28
Otis tarda UMMZ Birds 57476	IIPad1L: 15 IIPad2L: 9 IIClawL: 7 IIIPad1L: 17 IIIPad2L: 13 IIIPad3L: 12 IIIClawL: 8 Digit II length excluding claw: 24; digit II length: 31; digit III length excluding claw: 42; digit III length: 50; digit IV length excluding claw: 31; digit IV length: 38
Otis tarda UMMZ Birds 114965	IIPad1L: 9 IIPad2L: 8 IIClawL: 8 IIIPad1L: 12 IIIPad2L: 9 IIIPad3L: ——— IIIClawL: 9 Digit II length excluding claw: 17; digit II length: 25; digit III length excluding claw: ———; digit III length: ———; digit IV length excluding claw: 21; digit IV length: 28
Neotis denhami FMNH 23189	IIPad1L: 8 IIPad2L: 9 IIClawL: 8 IIIPad1L: 13 IIIPad2L: 10 IIIPad3L: 10 IIIClawL: 9 Digit II length excluding claw: 17; digit II length: 25; digit III length excluding claw: 33; digit III length: 42; digit IV length excluding claw: 23; digit IV length: 28
Neotis denhami FMNH 23190	IIPad1L: 9 IIPad2L: 7 IIClawL: 8 IIIPad1L: 15 IIIPad2L: 18 IIIPad3L: 0 IIIClawL: 8 Digit II length excluding claw: 16; digit II length: 24; digit III length excluding claw: 33; digit III length: 41; digit IV length excluding claw: 20; digit IV length: 25
Neotis heuglinii FMNH 7174	IIPad1L: 11 IIPad2L: 9 IIClawL: 8 IIIPad1L: 14 IIIPad2L: 10 IIIPad3L: 10 IIIClawL: 6 Digit II length excluding claw: 20; digit II length: 28; digit III length excluding claw: 34; digit III length: 40; digit IV length excluding claw: 24; digit IV length: 29

Table A4.1. *continued*

Specimen	Measurements
Galliforms	
Domestic chicken Berkeley embryo	IIPad1L: 2 IIPad2L: 2 IIClawL:2 IIIPad1L: 1 IIIPad2L: 2 IIIPad3L: 2 IIIClawL: 2 Digit II length excluding claw: 7; digit II length: 8; digit III length excluding claw: 10; digit III length: 12; digit IV length excluding claw: 10; digit IV length: 11
Domestic chicken Berkeley embryo	IIPad1L: 2 IIPad2L: 2 IIClawL: 2 IIIPad1L: 2 IIIPad2L: 2 IIIPad3L: 3 IIIClawL: 2 Digit II length excluding claw: 8; digit II length: 9; digit III length excluding claw: 12; digit III length: 13; digit IV length excluding claw: 10; digit IV length: 12
Domestic chicken Berkeley embryo	IIPad1L: 1 IIPad2L: 2 IIClawL: 2 IIIPad1L: 2 IIIPad2L: 2 IIIPad3L: 2 IIIClawL: 2 Digit II length excluding claw: 5; digit II length: 7; digit III length excluding claw: 9; digit III length: 11; digit IV length excluding claw: 8; digit IV length: 9
Domestic chicken Jed Freels farm, Auburn, Indiana	IIPad1L: 7 IIPad2L: 6 IIClawL: 9 IIIPad1L: 8 IIIPad2L: 8 IIIPad3L: 8 IIIClawL: 10 Digit II length excluding claw: 18; digit II length: 27; digit III length excluding claw: 33; digit III length: 43; digit IV length excluding claw: 27; digit IV length: 36
Lophophorus impejanus FMNH 210929	IIPad1L: 6 IIPad2L: 6 IIClawL: 14 IIIPad1L: 8 IIIPad2L: 10 IIIPad3L: 8 IIIClawL: 15 Digit II length excluding claw: 27; digit II length: 41; digit III length excluding claw: 42; digit III length: 57; digit IV length excluding claw: 36; digit IV length: 48
Phasianus colchicus FMNH 123275	IIPad1L: 6 IIPad2L: 6 IIClawL: 8 IIIPad1L: 7 IIIPad2L: 6 IIIPad3L: 6 IIIClawL: 7 Digit II length excluding claw: 21; digit II length: 29; digit III length excluding claw: 34; digit III length: 41; digit IV length excluding claw: 29; digit IV length: 35
Phasianus colchicus FMNH 12933	IIPad1L: 5 IIPad2L: 7 IIClawL: 7 IIIPad1L: 6 IIIPad2L: 6 IIIPad3L: 8 IIIClawL: 9 Digit II length excluding claw: 18; digit II length: 25; digit III length excluding claw: 33; digit III length: 42; digit IV length excluding claw: 24; digit IV length: 31
Phasianus versicolor FMNH Conover 26081	IIPad1L: 5 IIPad2L: 5 IIClawL: 10 IIIPad1L: 5 IIIPad2L: 6 IIIPad3L: 6 IIIClawL: 12 Digit II length excluding claw: 22; digit II length: 32; digit III length excluding claw: 37; digit III length: 49; digit IV length excluding claw: 32; digit IV length: 40
Pavo cristatus FMNH 65380	IIPad1L: 10 IIPad2L: 9 IIClawL: 11 IIIPad1L: 11 IIIPad2L: 13 IIIPad3L: 11 IIIClawL: 14 Digit II length excluding claw: 34; digit II length: 45; digit III length excluding claw: 59; digit III length: 73; digit IV length excluding claw: 43; digit IV length: 53
Pavo cristatus FMNH no number	IIPad1L: 10 IIPad2L: 9 IIClawL: 12 IIIPad1L: 13 IIIPad2L: 13 IIIPad3L: 11 IIIClawL: 13 Digit II length excluding claw: 35; digit II length: 47; digit III length excluding claw: 62; digit III length: 75; digit IV length excluding claw: 50; digit IV length: 61
Numida meleagris FMNH Conover 6216	IIPad1L: 6 IIPad2L: 5 IIClawL: 9 IIIPad1L: 8 IIIPad2L: 7 IIIPad3L: 6 IIIClawL: 11 Digit II length excluding claw: 18; digit II length: 27; digit III length excluding claw: 33; digit III length: 44; digit IV length excluding claw: 26; digit IV length: 34
Allen Haynes farm domestic turkey 1, Auburn, Indiana	TMT: 127; foot: 103 IIPad1L: 12 IIPad2L: 10 IIClawL: 12 IIIPad1L: 13 IIIPad2L: 13 IIIPad3L: 13 IIIClawL: 14 Digit II length excluding claw: 33; digit II length: 45; digit III length excluding claw: 57; digit III length: 71; digit IV length excluding claw: 46; digit IV length: 57
Allen Haynes farm domestic turkey 2, Auburn, Indiana	TMT: 145; foot: 123 IIPad1L: 16 IIPad2L: 12 IIClawL: 14 IIIPad1L: 16 IIIPad2L: 14 IIIPad3L: 14 IIIClawL: 18 Digit II length excluding claw: 39; digit II length: 53; digit III length excluding claw: 63; digit III length: 81; digit IV length excluding claw: 52; digit IV length: 66

Specimen	Measurements			
	TMT: 160; foot: 120			
Allen Haynes farm domestic turkey 3, Auburn, Indiana	IIPad1L: 15	IIPad2L: 10	IIClawL: 15	
	IIIPad1L: 16	IIIPad2L: 13	IIIPad3L: 14	IIIClawL: 17
	Digit II length excluding claw: 39; digit II length: 54; digit III length excluding claw: 62; digit III length: 79; digit IV length excluding claw: 55; digit IV length: 69			

Table A4.2. Summary measurements of mean footprint length and stride length in trackways consisting of three or more prints of bipedal or potentially bipedal dinosaurs. In most (but not all) cases, attribution of the trackmaker follows that of the author(s) of the cited study, but many such identifications are questionable, especially for smaller trackmakers. For sake of convenience, assignment of trackways to ichnotaxa follows that of the author(s) or the labels of museum specimens, and does not necessarily reflect current understanding of ichnotaxonomy. In cases in which there are multiple trackways assigned to the same ichnospecies, the specific name is indicated only in the first case. Data for trackways attributed to ornithopods or other ornithischians include measurements for both bipedally and quadrupedally moving animals. Trackmaker code: T = theropod; O = ornithischian (mostly ornithopod); UID = unidentified dinosaur or dinosauromorph; some early Mesozoic trackways (e.g., *Atreipus)* might have been made by non-dinosaurian dinosauromorphs. Fm = formation; MT = metatarsal. Measurements reported are those of the authors, wherever possible. In some cases, however, measurements were made from trackway diagrams or photographs. In trackways in which the dinosaur left an impression of the metatarsal region ("elongate," "metatarsal," or "semiplantigrade" trackways), footprint length excludes the metatarsal portion of the print. Some sites span the Jurassic-Cretaceous boundary; I assigned such sites to the Mid-Late Jurassic or Cretaceous on the basis of which time interval I thought better characterized the age of that site.

Site	Trackway	Code	Footprint mean length (cm)	Mean stride (cm)	Source of data and comments
Triassic–Early Jurassic trackways					
Newton Nottage, Wales, Mercia Mudstone Group	*Anchisauripus* sp. NMW G628	T	25.0	79.0	Lockley et al. (1996)
	Grallator	T	10.0	170.0	
	Grallator	T	10.0	167.0	
Severac-le-Chateau, Aveyron, France, Lias	Bipedal dinosaur	UID	9.0	73.0	Ellenberger and Fuchs (1965)
Largentiere (Ardeche), France	*Coelurosaurichnus largentierensis* piste VIII	T	13.0	107.0	Courel and Demathieu (1976)
	C. largentierensis piste IX	T	8.0	102.0	
Payzac, Ardeche, France	*Coelurosaurichnus grancieri*	T	11.0	82.0	Courel and Demathieu (2000: fig. 3*A*)
Le Veillon, France	*Grallator olonensis*	T	4.3	31.0	Lapparent and Montenat (1967)
	G. olonensis	T	3.8	40.0	
	G. variabilis	T	12.3	120.0	
	G. maximus	T	28.0	209.0	
	Eubrontes veillonensis	T	34.0	249.0	
	Saltopoides igalensis	T	15.5	349.0	
Assemblage BF2, Plateau d'Antully, France	*Anchisauripus bibractensis* piste A	T	4.1	52.5	Demathieu and Gand (1972)
	A. bibractensis piste B	T	4.5	61.8	
	A. bibractensis piste M	T	5.5	76.3	Gand (1979*b*)
	Coelurosaurichnus perriauxi piste D	T	7.3	86.5	Demathieu and Gand (1972)
	C. perriauxi piste E	T	8.7	126.6	
	C. perriauxi piste X	T	7.7	97.0	Gand (1979*b*)
	C. perriauxi piste Y1	T	8.5	119.5	
	C. perriauxi piste Y2	T	8.8	97.0	
	C. perriauxi piste O	T	7.1	77.9	
Assemblage CF2, Saint-Sernin-du-Bois, France	*Anchisauripus bibractensis*	T	7.3	96.1	Gand (1975)
Saint-Sernin-du-Bois, Antunois, France	*Coelurosaurichnus perriauxi* trackway A	T	7.6	92.0	Gand (1979*a*)
	C. perriauxi trackway B	T	8.2	114.0	
	C. perriauxi trackway D	T	7.8	82.0	
Sauclieres, Aveyron, France	*Grallator variabilis* piste P	T	13.0	139.0	Demathieu and Sciau (1994); authors don't report footprint lengths for individual trackways of this ichnospecies, but they all look about the same size
	G. variabilis piste Y	T	13.0	70.0	
	G. variabilis piste Z	T	13.0	137.0	
	G. variabilis piste AD	T	13.0	152.0	Demathieu and Sciau (1994) from authors' figures
	G. sauclierensis	T	16.1	177.0	Demathieu and Sciau (1994)
	G. minusculus piste T	T	30.0	187.0	
	G. minusculus piste P	T	30.0	220.0	
	Dilophosauripus williamsi	T	22.8	189.0	

Site	Trackway	Code	Footprint mean length (cm)	Mean stride (cm)	Source of data and comments
Gres superieurs et Argilites bariolees, La Grand-Combe France	*Grallator andeolensis* piste 1	T	19.0	189.0	Gand et al. (2000: fig. 10)
	G. andeolensis piste 3	T	20.0	193.0	
	G. andeolensis piste 4	T	20.0	204.0	
	G. andeolensis piste 6	T	19.0	197.0	
Saint-Leons, France	*Grallator minusculus* Pt-3	T	33.7	242.3	Demathieu et al. (2002)
	G. minusculus Pt-6	T	35.0	233.0	
	G. minusculus Pt-8	T	28.6	214.0	
	G. lescurei Pt-4	T	27.0	230.5	
	G. lescurei Pt-5	T	25.2	300.5	
	G. lescurei Pt-11	T	27.1	264.6	
	G. lescurei Pt-Ca-1	T	25.3	252.5	
	Eubrontes giganteus Pt-1	T	34.8	346.3	
	E. giganteus Pt-10	T	30.5	566.5	
Fondamente, France	*Grallator minusculus* Pt-2	T	30.0	180.0	
	G. minusculus Pt-3	T	28.3	262.0	
	G. minusculus Pt-4	T	24.7	165.5	
Camprieu, France	*Grallator minusculus* Pt-2	T	30.5	226.2	
	Eubrontes giganteus Pt-1	T	29.3	198.5	
Saint-Laurent de Treves, France	*Grallator minusculus* Pt-1	T	26.5	190.0	
	G. minusculus Pt-2	T	23.0	243.5	
	G. minusculus Pt-3	T	27.2	264.0	
Salsac, France	*Grallator lescurei* Pt-1	T	27.0	212.5	
	G. lescurei Pt-2	T	26.5	221.5	
	G. lescurei Pt-6	T	24.4	196.3	
	G. lescurei Pt-10	T	25.8	327.0	
	G. lescurei Pt-11	T	26.5	198.8	
	E. giganteus Pt-3	T	45.0	244.5	
	E. giganteus Pt-4	T	38.0	244.0	
	E. giganteus Pt-5	T	39.0	246.8	
	E. giganteus Pt-9	T	38.0	257.0	
Peyres, France	*E. giganteus* Pt-2	T	40.0	229.0	
Figeac, Lot, France	*Eubrontes* sp. trackway P1	T	29.0	163.0	Lange-Badré and Lafon (2000)
Dolomia Principale, Mt. Pelmetto Cadore, Italy	Theropod trackway A	T	6.5	65.0	Mietto (1985)
	Ornithopod trackway B	O	11.0	52.0	
Ciol de la Fratta A, Dolomia Principale, Italy	Theropod trackway CFA1	T	17.5	157.7	Dalla Vecchia and Mietto (1998)
Dolomia Principale, Friuli-Venezia Giulia, Italy	Theropod trackway AVG2-2	T	21.0	151.0	Marzola and Dalla Vecchia (2014)
	Theropod trackway VPE1-1	T	20.3	57.0	
Lavini di Marco, Calcari Grigi, Italy	Theropod ROLM 38	T	18.0	111.0	Leonardi and Mietto (2000: fig. 8.4)
	Theropod ROLM 12	T	31.0	161.0	
	Theropod ROLM 159	T	31.0	210.0	Leonardi and Mietto (2000: fig. 8.5)
	Theropod ROLM 193	T	17.0	122.0	Leonardi and Mietto (2000: fig 8.29)
	Ornithopod ROLM 181	O	25.0	111.0	

Table A4.2. *continued*

Site	Trackway	Code	Footprint mean length (cm)	Mean stride (cm)	Source of data and comments
Costa dell'Anglone, Rotzo Fm, Italy	CA1	T	21.0	195.3	Petti et al. (2011)
	CA2	T	22.3	125.9	
	CA3	T	17.0	162.0	
	CA4	T	23.0	261.4	
	CA5	T	25.3	142.0	
	CA6	T	21.3	138.4	
	CA7	T	19.2	176.0	
	CA8	T	19.2	157.0	
	CA9	T	22.3	204.0	
	CA10	T	21.3	169.0	
	CA11	T	14.5	148.7	
	CA12	T	17.6	153.3	
	CA13	T	26.0	167.1	
	CA14	T	25.0	234.8	
	CA15	T	18.0	258.4	
	CA16	T	14.5	124.3	
	CA17	T	22.2	145.2	
	CA18	T	21.1	134.6	
	CA19	T	20.0	115.8	
	CA20	T	24.0	271.0	
Mecsek Coal Fm, Hungary	PB1-B *Komlosaurus carbonis*	T	13.0	138.0	Ösi et al. (2011)
	PB1-C *K. carbonaris*	T	19.2	159.0	
	PB1-D *K. carbonaris*	T	18.2	148.5	
	PB1-E *Kcarbonaris*	T	17.5	203.0	
	PB2-A *K. carbonaris*	T	22.5	154.2	
	PB2-B *K. carbonaris*	T	21.7	140.8	
	PB2-D *K. carbonaris*	T	12.0	100.3	
	PB2-E *K. carbonaris*	T	20.4	112.7	
	PB2-F *K. carbonaris*	T	12.0	107.0	
	PB2-I *K. carbonaris*	T	23.1	134.3	
	PB2-K *K. carbonaris*	T	21.5	143.0	
	PB2-L *K. carbonaris*	T	16.0	161.0	
	PB2-N *K. carbonaris*	T	17.0	129.0	
Przysucha Ore-bearing series, Holy Cross Mountains, Poland	*Moyenisauripus karaszevskii*	O	24.0	93.0	Gierliński (1991)
Zapniów Mine and Clay Pit, Holy Cross Mountains, Poland	*Anchisauripus*	T	22.5	174	Niedźwiedzki and Pieńkowski (2016)
Ab-e-Haji Fm, Neizar Valley, Iran	Trackway UHUL-T1	T	39	253	Kellner et al. (2012)
Wulatezhongqi, China	"*Eubrontes glenrosensis*" trackway 1	T	43.5	167.0	Li et al. (2010)
	Kaypentapus hailiutuensis trackway 2	T	26.6	81.1	
	Kayentapus trackway 3	T	24.7	165.8	
	Kayentapus trackway 4	T	23.0	156.7	
	Kayentapus trackway 6	T	37.0	221.5	
	Kayentapus trackway 10	T	8.4	33.0	
	Kayentapus trackway 16	T	31.5	189.0	
	Kayentapus trackway 17	T	26.5	176.3	
Fuxian Fm, Shaanxi Province, China	*Shenmuichnus youngteilhardorum* quadrupedal trackway	O	16.1	88.4	Li J. et al. (2012)
Fuxian Fm, Shaanxi Province, China	*Shanbeipus caudatus* trackway ZJ	T	11	56.4	Wang B. et al. (2016)
	Shanbeipus caudatus trackway ZL	T	11	69	

Site	Trackway	Code	Footprint mean length (cm)	Mean stride (cm)	Source of data and comments
Lufeng Fm, Dalishu, Yunnan Province, China	*Shenmuichnus wangi* quadrupedal trackway DLS1	O	21.5	76.3	Xing et al. (2016*l*)
	Shenmuichnus wangi quadrupedal trackway DLS2	O	23.2	113.3	
	Shenmuichnus wangi quadrupedal trackway DLS3	O	23.1	109.8	
Lufeng Fm, Dalishu II, Yunnan Province, China	*Shenmuichnus wangi* quadrupedal trackway DLSII-O3	O	21.3	91.5	Xing et al. (2016*e*)
	Shenmuichnus wangi quadrupedal trackway	O	18.2	85.1	
	Shenmuichnus wangi quadrupedal trackway	O	25.7	79.9	
	Shenmuichnus wangi quadrupedal trackway	O	20.5	101.0	
	Shenmuichnus wangi quadrupedal trackway	O	20.5	107.5	
Lufeng Fm, Dalishu III, Yunnan Province, China	*Anomoepus*	O	9.0	57.0	Xing et al. (2016*c*)
Ziliujing Fm, Hejie tracksite, Zigong City, Sichuan, China	Theropod trackway T1-T3	T	14.5	124.0	Xing et al. (2014*j*)
Ziliujing Fm, Gulin, Sichuan, China	JYT1 type A *Eubrontes*	T	30.4	177	Xing et al. (2017*b*)
	JYT2 type A *Eubrontes*	T	29.2	190	
	JYT3 type B *Grallator*	T	14.1	82	
	JYT3 type B *Grallator*	T	13.0	100	
Zhenzhuchong F, Weiyuan County, Sichuan, China Lower Jurassic	*Eubrontes* (formerly *Weiyuanpus*) *zigongensis*	T	42.1	243.8	Xing et al. (2014*k*)
Otani, Japan, KrIII Fm	*Schizograllator otariensis* OTK-001 tracks 2, 4, and 6	T	21.7	183.0	Matsukawa et al. (2005)
Zone A/2 Lower Stormberg, South Africa	*Qemetrisauropus princeps*	T	35	282	Values of stride mostly represent midpoints of ranges given by author; Ellenberger (1972)
	Q. minor	T	24	280	
	Anatrisauropus ginsburgi	O	22.0	248	
	Pseudotrisauropus humilis	T	12	100	
	P. minusculus	T	7.3	60	
	Prototrisauropus crassidigitus	T	40	305	
	P. crassidigitus	T	32	210	
	P. angustidigitus	T	25	220	
	P. rectilineus	T	28	304	
	P. rectilineus	T	28	396	
	P. rectilineus	T	28	180	
	P. graciosus	T	15	200	
	Prototrisauropodiscus minimus	T	6.7	37	
	Seakatrisauropodus unguiferous	T	19	192	
	Bosiutrisauropus phuthiatsani	T	20	200	
	B. phuthiatsani	T	26.7	165.0	
	Paratriasauropus mendrezi	O	18	120	
	Deuterotrisauropus socialis	T	20	195	
	Mafatrisauropus errans	T	16	139	
	Mafatrisauropus errans	T	16	122	
	Psilotrisauropus subengensis	O	12.5	105.5	
	Trisauropodiscus aviforma	UID	3.5	24	
	T. aviforma	UID	3.2	16	
	T. aviforma	UID	2.7	13	
	Paratrisauropus equester	O	23	72	
	Pseudotrisauropus maserui	T	29	132	

Table A4.2. *continued*

Site	Trackway	Code	Footprint mean length (cm)	Mean stride (cm)	Source of data and comments
Zone A/2 Lower Stormberg, South Africa	*Tritotrisauropus medius*	T	17	124	Values of stride mostly represent midpoints of ranges given by author; Ellenberger (1972)
	Trisauropodiscus galliforma	UID	6.8	48	
	Trisauropodiscus phasianiforma	UID	4.4	32	
	Trisauropodiscus phasianiforma	UID	4.4	32	
	Trisauropodiscus phasianiforma	UID	4.4	24	
	T. levis	UID	2.5	20	
	T. popompoi	UID	12.5	76	
Upper Stormberg, L'Ile de Moyeni	*Neotrisauropus deambulator*	T	26	190	Ellenberger (1974); also see Wilson et al. (2009)
	Plastisauropus ingens	T	32	290	
	M. natator piste AH	O	18	110	
	Moyenisauropus natatilis piste D	O	13	82	
	M. natatilis piste E	O	14.5	64	
	M. natatilis piste I	O	13	120	
	M. dodai piste K	O	13.5	88	
	M. vermivorus piste AA	O	12	69	
	M. vermivorus piste AF	O	11.8	63	
	M. vermivorus piste AG	O	12	66	
	M. vermivorus piste AJ	O	12	67	
	M. minor	O	9.8	49	
	M. longicauda	O	11.5	67	
	Neotrisauropus mokanametsongensis piste A1	T	23.5	142	
	M. levicauda piste M1	O	10.5	82	
	M. levicauda piste N1	O	11.1	88	
	M. levicauda piste N3	O	11.0	90	
	Aetonychopus rapidus piste H1	T	18.4	175	
	A. rapidus piste H2	T	18.4	195	
	A. digitigradus piste J	T	15.0	120	
	A. digitigradus piste K	T	14.5	110	
	A. digitigradus piste L	T	13.0	112	
	Masitisisauropus palmipes piste AJ	T	10.5	61	
	M. palmipes piste AK1	T	10.3	60	
	M. palmipes piste AM	T	11.0	70.7	
	M. palmipes piste AO	T	10.5	78.	
	M. exiguus piste AR3	T	9.5	66	
	Masitisisauropezus minimus piste AT1	T	6.1	30	
	M. minutus piste AV	T	5.2	30	
	M. minutus piste AW2	T	4.8	35	
	Masitisisauropodiscus turdus piste AY1	T	3.9	21	
	M. turdus piste AY2	T	3.5	24	
Stormberg Group, Pot River, Rhodes, Cape Province, South Africa	Bipedal dinosaur	UID	19.5	181	Raath et al. (1990); trackmaker interpreted as a theropod by the authors, but the prints remind me of *Anomoepus*
Formation d'Agane, Haut Atlas Morocco	Groupe 1-1	T	30.0	146.0	Ishigaki (1988); cf. Ishigaki and Lockley (2010)
	Groupe 1-2	T	32.0	152.0	
	Groupe 1-3	T	28.0	168.0	
	Groupe 1-4	T	34.0	185.0	
	Groupe 1-5	T	32.0	133.0	
	Groupe 2-6	T	34.0	180.0	
	Groupe 2-7	T	41.0	189.0	
	Groupe 2-8	T	28.0	230.0	
	Groupe 3-9	T	38.0	197.0	
	Groupe 3-10	T	36.0	202.0	

Site	Trackway	Code	Footprint mean length (cm)	Mean stride (cm)	Source of data and comments
Site 7TAO, Taoudaat, Morocco	7TAO1 possible thyreophoran	O	38	192	Hadri et al. (2007)
	7TAO2 possible thyreophoran	O	34	140	
	7TAO3 *Eubrontes*	T	35	218	
Formacion Arganane, Ansous, Morocco	Theropod 3ANS1	T	30.0	249.0	Boutakiout et al. (2014)
Arganane Fm, Issil-n-Aït Arbi, Morocco	Theropod 1.4AAR1	T	28	204	Masrour et al. (2015)
Blackstone Fm, SE Queensland	cf. *Grallator*	T	7.2	43.2	Thulborn (1998)
Striped Bacon Coal Seam, Rhonda Colliery, Queensland	*Eubrontes*	T	43.0	191.0	Staines and Woods (1964); cf. Lucas et al. (2006)
Wood Bjerg, Fleming Fjord Fm, East Greenland	*Grallator* trackway A3.01	T	23.5	150.3	Jenkins et al. (1994); S. M. Gatesy (pers. comm.)
	Grallator trackway A3.02	T	7.1	65.8	
	Grallator trackway A3.03	T	5.3	36.5	
	Grallator trackway A3.04	T	4.3	40.6	
	Grallator trackway A3.06	T	20.7	198.1	
	Grallator trackway B.21	T	13.3	125.5	
	Grallator trackway F.01	T	25.4	171.1	
	Grallator trackway F.04	T	14.0	152.4	
	Grallator trackway L.02	T	20.3	157.5	
Dinosaur State Park, Rocky Hill, Connecticut: presently exposed trackway surface	1-A mean	T	34.0	194.5	This study; cf. Farlow and Galton (2003). Weems (1987, 1992, 2003) attributed such footprints to prosauropods
	1-B mean	T	31.0	190.3	
	1-C mean	T	39.2	246.4	
	1-D mean	T	36.5	303.8	
	1-F mean	T	38.1	230.1	
	1-G mean	T	34.8	215.7	
	1-H (1 Galt 22?) mean	T	34.8	188.0	
	1-I mean	T	30.8	195.7	
	1-J mean	T	31.5	233.0	
	1-K mean	T	38.4	224.1	
	1-L mean	T	38.6	224.5	
	1-M mean	T	33.3	186.5	
	1-N mean	T	38.7	232.5	
	1-O mean	T	34.3	202.5	
	1-P (1-S) mean	T	35.3	213.6	
	1-Q mean	T	35.3	229.5	
	1-R mean	T	36.0	212.5	
	1-T mean	T	37.3	201.5	
	1-W mean	T	32.3	201.3	
	1-Y mean	T	37.7	222.3	
	1-Z mean	T	32.5	214.7	
	1-BB mean	T	41.3	225.5	
	1-Hi mean	T	35.7	238.4	
	0-A mean	T	38.7	224.3	
	0-B mean	T	33.8	212.0	
	0-C mean	T	39.3	230.0	
	0-D mean	T	36.6	194.6	
Dinosaur State Park; presently covered trackway surface	Galton peel of buried print	T	36.5	193.8	
Pliny Moody home, South Hadley	AC 4/1a *Grallator cursorius*	T	7.7	119.0	This study
Turners Falls	AC 42/2 *Plectropterna lineans*	T	4.1	48.1	This study

Table A4.2. *continued*

Site	Trackway	Code	Footprint mean length (cm)	Mean stride (cm)	Source of data and comments
Turners Falls	AC 12/3 "left to right" *Grallator*	T	6.7	35.3	This study; trackways cross an *Anomoepus* trackway on the same slab; trackways labeled to reflect orientation on slab at the time they were measured
	AC 12/3 "right to left" *Grallator*	T	7.1	42.7	
Lily Pond	AC 30/1 *Grallator cuneatus* J-H-D	T	13.5	125.5	This study; individual footprint letter labels assigned by me
Walter Kidde Dinosaur Park, New Jersey; R. Salkin collection	YPM 5900 grallatorid undertracks	T	2.2	13.3	Olsen (1980: fig. 22*A*); Olsen (1995: fig. 13*E*)
Turners Falls, Massachusetts, Portland Fm	AC 14/4 *Anchisauripus minusculus*	T	26.0	168.5	This study
Below Smith's Ferry	AC 10/1 *A. minusculus*	T	33.0	220.5	
Horse Race, Gill	AC 22/1 *A. minusculus* first 3 prints	T	35.3	202.5	
Middlefield, Connecticut	AC 9/14 *A. sillimani* Somebody's toe-9-31	T	13.9	145.7	This study; individual footprint number or letter labels assigned by me
Middlefield, Connecticut	AC 9/14 *A. sillimani* 26-19-43	T	11.6	110.5	
Middlefield, Connecticut	AC 30/1 *A. tuberosus* B-G-I-L-O	T	20.8	130.9	
Connecticut Valley	AC 45/2 *A. minusculus*	T	32.3	201.5	
Turners Falls, south side	AC 49/1 *A. sillimani* (*tuberosus*) up (B) trail	T	19.0	222.0	This study; "up" and "down" refer to direction trails head in slab mounted on wall at the time the trackways were measured
	AC 49/1 *Anchisauripus* down or right (A) trail	T	18.0	165.7	
Smith's Ferry	AC 2/3 *Anchisauripus*	T	17.3	127.5	This study
Turners Falls	AC 64/2 *Anchisauripus* or *Eubrontes*	T	11.9	137.0	This study
Lily Pond, Gill, Massachusetts	AC 20/6 *Anchisauripus*	T	12.8	116.2	This study; trackway crosses *Anomoepus* on same slab
Connecticut Valley	AC 43/4 *A tuberosus*	T	21.9	128.3	This study
Below Smith's Ferry	AC 7/1 *Eubrontes giganteus*	T	42.0	261.5	
Smith's Ferry	AC 8/1 *E. giganteus*	T	37.4	201.8	
Connecticut Valley	AC 45/1 *E. giganteus* early trail	T	35.5	200.5	
	AC 45/1 *E. giganteus* later trail	T	37.4	212.5	
Lily Pond, Turners Falls	AC 9/9and 9/10 *Gigandipus caudatus*	T	43.3	195.3	
Turners Falls	YPM 2098 *Eubrontes giganteus*	T	32.5	173.3	This study
Nash Dinosaur Tracksite, Granby, MA	*Eubrontes* trackway	T	35.8	175.1	This study
Franklin Co., MA	AMNH FR 2246 *Eubrontes*	T	32	206	Measurements estimated from photographs supplied by Carl Mehling
Locality unknown	ACM.ICH 196 *Anchisauripus?*	T	17.8	121.0	This study
	ACM.ICH.1928-0 *Anchisauripus?*	T	20.6	190.2	
Lily Pond, Turners Falls	AC 24/1 *Anchisauripus exsertus*	T	22.0	187.5	This study
Lily Pond, Turners Falls	AC 52/10 right trail *Anomoepus curvatus*	O	9.8	47.5	This study
Field's orchard, Turners Falls	AC 26/10 *Anomoepus intermedius* left trail prints A-D	O	10.0	51.5	This study; individual letter labels for prints assigned by me
	AC 26/10 *Anomoepus* diagonal trail	O	9.3	39.0	This study
Ferry, Turners Falls	AC 48/1 1-13 *Anomoepus intermedius* right trail	O	9.9	40.8	
Turners Falls	AC 19/18 *Anomoepus* (*Plesiornis quadrupes*)	O	9.7	43.6	This study
Turners Falls	AC 16/5 *Anomoepus scambus* walking portion	O	7.5	36.0	(Olsen and Rainforth 2003: fig. 19.11)
Lily Pond, Gill, Massachusetts	AC 20/6 Anomoepus (*Apatichnus circumagens*)	O	8.2	37.0	This study
South Hadley	AC 16/2 *Anomoepus* sp. (Noah's Raven)	O	8.5	47.1	This study
Turners Falls	AC 26/21 *Anomoepus* (*Tridentipes* [*Steropoides*] *elegans*)	O	11.7	68.9	This study

Site	Trackway	Code	Footprint mean length (cm)	Mean stride (cm)	Source of data and comments
Marsh's quarry, Montague	AC 1/4 ?*Anomoepus* (*Steropoides loripes*)	O	16	93	This study
Lily Pond	AC 6/1 ?*Anomoepus* (*Tridentipes uncus*)	O	6.6	36.1	This study
Turners Falls	AC 52/1 *Anomoepus*	O	8.0	33.5	This study
Turners Falls, probably Lily Pond	AC 44/2 *Anomoepus curvatus*	O	7.8	41.6	This study
Lily Pond	AC 53/7 *Anomoepus* "left" trail	O	9.1	48.1	This study
	AC 53/7 *Anomoepus* "middle" trail	O	8.6	31.3	
	AC 53/7 *Anomoepus* "right" trail	O	9.3	45.9	
Turners Falls	AC 19/8 ?*Anomoepus*	O	6.7	35.5	This study
Lily Pond	AC 52/14 *Anomoepus*	O	7.6	45.4	This study
Locality unknown	ACM.ICH 8 *Anomoepus*	O	12.7	58.6	This study
Turners Falls	AC 12/3 *Anomoepus*	O	9.7	52.4	This study
Turners Falls? (removed from sidewalk in Greenfield)	AC 11/1 ?*Anomoepus* (*Platypterna concamerata*)	O	10.2	36.7	This study
Brunswick Fm, Essex County, New Jersey	YPM PU 021238 *Anomoepus* A trail	O	3.0	14.7	This study; trackway letter label assigned by me
Ferry, Gill	YPM 002045 *Anomoepus gracillimus*	O	7.8	37.3	This study
Towaco Fm, Essex County, New Jersey	YPM-PU 021722 *Anomoepus*	O	4.1	23.5	
Turners Falls	YPM-PU 002019 *Anomoepus minimus*	O	4.5	26.5	
	YPM-PU 002021 *Anomoepus curvatus*	O	5.9	33.8	
Upper unit 18 Walter Kidde Dinosaur Park Roseland New Jersey	*Anomoepus* A trail	O	1.8	13.3	Olsen (1980: fig. 22*D*); Olsen (1995: fig. 13*B*); trackway letter labels assigned by me
	Anomoepus B trail	O	2.0	13.1	
	Anomoepus C trail	O	3.9	25.2	
	Anomoepus D trail	O	1.9	15.6	
	Anomoepus E trail	O	1.8	14.4	
	Anomoepus F trail	O	1.8	8.8	
Newark Group R Salkin Collection	YPM VP 005910 *Anomoepus*	O	1.8	14.8	Olsen (1980: fig. 22*E*)
Upper unit 18 Walter Kidde Dinosaur Park Roseland NJ	*Anomoepus* lost specimen	O	6.4	38.2	Olsen (1995: fig. 12*I*)
Culpeper Quarry	*Kayentapus* K-11	T	25.0	218.0	Weems (1992)
	Kayentapus K-19	T	30.3	226.0	
Ferry, Gill	YPM VP 003473 "*Arachnichnus dehiscens*"	UID	2.7	6.8	This study
Turners Falls	YPM VP 002095B *Anchisauripus*and YPM VP 002102 *Anomoepus* slab trackway A	UID	10.1	49.0	This study; trackway letter labels assigned by me
Turners Falls	YPM VP 002095B and YPM VP 002102 slab trackway B	UID	8.7	43.0	
Turners Falls	YPM VP 002095B and YPM VP 002102 trackway C	UID	8.7	52.5	
Turners Falls	YPM VP 002030 *Argoides minimus*	UID	5.3	41.1	This study
Turners Falls	YPM VP 002090 *Steropoides diversus*	UID	13.5	95.8	
Ferry, Gill	YPM VP 002031 *Argoides minimus* final 5 prints in slab	UID	3.4	16.8	
Turners Falls	YPM VP 002132 *Argoides minimus*	UID	3.4	17.2	
Sugar Loaf Tunnel, North Branford, Connecticut	*Gigandipus caudatus* YPM 370	T	50.5	182.4	Thorpe (1929)
Murray Quarry, East Berlin Fm, Holyoke, Massachusetts	*Eubrontes* trackway on slab beneath carport at Wistariahurst Museum, Holyoke, MA	T	36.0	297	This study; probably the same as WM 13 of Getty (2005)
	FP 1 *Eubrontes* sp.	T	28.5	215	Getty (2005)
	FP 5 *Anchisauripus* sp.	T	20.6	176.2	
	FP 7 *Anchisauripus* sp	T	23.5	256	
	FP 10/WM 2 *Eubrontes giganteus*	T	35.4	242.2	
	FP 12 *Eubrontes* sp.	T	25.7	268	

Table A4.2. *continued*

Site	Trackway	Code	Footprint mean length (cm)	Mean stride (cm)	Source of data and comments
Murray Quarry, East Berlin Fm, Holyoke, Massachusetts	WM 13 *Eubrontes* sp.	T	35.5	297	Getty (2005)
	MQ 4 *Eubrontes* sp.	T	28.4	292.7	
	MQ 5 *Eubrontes* sp.	T	32.8	271.3	
	MQ 6 *Eubrontes* sp.	T	33.2	269.2	
	MQ 7 *Eubrontes* sp.	T	28.9	209.7	
	MQ 8 *Anchisauripus* sp.	T	24.6	177	
	MQ 9 *Anchisauripus* sp.	T	21.7	199.5	
	MQ 14 *Eubrontes* sp.	T	28.4	258.5	
Powder Hill Dinosaur Park, East Berlin Fm, Middlefield, Connecticut	Trackway 1 *Anchisauripus*	T	18.4	156	Getty et al. (2015); strides from fig. 3
	Trackway 4 *Eubrontes*	T	30.3	146	
	Trackway 8 *Eubrontes*	T	39.7	225	
	Trackway 10 *Eubrontes*	T	35.5	159	
	Trackway 11 *Eubrontes*	T	28.3	201	
	Trackway 26 *Eubrontes*	T	29.0	179	
	Trackway 33 *Eubrontes*	T	26	198	
	Trackway 35 *Anchisauripus*	T	15.3	121	
	Trackway 59 *Eubrontes*	T	38.0	218	
	Trackway 60 *Eubrontes*	T	28.0	162	
East Berlin Fm, Holyoke, Massachusetts Gaulin Site Upper Layer	*Anomoepus lacertoideus* quadrupedal trackway UT2 SSM2006/29-23	O	3.0	16.7	Dalman and Weems (2013)
	Anomoepus quadrupedal trackway GG2/5 LT1	O	4.7	21.3	
	Anomoepus quadrupedal trackway GG2/5 LT2	O	2.7	12.4	
	Anomoepus GG2/5 LT5	O	3.1	36.0	
	Anomoepus quadrupedal trackway SSM2006/29-16 LT12	O	3.6	15.7	
East Berlin Fm, Holyoke, Massachusetts Gaulin Site	*Eubrontes*	T	39	212.7	This study
Newark Supergroup	ANSP 9735	UID	5.3	26.2	This study
	ANSP 10021	UID	5.9	33.0	
Dry Fork Fm, Danville/Dan River Basin, Virginia	*Banisterobates boisseaui* VMNH 202	UID	1.8	12.2	Fraser and Olsen (1996)
Buckland Fm, Oak Hill, Culpeper Basin, Virginia	*Grallator cuneatus* trackway T1-T4	T	19	102.5	Pannell (1985); footprint length estimated from fig. 2
Aldie, Virginia	USNM 10912 *Eubrontes platypus*	T	27.5	282.0	This study
Maryland Science Center *Atreipus* block, Emmitsburg, Maryland	Bipedal dinosaur	UID	7.0	64.0	G. Paul (pers. comm.)
	Bipedal dinosaur	UID	8.0	56.0	
Bull Run Fm, Culpeper, Virginia	*Anchisauripus parallelus* trackway segments 1 and 2	T	18.0	145.0	Weems (1987); Weems (2006*a*) reinterpreted *Gregaripus* as a variant of *Anomoepus*, *Anchisauripus parallelus* as either *Grallator parallelus* or *Atreipus sulcatus*, and *Grallator* isp. possibly *Atreipus* cf. *milfordensis*
	Apatichnus minor trackway 1	T	18.4	154.0	
	Eubrontes divaricatus trackway 1	T	27.5	170.0	
	Grallator sp. trackway 1	T	12.5	173.0	
	Grallator sp. trackway 2	T	12.5	151.0	
	Grallator sp. trackway 3	T	12.6	154.0	
	Gregaripus bairdi trackway 1	O	8.9	106.0	
	G. bairdi trackway 2	O	8.9	97.0	
	G. bairdi trackway 3	O	8.8	99.0	
	G. bairdi trackway 4	O	8.8	87.0	
	G. bairdi trackway 5	O	8.8	114.0	
	G. bairdi trackway 6	O	8.7	93.0	
	G. bairdi trackway 7	O	8.8	105.0	
	G. bairdi trackway 8	O	8.7	96.0	

Site	Trackway	Code	Footprint mean length (cm)	Mean stride (cm)	Source of data and comments
	G. bairdi trackway 9	O	8.7	100.0	
	G. bairdi trackway 10	O	8.7	98.0	
	G. bairdi trackway 11	O	8.8	94.0	
	G. bairdi trackway 12	O	8.9	97.0	
	G. bairdi trackway 13	O	8.8	99.0	
	G. bairdi trackway 14	O	8.7	102.0	
	G. bairdi trackway 15	O	9.0	88.0	
Mt. Tom, Massachusetts	B	T	36.3	247.0	This study; trackway labels assigned by me; cf. Getty et al. (2012)
Mt. Tom	E	T	34.5	231.0	
Mt. Tom	H	T	38.8	240.5	
Mt. Tom	I	T	36.0	249.7	
Mt. Tom	K	T	36.7	225.0	
Mt. Tom	N	T	40.7	257.0	
Mt. Tom	Q	T	33.7	220.5	
Mt. Tom	R	T	34.9	226.4	
Mt. Tom	T	T	33.0	229.0	
Mt. Tom	U	T	32.9	199.7	
Mt. Tom	X	T	35.7	232.8	
Mt. Tom	Y	T	30.7	211.3	
Portland Fm, Granby, Massachusetts	Grallatorid GMS 1/1	T	2.2	24.6	Dalman (2012)
	Grallatorid GMS 1/2	T	2.2	17.7	
Sloan Canyon Fm, Peacock Canyon, New Mexico	*Coelurosaurichnus* sp.	T	7.7	58.0	Conrad et al. (1987: fig. 5*C*)
Sheep Pen Sandstone, Tucker Ranch, OK/NM boundary, upper track bed	*Grallator* sp.	T	5.5	45.0	Conrad et al. (1987: fig. 11)
	Grallator sp.	T	5.5	41.0	
Cub Creek South, near Dinosaur National Monument	Small theropod	T	7.7	42.8	This study
Chinle Fm, Four Mile Canyon, Glen Canyon National Recreation Area	*Atreipus*	UID	12.0	84.0	Lockley et al. (1998c: fig. 6*B*)
Chinle Group, Late Triassic, Gateway Area Colorado	CU-MWC 154.50 long *Grallator* trackway	T	6.0	69.0	Gaston et al. (2003) and this study
	Short *Grallator* trackway	T	7.0	49.0	Gaston et al. (2003: fig. 6)
Chinle Fm, Moab, Utah UCD 13 2004	*Grallator* trackway 1	T	7.1	72.0	Lockley and Gierliński (2009)
	Grallator trackway 4	T	11.5	98.8	
UCD Site 92, near Chew Ranch, just E of Dinosaur National Monument boundary	Small theropod	T	7.5	56.5	This study
Chinle Fm, Dirty Devil River Valley, Utah	*Grallator* trackway 15	T	11.0	110.0	Lockley and Eisenberg (2006)
	Grallator trackway 17	T	9.5	88.0	
	Grallator trackway 20	T	12.0	76.0	
Redonda Fm, Late Triassic NMMNH locality 5795	*Grallator*	T	8.5	117.0	Klein et al. (2006)
Moenave Fm, SGDS.568 slab	Trackway 1	T	22.0	208.5	Milner et al. (2006)
	Trackway 3	T	18.0	200.0	
	Trackway 5	T	17.0	135.0	
	Trackway 6	T	17.0	137.0	
	Trackway 7	T	12.0	110.0	
	Trackway 8	T	10.0	103.0	
	Trackway 9	T	9.0	81.0	
Moenave Fm, SGDS.18 Top Surface	*Eubrontes* SGDS.18.T1 walking portion	T	36.8	227.2	Milner et al. (2009)

Table A4.2. *continued*

Site	Trackway	Code	Footprint mean length (cm)	Mean stride (cm)	Source of data and comments
Moenave Fm, R.T. Bird Site	Q10	T	40.8	222.0	Irby (1996*b*)
	P11	T	26.5	215.0	
	N8	T	40.8	235.0	
	K12	T	33.0	216.5	
	K11	T	32.5	258.2	
	J11	T	26.3	240.0	
	J10	T	31.7	209.0	
	I6	T	32.0	218.0	
	G4	T	31.9	285.5	
	F4b	T	36.7	212.0	
	C8	T	36.8	246.0	
	C1	T	27.7	492.5	
	P12	T	18.3	167.0	
	N13	T	18.5	142.7	
	N11	T	21.0	132.0	
	L10	T	21.0	138.0	
	J9	T	20.5	162.0	
	J5	T	17.6	161.0	
	I9	T	19.9	140.7	
	H9	T	21.8	305.0	
	H5	T	18.0	159.0	
	G7	T	18.8	148.6	
	F4a	T	21.3	176.0	
	E2	T	16.1	108.3	
	D2	T	21.1	316.5	
	A6	T	23.3	196.0	
	G2	T	11.8	98.6	
	F7	T	11.7	124.0	
	F3	T	12.0	97.0	
	F2a	T	7.0	61.0	
	F2b	T	11.2	116.0	
	F2c	T	11.5	108.0	
	C5	T	10.7	159.1	
	C4	T	11.5	106.2	
Lisbon Valley Oilfield Site, Moab, UT, Wingate-Kayenta transition	*Anomoepus* CU-MWC 179.16	O	9.5	42.0	Lockley and Gierliński (2006)
	Grallator	T	20.0	156.0	Lockley and Gierliński (2006: fig. 5*A*)
	Bipedal dinosaur	UID	14.0	84.0	Lockley and Gierliński (2006: fig. 5*B*)
	Bipedal dinosaur	UID	15.0	110.0	Lockley and Gierliński (2006 fig. 5*C*)
Kayenta Fm	*Hopiichnus shingi* UCMP 83760	UID	10.0	382.0	Welles (1971); assigned to *Anomoepus* by Lockley and Gierliński (2006
Kayenta Fm, Flag Point Pictograph-Track Site, Kanab, Utah	*Kayentapus* trackway 1	T	44.0	335.0	Lockley et al. (2006*a*)
	Eubrontes trackway 6	T	33.0	254.0	
	Eubrontes trackway 11	T	37.0	245.0	

Site	Trackway	Code	Footprint mean length (cm)	Mean stride (cm)	Source of data and comments
Kayenta Fm, Desert Tortoise DT1	Trackway 5 *Eubrontes*	T	43.0	256.0	Lockley et al. (2006*f*)
	Trackway 8 *Eubrontes*	T	33.0	243.0	
	Trackway 9 *Eubrontes*	T	34.0	280.0	
Desert Tortoise DT2	Trackway 1 *Eubrontes*	T	50.0	245.0	
	Trackway 2 *Eubrontes*	T	36.0	209.0	
	Trackway 3 *Eubrontes*	T	38.0	223.0	
	Trackway 4 *Eubrontes*	T	36.0	198.0	
	Trackway 5 *Eubrontes*	T	35.0	172.0	
	Trackway 6 *Eubrontes*	T	49.0	254.0	
	Trackway 7 *Eubrontes*	T	38.0	323.0	
	Trackway 9 *Eubrontes*	T	45.0	176.0	
Desert Tortoise DT3	Trackway 1	T	30.0	272.0	
Desert Tortoise DT4	Trackway 20	T	27.0	235.0	
	Trackway 21	T	30.0	227.0	
	Trackway 22	T	26.0	210.0	
Desert Tortiose DT5	Trackway 25	T	30.0	226.0	
	Trackway 27	T	26.0	193.0	
	Trackway 28	T	30.0	233.0	
Kayenta Fm, Museum of Northern Arizona site 389 near Cameron	Large theropod	T	61.0	274.0	Morales and Bulkley (1996)
Kayenta-Navajo transition, Mile 68 Site, mouth of Escalante Canyon, Glen Canyon National Recreation Area	Small theropod	T	3.5	22	Lockley et al. (2014*g*: fig. 13)
Kayenta-Navajo transition, Slick Rock Canyon-Iceberg Canyon area, Glen Canyon National Recreation area	*Kayentapus* or *Eubrontes* trackway	T	40	169	Lockley et al. (2014*g*: fig. 8)
Kayenta-Navajo transition, north of Slick Rock mouth, Canyon, Glen Canyon National Recreation Area	Trackway 2	T	29	146	Lockley et al. (2014*g*: fig. 4*C*)
	Trackway 3	T	23	184	
Kayenta-Navajo transition, Choal Canyon, Glen Canyon National Recreation Area	*Eubrontes* trackway	T	42	239	Lockley et al. (2014*g*: fig. 18*B*)
	Eubrontes trackway	T	38	224	
Slick Rock Canyon area 2, Kayenta-Navajo transition	*Anomoepus* trackway 1	O	8.0	55.0	Lockley and Gierliński (2006: fig. 13)
Escalante Mouth, Navajo Fm	*Anomoepus*	O	20.0	126.0	Lockley and Gierliński (2006: fig. 17)
Kayenta-Navajo transition, Cedar Canyon, Glen Canyon National Recreation Area	*Grallator*	T	12.0	112.0	Lockley et al. (1998c: fig. 14*A*)
Navajo Sandstone, Moab, Utah	MWC 183.2 slab small theropod trackway	T	6.76	75	Lockley (2011)
Granite Creek, Grand County, Utah	MWC slab 180.1 *Grallator* trackway	T	13	121	Lockley et al. (2014*b*: fig. 5*C*)
Navajo Sandstone, Hole in the Rock lower level, Glen Canyon National Recreation Area	Theropod	T	39.0	321.0	Lockley et al. (1998*c*: fig. 20*A*)
Copper Mine, northern Arizona, Navajo Sandstone	Bipedal dinosaur	UID	3.5	45.5	Rainforth and Lockley (1996)
Navajo Fm, Buoy 62 Site, Slick Rock-Iceberg area, Glen Canyon National Recreation Area	*Grallator* trackway	T	27	131	Lockley et al. (2014*g*: fig. 17)
Nugget Sandstone, Lower Jurassic, Bear Lake, Idaho	*Grallator*	T	4	24	Lockley et al. (2011*b*)
	Grallator	T	4	33	

Table A4.2. *continued*

Site	Trackway	Code	Footprint mean length (cm)	Mean stride (cm)	Source of data and comments
Aztec Sandstone, Lower-Middle Jurassic, Tracksite UNLV-AZ-008, Red Rock Canyon National Conservation Area, Nevada	Grallatorid trackway	T	14	45	Rowland et al. (2014)
Mid-Late Jurassic trackways					
Ardley, U.K.	Theropod T80	T	66.0	310.0	Day et al. (2004); cf. Mossman et al. (2003)
	Theropod T13	T	72.0	270.0	
	Theropod T29	T	74.0	287.0	
Scalby Fm, Yorkshire	Bipedal dinosaur	UID	43.0	250.0	Romano and Whyte (2003: fig. 25)
Kilmaluag Fm, Isle of Skye Scotland	Trackway 22-12-5	T	9.5	55.7	Clark et al. (2005)
	Trackway 2-8-15	T	9.7	52.3	
	Trackway 3.2-3.3-3.4	T	12.7	53.0	
Gayrand, France	Theropod P3-P4-P5	T	38.0	301.0	Moreau et al. (2012*a*: fig. 4)
Louelle, France	LOU 05 *Carmelopodus*	T	27	177	Mazin et al. (2016); *Carmelopodus* measurements from fig. 13
	LOU 20 *Megalosauripus*	T	77	288	
Vale de Meios, Portugal	*Megalosauripus* VM1	T	63.28	321.81	Razzolini et al. (2016*a*)
	Megalosauripus VM2	T	71.77	303.66	
	Megalosauripus VM	T	38.27	182.30	
Chevez-Combe Ronde, Reuchenette Fm, Switzerland	T2 Bipedal dinosaur	UID	15	137	Marty (2008)
	T3 Morphotype I *Carmelopodus*	T	15	137	
	T4 Bipedal dinosaur	UID	15	80	
	T6 Bipedal dinosaur	UID	15	125	
	T7 Bipedal dinosaur	UID	14	148	
	T8 Morphotype I *Carmelopodus*	T	14	125	
	T9 Bipedal dinosaur	UID	16	136	
	T10 Morphotype I *Carmelopodus*	T	8	113	
	T11 Morphotype II Bipedal dinosaur	UID	24	220	
	T12 Bipedal dinosaur	UID	14	153	
	T13 Bipedal dinosaur	UID	14	141	
	T14 Morphotype I *Carmelopodus*	T	14	126	
	T15 Morphotype I *Carmelopodus*	T	12	125	
	T16 Morphotype I *Carmelopodus*	T	14	133	
	T17 Morphotype I *Carmelopodus*	T	12	131	
	T18 Morphotype I *Carmelopodus*	T	12	155	
	T19 Bipedal dinosaur	UID	10	165	
	T20 Bipedal dinosaur	UID	13	150	
	T21 Morphotype I *Carmelopodus*	T	13	170	
	T22 Bipedal dinosaur	UID	12	148	
	T23 Morphotype I *Carmelopodus*	T	10	124	
	T24 Morphotype I *Carmelopodus*	T	11	141	
	T25 Morphotype I *Carmelopodus*	T	15	176	
	T26 Morphotype I *Carmelopodus*	T	13	135	
	T27 Bipedal dinosaur	UID	11	148	
	T28 Morphotype I *Carmelopodus*	T	15	167	
	T29 Bipedal dinosaur	UID	14	190	
	T32 Bipedal dinosaur	UID	14	127	
	T33 Bipedal dinosaur	UID	10	130	
	T34 Bipedal dinosaur	UID	16	186	
	T36 Morphotype I *Carmelopodus*	T	11	122	
	T37 Bipedal dinosaur	UID	16	146	
	T38 Morphotype I *Carmelopodus*	T	12	130	
	T39 Morphotype I *Carmelopodus*	T	12	127	
	T40 Morphotype I *Carmelopodus*	T	14	144	
	T41 Morphotype I *Carmelopodus*	T	9	119	

Site	Trackway	Code	Footprint mean length (cm)	Mean stride (cm)	Source of data and comments
	T42 Morphotype I *Carmelopodus*	T	10	132	
	T43 Morphotype II ?Theropod	T	30	207	
Hemenkou, Shedian Fm, Yunnan Province, China	HMK-T1 Morphotype A	T	15.3	109.5	Xing et al. (2016*k*); Fujita et al. (2008)
	HMK-T2 Morphotype B	T	15.0	112.8	
	HMK-T3 Morphotype C cf. *Eubrontes*	T	43.9	211.2	
Nianpanshan, Xiashaximiao Fm, Zizhong, Sichuan, China	NPS-T1 *Anomoepus*	O	14.4	102.4	Xing et al. (2016*g*)
	NPS-T2 *Anomoepus*	O	6.8	45.9	
	NPS-T3 *Anomoepus*	O	7.6	51.9	
	NPS-T6 *Anomoepus*	O	14.1	119.5	
	NPS-T4 *Eubrontes nianpanshanensis*	T	34.3	197.4	
	NPS-T5 cf. *Eubrontes*	T	19.8	138.2	
Shemshak Group, Alborz Mountains, Iran	Royan section, Baladah Horizon 4 trackway a1	T	11.3	68.9	Abbassi and Madanipour (2014)
	Royan section, BaladahHorizon 4 a3	T	15	55.8	
	Royan section, Baladah Horizon 4 trackway b1	T	12.6	60.8	
	Royan section, Baladah Horizon 4 trackway e	T	11.6	59.7	
	Royan section, Baladah Horizon 4 trackway f	T	13.3	57.5	
	Royan section, Baladah Horizon 5 trackway a	T	18	85	
	Bol-Yasel section Horizon 2 Layer 2 trackway a	T	9	61	
	Bol-Yasel section Horizon 3 trackway d	T	22	77	
	Bol-Yael section Horizon 3 trackway e	T	11	87	
Hojedk Fm, Dashtkhak, Kerman, Iran	*Wildeichnus* trackway c	T	7.1	36.7	Abbassi et al. (2015)
El Hontanar, Alpuente, Valencia, Spain, Late Jurassic-Early Cretaceous	Theropod trackway	T	50.0	287.0	Santisteban et al. (2007)
Site AB-1, Ababuj, Teruel, Villar del Arzobispo Fm (Tithonian-Berriasian)	Trackway 1AB-1 theropod	T	49.0	299.0	Alcalá et al. (2012)
El Pozo 3CT-2, El Castellar, Villar del Arzobispo Fm	El Pozo 3CT-2 prints 2 to 5 quadrupedal ornithopod	O	12.4	50.7	Alcalá et al. (2014*a*)
	El Pozo 3CT-2 prints 13 to 19 ornithopod	O	17	94	
El Castellar, Villar del Arzobispo Fm, Teruel, Spain	*Iberosauripus grandis* 1CB	T	56.0	355.5	Cobos et al. (2014)
Corcollilla, Los Serranos, Valencia, Spain	Large theropod footprints 7, F,2 2	T	43.5	230.0	Santisteban et al. (2003)
Faros de Tazones, Asturias, Spain	cf *Anomoepus* trackway A	O	16.0	102.0	Lockley et al. (2008*a*: fig 14)
	cf *Anomoepus* trackway B	O	17.0	98.0	
	cf *Anomoepus* trackway c	O	15.0	73.0	
	cf *Anomoepus* trackway D	O	13.0	99.0	
Tereñes Fm, Tereñes, Asturias, Spain	Ornithopod OT1	O	28.0	159.3	Piñuela et al. (2016)
	Ornithopod OT2	O	30.3	130.8	
	Ornithopod OT3	O	28.8	126.4	
Barkhausen, Germany	*Megalosauropus teutonicus* trackway A	T	65.4	305.0	Diedrich (2011)
Sanjianfang Fm, Shanshan, China	*Changpeipus carbonicus* trackway TW1	T	47.4	269.0	Wings et al. (2007); Xing et al. (2014*a*)
Xincun Fm, Panxi, China	*Kayentapus* trackway SSA	T	28.0	156.4	Xing et al. (2013c)
	Kayentapus trackway SSB	T	28.2	170.9	
	Kayentapus trackway SSC	T	33.1	199.4	

Table A4.2. *continued*

Site	Trackway	Code	Footprint mean length (cm)	Mean stride (cm)	Source of data and comments
Yanan Fm, Zizhou County, China	Theropod NT2	T	22.2	107.5	Xing et al. (2015*c*)
	Anomoepus HT1	O	12.8	61.0	
	Anomoepus HT2	O	13.5	88.0	
Shangshaximiao Fm, Jinji, China	Jinji JJ trackway cf. *Therangospodus*	T	25.3	192.1	Xing et al. (2013*b*)
Houcheng Fm, Jurassic-Cretaceous boundary, Shangyi, northern Hebei, China	cf. *Therangospodus* SYT51-53	T	17.6	213.5	Xing et al. (2014*c*)
	Ornithopod SYO1	O	15.1	93.4	
Tuchengzi Fm, Jurassic-Cretaceous boundary, Qianjiadian, Yanqing, Beijing, China	Theropod QJDIUL-T1 morphotype A cf. *Eubrontes*	T	15.7	152.5	Xing et al. (2015*h*)
	Theropod QJDIUL-T2 morphotype A cf. *Eubrontes*	T	15.0	98.5	
	Theropod QJDIUL-T3 morphotype B *Grallator* isp.	T	15.3	185.5	
	Theropod QJDIUL-T4 morphotype B *Grallator* isp.	T	15.3	186.2	
	Theropod QJDIUL-T8 morphotype A cf. *Eubrontes*	T	11.3	125.1	
	Theropod QJDIUL-T10 morphotype B *Grallator* isp.	T	17.3	156.3	
	Theropod QJDIUL-T12 morphotype A cf. *Eubrontes*	T	14.2	186.2	
	Theropod QJDIUL-T13 morphotype A cf. *Eubrontes*	T	16.0	160.0	
	Theropod QJDIUL-T14 morphotype A cf. *Eubrontes*	T	13.7	103.0	
	Theropod QJDIUL-T15 morphotype B *Grallator* isp.	T	16.8	131.5	
Khodja-Pil Ata, Turkmenistan	*Megalosauripus uzbekistanicus* 7	T	50.0	310.0	Lockley et al. (1998e); cf. Fanti et al. (2013*b*)
	Megalosauripus 8	T	58.0	310.0	
	Megalosauripus 10	T	58.0	365.0	
	Megalosauripus 12	T	46.0	310.0	
	Megalosauripus 13	T	39.0	304.0	
	Megalosauripus 14	T	53.0	315.0	
	Therangospodus 16	T	26.0	198.0	
	Megalosauripus 17	T	58.0	321.0	
	Megalosauripus 18	T	52.0	300.0	
	Megalosauripus C	T	72.0	212.0	
	Megalosauripus 20	T	52.0	355.0	
	Megalosauripus 21	T	45.0	274.0	
	Megalosauripus 22	T	52.0	285.0	
	Megalosauripus 23	T	60.0	287.0	
	Megalosauripus 24	T	60.0	250.0	
	Therangospodus 25	T	27.0	198.0	
	Megalosauripus 26	T	50.0	250.0	
	Megalosauripus 28	T	55.0	238.0	
	Megalosauripus 29	T	62.0	217.0	
	Therangospodus 30	T	27.0	181.0	
	Megalosauripus 31	T	55.0	257.0	
	Megalosauripus 32	T	40.0	310.0	
Iouaridene, Demnat, Morocco	Deio I morphotype 2C	T	42.2	234.0	Belvedere (2008) and personal communication; cf. Boutakiout et al. (2009*a*, 2009*b*, 2010, 2012); Díaz-Martínez et al. (2009); Belvedere et al. (2010); Nouri et al. (2011)
	Deio II morphotype 2C	T	32.7	206.0	
	Deio III morphotype 2C	T	37.4	210.0	
	Deio IV morphotype 2C	T	44.0	247.0	
	Deio V morphotype 2C	T	39.5	232.0	
	Deio VI morphotype 2C	T	39.5	225.0	

Site	Trackway	Code	Footprint mean length (cm)	Mean stride (cm)	Source of data and comments
Iouaridene, Demnat, Morocco	Deio VII morphotype 2A	T	21.1	140.0	Belvedere (2008) and personal communication; cf. Boutakiout et al. (2009*a*, 2009*b*, 2010, 2012); Díaz-Martínez et al. (2009); Belvedere et al. (2010); Nouri et al. (2011)
	Deio VIII morphotype 2C	T	39.2	235.0	
	Deio IX morphotype 2C	T	22.3	124.0	
	Deio XIII morphotype 2A	T	17.2	102.0	
	Deio XIV morphotype 2C	T	38.9	216.0	
	Deio XV morphotype 2C	T	36.1	216.0	
	Deio XVI morphotype 2C	T	34.1	251.0	
	Deio XVII morphotype 2C	T	38.4	220.0	
	Deio XXVIII morphotype 2C	T	58.7	182.0	
	Deio XIX morphotype 2C	T	17.3	156.0	
	Deio XXX morphotype 2C	T	38.1	243.0	
	Deio XXXII morphotype 2C	T	34.0	206.0	
	Deio XXXIII morphotype 2C	T	34.5	202.0	
	Deio XXXIV morphotype 2C	T	38.5	217.0	
	Deio XXXV morphotype 2C	T	32.3	209.0	
	Deio XL morphotype 2A	T	17.3	105.0	
	Deio XLI morphotype 2A	T	14.9	247.0	
	Deio XLIII morphotype 2C	T	35.1	243.0	
	Deio XLIV morphotype 2C	T	37.3	196.0	
	Deio C morphotype 2C	T	37.8	212.0	
	Deio CII morphotype 2C	T	37.7	221.0	
	Deio CIII morphotype 2C	T	39.4	219.0	
	Deio CIV morphotype 2C	T	39.3	247.3	
	Deio CV morphotype 2C	T	36.9	224.2	
	Deio CVI morphotype 2B	UID	34.5	204.1	
	Deio CVII morphotype 2C	T	36.4	195.2	
	Deio CVIII morphotype 2C	T	29.0	192.2	
	Deio CIX morphotype 2C	T	26.5	181.9	
	Deio CX morphotype 2C	T	33.8	197.7	
	Deio CXI morphotype 2C	T	25.8	190.9	
	Deio CXII morphotype 2A	T	17.1	151.2	
	Deio CXIV morphotype 2C	T	37.7	210.2	
	Deio CXVII morphotype 2C	T	42.2	258.6	
	Deio CXVIII morphotype 2C	T	41.6	235.0	
	Deio CXIX morphotype 2A	T	24.0	193.9	
	Deio CXXI morphotype 2C	T	35.7	238.0	
	Deio CXXIII morphotype 2B	UID	21.2	203.3	
	Deio CXXV morphotype 2B	UID	29.7	199.5	
	Deio CXXVI morphotype 2C	T	37.5	239.9	
	Deio CXXVII morphotype 2C	T	35.1	230.3	
	Deio CXXVIII morphotype 2C	T	35.5	208.5	
	Deio CXXIX morphotype 2C	T	26.3	194.5	
	Deio CXXX morphotype UK	UID	31.5	216.5	
	Deio DI morphotype 2C	T	29.3	183.0	
	Deio DII morphotype 2C	T	39.7	238.0	
	Deio DIII morphotype 2C	T	39.5	256.0	
	Deio DIV morphotype 2C	T	36.7	234.9	
	Deio DV morphotype 2C	T	39.3	255.5	
	Deio DVI morphotype 2C	T	38.2	233.8	
	Deio DVII morphotype 2C	T	33.4	211.1	
	Deio DX morphotype 2E	T	63.3	305.5	
	Deio DXII morphotype 2A	T	19.8	112.4	

Table A4.2. *continued*

Site	Trackway	Code	Footprint mean length (cm)	Mean stride (cm)	Source of data and comments
Iouaridene, Demnat, Morocco	Deio DXIII morphotype 2C	T	35.0	199.9	Belvedere (2008) and personal communication; cf. Boutakiout et al. (2009*a*, 2009*b*, 2010, 2012); Díaz-Martínez et al. (2009); Belvedere et al. (2010); Nouri et al. (2011)
	Deio DXXII morphotype 2C	T	35.4	218.8	
	Deio DXXIV morphotype 2A	T	16.3	157.8	
	Deio MI morphotype 2C	T	38.5	248.8	
	Deio MIII morphotype 2C	T	42.8	216.8	
	Deio MIV morphotype 2E	T	48.3	287.2	
	Detk I morphotype 2C	T	49.4	297.4	
	Detk II morphotype 2C	T	23.2	200.0	
	Detk III morphotype 2C	T	35.5	213.0	
	Detk IV morphotype 2C	T	25.4	240.6	
	Detk VII morphotype 2C	T	42.6	218.0	
	Detk X morphotype 2C	T	21.9	190.0	
	Detk XIII morphotype 2C	T	37.7	225.3	
	Detk XIV morphotype 2C	T	31.0	206.6	
	Detk XV morphotype 2B	UID	17.2	135.0	
	Detk XVII morphotype 2C	T	32.9	180.0	
	Detk XVIII morphotype 2C	T	30.4	206.4	
	Detk XIX morphotype 2C	T	38.1	270.0	
	Detk M morphotype 2C	T	30.6	250.1	
	Detk MI morphotype 2C	T	29.0	220.5	
	Detk MVII morphotype 2C	T	40.8	288.0	
	Detk MVIII morphotype 2C	T	35.3	207.0	
	Detk MXIV morphotype 2D	T	28.1	220.0	
	Detk MXX morphotype UK	UID	46.5	211.1	
	Detk MXXI morphotype 2E	T	50.4	238.4	
	Detk MXXII morphotype 2C	T	33.5	215.3	
	Detk MXIII morphotype 2C	T	31.3	228.6	
	Detk MXXIV morphotype 2D	T	36.8	247.8	
	Detk MXXV morphotype 2C	T	40.5	241.1	
	Detk MXXVI morphotype 2A	T	23.9	180.8	
	Detk MXXIX morphotype 2C	T	40.1	242.0	
	Detk MXXXI morphotype 2D	T	20.8	157.0	
	Detk MXXXIV morphotype 2E	T	39.0	220.3	
	Detk MXXXV morphotype 2C	T	28.5	182.2	
	Detk MXXXVIII morphotype 2C	T	30.2	298.3	
	Detk MXLIV morphotype 2C	T	25.4	199.6	
	Detk MXLV morphotype 2C	T	43.2	256.7	
	Detk MXLVI morphotype 2D	T	33.2	216.0	
	Detk MXLVII morphotype 2D	T	37.5	200.0	
	Detk MXLIX morphotype 2C	T	28.4	207.3	
	Detk MLVI morphotype 2C	T	30.0	225.0	
	Detk MLXVII morphotype 2C	T	19.5	148.0	
	Detk MLXXIX morphotype 2E	T	50.9	320.4	
	Detk MLXXX morphotype 2A	T	17.4	152.7	
Dande Sandstone Fm, Ntumbe, Zimbabwe	First site UID *Eubrontes*	T	40	230.6	Broderick (1985); Munyika (1996)
	Unidentified quadrupedal juvenile dinosaur	U	5	45.8	Lingham-Soliar and Broderick (2000)
	Small theropod	T	7.0	80	
	New site theropod I	T	47.0	196.9	Lingham-Soliar et al. (2003)
Arhab, Mid-Jurassic to Early Cretaceous, Yemen	Ornithopod o1	O	56.0	207.0	Schulp et al. (2008); Schulp and Al-Wosabi (2012)

Site	Trackway	Code	Footprint mean length (cm)	Mean stride (cm)	Source of data and comments
Bemaraha Fm, Madagascar	Trackway A Tsl8–Tsl12	T	33.6	247.9	Wagensommer et al. (2010)
	Trackway B Tsl17-Tsl21	T	32.3	208.0	
	Trackway D Tsl24–Tsl29	T	36.6	200.0	
Banos del Flaco Fm, Rio Tinguiririca, Chile	*Camptosaurichnus fasolae* trackway 1	O	17.0	142.0	Moreno and Pino (2002)
	Camptosaurichnus trackway 2	O	24.0	118.2	
	Camptosaurichnus trackway 6A	O	20.0	140.0	
	Camptosaurichnus trackway 6C	O	21.0	182.0	
	Camptosaurichnus trackway 7A	O	24.0	141.0	
	Camptosaurichnus trackway 10	O	18.5	136.0	
	Ornithopod trackway 6D	O	20.0	157.5	
	Ornithopod trackway 12A	O	19.5	153.0	
	Ornithopod trackway 12B	O	21.5	150.0	
	Ornithopod trackway 6B	O	13.0	161.0	
	Theropod trackway 9	T	23.0	162.5	
	Theropod trackway 11	T	22.5	147.0	
San Salvador Fm, Calama, Chile	Theropod trackway 1 morphotype A	T	35.0	186.0	Moreno et al. (2004)
	Theropod trackway 3 morphotype A	T	52.0	220.0	
	Theropod trackway 4 Morphotype A	T	28.0	183.0	
	Theropod trackway 5 morphotype B	T	16.0	60.0	
	Theropod trackway 6 morphotype B	T	16.0	51.0	
	Theropod trackway 8 morphotype B	T	15.0	106.0	
	Theropod trackway 9 morphotype B	T	15.0	100.0	
Botucatu Fm, Araraquara, Late Jurassic-Early K, Brazil	Theropod	T	12.0	53.0	Leonardi (1980*b*); also see Francischini et al. (2015)
	"Group one" ornithischian	O	13.0	200.0	Leonardi and Godoy (1980)
	"Group one" ornithischian ARCE 6	O	6.0	33.0	
Guará Fm, Late Jurassic-Early K, Rio Grande do Sol, Brazil	Theropod RSCT-1	T	22.5	135.0	Francischini et al. (2015); Francischini (pers. comm.)
	Theropod RSCT-2	T	34.5	110.0	
	Theropod RSCT-5	T	20.0	193.0	
	Ornithopod RSCT-3	O	15.0	120.0	
	Ornithopod RSCT-9	O	43.0	248.0	
Laguna Manantiales, Santa Cruz, Argentina	*Delatorrichnus goyenechei* MLP 60-X-31-6 quadrupedal ornithopod	O	3.1	29.0	Casamiquela (1964); Calvo (2007); Novas (2009)
Gypsum Spring Fm, Bighorn Basin, Wyoming	BM3 trackway 2	T	25.3	157.4	Kvale et al. (2001)
	BME trackway 3	T	22.3	137.4	
	WSS1 trackway 1	T	25.1	161.6	
	WSS1 trackway 2	T	23.4	193.9	
Teec Nos Pos, Bluff Sandstone	*Megalosauripus*	T	42.0	190.0	Lockley and Mickelson (1997); note that Castanera et al. (2013*a*) regard Spanish *Therangospodus* as an ornithopod print
Summerville-Bluff Sequence, Butler Wash, Utah	*Therangospodus* track level 5	T	23.0	178.0	
Bull Canyon, US	*Megalosauripus* Bull 1	T	45.0	205.0	Lockley et al. (1998*b*)
	Megalosauripus Bull 4	T	42.0	217.0	
Moab Megatracksite, Utah	*Megalosauripus* MMA-1	T	48.0	235.0	Lockley et al. (1998e); Castanera et al. (2013*a*) regard Spanish *Therangospodus* as an ornithopod print
	Megalosauripus MMA-2	T	53.0	263.0	
	Megalosauripus MMA-6	T	43.0	225.0	
	Megalosauripus MMB-1	T	51.0	289.0	
	Megalosauripus MMB-4	T	51.0	266.0	
	Megalosauripus MMC-1	T	44.0	242.0	
	Megalosauripus MMC-3	T	44.0	270.0	
	Megalosauripus MMC-5	T	46.0	240.0	
	Megalosauripus MMC-6	T	45.0	244.0	
	Megalosauripus MME-6	T	46.0	245.0	
	Megalosauripus MMG-11	T	40.0	229.0	

Table A4.2. *continued*

Site	Trackway	Code	Footprint mean length (cm)	Mean stride (cm)	Source of data and comments
Moab Megatracksite, Utah	*Megalosauripus* MMG-6	T	50.0	238.0	Lockley et al. (1998*e*); Castanera et al. (2013*a*) regard Spanish *Therangospodus* as an ornithopod print
	Megalosauripus MMG-7	T	43.0	245.0	
	Megalosauripus MMH-3	T	40.0	214.0	
	Megalosauripus MMI-1	T	48.0	255.0	
	Megalosauripus MML-1	T	47.0	231.0	
	Megalosauripus MML-2	T	45.0	207.0	
	Megalosauripus MML-3	T	43.0	255.0	
	Megalosauripus MMO-1	T	40.0	243.0	
	Megalosauripus MMQ-1	T	48.0	297.0	
	Megalosauripus MMQ-2	T	44.0	290.0	
	Megalosauripus MMR-1	T	41.0	246.0	
	Megalosauripus MMR-2	T	42.0	249.0	
	Megalosauripus MMW-1	T	45.0	221.0	
	Megalosauripus MMW-2	T	49.0	219.0	
	Megalosauripus MMX-1	T	45.0	258.0	
	Megalosauripus MMX-2	T	48.0	250.0	
	Therangospodus MMA-3	T	32.6	213.0	
	Therangospodus MMA-5	T	31.0	193.0	
	Therangospodus MMA-7	T	30.0	211.0	
	Therangospodus MMB-2	T	26.0	172.0	
	Therangospodus MMB-3	T	26.0	202.0	
	Therangospodus MMC-1	T	26.0	183.0	
	Therangospodus MMC-4	T	24.0	163.5	
	Therangospodus MMC-7	T	28.0	188.0	
	Therangospodus MMD-1	T	31.0	230.0	
	Therangospodus MMD-2	T	27.0	225.0	
	Therangospodus MMD-3	T	22.0	180.0	
	Therangospodus MME-1	T	27.0	170.0	
	Therangospodus MME-2	T	33.0	184.0	
	Therangospodus MME-3	T	27.0	161.0	
	Therangospodus MME-4	T	30.0	212.0	
	Therangospodus MME-5	T	31.0	196.0	
	Therangospodus MMG-1	T	32.0	182.0	
	Therangospodus MMG-2	T	32.0	225.0	
	Therangospodus MMG-3	T	23.0	140.0	
	Therangospodus MMG-5	T	30.0	186.0	
	Therangospodus MMG-8	T	27.0	158.0	
	Therangospodus MMG-9	T	33.0	221.0	
	Therangospodus MMG-10	T	28.0	180.0	
	Therangospodus MMG-12	T	31.0	178.0	
	Therangospodus MMH-1	T	26.0	162.0	
	Therangospodus MMH-2	T	33.0	208.0	
	Therangospodus MMH-5	T	25.0	196.0	
	Therangospodus MMH-E	T	27.0	202.0	
Moab Member, Curtis Fm, Moab, Utah	Sovereign site 2 trackway	T	6	46	Lockley et al. (2007*b*)
	Sovereign site 3 trackway	T	6	30	
	Site J smaller dinosaur	T	20	112	
	Site J larger dinosaur	T	30	188	
Carmel Fm, Dinosaur National Park area	*Carmelopodus untermannorum* CU-MWC 184.20	T	5.2	48.0	Lockley et al. (1998*d*)
	Carmelopodus Jensen 1 and 2	T	15.0	130.0	
	Carmelopodus CU-MWC 184.12	T	12.0	81.0	

Site	Trackway	Code	Footprint mean length (cm)	Mean stride (cm)	Source of data and comments
Sundance Fm, Red Gulch Ballroom, Shell, Wyoming	Trackway B1	T	18.2	146.0	J. Sizemore (pers. comm.)
	Trackway B5	T	17.4	130.8	
	Trackway B6	T	18.8	161.9	
	Trackway B7	T	17.1	161.8	
	Trackway B15	T	8.5	74.1	
	Trackway B18	T	15.9	136.8	
	Crosstown trackway	T	21.9	192.2	
	Broadway trackway	T	22.4	170.8	
Sundance Fm, Discovery area	Trackway D2	T	17.6	126.6	
Sundance Fm, Karin's area	Trackway K1	T	15.5	121.5	
	Trackway K3	T	16.7	113.1	
Sundance Fm, Flitner site	Flitner 1	T	20.2	159.3	
Sundance Fm, Yellow Brick Road, Bighorn Basin, Shell, Wyoming	Theropod A	T	16.7	128	Adams et al. (2014); T. Adams (pers. comm.)
	Theropod E	T	17.3	145	
	Theropod F	T	17.4	153	
	Theropod G	T	18.3	120	
	Theropod H	T	20.6	194	
	Theropod K	T	18.8	148	
	Theropod L	T	16.7	120	
	Theropod M	T	15.7	121	
	Theropod O	T	16.6	138	
	Theropod P	T	18.5	165	
	Theropod U	T	19.0	151	
	Theropod V	T	16.2	143	
	Theropod W	T	8.0	102	
	Theropod X	T	7.4	79	
	Theropod AA	T	18.3	138	
	Theropod AB	T	19.8	150	
	Theropod AC	T	17.3	159	
	Theropod AD	T	17.1	178	
	Theropod AE	T	17.6	135	
	Theropod AF	T	14.1	157	
	Theropod AH	T	19.5	161	
	Theropod AJ	T	15.4	133	
Bell Ranch Fm, Carizzo Creek, Oklahoma	Ornithopod (*Gypsichnites*)	O	45.0	183.0	Conrad et al. (1987)
	Theropod	T	45.0	217.5	
Morrison Fm, Purgatoire, Colorado	Purgatoire 1	T	28.5	225.0	Lockley et al. (1986)
	Purgatoire 2	T	44.3	290.0	
	Purgatoire 3	T	42.6	220.0	
	Purgatoire 4	T	37.0	230.0	
	Purgatoire 5	T	42.4	232.0	
	Purgatoire 6	T	40.0	250.0	
	Purgatoire 7	T	33.5	245.0	
	Purgatoire 8	T	37.3	240.0	
	Purgatoire 9	T	42.0	258.0	
	Purgatoire 10	T	36.3	250.0	
	Purgatoire 12	T	28.0	206.0	
	Purgatoire 13	T	43.7	268.0	
	Purgatoire 14b	T	40.0	268.0	
	Purgatoire 16	T	37.0	224.0	
	Purgatoire 17	T	36.0	200.0	
	Purgatoire 18	T	40.0	248.0	

Table A4.2. *continued*

Site	Trackway	Code	Footprint mean length (cm)	Mean stride (cm)	Source of data and comments
Morrison Fm, Purgatoire, Colorado	Purgatoire 19	T	39.2	243.0	Lockley et al. (1986)
	Purgatoire 20	T	36.3	254.0	
	Purgatoire 22	T	44.7	327.0	
	Purgatoire 23	T	38.0	242.0	
	Purgatoire 26	T	13.5	129.0	
	Purgatoire 27	T	37.5	282.0	
	Purgatoire 29	T	37.0	248.0	
	Purgatoire 31	T	50.8	350.0	
	Purgatoire 32	T	36.0	248.0	
	Purgatoire 33	T	37.0	284.0	
	Purgatoire 33a	T	14.0	140.0	
	Purgatoire 34	T	35.0	220.0	
	Purgatoire 35	T	40.0	280.0	
	Purgatoire 36	T	42.0	264.0	
	Purgatoire 37	T	42.0	250.0	
	Purgatoire 39	T	30.0	240.0	
	Purgatoire 40	T	40.0	272.0	
Morrison Fm, Boundary Butte, Utah	CU-MWC 198.2 *Dinehichnus socialis*	O	11.0	67.0	Lockley et al. (1998*a*)
Morrison Fm, Black Dinah Mine, Uravan, Colorado	cf. *Anomoepus*	O	15.0	87.0	Lockley et al. (1998*b*)
Black Ridge Trail, Colorado National Monument	COLM 00203 ornithopod	O	9	42	Foster and Lockley (2006)
Morrison Fm, Copper Ridge, Utah	CR-T-1 *Hispanosauropus* isp.	T	47	287.5	Foster (2015)
Cretaceous trackways					
Purbeck Limestone Group, Suttle's Mutton Hole Quarry, Herston, England	Ornithopod trackway A	O	29.4	144.3	Wright (1999)
	Ornithopod trackway Cb	O	27.8	126.7	
Wealden, Paine's Quarry, Herston, England	*Iguanodontipus burreyi*	O	26.4	127.3	Sarjeant et al. (1998)
Wealden, Norman's Quarry, Langton Travers	*Iguanodontipus*	O	41.8	184.3	
Wealden, East Sussex, Cooden Beach	Ornithopod trackway 1	O	49.5	261.0	Woodhams and Hines (1989)
	Ornithopod trackway 2	O	56.0	247.0	
Wessex Fm, 200 m west of Hanover Point, Isle of Wight	Ornithopod	O	40	220	Lockwood et al. (2014)
Wessex Fm, Hanover Point, Isle of Wight	Ornithopod	O	52.2	406	Pond et al. (2014)
Wessex Fm, Chilton Chine, Isle of Wight	Theropod 1	T	23	138	Pond et al. (2014) from work by W. Blows
	Theropod 2	T	23	148	
	Theropod 3	T	23	148	
	Theropod 4	T	25	148	
	Theropod 5	T	36	178	
	Theropod 6	T	33	158	
	Ornithopod 7	O	64	294	
	Theropod X	T	25	168	
	Theropod XX	T	46	182	
Münchehagen, Germany, Bückeberg Fm	Bipedal dinosaur	UID	43.3	250.0	Lockley et al. (2004*b*)
Münchehagen, Germany, Obernkirchen Ss	Theropod T2	T	29.5	208	Lallensack et al. (2016)
	Theropod T3	T	40.4	228	
	Ornithopod I1	O	29.5	131	

Site	Trackway	Code	Footprint mean length (cm)	Mean stride (cm)	Source of data and comments
Bückeburg Fm, Obernkirchen Dobergmuseum-Geologisches Museum Ostwestfalen-Lippe Bunde	*Iguanodontipus* quadrupedal trackway 1	O	39.0	191.0	Diedrich (2004: fig. 4 and table 1)
	Iguanodontipus quadrupedal trackway 2	O	37.0	182.3	
Obernkirchen Sandstone, Chicken Yard, Bückeburg Fm	Ornithopod trackway A	O	31.0	149.0	A. Richter map and Richter et al. (2012); cf. Hornung et al. (2012); Hübner (2016); Richter and Böhme (2016)
	Ornithopod trackway B	O	39.0	183.0	
	Ornithopod trackway C	O	42.0	224.0	
	Ornithopod trackway D	O	33.0	157.0	
	Ornithopod trackway F	O	35.0	175.0	
	Ornithopod trackway G	O	36.0	206.0	
	Ornithopod trackway H	O	41.0	199.0	
	Theropod trackway I	T	43.0	271.0	
	Theropod trackway II	T	33.0	207.0	
	Theropod trackway III	T	43.0	302.0	
	Theropod trackway IV	T	35.0	229.0	
	Theropod trackway V	T	46.0	249.0	
	Theropod trackway VI	T	43.0	239.0	
	Theropod trackway VII	T	36.0	336.0	
	Theropod trackway VIII	T	32.0	223.0	
	Theropod trackway IX	T	39.0	233.0	
	Theropod trackway X	T	34.0	220.0	
Festningen, Svalbard, Helvetiafjellet Fm	Ornithopod trackway A	O	68.0	194.0	de Lapparent (1962: fig. 3); Lockley and Meyer (2000) identified these as theropod trackways
	Ornithopod trackway B	O	69.0	170.0	
Boltodden, Svalbard, Helvetiafjellet Fm	*Caririchnium billsarjeanti* quadrupedal trackway	O	28	125	Hurum et al. (2016)
Las Cerradicas, Villar del Arzobispo Fm, Galve, Spain	LCR1	O	22.0	116.0	Pérez-Lorente et al. (1997); also see Castanera et al. (2013*b*)
	LCR2	O	21.2	107.3	
	LCR3	O	25.3	120.6	
	LCR4	O	22.8	119.2	
Las Cerradicas	LCR8	O	24.4	103.0	Castanera et al. (2013*b*)
	LCR15	O	24.2	110.0	
	LCR16	O	23.3	108.5	
Del Cantalar, Villar del Arzobispo Fm, Galve	Theropod trackway 2ELC	T	24.0	255.0	Pérez-Lorente and Ortega (2003)
Las Sereas 7, Rupelo Fm, Tithonian-Berriasian, Burgos	Theropod trackway LS7D	T	43	192	Torcida Fernández-Baldor et al. (2015)
Costalomo, Pinilla de los Moros Fm, Burgos, Spain	Theropod rastrillada CLS.F	T	54.0	341.0	Torcida Fernández-Baldor et al. (2006); cf. Platt and Meyer (1991); Huerta et al. (2012)
	Theropod rastrillada CLS.G	T	64.0	367.0	
	Theropod rastrillada CLS.O	T	51.0	313.0	
	Theropod rastrillada CLS.P	T	46.0	326.0	
	Theropod rastrillada CLS.R	T	38.0	412.0	
	Theropod rastrillada CLS.S	T	49.0	386.0	
	Ornithopod rastrillada CLS.X	O	51.0	214.0	
	UID biped rastrillada CLS.H	UID	21.0	103.0	
	UID biped rastrillada CLS.I	UID	14.0	103.0	
El Camino a Treguajantes, Oncala Group Cigudosa, Aguilar del Río Fm, Oncala Group, Spain	Theropod TR1	T	67.0	286.0	Casanovas et al. (1995*j*)
	Theropod BDC1	T	26	292	Hernánde Medrano et al. (2014)
	Ornithopod BDC2	O	24	107	
	Theropod BDC3	T	34	226	
	Theropod BDC4	T	33	209	
	Theropod BDC6	T	32	240	
	Ornithopod BDC7	O	22	94	
	Theropod BDC7a	T	36	259	

Table A4.2. *continued*

Site	Trackway	Code	Footprint mean length (cm)	Mean stride (cm)	Source of data and comments
El Camino a Treguajantes, Oncala Group Cigudosa, Aguilar del Río Fm, Oncala Group, Spain	Theropod BDC9	T	49	242	Hernánde Medrano et al. (2014)
	Theropod BDC10	T	38	227	
	Theropod BDC11	T	50	249	
	Theropod BDC14	T	45	298	
	UID Dinosaur BDC17	UID	40	150	
	UID Dinosaur BDC19	UID	39	194	
	Theropod BDC23	T	26	160	
	Theropod BDC24	T	26	181	
	Theropod CAL1	T	50	258	
	Ornithopod CDC1	O	30	176	
Miraflores I, Huerteles Fm, Soria	Theropod Nivel V trackway1	T	33.0	198.0	Latorre Macarrón et al. (2006)
	Theropod Nivel IX-A trackway 1	T	23.0	152.0	
Juncal IV, Miraflores, Soria, Spain	Theropod	T	39.0	214.0	Pascual-Arribas et al. (2005)
Malvalvao II, Miraflores	Theropod	T	33.0	110.0	Pascual-Arribas et al. (2005)
La llaga, Oncala Group, La Rioja	Theropod trackway ILA1	T	28.0	131.0	Hernández Medrano and Pérez-Lorente (2003)
	Theropod trackway ILA2	T	28.0	118.0	
	Theropod trackway ILA3	T	33.0	135.0	
	Theropod trackway ILA4	T	30.0	150.0	
	Theropod trackway ILA5	T	25.0	154.0	
	Theropod trackway ILA6	T	27.0	143.0	
	Theropod trackway ILA8	T	33.0	175.0	
	Theropod trackway ILA9	T	28.0	207.0	
El Frontal, Huerteles Fm, Oncala Group	Trackway 1	T	7.8	44.9	Razzolini et al. (2014); also see Torcida et al. (2003)
	Trackway 2	T	42.4	210.0	
	Trackway 3	T	5.0	50.3	
	Trackway 4	T	25.4	203.0	
	Trackway 5	T	18.4	136.3	
	Trackway 7	T	23.0	187.5	
	Trackway 8	T	9.7	74.3	
	Trackway 9	T	13.1	92.7	
	Trackway 11	T	13.1	119.2	
	Trackway 12	T	18.5	93.1	
	Trackway 13	T	15.6	81.2	
	Trackway 16	T	11.9	62.1	
	Trackway 17	T	29.6	192.9	
	Trackway 19	T	4.5	34.8	
	Trackway 20	T	6.7	53.5	
	Trackway 23	T	5.4	29.5	
	Trackway 25	T	25.7	202.0	
	Trackway 31	T	23.4	131.0	
	Trackway 32	T	20.7	146.3	
	Trackway 33	T	23.5	163.7	
	Trackway 37	T	18.0	144.8	
	Trackway 41	T	29.4	133.0	
	Trackway 43	T	24.3	124.9	
	Trackway 44	T	24.6	111.9	
	Trackway 45	T	33.5	166.1	
Valdehijuelos, Soria, Oncala Group	Theropod MNS-2005/111-6	T	5.9	15.6	Pascual-Arribas and Hernandez-Medrano (2011)
Valdegen I, Villar del Rio, Soria, Huerteles Fm, Oncala Group	Theropod VDG-I-1	T	29.0	235.8	Hernandez Medrano et al. (2004)
	Theropod VDG-I-2	T	28.5	185.8	
	Theropod VDG-I-3	T	33.3	218.5	
	Theropod VDG-I-4	T	36.0	186.0	

Site	Trackway	Code	Footprint mean length (cm)	Mean stride (cm)	Source of data and comments
Valdegen I, Villar del Rio, Soria, Huerteles Fm, Oncala Group	Theropod VDG-I-5	T	29.3	190.0	Hernandez Medrano et al. (2004)
	Theropod VDG-I-6	T	30.3	231.5	
	Theropod VDG-I-7	T	32.3	192.0	
	Theropod VDG-I-8	T	30.3	204.0	
	Theropod VDG-I-9	T	33.0	238.5	
	Theropod VDG-I-10	T	24.0	177.6	
	Theropod VDG-I-11	T	35.8	223.5	
	Theropod VDG-I-12	T	32.0	188.5	
	Theropod VDG-I-13	T	32.0	190.0	
	Theropod VDG-I-14	T	30.6	219.0	
	Theropod VDG-I-15	T	34.5	238.0	
	Theropod VDG-I-16	T	33.0	260.0	
	Theropod VDG-I-18	T	31.8	208.0	
	Theropod VDG-I-19	T	32.0	232.0	
	Theropod VDG-I-20	T	30.0	215.0	
	Theropod VDG-I-21	T	34.8	234.0	
Santa Cristina, Villar del Rio, Soria, Huerteles Fm, Oncala Group	Rastro II *Kalohippus bretunensis*	T	6.9	49.7	Fuentes Vidarte and Meijide Calvo (1998); cf. Castanera et al. (2015)
	Rastro IV *Kalohippus bretunensis*	T	6.5	47.5	
Fuentesalvo, Oncala Group	*Therangospodus oncalensis* Tw1	O	22.2	110.0	Barco et al. (2006); *Therangospodus* was interpreted by them as a theropod but is considered an ornithopod by Castanera et al. (2013*a*)
	Therangospodus Tw2	O	23.0	130.3	
	Therangospodus Tw3	O	22.3	118.6	
	Therangospodus Tw4	O	21.8	110.1	
	Therangospodus Tw5	O	20.9	109.5	
	Therangospodus Tw6	O	21.6	121.0	
	Therangospodus Tw7	O	24.1	119.0	
	Therangospodus Tw8	O	26.0	126.1	
	Therangospodus Tw9	O	27.6	130.2	
	Therangospodus Tw10	O	21.6	111.9	
	Therangospodus Tw11	O	22.4	117.7	
	Therangospodus Tw12	O	21.3	112.9	
La Pena, Huerteles Fm, Oncala Group, Soria, Spain	?*Iguanodontipus oncalensis* ornithopod LP1	O	22.4	128.3	Castanera et al. (2013*a*)
	?*Iguanodontipus* LP2	O	28.4	161.6	
	?*Iguanodontipus* LP3	O	21.9	114.4	
La Losa, Huerteles Fm	?*Iguanodontipus* LLI1	O	26.0	139.8	Castanera et al. (2013*a*)
Valloria IV, Huerteles Fm	?*Iguanodontipus* VAIV.1	O	23.1	107.0	Castanera et al. (2013*a*)
Treguajantes, Oncala Fm	Theropod trackway TR1	T	66.0	281.0	Barco et al. (2005)
Valdelavilla, Soria, Spain, Huerteles Fm, Oncala Group	Nivel IV trackway VDV-IV5 to 8	T	28.8	173.0	Pascual-Arribas and Sanz Perez (2000)
	Nivel V trackway Tramo 1 VDV V1-1	T	23.6	125.0	
	Nivel V trackway Tramo 1 VDV V1-2	T	24.3	136.0	
	Nivel V trackway Tramo 1 VDV V1-4	T	26.0	187.5	
	Nivel V trackway Tramo 1 VDV V1-5	T	26.4	247.5	
	Nivel V trackway Tramo 2 VDV V2-1	T	21.0	152.0	
	Nivel VI trackway VDV VI-2	T	19.5	196.0	
	Nivel VI trackway VDV VII-1	T	22.8	118.0	
	Nivel VIII trackway VDV VIII-1	O	24.7	117.0	
Las Cuestas I, Soria, Spain	Ornithopod LCU-I-1 *Iguanodontipus*	O	28.5	163.4	Pascual-Arribas et al. (2009); Pascual-Arribas and Hernandez-Medrano (2010)
	LCU-I-4 *Iguanodontipus*	O	35.8	165.5	
	LCU-I-5 *Iguanodontipus*	O	37.8	184.3	
	LCU-I-10 *Iguanodontipus*	O	27.7	175.2	
	Ornithopod LCU-I-15	O	36.4	155.4	
	LCU-I-41 *Iguanodontipus*	O	28.1	166.8	
	LCU-I-43 *Iguanodontipus*	O	36.6	177.9	

Table A4.2. *continued*

Site	Trackway	Code	Footprint mean length (cm)	Mean stride (cm)	Source of data and comments
Las Cuestas I, Soria, Spain	LCU-I-47 *Iguanodontipus*	O	36.5	194.5	Pascual-Arribas et al. (2009); Pascual-Arribas and Hernandez-Medrano (2010)
	LCU-I-50 *Iguanodontipus*	O	32.2	116.0	
	LCU-I-51 *Iguanodontipus*	O	30.5	167.0	
	LCU-I-53 *Iguanodontipus*	O	41.5	209.0	
	Theropod LCU-I-2	T	37.1	175.0	
	Theropod LCU-I-3	T	44.3	237.0	
	Theropod LCU-I-7	T	44.3	234.0	
	Theropod LCU-I-8	T	37.0	176.0	
	Theropod LCU-I-9	T	35.5	247.7	
	Theropod LCU-I-11	T	19.7	100.5	
	Theropod LCU-I-12	T	35.6	214.0	
	Theropod LCU-I-13	T	36.0	230.0	
	Theropod LCU-I-21	T	47.6	227.0	
	Theropod LCU-I-24	T	46.5	201.3	
	Theropod LCU-I-26	T	22.4	147.1	
	Theropod LCU-I-27	T	41.3	220.0	
	Theropod LCU-I-36	T	25.5	210.5	
	Theropod LCU-I-40	T	31.0	173.0	
	Theropod LCU-I-42	T	37.4	165.0	
	Theropod LCU-I-44	T	38.7	219.5	
	Theropod LCU-I-46	T	43.8	189.5	
	Theropod LCU-I-52	T	38.3	213.5	
La Carretera de Soria, Urbion Group	Theropod 1	T	20.0	98.0	Casanovas et al. (1995*b*)
Barranco de Acrijos, La Rioja, Urbion Group, Early K	Bipedal dinosaur BAC1	UID	54.3	216.7	Casanovas et al. (1995*f*)
	Bipedal dinosaur BAC2	UID	53.5	202.0	
La Cela, Urbion Group	Ornithopod LACB4	O	42.0	196.0	Casanovas et al. (1995*i*)
	Theropod LACB5	T	30.0	191.0	
	Theropod LACB6	T	31.0	223.0	
	Theropod LACB7	T	26.0	212.0	
	Theropod LACB8	T	27.0	215.0	
	Theropod LACB9	T	27.0	161.0	
	Theropod LACB10	T	34.0	222.0	
	Theropod LACB11	T	38.0	239.0	
	Theropod LACB13	T	32.0	195.0	
Barranca de la Muga, Urbion Group, Lower K	Ornithopod trackway MUA	O	51.0	150.0	Caro et al. (1995)
Regumiel de la Sierra, Urbion Group	Theropod B	T	26.3	139.7	Moratalla et al. (1994)
	Ornithopod C	O	44.5	208.0	
	Ornithopod D	O	36.5	204.3	
	Ornithopod E	O	37.6	224.0	
	Ornithopod F	O	31.3	154.0	
	Ornithopod G	O	36.0	208.0	
La Rueda, Urbión Group	Ornithopod LRU2	O	13	85	Díaz-Martínez et al. (2015)
Las Obras de las Preso de Enciso, Urbion-Enciso Group	Semiplantigrade dinosaur 3.2PRE1	UID	17.0	157.0	Melero Rubio and Pérez-Lorente (2011)
	"Aviform" theropod 3.3PRE1	T	39.0	263.0	
	Semiplantigrade theropod 4.3PRE1	T	39.0	263.0	
Valdenocerillo, Enciso Group	Theropod	T	33.7	206.5	Casanovas et al. (1989*b*)

Site	Trackway	Code	Footprint mean length (cm)	Mean stride (cm)	Source of data and comments
Valdecevillo, Enciso Group	Theropod B	T	30.8	262.5	Casanovas et al. (1989*a*)
	Theropod C	T	55.6	306.3	
	Theropod E1	T	26.5	200.0	
	Theropod E2	T	37.5	275.0	
	Theropod G4	T	49.7	234.0	
	Theropod K1	T	31.0	201.5	
	Theropod K2	T	41.9	237.5	
	Theropod K4	T	18.8	187.5	
	Theropod L1	T	22.0	211.0	
	Theropod L2	T	30.0	197.0	
	Ornithopod G1	O	33.3	201.0	
	Ornithopod G2	O	22.5	100.0	
	Ornithopod G3	O	31.5	144.0	
	Ornithopod K9	O	33.0	176.0	
	Ornithopod L y K	O	50.9	249.0	
Barranco de Valdecevillo Este	Theropod VaeA	T	52.0	299.0	Casanovas et al. (1998)
	Theropod VaeB	T	52.0	252.0	
	Ornithopod VaeC	O	42.0	246.0	
La Senoba, Enciso Group	Theropod 1	T	40.8	282.0	Casanovas et al. (1989*a*)
	Theropod 2	T	47.5	275.0	
	Theropod 3	T	28.5	165.0	
	Theropod 4	T	36.9	196.0	
	Theropod 5	T	36.3	287.5	
	Theropod 6	T	31.3	262.5	
	Theropod 8	T	31.9	225.0	
	Theropod 9	T	24.0	181.0	
	Theropod 10	T	36.5	219.0	
	Theropod 12	T	40.0	229.0	
	Theropod 13	T	36.3	270.0	
	Theropod 14	T	23.1	158.8	
Camino de Igea a Valdebrajes, Spain, Enciso Group	Ornithopod 1	O	33.0	250.0	Casanovas et al. (1991*a*)
	Ornithopod 5	O	35.0	220.0	
	Ornithopod 14	O	45.0	229.0	
	Theropod 2	T	36.0	211.5	
	Theropod 3	T	46.0	264.0	
	Theropod 4	T	33.0	176.5	
	Theropod 15	T	31.5	185.0	
Icnitas 3, Enciso Group	Theropod A	T	33.0	164.0	Casanovas et al. (1992)
	Theropod B	T	16.0	97.0	
	Theropod C	T	29.0	179.0	
	Theropod D	T	28.0	143.0	
Valdevajes 1, Spain	Theropod CA-9	T	30.8	240.0	Casanovas et al. (1991*b*)
	Theropod CA-1	T	10.5	73.1	
	Theropod CA-2	T	10.6	70.0	
	Theropod CA-5	T	11.0	75.6	
	Theropod CA-6	T	10.9	84.7	
	Theropod CA-7	T	11.2	72.0	
	Theropod CA-8	T	10.9	68.0	

Table A4.2. *continued*

Site	Trackway	Code	Footprint mean length (cm)	Mean stride (cm)	Source of data and comments
Soto 1, Spain, Enciso Group	Theropod A	T	27.0	233.0	Casanovas et al. (1990*b*)
	Theropod H	T	44.0	214.0	
	Ornithopod G	O	22.0	268.0	
	Ornithopod K	O	27.0	141.0	
	Ornithopod Q	O	24.0	153.0	
Soto 3, Enciso Group	Ornithopod 1	O	23.0	170.0	Casanovas et al. (1995*d*)
	Ornithopod 2	O	25.0	173.0	
	Ornithopod 3	O	23.0	146.0	
	Ornithopod 6	O	30.0	167.0	
	Theropod 7	T	28.0	344.0	
San Vincente de Robres, Enciso Group	Ornithopod A	O	22.5	173.0	Casanovas et al. (1990*a*)
	Theropod B	T	29.5	171.0	
	Theropod C	T	24.0	178.0	
	Theropod D	T	20.0	147.0	
	Theropod E	T	30.0	185.0	
	Theropod F	T	29.0	196.0	
La Virgen del Campo, Enciso Group, Spain	Theropod 1	T	50.5	310.0	Casanovas et al. (1989*a*)
	Theropod 3	T	50.0	600.0	
	Theropod 5	T	52.5	335.0	
	Theropod 6sub1	T	30.0	186.0	
	Theropod 7	T	27.5	227.0	
	Theropod 8	T	40.0	254.0	
	Theropod 9	T	50.0	261.0	
	Theropod 16	T	54.6	292.0	
	Theropod Makoki	T	47.5	204.0	
	Theropod Tretas	T	50.0	285.0	
	Theropod Incordia	T	30.0	158.7	
	Theropod Ubres	T	38.8	209.0	
	Theropod Coco	T	31.5	221.0	
	Theropod Turo	T	33.3	350.0	
	Theropod Pedro	T	26.3	203.1	
	Theropod Enorme	T	56.3	216.3	
	Theropod Tote	T	27.3	188.0	
	Theropod Trepadora	T	33.5	215.0	
	Theropod Cascada	T	37.5	250.0	
	Theropod Doblada	T	47.9	263.0	
	Theropod Iguacentro	T	38.8	251.0	
	Theropod Owen	T	42.0	227.0	
	Theropod Lola	T	33.8	182.5	
	Theropod Hallux	T	30.0	180.0	
La Virgen del Campo II, Enciso Group, Spain	Theropod 2	T	42.0	313.0	Casanovas et al. (1991*c*)
	Theropod 5	T	36.8	250.0	
	Theropod 7	T	51.8	298.7	
La Cuesta de Andorra, Enciso Group	Ornithopod AND1	O	58	259	Pérez-Lorente (2015)
La Canal, Munilla, Spain, Enciso Group	Ornithopod 1	O	52.0	183.0	Casanovas et al. (1995*a*); also see Razzolini et al. (2016*b*)
	Ornithopod 2	O	37.0	197.0	
	Ornithopod 3	O	45.0	204.0	
	Ornithopod 4	O	26.0	142.0	
	Ornithopod 5	O	46.0	126.0	
	Ornithopod 6	O	44.0	183.0	
	Ornithopod 7	O	36.0	219.0	

Site	Trackway	Code	Footprint mean length (cm)	Mean stride (cm)	Source of data and comments
Munilla, Enciso Group	Theropod MN2	T	25	127	Casanovas et al. (1995*g*)
	Theropod MN3	T	26	160	
	Theropod MN5	T	25	99	
	Theropod MN6	T	32	220	
	Theropod MN7	T	25	165	
	Theropod MN8	T	33	168	
	Theropod MN9	T	27	165	
	Theropod MN11	T	22	137	
	Theropod MN12	T	50	334	
	Theropod MN13	T	24	137	
	Theropod MN14	T	22	150	
	Theropod MN15	T	31	228	
	Theropod MN17	T	26	148	
	Theropod MN18	T	36	212	
	Theropod MN19	T	37	192	
	Theropod MN21	T	27	160	
	Theropod MN22	T	27	172	
	Theropod MN23	T	25	147	
	Theropod MN25	T	31	207	
	Theropod MN26	T	27	195	
	Theropod MN27	T	27	171	
	Theropod MN28	T	33	200	
	Theropod MN31	T	26	158	
	Theropod MN32	T	32	214	
Las Navillas, Enciso Group	Theropod LNF	T	51.0	281.0	Casanovas et al. (1995*h*); Pérez-Lorente (2015)
	Theropod LNG	T	32.0	191.0	
	Theropod LNH	T	33.0	236.0	
	Theropod LNL	T	38.0	243.0	
	Theropod LNO	T	44.0	250.0	
	Theropod LNP	T	43.0	207.0	
	Theropod LNS (LN19)	T	51.9	255.9	
Era del Peladillo, Enciso Group	Theropod 1	T	30.0	196.5	Casanovas et al. (1993*a*)
	Theropod 3	T	28.0	225.0	
	Theropod 4	T	22.0	218.5	
	Theropod 6	T	36.0	212.0	
	Theropod 7	T	33.0	187.0	
	Theropod 8	T	26.0	214.0	
	Theropod 9	T	27.0	199.0	
	Theropod 10	T	31.0	186.5	
	Ornithopod 2	O	34.0	166.0	
	Ornithopod 11	O	38.0	176.0	
	Ornithopod 12	O	46.0	139.0	
Era del Peladillo 2, Enciso Group	Theropod 160	T	31.0	198.0	Casanovas et al. (1993*b*)
	Theropod 161	T	33.0	155.0	
	Theropod 162	T	34.0	179.0	
	Theropod 163	T	32.0	231.0	
	Theropod 164	T	23.0	160.0	
	Theropod 165	T	29.0	201.0	
	Theropod 166	T	34.0	244.0	
	Theropod 167	T	19.0	163.0	
	Theropod 172	T	31.0	247.0	
	Ornithopod 173	O	38.0	183.0	

Table A4.2. *continued*

Site	Trackway	Code	Footprint mean length (cm)	Mean stride (cm)	Source of data and comments
Era del Peladillo 3, Enciso Group	Theropod rastrillada 2	T	30.8	187.0	Casanovas et al. (1995c)
	Ornithopod rastrillada 5	O	25.7	106.0	
	Theropod rastrillada 6	T	34.8	211.0	
	Theropod rastrillada 7	T	34.2	193.3	
	Ornithopod rastrillada 8	O	36.0	148.5	
	Ornithopod rastrillada 9	O	29.0	140.0	
	Theropod rastrillada 12	T	23.0	180.3	
	Theropod rastrillada 13	T	14.3	106.0	
	Theropod rastrillada 14	T	26.4	170.0	
	Theropod rastrillada 15	T	23.4	181.2	
	Theropod rastrillada 16	T	18.0	160.0	
	Theropod rastrillada 17	T	23.6	177.7	
	Theropod rastrillada 19	T	34.0	96.7	
	Ornithopod rastrillada 21	O	37.2	161.3	
	Ornithopod rastrillada 22	O	28.3	141.3	
	Theropod rastrillada 24	T	37.5	235.0	
	Theropod rastrillada 25	T	24.5	170.4	
Era del Peladillo 4, Enciso Group	Theropod 4PLC	T	32.0	219.0	Casanovas et al. (1995*k*)
	Theropod 4PLL	T	29.0	163.0	
	Theropod 4PLN	T	23.0	187.0	
Era del Peladillo 5, Enciso Group	Ornithopod 5PL2	O	9.8	52.0	Casanovas et al. (1997)
	Ornithopod 5PL13	O	40.0	187.0	
	Theropod 5PL10	T	25.0	179.0	
Era del Peladillo 6, Enciso Group	UID bipedal dinosaur 6PL2	UID	36.0	98.0	Casanovas et al. (1998)
	Theropod 6PL4	T	20.0	110.0	
	Theropod 6PL5	T	25.0	108.0	
	Theropod 6PL8	T	20.0	147.0	
	Theropod 6PL9	T	24.0	155.0	
	UID bipedal dinosaur 6PF12	UID	36.0	239.0	
	UID bipedal dinosaur 6PL13	UID	44.0	175.0	
Era del Peladillo 7, Enciso Group	Theropod 7PL2	T	34.0	201.0	Casanovas et al. (1998)
	Theropod 7PL3	T	25.0	180.0	
	Theropod 7PL4	T	31.0	212.0	
	Theropod 7PL8	T	19.0	126.0	
	Theropod 7PL10	T	16.0	106.0	
	Theropod 7PL11	T	25.0	169.0	
	Theropod 7PL12	T	20.0	130.0	
	Theropod 7PL14	T	34.0	198.0	
	Theropod 7PL16	T	17.0	121.0	
Barranco de Valdefuentes	Ornithopod CVF1	O	49.0	181.0	Casanovas et al. (1998)
	Ornithopod CFV2	O	41.0	200.0	
Barranco de Valdeño, Enciso Group	UID bipedal dinosaur VDA	UID	59.0	147.0	Casanovas et al. (1998)
Umbria del Portillo, Enciso Group	Ornithopod UP1	O	48.0	218.0	Casanovas et al. (1998)
	Ornithopod UP2	O	40.0	239.0	
	Ornithopod UP4	O	37.0	194.0	
	Ornithopod UP5	O	51.0	272.0	
	Ornithopod UP6	O	46.0	229.0	

Site	Trackway	Code	Footprint mean length (cm)	Mean stride (cm)	Source of data and comments
Navalsaz, Enciso Group	Ornithopod	O	33.0	141.3	Casanovas et al. (1993*b*)
	Ornithopod 2	O	35.8	184.3	
	Ornithopod 3	O	29.0	191.7	
	Ornithopod 4	O	53.1	165.0	
	Ornithopod 8	O	38.6	208.7	
	Ornithopod 9	O	62.6	195.2	
	Ornithopod 10	O	25.0	162.0	
	Theropod 7	T	30.3	182.0	
El Villar-Poyales, Enciso Group	Theropod H	T	26.0	169.0	Casanovas et al. (1993*c*)
	Theropod I	T	28.0	168.0	
	Theropod plantigrade D length excluding metatarsus	T	26.0	120.0	
	Ornithopod B	O	28.0	212.0	
	Ornithopod C	O	28.0	156.0	
Las Mortajeres, Enciso Group	Theropod 23	T	28.7	218.0	Casanovas et al. (1993*b*)
	Theropod 24	T	40.0	274.0	
	Theropod 25	T	34.7	250.0	
	Theropod 26	T	31.0	216.0	
	Theropod 27	T	42.3	242.0	
	Theropod 28	T	44.3	274.0	
	Theropod 29	T	38.3	312.3	
	Theropod 30	T	32.3	227.0	
	Theropod 31	T	35.5	235.7	
	Theropod 32	T	27.0	196.3	
	Theropod 33	T	32.0	221.0	
	Theropod 34	T	33.8	192.3	
	Theropod 35	T	28.0	199.3	
	Theropod 36	T	32.0	193.0	
	Theropod 37	T	36.0	239.0	
	Theropod 38	T	40.0	255.0	
	Theropod 39	T	31.0	234.0	
	Theropod 40	T	32.0	254.0	
	Theropod 41	T	32.0	225.0	
	Theropod 42	T	25.0	175.0	
	Theropod 43	T	24.0	221.0	
	Theropod 44	T	32.0	234.0	
	Theropod 45	T	33.0	219.0	
	Theropod 46	T	32.0	204.0	
	Theropod 47	T	37.0	195.0	
	Theropod 48	T	35.0	214.0	
	Theropod 49	T	34.0	229.0	
	Theropod 50	T	39.0	256.0	
	Theropod 51	T	36.0	243.0	
	Theropod 52	T	29.0	190.0	
	Theropod 53	T	27.0	281.0	
Peñaportillo, Enciso Group	Theropod 144	T	30.0	168.0	Casanovas et al. (1993*b*)
	Theropod 145	T	42.0	222.0	
	Theropod 146	T	33.0	169.0	
	Theropod 147	T	14.0	72.0	
	Theropod 148	T	16.0	81.0	
	Theropod 149	T	13.0	68.0	

Table A4.2. *continued*

Site	Trackway	Code	Footprint mean length (cm)	Mean stride (cm)	Source of data and comments
Malvaciervo, Enciso Group	Ornithopod 151	O	53.0	149.0	Casanovas et al. (1993*b*)
	Theropod 152	T	27.0	211.0	
Hornillos de Cameros, Enciso Group	Theropod 1HR2	T	42.0	256.0	Álvarez et al. (2000)
	Theropod 2HR5	T	29.0	212.0	
	Ornithopod 2HR12	O	61.0	240.0	
	Theropod 3HR5	T	30.0	173.0	
	Theropod 4HR1	T	20.0	151.0	
Contadero, Torremuna, La Rioja	Ornithopod CT1	O	56.0	214.0	Pérez-Lorente et al. (2000)
	Ornithopod CT4	O	61.0	227.0	
	Ornithopod CT5	O	61.0	181.0	
	Ornithopod CT6	O	58.0	175.0	
Las Losas, Enciso Group	LL1	T	33.0	217.0	Blanco et al. (2000)
	LL2	T	24.0	185.0	
	LL3	T	27.0	201.0	
	LL4	T	35.0	203.0	
	LL5	T	38.0	179.0	
	LL6	T	25.0	176.0	
	LL7	T	33.0	195.0	
	LL8	T	30.0	199.0	
	LL9	T	31.0	188.0	
	LL10	T	21.0	165.0	
	LL11	T	25.0	164.0	
	LL12	T	35.0	150.0	
	LL13	T	35.0	282.0	
	LL14	T	27.0	199.0	
	LL15	T	28.0	135.0	
	LL17	T	28.0	221.0	
	LL18	T	32.0	151.0	
	LL19	T	32.0	214.0	
	LL21	T	39.0	225.0	
	LL22	T	28.0	168.0	
	LL23	T	24.0	136.0	
	LL24	T	31.0	213.0	
	LL25	T	31.0	191.0	
	LL26	T	29.0	152.0	
	LL27	T	28.0	175.0	
	LL28	T	37.0	209.0	
	LL29	T	25.0	161.0	
	LL30	T	29.0	168.0	
	LL31	T	31.0	346.0	
	LL32	T	24.0	309.0	
	LL33	T	26.0	262.0	
	LL34	T	29.0	200.0	
	LL35	T	29.0	225.0	
	LL36	T	36.0	209.0	
	LL37	T	33.0	263.0	
	LL38	T	32.0	202.0	
	LL39	T	21.0	157.0	
	LL40	T	30.0	238.0	
	LL41	T	31.0	202.0	
	LL42	T	24.0	154.0	
	LL43	T	30.0	307.0	

Site	Trackway	Code	Footprint mean length (cm)	Mean stride (cm)	Source of data and comments
Las Losas, Enciso Group	LL44	T	30.0	155.0	Blanco et al. (2000)
	LL45	T	27.0	196.0	
	LL46	T	37.0	204.0	
	LL47	T	33.0	284.0	
	LL48	T	22.0	207.0	
	LL49	T	22.0	237.0	
	LL50	T	32.0	222.0	
	LL51	T	37.0	278.0	
	LL52	T	36.0	304.0	
	LL53	T	38.0	279.0	
	LL54	T	22.0	210.0	
	LL55	T	41.0	191.0	
	LL56	T	34.0	207.0	
	LL57	T	30.0	137.0	
	LL58	T	43.0	245.0	
	LL59	T	33.0	220.0	
La Barguilla, Hornillos de Cameros, Enciso Group	Ornithopod LBG1	O	34.0	177.0	Pérez-Lorente (2003)
	Ornithopod LBG2	O	36.0	170.0	
	Theropod LBG3	T	31.0	240.0	
Santisol, Hornillos de Cameros, Enciso Group	Theropod STS3	T	25.0	169.0	Pérez-Lorente (2003)
	Theropod STS6	T	22.0	172.0	
	Theropod STS7	T	34.0	226.0	
	Theropod STS9	T	32.0	345.0	
	Theropod STS10	T	36.0	216.0	
	Theropod STS12	T	35.0	122.0	
	Theropod STS14	T	36.0	233.0	
	Theropod STS17	T	27.0	176.0	
	Theropod STS19	T	30.0	179.0	
	Ornithopod STS21	O	57.0	244.0	
Santa Juliana, Hornillos de Cameros, Enciso Group	Theropod SJA1	T	30.0	169.0	Pérez-Lorente (2003)
	Bipedal dinosaur SJA4	UID	29.0	192.0	
La Pellejera, Enciso Group	Theropod 1LP2	T	27.0	309.0	Requeta Loza et al. (2006–2007)
	Theropod 1LP5	T	33.0	186.0	
	Theropod 1LP6	T	34.0	176.0	
	Theropod 1LP8	T	29.0	163.0	
	Theropod 1LP9	T	51.0	235.0	
	Theropod 1LP10	T	25.0	131.0	
	Theropod 1LP11	T	30.0	170.0	
	Theropod 1LP12	T	24.0	148.0	
	Theropod 1LP13	T	49.0	197.0	
	Theropod 1LP17	T	23.0	141.0	
	Theropod 1LP18	T	32.0	190.0	
	Theropod 1LP19	T	24.0	163.0	
	Theropod 1LP20	T	29.0	179.0	
	Theropod 1LP21	T	34.0	191.0	
	Theropod 1LP22	T	32.0	202.0	
	Theropod 1LP24	T	34.0	233.0	
	Theropod 1LP25	T	20.0	140.0	
	Theropod 1LP26	T	30.0	179.0	
	Theropod 1LP27	T	31.0	302.0	
	Theropod 1LP28	T	29.0	149.0	
	Theropod 1LP29	T	35.0	214.0	

Table A4.2. *continued*

Site	Trackway	Code	Footprint mean length (cm)	Mean stride (cm)	Source of data and comments
La Pellejera, Enciso Group	Theropod 1LP30	T	41.0	267.0	Requeta Loza et al. (2006-2007)
	Theropod 1LP32	T	54.0	275.0	
	Theropod 1LP33	T	40.0	225.0	
	Theropod 1LP34	T	39.0	189.0	
	Theropod 1LP36	T	42.0	231.0	
	Theropod 1LP38	T	28.0	148.0	
	Theropod 1LP39	T	43.0	224.0	
	Theropod 1LP42	T	35.0	381.0	
	Theropod 1LP44	T	38.0	182.0	
	Semiplantigrade theropod 1LP46	T	37.0	192.0	
	Theropod 1LP47	T	30.0	196.0	
	Theropod 1LP51	T	45.0	199.0	
	Theropod 1LP52	T	46.0	300.0	
	Theropod 1LP56	T	27.0	231.0	
	Theropod 1LP57	T	42.0	157.0	
	Theropod 1LP58	T	44.0	275.0	
	Theropod 1LP59	T	28.0	193.0	
	Theropod 1LP61	T	40.0	185.0	
	Theropod 1LP64	T	31.0	129.0	
	Ornithopod 1LP3	O	53.0	149.0	
	Ornithopod 1LP4	O	55.0	225.0	
	Ornithopod 1LP14	O	63.0	176.0	
	Ornithopod 1LP53	O	48.0	191.0	
	Ornithopod 1LP54	O	62.0	175.0	
	Ornithopod 1LP55	O	46.0	175.0	
	Ornithopod 1LP66	O	55.0	179.0	
	Ornithopod 1LP67	O	52.0	170.0	
	Ornithopod 1LP68	O	50.0	186.0	
	Ornithopod 1LP69	O	50.0	166.0	
	Semiplantigrade ornithopod 1LP1	O	54.0	235.0	
	Semiplantigrade ornithopod 1LP7	O	55.0	162.0	
	Semiplantigrade ornithopd 1LP15	O	58.0	236.0	
	Semiplantigrade ornithopod 1LP16	O	51.0	212.0	
	Semiplantigrade ornithopod 1LP23	O	53.0	183.0	
	Semiplantigrade ornithopod 1LP31	O	44.0	203.0	
	Semiplantigrade ornithopod 1LP35	O	44.0	187.0	
	Semiplantigrade ornithopod 1LP37	O	48.0	201.0	
	Semiplantigrade ornithopod 1LP40	O	51.0	157.0	
	Semiplantigrade ornithopod 1LP41	O	47.0	213.0	
	Semiplantigrade ornithopod 1LP43	O	54.0	201.0	
	Semiplantigrade ornithopod 1LP45	O	53.0	195.0	
	Semiplantigrade ornithopod 1LP65	O	45.0	174.0	
El Corral del Totico 1, Enciso Group	Theropod 1TT1	T	45.9	245.6	Jiménez Vela and Pérez-Lorente (2006-2007)
	Theropod 1TT2	T	37.5	452.5	
	Ornithopod 1TT3	O	41.3	176.7	
Guilera, Enciso Group	1GI2 ornithopod	O	54.0	252.0	Álvarez Bermejo et al. (2013)
	2GI1 ornithopod	O	58.0	156.0	

Site	Trackway	Code	Footprint mean length (cm)	Mean stride (cm)	Source of data and comments
Valdete Afloriamente C, Prejano, Enciso Group, La Rioja	Ornithopod Valdete Afloriamente C	O	41.3	181.0	Moratalla et al. (1988*b*)
La Magdalena, Prejano, Enciso Group	Ornithopod La Magdalena	O	58.6	237.0	
Mura de Aguas, Barranco del Chorron del Saltadero, Enciso Group	Ornithopod Mura de Aguas B	O	35.0	136.5	
Los Cayos, Enciso Group	Theropod LCA-R3	T	42.9	201.0	Moratalla et al. (2003)
	Theropod Level B3	T	29.0	188.0	
Valdemayor, Cabezon de Cameros, La Rioja, Enciso Group	Quadrupedal ornithopod	O	58.4	227.5	Moratalla et al. (1992)
Rambla de Tambuc, Valencia Spain	Zone B theropod TB1-Rastro 9	T	35.0	157.0	Santisteban and Suñer (2003)
	Zone B theropod TB1-Rastro 10	T	33.0	130.0	
	Zone B theropod TB1-Rastro 11	T	30.0	145.0	
	Zone C theropod TB1-Rastro 13	T	25.0	142.0	
	Zone C theropod TB1-Rastro 14	T	33.0	78.1	
	Zone E theropod TB1-Rastro 17	T	18.0	152.0	
	Zone E theropod TB1-Rastro 18	T	20.0	138.0	
	Zone E theropod TB1-Rastro 19	T	20.0	138.0	
Cerrillares Fm, Maastrichtian, Rambla de los Gavilanes, Spain	Ornithopod 1RGV6	O	39	197	Herrero et al. (2016)
	Ornithopod 6RGV1	O	33	180	
	Ornithopod 6RGV2	O	37	181	
	Ornithopod 6RGV3	O	37	160	
	Ornithopod 6RGV4	O	29	177	
	Ornithopod 6RGV7	O	35	175	
	Ornithopod 6RGV9	O	31	170	
Salema Fm, Lower Cretaceous, Praia da Salema, Algarve Basin, Portugal	Theropod	T	19	123	Santos (2016: fig. 3)
Salema Fm, Lower Cretaceous, Algarve Basin, Portugal	Salema Beach ornithopod SAL1-T1	O	38	140	Santos et al. (2013)
	Santa Beach *Iguanodontipus* SAN2-T1	O	32	170	
	Santa Beach *Iguanodontipus* SAN2-T2	O	41	190	
Areniscas de Areny Fm, Areny, Huesca	Hadrosaurid trackway 1	O	50.2	217	Barco et al. (2001)
Schrattenkalk Fm, Beckenried, Switzerland	*Iguanodontipus billsarjeanti* trackway 1	O	28.6	118.3	Meyer and Thüring (2003)
	Iguanodontipus trackway 2	O	35.1	137.0	
	Iguanodontipus trackway 3	O	29.8	129.1	
Solaris I campground, Istria, Croatia	Theropod SOLI-T1	T	21.0	129.3	Dalla Vecchia and Tarlao (2000)
	Theropod SOLI-T2	T	20.8	155.5	
	Theropod SOLI-T3	T	21.0	141.3	
	Theropod SOLI-T4	T	20.0	141.3	
	Theropod SOLI-T5	T	20.9	180.5	
	Theropod SOLI-T6	T	16.5	149.0	
	Theropod SOLI-T7	T	21.1	120.0	
	Theropod SOLI-T8	T	20.2	167.0	
	Theropod SOLI-T9	T	17.0	138.0	
	Theropod SOLI-T10	T	17.8	165.0	
	Theropod SOLI-T11	T	22.0	155.5	
	Theropod SOLI-T12	T	19.2	220.0	
	Theropod SOLI-T13	T	20.5	159.0	

Table A4.2. *continued*

Site	Trackway	Code	Footprint mean length (cm)	Mean stride (cm)	Source of data and comments
Solaris II campground, Istria, Croatia	Theropod SOLII-T1	T	19.0	153.0	Dalla Vecchia and Tarlao (2000)
	Theropod SOLII-T2	T	18.3	160.0	
	Theropod SOLII-T3	T	18.9	171.3	
	Theropod SOLII-T4	T	21.0	153.0	
	Theropod SOLII-T5	T	18.7	138.7	
	Theropod SOLII-T6	T	20.5	176.7	
	Theropod SOLII-T7	T	18.5	131.0	
	Theropod SOLII-T8	T	20.0	190.0	
	Theropod SOLII-T9	T	17.0	123.0	
	Theropod SOLII-T10	T	20.0	140.0	
	Theropod SOLII-T11	T	20.0	138.0	
	Theropod SOLII-T12	T	20.5	140.0	
	Theropod SOLII-T13	T	21.8	145.5	
	Theropod SOLII-T14	T	18.5	188.0	
	Theropod SOLII-T15	T	20.8	157.0	
	Theropod SOLII-T16	T	19.3	173.0	
	Theropod SOLII-T17	T	19.8	130.3	
	Theropod SOLII-T18	T	20.3	138.0	
	Theropod SOLII-T19	T	18.8	173.0	
	Theropod SOLII-T20	T	20.7	160.5	
	Theropod SOLII-T21	T	19.3	152.0	
	Theropod SOLII-T22	T	19.5	158.0	
	Theropod SOLII-T24	T	16.5	140.0	
	Theropod SOLII-T25	T	18.8	177.0	
	Theropod SOLII-T26	T	21.0	175.0	
	Theropod SOLII-T27	T	18.5	175.0	
	Theropod SOLII-T28	T	20.5	142.0	
	Theropod SOLII-T30	T	19.5	149.0	
	Theropod SOLII-T31	T	20.0	140.0	
	Theropod SOLII-T32	T	19.5	145.0	
Puntisella, Istrian Peninsula, Croatia	Theropod? PUI-T1	T	14.0	112.7	Dalla Vecchia and Tarlao (2000)
	Theropod PUII-T1	T	21.0	138.5	
Istrian Peninsula, Croatia	SLD-T1	T	20.4	165.9	Dalla Vecchia et al. (2001)
	SLD-T2	T	15.7	118.8	
	SLD-T3	T	19.3	142.0	
	SLD-T4	T	20.1	161.7	
	SLD-T5	T	18.3	151.0	
	PR-T1	T	17.0	111.5	
	PR-T2	T	25.4	144.7	
	FEN-T1	T	16.8	142.7	
	FEN-T2	T	21.4	139.0	
	FEN-T3	T	17.5	182.0	
Pogledalo, Istrian Peninsula	T13	T	38.5	235.5	Dalla Vecchia et al. (2002)
	T24	T	45.0	260.0	
	T27	T	37.8	229.0	
	T31	T	31.5	129.0	
Ploce I	T1	T	21.6	134.1	
	T2	T	12.8	114.8	
Pljesivac	PLJ1-T1	O	28.0	149.0	
Fenoliga Island, Croatia; Cenomanian limestone	Theropod T and FG II	T	20.02	141.69	Mezga and Bajraktarević (1999)
	Theropod T and FG IV	T	16.77	144.70	
Altamura Limestone Fm, Altamura, Italy	*Apulisauripus federicianus* ACDL 99/3 quadrupedal ornithopod	O	25.0	105.0	Nicosia et al. (1999)

Site	Trackway	Code	Footprint mean length (cm)	Mean stride (cm)	Source of data and comments
Feitianshan Fm, Zhaojue, Sichuan, China	Theropod ZJI-T1	T	26.4	176.5	Xing et al. (2013*e*, 2014*h*, 2015*a*)
	Theropod ZJI-T2	T	12.7	66.4	
	Ornithopod ZJII-O1	O	21.6	111.5	
	Ornithopod ZJII-O2	O	22.6	148.7	
	Ornithopod ZJII-O3	O	27.2	151.8	
	Ornithopod ZJII-O4	O	20.7	86.8	
	Ornithopod ZJII-O6	O	21.3	139.6	
	Ornithopod ZJII-O7	O	24.7	156.2	
	Ornithopod ZJII-O8	O	19.7	80.4	
	Ornithopod ZJII-O9	O	23.9	180.0	
	Ornithopod ZJII-O10	O	24.8	168.7	
	Ornithopod ZJII-1O1	O	21.5	156.2	
	Ornithopod ZJIIN-O1	O	13.1	78.4	
	Ornithopod ZJIIN-O2	O	14.2	84.7	
	Ornithopod ZJIIN-O3	O	13.8	77.1	
	Ornithopod ZJINN-O8	O	24.2	143.0	
	Theropod ZJII-T1	T	22.5	166.3	
	Theropod ZJII-ST1	T	33.5	266.8	
	Theropod ZJIIN-T1	T	26.4	187.3	
	Theropod ZJIIN-T2	T	21.9	182.2	
	Theropod ZJIIN-T4 *Siamopodus xui*	T	7.1	72.4	
Feitianshan Fm, Baiju, Sichuan, China	BJB-T1 morphotype B *Dromaeopodus*	T	24.7	166.0	Xing et al. (2016*i*)
	BJA-T2 morphotype A	T	25.2	166.9	
	BJC-T5 morphotype A	T	25.6	173.0	
	BJC-T6 morphotype A	T	27.0	151.0	
Feitianshan Fm, Yangmozu, Sichuan, China	Morphotype A *Minisauripus zhenshuonani* YMZ-T11	T	2.6	40.4	Xing et al. (2016*h*)
	Morphotype A *Minisauripus zhenshuonani* YMZ-T12	T	2.5	75.0	
	Morphotype A *Minisauripus zhenshuonani* YMZ-T13	T	2.6	53.8	
	Morphotype B cf. *Jialingpus* YMZ-T2	T	19.4	228.1	
	Morphotype C YMZ-T1	T	12.8	120.0	
	Morphotype C YMZ-T3	T	13.3	126.4	
	Morphotype C YMZ-T4	T	13.8	119.5	
	Morphotype C YMZ-T5	T	16.5	180.1	
	Morphotype C YMZ-T7	T	17.6	163.2	
	Morphotype C YMZ-T8	T	19.6	241.8	
Xiaoba Fm, Mujiaowu, Sichuan, China	MJW-T1 Morphotype A cf. *Anomoepus*	O	13.3	70.0	Xing et al. (2015*d*)
Jiaguan Fm, Longjing, China	LJ-O1 *Caririchnium*	O	22.8	95.9	Xing et al. (2015*f*)
Jiaguan Fm, Hanxi, Sichuan, China	HX-T3 *Eubrontes*	T	30.0	172.7	Xing et al. (2015e)
	HX-T4	T	16.3	110.6	
	HX-T5	T	13.8	115.2	
	HX-T6	T	20	114.8	
	HX-O1 cf. *Ornithopodichnus*	O	15.5	147.0	
	HX-O2 cf. *Ornithopodichnus*	O	13.7	138.5	
	HX-O3 cf. *Ornithopodichnus*	O	13.4	65.1	
Jiaguan Fm, Xinyang, Sichuan, China	XY-T1 *Eubrontes*	T	26.4	155.2	Xing et al. (2016*m*)
	XY-T2 small theropod	T	12.1	84.6	

Table A4.2. *continued*

Site	Trackway	Code	Footprint mean length (cm)	Mean stride (cm)	Source of data and comments
Jiaguan Fm, Shimiaogou, Gulin, Sichuan, China	Theropod SMG-T1	T	11.6	81.4	Xing et al. (2016*f*)
	Theropod SMG-T2	T	10.4	72	
	Theropod SMG-T4	T	11.0	82.5	
	Theropod SMG-T5	T	12.5	111.7	
	Theropod SMG-T6	T	11.8	92.9	
	Dromaeosaurid SMG-D1	T	7.1	43.3	
	Caririchnium liucixini SMG-O1	O	29.9	126.6	
	Caririchnium liucixini SMG-O2	O	21.9	106.9	
	Caririchnium liucixini SMG-O3	O	22.3	121.3	
	Caririchnium liucixini SMG-O4	O	22.1	104.2	
	Caririchnium liucixini SMG-O5	O	21.7	111.6	
Jiaguan Fm (formerly Wotoushan Fm), Baoyuan, Chishui, Guizhou Province, China	cf. *Irenesauripus* MT trackway A	T	17.3	149.4	Xing et al. (2011*b*); print length excludes metatarsal impression
Jiaguan Fm, Emei Mountain, Sichuan, China	*Neograllator emeiensis* CFEC-C-1-3	T	2.7	18.4	Zhen et al. (1994)
Jiaguan Fm, Lebei, Gulin, Sichuan, China	Block 1 T4 *Eubrontes*	T	16.2	120	Xing et al. (2016*d*)
	Block 4 T14 cf. *Dromaeopodus*	T	29.8	163	
Jiaguan Fm, Lotus Fortress, Qijiang National Geological Park, Chongqing, Sichuan, China	*Caririchnium lotus* QII-O1	O	39.0	184.0	Xing et al. (2015*b*)
	Caririchnium lotus QII-O3	O	20.6	170.0	
	Caririchnium lotus QII-O5	O	26.5	142.7	
	Caririchnium lotus QII-O6	O	25.5	138.0	
	Caririchnium lotus QII-O7	O	39.5	228.0	
	Caririchnium lotus QII-O8	O	24.0	116.9	
	Caririchnium lotus QII-O9	O	36.8	162.6	
	Caririchnium lotus QII-O10	O	27.5	137.8	
	Caririchnium lotus QII-O11	O	35.0	153.5	
	Caririchnium lotus QII-O12	O	20.8	123.7	
	Caririchnium lotus QII-O13	O	39.7	175.0	
	Caririchnium lotus QII-O14	O	35.4	172.3	
	Caririchnium lotus QII-O15	O	25.8	128.0	
	Caririchnium lotus QII-O16	O	21.5	109.5	
	Caririchnium lotus QII-O17	O	34.1	132.4	
	Caririchnium lotus QII-O18	O	38.6	198.7	
	Caririchnium lotus QII-O19	O	33.4	179.8	
	Caririchnium lotus QII-O20	O	36.0	185.0	
	Caririchnium lotus QII-O21	O	36.5	190.0	
	Caririchnium lotus QII-O22	O	20.6	109.4	
	Caririchnium lotus QII-O23	O	20.7	122.5	
	Caririchnium lotus QII-O24	O	22.9	133.9	
	Caririchnium lotus QIII-O1	O	23.3	124.1	
	Caririchnium lotus QIII-O2	O	21.4	100.6	
	Caririchnium lotus QIII-O3	O	17.3	83.3	
Yangjiazhuang Fm, Huanglonggou, Shandong Province, China	Theropod morphotype A trackway T0	T	34.0	233.0	Lockley et al. (2015)
	Theropod morphotype A trackway T1	T	30.2	246.0	
	Theropod morphotype A trackway T2	T	24.7	211.0	
	Theropod morphotype A trackway T3	T	25.0	200.5	
	Theropod morphotype A trackway T6	T	26.2	221.5	
	Theropod morphotype B (*Grallator yangi*) trackway T4	T	20.5	160.0	
	Theropod morphotype B trackway T5	T	18.2	168.5	
	Theropod morphotype B trackway T7	T	12.5	121.5	
	Theropod morphotype B trackway T8	T	14.8	119.2	
	Theropod morphotype B trackway T9	T	13.0	127.5	

Site	Trackway	Code	Footprint mean length (cm)	Mean stride (cm)	Source of data and comments
Yangjiazhuang Fm, Huanglonggou, Shandong Province, China	Theropod morphotype B trackway T10	T	11.3	124.0	Lockley et al. (2015)
	Theropod morphotype B trackway T11	T	11.3	107.0	
	Theropod morphotype B trackway T12	T	11.8	124.0	
	Theropod morphotype B trackway T13	T	10.5	92.0	
	Theropod morphotype B trackway T15	T	13.2	123.0	
	Theropod morphotype B trackway T16	T	13.5	109.5	
	Theropod morphotype B trackway T17	T	10.8	123.0	
	Theropod morphotype B trackway T18	T	12.2	106.3	
	Theropod morphotype B trackway T19	T	11.5	102.0	
	Theropod morphotype B trackway T20	T	13.2	107.0	
	Theropod morphotype B trackway T21	T	12.5	136.0	
	*Corpulentapus lilasia*trackway C1	T	13.0	123.5	
	Corpulentapus trackway C2	T	12.2	121.0	
	Corpulentapus trackway C3	T	12.2	115.0	
	Corpulentapus trackway C4	T	11.7	140.0	
	Corpulentapus trackway C5	T	12.3	126.0	
	Corpulentapus trackway C6	T	11.7	120.0	
	Corpulentapus trackway C7	T	13.3	148.5	
	Corpulentapus trackway C8	T	12.5	132.0	
	Corpulentapus trackway C9	T	12.5	137.0	
	Corpulentapus trackway C12	T	11.7	94.0	
	Corpulentapus trackway C13	T	12.3	117.0	
	Corpulentapus trackway C14	T	12.7	121.0	
	Corpulentapus trackway C16	T	10.3	103.0	
	Corpulentapus trackway C17	T	13.2	145.0	
	Corpulentapus trackway C18	T	11.0	117.0	
Tianjialou Fm, Houzuoshan Dinosaur Park, Shandong Province, China	*Dromaeopodus shandongensis* trackway 7	T	28.5	186.0	Li et al. (2008)
	Level 0 *Ornithopodichnus*	O	26	141	Li et al. (2015: table 1 and fig. 5)
	Level 6 theropod	T	25	282	Li et al. (2015: fig. 10)
	Level 10 *Grallator*	T	10	112	Li et al. (2015: text and fig. 9)
	Level 15 *Grallator*	T	9.5	100	Li et al (2015: text and fig. 15)
	Level 18 *Ornithopodichnus*	O	31	139	Li et al. (2015: fig. 18)
Tianjialou Fm, Jishan tracksites, Shandong Province, China	LSI-D1 cf. *Dromaeosauripus*	T	19.5	113.0	Xing et al. (2013*e*)
	LSII-T1	T	26.4	180.0	
Hekou Group, Liujiaxia Dinosaur National Geopark, Gansu, China	*Dromaeosauripus yongjingensis* trackway GSLTZP-S2-TB	T	15.0	74.0	Xing et al. (2013*a*)
Hekou Fm, Liagou I, Zhongpu, Gansu, China	Theropod ZPI	T	41.9	121.4	Xing et al. (2014*b*)
Hekou Fm, Huang He, China	Theropod T1	T	32.0	200.0	Zhang et al. (2006)
Hekou Group, Yanguoxia, Gansu Province, China	Quadrupedal ornithopod OA	O	36	171	Fujita et al. (2012)
	Quadrupedal ornithopod OB	O	29	162	
Shenhuangshan Fm, Jiuquwan, Hunan, China	*Xiangxipus chenxiensis* JQWI-HV003	T	18.2	117.0	Xing et al. (2016*b*)
	Grallator jiuquwanensis JQWII-T1	T	19.5	129.5	
Zhutian Fm, Nanxiong Basin, Guangdong, China	*Hadrosauropodus* isp. NCBLR.F.MI prints F1–F4	O	36.22	148.7	Xing et al. (2009*a*)
Yijinholo Fm, Chabu, Inner Mongolia	*Chapus lockleyi*	T	58.0	233.0	Li et al. (2006)
Pianling Village, Luanping Co., Hebei Province, China, Xiguayuan Fm	Theropod A	T	19.8	189.0	You and Azuma (1995)
	Theropod B	T	32.8	232.0	
	Ornithopod C	O	49.2	314.0	
	Ornithopod D	O	47.2	266.0	
	Ornithopod E	O	50.0	230.0	

Table A4.2. *continued*

Site	Trackway	Code	Footprint mean length (cm)	Mean stride (cm)	Source of data and comments
Jingchuan Fm, Site I Chabu Sumu Touli	A	UID	22.0	202.5	Azuma et al. (2006)
	B	UID	27.1	305.0	
	C	UID	18.0	132.0	
	E Type 1 theropod	T	48.2	277.1	
	F	UID	29.2	226.1	
	G	UID	19.1	155.0	
	H	UID	38.0	210.7	
	I Type 1? theropod	T	40.0	279.8	
	J	UID	20.6	132.3	
	K	UID	30.7	278.4	
	L	UID	16.1	118.8	
	M	UID	17.0	141.0	
	N	UID	23.0	213.8	
	O	UID	15.8	112.5	
Site II Chabu Sumu Hadetu	A	UID	19.9	131.9	
	B	UID	30.3	281.3	
	C	UID	32.3	300.0	
	D	UID	13.1	150.9	
	E	UID	13.3	169.0	
	G	UID	25.1	157.5	
	H	UID	31.1	267.4	
	I	UID	14.1	106.1	
	J	UID	15.2	168.3	
	M	UID	29.1	310.4	
	N	UID	28.1	252.4	
	O	UID	33.0	218.3	
	P	UID	18.1	127.7	
	T Type 1 theropod	T	34.2	198.9	
	U	UID	12.1	102.2	
	W	UID	30.2	230.0	
	X	UID	16.1	227.5	
	Z	UID	27.0	285.0	
	AA	UID	40.2	248.7	
	CC	UID	25.0	485.0	
	HH	UID	12.2	104.8	
	II	UID	15.1	117.1	
	KK	UID	32.0	246.2	
	LL	UID	42.0	223.4	
	MM	UID	13.0	132.3	
	SS	UID	18.5	107.0	
	TT Type 1 theropod	T	40.5	227.9	
	UU	UID	40.1	245.4	
	VV	UID	34.1	230.2	
Site III Ahrubulage	F Type 2 theropod	T	38.0	255.0	
	G Type 1 theropod	T	47.0	270.0	
	H Type 3 ornithopod	O	11.2	93.3	
	K	UID	10.5	59.3	
	M Type 2 theropod	T	21.0	184.0	
Xiaoyan Fm, Xiaohutian, Huangshan, China	*Paracorpulentapus zhangsanfengi* XHT-28-31	T	16.0	83.7	Xing et al. (2014*g*)
Jierboshi I, Zhaojue, Liangshan, Sichuan, China	JBSI-T1; compared with *Gypsichnites pacensis*, but considered a theropod trackway	T	15.8	135.0	Xing et al. (2015*g*)

Site	Trackway	Code	Footprint mean length (cm)	Mean stride (cm)	Source of data and comments
Jindong Fm, Masan tracksite, Korea	*Ornithopodichnus masanensis* TW 1 (NHC 1002)	O	36.5	156.8	Kim et al. (2009)
	Ornithopodichnus TW 2 (NHC 1003)	O	42.7	189.8	
	Ornithopodichnus TW 3 (NHC 1004)	O	43.2	181.7	
	Ornithopodichnus TW 4 (NHC 1005)	O	36.3	159.7	
	Ornithopodichnus TW 5 (NHC 1001)	O	42.4	177.7	
Jindong Fm, Samcheonpo area South Korea Bongwhagol 1-3	Ornithopod 2	O	33.0	159.0	Lockley et al. (2006*b*)
	Ornithopod 3	O	34.0	155.0	
	Ornithopod 4	O	38.0	170.0	
	Ornithopod 5	O	34.0	172.0	
	Ornithopod 6	O	36.0	169.0	
	Ornithopod 7	O	37.0	177.0	
Bonwhagol 1-4	Ornithopod 1	O	34.0	150.0	
	Ornithopod 2	O	32.0	145.0	
Bonwhagol 1-6	Ornithopod 1	O	34.0	198.0	
	Ornithopod 2	O	32.0	172.0	
Bonwhagol 1-7	Ornithopod 1	O	33.0	118.0	
Bonwhagol 1-11	Ornithopod 2	O	38.0	180.0	
Dukmeyongri 1-5	Ornithopod 1	O	34.0	174.0	Lockley et al. (2006*b*)
Dukmeyongri 1-7	Ornithopod 1	O	31.0	151.0	
	Ornithopod 2	O	30.0	139.0	
	Ornithopod 3	O	29.0	153.0	
Dukmeyongri 1-15	Ornithopod 2	O	36.0	200.0	
	Ornithopod 3	O	33.0	173.0	
Dukmeyongri 1-16	Ornithopod 1	O	38.0	208.0	
Dukmeyongri 1-24	Ornithopod 1	O	35.0	166.0	
Dukmeyongri 1-27	Ornithopod 2	O	35.0	165.0	
	Ornithopod 3	O	35.0	165.0	
Dukmeyongri 1-28	Ornithopod 1	O	41.0	167.0	
Dukmeyongri 2-2	Ornithopod 1	O	35.0	161.0	
Dukmeyongri 2-3	Ornithopod 1	O	40.0	185.0	
Dukmeyongri 2-7	Ornithopod 1	O	38.0	203.0	
	Ornithopod 2	O	39.0	197.0	
Dukmeyongri 2-8	Ornithopod 1	O	31.0	167.0	
Dukmeyongri 6-6	Ornithopod 1	O	27.0	112.0	
Dukmeyongri 2-15	Ornithopod 1	O	37.0	162.0	
	Ornithopod 2	O	37.0	173.0	
	Ornithopod 3	O	37.0	171.0	
	Ornithopod 4	O	37.0	165.0	
Dukmeyongri 2-17	Ornithopod 1	O	28.0	172.0	
Dukmeyongri 2-18	Ornithopod 1	O	31.0	137.0	
Dukmeyongri 2-22	Ornithopod 1	O	36.0	190.0	
Dukmeyongri 2-23	Ornithopod 1	O	29.0	160.0	
Dukemongri 4-3	Ornithopod 1	O	30.0	146.0	
Dukmeongri 4-14	Ornithopod 1	O	30.0	106.0	
Dukmeongri 4-16	Ornithopod 1	O	39.0	205.0	
Dukmeongri 4-18	Ornithopod 1	O	35.0	171.0	
Sangjok 2A-9	Ornithopod 1	O	34.0	140.0	Lockley et al. (2006*b*)
	Ornithopod 3	O	36.0	203.0	
	Ornithopod 4	O	30.0	159.0	

Table A4.2. *continued*

Site	Trackway	Code	Footprint mean length (cm)	Mean stride (cm)	Source of data and comments
Sangjok 2A-13	Ornithopod 2	O	35.0	165.0	Lockley et al. (2006*b*)
	Ornithopod 3	O	39.0	170.0	
	Ornithopod 4	O	39.0	165.0	
	Ornithopod 5	O	34.0	194.0	
	Ornithopod 6	O	39.0	201.0	
	Ornithopod 7	O	36.0	167.0	
Sangjok 1B-3	Ornithopod 3	O	35.0	193.0	
Sangjok 1B-5	Ornithopod 2	O	34.0	176.0	
Sangjok 1B-12	Ornithopod 1	O	35.0	158.0	
Sangjok 1B-13	Ornithopod 2	O	28.0	142.0	
	Ornithopod 3	O	31.0	177.0	
	Ornithopod 4	O	32.0	178.0	
Sangjok 1B-15	Ornithopod 1	O	51.0	257.0	
Sangjok 1B-14	Ornithopod 2	O	33.0	173.0	
	Ornithopod 3	O	30.0	171.0	
Sangjok 2A-11	Ornithopod 1	O	32.0	186.0	
Sangjok 2B-4	Ornithopod 1	O	36.0	161.0	
	Ornithopod 2	O	27.0	150.0	
Sangjok 2B-5	Ornithopod 1	O	33.0	152.0	
	Ornithopod 2	O	30.0	131.0	
	Ornithopod 3	O	33.0	143.0	
	Ornithopod 4	O	32.0	115.0	
	Ornithopod 6	O	26.0	160.0	
	Ornithopod 7	O	36.0	224.0	
	Ornithopod 8	O	31.0	159.0	
	Ornithopod 9	O	31.0	139.0	
Sangjok 2B-6	Ornithopod 4	O	26.0	194.0	
	Ornithopod 5	O	30.0	171.0	
	Ornithopod 6	O	25.0	206.0	
	Ornithopod 7	O	32.0	181.0	
	Ornithopod 8	O	22.0	145.0	
	Ornithopod 9	O	30.0	175.0	
	Ornithopod 10	O	31.0	134.0	
	Ornithopod 11	O	35.0	171.0	
	Ornithopod 12	O	34.0	160.0	
	Ornithopod 13	O	36.0	169.0	
	Ornithopod 14	O	31.0	152.0	
	Ornithopod 15	O	32.0	169.0	
Sangjok 2B-7	Ornithopod 1	O	28.0	136.0	
	Ornithopod 2	O	29.0	126.0	
	Ornithopod 3	O	24.0	133.0	
	Ornithopod 4	O	24.0	143.0	
	Ornithopod 5	O	25.0	125.0	
	Ornithopod 6	O	27.0	130.0	
	Ornithopod 7	O	29.0	152.0	
	Ornithopod 8	O	44.0	176.0	
	Ornithopod 12	O	36.0	191.0	
	Ornithopod 13	O	23.0	119.0	
	Ornithopod 14	O	25.0	116.0	
	Ornithopod 16	O	24.0	127.0	
	Ornithopod 17	O	25.0	136.0	
	Ornithopod 18	O	27.0	133.0	

Site	Trackway	Code	Footprint mean length (cm)	Mean stride (cm)	Source of data and comments
Sangjok 2B-7	Ornithopod 19	O	27.0	138.0	Lockley et al. (2006*b*)
	Ornithopod 20	O	27.0	124.0	
	Ornithopod 21	O	34.0	180.0	
	Ornithopod 22	O	32.0	125.0	
	Ornithopod 23	O	27.0	128.0	
Sangjok 2B-8	Ornithopod 4	O	40.0	200.0	
	Ornithopod 5	O	34.0	185.0	
	Ornithopod 7	O	31.0	138.0	
	Ornithopod 8	O	34.0	186.0	
Sangjok 2B-10	Ornithopod 1	O	30.0	140.0	
	Ornithopod 2	O	37.0	202.0	
	Ornithopod 3	O	32.0	239.0	
	Ornithopod 4	O	30.0	141.0	
	Ornithopod 5	O	29.0	132.0	
	Ornithopod 6	O	31.0	124.0	
	Ornithopod 7	O	33.0	133.0	
	Ornithopod 8	O	33.0	125.0	
	Ornithopod 9	O	39.0	169.0	
	Ornithopod 10	O	29.0	126.0	
	Ornithopod 11	O	25.0	99.0	
	Ornithopod 12	O	30.0	184.0	
	Ornithopod 13	O	31.0	185.0	
Haman Fm, Sinsu Island, Korea	KML2 trackway	T	2.3	11.6	Lockley et al. (2008*b*)
Haman Fm, Namhae, Korea	*Dromaeosauripus hamanensis* KNUE 060801	T	15.5	204.0	Kim et al. (2008)
Jinju Fm, Bitu Island, Korea	*Dromaeosauripus jinjuensis* CU 214.243	T	9.3	80.4	Kim et al. (2012)
Saniri Fm, Yuli, Yeongdong, Korea	*Caririchnium yeongdongensis* KNUE 2011001	O	40.6	161.5	Kim et al. (2016)
Neungju Group, Seoyu-ri, South Korea Site L1	A	T	21.1	212.0	Huh et al. (2006)
	B	T	20.7	291.0	
	C	T	21.7	183.0	
	D	T	56.1	367.0	
	E	T	18.6	203.0	
	F	T	21.2	190.0	
	G	T	19.3	206.0	
Nejungju Group Site L2	T2	T	17.7	200.0	
	T5	T	35.0	223.0	
	T6	T	49.5	355.0	
	T7	T	17.9	169.0	
	T8	T	27.3	267.0	
	T10	T	20.8	241.0	
	T12	T	55.5	228.0	
	T14	T	26.8	238.0	
	T17	T	24.2	262.0	
	T18	T	29.4	170.0	
	T20	T	53.0	245.0	
	T21	T	27.0	234.0	
	T23	T	23.5	164.0	
	T24	T	27.8	228.0	
	T30	T	14.9	211.0	
	T31	T	19.3	182.0	
	T32	T	15.9	208.0	
	T33	T	25.0	221.0	

Table A4.2. *continued*

Site	Trackway	Code	Footprint mean length (cm)	Mean stride (cm)	Source of data and comments
Nejungju Group Site L2	T35	T	17.3	222.0	
	O1 *Ornithopodichnus*	O	12.7	73.8	Lockley et al. (2012*b*)
	O2 *Ornithopodichnus*	O	12.1	82.0	
	O5 *Ornithopodichnus*	O	12.1	75.8	
Nejungju Group Nejungju Group Site L4	LOT *Ornithopodichnus*	O	35.1	225.0	
Nejungju Group Site L5	T1	T	20.1	234.0	Huh et al. (2006)
	T2	T	17.0	267.0	
	T3	T	28.1	216.0	
	T5	T	25.1	125.0	
	T6	T	22.8	167.0	
	T7	T	15.9	165.0	
	T8	T	20.1	198.0	
	T10	T	21.6	206.0	
	T14	T	15.0	164.0	
	T15	T	24.8	212.0	
	T16	T	16.2	172.0	
	T17	T	20.5	198.0	
	T21	T	22.0	221.0	
	T24	T	23.5	211.0	
Babatag Range, Tadzhik Depression	Bipedal dinosaur	UID	37.0	153.0	Bulin et al. (1981)
Cenomanian, Tajikistan	*Macropodosaurus* (therizinosaur trackway)	T	50.0	150.0	Gierliński and Lockley (2013)
Nemegt, Mongolia	Hadrosaur trackway	O	60.0	239.0	Currie et al. (2003)
Shar Tsav, Mongolia	Theropod trackway fig. 3*A*	T	42.0	332.0	Ishigaki et al. (2009: figs. 3*A–D*, 5*A*, *B*); S. Ishigaki (pers.comm.)
	Theropod trackway fig. 3*B*	T	30.0	297.0	
	Theropod trackway fig. 3*C*	T	30.0	184.0	
	Theropod trackway fig. 3*D*	T	11.0	130.0	
	Dromaeosauripus	T	19	119	
Bugin Tsav	Ornithopod trackway fig. 5*A*	O	90.0	371.0	
	Ornithopod trackway fig. 5*B*	O	59.0	320.0	
Abdrant Nuru, Mongolia	Theropod A	T	25.5	186	Ishigaki (2010: table 1 and fig. 4)
	Theropod B	T	30.1	217	
	Theropod C	T	23.6	226	
	Theropod E	T	20.2	178	
	Theropod G	T	37.8	204	
	Theropod I	T	17.0	159	
	Theropod J (*Eubrontes*-like)	T	36.7	313	
	Theropod K	T	25.7	213	
	Theropod L	T	22.0	178	
	Theropod M	T	11.8	99	
	Theropod N	T	24.0	171	
Phu Phan Fm, Thailand	Theropod	T	36.0	280.0	Buffetaut et al. (1985)
Phra Wihan Fm, Thailand	Phu Faek theropod	T	40.5	178.0	Le Loeuff et al. (2002); cf. Lockley et al. (2009)
	Hin Lat Pa Chad quadrupedal ornithopod	O	11.0	55.8	
Khok Kruat Fm, Huai Dam Chum, Thailand	cf. *Asianopodus* T1	T	11.7	112.7	Kozu et al. (2017)
	cf. *Asianopodus* T2	T	15.3	137.5	
	cf. *Asianopodus* T3	T	16.7	162.5	
	cf. *Asianopodus* T4	T	12.4	121	
	cf. *Asianopodus* T7	T	13.9	142	
	cf. *Asianopodus* T8	T	15	111.6	
	cf. *Asianopodus* T9	T	16	154.8	

Site	Trackway	Code	Footprint mean length (cm)	Mean stride (cm)	Source of data and comments
Khok Kruat Fm, Huai Dam Chum, Thailand	cf. *Asianopodus* T10	T	16.1	159.5	Kozu et al. (2017)
	cf. *Asianopodus* T11	T	12.8	153.5	
	cf. *Asianopodus* T12	T	14.8	142	
	cf. *Asianopodus* T13	T	15.2	146.5	
	cf. *Asianopodus* T14	T	15.8	146	
	cf. *Asianopodus* T15	T	11.7	115.8	
	cf. *Asianopodus* T16	T	14.4	132.5	
	cf. *Asianopodus* T17	T	16	127.3	
	cf. *Asianopodus* T18	T	15.7	136	
	cf. *Asianopodus* T20	T	16.6	124.8	
	cf. *Asianopodus* T22	T	15.8	147.3	
	Theropod T23	T	17.4	186.5	
	cf. *Asianopodus* T24	T	11.6	107.5	
	cf. *Asianopodus* T27	T	12.8	110.3	
	cf. *Asianopodus* T28	T	14.3	127.2	
	cf. *Asianopodus* T29	T	12.8	115.5	
	cf. *Asianopodus* T30	T	14	120.4	
	cf. *Asianopodus* T31	T	14.8	134.6	
	cf. *Asianopodus* T32	T	14.1	132.4	
	cf. *Asianopodus* T33	T	15.5	126.9	
	cf. *Asianopodus* T34	T	12.3	109.4	
	Theropod T35	T	13.5	180.5	
	cf. *Asianopodus* T38	T	15.6	128.1	
	cf. *Asianopodus* T39	T	16.3	125	
	cf. *Asianopodus* T40	T	15	137.7	
	cf. *Asianopodus* T41	T	12.3	122.5	
	cf. *Asianopodus* T42	T	12.7	108.8	
	cf. *Asianopodus* T43	T	12.8	117.8	
	cf. *Asianopodus* T46	T	11.3	106.5	
	cf. *Asianopodus* T47	T	15.3	134.8	
	cf. *Asianopodus* T50	T	12.6	123.6	
	cf. *Asianopodus* T51	T	12.5	137.5	
	cf. *Asianopodus* T52	T	13.1	128.8	
	cf. *Asianopodus* T53	T	15.5	138.3	
	cf. *Asianopodus* T54	T	12.4	137.5	
	cf. *Asianopodus* T55	T	10.5	98	
	cf. *Asianopodus* T56	T	12.7	137	
	cf. *Asianopodus* T57	T	10.7	124	
	cf. *Asianopodus* T58	T	12.3	148	
	cf. *Asianopodus* T59	T	11	129	
	cf. *Asianopodus* T60	T	15.7	104.9	
	cf. *Asianopodus* T61	T	12.4	136.3	
	cf. *Asianopodus* T62	T	16.5	113.5	
	cf. *Asianopodus* T63	T	13	166	
	cf. *Asianopodus* T64	T	13.8	129	
	cf. *Asianopodus* T67	T	15.3	126	
	cf. *Asianopodus* T68	T	11	157	
	cf. *Asianopodus* T69	T	11.6	131.8	
	cf. *Asianopodus* T71	T	13.2	109	
	cf. *Asianopodus* T72	T	13.2	114	
	cf. *Asianopodus* T73	T	11	113.5	
	cf. *Asianopodus* T76	T	11.3	128.5	

Table A4.2. *continued*

Site	Trackway	Code	Footprint mean length (cm)	Mean stride (cm)	Source of data and comments
	cf. *Asianopodus* T78	T	12	114.8	
	cf. *Asianopodus* T79	T	14.8	127.2	
	cf. *Asianopodus* T80	T	13.8	145.8	
	cf. *Asianopodus* T81	T	12.7	141	
	cf. *Asianopodus* T82	T	11.8	123	
	cf. *Asianopodus* T83	T	16.1	144.8	
	cf. *Asianopodus* T84	T	15.2	171.5	
Gres superieurs Fm, Muong Phalane, Laos	Theropod	T	31.8	241.4	Allain et al. (1997)
	Ornithopod	O	32.7	238.0	
Tazought Fm, Imi-n-Tanout, Morocco	Theropod TML1	T	29	203	Masrour et al. (2013)
	Theropod TML2	T	30	202	
	Theropod? TML4	T	16	155	
	Theropod? TML6	T	18	172	
	Theropod? TML7	T	16	104	
	Theropod? TML8	T	16	137	
El Bayadh region, Brezina, Algeria	El Grarij theropod A	T	54.2	310	Bessedik et al. (2008)
	El Grarij theropod B	T	54.4	260	
	El Grarij theropod C	T	54.4	230	
	El Grarij theropod D	T	54.4	300	
	El Grarij theropod E	T	53.3	205	
	El Grarij theropod F	T	54.7	240	
	El Grarij theropod G	T	50.4	250	
	El Grarij theropod H	T	29.5	220	
	El Grarij theropod I	T	30.0	220	
	El Mouilah theropod	T	23.0	148	
	Abbatoirs theropod	T	66.0	291	
	Brézina (ElMezioued) theropod A	T	47.0	242	
	Brézina (El Mezioued) theropod B	T	47.0	340	
	Brézina (El Mezioued) theropod C	T	27.0	202	
	Brézina (El Mezioued) theropod D	T	47.0	310	
	Brézina (El Mezioued) theropod E	T	47.0	420	
Babouri-Figuil Basin, Cameroon	trackway B13-B7 ?ornithopod	O	27.0	132.0	Dejax et al. (1989); strides estimated from authors' figures
	Trackway C7-C10 theropod	T	26.0	170.0	
	Trackway C12-C15 theropod	T	24.5	139.0	
	Trackway C17-C19 theropod	T	24.5	132.0	
	Trackway J1-J5 ?ornithopod	O	38.0	230.0	
Koum Basin, Cameroon	Trackway 12 Class 2 theropod	T	17.0	139.0	Congleton (1990)
	Trackway 13 Class 3 theropod	T	25.0	271.0	
	Trackway 16 Class 3 theropod	T	31.0	176.0	
	Trackway 17 Class 3 theropod	T	29.0	228.0	
	Trackway 19 Class 2 theropod	T	24.0	145.0	
	Trackway 20 Class 3 theropod	T	28.0	195.0	
	Trackway 22 Class 3 theropod	T	48.0	240.0	
	Trackway 24 Class 4 ornithopod	O	34.0	186.0	
	Trackway 25 Class 4 ornithopod	O	39.0	184.0	
	Trackway 26 Class 3 theropod	T	18.0	160.0	
	Trackway 27 Class 3 theropod	T	39.0	212.0	
	Trackway 30 Class 3 theropod	T	25.0	197.0	
	Trackway 32 Class 4 ornithopod	O	46.0	225.0	
	Trackway 34 Class 3 theropod	T	30.0	205.0	
	Trackway 37 Class 4 ornithopod	O	42.0	164.0	
	Trackway 53 Class 3 theropod	T	39.0	255.0	

Site	Trackway	Code	Footprint mean length (cm)	Mean stride (cm)	Source of data and comments
Broome Sandstone, Gantheaume Point, Broome, Western Australia	*Megalosauropus broomensis*	T	32.5	194.5	Colbert and Merrilees (1967)
Eumerella Fm, Micronesia Beach, Victoria, Australia	Trackway A8-A10	T	6.8	43.5	Martin et al. (2012)
Lark Quarry, Winton Fm, Queensland	cf. *Tyrannosauropus*	UID	59.0	331.6	This study; identified as large theropod by Thulborn and Wade (1984) and Thulborn (2013), as large ornithopod by Romilio and Salisbury (2011, 2014)
Lark Quarry, Winton Fm	*Wintonopus latomorum* trackway 1	O	2.4	53.5	Thulborn (1984, 2016); Thulborn and Wade (1984). *Skartopus* was interpreted as a small theropod print by Thulborn and Wade (1984); Romilio et al. (2013) interpreted these as prints of swimming dinosaurs, and thought *Skartopus* a variant of *Wintonopus*
	Wintonopus 2	O	3.0	67.6	
	Wintonopus 3	O	3.2	65.0	
	Wintonopus 4	O	3.3	60.8	
	Wintonopus 5	O	3.5	65.3	
	Wintonopus 6	O	3.4	66.0	
	Wintonopus 7	O	3.7	73.8	
	Wintonopus 8	O	3.7	65.5	
	Wintonopus 9	O	3.7	83.5	
	Wintonopus 10	O	3.5	69.0	
	Wintonopus 11	O	3.6	82.5	
	Wintonopus 12	O	4.9	71.0	
	Wintonopus 13	O	4.5	103.7	
	Wintonopus 14	O	4.5	106.5	
	Wintonopus 15	O	4.8	117.0	
	Wintonopus 16	O	4.7	84.7	
	Wintonopus 17	O	5.3	105.0	
	Wintonopus 18	O	5.4	130.5	
	Wintonopus 19	O	5.4	111.5	
	Wintonopus 20	O	5.4	115.5	
	Wintonopus 21	O	5.2	147.0	
	Wintonopus 22	O	5.6	103.3	
	Wintonopus 23	O	6.1	107.5	
	Wintonopus 24	O	7.0	100.5	
	Wintonopus 25	O	6.0	126.8	
	Wintonopus 26	O	5.9	106.3	
	Wintonopus 27	O	6.1	115.0	
	Wintonopus 28	O	6.2	102.8	
	Wintonopus 29	O	6.3	120.4	
	Wintonopus 30	O	6.4	110.0	
	Wintonopus 31	O	6.6	118.0	
	Wintonopus 32	O	6.7	134.5	
	Wintonopus 33	O	6.6	89.0	
	Wintonopus 34	O	6.5	114.5	
	Wintonopus 36	O	7.0	117.9	
	Wintonopus 37	O	7.9	187.5	
	Wintonopus 38	O	6.9	175.0	
	Wintonopus 39	O	7.2	124.0	
	Wintonopus 40	O	7.7	132.5	
	Wintonopus 41	O	7.1	116.9	
	Wintonopus 42	O	7.8	121.3	
	Wintonopus 43	O	8.3	194.5	
	Wintonopus 44	O	8.6	103.6	
	Wintonopus 47	O	9.3	158.0	
	Wintonopus 48	O	9.1	245.0	

Table A4.2. *continued*

Site	Trackway	Code	Footprint mean length (cm)	Mean stride (cm)	Source of data and comments
Lark Quarry, Winton Fm	*Wintonopus* 49	O	9.1	214.0	Thulborn (1984, 2016); Thulborn and Wade (1984). *Skartopus* was interpreted as a small theropod print by Thulborn and Wade (1984); Romilio et al. (2013) interpreted these as prints of swimming dinosaurs, and thought *Skartopus* a variant of *Wintonopus*
	Wintonopus 50	O	9.4	195.4	
	Wintonopus 52	O	9.2	229.0	
	Wintonopus 53	O	9.7	263.0	
	Wintonopus 55	O	9.9	200.3	
	Wintonopus 56	O	10.9	249.3	
	Skartopus australis 1	O	3.6	60.0	
	Skartopus 2	O	3.6	65.8	
	Skartopus 3	O	3.9	55.0	
	Skartopus 4	O	3.9	60.4	
	Skartopus 5	O	4.0	65.3	
	Skartopus 6	O	4.0	54.7	
	Skartopus 7	O	4.0	56.3	
	Skartopus 8	O	4.1	61.7	
	Skartopus 9	O	4.1	61.3	
	Skartopus 10	O	4.1	56.0	
	Skartopus 11	O	4.2	61.5	
	Skartopus 13	O	4.2	68.7	
	Skartopus 14	O	4.2	65.0	
	Skartopus 15	O	4.3	67.5	
	Skartopus 16	O	4.3	57.0	
	Skartopus 17	O	4.4	61.0	
	Skartopus 18	O	4.4	63.7	
	Skartopus 19	O	4.4	61.0	
	Skartopus 20	O	4.6	62.1	
	Skartopus 21	O	4.7	66.0	
	Skartopus 22	O	4.7	66.3	
	Skartopus 23	O	4.7	58.0	
	Skartopus 24	O	4.8	55.7	
	Skartopus 25	O	4.8	54.0	
	Skartopus 26	O	4.8	66.0	
	Skartopus 27	O	4.8	57.3	
	Skartopus 28	O	5.0	66.7	
	Skartopus 29	O	5.0	56.0	
	Skartopus 30	O	5.1	61.5	
	Skartopus 31	O	5.2	73.3	
	Skartopus 32	O	5.2	62.0	
	Skartopus 33	O	5.6	66.3	
	Skartopus 34	O	5.6	68.3	
	Dinosaur AC1	O	3.2	64.8	
	Dinosaur AC2	O	3.7	72.4	
	Dinosaur AC3	O	3.4	59.3	
	Wintonopus 57	O	22.8	332.2	
New quarry near Lark Quarry	*Wintonopus*	O	9.2	73.0	
Areado Group, Minas Gerais, Brazil	Theropod JPAR 03–01	T	15.0	120.0	Carvalho and Kattah (1998)
Sousa Fm, Rio do Peixe, Paraiba, Brazil	Theropod SOPI1	T	26.3	141.0	Leonardi and Carvalho (in press)
	Theropod SOPM1	T	32.2	177.4	Leonardi et al. (1987*a*, 1987*b*)
	Theropod SOPM4	T	29.3	182.3	
	Sousaichnium pricei SOPP1	O	44.1	186.7	Leonardi (1979*b*, 1994); Leonardi et al. (1987*a*, 1987*b*); *Moraesichnium* footprint lengths exclude metatarsal impression
	Moraesichnium barberenae SOPP2	T	24.0	214.1	
	Moraesichnium SOPP3	T	27.0	482.8	

Site	Trackway	Code	Footprint mean length (cm)	Mean stride (cm)	Source of data and comments
Sousa Fm, Rio do Peixe, Paraiba, Brazil	SOPP6 ornithopod	O	29.9	169.5	Leonardi et al. (1987*a*, 1987*b*)
	SOPP12 theropod	T	24.2	217.5	
	SOPP13 theropod	T	26.0	169.0	
	Theropod SOPI 1	T	28.5	137.0	Leonardi (1980*a*, 1994)
	Varzea do Samos II theropod	T	30.0	180.0	Leonardi and dos Santos (2004)
	SOFP1 theropod	T	46.0	244.0	Leonardi and dos Santos (2004: fig. 22)
	SOCA 43 ornithopod	O	32.3	93.5	Leonardi et al. (1987*a*, 1987*b*)
	SOCA 44 theropod	T	30.9	201.5	
	SOCA 57 theropod	T	25.8	134.6	
	SOCA 97 theropod	T	33.7	116.7	
	SOCA 99 theropod	T	31.7	180.8	
	SOCA 991 UID dinosaur	UID	28.5	172.0	
	SOCA 993 theropod	T	31.0	212.5	
	SOCA 1323 theropod	T	11.9	103.1	
	SOCA 13236 theropod	T	19.4	120.7	
	SOCA 1331 theropod	T	31.1	261.5	
	SOCA 1350 theropod	T	23.9	178.7	
	SOCA 151 theropod	T	21.6	183.8	
	ANBD1 theropod	T	21.0	156.5	
	ANBD2 theropod	T	22.0	150.0	
	ANBD3 theropod	T	26.6	175.3	
	ANJU3 theropod	T	28.9	121.0	
	ANJU4 theropod	T	38.0	212.5	
	ANPV1 theropod	T	22.5	173.5	
	ANEN1 theropod	T	15.6	83.5	
	ANEN8 ornithopod	O	68.5	154.7	
	ANEN12 theropod	T	24.4	146.2	
	ANEN30 theropod	T	18.2	134.2	
Antenor Navarro Fm, Sousa Basin, Brazil	SOES 1 theropod	T	35.0	112.8	Leonardi et al. (1987*a*, 1987*b*)
	SOES 6 theropod	T	37.6	195.7	
	SOES 9 ornithopod *Caririchnium magnificum*?	O	51.9	257.7	
	SOES 10 theropod	T	28.8	213.4	
	SOES 11 theropod	T	30.5	197.3	
	SOES 12 theropod	T	28.8	161.7	
	SOES 17 theropod	T	38.4	445.3	
	SOES 18 theropod	T	27.7	172.2	
	SOES 22 theropod	T	25.1	124.3	
	SOSL2 theropod	T	21.9	96.0	
	SOSL5 theropod	T	20.5	90.0	
	SOSL8 ornithopod	O	68.2	158.0	
	SOSL41 theropod	T	19.0	81.5	
	SOSL42 theropod	T	25.6	128.0	
	SOSL44 theropod	T	24.7	125.0	
	ORSR3 theropod	T	42.7	311.0	Leonardi (1994); Leonardi and Spezzamonte (1992)
	ORSR4 ornithopod?	O	21.4	174.0	
	ORSR5 bipedal dinosaur	UID	10.2	66.0	
	ORSR6 bipedal dinosaur	UID	11.0	69.0	
	ORSR9 ornithopod	O	28.0	194.5	Leonardi and Spezzamonte (1992)
	ORSR12 bipedal dinosaur	UID	56.0	198.0	

Table A4.2. *continued*

Site	Trackway	Code	Footprint mean length (cm)	Mean stride (cm)	Source of data and comments
Yacoraite Fm, La Escalera, San Carlos, Salta, Argentina	*Hadrosaurichnus australis* trackway 1	O	37.0	198.0	Alonso (1989)
	Hadrosaurichnus trackway 2	O	34.0	200.0	
	Hadrosaurichnus trackway 3	O	30.0	193.0	
	Hadrosaurichnus trackway 4	O	25.0	150.0	
	Hadrosaurichnus trackway 5	O	27.0	170.0	
	Hadrosaurichnus trackway 6	O	27.0	120.0	
	Hadrosaurichnus trackway 7	O	19.0	120.0	
Yacoraite Fm, Quebrada del Tapon, Argentina	Theropod? trackway 1	T	40.0	164.0	Alonso (1989)
Playa Azul, Michoacan, Mexico	Theropod morfotipo F	T	30.5	260.0	Ferrusquía-Villafranca et al. (1978); Bravo Cuevas and Rodríguez-de la Rosa (2014)
	Theropod morfotipo G	T	16.5	98.0	
	Theropod morfotipo B	T	23.9	190.0	
Rio Magdalena, Puebla Mexico San Juan Raya Fm	Theropod ATRM-A	T	30.0	185.0	Rodríguez-de la Rosa et al. (2012)
	Theropod ATRM-B	T	29.3	179.0	
Cerro del Pueblo Fm, Las Aguilas, Coahuila, Mexico	Quadrupedal hadrosaur	O	38.6	225.0	Rodríguez-de la Rosa (2007)
Raritan Fm, Woodbridge, New Jersey	Theropod RUGM V/226	T	43.0	350.0	Baird (1989: text and fig. 6)
F6 Ranch, Texas, Fort Terrett Fm	CHHB–G48	T	47.5	322.6	Farlow (1981, 1987)
	E23–HOPE	T	36.5	269.1	
	Q94–Q98	T	29.0	565.6	
	BLV–A3	T	36.7	538.5	
	18D–PVW1	T	40.0	296.3	
	9B–JOF9	T	42.0	337.8	
	C17–RW	T	34.0	259.1	
	N77–MKR32	T	38.3	288.3	
	M73–M66	T	45.0	299.4	
	F90–P92	T	36.5	260.3	
	86–082	T	38.2	658.9	
	H52A–H53	T	37.5	322.6	
	J60–J64	T	37.1	214.1	
Nichols Site, Glen Rose Fm, Texas		T	33.9	211.7	Farlow (1987); cf. Farlow et al. (2012, 2015)
Ising Site, Glen Rose Fm, Texas	Small dinosaur	T	25.7	192.5	
Schmidt Site, Glen Rose Fm, Texas	"Lettered" trail	T	32.0	310.4	
Schmidt Site, Glen Rose Fm, Texas	19–22	T	47.0	346.0	
Crider Site, Glen Rose Fm, Texas	Crider 1–3	T	40.3	710.0	
	Crider 29–31	T	45.3	277.0	
Davenport Site, Glen Rose Fm, Texas	Davenport 1A–12	T	49.9	314.0	
Huckabay Site, Glen Rose Fm, Texas	TD1B–TD1E	T	45.7	306.0	
	TD2A–TD2C	T	51.3	315.0	
Rowlett Site, Glen Rose Fm, Sandy Creek, Leander, Travis Co, Texas	Sandy Creek trackway	T	47.0	260.3	
Glen Rose Fm, Branch of Live Oak Creek, Red Shed Ranch, Kinney County, Texas	Theropod	T	49.0	319.0	
Glen Rose Fm, Dinosaur Valley State Park (DVSP) Main Tracksite, Paluxy River, Glen Rose Fm, Texas	Prints 1–4	T	50.5	372.7	
DVSP Overlook, Main Tracklayer	Prints 1–"6"	T	49.5	276.4	

Site	Trackway	Code	Footprint mean length (cm)	Mean stride (cm)	Source of data and comments
DVSP Denio Branch Mouth, Main Tracklayer	Trail 1	T	46.3	273.5	
	Trail 2	T	43.8	245.3	
	Trail 4	T	48.0	297.0	
DVSP AMNH Bird Slab	Bird Slab carnosaur	T	53.7	320.5	
DVSP Park East site trackway	Park East theropod	T	51.3	354.6	
McFall Site, Paluxy River, *Diplocraterion* Bed, Glen Rose Fm	*Diplocraterion* bed trackway	T	40.5	253.0	This study; cf. Dattilo et al. (2014)
McFall Site, Paluxy River, Taylor Tracklayer	Trackway A	T	48.5	278.3	
	Trackway B	T	46.5	262.3	
	Trackway C	T	49.6	301.3	
	Trackway E	T	47.5	201.7	
	Trackway H	T	25.0	169.0	
	Trackway X	T	47.3	260.5	
	McFall upstream A	UID	53.0	228.7	
	McFall upstream B	UID	47.0	424.0	
DVSP Dattilo Station 754, Taylor Tracklayer, Paluxy River	Small dinosaur	T	31.0	233.0	
DVSP Main Tracklayer Highline, Paluxy River	Highline trackway	T	37.0	332.0	
Jeannie Mack Site, Taylor Tracklayer, Paluxy River	Dattilo 764 cluster A 9-9B	T	49.7	293.0	
	Dattilo 764 cluster B dinosaur 11	T	49.4	236.5	
	Dattilo 764 cluster C dinosaur 14	T	35.4	234.0	
	Dattilo 764 cluster D dinosaur 15	T	45.5	261.0	
	Dattilo 764 cluster E 18	T	49.5	264.0	
Bear Creek, Glen Rose Fm, Texas	Theropod	T	47.3	281.9	Farlow (1987)
South San Gabriel, Glen Rose Fm, Texas	Theropod	T	45.1	286.5	
Dry Cedar Creek, Glen Rose Fm, Texas	Trail A	T	34.0	218.0	
Girvin, Glen Rose Fm, Texas	Theropod	T	53.0	342.0	
Little Jimmys Creek, Glen Rose Fm (?), Texas	Big ornithopod	O	75.0	350.0	Farlow and Hawthorne (1989); this study
Glen Rose Fm, Branch of Live Oak Creek, Kinney County, Texas	Trackway 1	T	47.4	322.0	Rogers (2002)
	Trackway 2	T	56.3	322.0	
Glen Rose Fm, Boerne Lake Spillway, Texas	Theropod OA	T	49	242	Hawthorne et al.(2002)
Zilker Park, Austin, Edwards Limestone	A	T	23.5	255.6	This study
	B	T	26.8	241.3	
	C	T	29.0	277.0	
	D	T	26.0	165.5	
	E	T	23.2	173.5	
	F	T	25.0	140.5	
	G	T	24.3	243.0	
	H	T	23.7	170.7	
Heritage Museum (HMTHC), Glen Rose Fm, Canyon Lake, Texas	2001 01B	UID	52.3	243.0	Farlow et al. (2006)
	2001 01C	UID	52.0	318.5	
	2001 01A	T	37.1	236.0	
	2001 01E	O	55.2	281.0	
	2001 01F	UID	45.3	244.5	
	2001 B	UID	50.3	322.7	
	2001 JOF12	UID	44.5	260.5	
	2001 C	O	55.8	258.6	

Table A4.2. *continued*

Site	Trackway	Code	Footprint mean length (cm)	Mean stride (cm)	Source of data and comments
Heritage Museum (HMTHC), Glen Rose Fm, Canyon Lake, Texas	2001 F	O	48.8	265.4	Farlow et al. (2006)
	2001 J	UID	46.3	303.0	
	2001 I	UID	48.4	317.0	
	2001 K	UID	52.2	322.0	
	2001 H	UID	42.4	273.3	
	2001 G	T	54.2	329.0	
	2001 01K	UID	55.0	498.5	
	2001 D	UID	41.8	300.0	
	2001 O	T	45.8	287.4	
	2001 Q	UID	50.2	273.3	
	2001 R	UID	53.0	328.0	
	2001 T	UID	46.0	262.5	
	2001 01D	UID	40.7	236.5	
	2001 01G	UID	45.0	290.0	
	2001 01H	UID	49.5	276.0	
	2001 01J	T	44.7	311.0	
	2001 01L	O	54.0	298.0	
	2001 01O	UID	51.0	295.0	
	Mountain Creek (MC) 2000 trail	UID	49.3	241.5	
	MC SB	UID	44.7	245.0	
	MC SH	O	44.0	237.5	
	MC SI	UID	33.9	201.6	
	MC SJ	UID	41.5	229.0	
	MC SO	UID	40.2	258.3	
	MC SQ	UID	39.0	246.0	
De Queen Fm, Howard County, Arkansas	Theropod T1	T	56	285	Shell and Boss (2013); measurements estimated from their fig. 9
Woodbine Fm, Grapevine Lake, Texas	Magnoavipes lowei trackway 1	1	20.5	208.0	Lee (1997)
	Caririchnium protohadrosaurichnus trackway 3	O	40.2	169.0	
	Caririchnium trackway 4 quadrupedal	O	32.6	125.0	
	Caririchnium trackway 5	O	41.3	255.0	
	Caririchnium trackway 6	O	45.8	245.0	
	Caririchnium trackway 7	O	46.6	245.0	
	Caririchnium trackway 8	O	38.8	232.0	
Lakota Fm, Grace Coolidge Creek Site, Hermosa, South Dakota	Quadrupedal ornithopod	O	40.0	196.0	Lockley et al. (2001*a*: fig 30.4)
Cedar Mountain Fm, Arches National Park	Theropod trackway 1 CU-MWC 199.14	T	27.0	175.0	Lockley et al. (2004*a*)
Mill Canyon, Cedar Mountain Fm, Utah	T1 *Irenesauripus* UCM 199.74	T	55.0	312.0	
	T2 *Carmelopodus*-like	T	29.0	262.0	
	D2 *Dromaeosauripus* UCM 199.82	T	22.0	169.0	
	O1 *Caririchnium* UCM 199.69+73	O	39.0	226.0	
	O2 *Caririchnium*	O	42.0	264.0	
Gething Fm, Peace River, British Columbia	*Amblydactylus* sp. IIIa-B	O	38.2	209.1	P. J. Currie (1983) and personal communication; Harington et al. (2005)
	Amblydactylus sp. IIIe-A	O	43.0	247.0	
	Amblydactylus sp. IIIe-C	O	26.5	143.7	
	Amblydactylus sp. IV-A	O	54.3	225.1	
	Amblydactylus sp. IV-F	O	51.0	232.3	
	Amblydactylus sp. V-A	O	55.0	241.3	
	Amblydactylus sp. V-B	O	58.0	232.0	
	Amblydactylus sp. V-C	O	54.7	258.6	

Site	Trackway	Code	Footprint mean length (cm)	Mean stride (cm)	Source of data and comments
Gething Fm, Peace River, British Columbia	*Amblydactylus* sp. V-E	O	42.0	264.5	P. J. Currie (1983) and personal communication; Harington et al. (2005)
	Amblydactylus sp. VIII-B	O	62.8	221.8	
	Amblydactylus sp. VIII-D	O	43.0	214.3	
	Amblydactylus sp. VIII-F	O	41.3	167.0	
	Amblydactylus sp. VIII-N	O	50.0	199.5	
	Amblydactylus sp. XIII-B	O	43.4	214.0	
	Amblydactylus sp. XIV-A	O	59.4	283.5	
	Amblydactylus sp. XIV-B	O	73.0	346.0	
	?*Gypsichnites* III, K	O	28.3	173.7	
	cf. *Gypsichnites pacensis* III, P	O	24.6	172.4	
	cf. *Gypsichnites pacensis* III, O	O	29.5	237.8	
	cf. *Gypsichnites pacensis* III, S	O	29.2	197.9	
	cf. *Gypsichnites pacensis* III, T	O	29.4	205.5	
	Amblydactylus sp. III-J	O	14.0	69.7	
	Amblydactylus sp. IIIa-C	O	51.3	232.0	
	Amblydactylus sp. IIIa-E	O	56.0	218.0	
	Amblydactylus sp. IIId-B	O	46.6	252.6	
	Amblydactylus sp. IIId-C	O	50.5	225.0	
	Amblydactylus sp. VI-A	O	53.0	279.0	
	Amblydactylus sp. VI-B	O	46.6	172.7	
	Amblydactylus sp. VIa-A	O	41.0	217.0	
	Amblydactylus sp. VIa-B	O	55.7	263.3	
	Amblydactylus sp. IX-D	O	42.5	209.0	
	Amblydactylus sp. XII-A	O	64.0	243.3	
	Amblydactylus sp. (*Irenesauripus occidentalis* NMC 8550)	O	46.4	201.5	
	Amblydactylus sp. IIIc-D	O	27.5	173.8	
	Amblydactylus sp. IIIe-D	O	37.1	180.4	
	Irenesauripus mclearni III-M	T	36.5	205.2	
	Irenesauripus mclearni III-N	T	32.6	266.0	
	Irenesauripus mclearni III-Q	T	28.8	317.5	
	Irenesauripus mclearni III-R	T	33.4	256.0	
	Irenesauripus mclearni III-V	T	33.9	240.1	
	Irenesauripus mclearni III-W	T	33.4	194.6	
	Irenesauripus mclearni IIIc	T	30.0	175.0	
	Irenesauripus mclearni IIIc-A	T	29.1	216.3	
	Irenesauripus mclearni IIIc-B	T	33.3	188.3	
	Irenesauripus mclearni IIIc-F	T	26.6	181.7	
	Irenesauripus mclearni IIIc-G	T	19.7	192.3	
	Irenesauripus mclearni IX-B	T	23.7	128.0	
	Irenesauripus mclearni IX-C	T	31.0	215.6	
	Irenesauripus mclearni NMC 8561	T	28.1	193.6	
	Irenesauripus mclearni NMC 8562	T	29.7	210.2	
	Irenesauripus mclearni NMC 8563	T	25.7	169.0	
	Irenesauripus acutus III-L	T	41.9	239.2	
	Irenesauripus acutus III-DD	T	37.0	273.0	
	Irenesauripus acutus IIId-A	T	51.8	294.5	
	Irenesauripus acutus V-D	T	41.6	276.0	
	Irenesauripus acutus VIII-J	T	50.8	273.8	
	Irenesauripus acutus VIII-K	T	55.1	337.8	
	Irenesauripus acutus VIII-O	T	53.8	275.4	
	Irenesauripus acutus VIII-P	T	56.0	334.0	
	Irenesauripus acutus VIII-L	T	39.4	256.8	

Table A4.2. *continued*

Site	Trackway	Code	Footprint mean length (cm)	Mean stride (cm)	Source of data and comments
Gething Fm, Peace River, British Columbia	*Irenesauripus mclearni* III-I	T	31.8	199.1	P. J. Currie (1983) and personal communication; Harington et al. (2005)
	cf. *Irenesauripus mclearni* III-AA	T	27.4	193.6	
	Irenesauripus mclearni III-BB	T	36.5	249.0	
	Irenesauripus mclearni III-EE	T	27.5	240.0	
	Irenichnites gracilis NMC 8552	T	15.6	143.2	
	Irenichnites gracilis III-Z	T	14.7	116.9	
	Irenichnites gracilis III-CC	T	15.7	125.5	
	Irenichnites gracilis IIIc-E	T	16.0	139.9	
	cf. *Columbisauripus* III-B	T	30.7	218.2	
	cf. *Columbisauripus* III-D	T	25.9	197.0	
	?Theropod III-C	T	34.3	273.5	
	?Theropod III-E	T	28.5	208.0	
	?Theropod III-F	T	29.2	228.0	
	?Theropod III-H	T	29.3	203.5	
	?Theropod III-X	T	26.8	160.6	
	?Theropod VIIIa-U	T	33.2	201.1	
	?Theropod VIIIa-V	T	19.4	122.2	
	?Theropod VIIIa-W	T	24.7	162.4	
	?Theropod VIIIa-X	T	26.7	170.2	
	?Theropod VIIIa-Y	T	31.0	147.2	
	?Theropod VIIIa-Z	T	30.0	203.3	
	?Theropod IX-A	T	31.5	212.0	
	Irenesauripus mclearni CMN 8548	T	28.2	128.0	
9-Sting Site, Gething Fm, British Columbia	Trackway B	T	16.4	94.5	R. McCrea (pers. comm.)
W3 Main Site, Smoky River Coal Mine, British Columbia	*Irenesauripus mclearni* D1-I9	T	47.9	305.7	R. McCrea (pers. (comm.)
	Irenesauripus mclearni G6-bb6	T	45.1	302.6	
	Columbosauripus ungulatus cc16-gg15	T	26.5	264.0	
	Columbosauripus ungulatus A10-aa9	T	23.8	201.5	
	Columbosauripus ungulatus B27-D27	T	23.0	208.3	
	Columbosauripus ungulatus C26-C28	T	26.2	236.5	
	Columbosauripus ungulatus G23-E23	T	26.3	217.0	
	Gypsichnites pacensis aa5-E8	O	20.7	186.1	
	Gypsichnites pacensis cc8-E2	O	19.5	247.7	
	Gypsichnites pacensis hh11-ii13	O	24.5	219.1	
	Gypsichnites pacensis B4-aa8	O	25.5	206.2	
	Gypsichnites pacensis D9-C1	O	25.1	211.4	
	Gypsichnites pacensis E26-F24	O	18.5	227.0	
	Gypsichnites pacensis F2-aa4	O	23.8	225.0	
	Gypsichnites pacensis G22-D18	O	20.8	230.4	
	Gypsichnites pacensis H11-bb2	O	27.1	236.8	
	Gypsichnites pacensis J8-E26	O	23.0	195.1	
	Irenichnites gracilis aa21-C18	T	15.1	108.4	
	Irenichnites gracilis bb20-B22	T	17.5	149.8	
Fording River Coal Mine, Elk Valley Coal region, British Columbia; Mist Mountain Fm	*Neoanomoepus peregrinatus* trackway A	O	11.0	55.0	Lockley et al. (2009)
	Neoanomoepus trackway B	O	12.8	51.3	
	Neoanomoepus trackway C	O	15.3	42.0	
Barrett Wall tracksite, Gates Fm, Grande Cache, Alberta	*Irenesauripus mclearni* TMP 2001.35.01	T	47	269	McCrea et al. (2014*b*: fig. 49)
Kakwa Provincial Park, Alberta, Gorman Creek Fm	?Ornithopod	O	17.1	92.1	R. McCrea (pers. comm.)
Ross River Acess Road, Yukon Territory	*Amblydactylus kortmeyeri* F-5 to F-8	O	33.0	131.0	Gangloff et al. (2004)

Site	Trackway	Code	Footprint mean length (cm)	Mean stride (cm)	Source of data and comments
Mesaverde Fm, Colorado	AMNH 3650 big hadrosaur	O	86.0	457.0	Russell and Béland (1976); Thulborn (1981); Lockley et al. (1983)
Mesaverde Fm	"Dinosauropodes magrawii"# 1	O	85.0	366.0	Strevell (1932); F. L. DeCourten (pers. comm.); print length excludes MT impression
	"Dinosauropodes wilsonii"	O	52.0	376.0	
	"Dinosauropodes magrawii" # 2	O	63.0	310.0	
Wapiti Fm, British Columbia	*Bellatoripes fredlundi* PRPRC 2011.11.001 trackway A	T	57.9	346.0	McCrea et al. (2014*a*)
St. Mary River Fm, southern Alberta	*Ornithichnites* sp.	T	26	190	McCrea et al. (2014*b*: fig. 98)
Zerbst Ranch, Lance Fm, Wyoming	CU-MWC 224.3 ?ornithomimid morphotype 3	T	29.0	170.0	Lockley et al. (2003*b*)
Dakota Group, Lakes Quarry, Rule Canyon, El Paso County, Colorado	Lakes ornithopod trackway	O	48	170	Lockley et al. (2014*h*)
	New ornithopod trackway	O	40.5	190	
Dakota Group, Stone City, Fort Carson, Colorado	*Caririchnium* trackway	O	21.1	194.8	Lockley et al. (2014*e*)
Dakota Group, Dinosaur Ridge, Colorado	*Caririchnium leonardii* Alameda trackway A	O	34.0	156.0	Lockley (1987)
	Caririchnium Alameda trackway B	O	48.0	200.0	
	Caririchnium Alameda trackway C	O	43.0	204.0	
	Caririchnium N. Alameda trackway Nth 1	O	52.0	210.0	
	Caririchnium N. Alameda trackway Nth 2	O	36.0	158.0	
Dakota Group, Lamar, Colorado	*Caririchnium* DMNH 1608	O	41.0	204.0	
Dakota Group, Dinosaur Ridge	Markman 1 Alameda ?carnosaur	T	48.0	192.0	
	Markman 2 Alameda coelurosaur	T	25.0	214.0	
	small ornithopod	O	16.4	230.0	Lockley (2001)
	Magnoavipes caneeri trackway 1	T	21.4	277.3	Lockley et al. (2001*b*)
	Magnoavipes trackway 2	T	18.8	204.5	
	Magnoavipes trackway 3	T	18.3	184.0	
	Magnoavipes trackway 11	T	19.0	221.0	
Dakota Group, Turkey Creek, Jefferson County, Colorado	Ornithopod A my label	O	41.0	120.0	Lockley et al. (1992: fig. 5)
	Ornithopod B my label	O	42.0	147.0	
	Theropod C my label	T	17.0	93.0	
Dakota Group, Roxborough State Park, Douglas County, Colorado	*Caririchnium*	O	36.0	132.0	Lockley et al. (1992: fig. 6); Lockley et al. (2014*f*)
	Theropod	T	34.0	153.0	
	Caririchnium T1	O	30	141	Lockley et al. (2014*f*: fig. 4)
	Willow Creek locality L-00694 *Caririchnium*	O	28	125	Lockley et al. 2014*f* (fig. 5)
	Willow Creek locality L-00694 *Caririchnium*	O	29	127	
	Willow Creek locality L-00694 *Caririchnium*	O	25	110	
Dakota Group, Apshapa, Las Animas County, Colorado	Ornithopod T1	O	32.0	135.0	Lockley et al. (1992: fig. 7)
	Ornithopod T2	O	30.0	149.0	
	Ornithopod T3	O	47.0	175.0	
Dakota Group, Sand Canyon, Baca County, Colorado	Ornithopod	O	26.0	167.0	Lockley et al. (1992: fig. 10)
Dakota Group, Mosquero Creek, New Mexico	*Caririchnium leonardi* Group A trackway 19	O	21.4	126.2	Matsukawa et al. (2001); M. Matsukawa (pers. comm.); cf. Cotton et al. (1998)
	Caririchnium Group A trackway 26	O	20.8	124.5	
	Caririchnium Group A trackway 28	O	19.4	105.8	
	Caririchnium Group A trackway 29	O	19.2	106.6	
	Caririchnium Group A trackway 31	O	19.2	110.0	
	Caririchnium Group A trackway 32	O	18.3	97.7	
	Caririchnium Group A trackway 33	O	19.1	111.5	
	Caririchnium Group A trackway 35	O	21.0	64.0	

Table A4.2. *continued*

Site	Trackway	Code	Footprint mean length (cm)	Mean stride (cm)	Source of data and comments
Dakota Group, Mosquero Creek, New Mexico	*Caririchnium* Group A trackway 47	O	23.0	105.0	Matsukawa et al. (2001); M. Matsukawa (pers. comm.); cf. Cotton et al. (1998)
	Caririchnium Group A trackway 56	O	22.5	88.0	
	Caririchnium Group B trackway 3	O	22.7	105.3	
	Caririchnium Group B trackway 8	O	20.5	135.0	
	Caririchnium Group B trackway 11	O	20.2	132.0	
	Caririchnium Group B trackway 12	O	22.5	124.5	
	Caririchnium Group B trackway 17	O	22.3	128.7	
	Caririchnium Group B trackway 18	O	23.1	87.6	
	Caririchnium Group B trackway 20	O	17.0	101.0	
	Caririchnium Group B trackway 27	O	19.1	113.7	
	Caririchnium Group C trackway 7	O	20.5	108.3	
	Caririchnium Group C trackway 21	O	20.8	117.0	
	Caririchnium Group C trackway 46	O	22.0	119.0	
	Caririchnium Group C trackway 48	O	21.7	119.5	
	Caririchnium Group C trackway 50	O	20.3	52.0	
	Caririchnium Group C trackway 54	O	22.0	40.0	
	Caririchnium Group C trackway 58	O	22.8	124.5	
	Caririchnium Group C trackway 59	O	23.3	117.0	
	Caririchnium Group C trackway 60	O	19.8	105.5	
	Caririchnium Group C trackway 61	O	20.5	111.0	
	Caririchnium Group D trackway 13	O	23.2	97.0	
	Caririchnium Group D trackway 16	O	23.1	98.7	
	Caririchnium Group E trackway 23	O	28.0	180.0	
	Caririchnium Group E trackway 30	O	37.5	192.7	
	Caririchnium Group F trackway 37	O	37.5	193.0	
Dakota Group, Chatfield Site 1	*Caririchnium*	O	45.0	203.0	Lockley et al. (2006c: figs. 5, 6*D*)
Dakota Group, Witt Ranch, NE New Mexico	*Caririchnium* trackway 2	O	28.0	167.0	Lockley et al. (2006*c*)
	Caririchnium trackway 3	O	47.0	190.0	
	Theropod trackway 1	T	50.0	370.0	
Dakota Group, Dominguez-Escalante National Conservation Area, Colorado	*Irenesauripus* trackway T1	T	29.0	190.0	Lockley et al. (2014*c*, 2014*d*)
	Banner Road ornithopod trackway T2	O	43	160	Lockley et al. (2014*a*: fig. 3*A*)
Moore tracksite, Ferron Sandstone, Utah	Quadrupedal ornithopod	O	20.0	77.0	Gierliński and Lockley (2013: fig. 24.2 and text)
Oldman Fm, Milk River Natural Area Alberta	Hadrosaur trackway	O	58	268	Therrien et al. (2014)
Laramie Fm, Triceratops Trail	UCM 220.29 theropod morphotype A	T	9.5	82	Lockley and Tempel (2014: fig. 8*A*)
	UCM 220.29 theropod morphotype A	T	32	157	Lockley and Tempel (2014: fig. 8*B*)
Hell Creek Fm, Harding County, South Dakota	*Wakinyantanka styxi*	T	57.5	291	Lockley et al. (2014*i*); print length excludes metatarsal impression

Table A5.1. Comparison of lengths of horny claw sheaths of feet with the lengths of their underlying bony unguals. Such comparisons are possible because the claw sheaths of many specimens are transparent enough to see the tip of the ungual beneath the claw sheath. Specimens include skeletons, spirit-preserved specimens, and study skins. Mostly qualitative observations are reported for crocodylians and some birds; for these specimens, only a single entry of comments follows the cell identifying the specimen. For several bird specimens it was possible to compare the length of the claw sheath (measured from the joint of the underlying ungual to the tip of the claw) relative to the length of the underlying ungual (measured from the joint to the tip of the ungual); these data are reported as length of horny claw (ratio of claw length/underlying ungual length).

Specimen	Digit	Ungual length (mm)	Sheath length (mm); sheath length/ungual length
Crocodylians			
Alligator mississippiensis SREL captive XVI (Stump's baby in animal care facility); total length 47 cm	Unguals extend nearly to tips of horny claws	10.99	10.99
Alligator mississippiensis SREL captive XIX (in cattle tank 1 in experimental pond C); total length 55 cm	Unguals reach about ¾ of the way to the tips of the horny claws		
Alligator mississippiensis SREL captive XX (in cattle tank 1 in experimental pond C); total length 69 cm	Unguals extend from about ¾ of the way to the very tips of the horny claws		
Alligator mississippiensis SREL captive XXII; total length 67 cm	Unguals extend at least ¾ of the way to the tips of the horny claws		
Alligator mississippiensis SREL captive XXIII; total length 69 cm	Unguals extend almost to the tips of the horny claws		
Alligator mississippiensis USNM 211228	Tips of the ventral surfaces of the horny pedal claws are broken (?) away, revealing that the unguals extend nearly to the claw tips		
Alligator mississippiensis USNM 313409	The sharp, transparent claws on the foot of this individual extend a few mm beyond the tips of the unguals; ungual lengths ~ 8–10 mm long		
Alligator mississippiensis UF 115605	Digit I: sheath is transparent, and extends only a bit beyond tip of ungual		
	Digit II: sheath is transparent, and extends ~ 3 mm beyond tip of 11-mm ungual		
Alligator mississippiensis CITES 0041015	Claws are abraded or split at their tips, with unguals extending as far as edges of claw tips		
Alligator mississippiensis CITES 0056495	The claw tip of digit I is heavily worn and beveled; the ungual itself has been worn down, with the horny claw material around the worn terminal end of the ungual removed		
Caiman crocodilus UMMZ HERPS 130547	Claw sheaths transparent; unguals go nearly to tips of claws		
Caiman crocodilus UMMZ HERPS 134962	Claw sheath is present but transparent on digit I; ungual extends out to the tip of the sheath		
Caiman crocodilus FMNH 98960	Unguals of I1 and II2 extend nearly to tips of horny claw sheaths		
Paleosuchus trigonatus UMMZ HERPS 129764	Ungual of III4 extends right to tip of horny claw		
Paleosuchus trigonatus USNM 300660	Tip of claw of pedal digit I abraded and beveled; ungual extends to end of abraded claw		
Paleosuchus trigonatus USNM 302052	Claw tips are very abraded; unguals extend to tips of worn claws		
Paleosuchus trigonatus UF 56316	Claw sheath II3 is transparent; ungual extends nearly to tip of sheath		
Crocodylus acutus UF 99202	Claw sheaths transparent, and sheaths project very little beyond tips of unguals		
Crocodylus mindorensis USNM 252669	Claw tips are fairly worn; unguals go out nearly to claw tips		
Crocodylus mindorensis USNM 252670	Claw tips are worn, and unguals extend about to the ends of the claw tips		
Osteolaemus tetraspis UF 99204	Claw tips are abraded and extend only a little beyond tips of unguals		
Osteolaemus tetraspis UNSM ZM-15431	Horny claw of digit I is abraded and beveled; underlying ungual extends to abraded tip of claw. Claw on digit III of both feet is transparent; ungual extends to or close to tip of horny claw		

Table A5.1. *continued*

Specimen	Digit	Ungual length (mm)	Sheath length (mm); sheath length/ungual length
Tomistoma schlegelii UMMZ HERPS 128552	Claw sheaths transparent; unguals go nearly to tips of claws		
Tomistoma schlegelii UMMZ HERPS 174416	Claw sheaths transparent; unguals extend to the tips of the sheaths		
Gavialis gangeticus FMNH 82681	Claw sheaths nearly transparent; unguals extend nearly to tips of the claws. However, claws are heavily worn and beveled on ventral surfaces; possibly unworn claw sheaths would have projected further forward beyond tips of unguals		
Birds			
Archaeopteryx lithographica HMN 1880/81 Berlin specimen	III	7.8	11.1 (1.4)
	IV	6.5	8.6 (1.3)
	III	6.3	9.6 (1.5)
	IV	4.9	9.9 (2.0)
Struthionidae			
Struthio camelus Amherst College	Sheath is preserved on right digit III. Its dorsal edge passes proximally beyond the joint to end at about the proximal edge of the distal condyle of phalanx III3; thus the sheath extends almost halfway back along the length of III3. On the ventral edge of the sheath, in contrast, it doesn't get back as far as the joint between phalanges III3 and III4. The tip of the sheath is heavily worn, but even so it seems to extend twice the length of the bony ungual beyond the joint with III3		
Rheidae			
Rhea americana FMNH 101870	Horny claws are transparent; unguals reach nearly to the tip of the horny claw		
Dromaiidae			
Dromaius novaehollandiae FMNH 313620	II	29	31 (1.1)
	III	36	40 (1.1)
	IV	27	29 (1.1)
Dromaius novaehollandiae QM O12657	III	33	37 (1.1)
	IV	25	27 (1.1)
Casuariidae			
Casuarius casuarius FMNH 314889	II	61	85 (1.4)
	III	39	43 (1.1)
	IV	30	35 (1.2)
Apterygidae			
Apteryx australis FMNH 88262	Horny claws are transparent, and horny claw sheath extends about 4–6 mm beyond tips of unguals; claw lengths of digits II–IV range 14–18 mm (including sheath)		
Apteryx mantelli NMNZ S.000951	Sheath would have extended ~ 2 mm beyond the tip of the 13-mm long ungual of digit III		
Apteryx mantelli NMNH S540925 (spirit-preserved specimen)	Horny claw sheaths extend ~ 2 mm beyond tips of unguals II–IV; lengths of claws including sheaths are 10–13 mm		
Apteryx owenii NMNZ OR.023036	II	20	22 (1.1)
	III	22	24 (1.1)
	IV	17	19 (1.1)
Tinamidae			
Tinamus major FMNH Conover 12370	Horny claws are transparent; unguals seem to extend nearly to the tip of the horny claw		
Eudromia formosa FMNH Conover 17185	Horny claws are transparent; unguals extend nearly to tips of their sheaths		

Specimen	Digit	Ungual length (mm)	Sheath length (mm); sheath length/ungual length
Galliforms			
Bonasa umbellus UMA 1502	III	6.5	11.7 (1.8)
Bonasa umbellus UMA 3928	IV	4.3	6.9 (1.6)
Centrocercus urophasianus USNM 17974	II	10	16 (1.6)
Centrocercus urophasianus USNM 17985	II	7.5	12 (1.6)
Dendragapus canadensis USNM 288088	II	6	10.5 (1.7)
	III	7	13 (1.9)
	IV	6	10 (1.7)
Lagopus leucurus USNM 289726	II	5	7 (1.4)
Alectoris graeca MSU OR.3561	III	8	13 (1.6)
	IV	6.6	10.5 (1.6)
Francolinus afer AMNH SKEL 5050	Sheaths of claws are transparent, and underlying unguals are almost as long as sheaths		
Ithaginis cruentis AMNH SKEL 21985	II	9	13 (1.4)
	III	9	14 (1.6)
	IV	8	12 (1.5)
Lophura bulweri USNM 491472	II	10	8 (1.3)
Crossoptilon auritum YPM ORN 102045	III	16.5	19.2 (1.2)
Crossoptilon crossoptilon AMNH SKEL 11360	III	16	19 (1.2)
Phasianus colchicus UMMZ Birds 99579	III	9.3	11.6 (1.2)
Acryllium vulturinum AMNH SKEL 2567	IV	8	12 (1.5)
Cariamidae			
Cariama cristata uncatalogued Berkeley (J.R. Hutchinson) specimen	I	9	13 (1.4)
	II	14	17 (1.2)
	III	14	17 (1.2)
	IV	10	13 (1.3)
Cariama cristata AMNH SKEL 1722	III	12	16 (1.3)
Otididae			
Ardeotis kori FMNH 338449	III	19	20 (1.1)
Otis tarda several FMNH study skins	In many of these birds, the horny sheath is very little longer than the underlying ungual		

Table A5.2. Measurements of alligators. All measurements are in millimeters. TL = total length; SVL = snout-vent length (measured to anterior edge of cloaca); SHL = shoulder-hip length; Tail = tail length; UL = upper leg length; LL = lower leg length; Pes = overall pes length; IPd1L = length of first digital pad of digit I; IPd2L = length of second digital pad of digit I; IClawL = length of digit I claw; DigIL = digit I length; IHyp = digit I lateral hypex length; IIFreeNC = digit II free length excluding the claw; IIClawL = digit II claw length; IIMed = digit II medial hypex length; IILat = digit II lateral hypex length; IIIFreeNC = digit III free length excluding the claw; IIIClawL = digit III claw length; IIIMed = digit III medial hypex length; IIILat = digit III lateral hypex length; IVHyp = digit IV medial hypex length. See chapter 5 text for details of measurements. Claw lengths marked with an asterisk (*) indicate that the length of the claw was severely truncated by tip wear; note that worn claws will also reduce hypex lengths that include that claw. The sex of the animal is indicated, where known. SREL = Savannah River Ecology Laboratory (Aiken, SC); RWR = Rockefeller Wildlife Refuge (Grand Chenier, LA); CITES = central Florida alligator.

SREL I: female born in captivity of South Carolina stock	TL: 1,050	SVL: 490	SHL: 238	Tail: 560	
	UL: 54	LL: 66	PL: 88		
	IPd1L: 9.5	IPd2L: 10	IClawL: 12	DigIL: 32	IHyp: 22
	IIFreeNC: 40	IIClawL: 15	IIMed: 36	IILat: 34	
	IIIFreeNC: 47	IIIClawL: 10	IIIMed: 36	IIILat: 36	IVHyp: 21
SREL II: male born in captivity of South Carolina stock	TL:———	SVL: 570	SHL: 279	Tail:———	
	UL: 64	LL: 79	PL: 108		
	IPd1L: 13	IPd2L: 13	IClawL: 14	DigIL: 40	IHyp: 28
	IIFreeNC: 48	IIClawL: 14	IIMed: 43	IILat: 39	
	IIIFreeNC: 50	IIIClawL: 14	IIIMed: 46	IIILat: 45	IVHyp: 31
SREL III: female born in captivity of South Carolina stock	TL: 1,150	SVL: 560	SHL: 277	Tail: 590	
	UL: 56	LL: 72	PL: 99		
	IPd1L: 7.5	IPd2L: 12	IClawL: 14	DigIL: 32	IHyp:———
	IIFreeNC: 36	IIClawL: 13	IIMed: 40	IILat: 33	
	IIIFreeNC: 48	IIIClawL: 11	IIIMed: 40	IIILat: 38	IVHyp: 26
SREL IV: female born in captivity of South Carolina stock	TL: 1,162	SVL: 540	SHL: 259	Tail: 622	
	UL: 57	LL: 70	PL: 90		
	IPd1L: 9.5	IPd2L: 11	IClawL: 10	DigIL: 32	IHyp: 22
	IIFreeNC: 41	IIClawL: 13	IIMed: 43	IILat: 32	
	IIIFreeNC: 54	IIIClawL: 11	IIIMed: 37	IIILat: 36	IVHyp: 23
SREL Par Pond EEB wild female	TL: 2,246	SVL: 1,081	SHL: 530	Tail: 1,165	
	UL: 126	LL: 148	PL: 164		
	IPd1L: 13	IPd2L: 14	IClawL: 21	DigIL: 59	IHyp: 43
	IIFreeNC: 65	IIClawL: 22	IIMed: 71	IILat: 59	
	IIIFreeNC: 81	IIIClawL: 19	IIIMed: 66	IIILat: 65	IVHyp: 42
SREL Par Pond BFA wild male	TL: 3,230	SVL: 1,565	SHL: 798	Tail: 1,665	
	UL: 160	LL: 195	PL: 229		
	IPd1L: 18	IPd2L: 16	IClawL: 27	DigIL: 76	IHyp: 49
	IIFreeNC: 80	IIClawL: 25	IIMed: 103	IILat: 63	
	IIIFreeNC: 103	IIIClawL: 24	IIIMed: 81	IIILat: 80	IVHyp: 44
SREL Par Pond 00AC wild male	TL: 3,450	SVL:———	SHL: 885	Tail:———	
	UL: 168	LL: 181	PL: 215		
	IPd1L: 22	IPd2L: 23	IClawL: 27	DigIL: 84	IHyp: 57
	IIFreeNC: 83	IIClawL: 27	IIMed: 101	IILat: 78	
	IIIFreeNC: 104	IIIClawL: 24	IIIMed: 91	IIILat: 88	IVHyp: 53
SREL Par Pond wild juvenile 1	TL: 563	SVL: 263	SHL: 125	Tail: 300	
	UL: 31	LL: 38	PL: 47		
	IPd1L: 4.5	IPd2L: 6	IClawL: 6.5	DigIL: 18	IHyp: 12
	IIFreeNC: 18	IIClawL: 7	IIMed: 21	IILat: 16	
	IIIFreeNC: 23.5	IIIClawL: 6.5	IIIMed: 19	IIILat: 18	IVHyp: 10
SREL Par Pond wild juvenile 2	TL: 594	SVL: 279	SHL: 133	Tail: 315	
	UL: 32	LL: 40	PL: 47		
	IPd1L: 3.5	IPd2L: 6.5	IClawL: 7	DigIL: 16	IHyp: 11
	IIFreeNC: 18.5	IIClawL: 6.5	IIMed: 23	IILat: 17	
	IIIFreeNC: 23.5	IIIClawL: 6.5	IIIMed: 19	IIILat: 19	IVHyp: 10
SREL Par Pond wild juvenile 3	TL: 577	SVL: 276	SHL: 130	Tail: 301	
	UL: 32	LL: 38	PL: 45		
	IPd1L: 3.5	IPd2L: 7	IClawL: 6.5	DigIL: 16	IHyp: 11
	IIFreeNC: 17	IIClawL: 7	IIMed: 22	IILat: 17	
	IIIFreeNC: 20.5	IIIClawL: 6.5	IIIMed: 20	IIILat: 20	IVHyp: 12

SREL Par Pond wild juvenile 4	TL: 518 UL: 30 IPd1L: 4.5 IIFreeNC: 17.5 IIIFreeNC: 21	SVL: 243 LL: 33 IPd2L: 3 IIClawL: 6.5 IIIClawL: 6	SHL: 115 PL: 47 IClawL: 6 IIMed: 19 IIIMed: 19	Tail: 275 DigIL: 12 IILat: 15 IIILat: 18	IHyp: 9.5 IVHyp: 9
SREL Par Pond wild juvenile 5	TL: 557 UL: 36 IPd1L: 5.5 IIFreeNC: 16 IIIFreeNC: 23.5	SVL: 268 LL: 40 IPd2L: 3.5 IIClawL: 8 IIIClawL: 6.5	SHL: 123 PL: 47 IClawL: 7 IIMed: 23 IIIMed: 20	Tail: 289 DigIL: 16 IILat: 16 IIILat: 17	IHyp: 10 IVHyp: 10
SREL Par Pond wild juvenile 6	TL: 575 UL: 34 IPd1L: 5 IIFreeNC: 19 IIIFreeNC: 23	SVL: 267 LL: 37 IPd2L: 3 IIClawL: 6 IIIClawL: 6	SHL: 123 PL: 47 IClawL: 7 IIMed: 21 IIIMed: 20	Tail: 308 DigIL: 18 IILat: 16 IIILat: 19	IHyp: 13 IVHyp: 10
SREL V scute mark G captive Louisiana female	TL: 1,550 UL: 73 IPd1L: 12 IIFreeNC: 55 IIIFreeNC: 69	SVL: 720 LL: 98 IPd2L: 17 IIClawL: 12 IIIClawL: 11	SHL: 353 PL: 111 IClawL: 12 IIMed: 48 IIIMed: 46	Tail: 830 DigIL: 33 IILat: 35 IIILat: 42	IHyp: 27 IVHyp: 25
SREL VI scute mark EH captive Louisiana female	TL: 1,642 UL: 86 IPd1L: 10 IIFreeNC: 62 IIIFreeNC: 65	SVL: 790 LL: 98 IPd2L: 17 IIClawL: 13 IIIClawL: 12	SHL: 389 PL: 125 IClawL: 15 IIMed: 49 IIIMed: 52	Tail: 852 DigIL: 43 IILat: 38 IIILat: 43	IHyp: 32 IVHyp: 25
SREL VII scute mark AH captive Louisiana male	TL: 1,641 UL: 80 IPd1L: 13 IIFreeNC: 57 IIIFreeNC: 68	SVL: 771 LL: 94 IPd2L: 17 IIClawL: 12 IIIClawL: 10	SHL: 387 PL: 123 IClawL: 13 IIMed: 50 IIIMed: 52	Tail: 870 DigIL: 46 IILat: 38 IIILat: 48	IHyp: 28 IVHyp: 30
SREL VIII scute mark DF captive Louisiana female	TL: 1,692 UL: 85 IPd1L: 10 IIFreeNC: 67 IIIFreeNC: 67	SVL: 816 LL: 100 IPd2L: 18 IIClawL: 15 IIIClawL: 14	SHL: 394 PL: 118 IClawL: 16 IIMed: 51 IIIMed: 53	Tail: 876 DigIL: 45 IILat: 44 IIILat: 49	IHyp: 32 IVHyp: 27
SREL IX scute mark BH LDWF 6178 captive Louisiana female	TL: 1,714 UL: 94 IPd1L: 16 IIFreeNC: 50 IIIFreeNC: 65	SVL: 840 LL: 105 IPd2L: 13 IIClawL: 16 IIIClawL: 14	SHL: 420 PL: 123 IClawL: 15 IIMed: 50 IIIMed: 53	Tail: 874 DigIL: 44 IILat: 44 IIILat: 53	IHyp: 33 IVHyp: 28
SREL X scute mark CF captive Louisiana female	TL: 1,636 UL: 72 IPd1L: 15 IIFreeNC: 50 IIIFreeNC: 74	SVL: 792 LL: 106 IPd2L: 19 IIClawL: 15 IIIClawL: 13	SHL: 414 PL: 124 IClawL: 14 IIMed: 53 IIIMed: 53	Tail: 844 DigIL: 46 IILat: 42 IIILat: 49	IHyp: 32 IVHyp: 25
SREL XI animal care juvenile of South Carolina and Alabama parents	TL: 478 UL: 29 IPd1L: 4.5 IIFreeNC: 22 IIIFreeNC: 30	SVL: 236 LL: 30 IPd2L: 5 IIClawL: 6 IIIClawL: 5	SHL: 114 PL: 43 IClawL: 5 IIMed: 17 IIIMed: 17	Tail: 242 DigIL: 15 IILat: 14 IIILat: 17	IHyp: 11 IVHyp: 10

Table A5.2. *continued*

SREL XII animal care juvenile of South Carolina and Alabama parents	TL:———	SVL: 228	SHL: 112	Tail:———	
	UL: 29	LL: 27	PL: 41		
	IPd1L: 4.5	IPd2L: 5.5	IClawL: 5.5	DigIL: 16	IHyp: 11
	IIFreeNC: 20.5	IIClawL: 5.5	IIMed: 17	IILat: 15	
	IIIFreeNC: 23.5	IIIClawL: 4.5	IIIMed: 17	IIILat: 16	IVHyp: 9.5
SREL XIII animal care juvenile of South Carolina and Alabama parents	TL:———	SVL: 257	SHL: 118	Tail:———	
	UL: 33	LL: 31	PL: 46		
	IPd1L: 6	IPd2L: 5	IClawL: 5.5	DigIL: 18	IHyp: 11
	IIFreeNC: 21	IIClawL: 6	IIMed: 19	IILat: 17	
	IIIFreeNC: 30	IIIClawL: 5	IIIMed: 17	IIILat: 19	IVHyp: 10
SREL XIV animal care juvenile of South Carolina and Alabama parents	TL:———	SVL: 204	SHL: 94	Tail:———	
	UL: 24	LL: 23	PL: 32		
	IPd1L: 4	IPd2L: 4	IClawL: 4.5	DigIL: 13	IHyp: 9.5
	IIFreeNC: 16.5	IIClawL: 4.5	IIMed: 16	IILat: 13	
	IIIFreeNC: 19.5	IIIClawL: 3.5	IIIMed: 13	IIILat: 14	IVHyp: 8.5
SREL XV animal care juvenile of South Carolina and Alabama parents	TL: 375	SVL: 194	SHL: 95	Tail: 181	
	UL: 24	LL: 23	PL: 32		
	IPd1L: 5	IPd2L: 4.5	IClawL: 4.5	DigIL: 13	IHyp: 7.5
	IIFreeNC: 16.5	IIClawL: 4.5	IIMed: 16	IILat: 11	
	IIIFreeNC: 19	IIIClawL: 4	IIIMed: 14	IIILat: 14	IVHyp: 7
SREL XVI animal care juvenile of South Carolina and Alabama parents	TL: 467	SVL: 232	SHL: 112	Tail: 235	
	UL: 24	LL: 29	PL: 41		
	IPd1L: 3.5	IPd2L: 5	IClawL: 5.5	DigIL: 15	IHyp: 10
	IIFreeNC: 21.5	IIClawL: 5.5	IIMed: 17	IILat: 14	
	IIIFreeNC: 20.5	IIIClawL: 4.5	IIIMed: 16	IIILat: 16	IVHyp: 9
SREL XVII animal care juvenile of South Carolina and Alabama parents	TL: 469	SVL: 234	SHL: 109	Tail: 235	
	UL: 25	LL: 29	PL: 39		
	IPd1L: 5.5	IPd2L: 6	IClawL: 5.5	DigIL: 16	IHyp: 11
	IIFreeNC: 16	IIClawL: 5	IIMed: 18	IILat: 15	
	IIIFreeNC: 23.5	IIIClawL: 4.5	IIIMed: 17	IIILat: 17	IVHyp: 7.5
SREL XVIII alligator in cattle tank 1 in experimental pond C	TL: 609	SVL: 306	SHL: 147	Tail: 303	
	UL: 36	LL: 36	PL: 52		
	IPd1L: 5.5	IPd2L: 6.5	IClawL: 6.5	DigIL: 19	IHyp: 12
	IIFreeNC: 21.5	IIClawL: 7.5	IIMed: 24	IILat: 18	
	IIIFreeNC: 26	IIIClawL: 6	IIIMed: 23	IIILat:———	IVHyp:———
SREL XIX alligator in cattle tank 1 in experimental pond C	TL: 546	SVL: 275	SHL: 135	Tail: 271	
	UL: 30	LL: 33	PL: 49		
	IPd1L: 6.5	IPd2L: 5	IClawL: 8	DigIL: 20	IHyp: 14
	IIFreeNC: 21	IIClawL: 8	IIMed: 24	IILat: 18	
	IIIFreeNC: 29.5	IIIClawL: 6.5	IIIMed: 22	IIILat: 21	IVHyp: 11
SREL XX alligator in cattle tank 1 in experimental pond C	TL: 693	SVL: 339	SHL: 161	Tail: 354	
	UL: 35	LL: 42	PL: 60		
	IPd1L: 5.5	IPd2L: 7	IClawL: 8.5	DigIL: 20	IHyp: 15
	IIFreeNC: 22	IIClawL: 9	IIMed: 28	IILat: 20	
	IIIFreeNC: 29.5	IIIClawL: 7.5	IIIMed: 24	IIILat: 24	IVHyp: 14
SREL XXI alligator in cattle tank 1 in experimental pond C	TL: 697	SVL: 332	SHL: 154	Tail: 365	
	UL: 34	LL: 43	PL: 57		
	IPd1L: 4	IPd2L: 4.5	IClawL: 8.5	DigIL: 20	IHyp: 16
	IIFreeNC: 29	IIClawL: 8	IIMed: 26	IILat: 21	
	IIIFreeNC: 34	IIIClawL: 7	IIIMed: 25	IIILat: 23	IVHyp: 13
SREL XXII alligator in cattle tank 1 in experimental pond C	TL: 673	SVL: 323	SHL: 151	Tail: 350	
	UL: 34	LL: 42	PL: 59		
	IPd1L: 6	IPd2L: 7.5	IClawL: 8.5	DigIL: 22	IHyp: 16
	IIFreeNC: 25	IIClawL: 5	IIMed: 21	IILat: 18	
	IIIFreeNC: 33.5	IIIClawL: 6.5	IIIMed: 25	IIILat: 23	IVHyp: 12

SREL XXIII alligator in cattle tank 1 in experimental pond C	TL: 688	SVL: 335	SHL: 159	Tail: 353	
	UL: 41	LL: 43	PL: 58		
	IPd1L: 5.5	IPd2L: 7.5	IClawL: 8.5	DigIL: 22	IHyp: 18
	IIFreeNC: 25.5	IIClawL: 9.5	IIMed: 27	IILat: 22	
	IIIFreeNC: 33.5	IIIClawL: 7.5	IIIMed: 26	IIILat: 24	IVHyp: 12
SREL XXIV alligator in cattle tank 1 in experimental pond C	TL: 680	SVL: 324	SHL: 159	Tail: 356	
	UL: 36	LL: 40	PL: 56		
	IPd1L: 6	IPd2L: 7	IClawL: 8	DigIL: 21	IHyp: 15
	IIFreeNC: 22.5	IIClawL: 9.5	IIMed: 24	IILat: 20	
	IIIFreeNC: 31	IIIClawL: 7	IIIMed: 22	IIILat: 24	IVHyp: 12
SREL XXV cesium uptake alligator	TL: 1,583	SVL: 781	SHL: 415	Tail: 802	
	UL: 84	LL: 91	PL: 134		
	IPd1L: 14	IPd2L: 16	IClawL: 18	DigIL: 50	IHyp: 35
	IIFreeNC: 59	IIClawL: 17	IIMed: 48	IILat: 40	
	IIIFreeNC: 67	IIIClawL: 17	IIIMed: 56	IIILat: 51	IVHyp: 27
SREL XXVI cesium uptake alligator	TL: 1,924	SVL: 936	SHL: 470	Tail: 988	
	UL: 109	LL: 123	PL: 154		
	IPd1L: 16	IPd2L: 17	IClawL: 23	DigIL: 57	IHyp: 38
	IIFreeNC: 72	IIClawL: 22	IIMed: 61	IILat: 50	
	IIIFreeNC: 90	IIIClawL: 17	IIIMed: 61	IIILat: 50	IVHyp:32
RWR 1 captive Louisiana female	TL: 2,820	SVL:———	SHL:———	Tail:———	
	UL:———	LL:———	PL:———		
	IPd1L: 16	IPd2L: 22	IClawL: 25	DigIL: 72	IHyp: 55
	IIFreeNC: 82	IIClawL: 24	IIMed: 78	IILat: 65	
	IIIFreeNC: 94	IIIClawL: 20	IIIMed: 69	IIILat: 73	IVHyp: 52
RWR 2 captive Louisiana female	TL: 2,572	SVL:———	SHL:———	Tail:———	
	UL:———	LL:———	PL:———		
	IPd1L: 23	IPd2L: 28	IClawL: 22*	DigIL: 71	IHyp: 49
	IIFreeNC: 69	IIClawL: 25	IIMed: 81	IILat: 56	
	IIIFreeNC: 94	IIIClawL: 20	IIIMed: 86	IIILat: 74	IVHyp: 44
RWR 3 captive Louisiana female	TL: 2,769	SVL:———	SHL:———	Tail:———	
	UL:———	LL:———	PL:———		
	IPd1L: 17	IPd2L: 28	IClawL: 25	DigIL: 70	IHyp: 46
	IIFreeNC: 74	IIClawL: 29	IIMed: 89	IILat: 62	
	IIIFreeNC: 87	IIIClawL: 25	IIIMed: 86	IIILat: 75	IVHyp: 49
RWR 4 captive Louisiana male	TL: 3,823	SVL:———	SHL:———	Tail:———	
	UL:———	LL:———	PL:———		
	IPd1L: 43	IPd2L: 31	IClawL: 21*	DigIL: 104	IHyp: 60
	IIFreeNC: 96	IIClawL: 25	IIMed: 94	IILat: 74	
	IIIFreeNC: 131	IIIClawL: 23	IIIMed: 104	IIILat: 93	IVHyp: 67
RWR 5 captive Louisiana captive male	TL: 3,632	SVL:	SHL:	Tail:	
	UL:———	LL:———	PL:———		
	IPd1L: 19	IPd2L: 28	IClawL: 26	DigIL: 79	IHyp: 55
	IIFreeNC: 97	IIClawL: 20	IIMed: 103	IILat: 74	
	IIIFreeNC: 107	IIIClawL: 21	IIIMed: 99	IIILat: 89	IVHyp: 70
RWR 6 captive Louisiana female	TL: 2,845	SVL:———	SHL:———	Tail:———	
	UL:———	LL:———	PL:———		
	IPd1L: 21	IPd2L: 32	IClawL: 24	DigIL: 75	IHyp: 50
	IIFreeNC: 71	IIClawL: 23	IIMed: 84	IILat: 56	
	IIIFreeNC: 94	IIIClawL: 17	IIIMed: 78	IIILat: 94	IVHyp: 71
RWR 7 captive Louisiana female	TL: 2,807	SVL:———	SHL: 680	Tail:———	
	UL: 120	LL: 175	PL: 203		
	IPd1L: 37	IPd2L: 0	IClawL: 19*	DigIL: 54	IHyp: 42
	IIFreeNC: 78	IIClawL: 25	IIMed: 87	IILat: 74	
	IIIFreeNC: 99	IIIClawL: 22	IIIMed: 87	IIILat: 85	IVHyp: 42

Table A5.2. *continued*

RWR 8 captive Louisiana female	TL: 2,743	SVL: 1,350	SHL: 665	Tail: 1,393	
	UL: 138	LL: 184	PL:———		
	IPd1L: 24	IPd2L: 31	IClawL: 22	DigIL: 76	IHyp: 52
	IIFreeNC: 83	IIClawL: 17	IIMed: 81	IILat: 67	
	IIIFreeNC: 98	IIIClawL: 5*	IIIMed: 71	IIILat: 60	IVHyp: 49
RWR 9 captive Louisiana female	TL: 2,840	SVL:———	SHL: 695	Tail:———	
	UL: 160	LL: 180	PL: 192		
	IPd1L: 26	IPd2L: 20	IClawL: 23	DigIL: 71	IHyp: 45
	IIFreeNC: 83	IIClawL: 26	IIMed: 87	IILat: 58	
	IIIFreeNC: 103	IIIClawL: 24	IIIMed: 75	IIILat: 81	IVHyp: 50
RWR 10 captive Louisiana female	TL: 2,769	SVL: 1,385	SHL: 650	Tail: 1,384	
	UL: 131	LL: 171	PL: 179		
	IPd1L: 23	IPd2L: 25	IClawL: 25	DigIL: 74	IHyp: 50
	IIFreeNC: 79	IIClawL: 23	IIMed: 88	IILat: 68	
	IIIFreeNC: 103	IIIClawL: 21	IIIMed: 87	IIILat: 77	IVHyp: 53
RWR 11 captive Louisiana female	TL: 2,743	SVL: 1,397	SHL: 710	Tail: 1,346	
	UL: 141	LL: 183	PL: 194		
	IPd1L: 22	IPd2L: 32	IClawL: 29	DigIL: 78	IHyp: 55
	IIFreeNC: 87	IIClawL: 24	IIMed: 91	IILat: 74	
	IIIFreeNC: 96	IIIClawL: 23	IIIMed: 84	IIILat: 83	IVHyp: 59
RWR 12 captive Louisiana female	TL: 2,769	SVL: 1,375	SHL: 705	Tail: 1,394	
	UL: 133	LL: 183	PL: 184		
	IPd1L: 23	IPd2L: 29	IClawL: 32	DigIL: 81	IHyp: 58
	IIFreeNC:———	IIClawL:———	IIMed:———	IILat:———	
	IIIFreeNC:101	IIIClawL: 20	IIIMed: 81	IIILat: 88	IVHyp: 53
RWR 13 captive Louisiana male	TL: 3,759	SVL: 1,805	SHL: 870	Tail: 1,954	
	UL: 187	LL: 235	PL: 227		
	IPd1L: 33	IPd2L: 36	IClawL: 24	DigIL: 93	IHyp: 59
	IIFreeNC: 98	IIClawL: 23	IIMed: 113	IILat: 71	
	IIIFreeNC: 100	IIIClawL: 24	IIIMed: 109	IIILat: 92	IVHyp: 60
RWR 14 captive Louisiana female	TL: 2,845	SVL: 1,372	SHL: 690	Tail: 1,473	
	UL: 146	LL: 183	PL: 195		
	IPd1L: 15	IPd2L: 31	IClawL: 25	DigIL: 71	IHyp: 54
	IIFreeNC: 86	IIClawL: 21	IIMed: 92	IILat: 70	
	IIIFreeNC: 102	IIIClawL: 23	IIIMed: 90	IIILat: 80	IVHyp: 49
RWR 15 captive Louisiana female	TL: 2,679	SVL: 1,385	SHL: 700	Tail: 1,294	
	UL: 145	LL: 179	PL: 199		
	IPd1L: 21	IPd2L: 29	IClawL: 29	DigIL: 80	IHyp: 53
	IIFreeNC: 79	IIClawL: 24	IIMed: 85	IILat: 66	
	IIIFreeNC: 99	IIIClawL: 21	IIIMed: 95	IIILat: 85	IVHyp: 61
RWR 16 captive Louisiana female	TL: 2,565	SVL: 1,285	SHL: 640	Tail: 1,280	
	UL: 132	LL: 169	PL: 165		
	IPd1L: 26	IPd2L: 22	IClawL: 27	DigIL: 74	IHyp: 48
	IIFreeNC: 75	IIClawL: 25	IIMed: 83	IILat: 74	
	IIIFreeNC: 90	IIIClawL: 13*	IIIMed: 84	IIILat: 64	IVHyp: 45
RWR 17 wild Louisiana male	TL: 2,337	SVL: 1,102	SHL: 560	Tail: 1,235	
	UL: 121	LL: 144	PL: 153		
	IPd1L: 16	IPd2L: 20	IClawL: 18	DigIL: 54	IHyp: 33
	IIFreeNC: 59	IIClawL: 18	IIMed: 67	IILat: 48	
	IIIFreeNC: 81	IIIClawL: 17	IIIMed: 63	IIILat: 57	IVHyp: 42
RWR 18 wild Louisiana female	TL: 2,159	SVL: 1,050	SHL: 530	Tail: 1,109	
	UL: 111	LL: 124	PL: 164		
	IPd1L: 15	IPd2L: 24	IClawL: 22	DigIL: 63	IHyp: 40
	IIFreeNC: 65	IIClawL: 23	IIMed: 74	IILat: 52	
	IIIFreeNC: 79	IIIClawL: 23	IIIMed: 71	IIILat: 65	IVHyp: 40

RWR 19 wild Louisiana male	TL: 1,765	SVL: 859	SHL: 415	Tail: 906	
	UL: 97	LL: 104	PL: 118		
	IPd1L: 13	IPd2L: 14	IClawL: 18	DigIL: 45	IHyp: 30
	IIFreeNC: 47	IIClawL: 19	IIMed: 54	IILat: 43	
	IIIFreeNC: 59	IIIClawL: 17	IIIMed: 52	IIILat: 48	IVHyp: 29
RWR 20 captive Louisiana female	TL: 2,667	SVL: 1,335	SHL: 675	Tail: 1,332	
	UL: 146	LL: 156	PL: 173		
	IPd1L: 20	IPd2L: 31	IClawL: 23	DigIL: 75	IHyp: 49
	IIFreeNC: 77	IIClawL: 23	IIMed: 77	IILat: 63	
	IIIFreeNC: 92	IIIClawL: 22	IIIMed: 83	IIILat: 75	IVHyp: 50
RWR 22 captive Louisiana male	TL: 3,810	SVL: 1,915	SHL: 900	Tail: 1,895	
	UL: 214	LL: 208	PL: 252		
	IPd1L: 34	IPd2L: 39	IClawL: 29	DigIL: 100	IHyp: 59
	IIFreeNC: 95	IIClawL: 25	IIMed: 102	IILat: 78	
	IIIFreeNC: 127	IIIClawL: 22	IIIMed: 101	IIILat: 104	IVHyp: 74
RWR 23 wild Louisiana male	TL: 2,134	SVL: 1,021	SHL: 480	Tail: 1,113	
	UL: 110	LL: 132	PL: 150		
	IPd1L: 16	IPd2L: 16	IClawL: 20	DigIL: 53	IHyp: 38
	IIFreeNC: 60	IIClawL: 21	IIMed: 67	IILat: 51	
	IIIFreeNC: 72	IIIClawL: 20	IIIMed: 66	IIILat: 58	IVHyp: 33
RWR 24 wild Louisiana female	TL: 1,981	SVL: 932	SHL: 470	Tail: 1,049	
	UL: 107	LL: 121	PL: 142		
	IPd1L: 15	IPd2L: 16	IClawL: 19	DigIL: 54	IHyp: 35
	IIFreeNC: 60	IIClawL: 19	IIMed: 63	IILat: 47	
	IIIFreeNC: 72	IIIClawL: 19	IIIMed: 59	IIILat: 58	IVHyp: 33
RWR 25 wild Louisiana male	TL: 1,613	SVL: 773	SHL: 355	Tail: 840	
	UL: 90	LL: 96	PL: 109		
	IPd1L: 25	IPd2L: 0	IClawL: 17	DigIL: 41	IHyp: 30
	IIFreeNC: 45	IIClawL: 19	IIMed: 51	IILat: 39	
	IIIFreeNC: 58	IIIClawL: 16	IIIMed: 48	IIILat: 46	IVHyp: 26
RWR 26 wild Louisiana male	TL: 2,438	SVL: 1,220	SHL: 590	Tail: 1,218	
	UL: 136	LL: 136	PL: 172		
	IPd1L: 16	IPd2L: 23	IClawL: 20	DigIL: 57	IHyp: 39
	IIFreeNC: 65	IIClawL: 22	IIMed: 73	IILat: 54	
	IIIFreeNC: 86	IIIClawL: 20	IIIMed: 67	IIILat: 66	IVHyp: 39
RWR 27 captive Louisiana male	TL: 3,607	SVL: 1,740	SHL: 840	Tail: 1,867	
	UL: 182	LL: 232	PL: 237		
	IPd1L: 21	IPd2L: 30	IClawL: 28	DigIL: 84	IHyp: 61
	IIFreeNC: 98	IIClawL: 22	IIMed: 102	IILat: 84	
	IIIFreeNC: 128	IIIClawL: 19	IIIMed: 104	IIILat: 100	IVHyp: 66
RWR 28 captive Louisiana female	TL: 2,769	SVL: 1,370	SHL: 670	Tail: 1,399	
	UL: 137	LL: 147	PL: 198		
	IPd1L: 19	IPd2L: 28	IClawL: 31	DigIL: 80	IHyp: 61
	IIFreeNC: 75	IIClawL: 26	IIMed: 91	IILat: 70	
	IIIFreeNC: 99	IIIClawL: 26	IIIMed: 85	IIILat: 100	IVHyp: 70
RWR 29 captive Louisiana male	TL: 3,873	SVL: 1,823	SHL: 870	Tail: 2,050	
	UL: 235	LL: 190	PL: 250		
	IPd1L: 30	IPd2L: 36	IClawL: 27	DigIL: 94	IHyp: 60
	IIFreeNC: 103	IIClawL: 24	IIMed: 109	IILat: 82	
	IIIFreeNC: 112	IIIClawL: 24	IIIMed: 110	IIILat: 88	IVHyp: 72
RWR 30 captive Louisiana female	TL: 2,515	SVL: 1,233	SHL: 610	Tail: 1,282	
	UL: 139	LL: 154	PL: 185		
	IPd1L: 18	IPd2L: 26	IClawL: 22	DigIL: 69	IHyp: 55
	IIFreeNC:———	IIClawL:———	IIMed: 79	IILat: 66	
	IIIFreeNC: 75	IIIClawL: 19	IIIMed: 79	IIILat: 74	IVHyp: 47

Table A5.2. *continued*

RWR 31 captive Louisiana male	TL: 3,607	SVL: 1,787	SHL: 840	Tail: 1,820	
	UL: 197	LL: 211	PL: 229		
	IPd1L: 35	IPd2L: 40	IClawL: 25	DigIL: 98	IHyp: 58
	IIFreeNC: 100	IIClawL: 24	IIMed: 105	IILat: 84	
	IIIFreeNC: 124	IIIClawL: 22	IIIMed: 100	IIILat: 108	IVHyp: 72
RWR 32 captive Louisiana female	TL: 2,794	SVL: 1,378	SHL: 675	Tail: 1,416	
	UL: 150	LL: 163	PL: 189		
RWR 33 captive Louisiana female	TL: 2,686	SVL: 1,322	SHL: 670	Tail: 1,364	
	UL: 133	LL: 200	PL: 198		
	IPd1L: 28	IPd2L: 24	IClawL: 26	DigIL: 81	IHyp: 57
	IIFreeNC: 80	IIClawL: 24	IIMed: 91	IILat: 66	
	IIIFreeNC: 96	IIIClawL: 21	IIIMed: 85	IIILat: 82	IVHyp: 49
RWR captive Louisiana juvenile 1	TL: 440	SVL: 205	SHL: 92	Tail: 235	
	UL: 25	LL: 28	PL: 33		
	IPd1L: 4.5	IPd2L: 4	IClawL: 4.5	DigIL: 13	IHyp: 8.5
	IIFreeNC: 15	IIClawL: 5	IIMed: 17	IILat: 12	
	IIIFreeNC: 18	IIIClawL: 4	IIIMed: 15	IIILat: 13	IVHyp: 8
RWR captive Louisiana juvenile 2	TL: 766	SVL: 348	SHL: 173	Tail: 418	
	UL: 46	LL: 53	PL: 61		
	IPd1L: 6	IPd2L: 8	IClawL: 8.5	DigIL: 21	IHyp: 16
	IIFreeNC: 25	IIClawL: 10	IIMed: 28	IILat: 21	
	IIIFreeNC: 33.5	IIIClawL: 8.5	IIIMed: 28	IIILat: 24	IVHyp: 15
RWR roadkill wild female	TL: 1,651	SVL: 775	SHL: 380	Tail: 876	
	UL: 103	LL: 111	PL: 121		
	IPd1L: 9.5	IPd2L: 12	IClawL: 18	DigIL: 46	IHyp: 33
	IIFreeNC: 49	IIClawL: 17	IIMed: 52	IILat: 43	
	IIIFreeNC: 61	IIIClawL: 16	IIIMed: 49	IIILat: 54	IVHyp: 26
RWR roadkill wild male 1	TL: 1,715	SVL: 807	SHL: 400	Tail: 908	
	UL: 90	LL: 110	PL: 129		
	IPd1L: 10	IPd2L: 9.5	IClawL: 15	DigIL: 43	IHyp: 33
	IIFreeNC: 50	IIClawL: 18	IIMed: 50	IILat: 42	
	IIIFreeNC: 60	IIIClawL: 16	IIIMed: 50	IIILat: 51	IVHyp: 33
RWR roadkill wild male 2	TL: 3,073	SVL: 1,542	SHL: 767	Tail: 1,531	
	UL: 160	LL: 170	PL: 221		
	IPd1L: 17	IPd2L: 20	IClawL: 23	DigIL: 72	IHyp: 48
	IIFreeNC: 83	IIClawL: 25	IIMed: 93	IILat: 66	
	IIIFreeNC: 97	IIIClawL: 23	IIIMed: 89	IIILat: 79	IVHyp: 55
CITES 0011601 central Florida wild male	TL: 2,972	SVL:———	SHL:———	Tail:———	
	IPd1L: 19	IPd2L: 30	IClawL: 24	DigIL: 73	IHyp: 52
	IIFreeNC: 82	IIClawL: 23	IIMed: 88	IILat: 72	
	IIIFreeNC: 102	IIIClawL: 21	IIIMed: 83	IIILat: 78	IVHyp: 46
CITES 0027580 central Florida wild alligator	TL: 2,007	SVL:———	SHL:———	Tail:———	
	IPd1L: 14	IPd2L: 20	IClawL: 20	DigIL: 55	IHyp: 41
	IIFreeNC: 56	IIClawL: 20	IIMed: 60	IILat: 52	
	IIIFreeNC: 67	IIIClawL: 17	IIIMed: 59	IIILat: 57	IVHyp: 32
CITES 0040187 central Florida wild male	TL: 3,251	SVL:———	SHL:———	Tail:———	
	IPd1L: 22	IPd2L: 20	IClawL: 28	DigIL: 87	IHyp: 59
	IIFreeNC: 91	IIClawL: 27	IIMed: 101	IILat: 77	
	IIIFreeNC: 120	IIIClawL: 23	IIIMed: 90	IIILat: 89	IVHyp: 45
CITES 0040449 central Florida wild alligator	TL: 2,261	SVL:———	SHL:———	Tail:———	
	IPd1L: 12	IPd2L: 17	IClawL: 25	DigIL: 62	IHyp: 49
	IIFreeNC: 65	IIClawL: 27	IIMed: 78	IILat: 68	
	IIIFreeNC: 82	IIIClawL: 22	IIIMed: 73	IIILat: 75	IVHyp: 40
CITES 0041015 central Florida wild alligator	TL: 2,819	SVL:———	SHL:———	Tail:———	
	IPd1L: 17	IPd2L:	IClawL: 27	DigIL: 74	IHyp: 56
	IIFreeNC: 86	IIClawL: 28	IIMed: 96	IILat: 77	
	IIIFreeNC: 106	IIIClawL: 23*	IIIMed: 88	IIILat: 78	IVHyp: 52

CITES 0041602 central Florida wild male	TL: 3,404	SVL:———	SHL:———	Tail:———	
	IPd1L: 18	IPd2L: 23	IClawL: 25	DigIL: 75	IHyp: 58
	IIFreeNC: 98	IIClawL: 22	IIMed: 93	IILat: 73	
	IIIFreeNC: 118	IIIClawL: 22	IIIMed: 92	IIILat: 98	IVHyp: 45
CITES 0042092 central Florida wild alligator	TL: 2,337	SVL:———	SHL:———	Tail:———	
	IPd1L: 17	IPd2L: 19	IClawL: 26	DigIL: 62	IHyp: 46
	IIFreeNC: 66	IIClawL: 25	IIMed: 76	IILat: 59	
	IIIFreeNC: 82	IIIClawL: 22	IIIMed: 68	IIILat: 69	IVHyp: 30
CITES 0042177 central Florida wild male	TL: 3,353	SVL:———	SHL:———	Tail:———	
	IPd1L: 22	IPd2L: 26	IClawL: 26	DigIL: 74	IHyp: 57
	IIFreeNC: 84	IIClawL: 24	IIMed: 89	IILat: 75	
	IIIFreeNC: 112	IIIClawL: 20	IIIMed: 85	IIILat: 84	IVHyp: 54
CITES 0042178 central Florida wild male	TL: 3,150	SVL:———	SHL:———	Tail:———	
	IPd1L: 20	IPd2L: 24	IClawL: 24	DigIL: 78	IHyp: 54
	IIFreeNC: 82	IIClawL: 26	IIMed: 96	IILat: 77	
	IIIFreeNC: 104	IIIClawL: 24	IIIMed: 88	IIILat: 89	IVHyp: 58
CITES 0042233 central Florida wild alligator	TL: 2,464	SVL:———	SHL:———	Tail:———	
	IPd1L: 11	IPd2L: 18	IClawL: 22	DigIL: 61	IHyp: 47
	IIFreeNC: 65	IIClawL: 24	IIMed: 76	IILat: 65	
	IIIFreeNC: 81	IIIClawL: 21	IIIMed: 74	IIILat: 71	IVHyp: 41
CITES 0042410 central Florida wild alligator	TL: 1,651	SVL:———	SHL:———	Tail:———	
	IPd1L: 13	IPd2L: 14	IClawL: 18	DigIL: 46	IHyp: 36
	IIFreeNC: 49	IIClawL: 19	IIMed: 58	IILat: 44	
	IIIFreeNC: 62	IIIClawL: 16	IIIMed: 50	IIILat: 49	IVHyp: 29
CITES 0042448 central Florida wild alligator	TL: 2,616	SVL:———	SHL:———	Tail:———	
	IPd1L: 18	IPd2L: 17	IClawL: 24	DigIL: 68	IHyp: 51
	IIFreeNC: 78	IIClawL: 25	IIMed: 83	IILat: 65	
	IIIFreeNC: 95	IIIClawL: 20	IIIMed: 74	IIILat: 89	IVHyp: 59
CITES 0043066 central Florida wild alligator	TL: 2,540	SVL:———	SHL:———	Tail:———	
	IPd1L: 22	IPd2L: 19	IClawL: 28	DigIL: 70	IHyp: 50
	IIFreeNC: 71	IIClawL: 25	IIMed: 79	IILat: 65	
	IIIFreeNC: 85	IIIClawL: 23	IIIMed: 72	IIILat: 72	IVHyp: 41
CITES 0043153 Lake Okeechobee, Florida wild male	TL: 3,390	SVL: 1,694	SHL: 800	Tail: 1,696	
	UL: 175	LL: 217	PL: 234		
CITES 0043154 Lake Okeechobee, Florida wild male	TL: 3,470	SVL: 1,687	SHL: 790	Tail: 1,783	
	UL: 175	LL: 213	PL: 213		
	IPd1L: 18	IPd2L: 22	IClawL: 20*	DigIL: 74	IHyp: 49
	IIFreeNC: 76	IIClawL: 21*	IIMed: 94	IILat: 66	
	IIIFreeNC: 110	IIIClawL: 20*	IIIMed: 84	IIILat: 85	IVHyp: 58
CITES 0043258 Lake Okeechobee, Florida wild male	TL: 3,130	SVL: 1,530	SHL: 700	Tail: 1,600	
	UL: 165	LL: 195	PL: 212		
	IPd1L: 15	IPd2L: 19	IClawL: 24	DigIL: 73	IHyp: 51
	IIFreeNC: 87	IIClawL: 24	IIMed: 90	IILat: 69	
	IIIFreeNC: 101	IIIClawL: 22	IIIMed: 82	IIILat: 86	IVHyp: 45
CITES 0043361 central Florida wild alligator	TL: 2,718	SVL:———	SHL:———	Tail:———	
	IPd1L: 21	IPd2L: 23	IClawL: 21	DigIL: 66	IHyp: 48
	IIFreeNC: 78	IIClawL: 22	IIMed: 82	IILat: 60	
	IIIFreeNC: 100	IIIClawL: 17	IIIMed: 69	IIILat: 74	IVHyp: 48
CITES 0046862 central Florida wild alligator	TL: 2,235	SVL:———	SHL:———	Tail:———	
	IPd1L: 13	IPd2L: 15	IClawL: 20	DigIL: 57	IHyp: 41
	IIFreeNC: 63	IIClawL: 22	IIMed: 72	IILat: 56	
	IIIFreeNC: 81	IIIClawL: 18	IIIMed: 65	IIILat: 67	IVHyp: 40
CITES 0046863 central Florida wild alligator	TL: 1,981	SVL:———	SHL:———	Tail:———	
	IPd1L: 20	IPd2L: 19	IClawL: 16*	DigIL: 55	IHyp: 33
	IIFreeNC: 57	IIClawL: 17	IIMed: 60	IILat: 46	
	IIIFreeNC: 72	IIIClawL: 14	IIIMed: 53	IIILat: 53	IVHyp: 38

Table A5.2. *continued*

CITES 0054237 central Florida wild alligator	TL: 1549	SVL:———	SHL:———	Tail:———	
	IPd1L: 10	IPd2L: 7	IClawL: 15	DigIL: 37	IHyp: 27
	IIFreeNC: 36	IIClawL: 15	IIMed: 45	IILat: 36	
	IIIFreeNC: 45	IIIClawL: 13	IIIMed: 43	IIILat: 42	IVHyp: 24
CITES 0056493 central Florida wild alligator	TL: 2083	SVL:———	SHL:———	Tail:———	
	IPd1L: 15	IPd2L: 11	IClawL: 21	DigIL: 58	IHyp: 40
	IIFreeNC: 61	IIClawL: 21	IIMed: 69	IILat: 50	
	IIIFreeNC: 72	IIIClawL: 18	IIIMed: 62	IIILat: 57	IVHyp: 32
CITES 0056495 central Florida wild alligator	TL: 2,794	SVL:———	SHL:———	Tail:———	
	IPd1L: 19	IPd2L: 28	IClawL: 14*	DigIL: 57	IHyp: 39
	IIFreeNC: 81	IIClawL: 14*	IIMed: 77	IILat: 69	
	IIIFreeNC: 97	IIIClawL: 18	IIIMed: 84	IIILat: 95	IVHyp: 59
CITES 0059824 Lake Grassy, Florida wild male	TL: 2,060	SVL: 982	SHL: 465	Tail: 1,078	
	UL: 115	LL: 130	PL: 146		
	IPd1L: 21	IPd2L: 12	IClawL: 23	DigIL: 57	IHyp: 43
	IIFreeNC: 62	IIClawL: 22	IIMed: 72	IILat: 51	
	IIIFreeNC: 75	IIIClawL: 19	IIIMed: 61	IIILat: 59	IVHyp: 36
CITES 0059825 Lake Grassy, Florida wild male	TL: 2,400	SVL: 1,170	SHL: 580	Tail: 1,230	
	UL: 130	LL: 154	PL: 170		

Table A6.1. Measurements of overall footprint size in individual emus. All measurements are in centimeters. CV = coefficient of variation. Values for a given parameter are reported if the number of footprints for which that parameter was observed was 10 or more. If a juvenile bird grew detectably over the course of runs, data for that bird are broken down by run. Otherwise data are pooled across runs for individual birds. Data for individual birds are reported in the order of increasing footprint size (at time of first run for those juveniles run repeatedly). See text for description of abbreviation IIILGL.

Bird code	Run	Parameter	Minimum	Maximum	Mean (CV)	Number of footprints
1988 juvenile 2	6	Field length	4.2	5.1	4.80 (4.1)	39
		Thulborn length	4.2	5.1	4.71 (6.0)	13
		IIILGL	4.1	5.1	4.79 (4.3)	36
		Thulborn width	4.1	5.1	4.57 (6.7)	13
		Toetip width	3.6	5.2	4.68 (6.6)	39
		Clawbase width	3.8	4.6	4.15 (4.9)	24
		Size index	4.2	5.0	4.64 (4.7)	13
1988 juvenile 4	4	Field length	5.2	5.9	5.43 (3.2)	21
		IIILGL	5.0	5.5	5.26 (2.4)	19
		Toetip width	4.7	5.6	5.29 (4.2)	19
	10	Field length	7.1	7.4	7.23 (1.3)	12
		IIILGL	6.8	7.4	7.13 (2.0)	13
		Toetip width	6.6	7.4	7.15 (3.7)	12
	13	Field length	7.9	8.5	8.18 (2.1)	10
		Thulborn length	7.9	8.5	8.19 (2.1)	10
		IIILGL	7.6	8.6	8.15 (4.4)	10
		Thulborn width	7.6	8.4	8.02 (3.5)	10
		Clawbase width	6.6	7.0	6.86 (2.0)	10
		Size index	7.8	8.4	8.10 (2.4)	10
	18	Field length	9.5	10.4	9.95 (2.4)	18
		Thulborn length	9.5	10.4	9.97 (2.5)	14
		IIILGL	9.3	10.3	9.93 (2.9)	18
		Thulborn width	8.6	10.0	9.38 (4.7)	14
		Toetip width	9.0	10.0	9.50 (3.0)	17
		Clawbase width	8.0	8.7	8.37 (2.2)	20
		Size index	9.2	9.9	9.67 (2.7)	14
1988 juvenile 7		Field length	5.9	6.6	6.34 (4.2)	11
		IIILGL	5.5	6.9	6.30 (6.1)	11
1989 juvenile A		Field length	7.4	9.8	8.65 (6.8)	23
		IIILGL	7.3	10.0	8.48 (8.4)	24
		Toetip width	6.7	9.0	7.65 (9.1)	23
1988 juvenile 1		Field length	9.7	10.5	10.19 (2.2)	10
		IIILGL	9.0	10.2	9.85 (3.4)	10
		Toetip width	9.3	10.0	9.69 (2.3)	10
1989 juvenile C		Field length	8.2	12.0	10.88 (5.1)	48
		IIILGL	7.5	11.4	10.57 (6.9)	40
		Toetip width	7.0	10.9	9.90 (8.0)	38
		Clawbase width	8.2	9.6	9.11 (4.4)	10
1990 juvenile 12		Field length	15.8	18.0	17.04 (3.5)	19
		IIILGL	15.6	18.3	16.81 (4.8)	18
		Toetip width	12.3	15.9	13.96 (7.4)	19
1990 juvenile 13		Field length	15.5	18.4	17.26 (4.1)	13
		IIILGL	15.2	18.1	16.78 (6.3)	13
		Toetip width	11.5	14.7	13.51 (7.1)	13
1988 adult 6		Field length	16.0	23.0	18.90 (5.5)	81
		Thulborn length	16.0	23.0	18.92 (5.7)	72
		IIILGL	16.0	20.3	18.58 (4.8)	52
		Thulborn width	13.7	19.9	16.21 (7.8)	71
		Toetip width	13.5	18.4	16.14 (7.0)	45
		Clawbase width	13.5	15.7	14.66 (3.4)	18
		Size index	15.2	20.7	17.51 (5.9)	71

Table A6.1. *continued*

Bird code	Run	Parameter	Minimum	Maximum	Mean (CV)	Number of footprints
1988 adult 5		Field length	17.2	22.5	19.02 (5.5)	76
		Thulborn length	17.1	22.5	19.03 (5.6)	71
		IIILGL	17.0	22.1	18.63 (4.8)	54
		Thulborn width	12.3	21.1	15.80 (9.3)	69
		Toetip width	14.6	18.4	15.80 (6.0)	42
		Clawbase width	13.9	15.4	14.42 (3.5)	19
		Size index	15.0	20.9	17.34 (6.5)	69
1988 adult 3		Field length	17.9	22.4	19.51 (4.4)	64
		Thulborn length	17.8	22.4	19.55 (4.6)	56
		IIILGL	16.6	22.0	19.13 (4.6)	50
		Thulborn width	13.7	18.2	15.81 (6.7)	56
		Toetip width	13.0	17.7	16.01 (6.1)	53
		Clawbase width	11.9	15.6	14.30 (6.3)	14
		Size index	15.9	19.6	17.57 (4.8)	56

Table A6.2. Measurement of metatarsal pad and digit II parameters in individual emus. All measurements are in centimeters. Data reporting conventions as in table A6.1. See text for descriptions of abbreviations.

Bird code	Run	Parameter	Minimum	Maximum	Mean (CV)	Number of footprints
1988 juvenile 2	6	MtPdL	0.8	1.4	1.04 (16)	34
		MtPdW	0.5	1.0	0.80 (13)	32
		IIPdGp	0.9	1.4	1.14 (10)	30
		IIPd1L	0.4	0.7	0.56 (14)	27
		IIPd2L	0.4	0.8	0.58 (12)	26
		IIClwL	0.3	0.6	0.46 (20)	33
1988 juvenile 4	4	MtPdL	1.0	1.7	1.44 (14)	19
		MtPdW	0.7	1.3	1.05 (14)	19
		IIPdGp	1.0	1.4	1.18 (12)	18
		IIPd1L	0.5	0.8	0.64 (11)	16
	10	MtPdL	1.3	2.0	1.78 (13)	12
		MtPdW	1.1	1.9	1.46 (17)	11
		IIPdGp	1.3	2.2	1.63 (14)	12
		IIPd1L	0.7	0.9	0.79 (9.7)	10
		IIClwL	0.5	0.8	0.67 (13)	11
	13	MtPdL	1.6	2.5	2.14 (13)	10
		MtPdW	1.3	2.0	1.68 (14)	10
		IIPdGp	0.9	1.8	1.44 (19)	10
		IIPd1L	0.8	1.3	0.96 (15)	10
		IIPd2L	0.7	1.4	1.01 (21)	10
		IIClwL	0.6	1.1	0.79 (17)	10
	18	MtPdL	1.3	2.6	2.19 (13)	20
		MtPdW	1.7	2.6	2.15 (11)	20
		IIPdGp	1.5	2.8	2.11 (14)	19
		IIPd1L	0.8	1.2	1.01 (12)	19
		IIPd2L	1.1	1.6	1.30 (9.4)	19
		IIClwL	0.5	1.1	0.98 (14)	20
1988 juvenile 7		MtPdW	1.4	1.8	1.58 (7.8)	10
		IIPdGp	0.9	1.7	1.31 (22)	10
		IIClwL	0.3	0.7	0.45 (30)	10
1989 juvenile A		MtPdL	1.5	3.5	2.38 (20)	22
		MtPdW	1.5	2.6	2.12 (12)	20
		IIPdGp	0.9	2.1	1.48 (23)	19
1988 juvenile 1		MtPdW	1.5	2.4	2.11 (14)	10
1989 juvenile C		MtPdL	2.0	4.1	3.07 (17)	39
		MtPdW	2.1	3.7	2.89 (13)	34
		IIPdGp	0.8	3.1	2.05 (20)	35
		IIPd1L	1.2	1.8	1.44 (12)	10
		IIClwL	0.5	1.4	0.96 (29)	15
1990 juvenile 12		MtPdL	3.8	5.2	4.30 (10)	19
		MtPdW	3.7	4.7	4.25 (6.7)	19
		IIPdGp	2.1	4.1	3.08 (17)	19
1990 juvenile 13		MtPdL	3.9	5.5	4.63 (11)	13
		MtPdW	3.5	5.7	4.46 (13)	10
		IIPdGp	2.8	4.5	3.58 (15)	11
		IIPd1L	1.6	2.3	1.94 (14)	10
1988 adult 6		MtPdL	4.0	7.0	4.89 (11)	47
		MtPdW	3.9	7.2	4.47 (12)	46
		IIPdGp	2.1	4.6	3.36 (18)	42
		IIPd1L	1.6	2.8	2.20 (14)	19
		IIPd2L	1.9	3.0	2.35 (14)	15
		IIClwL	0.8	2.5	1.55 (23)	23

Table A6.2. *continued*

Bird code	Run	Parameter	Minimum	Maximum	Mean (CV)	Number of footprints
1988 adult 5		MtPdL	3.5	6.1	4.95 (12)	47
		MtPdW	4.0	5.8	4.77 (8.9)	47
		IIPdGp	1.8	4.7	3.50 (18)	47
		IIPd1L	1.7	2.7	2.13 (11)	31
		IIPd2L	1.3	2.6	2.02 (12)	25
		IIClwL	1.0	2.0	1.54 (18)	20
1988 adult 3		MtPdL	3.3	5.6	4.47 (10)	58
		MtPdW	3.6	5.2	4.48 (8.1)	57
		IIPdGp	2.7	4.7	3.66 (13)	53
		IIPd1L	1.5	2.5	2.10 (10)	32
		IIPd2L	1.3	2.3	1.81 (12)	22
		IIClwL	0.9	1.8	1.38 (19)	24

Table A6.3. Measurements of digit III parameters in individual emus. All measurements are in centimeters. Data reporting conventions as in table A6.1. See text for descriptions of abbreviations.

Bird code	Run	Parameter	Minimum	Maximum	Mean (CV)	Number of footprints
1988 juvenile 2	6	IIIPdGp	0.6	1.2	0.90 (17)	34
		IIIPd1L	0.7	1.0	0.84 (8.3)	27
		IIIPd2L	0.8	1.2	1.05 (8.1)	12
		IIIPd3L	0.5	0.9	0.68 (19)	14
		IIIClwL	0.2	0.8	0.42 (28)	37
1988 juvenile 4	4	IIIPdGp	0.3	1.1	0.64 (32)	19
		IIIPd1L	0.9	1.3	1.08 (8.7)	15
		IIIClwL	0.3	0.6	0.41 (25)	12
	10	IIIPdGp	0.6	1.4	0.87 (29)	12
	13	IIIPdGp	0.3	1.3	0.82 (42)	10
		IIIPd1L	1.4	1.7	1.52 (6.3)	10
		IIIClwL	0.6	1.3	0.86 (24)	10
	18	IIIPdGp	0.6	1.9	1.43 (27)	20
		IIIPd1L	1.5	2.2	1.83 (9.6)	20
		IIIPd2L	1.6	2.4	1.88 (8.3)	20
		IIIPd3L	1.5	1.9	1.65 (7.0)	19
		IIIClwL	0.9	1.2	1.07 (8.1)	17
1989 juvenile A		IIIPdGp	0.6	1.5	0.99 (28)	17
		IIIPd1L	1.5	2.0	1.76 (9.9)	13
1988 juvenile 1		IIIPd1L	1.5	2.1	1.83 (10)	10
		IIIPd2L	2.0	2.4	2.22 (5.5)	10
		IIIPd3L	1.0	1.6	1.40 (11)	10
		IIIClwL	0.5	0.8	0.62 (16)	10
1989 juvenile C		IIIPdGp	0.7	2.6	1.32 (34)	36
		IIIPd1L	1.7	2.8	2.27 (10)	39
		IIIPd2L	1.2	2.5	2.00 (13)	31
		IIIPd3L	1.0	1.9	1.39 (18)	22
		IIIClwL	0.4	1.2	0.83 (26)	27
1990 juvenile 12		IIIPdGp	1.7	2.7	2.32 (13)	19
		IIIPd1L	3.1	4.0	3.51 (6.3)	18
		IIIPd2L	2.7	3.6	3.19 (7.7)	17
1990 juvenile 13		IIIPdGp	1.3	2.7	2.20 (19)	13
		IIIPd1L	3.0	3.7	3.31 (5.4)	12
		IIIPd2L	2.8	3.7	3.21 (9.0)	12
1988 adult 6		IIIPdGp	1.2	3.0	2.04 (18)	49
		IIIPd1L	3.4	4.3	3.88 (5.8)	55
		IIIPd2L	2.9	3.8	3.41 (9.0)	24
		IIIPd3L	1.9	3.9	2.64 (17)	23
		IIIClwL	1.3	2.6	1.93 (16)	45
1988 adult 5		IIIPdGp	1.3	2.9	2.30 (18)	40
		IIIPd1L	3.0	4.1	3.53 (7.2)	42
		IIIPd2L	2.9	4.0	3.30 (9.0)	29
		IIIPd3L	2.0	3.5	2.56 (15)	25
		IIIClwL	1.3	2.6	1.85 (17)	32
1988 adult 3		IIIPdGp	1.0	3.4	2.42 (19)	48
		IIIPd1L	2.9	4.8	3.85 (11)	37
		IIIPd2L	2.8	3.6	3.21 (7.2)	20
		IIIPd3L	2.2	3.5	2.84 (12)	15
		IIIClwL	1.5	2.7	2.26 (17)	12

Table A6.4. Measurements of digit IV parameters in individual emus. All measurements in centimeters. Data reporting conventions as in table A6.1. See text for descriptions of abbreviations.

Bird code	Run	Parameter	Minimum	Maximum	Mean (CV)	Number of footprints
1988 juvenile 2	6	IVPdGp	0.6	1.2	0.85 (17)	31
		IVClwL	0.2	0.6	0.40 (24)	29
1988 juvenile 4	4	IVPdGp	0.7	1.4	0.97 (20)	19
	10	IVPdGp	0.9	1.6	1.19 (22)	11
	13	IVPdGp	1.1	1.6	1.28 (13)	10
	18	IVPdGp	0.9	1.8	1.38 (18)	19
		IVClwL	0.5	1.0	0.76 (17)	17
1989 juvenile A		IVPdGp	0.8	1.3	1.08 (16)	17
1988 juvenile 1		IVPdGp	0.8	2.0	1.52 (26)	10
1989 juvenile C		IVPdGp	0.8	2.1	1.42 (25)	35
		IVClwL	0.4	1.5	0.90 (29)	21
1990 juvenile 12		IVPdGp	1.7	2.6	2.13 (12)	19
1990 juvenile 13		IVPdGp	1.8	2.6	2.12 (13)	12
1988 adult 6		IVPdGp	2.2	3.9	2.93 (13)	44
		IVPd1L	2.3	3.6	2.74 (11)	18
		IVClwL	1.1	2.1	1.58 (17)	28
1988 adult 5		IVPdGp	1.6	3.2	2.51 (17)	43
		IVPd1L	1.9	3.3	2.39 (15)	12
		IVClwL	1.2	2.5	1.65 (22)	22
1988 adult 3		IVPdGp	1.4	3.7	2.27 (23)	43
		IVClwL	1.0	2.0	1.55 (17)	24

Table A6.5. Variability of measurements of overall digit lengths in individual emus. All measurements are in centimeters. Data reporting conventions as in table A6.1; for values of IIILGL, see table A6.1. See text for descriptions of abbreviations.

Bird code	Run	Parameter	Minimum	Maximum	Mean (CV)	Number of footprints
1988 juvenile 2	6	IIPdL	0.8	1.5	1.10 (14)	34
		IILgth	1.1	1.9	1.54 (11)	36
		IIFRNC	2.1	2.5	2.25 (4.0)	26
		IIFRL	1.9	2.9	2.66 (7.2)	30
		IILGNC	2.6	3.0	2.81 (3.3)	30
		IILGL	2.6	3.5	3.21 (5.3)	34
		IIIPdL	2.2	2.9	2.53 (6.9)	37
		IIILgth	2.5	3.3	2.89 (5.9)	41
		IIIFRNC	3.1	3.7	3.40 (5.2)	30
		IIIFRL	3.2	4.3	3.76 (5.6)	34
		IIILGNC	4.0	4.8	4.43 (4.0)	32
1988 juvenile 4	4	IILgth	1.3	2.1	1.74 (10)	20
		IIFRL	2.3	3.3	2.87 (8.1)	19
		IILGL	3.0	4.0	3.58 (7.0)	19
		IIIPdL	2.8	3.1	2.94 (3.0)	10
		IIILgth	3.0	3.4	3.25 (3.2)	19
		IIIFRNC	3.2	3.9	3.50 (6.4)	10
		IIIFRL	3.4	4.2	3.85 (5.8)	19
		IIILGNC	4.6	5.1	4.91 (3.0)	10
		IVLgth	1.7	2.3	1.93 (8.7)	18
		IVFRL	2.5	3.1	2.88 (4.8)	16
		IVLGL	3.4	3.8	3.62 (2.9)	16
	10	IIPdL	1.4	1.9	1.68 (7.2)	11
		IILgth	2.0	2.5	2.32 (5.1)	13
		IIFRNC	2.8	3.8	3.25 (8.7)	11
		IIFRL	3.5	4.4	3.90 (6.2)	12
		IILGNC	3.9	4.7	4.21 (5.8)	11
		IILGL	4.5	5.2	4.85 (4.6)	13
		IIILgth	4.3	4.8	4.54 (3.3)	13
		IIIFRL	5.1	6.1	5.39 (5.6)	12
		IVLgth	2.3	3.1	2.71 (10)	12
		IVFRL	3.7	4.4	3.85 (5.4)	10
		IVLGL	4.4	5.2	4.77 (6.1)	12
	13	IIPdL	1.8	2.3	1.96 (8.9)	10
		IILgth	2.5	3.0	2.75 (7.2)	10
		IIFRNC	3.1	3.8	3.35 (7.6)	10
		IIFRL	3.7	4.8	4.12 (8.1)	10
		IILGNC	4.2	4.9	4.58 (4.8)	10
		IILGL	4.9	6.0	5.35 (6.0)	10
		IIIPdL	4.1	4.8	4.42 (4.3)	10
		IIILgth	5.0	5.6	5.23 (3.2)	10
		IIIFRNC	4.4	5.8	5.22 (8.3)	10
		IIIFRL	5.2	6.5	6.03 (6.5)	10
		IIILGNC	6.7	7.8	7.34 (4.5)	10
		IVPdL	2.1	2.7	2.40 (9.6)	10
		IVFRNC	3.3	4.0	3.64 (5.3)	10
		IVLGNC	4.4	5.2	4.75 (5.5)	10
	18	IIPdL	1.9	2.5	2.30 (6.8)	19
		IILgth	2.4	3.5	3.25 (7.7)	19
		IIFRNC	3.8	4.9	4.32 (7.1)	20
		IIFRL	4.7	6.0	5.25 (6.7)	20
		IILGNC	5.0	6.3	5.63 (4.9)	20
		IILGL	6.0	7.3	6.56 (4.5)	20
		IIIPdL	5.0	5.7	5.35 (4.4)	19
		IIILgth	6.0	6.7	6.40 (2.9)	18

Table A6.5. *continued*

Bird code	Run	Parameter	Minimum	Maximum	Mean (CV)	Number of footprints
		IIIFRNC	6.2	7.2	6.75 (4.2)	19
		IIIFRL	7.2	8.2	7.76 (3.8)	18
		IIILGNC	8.2	9.4	8.91 (3.1)	19
		IVPdL	2.5	3.3	2.95 (7.1)	19
1988 juvenile 4	18	IVLgth	3.3	3.9	3.68 (5.8)	16
		IVFRNC	3.8	4.8	4.27 (6.1)	20
		IVFRL	4.3	5.5	4.99 (6.1)	17
		IVLGNC	5.0	6.1	5.61 (5.0)	20
		IVLGL	5.7	6.7	6.33 (4.9)	17
		IIPdL	0.8	1.7	1.34 (25)	10
		IILgth	1.1	2.1	1.74 (22)	10
		IIFRNC	2.3	3.0	2.61 (8.1)	10
1988 juvenile 7		IIFRL	2.5	3.7	3.05 (11)	11
		IILGNC	3.3	3.9	3.68 (5.4)	10
		IILGL	3.5	4.5	4.11 (7.2)	11
		IIILgth	3.1	4.2	3.84 (9.7)	10
		IILgth	2.2	3.6	2.92 (13)	19
		IIFRL	3.4	5.2	4.30 (11)	20
		IILGL	4.6	6.4	5.68 (9.6)	23
1989 juvenile A		IIILgth	4.2	6.5	4.97 (12)	17
		IIIFRL	5.0	7.5	6.03 (10)	22
		IVLgth	2.8	3.9	3.37 (8.8)	14
		IVFRL	3.2	5.5	4.40 (11)	18
		IVLGL	4.5	6.8	5.78 (8.7)	21
		IIFRL	4.8	5.9	5.33 (7.0)	10
		IILGL	6.3	7.1	6.67 (3.5)	10
		IIIPdL	5.0	5.7	5.42 (3.9)	10
1988 juvenile 1		IIILgth	5.4	6.2	5.87 (4.1)	10
		IIILGNC	8.7	9.7	9.41 (3.3)	10
		IVLgth	3.4	4.8	3.89 (9.9)	10
		IVFRL	4.9	5.6	5.35 (3.8)	10
		IVLGL	6.3	7.1	6.68 (3.8)	10
		IIPdL	2.2	3.4	2.58 (13)	17
		IILgth	2.3	4.6	3.47 (12)	37
		IIFRNC	3.0	5.3	4.56 (15)	15
		IIFRL	3.6	6.7	5.48 (11)	31
		IILGNC	4.2	7.1	6.28 (11)	16
		IILGL	4.7	8.1	7.04 (9.8)	34
		IIIPdL	4.9	6.4	5.72 (5.3)	21
		IIILgth	4.8	6.8	6.22 (7.5)	44
1989 juvenile C		IIIFRNC	6.5	8.7	7.36 (8.1)	15
		IIIFRL	5.5	9.0	7.55 (10)	37
		IIILGNC	9.4	10.8	10.21 (4.3)	17
		IVPdL	3.1	4.4	3.62 (8.7)	18
		IVLgth	2.8	4.7	4.24 (10)	37
		IVFRNC	4.3	5.7	5.02 (7.4)	17
		IVFRL	3.9	6.4	5.61 (9.9)	36
		IVLGNC	6.4	7.3	6.73 (3.6)	19
		IVLGL	5.5	8.0	7.35 (6.9)	38

Bird code	Run	Parameter	Minimum	Maximum	Mean (CV)	Number of footprints
1990 juvenile 12		IILgth	4.1	7.1	5.39 (15)	19
		IIFRL	7.5	9.6	8.35 (7.8)	19
		IILGL	9.3	12.1	10.54 (5.7)	19
		IIILgth	9.2	11.3	10.28 (6.7)	18
		IIIFRL	11.7	13.5	12.55 (4.8)	18
		IVLgth	5.8	8.0	6.86 (7.1)	19
		IVFRL	7.4	10.1	8.90 (6.8)	19
		IVLGL	10.2	12.2	11.40 (4.2)	19
1990 juvenile 13		IILgth	4.2	6.3	5.20 (13)	13
		IIFRL	7.1	9.5	8.82 (8.2)	11
		IILGL	9.6	12.2	11.01 (6.8)	13
		IIILgth	8.6	11.1	10.08 (8.3)	13
		IIIFRL	10.5	13.6	12.19 (7.8)	13
		IVLgth	6.9	7.6	7.26 (3.1)	13
		IVFRL	8.9	9.8	9.28 (3.1)	12
		IVLGL	10.8	12.4	11.82 (3.9)	13
1988 adult 6		IIPdL	3.6	5.5	4.34 (9.5)	32
		IILgth	3.7	7.6	5.84 (14)	35
		IIFRNC	6.1	8.7	7.80 (6.2)	28
		IIFRL	6.6	10.4	9.21 (9.1)	31
		IILGNC	8.9	11.3	10.34 (5.3)	30
		IILGL	9.4	14.7	11.89 (9.0)	34
		IIIPdL	9.2	11.0	9.93 (4.8)	41
		IIILgth	10.6	12.7	11.81 (4.9)	60
		IIIFRNC	10.4	13.4	12.02 (5.8)	32
		IIIFRL	11.9	15.0	13.89 (4.8)	47
		IIILGNC	14.6	18.2	16.77 (4.9)	38
		IVPdL	5.4	6.9	6.25 (6.3)	27
		IVLgth	6.5	8.6	7.82 (5.7)	51
		IVFRNC	7.5	9.7	8.97 (6.3)	23
		IVFRL	8.7	11.4	10.57 (5.6)	42
		IVLGNC	10.3	12.7	11.78 (4.8)	26
		IVLGL	11.9	14.3	13.22 (4.8)	49
1988 adult 5		IIPdL	3.5	4.9	4.08 (8.2)	30
		IILgth	4.2	7.9	5.79 (13)	38
		IIFRNC	6.8	9.0	7.65 (6.4)	29
		IIFRL	7.8	10.8	9.18 (9.0)	37
		IILGNC	9.7	11.6	10.52 (4.1)	27
		IILGL	10.3	14.3	12.07 (7.3)	35
		IIIPdL	8.6	10.8	9.41 (5.2)	33
		IIILgth	10.2	13.2	11.32 (6.2)	44
		IIIFRNC	10.0	12.8	11.78 (5.1)	27
		IIIFRL	11.9	16.3	13.71 (5.7)	50
		IIILGNC	15.2	17.5	16.43 (3.7)	29
		IVPdL	5.2	7.5	5.90 (9.7)	23
		IVLgth	6.2	9.5	7.48 (8.7)	45
		IVFRNC	7.6	9.2	8.47 (5.4)	22
		IVFRL	8.2	11.6	9.90 (7.0)	48
		IVLGNC	10.1	12.4	11.17 (5.2)	21
		IVLGL	10.8	14.4	12.76 (5.9)	51

Table A6.5. *continued*

Bird code	Run	Parameter	Minimum	Maximum	Mean (CV)	Number of footprints
		IIPdL	3.2	4.5	3.87 (6.8)	23
		IILgth	4.3	7.0	5.45 (12)	46
		IIFRNC	6.6	8.3	7.55 (6.0)	24
		IIFRL	7.1	10.2	8.96 (6.3)	53
		IILGNC	9.1	10.6	10.09 (3.8)	24
		IILGL	9.9	13.2	11.60 (5.0)	50
		IIIPdL	9.0	10.8	9.61 (4.8)	16
		IIILgth	10.7	14.1	12.31 (5.9)	45
1988 adult 3		IIIFRNC	10.9	13.1	12.04 (5.1)	14
		IIIFRL	12.3	16.7	14.73 (5.6)	52
		IIILGNC	15.2	17.3	16.34 (3.8)	15
		IVPdL	4.9	7.3	6.06 (9.1)	23
		IVLgth	6.2	8.9	7.60 (6.9)	43
		IVFRNC	7.2	9.3	8.12 (6.7)	22
		IVFRL	7.4	11.2	9.71 (7.2)	52
		IVLGNC	9.7	11.9	11.01 (5.3)	24
		IVLGL	10.8	14.5	12.62 (5.5)	54

Table A6.6. Values of best-fit interdigital angles (IDAs) in individual emus. All values are in degrees. Data reporting conventions as in table A6.1. See text for descriptions of abbreviations.

Bird code	Run	Parameter	Minimum	Maximum	Mean (CV)	Number of footprints
1988 juvenile 2	6	JOF IDA II–III	13	41	28.8 (23)	36
		MCC IDA II–III	32	48	38.0 (14)	11
		JOF IDA III–IV	25	75	58.6 (17)	35
		MCC IDA III–IV	36	76	51.5 (25)	11
		JOF IDA II–IV	65	101	88.3 (11)	31
		MCC IDA II–IV	69	108	89.3 (15)	10
1988 juvenile 4	4	JOF IDA II–III	35	51	42.7 (9.0)	17
		JOF IDA III–IV	43	70	55.5 (14)	15
		JOF IDA II–IV	88	115	97.9 (7.8)	18
	10	JOF IDA II–III	23	46	34.8 (20)	13
		JOF IDA III–IV	54	65	60.7 (6.1)	12
		JOF IDA II–IV	81	105	94.5 (8.0)	12
	13	JOF IDA II–III	22	49	36.8 (19)	10
		MCC IDA II–III	25	52	37.6 (22)	10
		JOF IDA III–IV	39	63	52.8 (14)	10
		MCC IDA III–IV	44	61	53.7 (10)	10
		JOF IDA II–IV	76	112	89.7 (11)	10
		JOF IDA II–IV	76	113	91.3 (11)	10
	18	JOF IDA II–III	25	41	35.0 (11)	19
		MCC IDA II–III	26	42	34.4 (14)	14
		JOF IDA III–IV	42	66	50.3 (12)	19
		MCC IDA III–IV	41	66	51.6 (15)	14
		JOF IDA II–IV	77	101	85.8 (7.3)	18
		MCC IDA II–IV	77	101	85.8 (7.3)	18
1989 juvenile A		JOF IDA II–III	21	47	38.6 (19)	14
		TOPO IDA II–III	25	52	36.4 (20)	14
		JOF IDA III–IV	29	60	44.0 (23)	11
		TOPO IDA III–IV	32	66	47.7 (21)	14
		JOF IDA II–IV	63	105	81.3 (14)	15
		TOPO IDA II–IV	57	109	82.1 (16)	14
1988 juvenile 1		JOF IDA III–IV	42	61	53.9 (8.9)	10
1989 juvenile C		JOF IDA II–III	27	53	37.2 (16)	33
		TOPO IDA II–III	25	48	34.9 (16)	15
		JOF IDA III–IV	40	72	49.5 (14)	35
		TOPO IDA III–IV	44	70	51.7 (13)	15
		JOF IDA II–IV	71	121	85.9 (10)	31
		TOPO IDA II–IV	73	103	86.7 (9.2)	15
1990 juvenile 12		JOF IDA II–III	8	55	27.8 (38)	18
		JOF IDA III–IV	28	59	42.7 (19)	18
		JOF IDA II–IV	45	96	69.9 (19)	19
1990 juvenile 13		JOF IDA II–III	20	32	25.4 (15)	13
		JOF IDA III–IV	25	47	38.0 (17)	13
		JOF IDA II–IV	46	74	63.4 (14)	13
1988 adult 6		JOF IDA II–III	15	51	33.1 (27)	45
		MCC IDA II–III	13	59	38.0 (27)	65
		TOPO IDA II–III	25	57	39.4 (25)	14
		JOF IDA III–IV	24	54	38.8 (14)	51
		MCC IDA III–IV	32	76	43.6 (18)	65
		TOPO IDA III–IV	35	62	46.3 (15)	20
		JOF IDA II–IV	49	90	72.0 (14)	38
		MCC IDA II–IV	55	118	81.8 (15)	70
		TOPO IDA II–IV	76	119	86.3 (13)	14

Table A6.6. *continued*

Bird code	Run	Parameter	Minimum	Maximum	Mean (CV)	Number of footprints
1988 adult 5		JOF IDA II–III	0	38	24.8 (29)	40
		MCC IDA II–III	2	61	31.2 (37)	68
		TOPO IDA II–III	11	43	29.2 (30)	20
		TRACE IDA II–III	19	30	23.6 (16)	10
		JOF IDA III–IV	23	52	43.3 (14)	41
		MCC IDA III–IV	15	61	43.9 (20)	70
		TOPO IDA III–IV	38	55	46.1 (12)	22
		TRACE IDA III–IV	43	51	47.5 (4.8)	10
		JOF IDA II–IV	51	89	68.6 (13)	39
		MCC IDA II–IV	43	116	75.3 (18)	70
		TOPO IDA II–IV	60	98	75.8 (13)	19
		TRACE IDA II–IV	65	78	71.2 (5.4)	10
1988 adult 3		JOF IDA II–III	16	41	30.9 (19)	34
		MCC IDA II–III	9	52	33.2 (27)	57
		TOPO IDA II–III	23	51	35.4 (24)	11
		JOF IDA III–IV	31	62	43.2 (17)	38
		MCC IDA III–IV	27	61	43.7 (17)	57
		TOPO IDA III–IV	32	54	42.2 (15)	13
		JOF IDA II–IV	56	93	74.4 (13)	32
		MCC IDA II–IV	45	99	76.9 (14)	57
		TOPO IDA II–IV	55	96	76.3 (15)	10

Table A6.7. Reference point–based interdigital angles (IDAs) and spread angles (SAs) in individual emus. All measurements are in degrees. Data reporting conventions as in table A6.1. See text for descriptions of abbreviations.

Bird code	Run	Parameter	Minimum	Maximum	Mean (CV)	Number of footprints
1988 juvenile 2	6	Toebase IDA	88	112	101.5 (6.3)	33
		Clawbase IDA	71	80	74.5 (3.1)	23
		Toetip IDA	61	89	79.1 (7.2)	39
		Clawbase SA	84	108	95.6 (6.4)	22
		Toetip SA	74	109	96.4 (8.3)	33
1988 juvenile 4	4	Toebase IDA	98	133	115.0 (8.3)	18
		Toetip IDA	79	90	85.3 (4.0)	19
		Toetip SA	89	102	95.3 (4.3)	16
	10	Toebase IDA	106	136	117.8 (7.1)	13
		Toetip IDA	82	90	84.4 (2.7)	12
		Toetip SA	85	105	95.8 (6.1)	12
	13	Toebase IDA	103	126	111.6 (7.7)	10
		Clawbase IDA	73	90	78.4 (6.2)	10
		Clawbase SA	88	108	95.1 (6.9)	10
	18	Toebase IDA	95	123	110.2 (6.6)	18
		Clawbase IDA	74	82	77.5 (3.1)	19
		Toetip IDA	75	85	80.0 (3.6)	15
		Clawbase SA	89	105	96.3 (4.4)	20
		Toetip SA	88	103	95.3 (4.7)	17
1989 juvenile A		Toebase IDA	86	115	102.1 (9.3)	12
		Toetip IDA	71	93	83.2 (6.1)	23
		Toetip SA	69	102	85.7 (10)	21
1988 juvenile 1		Toetip IDA	81	94	84.7 (4.6)	10
1989 juvenile C		Toebase IDA	84	118	103.3 (8.6)	30
		Toetip IDA	69	100	85.2 (8.2)	36
		Clawbase SA	85	100	90.1 (5.8)	10
		Toetip SA	78	100	88.2 (4.9)	31
1990 juvenile 12		Toebase IDA	85	107	98.4 (5.9)	19
		Toetip IDA	73	89	78.6 (5.5)	18
		Toetip SA	68	93	79.0 (8.9)	19
1990 juvenile 13		Toebase IDA	93	115	103.4 (5.7)	13
		Toetip IDA	74	97	82.7 (9.5)	13
		Toetip SA	62	80	72.5 (7.6)	13
1988 adult 6		Toebase IDA	86	130	111.6 (8.3)	42
		Clawbase IDA	72	88	79.7 (6.2)	12
		Toetip IDA	70	93	83.3 (6.4)	41
		Clawbase SA	77	91	82.5 (5.1)	16
		Toetip SA	67	91	78.9 (7.9)	31
1988 adult 5		Toebase IDA	90	121	107.3 (6.2)	33
		Clawbase IDA	77	83	79.4 (2.2)	13
		Toetip IDA	66	87	80.3 (5.3)	39
		Clawbase SA	74	94	82.9 (5.6)	17
		Toetip SA	70	93	79.2 (6.4)	32
1988 adult 3		Toebase IDA	77	132	108.1 (10)	36
		Toetip IDA	61	89	75.7 (7.6)	45
		Clawbase SA	76	92	86.1 (5.7)	14
		Toetip SA	72	92	82.6 (5.9)	45

Table A6.8. Reduced major axis (RMA) relationships between various linear dimensions of emu footprints and study skins. All measurements were log transformed prior to analysis. CI = confidence interval; SEE = standard error of estimate. For the "all footprints" treatment, individual footprints constitute the data cases, and particular birds will be represented by a variable number of prints. For the "emu means" treatment, each linear dimension represents the mean of all values for footprints of a given bird (where the trackmaker was a captive animal) or trackway (for footprints collected in the field), and the data cases constitute individual birds or trackways. Variable abbreviations as described in text. RMA slopes significantly different than 1 are indicated in **bold**. See text for descriptions of variable abbreviations.

Dependent variable	Treatment	r^2	SEE	Number of cases	Slope	95% CI of slope
Independent variable = length of first digital pad of digit III (IIIPd1L)						
MtPdL	All footprints	0.889	0.0666	364	0.969	0.934–1.006
	Emu means	0.906	0.0462	32	**0.865**[a]	0.766–0.976
MtPdW	All footprints	0.910	0.0676	360	**1.103**	1.064–1.140
	Emu means	0.909	0.0507	34	0.977	0.871–1.096
IIPdGp	All footprints	0.761	0.0872	357	**0.885**	0.834–0.939
	Emu means	0.837	0.0559	33	**0.795**	0.676–0.936
IIPd1L	All footprints	0.901	0.0652	217	0.956	0.914–1.000
	Emu means	0.963	0.0326	20	0.908	0.824–1.001
IIPd2L	All footprints	0.865	0.0704	187	**0.891**	0.841–0.944
	Emu means	0.890	0.0522	20	0.845	0.709–1.007
IIClwL	All footprints	0.720	0.117	238	1.018	0.939–1.104
	Emu means	0.864	0.0696	27	1.039	0.881–1.225
IIIPdGp	All footprints	0.590	0.136	376	1.035	0.950–1.127
	Emu means	0.785	0.0746	33	0.922	0.760–1.120
IIIPd2L	All footprints	0.846	0.0639	298	**0.877**	0.835–0.921
	Emu means	0.893	0.0500	31	0.895	0.784–1.021
IIIPd3L	All footprints	0.766	0.0876	217	0.928	0.861–1.000
	Emu means	0.831	0.0694	30	0.981	0.822–1.170
IIIClwL	All footprints	0.776	0.120	272	**1.163**	1.091–1.241
	Emu means	0.792	0.0971	34	**1.254**[a]	1.041–1.512
IVPdGp	All footprints	0.733	0.0969	354	**0.922**[a]	0.865–0.982
	Emu means	0.829	0.0704	33	0.987	0.834–1.167
IVPdL	All footprints	0.948	0.0469	223	0.989	0.959–1.020
	Emu means	0.968	0.0295	30	0.935	0.871–1.003
IVClwL	All footprints	0.701	0.134	213	**1.181**	1.081–1.292
	Emu means	0.793	0.105	29	**1.287**[a]	1.049–1.579
Proximal end digit II to proximal end digit III	All footprints	0.905	0.0541	391	**0.841**	0.814–0.869
	Emu means	0.884	0.0477	34	**0.808**	0.709–0.922
Proximal end digit III to proximal end digit IV	All footprints	0.887	0.0589	392	**0.854**	0.824–0.885
	Emu means	0.929	0.0357	34	**0.780**	0.706–0.862
Proximal end digit II to proximal end digit IV	All footprints	0.924	0.0473	361	**0.831**	0.807–0.856
	Emu means	0.916	0.0396	33	**0.786**	0.703–0.878
Clawbase II to clawbase III	All footprints	0.948	0.0441	193	**0.878**	0.849–0.908
	Emu means	0.960	0.0320	26	**0.904**[a]	0.829–0.985
Clawbase III to clawbase IV	All footprints	0.951	0.0374	176	**0.789**	0.762–0.816
	Emu means	0.974	0.0226	27	**0.791**	0.740–0.847
Clawbase II to clawbase IV	All footprints	0.955	0.0360	183	**0.825**	0.799–0.852
	Emu means	0.954	0.0315	26	**0.795**	0.725–0.873
Toetip II to toetip III	All footprints	0.913	0.0557	345	**0.890**	0.861–0.920
	Emu means	0.939	0.0384	31	**0.879**[a]	0.798–0.969
Toetip III to toetip IV	All footprints	0.897	0.0537	374	**0.804**	0.776–0.832
	Emu means	0.931	0.0366	35	**0.814**	0.739–0.897
Toetip II to toetip IV	All footprints	0.933	0.0443	337	**0.821**	0.797–0.845
	Emu means	0.946	0.0327	32	**0.810**	0.740–0.885
Independent variable = digit III length with claw (IIILgth)						
MtPdL	All footprints	0.886	0.0761	429	1.000	0.966–1.035
	Emu means	0.897	0.0484	32	0.888	0.782–1.008
MtPdW	All footprints	0.906	0.0774	415	**1.139**	1.103–1.175
	Emu means	0.884	0.0572	34	0.995	0.873–1.135

Dependent variable	Treatment	r^2	SEE	Number of cases	Slope	95% CI of slope
IIPdGp	All footprints	0.826	0.0861	411	**0.937**[a]	0.896–0.981
	Emu means	0.847	0.0542	33	**0.815**[a]	0.696–0.953
IILgth	All footprints	0.936	0.0562	396	0.978	0.953–1.004
	Emu means	0.972	0.0271	32	0.953	0.895–1.015
	Study skins	0.986	——	30	**0.929**[a]	0.887–0.973
IIIPdGp	All footprints	0.628	0.134	439	0.975	0.906–1.049
	Emu means	0.799	0.0720	33	0.949	0.788–1.143
IVPdGp	All footprints	0.792	0.0960	401	0.964	0.916–1.014
	Emu means	0.889	0.0568	33	1.003	0.880–1.142
IVLgth	All footprints	0.958	0.0467	409	**1.033**[a]	1.012–1.054
	Emu means	0.971	0.0288	34	1.007	0.946–1.072
	Study skins	0.991	——	30	0.986	0.951–1.023
Proximal end II to proximal end III	All footprints	0.930	0.0513	450	**0.866**	0.844–0.889
	Emu means	0.896	0.0452	34	**0.831**	0.735–0.940
Proximal end III to proximal end IV	All footprints	0.908	0.0574	447	**0.860**	0.834–0.886
	Emu means	0.936	0.0339	34	**0.796**	0.724–0.875
Proximal end II to proximal end IV	All footprints	0.958	0.0395	407	**0.880**	0.862–0.899
	Emu means	0.922	0.0382	33	**0.806**	0.724–0.897
Clawbase II to clawbase III	All footprints	0.968	0.0376	203	**0.971**[a]	0.947–0.996
	Emu means	0.970	0.0277	26	**0.914**[a]	0.849–0.985
Clawbase III to clawbase IV	All footprints	0.964	0.0347	186	**0.852**	0.828–0.877
	Emu means	0.969	0.0246	27	**0.810**	0.753–0.872
Clawbase II to clawbase IV	All footprints	0.968	0.0334	186	**0.893**	0.870–0.917
	Emu means	0.962	0.0284	26	**0.814**	0.749–0.886
Toetip II to toetip III	All footprints	0.960	0.0427	425	**0.946**[a]	0.928–0.965
	Emu means	0.965	0.0292	31	**0.910**[a]	0.846–0.978
Toetip III to toetip IV	All footprints	0.954	0.0402	457	**0.833**	0.817–0.851
	Emu means	0.967	0.0254	35	**0.832**	0.779–0.889
Toetip II to toetip IV	All footprints	0.966	0.0365	390	**0.866**	0.850–0.882
	Emu means	0.963	0.0273	32	**0.828**	0.770–0.891

[a]Only barely allometric (upper CI limit of RMA slope between 0.95 and 1.00, or lower CI limit between 1.00 and 1.05).

Table A6.9. Miscellaneous reduced major axis (RMA) relationships between various linear dimensions of emu footprints and study skins. Some of these relationships are counterparts or analogs of bivariate relationships examined in alligator feet. All measurements were log transformed prior to analysis. The "all footprints" and "emu means" treatments are as described in table A6.8. Variable abbreviations as described in text. RMA slopes significantly different than 1 are indicated in **bold**. See text for descriptions of variable abbreviations.

Independent variable	Dependent variable	Treatment	r^2	SEE	Number of cases	Slope	95% CI of slope
IILgth	Digit II lateral hypex length	All footprints	0.806	0.0488	262	0.990	0.932–1.051
		Emu means	0.887	0.0356	24	0.948	0.808–1.111
IIILgth	Digit III medial hypex length	All footprints	0.959	0.0271	276	**1.051**[a]	1.025–1.077
		Emu means	0.972	0.0227	24	1.053	0.977–1.136
	Digit III lateral hypex length	All footprints	0.957	0.0257	292	1.001	0.977–1.026
		Emu means	0.979	0.0180	28	0.985	0.929–1.045
IVLgth	Digit IV medial hypex length	All footprints	0.881	0.0401	286	0.961	0.921–1.003
		Emu means	0.952	0.0275	26	0.997	0.907–1.096
IILgth	IVLgth	All footprints	0.913	0.0670	373	**1.069**[a]	1.035–1.103
		Emu means	0.971	0.0298	31	1.064	0.997–1.137
		Study skins	0.992	———	29	**1.062**[a]	1.025–1.100
IIPdL	IIClwL	All footprints	0.723	0.115	277	1.060	0.985–1.142
		Emu means	0.831	0.0775	27	1.184	0.981–1.429
		Study skins	0.939	———	30	**1.163**	1.054–1.284
IIIPdL	IIIClwL	All footprints	0.791	0.121	304	**1.255**	1.183–1.330
		Emu means	0.800	0.0952	34	**1.367**	1.139–1.640
		Study skins	0.919	———	33	0.990	0.888–1.104
IVPdL	IVClwL	All footprints	0.674	0.147	240	**1.150**	1.052–1.258
		Emu means	0.743	0.117	29	**1.376**	1.086–1.744
		Study skins	0.954	———	30	**1.240**	1.139–1.350
IIIPdL	IIPdL	All footprints	0.951	0.0472	206	1.005	0.974–1.037
		Emu means	0.958	0.0325	26	0.968	0.886–1.057
	IVPdL	All footprints	0.963	0.0440	188	**1.083**	1.052–1.114
		Emu means	0.959	0.0337	29	1.040	0.958–1.129
IIPdL	IVPdL	All footprints	0.915	0.0641	196	**1.086**[a]	1.040–1.134
		Emu means	0.952	0.0375	27	1.074	0.979–1.178
IIIClwL	IIClwL	All footprints	0.697	0.128	186	**0.892**[a]	0.810–0.983
		Emu means	0.817	0.0776	26	0.837	0.684–1.025
		Study skins	0.962	———	32	1.040	0.965–1.120
	IVClwL	All footprints	0.670	0.153	174	1.044	0.939–1.161
		Emu means	0.691	0.129	28	1.084	0.822–1.429
		Study skins	0.976	———	33	**1.177**	1.111–1.247
IIClwL	IVClwL	All footprints	0.645	0.154	174	**1.169**[a]	1.045–1.308
		Emu means	0.768	0.110	27	1.210	0.961–1.523
		Study skins	0.971	———	32	**1.128**	1.057–1.203
Proximal end II to proximal end IV	IIIClwL	Emu means	0.774	0.100	31	**1.513**	1.239–1.864
Clawbase II to clawbase IV	IIIClwL	Emu means	0.791	0.101	25	**1.504**	1.200–1.884
Toetip II to toetip IV	IIIClwL	Emu means	0.789	0.0979	30	**1.484**	1.211–1.818
MtPdL	Toetip II to toetip IV	Emu means	0.912	0.0426	30	0.931	0.825–1.051
IIIPdGp	Toetip II to toetip IV	Emu means	0.765	0.0690	31	0.873	0.705–1.081
IIIPd2L	Toetip II to toetip IV	Emu means	0.928	0.0389	28	0.919	0.821–1.028
IIIPd3L	Toetip II to toetip IV	Emu means	0.870	0.0530	27	0.890	0.758–1.045
IIIPdL	Toetip II to toetip IV	Emu means	0.978	0.0214	30	**0.906**[a]	0.855–0.961

[a]Only barely allometric (upper CI limit of RMA slope between 0.95 and 1.00, or lower CI limit between 1.00 and 1.05).

Table A6.10. Coefficient of relative dispersion (D_d) about the reduced major axis for bivariate comparisons involving emu footprints. To make D_d values comparable to those calculated elsewhere in this book, all measurements were first converted from centimeters to millimeters before analysis. Values are calculated for raw measurements, and also for log-transformed data (in parentheses). See text for descriptions of variable abbreviations.

Variables		Treatment	D_d	N
MtPdL	IIIPd1L	All emu prints	16.2 (4.69)	364
	IIPd1L	All emu prints	18.1 (6.28)	250
	IIIPd1L	High-quality prints	15.2 (4.16)	124
MtPdW	IIIPd1L	All emu prints	14.8 (4.52)	360
	IIPd1L	All emu prints	17.4 (6.17)	240
	IIIPd1L	High-quality prints	13.5 (3.82)	129
IIPdGp	IIIPd1L	All emu prints	20.9 (6.86)	357
	IIPd1L	All emu prints	23.1 (8.31)	250
	IIIPd1L	High-quality prints	17.8 (5.32)	131
IIPd1L	IIIPd1L	All emu prints	15.9 (5.40)	217
	IIIPd1L	High-quality prints	13.0 (4.51)	76
IIPd2L	IIIPd1L	All emu prints	18.6 (6.03)	187
	IIPd1L	All emu prints	21.6 (8.07)	217
	IIIPd1L	High-quality prints	15.9 (5.11)	73
IIClwL	IIIPd1L	All emu prints	23.6 (10.0)	238
	IIPd1L	All emu prints	26.2 (11.7)	201
	IIIPd1L	High-quality prints	17.6 (7.41)	103
IIIPdGp	IIIPd1L	All emu prints	26.5 (10.9)	376
	IIPd1L	All emu prints	26.6 (12.5)	242
	IIIPd1L	High-quality prints	22.7 (7.95)	129
IIIPd2L	IIIPd1L	All emu prints	15.3 (5.00)	298
	IIPd1L	All emu prints	17.5 (6.44)	160
	IIIPd1L	High-quality prints	14.2 (4.60)	112
IIIPd3L	IIIPd1L	All emu prints	22.1 (7.05)	217
	IIPd1L	All emu prints	21.7 (8.09)	131
	IIIPd1L	High-quality prints	19.7 (6.57)	89
IIIClwL	IIIPd1L	All emu prints	22.5 (9.31)	272
	IIPd1L	All emu prints	27.4 (12.4)	160
	IIIPd1L	High-quality prints	18.4 (7.05)	97
IVPdGp	IIIPd1L	All emu prints	21.6 (7.89)	354
	IIPd1L	All emu prints	23.2 (9.24)	235
	IIIPd1L	High-quality prints	18.0 (5.83)	128
IVPdL	IIIPd1L	All emu prints	12.2 (3.16)	223
	IIPd1L	All emu prints	15.9 (4.98)	150
	IIIPd1L	High-quality prints	10.5 (2.78)	98
IVClwL	IIIPd1L	All emu prints	23.8 (10.7)	213
	IIPd1L	All emu prints	26.8 (12.9)	145
	IIIPd1L	High-quality prints	17.6 (8.90)	93
Digit III length	Digit II length	All emu prints	14.5 (3.36)	396
		High-quality prints	12.4 (2.91)	119
	Digit IV length	All emu prints	10.4 (2.60)	409
		High-quality prints	8.49 (2.19)	118
Digit II length	Digit IV length	All emu prints	15.3 (4.05)	373
		High-quality prints	12.8 (3.33)	119
Digit III length excluding claw	Digit II length excluding claw	All emu prints	10.8 (2.97)	206
		High-quality prints	9.48 (2.37)	81
	Digit IV length excluding claw	All emu prints	9.50 (2.54)	188
		High-quality prints	8.47 (2.06)	80
Digit II length excluding claw	Digit IV length excluding claw	All emu prints	14.1 (4.16)	196
		High-quality prints	10.7 (3.10)	92
Digit II length excluding claw	Digit II claw length	All emu prints	23.1 (9.55)	277
		High-quality prints	16.6 (7.07)	110
Digit III length excluding claw	Digit III claw length	All emu prints	21.4 (7.66)	304
		High-quality prints	16.2 (6.08)	100

Table A6.10. *continued*

Variables		Treatment	D_d	*N*
Digit IV length excluding claw	Digit IV claw length	All emu prints	26.3 (11.0)	240
		High-quality prints	20.0 (9.78)	98
Digit II length	Digit II lateral hypex length	All emu prints	11.3 (2.93)	262
		High-quality prints	9.61 (2.70)	108
Digit III length	Digit III medial hypex length	All emu prints	6.23 (1.33)	276
		High-quality prints	5.43 (1.14)	105
	Digit III lateral hypex length	All emu prints	6.10 (1.28)	292
		High-quality prints	5.52 (1.18)	110
Digit IV length	Digit IV medial hypex length	All emu prints	9.43 (2.27)	286
		High-quality prints	8.15 (2.08)	105
Digit III medial hypex length	Digit II lateral hypex length	All emu prints	13.1 (2.93)	290
		High-quality prints	11.0 (2.61)	109
	Digit III lateral hypex length	All emu prints	3.98 (0.85)	298
		High-quality prints	3.43 (0.72)	112
	Digit IV medial hypex length	All emu prints	11.4 (2.53)	285
		High-quality prints	9.16 (2.11)	106
Digit III claw length	Digit II claw length	All emu prints	32.2 (14.5)	186
		High-quality prints	31.8 (12.6)	73
	Digit IV claw length	All emu prints	31.8 (16.0)	174
		High-quality prints	34.1 (17.1)	72
Digit II claw length	Digit IV claw length	All emu prints	29.7 (15.6)	174
		High-quality prints	28.3 (15.0)	83
Digit III length	Toetip width	All emu prints	9.74 (2.03)	390
		High-quality prints	9.36 (2.13)	114

Table A6.11. Coefficient of relative dispersion (D_d) about the reduced major axis for selected bivariate comparisons involving emu footprints. To make D_d values comparable to those calculated elsewhere in this book, all measurements were first converted from centimeters to millimeters before analysis. In contrast to table A6.10, in which data cases are individual footprints, data cases in this table are means for particular birds ("emu means" treatment; see text for details), so N = the number of birds represented in the calculation. D_d calculations in this table were done only with raw data and not also with log-transformed data. Numbers in parentheses are based on high-quality footprints, but numbers outside parentheses are bird means without regard for footprint quality. See text for descriptions of variable abbreviations.

Variables		D_d	N
IIIPd1L	MtPdL	12.8 (12.7)	32 (22)
	MtPdW	13.1 (14.5)	34 (24)
	IIPdGp	15.6 (17.3)	33 (23)
	IIPd1L	7.26 (7.33)	20 (12)
	IIPd2L	15.2 (13.9)	20 (12)
	IIClwL	15.5 (12.4)	27 (19)
	IIIPdGp	15.9 (16.6)	33 (23)
	IIIPd2L	12.2 (12.6)	31 (21)
	IIIPd3L	14.8 (13.3)	30 (19)
	IIIClwL	17.9 (15.0)	34 (23)
	IVPdGp	17.2 (17.5)	33 (23)
	IVPdL	8.46 (7.60)	30 (19)
	IVClwL	19.3 (18.6)	29 (19)
Digit III length	Digit II length	6.42 (7.05)	32 (22)
	Digit IV length	7.10 (6.84)	34 (24)
Digit II length	Digit IV length	6.75 (8.36)	31 (21)
Digit III length excluding claw	Digit II length excluding claw	7.32 (6.64)	26 (17)
	Digit IV length excluding claw	8.33 (5.68)	29 (18)
Digit II length excluding claw	Digit IV length excluding claw	8.25 (6.37)	27 (18)
Digit II length excluding claw	Digit II claw length	15.6 (13.7)	27 (19)
Digit III length excluding claw	Digit III claw length	15.5 (12.7)	34 (23)
Digit IV length excluding claw	Digit IV claw length	19.5 (15.9)	29 (19)
Digit II length	Digit II lateral hypex length	8.26 (10.4)	24 (18)
Digit III length	Digit III medial hypex length	5.40 (4.87)	24 (18)
	Digit III lateral hypex length	4.29 (4.09)	28 (22)
Digit IV length	Digit IV medial hypex length	6.47 (5.50)	26 (20)
Digit III medial hypex length	Digit II lateral hypex length	8.08 (7.94)	24 (18)
	Digit III lateral hypex length	2.85 (2.80)	24 (18)
	Digit IV medial hypex length	8.66 (9.01)	23 (17)
Digit III claw length	Digit II claw length	19.6 (20.1)	26 (17)
	Digit IV claw length	27.8 (26.6)	28 (18)
Digit II claw length	Digit IV claw length	23.6 (21.4)	27 (18)
Digit III length	Toetip width	7.58 (7.29)	32 (22)

Table A6.12. Variability of emu footprints scaled to a common parameter. In each analysis, the variability of the test parameter is examined after being scaled to a common value of the scaling parameter for all footprints in the treatment. "High-quality prints" are particularly well preserved footprints of juvenile or adult birds. See text for descriptions of abbreviations used for parameters.

Test parameter	Scaling parameter	Treatment	Maximum/minimum	*N*
MtPdL	IIIPd1L	All emu prints	2.89	364
	IIPd1L	All emu prints	2.95	250
	IIIPd1L	High-quality prints	2.22	124
MtPdW	IIIPd1L	All emu prints	2.72	360
	IIPd1L	All emu prints	3.61	240
	IIIPd1L	High-quality prints	1.95	129
IIPdGp	IIIPd1L	All emu prints	5.66	357
	IIPd1L	All emu prints	4.62	250
	IIIPd1L	High-quality prints	2.36	131
IIPd1L	IIIPd1L	All emu prints	2.39	217
	IIIPd1L	High-quality prints	1.93	76
	IIIPdL	All emu prints	2.25	165
		High-quality prints	1.67	56
IIPd2L	IIIPd1L	All emu prints	2.51	187
	IIPd1L	All emu prints	3.45	217
	IIIPd1L	High-quality prints	2.07	73
	IIIPdL	All emu prints	2.37	150
		High-quality prints	2.28	56
IIClwL	IIIPd1L	All emu prints	4.32	238
	IIPd1L	All emu prints	5.00	201
	IIIPd1L	High-quality prints	4.27	103
IIIPdGp	IIIPd1L	All emu prints	7.95	376
	IIPd1L	All emu prints	8.49	242
	IIIPd1L	High-quality prints	4.37	129
IIIPd1L	IIPd1L	All emu prints	2.39	217
	IIIPdL	All emu prints	1.98	282
		High-quality prints	1.72	104
IIIPd2L	IIIPd1L	All emu prints	3.39	298
	IIPd1L	All emu prints	4.01	160
	IIIPd1L	High-quality prints	2.64	112
	IIIPdL	All emu prints	1.67	217
		High-quality prints	1.53	89
IIIPd3L	IIIPd1L	All emu prints	3.65	217
	IIPd1L	All emu prints	3.67	131
	IIIPd1L	High-quality prints	3.65	89
	IIIPdL	All emu prints	2.51	226
		High-quality prints	2.22	90
IIIClwL	IIIPd1L	All emu prints	5.41	272
	IIPd1L	All emu prints	8.26	160
	IIIPd1L	High-quality prints	3.11	97
IVPdGp	IIIPd1L	All emu prints	4.11	354
	IIPd1L	All emu prints	3.59	235
	IIIPd1L	High-quality prints	2.55	128
IVPdL	IIIPd1L	All emu prints	2.04	223
	IIPd1L	All emu prints	2.30	150
	IIIPd1L	High-quality prints	1.92	98
IVClwL	IIIPd1L	All emu prints	7.77	213
	IIPd1L	All emu prints	8.04	145
	IIIPd1L	High-quality prints	7.77	93
IILgth	IIILgth	All emu prints	2.20	396
	IIILgth	High-quality prints	2.02	119
IIILgth	IILgth	All emu prints	2.20	396

Test parameter	Scaling parameter	Treatment	Maximum/minimum	*N*
IVLgth	IIILgth	All emu prints	2.67	409
	IILgth	All emu prints	2.80	373
	IIILgth	High-quality prints	1.96	118
IILgth	IIIPdL	All emu prints	2.17	245
		High-quality prints	2.05	92
IIPdL	IIIPdL	All emu prints	1.83	206
		High-quality prints	1.75	81
IIClwL	IIIPdL	All emu prints	4.58	191
		High-quality prints	4.58	77
IIIClwL	IIIPdL	All emu prints	4.74	304
		High-quality prints	3.06	100
IVLgth	IIIPdL	All emu prints	1.68	256
		High-quality prints	1.53	92
IVPdL	IIIPdL	All emu prints	1.56	188
		High-quality prints	1.49	80
IVClwL	IIIPdL	All emu prints	6.72	176
		High-quality prints	6.72	76
II Lateral length	IIILgth	All emu prints	2.44	254
		High quality prints	1.77	100
III Medial length	IIILgth	All emu prints	1.50	276
		High-quality prints	1.50	105
III Lateral length	IIILgth	All emu prints	1.61	292
		High-quality prints	1.42	110
IV Medial length	IIILgth	All emu prints	2.41	285
		High-quality prints	1.91	104
Toetip II–toetip IV distance	IIILgth	All emu prints	2.06	390
		High-quality prints	1.74	114

Table A6.13. Variability of digital parameters of the feet of emu study skins scaled to a common parameter. I measured only one foot per bird, and so *N* = number of individual birds.

Test parameter	Scaling parameter	Maximum/minimum	*N*
Digit II first pad length	Digit III first pad length	3.82[a] (1.99)	31 (27)
	Digit III length excluding claw	3.57[a] (1.71)	31 (27)
Digit II second pad length	Digit III first pad length	——[a] (2.24)	32 (28)
	Digit III length excluding claw	——[a] (2.04)	32 (28)
Digit II claw length	Digit III first pad length	2.21	32
	Digit III length excluding claw	2.10	32
Digit III first pad length	Digit III length excluding claw	1.32	33
Digit III second pad length	Digit III first pad length	2.40[b] (1.83)	33 (29)
	Digit III length excluding claw	2.13[b] (1.42)	33 (29)
Digit III third pad length	Digit III first pad length	——[b] (1.65)	33 (29)
	Digit III length excluding claw	——[b] (1.49)	33 (29)
Digit III claw length	Digit III first pad length	3.75	33
	Digit III length excluding claw	3.60	33
Digit IV length excluding claw	Digit III first pad length	1.61	30
	Digit III length excluding claw	1.60	30
Digit II length excluding claw	Digit III length excluding claw	1.77	31
Digit IV length excluding claw	Digit II length excluding claw	1.43	30
Digit IV claw length	Digit III first pad length	2.78	33
	Digit III length excluding claw	2.86	33
Digit II length	Digit III length	1.56	30
Digit IV length	Digit III length	1.35	30
	Digit II length	1.34	29

[a]If the non-claw portion of digit II wasn't subdivided by a crease into two pads, the entire fleshy portion of the toe was considered to be equivalent to the first pad, and the second pad was assigned a value of zero. This makes determination of the maximum/minimum scaled values of the second pad impossible to compute, and makes maximum/minimum values of the first pad larger than in the analyses based on emu footprints. Numbers in parentheses represent "normal" individuals with the fleshy toe subdivided by a crease into two digital pads.

[b]Some individuals had the fleshy portion of digit III distal to the first pad undivided into two (second and third) pads. Consequently the entire distal fleshy portion of the toe was treated as the second digital pad, and the third digital pad was assigned a value of zero. Numbers in parentheses represent "normal" individuals with a subdivided distal fleshy portion of the toe.

Table A6.14. Variability of footprint measurements for birds other than emus. All lengths are in centimeters, and angular measurements are in degrees. CV = coefficient of variation. Values for a given parameter are reported if the number of footprints for which that parameter was observed was 10 or more. See text for descriptions of parameter abbreviations.

Birds	Parameter	Minimum	Maximum	Mean (CV)	Number of footprints
Evansville elegant crested-tinamous (unknown number of birds)	Field length	3.5	4.6	4.03 (6.3)	15
	Toetip width	3.8	4.5	4.15 (5.7)	16
	IILgth	1.1	2.1	1.47 (15)	16
	IILGL	2.3	3.0	2.61 (6.7)	15
	IIILgth	1.9	2.8	2.34 (11)	16
	IIILGL	3.3	4.3	3.82 (6.1)	15
	IVLgth	1.5	2.1	1.84 (11)	14
	IVLGL	2.7	3.5	2.99 (8.2)	15
	II Lateral length	1.3	1.9	1.63 (9.7)	15
	III Medial length	2.2	3.0	2.61 (9.8)	15
	III Lateral length	2.2	3.0	2.72 (7.6)	16
	IV Medial length	1.6	2.3	1.96 (9.5)	14
	JOF IDA II–III	23	62	48.2 (22)	15
	JOF IDA III–IV	27	60	49.7 (22)	13
	JOF IDA II–IV	69	122	99.9 (14)	14
	Toebase IDA	68	153	108.5 (22)	12
	Toetip IDA	82	113	94.0 (7.9)	16
	Toetip SA	80	108	95.2 (8.7)	15
Fort Wayne Zoo domestic chicken hen	Field length	7.6	8.2	7.89 (2.6)	15
	Toetip width	8.3	9.4	8.82 (3.3)	17
	MtPdL	1.2	1.9	1.55 (14)	11
	MtPdW	1.1	1.8	1.47 (14)	12
	IIPdGp	0.8	1.3	1.04 (19)	12
	IIClwL	0.4	1.2	0.82 (30)	10
	IIIPdGp	0.4	1.2	1.01 (24)	11
	IIIClwL	0.5	0.8	0.67 (17)	11
	IVPdGp	0.8	1.2	0.99 (16)	13
	IIPdL	2.2	3.0	2.66 (8.0)	10
	IILgth	2.8	3.8	3.42 (8.0)	18
	IIFRL	3.8	4.6	4.41 (4.9)	12
	IILGL	4.6	5.7	5.26 (4.8)	16
	IIIPdL	4.0	4.8	4.27 (5.0)	11
	IIILgth	4.5	5.3	4.87 (4.2)	17
	IIIFRL	5.3	6.2	5.84 (4.8)	11
	IIILGNC	6.5	7.4	6.82 (3.5)	10
	IIILGL	7.0	7.9	7.40 (3.6)	16
	IVLgth	3.2	4.4	3.90 (10)	17
	IVFRL	4.6	5.3	4.93 (4.6)	12
	IVLGL	4.8	6.2	5.67 (6.3)	15
	II Lateral length	1.3	1.9	1.63 (9.7)	15
	III Medial length	2.2	3.0	2.61 (9.8)	15
	III Lateral length	2.2	3.0	2.72 (7.6)	16
	IV Medial length	1.6	2.3	1.96 (9.5)	14
	JOF IDA II–III	43	67	53.2 (14)	17
	JOF IDA III–IV	50	79	62.1 (10)	16
	JOF IDA II–IV	101	145	115.6 (8.8)	17
	Toebase IDA	74	122	93.1 (13)	17
	Toetip IDA	80	102	92.3 (5.6)	17
	Toetip SA	98	124	108.2 (5.7)	15
Freels Cornish cross domestic chicken chick	Field length	3.5	3.7	3.56 (2.0)	10
	IIILgth	1.9	2.7	2.24 (12)	11
	IIILGL	3.0	3.6	3.39 (5.2)	11

Table A6.14. *continued*

Birds	Parameter	Minimum	Maximum	Mean (CV)	Number of footprints
Cassell common pheasants (unknown number of birds)	Toetip width	6.2	7.2	6.71 (5.1)	11
	IILgth	2.3	3.0	2.75 (8.4)	10
	IIILgth	3.5	4.5	4.31 (6.1)	13
	IVLgth	2.9	3.6	3.23 (6.6)	10
	JOF IDA III–IV	46	57	51.3 (8.3)	10
	Toebase IDA	82	139	102.2 (15)	11
	Toetip IDA	84	108	91.2 (7.6)	11
Fort Wayne Zoo domestic turkey hen	Field length	8.1	9.1	8.55 (3.1)	13
	Toetip width	9.9	11.3	10.74 (4.5)	14
	MtPdL	1.6	2.0	1.74 (8.3)	13
	MtPdW	1.2	1.7	1.36 (10)	14
	IIPdGp	1.1	1.7	1.32 (13)	14
	IIIPdGp	1.0	1.6	1.36 (15)	13
	IIIPd1L	1.2	1.7	1.41 (11)	12
	IVPdGp	1.1	2.5	1.69 (21)	14
	IILgth	2.8	4.1	3.64 (9.5)	15
	IIFRL	4.0	5.3	4.85 (7.1)	14
	IILGL	4.9	6.1	5.72 (5.8)	15
	IIILgth	4.8	6.0	5.42 (5.9)	15
	IIIFRL	6.1	7.4	6.77 (4.9)	13
	IIILGL	7.7	9.3	8.45 (4.7)	15
	IVLgth	3.2	5.1	4.30 (11)	14
	IVFRL	5.3	6.6	5.90 (5.5)	13
	IVLGL	6.2	7.2	6.66 (4.7)	14
	II Lateral length	3.6	4.7	4.21 (8.4)	14
	II Medial length	5.5	6.8	6.10 (5.4)	14
	III Lateral length	5.5	6.6	5.93 (6.2)	14
	IV Medial length	3.7	4.6	4.26 (6.5)	11
	JOF IDA II–III	54	82	70.3 (13)	15
	JOF IDA III–IV	51	90	66.9 (16)	14
	JOF IDA II–IV	122	155	137.1 (6.6)	14
	Toebase IDA	82	128	97.2 (12)	15
	Toetip IDA	83	99	89.1 (5.0)	14
	Toetip SA	113	127	120.6 (3.8)	14
Toledo male kori bustard	Field length	12.2	13.9	13.11 (4.1)	14
	Toetip width	9.4	12.6	11.10 (7.3)	13
	Clawbase width	8.1	10.6	9.43 (8.5)	11
	MtPdL	3.3	3.8	3.56 (6.0)	11
	MtPdW	2.5	3.1	2.75 (7.4)	13
	IIPdGp	1.2	2.3	1.62 (19)	15
	IIPd1L	1.5	2.2	1.77 (13)	13
	IIPd2L	1.4	2.0	1.71 (10)	12
	IIClwL	0.9	1.5	1.24 (18)	12
	IIIPdGp	0.6	1.0	0.69 (14)	12
	IIIPd1L	2.1	2.9	2.51 (7.0)	14
	IIIClwL	2.1	4.1	3.16 (24)	15
	IVPdGp	1.2	2.2	1.64 (16)	12
	IVClwL	0.8	1.7	1.28 (19)	12
	IIPdL	2.8	4.0	3.26 (9.4)	13
	IILgth	3.8	5.3	4.46 (9.4)	14
	IIFRNC	4.3	5.1	4.69 (5.1)	13
	IIFRL	5.1	6.5	5.95 (7.4)	14
	IILGNC	6.2	7.4	6.62 (5.7)	12
	IILGL	7.3	8.9	7.82 (6.5)	13

Birds	Parameter	Minimum	Maximum	Mean (CV)	Number of footprints
Toledo male kori bustard *continued*	IIIPdL	5.6	6.6	6.00 (4.4)	15
	IIILgth	7.9	9.6	8.93 (7.4)	15
	IIIFRNC	6.4	6.8	6.59 (1.6)	12
	IIIFRL	8.5	10.5	9.64 (7.1)	12
	IIILGNC	9.8	10.5	10.17 (2.6)	14
	IIILGL	12.2	14.1	13.11 (4.6)	14
	IVPdL	3.9	4.6	4.23 (4.7)	12
	IVLgth	4.9	5.9	5.41 (5.3)	12
	IVFRNC	5.3	6.2	5.81 (4.7)	12
	IVFRL	6.4	7.5	7.01 (4.6)	13
	IVLGNC	7.2	8.0	7.69 (3.1)	13
	IVLGL	8.2	9.5	8.92 (3.8)	13
	III Medial length	8.7	10.6	9.56 (8.0)	10
	III Lateral length	8.8	10.5	9.52 (5.5)	11
	IV Medial length	4.9	6.3	5.45 (7.4)	11
	JOF IDA II–III	13	42	32.9 (23)	15
	TOPO IDA II–III	7	41	27.2 (34)	11
	JOF IDA III–IV	32	53	40.4 (16)	12
	TOPO IDA III–IV	31	58	42.3 (23)	10
	JOF IDA II–IV	56	91	72.2 (15)	12
	Toebase IDA	85	121	108.7 (7.0)	12
	Clawbase IDA	80	92	87.9 (4.6)	11
	Toetip IDA	66	89	77.9 (9.2)	13
	Clawbase SA	72	92	81.9 (8.5)	11
	Toetip SA	72	90	82.6 (6.8)	12
Cleveland double-striped thick-knee	IILgth	1.8	2.3	2.02 (7.5)	10
	IIILgth	3.2	3.8	3.43 (6.9)	10
	JOF IDA II–III	18	34	26.9 (19)	10
St. Joseph River killdeer trail 1	IIILgth	1.9	2.2	2.08 (5.0)	10
	IVLgth	1.4	1.9	1.71 (11)	10
	III Lateral length	2.0	2.4	2.29 (5.2)	10
	JOF IDA II–III	51	75	59.1 (13)	10
	TRACE IDA II–III	41	66	59.0 (12)	11
	JOF IDA III–IV	25	57	48.7 (17)	11
	TRACE IDA III–IV	40	51	45.2 (7.9)	11
	JOF IDA II–IV	100	119	107.3 (5.4)	10
	TRACE IDA II–IV	92	114	104.4 (5.8)	11
	Toebase IDA	57	87	72.9 (13)	10
St. Joseph River killdeer trail 2	Toetip width	3.1	3.8	3.34 (6.9)	10

Table A6.15. Variability in size and shape of footprints of domestic chickens from a French farmyard. All linear measurements are in millimeters, and angles in degrees; number of footprints = 93. CV = coefficient of variation. Data from Gand (1976 and personal communication).

Parameter	Minimum	Maximum	Mean (CV)	Maximum/minimum
Digit II length	47	67	59.4 (7.5)	1.42
Digit III length	80	105	91.0 (5.8)	1.31
Digit IV length	57	80	69.3 (6.4)	1.40
Footprint width	72	107	93.9 (7.4)	1.49
Interdigital angle II–IV	80	125	98.2 (8.5)	1.56

Coefficient of dispersion about the reduced major axis (D_d) for bivariate comparisons[a]

Variables		D_d
Digit III length	Digit II length	8.67 (2.17)
	Digit IV length	7.15 (1.67)
Digit II length	Digit IV length	7.91 (1.93)
Digit III length	Footprint width	8.99 (2.03)

Maximum/minimum ratios of scaled parameters

Test parameter	Scaling parameter	Maximum/minimum
Digit II length	Digit III length	1.51
Digit IV length	Digit III length	1.54
	Digit II length	1.42
Footprint width	Digit III length	1.62
	Digit II length	1.56

[a]Values are reported for both raw and log-transformed (in parentheses) measurements.

Table A6.16. Selected summary statistics of size variability in footprints of adult emus. In the "all footprints" treatment, data cases are individual footprints, and individual birds are represented by varying numbers of prints. In the "emu means" treatment, data cases are means for individual adult birds. Compare with table A6.15 values for footprints of domestic chickens ("all footprints" treatment) and table 5.3 for hindfeet of mature alligators. All lengths are in centimeters; angles are in degrees. See text for descriptions of abbreviations.

Parameter	Treatment	Minimum	Maximum	Mean (CV)	Maximum/minimum	*N*
IILgth	All footprints	3.7	7.9	5.73 (13)	2.13	165
	Emu means	5.5	6.9	5.96 (6.8)	1.28	13
II lateral hypex length	All footprints	4.4	8.3	6.31 (12)	1.89	189
	Emu means	5.5	7.9	6.62 (10)	1.44	11
IIPdL	All footprints	3.2	5.5	4.16 (10)	1.70	119
	Emu means	3.9	4.7	4.27 (7.3)	1.22	10
IIClwL	All footprints	0.8	2.7	1.59 (23)	3.24	100
	Emu means	1.4	2.4	1.79 (18)	1.71	10
IIILgth	All footprints	9.7	14.1	11.8 (6.6)	1.46	204
	Emu means	11.3	13.9	12.1 (5.3)	1.23	17
III medial hypex length	All footprints	11.1	16.0	13.3 (6.3)	1.44	202
	Emu means	11.8	15.7	13.7 (8.1)	1.33	11
III lateral hypex length	All footprints	11.6	16.6	13.8 (6.0)	1.43	209
	Bird means	13.4	16.4	14.3 (6.4)	1.22	14
IIIPdL	All footprints	8.3	11.4	9.77 (6.2)	1.38	130
	Emu means	9.4	11.0	9.92 (4.6)	1.17	15
IIIClwL	All footprints	1.0	2.9	1.95 (19)	2.90	129
	Emu means	1.4	2.5	2.08 (15)	1.83	15
IVLgth	All footprints	6.1	9.5	7.62 (7.3)	1.54	182
	Emu means	6.7	8.3	7.52 (6.4)	1.24	15
IV medial hypex length	All footprints	5.3	9.9	7.94 (10)	1.87	213
	Emu means	7.5	9.7	8.44 (8.9)	1.29	14
IVPdL	All footprints	4.6	7.5	6.02 (9.0)	1.62	108
	Emu means	5.3	6.5	5.94 (7.1)	1.23	11
IVClwL	All footprints	1.0	2.8	1.67 (20)	2.68	105
	Emu means	1.5	2.4	1.82 (14)	1.55	11
Toetip width	All footprints	12.9	20.0	16.1 (7.1)	1.54	182
IIPd1L	All footprints	1.5	2.8	2.13 (12)	1.79	96
	Emu means	2.1	2.4	2.19 (6.2)	1.16	5
IIPd2L	All footprints	1.3	3.1	2.07 (17)	2.32	77
	Emu means	1.8	2.6	2.18 (15)	1.45	5
IIIPd1L	All footprints	2.9	5.3	3.80 (9.4)	1.82	186
	Emu means	3.3	5.0	3.86 (9.9)	1.49	17
IIIPd2L	All footprints	2.4	4.1	3.36 (9.6)	1.69	109
	Emu means	3.0	4.0	3.45 (8.7)	1.31	12
IIIPd3L	All footprints	1.4	3.9	2.66 (17)	2.69	95
	Emu means	2.0	3.4	2.71 (12)	1.71	12
JOF IDA II–IV	All footprints	49	104	73.1 (14)	2.12	154

Table A6.17. Summary comparison of intraspecific variability of equivalent or nearly equivalent parameters in the feet of mature alligators and footprints and study skin feet of adult or nearly adult emus (cf. table A6.18). CV = coefficient of variation; Mx/Mn = maximum/minimum ratio.

Parameter	Alligator feet		Emu footprints: Footprints = data cases		Emu footprints: "Emu means" treatment		Emu study skins	
	CV	Mx/Mn	CV	Mx/Mn	CV	Mx/Mn	CV	Mx/Mn
Digit II length[a]	13	1.72	13	2.13	6.8	1.28	8.9	1.40
Digit II lateral hypex length	15	1.83	12	1.89	10	1.44	——	——
Digit II length excluding claw[a]	16	1.84	10	1.70	7.3	1.22	9.7	1.35
Digit II claw length	12	2.07	23	3.24	18	1.71	14	1.69
Digit III length[a]	15	1.83	6.6	1.46	5.3	1.23	5.3	1.22
Digit III medial hypex length	17	2.08	6.3	1.44	8.1	1.33	——	——
Digit III lateral hypex length	18	2.16	6.0	1.43	6.4	1.22	——	——
Digit III length excluding claw[a]	16	1.96	6.2	1.38	4.6	1.17	5.6	1.23
Digit III claw length	17	5.20	19	2.90	15	1.83	11	1.47
Digit IV medial hypex length	24	2.47	10	1.87	8.9	1.29	——	——

[a]Digit lengths are free lengths for alligators, and "Lgth" or "PdL" (see text for descriptions of variable abbreviations) lengths for emus. Alligator data include animals with worn claws.

Table A6.18. Summary statistics of size variability in the feet of adult or nearly adult emus. Data cases are the feet of individual birds, with one foot measured per bird. Measurements were made on study skins; all measurements are in millimeters.

Parameter	Minimum	Maximum	Mean (CV)	Maximum/minimum	*N*
Digit II length	48	67	58.3 (8.9)	1.40	15
Digit II length excluding claw	31	42	37.0 (9.7)	1.35	16
Digit II claw length	16	27	21.2 (14)	1.69	15
Digit III length	103	126	112.4 (5.3)	1.22	16
Digit III length excluding claw	86	106	92.6 (5.6)	1.23	16
Digit III claw length	17	25	19.8 (11)	1.47	16
Digit IV length	64	78	72.0 (5.1)	1.22	15
Digit IV length excluding claw	50	59	54.7 (5.5)	1.18	15
Digit IV claw length	14	21	17.1 (14)	1.50	16
Digit II first pad length	15	22	19.1 (11)	1.47	16
Digit II second pad length	13	21	17.9 (13)	1.62	16
Digit III first pad length	32	43	35.7 (8.6)	1.34	16
Digit III second pad length[a]	26	52 (37)	31.9 [30.6] (19 [10])	2.00 (1.42)	16 (15)
Digit III third pad length[a]	0 (21)	32	24.9 [26.6] (29 [12])	—— (1.52)	16 (15)

[a]One of the adult emus failed to subdivide the distal fleshy portion of digit III into distinct second and third pads. Consequently all of the length of the combined pad was assigned to the second digital pad, and the third digital pad was assigned a value of zero. Values in brackets were calculated excluding this specimen.

Table A6.19. Summary comparison of variability of equivalent or nearly equivalent shape parameters of alligator hindfeet and emu footprints. D_d = coefficient of relative dispersion about the reduced major axis; values are for raw (not log-transformed) data. Digit lengths are free lengths for alligators, and Lgth or PdL (see text for abbreviation descriptions) lengths for emus. For emu footprints, numbers within parentheses are for high-quality footprints, while numbers outside parentheses are for all emu prints.

Variables		Alligator feet	Emu footprints	
			Footprints = data cases	"Emu means" treatment
		D_d	D_d	D_d
Digit III length[a]	Digit II length[a]	5.13	14.5 (12.4)	6.42 (7.05)
Digit III length excluding claw	Digit II length excluding claw	7.18	10.8 (9.48)	7.32 (6.64)
Digit II length[a]	Digit II lateral hypex length[a]	7.19	11.3 (9.61)	8.26 (10.4)
Digit III length[a]	Digit III medial hypex length[a]	8.57	6.23 (5.43)	5.40 (4.87)
	Digit III lateral hypex length[a]	8.46	6.10 (5.52)	4.29 (4.09)
Digit II length excluding claw	Digit II claw length[a]	18.7	23.1 (16.6)	15.6 (13.7)
Digit III length excluding claw	Digit III claw length[a]	17.3	21.4 (16.2)	15.5 (12.7)
Digit III medial hypex length	Digit II lateral hypex length	9.43	13.1 (11.0)	8.08 (7.94)
	Digit III lateral hypex length	10.7	3.98 (3.43)	2.85 (2.80)
	Digit IV medial hypex length	14.7	11.4 (9.16)	8.66 (9.01)

[a]In alligators, excluding individuals with heavily worn claws.

Table A6.20. Summary comparison of equivalent or nearly equivalent scaled footprint parameters in alligator feet and emu footprints and feet. In all comparisons, variables are scaled to digit III length, either free length (alligators) or "Lgth" (see text for abbreviation description) length (emus). For emu prints, values in parentheses are for high-quality prints, and values outside parentheses are for all emu prints, of whatever quality. For alligator feet, data include specimens with heavily worn claws.

Parameter	Alligator feet	Emu footprints		Emu study skins
		Footprints = data cases	"Emu means" treatment	
Digit II length with claw[a]	1.38	2.20 (2.02)	1.33 (1.46)	1.56
Digit II lateral hypex length	1.57	2.43 (1.77)	1.47 (1.44)	———
Digit II length without claw[a]	1.51	2.21 (1.85)	1.54 (1.38)	1.72
Digit II claw length	2.19	4.04 (4.04)	1.71 (1.67)	1.82
Digit III medial hypex length	1.49	1.50 (1.50)	1.26 (1.20)	———
Digit III lateral hypex length	1.49	1.61 (1.42)	1.15 (1.18)	———
Digit III length without claw[a]	1.25	1.26 (1.18)	1.18 (1.13)	1.20
Digit III claw length	4.96	4.00 (2.67)	1.95 (1.95)	3.01
Digit IV medial hypex length	2.22	2.41 (1.91)	1.34 (1.29)	———

[a]Digit lengths = free lengths for alligators, and "Lgth" lengths for emus.

Table A6.21. Variability of digital parameters of the feet of greater rhea (*Rhea americana*) study skins scaled to a common parameter. *N* = number of individual birds.

Test parameter	Scaling parameter	Maximum/minimum	*N*
Digit II first pad length	Digit III first pad length	3.67[a] (1.43)	24 (5)
	Digit III length excluding claw	3.00[a] (1.43)	24 (5)
Digit II second pad length	Digit III first pad length	——[a] (1.69)	24 (5)
	Digit III length excluding claw	——[a] (1.50)	24 (5)
Digit II claw length	Digit III first pad length	3.15	24
	Digit III length excluding claw	2.79	24
Digit III first pad length	Digit III length excluding claw	1.60	24
Digit III second pad length	Digit III first pad length	3.46[b] (1.60)	24 (11)
	Digit III length excluding claw	2.60[b] (1.51)	24 (11)
Digit III third pad length	Digit III first pad length	——[b] (1.90)	24 (11)
	Digit III length excluding claw	——[b] (1.75)	24 (11)
Digit III claw length	Digit III first pad length	3.63	24
	Digit III length excluding claw	3.22	24
Digit IV length excluding claw	Digit III first pad length	1.93	24
	Digit III length excluding claw	1.66	24
Digit II length excluding claw	Digit III length excluding claw	1.54	24
Digit IV length excluding claw	Digit II length excluding claw	1.38	24
Digit IV claw length	Digit III first pad length	3.03	24
	Digit III length excluding claw	2.68	24
Digit II length	Digit III length	1.43	24
Digit IV length	Digit III length	1.37	24
	Digit II length	1.29	24

[a]Most rheas do not subdivide the fleshy portion of digit II into two distinct pads. The entire length of the pad was therefore assigned to the first digital pad, and the second pad was assigned a value of zero. Numbers in parentheses represent unusual individuals in which two distinct pads could be recognized.
[b]About half of the rheas examined subdivided the fleshy portion of digit III distal to the first digital pad into two (second and third) pads. Numbers for these birds are in parentheses. For birds that did not subdivide this part of the toe, the entire length was assigned to the second pad, and the third pad was assigned a length of zero.

Table A6.22. Variability of digital parameters of the feet of the kori bustard (*Ardeotis kori*) study skins scaled to a common parameter. *N* = number of individual birds.

Test parameter	Scaling parameter	Maximum/minimum	*N*
Digit II first pad length	Digit III first pad length	1.59	15
	Digit III length excluding claw	1.49	15
Digit II second pad length	Digit III first pad length	1.44	15
	Digit III length excluding claw	1.43	15
Digit II claw length	Digit III first pad length	1.94	15
	Digit III length excluding claw	1.91	15
Digit III first pad length	Digit III length excluding claw	1.22	15
Digit III second pad length	Digit III first pad length	2.28[a] (1.53)	15 (14)
	Digit III length excluding claw	2.21[a] (1.31)	15 (14)
Digit III third pad length	Digit III first pad length	———[a] (1.40)	15 (14)
	Digit III length excluding claw	———[a] (1.29)	15 (14)
Digit III claw length	Digit III first pad length	2.00	14
	Digit III length excluding claw	2.08	14
Digit IV length excluding claw	Digit III first pad length	1.43	14
	Digit III length excluding claw	1.32	14
Digit II length excluding claw	Digit III length excluding claw	1.34	15
Digit IV length excluding claw	Digit II length excluding claw	1.28	14
Digit IV claw length	Digit III first pad length	1.77	15
	Digit III length excluding claw	1.74	15
Digit II length	Digit III length	1.24	14
Digit IV length	Digit III length	1.28	13
	Digit II length	1.23	14

[a]One individual failed to subdivide what in all other specimens are the second and third digital pads of digit III. In this specimen, the entire length of this undivided pad was assigned to the second digital pad, and the third digital pad was assigned a length of zero. Numbers in parentheses are for birds with normally subdivided digital pads.

Table A6.23. Variability of digital parameters of the feet of study skins of the great bustard (*Otis tarda*) scaled to a common parameter. *N* = number of individual birds.

Test parameter	Scaling parameter	Maximum/minimum	*N*
Digit II first pad length	Digit III first pad length	1.65	11
	Digit III length excluding claw	1.68	10
Digit II second pad length	Digit III first pad length	1.38	11
	Digit III length excluding claw	1.27	10
Digit III claw length	Digit III first pad length	2.31	11
	Digit III length excluding claw	2.36	10
Digit III first pad length	Digit III length excluding claw	1.14	10
Digit III second pad length	Digit III first pad length	1.45	11
	Digit III length excluding claw	1.32	10
Digit III third pad length	Digit III first pad length	1.43	10
	Digit III length excluding claw	1.33	10
Digit III claw length	Digit III first pad length	2.54	11
	Digit III length excluding claw	2.60	10
Digit IV length excluding claw	Digit III first pad length	1.46	11
	Digit III length excluding claw	1.43	10
Digit II length excluding claw	Digit III length excluding claw	1.29	10
Digit IV length excluding claw	Digit II length excluding claw	1.70	11
Digit IV claw length	Digit III first pad length	1.83	11
	Digit III length excluding claw	1.69	10
Digit II length	Digit III length	1.21	10
Digit IV length	Digit III length	1.41	10
	Digit II length	1.55	11

Table A6.24. Comparison of interspecific and intraspecific (emu) variability in bivariate comparisons of linear dimensions of bird footprints (bird means treatment). Footprints of all the species of birds (other than gastornithids) for which the various parameters could be measured are pooled in the interspecific analyses. Two versions of each interspecific analysis were run: one in which each bird, regardless of species, was considered a data case ("all birds" treatment), and the number of birds representing each species was highly variable, and one in which only one individual bird per species was that species' data case ("one per species" treatment). Data for ostrich were included in the all birds treatment, except where the comparison of interest could not be made (as with comparisons involving digit II), but ostrich data were excluded in the one per species treatment. Both kinds of bird interspecific analyses are compared with the comparable intraspecific emu results. Two measures of variability are compared in these analyses. First, values of the coefficient of relative dispersion around the reduced major axis (D_d) for interspecific and intraspecific (emus) relationships are reported. To make D_d values comparable to those calculated elsewhere in this book, all measurements were first converted from centimeters to millimeters before analysis. Values for D_d outside parentheses were calculated using raw data (in millimeters), and values inside parentheses were calculated using log-transformed data. The second measure of variability is the standard error of the estimate (SEE) of the regression involving the two variables, reported only for the log-transformed version of the relationship.

Independent variable	Dependent variable	Treatment	D_d	SEE of regression (log-transformed variables)	Number of cases
IIILgth	MtPdL	All birds	21.8 (6.18)	0.1095	70
		One per species	15.7 (6.54)	0.1081	29
		Emu means	12.1 (2.94)	0.0404	32
	MtPdW	All birds	26.9 (6.31)	0.1117	72
		One per species	13.8 (5.22)	0.0834	29
		Emu means	13.2 (3.29)	0.0572	34
	IIPdGp	All birds	28.0 (7.68)	0.1290	71
		One per species	26.5 (8.15)	0.1221	29
		Emu means	15.9 (3.58)	0.0542	33
	IILgth	All birds	23.5 (4.61)	0.0774	88
		One per species	24.6 (4.59)	0.0753	39
		Emu means	6.42 (1.52)	0.0271	32
	IIIPdGp	All birds	29.1 (8.14)	0.1222	69
		One per species	42.0 (11.2)	0.1374	28
		Emu means	16.1 (4.58)	0.0720	33
	IVPdGp	All birds	34.8 (6.29)	0.0997	70
		One per species	16.7 (5.14)	0.0770	29
		Emu means	12.6 (3.43)	0.0568	33
	IVLgth	All birds	19.3 (3.36)	0.0571	91
		One per species	12.1 (3.30)	0.0538	39
		Emu means	7.10 (1.56)	0.0288	34
	Proximal end II to proximal end III	All birds	17.8 (4.80)	0.0828	89
		One per species	23.6 (5.80)	0.0887	38
		Emu means	13.3 (2.82)	0.0452	34
	Proximal end III to proximal end IV	All birds	20.5 (5.37)	0.0943	90
		One per species	19.3 (5.35)	0.0829	38
		Emu means	10.9 (2.16)	0.0339	34
	Proximal end II to proximal end IV	All birds	20.0 (5.02)	0.0938	88
		One per species	20.8 (5.15)	0.0861	39
		Emu means	11.7 (2.29)	0.0382	33
	Toetip II to toetip III	All birds	19.1 (3.71)	0.0665	87
		One per species	26.2 (4.91)	0.0820	39
		Emu means	6.84 (1.57)	0.0292	31
	Toetip III to toetip IV	All birds	20.1 (4.61)	0.0838	92
		One per species	28.8 (6.46)	0.1107	38
		Emu means	6.64 (1.41)	0.0254	35
	Toetip II to toetip IV	All birds	21.6 (3.89)	0.0726	88
		One per species	25.7 (4.90)	0.0868	39
		Emu means	7.58 (1.48)	0.0273	32
IILgth	IILat	All birds	15.8 (3.27)	0.0521	75
		One per species	14.1 (2.98)	0.0466	34
		Emu means	8.08 (2.11)	0.0356	24

Independent variable	Dependent variable	Treatment	D_d	SEE of regression (log-transformed variables)	Number of cases
IIILgth	IIIMed	All birds	9.51 (1.74)	0.0312	72
		One per species	5.77 (1.57)	0.0260	33
		Emu means	5.40 (1.07)	0.0227	24
	IIILat	All birds	12.6 (2.33)	0.0437	78
		One per species	7.07 (2.07)	0.0361	34
		Emu means	4.30 (0.90)	0.0180	28
IVLgth	IVMed	All birds	17.4 (3.21)	0.0539	78
		One per species	7.86 (2.94)	0.0459	34
		Emu means	6.47 (1.52)	0.0275	26
IILgth	IVLgth	All birds	21.3 (3.93)	0.0640	88
		One per species	21.4 (4.36)	0.0659	39
		Emu means	6.75 (1.70)	0.0298	31

Table A6.25. Comparison of variability of selected footprint parameters (bird means treatment) scaled to a common dimension (IIILGL = 15 cm) in emus and other ground birds. "All birds" = all taxa for which footprints were collected (except gastornithids). Values outside the range for emus are indicated in **bold**.

Parameter	Taxon	Minimum	Maximum	Number of birds
Metatarsal pad length	*Dromaius*	3.2	4.3	32
	Rhea	3.8	3.9	2
	Casuarius	4.1	**4.5**	2
	Apteryx	4.5	**4.6**	2
	Pheasants	**2.0**	3.8	16
	Bustards	4.1	**4.4**	2
	All birds	**2.0**	**6.3**	71
Digit II length (IILgth)	*Dromaius*	4.1	5.2	31
	Rhea	5.1	**5.6**	2
	Casuarius	8.2	**9.5**	2
	Apteryx	6.6	**6.9**	2
	Pheasants	5.1	**8.1**	20
	Bustards	5.1	**5.3**	2
	All birds	4.1	**9.5**	86
Digit III length (IIILgth)	*Dromaius*	8.8	10.3	33
	Rhea	9.9	10.0	2
	Casuarius	9.3	9.8	2
	Apteryx	8.9	9.5	2
	Pheasants	**8.6**	**12.3**	20
	Bustards	9.1	10.2	2
	All birds	**7.1**	**12.3**	87
Digit IV length (IVLgth)	*Dromaius*	5.1	6.8	33
	Rhea	5.7	6.4	2
	Casuarius	6.4	**7.1**	2
	Apteryx	6.5	**7.0**	2
	Pheasants	6.7	**9.8**	20
	Bustards	**5.0**	6.2	2
	All birds	**4.5**	**9.8**	87
Digit II lateral hypex length	*Dromaius*	4.7	6.3	24
	Rhea	5.6	6.1	2
	Casuarius	7.7	**8.3**	2
	Apteryx	7.4	**7.7**	2
	Pheasants	5.5	**8.5**	20
	Bustards	4.7	5.6	2
	All birds	**3.7**	**9.9**	78
Digit III medial hypex length	*Dromaius*	9.4	12.0	24
	Rhea	10.4	11.4	2
	Casuarius	10.3	10.8	2
	Apteryx	9.5	10.8	2
	Pheasants	9.5	11.9	19
	Bustards	10.9	11.3	2
	All birds	**6.1**	**13.3**	77
Digit IV medial hypex length	*Dromaius*	5.8	8.0	27
	Rhea	6.6	6.8	2
	Casuarius	7.0	7.2	2
	Apteryx	6.9	8.0	2
	Pheasants	6.3	**9.5**	20
	Bustards	6.2	6.8	2
	All birds	**3.8**	**9.5**	81

Parameter	Taxon	Minimum	Maximum	Number of birds
Proximal end II to proximal end IV	*Dromaius*	5.0	7.7	32
	Rhea	5.1	5.6	2
	Casuarius	**4.6**	5.9	2
	Apteryx	5.0	5.4	2
	Pheasants	**4.3**	**10.1**	20
	Bustards	5.9	6.9	2
	All birds	**4.1**	**10.1**	86
Toetip II to toetip IV	*Dromaius*	11.8	15.4	31
	Rhea	**11.0**	11.1	2
	Casuarius	**10.7**	12.8	2
	Apteryx	**16.7**	**17.4**	2
	Pheasants	12.8	**23.3**	22
	Bustards	12.7	13.7	2
	All birds	**10.0**	**23.8**	91

Table A8.1. Selected summary measurements of Newark Supergroup tridactyl dinosaur trackways and footprints. For some trackways, only mean footprint length and mean stride length were used in this study; data for such trackways that are not reported here appear in table A4.2. The data reported here were used in the analysis of print shape and trackway dimensions of Newark Supergroup footprints. Measurements of footprint linear dimensions are in millimeters. Measurements of trackway paces and strides are in centimeters (in some cases to the nearest 0.1 cm). Measurements of interdigital angles, pace angulations, and footprint rotation are in degrees. It was not possible to measure all parameters on all footprints/trackways. Abbreviation isp. = ichnospecies. "Toebase" distances are the distances between the proximal ends of the first digital pads of the respective digits. "Toetip width" is the distance between the toetips of digits II and IV. For footprint rotation (reported as ROTFOLL; see text), the median is reported, rather than the mean, and marked by an asterisk (*).

Trackway	Parameter	Mean	Minimum	Maximum	N
AC 4/1a *Eubrontes cursorius* (*Grallator cursorius*)	Footprint field length	76.5	76	77	2
	Digit II first pad length	12.7	12	13	4
	Digit II first pad width	10.0	9	11	4
	Digit II second pad length	12.0	11	13	4
	Digit II second pad width	9.5	9	10	4
	Digit II length excluding claw	25.0	24	26	3
	Digit II claw length	7.7	6	9	3
	Digit II length	32.7	30	34	4
	Digit III first pad length	16.5	16	17	2
	Digit III first pad width	11.3	10	12	4
	Digit III second pad length	16.0	14	19	4
	Digit III second pad width	12.0	11	14	4
	Digit III third pad length	13.7	11	17	4
	Digit III third pad width	9.3	9	10	4
	Digit III length excluding claw	47.0	46	48	4
	Digit III claw length	7.7	6	9	4
	Digit III length	54.5	53	55	4
	Digit IV length	44.0	41	47	2
	Digit IV first pad width	9	——	——	1
	Toebase II–III	11.3	10	12	4
	Toebase III–IV	22.0	21	23	2
	Toebase II–IV	17.0	16	18	2
	Toetip II–III	35.5	35	36	4
	Toetip III–IV	38.5	36	41	4
	Toetip width	31.5	30	33	4
	Backfoot length	44.0	43	45	2
	Digit III projection	33.0	32	34	4
	Distance digit II proximal bend point to digit III midline	7.7	7	10	4
	Distance "heel" to toetip II	50.0	50	50	2
	Distance "heel" to toetip III	78.0	78	78	2
	Distance "heel" to toetip IV	44.0	42	46	2
	Interdigital angle II–III	12.0	11	13	4
	Interdigital angle III–IV	13.5	11	18	4
	Interdigital angle II–IV	25.5	23	31	4
	Pace	59.3	57	61	3
	Stride	119.0	117	121	2
	Pace angulation	180.0	180	180	2
AC 30/1 *Eubrontes josephbarratti* (*Grallator cuneatus*) possible sequence F–N	Footprint field length	133	——	——	1
	Digit II first pad length	28	——	——	1
	Digit II first pad width	21	——	——	1
	Digit II second pad length	27	——	——	1
	Digit II second pad width	20	——	——	1
	Digit II length excluding claw	55	——	——	1
	Digit II claw length	14	——	——	1
	Digit II length	67	——	——	1
	Digit III first pad length	27	——	——	1
	Digit III first pad width	25	——	——	1
	Digit III second pad pength	27	——	——	1
	Digit III second pad width	26	——	——	1
	Digit III third pad length	22	——	——	1
	Digit III third pad width	20	——	——	1

Trackway	Parameter	Mean	Minimum	Maximum	*N*
AC 30/1 *Eubrontes josephbarratti* (*Grallator cuneatus*) possible sequence F–N	Digit III length excluding claw	76	——	——	1
	Digit III claw length	16	——	——	1
	Digit III length	91	——	——	1
	Digit IV length	76	——	——	1
	Toebase II–III	30	——	——	1
	Toebase III–IV	38	——	——	1
	Toebase II–IV	22	——	——	1
	Toetip II–III	77	——	——	1
	Toetip III–IV	66	——	——	1
	Toetip width	80	——	——	1
	Backfoot length	74	——	——	1
	Digit III projection	58	——	——	1
	Distance digit II proximal bend point to digit III midline	15	——	——	1
	Distance "heel" to toetip II	95	——	——	1
	Distance "heel" to toetip III	133	——	——	1
	Distance "heel" to toetip IV	78	——	——	1
	Interdigital angle II–III	19	——	——	1
	Interdigital angle III–IV	18	——	——	1
	Interdigital angle II–IV	38	——	——	1
AC 30/1 *Eubrontes josephbarratti* (*Grallator cuneatus*) sequence E-Cw	Footprint field length	130	——	——	1
	Digit II first pad length	24	——	——	1
	Digit II first pad width	18	——	——	1
	Digit II length	57	——	——	—
	Digit III first pad length	34	——	——	1
	Digit III first pad width	25	——	——	1
	Digit III second pad length	26	——	——	1
	Digit III second pad width	24	——	——	1
	Digit III third pad length	20	——	——	1
	Digit III third pad width	16	——	——	1
	Digit III length excluding claw	80	——	——	1
	Digit III claw length	17	——	——	1
	Digit III length	97	——	——	1
	Digit IV length	81	——	——	1
	Digit IV first pad width	22	——	——	1
	Toebase II–III	17	——	——	1
	Toebase III–IV	33	——	——	1
	Toebase II–IV	25	——	——	1
	Toetip II–III	70	——	——	1
	Toetip III–IV	70	——	——	1
	Toetip width	76	——	——	1
	Backfoot length	72	——	——	1
	Digit III projection	59	——	——	1
	Distance digit II proximal bend point to digit III midline	17	——	——	1
	Distance "heel" to toetip II	85	——	——	1
	Distance "heel" to toetip III	130	——	——	1
	Distance "heel" to toetip IV	78	——	——	1
	Interdigital angle II–III	16	——	——	1
	Interdigital angle III–IV	23	——	——	1
	Interdigital angle II–IV	39	——	——	1

Table A8.1. *continued*

Trackway	Parameter	Mean	Minimum	Maximum	*N*
AC 30/1 *Eubrontes josephbarratti* (*Grallator cuneatus*) sequence J-H-D	Footprint field length	135	——	——	1
	Digit II first pad length	24	——	——	1
	Digit II first pad width	21	——	——	1
	Digit II second pad length	25	——	——	1
	Digit II second pad width	21	——	——	1
	Digit II length excluding claw	49	——	——	1
	Digit II claw length	15	——	——	1
AC 30/1 *Eubrontes josephbarratti* (*Grallator cuneatus*) sequence J-H-D	Digit II length	60	——	——	1
	Digit III first pad length	33	——	——	1
	Digit III first pad width	26	——	——	1
	Digit III second pad length	30	——	——	1
	Digit III second pad width	25	——	——	1
	Digit III third pad width	15	——	——	1
	Digit III length	94	——	——	1
	Digit IV length	86	——	——	1
	Toebase II–III	24	——	——	1
	Toebase III–IV	39	——	——	1
	Toebase II–IV	31	——	——	1
	Toetip II–III	71	——	——	1
	Toetip III–IV	73	——	——	1
	Toetip width	88	——	——	1
	Backfoot length	79	——	——	1
	Digit III projection	57	——	——	1
	Distance digit II proximal bend point to digit III midline	22	——	——	1
	Distance "heel" to toetip II	95	——	——	1
	Distance "heel" to toetip III	136	——	——	1
	Distance "heel" to toetip IV	86	——	——	1
	Interdigital angle II–III	9	——	——	1
	Interdigital angle III–IV	28	——	——	1
	Interdigital angle II–IV	38	——	——	1
AC 30/1 *Eubrontes loxonyx* (*Anchisauripus tuberosus*) trackway sequence B-G-I-L-O	Footprint field length	207.5	206	209	2
	Digit II first pad length	37.7	36	39	3
	Digit II first pad width	32.0	31	34	3
	Digit II second pad length	38.3	36	41	3
	Digit II second pad length	31.3	30	34	3
	Digit II length excluding claw	75.7	75	77	3
	Digit II claw length	23.3	19	30	3
	Digit II length	97.0	92	104	3
	Digit III first pad length	41.0	40	42	2
	Digit III first pad width	39.3	35	46	3
	Digit III second pad length	43.5	42	45	2
	Digit III second pad width	40.0	38	42	2
	Digit III third pad length	35.5	31	40	2
	Digit III third pad width	30.5	28	33	2
	Digit III length excluding claw	120.3	114	130	3
	Digit III claw length	21.7	17	27	2
	Digit III length	140.7	137	145	3
	Digit IV length	132.5	128	137	2
	Toebase II–III	37.0	30	42	3
	Toebase III–IV	68.5	65	72	3
	Toebase II–IV	59.5	56	63	2
	Toetip II–III	99.0	96	102	3
	Toetip III–IV	101.3	99	103	3
	Toetip width	127.3	120	137	3
	Backfoot length	127.5	125	130	2
	Digit III projection	77.7	77	79	3

Trackway	Parameter	Mean	Minimum	Maximum	N
	Distance digit II proximal bend point to digit III midline	26.7	18	33	3
	Distance "heel" to toetip II	154.5	153	156	2
	Distance "heel" to toetip III	207.5	206	209	2
	Distance "heel" to toetip IV	133.0	128	138	2
	Interdigital angle II–III	14.0	10	17	3
	Interdigital angle III–IV	16.3	12	20	3
	Interdigital angle II–IV	30.3	22	37	3
	Pace	65.1	64.2	66.1	2
	Stride	130.9	128.2	133.6	2
	Pace angulation	157.5	156	159	2
YPM VP 003314 *Eubrontes loxonyx* (*Anchisauripus tuberosus*)	Footprint field length	201	——	——	1
	Digit II first pad length	41	——	——	1
	Digit II first pad width	37	——	——	1
	Digit II second pad length	41	——	——	1
	Digit II second pad width	37	——	——	1
	Digit II length excluding claw	84	——	——	1
	Digit II claw length	20	——	——	1
	Digit II length	99	——	——	1
	Digit III first pad length	41	——	——	1
	Digit III first pad width	47	——	——	1
	Digit III second pad length	40	——	——	1
	Digit III second pad width	53	——	——	1
	Digit III third pad length	39	——	——	1
	Digit III third pad width	40	——	——	1
	Digit III length excluding claw	120	——	——	1
	Digit III claw length	21	——	——	1
	Digit III length	142	——	——	1
	Digit IV length	138	——	——	1
	Digit IV first pad width	41	——	——	1
	Toebase II–III	38	——	——	1
	Toebase III–IV	60	——	——	1
	Toebase II–IV	42	——	——	1
	Toetip II–III	112	——	——	1
	Toetip III–IV	93	——	——	1
	Toetip width	133	——	——	1
	Backfoot length	124	——	——	1
	Digit III projection	77	——	——	1
	Distance digit II proximal bend point to digit III midline	33	——	——	1
	Distance "heel" to toetip II	149	——	——	1
	Distance "heel" to toetip III	202	——	——	1
	Distance "heel" to toetip IV	134	——	——	1
	Interdigital angle II–III	24	——	——	1
	Interdigital angle III–IV	22	——	——	1
	Interdigital angle II–IV	46	——	——	1
YPM VP 001210 *Eubrontes loxonyx* (*Anchisauripus tuberosus*)	Footprint field length	176	——	——	1
	Digit II first pad length	35	——	——	1
	Digit II first pad width	32	——	——	1
	Digit II second pad length	35	——	——	1
	Digit II second pad width	26	——	——	1
	Digit II length excluding claw	69	——	——	1
	Digit III first pad length	31	——	——	1
	Digit III first pad width	32	——	——	1
	Digit III second pad length	32	——	——	1
	Digit III second pad width	43	——	——	1
	Digit III third pad length	36	——	——	1
	Digit III third pad width	35	——	——	1

Table A8.1. *continued*

Trackway	Parameter	Mean	Minimum	Maximum	*N*
	Digit III length excluding claw	100	——	——	1
	Digit III claw length	18	——	——	1
	Digit III length	116	——	——	1
	Digit IV length	132	——	——	1
	Digit IV first pad width	29	——	——	1
	Toebase II–III	40	——	——	1
	Toebase III–IV	58	——	——	1
	Toebase II–IV	37	——	——	1
	Toetip III–IV	76	——	——	1
	Distance digit II proximal bend point to digit III midline	35	——	——	1
	Distance "heel" to toetip III	176	——	——	1
	Distance "heel" to toetip IV	132	——	——	1
	Interdigital angle II–III	2	——	——	1
	Interdigital angle III–IV	32	——	——	1
	Interdigital angle II–IV	34	——	——	1
YPM VP 003315B *Eubrontes loxonyx* (*Anchisauripus tuberosus*)	Footprint field length	205	——	——	1
	Digit II second pad length	44	——	——	1
	Digit II second pad width	31	——	——	1
	Digit II claw length	25	——	——	1
	Digit III third pad length	39	——	——	1
	Digit III third pad width	39	——	——	1
	Digit III claw length	19	——	——	1
	Digit IV length	133	——	——	1
	Backfoot length	133	——	——	1
	Digit III projection	70	——	——	1
	Toetip II–III	79	——	——	1
	Toetip III–IV	92	——	——	1
	Toetip width	97	——	——	1
	Distance digit II proximal bend point to digit III midline	22	——	——	1
	Distance "heel" to toetip II	156	——	——	1
	Distance "heel" to toetip III	205	——	——	1
	Distance "heel" to toetip IV	131	——	——	1
	Interdigital angle II–III	8	——	——	1
	Interdigital angle III–IV	13	——	——	1
	Interdigital angle II–IV	20	——	——	1
AC 43/4 *Eubrontes loxonyx* (*Anchisauripus tuberosus*)	Footprint field length	218.7	215	225	4
	Digit II length excluding claw	95	——	——	1
	Digit II claw length	22	——	——	1
	Digit II length	116	——	——	1
	Digit III length	155	——	——	1
	Digit IV length	191	——	——	1
	Backfoot length	166	——	——	1
	Digit III projection	58	——	——	1
	Distance digit II proximal bend point to digit III midline	37	——	——	1
	Distance "heel" to toetip II	175	——	——	1
	Distance "heel" to toetip III	226	——	——	1
	Distance "heel" to toetip IV	191	——	——	1
	Interdigital angle II–III	18	——	——	1
	Interdigital angle III–IV	15	——	——	1
	Interdigital angle II–IV	33	——	——	1
	Pace	63.9	62	65.5	4
	Stride	128.3	126.5	130	3
	Pace angulation	180.0	180	180	3

Trackway	Parameter	Mean	Minimum	Maximum	N
YPM VP 003323 *Eubrontes sillimani* (*Anchisauripus sillimani*)	Footprint field length	137	——	——	1
	Digit II first pad length	28	——	——	1
	Digit II first pad width	22	——	——	1
	Digit II second pad length	25	——	——	1
	Digit II second pad width	17	——	——	1
	Digit II length excluding claw	53	——	——	1
	Digit II claw length	16	——	——	1
	Digit II length	68	——	——	1
	Digit III first pad length	31	——	——	1
	Digit III first pad width	23	——	——	1
	Digit III second pad length	28	——	——	1
	Digit III second pad width	33	——	——	1
	Digit III third pad length	23	——	——	1
	Digit III third pad width	24	——	——	1
	Digit III length excluding claw	83	——	——	1
	Digit III claw length	21	——	——	1
	Digit III length	103	——	——	1
	Digit IV length	89	——	——	1
	Digit IV first pad width	21	——	——	1
	Toebase II–III	20	——	——	1
	Toebase III–IV	34	——	——	1
	Toebase II–IV	31	——	——	1
	Toetip II–III	60	——	——	1
	Toetip III–IV	66	——	——	1
	Toetip width	71	——	——	1
	Backfoot length	86	——	——	1
	Digit III projection	53	——	——	1
	Distance digit II proximal bend point to digit III midline	19	——	——	1
	Distance "heel" to toetip II	98	——	——	1
	Distance "heel" to toetip III	140	——	——	1
	Distance "heel" to toetip IV	89	——	——	1
	Interdigital angle II–III	11	——	——	1
	Interdigital angle III–IV	20	——	——	1
	Interdigital angle II–IV	31	——	——	1
YPM VP 002054C *Eubrontes sillimani* (*Anchisauripus sillimani*)	Footprint field length	124	——	——	1
	Digit II first pad length	24	——	——	1
	Digit II first pad width	18	——	——	1
	Digit II second pad length	17	——	——	1
	Digit II second pad width	16	——	——	1
	Digit II length excluding claw	44	——	——	1
	Digit II claw length	14	——	——	1
	Digit II length	57	——	——	1
	Digit III first pad length	21	——	——	1
	Digit III first pad width	20	——	——	1
	Digit III second pad length	22	——	——	1
	Digit III second pad width	18	——	——	1
	Digit III third pad length	18	——	——	1
	Digit III third pad width	14	——	——	1
	Digit III length excluding claw	70	——	——	1
	Digit III claw length	15	——	——	1
	Digit III length	85	——	——	1
	Digit IV length	82	——	——	1
	Toebase II–III	23	——	——	1
	Toebase III–IV	40	——	——	1
	Toebase II–IV	28	——	——	1
	Toetip II–III	61	——	——	1

Table A8.1. *continued*

Trackway	Parameter	Mean	Minimum	Maximum	*N*
	Toetip III–IV	56	——	——	1
	Toetip width	64	——	——	1
	Backfoot length	78	——	——	1
	Digit III projection	48	——	——	1
	Distance digit II proximal bend point to digit III midline	15	——	——	1
	Distance "heel" to toetip II	85	——	——	1
	Distance "heel" to toetip III	127	——	——	1
	Distance "heel" to toetip IV	84	——	——	1
	Interdigital angle II–III	11	——	——	1
	Interdigital angle III–IV	14	——	——	1
	Interdigital angle II–IV	25	——	——	1
AC 9/14 *Eubrontes sillimani* (*Anchisauripus sillimani*) sequence 1-30	Footprint field length	136	——	——	1
	Digit II first pad length	26	——	——	1
	Digit II first pad width	18	——	——	1
	Digit II length excluding claw	50	——	——	1
	Digit II claw length	12	——	——	1
	Digit II length	60.5	60	61	2
	Digit III first pad length	29	——	——	1
	Digit III first pad width	22	——	——	1
	Digit III second pad length	27	——	——	1
	Digit III second pad width	23	——	——	1
	Digit III length excluding claw	83	——	——	1
	Digit III claw length	11	——	——	1
	Digit III length	100.0	94	106	2
	Digit IV length	89	——	——	1
	Toebase II–III	23.5	22	25	2
	Toebase III–IV	43	——	——	1
	Toebase II–IV	38	——	——	1
	Toetip II–III	66.0	59	73	2
	Toetip III–IV	64.5	62	67	2
	Toetip width	68.5	67	70	2
	Backfoot length	86	——	——	1
	Digit III projection	55.5	50	61	2
	Distance digit II proximal bend point to digit III midline	19.5	19	20	2
	Distance "heel" to toetip II	97	——	——	1
	Distance "heel" to toetip III	137	——	——	1
	Distance "heel" to toetip IV	88	——	——	1
	Interdigital angle II–III	5.5	5	6	2
	Interdigital angle III–IV	15.5	13	18	2
	Interdigital angle II–IV	21.5	20	23	2
	Pace	78.5	——	——	1
AC 9/14 *Eubrontes sillimani* (*Anchisauripus sillimani*) sequence 2-18-37	Digit II first pad length	24.0	19	27	3
	Digit II first pad width	20.0	20	20	2
	Digit II second pad length	22.3	15	26	3
	Digit II second pad width	16	——	——	1
	Digit II length excluding claw	51.7	50	53	3
	Digit II claw length	18.7	16	21	3
	Digit II length	68.7	65	71	3
	Digit III first pad length	28.0	26	30	2
	Digit III first pad width	19	——	——	1
	Digit III second pad length	28.5	28	29	2
	Digit III second pad width	20	——	——	1
	Toebase II–III	25.5	23	28	2
	Toetip II–III	70	——	——	1
	Toetip III–IV	68	——	——	2
	Toetip width	59.5	56	63	2

Trackway	Parameter	Mean	Minimum	Maximum	*N*
	Digit III projection	63	———	———	1
	Distance digit II proximal bend point to digit III midline	17.3	14	22	3
	Interdigital angle II–III	7.0	0	11	3
	Interdigital angle III–IV	9.5	6	13	2
	Interdigital angle II–IV	20.5	17	24	2
	Pace	59.2	59	59.4	2
	Stride	117.5	———	———	1
	Pace angulation	166	———	———	1
AC 9/14 *Eubrontes sillimani* (*Anchisauripus sillimani*) sequence 22-5	Footprint field length	156	———	———	1
	Digit II first pad length	24.5	24	25	2
	Digit II first pad width	21.0	20	22	2
	Digit II second pad length	23.5	22	25	2
	Digit II second pad width	17	———	———	1
	Digit II length excluding claw	49.0	48	50	2
	Digit II claw length	23.0	17	29	2
	Digit II length	71.5	68	75	2
	Digit III first pad length	32.0	31	33	2
	Digit III first pad width	20.0	19	21	2
	Digit III second pad length	29.5	29	30	2
	Digit III second pad width	23.5	23	24	2
	Digit III third pad length	24	———	———	1
	Digit III third pad width	17	———	———	1
	Digit III length excluding claw	87	———	———	1
	Digit III claw length	13	———	———	1
	Digit III length	107.5	97	118	2
	Digit IV length	95	———	———	1
	Toebase II–III	21.0	20	22	2
	Toebase III–IV	40	———	———	1
	Toebase II–IV	33	———	———	1
	Toetip II–III	69.0	62	76	2
	Toetip III–IV	61.5	50	73	2
	Toetip width	72.0	67	77	2
	Backfoot length	92	———	———	1
	Digit III projection	54	44	62	2
	Distance digit II proximal bend point to digit III midline	15.5	15	16	2
	Distance "heel" to toetip II	107	———	———	1
	Distance "heel" to toetip III	156	———	———	1
	Distance "heel" to toetip IV	94	———	———	1
	Interdigital angle II–III	15.5	10	21	2
	Interdigital angle II–III	8.5	3	14	2
	Interdigital angle II–III	24.0	23	25	2
	Pace	73.5	———	———	1
AC 9/14 *Eubrontes sillimani* (*Anchisauripus sillimani*) sequence 9-31	Footprint field length	139	———	———	1
	Digit II first pad length	23.5	23	24	2
	Digit II first pad width	20.0	18	22	2
	Digit II second pad length	22.0	22	22	2
	Digit II second pad width	18.0	17	19	2
	Digit II length excluding claw	44.5	43	46	2
	Digit II claw length	15.5	15	16	2
	Digit II length	57.5	56	59	2
	Digit III first pad length	23	———	———	1
	Digit III first pad width	17	———	———	1
	Digit III second pad length	28	———	———	1
	Digit III second pad width	19	———	———	1
	Digit III third pad length	24	———	———	1
	Digit III third pad length	14	———	———	1

Table A8.1. *continued*

Trackway	Parameter	Mean	Minimum	Maximum	*N*
AC 9/14 *Eubrontes sillimani* (*Anchisauripus sillimani*) sequence 9-31	Digit III length excluding claw	75	——	——	1
	Digit III claw length	21	——	——	1
	Digit III length	102	——	——	1
	Toebase II–III	22.5	22	23	2
	Toebase III–IV	36	——	——	1
	Toebase II–IV	27	——	——	1
	Toetip II–III	64.5	62	67	2
	Toetip III–IV	68	——	——	1
	Toetip width	52	——	——	1
	Digit III projection	59	——	——	1
	Distance digit II proximal bend point to digit III midline	21	——	——	1
	Distance "heel" to toetip II	86	——	——	1
	Distance "heel" to toetip III	138	138	——	1
	Interdigital angle II–III	12	——	——	1
	Interdigital angle III–IV	24	——	——	1
	Interdigital angle II–IV	32.0	28	36	2
	Pace	73.0	66	80	2
	Stride	145.7	——	——	1
	Pace angulation	173	——	——	1
AC 9/14 *Eubrontes sillimani* (*Anchisauripus sillimani*) sequence 32-12	Footprint field length	137	——	——	1
	Digit II first pad length	21.5	19	24	2
	Digit II first pad width	18.0	18	18	2
	Digit II second pad length	24.0	22	26	2
	Digit II second pad width	16.5	16	17	2
	Digit II length excluding claw	46.5	41	52	2
	Digit II claw length	19.5	19	20	2
	Digit II length	59.5	58	61	2
	Digit III first pad length	27.0	25	29	2
	Digit III first pad width	19.5	18	21	2
	Digit III second pad length	26.5	26	27	2
	Digit III second pad width	22.0	21	23	2
	Digit III third pad length	24.5	22	27	2
	Digit III third pad width	16.0	16	16	2
	Digit III length excluding claw	79.0	79	79	2
	Digit III claw length	18	——	——	1
	Digit III length	94	——	——	1
	Digit IV length	79.0	72	86	2
	Toebase II–III	19.5	19	20	2
	Toebase III–IV	38.0	33	43	2
	Toebase II–IV	33.0	30	36	2
	Toetip II–III	72	——	——	1
	Toetip III–IV	58	——	——	1
	Toetip width	68.0	63	73	2
	Backfoot length	76.5	72	81	2
	Digit III projection	53	——	——	1
	Distance digit II proximal bend point to digit III midline	16.0	16	16	2
	Distance "heel" to toetip II	91.5	86	97	2
	Distance "heel" to toetip III	135	——	——	1
	Distance "heel" to toetip IV	78.5	73	84	2
	Interdigital angle II–III	12.5	9	16	2
	Interdigital angle III–IV	9.5	6	13	2
	Interdigital angle II–IV	22.0	22	22	2
	Pace	50.5	——	——	1

Trackway	Parameter	Mean	Minimum	Maximum	N
AC 9/14 *Eubrontes sillimani* (*Anchisauripus sillimani*) sequence 17-38	Footprint field length	151.0	151	151	2
	Digit II first pad length	28.5	26	31	2
	Digit II first pad width	22.0	22	22	2
	Digit II second pad length	29.0	27	31	2
	Digit II second pad width	19.5	19	20	2
	Digit II length excluding claw	59.0	58	60	2
	Digit II claw length	20.0	17	23	2
	Digit II length	75.5	74	77	2
	Digit III first pad length	29.5	29	30	2
	Digit III first pad width	24.0	24	24	2
	Digit III second pad length	29.0	29	29	2
	Digit III second pad width	26.5	24	29	2
	Digit III third pad length	28	——	——	1
	Digit III third pad width	19	——	——	1
	Digit III length excluding claw	85	——	——	1
	Digit III claw length	26	——	——	1
	Digit III length	111.0	110	112	2
	Digit IV length	90.5	87	94	2
	Toebase II–III	23.5	23	24	2
	Toebase III–IV	40.0	38	42	2
	Toebase II–IV	32.0	31	33	2
	Toetip II–III	65.0	64	66	2
	Toetip III–IV	76.5	74	79	2
	Toetip width	74.5	73	76	2
	Backfoot length	91.0	89	93	2
	Digit III projection	61.0	59	63	2
	Distance digit II proximal bend point to digit III midline	18.5	18	19	2
	Distance "heel" to toetip II	107.5	106	109	2
	Distance "heel" to toetip III	152.5	152	153	2
	Distance "heel" to toetip IV	90.0	87	93	2
	Interdigital angle II–III	7.5	4	11	2
	Interdigital angle III–IV	17.5	16	19	2
	Interdigital angle III–IV	25.0	23	27	2
	Pace	70.5	——	——	1
AC 9/14 *Eubrontes sillimani* (*Anchisauripus sillimani*) sequence 26-19-43	Footprint field length	115.5	111	120	2
	Digit II first pad length	21.0	20	22	2
	Digit II first pad width	16.3	15	17	3
	Digit II second pad length	20.0	19	21	2
	Digit II second pad width	15	——	——	1
	Digit II length excluding claw	41.5	41	42	2
	Digit II claw length	17.5	16	19	2
	Digit II length	57.5	57	58	2
	Digit III first pad length	23.5	23	24	2
	Digit III first pad width	17.7	17	19	3
	Digit III second pad length	27.0	25	29	2
	Digit III second pad width	18.5	18	19	2
	Digit III third pad length	19.0	18	20	2
	Digit III third pad width	13.5	13	14	2
	Digit III length excluding claw	70.0	69	71	2
	Digit III claw length	13.5	11	16	2
	Digit III length	83.0	78	88	2
	Digit IV length	70.5	67	74	2
	Digit IV first pad width	12.0	9	15	2
	Toebase II–III	21.0	20	23	3
	Toebase III–IV	34.3	33	36	3
	Toebase II–IV	26.3	24	28	3

Table A8.1. *continued*

Trackway	Parameter	Mean	Minimum	Maximum	*N*
AC 9/14 *Eubrontes sillimani* (*Anchisauripus sillimani*) sequence 26-19-43	Toetip II–III	56.5	51	62	2
	Toetip III–IV	54.5	52	57	2
	Toetip width	59.5	55	64	2
	Distance digit II proximal bend point to digit III midline	14.0	14	14	2
	Distance "heel" to toetip II	82.0	82	82	2
	Distance "heel" to toetip III	116.5	112	121	2
	Distance "heel" to toetip IV	70.5	68	73	2
	Interdigital angle II–III	12.5	12	13	2
	Interdigital angle III–IV	11.0	7	15	2
	Interdigital angle II–IV	23.5	20	27	2
	Pace	56.3	55	57.5	2
	Stride	110.5	——	——	1
	Pace angulation	158	——	——	1
AC 9/14 *Eubrontes sillimani* (*Anchisauripus sillimani*) sequence 40-14-21	Footprint field length	142	——	——	1
	Digit II first pad length	21.0	21	21	2
	Digit II first pad width	16.0	15	17	2
	Digit II second pad length	23.5	22	25	2
	Digit II second pad width	16.0	14	18	2
	Digit II length excluding claw	45.0	44	46	2
	Digit II claw length	10	——	——	1
	Digit II length	54	——	——	1
	Digit III first pad length	26.0	23	29	2
	Digit III first pad width	19.5	19	20	2
	Digit III second pad length	27.0	27	27	2
	Digit III second pad width	21.5	20	23	2
	Digit III third pad length	21.5	21	22	2
	Digit III third pad width	15.5	15	16	2
	Digit III length excluding claw	75.0	71	79	2
	Digit III claw length	21.0	19	23	2
	Digit III length	92.0	91	93	2
	Toebase II–III	24.0	22	26	2
	Toebase III–IV	50	——	——	1
	Toebase II–IV	40	——	——	1
	Toetip II–III	62	——	——	1
	Toetip III–IV	88	——	——	1
	Distance digit II proximal bend point to digit III midline	16.5	16	17	2
	Distance "heel" to toetip II	95	——	——	1
	Distance "heel" to toetip III	142	——	——	1
	Interdigital angle II–III	11.5	11	12	2
	Pace	58	——	——	1
AC 9/14 *Eubrontes sillimani* (*Anchisauripus sillimani*) sequence 34-29	Digit II first pad length	24.0	23	25	2
	Digit II first pad width	19.0	19	19	2
	Digit II second pad length	23.0	22	24	2
	Digit II second pad width	16.5	16	17	2
	Digit II length excluding claw	47.0	45	49	2
	Digit III first pad length	26.5	26	27	2
	Digit III first pad width	21.5	21	22	2
	Digit III second pad length	27.0	26	28	2
	Digit III second pad width	23.0	22	24	2
	Digit III third pad length	21.5	21	22	2
	Digit III third pad width	16.5	16	17	2
	Digit III length excluding claw	75.0	75	75	2
	Digit III claw length	17	——	——	1
	Digit III length	89	——	——	1
	Toebase II–III	24.5	23	26	2
	Toebase III–IV	41	——	——	1

Trackway	Parameter	Mean	Minimum	Maximum	*N*
	Toebase II–IV	33	——	——	1
	Distance digit II proximal bend point to digit III midline	16.0	16	16	2
	Interdigital angle II–III	11.5	9	14	2
	Interdigital angle III–IV	12.5	11	14	2
	Interdigital angle II–IV	23.0	22	24	2
	Pace	59.7	——	——	1
AC 9/14 *Eubrontes sillimani* (*Anchisauripus sillimani*) sequence 33-8-20 (all measurements were made on print 20)	Digit II first pad length	21	——	——	1
	Digit II first pad width	17	——	——	1
	Digit II second pad length	23	——	——	1
	Digit II length excluding Claw	44	——	——	1
	Digit II length	54	——	——	1
	Digit III first pad length	30	——	——	1
	Digit III first pad width	20	——	——	1
	Digit III second pad length	23	——	——	1
	Digit III second pad width	22	——	——	1
	Digit III third pad length	19	——	——	1
	Digit III third pad width	19	——	——	1
	Digit III length excluding claw	72	——	——	1
	Digit III claw length	11	——	——	1
	Digit III length	83	——	——	1
	Toebase II–III	23	——	——	1
	Toetip II–III	78	——	——	1
	Toetip III–IV	51	——	——	1
	Toetip width	62	——	——	1
	Digit III projection	52	——	——	1
	Distance digit II proximal bend point to digit III midline	13	——	——	1
	Interdigital angle II–III	23	——	——	1
AC 9/14 *Eubrontes sillimani* (*Anchisauripus sillimani*) print 3	Footprint field length	147	——	——	1
	Digit II first pad length	25	——	——	1
	Digit II first pad width	20	——	——	1
	Digit II second pad length	24	——	——	1
	Digit II second pad width	21	——	——	1
	Digit II length excluding claw	51	——	——	1
	Digit II claw length	22	——	——	1
	Digit II length	67	——	——	1
	Digit III first pad length	29	——	——	1
	Digit III first pad width	21	——	——	1
	Digit III second pad length	27	——	——	1
	Digit III second pad width	22	——	——	1
	Digit III third pad length	22	——	——	1
	Digit III third pad width	17	——	——	1
	Digit III length excluding claw	76	——	——	1
	Digit III claw length	20	——	——	1
	Toebase II–III	25	——	——	1
	Toebase III–IV	52	——	——	1
	Toebase II–IV	41	——	——	1
	Toetip II–III	68	——	——	1
	Toetip III–IV	63	——	——	1
	Toetip width	74	——	——	1
	Digit III length	95	——	——	1
	Digit IV length	95	——	——	1
	Digit IV first pad width	19	——	——	1
	Backfoot length	93	——	——	1
	Digit III projection	53	——	——	1
	Distance digit II proximal bend point to digit III midline	15	——	——	1
	Distance "heel" to toetip II	110	——	——	1

Table A8.1. *continued*

Trackway	Parameter	Mean	Minimum	Maximum	*N*
	Distance "heel" to toetip III	148	——	——	1
	Distance "heel" to toetip IV	96	——	——	1
	Interdigital angle II–III	14	——	——	1
	Interdigital angle III–IV	7	——	——	1
	Interdigital angle II–IV	21	——	——	1
AC 9/14 *Eubrontes sillimani* (*Anchisauripus sillimani*) print 4	Footprint field length	161	——	——	1
	Digit II first pad length	31	——	——	1
	Digit II first pad width	21	——	——	1
	Digit II second pad length	27	——	——	1
	Digit II second pad width	20	——	——	1
	Digit II length excluding claw	55	——	——	1
	Digit II claw length	10	——	——	1
	Digit II length	66	——	——	1
	Digit III first pad length	34	——	——	1
	Digit III first pad width	26	——	——	1
	Digit III second pad length	32	——	——	1
	Digit III second pad width	25	——	——	1
	Digit III length	105	——	——	1
	Digit IV length	102	——	——	1
	Digit IV first pad width	22	——	——	1
	Toebase II–III	28	——	——	1
	Toebase III–IV	57	——	——	1
	Toebase II–IV	44	——	——	1
	Toetip II–III	77	——	——	1
	Toetip III–IV	64	——	——	1
	Toetip width	70	——	——	1
	Backfoot length	97	——	——	1
	Digit III projection	61	——	——	1
	Distance digit II proximal bend point to digit III midline	18	——	——	1
	Distance "heel" to toetip II	109	——	——	1
	Distance "heel" to toetip III	159	——	——	1
	Distance "heel" to toetip IV	101	——	——	1
	Interdigital angle II–III	9	——	——	1
	Interdigital angle III–IV	13	——	——	1
	Interdigital angle II–IV	22	——	——	1
AC 9/14 *Eubrontes sillimani* (*Anchisauripus sillimani*) print 6	Digit II first pad length	26	——	——	1
	Digit II second pad width	19	——	——	1
	Digit II length	69	——	——	1
	Digit III first pad length	30	——	——	1
	Digit III first pad width	20	——	——	1
	Digit III second pad length	29	——	——	1
	Digit III second pad width	24	——	——	1
	Digit III third pad length	25	——	——	1
	Digit III third pad width	18	——	——	1
	Digit III length excluding claw	85	——	——	1
	Toebase II–III	18	——	——	1
	Toetip width	61	——	——	1
	Distance digit II proximal bend point to digit III midline	10	——	——	1
	Interdigital angle II–III	23	——	——	1
AC 9/14 *Eubrontes sillimani* (*Anchisauripus sillimani*) print 10	Digit II first pad length	26	——	——	1
	Digit II first pad width	19	——	——	1
	Digit II second pad length	31	——	——	1
	Digit II second pad width	18	——	——	1
	Digit II length excluding claw	56	——	——	1
	Digit II length	69	——	——	1
	Digit III first pad length	29	——	——	1

Trackway	Parameter	Mean	Minimum	Maximum	N
	Digit III first pad width	22	——	——	1
	Digit III second pad length	30	——	——	1
	Digit III second pad width	24	——	——	1
	Digit III third pad length	28	——	——	1
	Digit III third pad width	20	——	——	1
	Digit III length excluding claw	86	——	——	1
	Digit III claw length	20	——	——	1
	Digit III length	105	——	——	1
	Toebase II–III	27	——	——	1
	Toetip II–III	65	——	——	1
	Toetip III–IV	66	——	——	1
	Toetip width	69	——	——	1
	Digit III projection	55	——	——	1
	Distance digit II proximal bend point to digit III midline	19	——	——	1
	Interdigital angle II–III	4	——	——	1
	Interdigital angle III–IV	11	——	——	1
	Interdigital angle II–IV	14	——	——	1
AC 9/14 *Eubrontes sillimani* (*Anchisauripus sillimani*) print 11	Digit II first pad length	23	——	——	1
	Digit II first pad width	21	——	——	1
	Digit II second pad length	27	——	——	1
	Digit II second pad width	17	——	——	1
	Digit II length excluding claw	50	——	——	1
	Digit II length	65	——	——	1
	Digit III first pad length	30	——	——	1
	Digit III first pad width	24	——	——	1
	Digit III second pad length	29	——	——	1
	Digit III second pad width	25	——	——	1
	Digit III third pad length	31	——	——	1
	Digit III third pad width	21	——	——	1
	Digit III length excluding claw	90	——	——	1
	Digit III claw length	17	——	——	1
	Digit III length	105	——	——	1
	Toebase II–III	22	——	——	1
	Toetip II–III	68	——	——	1
	Toetip III–IV	67	——	——	1
	Toetip width	68	——	——	1
	Digit III projection	59	——	——	1
	Distance digit II proximal bend point to digit III midline	23	——	——	1
	Interdigital angle II–III	8	——	——	1
	Interdigital angle III–IV	17	——	——	1
	Interdigital angle II–IV	25	——	——	1
AC 9/14 *Eubrontes sillimani* (*Anchisauripus sillimani*) print 13	Footprint field length	134	——	——	1
	Digit II first pad length	21	——	——	1
	Digit II first pad width	16	——	——	1
	Digit II second pad length	25	——	——	1
	Digit II second pad width	19	——	——	1
	Digit II length excluding claw	47	——	——	1
	Digit II claw length	11	——	——	1
	Digit II length	58	——	——	1
	Digit III first pad length	26	——	——	1
	Digit III first pad width	20	——	——	1
	Digit III second pad length	30	——	——	1
	Digit III second pad width	21	——	——	1
	Digit III third pad length	27	——	——	1
	Digit III third pad width	18	——	——	1
	Digit III length excluding claw	82	——	——	1

Table A8.1. *continued*

Trackway	Parameter	Mean	Minimum	Maximum	*N*
	Digit III claw length	12	——	——	1
	Digit III length	92	——	——	1
	Digit IV length	92	——	——	1
	Toebase II–III	21	——	——	1
	Toebase III–IV	41	——	——	1
	Toebase II–IV	35	——	——	1
	Toetip II–III	63	——	——	1
	Toetip III–IV	57	——	——	1
	Toetip width	72	——	——	1
	Backfoot length	83	——	——	1
	Digit III projection	48	——	——	1
	Distance digit II proximal bend point to digit III midline	12	——	——	1
	Distance "heel" to toetip II	92	——	——	1
	Distance "heel" to toetip III	133	——	——	1
	Distance "heel" to toetip IV	93	——	——	1
	Interdigital angle II–III	24	——	——	1
	Interdigital angle III–IV	12	——	——	1
	Interdigital angle II–IV	35	——	——	1
AC 9/14 *Eubrontes sillimani* (*Anchisauripus sillimani*) print 16	Footprint field length	146	——	——	1
	Digit II first pad length	22	——	——	1
	Digit II first pad width	20	——	——	1
	Digit II second pad length	20	——	——	1
	Digit II second pad width	18	——	——	1
	Digit II length excluding claw	43	——	——	1
	Digit II claw length	14	——	——	1
	Digit II length	55	——	——	1
	Digit III first pad width	23	——	——	1
	Digit III second pad length	25	——	——	1
	Digit III second pad width	20	——	——	1
	Digit III third pad length	23	——	——	1
	Digit III third pad width	17	——	——	1
	Digit III length excluding claw	86	——	——	1
	Digit III claw length	17	——	——	1
	Digit III length	101	——	——	1
	Digit IV length	92	——	——	1
	Toebase II–III	23	——	——	1
	Toebase III–IV	50	——	——	1
	Toebase II–IV	41	——	——	1
	Toetip II–III	90	——	——	1
	Toetip III–IV	59	——	——	1
	Toetip width	75	——	——	1
	Backfoot length	83	——	——	1
	Digit III projection	63	——	——	1
	Distance digit II proximal bend point to digit III midline	13	——	——	1
	Distance "heel" to toetip II	96	——	——	1
	Distance "heel" to toetip III	147	——	——	1
	Distance "heel" to toetip IV	92	——	——	1
	Interdigital angle II–III	25	——	——	1
	Interdigital angle III–IV	11	——	——	1
	Interdigital angle II–IV	35	——	——	1
AC 9/14 *Eubrontes sillimani* (*Anchisauripus sillimani*) print 23	Digit II length excluding claw	58	——	——	1
	Digit II claw length	15	——	——	1
	Digit II length	72	——	——	1
	Digit III first pad length	30	——	——	1
	Digit III first pad width	22	——	——	1
	Digit III second pad length	32	——	——	1

Trackway	Parameter	Mean	Minimum	Maximum	N
	Digit III second pad length	25	——	——	1
	Digit III third pad length	27	——	——	1
	Digit III third pad width	22	——	——	1
	Digit III length excluding claw	90	——	——	1
	Digit III claw length	19	——	——	1
	Digit III length	107	——	——	1
	Toebase II–III	24	——	——	1
	Toetip II–III	59	——	——	1
	Toetip III–IV	77	——	——	1
	Toetip II–IV	69	——	——	1
	Digit III projection	59	——	——	1
	Distance digit II proximal bend point to digit III midline	18	——	——	1
	Interdigital angle II–III	4	——	——	1
AC 9/14 *Eubrontes sillimani* (*Anchisauripus sillimani*) print 25	Digit II length excluding claw	57	——	——	1
	Digit III first pad length	25	——	——	1
	Digit III first pad width	26	——	——	1
	Digit III second pad length	28	——	——	1
	Digit III second pad width	27	——	——	1
	Digit III third pad length	33	——	——	1
	Digit III third pad width	25	——	——	1
	Digit III length excluding claw	86	——	——	1
	Digit IV first pad width	21	——	——	1
	Toebase II–III	27	——	——	1
	Toebase III–IV	45	——	——	1
	Toebase II–IV	35	——	——	1
	Distance digit II proximal bend point to digit III midline	22	——	——	1
	Interdigital angle II–III	9	——	——	1
	Interdigital angle III–IV	27	——	——	1
	Interdigital angle II–IV	36	——	——	1
AC 9/14 *Eubrontes sillimani* (*Anchisauripus sillimani*) print 27	Digit II first pad length	34	——	——	1
	Digit II first pad width	21	——	——	1
	Digit II second pad length	32	——	——	1
	Digit II second pad width	24	——	——	1
	Digit II length excluding claw	66	——	——	1
	Digit II claw length	25	——	——	1
	Digit II length	79	——	——	1
	Digit III first pad length	35	——	——	1
	Digit III first pad width	29	——	——	1
	Digit III second pad width	32	——	——	1
	Digit III third pad width	26	——	——	1
	Digit III length excluding claw	106	——	——	1
	Digit III claw length	12	——	——	1
	Digit III length	116	——	——	1
	Toebase II–III	30	——	——	1
	Toetip II–III	95	——	——	1
	Distance digit II proximal bend point to digit III midline	21	——	——	1
	Interdigital angle II–III	14	——	——	1
	Interdigital angle III–IV	18	——	——	1
	Interdigital angle II–IV	32	——	——	1
AC 9/14 *Eubrontes sillimani* (*Anchisauripus sillimani*) print 28	Digit II first pad length	28	——	——	1
	Digit II second pad length	15	——	——	1
	Digit II second pad width	20	——	——	1
	Digit II length excluding claw	46	——	——	1
	Digit II claw length	19	——	——	1
	Digit II length	65	——	——	1
	Digit III first pad length	25	——	——	1

Table A8.1. *continued*

Trackway	Parameter	Mean	Minimum	Maximum	*N*
AC 9/14 *Eubrontes sillimani* (*Anchisauripus sillimani*) print 28	Digit III first pad width	21	——	——	1
	Digit III second pad length	27	——	——	1
	Digit III second pad width	25	——	——	1
	Digit III third pad length	21	——	——	1
	Digit III third pad width	21	——	——	1
	Digit III length excluding claw	74	——	——	1
	Digit III claw length	21	——	——	1
	Digit III length	95	——	——	1
	Toebase II–III	19	——	——	1
	Toetip II–III	51	——	——	1
	Toetip III–IV	58	——	——	1
	Toetip width	55	——	——	1
	Digit III projection	47	——	——	1
	Distance digit II proximal bend point to digit III midline	16	——	——	1
	Interdigital angle II–III	9	——	——	1
	Interdigital angle III–IV	17	——	——	1
	Interdigital angle II–IV	26	——	——	1
AC 9/14 *Eubrontes sillimani* (*Anchisauripus sillimani*) print 33	Footprint field length	138	——	——	1
	Digit II first pad length	25	——	——	1
	Digit II first pad width	19	——	——	1
	Digit II second pad length	24	——	——	1
	Digit II length excluding claw	49	——	——	1
	Digit II claw length	12	——	——	1
	Digit II length	61	——	——	1
	Digit III second pad length	24	——	——	1
	Digit III third pad length	25	——	——	1
	Digit III third pad width	16	——	——	1
	Digit III claw length	18	——	——	1
	Digit IV length	88	——	——	1
	Toebase II–IV	33	——	——	1
	Toetip II–III	63	——	——	1
	Toetip III–IV	54	——	——	1
	Toetip width	51	——	——	1
	Backfoot length	86	——	——	1
	Digit III projection	51	——	——	1
	Distance digit II proximal bend point to digit III midline	15	——	——	1
	Distance "heel" to toetip II	90	——	——	1
	Distance "heel" to toetip III	137	——	——	1
	Distance "heel" to toetip IV	88	——	——	1
	Interdigital angle II–III	15	——	——	1
	Interdigital angle III–IV	3	——	——	1
	Interdigital angle II–IV	18	——	——	1
AC 9/14 *Eubrontes sillimani* (*Anchisauripus sillimani*) print 35	Digit III first pad length	27	——	——	1
	Digit III first pad width	22	——	——	1
	Digit III second pad length	24	——	——	1
	Digit III second pad width	25	——	——	1
	Digit IV length	86	——	——	1
	Digit IV first pad width	21	——	——	1
	Toebase III–IV	41	——	——	1
	Toetip width	71	——	——	1
	Backfoot length	84	——	——	1
	Distance digit II proximal bend point to digit III midline	26	——	——	1
	Distance "heel" to toetip II	96	——	——	1
	Distance "heel" to toetip IV	85	——	——	1
	Interdigital angle III–IV	18	——	——	1

Trackway	Parameter	Mean	Minimum	Maximum	N
AC 9/14 *Eubrontes sillimani* (*Anchisauripus sillimani*) print 36	Digit II first pad length	25	——	——	1
	Digit II first pad width	16	——	——	1
	Digit II second pad length	22	——	——	1
	Digit II second pad width	16	——	——	1
	Digit II length excluding claw	47	——	——	1
	Digit II claw length	14	——	——	1
	Digit II length	60	——	——	1
	Digit III first pad length	29	——	——	1
	Digit III length excluding claw	79	——	——	1
	Digit III claw length	16	——	——	1
	Digit III length	95	——	——	1
	Toebase II–III	23	——	——	1
	Toetip II–III	62	——	——	1
	Toetip III–IV	72	——	——	1
	Toetip width	69	——	——	1
	Digit III projection	57	——	——	1
	Interdigital angle II–III	9	——	——	1
	Interdigital angle III–IV	20	——	——	1
	Interdigital angle II–IV	29	——	——	1
AC 9/14 *Eubrontes sillimani* (*Anchisauripus sillimani*) print 39	Digit II first pad length	27	——	——	1
	Digit II first pad width	24	——	——	1
	Digit II second pad width	22	——	——	1
	Toebase II–III	25	——	——	1
	Toebase III–IV	39	——	——	1
	Toebase II–IV	37	——	——	1
	Interdigital angle II–IV	22	——	——	1
AC 9/14 *Eubrontes sillimani* (*Anchisauripus sillimani*) print 41	Digit II claw length	19	——	——	1
	Digit III length excluding claw	75	——	——	1
	Digit III claw length	21	——	——	1
	Digit III length	97	——	——	1
	Toetip II–III	65	——	——	1
	Interdigital angle III–IV	9	——	——	1
AC 9/14 *Eubrontes sillimani* (*Anchisauripus sillimani*) print 42	Digit II first pad length	29	——	——	1
	Digit II second pad width	23	——	——	1
	Digit II length	77	——	——	1
	Digit III first pad length	30	——	——	1
	Digit III first pad width	27	——	——	1
	Toebase II–III	27	——	——	1
	Distance digit II proximal bend point to digit III midline	23	——	——	1
	Interdigital angle II–III	7	——	——	1
	Interdigital angle III–IV	25	——	——	1
	Interdigital angle II–IV	32	——	——	1
AC 49/1 *Eubrontes sillimani*? (*Anchisauripus sillimani* or *tuberosus*) trackway A	Footprint field length	183.3	177	191	3
	Digit II first pad length	45	——	——	1
	Digit II first pad width	24	——	——	1
	Digit II second pad length	41	——	——	1
	Digit II second pad width	25.5	25	26	2
	Digit II length excluding claw	76.5	68	85	2
	Digit II claw length	19.0	19	19	2
	Digit II length	91.5	88	95	2
	Digit III first pad length	40.0	38	42	2
	Digit III first pad width	29	——	——	1
	Digit III second pad length	35.0	34	36	2
	Digit III second pad width	36.0	34	38	2
	Digit III third pad length	33.0	33	33	2
	Digit III third pad width	32.0	30	34	2

Table A8.1. *continued*

Trackway	Parameter	Mean	Minimum	Maximum	*N*
	Digit III length excluding claw	108.5	107	110	2
	Digit III claw length	22.5	20	25	2
	Digit III length	130.0	126	134	2
	Digit IV length	110.5	110	111	2
	Toebase II–III	44	———	———	1
	Toebase III–IV	69	———	———	1
	Toebase II–IV	64	———	———	1
	Toetip II–III	102	———	———	1
	Toetip III–IV	87	———	———	1
	Toetip width	149	———	———	1
AC 49/1 *Eubrontes sillimani*? (*Anchisauripus sillimani* or *tuberosus*) trackway A	Backfoot length	111.0	108	114	2
	Digit III projection	65.5	65	66	2
	Distance digit II proximal bend point to digit III midline	20.0	17	23	2
	Distance "heel" to toetip II	135.0	133	137	2
	Distance "heel" to toetip II	179.5	177	182	2
	Distance "heel" to toetip II	110.5	110	111	2
	Interdigital angle II–III	9.0	6	12	2
	Interdigital angle III–IV	3.5	3	4	2
	Interdigital angle II–IV	12.5	9	16	2
	Pace	82.7	80	84.5	3
	Stride	165.7	163.5	168	2
	Pace angulation	180.0	180	180	2
	Footprint field length	190.0	185	195	2
	Digit II first pad length	42.0	41	43	2
	Digit II first pad width	34	———	———	1
	Digit II second pad length	42.0	40	44	2
	Digit II second pad width	34.0	31	37	2
	Digit II length excluding claw	86.5	86	87	2
	Digit II claw length	15.5	15	16	2
	Digit II length	96.5	94	99	2
	Digit III first pad length	42.0	41	43	2
	Digit III first pad width	43	———	———	1
	Digit III second pad length	41	———	———	1
	Digit III second pad width	50	———	———	1
	Digit III third pad length	36	———	———	1
	Digit III third pad width	43	———	———	1
	Digit III length excluding claw	117.5	116	119	2
	Digit III claw length	19.5	19	20	2
AC 49/1 *Eubrontes sillimani*? (*Anchisauripus sillimani* or *tuberosus*) trackway B	Digit III length	136.5	136	137	2
	Digit IV length	121.0	116	126	2
	Toebase II–III	37.0	37	37	2
	Toebase III–IV	54.5	51	58	2
	Toebase II–IV	47.0	42	52	2
	Toetip II–III	85.0	85	85	2
	Toetip III–IV	88.0	87	89	2
	Toetip width	98.5	93	104	2
	Backfoot length	117.5	114	121	2
	Digit III projection	71.0	70	72	2
	Distance digit II proximal bend point to digit III midline	28.5	27	30	2
	Distance "heel" to toetip II	140.0	139	141	2
	Distance "heel" to toetip III	191.0	186	196	2
	Distance "heel" to toetip IV	119.0	114	124	2
	Interdigital angle II–III	11.0	10	12	2
	Interdigital angle III–IV	11.0	10	12	2
	Interdigital angle II–IV	22.5	20	25	2
	Pace	111.7	110	113.5	2

Trackway	Parameter	Mean	Minimum	Maximum	N
	Stride	222	——	——	1
	Pace angulation	167	——	——	1
YPM VP 003303 *Eubrontes exsertus* (*Anchisauripus exsertus*)	Footprint field length	223	——	——	1
	Digit II first pad length	51	——	——	1
	Digit II first pad width	46	——	——	1
	Digit II second pad length	39	——	——	1
	Digit II second pad width	39	——	——	1
	Digit II length excluding claw	91	——	——	1
	Digit II claw length	20	——	——	1
	Digit II length	111	——	——	1
	Digit III first pad length	46	——	——	1
	Digit III first pad width	50	——	——	1
	Digit III second pad length	51	——	——	1
	Digit III second pad width	53	——	——	1
	Digit III third pad length	34	——	——	1
	Digit III third pad width	53	——	——	1
	Digit III length excluding claw	131	——	——	1
	Digit III claw length	24	——	——	1
	Digit III length	156	——	——	1
	Digit IV length	145	——	——	1
	Digit IV first pad width	44	——	——	1
	Toebase II–III	39	——	——	1
	Toebase III–IV	69	——	——	1
	Toebase II–IV	47	——	——	1
	Toetip II–III	105	——	——	1
	Toetip III–IV	112	——	——	1
	Toetip width	134	——	——	1
	Backfoot length	135	——	——	1
	Digit III projection	85	——	——	1
	Distance digit II proximal bend point to digit III midline	25	——	——	1
	Distance "heel" to toetip II	160	——	——	1
	Distance "heel" to toetip III	221	——	——	1
	Distance "heel" to toetip IV	143	——	——	1
	Interdigital angle II–III	12	——	——	1
	Interdigital angle III–IV	30	——	——	1
	Interdigital angle II–IV	41	——	——	1
YPM VP 00361 *Eubrontes exsertus* (*Anchisauripus exsertus*)	Digit II first pad length	38	——	——	1
	Digit II first pad width	34	——	——	1
	Digit II second pad width	28	——	——	1
	Digit II length	102	——	——	1
	Digit III length excluding claw	116	——	——	1
	Digit III claw length	21	——	——	1
	Digit III length	135	——	——	1
	Toebase II–III	34	——	——	1
	Toetip II–III	83	——	——	1
	Toetip III–IV	107	——	——	1
	Toetip width	116	——	——	1
	Digit III projection	75	——	——	1
	Distance digit II proximal bend point to digit III midline	21	——	——	1
	Interdigital angle II–III	17	——	——	1
	Interdigital angle III–IV	22	——	——	1
	Interdigital angle II–IV	39	——	——	1
AC 10/1 *Eubrontes exsertus* (*Anchisauripus minusculus*)	Footprint field length	330.3	313	344	3
	Pace	110.3	108	112.5	2
	Stride	220.5	——	——	1
	Pace angulation	157	——	——	1

Table A8.1. *continued*

Trackway	Parameter	Mean	Minimum	Maximum	*N*
	Footprint rotation	0.5*	−7	8	2
YPM VP 003348 *Eubrontes hitchcocki* (*Anchisauripus hitchcocki*)	Digit II First Pad Length	23	——	——	1
	Digit II first pad width	14	——	——	1
	Digit II second pad length	17	——	——	1
	Digit II second pad width	14	——	——	1
	Digit II length excluding claw	41	——	——	1
	Digit II claw length	11	——	——	1
	Digit II length	51	——	——	1
	Digit III first pad length	18	——	——	1
	Digit III first pad width	19	——	——	1
	Digit III second pad length	17	——	——	1
	Digit III second pad width	21	——	——	1
	Digit III third pad length	16	——	——	1
	Digit III third pad width	17	——	——	1
	Digit III length excluding claw	51	——	——	1
	Digit III claw length	14	——	——	1
	Digit III length	65	——	——	1
	Digit III projection	37	——	——	1
	Toebase II–III	24	——	——	1
	Toetip II–III	49	——	——	1
	Toetip III–IV	46	——	——	1
	Toetip width	57	——	——	1
	Distance digit II proximal bend point to digit III midline	10	——	——	1
	Interdigital angle II–III	15	——	——	1
	Interdigital angle III–IV	15	——	——	1
	Interdigital angle II–IV	30	——	——	1
AC 56/1 *Eubrontes hitchcocki* (*Anchisauripus hitchcocki*) small print	Footprint field length	122	——	——	1
	Digit II first pad length	21	——	——	1
	Digit II first pad width	19	——	——	1
	Digit II second pad ength	22	——	——	1
	Digit II second pad width	16	——	——	1
	Digit II length excluding claw	44	——	——	1
	Digit II claw length	14	——	——	1
	Digit II length	57	——	——	1
	Digit III first pad length	25	——	——	1
	Digit III first pad width	19	——	——	1
	Digit III second pad length	24	——	——	1
	Digit III second pad width	27	——	——	1
	Digit III third pad length	26	——	——	1
	Digit III third pad width	21	——	——	1
	Digit III length excluding claw	76	——	——	1
	Digit III claw length	14	——	——	1
	Digit III length	89	——	——	1
	Digit IV length	80	——	——	1
	Digit IV first pad width	20	——	——	1
	Toebase II–III	19	——	——	1
	Toebase III–IV	34	——	——	1
	Toebase II–IV	25	——	——	1
	Toetip II–III	57	——	——	1
	Toetip III–IV	55	——	——	1
	Toetip width	65	——	——	1
	Backfoot length	72	——	——	1
	Digit III projection	45	——	——	1
	Distance digit II proximal bend point to digit III midline	13	——	——	1
	Distance "heel" to toetip II	79	——	——	1
	Distance "heel" to toetip III	119	——	——	1

Trackway	Parameter	Mean	Minimum	Maximum	N
	Distance "heel" to toetip IV	81	——	——	1
	Interdigital angle II–III	16	——	——	1
	Interdigital angle III–IV	7	——	——	1
	Interdigital angle II–IV	24	——	——	1
YPM VP 003337A *Eubrontes hitchcocki* (*Anchisauripus hitchcocki*)	Footprint field length	129	——	——	1
	Digit II first pad length	25	——	——	1
	Digit II second pad length	25	——	——	1
	Digit II length excluding claw	50	——	——	1
	Digit II claw length	13	——	——	1
	Digit II length	60	——	——	1
	Digit III first pad length	29	——	——	1
	Digit III first pad width	24	——	——	1
	Digit III second pad length	26	——	——	1
	Digit III second pad width	24	——	——	1
	Digit III third pad width	18	——	——	1
	Digit III length	94	——	——	1
	Digit IV length	83	——	——	1
	Toebase II–III	21	——	——	1
	Toebase III–IV	35	——	——	1
	Toebase II–IV	34	——	——	1
	Toetip II–III	61	——	——	1
	Toetip III–IV	64	——	——	1
	Toetip width	75	——	——	1
	Backfoot length	77	——	——	1
	Digit III projection	50	——	——	1
	Distance digit II proximal bend point to digit III midline	18	——	——	1
	Distance "heel" to toetip II	92	——	——	1
	Distance "heel" to toetip III	129	——	——	1
	Distance "heel" to toetip IV	82	——	——	1
	Interdigital angle II–III	9	——	——	1
	Interdigital angle III–IV	16	——	——	1
	Interdigital angle II–IV	25	——	——	1
AC 56/1 *Eubrontes minusculus* (*Anchisauripus minusculus*)	Footprint field length	276	——	——	1
	Digit II first pad length	57	——	——	1
	Digit II first pad width	51	——	——	1
	Digit II second pad length	53	——	——	1
	Digit II second pad width	45	——	——	1
	Digit II length excluding claw	113	——	——	1
	Digit II claw length	30	——	——	1
	Digit II length	143	——	——	1
	Digit III first pad length	61	——	——	1
	Digit III first pad width	56	——	——	1
	Digit III second pad length	49	——	——	1
	Digit III second pad width	52	——	——	1
	Digit III third pad length	51	——	——	1
	Digit III third pad width	45	——	——	1
	Digit III length excluding claw	161	——	——	1
	Digit III claw length	31	——	——	1
	Digit III length	192	——	——	1
	Digit IV length	203	——	——	1
	Toebase II–III	56	——	——	1
	Toebase III–IV	78	——	——	1
	Toebase II–IV	48	——	——	1
	Toetip II–III	145	——	——	1
	Toetip III–IV	146	——	——	1
	Toetip width	206	——	——	1

Table A8.1. *continued*

Trackway	Parameter	Mean	Minimum	Maximum	*N*
	Backfoot length	173	———	———	1
	Digit III projection	101	———	———	1
	Distance digit II proximal bend point to digit III midline	40	———	———	1
	Distance "heel" to toetip II	202	———	———	1
	Distance "heel" to toetip III	278	———	———	1
	Distance "heel" to toetip IV	203	———	———	1
	Interdigital angle II–III	25	———	———	1
	Interdigital angle III–IV	29	———	———	1
	Interdigital angle II–IV	54	———	———	1
AC 16/1 *Eubrontes minusculus* (*Anchisauripus minusculus*)	Footprint field length	304.5	299	310	2
	Digit II first pad length	63.5	63	64	2
	Digit II first pad width	56.5	56	57	2
	Digit II second pad length	63.0	62	64	2
	Digit II second pad width	45.0	45	45	2
	Digit II length excluding claw	124.0	121	127	2
	Digit II claw length	26.0	26	26	2
	Digit II length	145.0	144	146	2
	Digit III first pad length	69.0	63	75	2
	Digit III first pad width	59.5	59	60	2
	Digit III second pad length	52.5	51	54	2
	Digit III second pad width	53.0	51	55	2
	Digit III third pad length	52.5	45	60	2
	Digit III third pad width	46.0	45	47	2
	Digit III length excluding claw	169	———	———	1
	Digit III claw length	25	———	———	1
	Digit III length	169	———	———	1
	Digit IV length	227.0	219	235	2
	Digit IV first pad width	60	———	———	1
	Toebase II–III	58.5	56	61	2
	Toebase III–IV	106.0	101	111	2
	Toebase II–IV	70.0	69	71	2
	Toetip II–III	144.5	141	148	2
	Toetip III–IV	140.0	137	145	2
	Toetip width	196.0	193	199	2
	Backfoot length	200.5	196	205	2
	Digit III projection	101.5	98	105	2
	Distance digit II proximal bend point to digit III midline	45.0	41	49	2
	Distance "heel" to toetip II	227.0	225	229	2
	Distance "heel" to toetip III	305.0	299	311	2
	Distance "heel" to toetip IV	222.0	213	231	2
	Interdigital angle II–III	21.5	19	24	2
	Interdigital angle III–IV	19.0	18	20	2
	Interdigital angle II–IV	40.0	37	43	2
	Pace	98.2	———	———	1
	Footprint rotation	−5	———	———	1
AC 22/1 *Eubrontes tuberatus* (*Anchisauripus minusculus*)	Footprint field length	352.7	333	375	3
	Pace	101.0	99	103	2
	Stride	202.5	———	———	1
	Pace angulation	180	———	———	1
	Footprint rotation	11.5*	7	16	2
AC 45/2 *Eubrontes minusculus* (*Anchisauripus minusculus*)	Footprint field length	323.3	320	325	3
	Digit II first pad length	68	———	———	1
	Digit II first pad width	53	———	———	1
	Digit II second pad length	68	———	———	1
	Digit II second pad width	44	———	———	1
	Digit II length excluding claw	140	———	———	1

Trackway	Parameter	Mean	Minimum	Maximum	*N*
	Digit II claw length	30	——	——	1
	Digit II length	171	——	——	1
	Digit III first pad length	77	——	——	1
	Digit III first pad width	67	——	——	1
	Digit III length	224	——	——	1
	Digit IV length	232	——	——	1
	Toebase II–III	63	——	——	1
	Toebase III–IV	98	——	——	1
	Toebase II–IV	90	——	——	1
	Toetip II–III	131	——	——	1
	Toetip III–IV	150	——	——	1
	Toetip width	197	——	——	1
	Backfoot length	224	——	——	1
	Digit III projection	100	——	——	1
	Distance digit II proximal bend point to digit III midline	53	——	——	1
	Distance "heel" to toetip II	259	——	——	1
	Distance "heel" to toetip III	326	——	——	1
	Distance "heel" to toetip IV	232	——	——	1
	Interdigital angle II–III	6	——	——	1
	Interdigital angle III–IV	23	——	——	1
	Interdigital angle II–IV	29	——	——	1
	Pace	101.3	99	103.5	2
	Stride	201.5	——	——	1
	Pace angulation	169	——	——	1
AC 14/4 *Eubrontes divaricatus* (*Anchisauripus divaricatus*)	Footprint field length	260.0	260	260	2
	Pace	86.0	84	88	2
	Stride	168.5	——	——	1
	Pace angulation	157	——	——	1
AC 58/1 *Eubrontes divaricatus*	Footprint field length	333.5	331	336	2
	Digit II length	177	——	——	1
	Digit IV length	239	——	——	1
	Toetip II–III	160.5	160	161	2
	Toetip III–IV	185.0	185	185	2
	Toetip width	268.0	263	273	2
	Backfoot length	220.5	220	221	2
	Digit III projection	109.0	105	113	2
	Distance digit II proximal bend point to digit III midline	55.5	54	57	2
	Distance "heel" to toetip II	277.5	277	278	2
	Distance "heel" to toetip III	333.5	331	336	2
	Distance "heel" to toetip IV	241.5	237	246	2
	Interdigital angle II–III	15.5	12	19	2
	Interdigital angle III–IV	31.0	28	34	2
	Interdigital angle II–IV	46.5	46	47	2
AC 15/3 *Eubrontes giganteus*	Digit II first pad length	90	——	——	1
	Digit II first pad width	67	——	——	1
	Digit II second pad length	88	——	——	1
	Digit II second pad width	54	——	——	1
	Digit II length	188	——	——	1
	Digit III first pad length	109	——	——	1
	Digit III first pad width	74	——	——	1
	Digit III second pad length	68	——	——	1
	Digit III second pad width	57	——	——	1
	Digit III third pad length	80	——	——	1
	Digit III third pad width	53	——	——	1
	Digit III length	286	——	——	1
	Digit IV length	274	——	——	1

Table A8.1. *continued*

Trackway	Parameter	Mean	Minimum	Maximum	*N*
	Toebase II–III	45	———	———	1
	Toebase III–IV	91	———	———	1
	Toebase II–IV	109	———	———	1
	Toetip II–III	181	———	———	1
	Toetip III–IV	175	———	———	1
	Toetip width	260	———	———	1
	Backfoot length	255	———	———	1
	Digit III projection	120	———	———	1
	Distance digit II proximal bend point to digit III midline	64	———	———	1
	Distance "heel" to toetip II	305	———	———	1
	Distance "heel" to toetip III	379	———	———	1
	Distance "heel" to toetip IV	274	———	———	1
	Interdigital angle II–III	25	———	———	1
	Interdigital angle III–IV	20	———	———	1
	Interdigital angle II–IV	45	———	———	1
AC 45/8 *Eubrontes giganteus*	Footprint field length	423	———	———	1
	Toetip II–III	260	———	———	1
	Toetip III–IV	175	———	———	1
	Toetip II–IV	346	———	———	1
	Backfoot length	290	———	———	1
	Digit III projection	132	———	———	1
	Distance digit II proximal bend point to digit III midline	81	———	———	1
	Distance "heel" to toetip II	342	———	———	1
	Distance "heel" to toetip III	424	———	———	1
	Distance "heel" to toetip IV	335	———	———	1
	Interdigital angle II–III	33	———	———	1
AC 7/1 *Eubrontes giganteus*	Footprint field length	42.0	———	———	1
	Pace	132.0	127.5	136.5	2
	Stride	261.5	———	———	1
	Pace angulation	164	———	———	1
AC 45/1 *Eubrontes giganteus* "early" trackway	Footprint field length	354.5	349	360	2
	Pace	102.0	101.5	102.5	2
	Stride	200.5	———	———	1
	Pace angulation	159	———	———	1
AC 45/1 *Eubrontes giganteus* "late" trackway	Footprint field length	373.7	366	388	3
	Pace	108.0	105	111	2
	Stride	212.5	———	———	1
	Pace angulation	159	———	———	1
	Footprint rotation	–2.5*	–8	3	2
AC 43/2 *Eubrontes giganteus*	Footprint field length	405	———	———	1
	Pace	120.5	———	———	1
YPM VP 002098 *Eubrontes giganteus*	Footprint field length	322.5	310	330	4
	Digit II first pad length	83	———	———	1
	Digit II first pad width	63	———	———	1
	Digit II second pad length	49	———	———	1
	Digit II second pad width	48	———	———	1
	Digit II length excluding claw	133	———	———	1
	Digit II claw length	28	———	———	1
	Digit II length	161	———	———	1
	Digit III first pad length	72	———	———	1
	Digit III first pad width	65	———	———	1
	Digit III second pad length	48	———	———	1
	Digit III second pad width	62	———	———	1
	Digit III third pad length	57	———	———	1
	Digit III third pad width	51	———	———	1

Trackway	Parameter	Mean	Minimum	Maximum	N
	Digit III length excluding claw	177	———	———	1
	Digit III claw length	23	———	———	1
	Digit III length	199	———	———	1
	Digit IV length	219	———	———	1
	Toebase II–III	73	———	———	1
	Toebase III–IV	105	———	———	1
	Toebase II–IV	74	———	———	1
	Toetip II–III	136	———	———	1
	Toetip III–IV	151	———	———	1
	Toetip width	194	———	———	1
	Backfoot length	230	———	———	1
	Digit III projection	105	———	———	1
	Distance digit II proximal bend point to digit III midline	49	———	———	1
	"Heel" to toetip II	236	———	———	1
	"Heel" to toetip III	308	———	———	1
	"Heel" to toetip IV	213	———	———	1
	Interdigital angle II–III	11	———	———	1
	Interdigital angle III–IV	21	———	———	1
	Interdigital angle II–IV	33	———	———	1
	Pace	86.8	85	88.3	3
	Stride	173.3	171.6	174.9	2
	Pace angulation	168.1	163.9	172.3	2
AC 8/1 *Eubrontes giganteus*	Footprint field length	373.5	364	385	6
	Pace	103.6	92	110	5
	Stride	201.8	192.5	211.5	3
	Pace angulation	164.0	162	166	2
	Footprint rotation	0.5*	−2	3	2
AC 13/4 *Eubrontes platypus* sequence B1 and B2	Digit II first pad length	51.0	47	55	2
	Digit II first pad width	68	———	———	1
	Digit II second pad length	66.0	64	68	2
	Digit II second pad width	57.5	57	58	2
	Digit II length excluding claw	122.0	121	123	2
	Digit II claw length	31.5	25	38	2
	Digit II length	140.0	136	144	2
	Digit III first pad length	62	———	———	1
	Digit III first pad width	62	———	———	1
	Digit III second pad length	54.5	49	60	2
	Digit III second pad width	58	———	———	1
	Digit III third pad length	53.5	53	54	2
	Digit III third pad width	63	———	———	1
	Digit III length excluding claw	162.0	———	———	1
	Digit III claw length	33.0	32	34	2
	Digit III length	194	———	———	1
	Toebase II–III	58	———	———	1
	Toetip II–III	117.5	116	119	2
	Toetip III–IV	138.0	127	149	2
	Toetip width	191.5	187	196	2
	Digit III projection	80.5	73	88	2
	Distance digit II proximal bend point to digit III midline	61.5	57	66	2
	Interdigital angle II–III	12.0	11	13	2
	Interdigital angle III–IV	15.0	15	15	2
	Interdigital angle II–IV	26.5	25	28	2
	Pace	108	———	———	1

Table A8.1. *continued*

Trackway	Parameter	Mean	Minimum	Maximum	*N*
AC 9/9 and 9/10 *Eubrontes caudatus* (*Gigandipus caudatus*)	Footprint field length	433.3	420	450	3
	Toetip II–III	172	——	——	1
	Toetip III–IV	204	——	——	1
	Toetip width	268	——	——	1
	Digit III projection	131	——	——	1
	Interdigital angle II–III	37	——	——	1
	Interdigital angle III–IV	20	——	——	1
	Interdigital angle II–IV	56	——	——	1
	Pace	97.2	88	104.5	5
	Stride	195.2	178.5	205	3
	Pace angulation	172.3	166	180	3
Gaulin site large *Eubrontes*	Footprint field length	390	——	——	1
	Pace	106.4	79	117	9
	Stride	212.7	195	220.5	8
	Pace angulation	162.3	146.9	180	8
	Footprint rotation	−2.5*	−5	5	8
Murray Quarry Wistariahurst Museum slab large *Eubrontes*	Footprint field length	360.0	345	380	3
	Pace	150.3	149.5	151	2
	Stride	297	——	——	1
	Pace angulation	162.5	——	——	1
	Footprint rotation	−0.5*	−3	2	2
Nash site large *Eubrontes*	Footprint field length	358.3	350	370	6
	Pace	88.1	83.6	91.5	5
	Stride	175.1	171	179.5	4
	Pace angulation	173.1	165.6	180	4
	Footprint rotation	0.0*	−4	1	5
AC 13/4 *Eubrontes* isp. (*Anchisauripus* isp.?) print A	Footprint field length	152	——	——	1
	Digit II length excluding claw	61	——	——	1
	Digit II claw length	19	——	——	1
	Digit II length	76	——	——	1
	Digit III claw length	18	——	——	1
	Toebase II–IV	39	——	——	1
	Toetip II–III	69	——	——	1
	Toetip III–IV	71	——	——	1
	Toetip width	86	——	——	1
	Backfoot length	92	——	——	1
	Digit III projection	59	——	——	1
	Distance digit II proximal bend point to digit III midline	25	——	——	1
	Distance "heel" to toetip II	108	——	——	1
	Distance "heel" to toetip III	152	——	——	1
	Distance "heel" to toetip IV	97	——	——	1
	Interdigital angle II–III	4	——	——	1
	Interdigital angle III–IV	20	——	——	1
	Interdigital angle II–IV	24	——	——	1
AC 13/4 *Eubrontes* isp. (*Anchisauripus* isp.?) print C	Digit II first pad length	40	——	——	1
	Digit II first pad width	42	——	——	1
	Digit II second pad width	34	——	——	1
	Digit II length	91	——	——	1
	Toebase II–IV	51	——	——	1
	Distance digit II proximal bend point to digit III midline	32	——	——	1
	Distance "heel" to toetip II	142	——	——	1
	Interdigital angle II–III	16	——	——	1
	Interdigital angle III–IV	22	——	——	1
	Interdigital angle II–IV	38	——	——	1

Trackway	Parameter	Mean	Minimum	Maximum	*N*
	Digit II first pad length	51	——	——	1
	Digit II second pad length	47	——	——	1
	Digit II second pad width	36	——	——	1
	Digit II length excluding claw	100	——	——	1
	Digit II claw length	29	——	——	1
	Digit II length	119	——	——	1
	Digit III claw length	28	——	——	1
AC 13/4 *Eubrontes* isp. (*Anchisauripus* isp.?) print G	Digit III projection	81	——	——	1
	Toetip II–III	102	——	——	1
	Toetip II–IV	112	——	——	1
	Toetip width	132	——	——	1
	Distance digit II proximal bend point to digit III midline	30	——	——	1
	Interdigital angle II–III	15	——	——	1
	Interdigital angle III–IV	4	——	——	1
	Interdigital angle II–IV	19	——	——	1
	Footprint field length	178	——	——	1
	Digit II first pad length	34	——	——	1
	Digit II first pad width	32	——	——	1
	Digit II second pad length	41	——	——	1
	Digit II second pad width	31	——	——	1
	Digit II length excluding claw	75	——	——	1
	Digit II claw length	21	——	——	1
	Digit II length	92	——	——	1
	Digit III length	126	——	——	1
	Digit IV length	126	——	——	1
	Toebase II–III	36	——	——	1
	Toebase III–IV	50	——	——	1
AC 13/4 *Eubrontes* isp. (*Anchisauripus* isp.?) print H	Toebase II–IV	47	——	——	1
	Toetip II–III	80	——	——	1
	Toetip III–IV	87	——	——	1
	Toetip width	118	——	——	1
	Backfoot length	118	——	——	1
	Digit III projection	59	——	——	1
	Distance digit II proximal bend point to digit III midline	30	——	——	1
	Distance "heel" to toetip II	141	——	——	1
	Distance "heel" to toetip III	178	——	——	1
	Distance "heel" to toetip IV	123	——	——	1
	Interdigital angle II–III	6	——	——	1
	Interdigital angle III–IV	20	——	——	1
	Interdigital angle II–IV	26	——	——	1
	Footprint field length	186	——	——	1
	Digit II first pad length	39	——	——	1
	Digit II first pad width	35	——	——	1
	Digit II second pad length	45	——	——	1
	Digit II second pad width	31	——	——	1
	Digit II length excluding claw	84	——	——	1
	Digit II claw length	21	——	——	1
AC 13/4 *Eubrontes* isp. (*Anchisauripus* isp.?) print J	Digit II length	99	——	——	1
	Digit III first pad length	43	——	——	1
	Digit III first pad width	36	——	——	1
	Digit III second pad length	38	——	——	1
	Digit III second pad width	42	——	——	1
	Digit III third pad length	36	——	——	1
	Digit III third pad width	38	——	——	1
	Digit III length excluding claw	116	——	——	1
	Digit III claw length	20	——	——	1

Table A8.1. *continued*

Trackway	Parameter	Mean	Minimum	Maximum	*N*
	Digit III length	135	———	———	1
	Digit IV length	134	———	———	1
	Toebase II–III	38	———	———	1
	Toebase III–IV	50	———	———	1
	Toebase II–IV	44	———	———	1
	Toetip II–III	83	———	———	1
	Toetip III–IV	71	———	———	1
	Toetip width	95	———	———	1
	Backfoot length	125	———	———	1
	Digit III projection	61	———	———	1
	Distance digit II proximal bend point to digit III midline	28	———	———	1
	Distance "heel" to toetip II	135	———	———	1
	Distance "heel" to toetip III	187	———	———	1
	Distance "heel" to toetip IV	134	———	———	1
	Interdigital angle II–III	3	———	———	1
	Interdigital angle III–IV	10	———	———	1
	Interdigital angle II–IV	13	———	———	1
AC 13/4 *Eubrontes* isp. (*Anchisauripus* isp.?) print K	Footprint field length	179	———	———	1
	Digit II first pad length	43	———	———	1
	Digit II first pad width	28	———	———	1
	Digit II second pad length	40	———	———	1
	Digit II second pad width	31	———	———	1
	Digit II length excluding claw	84	———	———	1
	Digit II claw length	17	———	———	1
	Digit II length	96	———	———	1
	Digit III first pad width	38	———	———	1
	Digit III second pad width	46	———	———	1
	Digit III length excluding claw	111	———	———	1
	Digit III claw length	16	———	———	1
	Digit III length	125	———	———	1
	Digit IV length	114	———	———	1
	Backfoot length	112	———	———	1
	Digit III projection	64	———	———	1
	Toebase II–III	32	———	———	1
	Toebase III–IV	53	———	———	1
	Toebase II–IV	46	———	———	1
	Toetip II–III	79	———	———	1
	Toetip III–IV	91	———	———	1
	Toetip width	113	———	———	1
	Distance digit II proximal bend point to digit III midline	23	———	———	1
	Distance "heel" to toetip II	143	———	———	1
	Distance "heel" to toetip III	179	———	———	1
	Distance "heel" to toetip IV	112	———	———	1
	Interdigital angle II–III	15	———	———	1
	Interdigital angle III–IV	14	———	———	1
	Interdigital angle II–IV	29	———	———	1
AC 13/4 *Eubrontes* isp. (*Anchisauripus* isp.?) print R	Digit II first pad length	27	———	———	1
	Digit II second pad length	29	———	———	1
	Digit II second pad width	19	———	———	1
	Digit II length excluding claw	57	———	———	1
	Digit II claw length	11	———	———	1
	Digit II length	67	———	———	1
	Digit III length excluding claw	83	———	———	1
	Digit III claw length	18	———	———	1
	Digit III length	101	———	———	1

Trackway	Parameter	Mean	Minimum	Maximum	*N*
	Toebase II–III	27	——	——	1
	Toetip II–III	68	——	——	1
	Toetip III–IV	60	——	——	1
	Toetip width	65	——	——	1
	Digit III projection	54	——	——	1
	Distance digit II proximal bend point to digit III midline	18	——	——	1
	Interdigital angle II–III	19	——	——	1
	Interdigital angle III–IV	10	——	——	1
	Interdigital angle II–IV	29	——	——	1
AC 60/33 *Eubrontes* isp.	Footprint field length	155	——	——	1
	Digit II first pad length	29	——	——	1
	Digit II first pad width	21	——	——	1
	Digit II second pad length	24	——	——	1
	Digit II second pad width	20	——	——	1
	Digit II length excluding claw	54	——	——	1
	Digit II claw length	18	——	——	1
	Digit II length	71	——	——	1
	Digit III first pad length	38	——	——	1
	Digit III first pad width	28	——	——	1
	Digit III length excluding claw	95	——	——	1
	Digit III claw length	18	——	——	1
	Digit III length	114	——	——	1
	Digit IV length	99	——	——	1
	Toebase II–III	19	——	——	1
	Toebase III–IV	40	——	——	1
	Toebase II–IV	29	——	——	1
	Toetip II–III	79	——	——	1
	Toetip III–IV	76	——	——	1
	Toetip width	85	——	——	1
	Backfoot length	90	——	——	1
	Digit III projection	65	——	——	1
	Distance digit II proximal bend point to digit III midline	14	——	——	1
	Distance "heel" to toetip II	102	——	——	1
	Distance "heel" to toetip III	154	——	——	1
	Distance "heel" to toetip IV	97	——	——	1
	Interdigital angle II–III	21	——	——	1
	Interdigital angle III–IV	20	——	——	1
	Interdigital angle II–IV	41	——	——	1
AC 16/1 *Eubrontes* isp. (*Anchisauripus* isp.?) small print	Footprint field length	192	——	——	1
	Digit II first pad length	34	——	——	1
	Digit II first pad width	34	——	——	1
	Digit II second pad length	41	——	——	1
	Digit II second pad width	29	——	——	1
	Digit II length excluding claw	75	——	——	1
	Digit II claw length	23	——	——	1
	Digit II length	92	——	——	1
	Digit III first pad length	41	——	——	1
	Digit III first pad width	36	——	——	1
	Digit III second pad length	40	——	——	1
	Digit III second pad width	41	——	——	1
	Digit III third pad length	35	——	——	1
	Digit III third pad width	32	——	——	1
	Digit III length excluding claw	116	——	——	1
	Digit III claw length	21	——	——	1
	Digit III length	137	——	——	1

Table A8.1. *continued*

Trackway	Parameter	Mean	Minimum	Maximum	*N*
	Digit IV length	127	———	———	1
	Toebase II–III	34	———	———	1
	Toebase III–IV	55	———	———	1
	Toebase II–IV	49	———	———	1
	Toetip II–III	101	———	———	1
	Toetip III–IV	90	———	———	1
	Toetip width	109	———	———	1
	Backfoot length	121	———	———	1
	Digit III projection	72	———	———	1
	Distance digit II proximal bend point to digit III midline	35	———	———	1
	Distance "heel" to toetip II	137	———	———	1
	Distance "heel" to toetip III	193	———	———	1
	Distance "heel" to toetip IV	128	———	———	1
	Interdigital angle II–III	1	———	———	1
	Interdigital angle III–IV	19	———	———	1
	Interdigital angle II–IV	20	———	———	1
USNM 481087 *Eubrontes* isp. (*Anchisauripus* isp.?) large print	Footprint field length	222	———	———	1
	Digit II first pad length	53	———	———	1
	Digit II second pad length	46	———	———	1
	Digit II second pad width	43	———	———	1
	Digit II length excluding claw	98	———	———	1
	Digit II claw length	27	———	———	1
	Digit II length	118	———	———	1
	Digit III second pad width	55	———	———	1
	Digit III third pad length	42	———	———	1
	Digit III third pad width	43	———	———	1
	Digit III length excluding claw	136	———	———	1
	Digit III claw length	22	———	———	1
	Digit III length	154	———	———	1
	Digit IV length	160	———	———	1
	Toebase II–III	39	———	———	1
	Toebase III–IV	67	———	———	1
	Toebase II–IV	56	———	———	1
	Toetip II–III	95	———	———	1
	Toetip III–IV	96	———	———	1
	Toetip width	130	———	———	1
	Digit IV first pad width	35	———	———	1
	Backfoot length	152	———	———	1
	Digit III projection	70	———	———	1
	Distance digit II proximal bend point to digit III midline	34	———	———	1
	Distance "heel" to toetip II	172	———	———	1
	Distance "heel" to toetip III	223	———	———	1
	Distance "heel" to toetip IV	160	———	———	1
	Interdigital angle II–III	7	———	———	1
	Interdigital angle III–IV	25	———	———	1
	Interdigital angle II–IV	32	———	———	1
AC 16/2 *Anomoepus scambus*	Footprint field length	85.3	82	87	3
	Pace	23.6	22.9	24.6	3
	Stride	47.1	47	47.2	2
	Pace angulation	162.4	157.7	167.1	2
AC 52/10 *Anomoepus scambus* (*Anomoepus curvatus*) left trail	Footprint field length	84.5	84	85	2
	Pace	24.8	———	———	1
AC 52/10 *Anomoepus scambus* (*Anomoepus curvatus*) right trail	Footprint field length	98.0	97	99	2
	Digit II first pad length	20.5	20	21	2
	Digit II first pad width	16.0	15	17	2
	Digit II second pad length	18	———	———	1

Trackway	Parameter	Mean	Minimum	Maximum	N
	Digit II second pad width	16	———	———	1
	Digit II length excluding claw	35.5	34	37	2
	Digit II claw length	15.5	14	17	2
	Digit II length	52.3	51	55	3
	Digit III first pad length	19.5	19	20	2
	Digit III first pad width	17.0	16	18	2
	Digit III second pad length	15.5	15	16	2
	Digit III second pad width	16.5	15	18	2
	Digit III third pad length	14.0	14	14	2
	Digit III third pad width	14.0	13	15	2
	Digit III length excluding claw	48.3	48	49	3
	Digit III claw length	17.0	17	17	2
	Digit III length	65.3	64	67	3
	Digit IV length	72.5	71	74	2
	Toebase II–III	15.0	13	18	3
	Toebase III–IV	32.5	32	33	2
	Toebase II–IV	29.0	28	30	2
	Toetip II–III	39.7	39	41	3
	Toetip III–IV	55.7	52	61	3
	Toetip width	74.7	71	81	3
	Backfoot length	67.0	67	67	2
	Digit III projection	30.0	29	31	3
	Distance digit II proximal bend point to digit III midline	15.3	14	16	3
	"Heel" to toetip II	83.0	82	84	2
	"Heel" to toetip III	98.0	97	99	2
	"Heel" to toetip IV	72.0	71	73	2
	Interdigital angle II–III	22.3	20	26	3
	Interdigital angle III–IV	28.3	23	31	3
	Interdigital angle II–IV	51.0	43	57	3
	Pace	24.7	23.3	25.9	4
	Stride	47.5	46.9	48.5	3
	Pace angulation	153.0	148	159	3
	Footprint rotation	–4.5*	–13	–1	4
AC 26/10 *Anomoepus scambus* (*Anomoepus intermedius*) left trail prints A–D	Footprint field length	99.7	90	109	4
	Digit II first pad length	18.5	18	19	2
	Digit II first pad width	16.5	15	18	2
	Digit II second pad length	15	———	———	1
	Digit II second pad width	17	———	———	1
	Digit II length Excluding claw	36.0	34	40	3
	Digit II claw length	12.7	11	15	3
	Digit II length	48.3	42	51	4
	Digit III first pad length	19	———	———	1
	Digit III first pad width	19	———	———	1
	Digit III second pad length	14	———	———	1
	Digit III second pad width	18	———	———	1
	Digit III length excluding claw	45	———	———	1
	Digit III claw length	15	———	———	1
	Digit III length	61.3	60	64	4
	Digit IV length	79.5	69	84	4
	Digit IV first pad width	17	———	———	1
	Toebase II–III	19.3	18	22	4
	Toebase III–IV	38.3	29	50	4
	Toebase II–IV	31.3	28	34	4
	Toetip II–III	39.5	37	42	4
	Toetip III–IV	52.5	40	58	4
	Toetip width	71.7	66	78	4

Table A8.1. *continued*

Trackway	Parameter	Mean	Minimum	Maximum	*N*
	Backfoot length	71.0	65	75	4
	Digit III projection	28.7	25	34	4
	Distance digit II proximal bend point to digit III midline	17.7	12	21	4
	"Heel" to toetip II	80.3	78	83	4
	"Heel" to toetip III	99.5	90	109	4
	"Heel" to toetip IV	79.5	68	84	4
	Interdigital angle II–III	12.0	6	23	4
	Interdigital angle III–IV	34.3	26	41	4
	Interdigital angle II–IV	46.0	33	54	4
	Pace	25.7	22.9	27.5	3
	Stride	51.5	48.8	54.2	2
	Pace angulation	170.5	161	180	2
	Footprint rotation	–24.0*	–27	–20	3
AC 26/10 *Anomoepus scambus* (*Anomoepus intermedius*) "diagonal" trail prints 1–5	Footprint field length	92.7	91	94	3
	Pace	19.7	14.4	23.2	4
	Stride	39.0	30.4	44.9	3
	Pace angulation	151.7	132	164	3
AC 48/1 *Anomoepus scambus* (*Anomoepus intermedius*) right trail	Footprint field length	98.5	89	104	4
	Digit II first pad length	19.7	19	21	4
	Digit II first pad width	16.3	14	19	4
	Digit II second pad length	17.3	16	19	4
	Digit II second pad width	13.7	12	15	4
	Digit II length excluding claw	38.3	36	42	4
	Digit II claw length	12.5	9	15	4
	Digit II length	51.0	46	54	4
	Digit III first pad length	17.0	15	21	3
	Digit III first pad width	16.7	15	18	3
	Digit III second pad length	15.3	13	19	3
	Digit III second pad width	16.7	16	17	3
	Digit III third pad length	17.3	15	18	4
	Digit III third pad width	13.7	12	15	4
	Digit III length excluding claw	51.5	50	54	4
	Digit III claw length	15.0	12	17	4
	Digit III length	66.0	61	70	4
	Digit IV length	72.7	72	73	4
	Digit IV first pad width	19	———	———	1
	Toebase II–III	19.5	18	21	4
	Toebase III–IV	30.5	24	34	4
	Toebase II–IV	24.5	15	29	4
	Toetip II–III	50.0	48	52	4
	Toetip III–IV	61.3	53	69	4
	Toetip width	85.5	75	91	4
	Backfoot length	62.7	60	65	4
	Digit III projection	35.7	29	40	4
	Distance digit II proximal bend point to digit III midline	21.0	19	22	4
	"Heel" to toetip II	80.0	75	84	4
	"Heel" to toetip III	98.5	89	104	4
	"Heel" to toetip IV	71.7	71	73	4
	Interdigital angle II–III	25.0	22	27	4
	Interdigital angle III–IV	40.3	36	47	4
	Interdigital angle II–IV	64.5	58	71	4
	Pace	22.5	20.9	24.2	9
	Stride	40.8	31.5	45.1	8
	Pace angulation	132.2	96	155	8
	Footprint rotation	–1.0*	–14	7	9

Trackway	Parameter	Mean	Minimum	Maximum	*N*
AC 48/1 *Anomoepus scambus* (*Anomoepus intermedius*) left (sitting or P) trail	Footprint field length	99.0	96	102	3
	Digit II first pad length	20	———	———	1
	Digit II first pad width	16	———	———	1
	Digit II second pad length	20	———	———	1
	Digit II second pad width	15	———	———	1
	Digit II length excluding claw	38.7	38	40	3
	Digit II claw length	14.3	12	18	3
	Digit II length	51.3	49	54	3
	Digit III first pad length	16	———	———	1
	Digit III first pad width	15	———	———	1
	Digit III length excluding claw	55	———	———	1
	Digit III claw length	17	———	———	1
	Digit III length	67.0	62	73	3
	Digit IV length	72.0	70	73	3
	Digit IV first pad length	16	———	———	1
	Toebase II–III	19.0	16	22	3
	Toebase III–IV	30.0	24	34	3
	Toebase II–IV	27.3	25	29	3
	Toetip II–III	50.3	48	52	3
	Toetip III–IV	60.7	50	68	3
	Toetip II–IV	86.7	78	93	3
	Backfoot length	62.7	61	65	3
	Digit III projection	35.3	33	39	3
	Distance digit II proximal bend point to digit III midline	18.7	17	20	3
	"Heel" to toetip II	83.7	82	86	3
	"Heel" to toetip III	99.0	96	102	3
	"Heel" to toetip IV	70.3	68	72	3
	Interdigital angle II–III	17.7	16	19	3
	Interdigital angle III–IV	31.7	27	39	3
	Interdigital angle II–IV	49.0	45	54	3
AC 19/18 *Anomoepus scambus* (*Plesiornis quadrupes*)	Footprint field length	97.5	95	100	2
	Pace	22.5	19.5	25.6	2
	Stride	43.6	———	———	1
	Pace angulation	150	———	———	1
YPM VP 003397A *Anomoepus scambus* (*Anomoepus intermedius*)	Footprint field length	104	———	———	1
	Digit II first pad length	20	———	———	1
	Digit II second pad length	13	———	———	1
	Digit II second pad width	16	———	———	1
	Digit II length excluding claw	32	———	———	1
	Digit II claw length	9	———	———	1
	Digit II length	41	———	———	1
	Digit III length excluding claw	51	———	———	1
	Digit III claw length	13	———	———	1
	Digit III length	64	———	———	1
	Digit IV length	89	———	———	1
	Toebase II–III	16	———	———	1
	Toebase III–IV	38	———	———	1
	Toebase II–IV	30	———	———	1
	Toetip II–III	51	———	———	1
	Toetip III–IV	53	———	———	1
	Toetip width	79	———	———	1
	Backfoot length	71	———	———	1
	Digit III projection	33	———	———	1
	Distance digit II proximal bend point to digit III midline	15	———	———	1
	Distance "heel" to toetip II	75	———	———	1
	Distance "heel" to toetip III	105	———	———	1

Table A8.1. *continued*

Trackway	Parameter	Mean	Minimum	Maximum	*N*
	Distance "heel" to toetip IV	88	———	———	1
	Interdigital angle II–III	30	———	———	1
	Interdigital angle III–IV	28	———	———	1
	Interdigital angle II–IV	57	———	———	1
YPM VP 003398 *Anomoepus scambus* (*Anomoepus intermedius*)	Footprint field length	206	———	———	1
	Digit II claw length	25	———	———	1
	Digit III claw length	24	———	———	1
	Toetip II–III	106	———	———	1
	Toetip III–IV	114	———	———	1
	Toetip width	153	———	———	1
	Backfoot length	125	———	———	1
	Digit III projection	80	———	———	1
	Distance "heel" to toetip II	151	———	———	1
	Distance "heel" to toetip III	205	———	———	1
	Distance "heel" to toetip IV	143	———	———	1
	Interdigital angle III–IV	23	———	———	1
YPM VP 003384 *Anomoepus scambus* (*Anomoepus* cf. *curvatus*)	Footprint field length	90.0	85	95	2
	Pace	25.2	———	———	1
YPM VP 002045 *Anomoepus scambus* (*Anomoepus gracillimus*)	Footprint field length	77.5	75	80	2
	Digit II first pad length	14	———	———	1
	Digit II first pad width	13	———	———	1
	Digit II second pad length	9	———	———	1
	Digit II second pad width	10	———	———	1
	Digit II length excluding claw	27	———	———	1
	Digit II claw length	13	———	———	1
	Digit II length	40	———	———	1
	Digit III length excluding claw	36	———	———	1
	Digit III claw length	14	———	———	1
	Digit III length	50	———	———	1
	Digit IV length	53	———	———	1
	Toebase II–III	17	———	———	1
	Toebase III–IV	24	———	———	1
	Toebase II–IV	21	———	———	1
	Toetip II–III	44	———	———	1
	Toetip III–IV	40	———	———	1
	Toetip width	64	———	———	1
	Backfoot length	47	———	———	1
	Digit III projection	27	———	———	1
	Distance digit II proximal bend point to digit III midline	16	———	———	1
	"Heel" to toetip II	65	———	———	1
	"Heel" to toetip III	75	———	———	1
	"Heel" to toetip IV	52	———	———	1
	Interdigital angle II–III	25	———	———	1
	Interdigital angle II–III	33	———	———	1
	Interdigital angle II–III	58	———	———	1
	Pace	18.9	18.8	18.9	2
	Stride	37.3	———	———	1
	Pace angulation	163	———	———	1
YPM VPPU 021238 *Anomoepus* isp. A trail	Footprint field length	29.5	29	30	2
	Digit II length excluding claw	10	———	———	1
	Digit II claw length	3	———	———	1
	Digit II length	12.5	12	13	4
	Digit III length	20.5	20	21	2
	Digit IV length	20	———	———	1
	Digit IV first pad width	4.5	4	5	2
	Toebase II–III	4.0	4	4	2

Trackway	Parameter	Mean	Minimum	Maximum	*N*
	Toebase III–IV	8.0	8	8	2
	Toebase II–IV	6.5	6	7	2
	Toetip II–III	15.7	14	17	4
	Toetip III–IV	16.3	16	17	3
	Toetip width	20.3	19	21	3
	Backfoot length	17	——	——	1
	Digit III projection	12.3	11	13	3
	Distance digit II proximal bend point to digit III midline	4.3	4	5	4
	"Heel" to toetip II	20.5	20	21	2
	"Heel" to toetip III	29.5	29	30	2
	"Heel" to toetip IV	20	——	——	1
	Interdigital angle II–III	23.5	15	31	4
	Interdigital angle III–IV	29.7	28	31	3
	Interdigital angle II–IV	50.7	46	57	3
	Pace	7.4	7.3	7.5	3
	Stride	14.7	14.7	14.8	2
	Pace angulation	164.0	161	167	2
YPM VPPU 021238 *Anomoepus* isp. print C	Footprint field length	32	——	——	1
	Digit II length	13	——	——	1
	Digit IV length	21	——	——	1
	Toebase II–IV	9	——	——	1
	Toetip II–III	18	——	——	1
	Toetip III–IV	16	——	——	1
	Toetip width	23	——	——	1
	Backfoot length	20	——	——	1
	Digit III projection	12	——	——	1
	Distance digit II proximal bend point to digit III midline	6	——	——	1
	Distance "heel" to toetip II	23	——	——	1
	Distance "heel" to toetip III	32	——	——	1
	Distance "heel" to toetip IV	21	——	——	1
	Interdigital angle II–III	25	——	——	1
	Interdigital angle III–IV	38	——	——	1
	Interdigital angle II–IV	63	——	——	1
YPM VPPU 021238 *Anomoepus* isp. print D	Toetip width	30	——	——	1
	Digit III projection	15	——	——	1
	Interdigital angle II–IV	62	——	——	1
	Digit III projection	15	——	——	1
YPM VPPU 021722 *Anomoepus* isp.	Footprint field length	40.5	40	42	4
	Pace	10.9	10	12.5	3
	Stride	22.7	22.4	23	2
	Page angulation	174.5	169	180	2
YPM VPPU 023665 *Anomoepus scambus*	Footprint field length	100.0	100	100	2
	Pace	32.5	——	——	1
YPM VPPU 002019 *Anomoepus scambus* (cf. *Anomoepus minimus*)	Footprint field length	45.0	43	47	3
	Pace	13.6	13.3	14.6	5
	Stride	26.5	26.2	27.1	4
	Pace angulation	158.7	151	170	4
YPM VPPU 002021 *Anomoepus scambus* (*Anomoepus curvatus*)	Footprint field length	58.7	56	62	5
	Pace	17.8	16.9	19.4	5
	Stride	33.8	32.9	34.3	4
	Pace angulation	148.3	138	165	4
	Footprint rotation	−6.0*	−11	−3	6
ANSP 15641 *Anomoepus* isp.	Footprint field length	96.5	95	98	2
	Pace	34.7	——	——	1
YPM VP 007732 unidentified bipedal dinosaur (*Anchisauripus* or *Anomoepus*)	Digit II first pad length	17	——	——	1
	Digit II first pad width	15	——	——	1

Table A8.1. *continued*

Trackway	Parameter	Mean	Minimum	Maximum	N
	Digit II second pad length	15	———	———	1
	Digit II second pad width	12	———	———	1
	Digit II length excluding claw	33	———	———	1
	Digit II claw length	14	———	———	1
	Digit II length	46	———	———	1
	Digit III first pad length	21	———	———	1
	Digit III first pad width	18	———	———	1
	Digit III length excluding claw	53	———	———	1
	Digit III claw length	16	———	———	1
	Digit III length	68	———	———	1
	Toebase II–III	14	———	———	1
	Toetip II–III	49	———	———	1
	Toetip III–IV	46	———	———	1
	Toetip width	59	———	———	1
	Digit III projection	38	———	———	1
	Distance digit II proximal bend point to digit III midline	14	———	———	1
	Interdigital angle II–III	15	———	———	1
	Interdigital angle III–IV	23	———	———	1
	Interdigital angle II–IV	38	———	———	1
YPM VP 003473 *Arachnichnus dehiscens*	Footprint field length	27.0	———	———	1
	Pace	3.8	2.9	4.8	5
	Stride	6.8	5.4	8.2	4
	Pace angulation	133.3	107	164	4
YPM VP 002095B / 002102 bipedal dinosaur A trail	Footprint field length	101.1	95	106	7
	Pace	25.6	23.9	26.8	6
	Stride	49.0	48	50.3	5
	Pace angulation	146.2	141	160	5
YPM VP 002095B / 002102 bipedal dinosaur B trail	Footprint field length	87.1	82	91	8
	Pace	21.7	21	23.1	7
	Stride	43.0	42.5	43.6	6
	Pace angulation	161.5	157	167	6
YPM VP 002095B / 002102 bipedal dinosaur C trail	Footprint field length	86.8	79	91	5
	Pace	26.6	26	27.3	4
	Stride	52.5	51.6	53.4	3
	Pace angulation	161.0	159	163	3
YPM VP 002030 *Argoides minimus*	Footprint field length	52.8	48	57	12
	Pace	20.5	17.6	22.9	13
	Stride	41.1	37.5	44.3	9
	Pace angulation	174.0	161	180	9
YPM VP 002090 *Eubrontes diversus* (*Steropoides diversus*) unidentified bipedal dinosaur	Footprint field length	135.0	———	———	1
	Pace	47.9	44	51.5	4
	Stride	95.8	94	98	3
	Pace angulation	165.3	160	168	3
YPM VP 002031 unidentified bipedal dinosaur (*Argoides minimus*)	Footprint field length	34	———	———	1
	Pace	8.4	7.8	9	4
	Stride	16.8	16.4	17	3
	Pace angulation	164.7	155	180	3
YPM VP 002132 unidentified bipedal dinosaur (*Argoides minimus*)	Footprint field length	34.4	30	37	4
	Pace	9.3	9.3	9.3	2
	Stride	17.2	16.7	17.7	2
	Pace angulation	144	———	———	1
ANSP 9735 unidentified bipedal dinosaur	Footprint field length	53.3	49	61	3
ANSP 10021 unidentified bipedal dinosaur	Footprint field length	59.0	55	63	2
	Pace	16.9	16.1	17.7	2
	Stride	33	———	———	1

Trackway	Parameter	Mean	Minimum	Maximum	*N*
YPM VPPU 021808 *Atreipus acadianus*	Footprint field length	129	——	——	1
	Pace	50.8	——	——	1
YPM VPPU 018581 *Atreipus acadianus*	Footprint field length	87	87	87	2
	Pace	39.4	——	——	1
YPM VPPU 021713 *Atreipus acadianus*	Footprint field length	136	——	——	1
	Pace	35.6	——	——	1
Kayentapus minor A trackway	Footprint field length	255.3	249	261	4
	Digit II first pad length	51.0	43	57	5
	Digit II first pad width	39.6	34	49	5
	Digit II second pad length	50.7	44	57	3
	Digit II second pad width	39.0	37	42	3
	Digit II length excluding claw	103.0	97	108	3
	Digit II claw length	30.3	29	32	3
	Digit II length	130.6	123	138	5
	Digit III first pad length	49	——	——	1
	Digit III first pad width	41	——	——	1
	Digit III second pad length	51	——	——	1
	Digit III second pad width	48	——	——	1
	Digit III length excluding claw	154	——	——	1
	Digit III claw length	24	——	——	1
	Digit III length	177.0	172	182	2
	Digit IV length	160.2	155	166	5
	Digit IV first pad width	30.3	28	33	4
	Toebase II–III	46.7	44	49	3
	Toebase III–IV	81.7	79	85	3
	Toebase II–IV	75.6	69	79	5
	Toetip II–III	119.7	113	127	4
	Toetip III–IV	167.0	159	172	4
	Toetip width	205.4	205	206	5
	Backfoot length	150.4	142	157	5
	Digit III projection	101.0	93	111	4
	Distance digit II proximal bend point to digit III midline	45.2	38	59	5
	"Heel" to toetip II	206.6	199	215	5
	"Heel" to toetip III	255.5	249	261	4
	"Heel" to toetip IV	157.8	152	165	5
	Interdigital angle II–III	18.6	10	25	5
	Interdigital angle III–IV	28.2	27	30	5
	Interdigital angle II–IV	46.8	38	53	5
Kayentapus minor B trackway	Footprint field length	223.4	204	239	5
	Digit II first pad length	47.6	43	53	5
	Digit II first pad width	36.6	30	46	5
	Digit II second pad length	44.7	43	48	4
	Digit II second pad width	34.5	32	39	4
	Digit II length excluding claw	92.7	91	96	4
	Digit II claw length	33.3	31	36	4
	Digit II length	121.6	117	127	5
	Digit III first pad length	50	——	——	1
	Digit III first pad width	41	——	——	1
	Digit III second pad length	49	——	——	1
	Digit III second pad width	40	——	——	1
	Digit III third pad length	44	——	——	1
	Digit III third pad width	41	——	——	1
	Digit III length excluding claw	139.5	136	143	2
	Digit III claw length	27.0	26	28	2
	Digit III length	152.5	140	167	4
	Digit IV length	141.0	129	149	5

Table A8.1. *continued*

Trackway	Parameter	Mean	Minimum	Maximum	*N*
	Digit IV first pad width	22.0	20	24	2
	Toebase II–III	51.3	47	57	3
	Toebase III–IV	69.7	58	83	4
	Toebase II–IV	69.0	64	77	4
	Toetip II–III	95.5	86	101	6
	Toetip III–IV	154.5	134	174	6
	Toetip width	179.0	166	188	6
	Backfoot length	136.2	125	144	5
	Digit III projection	84.6	68	95	5
	Distance digit II proximal bend point to digit III midline	41.0	39	46	4
	"Heel" to toetip II	180.4	171	191	5
	"Heel" to toetip III	223.6	204	240	5
	"Heel" to toetip IV	139.2	128	147	5
	Interdigital angle II–III	9.2	3	15	5
	Interdigital angle III–IV	32.0	26	40	6
	Interdigital angle II–IV	41.2	32	47	5
Kayentapus minor C trackway	Footprint field length	243.0	234	252	2
	Digit II first pad length	49.3	44	55	3
	Digit II first pad width	44.0	44	44	2
	Digit II second pad length	50.0	49	51	2
	Digit II second pad width	32	—	—	1
	Digit II length excluding claw	102.0	97	107	2
	Digit II claw length	36.5	34	39	2
	Digit II length	131.5	129	134	2
	Digit III second pad length	52	—	—	1
	Digit III second pad width	37	—	—	1
	Digit III third pad length	48	—	—	1
	Digit III third pad width	34	—	—	1
	Digit III claw length	22	—	—	1
	Digit III length	167	—	—	1
	Digit IV length	144.0	140	148	2
	Toebase II–III	48	—	—	1
	Toebase III–IV	84	—	—	1
	Toebase II–IV	68.3	62	74	3
	Toetip II–III	111.5	109	114	2
	Toetip III–IV	149.5	149	150	2
	Toetip width	184.5	181	188	2
	Backfoot length	138	—	—	1
	Digit III projection	94	—	—	1
	Distance digit II proximal bend point to digit III midline	46.0	45	47	2
	"Heel" to toetip II	194	—	—	1
	"Heel" to toetip III	244.0	235	253	2
	"Heel" to toetip IV	140	—	—	1
	Interdigital angle II–III	18.0	16	20	2
	Interdigital angle III–IV	29.5	25	34	2
	Interdigital angle II–IV	47.5	41	54	2

Table A8.2. Variability of digit II footprint shape parameters among individuals and species of ground birds, and within trackways of non-avian dinosaurs. All parameters are scaled as a percentage of the length of the first digital pad of digit III. Values reported in most cells are the ratio of the maximum/minimum value of the scaled parameter (number of footprints or trackways); for samples in which the number of cases is six or more, the coefficient of variation of the scaled parameter is reported in a second line below the data for the maximum/minimum ratio. Emu data are for birds characterized as adults at the time their footprints were collected. For individual emus, data cases represent individual footprints. For pooled samples of more than one individual emu, results are reported for treatments in which the data cases are individual footprints, but also treatments in which the data cases represent means for each emu. For the three 1988 Fort Wayne Zoo adult emus, results are reported both for the best-preserved footprints, and for all footprints, of those birds. For dinosaur trackways, the data cases for which the maximum/minimum ratio of the scaled parameter is calculated are the individual footprints within the trackway. In all treatments, empty cells mean that there were fewer than three (ratio of the maximum/minimum ratio of the scaled parameter) or six (coefficient of variation of the scaled parameter) measurements of that scaled parameter. Digital pad widths were not measured in the bird footprint study, but were measured for dinosaur trackways.

Bird, trackway or group				Digit II		
		First pad length	**First pad width**	**Second pad length**	**Second pad width**	**Claw length**
1988 adult emu 3	Preservation class ≥ 3	1.46 (10)	—	1.40 (9)	—	1.74 (10)
		10.6	—	10.7	—	17.2
	All footprints	1.46 (16)	—	1.40 (13)	—	1.74 (14)
		9.77	—	10.3	—	16.5
1988 adult emu 5	Preservation class ≥ 3	1.40 (9)	—	1.27 (9)	—	1.58 (11)
		11.2	—	8.20	—	15.0
	All footprints	1.71 (25)	—	1.89 (23)	—	1.99 (18)
		14.3	—	12.3	—	19.5
1988 adult emu 6	Preservation class ≥ 3	1.37 (4)	—	1.24 (4)	—	1.89 (8)
		—	—	—	—	17.4
	All footprints	1.62 (17)	—	1.65 (14)	—	2.63 (20)
		13.9	—	15.9	—	24.4
Pooled adult emus 3, 5, and 6	Individual footprint cases (preservation class ≥ 3)	1.70 (23)	—	1.91 (22)	—	2.03 (29)
		11.3	—	16.1	—	16.6
	Emu means treatment	1.11 (3)	—	1.29 (3)	—	1.22 (3)
Radandt adult male emu		1.50 (3)	—	1.61 (3)	—	1.55 (6)
		—	—	—	—	17.8
Outback emu farm female		1.07 (3)	—	1.17 (3)	—	1.12 (3)
All adult emus	Individual footprint cases	1.96 (70)	—	2.09 (62)	—	3.04 (74)
		13.6	—	16.3	—	21.0
	Emu means treatment	1.11 (5)	—	1.45 (5)	—	1.64 (10)
		—	—	—	—	15.5
Cleveland Zoo female greater rhea		—	—	—	—	1.25 (3)
Oklahoma City Zoo male lesser rhea		—	—	—	—	1.77 (5)
Fort Wayne Zoo turkey hen		1.51 (8)	—	1.77 (5)	—	1.69 (8)
		17.0	—	—	—	19.4
Toledo Zoo kori bustard		1.48 (13)	—	1.75 (12)	—	1.88 (12)
		11.1	—	16.1	—	19.2
Toledo Zoo secretarybird		1.22 (3)	—	1.24 (2)	—	1.27 (3)
San Diego Zoo ground-hornbill		1.19 (4)	—	—	—	—
AC 4/1 *Eubrontes cursorius*		1.08 (4)	1.15 (4)	1.18 (4)	1.18 (4)	1.41 (3)
AC 48/1 *Anomoepus scambus* (*A. intermedius*) prints 1–13		1.55 (3)	1.70 (3)	1.25 (3)	1.75 (3)	1.33 (3

Table A8.3. Variability of digit III footprint shape parameters among individuals and species of ground birds, and within trackways of non-avian dinosaurs. All parameters are scaled as a percentage of the length of the first digital pad of digit III. Values reported in most cells are the ratio of the maximum/minimum value of the scaled parameter (number of footprints or trackways); for samples in which the number of cases is six or more, the coefficient of variation of the scaled parameter is reported in a second line below the data for the maximum/minimum ratio. Emu data are for birds characterized as adults at the time their footprints were collected. For individual emus, data cases represent individual footprints. For pooled samples of more than one individual emu, results are reported for treatments in which the data cases are individual footprints, but also treatments in which the data cases represent means for each emu. For the three 1988 Fort Wayne Zoo adult emus, results are reported both for the best-preserved footprints, and for all footprints, of those birds. For dinosaur trackways, the data cases for which the maximum/minimum ratio of the scaled parameter is calculated are the individual footprints within the trackway. In all treatments, empty cells mean that there were fewer than three (ratio of the maximum/minimum ratio of the scaled parameter) or six (coefficient of variation of the scaled parameter) measurements of that scaled parameter. Digital pad widths were not measured in the bird footprint study, but were measured for dinosaur trackways.

		Digit III					
Bird, trackway or group		**First pad width**	**Second pad length**	**Second pad width**	**Third pad length**	**Third pad width**	**Claw length**
1988 adult emu 3	Preservation class ≥ 3	—	1.38 (13)	—	1.46 (12)	—	2.01 (9)
		—	9.55	—	11.7	—	18.0
	All footprints	—	1.69 (20)	—	1.68 (15)	—	2.19 (12)
		—	12.1	—	14.0	—	18.8
1988 adult emu 5	Preservation class ≥ 3	—	1.24 (10)	—	1.73 (10)	—	1.57 (8)
		—	6.34	—	17.1	—	15.2
	All footprints	—	1.67 (25)	—	1.98 (22)	—	2.18 (28)
		—	14.9	—	16.2	—	17.5
1988 adult emu 6	Preservation class ≥ 3	—	1.31 (7)	—	1.13 (4)	—	1.66 (6)
		—	8.58	—	—	—	17.4
	All footprints	—	1.52 (24)	—	2.07 (20)	—	1.94 (38)
		—	10.7	—	18.0	—	16.6
Pooled adult emus 3, 5, and 6	Individual footprint cases (preservation class ≥ 3)	—	1.38 (30)	—	1.73 (26)	—	2.12 (23)
		—	8.81	—	13.1	—	19.3
	Emus means treatment	—	1.12 (3)	—	1.08 (3)	—	1.18 (3)
Radandt adult male emu		—	1.64 (9)	—	2.17 (8)	—	1.56 (8)
		—	13.9	—	22.4	—	12.8
All adult emus	Individual footprint cases	—	2.19 (96)	—	2.77 (80)	—	2.50 (105)
		—	13.1	—	16.7	—	19.0
	Emu means treatment	—	1.41 (12)	—	1.60 (12)	—	1.92 (15)
		—	11.3	—	10.9	—	16.3
Cleveland Zoo female greater rhea		—	—	—	—	—	1.45 (6)
		—	—	—	—	—	15.8
Fort Wayne Zoo male ostrich		—	—	—	—	—	1.45 (6)
		—	—	—	—	—	18.1
Fort Wayne Zoo turkey hen		—	1.63 (6)	—	—	—	1.37 (3)
		—	18.3	—	—	—	—
Toledo Zoo kori bustard		—	—	—	—	—	1.98 (14)
		—	—	—	—	—	23.2
Toledo Zoo secretarybird		—	1.41 (3)	—	—	—	1.47 (3)
San Diego Zoo ground-hornbill		—	1.48 (5)	—	1.11 (3)	—	1.13 (3)
AC 4/1 *Eubrontes cursorius*		1.28 (4)	1.44 (4)	1.27 (4)	1.55 (4)	1.11 (4)	1.50 (4)
AC 48/1 *Anomoepus scambus* (*A. intermedius*) prints 1–13		1.32 (3)	1.08 (3)	1.49 (3)	1.68 (3)	1.63 (3)	1.59 (3)

Table A8.4. Variability of digit IV footprint shape parameters among individuals and species of ground birds, and within trackways of non-avian dinosaurs. All parameters are scaled as a percentage of the length of the first digital pad of digit III. Values reported in most cells are the ratio of the maximum/minimum value of the scaled parameter (number of footprints or trackways); for samples in which the number of cases is six or more, the coefficient of variation of the scaled parameter is reported in a second line below the data for the maximum/minimum ratio. Emu data are for birds characterized as adults at the time their footprints were collected. For individual emus, data cases represent individual footprints. For pooled samples of more than one individual emu, results are reported for treatments in which the data cases are individual footprints, but also treatments in which the data cases represent means for each emu. For the three 1988 Fort Wayne Zoo adult emus, results are reported both for the best-preserved footprints, and for all footprints, of those birds. For individual dinosaur trackways, the data cases for which the maximum/minimum ratio of the scaled parameter is calculated are the individual footprints within the trackway. In all treatments, empty cells mean that there were fewer than three (ratio of the maximum/minimum ratio of the scaled parameter) or six (coefficient of variation of the scaled parameter) measurements of that scaled parameter. Digital pad widths were not measured in the bird footprint study, but were measured for dinosaur trackways.

Bird, trackway or group		Digit IV First pad length	Digit IV Aggregate distal pad length	Digit IV Claw length
1988 adult emu 3	Preservation class ≥ 3	—	—	1.62 (10)
		—	—	15.4
	All footprints	—	—	2.49 (18)
		—	—	22.3
1988 adult emu 5	Preservation class ≥ 3	1.28 (7)	1.31 (6)	1.55 (9)
		10.2	10.3	15.1
	All footprints	1.49 (11)	1.35 (10)	2.16 (20)
		12.3	9.12	24.6
1988 adult emu 6	Preservation class ≥ 3	1.19 (5)	1.24 (5)	1.62 (7)
		—	—	16.9
	All footprints	1.64 (18)	1.34 (10)	2.18 (22)
		13.4	9.64	19.0
Pooled adult emus 3, 5, and 6 (preservation class ≥ 3)	Individual footprint cases	1.43 (13)	1.31 (12)	1.78 (26)
		9.69	8.64	15.6
	Emu means treatment	1.18 (3)	1.10 (3)	1.16 (3)
Radandt adult male emu		—	—	1.86 (6)
		—	—	19.7
All adult emus	Individual footprint cases	2.24 (38)	1.35 (27)	3.37 (82)
		15.0	8.83	23.2
	Emu means treatment	1.68 (6)	1.12 (5)	1.95 (11)
		18.6	—	18.6
Cleveland Zoo female greater rhea		—	—	1.26 (5)
Oklahoma City Zoo male lesser rhea		—	—	1.10 (3)
Fort Wayne Zoo turkey hen		1.54 (6)	1.24 (3)	2.31 (7)
		14.2	—	33.3
Toledo Zoo kori bustard		—	—	1.90 (12)
		—	—	18.2
Toledo Zoo secretarybird		—	—	1.55 (3)
San Diego Zoo ground-hornbill		1.57 (4)	1.15 (3)	1.57 (3)
AC 4/1 *Eubrontes cursorius*		—	—	1.50 (3)
AC 48/1 *Anomoepus scambus* (*A. intermedius*) prints 1–13		1.25 (3)	1.58 (3)	1.40 (3)

Table A8.5. Variability of footprint width across the proximal ends of digits II–IV among individuals and species of ground birds and within trackways of non-avian dinosaurs. All parameters are scaled as a percentage of backfoot length. Values reported in most cells are the ratio of the maximum/minimum value of the scaled parameter (number of footprints or trackways); for samples in which the number of cases is six or more, the coefficient of variation of the scaled parameter is reported in a second line below the data for the maximum/minimum ratio. Emu data are for birds characterized as adults at the time their footprints were collected. For individual emus, data cases represent individual footprints. For pooled samples of more than one individual emu, results are reported for treatments in which the data cases are individual footprints, but also treatments in which the data cases represent means for each emu. For the three 1988 Fort Wayne Zoo adult emus, results are reported both for the best-preserved footprints, and for all footprints, of those birds. For individual dinosaur trackways, the data cases for which the maximum/minimum ratio of the scaled parameter is calculated are the individual footprints within the trackway. In all treatments, empty cells mean that there were fewer than three (ratio of the maximum/minimum ratio of the scaled parameter) or six (coefficient of variation of the scaled parameter) measurements of that scaled parameter.

Bird, trackway, or group		Distance between proximal ends of:		
		Digit II–digit III	Digit III–digit IV	Digit II–digit IV
1988 adult emu 3	Preservation class ≥ 3	1.36 (10) 9.84	1.25 (9) 7.20	1.19 (9) 5.99
	All footprints	1.50 (23) 9.65	1.39 (23) 10.1	1.20 (23) 5.64
1988 adult emu 5	Preservation class ≥ 3	1.35 (8) 9.98	1.35 (8) 12.0	1.34 (8) 9.44
	All footprints	1.45 (21) 10.7	1.58 (20) 11.6	1.69 (27) 12.3
1988 adult emu 6	Preservation class ≥ 3	1.21 (4)	1.19 (3)	1.16 (3)
	All footprints	1.69 (22) 15.2	1.64 (23) 12.3	1.60 (20) 12.7
Pooled adult emus 3, 5, and 6	Individual footprint cases (preservation class ≥ 3)	1.37 (22) 9.83	1.61 (20) 11.5	1.34 (20) 8.13
	Emu means treatment	1.12 (3)	1.12 (3)	1.10 (3)
Emu 00-0021-26ED		1.27 (4)	1.16 (4)	1.14 (4)
Radandt adult male emu		1.34 (3)	1.19 (4)	1.12 (3)
Outback emu farm female		1.01 (3)	1.09 (3)	1.06 (3)
Outback emu farm male		1.20 (3)	1.05 (3)	1.13 (3)
All adult emus	Individual footprint cases	1.71 (91) 13.3	1.93 (92) 13.0	1.69 (95) 11.2
	Emu means treatment	1.40 (12) 9.99	1.34 (12) 9.99	1.34 (12) 7.73
Cleveland Zoo female greater rhea		1.47 (7) 12.0	1.28 (7) 9.57	1.21 (7) 7.07
Oklahoma City Zoo male lesser rhea		1.61 (5)	1.15 (5)	1.31 (5)
San Antonio Zoo male cassowary		1.18 (3)	1.09 (3)	1.22 (3)
San Diego Zoo kiwi		1.25 (3)	1.23 (3)	1.30 (3)
Fort Wayne Zoo turkey hen		1.66 (14) 11.7	1.46 (14) 10.4	1.49 (14) 12.9
Fort Wayne Zoo chicken hen		1.62 (14) 14.7	1.73 (14) 15.8	1.69 (14) 15.4
Freels Farm Plymouth barred rock hen		1.38 (6) 15.1	1.36 (6) 12.4	1.81 (6) 20.7
Freels Farm Cornish cross chick		1.28 (6) 9.82	1.64 (6) 21.3	1.26 (6) 10.1
Toledo Zoo male golden pheasant		1.52 (5)	1.42 (4)	1.74 (4)
Bob Cassell farm male silver pheasant		1.17 (4)	1.23 (4)	1.58 (5)
Cincinnati Zoo Congo peahen		1.51 (5)	2.46 (5)	1.87 (5)
Cleveland Zoo black-legged seriema		1.09 (4)	1.21 (4)	1.43 (4)
San Diego Zoo Abyssinian ground hornbill		1.20 (3)	1.27 (3)	1.20 (3)
Cleveland Zoo double-striped thick-knee		1.35 (7) 13.9	1.40 (7) 10.1	1.47 (7) 11.7
St. Joseph River, Fort Wayne killdeer		1.43 (6) 14.2	1.35 (6) 10.0	1.34 (6) 13.2
St. Joseph River, Fort Wayne killdeer		1.95 (3)	2.02 (3)	2.50 (3)
St. Joseph River, Fort Wayne killdeer		2.05 (6) 25.4	1.81 (6) 19.6	1.59 (6) 17.0

Bird, trackway, or group	Distance between proximal ends of:		
	Digit II–digit III	Digit III–digit IV	Digit II–digit IV
San Diego Zoo weka	1.31 (4)	1.16 (4)	1.37 (4)
Toledo Zoo kori bustard	1.64 (12)	1.25 (11)	1.33 (11)
	15.2	7.40	9.17
Kayentapus minor trackway A	1.06 (3)	1.08 (3)	1.21 (3)
AC 26/10 *Anomoepus scambus* (*A. intermedius*) left trackway	1.26 (4)	1.49 (4)	1.36 (4)
AC 48/1 *Anomoepus scambus* (*A. intermedius*) prints 1–13	1.17 (4)	1.31 (4)	1.84 (4)
AC 48/1 *Anomoepus scambus* (*A. intermedius*) sitting animal	1.47 (3)	1.49 (3)	1.18 (3)

Table A8.6. Variability of footprint width across the tips of digits II–IV, and the projection of digit III beyond the tips of digits II and IV, among individuals and species of ground birds and within trackways of non-avian dinosaurs. All parameters are scaled as a percentage of backfoot length. Values reported in most cells are the ratio of the maximum/minimum value of the scaled parameter (number of footprints or trackways); for samples in which the number of cases is six or more, the coefficient of variation of the scaled parameter is reported in a second line below the data for the maximum/minimum ratio. Emu data are for birds characterized as adults at the time their footprints were collected. For individual emus, data cases represent individual footprints. For pooled samples of more than one individual emu, results are reported for treatments in which the data cases are individual footprints, but also treatments in which the data cases represent means for each emu. For the three 1988 Fort Wayne Zoo adult emus, results are reported both for the best-preserved footprints, and for all footprints, of those birds. For individual dinosaur trackways, the data cases for which the maximum/minimum ratio of the scaled parameter is calculated are the individual footprints within the trackway. In all treatments, empty cells mean that there were fewer than three (ratio of the maximum/minimum ratio of the scaled parameter) or six (coefficient of variation of the scaled parameter) measurements of that scaled parameter.

Bird, trackway, or group		Distance between tips of: Digit II–digit III	Digit III–digit IV	Digit II–digit IV	Digit III projection
1988 adult emu 3	Preservation class ≥ 3	1.34 (10) 7.57	1.57 (10) 14.5	1.42 (10) 10.1	1.39 (10) 10.1
	All footprints	1.43 (38) 9.23	1.70 (38) 12.4	1.42 (38) 8.60	1.72 (38) 11.4
1988 adult emu 5	Preservation class ≥ 3	1.48 (8) 15.2	1.34 (8) 10.5	1.37 (8) 11.8	1.39 (8) 12.8
	All footprints	1.83 (31) 14.0	1.59 (31) 11.0	1.60 (31) 10.8	1.62 (31) 13.6
1988 adult emu 6	Preservation class ≥ 3	1.27 (4)	1.13 (4)	1.12 (4)	1.19 (4)
	All footprints	2.19 (28) 20.9	1.72 (28) 12.4	1.58 (28) 11.8	1.92 (28) 18.8
Pooled adult emus 3, 5, and 6	Individual footprint cases (preservation class ≥ 3)	1.63 (22) 12.8	1.69 (22) 13.0	1.50 (22) 10.7	1.61 (22) 12.8
	Emu means treatment	1.17 (3)	1.18 (3)	1.06 (3)	1.22 (3)
Emu 00-0021-26ED		1.31 (4)	1.16 (4)	1.22 (4)	1.24 (4)
Radandt adult male emu		1.40 (4)	1.25 (4)	1.22 (4)	1.42 (4)
Outback emu farm female		1.07 (3)	1.19 (3)	1.10 (3)	1.10 (3)
Outback emu farm male		1.08 (3)	1.06 (3)	1.10 (3)	1.04 (3)
All adult emus	Individual footprint cases	2.19 (123) 15.2	2.00 (123) 13.8	1.71 (123) 11.2	2.10 (123) 15.7
	Emu means treatment	1.31 (11) 9.90	1.41 (12) 11.3	1.28 (12) 8.65	1.44 (12) 12.0
Cleveland Zoo female greater rhea		1.31 (7) 8.85	1.35 (7) 12.1	1.22 (7) 7.08	1.39 (7) 13.2
Oklahoma City Zoo lesser rhea		1.25 (5)	1.19 (5)	1.14 (5)	1.24 (5)
San Diego Zoo kiwi		1.25 (3)	1.71 (3)	1.49 (3)	1.46 (3)
Fort Wayne Zoo turkey hen		1.52 (14) 11.8	1.39 (14) 11.0	1.28 (14) 8.85	1.47 (14) 10.3
Fort Wayne Zoo chicken hen		1.91 (15) 18.0	1.91 (15) 15.7	1.62 (15) 12.3	2.20 (15) 21.2
Freels Farm Plymouth barred rock hen		1.34 (6) 13.3	1.26 (6) 9.94	1.28 (6) 10.9	1.28 (6) 11.5
Freels Farm Cornish cross chick		1.44 (6) 12.2	1.24 (6) 7.79	1.30 (6) 9.92	1.46 (6) 12.4
San Diego Zoo female Philby's partridge		1.58 (3)	1.74 (3)	1.37 (3)	2.04 (3)
Cincinnati Zoo Congo peahen		1.61 (5)	1.72 (5)	1.50 (5)	1.81 (5)
Cincinnati Zoo male crested fireback		1.34 (3)	1.15 (3)	1.23 (3)	1.11 (3)
Bob Cassell Farm male silver pheasant		1.48 (5)	1.43 (5)	1.37 (5)	1.63 (5)
Toledo Zoo male golden pheasant		1.87 (5)	1.71 (5)	1.78 (5)	1.83 (5)
San Diego female Himalayan monal		1.16 (3)	1.47 (3)	1.17 (3)	1.47 (3)
Toledo Zoo secretarybird		1.11 (4)	1.05 (4)	1.18 (4)	1.12 (4)
Cleveland Zoo black-legged seriema		1.47 (4)	1.39 (4)	1.32 (4)	1.39 (4)
San Diego Zoo Abyssinian ground hornbill		1.25 (3)	1.28 (3)	1.22 (3)	1.55 (3)
Cleveland Zoo double-striped thick-knee		1.36 (7) 12.2	1.27 (7) 7.46	1.32 (7) 11.6	1.26 (7) 9.07
St. Joseph River, Fort Wayne killdeer		1.33 (6) 11.6	1.33 (6) 10.6	1.35 (6) 11.0	1.35 (6) 12.6

Bird, trackway, or group	Distance between tips of: Digit II–digit III	Digit III–digit IV	Digit II–digit IV	Digit III projection
St. Joseph River, Fort Wayne killdeer	1.87 (3)	1.83 (3)	1.67 (3)	2.05 (3)
St. Joseph River, Fort Wayne killdeer	1.71 (6) 20.3	1.57 (6) 14.6	1.50 (6) 16.0	1.68 (6) 19.7
Louisville Zoo juvenile sarus crane	1.28 (3)	1.48 (3)	1.39 (3)	1.35 (3)
San Diego Zoo weka	1.15 (4)	1.29 (4)	1.10 (4)	1.35 (4)
San Diego Zoo buff-crested bustard	1.54 (3)	1.56 (3)	1.38 (3)	1.76 (3)
Toledo Zoo kori bustard	1.34 (12) 10.1	1.27 (12) 7.52	1.35 (12) 9.02	1.37 (12) 9.22
San Diego Zoo pheasant pigeon	1.48 (3)	1.15 (3)	1.26 (3)	1.36 (3)
Kayentapus minor trackway A	1.21 (4)	1.14 (4)	1.11 (5)	1.22 (4)
Kayentapus minor trackway B	1.33 (5)	1.34 (5)	1.21 (5)	1.44 (5)
AC 26/10 *Anomoepus scambus* (*A. intermedius*) left trackway	1.24 (4)	1.33 (4)	1.19 (4)	1.19 (4)
AC 48/1 *Anomoepus scambus* (*A. intermedius*) prints 1–13	1.13 (4)	1.24 (4)	1.22 (4)	1.31 (4)
AC 48/1 *Anomoepus scambus* (*A. intermedius*) sitting animal	1.08 (3)	1.43 (3)	1.27 (3)	1.24 (3)

Table A8.7. Variability of digit II footprint shape parameters among groups of trackways of non-avian dinosaurs. All parameters are scaled as a percentage of the length of the first digital pad of digit III. Data cases are mean values for trackways, or measurements of singleton footprints. For potentially homogeneous groups, the cell contents of the first (and sometimes only) row of data are reported as mean (maximum/minimum ratio; number of footprints or trackways). For more inclusive categories composed of one or more homogeneous groups, the cell contents of the first (and sometimes only) row of data are reported as maximum/minimum ratio (number of footprints or trackways). For samples in which the number of cases is six or more, the coefficient of variation of the scaled parameter is reported in a second line below the data for the mean and/or maximum/minimum ratio. In all treatments, empty cells mean that there were fewer than three (ratio of the maximum/minimum ratio of the scaled parameter) or six (coefficient of variation of the scaled parameter) measurements of that scaled parameter. Parameters for which the more inclusive intergroup comparison results in larger values (greater variability) than for at least one of the less inclusive groups with which the comparison is made are indicated in **bold**; if the more inclusive group values are larger than all of the less inclusive group values of the same parameter, the more inclusive group value is also *italicized*.

	Digit II				
Group	**First pad length**	**First pad width**	**Second pad length**	**Second pad width**	**Claw length**
Homogeneous groups:					
"*Anchisauripus*" cluster	89.3 (1.49; 13) 13.6	70.8 (1.43; 13) 11.7	87.9 (1.41; 13) 11.1	65.6 (1.44; 13) 11.8	60.5 (1.79; 13) 19.2
All AC 9/14	88.2 (1.60; 20) 11.0	66.4 (1.58; 17) 11.4	85.7 (1.78; 17) 13.0	65.0 (1.51; 18) 12.5	59.6 (2.58; 15) 26.9
"*Eubrontes*" cluster	100.8 (1.27; 8) 8.83	85.4 (1.29; 8) 8.28	92.0 (1.54; 8) 13.0	76.0 (1.38; 8) 11.5	44.5 (1.41; 8) 13.4
Large *Eubrontes*[a]	92.3 (1.40; 6) 13.2	82.2 (1.78; 6) 20.3	87.0 (1.56; 6) 14.5	67.5 (1.87; 6) 22.1	43.1 (1.35; 5) ———
All *Anomoepus*	110.8 (1.28; 4)	91.2 (1.22; 4)	99.5 (1.58; 4)	86.5 (1.16; 4)	77.3 (1.34; 4)
More inclusive groups:					
"*Anchisauripus*" + "*Eubrontes*" cluster	***1.51*** (21) **13.0**	***1.67*** (21) ***13.8***	**1.54** (21) **11.8**	***1.70*** (21) ***13.6***	***2.06*** (21) ***23.0***
"*Anchisauripus*" + *Anomoepus* cluster[b]	**1.51** (15) **14.8**	**1.60** (15) **14.1**	1.41 (15) 10.9	**1.54** (15) **13.4**	**1.88** (15) **19.5**
All AC 9/14 + all *Anomoepus*[c]	***1.79*** (24) **14.3**	***1.81*** (21) **17.5**	***2.08*** (21) **15.4**	***1.76*** (22) **16.6**	***3.04*** (19) 26.0
"*Eubrontes*" + *Anomoepus* cluster[b]	**1.28** (10) **8.97**	1.29 (10) **8.36**	1.54 (10) 11.8	1.38 (10) 10.4	**2.15** (10) **28.7**
"*Anchisauripus*" + "*Eubrontes*" + *Anomoepus* cluster[b]	***1.51*** (23) **13.3**	**1.67** (23) ***14.0***	**1.54** (23) **11.5**	***1.70*** (23) ***13.6***	***2.15*** (23) ***24.1***
All Newark Supergroup except Dinosaur State Park	***1.83*** (51) ***14.9***	***2.07*** (47) **18.3**	***2.08*** (47) ***14.7***	***1.89*** (47) **16.5**	***3.04*** (43) **25.1**

[a]Large *Eubrontes* potentially include the ichnospecies *E. minusculus*, *E. tuberatus*, *E. platypus*, *E. divaricatus*, *E. giganteus*, and *E. caudatus*, but not necessarily for all parameters.
[b]There were only two data cases in the *Anomoepus* cluster, and so comparisons of variability between that cluster and more inclusive intergroup combinations that include the *Anomoepus* cluster were not made.
[c]There were only four data cases in the all *Anomoepus* sample, and so comparisons of the coefficient of variation between that cluster and more inclusive intergroup combinations that include the all *Anomoepus* sample were not made.

Table A8.8. Variability of digit III footprint shape parameters among groups of trackways of non-avian dinosaurs. All parameters are scaled as a percentage of the length of the first digital pad of digit III. Data cases are mean values for trackways or measurements of singleton footprints. For potentially homogeneous groups, the cell contents of the first (and sometimes only) row of data are reported as mean (maximum/minimum ratio; number of footprints or trackways). For more inclusive categories composed of one or more homogeneous groups, the cell contents of the first (and sometimes only) row of data are reported as maximum/minimum ratio (number of footprints or trackways). For samples in which the number of cases is six or more, the coefficient of variation of the scaled parameter is reported in a second line below the data for the mean and/or maximum/minimum ratio. In all treatments, empty cells mean that there were fewer than three (ratio of the maximum/minimum ratio of the scaled parameter) or six (coefficient of variation of the scaled parameter) measurements of that scaled parameter. Parameters for which the more inclusive intergroup comparison results in larger values (greater variability) than for at least one of the less inclusive groups with which the comparison is made are indicated in **bold**; if the more inclusive group values are larger than all of the less inclusive group values of the same parameter, the more inclusive group value is also *italicized.*

	Digit III					
Group	**First pad width**	**Second pad length**	**Second pad width**	**Third pad length**	**Third pad width**	**Claw length**
Homogeneous groups:						
"*Anchisauripus*" cluster	77.8 (1.53; 13) 12.7	99.3 (1.32; 13) 8.76	87.7 (1.48; 13) 13.8	86.4 (1.40; 13) 11.4	67.6 (1.58; 13) 15.3	59.2 (2.17; 13) 21.8
All AC 9/14	77.1 (1.66; 22) 11.4	100.9 (1.59; 20) 10.4	83.4 (1.51; 21) 10.9	90.1 (2.08; 16) 17.8	67.0 (1.88; 17) 17.0	61.2 (2.66; 17) 29.7
"*Eubrontes*" cluster	96.3 (1.37; 8) 11.5	89.7 (1.66; 8) 16.2	100.7 (1.68; 8) 18.4	82.4 (1.29; 8) 7.94	86.1 (1.73; 8) 20.4	46.8 (1.86; 8) 19.0
Large *Eubrontes*[a]	87.2 (1.47; 6) 12.2	74.7 (1.41; 5) ———	78.8 (1.79; 5) ———	79.7 (1.18; 5) ———	72.3 (2.09; 5) ———	43.1 (1.67; 4) ———
All *Anomoepus*	94.8 (1.15; 4)	81.1 (1.22; 3)	92.5 (1.16; 3)	———	———	90.2 (1.35; 4)
More inclusive groups:						
"*Anchisauripus*" + "*Eubrontes*" cluster	***1.83*** (21) ***16.1***	***1.73*** (21) **12.5**	***1.78*** (21) **17.1**	***1.41*** (21) **10.3**	***2.17*** (21) ***21.4***	***2.76*** (21) ***23.7***
"*Anchisauripus*" + *Anomoepus* cluster[b]	**1.57** (15) **13.5**	**1.45** (15) **10.0**	1.48 (15) 13.1	**1.45** (15) **12.4**	1.58 (15) 14.8	2.17 (15) **24.8**
All AC 9/14 + all *Anomoepus*[c]	**1.66** (26) **13.2**	***1.65*** (23) **12.4**	**1.51** (24) 10.9	2.08 (18) 17.7	1.88 (19) 16.6	***3.10*** (21) **30.7**
"*Eubrontes*" + *Anomoepus* cluster[b]	1.37 (10) 10.7	1.66 (10) 14.9	1.68 (10) 17.3	**1.42** (10) **11.2**	1.73 (10) 19.2	**2.76** (10) **34.4**
"*Anchisauripus*" + "*Eubrontes*" + *Anomoepus* cluster[b]	***1.83*** (23) ***15.6***	***1.73*** (23) **12.6**	***1.78*** (23) **16.5**	***1.45*** (23) **11.2**	***2.17*** (23) ***20.5***	***2.76*** (23) ***27.1***
All Newark Supergroup except Dinosaur State Park	***1.83*** (53) ***14.6***	***1.95*** (47) **13.2**	***2.65*** (48) **18.1**	***2.24*** (39) **15.9**	***2.54*** (42) ***23.2***	***3.33*** (44) **28.8**

[a]Large *Eubrontes* potentially include the ichnospecies *E. minusculus*, *E. tuberatus*, *E. platypus*, *E. divaricatus*, *E. giganteus*, and *E. caudatus*, but not necessarily for all parameters.
[b]There were only two data cases in the *Anomoepus* cluster, and so comparisons of variability between that cluster and more inclusive intergroup combinations that include the *Anomoepus* cluster were not made.
[c]There were only up to four data cases in the all *Anomoepus* sample, and so comparisons of the coefficient of variation between that cluster and more inclusive intergroup combinations that include the all *Anomoepus* sample were not made.

Table A8.9. Variability of digit IV footprint shape parameters among groups of trackways of non-avian dinosaurs. All parameters are scaled as a percentage of the length of the first digital pad of digit III. Data cases are mean values for trackways, or measurements of singleton footprints. For potentially homogeneous groups, the cell contents of the first (and sometimes only) row of data are reported as mean (maximum/minimum ratio; number of footprints or trackways). For more inclusive categories composed of one or more homogeneous groups, the cell contents of the first (and sometimes only) row of data are reported as maximum/minimum ratio (number of footprints or trackways). For samples in which the number of cases is six or more, the coefficient of variation of the scaled parameter is reported in a second line below the data for the mean and/or maximum/minimum ratio. In all treatments, empty cells mean that there were fewer than three (ratio of the maximum/minimum ratio of the scaled parameter) or six (coefficient of variation of the scaled parameter) measurements of that scaled parameter. Parameters for which the more inclusive intergroup comparison results in larger values (greater variability) than for at least one of the less inclusive groups with which the comparison is made are indicated in **bold**; if the more inclusive group values are larger than all of the less inclusive group values of the same parameter, the more inclusive group value is also *italicized*.

	Digit IV			
Group	**First pad length**	**First pad width**	**Aggregate distal pad length**	**Claw length**
Homogeneous groups:				
"*Anchisauripus*" cluster	64.5 (2.18; 12)	63.8 (1.57; 5)	197.6 (1.51; 10)	53.2 (2.03; 12)
	26.7	———	12.6	21.4
All AC 9/14	66.6 (1.95; 11)	68.6 (1.65; 5)	202.4 (1.27; 8)	48.5 (3.74; 14)
	23.7	———	8.53	32.6
"*Eubrontes*" cluster	77.6 (1.26; 6)	94.2 (1.15; 3)	203.5 (1.30; 4)	39.8 (1.44; 6)
	8.35	———	———	14.2
Large *Eubrontes*[a]	74.0 (1.63; 5)	———	———	40.4 (1.35; 3)
All *Anomoepus*	90.5 (1.14; 4)	100.4 (1.25; 3)	262.7 (1.23; 3)	61.1 (1.82; 3)
More inclusive groups:				
"*Anchisauripus*" + "*Eubrontes*" cluster	**2.18** (18)	***1.96*** (8)	**1.51** (14)	***2.31*** (18)
	22.7	24.4	11.8	***23.9***
"*Anchisauripus*" + *Anomoepus* cluster[b]	2.18 (14)	**2.19** (6)	**1.75** (12)	2.03 (14)
	27.9	30.8	**16.8**	20.2
All AC 9/14 + all *Anomoepus*[c]	**1.95** (15)	***2.19*** (8)	***1.56*** (11)	***4.22*** (17)
	23.9	24.8	**15.5**	**33.5**
"*Eubrontes*" + *Anomoepus* cluster[b]	**1.39** (8)	**1.29** (4)	**1.68** (6)	**1.46** (8)
	11.5	———	17.1	**14.8**
"*Anchisauripus*" + "*Eubrontes*" + *Anomoepus* cluster[b]	**2.18** (20)	***2.19*** (9)	***1.75*** (16)	***2.31*** (20)
	23.4	26.6	**15.3**	***22.6***
All Newark Supergroup except Dinosaur State Park	***2.31*** (36)	***2.54*** (18)	***2.03*** (24)	***4.22*** (35)
	24.2	24.7	***16.2***	**28.3**

[a]Large *Eubrontes* potentially include the ichnospecies *E. minusculus*, *E. tuberatus*, *E. platypus*, *E. divaricatus*, *E. giganteus*, and *E. caudatus*, but not necessarily for all parameters.
[b]There were only two data cases in the *Anomoepus* cluster, and so comparisons of variability between that cluster and more inclusive intergroup combinations that include the *Anomoepus* cluster were not made.
[c]There were only up to four data cases in the all *Anomoepus* sample, and so comparisons of the coefficient of variation between that cluster and more inclusive intergroup combinations that include the all *Anomoepus* sample were not made.

Table A8.10. Variability of shape parameters associated with distances between the proximal ends of digital impressions among groups of trackways of non-avian dinosaurs. All parameters are scaled as a percentage of backfoot length. Data cases are mean values for trackways, or measurements of singleton footprints. For potentially homogeneous groups, the cell contents of the first (and sometimes only) row of data are reported as mean (maximum/minimum ratio; number of footprints or trackways). For more inclusive categories composed of one or more homogeneous groups, the cell contents of the first (and sometimes only) row of data are reported as maximum/minimum ratio (number of footprints or trackways). For samples in which the number of cases is six or more, the coefficient of variation of the scaled parameter is reported in a second line below the data for the mean and/or maximum/minimum ratio. In all treatments, empty cells mean that there were fewer than three (ratio of the maximum/minimum ratio of the scaled parameter) or six (coefficient of variation of the scaled parameter) measurements of that scaled parameter. Parameters for which the more inclusive intergroup comparison results in larger values (greater variability) than for at least one of the less inclusive groups with which the comparison is made are indicated in **bold**; if the more inclusive group values are larger than all of the less inclusive group values of the same parameter, the more inclusive group value is also *italicized.*

Group	Distance between proximal ends of first digital pad of digits		
	II–III	III–IV	II–IV
Homogeneous groups:			
"*Anchisauripus*" cluster	26.5 (1.32; 13) 8.45	47.9 (1.41; 13) 9.54	39.4 (1.34; 13) 9.85
All AC 9/14	26.7 (1.32; 9) 8.12	51.0 (1.39; 10) 11.1	41.6 (1.41; 10) 11.0
"*Eubrontes*" cluster	32.4 (1.40; 8) 12.2	48.4 (1.32; 8) 9.00	34.1 (1.44; 8) 11.2
Large *Eubrontes*[a]	28.7 (2.04; 5)	45.8 (1.48; 5)	36.4 (1.54; 5)
All *Anomoepus*	27.6 (1.62; 7) 18.9	50.1 (1.15; 7) 5.55	42.5 (1.18; 8) 5.98
More inclusive groups:			
"*Anchisauripus*" + "*Eubrontes*" cluster	***1.78*** (21) ***14.5***	**1.41** (21) **9.12**	***1.68*** (21) ***12.3***
"*Anchisauripus*" + *Anomoepus* cluster[b]	**1.39** (15) **9.99**	1.41 (15) 8.83	1.34 (15) 9.41
All AC 9/14 + all *Anomoepus*	**1.62** (16) **13.6**	**1.39** (17) **9.09**	**1.41** (18) **8.92**
"*Eubrontes*" + *Anomoepus* cluster[b]	**1.81** (10) **15.0**	1.32 (10) 7.93	**1.56** (10) **13.0**
"*Anchisauripus*" + "*Eubrontes*" + *Anomoepus* cluster[b]	***1.81*** (23) ***14.8***	**1.41** (23) 8.69	***1.68*** (23) ***12.1***
All Newark Supergroup except Dinosaur State Park	***2.30*** (44) **15.8**	***1.71*** (45) **10.8**	***1.83*** (47) ***13.0***

[a]Large *Eubrontes* potentially include the ichnospecies *E. minusculus*, *E. tuberatus*, *E. platypus*, *E. divaricatus*, *E. giganteus*, and *E. caudatus*, but not necessarily for all parameters.
[b]There were only two data cases in the *Anomoepus* cluster, and so comparisons of variability between that cluster and more inclusive intergroup combinations that include the *Anomoepus* cluster were not made.

Table A8.11. Variability of shape parameters associated with toetip distances, the projection of digit III beyond the tips of digits II and IV, and the length of digit IV, among groups of trackways of non-avian dinosaurs. All parameters are scaled as a percentage of backfoot length. Data cases are mean values for trackways, or measurements of singleton footprints. For potentially homogeneous groups, the cell contents of the first (and sometimes only) row of data are reported as mean (maximum/minimum ratio; number of footprints or trackways). For more inclusive categories composed of one or more homogeneous groups, the cell contents of the first (and sometimes only) row of data are reported as maximum/minimum ratio (number of footprints or trackways). For samples in which the number of cases is six or more, the coefficient of variation of the scaled parameter is reported in a second line below the data for the mean and/or maximum/minimum ratio. In all treatments, empty cells mean that there were fewer than three (ratio of the maximum/minimum ratio of the scaled parameter) or six (coefficient of variation of the scaled parameter) measurements of that scaled parameter. Parameters for which the more inclusive intergroup comparison results in larger values (greater variability) than for at least one of the less inclusive groups with which the comparison is made are indicated in **bold**; if the more inclusive group values are larger than all of the less inclusive group values of the same parameter, the more inclusive group value is also *italicized*.

Group	Toetip distance II–III	 III–IV	 II–IV	Projection of digit III beyond the tips of digits II and IV	Digit IV length
Homogeneous groups:					
"*Anchisauripus*" cluster	77.5 (1.38; 13) 8.77	75.4 (1.31; 13) 8.14	84.2 (1.39; 13) 8.54	62.8 (1.32; 13) 8.38	103.7 (1.12; 13) 3.59
All AC 9/14	80.9 (1.52; 10) 14.4	71.6 (1.34; 10) 9.15	80.6 (1.52; 11) 10.9	63.9 (1.33; 10) 9.42	104.1 (1.11; 11) 3.50
"*Eubrontes*" cluster	79.2 (1.57; 8) 16.4	75.9 (1.57; 8) 13.2	98.4 (1.57; 8) 14.0	59.2 (1.61; 8) 16.0	108.8 (1.14; 8) 4.60
Large *Eubrontes*[a]	73.5 (1.53; 7) 14.1	72.6 (1.40; 7) 12.3	106.2 (1.38; 7) 12.8	49.6 (1.31; 7) 9.39	109.6 (1.13; 6) 4.44
All *Anomoepus*	78.6 (1.68; 9) 17.6	86.5 (1.32; 9) 10.8	121.3 (1.37; 9) 10.9	55.4 (1.79; 9) 18.1	114.0 (1.19; 8) 5.41
More inclusive groups:					
"*Anchisauripus*" + "*Eubrontes*" cluster	**1.57** (21) **11.9**	**1.57** (21) **10.1**	***1.66*** (21) **13.5**	**1.61** (21) **11.7**	***1.18*** (21) **4.57**
"*Anchisauripus*" + *Anomoepus* cluster[b]	**1.59** (15) **10.3**	**1.46** (15) **10.7**	**1.90** (15) **18.1**	**1.67** (15) **11.2**	**1.17** (15) **4.54**
All AC 9/14 + all *Anomoepus*	***1.95*** (19) **15.6**	***1.56*** (19) ***13.8***	**2.33** (20) ***23.6***	***1.88*** (19) **15.1**	***1.26*** (19) ***6.37***
"*Eubrontes*" + "*Anomoepus*" cluster[b]	**1.76** (10) **16.9**	**1.72** (10) **14.3**	**1.79** (10) **16.7**	**1.75** (10) **16.5**	1.14 (10) 4.55
"*Anchisauripus*" + "*Eubrontes*" + *Anomoepus* cluster[b]	***1.76*** (23) **12.6**	***1.72*** (23) **11.3**	***1.90*** (23) ***16.9***	***1.75*** (23) **12.7**	***1.18*** (23) ***4.80***
All Newark Supergroup except Dinosaur State Park	***1.95*** (51) **14.7**	***2.16*** (51) ***16.5***	**2.33** (52) ***19.8***	**2.35** (51) **16.8**	***1.26*** (49) **5.13**

[a]Large *Eubrontes* potentially include the ichnospecies *E. minusculus*, *E. tuberatus*, *E. platypus*, *E. divaricatus*, *E. giganteus*, and *E. caudatus*, but not necessarily for all parameters.
[b]There were only two data cases in the "*Anomoepus*" cluster, and so comparisons of variability between that cluster and more inclusive intergroup combinations that include the "*Anomoepus*" cluster were not made.

Table A8.12. Variability of selected parameters (mostly those from table 8.3) of footprint shape in potentially homogeneous groups of Newark Supergroup dinosaur footprints ("*Anchisauripus*" and "*Eubrontes*" clusters, all footprints on slab AC 9/14, and all *Anomoepus* footprints), combinations of those homogeneous clusters, and (for comparative purposes) adult emus (emu means treatment). Variability is expressed in terms of the maximum/minimum ratio and coefficient of variation of parameters scaled either (1) as a percentage of the length of the first digital pad of digit III (width of the first digital pad of digit III) or (2) as a percentage of backfoot length (all other parameters), and in terms of the range and standard deviation of GM-scaled variables. Variability in terms of GM scaling could not be examined for comparisons involving the "all AC 9/14" or the "all *Anomoepus*" samples because not all of the data cases had measurable values of all of the 19 variables (table 8.2) used to create the scaling factor (mean of log-transformed values of all 19 variables from table 8.2). Variability of more inclusive groups is calculated in two ways: (1) based on all of the cases in the combined samples (tables A8.3–A8.11) and (2) as the median of 15 trials of randomly selected cases from the combined sample; the number of cases randomly selected corresponds to the number of data cases in the smaller of the homogeneous groups in the combined sample. Parameters for which the more inclusive intergroup comparison results in larger values (greater variability) than for at least one of the less inclusive groups with which the comparison is made are indicated in **bold**; if the more inclusive group values are larger than all of the less inclusive group values of the same parameter, the more inclusive group value is also *italicized*.

Group	Minimum/maximum ratio of % scaled length	Coefficient of variation of % scaled length	Range of GM-scaled length	Standard deviation of GM-scaled length
Distance between proximal ends of first digital pads of digits III–IV (toebase III–IV)				
Homogeneous groups:				
"*Anchisauripus*" cluster (*N* = 13)	1.41	9.54	0.176	0.0497
All AC 9/14 (*N* = 10)	1.39	11.1	—	—
"*Eubrontes*" cluster (*N* = 8)	1.32	9.00	0.136	0.0464
All *Anomoepus* (*N* = 7)	1.15	5.55	—	—
All adult emus[a] (emu means treatment: *N* = 12)	1.34	9.99	—	—
More inclusive groups:				
"*Anchisauripus*" + "*Eubrontes*" cluster (*N* = 21): entire sample	**1.41**	**9.12**	**0.176**	**0.0486**
"*Anchisauripus*" + "*Eubrontes*" cluster (*N* = 21): median value of 15 trials of 8 randomly selected cases	**1.34**	8.79	**0.142**	**0.0487**
All AC 9/14 + all *Anomoepus* (*N* = 17)	**1.39**	**9.09**	—	—
All AC 9/14 + all *Anomoepus*: median value of 15 trials of 7 randomly selected cases from 17 possible cases	**1.26**	**8.10**	—	—
"*Eubrontes*" cluster + all *Anomoepus* (*N* = 15)	***1.35***	**7.49**	—	—
"*Eubrontes*" cluster + all *Anomoepus*: median value of 15 trials of 7 randomly selected cases from 15 possible cases	**1.20**	**6.57**	—	—
Distance between proximal ends of first digital pads of digits II–IV (toebase II–IV)				
Homogeneous groups:				
"*Anchisauripus*" cluster (*N* = 13)	1.34	9.85	0.147	0.0436
All AC 9/14 (*N* = 10)	1.41	11.0	—	—
"*Eubrontes*" cluster (*N* = 8)	1.44	11.2	0.151	0.0630
All *Anomoepus* (*N* = 8)	1.18	5.98	—	—
All adult emus[a] (emu means treatment: *N* = 12)	1.34	7.73	—	—
More inclusive groups:				
"*Anchisauripus*" + "*Eubrontes*" cluster (*N* = 21)	***1.68***	***12.3***	***0.253***	***0.0675***
"*Anchisauripus*" + "*Eubrontes*" cluster (*N* = 21): median value of 15 trials of 8 randomly selected cases	***1.45***	***11.9***	***0.189***	**0.0627**
All AC 9/14 + all *Anomoepus* (*N* = 18)	**1.41**	**8.92**	—	—
All AC 9/14 + all *Anomoepus*: median value of 15 trials of 8 randomly selected cases from 16 possible cases	**1.29**	**8.20**	—	—
"*Eubrontes*" cluster + all *Anomoepus* (*N* = 16)	***1.62***	***14.0***	—	—
"*Eubrontes*" cluster + all *Anomoepus*: median value of 15 trials of 8 randomly selected cases from 16 possible cases	**1.57**	***13.8***	—	—
Distance between toetip II–toetip III				
Homogeneous groups:				
"*Anchisauripus*" cluster (*N* = 13)	1.38	8.77	0.131	0.0373
All AC 9/14 (*N* = 10)	1.52	14.4	—	—
"*Eubrontes*" cluster (*N* = 8)	1.57	16.4	0.137	0.0494
All *Anomoepus* (*N* = 9)	1.68	17.6	—	—
All adult emus (emu means treatment: *N* = 11)	1.31	9.90	—	—

Table A8.12. *continued*

Group	Minimum/maximum ratio of % scaled length	Coefficient of variation of % scaled length	Range of GM-scaled length	Standard deviation of GM-scaled length
More inclusive groups:				
"Anchisauripus" + *"Eubrontes"* cluster (*N* = 21)	**1.57**	**11.9**	**0.137**	**0.0424**
"Anchisauripus" + *"Eubrontes"* cluster (*N* = 21): median value of 15 trials of 8 randomly selected cases	1.38	**10.3**	0.117	**0.0390**
All AC 9/15 + all *Anomoepus* (*N* = 19)	***1.95***	**15.6**	——	——
All AC 9/14 + all *Anomoepus*: median value of 15 trials of 9 randomly selected cases from 19 possible cases	***1.83***	**16.6**	——	——
"Eubrontes" cluster + all *Anomoepus* (*N* = 17)	***1.87***	**16.5**	——	——
"Eubrontes" cluster + all *Anomoepus*: median value of 15 trials of 8 randomly selected cases from 17 possible cases	**1.68**	**16.6**	——	——
Distance between toetip III–toetip IV				
Homogeneous groups:				
"Anchisauripus" cluster (*N* = 13)	1.31	8.14	0.100	0.0290
All AC 9/14 (*N* = 10)	1.34	9.15	——	——
"Eubrontes" cluster (*N* = 8)	1.57	13.2	0.147	0.0466
All *Anomoepus* (*N* = 9)	1.32	10.8	——	——
All adult emus (emu means treatment: *N* = 12)	1.41	11.3	——	——
More inclusive groups:				
"Anchisauripus" + *"Eubrontes"* cluster (*N* = 21)	**1.57**	**10.1**	***0.194***	**0.0378**
"Anchisauripus" + *"Eubrontes"* cluster (*N* = 21): median value of 15 trials of 8 randomly selected cases	**1.32**	**9.20**	**0.113**	**0.0323**
All AC 9/14 + all *Anomoepus* (*N* = 19)	***1.56***	***13.8***	——	——
All AC 9/14 + all *Anomoepus*: median value of 15 trials of 9 randomly selected cases from 19 possible cases	***1.53***	***14.7***	——	——
"Eubrontes" cluster + all *Anomoepus* (*N* = 17)	***1.72***	***13.3***	——	——
"Eubrontes" cluster + all *Anomoepus*: median value of 15 trials of 8 randomly selected cases from 17 possible cases	**1.40**	**12.1**	——	——
Distance between toetip II–toetip IV				
Homogeneous groups:				
"Anchisauripus" cluster (*N* = 13)	1.39	8.54	0.119	0.0316
All AC 9/14 (*N* = 11)	1.52	10.9	——	——
"Eubrontes" cluster (*N* = 8)	1.57	14.0	0.170	0.0581
All *Anomoepus* (*N* = 9)	1.37	10.9	——	——
All adult emus (emu means treatment: *N* = 12)	1.28	8.65	——	——
More inclusive groups:				
"Anchisauripus" + *"Eubrontes"* cluster (*N* = 21)	***1.66***	**13.5**	***0.190***	**0.0463**
"Anchisauripus" + *"Eubrontes"* cluster (*N* = 21): median value of 15 trials of 8 randomly selected cases	1.39	**11.7**	**0.125**	**0.0443**
All AC 9/14 + all *Anomoepus* (*N* = 20)	***2.33***	***23.6***	——	——
All AC 9/14 + all *Anomoepus*: median value of 15 trials of 9 randomly selected cases from 20 possible cases	***1.92***	***23.7***	——	——
"Eubrontes" cluster + all *Anomoepus* (*N* = 17)	***1.82***	***15.9***	——	——
"Eubrontes" cluster + all *Anomoepus*: median value of 15 trials of 8 randomly selected cases from 17 possible cases	***1.63***	***15.3***	——	——
Digit III first pad width				
Homogeneous groups:				
"Anchisauripus" cluster (*N* = 13)	1.53	12.7	0.0947	0.0287
All AC 9/14 (*N* = 22)	1.66	11.4	——	——
"Eubrontes" cluster (*N* = 8)	1.37	11.5	0.0725	0.0271
All *Anomoepus* (*N* = 4)	1.15	——	——	——

Group	Minimum/maximum ratio of % scaled length	Coefficient of variation of % scaled length	Range of GM-scaled length	Standard deviation of GM-scaled length
More inclusive groups:				
"Anchisauripus" + *"Eubrontes"* cluster (*N* = 21)	***1.83***	***16.1***	***0.157***	***0.0503***
"Anchisauripus" + *"Eubrontes"* cluster: median value of 15 trials of 8 randomly selected cases from 21 possible cases	***1.59***	***16.3***	***0.147***	***0.0513***
All AC 9/14 + all *Anomoepus* (*N* = 26)	**1.66**	**13.2**	——	——
Digit III projection				
Homogeneous groups:				
"Anchisauripus" cluster (*N* = 13)	1.32	8.38	——	——
All AC 9/14 (*N* = 10)	1.33	9.42	——	——
"Eubrontes" cluster (*N* = 8)	1.61	16.0	——	——
All *Anomoepus* (*N* = 9)	1.79	18.1	——	——
All adult emus (emu means treatment: *N* = 12)	1.44	12.0	——	——
More inclusive groups:				
"Anchisauripus" + *"Eubrontes"* cluster (*N* = 21)	**1.61**	**11.7**	——	——
"Anchisauripus" + *"Eubrontes"* cluster (*N* = 21): median value of 15 trials of 8 randomly selected cases	**1.37**	**10.9**	——	——
All AC 9/14 + all *Anomoepus* (*N* = 19)	***1.88***	**15.1**	——	——
All AC 9/14 + all *Anomoepus*: median value of 15 trials of 9 randomly selected cases from 19 possible cases	**1.71**	**15.2**	——	——
"Eubrontes" cluster + all *Anomoepus* (*N* = 17)	***1.94***	**16.9**	——	——
"Eubrontes" cluster + all *Anomoepus*: median value of 15 trials of 8 randomly selected cases from 17 possible cases	**1.75**	**16.9**	——	——

[a]Because of the different architecture of the proximal end of footprints in birds and non-avian dinosaurs, this measurement is not strictly comparable between emus and non-avian dinosaurs.

Table A8.13. Variability of "simple" footprint parameters scaled to a common dimension (overall footprint length) of non-avian dinosaurs ("mean or single" treatment) and ground birds (bird means treatment; excludes ostriches [both the "all birds" and "one per species" because the metatarsal pad does not register as part of overall footprint length]). For each parameter, the various groups are ranked in order of decreasing variability (1 = the group with the greatest variability), allowing for ties.

Parameter	Group	Maximum/minimum *(rank)*	Number of cases
Proximal end (toebase) II–proximal end (toebase) III	*Eubrontes*	1.94 ***(3)***	36
	AC 9/14	1.35 ***(6)***	11
	Anomoepus	1.62 ***(5)***	7
	Emus	1.71 ***(4)***	32
	All birds	2.48 ***(1)***	85
	Birds, one per species	2.48 ***(1)***	36
	Phasianids	2.41 ***(2)***	19
Proximal end III–proximal end IV	*Eubrontes*	1.43 ***(5)***	36
	AC 9/14	1.38 ***(7)***	11
	Anomoepus	1.42 ***(6)***	7
	Emus[a]	1.53 ***(4)***	33
	All birds[a]	2.09 ***(1)***	85
	Birds, one per species[a]	1.87 ***(2)***	36
	Phasianids[a]	1.58 ***(3)***	19
Proximal end II–proximal end IV	*Eubrontes*	1.77 ***(3)***	38
	AC 9/14	1.45 ***(5)***	12
	Anomoepus	1.42 ***(6)***	8
	Emus[a]	1.54 ***(4)***	32
	All birds[a]	2.47 ***(1)***	86
	Birds, one per species[a]	2.47 ***(1)***	37
	Phasianids[a]	2.34 ***(2)***	20
Toetip II–toetip III	*Eubrontes*	1.60 ***(2)***	41
	AC 9/14	1.43 ***(4)***	12
	Anomoepus	1.43 ***(4)***	9
	Emus	1.28 ***(5)***	30
	All birds	2.60 ***(1)***	90
	Birds, one per species	2.60 ***(1)***	37
	Phasianids	1.45 ***(3)***	22
Toetip III–toetip IV	*Eubrontes*	1.62 ***(3)***	42
	AC 9/14	1.58 ***(4)***	12
	Anomoepus	1.24 ***(6)***	9
	Emus	1.37 ***(5)***	33
	All birds	4.44 ***(1)***	92
	Birds, one per species	4.44 ***(1)***	36
	Phasianids	2.16 ***(2)***	21
Toetip II–toetip IV	*Eubrontes*	2.21 ***(3)***	40
	AC 9/14	1.45 ***(5)***	11
	Anomoepus	1.27 ***(7)***	9
	Emus	1.30 ***(6)***	31
	All birds	2.38 ***(1)***	91
	Birds, one per species	2.25 ***(2)***	37
	Phasianids	1.82 ***(4)***	22
Digit II length	*Eubrontes*	1.42 ***(4)***	38
	AC 9/14	1.33 ***(6)***	12
	Anomoepus	1.35 ***(5)***	8
	Emus	1.25 ***(7)***	31
	All birds	2.29 ***(1)***	86
	Birds, one per species	2.02 ***(2)***	37
	Phasianids	1.59 ***(3)***	20

Parameter	Group	Maximum/minimum *(rank)*	Number of cases
Digit III projection	*Eubrontes*	1.71 ***(2)***	40
	AC 9/14	1.25 ***(6)***	11
	Anomoepus	1.45 ***(4)***	9
	Emus	1.34 ***(5)***	33
	All birds	4.67 ***(1)***	92
	Birds, one per species	4.67 ***(1)***	36
	Phasianids	1.47 ***(3)***	21
Backfoot length	*Eubrontes*	1.37 ***(4)***	38
	AC 9/14	1.13 ***(6)***	10
	Anomoepus	1.24 ***(5)***	9
	Emus	1.45 ***(3)***	33
	All birds	4.61 ***(1)***	92
	Birds, one per species	4.61 ***(1)***	36
	Phasianids	1.68 ***(2)***	21
Digit IV length	*Eubrontes*	1.53 ***(3)***	37
	AC 9/14	1.19 ***(7)***	10
	Anomoepus	1.30 ***(6)***	8
	Emus[a]	1.34 ***(5)***	33
	All birds[a]	2.18 ***(1)***	87
	Birds, one per species[a]	1.98 ***(2)***	37
	Phasianids[a]	1.47 ***(4)***	20

[a]Because of differences in the anatomical configuration of the "heel" between non-avian dinosaurs and birds, this comparison is only analogous between emu and non-avian dinosaur footprints.

Table A8.14. Variability of "complex" footprint parameters scaled to a common dimension (digit III length excluding the claw) of non-avian dinosaurs ("mean or single" treatment) and emus. For each parameter, the various groups are ranked in order of decreasing variability (1 = the group with the greatest variability), allowing for ties.

Parameter	Group	Maximum/minimum *(rank)*	Number of cases
Digit II first pad length	*Eubrontes*	1.83 ***(1)***	42
	AC 9/14	1.48 ***(3)***	18
	Anomoepus	1.17 ***(4)***	6
	Emus	1.56 ***(2)***	20
Digit II second pad length	*Eubrontes*	2.01 ***(1)***	38
	AC 9/14	1.78 ***(2)***	16
	Anomoepus	1.49 ***(3)***	6
	Emus	1.49 ***(3)***	20
Digit III first pad length	*Eubrontes*	1.46 ***(1)***	39
	AC 9/14	1.43 ***(3)***	19
	Anomoepus	1.45 ***(2)***	4
	Emus	1.27 ***(4)***	34
Digit III second pad length	*Eubrontes*	1.44 ***(2)***	37
	AC 9/14	1.33 ***(3)***	18
	Anomoepus	1.08 ***(4)***	3
	Emus	1.48 ***(1)***	31
Digit III third pad length	*Eubrontes*	1.53 ***(2)***	37
	AC 9/14	1.45 ***(3)***	17
	Anomoepus	1.16 ***(4)***	2
	Emus	1.73 ***(1)***	30
Digit IV length excluding claw	*Eubrontes*	1.56 ***(1)***	23
	AC 9/14	1.28 ***(3)***	7
	Anomoepus	1.19 ***(4)***	5
	Emus[a]	1.43 ***(2)***	29
Proximal end (toebase) II–proximal end (toebase) III	*Eubrontes*	2.35 ***(1)***	43
	AC 9/14	1.46 ***(4)***	20
	Anomoepus	1.52 ***(3)***	6
	Emus	1.77 ***(2)***	32
Proximal end III–proximal end IV	*Eubrontes*	1.63 ***(1)***	32
	AC 9/14	1.45 ***(4)***	12
	Anomoepus	1.56 ***(3)***	6
	Emus[a]	1.61 ***(2)***	32
Proximal end II–proximal end IV	*Eubrontes*	1.84 ***(1)***	32
	AC 9/14	1.48 ***(3)***	12
	Anomoepus	1.46 ***(4)***	6
	Emus[a]	1.63 ***(2)***	31
Clawbase II–clawbase III	*Eubrontes*	1.52 ***(1)***	39
	AC 9/14	1.52 (***1***)	17
	Anomoepus	1.36 ***(2)***	6
	Emus	1.21 ***(3)***	26
Clawbase III–clawbase IV	*Eubrontes*	1.57 ***(1)***	32
	AC 9/14	1.44 ***(2)***	13
	Anomoepus	1.31 ***(4)***	5
	Emus	1.34 ***(3)***	27
Clawbase II–clawbase IV	*Eubrontes*	1.94 ***(1)***	30
	AC 9/14	1.76 ***(2)***	12
	Anomoepus	1.15 ***(4)***	5
	Emus	1.30 ***(3)***	25
Heel to clawbase II	*Eubrontes*	1.37 ***(2)***	30
	AC 9/14	1.27 ***(3)***	11
	Anomoepus	1.19 ***(4)***	6
	Emus[a]	1.38 ***(1)***	26

[a]Because of differences in the anatomical configuration of the "heel" between non-avian dinosaurs and birds, this comparison is only analogous between emu and non-avian dinosaur footprints.

Table A8.15. Variability of scaled "simple" footprint parameters among footprints of non-avian dinosaurs and ground birds ("mean or single" treatment for dinosaurs, and "bird means" treatment for ground birds). Each parameter was "GM-scaled" by subtracting from its log-transformed version the mean of all the log-transformed parameters used in the comparison (all the parameters listed in this table). Variability is expressed in terms of the range of scaled values of each parameter, and the standard deviation of those scaled values. For each parameter, the various groups are ranked in order of decreasing variability (1 = the group with the greatest variability), allowing for ties. The number of cases is uniform across parameters for the same group: *Eubrontes* = 34; AC 9/14 (footprints on the Amherst College Slab of that catalog number) = 9; *Anomoepus* = 7; emus = 30; all birds = 81; birds (one per species) = 35; phasianids = 18.

Parameter	Group	Mean	Range *(rank)*	Standard deviation *(rank)*
	Eubrontes	−0.370	0.265 ***(2)***	0.0571 ***(4)***
	AC 9/14	−0.388	0.080 ***(5)***	0.0288 ***(6)***
	Anomoepus	−0.396	0.162 ***(3)***	0.0605 ***(2)***
Proximal end (toebase) II–proximal end (toebase) III	Emus	−0.223	0.121 ***(4)***	0.0273 ***(7)***
	All birds	−0.257	0.282 ***(1)***	0.0538 ***(5)***
	Birds, one per species	−0.284	0.282 ***(1)***	0.0609 ***(1)***
	Phasianids	−0.304	0.282 ***(1)***	0.0595 ***(3)***
	Eubrontes	−0.136	0.175 ***(3)***	0.0384 ***(6)***
	AC 9/14	−0.106	0.118 ***(6)***	0.0410 ***(5)***
	Anomoepus	−0.131	0.116 ***(7)***	0.0494 ***(4)***
Proximal end III–proximal end IV	Emus[a]	−0.210	0.135 ***(5)***	0.0271 ***(7)***
	All birds[a]	−0.268	0.298 ***(1)***	0.0716 ***(1)***
	Birds, one per species[a]	−0.314	0.246 ***(2)***	0.0568 ***(2)***
	Phasianids[a]	−0.337	0.174 ***(4)***	0.0505 ***(3)***
	Eubrontes	−0.227	0.253 ***(4)***	0.0618 ***(3)***
	AC 9/14	−0.193	0.103 ***(6)***	0.0402 ***(6)***
	Anomoepus	−0.206	0.115 ***(5)***	0.0461 ***(5)***
Proximal end II–proximal end IV	Emus[a]	−0.013	0.085 ***(7)***	0.0227 ***(7)***
	All birds[a]	−0.069	0.319 ***(1)***	0.0762 ***(1)***
	Birds, one per species[a]	−0.118	0.277 ***(2)***	0.0727 ***(2)***
	Phasianids[a]	−0.141	0.254 ***(3)***	0.0603 ***(4)***
	Eubrontes	0.077	0.195 ***(3)***	0.0432 ***(4)***
	AC 9/14	0.095	0.134 ***(5)***	0.0428 ***(5)***
	Anomoepus	0.043	0.167 ***(4)***	0.0613 ***(2)***
Toetip II–toetip III	Emus	0.137	0.097 ***(7)***	0.0253 ***(7)***
	All birds	0.164	0.299 ***(1)***	0.0571 ***(3)***
	Birds, one per species	0.192	0.284 ***(2)***	0.0630 ***(1)***
	Phasianids	0.210	0.133 ***(6)***	0.0409 ***(6)***
	Eubrontes	0.063	0.178 ***(3)***	0.0400 ***(4)***
	AC 9/14	0.047	0.120 ***(4)***	0.0363 ***(5)***
	Anomoepus	0.105	0.077 ***(6)***	0.0337 ***(6)***
Toetip III–toetip IV	Emus	0.206	0.101 ***(5)***	0.0274 ***(7)***
	All birds	0.198	0.511 ***(1)***	0.0676 ***(2)***
	Birds, one per species	0.203	0.511 ***(1)***	0.0873 ***(1)***
	Phasianids	0.241	0.195 ***(2)***	0.0459 ***(3)***
	Eubrontes	0.146	0.207 ***(1)***	0.0482 ***(3)***
	AC 9/14	0.103	0.089 ***(5)***	0.0247 ***(6)***
	Anomoepus	0.253	0.073 ***(6)***	0.0263 ***(5)***
Toetip II–toetip IV	Emus	0.287	0.066 ***(7)***	0.0187 ***(7)***
	All birds	0.320	0.197 ***(2)***	0.0508 ***(2)***
	Birds, one per species	0.343	0.187 ***(3)***	0.0540 ***(1)***
	Phasianids	0.370	0.110 ***(4)***	0.0308 ***(4)***
	Eubrontes	0.077	0.157 ***(4)***	0.0339 ***(6)***
	AC 9/14	0.056	0.145 ***(5)***	0.0459 ***(4)***
	Anomoepus	0.042	0.133 ***(7)***	0.0419 ***(5)***
Digit II length	Emus	−0.166	0.137 ***(6)***	0.0257 ***(7)***
	All birds	−0.089	0.427 ***(1)***	0.0845 ***(1)***
	Birds, one per species	−0.036	0.386 ***(2)***	0.0770 ***(2)***
	Phasianids	−0.052	0.242 ***(3)***	0.0604 ***(3)***

Table A8.15. *continued*

Parameter	Group	Mean	Range *(rank)*	Standard deviation *(rank)*
Digit III projection	*Eubrontes*	−0.034	0.280 ***(2)***	0.0590 ***(4)***
	AC 9/14	−0.006	0.078 ***(6)***	0.0264 ***(6)***
	Anomoepus	−0.109	0.200 ***(3)***	0.0624 ***(3)***
	Emus	0.050	0.102 ***(5)***	0.0262 ***(7)***
	All birds	0.035	0.541 ***(1)***	0.0809 ***(2)***
	Birds, one per species	0.044	0.541 ***(1)***	0.0980 ***(1)***
	Phasianids	0.080	0.155 ***(4)***	0.0358 ***(5)***
Backfoot length	*Eubrontes*	0.189	0.112 ***(4)***	0.0251 ***(6)***
	AC 9/14	0.187	0.069 ***(6)***	0.0194 ***(7)***
	Anomoepus	0.169	0.078 ***(5)***	0.0321 ***(5)***
	Emus	0.006	0.197 ***(3)***	0.0397 ***(4)***
	All birds	−0.007	0.800 ***(1)***	0.1196 ***(2)***
	Birds, one per species	−0.042	0.800 ***(1)***	0.1579 ***(1)***
	Phasianids	−0.089	0.282 ***(2)***	0.0777 ***(3)***
Digit IV length	*Eubrontes*	0.216	0.161 ***(2)***	0.0302 ***(5)***
	AC 9/14	0.205	0.059 ***(6)***	0.0202 ***(6)***
	Anomoepus	0.231	0.114 ***(5)***	0.0378 ***(3)***
	Emus[a]	−0.074	0.120 ***(4)***	0.0308 ***(4)***
	All birds[a]	−0.026	0.285 ***(1)***	0.0613 ***(1)***
	Birds, one per species[a]	0.013	0.285 ***(1)***	0.0524 ***(2)***
	Phasianids[a]	0.023	0.123 ***(3)***	0.0308 ***(4)***

[a]Because of differences in the anatomical configuration of the "heel" between non-avian dinosaurs and birds, this comparison is only analogous between bird and non-avian dinosaur footprints.

Table A8.16. Variability of scaled "complex" footprint parameters among footprints of non-avian dinosaurs (all *Eubrontes,* and also those trackways on AC 9/14; "mean or single" treatment) and emus ("bird means" treatment). Each parameter was "GM scaled" by subtracting from its log-transformed version the mean of all the log-transformed parameters used in the comparison (all the parameters listed in this table). Variability is expressed in terms of the range of scaled values of each parameter, and the standard deviation of those scaled values. The number of cases is uniform across parameters for the same group: all *Eubrontes* = 17; AC 9/14 = 5; emus = 19. Parameters for which the emu measure of variability is greater than that for the all *Eubrontes* sample are indicated in **bold**.

Parameter	Group	Mean	Range	Standard deviation
Digit II first pad length	*Eubrontes*	−0.166	0.205	0.0509
	AC 9/14	−0.207	0.107	0.0405
	Emus	−0.421	0.171	0.0409
Digit II second pad length	*Eubrontes*	−0.177	0.195	0.0502
	AC 9/14	−0.188	0.075	0.0284
	Emus	−0.397	0.152	0.0412
Digit III first pad length	*Eubrontes*	−0.134	0.116	0.0285
	AC 9/14	−0.142	0.045	0.0169
	Emus	−0.186	0.110	**0.0320**
Digit III second pad length	*Eubrontes*	−0.148	0.173	0.0428
	AC 9/14	−0.127	0.098	0.0378
	Emus	−0.199	0.145	0.0362
Digit III third pad length	*Eubrontes*	−0.212	0.127	0.0376
	AC 9/14	−0.194	0.107	0.0470
	Emus	−0.337	**0.211**	**0.0522**
Digit IV length excluding claw	*Eubrontes*	0.291	0.126	0.0349
	AC 9/14	0.284	0.054	0.0233
	Emus[a]	0.019	0.087	0.0261
Proximal end (toebase) II–proximal end (toebase) III	*Eubrontes*	−0.209	0.164	0.0466
	AC 9/14	−0.232	0.083	0.0363
	Emus	−0.028	0.085	0.0211
Proximal end III–proximal end IV	*Eubrontes*	0.017	0.157	0.0434
	AC 9/14	0.037	0.121	0.0450
	Emus[a]	−0.013	0.119	0.0249
Proximal end II–proximal end IV	*Eubrontes*	−0.089	0.248	0.0671
	AC 9/14	−0.052	0.115	0.0450
	Emus[a]	0.186	0.051	0.0156
Clawbase II–clawbase III	*Eubrontes*	0.172	0.119	0.0388
	AC 9/14	0.197	0.047	0.0214
	Emus	0.299	0.080	0.0231
Clawbase III–clawbase IV	*Eubrontes*	0.152	0.150	0.0356
	AC 9/14	0.150	0.074	0.0298
	Emus	0.367	0.098	0.0243
Clawbase II–clawbase IV	*Eubrontes*	0.173	0.218	0.0594
	AC 9/14	0.138	0.063	0.0240
	Emus	0.436	0.062	0.0165
"Heel" to clawbase II	*Eubrontes*	0.331	0.067	0.0176
	AC 9/14	0.335	0.066	0.0254
	Emus[a]	0.274	**0.075**	0.0164

[a]Because of differences in the anatomical configuration of the "heel" between non-avian dinosaurs and birds, this comparison is only analogous between emu and non-avian dinosaur footprints.

Table A8.17. Summary of comparisons of variability of scaled footprint parameters in *Eubrontes* with emu and potentially homogeneous dinosaur footprint samples. AA = all *Anomoepus* prints; AC = AC 9/14 footprints; E = *Eubrontes*; Em = emus. Detailed results reported in tables A8.13–A8.16. Mx/Mn = maximum/minimum ratio of parameters either "simple" scaled to the same value of footprint length, or "complex" scaled to the same value of digit III length excluding the claw; SD = standard deviation.

Parameter	Measure of variability	***Eubrontes* and emus**		***Eubrontes* and AC 9/14**		***Eubrontes* and all *Anomoepus***	
		E > Em	**E < Em**	**E > AC**	**E < AC**	**E > AA**	**E < AA**
Toebase (proximal end) II–toebase (proximal end) III: "simple" scaled	Mx/Mn	x		x		x	
Toebase (proximal end) II–toebase (proximal end) III: "complex" scaled	Mx/Mn	x		x		x	
Toebase II–toebase III: "simple" GM scaled	Range	x		x		x	
	SD	x		x			x
Toebase II–toebase III: "complex" GM scaled	Range	x		x			
	SD	x		x			
Toebase III–toebase IV: "simple" scaled	Mx/Mn		x[a]	x		x	
Toebase III–toebase IV: "complex" scaled	Mx/Mn		x[a]	x		x	
Toebase III–toebase IV: "simple" GM scaled	Range	x[a]		x		x	
	SD	x[a]			x		x
Toebase III–toebase IV: "complex" GM scaled	Range	x[a]		x			
	SD	x[a]			x		
Toebase II–toebase IV: "simple" scaled	Mx/Mn	x[a]		x		x	
Toebase II–toebase IV: "complex" scaled	Mx/Mn	x[a]		x		x	
Toebase II–toebase IV: "simple" GM scaled	Range	x[a]		x		x	
	SD	x[a]		x		x	
Toebase II–toebase IV: "complex" GM scaled	Range	x[a]		x			
	SD	x[a]		x			
Clawbase II–clawbase III: "complex" scaled	Mx/Mn	x		Tie		x	
Clawbase II–clawbase III: "complex" GM scaled	Range	x		x			
	SD	x		x			
Clawbase III–clawbase IV: "complex" scaled	Mx/Mn	x		x		x	
Clawbase III–clawbase IV: "complex" GM scaled	Range	x		x			
	SD	x		x			
Clawbase II–clawbase IV: "complex" scaled	Mx/Mn	x		x		x	
Clawbase II–clawbase IV: "complex" GM scaled	Range	x		x			
	SD	x		x			
"Heel" to clawbase II[a]: "complex" scaled	Mx/Mn		x	x		x	
"Heel" to clawbase II[a]: "complex" GM scaled	Range		x	x			
	SD	x			x		
Toetip II–toetip III: "simple" scaled	Mx/Mn	x		x		x	
Toetip II–toetip III: "simple" GM scaled	Range	x		x		x	
	SD	x		x			x
Toetip III–toetip IV: "simple" scaled	Mx/Mn	x		x		x	
Toetip III–toetip IV: "simple" GM scaled	Range	x		x		x	
	SD	x		x		x	
Toetip II–toetip IV: "simple" scaled	Mx/Mn	x		x		x	
Toetip II–toetip IV: "simple" GM scaled	Range	x		x		x	
	SD	x		x		x	
Digit II length: "simple" scaled	Mx/Mn	x		x		x	
Digit II length: "simple" GM scaled	Range	x		x		x	
	SD	x			x		x
Digit III projection: "simple" scaled	Mx/Mn	x		x		x	
Digit III projection: "simple" GM scaled	Range	x		x		x	
	SD	x		x			x
Backfoot length: "simple" scaled	Mx/Mn		x	x		x	
Backfoot length: "simple" GM scaled	Range		x	x		x	
	SD		x	x			x
Digit IV length: "simple" scaled	Mx/Mn	x[a]		x		x	

Parameter	Measure of variability	*Eubrontes* and emus		*Eubrontes* and AC 9/14		*Eubrontes* and all *Anomoepus*	
		E > Em	**E < Em**	**E > AC**	**E < AC**	**E > AA**	**E < AA**
Digit IV length: "simple" GM scaled	Range	x[a]		x		x	
	SD		x[a]	x			x
Digit IV length excluding claw: "complex" scaled	Mx/Mn	x[a]		x		x	
Digit IV length excluding claw: "complex" GM scaled	Range	x[a]		x			
	SD	x[a]		x			
Digit II first pad length: "complex" scaled	Mx/Mn	x		x		x	
Digit II first pad length: "complex" GM scaled	Range	x		x			
	SD	x		x			
Digit II second pad: "complex" scaled	Mx/Mn	x		x		x	
Digit II second pad length: "complex" GM scaled	Range	x		x			
	SD	x		x			
Digit III first pad length: "complex" scaled	Mx/Mn	x		x		x	
Digit III first pad length: "complex" GM scaled	Range	x		x			
	SD		x	x			
Digit III second pad length: "complex" scaled	Mx/Mn		x	x		x	
Digit III second pad length: "complex" GM scaled	Range	x		x			
	SD	x		x			
Digit III third pad length: "complex" scaled	Mx/Mn		x	x		x	
Digit III third pad length: "complex" GM scaled	Range		x	x			
	SD		x		x		

[a]Because of differences in the anatomical configuration of the "heel" between non-avian dinosaurs and birds, this comparison is only analogous between emu and non-avian dinosaur footprints.

Table A8.18. Summary of comparisons of variability of scaled footprint parameters between the emu and the potentially homogeneous AC 9/14 and all *Anomoepus* samples. AA = all *Anomoepus* prints; AC = AC 9/14 footprints. Detailed results reported in tables A8.13–A8.16. Mx/Mn = maximum/minimum ratio either "simple" scaled to the same value of footprint length, or "complex" scaled to the same value of digit III length excluding the claw; SD = standard deviation.

Parameter	Measure of variability	*AC 9/14* and emus		All *Anomoepus* and emus	
		AC > Emus	AC < Emus	AA > Emus	AA < Emus
Toebase (proximal end) II–toebase (proximal end) III: "simple" scaled	Mx/Mn		x		x
Toebase II–toebase III: "complex" scaled	Mx/Mn		x		x
Toebase II–toebase III: "simple" GM scaled	Range		x	x	
	SD	x		x	
Toebase II–toebase III: "complex" GM scaled	Range		x		
	SD	x			
Toebase III–toebase IV[a]: "simple" scaled	Mx/Mn		x		x
Toebase III–toebase IV[a]: "complex" scaled	Mx/Mn		x		x
Toebase III–toebase IV[a]: "simple" GM scaled	Range		x		x
	SD	x		x	
Toebase III–toebase IV[a]: "complex" GM scaled	Range	x			
	SD	x			
Toebase II–toebase IV[a]: "simple" scaled	Mx/Mn		x		x
Toebase II–toebase IV[a]: "complex" scaled	Mx/Mn		x		x
Toebase II–toebase IV[a]: "simple" GM scaled	Range	x		x	
	SD	x		x	
Toebase II–toebase IV[a]: "complex" GM scaled	Range	x			
	SD	x			
Clawbase II–clawbase III: "complex" scaled	Mx/Mn	x		x	
Clawbase II–clawbase III: "complex" GM scaled	Range		x		
	SD		x		
Clawbase III–clawbase IV: "complex" scaled	Mx/Mn	x			x
Clawbase III–clawbase IV: "complex" GM scaled	Range		x		
	SD	x			
Clawbase II–clawbase IV: "complex" scaled	Mx/Mn	x			x
Clawbase II–clawbase IV: "complex" GM scaled	Range	x			
	SD	x			
"Heel" to clawbase II[a]: "complex" scaled	Mx/Mn		x		x
"Heel" to clawbase II[a]: "complex" GM scaled	Range		x		
	SD	x			
Toetip II–toetip III: "simple" scaled	MxMn	x		x	
Toetip II–toetip III: "simple" GM scaled	Range	x		x	
	SD	x		x	
Toetip III–toetip IV: "simple" scaled	Mx/Mn	x			x
Toetip III–toetip IV: "simple" GM scaled	Range	x			x
	SD	x		x	
Toetip II–toetip IV: "simple" scaled	Mx/Mn	x			x
Toetip II–toetip IV: "simple" GM scaled	Range	x		x	
	SD	x		x	
Digit II length: "simple" scaled	Mx/Mn	x		x	
Digit II length: "simple" GM scaled	Range	x			x
	SD	x		x	
Digit III projection: "simple" scaled	Mx/Mn		x	x	
Digit III projection: "simple" GM scaled	Range		x	x	
	SD	x		x	
Backfoot length: "simple" scaled	Mx/Mn		x		x
Backfoot length: "simple" GM scaled	Range		x		x
	SD		x		x
Digit IV length[a]: "simple" scaled	Mx/Mn		x		x
Digit IV length[a]: "simple" GM scaled	Range		x		x
	SD		x	x	
Digit IV length excluding claw[a]: "complex" scaled	Mx/Mn		x		x

Parameter	Measure of variability	*AC 9/14* and emus		All *Anomoepus* and emus	
		AC > Emus	AC < Emus	AA > Emus	AA < Emus
Digit IV length excluding claw[a]: "complex" GM scaled	Range		x		
	SD		x		
Digit II first pad length: "complex" scaled	Mx/Mn		x		x
Digit II first pad length: "complex" GM scaled	Range		x		
	SD		x		
Digit II second pad length: "complex" scaled	Mx/Mn	x		Tie	
Digit II second pad length: "complex" GM scaled	Range		x		
	SD		x		
Digit III first pad length: "complex" scaled	Mx/Mn	x		x	
Digit III first pad length: "complex" GM scaled	Range		x		
	SD		x		
Digit III second pad length: "complex" scaled	Mx/Mn		x		x
Digit III second pad length: "complex" GM scaled	Range		x		
	SD	x			
Digit III third pad length: "complex" scaled	Mx/Mn		x		x
Digit III third pad length: "complex" GM scaled	Range		x		
	SD		x		

[a]Because of differences in the anatomical configuration of the "heel" between non-avian dinosaurs and birds, this comparison is only analogous between emu and non-avian dinosaur footprints.

Table A8.19. Comparison of regression and reduced major axis (RMA) relationships between various linear dimensions of footprints and feet. Known intraspecific comparisons are for emu footprints (emu means treatment) and alligator feet. For alligators, digit lengths = free lengths. Alligator cases are excluded from the analysis when one or both of the parameters involves claw lengths and when one or both of the claws were heavily worn. Known interspecific comparisons are for all ground bird species combined ("all birds": bird means treatment, with variable numbers of individual birds represented per species [and so dominated by emu cases]) and all ground bird species (bird means treatment, but with only one individual represented per species ["one per species"). Data for ostrich were included in the "all birds" treatment, except where the comparison of interest could not be made (as with comparisons involving digit II), but ostrich data were excluded in the "one per species" ground bird treatment. The intraspecific and interspecific shape comparisons are compared with the same or analogous comparisons of dinosaur footprints assigned to the ichnogenus *Eubrontes* ("mean or single" treatment), for which the nature of the sample (intraspecific and/or interspecific) is unknown. All measurements in are in millimeters, and all measurements were log transformed prior to analysis. SEE = standard error of estimate of the regression equation between the two variables; where results for more than one group are reported in a bivariate comparison, SEE is ranked in decreasing order (1 = largest values [most variability]). Comparisons in which the 95% confidence interval (CI) of the RMA slope excluded 1 are indicated in **bold**.

Dependent variable	Group	r^2	SEE of regression *(rank)*	Number of cases	Slope	95% CI of slope
Independent variable = footprint length from heel to toetip III						
Toetip II to toetip IV	*Eubrontes*	0.934	0.0573 ***(3)***	40	**1.386**	1.270–1.513
	Emus	0.971	0.0244 ***(4)***	31	**0.882**	0.826–0.942
	All birds	0.917	0.0723 ***(2)***	91	**0.915**[a]	0.858–0.975
	Birds, one per species	0.795	0.0904 ***(1)***	37	0.959	0.804–1.143
Distance from digit II proximal bend point to digit III long axis	*Eubrontes*	0.881	0.0775	42	**1.431**	1.272–1.610
Distance from digit II proximal bend point to "heel"	*Eubrontes*	0.943	0.0469	41	**1.237**	1.142–1.340
Independent variable = backfoot length						
Digit III projection	*Eubrontes*	0.866	0.0482 ***(4)***	39	**0.719**	0.630–0.820
	Emus	0.894	0.0504 ***(3)***	31	**0.871**[a]	0.763–0.993
	All birds	0.705	0.1367 ***(2)***	92	**0.772**	0.673–0.885
	Birds, one per species	0.272	0.1829 ***(1)***	36	0.822	0.430–1.571
Independent variable = digit III length						
Digit II length	*Eubrontes*	0.957	0.0336 ***(3)***	47	**1.161**	1.090–1.238
	Emus	0.972	0.0271 ***(4)***	32	0.953	0.895–1.015
	Alligators	0.988	0.0276 ***(5)***	85	1.001	0.977–1.025
	All birds	0.896	0.0774 ***(1)***	88	0.968	0.899–1.041
	Birds, one per species	0.878	0.0753 ***(2)***	39	1.050	0.927–1.189
Digit IV length	*Eubrontes*	0.955	0.0372 ***(3)***	35	**1.212**	1.122–1.309
	Emus[b]	0.971	0.0288 ***(4)***	34	1.007	0.946–1.072
	All birds[b]	0.943	0.0571 ***(1)***	91	**0.937**[a]	0.890–0.987
	Birds, one per species[b]	0.932	0.0538 ***(2)***	39	1.004	0.918–1.099
Proximal end (toebase) digit II to proximal end (toebase) digit III	*Eubrontes*	0.760	0.0796 ***(3)***	48	**1.200**[a]	1.014–1.420
	Emus	0.896	0.0452 ***(4)***	34	**0.831**	0.735–0.940
	All birds	0.910	0.0828 ***(2)***	89	**1.114**[a]	1.042–1.191
	Birds, one per species	0.838	0.0887 ***(1)***	38	1.065	0.917–1.237
Proximal end digit III to proximal end digit IV	*Eubrontes*	0.834	0.0635 ***(3)***	38	1.095	0.941–1.275
	Emus[b]	0.936	0.0339 ***(4)***	34	**0.796**	0.724–0.875
	All birds[b]	0.899	0.0943 ***(1)***	90	**1.167**	1.087–1.253
	Birds, one per species[b]	0.866	0.0829 ***(2)***	38	1.097	0.960–1.254
Proximal end digit II to proximal end digit IV	*Eubrontes*	0.832	0.0695 ***(3)***	38	**1.191**[a]	1.022–1.388
	Emus[b]	0.922	0.0382 ***(4)***	33	**0.806**	0.724–0.897
	All birds[b]	0.903	0.0938 ***(1)***	88	**1.216**	1.133–1.304
	Birds, one per species[b]	0.871	0.0861 ***(2)***	39	**1.168**[a]	1.027–1.329
Clawbase II to clawbase III	*Eubrontes*	0.895	0.0444 ***(1)***	38	1.075	0.957–1.207
	Emus	0.970	0.0277 ***(2)***	26	**0.914**[a]	0.849–0.985
Clawbase III to clawbase IV	*Eubrontes*	0.900	0.0427 ***(1)***	31	1.096	0.965–1.245
	Emus	0.969	0.0246 ***(2)***	27	**0.810**	0.753–0.872
Clawbase II to clawbase IV	*Eubrontes*	0.908	0.0584 ***(1)***	33	**1.506**	1.339–1.693
	Emus	0.962	0.0284 ***(2)***	26	**0.814**	0.749–0.886
Toetip II to toetip III	*Eubrontes*	0.867	0.0504 ***(3)***	48	0.994	0.884–1.117
	Emus	0.965	0.0292 ***(4)***	31	**0.910**[a]	0.846–0.978
	All birds	0.921	0.0665 ***(2)***	87	0.956	0.897–1.018
	Birds, one per species	0.831	0.0820 ***(1)***	39	0.970	0.834–1.129

Dependent variable	Group	r^2	SEE of regression *(rank)*	Number of cases	Slope	95% CI of slope
Toetip III to toetip IV	*Eubrontes*	0.903	0.0458 ***(3)***	47	1.050	0.952–1.159
	Emus	0.967	0.0254 ***(4)***	35	**0.832**	0.779–0.889
	All birds	0.881	0.0838 ***(2)***	92	0.953	0.882–1.029
	Birds, one per species	0.733	0.1107 ***(1)***	38	1.035	0.841–1.272
Toetip II to toetip IV	*Eubrontes*	0.910	0.0583 ***(3)***	45	**1.363**	1.237–1.502
	Emus	0.963	0.0273 ***(4)***	32	**0.828**	0.770–0.891
	All birds	0.906	0.0726 ***(2)***	88	0.953	0.889–1.021
	Birds, one per species	0.819	0.0868 ***(1)***	39	0.993	0.848–1.163
Independent variable = digit II length						
Digit IV length	*Eubrontes*	0.955	0.0386 ***(3)***	36	1.022	0.947–1.102
	Emus[b]	0.971	0.0298 ***(4)***	31	1.064	0.997–1.137
	All birds[b]	0.929	0.0640 ***(2)***	88	0.990	0.933–1.050
	Birds, one per species[b]	0.898	0.0659 ***(1)***	39	0.957	0.855–1.071
Independent variable = digit II length excluding claw						
Digit II claw length	*Eubrontes*	0.617	0.0918 ***(1)***	45	0.890	0.695–1.140
	Emus	0.831	0.0775 ***(3)***	27	1.184	0.981–1.429
	Alligators	0.889	0.0799 ***(2)***	89	0.939	0.871–1.012
Independent variable = digit III length excluding claw						
Digit III claw length	*Eubrontes*	0.456	0.0915 ***(2)***	43	0.962	0.672–1.378
	Emus	0.800	0.0952 ***(1)***	34	**1.367**	1.139–1.640
	Alligators	0.912	0.0732 ***(3)***	88	0.981	0.918–1.049
Independent variable = digit IV length excluding claw						
Digit IV claw length	*Eubrontes*	0.468	0.1204 ***(1)***	26	0.977	0.602–1.585
	Emus[b]	0.743	0.1167 ***(2)***	29	**1.376**	1.086–1.744
Independent variable = digit III length excluding claw						
Digit II length excluding claw4	*Eubrontes*	0.934	0.0411 ***(1)***	41	**1.225**	1.124–1.336
	Emus	0.958	0.0325 ***(3)***	26	0.968	0.886–1.057
	Alligators	0.980	0.0356 ***(2)***	90	1.015	0.985–1.046
Digit IV length excluding claw	*Eubrontes*	0.960	0.0344 ***(1)***	23	**1.285**	1.171–1.411
	Emus[b]	0.959	0.0337 ***(2)***	29	1.040	0.958–1.129
Independent variable = digit II length excluding claw						
Digit IV length excluding claw	*Eubrontes*	0.922	0.0463 ***(1)***	27	1.038	0.920–1.171
	Emus[b]	0.952	0.0375 ***(2)***	27	1.074	0.979–1.178
Independent variable = digit III claw length						
Digit II claw length	*Eubrontes*	0.440	0.1099 ***(1)***	40	1.160	0.786–1.711
	Emus	0.817	0.0776 ***(2)***	26	0.837	0.684–1.025
	Alligators	0.972	0.0407 ***(3)***	85	0.971	0.936–1.008
Digit IV claw length	*Eubrontes*	0.467	0.1046 ***(2)***	34	1.115	0.743–1.672
	Emus	0.691	0.1289 ***(1)***	28	1.084	0.822–1.429
Independent variable = digit II claw length						
Digit IV claw length	*Eubrontes*	0.496	0.1136 ***(1)***	34	1.114	0.761–1.629
	Emus	0.768	0.1100 ***(2)***	27	1.210	0.961–1.523

Table A8.19. *continued*

Dependent variable	Group	r^2	SEE of regression *(rank)*	Number of cases	Slope	95% CI of slope
Independent variable = digit III length excluding claw						
Clawbase II to clawbase IV	*Eubrontes*	0.897	0.0625 ***(1)***	30	**1.419**	1.243–1.619
	Emus	0.979	0.0211 **(2)**	25	**0.889**	0.783–0.947
Independent variable = digit III first pad length						
Digit II first pad length	*Eubrontes*	0.880	0.0620 ***(1)***	44	1.060	0.945–1.190
	Emus	0.963	0.0326 **(2)**	20	0.908	0.824–1.001
Digit II second pad length	*Eubrontes*	0.893	0.0627 ***(1)***	40	1.083	0.966–1.214
	Emus	0.890	0.0521 **(2)**	20	0.845	0.709–1.007
Digit II claw length	*Eubrontes*	0.597	0.0955 ***(1)***	36	0.918	0.684–1.232
	Emus	0.864	0.0696 **(2)**	27	1.039	0.881–1.225
Digit III second pad length	*Eubrontes*	0.884	0.0448 **(2)**	42	**0.798**	0.711–0.897
	Emus	0.893	0.0500 ***(1)***	31	0.895	0.784–1.021
Digit III third pad length	*Eubrontes*	0.852	0.0650 **(2)**	36	0.958	0.828–1.109
	Emus	0.831	0.0694 ***(1)***	30	0.981	0.822–1.170
Digit III claw length	*Eubrontes*	0.448	0.0981 ***(1)***	37	0.881	0.590–1.316
	Emus	0.792	0.0971 **(2)**	34	**1.254**[a]	1.041–1.512
Digit IV length excluding claw	*Eubrontes*	0.885	0.0565 ***(1)***	23	1.127	0.955–1.329
	Emus[b]	0.968	0.0295 **(2)**	30	0.935	0.871–1.003
Digit IV claw length	*Eubrontes*	0.439	0.1170 ***(1)***	30	1.105	0.691–1.767
	Emus	0.793	0.1047 **(2)**	29	**1.287**[a]	1.049–1.579
Proximal end digit II to proximal end digit III	*Eubrontes*	0.676	0.0930 ***(1)***	45	1.001	0.806–1.243
	Emus	0.884	0.0477 **(2)**	34	**0.808**	0.709–0.922
Proximal end digit III to proximal end digit IV	*Eubrontes*	0.777	0.0755 ***(1)***	35	0.904	0.746–1.095
	Emus[b]	0.929	0.0357 **(2)**	34	**0.780**	0.706–0.862
Proximal end digit II to proximal end digit IV	*Eubrontes*	0.805	0.0765 ***(1)***	34	0.970	0.811–1.160
	Emus[b]	0.916	0.0396 **(2)**	33	**0.786**	0.703–0.878
Clawbase II to clawbase III	*Eubrontes*	0.879	0.0491 ***(1)***	34	0.896	0.783–1.025
	Emus	0.960	0.0320 **(2)**	26	**0.904**[a]	0.829–0.985
Clawbase III to clawbase IV	*Eubrontes*	0.795	0.0625 ***(1)***	28	0.935	0.760–1.151
	Emus	0.974	0.0226 **(2)**	27	**0.791**	0.740–0.847
Clawbase II to clawbase IV	*Eubrontes*	0.839	0.0769 ***(1)***	29	**1.252**	1.051–1.491
	Emus	0.954	0.0315 **(2)**	26	**0.795**	0.725–0.873
Toetip II to toetip III	*Eubrontes*	0.890	0.0488 ***(1)***	41	**0.845**	0.754–0.947
	Emus	0.939	0.0384 **(2)**	31	**0.879**[a]	0.798–0.969
Toetip III to toetip IV	*Eubrontes*	0.875	0.0540 ***(1)***	41	**0.877**[a]	0.776–0.992
	Emus	0.931	0.0366 **(2)**	35	**0.814**	0.739–0.897
Toetip II to toetip IV	*Eubrontes*	0.908	0.0603 ***(1)***	41	**1.141**[a]	1.029–1.265
	Emus	0.946	0.0327 **(2)**	32	**0.810**	0.740–0.885

[a]Only barely allometric (upper CI limit of RMA slope between 0.95 and 1.00, or lower CI limit between 1.00 and 1.05).
[b]Because of differences in the anatomical configuration of the "heel" between non-avian dinosaurs and birds, this comparison is only analogous between bird and non-avian dinosaur footprints.

Table A8.20. Comparison of intraspecific and interspecific variability in bivariate comparisons of linear dimensions of archosaur feet and footprints. Known intraspecific comparisons are for emu footprints (emu means treatment) and alligator feet. For alligators, digit lengths = free lengths. Alligator cases are excluded from the analysis when one or both of the parameters involves claw lengths, and also when one or both of the claws were heavily worn. Known interspecific comparisons are for all ground bird species combined ("all birds": bird means treatment, with variable numbers of individual birds represented per species [and so dominated by emu cases]), and all ground bird species (bird means treatment, but with only one individual represented per species ["one per species"]). Data for ostrich were included in the "all birds" treatment, except where the comparison of interest could not be made (as with comparisons involving digit II), but ostrich data were excluded in the "one per species" ground bird treatment. The intraspecific and interspecific shape comparisons are compared with the same or analogous comparisons of dinosaur footprints assigned to the ichnogenus *Eubrontes* ("mean or single" treatment), for which the nature of the sample (intraspecific or interspecific) is unknown. All measurements are in millimeters. Variability is expressed in terms of the coefficient of relative dispersion around the reduced major axis (D_d); values outside parentheses were calculated using raw data, and values inside parentheses were calculated using log-transformed data. For each bivariate comparison between or among groups, the groups are ranked in order of decreasing variability (1 = the group with the greatest variability), allowing for ties (ties include instances where rounding causes the rank order of log-transformed variables to be slightly different from that of raw data).

Independent variable	Dependent variable	Group	D_d *(rank)*	Number of cases
Footprint length (heel to toetip III)	Toetip II–IV	*Eubrontes*	10.8 (2.35) ***(3)***	40
		Emus	6.44 (1.22) ***(4)***	31
		All birds	20.8 (3.76) ***(2)***	91
		Birds, one per species	28.0 (5.00) ***(1)***	37
	Distance from digit II bend point to digit III long axis	*Eubrontes*	11.3 (3.60)	42
	Distance from digit II bend point to "heel"	*Eubrontes*	8.46 (2.13)	43
Backfoot length	Digit III projection	*Eubrontes*	14.6 (3.04) ***(3)***	39
		Emus	12.4 (2.95) ***(4)***	31
		All birds	35.2 (9.52) ***(2)***	92
		Birds, one per species	61.5 (14.9) ***(1)***	36
Digit III length	Digit II length	*Eubrontes*	7.16 (1.59) **(2)**	47
		Emus	6.42 (1.52) **(3)**	32
		Alligators	6.03 (1.48) **(4)**	85
		All birds	23.5 (4.61) **(1)**	88
		Birds, one per species	24.6 (4.59) **(1)**	39
	Digit IV length	*Eubrontes*	9.12 (1.65) **(3)**	35
		Emus[a]	7.10 (1.56) **(4)**	34
		All birds[a]	19.3 (3.36) **(1)**	91
		Birds, one per species[a]	12.1 (3.30) **(2)**	39
	Proximal end (toebase) II to proximal end (toebase) III	*Eubrontes*	19.9 (4.21) **(3)**	48
		Emus	13.3 (2.82) **(4)**	34
		All birds	17.8 (4.80) **(2)**	89
		Birds, one per species	23.6 (5.80) **(1)**	38
	Proximal end III to proximal end IV	*Eubrontes*	16.8 (3.25) **(3)**	38
		Emus[a]	10.9 (2.16) **(4)**	34
		All birds[a]	20.6 (5.37) **(1)**	90
		Birds, one per species[a]	19.3 (5.35) **(2)**	38
	Proximal end II to proximal end IV	*Eubrontes*	13.3 (3.50) **(3)**	38
		Emus[a]	11.7 (2.29) **(4)**	33
		All birds[a]	19.7 (5.02) **(2)**	88
		Birds, one per species[a]	20.8 (5.15) **(1)**	39
	Clawbase II to clawbase III	*Eubrontes*	9.59 (2.21) **(1)**	38
		Emus	6.37 (1.49) **(2)**	26
	Clawbase III to clawbase IV	*Eubrontes*	8.81 (2.10) **(1)**	31
		Emus	6.40 (1.42) **(2)**	27
	Clawbase II to clawbase IV	*Eubrontes*	10.9 (2.53) **(1)**	33
		Emus	7.58 (1.59) **(2)**	26
	Toetip II to toetip III	*Eubrontes*	11.7 (2.57) **(3)**	48
		Emus	6.84 (1.57) **(4)**	31
		All birds	19.1 (3.71) **(2)**	87
		Birds, one per species	26.2 (4.91) **(1)**	39
	Toetip III to toetip IV	*Eubrontes*	10.7 (2.27) **(3)**	47
		Emus	6.64 (1.41) **(4)**	35
		All birds	20.1 (4.61) **(2)**	92
		Birds, one per species	28.8 (6.46) **(1)**	38
	Toetip II to toetip IV	*Eubrontes*	10.7 (2.54) **(3)**	45
		Emus	7.58 (1.49) **(4)**	32
		All birds	21.6 (3.89) **(2)**	88
		Birds, one per species	25.7 (4.90) **(1)**	39

Table A8.20. *continued*

Independent variable	Dependent variable	Group	D_d *(rank)*	Number of cases
Digit II length	Digit IV length	*Eubrontes*	8.47 (1.92) **(3)**	36
		Emus[a]	6.75 (1.70) **(4)**	31
		All birds[a]	21.3 (3.93) **(2)**	88
		Birds, one per species[a]	21.4 (4.36) **(1)**	39
Digit II length excluding claw	Digit II claw length	*Eubrontes*	24.8 (6.60) **(1)**	45
		Emus	15.6 (5.55) **(3)**	27
		Alligators	19.8 (5.66) **(2)**	89
Digit III length excluding claw	Digit III claw length	*Eubrontes*	24.1 (6.10) **(1)**	43
		Emus	15.5 (5.47) **(3)**	34
		Alligators	18.0 (4.94) **(2)**	88
Digit IV length excluding claw	Digit IV claw length	*Eubrontes*	31.3 (7.95) **(1)**	26
		Emus[a]	19.5 (7.58) **(2)**	29
Digit III length excluding claw	Digit II length excluding claw	*Eubrontes*	6.82 (2.00) **(2)**	41
		Emus	7.32 (1.93) **(2)**	26
		Alligators	8.03 (2.03) **(1)**	90
	Digit IV length excluding claw	*Eubrontes*	7.58 (1.54) **(2)**	23
		Emus[a]	8.33 (1.87) **(1)**	29
Digit II length excluding claw	Digit IV length excluding claw	*Eubrontes*	10.8 (2.40) **(1)**	27
		Emus[a]	8.25 (2.28) **(2)**	27
Digit III claw length	Digit II claw length	*Eubrontes*	23.4 (8.83) **(1)**	40
		Emus	19.6 (8.06) **(2)**	28
		Alligators	9.02 (3.52) **(3)**	85
	Digit IV claw length	*Eubrontes*	22.6 (8.60) **(2)**	34
		Emus	27.8 (12.0) **(1)**	28
Digit II claw length	Digit IV claw length	*Eubrontes*	26.3 (9.28) **(1)**	34
		Emus	23.6 (9.80) **(1)**	27
Digit III length excluding claw	Clawbase width	*Eubrontes*	12.7 (2.83) **(1)**	30
		Emus	5.37 (2.31) **(2)**	25
Digit III first pad length	Digit II first pad length	*Eubrontes*	14.5 (4.01) **(1)**	44
		Emus	7.26 (2.65) **(2)**	20
	Digit II second pad length	*Eubrontes*	14.8 (4.01) **(2)**	40
		Emus	15.2 (4.41) **(1)**	20
	Digit II claw length	*Eubrontes*	24.9 (7.53) **(1)**	36
		Emus	15.5 (5.44) **(2)**	27
	Digit III second pad length	*Eubrontes*	15.4 (3.38) **(1)**	42
		Emus	12.2 (3.70) **(1)**	31
	Digit III third pad length	*Eubrontes*	14.0 (4.49) **(2)**	36
		Emus	14.8 (5.17) **(1)**	30
	Digit III claw length	*Eubrontes*	28.5 (8.15) **(1)**	37
		Emus	17.9 (6.84) **(2)**	34
	Digit IV length excluding claw	*Eubrontes*	10.9 (3.03) **(1)**	23
		Emus[a]	8.46 (1.95) **(2)**	30
	Digit IV claw length	*Eubrontes*	25.8 (8.88) **(1)**	30
		Emus	19.3 (7.57) **(2)**	29

[a]Because of differences in the anatomical configuration of the "heel" between non-avian dinosaurs and birds, this comparison is only analogous between bird and non-avian dinosaur footprints.

Table A8.21. Summary of comparisons of variability of footprint or foot bivariate relationships in *Eubrontes* with known intraspecific samples of footprints (emus) and feet (alligators). Detailed results reported in tables A8.19 and A8.20. All = alligators; E = *Eubrontes*; Em = emus. D_d = coefficient of relative dispersion around the reduced major axis; SEE = standard error of estimate of regression equation.

Comparison	**Measure of variability**	***Eubrontes* and emus**		***Eubrontes* and alligators**	
		E > Em	**E < Em**	**E > All**	**E < All**
Toetip II–toetip IV vs. footprint length	SEE of regression	x			
	D_d	x			
Digit III projection vs. backfoot length	SEE of regression		x		
	D_d	x			
Digit II length vs. digit III length	SEE of regression	x		x	
	D_d	x		x	
Digit IV length vs. digit III length	SEE of regression	x			
	D_d	x			
Toebase (proximal end) II–toebase (proximal end) III vs. digit III length	SEE of regression	x			
	D_d	x			
Toebase III–toebase IV vs. digit III length	SEE of regression	x			
	D_d	x			
Toebase II–toebase IV vs. digit III length	SEE of regression	x			
	D_d	x			
Clawbase II–clawbase III vs. digit III length	SEE of regression	x			
	D_d	x			
Clawbase III–clawbase IV vs. digit III length	SEE of regression	x			
	D_d	x			
Clawbase II–clawbase IV vs. digit III length	SEE of regression	x			
	D_d	x			
Toetip II–toetip III vs. digit III length	SEE of regression	x			
	D_d	x			
Toetip III–toetip IV vs. digit III length	SEE of regression	x			
	D_d	x			
Toetip II–toetip IV vs. digit III length	SEE of regression	x			
	D_d	x			
Digit II length vs. digit IV length	SEE of regression	x			
	D_d	x			
Digit II claw length vs. digit II length excluding claw	SEE of regression	x		x	
	D_d	x		x	
Digit III claw length vs. digit III length excluding claw	SEE of regression		x	x	
	D_d	x		x	
Digit IV claw length vs. digit IV length excluding claw	SEE of regression	x			
	D_d	x			
Digit II length excluding claw vs. digit III length excluding claw	SEE of regression	x		x	
	D_d	Tie			x
Digit IV length excluding claw vs. digit III length excluding claw	SEE of regression	x			
	D_d		x		
Digit IV length excluding claw vs. digit II length excluding claw	SEE of regression	x			
	D_d	x			
Digit II claw length vs. digit III claw length	SEE of regression	x		x	
	D_d	x		x	
Digit IV claw length vs. digit III claw length	SEE of regression		x		
	D_d		x		
Digit IV claw length vs. digit II claw length	SEE of regression	x			
	D_d	Tie			
Clawbase II–clawbase IV vs. digit III length excluding claw	SEE of regression	x			
	D_d	x			
Digit II first pad length vs. digit III first pad length	SEE of regression	x			
	D_d	x			
Digit II second pad length vs. digit III first pad length	SEE of regression	x			
	D_d		x		

Table A8.21. *continued*

Comparison	Measure of variability	*Eubrontes* and emus		*Eubrontes* and alligators	
		E > Em	E < Em	E > All	E < All
Digit II claw length vs. digit III first pad length	SEE of regression	x			
	D_d	x			
Digit III second pad length vs. digit III first pad length	SEE of regression		x		
	D_d	Tie			
Digit III third pad length vs. digit III first pad length	SEE of regression		x		
	D_d		x		
Digit III claw length vs. digit III first pad length	SEE of regression	x			
	D_d	x			
Digit IV length excluding claw vs. digit III first pad length	SEE of regression	x			
	D_d	x			
Digit IV claw length vs. digit III first pad length	SEE of regression	x			
	D_d	x			
Toebase II–toebase III vs. digit III first pad length	SEE of regression	x			
Toebase III–toebase IV vs. digit III first pad length	SEE of regression	x			
Toebase II–toebase IV vs. digit III first pad length	SEE of regression	x			
Clawbase II–clawbase III vs. digit III first pad length	SEE of regression	x			
Clawbase III–clawbase IV vs. digit III first pad length	SEE of regression	x			
Clawbase II–clawbase IV vs. digit III first pad length	SEE of regression	x			
Toetip II–toetip III vs. digit III first pad length	SEE of regression	x			
Toetip III–toetip IV vs. digit III first pad length	SEE of regression	x			
Toetip II–toetip IV vs. digit III first pad length	SEE of regression	x			

Table A8.22. Multivariate relationships between parameters of footprint size in *Eubrontes*. Data were log transformed prior to analysis. Each log-transformed parameter was examined as a dependent variable in a RMA analysis with respect to a footprint size factor, the latter calculated as the mean of all log-transformed parameters in the analysis ("simple" or "complex" version) except the one being examined. CI = confidence interval; RMA = reduced major axis; SEE = standard error of the estimate of the regression version of the relationship. Slopes significantly different than 1 are indicated in **bold**.

Dependent variable	r^2	SEE	RMA slope	95% CI of slope
"Simple" version of the analysis: number of cases = 34				
Digit II length	0.954	0.0373	1.079	0.997–1.169
Backfoot length	0.976	0.0268	**1.070**[a]	1.011–1.132
Digit III projection	0.881	0.0423	**0.737**	0.645–0.841
Digit IV length	0.972	0.0301	**1.117**	1.051–1.187
Proximal end digit II to proximal end digit III	0.883	0.0630	**1.154**[a]	1.011–1.316
Proximal end digit III to proximal end digit IV	0.930	0.0430	1.005	0.910–1.109
Proximal end digit II to proximal end digit IV	0.844	0.0697	1.096	0.938–1.282
Toetip II–III	0.915	0.0426	**0.894**[a]	0.801–0.998
Toetip III–IV	0.923	0.0426	0.946	0.852–1.050
Toetip II–IV	0.952	0.0429	**1.242**	1.145–1.347
"Complex" version of the analysis: number of cases = 17				
Digit II first pad length	0.914	0.0554	1.127	0.951–1.337
Digit II second pad length	0.925	0.0528	1.161	0.991–1.360
Digit III first pad length	0.964	0.0317	0.997	0.896–1.110
Digit III second pad length	0.937	0.0362	**0.844**[a]	0.731–0.974
Digit III third pad length	0.938	0.0401	0.955	0.828–1.101
Digit IV length excluding claw	0.968	0.0339	**1.135**[a]	1.027–1.255
Proximal end digit II to proximal end digit III	0.924	0.0510	1.107	0.944–1.297
Proximal end digit III to proximal end digit IV	0.918	0.0476	0.984	0.834–1.162
Proximal end digit II to proximal end digit IV	0.809	0.0713	0.958	0.729–1.261
Clawbase II to clawbase III	0.944	0.0353	0.877	0.767–1.004
Clawbase III to clawbase IV	0.950	0.0344	0.904	0.796–1.026
Clawbase II to clawbase IV	0.954	0.0465	**1.319**	1.168–1.489
"Heel" to clawbase II	0.988	0.0192	1.033	0.972–1.097

[a]Only barely allometric (upper CI limit of RMA slope between 0.95 and 1.00, or lower CI limit between 1.00 and 1.05).

Table A8.23. Comparison of comparable or analogous multivariate relationships between parameters of footprint size in emus (emu means treatment) and *Eubrontes* (mean or single treatment). The parameters examined are those used in the principal component analysis (table 8.2). Data (all measurements in millimeters) were log transformed prior to analysis. Each log-transformed parameter was examined as a dependent variable in a RMA analysis with respect to a footprint size factor, the latter calculated as the mean of all log-transformed parameters in the analysis except the one being examined. Two versions of the *Eubrontes* analyses were run, one of them excluding the widths of the digital pads of digits II and III (no pw), to make the analysis more comparable to that of emu footprints, and one including the digital pad widths (pw). CI = confidence interval; RMA = reduced major axis; SEE = standard error of the estimate of the regression version of the relationship. Slopes significantly different than 1 indicated in **bold**. Number of data cases: emus: $N = 20$; *Eubrontes*: $N = 21$ (both versions).

Dependent variable	Group	r^2	SEE	RMA slope	95% CI of slope
Digit II first pad length	Emus	0.926	0.0457	1.016	0.882–1.170
	Eubrontes no pw	0.925	0.0551	**1.255**	1.093–1.440
	Eubrontes pw	0.940	0.0490	**1.169**[a]	1.035–1.321
Digit II first pad width	*Eubrontes* pw	0.968	0.0379	**1.239**	1.136–1.353
Digit II second pad length	Emus	0.905	0.0484	0.940	0.799–1.105
	Eubrontes no pw	0.907	0.0566	1.150	0.985–1.343
	Eubrontes pw	0.921	0.0520	1.074	0.932–1.237
Digit II second pad width	*Eubrontes* pw	0.970	0.0340	**1.150**	1.057–1.251
Digit II claw length	Emus	0.907	0.0553	1.100	0.938–1.291
	Eubrontes no pw	0.663	0.0831	0.864	0.605–1.234
	Eubrontes pw	0.624	0.0877	0.813	0.549–1.202
Digit III first pad length	Emus	0.977	0.0279	**1.130**[a]	1.047–1.219
	Eubrontes no pw	0.960	0.0342	1.055	0.957–1.165
	Eubrontes pw	0.962	0.0333	0.987	0.897–1.086
Digit III first pad width	*Eubrontes* pw	0.968	0.0383	**1.247**	1.142–1.361
Digit III second pad length	Emus	0.934	0.0430	1.008	0.883–1.150
	Eubrontes no pw	0.919	0.0395	**0.845**[a]	0.732–0.976
	Eubrontes pw	0.935	0.0355	**0.794**	0.699–0.902
Digit III second pad width	*Eubrontes* pw	0.903	0.0626	**1.173**[a]	1.001–1.374
Digit III third pad length	Emus	0.856	0.0653	1.035	0.842–1.272
	Eubrontes no pw	0.919	0.0466	1.003	0.869–1.158
	Eubrontes pw	0.926	0.0443	0.939	0.819–1.077
Digit III third pad width	*Eubrontes* pw	0.888	0.0753	**1.317**	1.109–1.565
Digit III claw length	Emus	0.853	0.0827	**1.324**	1.075–1.631
	Eubrontes no pw	0.682	0.0803	0.859	0.611–1.207
	Eubrontes pw	0.680	0.0805	0.808	0.574–1.138
Digit IV length[b]	Emus	0.976	0.0296	**1.161**	1.074–1.255
	Eubrontes no pw	0.969	0.0316	**1.110**[a]	1.019–1.210
	Eubrontes pw	0.963	0.0344	1.036	0.943–1.139
Proximal end (toebase) of digit II to proximal end (toebase) of digit III	Emus	0.970	0.0277	0.966	0.885–1.054
	Eubrontes no pw	0.924	0.0544	**1.237**	1.077–1.421
	Eubrontes pw	0.931	0.0519	**1.153**[a]	1.011–1.315
Proximal end of digit III to proximal end of digit IV[b]	Emus	0.978	0.0224	**0.903**[a]	0.839–0.973
	Eubrontes no pw	0.925	0.0473	1.068	0.931–1.225
	Eubrontes pw	0.910	0.0519	0.998	0.857–1.162
Proximal end of digit II to proximal end of digit IV[b]	Emus	0.988	0.0170	**0.945**[a]	0.895–0.999
	Eubrontes no pw	0.841	0.0662	1.017	0.823–1.257
	Eubrontes pw	0.826	0.0692	0.952	0.761–1.192
Toetip II to toetip III	Emus	0.965	0.0313	1.016	0.924–1.117
	Eubrontes no pw	0.939	0.0390	0.965	0.853–1.091
	Eubrontes pw	0.934	0.0403	0.904	0.795–1.028
Toetip III to toetip IV	Emus	0.972	0.0251	**0.901**[a]	0.829–0.981
	Eubrontes no pw	0.947	0.0374	0.997	0.889–1.117
	Eubrontes pw	0.946	0.0376	0.933	0.831–1.047
Toetip II to toetip IV	Emus	0.985	0.0189	**0.939**[a]	0.883–0.998
	Eubrontes no pw	0.961	0.0401	**1.278**	1.160–1.408
	Eubrontes pw	0.958	0.0416	**1.188**	1.074–1.314

[a]Only barely allometric (upper CI limit of RMA slope between 0.95 and 1.00, or lower CI limit between 1.00 and 1.05).

[b]Because the proximal end of digit IV is not defined the same way in emu as in non-avian dinosaur footprints, this comparison is only analogous between them.

Table A8.24. Analysis of variance of hindfoot size parameters in three size classes of *Alligator mississippiensis*. Parameters were "GM scaled" by subtracting the mean of all the log-transformed parameters from each log-transferred parameter. Alligator size classes (SC) were subdivided on the basis of animal total length: SC1 = 375–1,541 mm (*N* = 25); SC2 = 1,542–2,707 mm (*N* = 31); SC3 = 2,708–3,873 mm (*N* = 23). In all tests, the *P* value for Levene's test of equality of error variances >.05.

Parameter	*P* of *F*-test	Result of comparison (Bonferroni test)
Digit I first pad length	.634	No significant differences between groups
Digit I second pad length	.028	No significant difference between SC1 and SC2, or between SC2 and SC3; SC3 significantly different than SC1 (SC3 > SC1): pad length becomes relatively longer in bigger alligators
Digit I claw length	.010	No significant difference between SC1 and SC2; SC3 significantly different than SC1 (SC3 < SC1) and SC2 (SC3 < SC2); claw length becomes relatively shorter in bigger alligators
Digit II length excluding claw	.969	No significant differences between groups
Digit II claw length	<.001	No significant difference between SC1 and SC2; SC3 significantly different than SC1 (SC3 < SC1) and SC2 (SC3 < SC2); claw length becomes relatively shorter in bigger alligators
Digit III length excluding claw	.333	No significant differences between groups
Digit III claw length	.002	No significant difference between SC1 and SC2; SC3 significantly different than SC1 (SC3 < SC1) and SC2 (SC3 < SC2); claw length becomes relatively shorter in bigger alligators
Digit IV medial hypex length	<.001	No significant difference between SC1 and SC2; SC3 significantly different than SC1 (SC3 > SC1) and SC2 (SC3 > SC2); digit IV hypex length becomes relatively longer in bigger alligators

References

R

Note: The format used to order references here may differ somewhat from usual practices, but seems more reasonable to me (and this is, after all, my book). References are first arranged in alphabetical order in terms of the first letter of the first (or single) author's surname, as usual, beginning with the oldest publications and proceeding from there to subsequent publications. When there are two or more authors with the same surname, their publications are listed in alphabetical order of the first letter of the first author's given name.

When there is a series of papers by the same set of multiple authors (as is most notably the case with many of the Spanish publications), presented in the same order from first author to last, their publications are listed in order of date of publication. If there are multiple publications in the same year by the same set of authors with names in the same order, these are presented in alphabetical order of the first letter of the title of the publication.

When a particular author (e.g., M. L. Casanovas Cladellas, Lida Xing, and M. G. Lockley) has published many articles or books with many different coauthors, their publications are presented in alphabetical order of the first letter of the surname of the second author, regardless of whether or not there are any subsequent authors, or how many there are, after that second author. That is, I do not organize the references by first presenting publications by the first author and one other coauthor, then the same first author and two other coauthors, and so on. If the first two authors of a publication are the same, publications are then presented in alphabetical order of the surname of the third author, and so on.

Some author names require explanation. Some authors publish under different versions of their name; the most striking example of this is the incredibly prolific Maria Lourdes Casanovas Cladellas, whose papers variously appear under the names L. Casanovas, M. L. Casanovas, and M. L. Casanovas Cladellas. In the references I present the names following the format used in each individual publication, but I arrange the references in alphabetical order of the first letter of the first surname (e.g., Casanovas rather than Cladellas).

The names of East Asian authors may also be confusing. The standard format for such names is for the surname to precede the given name, unlike the Western format, in which the order is reversed (e.g., Lincoln Abraham vs. Abraham Lincoln). I present Asian names following the format employed in the cited publication. Thus publications by Lida (or Li-Da) Xing sometimes are cited as Xing L., sometimes as Xing, L., and sometimes as Xing, L.-D.

As in God's Final Message to His Creation (as related by the late and great Douglas Adams), we apologize for any inconvenience.

Abbassi, N., S. d'Orazi Porchetti, A. Wagensommer, and M. G. Dehnavi. 2015. Dinosaur and crocodylomorph footprints from the Hojedk Formation (Bajocian, Middle Jurassic) of north Kerman, central Iran. *Italian Journal of Geoscience* 134: 86–94.

Abbassi, N., and S. Madanipour. 2014. Dinosaur tracks from the Jurassic Shemshak Group in the Central Alborz Mountains (northern Iran). *Geologica Carpathica* 65: 99–115.

Abourachid, A., and S. Renous. 2000. Bipedal locomotion in ratites (Paleognatiform): examples of cursorial birds. *Ibis* 142: 538–549.

Adams, T. L., B. H. Breithaupt, and N. A. Matthews. 2014. Paleoecology and paleoenvironment of Middle Jurassic dinosaurs (Yellow Brick Road Dinosaur Tracksite, Bighorn Basin, northern Wyoming). In M. G. Lockley and S. G. Lucas (eds.), *Fossil Footprints of Western North America*, 225–233. Albuquerque, NM: New Mexico Museum of Natural History and Science Bulletin 62.

Agirrezabala, L. M., J. A. Torres, and L. I. Viera. 1985. El Weald de Igea (Cameros-La Rioja). Sedimentología, bioestratigrafía, y paleoicnoloía de grandes reptiles (Dinosaurios). *Munibe* 37: 111–138.

Agnolin, F. L. 2017. Unexpected diversity of ratites (Aves, Palaeognathae) in the early Cenozoic of South America: palaeobiogeographical implications. *Alcheringa* 41: 101–111.

Agurirre, J., E. Aramburu, E. Campos, M. Gallego, A. Garritz, L. Gómez Cubillo, R. Gómez Díez, A. Gonzáles Díaz, C. Gonzáles Ferrero, P. Gonzáles Martín, C. Herreros, M. Izco, L. Juániz, S. López Fernández, C. Marca, N. Marín, J. Monreal, M. Ochoa, E. Osés, A. Ramos, A. Rodrígez Ochagavía, F. Sánchez Blanco, C. Suescun, L. Vega, and I. Viteri. 2001. Las anomalías de una rastrillada terópoda (6BLT16). Afloramiento de la Torre 6, Igea (La Rioja, España). *Zubía* 19: 97–111.

Alcalá, L., A. Cobos, E. Espílez, F. Gascó, L. Mampel, C. Martín Escorza, and R. Royo-Torres. 2012. Icnitas de dinosaurios en la Formación Villar del Arzobispo de Ababuj (Teruel, España). *Geogaceta* 51: 35–38.

Alcalá, L., L. Mampel, R. Royo-Torres, and A. Cobos. 2014*a*. On small quadrupedal ornithopod tracks in Jurassic-Cretaceous transition intertidal deposits (El Castellar, Teruel, Spain). *Spanish Journal of Palaeontology* 29: 183–190.

Alcalá, L., F. Pérez-Lorente, L. Luque, A. Cobos, R. Royo-Torres, and L. Mampel. 2014*b*. Preservation of dinosaur footprints in shallow intertidal deposits of the

Jurassic-Cretaceous transition in the Iberian Range (Teruel, Spain). *Ichnos* 21: 19–31.
Alexander, R. McN. 1985. Mechanics of posture and gait of some large dinosaurs. *Zoological Journal of the Linnean Society* 83: 1–25.
Allain, R., P. Taquet, B. Battail, J. Dejax, P. Richir, M. Veran, P. Sayarath, B. Khenthavong, P. Thamvirith, and B. Hom. 1997. Pistes de dinosaures dans les niveaux du Crétacé inférieur de Muong Phalane, province de Savannakhet (Laos). *Comptes Rendus Académie des Sciences* 325: 815–821.
Allen, J. R. L. 1989. Fossil vertebrate tracks and indenter mechanics. *Journal of the Geological Society* 146: 600–602.
Allen, J. R. L. 1997. Subfossil mammalian tracks (Flandrian) in the Severn Estuary, S.W. Britain: the mechanics of formation, preservation and distribution. *Philosophical Transactions of the Royal Society B: Biological Sciences* 352: 481–518.
Allen, V., R. M. Elsey, N. Jones, J. Wright, and J. R. Hutchinson. 2010. Functional specialization and ontogenetic scaling of limb anatomy in *Alligator mississippiensis*. *Journal of Anatomy* 216: 423–445.
Allentoft, M. E., M. Bunce, R. P. Scofield, M. L. Hale, and R. N. Holdaway. 2010. Highly skewed sex ratios and biased fossil deposition of moa: ancient DNA provides new insight on New Zealand's extinct megafauna. *Quaternary Science Reviews* 29: 753–762.
Allentoft, M. E., and N. J. Rawlence. 2012. Moa's Ark or volant ghosts of Gondwana? Insights from nineteen years of ancient DNA research on the extinct moa (Aves: Dinornithiformes) of New Zealand. *Annals of Anatomy* 194: 36-51.
Alonso, R. N. 1989. Late Cretaceous dinosaur trackways in northern Argentina. In D. D. Gillette and M. G. Lockley (eds.), *Dinosaur Tracks and Traces*, 223–228. Cambridge: Cambridge University Press.
Alter, Robert. 2004. *The Five Books of Moses: A Translation with Commentary*. New York: W. W. Norton, 1064 pp.
Altrichter, M., and P. M. Sherman. 1999. Distribution and abundance of the American alligator (*Alligator mississippiensis*) in the Welder Wildlife Refuge, Texas. *Texas Journal of Science* 51: 139–146.
Alvarenga, H., L. Chiappe, and S. Bertelli. 2011. Phorusrhacids: the terror birds. In G. Dyke and G. Kaiser (eds.), *Living Dinosaurs: The Evolutionary History of Modern Birds*, 187–208. West Sussex, U.K.: Wiley-Blackwell.
Álvarez Bermejo, J. M., O. Argaíz Gil, I. Díaz-Martínez, I. Eguiluz Garcia, M. Herreros Martínez, J. Lucas López, C. Martínez García, J. Martínez Pérez, A. Pérez Llanos, F. Pérez-Lorente, and P. Vega García. 2013. El yacimiento paleoicnológico de Guilera redescubrimiento, situación, definición y anális de sus icnitas. *Zubía* 31: 199–210.
Álvarez, S., E. Aragón, S. Azofra, J. Barrau, M. P. Benzal, S. J. Berger, J. M. Camarón, I. Catalán, A. Coda, L. Colomé, E. Gallago, A. Garrido, M. Gonzalo, A. Hernáez, P. Del Hoyo, S. López, N. Lorente, F. J. Madariaga, M. E. Marín, D. Martínez, J. Menéndez, A. Moneo, J. Pérez, A. Rodríguez, V. Sáenz, M. E. Sáenz, M. Sáenz, I. Torróntegui, M. Yanguas, and C. Zorzano. 2000. El yacimiento de huellas de dinosaurios de Hornillos de Cameros (La Rioja, España). *Zubía* 18: 73–95.
Andors, A. 1992. Reappraisal of the Eocene groundbird *Diatryma* (Aves: Anserimorphae). *Natural History Museum of Los Angeles County, Science Series* 36: 109–125.
Angst, D., and E. Buffetaut. 2017. *Paleobiology of Giant Flightless Birds*. London: ISTE Press, Elsevier, 281 pp.
Ansorena, P., I. Díaz-Martínez, and F. Pérez-Lorente. 2007–2008. Mina Victoria (Navajún) y Valdeperillo (Cornago). Nuevos yacimientos de icnitas de dinosaurio en el Grupo de Urbión (Cuenca de Cameros. La Rioja. España). *Zubía* 25–26: 75–96.
Apesteguía, S., N. D. Smith, R. Juárez Valleri, and P. J. Makovicky. 2016. An unusual new theropod with a didactyl manus from the Upper Cretaceous of Patagonia, Argentina. *PloS One* 11(7): e0157793. doi:10.1371/journal.pone.0157793.
ASTM D422-63. 1972. Standard method for particle-science analysis of soils., West Conshohocken, PA: ASTM, pp. 116–126.
Avanzini, M. 2000. Tetrapod footprints from the Mesozoic carbonate platforms of the Italian Alps. *Zubía* 21: 175–186.
Avanzini, M., J. C. García-Ramos, J. Lires, L. Piñuela, and M. G. Lockley. 2007. Crocodylomorph tracks from the Late Jurassic of Asturias (Spain). *Ichnos* 14: 143–153.
Avanzini, M., and F. M. Petti. 2008. Updating the dinosaur tracksites from the Lower Jurassic Calcari Grigi Group (southern Alps, northern Italy). *Studi Trentini di Scienze Naturali Acta Geologica* 83: 289–301.
Avanzini, M., L. Piñuela, and J. García-Ramos. 2012. Late Jurassic footprints reveal walking kinematics of theropod dinosaurs. *Lethaia* 45: 238–252.
Avanzini, M., D. Piubelli, P. Mietto, G. Roghi, R. Romano, and D. Masetti. 2006. Lower Jurassic (Hettangian-Sinemurian) dinosaur track megasites, southern Alps, northern Italy. In J. D. Harris, S. G. Lucas, J. A. Spielmann, M. G. Lockley, A. R. C. Milner, and J. I. Kirkland (eds.), *The Triassic-Jurassic Terrestrial Transition*, 207–216. Albuquerque, NM: New Mexico Museum of Natural History and Science Bulletin 37.
Azuma, Y., R. Li, P. J. Currie, Z. Dong, M. Shibata, and J. Li. 2006. Dinosaur footprints from the Lower Cretaceous of Inner Mongolia, China. *Memoir of the Fukui Prefecture Dinosaur Museum* 5: 1–14.
Baele, J.-M., P. Godefroit, P. Spagna, and C. Dupuis. 2012. Geological model and cyclic mass mortality scenarios for the Lower Cretaceous Bernissart *Iguanodon* bonebeds. In P. Godefroit (ed.), *Bernissart Dinosaurs and Early Cretaceous Terrestrial Ecosystems*, 154–171. Bloomington: Indiana University Press.
Baird, D. 1957. Triassic footprint faunules from Milford, New Jersey. *Bulletin of the Museum of Comparative Zoology, Harvard University* 117(5): 449–520.
Baird, D. 1980. A prosauropod dinosaur trackway from the Navajo Sandstone (Lower Jurassic) of Arizona. In L. L. Jacobs (ed.), *Aspects of Vertebrate History: Essays in Honor of Edwin Harris Colbert*, 219–230. Flagstaff, AZ: Museum of Northern Arizona Press.
Baird, D. 1989. Medial Cretaceous carnivorous dinosaur and footprints from New Jersey. *The Mosasaur* 4: 53–63.
Baker, A. J., O. Haddrath, J. D. McPherson, and A. Cloutier. 2014. Genomic support for a moa-tinamou clade and adaptive morphological convergence in flightless ratites. *Molecular Biology and Evolution* 31: 1686–1696.
Baker, A. J., L. J. Huynen, O. Haddrath, C. D. Millar, and D. M. Lambert. 2005. Reconstructing the tempo and mode of evolution in an extinct clade of birds with ancient DNA: the giant moas of New Zealand. *Proceedings of the National Academy of Science USA* 102: 8257–8262.
Bakker, R. T. 2004. Dinosaurs acting like birds, and vice versa: an homage to the Reverend Edward Hitchcock, first director of the Massachusetts Geological Survery. In P. J. Currie, E. B. Koppelhus, M. A Shugar, and J. L. Wright (eds.), *Feathered Dragons: Studies on the Transition from Dinosaurs to Birds*, 1–11. Bloomington: Indiana University Press.
Bakker, R. T., M. Williams, and P. Currie. 1988. *Nanotyrannus*, a new genus of pygmy tyrannosaur, from the latest Cretaceous of Montana. *Hunteria* 1: 1–30.
Barco, J. L., L. I. Ardévol, and J. I. Canudo. 2001. Descripción de los primeros rastros asignados a Hadrosauridae (Ornithopoda, Dinosauria) del Maatrichtiense de la Península Ibérica (Areny, Huesca). *Geogaceta* 20: 235–238.
Barco, J. L., J. I. Canudo, and J. Ignacio Ruiz-Omenaca. 2006. New data on *Therangospodus oncalensis* from the Berriasian Fuentesalvo Tracksite (Villar del Río, Soria, Spain): an example of gregarious behaviour in theropod dinosaurs. *Ichnos* 13: 237–248.
Barco, J. L., J. I. Canudo, J. Ignacio Ruiz-Omenaca, and J. L. Rubio. 2005. Evidencia icnológica de un dinosaurio terópodo gigante en el Berriasiense (Cretácico Inferior) de Laurasia (Las Villasecas, Soria, España). *Revista Española de Paleontología* 10: 59–71.
Baron, M. G., D. B. Norman, and P. M. Barrett. 2017*a*. A new hypothesis of dinosaur relationships and early dinosaur evolution. *Nature* 543: 501–506.
Baron, M. G., D. B. Norman, and P. M. Barrett. 2017*a1*. Baron *et al.* reply. *Nature* 551: E4–E5.
Baron, M. G., D. B. Norman, and P. M. Barrett. 2017*b*. Postcranial anatomy of *Lesothosaurus diagnosticus* (Dinosauria: Ornithischia) from the Lower Jurassic of southern Africa: implications for basal ornithischian taxonomy and systematics. *Zoological Journal of the Linnean Society* 179: 125–168.
Barrett, P. M., and S. C. R. Maidment. 2017. The evolution of ornithischian quadrupedality. *Journal of Iberian Geology* 43: 363–377. doi:10.1007/s41513-017-0036-0.
Bates, K. T., P. L. Manning, L. Margetts, and W. I. Sellers. 2010. Sensitivity analysis in evolutionary robotic simulations of bipedal dinosaur running. *Journal of Vertebrate Paleontology* 30: 458–466.
Bates, K. T., R. B. J. Benson, and P. L. Falkingham. 2012. A computational analysis of locomotor anatomy and body mass evolution in Allosauroidea (Dinosauria: Theropoda). *Paleobiology* 38: 486–507.

Bates, K. T., and E. R. Schachner. 2012. Disparity and convergence in bipedal archosaur locomotion. *Journal of the Royal Society Interface* 9: 1339–1353.
Baumel, J. J., A. S. King, A. M. Lucas, J. E. Breazile, and H. E. Evans (eds.). 1979. *Nomina Anatomica Avium: An Annotated Dictionary of Birds*. London: Academic Press, 637 pp.
Becerra, M. G., D. Pol, O. W. M. Rauhut, and I. A. Cerda. 2016. New heterodontosaurid remains from the Cañadón Asfalto Formation: cursoriality and the functional importance of the pes in small heterodontosaurids. *Journal of Paleontology* 50: 555–577.
Belvedere, M. 2008. *Ichnological Researches on the Upper Jurassic Dinosaur Tracks in the Iouaridène Area (Demnat, Central High Atlas, Morocco)*. PhD Thesis. Padua, Italy: University of Padua, 121 pp.
Belvedere, M., and J. O. Farlow. 2016. A numerical scale for quantifying the quality of preservation of vertebrate tracks. In P. L. Falkingham, D. Marty, and A. Richter (eds.), *Dinosaur Tracks: The Next Steps*, 92–98. Bloomington: Indiana University Press.
Belvedere, M., N.-E. Jalil, A. Breda, G. Gattolin, H. Bourget, F. Khaldoune, and G. J. Dyke. 2013. Vertebrate footprints from the Kem Kem beds (Morocco): a novel ichnological approach to faunal reconstruction. *Palaeogeography, Palaeoclimatology, Palaeoecology* 383–384: 52–58.
Belvedere, M., P. Mietto, and S. Ishigaki. 2010. A Late Jurassic diverse ichnocoenosis from the siliciclastic Iouaridène Formation (Central High Atlas, Morocco). *Geological Quarterly* 54: 367–380.
Bengoechea, A., L. A. Izquierdo, J. M. Martinez, J. L. Molinero, D. Montero, T. Torcida, and V. Urien. 1993. Icnitas de dinosaurios en el sureste de la provincia de Burgos. *Boletín Geológico y Minero* 104(3): 19–34.
Bennett, S. C. 1992. Reinterpretation of problematical tracks at Clayton Lake State Park, New Mexico: not one pterosaur, but several crocodiles. *Ichnos* 2: 37–42.
Bessedik, M., C. Mammeri, L. Belkebir, M. Mahboubi, M. Adaci, H. Hebib, M. Bensalah, B. Mansour, and M. El Habib Mansouri. 2008. Nouvelles donnes sur les ichnites de dinosaures de la region d'El Bayadh (Crétacé Inférieur, Algéie). *Palaeovertebrata* 36: 7–35.
Bird, R. T. 1939. Thunder in his footsteps. *Natural History* 43: 254–261.
Bird, R. T. 1941. A dinosaur walks into the museum. *Natural History* 47: 74–81.
Bird, R. T. 1944. Did *Brontosaurus* ever walk on land? *Natural History* 53: 61–67.
Bird, R. T. 1954. We captured a "live" brontosaur. *National Geographic* 105: 707–722.
Bird, R. T. 1985. *Bones for Barnum Brown: Adventures of a Dinosaur Hunter*. Fort Worth: Texas Christian University Press, 225 pp.
Bishop, P. J. 2015. A critical re-evaluation of the hindlimb myology of moa (Aves: Dinornithiformes). *Memoirs of the Queensland Museum-Nature* 59: 187–246.
Blanco, M., S. Caro, A. López, F. Pérez-Lorente, E. Requeta, and M. Romero. 2000. El yacimiento de icnitas de dinosaurio del Cretácico Inferior de Las Losas (Enciso, La Rioja, España). *Zubía* 18: 97–138.
Bock, W. 1952. Triassic reptilian tracks and trends of locomotive evolution. *Journal of Paleontology* 26: 395–453.
Bonaparte, J. F. 1971. Los tetrapodos del sector superior de la Formacion Los Colorados, La Rioja, Argentina (Triásico Superior). *Opera Lilloana* 22: 1–183.
Bonaparte, J. F. 1991. Los vertebrados fosiles de la Formacion Rio Colorado, de la Ciudad de Neuquen y Cercanias, Cretácico Superior, Argentina. *Revista del Museo Argentino de Ciencias Naturales "Bernardino Rivadavia" e Instituto Nacional de Investigación de las Ciencias Naturales, Paleontología* 4: 15–123.
Bonnan, M. F., J. O. Farlow, and S. L. Masters. 2008. Using linear and geometric morphometrics to detect intraspecific variability and sexual dimorphism in femoral shape in *Alligator mississippiensis* and its implications for sexing fossil archosaurs. *Journal of Vertebrate Paleontology* 28: 422–431.
Bonnan, M. F., J. L. Sandrick, T. Nishiwaki, D. R. Wilhite, R. M. Elsey, and C. Vittore. 2010. Calcified cartilage shape in archosaur long bones reflects overlying joint shape in stress-bearing elements: implications for nonavian dinosaur locomotion. *The Anatomical Record* 293: 2044–2055.
Bonnan, M. F., and P. Senter. 2007. Were the basal sauropodomorph dinosaurs *Plateosaurus* and *Massospondylus* habitual quadrupeds? *Special Papers in Paleontology* 77: 139–155.
Bordy, E. M., M. Abrahams, and L. Sciscio. 2017. The Subeng vertebrate tracks: stratigraphy, sedimentology and a digital archive of a historic Upper Triassic palaeosurface (lower Elliot Formation), Leribe, Lesotho (southern Africa). *Bollettino della Società Paleontologica Italiana* 56: 181–198.
Bourdon, E., A. De Ricqlès, and J. Cubo. 2009. A new Transantarctic relationship: morphological evidence for a Rheidae-Dromaiidae-Casuariidae clade (Aves, Palaeognathae, Ratitae). *Zoological Journal of the Linnean Society* 156: 641–663.
Boutakiout, M., M. Hadri, J. Nouri, I. Díaz-Martínez, and F. Pérez-Lorente. 2009*a*. Rastrilladas de icnitas terópodas gigantes del Jurásico Superior (Sinclinal de Iouaridène, Marruecos). *Revista Española de Paleontología* 24: 31–46.
Boutakiout, M., L. Ladel, I. Díaz-Martínez, and F. Pérez-Lorente. 2009*b*. Prospecciones paleoicnológicas en el sinclinal de Iouaridène (Alto Atlas, Marruecos). 2: Parte oriental. *Geogaceta* 47: 33–36.
Boutakiout, M., M. Masrour, L. Ladel, I. Díaz-Martínez, and F. Pérez-Lorente. 2010. Nuevos yacimientos de icnitas domerienses en Ibaqalliwan (Aït Bou Guemez, Alto Atlas Central, Marruecos). *Geogaceta* 48: 91–94.
Boutakiout, M., M. Masrour, and F. Pérez-Lorente. 2014. Icnitas tireóforas y terópodos en Ansous (Pliensbachiense, Alto Atlas Central, Marruecos). *Geogaceta* 55: 75–78.
Boutakiout, M., J. Nouri, L. Ladel, I. Díaz-Martínez, and F. Pérez-Lorente. 2012. Contenido anómalo de icnitas semiplantígradas terópodas en el yacimiento de Oumzawrou (36IGR) del Atlas Marroquí. *Geogaceta* 52: 173–176.
Bouyoucos, G. J. 1936. Directions for making mechanical analysis of soils by the hydrometer method. *Soil Science* 42: 225–228.
Brand, L. R. 1996. Variations in salamander trackways resulting from substrate differences. *Journal of Paleontology* 70: 1004–1010.
Brandt, L. A. 1991*a*. Growth of juvenile alligators in Par Pond, Savannah River Site, South Carolina. *Copeia* 1881: 1123–1129.
Brandt, L. A. 1991*b*. Long term changes in a population of *Alligator mississippiensis* in South Carolina. *Journal of Herpetology* 25: 419–424.
Brassey, C. A., R. N. Holdaway, A. G. Packham, J. Anné, P. L. Manning, and W. I. Sellers. 2013. More than one way of being a moa: differences in leg bone robustness map divergent evolutionary trajectories in Dinornithidae and Emeidae (Dinornithiformes). *PloS One* 8(12): e82668. doi:10.1371/journal/pone.0082668.
Bravo-Cuevas, V. M., and R. Rodríguez-de la Rosa. 2014. A summary of the Mesozoic vertebrate tracks of Mexico. In H. Rivera-Sylva, K. Carpenter, and E. Frey (eds.), *Dinosaurs and Other Reptiles from the Mesozoic of Mexico*, 181–192. Bloomington: Indiana University Press.
Brazaitis, P., and M. E. Watanabe. 2011. Crocodilian behaviour: a window to dinosaur behaviour? *Historical Biology* 23: 73–90.
Brinkman, D. L., R. L. Cifelli, and N. J. Czaplewski. 1998. First occurrence of *Deinonychus antirrhopus* (Dinosauria: Theropoda) from the Antlers Formation (Lower Cretaceous: Aptian-Albian) of Oklahoma. *Oklahoma Geological Survey Bulletin* 146: 1–27.
Brisbin, I. L., Jr., K. F. Gaines, C. H. Jagoe, and P. A. Consolie. 1996. Population studies of American alligators (*Alligator mississippiensis*) inhabiting a reservoir: responses to long-term drawdown and subsequent refill. In *Crocodiles: Proceedings of the 13th Working Meeting of the Crocodile Specialist Group, IUCN*, 446–477 (editor not specified). Gland, Switzerland.
Brochu, C. A. 2003. Phylogenetic approaches toward crocodylian history. *Annual Review of Earth and Planetary Sciences* 31: 357–397.
Brochu, C. A., D. C. Parris, B. S. Grandstaff, R. K. Denton, Jr., and W. B. Gallagher. 2012. A new species of *Borealosuchus* (Crocodyliformes, Eusuchia) from the Late Cretaceous-Early Paleogene of New Jersey. *Journal of Vertebrate Paleontology* 32: 105–116.
Broderick, T. J. 1985. Zambezi Valley notes. *Annals of the Zimbabwe Geological Survey* 10: 48–63.
Broom, R. 1911. On the dinosaurs of the Stormberg, South Africa. *Annals of the South African Museum* 7: 291–308.
Brown, B. 1916. *Corythosaurus casuarius*: skeleton, musculature and epidermis. *Bulletin of the American Museum of Natural History* 35: 709–716.
Brown, J. W. 2005. Megapodiidae. Megapodes. Version 20 December 2005 (temporary); http://tolweb.org/Megapodiidae/26297/2005.12.20 in the Tree of Life Project.
Brownstein, C. D. 2017. Description of Arundel Clay ornithomimosaur material

and a reinterpretation of *Nedocolbertia junstinhofmanni* as an "ostrich dinosaur"; biogeographic implications. *PeerJ* 5: e3110. doi:10.7717/peerj.3110.
Brusatte, S. L., T. D. Carr, T. E. Williamson, T. R. Holtz, Jr., D. W. E. Hone, and S. A. Williams. 2016. Dentary groove morphology does not distinguish *'Nanotyrannus'* as a valid taxon of tyrannosauroid dinosaur. Comments on "Distribution of the dentary groove of theropod dinosaurs: implications for theropod phylogeny and the validity of the genus *Nanotyrannus* Bakker et al., 1988." *Cretaceous Research* 65: 232–237.
Bryant, H. N., and A. P. Russell. 1992. The role of phylogenetic analysis in the inference of unpreserved attributes of extinct taxa. *Philosophical Transactions of the Royal Society of London* B 337: 405–418.
Buckley, L. G. 2015. *Bare-Bones Paleontology: An Examination of the SystematicMethods Used in Vertebrate Paleontology and Their Congruence with Avian Ichnotaxonomy.* PhD Thesis. Edmonton: University of Alberta, 519 pp.
Buckley, L. G., R. T. McCrea, and M. G. Lockley. 2016. Analyzing and resolving Cretaceous avian ichnotaxonomy using multivariate statistical analyses: approaches and results. In P. L. Falkingham, D. Marty, and A. Richter (eds.), *Dinosaur Tracks: The Next Steps*, 258–308. Bloomington: Indiana University Press.
Buffetaut, E. 2004. Footprints of giant birds from the Upper Eocene of the Paris Basin: an ichnological enigma. *Ichnos* 11: 357–362.
Buffetaut, E., R. Ingavat, N. Sattayarak, and V. Suteetorn. 1985. Les premières empreintes de pas de Dinosaures du Sud-Est asiatique: pistes de Carnosaures du Crétacé inférieur de Thaïlande. *Comptes Rendus de l'Académie des Sciences* Serie 2 301: 643–648.
Bulin, V. P., M. R. Dzhalilov, V. I. Dronov, A. B. Makhkamov, G. K. Melnikova, and G. K. Salibayev. 1981. (Finds of tracks of dinosaurs in Lower Cretaceous sediments of the Babatag Range (Tadzhik Depression). *Izvestiya Akademiy Nauk Tadzikskoy SSR, Otdeleniye Fiziko-Matematicheskikh, Khimicheskikh, Geologicheskikh Nauk* 2: 105–108 (in Russian; translation by G. W. Storrs).
Bunce, M., T. H. Worthy, T. Ford, W. Hoppitt, E. Willerslev, A. Drummond, and A. Cooper. 2003. Extreme reversed sexual size dimorphism in the extinct New Zealand moa *Dinornis*. *Nature* 425: 172–175.
Bunce, M., T. H. Worthy, M. J. Phillips, R. N. Holdaway, E. Willerslev, J. Haile, B. Shapiro, R. P. Scofield, A. Drummond, P. J. J. Kamp, and A. Cooper. 2009. The evolution of the extinct ratite moa and New Zealand Neogene paleogeography. *Proceedings of the National Academy of Sciences USA* 106: 20646–20651.
Bybee, P. J., A. H. Lee, and E.-T. Lamm. 2006. Sizing the Jurassic theropod *Allosaurus*: assessing growth strategy and evolution of ontogenetic scaling of limbs. *Journal of Morphology* 267: 347–359.
Cabreira, S. F., A. W. A. Kellner, S. Dias-da-Silva, L. R. da Silva, M. Bronzati, J. C. de Almeida Marsola, R. T. Müller, J. de Souza Bittencourt, B. J. Batista, T. Raugust, R. Carrilho, A. Brodt, and M. C. Langer. 2016. A unique Late Triassic dinosauromorph assemblage reveals dinosaur ancestral anatomy and diet. *Current Biology* 26: 1–6.
Calvo, J. O. 1991. Huellas de dinosaurios en la Formación Río Limay (Albiano-Cenomaniano?), Picún Leufú, Provincia de Neuquén, Republica Argentina. (Ornithischia-Saurischia: Sauropoda-Theropoda). *Ameghiniana* 28: 241–258.
Calvo, J. O. 2007. Ichnology. In Z. Gasparini, L. Salgado, and R. A. Coria (eds.), *Patagonian Mesozoic Reptiles*, 314–334. Bloomington: Indiana University Press.
Calvo, J. O., and G. V. Mazzetta. 2004. Nuevos hallazgos de huellas de dinosaurios en la Formación Candeleros (Albiano-Cenomaniano), Picún Leufú, Neuquén, Argentina. *Ameghiniana* 41: 545–554.
Campione, N. E., and D. C. Evans. 2011. Cranial growth and variation in edmontosaurs (Dinosauria: Hadrosauridae): implications for latest Cretaceous megaherbivore diversity in North America. *PloS One* 6(9): e25186. doi:10.1371/journal.pone.0025186.
Cariou, E., N. Olivier, B. Pittet, J.-M. Mazin, and P. Hantzpergue. 2014. Dinosaur track record on a shallow carbonate-dominated ramp (Loulle section, Late Jurassic, French Jura). *Facies* 60: 229–253.
Caro, S., C. Fuentes, M. Meijide, and F. Pérez-Lorente. 1995. Una rastrillada nueva de ornitópodo encontrada en el Barranco de la Muga (Soria-La Rioja, España). In F. Pérez-Lorente (ed.), *Huellas Fósiles de Dinosaurios de La Rioja: Nuevos Yacimientos*, 25–26. Logroño, España: Gobierno de La Rioja, Instituto de Estudios Riojanos.
Carpenter, K. 1999. *Eggs, Nests, and Baby Dinosaurs: A Look at Dinosaur Reproduction.* Bloomington: Indiana University Press, 336 pp.
Carpenter, K. 2009. Role of lateral body bending in crocodylian track making. *Ichnos* 16: 202–207.
Carpenter, K. 2016. Acrocanthosaurus *Inside and Out.* Norman, OK: University of Oklahoma Press, 137 pp.
Carr, T. D. 1999. Craniofacial ontogeny in Tyrannosauridae (Dinosauria, Coelurosauria). *Journal of Vertebrate Paleontology* 19: 497–520.
Carr, T. D., and T. E. Williamson. 2004. Diversity of late Maastrichtian Tyrannosauridae (Dinosauria: Theropoda) from western North America. *Zoological Journal of the Linnean Society* 142: 479–523.
Carrano, M. T., and J. Choiniere. 2016. New information on the forearm and manus of *Ceratosaurus nasicornis* Marsh, 1884 (Dinosauria, Theropoda), with implications for theropod forelimb evolution. *Journal of Vertebrate Paleontology* 36(3): e1054497 (14 pages).
Carvalho, I. de S. 1995. As pistas de dinossauros da Ponta da Guia (Bacia de São Luís, Cretáceo Superior—Maranhão, Brasil. *Anais da Academia Brasileira de Ciências* 67: 413–431.
Carvalho, I. de S., L. Borghi, and G. Leonardi. 2013. Preservation of dinosaur tracks induced by microbial mats in the Sousa Basin (Lower Cretaceous), Brazil. *Cretaceous Research* 44: 112–121.
Carvalho, I. de S., and S. da S. Kattah. 1998. As pegadas fósseis do paleodeserto da Bacia Sanfranciscana (Jurássico Superior-Cretáceo Inferior, Minas Gerais. *Anais da Academia Brasileira de Ciências* 70: 53–67.
Casamiquela, R. M. 1964. *Estudios Icnológicos: Problemas y Métodos de la Icnología con Aplicación al Estudio de Pisadas Mesozoicas (Reptilia, Mammalia) de la Patagonia.* Buenos Aires, Argentina: Colegio Industrial Pío 9, 229 pp.
Casamiquela, R. M. 1967. Un nuevo dinosaurio ornitisquio Triásico (*Pisanosaurus mertii*; Ornithopoda) de la Formación Ischigualasto, Argentina. *Ameghiniana* 5: 47–64.
Casanovas, L., R. Ezquerra, A. Fernández, D. Montero, F. Pérez-Lorente, J. V. Santafé, F. Torcida, and L. I. Viera. 1995*a*. El yacimiento de La Canal (Munilla, La Rioja, España). La variación de velocidad en función del tamaño del pie de los ornitópodos. *Zubía* 13: 55–81.
Casanovas, M. L., R. Ezquerra, A. Fernández, F. Pérez-Lorente, J. V. Santafé, and F. Torcida. 1995*b*. Dos nuevos yacimientos de icnitas de dinosaurios en La Rioja y en la provincia de Soria (España). *Coloquios de Paleontología* 47: 9–23.
Casanovas, M. L., R. Ezquerra, A. Fernández, F. Pérez-Lorente, J. V. Santafé, and F. Torcida. 1995*c*. Huellas de dinosaurio en la Era del Peladillo 3 (La Rioja, España). Primera nota. *Zubía* 13: 83–101.
Casanovas, M. L., R. Ezquerra, A. Fernández, F. Pérez-Lorente, J. V. Santafé, and F. Torcida. 1995*d*. Pisadas de dinosaurios en el Yacimiento Soto 3 (La Rioja, España). In F. Pérez-Lorente (ed.), *Huellas Fósiles de Dinosaurios de La Rioja: Nuevos Yacimientos*, 27–31. Logroño, España: Gobierno de La Rioja, Instituto de Estudios Riojanos.
Casanovas Cladellas, M. L., R. Ezquerra Miguel, A. Fernández Ortega, F. Pérez-Lorente, and J. V. Santafé Llopis. 1990*a*. Huellas de dinosaurio en San Vincente de Robres (La Rioja, España). *Zubía* 8: 33–47.
Casanovas Cladellas, M. L., R. Ezquerra Miguel, A. Fernández Ortega, F. Pérez-Lorente, and J. V. Santafé Llopis. 1990*b*. Huellas de dinosaurio en Soto de Cameros (La Rioja, España). *Zubía* 8: 49–71.
Casanovas Cladellas, M. L., R. Ezquerra Miguel, A. Fernández Ortega, F. Pérez-Lorente, and J. V. Santafé Llopis. 1991*a*. Huellas de dinosaurio en el Camino de Igea a Valdebrajes (La Rioja, España). *Zubía* 9: 89–111.
Casanovas, M. L., R. Ezquerra, A. Fernández, F. Pérez-Lorente, J. V. Santafé, and F. Torcida. 1993*a*. Huellas de dinosaurios palmeados y de terópodos den la "Era del Peladillo" Igea. (La Rioja). *Zubía* 11: 11–53.
Casanovas Cladellas, M. L., R. Ezquerra Miguel, A. Fernández Ortega, F. Pérez-Lorente, J. V. Santafé Llopis, and F. Torcida Fernández. 1993*b*. Icnitas de dinosaurios. Yacimientos de Navalsaz, Las Mortajeras, Peñaportillo, Malvaciervo y La Era del Peladillo 2 (La Rioja, España). *Zubía* 5: 9–133.
Casanovas, M. L., E. Ezquerra, A. Fernández, F. Pérez-Lorente, J. V. Santafé, and F. Torcida. 1995e. Pisadas de dinosaurios en el yacimiento Soto 3 (La Rioja, España). In F. Pérez-Lorente (ed.), *Huellas Fósiles*

de Dinosaurios de La Rioja: Nuevos Yacimientos. Logroño, España: Gobierno de La Rioja, Instituto de Estudios Riojanos.

Casanovas Cladellas, M. L., R. Ezquerra Miguel, A. Fernández Ortega, F. Pérez-Lorente, J. V. Santafé Llopis, and F. Torcida Fernández. 1993*c*. Icnitas digitígradas de dinosaurios en el afloramiento de El Villar-Poyales (La Rioja, España). *Zubía* 5: 135–163.

Casanovas, M. L., R. Ezquerra, F. Pérez-Lorente, and J. V. Santafé. 1992. Revisión del Yacimiento "Icnitas 3" de huellas de dinosaurio (Enciso, La Rioja, España). *Zubía* 10: 31–44.

Casanovas, M. L., A. Fernandez, M. C. Ondiviela, F. Pérez-Lorente, J. V. Santafé, and R. Serrano. 1995*f*. El rastro del barranco de Acrijos (Cornago, La Rioja, España). In F. Pérez-Lorente (ed.), *Huellas Fósiles de Dinosaurios de La Rioja: Nuevos Yacimientos*, 15–16. Logroño, España: Gobierno de La Rioja, Instituto de Estudios Riojanos.

Casanovas Cladellas, M. L., A. Fernández Ortega, F. Pérez-Lorente, and J. V. Santafé Llopis. 1989*a*. *Huellas Fosiles de Dinosaurios en La Rioja Yacimientos de Valdecevillo, La Senoba, y de La Virgen del Campo*. Logroño, Spain: Gobierno de La Rioja, Instituto de Estudios Riojanos, 190 pp.

Casanovas Cladellas, M. L., A. Fernández Ortega, F. Pérez-Lorente, and J. V. Santafé Llopis. 1991*b*. Dinosaurios coelúridos gregarios en el yacimiento de Valdevajes (La Rioja, España). *Revista Española de Paleontología* 6: 177–189.

Casanovas, M. L., A. Fernández, F. Pérez-Lorente, and J. V. Santafé. 1995*g*. Dinosaurios terópodos del afloramiento de Munilla (La Rioja, España). In F. Pérez-Lorente (ed.), *Huellas Fósiles de Dinosaurios de La Rioja: Nuevos Yacimientos*, 53–57. Logroño, España: Gobierno de La Rioja, Instituto de Estudios Riojanos.

Casanovas, M. L., A. Fernández, F. Pérez-Lorente, and J. V. Santafé. 1995*h*. Icnitas de terópodos y saurópodos del yacimiento de Las Navillas (La Rioja, España). In F. Pérez-Lorente (ed.), *Huellas Fósiles de Dinosaurios de La Rioja: Nuevos Yacimientos*, 33–44. Logroño, España: Gobierno de La Rioja, Instituto de Estudios Riojanos.

Casanovas, M. L., A. Fernández, F. Pérez-Lorente, and J. V. Santafé. 1995*i*. Icnitas terópodas y saurópodas en La Cela, Mura en Cameros (La Rioja, España). In F. Pérez-Lorente (ed.), *Huellas Fósiles de Dinosaurios de La Rioja: Nuevos Yacimientos*, 17–24. Logroño, España: Gobierno de La Rioja, Instituto de Estudios Riojanos.

Casanovas, M. L., A. Fernández, F. Pérez-Lorente, and J. V. Santafé. 1995*j*. Un terópodo carnosaurio en el Camino a Treguajantes (La Rioja, España). In F. Pérez-Lorente (ed.), *Huellas Fósiles de Dinosaurios de La Rioja: Nuevos Yacimientos*, 13–14. Logroño, España: Gobierno de La Rioja, Instituto de Estudios Riojanos.

Casanovas, M. L., A. Fernández, F. Pérez-Lorente, and J. V. Santafé. 1998. Ocho nuevos yacimientos de huellas de dinosaurio. *Zubía* 16: 117–152.

Casanovas, M. L., A. Fernández, F. Pérez-Lorente, J. V. Santafé, and F. Torcida. 1995*k*. La Era del Peladillo 4 (La Rioja, España). In F. Pérez-Lorente (ed.), *Huellas Fósiles de Dinosaurios de La Rioja: Nuevos Yacimientos*, 45–52. Logroño, España: Gobierno de La Rioja, Instituto de Estudios Riojanos.

Casanovas, M. L., A. Fernández, F. Pérez-Lorente, J. V. Santafé, and F. Torcida. 1997. Pisadas de ornitópodos, terópodos y saurópodos en La Era del Peladillo 5 (La Rioja, España). *Zubía* 15: 229–246.

Casanovas Cladellas, M. L., F. Pérez-Lorente, M. Ruiz de Arcaute, J. V. Santafé Llopis, and F. Torcida Fernández. 1991*c*. Terópodos carnosaurios en La Virgen del Campo II. Enciso (La Rioja, España). *Zubía* 9: 113–126.

Casanovas Cladellas, M. L., F. Pérez-Lorente, and J. V. Santafé Llopis. 1989*b*. Huellas de dinosaurio en Valdenocerillo (Cornago, La Rioja, España). *Zubía* 7: 29–35.

Castanera, D., J. Colmenar, V. Sauqué, and J. L. Canudo. 2015. Geometric morphometric analysis applied to theropod tracks from the Lower Cretaceous (Berriasian) of Spain. *Palaeontology* 58: 183–200.

Castanera, D., C. Pascual, N. L. Razzolini, B. Vila, J. L. Barco, and J. I. Canudo. 2013*a*. Discriminating between medium-sized tridactyl trackmakers: tracking ornithopod tracks in the base of the Cretaceous (Berriasian, Spain). *PloS One* 8(11): e81830. doi: 10.1371/journal.pone.0081830.

Castanera, D., L. Piñuela, and J. C. García-Ramos. 2016. *Grallator* theropod tracks from the Late Jurassic of Asturias (Spain): ichnotaxonomic implications. *Spanish Journal of Paleontology* 31: 283–296.

Castanera, D., B. Vila, N. L. Razzolini, P. L. Falkingham, J. I. Canudo, P. L. Manning, and À. Galobart. 2013*b*. Manus track perservation bias as a key factor for assessing trackmaker identity and quadrupedalism in basal ornithopods. *PloS One* 8(1): e54177. doi:10.1371/journal.pone.0054177.

Chabreck, R. H. 1963. Methods of capturing, marking and sexing alligators. *Proceedings of the 17th Annual Conference of the Southeastern Association of Game and Fish Commissioners* 17: 47–50.

Chabreck, R. H. 1966. Methods of determining the size and composition of alligator populations in Louisiana. *Proceedings of the 20th Annual Conference of the Southeastern Association of Game and Fish Commissioners* 20: 105–112.

Chabreck, R. H., and T. Joanen. 1979. Growth rates of American alligators in Louisiana. *Herpetologica* 35: 51–57.

Charig, A. J., and S. D. Chapman. 1998. Case 3037. *Bulletin of Zoological Nomenclature* 55: 99–104.

Charig, A. J., and S. D. Chapman. 2000. Opinion 1947. *Bulletin of Zoological Nomenclature* 57: 61–62.

Chiappe, L. M., and U. B. Göhlich. 2010. Anatomy of *Juravenator starki* (Theropoda: Coelurosauria) from the Late Jurassic of Germany. *Neues Jahrbuch für Geologie und Paläontologie Abhandlungen* 258: 257–296.

Choiniere, J. N., J. M. Clark, C. A. Forster, M. A. Norell, D. A. Eberth, G. M. Erickson, H. Chu, and X. Xu. 2014. A juvenile specimen of a new coelurosaur (Dinosauria: Theropoda) from the Middle-Late Jurassic Shishugou Formation of Xinjiang, People's Republic of China. *Journal of Systematic Paleontology* 12: 177–215.

Chojnowski, J. L., R. T. Kimball, and E. L. Braun. 2008. Introns outperform exons in analyses of basal avian phylogeny using clathrin heavy chain genes. *Gene* 410: 89–96.

Christian, A., and H. Preuschoft. 1996. Deducing the body posture of extinct large vertebrates from the shape of the vertebral column. *Palaeontology* 39: 801–812.

Christiansen, P. 1999. Long bone scaling and limb posture in non-avian theropods: evidence for differential allometry. *Journal of Vertebrate Paleontology* 19: 666–680.

Chubb, A. L. 2004. New nuclear evidence for the oldest divergence among neognath birds: the phylogenetic utility of ZENK (I). *Molecular Phylogenetics and Evolution* 30: 140–151.

Claessens, L. P. A. M., H. J. M. Meijer, J. P. Hume, and K. F. Rijsdijk (eds.). 2015. Anatomy of the dodo (*Raphus cucullatus* L., 1758): an osteological study of the Thirioux specimens. *Journal of Vertebrate Paleontology* Memoir 15, 35 (Supplement to No. 6): 1–187.

Clark, N. D. L., and J. L. Barco Rodríguez. 1998. The first dinosaur trackway from the Valtos Sandstone Formation (Bathonian, Jurassic) of the Isle of Skye, Scotland, UK. *Geogaceta* 24: 70–82.

Clark, N. D. L., P. Booth, C. Booth, and D. A. Ross. 2004. Dinosaur footprints from the Duntulm Formation (Bathonian, Jurassic) of the Isle of Skye. *Scottish Journal of Geology* 40: 13–21.

Clark, N. D. L., and M. K. Brett-Surman. 2008. A comparison between dinosaur footprints from the Middle Jurassic of the Isle of Skye, Scotland, UK, and Shell, Wyoming, USA. *Scottish Journal of Geology* 44: 139–150.

Clark, N. D. L., D. A. Ross, and P. Booth. 2005. Dinosaur tracks from the Kilmaluag Formation (Bathonian, Middle Jurassic) of Score Bay, Isle of Skye, Scotland, UK. *Ichnos* 12: 93–104.

Clarke, J. A., Z. Zhou, and F. Zhang. 2006. Insight into the evolution of avian flight from a new clade of Early Cretaceous ornithurines from China and the morphology of *Yixianornis grabaui*. *Journal of Anatomy* 208: 287–308.

Clements, J. F. 2007. *The Clements Checklist of Birds of the World*, 6th ed. Ithaca, NY: Cornell University Press, 843 pp.

Cobos, A., F. Gascó, R. Royo-Torres, M. G. Lockley, and L. Alcalá. 2016. Dinosaur tracks as "four-dimensional phenomena" reveal how different species moved. In P. L. Falkingham, D. Marty, and A. Richter (eds.), *Dinosaur Tracks: The Next Steps*, 244–254. Bloomington: Indiana University Press.

Cobos, A., M. G. Lockley, F. Gascó, R. Royo-Torres, and Luis Alcalá. 2014. Megatheropods as apex predators in the typically Jurassic ecosystems of the Villar del Arzobispo Formation (Iberian Range, Spain). *Palaeogeography, Palaeoclimatology, Palaeoecology* 399: 31–41.

Colbert, E. H. 1968. *Men and Dinosaurs: The Search in Field and Laboratory.* New York: E.P. Dutton, 283 pp.

Colbert, E. H., and D. Merrilees. 1967. Cretaceous dinosaur footprints from Western Australia. *Journal of the Royal Society of Western Australia* 50: 21–25.

Congleton, J. D. 1990. *Vertebrate Paleontology of the Koum Basin, Northern Cameroon, and Archosaurian Paleobiogeography in the Early Cretaceous*. Master's Thesis. Dallas, TX: Southern Methodist University, 244 pp.

Conrad, K. P., M. G. Lockley, and N. K. Prince. 1987. Triassic and Jurassic vertebrate-dominated trace fossil assemblages of the Cimarron Valley region: implications for paleoecology and biostratigraphy. *New Mexico Geological Society Guidebook, 38th Field Conference, Northeastern New Mexico*, pp. 127–138.

Contessi, M., and F. Fanti. 2012. Vertebrate tracksites in the Middle Jurassic-Upper Cretaceous of South Tunisia. *Ichnos* 19: 211–227.

Cook, A. G., N. Saini, and S. A. Hocknull. 2010. Dinosaur footprints from the lower Jurassic of Mount Morgan, Queensland. *Memoirs of the Queensland Museum / Nature* 55: 135–146.

Coombs, W. P., Jr. 1978. Theoretical aspects of cursorial adaptations in dinosaurs. *Quarterly Review of Biology* 53: 393–418.

Cooper, A., C. Lalueza-Fox, S. Anderson, A. Rambaut, J. Austin, and R. Ward. 2001. Complete mitochondrial genome sequences of two extinct moas clarify ratite evolution. *Nature* 409: 704–707.

Cooper, A., C. Mourer-Chauviré, G. K. Chambers, A. von Haeseler, A. C. Wilson, and S. Pääbo 1992. Independent origins of New Zealand moas and kiwis. *Proceedings of the National Academy of Sciences, U.S.A.* 89: 8741–8744.

Cooper, M. R. 1981. The prosauropod dinosaur *Massospondylus carinatus* Owen from Zimbabwe: its biology, mode of life and phylogenetic significance. *Occasional Papers of the National Museums and Monuments of Rhodesia Series B* 6: 689–840.

Cotton, W. D., J. E. Cotton, and A. P. Hunt. 1998. Evidence for social behavior in ornithopod dinosaurs from the Dakota Group of northeastern New Mexico, U.S.A. *Ichnos* 6: 141–149.

Coulson, R. A., and T. Hernandez. 1983. *Alligator metabolism: Studies on Chemical Reactions in vivo.* Oxford, U.K.: Pergamon Press, 182 pp.

Coulson, T. D., R. A. Coulson, and T. Hernandez. 1973. Some observations on the growth of captive alligators. *Zoologica* 58: 47–52.

Courel, L., and G. Demathieu. 1976. Une ichnofaune reptilienne remarquable dans les grès Triasiques de Largentière (Ardèche, France). *Palaeontographica A* 151: 194–216.

Courel, L., and G. Demathieu. 2000. Une nouvelle ichnoespèce *Coelurosaurichnus grancieri* du Trias supérieur de l'Ardèche, France. *Geodiversitas* 22: 35–46.

Covacevich, V., and P. V. Rich. 1982. New bird ichnites from Fildes Peninsula, King George Island, West Antarctica. In C. Craddock (ed.), *Antarctic Geoscience*, 245–254. Madison: University of Wisconsin Press.

Cracraft, J., F. K. Barker, M. J. Braun, J. Harshman, G. Dyke, J. Feinstein, S. Stanley, A. Cibois, P. Schikler, P. Beresford, J. García-Moreno, M. D. Sorenson, T. Yuri, and D. P. Mindell. 2004. Phylogenetic relationships among modern birds (Neornithes): toward an avian tree of life. In J. Cracraft and M. J. Donoghue (eds.), *Assembling the Tree of Life*, 468–489. New York: Oxford University Press.

Cramp, S. (ed.). 1977. *Handbook of the Birds of Europe, the Middle East and North Africa: The Birds of the Western Palearctic. Volume I: Ostrich to Ducks.* Oxford, U.K.: Oxford University Press, 722 pp.

Cramp, S. (ed.). 1980. *Handbook of the Birds of Europe, the Middle East and North Africa: The Birds of the Western Palearctic. Volume II: Hawks to Bustards.* Oxford, U.K.: Oxford University Press, 695 pp.

Cramp, S. (ed.). 1983. *Handbook of the Birds of Europe, the Middle East and North Africa: The Birds of the Western Palearctic. Volume III: Waders to Gulls.* Oxford, U.K.: Oxford University Press, 913 pp.

Cuesta, E., I. Díaz-Martínez, F. Ortega, and J. L. Sanz. 2015. Did all theropods have chicken-like feet? First evidence of a non-avian dinosaur podotheca. *Cretaceous Research* 56: 53–59.

Cullen, T. M., M. J. Ryan, C. Schröder-Adams, P. J. Currie, and Y. Kobayashi. 2013. An ornithomimid (Dinosauria) bonebed from the Late Cretaceous of Alberta, with implications for the behavior, classification, and stratigraphy of North American ornithomimids. *PloS One* 8(3): e58853. doi:10.1371/journal.pone.0058853.

Currie, P. J. 1983. Hadrosaur trackways from the Lower Cretaceous of Canada. *Acta Palaeontologica Polonica* 28: 63–73.

Currie, P. J. 2003*a*. Allometric growth in tyrannosaurids (Dinosauria: Theropoda) from the Upper Cretaceous of North America and Asia. *Canadian Journal of Earth Sciences* 40: 651–665.

Currie, P. J. 2003*b*. Cranial anatomy of tyrannosaurid dinosaurs from the late Cretaceous of Alberta, Canada. *Acta Palaeontologica Polonica* 48: 191–226.

Currie, P. J., D. Badamgarav, and E. B. Koppelhus. 2003. The first Late Cretaceous footprints from the Nemegt locality in the Gobi of Mongolia. *Ichnos* 10: 1–13.

Dai, H., L. Xing, D. Marty, J. Zhang, W. S. Persons, IV, H. Hu, and F. Wang. 2015. Microbially-induced sedimentary wrinkle structures and possible impact of microbial mats for the enhanced preservation of dinosaur tracks from the Lower Cretaceous Jiguan Formation near Qijiang (Chongqing, China). *Cretaceous Research* 53: 98–109.

Dalla Vecchia, F. M. 2009. *Tethyshadros insularis*, a new hadrosauroid dinosaur (Ornithischia) from the Upper Cretaceous of Italy. *Journal of Vertebrate Paleontology* 29: 1100–1116.

Dalla Vecchia, F. M., and P. Mietto. 1998. Impronte de rettili terrestri nella Dolomia Principale (Triassico Superiore) delle Prealpi Carniche (Pordenone, Friuli). *Atti Ticinensi de Scienze della Terra* 7: 87–107.

Dalla Vecchia, F. M., and A. Tarlao. 2000. New dinosaur track sites in the Albian (Early Cretaceous) of the Istrian Peninsula (Croatia)—Part II—paleontology. *Memorie di Scienze Geologiche* 52: 227–292.

Dalla Vecchia, F. M., G. Tunis, S. Venturini, and A. Tarlao. 2001. Dinosaur tracks in the upper Cenomanian (Late Cretaceous) of Istrian Peninsula (Croatia). *Bollettino della Società Paleontologica Italiana* 40: 25–53.

Dalla Vecchia, F. M., I. Vlahovi☒, L. Posocco, A. Tarlao, and M. Tentor. 2002. Late Barremian and Late Albian (Early Cretaceous) dinosaur track sites in the main Brioni/Brijun Island (SW Istria, Croatia). *Natura Nascosta* 25: 1–36.

Dalman, S. G. 2012. New data on small theropod footprints from the Early Jurassic (Hettangian) Hartford Basin of Massachusetts, United States. *Bulletin of the Peabody Museum of Natural History* 53: 333–353.

Dalman, S. G., and R. E. Weems. 2013. A new look at morphological variation in the ichnogenus *Anomoepus*, with special reference to material from the Lower Jurassic Newark Supergroup: implications for ichnotaxonomy and ichnodiversity. *Bulletin of the Peabody Museum of Natural History* 54: 67–124.

Dattilo, B. F., S. C. Howald, R. Bonem, J. Farlow, A. J. Martin, M. O'Brien, M. G. Blair, G. Kuban, L. K. Mark, A. R. Knox, W. N. Ward, and T. Joyce. 2014. Stratigraphy of the Paluxy River tracksites in and around Dinosaur Valley State Park, Lower Cretaceous Glen Rose Formation, Somervell County, Texas. In M. G. Lockley and S. G. Lucas (eds.), *Fossil Footprints of Western North America*, 307–338. Albuquerque, NM: New Mexico Museum of Natural History and Science Bulletin 62.

Davies, S. J. J. F. 2002. *Ratites and Tinamous (Tinamidae, Rheidae, Dromaiidae, Casuariidae, Apterygidae, Struthionidae) (Bird Families of the World).* Oxford, U.K.: Oxford University Press, 310 pp.

Davis, L. M., T. C. Glenn, R. M. Elsey, H. C. Dessauer, and R. H. Sawyer. 2001. Multiple paternity and mating patterns in the American alligator, *Alligator mississippiensis*. *Molecular Ecology* 10: 1011–1024.

Davis, L. M., T. C. Glenn, D. C. Strickland, L. J. Guillette, Jr., R. M. Elsey, W. E. Rhodes, H. C. Dessauer, and R. H. Sawyer. 2002. Microsatellite DNA analyses support an east-west phylogeographic split of American alligator populations. *Journal of Experimental Zoology (Mol Dev Evol)* 294: 352–372.

Day, J. J., D. B. Norman, A. S. Gale, P. Upchurch, and H. P. Powell. 2004. A Middle Jurassic dinosaur trackway site from Oxfordshire, UK. *Palaeontology* 47: 319–348.

Dean, D. R. 1969. Hitchcock's dinosaur tracks. *American Quarterly* 21: 639–644.

Degrange, F. J., C. P. Tambussi, M. L. Taglioretti, A. Dondas, and F. Scaglia. 2015. A new Mesembriornithinae (Aves, Phorusrhacidae) provides new insights into the phylogeny and sensory capabilities of terror birds. *Journal of Vertebrate Paleontology* 35(2) e912656. doi:10.1080/02724634.2014.912656.

Dejax, J., J.-G. Michard, M. Brunet, and J. Hell. 1989. Empreintes de pas de dinosauriens datées du Crétacé inférieur dans le Bassin de Babouri-Figuil (Fossé de la Bénoué, Cameroun). *Neues Jahrbuch für Geologie und Paläontologie Abhandlungen* 178: 85–108.

Delany, M. F., and C. L. Abercrombie. 1986. American alligator food habits in northcentral Florida. *Journal of Wildlife Management* 50: 348–353.

Demathieu, G. 1990. Problems of discrimination of tridactyl dinosaur footprints exemplified by the Hettangian trackways, the Causses, France. *Ichnos* 1: 97–110.

Demathieu, G. 1993. Empreintes de pas de dinosaures dans les Causses (France). *Zubía Monográfico* 5: 229–252.

Demathieu, G., and G. Gand. 1972. *Coelurosaurichnus perriauxi*: empreinte dinosauroide nouvelle du Trias du plateau d'Antully (Saône-et-Loire). *Bulletin Trimestriel de la Société d'Histoire Naturelle et des Amis du Musèum d'Autun* 62: 2–18.

Demathieu, G., G. Gand, J. Sciau, P. Freytet, and J. Garric. 2002. Les traces de pas de dinosaures et autres archosaures du Lias Inferieur des Grands Causses, sud de la France. *Palaeovertebrata* 31: 1–143.

Demathieu, G., and J. Sciau. 1994. L'ichnofaune Hettangienne d'archosauriens de Sauclières, Aveyron, France. *Bulletin de la Société d'Histoire Naturelle Autun* 151: 5–46.

De Velais, S. 2011. Revision of dinosaur ichnotaxa from the La Matilde Formation (Middle Jurassic), Santa Cruz Province, Argentina. *Ameghiniana* 48: 28–42.

Díaz-Martínez, I., E. García-Ortiz, and F. Pérez-Lorente. 2015. A new dinosaur tracksite with small footprints in the Urbión Group (Cameros Basin, Lower Cretaceous, La Rioja, Spain). *Journal of Iberian Geology* 41: 167–175.

Díaz-Martínez, I., L. Ladel, D. Martín, Saida Saadani, J. Buzzi, S. Ghadbane, A. H. Luján, D. Moussa, V. Martínez, Y. Elazzouzi, and E. García-Ortiz de Landaluce. 2009. Nuevas huellas terópodas en el yacimiento 4AM en Aït Mimoun (Sinclinal de Iouaridène, Alto Atlas, Marruecos). *Paleolusitana* 1: 151–159.

Dickinson, E. C. 2003. *The Howard and Moore Complete Checklist of the Birds of the World*. Princeton, NJ: Princeton University Press, 1039 pp.

Diedrich, C. 2004. New important iguanodontid and theropod trackways of the Tracksite Obernkirchen in the Berriasian of NW Germany and megatracksite concept of central Europe. *Ichnos* 11: 215–228.

Diedrich, C. 2011. Upper Jurassic tidal flat megatracksites of Germany—coastal dinosaur migration highways between European islands, and a review of the dinosaur footprints. *Palaeobiodiversity and Palaeoenvironments* 91: 129–155.

Dodson, P. 2003. Allure of El Lagarto—Why do dinosaur paleontologists love alligators, crocodiles, and their kin? *The Anatomical Record* 274A: 887–890.

D'Orazi Porchetti, S., M. Bernardi, A. Cinquegranelli, V. F. dos Santos, D. Marty, F. M. Petti, P. Sá Caetano, and A. Wagensommer. 2016. A review of the dinosaur track record from Jurassic and Cretaceous shallow marine carbonate depositional environments. In P. L. Falkingham, D. Marty, and A. Richter (eds.), *Dinosaur Tracks: The Next Steps*, 380–390. Bloomington: Indiana University Press.

D'Orazi Porchetti, S., and U. Nicosia. 2007. Re-examination of some large early Mesozoic tetrapod footprints from the African collection of Paul Ellenberger. *Ichnos* 14: 219–245.

Dyke, G. J., and M. Van Tuinen. 2004. The evolutionary radiation of modern birds (Neornithes): reconciling molecules, morphology and the fossil record. *Zoological Journal of the Linnean Society* 141: 153–177.

Edwards, S.V., B. Furtil, A. Giron, and P. J. Deschavanne. 2002. A genomic schism in birds revealed by phylogenetic analysis of DNA strings. *Systematic Biology* 5: 599–613.

Elbroch, M., and E. Marks. 2001. *Bird Tracks and Sign: A Guide to North American Species*. Mechanicsburg, PA: Stackpole Books, 456 pp.

Ellenberger, P. 1972. Contribution à la classification des Pistes de Vertébrés du Trias: Les Types du Stormberg d'Afrique du Sud (I). *Palaeovertebrata Mémoire Éxtraordinaire,* Montpellier, France, 123 pp.

Ellenberger, P. 1974. Contribution à la classification des Pistes de Vertébrés du Trias: Les Types du Stormberg d'Afrique du Sud (IIème Partie: Le Stormberg Superieur-I. Le Biome de la Zone B/1 ou Niveau de Moyeni: ses Biocènoses). *Palaeovertebrata Mémoire Éxtraordinaire,* Montpellier, France, 153 pp.

Ellenberger, F., & Y. Fuchs. 1965. Sur la présence de pistes de Vertébrés dans le Lotharingien marin de la région de Sévérac-le-Châtea (Aveyron). *Comptes Rendus de la Société Géolgique de France* 2: 39–40.

Ellis, R. G., and S. M. Gatesy. 2013. A biplanar X-ray method for three-dimensional analysis of track formation. *Palaeontologia Electronica* 16(1): 16 pp. palaeo-electronica.org/content/2013/371-x-ray-track-analysis.

Erickson, B. R. 2005. Crocodile and arthropod tracks from the Late Paleocene Wannagan Creek fauna of North Dakota, USA. *Ichnos* 12: 303–308.

Erickson, G. M., K. C. Curry-Rogers, D. J. Varricchio, M. A. Norell, and Xing Xu. 2007. Growth patterns in brooding dinosaurs reveals the timing of sexual maturity in non-avian dinosaurs and genesis of the avian condition. *Biology Letters* 3: 558–561.

Erickson, G. M., P. J. Makovicky, B. D. Inouye, C.-F. Zhoum, and K.-Q. Gao. 2009. A life table for *Psittacosaurus lujiatunensis*: initial insights into ornithischian dinosaur population biology. *The Anatomical Record* 292: 1514–1521.

Ericson, P. G. P., C. L. Anderson, T. Britton, A. Elzanowski, U. S. Johansson, M. Källersjö, J. I. Ohlson, T. J. Parsons, D. Zuccon, and G. Mayr. 2006. Diversification of Neoaves: integration of molecular sequence data and fossils. *Biology Letters* 2: 543–547.

Escaso, F., F. Ortega, P. Dantas, E. Malafaia, B. Silva, J. M. Gasulla, P. Mocho, I. Narváez, and J. L. Sanz. 2014. A new dryosaurid ornithopod (Dinosauria, Ornithischia) from the late Jurassic of Portugal. *Journal of Vertebrate Paleontology* 34: 1102–1112.

Fain, M. G., and P. Houde. 2004. Parallel radiations in the primary clades of birds. *Evolution* 58: 2558–2573.

Falk, A. R., S. T. Hasiotis, E. Gong, J.-D. Lim, and E. D. Brewer. 2017. A new experimental setup for studying avian neoichnology and the effects of grain size and moisture content on tracks: trials using the domestic chicken (*Gallus gallus*). *Palaios* 32: 689–707.

Falk, A. R., L. D. Martin, and S. T. Hasiotis. 2011. A morphologic criterion to distinguish bird tracks. *Journal of Ornithology* 152: 701–716.

Falkingham, P. L. 2014. Interpreting ecology and behavior from the vertebrate fossil track record. *Journal of Zoology* 292: 222–229.

Falkingham, P. L. 2016. Applying objective methods to subjective track outlines. In P. L. Falkingham, D. Marty, and A. Richter (eds.), *Dinosaur Tracks: The Next Steps*, 72–80. Bloomington: Indiana University Press.

Falkingham, P. L., K. T. Bates, L. Margetts, and P. L. Manning. 2011*a*. Simulating sauropod manus-only trackway formation using finite-element analysis. *Biology Letters* 7: 142–145.

Falkingham, P. L., K. T. Bates, L. Margetts, and P. L. Manning. 2011*b*. The "Goldilocks" effect: preservation bias in vertebrate track assemblages. *Journal of the Royal Society Interface* 8: 1142–1154.

Falkingham, P. L., and S. M. Gatesy. 2014. The birth of a dinosaur footprint: subsurface 3D motion reconstruction and discrete element simulation reveal track ontogeny. *Proceedings of the National Academy of Science USA* 111: 18279–18284.

Falkingham, P. L., L. Margetts, and P. L. Manning. 2010. Fossil vertebrate tracks as paleopenetrometers: confounding effects of foot morphology. *Palaios* 25: 356–360.

Fanti, F. M., P. R. Bell, and R. L. Sissons. 2013*a*. A diverse, high-latitude ichnofauna from the Late Cretaceous Wapiti Formation, Alberta, Canada. *Cretaceous Research* 41: 256–269.

Fanti, F., M. Contessi, A. Nigarov, and P. Esenov. 2013*b*. New data on two large dinosaur tracksites from the Upper Jurassic of eastern Turkmenistan (central Asia). *Ichnos* 20: 54–71.

Farlow, J. O. 1981. Estimates of dinosaur speeds from a new trackway site in Texas. *Nature* 294: 747–748.

Farlow, J. O. 1987. *A Guide to Lower Cretaceous Dinosaur Footprints and Tracksites of the Paluxy River, Somevell County, Texas.* Field Trip Guidebook, South Central Section. Waco, Texas: Geological Society of America, Baylor University, 50 pp.

Farlow, J. O. 1989. Ostrich footprints and trackways: implications for dinosaur ichnology. In D. D. Gillette and M. G. Lockley (eds.), *Dinosaur Tracks and Traces*, 243–248. Cambridge: Cambridge University Press.

Farlow, J. O. 2001. *Acrocanthosaurus* and the maker of Comanchean large-theropod footprints. In D.H. Tanke and K. Carpenter (eds.), *Mesozoic Vertebrate Life: New Research Inspired by the Paleontology of Philip J. Currie*, 408–427. Bloomington: Indiana University Press.

Farlow, J. O., K. T. Bates, R. M. Bonem, B. F. Dattilo, P. L. Falkingham, R. Gildner, J. Jacene, G. J. Kuban, A. J. Martin, M. O'Brien, and J. Whitcraft. 2015. Dinosaur footprints from the Glen Rose Formation (Paluxy River, Dinosaur Valley State Park, Somervell County, Texas). In C. Noto (ed.), *Early- and Middle-Cretaceous Archosaur Localities of North-Central Texas*, 14–37. Field Trip Guidebook, 75th Annual Meeting. Dallas, TX: Society of Vertebrate Paleontology.

Farlow, J. O., and A. Britton. 2000. Size and body proportions in *Alligator mississippiensis*: implications for archosaurian ichnology. *Paleontological Society of Korea Special Publication* 4: 189–206.

Farlow, J. O., and R. E. Chapman. 1997. The scientific study of dinosaur footprints. In J. O. Farlow and M. K. Brett-Surman (eds.), *The Complete Dinosaur*, 519–553. Bloomington: Indiana University Press.

Farlow, J. O., R. E. Chapman, B. Breithaupt, and N. Matthews. 2012. The scientific study of dinosaur footprints. In M. K. Brett-Surman, T. R. Holtz, Jr., and J. O. Farlow (eds.), *The Complete Dinosaur*, 2nd ed., 712–759. Bloomington: Indiana University Press.

Farlow, J. O., and R. M. Elsey. 2004. Femoral dimensions and mid-thigh circumference in *Alligator mississippiensis*. *Lethaia* 37: 401–405.

Farlow, J. O., and R. M. Elsey. 2010. Footprints and trackways of the American alligator, Rockefeller Wildlife Refuge, Louisiana. In J. Milàn, S. G. Lucas, M. G. Lockley, and J. A. Spielman (eds.), *Crocodyle Tracks and Traces*, 31–39. Albuquerque, NM: New Mexico Museum of Natural History and Science Bulletin 51.

Farlow, J. O., and P. M. Galton. 2003. Dinosaur trackways of Dinosaur State Park, Rocky Hill, Connecticut. In P. M. Letourneau and P. E. Olsen (eds.), *The Great Rift Valleys of Pangea in Eastern North America. Volume Two: Sedimentology, Stratigraphy, and Paleontology*, 248–263. New York: Columbia University Press.

Farlow, J. O., S. M. Gatesy, T. R. Holtz, Jr., J. R. Hutchinson, and J. M. Robinson. 2000. Theropod locomotion. *American Zoologist* 40: 640–663.

Farlow, J. O., and J. M. Hawthorne. 1989. Comanchean dinosaur footprints. In D. A. Winkler, P. A. Murry, and L. L. Jacobs (eds.), *Field Guide to the Vertebrate Paleontology of the Trinity Group, Lower Cretaceous of Central Texas*, 23–30. Dallas, TX: Institute for the Study of Earth and Man, Southern Methodist University.

Farlow, J. O., T. R. Holtz, Jr., T. H. Worthy, and R. E. Chapman. 2013. Feet of the fierce (and not so fierce): pedal proportions in large theropods, other non-avian dinosaurs, and large ground birds. In J. M. Parrish, R. E. Molnar, P. J. Currie, and E. B. Koppelhus (eds.), *Tyrannosaurid Paleobiology*, 88–132. Bloomington: Indiana University Press.

Farlow, J. O., G. R. Hurlburt, R. M. Elsey, A. R. C. Britton, and W. Langston, Jr. 2005. Femoral dimensions and body size of *Alligator mississippiensis*: estimating the size of extinct mesoeucrocodylians. *Journal of Vertebrate Paleontology* 25: 354–369.

Farlow, J. O., W. Langston, Jr., E. E. Deschner, R. Solis, W. Ward, B. L. Kirkland, S. Hovorka, T. L. Reece, and J. Whitcraft. 2006. *Texas Giants: Dinosaurs of the Heritage Museum of the Texas Hill Country.* Canyon Lake, TX: Heritage Museum of the Texas Hill Country, 105 pp.

Farlow, J. O., and M. G. Lockley. 1993. An osteometric approach to the identification of the makers of early Mesozoic tridactyl dinosaur footprints. In S. G. Lucas and M. Morales (eds.), *The Nonmarine Triassic*, 123–131. Albuquerque, NM: New Mexico Museum of Natural History and Science Bulletin 3.

Farlow, J. O., N. J. Robinson, C. J. Kumagai, F. V. Paladino, P. L. Falkingham, R. M. Elsey, and A. J. Martin. 2018. Trackways of the American crocodile (*Crocodylus acutus*) in northwestern Costa Rica: implications for crocodylian ichnology. *Ichnos* 25: 30–65. doi.org/10.1080/10420940.2017.1350856.

Farlow, J. O., E. R. Schachner, J. C. Sarrazin, H. Klein, and P. J. Currie. 2014. Pedal proportions in *Poposaurus gracilis*: convergence and divergence in the feet of archosaurs. *The Anatomical Record* 297: 1022–1046.

Ferrusquía-Villafranca, I., S. P. Applegate, and L. Espinosa Arrubarrena. 1978. Las huellas más australes de dinosaurios en Norteamérica y su significación geobiológica. *Actas del Segundo Congreso Argentino de Paleontología y Biostratigrafía y Primer Congreso Latinoamericano de Paleontología* 1: 249–263.

Finkel, I. 2014. *The Ark Before Noah: Decoding the Story of the Flood.* New York: Doubleday, 421 pp.

Fiorillo, A. R., and T. L. Adams. 2012. A therizinosaur track from the lower Cantwell Formation (Upper Cretaceous) of Denali National Park, Alaska. *Palaios* 27: 395–400.

Fiorillo, A. R., M. Contessi, Y. Kobayashi, and P. J. McCarthy. 2014*a*. Theropod tracks from the lower Cantwell Formation (Upper Cretaceous) of Denali National Park, Alaska, USA with comments on theropod diversity in an ancient, high-latitude terrestrial ecosystem. In M. G. Lockley and S. G. Lucas (eds.), *Fossil Footprints of Western North America*, 429–439. Albuquerque, NM: New Mexico Museum of Natural History and Science Bulletin 62.

Fiorillo, A. R., S. T. Hasiotis, and Y. Kobayashi. 2014*b*. Herd structure in Late Cretaceous polar dinosaurs: a remarkable new dinosaur tracksite, Denali National Park, Alaska, USA. *Geology* 42: 719–722.

Foster, J. R. 2015. Theropod dinosaur ichnogenus *Hispanosauropus* identified from the Morrison Formation (Upper Jurassic), western North America. *Ichnos* 22: 183–191.

Foster, J. R., and D. J. Chure. 2006. Hindlimb allometry in the Late Jurassic theropod dinosaur *Allosaurus*, with comments on its abundance and distribution. In J. R. Foster and S. G. Lucas (eds.), *Paleontology and Geology of the Upper Jurassic Morrison Formation*, 119–122. Albuquerque, NM: New Mexico Museum of Natural History and Science Bulletin 36.

Foster, J. R., and M. G. Lockley. 1997. Probable crocodilian tracks and traces from the Morrison Formation (Upper Jurassic) of eastern Utah. *Ichnos* 5: 121–129.

Foster, J. R., and M. G. Lockley. 2006. The vertebrate ichnological record of the Morrison Formation (Upper Jurassic, North America). In J. R. Foster and S. G. Lucas (eds.), *Paleontology and Geology of the Upper Jurassic Morrison Formation*, 203–216. Albuquerque, NM: New Mexico Museum of Natural History and Science Bulletin 36.

Foth, C., H. Tischlinger, and O. W. M. Rauhut. 2014. New specimen of *Archaeopteryx* provides insights into the evolution of pennaceous feathers. *Nature* 511: 79–82.

Francischini, H., P. C. Dentzien-Dias, M. A. Fernandes, and C. L. Schultz. 2015. Dinosaur ichnofauna of the Upper Jurassic/Lower Cretaceous of the Paraná Basin (Brazil and Uruguay). *Journal of South American Earth Sciences* 63: 180–190.

Fraser, N. C., and P. E. Olsen. 1996. A new dinosauromorph ichnogenus from the Triassic of Virginia. *Jeffersoniana* 7: 1–17.

Fuentes Vidarte, C., and M. Meijide Calvo. 1998. Icnitas de dinosaurios teropodos en el Weald de Soria (España). Nuevo icnogenero *Kalohippus*. *Estudios Geológicos* 54: 147–152.

Fuentes Vidarte, C., and M. Meijide Calvo. 2001. Primeras huellas de cocodrilo en el Weald de Cameros (Soria, España). Nueva familia: Crocodylopodidae, nuevo icnogénero: *Crocodylopodus*, nueva icnoespecie: *C. meijidei*. In Colectivo Arqueolóico-Paleontológico (eds.), *Actas de las I Jornadas Internacionales sobre Paleontología de Dinosaurios y su Entorno*, 329–337. Burgos, España: Salas de los Infantes.

Fujita, M., Y.-N. Lee, Y. Azuma, and D. Li. 2012. Unusual tridactyl trackways with tail traces from the Lower Cretaceous Hekou Group, Gansu Province, China. *Palaios* 27: 560–570.

Fujita, M., Z. Wang, Y. Azuma, M. Shibata, and Z. Dong. 2008. First dinosaur track site from the Lower Cretaceous of Yunnan Province, China. *Memoir of the Fukui Prefectural Dinosaur Museum* 7: 33–43.

Fuller, M. K. 1981. *Characteristics of an American Alligator (*Alligator mississippiensis*) Population in the Vicinity of Lake Ellis Simon, North Carolina.* Master's Thesis. Raleigh, NC: North Carolina State University, 136 pp.

Galton, P. M. 1971. The prosauropod dinosaur *Ammosaurus*, the crocodile *Protosuchus*, and their bearing on the age of the Navajo Sandstone of northeastern Arizona. *Journal of Paleontology* 45: 781–795.

Galton, P. M. 1973. On the anatomy and relationships of *Efraasia diagnostica* (Huene), n. gen., a prosauropod dinosaur (Reptilia: Saurischia) from the Upper Triassic of Germany. *Paläontologische Zeitschrift* 47: 229–255.

Galton, P. M. 1974. The ornithischian dinosaur *Hypsilophodon* from the Wealden of the Isle of Wight. *Bulletin of the British Museum of Natural History (Geology)* 25: 1–152.

Galton, P. M. 1976. Prosauropod dinosaurs (Reptilia: Saurischia) of North America. Yale Peabody Museum *Postilla* 169: 1–98.

Galton, P. M. 1981. *Dryosaurus*, a hypsilophodontid dinosaur from the Upper Jurassic of North America and Africa. Postcranial skeleton. *Paläontologische Zeitschrift* 55: 271–312.

Galton, P. M. 1983. The cranial anatomy of *Dryosaurus*, a hypsilophodontid dinosaur from the Upper Jurassic of North America and East Africa, with a review of the hypsilophodontids from the Upper Jurassic of North America. *Geologica et Palaeontologica* 17: 207–243.

Galton, P. M. 1984*a*. An early prosauropod dinosaur from the Upper Triassic of Nordwürttemberg, West Germany. *Stuttgarter Beiträge zur Naturkunde Serie B (Geologie und Paläontologie)* 106: 25.

Galton, P. M. 1984*b*. Cranial anatomy of the prosauropod dinosaur *Plateosaurus* from the Knollenmergel (Middle Keuper, Upper Triassic) of Germany. *Geologica et Palaeontologica* 18: 139–171.

Galton, P. M. 1986. Prosauropod dinosaur *Plateosaurus* (=*Gresslyosaurus*) (Saurischia: Sauropodomorpha) from the Upper Triassic of Switzerland. *Geologica et Palaeontologica* 20: 167–183.
Galton, P. M. 1998. The prosauropod dinosaur *Plateosaurus (Dimodosaurus) poligniensis* (Pidancet & Chopard, 1862) (Upper Triassic, Poligny, France). *Neues Jahrbuch für Geologie und Paläontologie Abhandlungen* 207: 255–288.
Galton, P. M. 2015. The holotype pes of the Morrison dinosaur *Camptonotus amplus* Marsh, 1879 (Upper Jurassic, western USA)–is it *Camptosaurus*, Sauropoda, or *Allosaurus*? *Neues Jahrbuch für Geologie und Paläontologie Abhandlungen* 275: 317–335.
Galton, P. M., and M. A. Cluver. 1976. *Anchisaurus capensis* (Broom) and a revision of the Anchisauridae (Reptilia, Saurischia). *Annals of the South African Museum* 69: 121–159.
Galton, P. M., and D. Kermack. 2010. The anatomy of *Pantydraco caducus*, a very basal sauropodomorph dinosaur from the Rhaetian (Upper Triassic) of south Wales, UK. *Revue de Paléontologie, Genève* 29: 341–404.
Galton, P. M., and H. P. Powell. 1980. The ornithischian dinosaur *Camptosaurus prestwichii* from the Upper Jurassic of England. *Palaeontology* 23: 411–443.
Galton, P. M., and P. Upchurch. 2004. Prosauropoda. In D. B. Weishampel, P. Dodson, and H. Osmólska (eds.). *The Dinosauria*, 2nd ed., 232–258. Berkeley: University of California Press.
Galton, P. M., and J. VanHeerden. 1985. Partial hindlimb of *Blikanasaurus cromptoni* N. gen. and N. sp., representing a new family of prosauropod dinosaurs from the Upper Triassic of South Africa. *Geobios* 18: 509–516.
Gand, G. 1975. Sur l'interprétation paléozoologique d'un nouvel assemblage ichnologique des carrières de Saint-Sernin-du-Bois (Saône-et-Loire). *Bulletin Trimestriel de la Société d'Histoire Naturelle et des Amis du Musèum d'Autun* 73: 6–23.
Gand, G. 1976. Étude biométrique de traces de pas laissées par *Gallus gallus*, variété Sussex. *Bulletin de la Société d'Histoire Naturelle du Cruesot* 34: 19–23.
Gand, G. 1979*a*. Interprétations paléontologique et paléoecologique d'un sixième assemblage à traces de Reptiles des carrières triasiques de Saint-Sernin-du-Bois (Autunois-France); conclusions générales à l'étude du fisement fossilifère. *Bulletin Trimestriel de la Société d'Histoire Naturelle et des Amis du Musèum d'Autun* 87: 9–29.
Gand, G. 1979*b*. Note sur quelques nouvelles pistes de Reptiles observées dans le Trias moyen du Plateau d'Antully (Saône-et-Loire–France). *Bulletin Trimestriel de la Société d'Histoire Naturelle et des Amis du Musèum d'Autun* 92: 7–20.
Gand, G., and G. Demathieu. 2005. Les pistes dinosauroïdes du Trias moyen Français: interprétation et réévaluation de la nomenclature. *Geobios* 38: 725–749.
Gand, G., G. Demathieu, and C. Montenat. 2007. Les traces de pas d'amphibiens, de dinosaures et autres reptiles du Mesozoïque Français: inventaire et interpretations. *Palaeovertebrata* 2007: 1–149.
Gand, G., M. Vianey-Liaud, G. Demathieu, and J. Garric. 2000. Deux novelles traces de pas de dinosaures du Trias Supérieur de la Bordure Cévenole (La Grand-Comb, sud-est de la France). *Geobios* 33: 599–624.
Gangloff, R. A., K. C. May, and J. E. Storer. 2004. An early Late Cretaceous dinosaur tracksite in central Yukon Territory, Canada. *Ichnos* 11: 299–309.
Gao, C., E. M. Morschhauser, D. J. Varrichio, J. Liu, and B. Zhao. 2012. A second soundly sleeping dragon: new anatomical details of the Chinese troodontid *Mei long* with implications for phylogeny and taphonomy. *PloS One* 7(9): e45203. doi:10.1371/journal.pone.0045203.
García-Moreno, J., M. D. Sorenson, and D. P. Mindell. 2003. Congruent avian phylogenies inferred from mitochondrial and nuclear DNA sequences. *Journal of Molecular Evolution* 57: 27–37.
Gaston, R., M. G. Lockley, S. G. Lucas, and A. P. Hunt. 2003. *Grallator*-dominated fossil footprint assemblages and associated enigmatic footprints from the Chinle Group (Upper Triassic, Gateway Area, Colorado. *Ichnos* 10: 153–163.
Gatesy, S. M. 2001. Skin impressions of Triassic theropods as records of foot movement. *Bulletin of the Museum of Comparative Zoology, Harvard University* 156: 137–149.
Gatesy, S. M. 2003. Direct and indirect track features: what sediment did a dinosaur touch? *Ichnos* 10: 91–98.
Gatesy, S. M., M. Bäker, and J. R. Hutchinson. 2009. Constraint-based exclusion of limb poses for reconstructing theropod dinosaur locomotion. *Journal of Vertebrate Paleontology* 29: 535–544.
Gatesy, S. M., and R. G. Ellis. 2016. Beyond surfaces: a particle-based perspective on track formation. In P. L. Falkingham, D. Marty, and A. Richter (eds.), *Dinosaur Tracks: The Next Steps*, 82–91. Bloomington: Indiana University Press.
Gatesy, S. M., K. M. Middleton, F. A. Jenkins, Jr., and N. H. Shubin. 1999. Three-dimensional preservation of foot movements in Triassic theropod dinosaurs. *Nature* 399: 141–144.
Gauthier, J. A., S. J. Nesbitt, E. R. Schachner, G. S. Bever, and W. G. Joyce. 2011. The bipedal stem crocodilian *Poposaurus gracilis*: inferring function in fossils and innovation in archosaur locomotion. *Bulletin of the Peabody Museum of Natural History* 52: 107–126.
Getty, P. R. 2005. Excavated and *in situ* dinosaur footprints form the Murray Quarry (Early Jurassic East Berlin Formation), Holyoke, Massachusetts, USA. *Ichnos* 12: 163–178.
Getty, P. R., C. Aucoin, N. Fox, A. Judge, L. Hardy, and A. M. Bush. 2017. Perennial lakes as an environmental control on theropod movement in the Jurassic of the Hartford Basin. *Geosciences* 7: 13. doi:10.3390/geosciences 7010013.
Getty, P. R., L. Hardy, and A. M. Bush. 2015. Was the *Eubrontes* track maker gregarious? Testing the herding hypothesis at Powder Hill Dinosaur Park, Middlefield, Connecticut. *Bulletin of the Peabody Museum of Natural History* 56: 95–106.
Getty, P. R., A. Judge, J. Csonka, and A. M. Bush. 2012. Were Early Jurassic dinosaurs gregarious? Reexamining the evidence from Dinosaur Footprint Reservation in Holyoke, Massachusetts. In M. A. Thomas (ed.), *Guidebook for Fieldtrips in Connecticut and Massachusetts*, A1–A18. Geological Society of America Northeast Section and 47th Annual Meeting, Hartford, Connecticut: State Geological and Natural History of Connecticut Guidebook 9.
Giebeler, E. M., and J. Werner. 2011. The life cycle of sauropod dinosaurs. In N. Klein, K. Remes, C. T. Gee, and P. M. Sander (eds.), *Biology of the Sauropod Dinosaurs: Understanding the Life of Giants*, 263–275. Bloomington: Indiana University Press.
Gierliński, G. 1991. New dinosaur ichnotaxa from the Early Jurassic of the Holy Cross Mountains, Poland. *Palaeogeography, Palaeoclimatology, Palaeoecology* 85: 137–148.
Gierliński, G. 1994. Early Jurassic theropod tracks with the metatarsal impressions. *Przeglad Geologiczny* 42: 280–284.
Gierliński, G. 1995. New theropod tracks from the Early Jurassic strata of Poland. *Przeglad Geologiczny* 43: 931–934.
Gierliński, G. 1996*a*. Avialian theropod tracks from the Early Jurassic strata of Poland. *Zubía* 14: 79–87.
Gierliński, G. 1996*b*. Dinosaur ichnotaxa from the Lower Jurassic of Hungary. *Geological Quarterly* 40: 119–128.
Gierliński, G., A. Lagnaoui, H. Klein, H. Saber, M. Oukassou, and A. Charrière. 2017. Bird-like tracks from the Imchil Formation (Middle Jurassic, Bajocian-Bathonian) of the Central High Atlas, Morocco, in comparison with similar Mesozoic tridactylous ichnotaxa. *Bollettino della Società Paleontologica Italiana* 56: 207–215.
Gierliński, G., and M. G. Lockley. 2013. First report of probable therizinosaur (cf. *Macropodosaurus*) tracks from North America, with notes on the neglected vertebrate ichnofauna of the Ferron Sandstone (Late Cretaceous) of central Utah. In A. L. Titus and M. A. Loewen (eds.), *At the Top of the Grand Staircase: The Late Cretaceous of Southern Utah*, 530–535. Bloomington: Indiana University Press.
Gierliński, G. D., M. G. Lockley, and G. Niedźwiedzki. 2009. A distinctive crouching theropod trace from the Lower Jurassic of Poland. *Geological Quarterly* 53: 471–476.
Gierliński, G., and G. Niedźwiedzki. 2002. Dinosaur footprints from the Upper Jurassic of Blaziny, Poland. *Geological Quarterly* 46: 463–465.
Gierliński, G., and G. Niedźwiedzki. 2005. New saurischian dinosaur footprints from the Lower Jurassic of Poland. *Geological Quarterly* 49: 99–104.
Gierliński, G., G. Niedźwiedzki, and P. Nowacki. 2009. Small theropod and ornithopod footprints in the Late Jurassic of Poland. *Acta Geologica Polonica* 59: 221–234.
Gierliński, G., G. Niedźwiedzki, and G. Pieńkowski. 2001. Gigantic footprint of a theropod dinosaur in the Early Jurassic of Poland. *Acta Palaeontologica Polonica* 46: 441–446.

Gierliński, G., and G. Pieńkowski. 1999. Dinosaur track assemblages from the Hettangian of Poland. *Geological Quarterly* 43: 329–346.

Gierliński, G., G. Pieńkowski, and G. Niedźwiedzki. 2004. Tetrapod track assemblage in the Hettangian of Sołtyków, Poland, and its paleoenvironmental background. *Ichnos* 11: 195–213.

Gierliński, G., and A. Potemska. 1987. Lower Jurassic dinosaur footprints from Gliniany Las, northern slope of the Holy Cross Mountains, Poland. *Neues Jahrbuch für Geologie und Paläontologie Abhandlungen* 175: 107–120.

Gilmore, C. W. 1915. Osteology of *Thescelosaurus*, an orthopodous dinosaur from the Lance Formation of Wyoming. *Proceedings of the United States National Museum* 49: 591–616.

Gilmore, C. W. 1920. Osteology of the Carnivorous Dinosauria in the United States National Museum, with Special Reference to the Genera *Antrodemus* (*Allosaurus*) and *Ceratosaurus*. Bulletin 110. Washington, DC: United States National Museum, Smithsonian Institution, 159 pp.

Godefroit, P., A. Cau, H. Dong-Yu, F. Escuillié, W. Wenhao, and G. Dyke. 2013. A Jurassic avialan dinosaur from China resolves the early phylogenetic history of birds. *Nature* 498: 359–362.

Goodwin, T. M., and W. R. Marion. 1978. Aspects of the nesting ecology of American alligators (*Alligator mississippiensis*) in north-central Florida. *Herpetologica* 34: 43–47.

Graversen, O., J. Milàn, and D. B. Loope. 2007. Dinosaur tectonics: a structural analysis of theropod undertracks with a reconstruction of theropod walking dynamics. *Journal of Geology* 115: 641–654.

Grealy, A., M. Phillips, G. Miller, M. T. P. Gilbert, J.-M. Rouillard, D. Lambert, M. Bunce, and J. Halle. 2017. Eggshell palaeogenomics: palaeognath evolutionary history revealed through ancient nuclear and mitochondrial DNA from Madagascan elephant bird (*Aepyornis* sp.) eggshell. *Molecular Phylogenetics and Evolution* 109: 151–163.

Green, R. E., E. L. Braun, J. Armstrong, D. Earl, N. Nguyen, G. Hickey, M. W. Vandewege, J. A. St. John, S. Capella-Gutiérrez, T. A. Castoe, C. Kern, M. K. Fujita, J. C. Opazo, J. Jurka, K. K. Kojima, J. Caballero, R. M. Hubley, A. F. Smit, R. N. Platt, C. A. Lavoie, M. P. Ramakodi, J. W. Finger Jr., A. Suh, S. R. Isberg, L. Miles, A. Y. Chong, W. Jaratlerdsiri, J. Gongora, C. Moran, A. Iriarte, J. McCormack, S. C. Burgess, S. V. Edwards, E. Lyons, C. Williams, M. Breen, J. T. Howard, C. R. Gresham, D. G. Peterson, J. Schmitz, D. D. Pollock, D. Haussler, E. W. Triplett, G. Zhang, N. Irie, E. D. Jarvis, C. A. Brochu, C. J. Schmidt, F. M. McCarthy, B. C. Faircloth, F. G. Hoffman, T. C. Glen, T. Gabaldón, B. Paten, and D. A. Ray. 2014. Three crocodilian genomes reveal ancestral patterns of evolution among archosaurs. *Science* 346, 1254449. DOI:10.1126/science.1254449.

Grigg, G., and D. Kirshner. 2015. *Biology and Evolution of Crocodylians*. Ithaca, NY: Cornell University Press, 649 pp.

Gulka, J., and D. P. Mindell. 2010. Phasianidae. Chickens, grouse, partridges, quail, pheasants, etc. Version 23 July 2011 (temporary). http://tolweb.org/Phasianidae/26302/2011.07.23.

Hackett, S. J., R. T. Kimball, S. Reddy, R. C. K. Bowie, E. L. Braun, M. J. Braun, J. L. Chojnowski, W. A. Cox, K.-L. Han, J. Harshman, C. J. Huddleston, B. D. Marks, K. J. Miglia, W. S. Moore, F. H. Sheldon, D. W. Steadman, C. C. Witt, and T. Yuri. 2008. A phylogenomic study of birds reveals their evolutionary history. *Science* 320: 1763–1768.

Haddrath, O., and A. J. Baker. 2012. Multiple nuclear genes and retroposons support vicariance and dispersal of the palaeognaths, and an Early Cretaceous origin of modern birds. *Proceedings of the Royal Society B* 279: 4617–4625.

Hadri, M. H., M. Boutakiout, F. Gómez, and F. Pérez-Lorente. 2015. Crocodyliform footprints from "les couches rouges" of the Middle Jurassic of Msemrir, High Atlas, Morocco. *Geogaceta* 58: 39–42.

Hadri, M., X. Pereda Suberbiola, M. Boutakiout, and F. Pérez-Lorente. 2007. Icnitas de posibles dinosaurios tireóforos del Jurásico Inferior (Alto Atlas, Goulmima, Marruecos). *Revista Española de Paleontología* 22: 147–156.

Hamblin, A. H., M. G. Lockley, and A. R. C. Milner. 2006. More reports of theropod dinosaur tracksites from the Kayenta Formation (Lower Jurassic), Washington County, Utah: implications for describing the Springdale Megatracksite. In J. D. Harris, S. G. Lucas, J. A. Spielmann, M. G. Lockley, and A. R. C. Milner (eds.), *The Triassic-Jurassic Transition*, 276–281. Albuquerque, NM: New Mexico Museum of Natural History and Science Bulletin 37.

Hammer, Ø., and D. Harper. 2006. *Paleontological Data Analysis*. Malden, MA: Blackwell Publishing, 351 pp.

Han, F.-L., P. M. Barrett, R. J. Butler, and X. Xu. 2012. Postcranial anatomy of *Jeholosaurus shangyuanensis* (Dinosauria, Ornithischia) from the Lower Cretaceous Yixian Formation of China. *Journal of Vertebrate Paleontology* 32: 1370–1395.

Harington, C. R., H. Foster, R. Holmes, and P. J. Currie. 2005. *Photographic Catalogue of Trackways in the Canadian Museum of Nature*. *Syllogeus* 74. Ottawa, Canada: Canadian Museum of Nature, 151 pp.

Harrison, G. L., P. A. McLenachan, M. J. Phillips, K. E. Slack, A. Cooper, and D. Penny. 2004. Four new avian mitochondrial genomes help get to basic evolutionary questions in the Late Cretaceous. *Molecular Biology and Evolution* 21: 974–983.

Harshman, J. 2008*a*. Gruiformes. Cranes, Rails, Coots, and allies. Version 27 June 2008 (under construction); http://tolweb.org/Gruiformes/26307/2008.06.27 in the Tree of Life Project.

Harshman, J. 2008*b*. 'Land Birds.' Version 27 June 2008 (under construction); http://tolweb.org/%27Land_Birds%27/26410/2008.06.27 in the Tree of Life Project.

Harshman, J., E. L. Braun, M. J. Braun, C. J. Huddleston, R. C. K. Bowie, J. L. Chojnowski, S. J. Hackett, K.-L. Han, R. T. Kimball, B. D. Marks, K. J. Miglia, W. S. Moore, S. Reddy, F. H. Sheldon, D. W. Steadman, S. J. Steppan, C. C. Witt, and T. Yuri. 2008. Phylogenomic evidence for multiple losses of flight in ratite birds. *Proceedings of the National Academy of Science USA* 105: 13462–13467.

Harshman, J., and J. W. Brown. 2008. Charadriiformes. Shorebirds and relatives. Version 24 June 2008 (temporary); http://tolweb.org/Charadriiformes/26342/2008.06.24 in the Tree of Life Project.

Harshman, J., and J. W. Brown. 2010. Palaeognathae. Version 13 May 2010 (under construction); http://tolweb.org/Palaeognathae/15837/2010.05.13 in the Tree of Life Project.

Haubold, H., and H. Klein. 2000. Die dinosauroiden Fährten *Parachirotherium-Atreipus-Grallator* aus dem unteren Mittelkeuper (Obere Trias: Ladin, Karn, ?Nor) in Franken. *Hallesches Jahrbuch für Geowissenschaften* B22: 59–85.

Haubold, H., and H. Klein. 2002. Chirotherien und Grallatoriden aus der Unteren bis Oberen Trias Mitteleuropas und die Enstehung der Dinosauria. *Hallesches Jahrbuch für Geowissenschaften* B24: 1–22.

Haughton, S. H. 1924. The fauna and stratigraphy of the Stormberg Series. *Annals of the South African Museum* 12: 323–497.

Hawthorne, J. M., R. M. Bonem, J. O. Farlow, and J. O. Jones. 2002. Ichnology, stratigraphy and paleoenvironment of the Boerne Lake Spillway dinosaur tracksite, south-central Texas. *Texas Journal of Science* 54: 309–324.

Henderson, D. M. 2006. Simulated weathering of dinosaur tracks and the implications for their characterization. *Canadian Journal of Earth Sciences* 43: 691–704.

Hernández Medrano, N., C. Pascual Arribas, P. Latorre Macarrón, and E. Sanz Pérez. 2004. Huellas de terópodos y pterosaurios en Valdegen I (Villar del Río, Soria, España). In Colectivo Arqueológico-Paleontológico Salense (eds.), *Actas de las III Jornadas sobre Dinosauros y su Entorno*, 253–271. Burgos, Spain: Salas de los Infantes.

Hernández Medrano, N., C. Pascual Arribas, F. Pérez-Lorente, and R. Sesma. 2014. Icnitas terópodas, ornitópodas y de pterosaurio en la formación Aguilar del Río Alhama, Grupo de Oncala, Cuenca de Cameros. *Zubía* 32: 33–71.

Hernández Medrano, N., and F. Pérez-Lorente. 2003. Un nuevo yacimiento de icnitas de dinosaurio La llaga (Terroba, La Rioja. España). Su flexible y rastrilladas paralelas. In F. Pérez (ed.), *Dinosaurios y Otros Reptiles Mesozoicos en España*, 195–214. Logroño, Spain: Gobierno de La Rioja, Fundación Patrimonio Paleontológico de La Rioja, Instituto de Estudios Riojanos, Universidad de La Rioja.

Herrero, C., E. Herrero, J. Martín-Chivelet, and F. Pérez-Lorente. 2016. Contribution to knowledge of the last dinosaur footprints in Europe. Persistence of ornithopods in the upper Maastrichtian of SE Spain. *Cretaceous Research* 57: 490–507.

Higgins, P. J., and S. J. J. F. Davies (eds.). 1996. *Handbook of Australian, New Zealand and Antarctic Birds. Volume 3: Snipe to Pigeons*. Melbourne: Oxford University Press Australia, 1028 pp.

Hines, T. C., M. J. Fogarty, and L. C. Chappell. 1968. Alligator research in Florida: a progress report. *Proceedings of the 22nd Annual*

Conference of the Southeastern Association of Game and Fish Commissioners 22: 166–180.
Hitchcock, E. 1844. Report on ichnolithology, or fossil footmarks, with a description of several new species, and the coprolites of birds, from the valley of Connecticut River, and of a supposed footmark from the Valley of Hudson River. *American Journal of Science* (second series) 47: 292–322.
Hitchcock, E. 1858. *Ichnology of New England: A Report on the Sandstone of the Connecticut Valley, Especially Its Fossil Footmarks, Made to the Commonwealth of Massachusetts.* Boston: William White (Reprint edition, Arno Press, New York), 199 pp.
Hitchcock, E. 1863. *Supplement to the Ichnology of New England. A Report to the Government of Massachusetts.* Boston: Wright and Potter, 96 pp.
van Hoepen, E. C. N. 1920*a*. Contributions to the knowledge of the reptiles of the Karroo Formation. 5. A new dinosaur from the Stormberg beds. *Annals of the Transvaal Museum* 7: 77–92.
van Hoepen, E. C. N. 1920*b*. Contributions to the knowledge of the reptiles of the Karroo Formation. 6. Further dinosaurian material in the Transvaal Museum. *Annals of the Transvaal Museum* 7: 93–140.
Holliday, C. M., R. C. Ridgely, J. C.Sedlmayr, and L.M. Witmer. 2010. Cartilaginous epiphyses in extant archosaurs and their implications for reconstructing limb function in dinosaurs. *PLoS One* 5(9): e13120. doi:10.1371/journal.pone.0013120.
Holtz, T. R., Jr. 1994. The arctometatarsalian pes, an unusual structure of the metatarsus of Cretaceous Theropoda (Dinosauria: Saurischia). *Journal of Vertebrate Paleontology* 14: 480–519.
Hone, D. W. E., A. A. Farke, and M. J. Wedel. 2016. Ontogeny and the fossil record: what, if anything, is an adult dinosaur? *Biology Letters* 12: 20150947. dx.doi.org/10.1098/rsbl.2015.0947.
Horner, J. R., A. de Ricqlès, and K. Padian. 1999. Variation in dinosaur skeletochronology indicators: implications for age assessment and physiology. *Paleobiology* 25: 295–304.
Horner, J. R., A. de Ricqlès, and K. Padian. 2000. Long bone histology of the hadrosaurid dinosaur *Maiasaura peeblesorum*: growth dynamics and physiology based on an ontogenetic series of skeletal elements. *Journal of Vertebrate Paleontology* 20: 115–129.
Hornung, J. J., A. Böhme, N. Schlüter, and M. Reich. 2016. Diversity, ontogeny, or both? A morphometric approach to iguanodontian ornithopod (Dinosauria: Ornithischia) track assemblages from the Berriasian (Lower Cretaceous) of northwestern Germany. In P. L. Falkingham, D. Marty, and A. Richter (eds.), *Dinosaur Tracks: The Next Steps*, 202–225. Bloomington: Indiana University Press.
Hornung, J. J., A. Böhme, T. Van der Lubbe, M. Reich, and A. Richter. 2012. Vertebrate tracksites in the Obernkirchen Sandstone (Late Berriasian, Early Cretaceous) of northwest Germany—their stratigraphical, palaeogeographical, palaeoecological, and historical context. *Paläontologische Zeitschrift* 86: 231–267.
Houck, M. A., J. A. Gauthier, and R. E. Strauss. 1990. Allometric scaling in the earliest fossil bird, *Archaeopteryx lithographica*. *Science* 247: 195–198.
Hu, D., L. Hou, L. Zhang, and X. Xu. 2009. A pre-*Archaeopteryx* troodontid theropod from China with long feathers on the metatarsus. *Nature* 461: 640–643.
Huang, J., X. Wang, Y. Hu, J. Liu, J. A. Peteya, and J. A. Clarke. 2016. A new ornithurine from the Early Cretaceous of China sheds light on the evolution of early ecological and cranial diversity in birds. *PeerJ* 4: e1765. doi:10.7717/peerj.1765.
Hübner, T. 2016. Elusive ornithischian tracks in the famous Berriasian (Lower Cretaceous) "Chicken Yard" tracksite of northern Germany: quantitative differentiation between small tridactyl trackmakers. In P. L. Falkingham, D. Marty, and A. Richter (eds.), *Dinosaur Tracks: The Next Steps*, 310–330. Bloomington: Indiana University Press.
von Huene, F. 1907–1908. Die Dinosaurier der europäischen Triasformation mit Berucksichtigung der aussereuropäischen Vorkommnisse. *Geologische und Paläontologische Abhandlungen Supplement 1*, 419 pp.
von Huene, F. 1913. Beobachtungen über die Bewegungsart der Extremitäten bei Krokodilen. *Biologisches Centralblatt* 33: 468–472.
von Huene, F. 1932. Die fossile Reptil-Ordnung Saurischia, ihre Entwicklung und Geschichte. *Monographien zur Geologie und Palaeontologie (serie 1)*. Leipzig: Verlag von Gebrüder Bornträger, 361 pp.
von Huene, F. 1934. Ein neuer Coelurosaurier in der thüringischen Trias. *Palaeontologische Zeitschrift* 16: 145–170.
Huerta, P., F. T. Fernández-Baldor, J. O. Farlow, and D. Montero. 2012. Exceptional preservation processes of 3D dinosaur footprint casts in Costalomo (Lower Cretaceous, Cameros Basin, Spain). *Terra Nova* 24: 136–141.
Huh, M., I. S. Paik, M. G. Lockley, K. G. Hwang, B. S. Kim, and S. K. Kwak. 2006. Well-preserved theropod tracks from the Upper Cretaceous of Hwasun County, southwestern Korea, and their paleobiological implications. *Cretaceous Research* 27: 123–138.
Hulke, J. W. 1882. Description of some *Iguanodon*-remains indicating a new species, *I. seeleyi*. *Quarterly Journal of the Geological Society of London* 38: 135–144.
Hunt, A. P., and S. G. Lucas. 2006. The taxonomic status of *Navahopus falcipollex* and the ichnofauna and ichnofacies of the Navajo Lithosome (Lower Jurassic) of western North America. In J. D. Harris, S. G. Lucas, J. A. Spielmann, M. G. Lockley, A. R. C. Milner, and J. I. Kirkland (eds.), *The Triassic-Jurassic Terrestrial Transition*, 164–169. Albuquerque, NM: New Mexico Museum of Natural History and Science Bulletin 37.
Hunt, A. P., and S. G. Lucas. 2007. Late Triassic tetrapod tracks of western North America. In S. G. Lucas and J. A. Spielmann (eds.), *Triassic of the American West*, 215–230. Albuquerque, NM: New Mexico Museum of Natural History and Science Bulletin 40.
Hurum, J. H., P. S. Druckenmiller, O. Hammer, H. A. Nakrem, and S. Olaussen. 2016. The theropod that wasn't: an ornithopod tracksite from the Helvetiafjellet Formation (Lower Cretaceous) of Boltodden, Svalbard. In B. P. Kear, J. Lindgren, J. H. Hurum, J. Milàn, and V. Vajda (eds.), *Mesozoic Biotas of Scandinavia and Its Arctic Territories*. Geological Society of London Special Publication 434. http://doi.org/10.1144/SP434.10.
Hutchinson, J. R. 2004. Biomechanical modeling and sensitivity analysis of bipedal running ability. II. Extinct taxa. *Journal of Morphology* 262: 441–461.
Hutchinson, J. R., F. C. Anderson, S. S. Blemker, and S. L. Delp. 2005. Analysis of hindlimb muscle moment arms in *Tyrannosaurus rex* using a three-dimensional musculoskeletal computer model: implications for stance, gait, and speed. *Paleobiology* 31: 676–701.
Hutchinson, J. R., K. T. Bates, J. Molnar, V. Allen, and P. J. Makovicky. 2011. A computational analysis of limb and body proportions in *Tyrannosaurus rex* with implications for locomotion, ontogeny, and growth. *PLoS One* 6(10): e326037. doi: 10.1371/journal.pone.0026037.
Hutchinson, J. R., and M. Garcia. 2002. *Tyrannosaurus* was not a fast runner. *Nature* 415: 1018–1021.
Hutchinson, J. R., V. Ng-Thow-Hing, and F. C. Anderson. 2007. A 3D interactive method for estimating body segmental parameters in animals: application to the turning and running performance of *Tyrannosaurus rex*. *Journal of Theoretical Biology* 246: 660–680.
Huynen, L., B. J. Gill, A. Doyle, C. D. Millar, and D. M. Lambert. 2014. Identification, classification, and growth of moa chicks (Aves: Dinornithiformes) from the genus *Euryapteryx*. *PLoS One* 9(6): e99929. doi:10.1371/journal.pone.0099929.
Huynen, L., and D. M. Lambert. 2014. Complex species status for extinct moa (Aves: Dinornithiformes) from the genus *Euryapteryx*. *PLoS One* 9(3): e90212. doi:10.1371/journal.pone.0090212.
Huynen, L., C. D. Millar, R. P. Scofield, and D. M. Lambert. 2003. Nuclear DNA sequences detect species limits in ancient moa. *Nature* 425: 175–178.
Ibrahim, N., P. C. Sereno, C. Dal Sasso, S. Manganuco, M. Fabbri, D. M. Martill, S. Zouhri, N. Myhrvold, and D. A. Iurino. 2014. Semiaquatic adaptations in a giant predatory dinosaur. *Science* 345: 1613–1616.
Imbrie, J. 1956. Biometrical methods in the study of invertebrate fossils. *Bulletin of the American Museum of Natural History* 108: 214–252.
Irby, G. V. 1996*a*. Paleoichnological evidence for running dinosaurs worldwide. In M. Morales (ed.), *The Continental Jurassic*, 109–112. Flagstaff, AZ: Museum of Northern Arizona Bulletin 60.
Irby, G. V. 1996*b*. Paleoichnology of the Cameron Dinosaur Tracksite, Lower Jurassic Moenave Formation, Northeastern Arizona. In M. Morales (ed.), *The Continental Jurassic*, 147–166. Flagstaff, AZ: Museum of Northern Arizona Bulletin 60.

Irmis, R. B. 2005. A review of the vertebrate fauna of the Lower Jurassic Navajo Sandstone in Arizona. In R. D. McCord (ed.), *Vertebrate Paleontology of Arizona*, 55–71. Mesa, AZ: Mesa Southwest Museum Bulletin 11.

Ishigaki, S. 1988. Les empreintes de dinosaures du Jurassique inférieur du Haut Atlas central marocain. *Notes du Service Géologique du Maroc* 44: 79–86.

Ishigaki, S. 2010. Theropod trampled bedding plane with laboring trackways from the Upper Cretaceous Abdrant Nuru fossil site, Mongolia. *Hayashibara Museum of Natural Sciences Research Bulletin* 3: 133–141.

Ishigaki, S., and M. G. Lockley. 2010. Didactyl, tridactyl and tetradactyl theropod trackways from the Lower Jurassic of Morocco: evidence of limping, labouring and other irregular gaits. *Historical Biology* 22: 100–108.

Ishigaki, S., M. Watabe, K. Tsogtbaatar, and M. Saneyoshi. 2009. Dinosaur footprints from the Upper Cretaceous of Mongolia. *Geological Quarterly* 53: 449–460.

Jackson, S. J., M. A. Whyte, and M. Romano. 2009. Laboratory-controlled simulations of dinosaur footprints in sand: a key to understanding vertebrate track formation and preservation. *Palaios* 24: 222–238.

Jackson, S. J., M. A. Whyte, and M. Romano. 2010. Range of experimental dinosaur (*Hypsilophodon foxii*) footprints due to variations in sand consistency: how wet was the track? *Ichnos* 17: 197–214.

Jarvis, E. D., S. Mirarab, A. J. Aberer, B. Li, P. Houde, C. Li, S. Y. W. Ho, B. C. Faircloth, B. Nabholz, J. T. Howard, A. Suh, C. C. Weber, R. R. da Fonseca, J. Li, F. Zhang, H. Li, L. Zhou, N. Narula, L. Liu, G. Ganapathy, B. Boussau, M. S. Bayzid, V. Zavidovych, S. Subramanian, T. Gabaldón, S. Capella-Gutiérrez, J. Huerta-Cepas, B. Rekepalli, K. Munch, M. Schierup, B. Lindow, W. C. Warren, D. Ray, R. E. Green, M. W. Bruford, X. Zhan, A. Dixon, S. Li, N. Li, Y. Huang, E. P. Derryberry, M. F. Bertelsen, F. H. Sheldon, R. T. Brumfield, C. V. Mello, P. V. Lovell, M. Wirthlin, M. P. C. Schneider, F. Prosdocimi, J. A. Samaniego, A. M. V. Velazquez, A. Alfaro-Núñez, P. F. Campos, B. Petersen, T. Sicheritz-Ponten, A. Pas, T. Bailey, P. Scofield, M. Bunce, D. M. Lambert, Q. Zhou, P. Perelman, A. C. Driskell, B. Shapiro, Z. Xiong, Y. Zeng, S. Liu, Z. Li, B. Liu, K. Wu, J. Xiao, X. Yinqi, Q. Zheng, Y. Zhang, H. Yang, J. Wang, L. Smeds, F. E. Rheindt, M. Braun, J. Fjeldsa, L. Orlando, F. K. Barker, K. A. Jansson, W. Johnson, K.-P. Koepfli, S. O'Brien, D. Haussler, O. A. Ryder, C. Rahbek, E. Willerslev, G. R. Graves, T. C. Glenn, J. McCormack, D. Burt, H. Ellegren, P. Alström, S. V. Edwards, A. Stamatakis, D. P. Mindell, J. Cracraft, E. L. Braun, T. Warnov, W. Jun, M. T. P. Gilbert, and G. Zhang. 2014. Whole-genome analyses resolve early branches in the tree of life of modern birds. *Science* 346: 1320–1331.

Jasinski, L. E. 2008. *Dinosaur Highway: A History of Dinosaur Valley State Park*. Fort Worth: Texas Christian University Press, 212 pp.

Jenkins, F. A., Jr., N. H. Shubin, W. W. Amaral, S. M. Gatesy, C. R. Schaff, L. B. Clemmensen, W. R. Downs, A. R. Davidson, N. Bonde, and F. Osbæck. 1994. Late Triassic continental vertebrates and depositional environments of the Fleming Fjord Formation, Jameson Land, East Greenland. *Meddelelser om Grønland, Geoscience* 32: 1–25.

Ji, S., J. Atterholt, J. K. O'Connor, M. C. Lamanna, J. D. Harris, D.-Q. Li, H.-L. You, and P. Dodson. 2011. A new, three-dimensionally preserved enantiornithine bird (Aves: Ornithothoraces) from Gansu Province, north-western China. *Zoological Journal of the Linnean Society* 162: 201–219.

Jiménez Vela, A., and F. Pérez-Lorente. 2006–2007. El Corral del Totico. Dos nuevos yacimientos con pistas singulares. (Enciso, La Rioja, España). *Zubía Monográfico* 18–19: 115–144.

Joanen, T. 1969. Nesting ecology of alligators in Louisiana. *Proceedings of the 23rd Annual Conference of the Southeastern Association of Game and Fish Commissioners* 23: 141–151.

Joanen, T., and L. McNease. 1970. A telemetric study of nesting female alligators on Rockefeller Refuge, Louisiana. *Proceedings of the 24th Annual Conference of the Southeastern Association of Game and Fish Commissioners* 24: 175–193.

Joanen, T., and L. McNease. 1972. A telemetric study of adult male alligators on Rockefeller Refuge, Louisiana. *Proceedings of the 26th Annual Conference of the Southeastern Association of Game and Fish Commissioners* 26: 252–275.

Joanen, T., and L. McNease. 1980. Reproductive biology of the American alligator in southwest Louisiana. In J. B. Murphy and J. T. Collins (eds.), *Reproductive Biology and Diseases of Captive Reptiles*, 153–159. Lawrence, KS: Contributions to Herpetology 1, Society for the Study of Amphibians and Reptiles.

Joanen, T., and L. McNease. 1987. The management of alligators in Louisiana, USA. In G. J. W. Webb, S. C. Manolis, and P. J. Whitehead (eds.), *Wildlife Management, Crocodiles and Alligators*, 33–42. Northern Territory, Australia: Surrey Beatty and Sons.

Johnsgard, P.A. 1991. *Bustards, Hemipodes, and Sandgrouse: Birds of Dry Places*. Oxford, U.K.: Oxford University Press, 276 pp.

Johnston, P. 2011. New morphological evidence supports congruent phylogenies and Gondwana vicariance for palaeognathous birds. *Zoological Journal of the Linnean Society* 163: 959–982.

Kaiser, G. W. 2007. *The Inner Bird: Anatomy and Evolution*. Vancouver, B.C., Canada: University of British Columbia Press, 386 pp.

Kellner, A. W. A., F. M. Dalla Vecchia, M. M. Ataabadi, H. De Paula Silva, and E. Khosravi. 2012. Review of dinosaur record from Iran with the description of new material. *Rivista Italiana di Paleontologia e Stratigrafia* 118: 261–275.

Kellogg, R. 1929. *The Habits and Economic Importance of Alligators*. Technical Bulletin 147. Washington, DC: U.S. Department of Agriculture, 36 pp.

Kim, J. Y., K. S. Kim, M. G. Lockley, S. Y. Yang, S. J. Seo, H. I. Choi, and J. D. Lim. 2008. New didactyl dinosaur footprints (*Dromaeosauripus hamanensis* ichnogen. et ichnosp. nov.) from the Early Cretaceous Haman Formation, south coast of Korea. *Palaeogeography, Palaeoclimatology, Palaeoecology* 262: 72–78.

Kim, J. Y., M. G. Lockley, and H. Y. Cun. 2016. New dinosaur tracks from the Lower Cretaceous (Valanginian-Hauterivian) Saniri Formation of Yeongdong area, central Korea: implications for quadrupedal ornithopod locomotion. *Cretaceous Research* 61: 5–16.

Kim, J. Y., M. G. Lockley, H. M. Kim, J.-D. Lim, and K. S. Kim. 2009. New dinosaur tracks from Korea, *Ornithopodichnus masanensis* ichnogen. et ichnosp. nov. (Jindong Formation, Lower Cretaceous): implications for polarities in ornithopod foot morphology. *Cretaceous Research* 30: 1387–1397.

Kim, J. Y., M. G. Lockley, J. O. Woo, and S. H. Kim. 2012. Unusual didactyl traces from the Jinju Formation (Early Cretaceous, South Korea) indicate a new ichnospecies of *Dromaeosauripus*. *Ichnos* 19: 75–83.

Klein, H., S. G. Lucas, and H. Haubold. 2006. Tetrapod track assemblage of the Redonda Formation (Upper Triassic, Chinle Group), in east-central New Mexico—re-evaluation of ichnofaunal diversity from studies of new material. In J. D. Harris, S. G. Lucas, J. A. Spielmann, M. G. Lockley, A. R. C. Milner, and J. I. Kirkland (eds.), *The Triassic-Jurassic Terrestrial Transition*, 241–250. Albuquerque, NM: New Mexico Museum of Natural History and Science Bulletin 37.

Klein, H., S. Voigt, H. Saber, J. W. Schneider, A. Hminna, J. Fischer, A. Lagnaoui, and A. Brosig. 2011. First occurrence of a Middle Triassic tetrapod ichnofauna from the Argana Basin (western High Atlas, Morocco). *Palaeogeography, Palaeoclimatology, Palaeoecology* 307: 218–231.

Klein, H., A. Lagnaoui, G. D. Gierliński, H. Saber, J. N. Hallensack, M. Oukassou, and A. Charrière. 2018. Crocodylomorph, turtle and mammal tracks in dinosaur-dominated Middle-? Upper Jurassic and mid-Cretaceous ichnoassemblages of Morocco. *Palaeogeography, Palaeoclimatology, Palaeoecology*. 498: 39–52. https://doi.org/10.1016/j.palaeo.2018.02.028.

Klein, N., and P. M. Sander. 2008. Ontogenetic changes in the long bone histology of sauropod dinosaurs. *Paleobiology* 34: 247–263.

Kobayashi, Y., and J.-C. Lü. 2003. A new ornithomimid dinosaur with gregarious habits from the Late Cretaceous of China. *Acta Palaeontologica Polonica* 48: 235–259.

Kowalewski, M., E. Dyreson, J. D. Marcot, J. A. Vargas, K. W. Flessa, and D. P. Hallman. 1997. Phenetic discrimination of biometric simpletons: paleobiological implications of morphospecies in the lingulide brachiopod *Glottidia*. *Paleobiology* 23: 444–469.

Kozu, S., A. Sardsud, D. Saesaengseerung, C. Pothichaiya, S. Agematsu, and K. Sashida. 2017. Dinosaur footprint assemblage from the Lower Cretaceous Khok Kruat Formation, Khorat Group, northeastern Thailand. *Geoscience Frontiers* 8: 1479–1493.

Kuban, G. J. 1989. Elongate dinosaur tracks. In D. D. Gillette and M. G. Lockley (eds.), *Dinosaur Tracks and Traces*, 57–72. Cambridge: Cambridge University Press.

Kubo, T. 2008. In quest of the *Pteraichnus* trackmaker: comparisons to modern

crocodilians. *Acta Palaeontologica Polonica* 53: 405–412.
Kubo, T., and M. O. Kubo. 2012. Associated evolution of bipedality and cursoriality among Triassic archosaurs: a phylogenetically controlled evaluation. *Paleobiology* 38: 474–485.
Kumagai, C. J. 2010. *A Morphometric Analysis of the Autopodia of the American Crocodile (*Crocodylus acutus*) in Costa Rica*. M.S. Thesis. West Lafayette, IN: Purdue University, 95 pp.
Kvale, E. P., G. D. Johnson, D. L. Mickelson, K. Keller, L. C. Furer, and A. W. Archer. 2001. Middle Jurassic (Bajocian and Bathonian) dinosaur megatracksites, Bighorn Basin, Wyoming. *Palaios* 16: 233–254.
Lagnaoui, A., H. Klein, S. Voigt, A. Hminna, H. Saber, J. W. Schneider, and R. Werneburg. 2012. Late Triassic tetrapod-dominated ichnoassemblages from the Argana Basin (western High Atlas, Morocco). *Ichnos* 19: 238–253.
Lallensack, J. N., A. K. von Heteren, and O. Wings. 2016. Geometric morphometric analysis of intratrackway variability: a case study on theropod and ornithopod dinosaur trackways from Münchehagen (Lower Cretaceous, Germany). *PeerJ* 4:e2059. doi:10.7717/peerj.2059.
Lande, R. 1977. On comparing coefficients of variation. *Systematic Zoology* 26: 214–217.
Lange-Badré, B., and J.-P. Lafon. 2000. Découverte de pistes de dinosaures théropodes dans le Lias inférieur des environs de Figeac (Lot). *Comptes Rendus de l'Académie des Sciences de Paris, Sciences de la Terre et des planètes* 330: 379–384.
Langer, M. C., M. D. Ezcurra, O. W. M. Rauhut, M. J. Benton, F. Knoll, B. W. McPhee, F. E. Novas, D. Pol, and S. L. Brusatte. 2017. Untangling the dinosaur family tree. *Nature* 551: E1–E3.
Langston, W., Jr. 1974. Nonmammalian Comanchean tetrapods. *Geoscience and Man* 8: 77–102.
Lapparent, A. F. de. 1962. Footprints of dinosaur in the Lower Cretaceous of Vestspitsbergen-Svalbard. *Norsk Polarinstitutt Årbok* 1960: 14–21.
Lapparent, A. F. de, and C. Montenat. 1967. Les empreintes de pas de reptiles de l'Infralias du Veillon (Vendée). *Mémoires de la Société Géologique de France (Nouvelle Serie)* 107: 1–41.
Larson, P. 2013. The case for *Nanotyrannus*. In J. M. Parrish, R. E. Molnar, P. J. Currie, and E. B. Koppelhus (eds.), *Tyrannosaurid Paleobiology*, 15–53. Bloomington: Indiana University Press.
Latorre Macarrón, P., C. Pascual Arribas, E. Sanz Pérez, and N. Hernández Medrano. 2006. El yacimiento con huellas de saurópodos de Miraflores I, Fuentes de Magaña (Soria, España). In Colectivo Arqueológico-Paleontológico Salense (eds.), *Actas de las III Jornadas sobre Dinosauros y su Entorno*, 273–296. Burgos, Spain: Salas de los Infantes.
Leduc, D. J. 1987. A comparative analysis of the reduced major axis technique of fitting lines to bivariate data. *Canadian Journal of Forestry Research* 17: 654–659.
Lee, A. H., A. K. Huttenlocker, K. Padian, and H. N. Woodward. 2013. Analysis of growth rates. In K. Padian and E.-T. Lamm (eds.), *Bone Histology of Fossil Tetrapods: Advancing Methods, Analysis, and Interpretation*, 217–521. Berkeley, CA: University of California Press.
Lee, A. H., and S. Werning. 2008. Sexual maturity in growing dinosaurs does not fit reptilian growth models. *Proceedings of the National Academy of Sciences USA* 105: 582–587.
Lee, Y.-N. 1997. Bird and dinosaur footprints in the Woodbine Formation (Cenomanian), Texas. *Cretaceous Research* 18: 849–864.
Lee, Y.-N., R. Barsbold, P. J. Currie, Y. Kobayashi, H.-J. Lee, P. Godefroit, F. Escuillié, and T. Chinzorig. 2014. Resolving the long-standing enigmas of a giant ornithomimosaur *Deinocheirus mirificus*. *Nature* 515: 257–260.
Lee, Y.-N., M. J. Ryan, and Y. Kobayashi. 2011. The first ceratopsian dinosaur from South Korea. *Naturwissenschaften* 98: 39–49.
Le Loeuff, J., S. Khansubha, E. Buffetaut, V. Suteethorn, H. Tong, and C. Souillat. 2002. Dinosaur footprints from the Phra Wihan Formation (Early Cretaceous of Thailand). *Comptes Rendus Palevol* 1: 287–292.
Leonardi, G. 1979*a*. New archosaurian trackways from the Rio do Peixe Basin, Paraíba, Brasil. *Annali dell' Univesita di Ferrara (N.S.)*, S. 9, 5(14): 239–249.
Leonardi, G. 1979*b*. Nota preliminar dobre seis pistas de dinossaurios Ornithischia da Bacia do Rio do Peixe, em Sousa, Paraiba, Brasil. *Anais da Academia Brasileria de Ciencias* 51: 501–516.
Leonardi, G. 1980*a*. Dez novas pistas de dinossaurios (Theropoda Marsh, 1881) na Bacia do Rio do Peixe, Paraiba, Brasil. *Actas I Congreso Latínoamericano Paleontología, Buenas Aires* 1: 243–248.
Leonardi, G. 1980*b*. On the discovery of an abundant ichno-fauna (vertebrates and invertebrates) in the Botucatu Formation s.s. in Araraquara, São Paulo, Brazil. *Anais de al Academia Brasileira de Ciencias* 52: 559–567.
Leonardi, G. 1984. Dimensions and weight of some modern Brazilian reptiles, for the use of paleoichnology. *Dusenia* 14: 45–49.
Leonardi, G. (ed.). 1987. *Glossary and Manual of Tetrapod Footprint Palaeoichnology*. Brasília, Brazil: República Federativa do Brasil, Ministério das Minas e Energia, Departamento Nacional da Produção Mineral, 75 pp.
Leonardi, G. 1989. Inventory and statistics of the South American dinosaurian ichnofauna and its paleobiological interpretation. In D. D. Gillette and M. G. Lockley (eds.), *Dinosaur Tracks and Traces*, 165–178. Cambridge: Cambridge University Press.
Leonardi, G. 1994. *Annotated Atlas of South America Tetrapod Footprints (Devonian to Holocene) with An Appendix on Mexico and Central America*. Brasilia, Brazil: República Federativa do Brasil, Ministério de Minas e Energia, Secretaria de Minas e Metalurgia, Companhia de Pesuisa de Recursos Minerals, 247 pp.
Leonardi, G., and I. de Souza Carvalho, in press. *Steps Across Gondwana: Dinosaur Tracks of the Early Cretaceous, West Paraiba, Brazil*. Bloomington: Indiana University Press.
Leonardi, G., and M. de Fátima C. F. dos Santos. 2004. New dinosaur tracksites from the Sousa Lower Cretaceous basin (Paraíba, Brasil). *Studi Trentini di Scienze Naturali Acta Geologica* 81: 5–21.
Leonardi, G., and L. C. Godoy. 1980. Novas pistas de tetrápodes da Formação Botucatu no Estado de São Paulo. *Anais XXXI Congresso Brasileiro de Geologia* 5: 3080–3089.
Leonardi, G., C. V. de Lima, and F. H. De Oliveira Lima. 1987*a*. Os dados numéricos relativos às pistas (e suas pegadas) das icnofaunas dinossaurianas do Cretáceo Inferior da Paraíba, e sua interpretação estatística; I—parâmetros das pistas. *Anais do X Congresso Brasileiro de Paleontologia, Rio de Janeiro*: 377–394.
Leonardi, G., C. V. de Lima, and F. H. De Oliveira Lima. 1987*b*. Os dados numéricos relativos às pistas (e suas pegadas) das icnofaunas dinossaurianas do Cretáceo Inferior da Paraíba, e sua interpretação estatística; II—parâmetros das pegadas. *Anais do X Congresso Brasileiro de Paleontologia, Rio de Janeiro*: 395–416.
Leonardi, G., and P. Mietto. 2000. Le piste liassiche di dinosauri dei Lavini di Marco. In G. Leonardi, P. Mietto and M. Avanzini (eds.), *Dinosauri in Italia: Le Orme Giurassiche dei Lavini di Marco (Trentino) e Gli Altri Resti Fossili Italiani*, 169–246. Pisa, Italy: Accademia Editoriale.
Leonardi, G., and M. Spezzamonte. 1992. New tracksites (Dinosauria: Theropoda and Ornithopoda) from the Lower Cretaceous of the Ceará, Brasil. *Studi Trentini Scienze Naturali—Acta Geologica* 69: 61–70.
Li, J., Z. Bai, M. Lockley, B. Zhou, J. Liu, and Y. Song. 2010. Dinosaur tracks in Wulatezhongqi, Inner Mongolia. *Acta Geologica Sinica* 84: 723–742 (in Chinese with English summary).
Li, J.-J., M. Bater, W.-H. Zhang, B.-L. Hu, and L.-H. Gao. 2006. A new type of dinosaur tracks from Lower Cretaceous of Chabu, Otog Qi, Inner Mongolia. *Acta Palaeontologica Sinica* 45: 221–234 (in Chinese with English summary).
Li, J., M. G. Lockley, Y. Zhang, S. Hu, M. Matsukawa, and Z. Bai. 2012. An important ornithischian tracksite in the Early Jurassic of the Shenmu Region, Shaanxi, China. *Acta Geologica Sinica* 86: 1–10.
Li, L., J. Wang, X. Zhang, and S. Hou. 2012. A new enantiornithine bird from the Lower Cretaceous Jiufotang Formation in Jinzhou area, western Liaoning Province, China. *Acta Geologica Sinica* 86: 1039–1044.
Li, R., M. G. Lockley, P. J. Makovicky, M. Matsukawa, M. A. Norell, J. D. Harris, and M. Liu. 2008. Behavioral and faunal implications of Early Cretaceous deinonychosaur trackways from China. *Naturwissenschaften* 95: 185–191.
Li, R., M. G. Lockley, M. Matsukawa, and M. Liu. 2015. Important dinosaur-dominated footprint assemblages from the Lower Cretaceous Tianjialou Formation at the Houzuoshan Dinosaur Park, Junan County, Shandong Province, China. *Cretaceous Research* 52: 83–100.
Lingham-Soliar, T., and T. Broderick. 2000. An enigmatic early Mesozoic dinosaur trackway from Zimbabwe. *Ichnos* 7: 135–148.

Lingham-Soliar, T., T. Broderick, and A. A. K. Ahmed. 2003. Closely associated theropod trackways from the Jurassic of Zimbabwe. *Naturwissenschaften* 90: 572–576.
Livezey, B. C., and R. L. Zusi. 2007. Higher-order phylogeny of modern birds (Theropoda, Aves: Neornithes) based on comparative anatomy. *Zoological Journal of the Linnean Society* 149: 1–95.
Livingston, V. J., M. F. Bonnan, R. M. Elsey, J. L. Sandrik, and D. R. Wilhite. 2009. Differential limb scaling in the American alligator (*Alligator mississippiensis*) and its implications for archosaur locomotor evolution. *Anatomical Record* 292: 787–797.
Llandres Serrano, M., R. Vullo, J. Marugán-Lobón, F. Ortega, and A. D. Buscalioni. 2013. An articulated limb of a basal iguanodont (Dinosauria, Ornithopoda) from the Early Cretaceous Las Hoyas Lagerstätte (Spain). *Geological Magazine* 150: 572–576.
Lockley, M. G. 1987. Dinosaur footprints from the Dakota Group of eastern Colorado. *The Mountain Geologist* 24: 107–122.
Lockley, M. G. 2001. A decade of dinosaur tracking at Dinosaur Ridge. *The Mountain Geologist* 38: 119–135.
Lockley, M. G. 2009. New perspectives on morphological variation in tridactyl footprints: clues to widespread convergence in developmental dynamics. *Geological Quarterly* 53: 415–432.
Lockley, M. G. 2011. Putting the best foot forward: a single case of "toe extension" has general implications for the toe extension concept in theropod dinosaur feet and footprints. In R. M. Sullivan, S. G. Lucas, and J. A. Spielmann (eds.), *Fossil Record 3*, 301–305. Albuquerque, NM: New Mexico Museum of Natural History and Science Bulletin 53.
Lockley, M. G., K. Cart, J. Martin, R. Prunty, K. Houck, K. Hups, J.-D. Lim, K. S. Kim, and G. Gierliński. 2014*a*. A bonanza of new tetrapod tracksites from the Cretaceous Dakota Group, western Colorado: implications for paleoecology. In M. G. Lockley and S. G. Lucas (eds.), *Fossil Footprints of Western North America*, 393–409. Albuquerque, NM: New Mexico Museum of Natural History and Science Bulletin 62.
Lockley, M. G., V. F. dos Santos, C. Meyer, and A. Hunt. 1998*a*. A new dinosaur tracksite in the Morrison Formation, Boundary Butte, southeastern Utah. *Modern Geology* 23: 317–330.
Lockley, M. G., and L. Eisenberg. 2006. A preliminary report on a spectacular dinosaur tracksite in the Chinle Group, Dirty Devil River valley, Wayne County, Utah. In J. D. Harris, S. G. Lucas, J. A. Spielmann, M. G. Lockley, A. R. C. Milner, and J. I. Kirkland (eds.), *The Triassic-Jurassic Terrestrial Transition*, 263–268. Albuquerque, NM: New Mexico Museum of Natural History and Science Bulletin 37.
Lockley, M. G., R. H. Foster, J. Foster, and S. Gerwe. 2014*b*. Early Jurassic track assemblages from the Granite Creek area of eastern Utah. In M. G. Lockley and S. G. Lucas (eds.), *Fossil Footprints of Western North America*, 205–210. Albuquerque, NM: New Mexico Museum of Natural History and Science Bulletin 62.
Lockley, M. G., J. Foster, and A. Hunt. 1998*b*. A short summary of dinosaur tracks and other fossil footprints from the Morrison Formation. *Modern Geology* 23: 277–290.
Lockley, M. G., J. C. Garcia-Ramos, L. Piñuela, and M. Avanzini. 2008*a*. A review of vertebrate track assemblages from the Late Jurassic of Asturias, Spain with comparative notes on coeval ichnofaunas from the western USA: implications for faunal diversity in siliciclastic assemblages. *Oryctos* 8: 53–70.
Lockley, M. G., and G. D. Gierliński. 2006. Diverse vertebrate ichnofaunas containing *Anomoepus* and other unusual trace fossils from the Lower Jurassic of the western United States: implications for paleoecology and palichnostratigraphy. In J. D. Harris, S. G. Lucas, J. A. Spielmann, M. G. Lockley, A. R. C. Milner, and J. I. Kirkland (eds.), *The Triassic-Jurassic Terrestrial Transition*, 176–191. Albuquerque, NM: New Mexico Museum of Natural History and Science Bulletin 37.
Lockley, M. G., and G. D. Gierliński. 2009. A *Grallator*-dominated tracksite from the Chile Group (Late Triassic), Moab, Utah. *Geological Quarterly* 53: 433–440.
Lockley, M. G., G. D. Gierliński, K. Houk, J.-D. Lim, K. S. Kim, D.-Y. Kim, T. H. Kim, S.-H. Kang, R. H. Foster, R. Li, C. Chessser, R. Gay, Z. Dubicka, K. Cart, and C. Wright. 2014*c*. New excavations at the Mill Canyon Dinosaur Track Site (Cedar Mountain Formation, Lower Cretaceous) of eastern Utah. In M. G. Lockley and S. G. Lucas (eds.), *Fossil Footprints of Western North America*, 287–300. Albuquerque, NM: New Mexico Museum of Natural History and Science Bulletin 62.
Lockley, M. G., G. D. Gierliński, and S. G. Lucas. 2011*a*. *Kayentapus* revisited: notes on the type material and the importance of this theropod footprint ichnogenus. In R. M. Sullivan, S. G. Lucas, and J. A. Spielmann (eds.), *Fossil Record 3*, 330–336. Albuquerque, NM: New Mexico Museum of Natural History and Science Bulletin 62.
Lockley, M. G., G. D. Gierliński, J. Martin, and K. Cart. 2014*d*. An unusual theropod tracksite in the Cretaceous Dakota Group, western Colorado: implications for ichnodiversity. In M. G. Lockley and S. G. Lucas (eds.), *Fossil Footprints of Western North America*, 411–415. Albuquerque, NM: New Mexico Museum of Natural History and Science Bulletin 62.
Lockley, M. G., G. D. Gierliński, A. L. Titus, and B. Albright. 2006*a*. An introduction to thunderbird footprints at the Flag Point Pictograph-Tracksite: preliminary observations on Lower Jurassic theropod tracks from the Vermillion Cliffs area, southwestern Utah. In J. D. Harris, S. G. Lucas, J. A. Spielmann, M. G. Lockley, A. R. C. Milner, and J. I. Kirkland (eds.), *The Triassic-Jurassic Terrestrial Transition*, 310–314. Albuquerque, NM: New Mexico Museum of Natural History and Science Bulletin 37.
Lockley, M. G., M. R. Gregory, and B. J. Gill. 2007*a*. The ichnological record of New Zealand's moas: a preliminary survey. In S. G. Lucas, J. A. Spielmann, and M. G. Lockley (eds.), *Cenozoic Tracks and Traces*, 73–78. Albuquerque, NM: New Mexico Museum of Natural History and Science Bulletin 42.
Lockley, M. G., K. K. Honda, and B. Simmons. 2014*e*. A new dinosaur tracksite in the Dakota Group (Cretaceous) of the historic Stone City area, Colorado. In M. G. Lockley and S. G. Lucas (eds.), *Fossil Footprints of Western North America*, 355–359. Albuquerque, NM: New Mexico Museum of Natural History and Science Bulletin 62.
Lockley, M. G., K. Houck, C. Green, and M. Caldwell. 2014*f*. New fossil footprints from the Dakota Group (Cretaceous) Roxborough State Park, Colorado. In M. G. Lockley and S. G. Lucas (eds.), *Fossil Footprints of Western North America*, 373–384. Albuquerque, NM: New Mexico Museum of Natural History and Science Bulletin 62.
Lockley, M. G., K. J. Houck, and N. K. Prince. 1986. North America's largest dinosaur trackway site: implications for Morrison Formation paleoecology. *Geological Society of America Bulletin* 97: 1163–1176.
Lockley, M. G., K. Houck, S.-Y. Yang, M. Matsukawa, and S.-K. Lim. 2006*b*. Dinosaur-dominated footprint assemblages from the Cretaceous Jindong Formation, Hallyo Haesang National Park area, Goseong County, South Korea: evidence and implications. *Cretaceous Research* 27: 70–101.
Lockley, M. G., M. Huh, S.-G. Gwak, K. G. Hwang, and I. S. Paik. 2012*a*. Multiple tracksites with parallel trackways from the Cretaceous of Yeosu City area Korea: implications for gregarious behavior in ornithopod and sauropod dinosaurs. *Ichnos* 19: 105–114.
Lockley, M. G., M. Huh, and B. S. Kim. 2012*b*. *Ornithopodichnus* and pes-only sauropod trackways from the Hwasun Tracksite, Cretaceous of Korea. *Ichnos* 19: 93–100.
Lockley, M. G., and A. P. Hunt. 1995. *Dinosaur Tracks and Other Fossil Footprints of the Western United States*. New York: Columbia University Press, 338 pp.
Lockley, M. G., A. Hunt, J. Holbrook, M. Matsukawa, and C. Meyer. 1992. The Dinosaur Freeway: a preliminary report on the Cretaceous megatracksite, Dakota Group, Rocky Mountain Front Range, and High Plains, Colorado, Oklahoma, and New Mexico. In R. Flores (ed.), *Mesozoic of the Western Interior*, 39–54. Tulsa, OK: Field Trip Guidebook, Rocky Mountain Section, American Association of Petroleum Geologists.
Lockley, M. G., A. P. Hunt, C. Meyer, E. C. Rainforth, and R. J. Schultz. 1998*c*. A survey of fossil footprint sites at Glen Canyon National Recreation Area (western USA): a case study in documentation of trace fossil resources at a national preserve. *Ichnos* 5: 177–211.
Lockley, M. G., A. Hunt, M. Paquette, S.-A. Bilbey, and A. Hamblin. 1998*d*. Dinosaur tracks from the Carmel Formation, northeastern Utah: implications for Middle Jurassic paleoecology. *Ichnos* 5: 255–267.
Lockley, M. G., P. Janke, and L. Thiesen. 2001*a*. First reports of bird and ornithopod tracks from the Lakota Formation (Early Cretaceous), Black Hills, South Dakota. in D. H. Tanke and K. Carpenter (eds.), *Mesozoic Vertebrate Life: New Research Inspired by the Paleontology of Philip J. Currie*, 443–452. Bloomington: Indiana University Press.

Lockley, M. G., J. Y. Kim, K. S. Kim, S. H. Kim, M. Matsukawa, L. Rihui, L. Jianjun, and S.-Y. Yang. 2008*b*. *Minisauripus*—the track of a diminutive dinosaur from the Cretaceous of China and South Korea: implications for stratigraphic correlation and theropod foot morphodynamics. *Cretaceous Research* 29: 115–130.

Lockley, M. G., M. King, S. Howe, and T. Sharp. 1996. Dinosaur tracks and other footprints from the Triassic of south Wales. *Ichnos* 5: 23–41.

Lockley, M. G., R. Kukihara, and L. Mitchell. 2006*c*. New dinosaur and crocodile tracksites from the Cretaceous Dakota Group of the Colorado Front Range, Boulder and Jefferson Counties, Colorado. In S. G. Lucas and R. M. Sullivan (eds.), *Late Cretaceous Vertebrates from the Western Interior*. Albuquerque, NM: New Mexico Museum of Natural History and Science Bulletin 35.

Lockley, M. G., R. Kukihara, L. Pionek, and A. Delgalvis. 2014*g*. A survey of new fossil footprint sites from Glen Canyon National Recreation Area (western USA), with special reference to the Kayenta-Navajo transition zone (Glen Canyon Group, Lower Jurassic). In M. G. Lockley and S. G. Lucas (eds.), *Fossil Footprints of Western North America*, 157–179. Albuquerque, NM: New Mexico Museum of Natural History and Science Bulletin 62.

Lockley, M. G., J. Li, R. Li, M. Matsukawa, J. D. Harris, and L. Xing. 2013. A review of the tetrapod track record in China, with special reference to type ichnospecies: implications for ichnotaxonomy and paleobiology. *Acta Geologica Sinica* 87: 1–20.

Lockley, M. G., R. Li, M. Matsukawa, L. Xing, J. Li, M. Liu, and X. Xu. 2015. Tracking the yellow dragons: implications of China's largest dinosaur tracksite (Cretaceous of the Zhugheng area, Shandong Province, China). *Palaeogeography, Palaeoclimatology, Palaeoecology* 423: 62–79.

Lockley, M. G., M. Matsukawa, and J. Li. 2003*a*. Crouching theropods in taxonomic jungles: ichnological and ichnotaxonomic investigations of footprints with metatarsal and ischial impressions. *Ichnos* 10: 169–177.

Lockley, M. G., M. Matsukawa, Y. Sato, M. Polahan, and V. Daorerk. 2006*d*. A distinctive new theropod dinosaur track from the Cretaceous of Thailand: implications for theropod track diversity. *Cretaceous Research* 27: 139–145.

Lockley, M. G., M. Matsukawa, and D. Witt. 2006*e*. Giant theropod tracks from the Cretaceous Dakota Group of northeastern New Mexico. In S. G. Lucas and R. M. Sullivan (eds.), *Late Cretaceous Vertebrates from the Western Interior*, 83–87. Albuquerque, NM: New Mexico Museum of Natural History and Science Bulletin 35.

Lockley, M. G., R. T. McCrea, and M. Matsukawa. 2009. Ichnological evidence for small quadrupedal ornithischians from the basal Cretaceous of SE Asia and North America: implications for a global radiation. *The Geological Society, London, Special Publications* 315: 255–269.

Lockley, M. G., and C. A. Meyer. 2000. *Dinosaur Tracks and Other Fossil Footprints of Europe*. New York: Columbia University Press, 323 pp.

Lockley, M. G., and C. A. Meyer. 2004. Crocodylomorph trackways from the Jurassic to Early Cretaceous of North America and Europe: implications for ichnotaxonomy. *Ichnos* 11: 167–178.

Lockley, M. G., C. A. Meyer, and J. J. Moratalla. 1998*e*. *Therangospodus*: trackway evidence for the widespread distribution of a Late Jurassic theropod with well-padded feet. *Gaia* 15: 339–353.

Lockley, M. G., C. A. Meyer, and V. F. dos Santos. 1998*f*. *Megalosauripus* and the problematic concept of megalosaur footprints. *Gaia* 15: 313–337.

Lockley, M G., and D. L. Mickelson. 1997. Dinosaur and pterosaur tracks in the Summerville and Bluff (Jurassic) beds of eastern Utah and northeastern Arizona. In *Mesozoic Geology and Paleontology of the Four Corners Region*, 133–138. New Mexico Geological Society Guidebook, 48th Field Conference.

Lockley, M. G., A. R. C. Milner, D. Slauf, and A. H. Hamblin. 2006*f*. Dinosaur tracksites from the Kayenta Formation (Lower Jurassic) "Desert Tortoise Site," Washington County, Utah. In J. D. Harris, S. G. Lucas, J. A. Spielmann, M. G. Lockley, and A. R. C. Milner (eds.), *The Triassic-Jurassic Transition*, 269–281. Albuquerque, NM: New Mexico Museum of Natural History and Science Bulletin 37.

Lockley, M. G., L. Mitchell, and G. P. Odier. 2007*b*. Small theropod track assemblages from Middle Jurassic eolianites of eastern Utah: paleoecological insights from dune ichnofacies in a transgressive sequence. *Ichnos* 14: 131–142.

Lockley, M. G., G. Nadon, and P. J. Currie. 2003*b*. A diverse dinosaur-bird footprint assemblage from the Lance Formation, Upper Cretaceous, eastern Wyoming: implications for ichnotaxonomy. *Ichnos* 11: 229–249.

Lockley, M. G., and E. C. Rainforth. 2002. The track record of Mesozoic birds and pterosaurs: an ichnological and paleoecological perspective. In L. M. Chiappe and L. M. Witmer (eds.), *Mesozoic Birds: Above the Heads of Dinosaurs*, 405–418. Berkeley: University of California Press.

Lockley, M. G., A. S. Schulp, C. A. Meyer, G. Leonardi, and D. K. Mamani. 2002. Titanosaurid trackways from the Upper Cretaceous of Bolivia: evidence for large manus, wide-gauge locomotion and gregarious behaviour. *Cretaceous Research* 23: 383–400.

Lockley, M. G., B. Simmons, and K. K. Honda. 2014*h*. The rediscovery of Arthur Lakes' 1902 dinosaur tracks quarry: what he found, how he excavated it, where it went and what it means. In M. G. Lockley and S. G. Lucas (eds.), *Fossil Footprints of Western North America*, 345–354. Albuquerque, NM: New Mexico Museum of Natural History and Science Bulletin 62.

Lockley, M. G., A. R. Tedrow, K. C. Chamberlain, N. J. Minter, and J.-D. Lim. 2011*b*. Footprints and invertebrate traces from a new site in the Nugget Sandstone (Lower Jurassic) of Idaho: implications for life in the northern reaches of the great Navajo-Nugget erg system in the western USA. In R. M. Sullivan, S. G. Lucas, and J. A. Spielmann (eds.), *Fossil Record 3*, 344–356. Albuquerque, NM: New Mexico Museum of Natural History and Science Bulletin 53.

Lockley, M. G., and J. Tempel. 2014. "Fossil Trace" trace fossils: the historic, scientific and educational significance of the Triceratops Trail.—A controversial Upper Cretaceous tracksite complex in the Laramie Formation, Golden, Colorado. In M. G. Lockley and S. G. Lucas (eds.), *Fossil Footprints of Western North America*, 441–457. Albuquerque, NM: New Mexico Museum of Natural History and Science Bulletin 62.

Lockley, M. G., M. Triebold, and P. R. Janke. 2014*i*. Dinosaur tracks from the Hell Creek Formation (Upper Cretaceous, Maastrichtian), South Dakota. In M. G. Lockley and S. G. Lucas (eds.), *Fossil Footprints of Western North America*, 459–468. Albuquerque, NM: New Mexico Museum of Natural History and Science Bulletin 62.

Lockley, M. G., D. White, J. Kirkland, and V. Santucci. 2004*a*. Dinosaur tracks from the Cedar Mountain Formation (Lower Cretaceous), Arches National Park, Utah. *Ichnos* 11: 285–293.

Lockley, M. G., J. L. Wright, and M. Matsukawa. 2001*b*. A new look at *Magnoavipes* and so-called "big bird" tracks from Dinosaur Ridge, Colorado. *The Mountain Geologist* 38: 137–146.

Lockley, M. G., J. L. Wright, and D. Thies. 2004*b*. Some observations on the dinosaur tracks at Münchehagen (Lower Cretaceous), Germany. *Ichnos* 11: 261–274.

Lockley, M. G., and L. Xing. 2015. Flattened fossil footprints: implications for paleobiology. *Palaeogeography, Palaeoclimatology, Palaeoecology* 426: 85–94.

Lockley, M. G., L. Xing, J. A. F. Lockwood, and S. Pond. 2014*j*. A review of large Cretaceous ornithopod tracks, with special reference to their ichnotaxonomy. *Biological Journal of the Linnean Society* 113: 721–736.

Lockley, M. G., B. H. Young, and K. Carpenter. 1983. Hadrosaur locomotion and herding behavior: evidence from footprints in the Mesaverde Formation, Grand Mesa Coal Field, Colorado. *The Mountain Geologist* 20: 5–14.

Lockwood, J. A. F., M. G. Lockley, and S. Pond. 2014. A review of footprints from the Wessex Formation (Wealden Group, Lower Cretaceous) at Hanover Point, the Isle of Wight, southern England. *Biological Journal of the Linnean Society* 113: 707–720.

Lomax, D. R., and N. Tamura. 2014. *Dinosaurs of the British Isles*. Manchester, U.K.: Siri Scientific Press, 424 pp.

Loope, D. B., and J. Milàn. 2016. Dinosaur tracks in eolian strata: new insights into track formation, walking kinetics, and trackmaker behavior. In P. L. Falkingham, D. Marty, and A. Richter (eds.), *Dinosaur Tracks: The Next Steps*, 358–364. Bloomington: Indiana University Press.

Lucas, A. M., and P. R. Stettenheim. 1972. *Avian Anatomy: Integument*. Part I. Agricultural Handbook 362. Washington, DC: U.S. Department of Agriculture, 340 pp.

Lucas, S. G., H. Klein, M. G. Lockley, J. A. Spielmann, G. D. Gierlinski, A. P. Hunt, and L. H. Tanner. 2006. Triassic-Jurassic stratigraphic distribution of the theropod footprint ichnogenus *Eubrontes*. In J. D. Harris, S. G. Lucas, J. A. Spielmann, M. G. Lockley, A. R. C. Milner, and J. I. Kirkland (eds.), *The Triassic-Jurassic Terrestrial Transition*, 86–93. Albuquerque, NM: New Mexico Museum of Natural History and Science Bulletin 37.

Lull, R. S. 1953. *Triassic Life of the Connecticut Valley.* Connecticut State Geological and Natural History Survey Bulletin 81, 331 pp.

Maidment, S. C. R., and P. M. Barrett. 2014. Osteological correlates for quadrupedality in ornithischian dinosaurs. *Acta Palaeontologica Polonica* 59: 53–70.

Mallison, H. 2010*a*. The digital *Plateosaurus* I: body mass, mass distribution and posture assessed using CAD and CAE on a digitally mounted skeleton. *Palaeontologia Electronica* 13: 1–26.

Mallison, H. 2010*b*. The digital *Plateosaurus* II: an assessment of the range of motion of the limbs and vertebral column and of previous reconstructions using a digital skeleton mount. *Acta Palaeontologica Polonica* 55: 433–458.

Mallison, H. 2011. *Plateosaurus* in 3D: how CAD models and kinetic-dynamic modeling bring an extinct animal to life. In N. Klein, K. Remes, C. T. Gee and P. M. Sander (eds.), *Biology of the Sauropod Dinosaurs: Understanding the Life of Giants*, 219–236. Bloomington: Indiana University Press.

Manion, B. L. 1984. *The Effects of Size and Growth on Hindlimb Locomotion in the Chicken.* PhD Thesis. Chicago: University of Illinois at Chicago, 206 pp.

Manning, P. L. 2004. A new approach to the analysis and interpretation of tracks: examples from the Dinosauria. *Geological Society of London Special Publications* 228: 93–123.

Manning, P. L. 2008. *T. rex* speed trap. In P. Larson and K. Carpenter (eds.). *Tyrannosaurus rex, the Tyrant King*, 204–231. Bloomington: Indiana University Press.

Marchant, S., and P. J. Higgins (co-ordinators). 1990*a*. *Handbook of Australian, New Zealand and Antarctic Birds. Volume 1: Ratites to Ducks, Part A: Ratites to Petrels.* Melbourne: Oxford University Press Australia, 735 pp.

Marchant, S., and P. J. Higgins (co-ordinators). 1990*b*. *Handbook of Australian, New Zealand and Antarctic Birds. Volume 1: Ratites to Ducks, Part B: Australian Pelican to Ducks.* Melbourne: Oxford University Press Australia, 1400 pp.

Marchant, S., and P. J. Higgins (co-ordinators). 1993. *Handbook of Australian, New Zealand and Antarctic Birds. Volume 2: Raptors to Lapwings.* Melbourne: Oxford University Press Australia, 984 pp.

Martin, A. J. 2013. *Life Traces of the Georgia Coast: Revealing the Unseen Lives of Plants and Animals.* Bloomington: Indiana University Press, 670 pp.

Martin, A. J. 2014. *Dinosaurs Without Bones: Dinosaur Lives Revealed by Their Trace Fossils.* New York: Pegasus Books, 460 pp.

Martin, A. J. 2016. A close look at Victoria's first known dinosaur tracks. *Memoirs of Museum Victoria* 74: 63–71.

Martin, A. J., T. H. Rich, M. Hall, P. Vickers-Rich, and G. Vazquez-Prokopec. 2012. A polar dinosaur-track assemblage from the Eumerella Formation (Albian), Victoria, Australia. *Alcheringa* 36: 171–188.

Marty, D. 2008. Sedimentology, taphonomy, and ichnology of Late Jurassic dinosaur tracks from the Jura carbonate platform (Chevenez-Combe Ronde tracksite, NW Switzerland): insights into the tidal-flat palaeoenvironment and dinosaur diversity, locomotion, and palaeoecology. *Geofocus* 21: 278.

Marty, D., P. L. Falkingham, and A. Richter. 2016. Dinosaur track terminology: a glossary of terms. In P. L. Falkingham, D. Marty, and A. Richter (eds.), *Dinosaur Tracks: The Next Steps*, 399–402. Bloomington: Indiana University Press.

Marty, D., A. Strasser, and C. A. Meyer. 2009. Formation and taphonomy of human footprints in microbial mats of present-day tidal-flat environments: implications for the study of fossil footprints. *Ichnos* 16: 127–142.

Marzola, M., and F. Dalla Vecchia. 2014. New dinosaur tracks from the Dolomia Principale (Upper Triassic) of the Carnic Prealps (Friuli-Venezia Giulia, NE Italy). *Bolletino della Società Paleontologica Italiana* 53(1): 1–18. doi:10.4435/BSPI.2014.01.

Masrour, M., L. Ladel, and F. Pérez-Lorente. 2015. New theropod and prosauropod ichnites from Issil-n-Aït Arbi (Lower Jurassic, Central High Atlas, Morocco). *Geogaceta* 57: 55–58.

Masrour, M., F. Pérez-Lorente, S. Ferry, N. Içame, and D. Grosheny. 2013. First dinosaur tracks from the Lower Cretaceous of the Western High Atlas (Morocco). *Geogaceta* 53: 33–36.

Mateus, O., M. Marzola, A. S. Sculp, L. L. Jacobs, M. J. Polcyn, V. Pervov, A. O. Gonçalves, and M. L. Morais. 2017. Angolan ichnosite in a diamond mine shows the presence of a larger terrestrial mammaliamorph, a crocodylomorph, and sauropod dinosaurs in the Early Cretaceous of Africa. *Palaeogeography, Palaeoclimatology, Palaeoecology* 471: 220–232.

Matsukawa, M., M. Futakami, M. G. Lockley, C. Peiji, C. Jinhua, C. Zhenyao, and U. L. Bolotsky. 1995. Dinosaur footprints from the Lower Cretaceous of eastern Manchuria, northeastern China: implications for the recognition of an ornithopod ichnofacies in East Asia. *Palaios* 10: 3–15.

Matsukawa, M., M. G. Lockley, and A. P. Hunt. 1999. Three groups of ornithopods inferred from footprints in the mid-Cretaceous Dakota Group, eastern Colorado, North America. *Palaeogeography, Palaeoclimatology, Palaeoecology* 147: 39–51.

Matsukawa, M., M. Lockley, and J. Li. 2006. Cretaceous terrestrial biotas of east Asia, with special reference to dinosaur-dominated ichnofaunas: towards a synthesis. *Cretaceous Research* 27: 3–21.

Matsukawa, M., T. Matsui, and M. G. Lockley. 2001. Trackway evidence of herd structure among ornithopod dinosaurs from the Cretaceous Dakota Group of northeastern New Mexico, USA. *Ichnos* 9: 197–206.

Matsukawa, M., K. Shibata, R. Kukihara, K. Koarai, and M. G. Lockley. 2005. Review of Japanese dinosaur track localities: implications for ichnotaxonomy, paleogeography and stratigraphic correlation. *Ichnos* 12: 201–222.

Matthew, W. D., and W. Granger. 1917. The skeleton of *Diatryma*, a gigantic bird from the Lower Eocene of Wyoming. *Bulletin of the American Museum of Natural History* 37: 307–326.

Mayr, G. 2008. Avian higher-level phylogeny: well-supported clades and what we can learn from a phylogenetic analysis of 2954 morphological characters. *Journal of Zoological Systematics and Evolutionary Research* 46: 63–72.

Mayr, G. 2012. The origins of crown group birds: molecules and fossils. *Palaeontology* 57: 231–242.

Mayr. G. 2017. *Avian Evolution: The Fossil Record of Birds and Its Palaeobiological Significance.* Chichester, UK: Wiley Blackwell, 293 pp.

Mayr, G., and J. Clarke. 2003. The deep divergences of neornithine birds: a phylogenetic analysis of morphological characters. *Cladistics* 19: 527–553.

Mazin, J.-M., J. P. Bilon-Bruyat, P. Hantzpergue, and G. Lafaurie. 2003. Ichnological evidence for quadrupedal locomotion in pterodactyloid pterosaurs: trackways from the Late Jurassic of Crayssac (southwestern France). In E. Buffetaut and J. M. Mazin (eds.), *Evolution and Palaeobiology of Pterosaurs*, 283–296. London, U.K.: Geological Society of London, Special Publication 217.

Mazin, J.-M., P. Hantzpergue, and J. Pouech. 2016. The dinosaur tracksite of Louelle (early Kimmeridgian; Jura, France). *Geobios* 49: 211–228.

Mazzetta, G. V., and R. E. Blanco. 2004. Speeds of dinosaurs from the Albian-Cenomanian of Patagonia and sauropod stance and gait. *Acta Palaeontologica Polonica* 46: 235–246.

McCormack, J. E., M. G. Harvey, B. C. Faircloth, N. G. Crawford, T. C. Glenn, and R. T. Brumfield. 2013. A phylogeny of birds based on over 1,500 loci collected by target enrichment and high-throughput sequencing. *PloS One* 8(1): e54848. doi:10.1371/journal.pone.0054848.

McCrea, R. T., L. G. Buckley, J. O. Farlow, M. G. Lockley, P. J. Currie, N. A. Matthews, and S. G. Pemberton. 2014*a*. A 'terror of tyrannosaurs': the first trackways of tyrannosaurids and evidence of gregariousness and pathology in Tyrannosauridae. *PloS One* 9(7): e103613. doi:10.1371/journal.pone.0103613.

McCrea, R. T., L. G. Buckley, A. G. Plint, P. J. Currie, J. W. Haggart, C. W. Helm, and S. G. Plimpton. 2014*b*. A review of vertebrate track-bearing formations from the Mesozoic and earliest Cenozoic of western Canada with a description of a new theropod ichnospecies and reassignment of an avian ichnogenus. In M. G. Lockley and S. G. Lucas (eds.), *Fossil Footprints of Western North America*, 5–93. Albuquerque, NM: New Mexico Museum of Natural History and Science Bulletin 62.

McCrea, R. T., S. G. Pemberton, and P. J. Currie. 2004. New ichnotaxa of mammal and reptile tracks from the Upper Paleocene of Alberta. *Ichnos* 11: 323–339.

McCrea, R. T., and T. S. Pigeon. 2014. Replication and description of a large theropod and large ornithopod trackway from the Upper Minnes Group (Lower Cretaceous: Valanginian) of the Peace Region of northeastern British

Columbia, Canada. In M. G. Lockley and S. G. Lucas (eds.), *Fossil Footprints of Western North America*, 269–277. Albuquerque, NM: New Mexico Museum of Natural History and Science Bulletin 62.

McIlhenny, E. A. 1935. *The Alligator's Life History*. Boston, MA: Christopher Publishing House, 117 pp. (1976 reprint edition, Society for the Study of Amphibians and Reptiles).

McIntosh, J. S. 1981. Annotated catalogue of the dinosaurs (Reptilia, Archosauria) in the collections of the Carnegie Museum of Natural History. *Bulletin of the Carnegie Museum of Natural History* 18: 67.

Melero Rubio, M., and F. Pérez-Lorente. 2011. Huellas en Las Obras. Reconocimiento y estudio de huellas fósiles de dinosaurio en Las Obras de La Presa de Enciso (La Rioja, España). *Zubía* 29: 31–60.

Meyer, C. A., and B. Thüring. 2003. The first iguanodontid dinosaur tracks from the Swiss Alps (Schrattenkalk Formation, Aptian). *Ichnos* 10: 221–228.

Mezga, A., and Z. Bajraktarevi[illegible]. 1999. Cenomanian dinosaur tracks on the islet of Fenoliga in southern Istria, Croatia. *Cretaceous Research* 20: 735–746.

Mietto, P. 1985. Piste di dinosauri nella Dolomia Principale (Triassico Superiore) del Monte Pelmetto (Cadore). *Memorie della Società Geologia Italiana* 30: 307–310.

Mietto, P., G. Roghi, and R. Zorzin. 2000. Le impronte di dinosauri liassici dei Monti Lessini Veronesi. *Bolletino del Museo Civico de Storia Naturale di Verona* 24: 55–72.

Milàn, J. 2006. Variations in the morphology of emu (*Dromaius novaehollandiae*) tracks reflecting differences in walking pattern and substrate consistency: ichnotaxonomic implications. *Palaeontology* 49: 405–420.

Milàn, J., M. Avanzini, L. B. Clemmensen, J. C. García-Ramos, and L. Piñuela. 2006. Theropod foot movement recorded from Late Triassic, Early Jurassic, and Late Jurassic fossil footprints. In J. D. Harris, S. G. Lucas, J. A. Spielmann, M. G. Lockley, A. R. C. Milner, and J. I. Kirkland (eds.), *The Triassic-Jurassic Terrestrial Transition*, 352–364. Albuquerque, NM: New Mexico Museum of Natural History and Science Bulletin 37.

Milàn, J., and R. G. Bromley. 2006. True tracks, undertracks and eroded tracks, experimental work with tetrapod tracks in laboratory and field. *Palaeogeography, Palaeoclimatology, Palaeoecology* 231: 253–264.

Milàn, J., and R. G. Bromley. 2008. The impact of sediment consistency on track and undertrack morphology: experiments with emu tracks in layered cement. *Ichnos* 15: 19–27.

Milàn, J., P. Christiansen, and O. Mateus. 2005. A three-dimensionally preserved sauropod manus impression from the Upper Jurassic of Portugal: implications for sauropod manus shape and locomotor mechanics. *Kaupia* 14: 47–52.

Milàn, J., L. B. Clemmensen, and N. Bonde. 2004. Vertical sections through dinosaur tracks (Late Triassic lake deposits, East Greenland): undertracks and other subsurface deformation structures revealed. *Lethaia* 37: 285–296.

Milàn, J., and P. L. Falkingham. 2016. Experimental and comparative ichnology. In P. L. Falkingham, D. Marty, and A. Richter (eds.), *Dinosaur Tracks: The Next Steps*, 14–27. Bloomington: Indiana University Press.

Milàn, J., and A. Hedegaard. 2010. Interspecific variation in tracks and trackways from extant crocodylians. In J. Milàn, S. G. Lucas, M. G. Lockley, and J. A. Spielmann (eds.), *Crocodyle Tracks and Traces*, 15–29. Albuquerque, NM: New Mexico Museum of Natural History and Science Bulletin 51.

Milàn, J., D. B. Loope, and R. G. Bromley. 2008. Crouching theropod and *Navahopus* sauropodomorph tracks from the Early Jurassic Navajo Sandstone of USA. *Acta Palaeontologica Polonica* 53: 197–205.

Milàn, J., S. G. Lucas, M. G. Lockley, and J. A. Spielmann (eds.). 2010. *Crocodyle Tracks and Traces*. Albuquerque, NM: New Mexico Museum of Natural History and Science Bulletin 51, 244 pp.

Miller, W. E., B. B. Britt, and K. L. Stadtman. 1989. Tridactyl trackways from the Moenave Formation of southwestern Utah. In D. D. Gillette and M. G. Lockley (eds.), *Dinosaur Tracks and Traces*, 209–215. Cambridge: Cambridge University Press.

Milner, A. R. C., J. D. Harris, M. G. Lockley, J. I. Kirkland, and N. A. Matthews. 2009. Bird-like anatomy, posture, and behavior revealed by an Early Jurassic theropod dinosaur resting trace. *PloS One* 4(3): e4591. doi:10.1371/journal.pone.0004591.

Milner, A. R. C., M. G. Lockley, and S. B. Johnson. 2006. The story of the St. George Dinosaur Discovery Site at Johnson Farm: an important new Lower Jurassic Dinosaur Tracksite from the Moenave Formation of southwestern Utah. In J. D. Harris, S. G. Lucas, J. A. Spielmann, M. G. Lockley, and A. R. C. Milner (eds.), *The Triassic-Jurassic Transition*, 329–345. Albuquerque, NM: New Mexico Museum of Natural History and Science Bulletin 37.

Mindell, D. P., and J. W. Brown. 2005. Neornithes. Modern Birds. Version 14 December 2005 (under construction); http://tolweb.org/Neornithes/15834/2005.12.14 in the Tree of Life Web Project.

Mindell, D. P., and J. W. Brown. 2007. Galloanserae. Version 07 March 2007 (temporary); http://tolweb.org/Galloanserae/15840/2007.03.07 in the Tree of Life Project.

Mindell, D. P., J. W. Brown, and J. Harshman. 2008. Neoaves. Version 27 June 2008 (under construction); http://tolweb.org/Neoaves/26305/2008.06.27 in the Tree of Life Project.

Mitchell, K. J., J. S. Bastien Llamas, N. J. Rawlence, T. H.Worthy, J. Wood, M. S. Y. Lee, and A. Cooper. 2014. Ancient DNA reveals elephant birds and kiwi are sister taxa and clarifies ratite bird evolution. *Science* 344: 898–900.

Montgomery, D. R. 2012. *The Rocks Don't Lie: A Geologist Investigates Noah's Flood*. New York: W. W. Norton, 302 pp.

Morales, M., and S. Bulkley. 1996. Paleoichnological evidence for a theropod dinosaur larger than *Dilophosaurus* in the Lower Jurassic Kayenta Formation. In M. Morales (ed.), *The Continental Jurassic*, 143–145. Flagstaff: Museum of Northern Arizona Bulletin 60.

Moratalla, J. J., J. Hernán, and S. Jiménez. 2003. Los Cayos dinosaur tracksite: an overview on the Lower Cretaceous ichno-diversity of the Cameros Basin (Cornago, La Rioja Province, Spain). *Ichnos* 10: 229–240.

Moratalla, J. J., M. G. Lockley, A. D. Buscalioni, M. A. Frenegal-Martínez, N. Meléndez, F. Ortega, B. P. Pérez-Moreno, E. Pérez-Asensio, J. L. Sanz, and R. J. Schultz. 1995. A preliminary note on the first tetrapod trackways from the lithographic limestones of Las Hoyas (Lower Cretaceous, Cuenca, Spain). *Geobios* 28: 777–782.

Moratalla, J. J., J. L. Sanz, and S. Jiménez. 1988*a*. Multivariate analysis on Lower Cretaceous dinosaur footprints: discrimination between ornithopods and theropods. *Geobios* 21: 395–408.

Moratalla, J. J., J. L. Sanz, and S. Jiménez. 1988*b*. Nueva evidencia icnologica de dinosaurios el el Cretácico Inferior de la Rioja (España). *Estudios Geológicos* 44: 119–131.

Moratalla, J. J., J. L. Sanz, S. Jiménez, and M. G. Lockley. 1992. A quadrupedal ornithopod trackway from the Lower Cretaceous of La Rioja (Spain): inferences on gait and hand structure. *Journal of Vertebrate Paleontology* 12: 150–157.

Moratalla, J. J., J. L. Sanz, and S. Jiménez. 1994. Dinosaur tracks from the Lower Cretaceous of Regumiel de la Sierra (province of Burgos, Soria): inferences on a new quadrupedal ornithopod trackway. *Ichnos* 3: 89–97.

Moreau, J. D., L. Baret, G. Gand, E. Fara, C. Durlet, and G. Caravaca. 2012*a*. Dècouverte d'un nouveau site à traces de pas de Dinosaures dans le Bathonien des Causses (Le Gayrand, Gorges de la Jonte, Lozère, France). In *Ichnologie Dinosaurenne du Jurassique de Meyrueis*, 13–19. Mende, Lozère, France: Association Paléontologique des Hauts Plateaux du Languedoc.

Moreau, J.-D., G. Gand, E. Fara, and A. Michelin. 2012*b*. Biometric and morphometric approaches on Lower Hettangian dinosaur footprints from the Rodez Strait (Aveyron, France). *Comptes Rendus Palevol* 11: 231–239.

Moreau, J.-D., V. Trincal, G. Gand, D. Néraudeau, G. Bessière, and B. Bourel. 2014. Deux nouveaux sites à traces de pas dinosauroïdes dans la Formation Dolomitique de l'Hettangien de Lozère, Languedoc-Roussillon, France. *Annales de Paléontologie* 100: 361–369.

Moreno, K., N. Blanco, and A. Tomlinson. 2004. Nuevas huellas de dinosaurios del Jurásico Superior en el norte de Chile. *Ameghiniana* 41: 535–544.

Moreno, K., M. T. Carrano, and R. Snyder. 2007. Morphological changes in pedal phalanges through ornithopod dinosaur evolution: a biomechanical approach. *Journal of Morphology* 268: 50–63.

Moreno, K., S. De Valais, N. Blanco, A. J. Tomlinson, J. Jacay, and J. O. Calvo. 2012. Large theropod dinosaur footprint associations in western Gondwana: behavioural and palaeogeographic implications. *Acta Palaeontologica Polonica* 57: 73–83.

Moreno, K., and M. Pino. 2002. Huellas de dinosaurios en la Formación Baños del Flaco (Titoniano-Jurásico Superior), VI Región,

Chile: paleoetología y paleoambiente. *Revista Geológia de Chile* 29: 191–206.
Morgan-Richards, M., S. A. Trewick, A. Bartosch-Härlid, O. Kardailsky, M. J. Phillips, P. A. McLenachan, and D. Penny. 2008. Bird evolution: testing the metaves clade with six new mitochondrial genomes. *BMC Evolutionary Biology* 8: 20. doi:10.1186/1471-2148-8-20.
Morris, W. J. 1976. Hypsilophodont dinosaurs: a new species and comments on their systematics. In C. S. Churcher (ed.), *Athlon: Essays on Palaeontology in Honour of Loris Shano Russell*, 93–113. Toronto, Ontario: Royal Ontario Museum Life Sciences Miscellaneous Publication.
Moser, M. 2003. *Plateosaurus engelhardti* Meyer, 1837 (Dinosauria: Sauropodomorpha) aus dem Feuerletten (Mittelkeuper; Obertrias) von Bayern. *Zitteliana B* 24: 3–186.
Mossman, D. J., R. Brüning, and H. P. Powell. 2003. Anatomy of a Jurassic theropod trackway from Ardley, Oxfordshire, U.K. *Ichnos* 10: 195–207.
Müller, R. T., M. C. Langer, M. Bronzati, C. P. Pacheco, S. F. Cabreira, and S. Dias-Da-Silva. May 15, 2018. Early evolution of sauropodomorphs: anatomy and phylogenetic relationships of a remarkably well-preserved dinosaur from the Upper Triassic of southern Brazil. *Zoological Journal of the Linnean Society.*
Munyika, D. 1996. Description of the first dinosaur trackway found in Zimbabwe. *Arnoldia Zimbabwe* 10: 36–45.
Murphy, T. M. 1981. *The Population Status of the American Alligator on the Savannah River Plant, South Carolina.* U.S. Department of Energy Savannah River National Environmental Research Park Publication 4, 20 pp.
Murray, P. F., and P. Vickers-Rich. 2004. *Magnificent Mihirungs; The Colossal Flightless Birds of the Australian Dreamtime.* Bloomington: Indiana University Press, 410 pp.
Mustoe, G. E., D. S. Tucker, and K. L. Kemplin. 2012. Giant Eocene bird footprints from northwest Washington USA. *Palaeontology* 55: 1293–1305.
Myers, T. S., and A. R. Fiorillo. 2009. Evidence for gregarious behavior and age segregation in sauropod dinosaurs. *Palaeogeography, Palaeoclimatology, Palaeoecology* 274: 96–104.
Nadon, G. C. 2001. The impact of sedimentology on vertebrate track studies. In D. H. Tanke and K. Carpenter (eds.), *Mesozoic Vertebrate Life: New Research Inspired by the Paleontology of Philip J. Currie*, 395–407. Bloomington Indiana University Press.
Naish, D. 2012. Birds. In M. K. Brett-Surman, T. R. Holtz, Jr., and J. O. Farlow (eds.), *The Complete Dinosaur*, 379–424. Bloomington: Indiana University Press.
Nesbitt, S. J. 2011. The early evolution of archosaurs: relationships and the origin of major clades. *Bulletin of the American Museum of Natural History* 352: 1–292.
Nesbitt, S. J., S. L. Brusatte, J. B. Desojo, A. Liparini, M. A. G. De França, J. C. Weinbaum, and D. J. Gower. 2013*a*. Rauisuchia. In S. J. Nesbitt, J. B. Desojo, and R. B. Irmis (eds.). 2013. *Anatomy, Phylogeny and Palaeobiology of Early Archosaurs and Their Kin*, 241–274. London: The Geological Society, Special Publication 379.
Nesbitt, S. J., and J. A. Clarke. 2016. The anatomy and taxonomy of the exquisitely preserved Green River Formation (early Eocene) lithornithids (Aves) and the relationships of Lithornithidae. *American Museum Novitates* 406: 91 pp.
Nesbitt, S. J., J. B. Desojo, and R. B. Irmis (eds.). 2013*b*. *Anatomy, Phylogeny and Palaeobiology of Early Archosaurs and Their Kin.* London: The Geological Society, Special Publication 379, 608 pp.
Nicosia, U., M. Marino, N. Mariotti, C. Muraro, S. Panigutti, F. M. Petti, and E. Sacchi. 1999. The Late Cretaceous dinosaur tracksite near Altamura (Bari, southern Italy). II-*Apulisauripus federicianus* new ichnogen. and new ichnosp. *Geologica Romana* 35: 237–247.
Niedźwiedzki, G. 2011. A Late Triassic dinosaur-dominated ichnofauna from the Tomanová Formation of the Tatra Mountains, central Europe. *Acta Palaeontologica Polonica* 56: 291–300.
Niedźwiedzki, G., and G. Pieńkowski. 2004. A dinosaur track association from the Early Jurassic deltaic deposits of Podole near Opatów, Poland. *Geological Quarterly* 48: 333–338.
Niedźwiedzki, G., and G. Pieńkowski. 2016. A dinosaur track assemblage from the Upper Hettangian (Lower Jurassic) marginal-marine deposits of Zapniów, Holy Cross Mountains, Poland. *Geological Quarterly* 60: 840–856.
Noblet, C., G. Leonardi, P. Taquet, R. Marocco, and E. Cordova. 1995. Nouvelle découverte d'empreintes laissées par des dinosaures dans la Formation des Couches Rouges (bassin de Cuzco-Sicuani, Sud du Pérou: conséquenes stratigraphiques et tectoniques. *Comptes Rendus de l'Academie des Sciénces* 320: 785–791.
Norman, D. B. 1980. On the ornithischian dinosaur *Iguanodon bernissartensis* of Bernissart (Belgium). *Institut Royal des Sciences Naturelles de Belgique Mémoire* 178: 1–104.
Norman, D. B. 1986. On the anatomy of *Iguanodon atherfieldensis* (Ornithschia: Ornithopoda). *Bulletin de l'Institut Royal des Sciences Naturelles de Belgique* 56: 281–372.
Norman, D. B. 1987*a*. A mass-accumulation of vertebrates from the Lower Cretaceous of Nehden (Sauerland), West Germany. *Proceedings of the Royal Society of London B* 230: 215–255.
Norman, D. B. 1987*b*. On the history of the discovery of fossils at Bernissart in Belgium. *Archives of Natural History* 14: 59–75.
Norman, D. B. 2010. A taxonomy of iguanodontians (Dinosauria: Ornithopoda) from the lower Wealden Group (Cretaceous: Valanginian) of southern England. *Zootaxa* 2489: 47–66.
Norman, D. B. 2012. Iguanodontian taxa (Dinosauria: Ornithischia) from the Lower Cretaceous of England and Belgium. In P. Godefroit (ed.), *Bernissart Dinosaurs and Early Cretaceous Terrestrial Ecosystems*, 174–212. Bloomington: Indiana University Press.
Norman, D. B. 2014. Iguanodonts from the Wealden of England: do they contribute to the discussion of hadrosaur origins? In D. A. Eberth and D. C. Evans (eds.), *Hadrosaurs*, 10–43. Bloomington: Indiana University Press.
Nouri, J., I. Díaz-Martínez, and F. Pérez-Lorente. 2011. Tetradactyl footprints of an unknown affinity theropod dinosaur from the Upper Jurassic of Morocco. *PloS One* 6(12): e26882. doi:10.1371/journal.pone.0026882.
Novas, F. E. 2009. *The Age of Dinosaurs in South America.* Bloomington: Indiana University Press, 453 pp.
Oaks, J. R. 2011. A time-calibrated species tree of Crocodylia reveals a recent radiation of the true crocodiles. *Evolution* 65: 3285–3297.
O'Connor, J. K., K.-G. Gao, and L. M. Chiappe. 2010. A new ornithuromorph (Aves: Ornithothoraces) bird from the Jehol Group indicative of higher-level diversity. *Journal of Vertebrate Paleontology* 30: 311–321.
O'Connor, J. K., D.-Q. Li, M. C. Lamanna, M. Wang, J. D. Harris, J. Atterholt, and H.-L. You. 2015. A new Early Cretaceous enantiornithine (Aves: Ornithothoraces) from northwestern China with elaborate tail ornamentation. *Journal of Vertebrate Paleontology* (published online). doi:10.1080/02724634.2015.1054035.
Olsen, P. E. 1980. Fossil great lakes of the Newark Supergroup in New Jersey. In W. Manspeizer (ed.), *Field Studies of New Jersey Geology and Guide to Field Trips*, 352–398. Newark, NJ: 52nd Annual Meeting of the New York State Geological Association, Rutgers University.
Olsen, P. E. 1995. Paleontology and paleoenvironments of Early Jurassic age strata in the Walter Kidde Dinosaur Park (New Jersey, USA). In J. E. B. Baker (ed.), *Field Guide and Proceedings of the Twelfth Annual Meeting of the Geological Association of New Jersey*, 156–190. Patterson, NJ: William Patterson College.
Olsen, P. E., and E. C. Rainforth. 2003. The Early Jurassic ornithischian ichnogenus *Anomoepus*. In P. M. Letourneau and P. E. Olsen (eds.), *The Great Rift Valleys of Pangea in Eastern North America. Volume Two: Sedimentology, Stratigraphy, and Paleontology*, 314–368. New York: Columbia University Press.
Olsen, P. E., J. B. Smith, and N. G. McDonald. 1998. Type material of the type species of the classic theropod footprint genera *Eubrontes*, *Anchisauripus*, and *Grallator* (Early Jurassic, Hartford and Deerfield Basins, Connecticut and Massachusetts, U.S.A. *Journal of Vertebrate Paleontology* 18: 586–601.
Olson, V. A., and S. T. Turvey. 2013. The evolution of sexual dimorphism in New Zealand giant moa (*Dinornis*) and other ratites. *Proceedings of the Royal Society B* 280: 20130401. dx.doi.org/10.1098/rspb.2013.0401.
Osborn, H. F. 1899. Fore and hind limbs of carnivorous and herbivorous dinosaurs from the Jurassic of Wyoming. *Bulletin of the American Museum of Natural History* 12: 161–172.
Ősi, A., J. Pálfy, L. Makádi, Z. Szentesi, P. Gulyás, M. Rabi, G. Botfalvai, and K. Hips. 2011. Hettangian (Early Jurassic) dinosaur tracksites from the Mecsek Mountains, Hungary. *Ichnos* 18: 79–94.
Ostrom, J. H. 1970. *Stratigraphy and Paleontology of the Cloverly Formation (Lower Cretaceous) of the Bighorn Basin*

area, Wyoming and Montana. Bulletin 35. New Haven, CT: Peabody Museum of Natural History, Yale University, 234 pp.

Ostrom, J. H. 1976. On a new specimen of the Lower Cretaceous theropod dinosaur *Deinonychus antirrhopus*. *Breviora* 439: 1–21.

Ostrom, J. H. 1978. The osteology of *Compsognathus longipes* Wagner. *Zitteliana* 4: 73–118.

Owen, R. 1879. *Memoirs on the Extinct Wingless Birds of New Zealand, with an Appendix on Those of England, Australia, Newfoundland, Mauritius and Rodriguez*. London: John van Voorst.

Padian, K., and E.-T. Lamm. 2013. *Bone Histology of Fossil Tetrapods: Advancing Methods, Analysis, and Interpretation*. Berkeley, CA: University of California Press, 285 pp.

Padian, K., and P. E. Olsen. 1984. The fossil trackway *Pteraichnus*; not pterosaurian, but crocodilian. *Journal of Paleontology* 58: 178–184.

Padian, K., and P. E. Olsen. 1989. Ratite footprints and the stance and gait of Mesozoic theropods. In D. D. Gillette and M. G. Lockley (eds.), *Dinosaur Tracks and Traces*, 231–241. Cambridge: Cambridge University Press.

Padian, K., and K. Stein. 2013. Evolution of growth rates and their implications. In K. Padian and E.-T. Lamm (eds.), *Bone Histology of Fossil Tetrapods: Advancing Methods, Analysis, and Interpretation*, 253–264. Berkeley, CA: University of California Press.

Pannell, N. K. 1985. *Dinosaur Footprints at Oak Hill, Virginia*. Master's Thesis. Washington, DC: George Washington University, 30 pp.

Parish, J. C. 2013. *The Dodo and the Solitaire: A Natural History*. Bloomington: Indiana University Press, 407 pp.

Parkes, A. S. 1993. Dinosaur footprints in the Wealden at Fairlight, East Sussex. *Proceedings of the Geologists' Association* 104: 15–21.

Parks, W. A. 1920. The osteology of the trachodont dinosaur *Kritosaurus incurvimanus*. *University of Toronto Studies Geological Series* 11: 1–75.

Parks, W. A. 1933. New species of dinosaurs and turtles from the Upper Cretaceous formations of Alberta. *University of Toronto Studies Geological Series* 34: 1–33.

Parks, W. A. 1935. New species of trachodont dinosaurs from the Cretaceous formations of Alberta with notes on other species. *University of Toronto Studies Geological Series* 37: 1–45.

Pascual-Arribas, C., and N. Hernández-Medrano. 2010. Nuevos datos sobre el yacimiento icnítico de las Cuestas I (Santa Cruz de Yanguas, Soria, España). *Studia Geologica Salmanticensia* 46: 121–157.

Pascual-Arribas, C., and N. Hernández-Medrano. 2011. Posibles huellas de crías de terópodo en el yacimiento de Valdehijuelos (Soria, España). *Studia Geologica Salmanticensia* 47: 77–110.

Pascual-Arribas, C., N. Hernández-Medrano, P. Latorre-Macarrón, and E. Sanz-Pérez. 2005. Nuevo rastro de icnitas de cocodrilo en la Formación Huérteles de la Cuenca de Cameros, Yacimiento del Barranco de Valdelavilla (Valdelavilla, Soria, España). *Studio Geologica Salmanticensia* 41: 77–91.

Pascual-Arribas, C., N. Hernández-Medrano, P. Latorre-Macarrón, and E. Sanz-Pérez. 2009. El icnogénero *Iguanodontipus* en el yacimiento de "Las Cuestas I" (Santa Cruz de Yanguas, Soria, España). *Studia Geologica Salmanticensia* 45: 105–128.

Pascual-Arribas, C., P. Latorre Macarrón, N. Hernández Medrano, and E. Sanz Pérez. 2005. Las huellas de dinosaurios de los yacimientos del Arroyo Miraflores (Fuentes de Magaña-Cerbón-Magaña, Soria). *Celtiberia* 55: 413–442.

Pascual-Arribas, C., and E. Sanz Pérez. 2000. Icnitas de dinosaurios en Valdelavilla (Soria, España). *Estudios Geológicos* 56: 41–61.

Paul, G. S. 2007. Turning the old into the new: a separate genus for the gracile iguanodont from the Wealden of England. In K. Carpenter (ed.), *Horns and Beaks: Ceratopsian and Ornithopod Dinosaurs*, 69–77. Bloomington: Indiana University Press.

Paul, G. S. 2008*a*. The extreme lifestyles and habits of the gigantic tyrannosaurid superpredators of the Late Cretaceous of North America and Asia. In P. Larson and K. Carpenter (eds.), *Tyrannosaurus rex, the Tyrant King*, 307–352. Bloomington: Indiana University Press.

Paul, G. S. 2008*b*. A revised taxonomy of the iguanodont dinosaur genera and species. *Cretaceous Research* 29: 192–216.

Paul, G. S. 2010. *The Princeton Field Guide to Dinosaurs*. Princeton, NJ: Princeton University Press, 320 pp.

Pei, R., Q. Li, Q. Meng, K.-Q. Gao, and M. A. Norell. 2014. A new specimen of *Microraptor* (Theropoda: Dromaeosauridae) from the Lower Cretaceous of western Liaoning, China. *American Museum Novitates* 3821: 1–28.

Pemberton, S. G., M. K. Gingras, and J. A. MacEachern. 2007. Edward Hitchcock and Roland Bird: two early titans of vertebrate ichnology in North America. In W. Miller III (ed.), *Trace Fossils: Concepts, Problems, Prospects*, 32–51. Amsterdam: Elsevier Science.

Peng G. 1992. Jurassic ornithopod *Agilisaurus louderbacki* (Ornithopoda: Fabrosauridae) from Zigong, Sichuan, China. *Vertebrata PalAsiatica* 4: 39–53 (in Chinese with English summary).

Pérez-Lorente, F. 1993. Dinosaurios plantígrados en La Rioja. *Zubía* 5: 189–228.

Pérez-Lorente, F. 1996. Pistas terópodos en cifras. *Zubía* 14: 37–55.

Pérez-Lorente, F. 2003. Aportaciones de los yacimientos de la Barguilla, Santisol, y Santa Juliana (Hornillos de Cameros, La Rioja, España). In F. Pérez (ed.), *Dinosaurios y Otros Reptiles Mesozoicos en España*, 161–194. Logroño, Spain: Gobierno de La Rioja, Fundación Patrimonio Paleontológico de La Rioja, Instituto de Estudios Riojanos, Universidad de La Rioja.

Pérez-Lorente, F. 2013. Las huellas de dinosaurio más modernas de Enciso (La Rioja, España). Yacimiento de las Riscas. *Zubía* 31: 211–222.

Pérez-Lorente, F. 2015. *Dinosaur Footprints and Trackways of La Rioja*. Bloomington: Indiana University Press, 363 pp.

Pérez-Lorente, F., G. Cuenca-Bescós, M. Aurell, J. I. Canudo, A. R. Soria, and J. I. Ruiz-Omeñaca. 1997. Las Cerradicas tracksite (Berriasian, Galve, Spain): growing evidence for quadrupedal ornithopods. *Ichnos* 5: 109–120.

Pérez-Lorente, F., M. M. Romero-Medina, and J. C. Pereda Olásolo. 2000. Icnitas ornitópodas del Contadero (Torremuña, La Rioja, España). Estructuras de barro y baja velocidad de marcha. In *Investigación Humanistica y Cientifica en La Rioja: Homenaje a Julio Luis Fernández*, 29–41. Spain: Sevilla y Mayela Balmaseda Aróspide.

Pérez-Lorente, F., and M. M. Romero Molina. 2001. Icnitas terópodas del Cretácico Inferior de La Rioja (España). *Zubía* 19: 115–138.

Pérez-Lorente, F., and F. Ortega. 2003. Dos nuevos rastros de arcosaurios en el Cretácico Inferior de Galve (Teruel, España). El yacimiento del Cantalar. In F. Pérez (ed.), *Dinosaurios y Otros Reptiles Mesozoicos en España*, 129–136. Logroño, Spain: Gobierno de La Rioja, Fundación Patrimonio Paleontológico de La Rioja, Instituto de Estudios Riojanos, Universidad de La Rioja.

Persons, W. S., IV, and P. J. Currie. 2016. An approach to scoring cursorial limb proportions in carnivorous dinosaurs and an attempt to account for allometry. *Nature Scientific Reports* 6: 19828. doi:10.1038/SREP19828.

Petti, F. M., M. Bernardi, P. Ferretti, R. Tomasoni, and M. Avanzini. 2011. Dinosaur tracks in a marginal marine environment: the Coste dell'Anglone ichnosite (Early Jurassic, Trento Platform, NE Italy). *Italian Journal of Geoscience* 130: 27–41.

Petti, F. M., M. A. Conti, S. D'Orazi Porchetti, M. Morsilli, U. Nicosia, and P. Gianolla. 2008. A theropod dominated ichnocoenosis from late Hauterivian-early Barremian of Borgo Celano (Gargano Promontory, Apulia, southern Italy). *Rivista Italiana de Paleontologia e Stratigrafia* 114: 3–17.

Phillips, M. J., G. C. Gibbs, E. A. Crimp, and D. Penny. 2010. Tinamous and moas flock together: mitochondrial genome sequence analysis reveals independent losses of flight among ratites. *Systematic Biology* 59: 90–107.

Pieńkowski, G., P. Brański, D. K. Pandey, J. Schlögel, M. Alberti, and F. T. Fürsich. 2015. Dinosaur footprints from the Thaiat ridge and their palaeoenvironmental background, Jaisalmer Basin, Rajastan, India. *Volumina Jurassica* 13: 17–26.

Piñuela, L., J. C. García Ramos, M. Romano, and J. I. Ruiz-Omeñaca. 2016. First record of gregarious behavior in robust medium-sized Jurassic ornithopods: evidence from the Kimmeridgian trackways of Asturias (N. Spain) and some general considerations on other medium-large ornithopod tracks in the Mesozoic record. *Ichnos* 23: 298–311.

Piubelli, D., M. Avanzini, and P. Mietto. 2005. The Early Jurassic ichnogenus *Kayentapus* at Lavini di Marco ichnosite (NE Italy). Global distribution and palaeogeographic implications. *Bollettino della Società Geologica Italiana* 124: 259–267.

Platt, B. F., S. T. Hasiotis, and D. R. Hirmas. 2012. Empirical determination of physical controls on megafaunal footprint formation through neoichnological experiments with elephants. *Palaios* 27: 725–737.

Platt, N. H., and C. A. Meyer. 1991. Dinosaur footprints from the Lower Cretaceous of northern Spain: their sedimentological and paleoecological context. *Palaeogeography, Palaeoclimatology, Palaeoecology* 86: 321–333.

Poe, S., and A. L. Chubb. 2004. Birds in a bush: five genes indicate explosive evolution of avian orders. *Evolution* 58: 404–415.

Pond, S., M. G. Lockley, J. A. F. Lockwood, B. H. Breithaupt, and N. A. Matthews. 2014. Tracking dinosaurs on the Isle of Wight: a review of tracks, sites, and current research. *Biological Journal of the Linnean Society* 113: 737–757.

Prieto-Márques, A., and M. A. Norell. 2011. Redescription of a nearly complete skull of *Plateosaurus* (Dinosauria: Sauropodomorpha) from the Late Triassic of Trossingen (Germany). *American Museum Novitates* 3277: 1–58.

Prum, R. O., J. S. Berv, A. Dornburg, D. J. Field, J. P. Townsend, E. M. Lemmon, and A. R. Lemmon. 2015. A comprehensive phylogeny of birds (Aves) using targeted next-generation DNA sequencing. *Nature* 526: 569–573.

Raath, M. A., J. W. Kitching, R. W. Shone, and G. J. Rossouw. 1990. Dinosaur tracks in Triassic Molteno sediments: the earliest evidence of dinosaurs in South Africa? *Palaeontologia africana* 27: 89–95.

Rainforth, E. C. 2003. Revision and re-evaluation of the Early Jurassic dinosaurian ichnogenus *Otozoum*. *Palaeontology* 46: 803–838.

Rainforth, E. C. 2005. *Ichnotaxonomy of the fossil footprints of the Connecticut Valley (Early Jurassic, Newark Supergroup, Connecticut and Massachusetts*. PhD Thesis. New York: Columbia University, 1301 pp.

Rainforth, E. C. 2007. Ichnotaxonomic updates from the Newark Supergroup. In E. C. Rainforth (ed.), *Contributions to the Paleontology of New Jersey II): Field Guide and Proceedings*, 49–59. Geological Association of New Jersey, XXIV annual conference and field trip. East Stroudsburg, PA: East Stroudsburg University.

Rainforth, E. C., and M. G. Lockley. 1996. Tracks of diminutive dinosaurs and hopping mammals from the Jurassic of North and South America. In M. Morales (ed.), *The Continental Jurassic*, 265–269. Flagstaff, AZ: Museum of Northern Arizona Bulletin 60.

Rainforth, E. C., and M. Manzella. 2007. Estimating speeds of dinosaurs from trackways: a re-evaluation of assumptions. In E. C. Rainforth (ed.), *Contributions to the Paleontology of New Jersey II): Field Guide and Proceedings*, 41–48. Geological Association of New Jersey, XXIV annual conference and field trip. East Stroudsburg, PA: East Stroudsburg University.

Rajkumar, H. S., G. E. Mustoe, K. S. Khaidem, and I. Soiban. 2015. Crocodylian tracks from the Lower Oligocene flysch deposits of the Barail Group, Manipur, India. *Ichnos* 22: 122–131.

Rasskin-Gutman, D., G. Hunt, R. E. Chapman, J. L. Sanz, and J. J. Moratalla. 1997. The shapes of tridactyl dinosaur footprints: procedures, problems, and potentials. In D. L. Wolberg, E. Stump, and G. D. Roseberg (eds.), *Dinofest International*, 377–383. Philadelphia: Arizona State University and Academy of Natural Science.

Rauhut, O. W. M., C. Foth, H. Tischlinger, and M. A. Norell. 2012. Exceptionally preserved juvenile megalosaurid theropod dinosaur with filamentous integument from the Late Jurassic of Germany. *Proceedings of the National Academy of Sciences USA* 109: 11746–11751.

Rayner, J. M. V. 1985. Linear relations in biomechanics: the statistics of scaling functions. *Journal of Zoology, London* 206: 415–439.

Razzolini, N. L., M. Belvedere, D. Marty, G. Paratte, C. Lovis, M. Cattin, and C. A. Meyer. 2017. *Megalosauripus transjuranicus* ichnosp. nov. A new Late Jurassic theropod ichnotaxon from NW Switzerland and implications for tridactyl dinosaur ichnology and ichnotaxonomy. *PLoS One* 12(7): e0180289. https://doi.org/10.1371/journal.pone.0180289.

Razzolini, N. L., O. Oms, D. Castanera, B. Vila, V. F. dos Santos, and À. Galobart. 2016*a*. Ichnological evidence of megalosaurid dinosaurs crossing Middle Jurassic tidal flats. *Scientific Reports* 6: 31494. doi:10.1038/srep31494.

Razzolini, N. L., B. Vila, D. Castanera, P. L. Falkingham, J. L. Barco, J. I. Canudo, P. L. Manning, and À. Galobart. 2014. Intra-trackway morphological variations due to substrate consistency; the El Frontal dinosaur tracksite (Lower Cretaceous, Spain). *PloS One* 9(4): e93708. doi:10.1371/journal.pone.0093708.

Razzolini, N. L., B. Vila, I. Díaz-Martínez, and P. L. Manning. 2016*b*. Pes shape variation in an ornithopod dinosaur trackway (Lower Cretaceous, NW Spain): new evidence of an antalgic gait in the fossil track record. *Cretaceous Research* 58: 125–134.

Reid, R. E. H. 1997. How dinosaurs grew. In J. O. Farlow and M. K. Brett-Surman (eds.), *The Complete Dinosaur*, 403–413. Bloomington: Indiana University Press.

Reilly, S. M., and J. A. Elias. 1998. Locomotion in *Alligator mississippiensis*: kinematic effects of speed and posture and their relevance to the sprawling-to-erect paradigm. *Journal of Experimental Biology* 201: 2559–2574.

Reineck, H.-E., and J. D. Howard. 1978. Alligatorfährten. *Natur und Museum* 108: 10–15.

Reisz, R. R., D. Scott, H.-D. Sues, D. C. Evans, and M. A. Raath. 2005. Embryos of an Early Jurassic prosauropod dinosaur and their evolutionary significance. *Science* 309: 761–764.

Requeta Loza, L. E., N. Hernández Medrano, and F. Pérez-Lorente. 2006–2007. La Pellejera: descripción y aportaciones. Heterocronía y variabilidad de un yacimiento con huellas de dinosaurio de La Rioja (España). *Zubía Monografico* 18–19: 21–114.

Rich, P. V. 1979. *The Dromornithidae, an extinct family of large ground birds endemic to Australia*. Department of National Development, Bureau of National Resources, Geology, and Geophysics Bulletin 184. Canberra: Australia Government Publishing Service, 196 pp.

Rich, P. V., and E. Gill. 1976. Possible dromornithid footprints from Pleistocene dune sands of southern Victoria, Australia. *Emu* 76: 221–223.

Rich, P. V., and R. H. Green. 1974. Bird footprints at South Mt. Cameron, Tasmania. *Emu* 74: 245–248.

Rich, P. V., and R. E. Molnar. 1996. The foot of a bird from the Eocene redbank Plains Formation of Queensland, Australia. *Alcheringa* 20: 21–29.

Richter, A., and A. Böhme. 2016. Too many tracks: preliminary description and interpretation of the diverse and heavily dinoturbated Early Cretaceous "Chicken Yard" ichnoassemblage (Obernkirchen Tracksite, northern Germany). In P. L. Falkingham, D. Marty, and A. Richter (eds.), *Dinosaur Tracks: The Next Steps*, 334–357. Bloomington: Indiana University Press.

Richter, A., J. J. Hornung, A. Böhme, and U. Stratmann. 2012. Excursion guide A1: Obernkirchen Sandstone quarries—a natural workstone lagerstaette and a dinosaur tracksite. In A. Richter, M. Reich (eds.), *Dinosaur Tracks 2011: An International Symposium, Obernkirchen, April 14–17, 2011 Abstract Volume and Field Guide to Excursions*, 73–99. Germany: Universitätsverlag Göttingen.

Rinehart, L. F., S. G. Lucas, A. B. Heckert, J. A. Spielmann, and M. D. Celeskey. 2009. The paleobiology of *Coelophysis bauri* (Cope) from the Upper Triassic (Apachean) Whitaker quarry, New Mexico, with detailed analysis of a single quarry block. Albuquerque, NM: New Mexico Museum of Natural History and Science Bulletin 45, 260 pp.

Rodrigues, L. A., and V. F. Santos. 2002. Sauropod tracks—a geometric morphometric study. In A. Elewa (ed.), *Morphometrics: Applications in Biology and Paleontology*, 129–142. Berlin: Springer-Verlag.

Rodríguez-de la Rosa, R. A. 2007. Hadrosaurian footprints from the Late Cretaceous Cerro del Pueblo Formation of Coahuila, Mexico. In E. Díaz-Martínez and I. Rábano (eds.), *4th European Meeting on the Palaeontology and Stratigraphy of Latin America*, 339–343. Madrid, Spain: Cuademos del Museo Geominero 8, Instituto Geológico y Minero de España.

Rodríguez-de la Rosa, R. A., M. C. Aguillón-Martínez, J. López-Espinoza, and D. A. Eberth. 2004. The fossil record of vertebrate tracks in Mexico. *Ichnos* 11: 27–37.

Rodríguez-de la Rosa, R. A., V. M. Bravo-Cuevas, E. Carillo-Montiel, and A. Ortiz-Ubilla. 2012. Lower Cretaceous dinosaur tracks from Puebla, Mexico. *Journal of Geological Research* 2012: 1–7. doi:10.1155/2012/808729.

Rogers, J. V., II. 2002. Theropod trackways in the Lower Cretaceous (Albian) Glen Rose Formation, Kinney County, Texas. *Texas Journal of Science* 54: 133–142.

Romano, M., and P. Citton. 2017. Crouching theropods at the seaside. Matching footprints with metatarsal impressions and theropod autopods: a morphometric approach. *Geological Magazine* 154: 946–962.

Romano, M., and M. A. Whyte. 2003. Jurassic dinosaur tracks and trackways of the Cleveland Basin, Yorkshire: preservation, diversity and distribution. *Proceedings of the Yorkshire Geological Society* 54: 185–215.

Romero Molina, M. M., F. Pérez-Lorente, and P. Rivas Carrera. 2003*a*. Análisis de la parataxonomía utilizada con las huellas de dinosaurio. In F. Pérez (ed.), *Dinosaurios y Otros Reptiles Mesozoicos en España*, 13–32. Logroño, Spain: Gobierno de La Rioja, Fundación Patrimonio Paleontológico de La Rioja, Instituto de Estudios Riojanos, Universidad de La Rioja.

Romero Molina, M. M., W. A. S. Sarjeant, F. Pérez-Lorente, A. López, and E. Requeta. 2003*b*. Orientation and characteristics of theropod trackways from the Las Losas palaeoichnological site (La Rioja, Spain). *Ichnos* 10: 241–254.

Romilio, A., and S. W. Salisbury. 2011. A reassessment of large theropod dinosaur tracks from the mid-Cretaceous (Late Albian-Cenomanian) Winton Formation of Lark Quarry, central-western Queensland, Australia: a case for mistaken identity. *Cretaceous Research* 32: 135–142.

Romilio, A., and S. W. Salisbury. 2014. Large dinosaurian tracks from the Upper Cretaceous (Cenomanian-Turonian) portion of the Winton Formation, Lark Quarry, central-western Queensland, Australia: 3D photogrammetric analysis renders the 'stampede trigger' scenario unlikely. *Cretaceous Research* 51: 186–207.

Romilio, A., R. T. Tucker, and S. W. Salisbury. 2013. Reevaluation of the Lark Quarry dinosaur tracksite (late Albian-Cenomanian Winton Formation, central-western Queensland): no longer a stampede? *Journal of Vertebrate Paleontology* 33: 102–120.

Rowland, S. M., B. H. Breithaupt, H. M. Stoller, N. A. Matthews, and M. Saines. 2014. First report of dinosaur, synapsid, and arthropod tracks in the Aztec Sandstone (Lower-Middle Jurassic) of Red Rock Canyon National Conservation Area, southern Nevada. In M. G. Lockley and S. G. Lucas (eds.), *Fossil Footprints of Western North America*, 249–259. Albuquerque, NM: New Mexico Museum of Natural History and Science Bulletin 62.

Ruben, J. 1995. The evolution of endothermy in mammals and birds: from physiology to fossils. *Annual Review of Physiology* 57: 69–95.

Rubilar-Rogers, D., K. Moreno, N. Blanco, and J. O. Calvo. 2008. Theropod dinosaur trackways from the Lower Cretaceous of the Chacarilla Formation, Chile. *Revista Geológica de Chile* 35: 175–184.

Russell, D. A., and P. Béland. 1976. Running dinosaurs. *Nature* 264: 486.

Ryan, W., and W. Pittman. 1998. *Noah's Flood: The New Scientific Discoveries about the Event that Changed History.* New York: Simon and Schuster, 319 pp.

Salisbury, S. W., A. Romilio, M. C. Herne, R. T. Tucker, and J. P. Nair. 2016. The dinosaurian ichnofauna of the Lower Cretaceous (Valanginian-Barremian) Broome Sandstone of the Walmadany Area (James Price Point), Dampier Peninsula, Western Australia. *Journal of Vertebrate Paleontology* Memoir 16: 1–152; supplement to *Journal of Vertebrate Paleontology* 36(1).

Sander, P. M. 1992. The Norian *Plateosaurus* bonebeds of central Europe and their taphonomy. *Palaeogeography, Palaeoclimatology, Palaeoecology* 93: 255–299.

Sander, P. M. 2000. Long bone histology of the Tendaguru sauropods: implications for growth and biology. *Paleobiology* 26: 466–488.

Sander, P. M., N. Klein, K. Stein, and O. Wings. 2011. Sauropod bone histology and its implications for sauropod biology. In N. Klein, K. Remes, C. T. Gee, and P. M. Sander (eds.), *Biology of the Sauropod Dinosaurs: Understanding the Life of Giants*, 276–302. Bloomington: Indiana University Press.

Santa Luca, A. P. 1980. The postcranial skeleton of *Heterodontosaurus tucki* (Reptilia, Ornithischia) from the Stormberg of South Africa. *Annals of the South African Museum* 79: 159–211.

Santa Luca, A. P. 1984. Postcranial remains of Fabrosauridae (Reptilia: Ornithischia) from the Stormberg of southern Africa. *Palaeontologia Africana* 25: 151–180.

Santisteban, C., R. Gaete, A. Galobart, and M. Suñer. 2003. Rastros de dinosaurios en el Jurásico terminal (facies Purbeck) de Corcolilla (Los Serranos, Valencia). In F. Pérez (ed.), *Dinosaurios y Otros Reptiles Mesozoicos en España*, 33–40. Logroño, Spain: Gobierno de La Rioja, Fundación Patrimonio Paleontológico de La Rioja, Instituto de Estudios Riojanos, Universidad de La Rioja.

Santisteban, C., and M. Suñer. 2003. Rastros de huellas de dinosaurios en carbonatos inter-supramareales del Cretácico Superior, Cuenca Ibérica Suroccidental, Valencia, España. In F. Pérez (ed.), *Dinosaurios y Otros Reptiles Mesozoicos en España*, 147–160. Logroño, Spain: Gobierno de La Rioja, Fundación Patrimonio Paleontológico de La Rioja, Instituto de Estudios Riojanos, Universidad de La Rioja.

Santisteban, C. de, B. Vila, and M. Suñer. 2007. Huellas de dinosaurios conservadas en materiales del cortejo transgresivo en cauces encajados, Jurásico superior y Cretácico inferior de Alpuente, Valencia. *Geogaceta* 42: 79–82.

Santos, V. F. 2016. Dinosaur tracks in the southwest Algarve Basin. *Comunicações Geológicas* 103: 117–120.

Santos, V. F., P. M. Callapez, and N. P. C. Rodrigues. 2013. Dinosaur footprints from the Lower Cretaceous of the Algarve Basin (Portugal): new data on the ornithopod palaeoecology and palaeobiogeography of the Iberian Peninsula. *Cretaceous Research* 40: 158–169.

Sarjeant, W. A. S. 1983–1984. A mélange of geological clerihews. *Geolog Canada* 13(4): 34–36.

Sarjeant, W. A. S., J. B. Delair, and M. G. Lockley. 1998. The footprints of *Iguanodon*: a history and taxonomic study. *Ichnos* 6: 183–202.

Scannella, J. B., D. W. Fowler, M. B. Goodwin, and J. R. Horner. 2014. Evolutionary trends in *Triceratops* from the Hell Creek Formation, Montana. *Proceedings of the National Academy of Sciences USA* 111: 10245–10250.

Schachner, E. R., P. L. Manning, and P. Dodson. 2011. Pelvic and hindlimb myology of the basal archosaur *Poposaurus gracilis* (Archosauria: Poposauroidea). *Journal of Morphology* 272: 1464–1491.

Schanz, T., Y. Lins, H. Viefhaus, T. Barciaga, S. Läbe, H. Preuschoft, U. Witzel, and P. M. Sander. 2013. Quantitative interpretation of tracks for determination of body mass. *PloS One* 8(10): e77606. doi:10.1371/journal.pone.0077606.

Schmerge, J. D., and B. M. Rothschild. 2016. Distribution of the dentary groove of theropod dinosaurs: implications for theropod phylogeny and the validity of the genus *Nanotyrannus* Bakker et al., 1988. *Cretaceous Research* 61: 26–33.

Schulp, A. S., and M. Al-Wosabi. 2012. Telling apart ornithopod and theropod trackways: a closer look at a large, Late Jurassic tridactyl dinosaur trackway at Serwah, Republic of Yemen. *Ichnos* 19: 194–198.

Schulp, A. S., M. Al-Wosabi, and N. J. Stevens. 2008. First dinosaur tracks from the Arabian Peninsula. *PLoS One* 3(5): 32243. doi:10.1371/journal.pone.0002243.

Schweitzer, M. H., J. L. Wittmeyer, and J. R. Horner. 2005. Gender-specific reproductive tissue in ratites and *Tyrannosaurus rex*. *Science* 308: 1456–1460.

Sciscio, L., E. M. Bordy, M. Abrahams, F. Knoll, and B. W. McPhee. 2017. The first megatheropod tracks from the Lower Jurassic upper Elliot Formaiton, Karoo Basin, Lesotho. *PLoS One* 12(10): e0185941. https://doi.org/10.1371/journal.pone.0185941.

Sciscio, L., E. M. Bordy, M. Reid, and M. Abrahams. 2016. Sedimentology and ichnology of the Mafube dinosaur track site (Lower Jurassic, eastern Free State, South Africa): a report on footprint preservation and palaeoenvironment. *PeerJ* 4:e2285. doi:10.7717/peerj.2285.

Seebacher, F., H. Guderley, R. M. Elsey, and P. L. Trosclair, III. 2003. Seasonal acclimatisation of muscle metabolic enzymes in a reptile (*Alligator mississippiensis*). *Journal of Experimental Biology* 206: 1193–1200.

Segura, M., F. Barroso-Barcenilla, M. Berrocal-Casero, D. Castanera, J. García-Hidalgo, and V. F. Santos. 2016. A new Cenomanian vertebrate tracksite at Tamajón (Guadalajara, Spain). *Cretaceous Research* 57: 508–518.

Sellers, W. L., and P. L. Manning. 2007. Estimating dinosaur maximum running speeds using evolutionary robotics. *Proceedings of the Royal Society of London B* 274: 2711–2716.

Sereno, P. C. 2012. Taxonomy, morphology, masticatory function and phylogeny of heterodontosaurid dinosaurs. *ZooKeys* 226: 1–225.

Sereno, P. C., R. N. Martinez, and O. A. Alcobar. 2013. Osteology of *Eoraptor lunensis* (Dinosauria, Sauropodomorpha). *Journal of Vertebrate Paleontology* 32(supplement 1): 83–179.

Sereno, P. C., L. Tan, S. L. Brusatte, H. J. Kriegstein, X. Zhao, and K. Cloward. 2009. Tyrannosaurid skeletal design first evolved at small body size. *Science* 326: 418–422.

Sertich, J. J. W., and M. A. Loewen. 2010. A new basal sauropodomorph dinosaur from the Lower Jurassic Navajo Sandstone of southern Utah. *PloS One* 5(3): 39789. doi:10.1371/journal.pone.0009789.

Shell, R., and S. K. Boss. 2013. Morphometric analysis of dinosaur tracks from southwest Arkansas. *Journal of the Arkansas Academy of Science* 67: 121–130.

Shuler, E. W. 1917. Dinosaur tracks in the Glen Rose Limestone near Glen Rose, Texas. *American Journal of Science* 44: 294–298.
Shuler, E. W. 1935. Dinosaur track mounted in the band stand at Glen Rose, Texas. *Field and Laboratory* 4: 9–13.
Shuler, E. W. 1937. Dinosaur tracks at the fourth crossing of the Paluxy River near Glen Rose, Texas. *Field and Laboratory* 5: 33–36.
Sibley, C. G., and B. L. Monroe, Jr. 1990. *Distribution and Taxonomy of Birds of the World*. New Haven, CT: Yale University Press, 1111 pp.
Sibley, C. G., and B. L. Monroe, Jr. 1993. *A Supplement to Distribution and Taxonomy of Birds of the World*. New Haven, CT: Yale University Press, 108 pp.
da Silva, R. C., R. Barboni, T. Dutra, M. M. Godoy, and R. B. Binotto. 2012. Footprints of large theropod dinosaurs and implications on the age of Triassic biotas from southern Brazil. *Journal of South American Earth Sciences* 39: 16–23.
Slack, K. E., F. Delsuc, P. A. McLenachan, U. Arnason, and D. Penny. 2007. Resolving the root of the avian mitogenic tree by breaking up long branches. *Molecular Phylogenetics and Evolution* 42: 1–13.
Smith, J. B., and J. O. Farlow. 2003. Osteometric approaches to trackmaker assignment for the Newark Supergroup ichnogenera *Grallator, Anchisauripus*, and *Eubrontes*. In P. M. LeTourneau and P. E. Olsen (eds.), *The Great Rift Valleys of Pangea in Eastern North America*, vol. 2: *Sedimentology, Stratigraphy, and Paleontology*, 273–292. New York: Columbia University Press.
Smith, J. V., E. L. Braun, and R. T. Kimball. 2013. Ratite nonmonophyly: independent evidence from 40 novel loci. *Systematic Biology* 62: 35–49.
Sokal, R R., and F. J. Rohlf. 1995. *Biometry: The Principles and Practice of Statistics in Biological Research*. New York: W. H. Freeman, 887 pp.
Stains, H. R. E., and J. T. Woods. 1964. Recent discovery of Triassic dinosaur footprints in Queensland. *Australian Journal of Science* 27: 55.
Stanford, R., R. E. Weems, and M. G. Lockley. 2004. A new dinosaur ichnotaxon from the Lower Cretaceous Patuxent Formation of Maryland and Virginia. *Ichnos* 11: 251–259.
Steinbock, R. T. 1989. Ichnology of the Connecticut Valley: a vignette of American science in the early Nineteenth Century. In D. D. Gillette and M. G. Lockley (eds.), *Dinosaur Tracks and Traces*, 27–32. Cambridge: Cambridge University Press.
Sternberg, C. M. 1932. Dinosaur tracks from Peace River, British Columbia. *National Museum of Canada Annual Report* 1930: 59–85.
Strevell, C. N. 1932. *Dinosauropodes*. Salt Lake City, UT: Deseret News.
Strickland, H. E., and A. G. Melville. 1848. *The Dodo and Its Kindred; or the History, Affinities, and Osteology of the Dodo, Solitaire, and Other Extinct Birds of the Islands Mauritius, Rodriguez, and Bourbon*. London: Reeve, Benham, and Reeve, 141 pp.
Sullivan, C., D. W. E. Hone, T. D. Cope, L. Yang, and L. Jun. 2009. A new occurrence of small theropod tracks in the Houcheng (Tuchengzi) Formation of Hebei Province, China. *Vertebrata PalAsiatica* 47: 33–52.
Taylor, D., and W. Neal. 1984. Management implications of size-class frequency distributions in Louisiana alligator populations. *Wildlife Society Bulletin* 12: 312–319.
Tennyson, A. 1998. List of moa footprints. Wellington, NZ: Internal document, National Museum of New Zealand, Te Papa Tongarewa, 9 pp.
Therrien, F., D. K. Zelenitsky, A. Quinney, and K. Tanaka. 2015. Dinosaur trackways from the Upper Cretaceous Oldman and Dinosaur Park formations (Belly River Group) of southern Alberta, Canada reveal novel ichnofossil preservation style. *Canadian Journal of Earth Sciences* 52: 530–641.
Therrien, F., D. K. Zelenitsky, K. Tanaka, and W. J. Sloboda. 2014. First hadrosaur trackway from the Upper Cretaceous (Late Campanian) Oldman Formation, southeastern Alberta. In D. A. Eberth and D. C. Evans (eds.), *Hadrosaurs*, 532–539. Bloomington: Indiana University Press.
Thorpe, M. R. 1929. A new Triassic fossil field. *American Journal of Science* Series 5 18: 277–300.
Thulborn, R. A. 1972. The post-cranial skeleton of the Triassic ornithischian dinosaur *Fabrosaurus australis*. *Palaeontology* 15: 29–60.
Thulborn, R. A. 1981. Estimated speed of a giant bipedal dinosaur. *Nature* 292: 273–274.
Thulborn, R. A. 1984. Preferred gaits of bipedal dinosaurs. *Alcheringa* 8: 243–252.
Thulborn, R.A. 1994. Ornithopod dinosaur tracks from the Lower Jurassic of Queensland. *Alcheringa* 18: 247–258.
Thulborn, R. A. 2013. Lark Quarry revisited: a critique of methods used to identify a large dinosaurian track-maker in the Winton Formation (Albian-Cenomanian), western Queensland, Australia. *Alcheringa* 37: 312–330.
Thulborn, R. A., and M. Wade. 1984. Dinosaur trackways in the Winton Formation (Mid-Cretaceous) of Queensland. *Memoirs of the Queensland Museum* 21: 413–517.
Thulborn, R. A., and M. Wade. 1989. A footprint as a history of movement. In D. D. Gillette and M. G. Lockley (eds.), *Dinosaur Tracks and Traces*, 51–56. Cambridge: Cambridge University Press.
Thulborn, T. 1990. *Dinosaur Tracks*. London: Chapman and Hall, 410 pp.
Thulborn, T. 1998. Australia's earliest theropods: footprint evidence in the Ipswich Coal Measures (Upper Triassic) of Queensland. *Gaia* 15: 301–311.
Thulborn, T. 2016. Behaviour of dinosaurian track-makers in the Winton Formation (Cretaceous, Albian-Cenomanian) at Lark Quarry, western Queensland, Australia: running or swimming? *Ichnos* 24(1): 1–18. http://doi.org/10.1080/10420940.2015.1129326.
Torcida Fernández-Baldor, F., I. Díaz-Martínez, R. Contreras, P. Huerta, D. Montero, and V. Urién. 2015. Unusual sauropod tracks in the Jurassic-Cretaceous interval of the Cameros Basin (Burgos, Spain). *Journal of Iberian Geology* 41: 141–154.
Torcida Fernández-Baldor, F., L. A. Izquierdo Montero, P. Huerta, D. Montero Huerta, G. Pérez Martínez, and V. Urién Montero. 2006. El yacimiento de icnitas de dinosaurios de Costalomo (Salas de los Infantes, Burgos, España): nuevos datos. In Colectivo Arqueológico-Paleontológico Salense (eds.), *Actas de las III Jornadas sobre Dinosauros y su Entorno*, 313–347. Burgos, Spain: Salas de los Infantes.
Torcida, F., D. Montero, P. Huerta, L. A. Izquierdo, G. Pérez, F. Pérez-Lorente, and V. Urbién. 2003. Rastro ornitópodo de andar cuadrúpedo con marca de cola. Cretácico Inferior (Burgos, España). In F. Pérez (ed.), *Dinosaurios y Otros Reptiles Mesozoicos en España*, 109–118. Logroño, Spain: Gobierno de La Rioja, Fundación Patrimonio Paleontológico de La Rioja, Instituto de Estudios Riojanos, Universidad de La Rioja.
Tree of Life Project. 2007*a*. Anseriformes. Ducks, Geese and Relatives. Version 07 March 2007 (temporary); http://tolweb.org/Anseriformes/15841/2007.03.07.
Tree of Life Project. 2007*b*. Galliformes. Fowl, Quail, Guans, Currasows, Megapodes. Version 07 March 2007 (temporary); http://tolweb.org/Galliformes/26296/2007.03.07.
Tree of Life Project. 2007*c*. Odontophoridae. New World Quail. Version 07 March 2007 (temporary); http://tolweb.org/Odontophoridae/26303/2007.03.07.
Tree of Life Project. 2009. Phasianidae. Chickens, Grouse, Partridges, Quail, Pheasants, etc. Version 16 February 2009 (temporary); http://tolweb.org/Phasianidae/26302/2009.02.16.
Tsai, H. P., K. M. Middleton, J. R. Hutchinson, and C. M. Holliday. 2018. Hip joint articular soft tissues of non-dinosaurian Dinosauromorpha and early Dinosauria: evolutionary and biomechanical implications for Saurischia. *Journal of Vertebrate Paleontology*. doi: 10.1080/02724634.2017.1427593.
Van Heerden, J. 1979. The morphology and taxonomy of *Euskelosaurus* (Reptilia: Saurischia; Late Triassic) from South Africa. *Navorsinge van die Nasionale Museum* 4: 21–84.
Van Heerden, J. 1997. Prosauropods. In J. O. Farlow and M. K. Brett-Surman (eds.), *The Complete Dinosaur*, 242–263. Bloomington: Indiana University Press.
Varricchio, D. J., P. C. Sereno, Z. Xijin, L. Tan, J. A. Wilson, and G. H. Lyon. 2008. Mud-trapped herd captures evidence of distinctive dinosaur socialty. *Acta Palaeontologica Polonica* 53: 567–578.
Verdú, F. J., P. Godefroit, R. Royo-Torres, A. Cobos, and L. Alcalá. 2017. Individual variation in the postcranial skeleton of the Early Cretaceous *Iguanodon bernissartensis* (Dinosauria, Ornithopoda). *Cretaceous Research* 74: 65–86.
Vierra, L. I., and J. A. Torres. 1995. Anális comparativo sobre dos rastros de dinosaurios terópodos: forma de marcha y velocidad. *Munibe* 47: 53–56.
Vila, B., D. Castanera, J. Marmi, J. I. Canudo, and À. Galobart. 2015. Crocodile swim tracks from the latest Cretaceous of Europe. *Lethaia* 48: 256–266.
Wagensommer, A., M. Latiano, G. Leroux, G. Cassano, and S. D'Orazi Porchetti. 2012. New dinosaur tracksites from the Middle

Jurassic of Madagascar: ichnotaxonomical, behavioural and palaeoenvironmental implications. *Palaeontology* 55: 109–126.
Wagensommer, A., M. Latiano, H. B. Mocke, S. D'Orazi Porchetti, and A. Wanke. 2016. A dinosaur ichnocoenosis from the Waterberg Plateau (Etjo Formation, Lower Jurassic), Namibia. *Ichnos* 23: 312–321.
Wagensommer, A., M. Latiano, and U. Nicosia. 2010. First report of dinosaur footprints from Madagascar: two tracksites from the Middle Jurassic Bemaraha Formation. *Ichnos* 17: 127–136.
Wang, B., J. Li, Z. Bai, J. Gao, S. Dong, B. Hu, S. Zhao, and J. Chang. 2016. Research on dinoasur (sic) footprints in Zizhou, Shaanxi Province, China. *Acta Geologica Sinica* (English edition) 90: 1–18.
Wang, M. 2014. *Taxonomical Revision, Ontogenetic, Ecological and Phylogenetic Analyses of Enantiornithes (Aves: Ornithothoraces) of China.* PhD Thesis. Beijing: University of Chinese Academy of Sciences, 464 pp.
Wang, M., X. Wang, Y. Wang, and Z. Zhou. 2016*a*. A new basal bird from China with implications for morphological diversity in early birds. *Scientific Reports* 6: 19700. doi:10.1038/srep19700.
Wang, M., Zhou Z., J. K. O'Connor, and N. V. Zelenkov. 2014*a*. A new diverse enantiornithine family (Bohaiornithidae fam. nov.) from the Lower Cretaceous of China with information from two new species. *Vertebrata PalAsiatica* 52: 31–76.
Wang, M., Z. Zhou, and G. Xu. 2014*b*. The first enantiornithine bird from the Upper Cretaceous of China. *Journal of Vertebrate Paleontology* 34: 135–145.
Wang, M., Z. Zhou, and S. Zhou. 2016*b*. A new basal ornithuromorph bird (Aves: Ornithothoraces) from the Early Cretaceous of China with implications for morphology of early Ornithuromorpha. *Zoological Journal of the Linnean Society* 176: 207–233.
Wang, X., M. Pittman, X. Zheng, T. G. Kaye, A. R. Falk, S. A. Hartman, and X. Xu. 2017. Basal paravian functional anatomy illuminated by high-detail body outline. *Nature Communications* 8: 14576. doi:10.1038, ncomms14576.
Wang, Y.-M., H.-L. You, and T. Wang. 2017. A new basal sauropodiform dinosaur from the Lower Jurassic of Yunnan Province, China. *Scientific Reports* 7: 41881. doi: 10.1038/srep41881.
Webb, G. J. W., and H. Messel. 1978. Morphometric analysis of *Crocodylus porosus* from the North Coast of Arnhem Land, Northern Australia. *Australian Journal of Zoology* 26: 1–27.
Weems, R. E. 1987. A Late Triassic footprint fauna from the Culpeper Basin Northern Virginia. *Transactions of the American Philosophical Society* 77(1): 1–79.
Weems, R. E. 1992. A re-evaluation of the taxonomy of Newark Supergroup saurischian dinosaur tracks, using extensive statistical data from a recently exposed tracksite near Culpeper, Virginia. In P. C. Sweet (ed.), *Proceedings of the 26th Forum on the Geology of Industrial Minerals*, 113–127. May 14–18, 1990, Charlottesville, Virginia: Publication 119, Virginia Division of Mineral Resources.
Weems, R. E. 2003. *Plateosaurus* foot structure suggests a single trackmaker for *Eubrontes* and *Gigandipus* footprints. In P. M. Letourneau and P. E. Olsen (eds.), *The Great Rift Valleys of Pangea in Eastern North America. Volume Two: Sedimentology, Stratigraphy, and Paleontology*, 293–313. New York: Columbia University Press.
Weems, R. E. 2006*a*. The manus print of *Kayentapus minor*: its bearing on the biomechanics and ichnotaxonomy of early Mesozoic saurischian dinosaurs. In J. D. Harris, S. G. Lucas, J. A. Spielmann, M. G. Lockley, A. R. C. Milner, and J. I. Kirkland (eds.), *The Triassic-Jurassic Terrestrial Transition*, 369–378. Albuquerque, NM: New Mexico Museum of Natural History and Science Bulletin 37.
Weems, R. E. 2006*b*. Locomotor speeds and patterns of running behavior in non-maniraptoriform theropod dinosaurs. In J. D. Harris, S. G. Lucas, J. A. Spielmann, M. G. Lockley, A. R. C. Milner, and J. I. Kirkland (eds.), *The Triassic-Jurassic Terrestrial Transition*, 379–389. Albuquerque, NM: New Mexico Museum of Natural History and Science Bulletin 37.
Weems, R. E., and P. G. Kimmel. 1993. Upper Triassic reptile footprints and a coelacanth fish scale from the Culpeper Basin, Virginia. *Proceedings of the Biological Society of Washington* 106: 390–401.
Weinbaum, J. C., and A. Hungerbühler. 2007. A revision of *Poposaurus gracilis* (Archosauria: Suchia) based on two new specimens from the Late Triassic of the southwestern U.S.A. *Palaontologische Zeitschrift* 81: 131–145.
Weishampel, D. B., and F. Westphal. 1986. *Die Plateosaurier von Trossingen in Geologischen Institut der Eberhard-Karls-Universität Tübingen*. Tübingen, Germany: Attempto Verlag, 27 pp.
Welles, S. P. 1971. Dinosaur footprints from the Kayenta Formation of Northern Arizona. *Plateau* 44: 27–38.
Welles, S. P. 1984. *Dilophosaurus wetherilli* (Dinosauria: Theropoda) osteology and comparisons. *Palaeontographica A* 185: 830–180.
Wellnhofer, P. 1974. Das fünfte Skelettexemplar von *Archaeopteryx*. *Palaeontographica* A 147: 169–216.
Wellnhofer, P. 1993*a*. Das siebte Exemplar von *Archaeopteryx* aus den Solnhofen Schichten. *Archaeopteryx* 11: 1–47.
Wellnhofer, P. 1993*b*. Prosauropod dinosaurs from the Feuerletten (Middle Norian) of Ellingen near Weissenburg in Bavaria. *Review de Paléobiologie* Volume Special 7: 263–271.
White, M. A., R. B. J. Benson, T. R. Tischler, S. A. Hocknull, A. G. Cook, D. G. Barnes, S. F. Poropat, S. J. Wooldridge, T. Sloan, G. H. K. Sinaplus, and D. A. Elliott. 2013. New *Australovenator* hind limb elements pertaining to the holotype reveal the most complete neovenatorid leg. *PloS One* 8(7): e68649. doi:10.1371/journal.pone.0068649.
White, M. A., A. G. Cook, A. J. Klinkhamer, and D. A. Elliott. 2016. The pes of *Australovenator wintonensis* (Theropoda: Megaraptoridae): analysis of the pedal range of motion and biological restoration. *PeerJ* 4: e2312. doi:10.7717/peerj.2312.
Wilkinson, P. M. 1983. *Nesting Ecology of the American Alligator in Coastal South Carolina. Study Completion Report August 1978-September 1983*. Charleston, SC: South Carolina Wildlife and Marine Resources Department, 113 pp.
Wilkinson, P. M., T. R. Rainwater, A. R. Woodward, E. H. Leone, and C. Carter. 2016. Determinate growth and reproductive lifespan in the American alligator (*Alligator mississippiensis*): evidence from long-term recaptures. *Copeia* 104: 843–852.
Wilkinson, P. M., and W. E. Rhodes. 1997. Growth rates of American alligators in coastal South Carolina. *Journal of Wildlife Management* 61: 397–402.
Wilkinson, P. M., and K. G. Rice. 1996. Hind-foot track length: a method for determining the size of American alligators. In *Proceedings of the 13th Working Meeting of the Crocodile Specialist Group of the Species Survival Commission of IUCN* (unedited), 429–436. Santa Fe, Argentina.
Wilson, J. A., C. A. Marsicano, and R. M. H. Smith. 2009. Dynamic locomotor capabilities revealed by early dinosaur trackmakers from southern Africa. *PloS One* 4(10): e7331. doi:10.1371/journal.pone.0007331.
Wings, O., J. N. Lallensack, and H. Mallison. 2016. The Early Cretaceous dinosaur trackways in Münchehagen (Lower Saxony, Germany): 3-D photogrammetry as basis for geometric morphometric analysis of shape variation and evaluation of material loss during excavation. In P. L. Falkingham, D. Marty, and A. Richter (eds.), *Dinosaur Tracks: The Next Steps*, 56–70. Bloomington: Indiana University Press.
Wings, O., R. Schellhorn, H. Mallison, B. Thuy, W. Wu, and G. Sun. 2007. The first dinosaur tracksite from Xinjiang, NW China (Middle Jurassic Sanjianfang Formation, Turpan Basin)—a preliminary report. *Global Geology* 10: 113–129.
Witmer, L. M. 1995. The extinct phylogenetic bracket and the importance of reconstructing soft tissues in fossils. In J. Thomason (ed.), *Functional Morphology in Vertebrate Paleontology*, 19–33. Cambridge: Cambridge University Press.
Woodhams, K. E., and J. S. Hines. 1989. Dinosaur footprints from the Lower Cretaceous of East Sussex, England. In D. D. Gillette and M. G. Lockley (eds.), *Dinosaur Tracks and Traces*, 301–307. Cambridge: Cambridge University Press.
Woodward, A. R., J. H. White, and S. B. Linda. 1995. Maximum size of the American alligator (*Alligator mississippiensis*). *Journal of Herpetology* 29: 507–513.
Woodward, H. N., E. A. Freedman Fowler, J. O. Farlow, and J. R. Horner. 2015. *Maiasaura*, a model organism for extinct vertebrate population biology: a large sample statistical assessment of growth dynamics and survivorship. *Paleobiology* 41: 503–527.
Woodward, H. N., J. R. Horner, and J. O. Farlow. 2011. Osteohistological evidence for determinate growth in the American alligator. *Journal of Herpetology* 45: 339–342.

Woodward, H. N., J. R. Horner, and J. O. Farlow. 2014. Quantification of intraskeletal histovariability in *Alligator mississippiensis* and implications for vertebrate osteohistology. *PeerJ* 2: 3422. doi:10.7717/peerj.422.
Worthy, T. H. 2005. Rediscovery of the types of *Dinornis curtus* Owen and *Palapteryx geranoides* Owen, with a new synonymy (Aves: Dinornithiformes). *Tuhinga* 16: 33–43.
Worthy, T. H., F. J. Degrange, W. D. Handley, and M. S. Y. Lee. 2017. The evolution of giant flightless birds and novel phylogenetic relationships for extinct fowl (Aves, Galloanseres). *Royal Society Open Science* 4: 270975. http://dx.doi.org/10.1098/rsos.170975.
Worthy, T. H., and R. N. Holdaway. 2002. *The Lost World of the Moa: Prehistoric Life of New Zealand*. Bloomington: Indiana University Press, 718 pp.
Worthy, T. H., and R. P. Scofield. 2012. Twenty-first century advances in knowledge of the biology of moa (Aves: Dinornithiformes): a new morphological analysis and moa diagnoses revised. *New Zealand Journal of Zoology* 39: 87–153.
Wosik, M., M. B. Goodwin, and D. C. Evans. 2018. A nestling-sized skeleton of *Edmontosaurus* (Ornithischia, Hadrosauridae) from the Hell Creek Formation of northeastern Montana, U.S.A., with an analysis of ontogenetic limb ontogeny. *Journal of Vertebrate Paleontology*. doi:10.1080/02724634.2017.1398168.
Wright, J. L. 1996. *Fossil Terrestrial Trackways: Function, Taphonomy, and Palaeoecological Significance*. PhD Thesis. Bristol, U.K.: University of Bristol, 300 pp.
Wright, J. L. 1999. Ichnological evidence for the use of the forelimb in iguanodontid locomotion. *Special Papers in Palaeontology* 60: 209–219.
Wright, J. L. 2004. Bird-like features of dinosaur footprints. In P. J. Currie, E. B. Koppelhus, M. A. Shugar, and J. L. Wright (eds.), *Feathered Dragons: Studies on the Transition from Dinosaurs to Birds*, 167–181. Bloomington: Indiana University Press.
Xing, L. D., N. Abbassi, M. G. Lockley, H. Klein, S. Jia, R. T. McCrea, and W. S. Persons, IV. 2017*a*. The first record of *Anomoepus* from the Middle Jurassic of Henan Province, central China. *Historical Biology* 29: 223–229.
Xing, L. D., J. D. Harris, and G. D. Gierliński. 2011*a*. *Therangospodus* and *Megalosauripus* track assemblage from the Upper Jurassic-Lower Cretaceous Tuchengzi Formation of Chicheng County, Hebei Province, China and their paleoecological implications. *Vertebrata PalAsiatica* 49: 423–434.
Xing, L., J. D. Harris, Z. Dong, Y. Lin, W. Chen, S. Guo, and Q. Ji. 2009*a*. Ornithopod (Dinosauria: Ornithischia) tracks from the Upper Cretaceous Zhutian Formation in the Nanxiong basin, Guangdong, China and general observations on large Chinese ornithopod footprints. *Geological Bulletin of China* 28: 829–843.
Xing, L., J. D. Harris, S. Toru, F. Masato, and Z. Dong. 2009*b*. Discovery of dinosaur footprints from the Lower Jurassic Lufeng Formation of Yunnan Province, China and new observations of *Changpeipus*. *Geological Bulletin of China* 28: 16–29.
Xing, L. D., J. D. Harris, G. D. Gierliński, W. M. Wang, Z. Wang, and D. Li. 2011*b*. Mid-Cretaceous non-avian theropod trackways from the southern margin of the Sichuan Basin, China. *Acta Palaeontologica Sinica* 50: 470–480.
Xing, L.-D., H. Klein, M. G. Lockley, A. Wetzel, Z.-D. Li, J.-J. Li, G. D. Gierliński, J.-P. Zhang, M. Matsukawa, J. D. Divay, and L. Zhou. 2014*a*. *Changpeipus* (theropod) tracks from the Middle Jurassic of the Turpan Basin, Xinjiang, northwest China: review, new discoveries, ichnotaxonomy, preservation and paleoecology. *Vertebrata PalAsiatica* 52: 233–259.
Xing, L., D. Li, J. D. Harris, P. R. Bell, Y. Azuma, M. Fujita, Y.-N. Lee, and P. J. Currie. 2013*a*. A new deinonychosaurian track from the Lower Cretaceous Hekou Group, Gansu Province, China. *Acta Palaeontologica Polonica* 58: 723–730.
Xing, L., D. Li, M. G. Lockley, J. Zhang, H. You, H. Klein, D. Marty, W. S. Persons, IV, and C. Peng. 2014*b*. Theropod and sauropod track assemblages from the Lower Cretaceous Hekou Group of Zhongpu, Gansu Province, China. *Acta Palaeontologica Sinica* 53: 381–391.
Xing, L. D., Y. Q. Liu, H. W. Kuang, H. Klein, J. P. Zhang, M. E. Burns, J. Chen, M. W. Wang, and J. Hu. 2014*c*. Theropod and possible ornithopod track assemblages from the Jurassic-Cretaceous boundary Houcheng Formation, Shangyi, northern Hebei, China. *Palaeoworld* 23: 200–208.
Xing, L., and M. G. Lockley. 2016. Early Cretaceous dinosaur and other tetrapod tracks of southwestern China. *Science Bulletin* 61: 1044–1051.
Xing, L.-D., M. G. Lockley, W. Chen, G. D. Gierliński, J.-J. Li, W. S. Persons, IV, M. Matsukawa, Y. Yong, M. K. Gingras, and C.-W. Wang. 2013*b*. Two theropod track assemblages from the Jurassic of Chongqing, China, and the Jurassic stratigraphy of Sichuan Basin. *Vertebrata PalAsiatica* 51: 107–130.
Xing, L., M. G. Lockley, S. Hu, Q. Li, and W. S. Persons, IV. 2016*a*. Early Jurassic *Anomoepus* track from the Fengjiahe Formation of northern central Yunnan, China. In R. M. Sullivan and S. G. Lucas (eds.), *Fossil Record 5*, 327–330. Albuquerque, NM: New Mexico Museum of Natural History and Science Bulletin 74.
Xing, L., M. G. Lockley, N. Hu, G. Li, G. Tong, M. Matsukawa, H. Klein, Y. Ye, J. Zhang, and W. S. Persons, IV. 2016*b*. Saurischian track assemblages from the Lower Cretaceous Shenhuangshan Formation in the Yuanma Basin, southern China. *Cretaceous Research* 65: 1–9.
Xing, L., M. G. Lockley, H. Klein, P. L. Falkingham, J. Y. Kim, R. T. McCrea, J. Zhang, W. S. Persons, IV, T. Wang, and Z. Wang. 2016*c*. First Early Jurassic small ornithischian tracks from Yunnan Province, southwestern China. *Palaios* 31: 516–524.
Xing, L.-D., M. G. Lockley, H. Klein, G. D. Gierliński, J. D. Divay, S.-M. Hu, J.-P. Zhang, Y. Ye, and Y.-P. He. 2014*d*. The non-avian track *Jialingpus* from the Cretaceous of the Ordos Basin, China, with a revision of the type material: implications for ichnotaxonomy and trackmaker morphology. *Palaeoworld* 23: 187–199.
Xing, L., M. G. Lockley, H. Klein, G. Peng, Y. Ye, S. Jiang, J. Zhang, W. S. Persons, IV, and T. Xu. 2016*d*. A theropod track assemblage including large deinonychosaur tracks from the Lower Cretaceous of China. *Cretaceous Research* 65: 213–222.
Xing, L., M. G. Lockley, H. Klein, J. Zhang, and W. S. Persons, IV. 2016*e*. A new ornithischian-dominated and theropod footprint assemblage from the Lower Jurassic Lufeng Formation of Yunnan Province, China. In R. M. Sullivan and S. G. Lucas (eds.), *Fossil Record 5*, 331–338. Albuquerque, NM: New Mexico Museum of Natural History and Science Bulletin 74.
Xing, L.-D., M. G. Lockley, Z.-D. Li, H. Klein, J.-P. Zhang, G. D. Gierliński, Y. Ye, W. S. Persons, IV, and L. Zhou. 2013*c*. Middle Jurassic theropod trackways from the Panxi region, southwest China and a consideration of their geologic age. *Palaeoworld* 22: 36–41.
Xing, L., M. G. Lockley, D. Marty, H. Klein, L. G. Buckley, R. T. McCrea, J. Zhang, G. D. Gierliński, J. D. Divay, and Q. Wu. 2013*d*. Diverse dinosaur ichnoassemblages from the Lower Cretaceous Dasheng Group in the Yishu fault zone, Shandong Province, China. *Cretaceous Research* 45: 114–134.
Xing, L., M. G. Lockley, D. Marty, H. Klein, G. Yang, J. Zhang, G. Peng, Y. Ye, W. S. Persons, IV, X. Yin, and T. Xu. 2016*f*. A diverse saurischian (theropod-sauropod) dominated footprint assemblage from the Lower Cretaceous Jiaguan Formation in the Sichuan Basin, southwestern China: a new ornithischian ichnotaxon, pterosaur tracks and an unusual sauropod walking pattern. *Cretaceous Research* 60: 176–193.
Xing, L., M. G. Lockley, D. Marty, L. Piñuela, H. Klein, J. Zhang, and W. Scott Persons. 2015*a*. Re-description of the partially collapsed Early Cretaceous Zhaojue dinosaur tracksite (Sichuan Province, China) by using previously registered video coverage. *Cretaceous Research* 52: 138–152.
Xing, L., M. G. Lockley, D. Marty, J. Zhang, Y. Wang, H. Klein, R. T. McCrea, L. G. Buckley, M. Belvedere, O. Mateus, G. D. Gierliński, L. Piñuela, W. S. Persons, IV, F. Wang, H. Ran, H. Dai, and X. Xie. 2015*b*. An ornithopod-dominated tracksite from the Lower Cretaceous Jiaguan Formation (Barremian-Albian) of Qijiang, south-central China: new discoveries, ichnotaxonomy, preservation and palaeoecology. *PloS One* 10(10): e0141059. doi:10.1371/journal.pone.0141059.
Xing, L.-D., M. G. Lockley, T. Miyashita, H. Klein, T. Wang, W. S. Persons, IV, S.-G. Pan, J.-P. Zhang, and Z.-M. Dong. 2014*e*. Large sauropod and theropod tracks from the Middle Jurassic Chuanjie Formation of Lufeng County, Yunnan Province and palaeobiogeography of the Middle Jurassic sauropod tracks from southwestern China. *Palaeoworld* 23: 294–303.
Xing, L., M. G. Lockley, G. Peng, Y. Ye, J. Zhang, M. Matsukawa, H. Klein, R. T. McCrea,

and W. S. Persons, IV. 2016*g*. *Eubrontes* and *Anomoepus* track assemblages from the Middle Jurassic Xiashaximiao Formation of Zizhong County, China: review, ichnotaxonomy and notes on preserved tail tracks. In R. M. Sullivan and S. G. Lucas (eds.), *Fossil Record 5*, 345–352. Albuquerque, NM: New Mexico Museum of Natural History and Science Bulletin 74.

Xing, L., M. G. Lockley, Y. Tang, H. Klein, J. Zhang, W. S. Persons, IV, H. Dai, and Y. Ye. 2015*c*. Theropod and ornithischian footprints from the Middle Jurassic Yanan Formation of Zizhou County, Shaanxi, China. *Ichnos* 22: 1–11.

Xing, L.-D., M. G. Lockley, Q.-F. Wang, Z.-D. Li, H. Klein, W. S. Persons, IV, Y. Yong, and M. Matsukawa. 2014*f*. Earliest records of dinosaur footprints in Xinjiang, China. *Vertebrata PalAsiatica* 52: 340–348.

Xing, L., M. G. Lockley, G. Yang, J. Cao, M. Benton, X. Xu, J. Zhang, H. Klein, W. S. Persons, IV, J. Y. Kim, G. Peng, Y. Ye, and H. Ran. 2016*h*. A new *Minisauripus* site from the Lower Cretaceous of China: tracks of small adults or juveniles? *Palaeogeography, Palaeoclimatology, Palaeoecology* 452: 28–39.

Xing, L., M. G. Lockley, G. Yang, J. Cao, R. T. McCrea, H. Klein, J. Zhang, W. S. Persons, IV, and H. Dai. 2016*i*. A diversified vertebrate ichnite fauna from the Feitianshan Formation (Lower Cretaceous) of southwestern Sichuan, China. *Cretaceous Research* 57: 79–89.

Xing, L.-D., M. G. Lockley, G. Yang, X. Xu, J. Cao, H. Klein, W. S. Persons, IV, H.-J. Shen, and X.-M. Zheng. 2015*d*. Unusual deinonychosaurian track morphology (*Velociraptorichnus zhangi* n. ichnosp.) from the Lower Cretaceous Xiaoba Formation, Sichuan Province, China. *Palaeoworld* 24: 283–292.

Xing, L., M. G. Lockley, J. Zhang, H. Klein, J. Y. Kim, W. S. Persons, IV, Masaki Matuskawa, X. You, J. Li, G. Chen, and Y. Hu. 2014*g*. Upper Cretaceous dinosaur track assemblages and a new theropod ichnotaxon from Anhui Province, eastern China. *Cretaceous Research* 49: 190–204.

Xing, L.-D., M. G. Lockley, J.-P. Zhang, H. Klein, S. B. Kümmel, W. S. Persons, IV, and H. Dai. 2017*b*. Theropod tracks from the Lower Jurassic of Gulin area, Sichuan Province, China. *Palaeoworld* 26: 115–123.

Xing, L., M. G. Lockley, J. Zhang, H. Klein, D. Li, T. Miyashita, Z. Li, and S. B. Kümmel. 2016*j*. A new sauropodomorph ichnogenus from the Lower Jurassic of Sichuan, China fills a gap in the track record. *Historical Biology* 7: 881–895.

Xing, L., M. G. Lockley, J. Zhang, H. Klein, D. Marty, G. Peng, Y. Ye, R. T. McCrea, W. S. Persons, IV, and T. Xu. 2015*e*. The longest theropod trackway from east Asia, and a diverse sauropod-, theropod-, and ornithopod-track assemblage from the Lower Cretaceous Jiaguan Formation, southwest China. *Cretaceous Research* 56: 345–362.

Xing, L., M. G. Lockley, J. Zhang, H. Klein, W. S. Persons, IV, and H. Dai. 2014*h*. Diverse sauropod-, theropod-, and ornithopod-track assemblages and a new ichnotaxon *Siamopodus xui* ichnosp. nov. from the Feitianshan Formation, Lower Cretaceous of Sichuan Province, southwest China. *Palaeogeography, Palaeoclimatology, Palaeoecology* 414: 79–97.

Xing, L., M. G. Lockley, J.-P. Zhang, H. Klein, T. Wang, S. B. Kümmel, and M. E. Burns. 2016*k*. A theropod-sauropod track assemblage from the ?Middle-Upper Jurassic Shedian Formation at Shuangbai, Yunnan Province, China, reflecting sizes of trackmakers: review and new observations. *Palaeoworld* 25: 84–94.

Xing, L., M. G. Lockley, J. Zhang, A. R. C. Milner, H. Klein, D. Li, W. S. Persons, IV, and J. Ebi. 2013*e*. A new Early Cretaceous dinosaur track assemblage and the first definite non-avian swim trackway from China. *Chinese Science Bulletin* 58: 2370–2378.

Xing, L., M. G. Lockley, J. Zhang, H. You, H. Klein, W. S. Persons, IV, H. Dai, and Z. Dong. 2016*l*. First Early Jurassic ornithischian and theropod footprint assemblage and a new ichnotaxon *Shenmuichnus wangi* ichnosp. nov. from Yunnan Province, southwestern China. *Historical Biology* 28: 721–733.

Xing, L.-D., G. Niedźwiedzki, M. G. Lockley, J.-P. Zhang, X.-F. Cai, W. S. Persons, IV, and Y. Ye. 2014*i*. *Asianopodus*-type footprints from the Hekou Group of Honggu District, Lanzhou City, Gansu, China and the "heel" of large theropod tracks. *Palaeworld* 23: 304–313.

Xing, L., G. Peng, M. G. Lockley, Y. Ye, H. Klein, R. T. McCrea, J. Zhang, and W. S. Persons, IV. 2016*m*. Saurischian (theropod-sauropod) track assemblages from the Jiaguan Formation in the Sichuan Basin, southwest China: ichnology and indications to differential track preservation. *Historical Biology* 28: 1003–1013.

Xing, L., G. Peng, M. G. Lockley, Y. Ye, H. Klein, J. Zhang, and W. S. Persons, IV. 2015*f*. Early Cretaceous sauropod and ornithopod trackways from a stream course in Sichuan Basin, southwest China. In R. M. Sullivan and S. G. Lucas (eds.), *Fossil Record 4*, 319–325. Albuquerque, NM: New Mexico Museum of Natural History and Science.

Xing, L., G. Peng, Y. Ye, M. G. Lockley, H. Klein, W. S. Persons, IV, J. Zhang, C. Shu, and B. Hao. 2014*j*. Sauropod and small theropod tracks from the Lower Jurassic Ziliujing Formation of Zigong City, Sichuan, China, with an overview of Triassic-Jurassic dinosaur fossils and footprints of the Sichuan Basin. *Ichnos* 21: 119–130.

Xing, L., G. Peng, Y. Ye, M. G. Lockley, R. T. McCrea, P. J. Currie, J. Zhang, and M. E. Burns. 2014*k*. Large theropod trackway from the Lower Jurassic Zhenzhuchong Formation of Weiyuan County, Sichuan Province, China: review, new observations and special preservation. *Palaeoworld* 23: 285–293.

Xing, L., G. Yang, J. Cao, M. G. Lockley, H. Klein, J. Zhang, W. S. Persons, IV, H. Hu, H. Shen, X. Zheng, and Y. Qin. 2015*g*. Cretaceous saurischian tracksites from southwest Sichuan Province and overview of Late Cretaceous dinosaur track assemblages of China. *Cretaceous Research* 56: 458–469.

Xing, L., J. Zhang, M. G. Lockley, R. T. McCrea, H. Klein, L. Alcalá, L. G. Buckley, M. E. Burns, S. B. Kümmel, and Q. He. 2015*h*. Hints of the early Jehol Biota: important dinosaur footprint assemblages from the Jurassic-Cretaceous boundary Tuchengzi Formation in Beijing, China. *PloS One* 10(4): e0122715. doi:10.1371/journal. pone.0122715.

Xu, L., Y. Kobayashi, J. Lü, Y.-N. Lee, Y. Liu, K. Tanaka, X. Zhang, S. Jia, and J. Zhang. 2011. A new ornithomimid dinosaur with North American affinities from the Late Cretaceous Qiupa Formation in Henan Province of China. *Cretaceous Research* 32: 213–222.

Xu, X., J. M. Clark, J. Mo, J. Choiniere, C. A. Forster, G. M. Erickson, D. W. E. Hone, C. Sullivan, D. A. Eberth, S. Nesbitt, Q. Zhao, R. Hernandez, C.-K. Jia, F.-L Han, and Y. Guo. 2009. A Jurassic ceratosaur from China helps clarify avian digital homologies. *Nature* 459: 940–944.

Xu, X., Q.-W. Tan, S. Wang, C. Sullivan, D. W. E. Hone, F.-L. Han, Q.-Y. Ma, L. Tan, and D. Xiao. 2013. A new oviraptorid from the Upper Cretaceous of Nei Mongol, China, and its stratigraphic implications. *Vertebrata PalAsiatica* 51: 85–101.

Xu, X., and F. Zhang. 2005. A new maniraptoran dinosaur from China with long feathers on the metatarsus. *Naturwissenschaften* 92: 173–177.

Yablokov, A. V. 1974. *Variability of Mammals*. New Delhi: Amerind Publishing Company, 299 pp.

Yates, A. M. 2003. A new species of the primitive dinosaur *Thecodontosaurus* (Saurischia: Saruopodomorpha) and its implications for the systematics of early dinosaurs. *Journal of Systematic Paleontology* 1: 1–42.

Yates, A. M. 2004. *Anchisaurus polyzelus* (Hitchcock): the smallest known sauropod dinosaur and the evolution of gigantism among sauropodomorph dinosaurs. *Postilla* 230: 58.

Yates, A. M., M. F. Bonnan, J. Neveling, A. Chinsamy, and M. G. Blackbeard. 2010. A new transitional sauropodomorph dinosaur from the Early Jurassic of South Africa and the evolution of sauropod feeding and quadrupedalism. *Proceedings of the Royal Society B* 277: 787–794.

Yonezawa, T., T. Segawa, H. Mori, P. F. Campos, Y. Hongoh, H. Endo, A. Akiyoshi, N. Kohno, S. Nishida, J. Wu, H. Jin, J. Adachi, H. Kishino, K. Kurokawa, Y. Nogi, H. Tanabe, H. Mukoyama, K. Yoshida, A. Rasoamiaramanana, S. Yamagishi, Y. Hayashi, A. Yoshida, H. Koike, F. Akishinonomiya, E. Willerslev, and M. Hasegawa. 2017. Phylogenomics and morphology of extinct ratites reveal the origin and evolution of the ratites. *Current Biology* 27: 68–77.

You, H., and Y. Azuma. 1995. Early Cretaceous dinosaur footprints from Luanping, Hebei Province, China. In A. Sun and Y. Wang (eds.), *Sixth Symposium on Mesozoic Terrestrial Ecosystems and Biota, Short Papers*, 151–156. Beijing: China Ocean Press.

You, H., M. C. Lamanna, J. D. Harris, L. M. Chiappe, J. O'Connor, J. Ji, J. Lü, C. Yuan, D. Li, X. Zhang, K. J. Lacovara, P. Dodson, and Q. Ji. 2006. A nearly modern amphibious bird from the Early Cretaceous of northwestern China. *Science* 312: 1640–1643.

Young, C.-C. 1941*a*. A complete osteology of *Lufengosaurus huenei* Young (gen. et. sp. nov.) from Lufeng, China. *Paleontologia Sinica* New Series C 7: 1–53.

Young, C.-C. 1941*b*. *Gyposaurus sinensis* Young (sp. nov.), a new Prosauropoda

from the Upper Triassic beds at Lufeng, Yunnan. *Bulletin of the Geological Society of China* 21: 205–253.
Young, C.-C. 1958. The dinosaurian remains of Laiyang, Shantung. *Palaeontologia Sinica* New Series C 16: 1–138.
Zanno, L. E., D. J. Varricchio, P. M. O'Connor, A. L. Titus, and M. J. Knell. 2011. A new troodontid theropod, *Talos sampsoni* gen et. sp. nov., from the Upper Cretaceous Western Interior Basin of North America. *PloS One* 6(9): e24487. doi: 10.1371/journal.pone.0024487.
Zelenitsky, D., and S. P. Modesto. 2003. New information on the eggshell of ratites (Avens) and its phylogenetic implications. *Canadian Journal of Zoology* 81: 962–970.
Zhang, J., D. Li, M. Li, M. G. Lockley, and Z. Bai. 2006. Diverse dinosaur-, pterosaur-, and bird-track assemblages from the Hakou Formation, Lower Cretaceous of Gansu Province, northwest China. *Cretaceous Research* 27: 44–55.
Zhang, M., Y. Wang, P. Yan, and X. Wu. 2011. Crocodilian phylogeny inferred from twelve mitochondrial protein-coding genes, with new complete mitochondrial genomic sequences for *Crocodylus acutus* and *Crocodylus novaeguineae*. *Molecular Phylogenetics and Evolution* 60: 62–67.
Zhang, Z., L. M. Chiappe, G. Han, and A. Chinsamy. 2013. A large bird from the Early Cretaceous of China: new information on the skull of enantiornithines. *Journal of Vertebrate Paleontology* 33: 1176–1189.
Zhen, S., J. Li, and B. Zhang. 1994. Dinosaur and bird footprints from the Lower Cretaceous of Emei County, Sichuan, China. *Memoirs of the Beijing Natural History Museum* 54: 105–120.
Zhou, S., Z. Zhou, and J. O'Connor. 2014. A new piscivorous ornithuromorph from the Jehol biota. *Historical Biology* 5: 608–618.
Zhou, Z., J. Clarke, and F. Zhang. 2008. Insight into diversity, body size and morphological evolution from the largest Early Cretaceous enantiornithine bird. *Journal of Anatomy* 212: 565–577.
Zhou, Z., and F. Zhang. 2001. Two new ornithurine birds from the Early Cretaceous of western Lianoning, China. *Chinese Science Bulletin* 46: 1258–1260.

Index

Note: The appendix tables have not been indexed. References to these are included in the text. Tables and figures appear in italic type. An italic t *following the page number indicates a table; italic* f *indicates illustrations, photos, or graphics and their captions. Species receiving heavy treatment are main entries; otherwise specific species are listed at the end of the subheadings for genera (moa and dinosaurs) or common names (birds).*

JAMES O. FARLOW (the one without feathers) with Pecky the emu, Black Pine Animal Sanctuary, Albion, Indiana. Dr. Farlow received an undergraduate degree from Indiana University, and a PhD from Yale University. His research interests include dinosaur paleobiology and ichnology, and pre-glacial Cenozoic fossil vertebrates of northern Indiana. He is presently Emeritus Professor of Geology in the Department of Biology, Purdue University Fort Wayne, and is editor of the Life of the Past series for Indiana University Press. *Photograph by James Whitcraft.*